STUDENT'S SOLUTIONS MANUAL

BEVERLY FUSFIELD

COLLEGE ALGEBRA AND TRIGONOMETRY AND PRECALCULUS
A RIGHT TRIANGLE APPROACH
FOURTH EDITIONS

J. S. Ratti
University of South Florida

Marcus McWaters
University of South Florida

Leslaw A. Skrzypek
University of South Florida

The author and publisher of this book have used their best efforts in preparing this book. These efforts include the development, research, and testing of the theories and programs to determine their effectiveness. The author and publisher make no warranty of any kind, expressed or implied, with regard to these programs or the documentation contained in this book. The author and publisher shall not be liable in any event for incidental or consequential damages in connection with, or arising out of, the furnishing, performance, or use of these programs.

Reproduced by Pearson from electronic files supplied by the author.

Copyright © 2019, 2015, 2011 Pearson Education, Inc.
Publishing as Pearson, 330 Hudson Street, NY, NY 10013

All rights reserved. No part of this publication may be reproduced, stored in a retrieval system, or transmitted, in any form or by any means, electronic, mechanical, photocopying, recording, or otherwise, without the prior written permission of the publisher. Printed in the United States of America.
   3 2020

ISBN-13: 978-0-13-469907-3
ISBN-10: 0-13-469907-6

CONTENTS

Chapter P   **Basic Concepts of Algebra**
- P.1   The Real Numbers and Their Properties 1
- P.2   Integer Exponents and Scientific Notation 5
- P.3   Polynomials 8
- P.4   Factoring Polynomials 12
- P.5   Rational Expressions 16
- P.6   Rational Exponents and Radicals 21
- Chapter P Review Exercises 27
- Chapter P Practice Test 29

Chapter 1   **Equations and Inequalities**
- 1.1   Linear Equations in One Variable 31
- 1.2   Applications of Linear Equations: Modeling 36
- 1.3   Quadratic Equations 43
- 1.4   Complex Numbers: Quadratic Equations with Complex Solutions 50
- 1.5   Solving Other Types of Equations 54
- 1.6   Inequalities 65
- 1.7   Equations and Inequalities Involving Absolute Value 74
- Chapter 1 Review Exercises 83
- Chapter 1 Practice Test A 88
- Chapter 1 Practice Test B 90

Chapter 2   **Graphs and Functions**
- 2.1   The Coordinate Plane 92
- 2.2   Graphs of Equations 97
- 2.3   Lines 105
- 2.4   Functions 114
- 2.5   Properties of Functions 120
- 2.6   A Library of Functions 125
- 2.7   Transformations of Functions 131
- 2.8   Combining Functions; Composite Functions 141
- 2.9   Inverse Functions 150
- Chapter 2 Review Exercises 156
- Chapter 2 Practice Test A 162
- Chapter 2 Practice Test B 163
- Cumulative Review Exercises (Chapters P–2) 164

Chapter 3   **Polynomial and Rational Functions**
- 3.1   Quadratic Functions 168
- 3.2   Polynomial Functions 179
- 3.3   Dividing Polynomials 187
- 3.4   The Real Zeros of a Polynomial Function 194
- 3.5   The Complex Zeros of a Polynomial Function 205
- 3.6   Rational Functions 211
- 3.7   Variation 223
- Chapter 3 Review Exercises 226
- Chapter 3 Practice Test A 234
- Chapter 3 Practice Test B 236
- Cumulative Review Exercises (Chapters P–3) 237

Chapter 4   Exponential and Logarithmic Functions

- 4.1 Exponential Functions 240
- 4.2 Logarithmic Functions 248
- 4.3 Rules of Logarithms 255
- 4.4 Exponential and Logarithmic Equations and Inequalities 262
- 4.5 Logarithmic Scales; Modeling 269
- Chapter 4 Review Exercises 275
- Chapter 4 Practice Test A 279
- Chapter 4 Practice Test B 280
- Cumulative Review Exercises (Chapters P–4) 281

Chapter 5   Trigonometric Functions

- 5.1 Angles and Their Measure 284
- 5.2 Right-Triangle Trigonometry 288
- 5.3 Trigonometric Functions of Any Angle; The Unit Circle 295
- 5.4 Graphs of the Sine and Cosine Functions 302
- 5.5 Graphs of the Other Trigonometric Functions 309
- 5.6 Inverse Trigonometric Functions 316
- Chapter 5 Review Exercises 321
- Chapter 5 Practice Test A 323
- Chapter 5 Practice Test B 324
- Cumulative Review Exercises (Chapters P–5) 325

Chapter 6   Trigonometric Identities and Equations

- 6.1 Verifying Identities 328
- 6.2 Sum and Difference Formulas 335
- 6.3 Double-Angle and Half-Angle Formulas 344
- 6.4 Product-to-Sum and Sum-to-Product Formulas 353
- 6.5 Trigonometric Equations I 358
- 6.6 Trigonometric Equations II 364
- Chapter 6 Review Exercises 376
- Chapter 6 Practice Test A 380
- Chapter 6 Practice Test B 381
- Cumulative Review Exercises (Chapters 1–6) 383

Chapter 7   Applications of Trigonometric Functions

- 7.1 The Law of Sines 386
- 7.2 The Law of Cosines 394
- 7.3 Areas of Polygons Using Trigonometry 402
- 7.4 Vectors 413
- 7.5 The Dot Product 418
- 7.6 Polar Coordinates 424
- 7.7 Polar Form of Complex Numbers; DeMoivre's Theorem 434
- Chapter 7 Review Exercises 441
- Chapter 7 Practice Test A 445
- Chapter 7 Practice Test B 447
- Cumulative Review Exercises (Chapters 1–7) 449

Chapter 8 **Systems of Equations and Inequalities**
    8.1   Systems of Linear Equations in Two Variables 452
    8.2   Systems of Linear Equations in Three Variables 463
    8.3   Partial-Fraction Decomposition 474
    8.4   Systems of Nonlinear Equations 486
    8.5   Systems of Inequalities 493
    8.6   Linear Programming 501
    Chapter 8 Review Exercises 509
    Chapter 8 Practice Test A 515
    Chapter 8 Practice Test B 518
    Cumulative Review Exercises (Chapters P–8) 520

Chapter 9 **Matrices and Determinants**
    9.1   Matrices and Systems of Equations 523
    9.2   Matrix Algebra 534
    9.3   The Matrix Inverse 545
    9.4   Determinants and Cramer's Rule 557
    Chapter 9 Review Exercises 564
    Chapter 9 Practice Test A 569
    Chapter 9 Practice Test B 571
    Cumulative Review Exercises (Chapters P–9) 572

Chapter 10 **The Conic Sections**
    10.2   The Parabola 575
    10.3   The Ellipse 583
    10.4   The Hyperbola 593
    Chapter 10 Review Exercises 607
    Chapter 10 Practice Test A 611
    Chapter 10 Practice Test B 613
    Cumulative Review Exercises (Chapters P–10) 614

Chapter 11 **Further Topics in Algebra**
    11.1   Sequences and Series 617
    11.2   Arithmetic Sequences; Partial Sums 623
    11.3   Geometric Sequences and Series 627
    11.4   Mathematical Induction 633
    11.5   The Binomial Theorem 639
    11.6   Counting Principles 643
    11.7   Probability 647
    Chapter 11 Review Exercises 650
    Chapter 11 Practice Test A 652
    Chapter 11 Practice Test B 652
    Cumulative Review Exercises (Chapters P–11) 653

Chapter P Basic Concepts of Algebra

P.1 The Real Numbers and Their Properties

P.1 Practice Problems

1. Let $x = 2.132132132\ldots$
   Then, $1000x = 2132.132132\ldots$
   $$1000x = 2132.132132\ldots$$
   $$x = 2.132132\ldots$$
   $$999x = 2130$$
   $$x = \frac{2130}{999} = \frac{710}{333}$$

2. a. Natural numbers: 2, 7
   b. Whole numbers: 0, 2, 7
   c. Integers: $-6, -\frac{21}{7} = -3, 0, 2, 7$
   d. Rational numbers:
      $-6, -\frac{21}{7} = -3, -\frac{1}{2}, 0, \frac{4}{3}, 2, 7$
   e. Irrational numbers: $\sqrt{3}, \sqrt{17}$
   f. Real numbers: the set B

3. a. $2^3 = 8$
   b. $(3a)^2 = (3a)(3a) = 9a^2$
   c. $\left(\frac{1}{2}\right)^4 = \frac{1}{2} \cdot \frac{1}{2} \cdot \frac{1}{2} \cdot \frac{1}{2} = \frac{1}{16}$

4. a. True b. True
   c. False

5. $A = \{-3, -1, 0, 1, 3\}, B = \{-4, -2, 0, 2, 4\}$
   $A \cap B = \{0\}$,
   $A \cup B = \{-4, -3, -2, -1, 0, 1, 2, 3, 4\}$

6. a. $I_1 \cup I_2 = (-\infty, \infty)$
   b. $I_1 \cap I_2 = [-2, 5)$

7. a. $|-10| = 10$
   b. $|3 - 4| = |-1| = 1$
   c. $|2(-3) + 7| = |1| = 1$

8. [number line from -7 to 2 showing 7 units from -7 to 0 and 2 units from 0 to 2; $7 + 2 = 9$ units]
   $d(-7, 2) = |-7 - 2| = |-9| = 9$

9. a. $(-3) \cdot 5 + 20 = -15 + 20 = 5$
   b. $5 - 12 \div 6 \cdot 2 = 5 - 2 \cdot 2 = 5 - 4 = 1$
   c. $\frac{9-1}{4} - 5 \cdot 7 = \frac{8}{4} - 5 \cdot 7 = 2 - 35 = -33$
   d. $-3 + (x-4)^2$ for $x = 6$
      $-3 + (6-4)^2 = -3 + 2^2 = -3 + 4 = 1$

10. a. $\frac{7}{4} + \frac{3}{8} = \frac{14}{8} + \frac{3}{8} = \frac{17}{8}$
    b. $\frac{8}{3} - \frac{2}{5} = \frac{40}{15} - \frac{6}{15} = \frac{34}{15}$
    c. $\frac{9}{14} \cdot \frac{7}{3} = \frac{\cancel{9}^3}{\cancel{14}_2} \cdot \frac{\cancel{7}^1}{\cancel{3}_1} = \frac{3}{2}$
    d. $\dfrac{\frac{5}{8}}{\frac{15}{16}} = \frac{5}{8} \div \frac{15}{16} = \frac{5}{8} \cdot \frac{16}{15} = \frac{\cancel{5}^1}{\cancel{8}_1} \cdot \frac{\cancel{16}^2}{\cancel{15}_3} = \frac{2}{3}$

11. a. $(x-2) \div 3 + x = (3-2) \div 3 + 3$
       $= 1 \div 3 + 3 = \frac{1}{3} + 3 = \frac{10}{3}$
    b. $7 - \frac{x}{|y|} = 7 - \frac{-1}{|3|} = 7 + \frac{1}{3} = \frac{22}{3}$

12. $°C = \frac{39}{3} + 4 = 13 + 4 = 17°C$

P.1 Concepts and Vocabulary

1. Whole numbers are formed by adding the number <u>zero</u> to the set of natural numbers.

3. If $a < b$, then a is to the <u>left</u> of b on the number line.

5. True

7. True

Chapter P Basic Concepts of Algebra

P.1 Building Skills

9. $0.\overline{3}$, repeating

11. -0.8, terminating

13. $0.\overline{27}$, repeating

15. $3.1\overline{6}$, repeating

17. $3.75 = \dfrac{375}{100} = \dfrac{15}{4}$

19. $10x = 53.\overline{3}$
    $x = 5.\overline{3}$
    $\overline{9x = 48}$
    $x = \dfrac{48}{9} = \dfrac{16}{3}$
    $-5.\overline{3} = -\dfrac{16}{3}$

21. $100x = 213.\overline{13}$
    $x = 2.\overline{13}$
    $\overline{99x = 211}$
    $x = \dfrac{211}{99}$

23. $100x = 452.3\overline{23}$
    $x = 4.5\overline{23}$
    $\overline{99x = 447.8}$
    $x = \dfrac{447.8}{99} = \dfrac{4478}{990} = \dfrac{2239}{495}$

25. Rational   27. Rational

29. Rational   31. Irrational

Exercises 33-38 refer to the set
$A = \left\{-19, -\dfrac{12}{3}, -\sqrt{3}, 0, 2, \sqrt{10}, \dfrac{17}{4}, 11\right\}$

33. Natural numbers: 2, 11

35. Integers: $-19, -\dfrac{12}{3} = -4, 0, 2, 11$

37. Irrational numbers: $-\sqrt{3}, \sqrt{10}$

39. 10^3
    base: 10; exponent: 3
    $10^3 = 10 \cdot 10 \cdot 10 = 1000$

41. $\left(\dfrac{2}{3}\right)^3$
    base: $\dfrac{2}{3}$; exponent: 3
    $\left(\dfrac{2}{3}\right)^3 = \dfrac{2}{3} \cdot \dfrac{2}{3} \cdot \dfrac{2}{3} = \dfrac{8}{27}$

43. $(-2)^3$
    base: -2; exponent: 3
    $(-2)^3 = (-2) \cdot (-2) \cdot (-2) = -8$

45. $2 \cdot 3^4$
    base: 3; exponent: 4
    $2 \cdot 3^4 = 2 \cdot 3 \cdot 3 \cdot 3 \cdot 3 = 2 \cdot 81 = 162$

47. $-2 \cdot 3^4$
    base: 3; exponent: 4
    $-2 \cdot 3^4 = -2 \cdot 3 \cdot 3 \cdot 3 \cdot 3 = -2 \cdot 81 = -162$

49. $-3 \cdot (-2)^5$
    base: -2; exponent: 5
    $-3 \cdot (-2)^5 = -3(-2)(-2)(-2)(-2)(-2)$
    $ = -3 \cdot (-32) = 96$

51. $3 > -2$ 53. $\dfrac{1}{2} \geq \dfrac{1}{2}$

55. $5 \leq 2x$ 57. $-x > 0$

59. $2x + 7 \leq 14$ 61. $4 = \dfrac{24}{6}$

63. $-4 < 0$

65. $A \cup B = \{-4, -3, -2, 0, 1, 2, 3, 4\}$

67. $A \cap C = \{-4, -2, 0, 2\}$

69. $(B \cap C) \cup A = \{-3, 0, 2\} \cup A$
    $ = \{-4, -3, -2, 0, 2, 4\}$

71. $(A \cup B) \cap C = \{-4, -3, -2, 0, 1, 2, 3, 4\} \cap C$
    $ = \{-4, -3, -2, 0, 2\}$

73. $I_1 \cup I_2 = (-2, 5);\ I_1 \cap I_2 = [1, 3]$

75. $I_1 \cup I_2 = (-6, 10);\ I_1 \cap I_2 = \varnothing$

77. $I_1 \cup I_2 = (-\infty, \infty);\ I_1 \cap I_2 = [2, 5)$

79. $I_1 \cup I_2 = (-\infty, \infty);\ I_1 \cap I_2 = [-1, 3) \cup [5, 7]$

81. $|-4| = 4$

83. $\left|\dfrac{5}{-7}\right| = \dfrac{5}{7}$

85. $|5-\sqrt{2}| = 5-\sqrt{2}$

87. $|\sqrt{3}-2| = 2-\sqrt{3}$

89. $\dfrac{8}{|-8|} = \dfrac{8}{8} = 1$

91. $|5-|-7|| = |5-7| = |-2| = 2$

93.
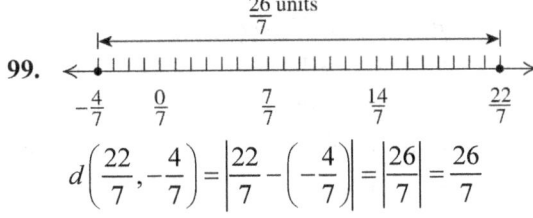
$d(3,8) = |3-8| = |-5| = 5$

95.
$d(-6,9) = |-6-9| = |-15| = 15$

97.
$d(-20,-6) = |-20-(-6)| = |-14| = 14$

99.
$d\left(\dfrac{22}{7}, -\dfrac{4}{7}\right) = \left|\dfrac{22}{7} - \left(-\dfrac{4}{7}\right)\right| = \left|\dfrac{26}{7}\right| = \dfrac{26}{7}$

101.
$-3 < x \le 1$

103.
$x \ge -3$

105.
$x \le 5$

107.
$-\dfrac{3}{4} < x < \dfrac{9}{4}$

109. $4(x+1) = 4x+4$

111. $5(x-y+1) = 5x-5y+5$

113. Additive inverse: -5; reciprocal: $\dfrac{1}{5}$

115. Additive inverse: 0; no reciprocal

117. Additive inverse

119. Multiplicative identity

121. Associative property of multiplication

123. Multiplicative inverse

125. Additive identity

127. Associative property of addition

129. $\dfrac{3}{5} + \dfrac{4}{3} = \dfrac{9}{15} + \dfrac{20}{15} = \dfrac{29}{15}$

131. $\dfrac{6}{5} + \dfrac{5}{7} = \dfrac{42}{35} + \dfrac{25}{35} = \dfrac{67}{35}$

133. $\dfrac{5}{6} + \dfrac{3}{10} = \dfrac{25}{30} + \dfrac{9}{30} = \dfrac{34}{30} = \dfrac{17}{15}$

135. $\dfrac{5}{8} - \dfrac{9}{10} = \dfrac{25}{40} - \dfrac{36}{40} = -\dfrac{11}{40}$

137. $\dfrac{5}{9} - \dfrac{7}{11} = \dfrac{55}{99} - \dfrac{63}{99} = -\dfrac{8}{99}$

139. $\dfrac{2}{5} - \dfrac{1}{2} = \dfrac{4}{10} - \dfrac{5}{10} = -\dfrac{1}{10}$

141. $\dfrac{3}{4} \cdot \dfrac{8}{27} = \dfrac{2}{9}$

143. $\dfrac{\frac{8}{5}}{\frac{16}{15}} = \dfrac{8}{5} \div \dfrac{16}{15} = \dfrac{8}{5} \cdot \dfrac{15}{16} = \dfrac{3}{2}$

145. $\dfrac{\frac{7}{8}}{\frac{21}{16}} = \dfrac{7}{8} \div \dfrac{21}{16} = \dfrac{7}{8} \cdot \dfrac{16}{21} = \dfrac{2}{3}$

147. $5 \cdot \dfrac{3}{10} - \dfrac{1}{2} = \dfrac{3}{2} - \dfrac{1}{2} = \dfrac{2}{2} = 1$

149. $3 \cdot \dfrac{2}{15} - \dfrac{1}{3} = \dfrac{2}{5} - \dfrac{1}{3} = \dfrac{6}{15} - \dfrac{5}{15} = \dfrac{1}{15}$

151. $2(x+y) - 3y = 2(3+(-5)) - 3(-5)$
$= 2(-2) - (-15) = -4+15 = 11$

153. $3|x| - 2|y| = 3|3| - 2|-5| = 3(3) - 2(5)$
$= 9 - 10 = -1$

155. $\dfrac{x-3y}{2}+xy = \dfrac{3-3(-5)}{2}+3(-5)$
$= \dfrac{3-(-15)}{2}+(-15)$
$= \dfrac{18}{2}+(-15) = 9+(-15) = -6$

157. $\dfrac{2(1-2x)}{y}-(-x)y = \dfrac{2(1-2(3))}{-5}-(-3)(-5)$
$= \dfrac{2(-5)}{-5}-15 = 2-15 = -13$

159. $\dfrac{\dfrac{14}{x}+\dfrac{1}{2}}{\dfrac{-y}{4}} = \dfrac{\dfrac{14}{3}+\dfrac{1}{2}}{\dfrac{-(-5)}{4}} = \dfrac{31}{6}\cdot\dfrac{4}{5} = \dfrac{62}{15}$

161. $\dfrac{x}{y}+\dfrac{x}{3} = \dfrac{3x+xy}{3y}$

163. $5(x+3) = 5x+5(3) = 5x+15$.

165. $x-(3y+2) = x-3y-2$

167. $\dfrac{x+y}{x} = \dfrac{x}{x}+\dfrac{y}{x} = 1+\dfrac{y}{x}$

169. $(x+1)(y+1) = xy+x+y+1$

171. $d(P, M) = \left|\dfrac{a+b}{2}-a\right| = \left|\dfrac{a+b-2a}{2}\right| = \left|\dfrac{-a+b}{2}\right|$
$d(Q, M) = \left|b-\dfrac{a+b}{2}\right| = \left|\dfrac{2b-a+b}{2}\right| = \left|\dfrac{b-a}{2}\right|$
Since
$d(P, M) = \left|\dfrac{-a+b}{2}\right| = \left|\dfrac{b-a}{2}\right| = d(Q, M)$, M is the midpoint of the line segment PQ.

P.1 Applying the Concepts

173. a. people who own either MP3 players or people who own DVD players.

 b. people who own both MP3 players and DVD players.

175. $119.5 \le x \le 134.5$

 [number line from 119.5 to 134.5]

176. $30 \le x \le 107$

 [number line from 30 to 107]

177. a. $|124-120| = 4$

 b. $|137-120| = 17$

 c. $|114-120| = |-6| = 6$

179. Let x = the number of calories from broccoli. Then we have
$522.5 - 55x = 0 \Rightarrow 522.5 = 55x \Rightarrow 9.5 = x$
The number of grams of broccoli is $9.5 \times 100 = 950$ grams.

P.1 Beyond the Basics

181. True

183. a. False. For example, $0\cdot\sqrt{2} = 0$.

 b. The products of a nonzero rational number and an irrational number is an irrational number.

185. False. For example, $(2+\sqrt{3})-(6+\sqrt{3}) = -4$.

187. True

For exercises 189–190, use the following definition: An integer P is *even* if $p = 2n$ for some integer n. An integer q is odd if $q = 2k+1$ for some integer k.

189. a. If a is odd, then $a = 2m+1$ for some integer m.
$a^2 = (2m+1)^2 = (2m+1)(2m+1)$
$= 4m^2+4m+1 = 4m(m+1)+1$
$= 2(2m(m+1))+1$,
which is of the format $2k+1$. Therefore, a^2 is an odd integer.

 b. Assume that b is odd. Then, from part (a), b^2 is also odd. However, b^2 is an even integer, so, by contradiction, b must be an even integer.

P.1 Getting Ready for the Next Section

191. a. $a^2\cdot a^3 = a^{\underline{5}}$

 b. $a^4\cdot a^7 = a^{\underline{11}}$

 c. $a^m\cdot a^n = a^{\underline{m+n}}$

193. a. $(a^2)^3 = a^{\underline{6}}$

 b. $(a^4)^2 = a^{\underline{8}}$

 c. $(a^m)^n = a^{\underline{mn}}$

P.2 Integer Exponents and Scientific Notation

P.2 Practice Exercises

1. a. $2^{-1} = \dfrac{1}{2}$

   b. $\left(\dfrac{4}{5}\right)^0 = 1$

   c. $\left(\dfrac{3}{2}\right)^{-2} = \left(\dfrac{2}{3}\right)^2 = \dfrac{4}{9}$

2. a. $x^2 \cdot 3x^7 = 3x^9$

   b. $(2^2 x^3)(4x^{-3}) = 4x^3 \cdot 4x^{-3} = 16x^{3+(-3)}$
   $= 16x^0 = 16 \cdot 1 = 16$

3. a. $\dfrac{3^4}{3^0} = 3^{4-0} = 3^4 = 81$

   b. $\dfrac{5}{5^{-2}} = 5^{1-(-2)} = 5^3 = 125$

   c. $\dfrac{2x^3}{3x^{-4}} = \dfrac{2x^{3-(-4)}}{3} = \dfrac{2x^7}{3}$

4. a. $\left(7^{-5}\right)^0 = 7^{(-5)(0)} = 7^0 = 1$

   b. $\left(7^0\right)^{-5} = 7^{(0)(-5)} = 7^0 = 1$

   c. $\left(x^{-1}\right)^8 = x^{(-1)(8)} = x^{-8} = \dfrac{1}{x^8}$

   d. $\left(x^{-2}\right)^{-5} = x^{(-2)(-5)} = x^{10}$

5. a. $\left(\dfrac{1}{2}x\right)^{-1} = \left(\dfrac{1}{2}\right)^{-1}(x)^{-1} = 2\left(\dfrac{1}{x}\right) = \dfrac{2}{x}$

   b. $(5x^{-1})^2 = 5^2(x^{-1})^2 = 5^2 x^{(-1)(2)}$
   $= 25x^{-2} = \dfrac{25}{x^2}$

   c. $(xy^2)^3 = x^3 y^{2(3)} = x^3 y^6$

   d. $(x^{-2} y)^{-3} = x^{(-2)(-3)} y^{-3} = \dfrac{x^6}{y^3}$

6. a. $\left(\dfrac{1}{3}\right)^2 = \dfrac{1^2}{3^2} = \dfrac{1}{9}$

   b. $\left(\dfrac{10}{7}\right)^{-2} = \left(\dfrac{7}{10}\right)^2 = \dfrac{7^2}{10^2} = \dfrac{49}{100}$

7. a. $(2x^4)^{-2} = \left(\dfrac{1}{2x^4}\right)^2 = \dfrac{1}{4x^8}$

   b. $\dfrac{x^2(-y)^3}{(xy^2)^3} = \dfrac{-x^2 y^3}{x^3 y^6} = -\dfrac{1}{xy^3}$

8. $732{,}000 = 7.32 \times 10^5$

9. $\dfrac{2.9 \times 10^{10}}{3.25 \times 10^8} = \dfrac{290}{325} \times 10^2 \approx 0.89 \times 10^2 \approx \$89$ per person

P.2 Concepts and Vocabulary

1. In the expression 7^2, the number 2 is called the <u>exponent</u>.

3. The number $\dfrac{1}{4^{-2}}$ simplifies to be the positive integer <u>16</u>.

5. False. $(-11)^{10} = 11^{10}$

7. True

P.2 Building Skills

9. $3^{-2} = \left(\dfrac{1}{3}\right)^2 = \left(\dfrac{1}{3}\right)\left(\dfrac{1}{3}\right) = \dfrac{1}{9}$

11. $\left(\dfrac{1}{2}\right)^{-4} = 2^4 = 16$

13. $7^0 = 1$

15. $-(-7)^0 = -1$

17. $(2^3)^2 = 2^{3 \cdot 2} = 2^6 = 64$

19. $(3^2)^{-2} = 3^{2(-2)} = 3^{-4} = \dfrac{1}{3^4} = \dfrac{1}{81}$

21. $(5^{-2})^3 = \left(\dfrac{1}{5^2}\right)^3 = \left(\dfrac{1}{25}\right)^3 = \dfrac{1}{15{,}625}$

23. $(4^{-3}) \cdot (4^5) = 4^{(-3+5)} = 4^2 = 16$

25. $\left(\sqrt{3}\right)^0 + 10^0 = 1 + 1 = 2$

27. $3^{-2} + \left(\dfrac{1}{3}\right)^2 = \left(\dfrac{1}{3}\right)^2 + \left(\dfrac{1}{3}\right)^2 = \dfrac{1}{9} + \dfrac{1}{9} = \dfrac{2}{9}$

29. $-2^{-3} = -\dfrac{1}{2^3} = -\dfrac{1}{8}$

31. $(-3)^{-2} = \dfrac{1}{(-3)^2} = \dfrac{1}{9}$

33. $\dfrac{2^{11}}{2^{10}} = 2^{(11-10)} = 2^1 = 2$

35. $\dfrac{\left(5^3\right)^4}{5^{12}} = \dfrac{5^{12}}{5^{12}} = 1$

37. $\dfrac{2^5 \cdot 3^{-2}}{2^4 \cdot 3^{-3}} = 2^{(5-4)} \cdot 3^{(-2-(-3))} = 2^1 \cdot 3^1 = 6$

39. $\dfrac{-5^{-2}}{2^{-1}} = -\dfrac{1}{5^2} \cdot 2^1 = -\dfrac{1}{25} \cdot 2 = -\dfrac{2}{25}$

41. $\left(\dfrac{2}{3}\right)^{-1} = \dfrac{3}{2}$

43. $\left(\dfrac{2}{3}\right)^{-2} = \left(\dfrac{3}{2}\right)^2 = \dfrac{3^2}{2^2} = \dfrac{9}{4}$

45. $\left(\dfrac{11}{7}\right)^{-2} = \dfrac{1}{\left(\dfrac{11}{7}\right)^2} = \dfrac{1}{\dfrac{121}{49}} = \dfrac{49}{121}$

47. $x^4 y^0 = x^4 \cdot 1 = x^4$

49. $x^{-1} y = \dfrac{1}{x} \cdot y = \dfrac{y}{x}$

51. $x^{-1} y^{-2} = \dfrac{1}{xy^2}$

53. $\left(x^{-3}\right)^4 = x^{-12} = \dfrac{1}{x^{12}}$

55. $\left(x^{-11}\right)^{-3} = x^{(-11)\cdot(-3)} = x^{33}$

57. $-3(xy)^5 = -3x^5 y^5$

59. $4\left(xy^{-1}\right)^2 = 4x^2 y^{-2} = \dfrac{4x^2}{y^2}$

61. $3\left(x^{-1} y\right)^{-5} = 3x^{(-1)\cdot(-5)} y^{-5} = \dfrac{3x^5}{y^5}$

63. $\dfrac{\left(x^3\right)^2}{\left(x^2\right)^5} = \dfrac{x^{3\cdot 2}}{x^{2\cdot 5}} = \dfrac{x^6}{x^{10}} = x^{6-10} = x^{-4} = \dfrac{1}{x^4}$

65. $\left(\dfrac{2xy}{x^2}\right)^3 = \dfrac{2^3 x^3 y^3}{x^{2\cdot 3}} = \dfrac{8x^3 y^3}{x^6} = 8x^{3-6} y^3$
$= 8x^{-3} y^3 = \dfrac{8y^3}{x^3}$

67. $\left(\dfrac{-3x^2 y}{x}\right)^5 = \dfrac{(-3)^5 x^{2\cdot 5} y^5}{x^5} = \dfrac{-243 x^{10} y^5}{x^5}$
$= -243 x^{10-5} y^5 = -243 x^5 y^5$

69. $\left(\dfrac{-3x}{5}\right)^{-2} = \dfrac{(-3)^{-2} x^{-2}}{5^{-2}} = \dfrac{\dfrac{1}{9} \cdot \dfrac{1}{x^2}}{\dfrac{1}{25}} = \dfrac{\dfrac{1}{9x^2}}{\dfrac{1}{25}} = \dfrac{25}{9x^2}$

71. $\left(\dfrac{4x^{-2}}{xy^5}\right)^3 = \dfrac{4^3 x^{-2\cdot 3}}{x^3 y^{5\cdot 3}} = \dfrac{64 x^{-6}}{x^3 y^{15}} = \dfrac{64}{x^9 y^{15}}$

73. $\dfrac{x^3 y^{-3}}{x^{-2} y} = x^{3-(-2)} y^{-3-1} = x^5 y^{-4} = \dfrac{x^5}{y^4}$

75. $\dfrac{27 x^{-3} y^5}{9 x^{-4} y^7} = 3 x^{-3-(-4)} y^{5-7} = 3 x^1 y^{-2} = \dfrac{3x}{y^2}$

77. $\dfrac{1}{x^3}\left(x^2\right)^3 x^{-4} = x^{-3}\left(x^{2\cdot 3}\right) x^{-4} = x^{-3}\left(x^6\right) x^{-4}$
$= x^{-3+6-4} = x^{-1} = \dfrac{1}{x}$

79. $\left(-xy^2\right)^3 \left(-2x^2 y^2\right)^{-4} = \dfrac{\left(-xy^2\right)^3}{\left(-2x^2 y^2\right)^4}$
$= \dfrac{(-x)^3 \left(y^2\right)^3}{(-2)^4 \left(x^2\right)^4 \left(y^2\right)^4}$
$= \dfrac{-x^3 \cdot y^{2\cdot 3}}{16 x^{2\cdot 4} y^{2\cdot 4}} = \dfrac{-x^3 y^6}{16 x^8 y^8}$
$= -\dfrac{1}{16}\left(x^{3-8}\right)\left(y^{6-8}\right)$
$= -\dfrac{1}{16} x^{-5} y^{-2} = -\dfrac{1}{16 x^5 y^2}$

Section P.2 Integer Exponents and Scientific Notation

81. $(4x^3y^2z)^2(x^3y^2z)^{-7} = \dfrac{(4x^3y^2z)^2}{(x^3y^2z)^7}$

$= \dfrac{4^2(x^3)^2(y^2)^2z^2}{(x^3)^7(y^2)^7z^7}$

$= \dfrac{16x^{3\cdot 2}y^{2\cdot 2}z^2}{x^{3\cdot 7}y^{2\cdot 7}z^7}$

$= \dfrac{16x^6y^4z^2}{x^{21}y^{14}z^7}$

$= 16x^{6-21}y^{4-14}z^{2-7}$

$= 16x^{-15}y^{-10}z^{-5}$

$= \dfrac{16}{x^{15}y^{10}z^5}$

83. $\dfrac{5a^{-2}bc^2}{a^4b^{-3}c^2} = 5a^{-2-4}b^{1-(-3)}c^{2-2}$

$= 5a^{-6}b^4 \cdot 1 = \dfrac{5b^4}{a^6}$

85. $\left(\dfrac{xy^{-3}z^{-2}}{x^2y^{-4}z^3}\right)^{-3} = \dfrac{x^{-3}y^{(-3)(-3)}z^{(-2)(-3)}}{x^{2(-3)}y^{(-4)(-3)}z^{3(-3)}}$

$= \dfrac{x^{-3}y^9z^6}{x^{-6}y^{12}z^{-9}}$

$= x^{-3-(-6)}y^{9-12}z^{6-(-9)}$

$= x^3y^{-3}z^{15} = \dfrac{x^3z^{15}}{y^3}$

87. $125 = 1.25 \times 10^2$

89. $850,000 = 8.5 \times 10^5$

91. $0.007 = 7.0 \times 10^{-3}$

93. $0.00000275 = 2.75 \times 10^{-6}$

P.2 Applying the Concepts

95. $135 \text{ ft}^3 \times 2^3 = 1080 \text{ ft}^3$

97. a. $(2x)^2 = 4x^2 = 4A = 2^2 A$

b. $(3x)^2 = 9x^2 = 9A = 3^2 A$

99. $F = Sw^2 = 25,000(0.25)^2 = 1562.5 \text{ lb}$

101. $365.25 \text{ days} \times \dfrac{24 \text{ hr}}{\text{day}} \times \dfrac{60 \text{ min}}{\text{hr}} \times \dfrac{60 \text{ sec}}{\text{min}}$

$= 365.25 \times 24 \times 60 \times 60 \text{ sec}$

$= 31,557,600 \text{ sec}$

$= 3.15576 \times 10^7 \text{ sec}$

103.

Celestial Body	Equatorial Diameter (km)	Scientific Notation
Earth	12,700	1.27×10^4 km
Moon	3480	3.48×10^3 km
Sun	1,390,000	1.39×10^6 km
Jupiter	134,000	1.34×10^5 km
Mercury	4800	4.8×10^3 km

105. $602,000,000,000,000,000,000 = 6.02 \times 10^{20}$ atoms

107. $0.00000000000000000000167 = 1.67 \times 10^{-21}$ kg

109. $5,980,000,000,000,000,000,000,000 = 5.98 \times 10^{24}$ kg

P.2 Beyond the Basics

111. $2^x = 32$

a. $2^{x+2} = 2^x \cdot 2^2 = 32 \cdot 4 = 128$

b. $2^{x-1} = 2^x \cdot 2^{-1} = 32 \cdot \dfrac{1}{2} = 16$

113. $5^x = 11$

a. $5^{x+1} = 5^x \cdot 5^1 = 11 \cdot 5 = 55$

b. $5^{x-2} = 5^x \cdot 5^{-2} = 11 \cdot \left(\dfrac{1}{5}\right)^2 = 11 \cdot \dfrac{1}{25} = \dfrac{11}{25}$

115. $\dfrac{3^{2-n} \cdot 9^{2n-2}}{3^{3n}} = \dfrac{3^{2-n} \cdot (3^2)^{2n-2}}{3^{3n}}$

$= \dfrac{3^{2-n} \cdot 3^{2(2n-2)}}{3^{3n}}$

$= \dfrac{3^{3n-2}}{3^{3n}} = 3^{(3n-2)-3n}$

$= 3^{-2} = \dfrac{1}{3^2} = \dfrac{1}{9}$

117. $\dfrac{2^{x(y-z)}}{2^{y(x-z)}} \div \left(\dfrac{2^y}{2^x}\right)^z = \dfrac{2^{xy-xz}}{2^{xy-yz}} \div \dfrac{2^{yz}}{2^{xz}}$

$= \dfrac{\dfrac{2^{xy}}{2^{xz}}}{\dfrac{2^{xy}}{2^{yz}}} \cdot \dfrac{2^{xz}}{2^{yz}}$

$= \dfrac{2^{yz}}{2^{xz}} \cdot \dfrac{2^{xz}}{2^{yz}} = 1$

P.2 Getting Ready for the Next Section

119. a. $x^2 \cdot x^5 = x^{2+5} = x^7$

   b. $(2x)(-5x^2) = -10x^3$

   c. $(2y^2)(3y^3)(4y^5) = 24y^{2+3+5} = 24y^{10}$

121. False. $5x^2$ and $3x^3$ are not like terms, so they cannot be combined.

P.3 Polynomials

P.3 Practice Problems

1. $16(7)^2 + 15(7) = 889$ ft

2. $(7x^3 + 2x^2 - 5) + (-2x^3 + 3x^2 + 2x + 1)$
   $= 5x^3 + 5x^2 + 2x - 4$

3. $(3x^4 - 5x^3 + 2x^2 + 7) - (-2x^4 + 3x^2 + x - 5)$
   $= 5x^4 - 5x^3 - x^2 - x + 12$

4. $-2x^3(4x^2 + 2x - 5) = -8x^5 - 4x^4 + 10x^3$

5. $(5x^2 + 2x)(-2x^2 + x - 7)$
   $= 5x^2(-2x^2 + x - 7) + 2x(-2x^2 + x - 7)$
   $= -10x^4 + 5x^3 - 35x^2 - 4x^3 + 2x^2 - 14x$
   $= -10x^4 + x^3 - 33x^2 - 14x$

6. a. $(4x - 1)(x + 7) = 4x^2 + 28x - x - 7$
      $= 4x^2 + 27x - 7$

   b. $(3x - 2)(2x - 5) = 6x^2 - 15x - 4x + 10$
      $= 6x^2 - 19x + 10$

7. $(3x + 2)^2 = (3x)^2 + 2(3x)(2) + 2^2$
   $= 9x^2 + 12x + 4$

8. $(1 - 2x)(1 + 2x) = 1^2 - (2x)^2 = 1 - 4x^2$

9. a. $(x + 2y)(x - 2y) = x^2 - (2y)^2 = x^2 - 4y^2$

   b. $(2x - y)^3 = (2x)^3 - 3(2x)^2 + 3(2x)y^2 - y^3$
      $= 8x^3 - 3(4x^2) + 6xy^2 - y^3$
      $= 8x^3 - 12x^2 + 6xy^2 - y^3$

P.3 Concepts and Vocabulary

1. The polynomial $-3x^7 + 2x^2 - 9x + 4$ has leading coefficient <u>−3</u> and degree <u>7</u>.

3. When a polynomial in x of degree 3 is added to a polynomial in x of degree 4, the resulting polynomial has degree <u>4</u>.

5. True

7. False. This is not a polynomial because the term $-3\sqrt{x}$ does not have an exponent that is either a positive integer or zero.

P.3 Building Skills

9. A polynomial; $x^2 + 2x + 1$

11. Not a polynomial

13. Not a polynomial

15. A polynomial; in standard form

17. Degree: 1; terms: $7x$, 3

19. Degree: 4; terms: $-x^4, x^2, 2x, -9$

21. $(x^3 + 2x^2 - 5x + 3) + (-x^3 + 2x - 4)$
    $= (x^3 + (-x^3)) + 2x^2 + (-5x + 2x) + (3 - 4)$
    $= 2x^2 - 3x - 1$

23. $(2x^3 - x^2 + x - 5) - (x^3 - 4x + 3)$
    $= (2x^3 - x^3) - x^2 + (x - (-4x)) + (-5 - 3)$
    $= x^3 - x^2 + 5x - 8$

25. $(-2x^4 + 3x^2 - 7x) - (8x^4 + 6x^3 - 9x^2 - 17)$
    $= (-2x^4 - 8x^4) - 7x - 6x^3$
    $\quad + (3x^2 - (-9x^2)) - (-17)$
    $= -10x^4 - 6x^3 + 12x^2 - 7x + 17$

27. $-2(3x^2 + x + 1) + 6(-3x^2 - 2x - 2)$
    $= -6x^2 - 2x - 2 - 18x^2 - 12x - 12$
    $= -24x^2 - 14x - 14$

29. $(3y^3 - 4y + 2) + (2y + 1) - (y^3 - y^2 + 4)$
$= 3y^3 - 4y + 2 + 2y + 1 - y^3 + y^2 - 4$
$= 2y^3 + y^2 - 2y - 1$

31. $6x(2x + 3) = 12x^2 + 18x$

33. $(x+1)(x^2 + 2x + 2)$
$= x(x^2 + 2x + 2) + 1(x^2 + 2x + 2)$
$= x^3 + 2x^2 + 2x + x^2 + 2x + 2$
$= x^3 + 3x^2 + 4x + 2$

35. $(3x - 2)(x^2 - x + 1)$
$= 3x(x^2 - x + 1) - 2(x^2 - x + 1)$
$= 3x^3 - 3x^2 + 3x - 2x^2 + 2x - 2$
$= 3x^3 - 5x^2 + 5x - 2$

37. $(x+1)(x+2) = x(x+2) + 1(x+2)$
$= x^2 + 2x + x + 2 = x^2 + 3x + 2$

39. $(3x + 2)(3x + 1) = 9x^2 + 3x + 6x + 2$
$= 9x^2 + 9x + 2$

41. $(-4x + 5)(x + 3) = -4x^2 - 12x + 5x + 15$
$= -4x^2 - 7x + 15$

43. $(3x - 2)(2x - 1) = 6x^2 - 3x - 4x + 2$
$= 6x^2 - 7x + 2$

45. $(2x - 3a)(2x + 5a) = 4x^2 + 10ax - 6ax - 15a^2$
$= 4x^2 + 4ax - 15a^2$

47. $(x + 2)^2 - x^2 = x^2 + 4x + 4 - x^2 = 4x + 4$

49. $(4x + 1)^2 = 16x^2 + 8x + 1$

51. $\left(x + \dfrac{3}{4}\right)^2 = x^2 + 2\left(\dfrac{3}{4}\right)x + \left(\dfrac{3}{4}\right)^2$
$= x^2 + \dfrac{3}{2}x + \dfrac{9}{16}$

53. $(3x - 4y)^2 = (3x)^2 - 2(3x)(4y) + (4y)^2$
$= 9x^2 - 24xy + 16y^2$

55. $(x^2 + 2)^2 = (x^2)^2 + 2(x^2)(2) + 2^2$
$= x^4 + 4x^2 + 4$

57. $\left(\dfrac{x}{2} + \dfrac{2}{x}\right)^2 = \left(\dfrac{x}{2}\right)^2 + 2\left(\dfrac{x}{2}\right)\left(\dfrac{2}{x}\right) + \left(\dfrac{2}{x}\right)^2$
$= \dfrac{x^2}{4} + 2 + \dfrac{4}{x^2} = \dfrac{x^2}{4} + \dfrac{4}{x^2} + 2$

59. $(x + 2)^3 = x^3 + 3x^2(2) + 3x(2)^2 + 2^3$
$= x^3 + 6x^2 + 12x + 8$

61. $(x + 3)^3 - x^3$
Note that this is the difference of cubes, with $A = x + 3$ and $B = x$.
$(x + 3)^3 - x^3$
$= \left[(x+3) - x\right]\left[(x+3)^2 + (x+3)x + x^2\right]$
$= 3\left(x^2 + 6x + 9 + x^2 + 3x + x^2\right)$
$= 3\left(3x^2 + 9x + 9\right)$
$= 9x^2 + 27x + 27$

63. $(x + 2y)^3 = x^3 + 3x^2(2y) + 3x(2y)^2 + 2^3$
$= x^3 + 6x^2y + 12xy^2 + 8$

65. $(5 + 2x)(5 - 2x) = 5^2 - (2x)^2 = 25 - 4x^2$

67. $(2x + 3y)(2x - 3y) = (2x)^2 - (3y)^2$
$= 4x^2 - 9y^2$

69. $\left(x + \dfrac{1}{x}\right)\left(x - \dfrac{1}{x}\right) = x^2 - \left(\dfrac{1}{x}\right)^2 = x^2 - \dfrac{1}{x^2}$

71. $(x^2 + 3)(x^2 - 3) = (x^2)^2 - 3^2 = x^4 - 9$

73. $(2x - 3)(x^2 - 3x + 5)$
$= 2x(x^2 - 3x + 5) - 3(x^2 - 3x + 5)$
$= 2x^3 - 6x^2 + 10x - 3x^2 + 9x - 15$
$= 2x^3 - 9x^2 + 19x - 15$

75. $(1 + y)(1 - y + y^2)$
$= 1(1 - y + y^2) + y(1 - y + y^2)$
$= 1 - y + y^2 + y - y^2 + y^3 = y^3 + 1$

77. $(x - 6)(x^2 + 6x + 36)$
$= x(x^2 + 6x + 36) - 6(x^2 + 6x + 36)$
$= x^3 + 6x^2 + 36x - 6x^2 - 36x - 216$
$= x^3 - 216$

79. $(x+2y)(3x+5y) = 3x^2 + 5xy + 6xy + 10y^2 = 3x^2 + 11xy + 10y^2$

81. $(2x-y)(3x+7y) = 6x^2 + 14xy - 3xy - 7y^2 = 6x^2 + 11xy - 7y^2$

83. $(x-y)^2(x+y)^2 = (x-y)(x-y)(x+y)(x+y) = [(x-y)(x+y)][(x-y)(x+y)]$
$= (x^2-y^2)(x^2-y^2) = x^4 - 2x^2y^2 + y^4$

85. $(x+y)(x-2y)^2 = (x+y)(x^2 - 4xy + 4y^2) = x(x^2 - 4xy + 4y^2) + y(x^2 - 4xy + 4y^2)$
$= x^3 - 4x^2y + 4xy^2 + x^2y - 4xy^2 + 4y^3 = x^3 - 3x^2y + 4y^3$

87. $(x-2y)^3(x+2y) = (x^3 + 3x^2(-2y) + 3x(-2y)^2 + (-2y)^3) \cdot (x+2y)$
$= (x^3 - 6x^2y + 12xy^2 - 8y^3)(x+2y)$
$= x(x^3 - 6x^2y + 12xy^2 - 8y^3) + 2y(x^3 - 6x^2y + 12xy^2 - 8y^3)$
$= x^4 - 6x^3y + 12x^2y^2 - 8xy^3 + 2x^3y - 12x^2y^2 + 24xy^3 - 16y^4$
$= x^4 - 4x^3y + 16xy^3 - 16y^4$

For exercises 89–93, use $a^2 + b^2 = (a+b)^2 - 2ab = (a-b)^2 + 2ab$.

89. $x+y=4,\ xy=3$
$x^2 + y^2 = (x+y)^2 - 2xy = 4^2 - 2 \cdot 3 = 16 - 6 = 10$

91. a. $x + \dfrac{1}{x} = 3$

   Let $a = x$ and $b = \dfrac{1}{x}$. Then $ab = x\left(\dfrac{1}{x}\right) = 1$.

   $x^2 + \dfrac{1}{x^2} = \left(x + \dfrac{1}{x}\right)^2 - 2x\left(\dfrac{1}{x}\right) = 3^2 - 2 = 9 - 2 = 7$

   b. Let $a = x^2$ and $b = \dfrac{1}{x^2}$. Using the result from part a, we have

   $x^4 + \dfrac{1}{x^4} = \left(x^2 + \dfrac{1}{x^2}\right)^2 - 2x^2\left(\dfrac{1}{x^2}\right) = 7^2 - 2 = 49 - 2 = 47$

93. Let $a = 3x$ and $b = 2y$. Then $ab = 6xy$.
$9x^2 + 4y^2 = (3x+2y)^2 - 2(6xy)$
$= 12^2 - 2 \cdot 6 \cdot 6 = 144 - 72 = 72$

P.3 Applying the Concepts

95. $-0.025(6^2) + 0.44(6) + 4.28 = \$6.02$ in 2012 (six years after 2006)

97. $0.1(40^2) + 40 + 50 = \$250.00$

99. $d = 16(5^2) + 20(5) = 16(25) + 100 = 500$ feet

101. a. $-x + 22.50$

   b. $(30 + 10x)(22.50 - x) = 30(22.50) - 30x + 225x - 10x^2 = 675 + 195x - 10x^2 = -10x^2 + 195x + 675$

P.3 Beyond the Basics

103. $(a+b+c)^2 = (a+b+c)(a+b+c)$
$ = a(a+b+c) + b(a+b+c) + c(a+b+c)$
$ = a^2 + ab + ac + ab + b^2 + bc + ac + bc + c^2$
$ = a^2 + b^2 + c^2 + 2ab + 2bc + 2ac$

105. $(2x+y-z)^2 = (2x)^2 + y^2 + (-z)^2 + 2(2x)y + 2y(-z) + 2(2x)(-z)$
$ = 4x^2 + y^2 + z^2 + 4xy - 2yz - 4xz$
$(2x-y+z)^2 = (2x)^2 + (-y)^2 + z^2 + 2(2x)(-y) + 2(-y)z + 2(2x)z$
$ = 4x^2 + y^2 + z^2 - 4xy - 2yz + 4xz$
$(2x+y-z)^2 - (2x-y+z)^2 = (4x^2 + y^2 + z^2 + 4xy - 2yz - 4xz) - (4x^2 + y^2 + z^2 - 4xy - 2yz + 4xz)$
$ = 8xy - 8xz$

Alternatively, recognize the difference of two squares.
$(2x+y-z)^2 - (2x-y+z)^2 = [(2x+y-z) + (2x-y+z)][(2x+y-z) - (2x-y+z)]$
$ = (4x)(2y-2z) = 8xy - 8xz$

107. $x+y+z = 12$; $x^2 + y^2 + z^2 = 44$
$(x+y+z)^2 = x^2 + y^2 + z^2 + 2xy + 2yz + 2xz = x^2 + y^2 + z^2 + 2(xy + yz + xz)$
$12^2 = 44 + 2(xy + yz + xz)$
$144 = 44 + 2(xy + yz + xz)$
$100 = 2(xy + yz + xz)$
$50 = xy + yz + xz$

109. $(a+b+c)(a^2 + b^2 + c^2 - ab - bc - ca)$
$= a(a^2 + b^2 + c^2 - ab - bc - ca) + b(a^2 + b^2 + c^2 - ab - bc - ca) + c(a^2 + b^2 + c^2 - ab - bc - ca)$
$= a^3 + ab^2 + ac^2 - a^2b - abc - ca^2 + a^2b + b^3 + bc^2 - ab^2 - b^2c - abc + a^2c + b^2c + c^3 - abc - bc^2 - c^2a$
$= a^3 + b^3 + c^3 - 3abc$

111. Let $a = x - y$, $b = y - z$, and $c = z - x$. Then $a + b + c = (x-y) + (y-z) + (z-x) = 0$.

From exercise 110, if $a + b + c = 0$, then $a^3 + b^3 + c^3 = 3abc$. So,
$(x-y)^3 + (y-z)^3 + (z-x)^3 = 3(x-y)(y-z)(z-x)$.

113. From exercise 109, $(a+b+c)(a^2 + b^2 + c^2 - ab - bc - ca) = a^3 + b^3 + c^3 - 3abc$.

If $a + b + c = 8$ and $ab + bc + ca = 19$, we have
$(a+b+c)(a^2 + b^2 + c^2 - ab - bc - ca) = (a+b+c)(a^2 + b^2 + c^2 - (ab+bc+ca)) = a^3 + b^3 + c^3 - 3abc \Rightarrow$
$8(a^2 + b^2 + c^2 - 19) = a^3 + b^3 + c^3 - 3abc$ (1)

From exercise 103,
$(a+b+c)^2 = a^2 + b^2 + c^2 + 2ab + 2bc + 2ac = a^2 + b^2 + c^2 + 2(ab + bc + ca) \Rightarrow$
$8^2 = a^2 + b^2 + c^2 + 2(19) \Rightarrow 64 = a^2 + b^2 + c^2 + 38 \Rightarrow a^2 + b^2 + c^2 = 26$

Substituting into equation (1) gives us $8(26 - 19) = a^3 + b^3 + c^3 - 3abc \Rightarrow 56 = a^3 + b^3 + c^3 - 3abc$.

P.3 Getting Ready for the Next Section

	a	b	$a+b$	ab
115.	3	4	7	12
117.	4	2	6	8
119.	2	5	7	10
121.	−5	7	2	−35
123.	−2	−3	−5	6

P.4 Factoring Polynomials

P.4 Practice Problems

1. a. $6x^5 + 14x^3 = 2x^3(3x^2 + 7)$

   b. $7x^5 + 21x^4 + 35x^2 = 7x^2(x^3 + 3x^2 + 5)$

   c. $5x^2(x-y) + 2(x-y) = (x-y)(5x^2 + 2)$

2. a. $x^2 + 6x + 8 = (x+4)(x+2)$

   b. $x^2 - 3x - 10 = (x-5)(x+2)$

3. a. $x^2 + 4x + 4 = (x+2)^2$

   b. $9x^2 - 6x + 1 = (3x-1)^2$

4. a. $x^2 - 16 = (x-4)(x+4)$

   b. $4x^2 - 25 = (2x-5)(2x+5)$

5. $x^4 - 81 = (x^2 - 9)(x^2 + 9)$
   $= (x-3)(x+3)(x^2 + 9)$

6. a. $x^3 - 125 = (x-5)(x^2 + 5x + 25)$

   b. $27x^3 + 8 = (3x+2)(9x^2 - 6x + 4)$

7. Following the reasoning in Example 7, in four years, the company will have invested the initial 12 million dollars, plus an additional 4 million dollars. Thus, the total investment is 16 million dollars. To find the profit or loss with 16 million dollars invested, let $x = 16$ in the profit-loss polynomial:

   $(0.012)(x-14)(x^2 + 14x + 196)$
   $= (0.012)(16-14)(16^2 + 14 \cdot 16 + 196)$
   $= 16.224$

   The company will have made a profit of 16.224 million dollars in four years.

8. a. $x^3 + 3x^2 + x + 3 = x^2(x+3) + 1(x+3)$
   $= (x+3)(x^2 + 1)$

   b. $28x^3 - 20x^2 - 7x + 5$
   $= 4x^2(7x-5) - (7x-5)$
   $= (7x-5)(4x^2 - 1)$
   $= (7x-5)(2x+1)(2x-1)$

   c. $x^2 - y^2 - 2y - 1 = x^2 - (y^2 + 2y + 1)$
   $= x^2 - (y+1)^2$
   $= (x-(y+1))(x+(y+1))$
   $= (x-y-1)(x+y+1)$

9. $x^4 + 5x^2 + 9 = x^4 + 6x^2 - x^2 + 9$
   $= x^4 + 6x^2 + 9 - x^2$
   $= (x^2 + 3)^2 - x^2$
   $= [(x^2 + 3) + x][(x^2 + 3) - x]$
   $= (x^2 + x + 3)(x^2 - x + 3)$

10. a. $5x^2 + 11x + 2 = 5x^2 + 10x + x + 2$
    $= (5x^2 + 10x) + (x+2)$
    $= 5x(x+2) + 1(x+2)$
    $= (5x+1)(x+2)$

    b. $4x^2 + x - 3 = 4x^2 + 4x - 3x - 3$
    $= (4x^2 + 4x) - (3x+3)$
    $= 4x(x+1) - 3(x+1)$
    $= (4x-3)(x+1)$

P.4 Concepts and Vocabulary

1. The polynomials $x + 2$ and $x - 2$ are called <u>factors</u> of the polynomial $x^2 - 4$.

3. The GCF of the polynomial $10x^3 + 30x^2$ is <u>$10x^2$</u>.

5. True

7. False. The polynomial $x^2 - 4$ can be factored as $(x+2)(x-2)$. Therefore, it is not irreducible.

P.4 Building Skills

9. $8x - 24 = 8(x-3)$

11. $-6x^2 + 12x = -6x(x-2)$

13. $7x^2 + 14x^3 = 7x^2(1+2x)$

15. $x^4 + 2x^3 + x^2 = x^2(x^2 + 2x + 1)$

17. $3x^3 - x^2 = x^2(3x-1)$

19. $8ax^3 + 4ax^2 = 4ax^2(2x+1)$

21. $x(x-y) + 3(x-y) = (x+3)(x-y)$

23. $3x(2x+y) + 5y(2x+y) = (3x+5y)(2x+y)$

25. $(y+2)^2 - 3(y+2) = (y+2)(y+2-3)$
$= (y+2)(y-1)$

27. $x^3 + 3x^2 + x + 3 = x^2(x+3) + 1(x+3)$
$= (x+3)(x^2+1)$

29. $x^3 - 5x^2 + x - 5 = x^2(x-5) + 1(x-5)$
$= (x-5)(x^2+1)$

31. $6x^3 + 4x^2 + 3x + 2 = 2x^2(3x+2) + 1(3x+2)$
$= (3x+2)(2x^2+1)$

33. $12x^7 + 4x^5 + 3x^4 + x^2$
$= 4x^5(3x^2+1) + x^2(3x^2+1)$
$= (3x^2+1)(4x^5 + x^2)$
$= x^2(3x^2+1)(4x^3+1)$

35. $xy + ab + bx + ay = (xy + bx) + (ab + ay)$
$= x(y+b) + a(b+y)$
$= (x+a)(b+y)$

37. $x^2 + 7x + 12 = (x+3)(x+4)$

39. $x^2 - 6x + 8 = (x-4)(x-2)$

41. $x^2 - 3x - 4 = (x-4)(x+1)$

43. Irreducible

45. $2x^2 + x - 36 = (2x+9)(x-4)$

47. $6x^2 + 17x + 12 = (2x+3)(3x+4)$

49. $3x^2 - 11x - 4 = (3x+1)(x-4)$

51. Irreducible

53. $x^2 - xy - 20y^2 = (x+4y)(x-5y)$

55. $15x^2 - 28 + x = 15x^2 + x - 28$
$= (3x-4)(5x+7)$

57. $x^2 + 6x + 9 = (x+3)^2$

59. $9x^2 + 6x + 1 = (3x+1)^2$

61. $25x^2 - 20x + 4 = (5x-2)^2$

63. $49x^2 + 42x + 9 = (7x+3)^2$

65. $x^2 - 64 = (x-8)(x+8)$

67. $4x^2 - 1 = (2x-1)(2x+1)$

69. $16x^2 - 9 = (4x-3)(4x+3)$

71. $x^4 - 1 = (x^2-1)(x^2+1) = (x-1)(x+1)(x^2+1)$

73. $20x^4 - 5 = 5(4x^4 - 1) = 5(2x^2-1)(2x^2+1)$

75. $x^3 + 64 = x^3 + 4^3 = (x+4)(x^2 - 4x + 16)$

77. $x^3 - 27 = x^3 - 3^3 = (x-3)(x^2 + 3x + 9)$

79. $8 - x^3 = 2^3 - x^3 = (2-x)(4 + 2x + x^2)$

81. $8x^3 - 27 = (2x)^3 - 3^3 = (2x-3)(4x^2 + 6x + 9)$

83. $40x^3 + 5 = 5(8x^3+1) = 5((2x)^3 + 1^3)$
$= 5(2x+1)(4x^2 - 2x + 1)$

85. $x^4 + x^2 + 25 = x^4 + 10x^2 - 9x^2 + 25$
$= (x^4 + 10x^2 + 25) - 9x^2$
$= (x^2+5)^2 - 9x^2$
$= [(x^2+5) + 3x][(x^2+5) - 3x]$
$= (x^2 + 3x + 5)(x^2 - 3x + 5)$

87. $x^4 + 15x^2 + 64 = x^4 + 16x^2 - x^2 + 64$
$= (x^4 + 16x^2 + 64) - x^2$
$= (x^2 + 8)^2 - x^2$
$= [(x^2 + 8) + x][(x^2 + 8) - x]$
$= (x^2 + x + 8)(x^2 - x + 8)$

89. $x^4 - x^2 + 16 = x^4 + 8x^2 - 9x^2 + 16$
$= (x^4 + 8x^2 + 16) - 9x^2$
$= (x^2 + 4)^2 - 9x^2$
$= [(x^2 + 4) + 3x][(x^2 + 4) - 3x]$
$= (x^2 + 3x + 4)(x^2 - 3x + 4)$

91. $x^4 + 4 = x^4 + 4x^2 - 4x^2 + 4$
$= (x^4 + 4x^2 + 4) - 4x^2$
$= (x^2 + 2)^2 - 4x^2$
$= [(x^2 + 2) + 2x][(x^2 + 2) - 2x]$
$= (x^2 + 2x + 2)(x^2 - 2x + 2)$

93. $1 - 16x^2 = (1 + 4x)(1 - 4x)$

95. $x^2 - 6x + 9 = (x - 3)^2$

97. $4x^2 + 4x + 1 = (2x + 1)^2$

99. $2x^2 - 8x - 10 = 2(x^2 - 4x - 5)$
$= 2(x - 5)(x + 1)$

101. $2x^2 + 3x - 20 = (2x - 5)(x + 4)$

103. $x^2 - 24x + 36$ is irreducible.

105. $3x^5 + 12x^4 + 12x^3 = 3x^3(x^2 + 4x + 4)$
$= 3x^3(x + 2)^2$

107. $9x^2 - 1 = (3x - 1)(3x + 1)$

109. $16x^2 + 24x + 9 = (4x + 3)^2$

111. $x^2 + 15$ is irreducible.

113. $45x^3 + 8x^2 - 4x = x(45x^2 + 8x - 4)$
$= x(5x + 2)(9x - 2)$

115. $ax^2 - 7a^2x - 8a^3 = a(x^2 - 7ax - 8a^2)$
$= a(x - 8a)(x + a)$

117. $x^2 - 16a^2 + 8x + 16 = (x^2 + 8x + 16) - 16a^2$
$= (x + 4)^2 - 16a^2$
$= (x + 4 - 4a)(x + 4 + 4a)$

119. $3x^5 + 12x^4y + 12x^3y^2 = 3x^3(x^2 + 4xy + 4y^2)$
$= 3x^3(x + 2y)^2$

121. $x^2 - 25a^2 + 4x + 4 = (x^2 + 4x + 4) - 25a^2$
$= (x + 2)^2 - 25a^2$
$= (x + 2 - 5a)(x + 2 + 5a)$

123. $18x^6 + 12x^5y + 2x^4y^2 = 2x^4(9x^2 + 6xy + y^2)$
$= 2x^4(3x + y)^2$

P.4 Applying the Concepts

125.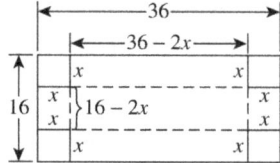

If one side of the garden is x feet, and the perimeter is 16 feet, then $\frac{16 - 2x}{2} = 8 - x$ gives the other dimension of the rectangle. So, the area of the garden is $x(8 - x)$.

Use the figure below for exercises 127 and 128.

127. If $x =$ the length of the cut corner, then $36 - 2x =$ the length of the box, and $16 - 2x =$ the width of the box. The height of the box is x. So,
$v = x(36 - 2x)(16 - 2x) = x(2(18 - x))(2(8 - x))$
$= 4x(18 - x)(8 - x)$

129. The area of the outside circle $= 4\pi \, \text{cm}^2$. The area of the inside circle $= \pi x^2 \, \text{cm}^2$. So the area of the disk = area of outside circle − area of inside circle $= 4\pi - \pi x^2 = \pi(4 - x^2) = \pi(2 - x)(2 + x) \, \text{cm}^2$.

131. If one side of the fence is x feet, and the rancher needs a total of 2800 feet of fencing, then the width of the fence is $2800 - 2x$. So, the area of the pen is $x(2800 - 2x) = 2800x - 2x^2 = 2x(1400 - x) \text{ ft}^2$.

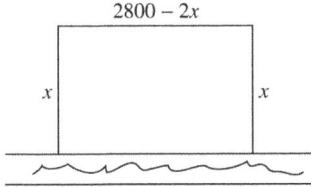

P.4 Beyond the Basics

133. $x^3 + y^3 + x + y = (x+y)(x^2 - xy + y^2) + x + y$
$= (x+y)(x^2 - xy + y^2 + 1)$

135. $x^4 + y^4 + x^2 y^2 = x^4 + 2x^2 y^2 + y^4 - x^2 y^2 = (x^2 + y^2)^2 - x^2 y^2$
$= (x^2 + y^2 - xy)(x^2 + y^2 + xy) = (x^2 - xy + y^2)(x^2 + xy + y^2)$

137. $x^6 - 64 = (x^3)^2 - 8^2 = (x^3 - 8)(x^3 + 8)$
$= (x - 2)(x + 2x + 4)(x + 2)(x - 2x + 4)$
$= (x - 2)(x + 2)(x + 2x + 4)(x - 2x + 4)$

139. $(x^2 - 3x)^2 - 38(x^2 - 3x) - 80 = \left[(x^2 - 3x) - 40\right]\left[(x^2 - 3x) + 2\right]$
$= (x^2 - 3x - 40)(x^2 - 3x + 2)$
$= (x - 8)(x + 5)(x - 2)(x - 1)$

141. From exercise 109 in Section P.3, we have $(a + b + c)(a^2 + b^2 + c^2 - ab - bc - ca) = a^3 + b^3 + c^3 - 3abc$.

Then, $a^3 + b^3 + c^3 - 3abc = (a + b + c)(a^2 + b^2 + c^2 - ab - bc - ca)$
$= \frac{1}{2}(a + b + c) \cdot 2(a^2 + b^2 + c^2 - ab - bc - ca)$
$= \frac{1}{2}(a + b + c)(2a^2 + 2b^2 + 2c^2 - 2ab - 2bc - 2ca)$
$= \frac{1}{2}(a + b + c)\left((a^2 - 2ab + b^2) + (b^2 - 2bc + c^2) + (c^2 - 2ca + a^2)\right)$
$= \frac{1}{2}(a + b + c)\left[(a - b)^2 + (b - c)^2 + (c - a)^2\right]$

If $a^3 + b^3 + c^3 - 3abc = 0$, then either $a + b + c = 0$ or $(a - b)^2 + (b - c)^2 + (c - a)^2 = 0 \Rightarrow a = b = c$.

143. Let $a = z(x - y) = xz - yz$, $b = x(y - z) = xy - xz$, and $c = y(z - x) = yz - xy$. Then
$a + b + c = xz - yz + xy - xz + yz - xy = 0$, and we can apply Section P.3 exercise 110.
$z^3(x - y)^3 + x^3(y - z)^3 + y^3(z - x)^3 = 3xyz(x - y)(y - z)(z - x)$

P.4 Getting Ready for the Next Section

145. $\dfrac{1}{4}+\dfrac{1}{6}=\dfrac{3}{12}+\dfrac{2}{12}=\dfrac{5}{12}$

147. $\dfrac{\cancel{8}^{1}}{\cancel{20}_{5}} \cdot \dfrac{\cancel{16}^{4}}{\cancel{27}_{3}} = \dfrac{4}{15}$

149. $\dfrac{2x}{4x+6}=\dfrac{2x}{2(2x+3)}=\dfrac{x}{2x+3}$

151. $\dfrac{x^2+3x+2}{x^2+4x+3}=\dfrac{(x+1)(x+2)}{(x+1)(x+3)}=\dfrac{x+2}{x+3}$

P.5 Rational Expressions

P.5 Practice Problems

1. $\dfrac{0.05(1-0.1)}{0.95(0.1)+(0.05)(1-0.1)} \approx 0.32$

The likelihood that a student who tests positive is a nonuser when the test that is used is 95% accurate is about 32%.

2. a. $\dfrac{2x^3+8x^2}{3x^2+12x}=\dfrac{2x^2(x+4)}{3x(x+4)}=\dfrac{2x}{3}$

b. $\dfrac{x^2-4}{x^2+4x+4}=\dfrac{(x-2)(x+2)}{(x+2)^2}=\dfrac{x-2}{x+2}$

3. $\dfrac{\dfrac{x^2-2x-3}{7x^3+28x^2}}{\dfrac{4x+4}{2x^4+8x^3}}=\dfrac{x^2-2x-3}{7x^3+28x^2} \div \dfrac{4x+4}{2x^4+8x^3}$

$=\dfrac{x^2-2x-3}{7x^3+28x^2} \cdot \dfrac{2x^4+8x^3}{4x+4}$

$=\dfrac{(x-3)(x+1)}{7x^2(x+4)} \cdot \dfrac{2x^3(x+4)}{4(x+1)}$

$=\dfrac{x(x-3)}{14}$

4. a. $\dfrac{5x+22}{x^2-36}+\dfrac{2(x+10)}{x^2-36}=\dfrac{5x+22+2x+20}{x^2-36}$

$=\dfrac{7x+42}{x^2-36}$

$=\dfrac{7(x+6)}{(x-6)(x+6)}=\dfrac{7}{x-6}$

b. $\dfrac{4x+1}{x^2+x-12}-\dfrac{3x+4}{x^2+x-12}=\dfrac{(4x+1)-(3x+4)}{x^2+x-12}$

$=\dfrac{x-3}{x^2+x-12}$

$=\dfrac{x-3}{(x+4)(x-3)}$

$=\dfrac{1}{x+4}$

5. a. $\dfrac{2x}{x+2}+\dfrac{3x}{x-5}$

$=\dfrac{2x(x-5)}{(x+2)(x-5)}+\dfrac{3x(x+2)}{(x+2)(x-5)}$

$=\dfrac{2x^2-10x+3x^2+6x}{(x+2)(x-5)}=\dfrac{5x^2-4x}{(x+2)(x-5)}$

$=\dfrac{x(5x-4)}{(x+2)(x-5)}$

b. $\dfrac{5x}{x-4}-\dfrac{2x}{x+3}$

$=\dfrac{5x(x+3)}{(x-4)(x+3)}-\dfrac{2x(x-4)}{(x-4)(x+3)}$

$=\dfrac{5x^2+15x-2x^2+8x}{(x-4)(x+3)}=\dfrac{3x^2+23x}{(x-4)(x+3)}$

$=\dfrac{x(3x+23)}{(x-4)(x+3)}$

6. a. $\dfrac{x^2+3x}{x^2(x+2)^2(x-2)}, \dfrac{4x^2+1}{3x(x-2)^2}$

LCD $=3x^2(x+2)^2(x-2)^2$

b. $\dfrac{2x-1}{x^2-25}, \dfrac{3-7x^2}{x^2+4x-5}$

$x^2-25=(x-5)(x+5)$

$x^2+4x-5=(x-1)(x+5)$

LCD $=(x-1)(x-5)(x+5)$

7. a. $\dfrac{4}{x^2-4x+4}+\dfrac{x}{x^2-4}$

$=\dfrac{4}{(x-2)^2}+\dfrac{x}{(x-2)(x+2)}$

$=\dfrac{4(x+2)}{(x-2)^2(x+2)}+\dfrac{x(x-2)}{(x-2)^2(x+2)}$

$=\dfrac{(4x+8)+(x^2-2x)}{(x-2)^2(x+2)}=\dfrac{x^2+2x+8}{(x-2)^2(x+2)}$

b. $\dfrac{2x}{3(x-5)^2} - \dfrac{6x}{2(x^2-5x)}$

$= \dfrac{2x}{3(x-5)^2} - \dfrac{6x}{2x(x-5)}$

$= \dfrac{2x(2x)}{3\cdot 2x(x-5)^2} - \dfrac{6x(3)(x-5)}{3\cdot 2x(x-5)^2}$

$= \dfrac{4x^2 - 18x^2 + 90x}{6x(x-5)^2} = \dfrac{-14x^2 + 90x}{6x(x-5)^2}$

$= \dfrac{2x(-7x+45)}{6x(x-5)^2} = \dfrac{-7x+45}{3(x-5)^2}$

8. $\dfrac{\dfrac{5}{3x} + \dfrac{1}{3}}{\dfrac{x^2-25}{3x}} = \dfrac{\left(\dfrac{5}{3x} + \dfrac{1}{3}\right)\cdot 3x}{\left(\dfrac{x^2-25}{3x}\right)\cdot 3x} = \dfrac{5+x}{x^2-25}$

$= \dfrac{x+5}{(x-5)(x+5)} = \dfrac{1}{x-5}$

9. $\dfrac{5x}{3x - \dfrac{4}{2 - \dfrac{1}{3}}} = \dfrac{5x}{3x - \dfrac{4}{\dfrac{5}{3}}} = \dfrac{5x}{3x - 4\left(\dfrac{3}{5}\right)} = \dfrac{5x}{3x - \dfrac{12}{5}}$

$= \dfrac{5x \cdot 5}{\left(3x - \dfrac{12}{5}\right)\cdot 5} = \dfrac{25x}{15x - 12}$

P.5 Concepts and Vocabulary

1. The least common denominator of two rational expressions is the polynomial of least degree that contains <u>each denominator</u> as a factor.

3. If the denominators of two rational expressions are $x^2 - 2x$ and $x^2 - x - 2$, then the LCD is $\underline{x(x-2)(x+1)}$.

5. False

7. False

P.5 Building Skills

9. $\dfrac{2x+2}{x^2+2x+1} = \dfrac{2(x+1)}{(x+1)^2} = \dfrac{2}{x+1}, x \neq -1$

11. $\dfrac{3x+3}{x^2-1} = \dfrac{3(x+1)}{(x-1)(x+1)} = \dfrac{3}{x-1}, x \neq -1, x \neq 1$

13. $\dfrac{2x-6}{9-x^2} = \dfrac{2(x-3)}{(3-x)(3+x)} = \dfrac{-2(3-x)}{(3-x)(3+x)}$

$= -\dfrac{2}{3+x}, x \neq -3, x \neq 3$

15. $\dfrac{2x-1}{1-2x} = \dfrac{-1(1-2x)}{1-2x} = -1, x \neq \dfrac{1}{2}$

17. $\dfrac{x^2-6x+9}{4x-12} = \dfrac{(x-3)^2}{4(x-3)} = \dfrac{x-3}{4}, x \neq 3$

19. $\dfrac{7x^2+7x}{x^2+2x+1} = \dfrac{7x(x+1)}{(x+1)^2} = \dfrac{7x}{x+1}, x \neq -1$

21. $\dfrac{x^2-11x+10}{x^2+6x-7} = \dfrac{(x-10)(x-1)}{(x+7)(x-1)}$

$= \dfrac{x-10}{x+7}, x \neq -7, x \neq 1$

23. $\dfrac{6x^4+14x^3+4x^2}{6x^4-10x^3-4x^2} = \dfrac{2x^2(3x^2+7x+2)}{2x^2(3x^2-5x-2)}$

$= \dfrac{2x^2(3x+1)(x+2)}{2x^2(3x+1)(x-2)}$

$= \dfrac{x+2}{x-2}, x \neq -\dfrac{1}{3}, x \neq 0, x \neq 2$

25. $\dfrac{x-3}{2x+4} \cdot \dfrac{10x+20}{5x-15} = \dfrac{x-3}{2(x+2)} \cdot \dfrac{10(x+2)}{5(x-3)} = 1$

27. $\dfrac{2x+6}{4x-8} \cdot \dfrac{x^2+x-6}{x^2-9} = \dfrac{2(x+3)}{4(x-2)} \cdot \dfrac{(x+3)(x-2)}{(x+3)(x-3)}$

$= \dfrac{x+3}{2(x-3)}$

29. $\dfrac{x^2-7x}{x^2-6x-7} \cdot \dfrac{x^2-1}{x^2}$

$= \dfrac{x(x-7)}{(x+1)(x-7)} \cdot \dfrac{(x+1)(x-1)}{x^2} = \dfrac{x-1}{x}$

31. $\dfrac{x^2-x-6}{x^2+3x+2} \cdot \dfrac{x^2-1}{x^2-9}$

$= \dfrac{(x-3)(x+2)}{(x+1)(x+2)} \cdot \dfrac{(x-1)(x+1)}{(x-3)(x+3)} = \dfrac{x-1}{x+3}$

33. $\dfrac{2-x}{x+1} \cdot \dfrac{x^2+3x+2}{x^2-4}$

$= \dfrac{-1(x-2)}{x+1} \cdot \dfrac{(x+1)(x+2)}{(x-2)(x+2)} = -1$

35. $\dfrac{x+2}{6} \div \dfrac{4x+8}{9} = \dfrac{x+2}{6} \cdot \dfrac{9}{4(x+2)} = \dfrac{3}{8}$

37. $\dfrac{x^2-9}{x} \div \dfrac{2x+6}{5x^2} = \dfrac{(x-3)(x+3)}{x} \cdot \dfrac{5x^2}{2(x+3)} = \dfrac{5x(x-3)}{2}$

39. $\dfrac{x^2+2x-3}{x^2+8x+16} \div \dfrac{x-1}{3x+12} = \dfrac{(x-1)(x+3)}{(x+4)^2} \cdot \dfrac{3(x+4)}{x-1} = \dfrac{3(x+3)}{x+4}$

41. $\left(\dfrac{x^2-9}{x^3+8} \div \dfrac{x+3}{x^3+2x^2-x-2} \right)\left(\dfrac{1}{x^2-1} \right) = \dfrac{(x-3)(x+3)}{(x+2)(x^2-2x+4)} \cdot \dfrac{(x^2-1)(x+2)}{x+3} \cdot \dfrac{1}{x^2-1} = \dfrac{x-3}{x^2-2x+4}$

43. $\dfrac{x}{5} + \dfrac{3}{5} = \dfrac{x+3}{5}$

45. $\dfrac{x}{2x+1} + \dfrac{4}{2x+1} = \dfrac{x+4}{2x+1}$

47. $\dfrac{x^2}{x+1} - \dfrac{x^2-1}{x+1} = \dfrac{x^2-(x^2-1)}{x+1} = \dfrac{1}{x+1}$

49. $\dfrac{4}{3-x} + \dfrac{2x}{x-3} = \dfrac{4}{-1(x-3)} + \dfrac{2x}{x-3} = \dfrac{-4+2x}{x-3} = \dfrac{2(x-2)}{x-3}$

51. $\dfrac{5x}{x^2+1} + \dfrac{2x}{x^2+1} = \dfrac{7x}{x^2+1}$

53. $\dfrac{7x}{2(x-3)} + \dfrac{x}{2(x-3)} = \dfrac{8x}{2(x-3)} = \dfrac{4x}{x-3}$

55. $\dfrac{x}{x^2-4} - \dfrac{2}{x^2-4} = \dfrac{x-2}{(x-2)(x+2)} = \dfrac{1}{x+2}$

57. $\dfrac{x-2}{2x+1} - \dfrac{x}{2x-1} = \dfrac{(x-2)(2x-1)}{(2x+1)(2x-1)} - \dfrac{x(2x+1)}{(2x+1)(2x-1)} = \dfrac{(2x^2-5x+2)-(2x^2+x)}{(2x+1)(2x-1)}$
$= \dfrac{-6x+2}{(2x+1)(2x-1)} = \dfrac{2(1-3x)}{(2x+1)(2x-1)}$

59. $\dfrac{-x}{x+2} + \dfrac{x-2}{x} - \dfrac{x}{x-2} = \dfrac{-x^2(x-2)}{x(x+2)(x-2)} + \dfrac{(x-2)^2(x+2)}{x(x+2)(x-2)} - \dfrac{x^2(x+2)}{x(x+2)(x-2)}$
$= \dfrac{(-x^3+2x^2)}{x(x+2)(x-2)} + \dfrac{(x^3-2x^2-4x+8)}{x(x+2)(x-2)} + \dfrac{-(x^3+2x^2)}{x(x+2)(x-2)}$
$= \dfrac{-x^3-2x^2-4x+8}{x(x+2)(x-2)} = -\dfrac{x^3+2x^2+4x-8}{x(x+2)(x-2)}$

In exercises 61–68, to find the LCD, first factor each denominator and then multiply each prime factor the greatest number of times it appears as a factor.

61. The denominators are $3x-6 = 3(x-2)$ and $4x-8 = 4(x-2) \Rightarrow$ LCD $= 3 \cdot 4(x-2) = 12(x-2)$

63. The denominators are $4x^2-1 = (2x+1)(2x-1)$ and $(2x+1)^2 = (2x+1)(2x+1) \Rightarrow$ LCD $= (2x-1)(2x+1)^2$

65. The denominators are $x^2+3x+2 = (x+1)(x+2)$ and $x^2-1 = (x-1)(x+1) \Rightarrow$ LCD $= (x-1)(x+1)(x+2)$

67. The denominators are $x^2-5x+4 = (x-1)(x-4)$ and
$x^2+x-2 = (x-1)(x+2) \Rightarrow$ LCD $= (x-1)(x-4)(x+2)$

Section P.5 Rational Expressions

69. $\dfrac{5}{x-3}+\dfrac{2x}{x^2-9}=\dfrac{5}{x-3}+\dfrac{2x}{(x-3)(x+3)}=\dfrac{5(x+3)}{(x-3)(x+3)}+\dfrac{2x}{(x-3)(x+3)}=\dfrac{5x+15+2x}{(x-3)(x+3)}=\dfrac{7x+15}{(x-3)(x+3)}$

71. $\dfrac{2x}{x^2-4}-\dfrac{x}{x+2}=\dfrac{2x}{(x-2)(x+2)}-\dfrac{x}{x+2}=\dfrac{2x}{(x-2)(x+2)}-\dfrac{x(x-2)}{(x+2)(x-2)}$

$\quad=\dfrac{2x-(x^2-2x)}{(x-2)(x+2)}=\dfrac{-x^2+4x}{(x-2)(x+2)}=\dfrac{x(4-x)}{(x-2)(x+2)}$

73. $\dfrac{x-2}{x^2+3x-10}+\dfrac{x+3}{x^2+x-6}=\dfrac{x-2}{(x-2)(x+5)}+\dfrac{x+3}{(x-2)(x+3)}=\dfrac{1}{x+5}+\dfrac{1}{x-2}$

$\quad=\dfrac{x-2}{(x+5)(x-2)}+\dfrac{x+5}{(x+5)(x-2)}=\dfrac{2x+3}{(x-2)(x+5)}$

75. $\dfrac{2x-3}{9x^2-1}+\dfrac{4x-1}{(3x-1)^2}=\dfrac{2x-3}{(3x-1)(3x+1)}+\dfrac{4x-1}{(3x-1)^2}=\dfrac{(2x-3)(3x-1)}{(3x+1)(3x-1)^2}+\dfrac{(4x-1)(3x+1)}{(3x+1)(3x-1)^2}$

$\quad=\dfrac{(6x^2-11x+3)+(12x^2+x-1)}{(3x+1)(3x-1)^2}=\dfrac{18x^2-10x+2}{(3x+1)(3x-1)^2}=\dfrac{2(9x^2-5x+1)}{(3x+1)(3x-1)^2}$

77. $\dfrac{x-3}{x^2-25}-\dfrac{x-3}{x^2+9x+20}=\dfrac{x-3}{(x-5)(x+5)}-\dfrac{x-3}{(x+4)(x+5)}=\dfrac{(x-3)(x+4)}{(x-5)(x+5)(x+4)}-\dfrac{(x-3)(x-5)}{(x+4)(x+5)(x-5)}$

$\quad=\dfrac{(x^2+x-12)-(x^2-8x+15)}{(x-5)(x+5)(x+4)}=\dfrac{9x-27}{(x-5)(x+5)(x+4)}=\dfrac{9(x-3)}{(x-5)(x+5)(x+4)}$

79. $\dfrac{3}{x^2-4}+\dfrac{1}{2-x}-\dfrac{1}{2+x}=-\dfrac{3}{4-x^2}+\dfrac{1}{2-x}-\dfrac{1}{2+x}=-\dfrac{3}{(2-x)(2+x)}+\dfrac{1}{2-x}-\dfrac{1}{2+x}$

$\quad=\dfrac{-3}{(2-x)(2+x)}+\dfrac{1(2+x)}{(2-x)(2+x)}-\dfrac{1(2-x)}{(2+x)(2-x)}=\dfrac{-3+(2+x)-(2-x)}{(2-x)(2+x)}=\dfrac{2x-3}{(2-x)(2+x)}$

81. $\dfrac{x+3a}{x-5a}-\dfrac{x+5a}{x-3a}=\dfrac{(x+3a)(x-3a)}{(x-5a)(x-3a)}-\dfrac{(x+5a)(x-5a)}{(x-3a)(x-5a)}=\dfrac{(x^2-9a^2)-(x^2-25a^2)}{(x-5a)(x-3a)}=\dfrac{16a^2}{(x-5a)(x-3a)}$

83. $\dfrac{1}{x+h}-\dfrac{1}{x}=\dfrac{x}{x(x+h)}-\dfrac{1(x+h)}{x(x+h)}$

$\quad=\dfrac{x-(x+h)}{x(x+h)}=-\dfrac{h}{x(x+h)}$

85. $\dfrac{\dfrac{2}{x}}{\dfrac{3}{x^2}}=\dfrac{2}{x}\cdot\dfrac{x^2}{3}=\dfrac{2x}{3}$

87. $\dfrac{\dfrac{1}{x}}{1-\dfrac{1}{x}}=\dfrac{1}{x}\div\left(1-\dfrac{1}{x}\right)=\dfrac{1}{x}\div\left(\dfrac{x-1}{x}\right)$

$\quad=\dfrac{1}{x}\cdot\dfrac{x}{x-1}=\dfrac{1}{x-1}$

We use Method 2 for exercises 89–91.

89. $\dfrac{\dfrac{1}{x}-1}{\dfrac{1}{x}+1}=\dfrac{\left(\dfrac{1}{x}-1\right)x}{\left(\dfrac{1}{x}+1\right)x}=\dfrac{1-x}{1+x}$

91. $\dfrac{\dfrac{1}{x}-x}{1-\dfrac{1}{x^2}}=\dfrac{\left(\dfrac{1}{x}-x\right)x^2}{\left(1-\dfrac{1}{x^2}\right)x^2}=\dfrac{x-x^3}{x^2-1}=\dfrac{x(1-x^2)}{x^2-1}$

$\quad=-x,\ x\neq 0$

93. $x - \dfrac{x}{x+\dfrac{1}{2}} = x - \dfrac{x}{\dfrac{2x+1}{2}} = x - \left(x \cdot \dfrac{2}{2x+1}\right) = x - \dfrac{2x}{2x+1} = \dfrac{x(2x+1)}{2x+1} - \dfrac{2x}{2x+1}$

$= \dfrac{2x^2 + x - 2x}{2x+1} = \dfrac{2x^2 - x}{2x+1} = \dfrac{x(2x-1)}{2x+1}$

95. $\dfrac{\dfrac{1}{x+h} - \dfrac{1}{x}}{h} = \left(\dfrac{1}{x+h} - \dfrac{1}{x}\right) \div h = \left(\dfrac{x}{x(x+h)} - \dfrac{x+h}{x(x+h)}\right) \div h$

$= \dfrac{x - (x+h)}{x(x+h)} \div h = \dfrac{-h}{x(x+h)} \div h = \dfrac{-h}{x(x+h)} \cdot \dfrac{1}{h} = -\dfrac{1}{x(x+h)}$

97. $\dfrac{\dfrac{1}{x-a} + \dfrac{1}{x+a}}{\dfrac{1}{x-a} - \dfrac{1}{x+a}} = \left(\dfrac{1}{x-a} + \dfrac{1}{x+a}\right) \div \left(\dfrac{1}{x-a} - \dfrac{1}{x+a}\right) = \dfrac{(x+a)+(x-a)}{(x-a)(x+a)} \div \dfrac{(x+a)-(x-a)}{(x-a)(x+a)}$

$= \dfrac{2x}{(x-a)(x+a)} \cdot \dfrac{(x-a)(x+a)}{2a} = \dfrac{x}{a}$

P.5 Applying the Concepts

99. $d = 4 \Rightarrow r = 2$. Substituting 2 for r in the formula gives $\dfrac{125.6}{(3.14)(2^2)} = \dfrac{125.6}{12.56} = 10$ cm.

101. a. Originally the citrus extract is 3 out of 100 gallons or $\dfrac{3}{100}$. When x gallons of water are added to the mixture, the total number of gallons in the mixture is $100 + x$. There are still 3 gallons of citrus extract, so the fraction is $\dfrac{3}{100+x}$.

b. If $x = 50$, then $\dfrac{3}{100+x} = \dfrac{3}{150} = 0.02 = 2\%$.

103. a. The volume of a cylinder is given by $V = 120 = \pi r^2 h \Rightarrow h = \dfrac{120}{\pi r^2}$. The cost of each base $= 0.05\pi x^2$; since there are two bases, the total cost of the bases is $0.10\pi x^2$.

The cost of the side is given by $0.01(2\pi rh) = 0.02\pi rh$. Since $r = x$ and $h = \dfrac{120}{\pi r^2}$, this gives

$0.02\pi x\left(\dfrac{120}{\pi x^2}\right) = \dfrac{2.4}{x}$. So the total cost for the container is $0.1\pi x^2 + \dfrac{2.4}{x} = \dfrac{0.1\pi x^3 + 2.4}{x}$.

b. $\dfrac{0.1(3.14)(2^3 + 240)}{2} \approx \$2.46$

P.5 Beyond the Basics

105. $\dfrac{(3x+4)^2 - (2x+3)^2}{(3x+4)^2} = \dfrac{(9x^2 + 24x + 16) - (4x^2 + 12x + 9)}{(3x+4)^2} = \dfrac{5x^2 + 12x + 7}{(3x+4)^2} = \dfrac{(5x+7)(x+1)}{(3x+4)^2}$

107. $\dfrac{(x^2-2x+1)(2x+2)-(x^2+2x+1)(2x-2)}{(x^2-2x+1)^2} = \dfrac{(x-1)^2(2)(x+1)-(x+1)^2(2)(x-1)}{\left[(x-1)^2\right]^2}$

$= \dfrac{2(x-1)(x+1)\left[(x-1)-(x+1)\right]}{(x-1)^4}$

$= \dfrac{2(x+1)(-2)}{(x-1)^3} = -\dfrac{4(x+1)}{(x-1)^3}$

109. $\dfrac{x^2-x-20}{x^2-25} \cdot \dfrac{x^2-x-2}{x^2+2x-8} \div \dfrac{x+1}{x^2+5x} = \dfrac{(x-5)(x+4)}{(x-5)(x+5)} \cdot \dfrac{(x-2)(x+1)}{(x+4)(x-2)} \cdot \dfrac{x(x+5)}{x+1} = x$

111. $\dfrac{1}{a-\dfrac{1}{a+\dfrac{1}{a}}} - \dfrac{1}{a+\dfrac{1}{a-\dfrac{1}{a}}} = \dfrac{1}{a-\dfrac{1}{\dfrac{a^2+1}{a}}} - \dfrac{1}{a+\dfrac{1}{\dfrac{a^2-1}{a}}} = \dfrac{1}{a-\dfrac{a}{a^2+1}} - \dfrac{1}{a+\dfrac{a}{a^2-1}}$

$= \dfrac{1}{\dfrac{a^3+a-a}{a^2+1}} - \dfrac{1}{\dfrac{a^3-a+a}{a^2-1}} = \dfrac{1}{\dfrac{a^3}{a^2+1}} - \dfrac{1}{\dfrac{a^3}{a^2-1}} = \dfrac{a^2+1}{a^3} - \dfrac{a^2-1}{a^3} = \dfrac{2}{a^3}$

113. $\left(\dfrac{x^2}{1-x^4} + \dfrac{2x^4}{1-x^8}\right) \div \dfrac{x^2+1}{x} = \left(\dfrac{x^2}{1-x^4} + \dfrac{2x^4}{(1-x^4)(1+x^4)}\right) \div \dfrac{x^2+1}{x} = \dfrac{x^2(1+x^4)+2x^4}{(1-x^4)(1+x^4)} \div \dfrac{x^2+1}{x}$

$= \dfrac{x^6+2x^4+x^2}{(1-x^2)(1+x^2)(1+x^4)} \cdot \dfrac{x}{x^2+1} = \dfrac{x^2(x^4+2x^2+1)}{(1-x^2)(1+x^2)(1+x^4)} \cdot \dfrac{x}{x^2+1}$

$= \dfrac{x^2(x^2+1)^2}{(1-x^2)(1+x^2)(1+x^4)} \cdot \dfrac{x}{x^2+1} = \dfrac{x^3}{(1-x^2)(1+x^4)} = -\dfrac{x^3}{(x^2-1)(1+x^4)}$

P.5 Getting Ready for the Next Section

115. True

117. False

$\sqrt{(-4)^2} = \sqrt{16} = 4$

119. True

P.6 Rational Exponents and Radicals

P.6 Practice Problems

1. a. $\sqrt{144} = \sqrt{12^2} = 12$

b. $\sqrt{\dfrac{1}{49}} = \sqrt{\left(\dfrac{1}{7}\right)^2} = \dfrac{1}{7}$

c. $\sqrt{\dfrac{4}{64}} = \sqrt{\dfrac{2^2}{8^2}} = \sqrt{\left(\dfrac{2}{8}\right)^2} = \dfrac{2}{8} = \dfrac{1}{4}$

2. a. $\sqrt{20} = \sqrt{4 \cdot 5} = \sqrt{4}\sqrt{5} = 2\sqrt{5}$

b. $\sqrt{6}\sqrt{8} = \sqrt{6 \cdot 8} = \sqrt{48} = \sqrt{16 \cdot 3}$
$= \sqrt{16}\sqrt{3} = 4\sqrt{3}$

c. $\sqrt{12x^2} = \sqrt{4 \cdot 3 \cdot x^2} = \sqrt{4}\sqrt{3}\sqrt{x^2} = 2|x|\sqrt{3}$

d. $\sqrt{\dfrac{20y^3}{27x^2}} = \dfrac{\sqrt{20y^3}}{\sqrt{27x^2}} = \dfrac{\sqrt{4 \cdot 5 \cdot y^2 \cdot y}}{\sqrt{9 \cdot 3 \cdot x^2}}$

$= \dfrac{\sqrt{4}\sqrt{5}\sqrt{y^2}\sqrt{y}}{\sqrt{9}\sqrt{3}\sqrt{x^2}} = \dfrac{2y\sqrt{5y}}{3|x|\sqrt{3}}$

3. Using Bernoulli's equation, the velocity of the water is

$v = \sqrt{2gh} \approx \sqrt{2(10\,\text{m/s}^2)(45\,\text{m})}$
$= \sqrt{900\,\text{m}^2/\text{s}^2} = 30\,\text{m/s}$

$30\,\text{m/s} \times 60\,\text{s/min} \times 60\,\text{min/hr} = 108{,}000\,\text{m/hr}$
$108\,\text{km/hr} \times 0.6214\,\text{mi/km} \approx 67\,\text{mi/hr}$

4. a. $\sqrt[3]{-8} = -2$ b. $\sqrt[5]{32} = 2$

   c. $\sqrt[4]{81} = 3$

   d. $\sqrt[6]{-4}$ is not a real number

5. a. $\sqrt[3]{72} = \sqrt[3]{8 \cdot 9} = \sqrt[3]{8} \cdot \sqrt[3]{9} = 2\sqrt[3]{9}$

   b. $\sqrt[4]{48a^2} = \sqrt[4]{16 \cdot 3a^2} = 2\sqrt[4]{3a^2}$

   c. $\sqrt[3]{\sqrt{64}} = \sqrt[3]{8} = 2$

6. a. $3\sqrt{12} + 7\sqrt{3} = 3\sqrt{4 \cdot 3} + 7\sqrt{3}$
   $= 6\sqrt{3} + 7\sqrt{3} = 13\sqrt{3}$

   b. $2\sqrt[3]{135x} - 3\sqrt[3]{40x} = 2\sqrt[3]{27 \cdot 5x} - 3\sqrt[3]{8 \cdot 5x}$
   $= 2 \cdot 3\sqrt[3]{5x} - 3 \cdot 2\sqrt[3]{5x}$
   $= 6\sqrt[3]{5x} - 6\sqrt[3]{5x} = 0$

7. a. $\sqrt{3}\sqrt[5]{2} = \sqrt[2]{3}\sqrt[5]{2} = \sqrt[2 \cdot 5]{3^5} \cdot \sqrt[5 \cdot 2]{2^2} = \sqrt[10]{3^5}\sqrt[10]{2^2}$
   $= \sqrt[10]{2^2 \cdot 3^5} = \sqrt[10]{972}$

   b. The least common multiple of the indices, 6 and 8, is 24.
   $\sqrt[6]{x^5} = \sqrt[6 \cdot 4]{x^{5 \cdot 4}} = \sqrt[24]{x^{20}}$
   $\sqrt[8]{y^7} = \sqrt[8 \cdot 3]{y^{7 \cdot 3}} = \sqrt[24]{y^{21}}$

8. a. $\dfrac{7}{\sqrt{8}} = \dfrac{7}{2\sqrt{2}} = \dfrac{7\sqrt{2}}{2\sqrt{2}\sqrt{2}} = \dfrac{7\sqrt{2}}{2 \cdot 2} = \dfrac{7\sqrt{2}}{4}$

   b. $\dfrac{\sqrt[3]{3}}{\sqrt[3]{4}} = \dfrac{\sqrt[3]{3} \cdot \sqrt[3]{2}}{\sqrt[3]{4} \cdot \sqrt[3]{2}} = \dfrac{\sqrt[3]{6}}{\sqrt[3]{8}} = \dfrac{\sqrt[3]{6}}{2}$

9. a. $\dfrac{2}{\sqrt{7} - \sqrt{5}} = \dfrac{2(\sqrt{7} + \sqrt{5})}{(\sqrt{7} - \sqrt{5})(\sqrt{7} + \sqrt{5})}$
   $= \dfrac{2(\sqrt{7} + \sqrt{5})}{7 - 5} = \dfrac{2(\sqrt{7} + \sqrt{5})}{2}$
   $= \sqrt{7} + \sqrt{5}$

   b. $\dfrac{x - 2}{\sqrt{x} + \sqrt{2}} = \dfrac{(x - 2)(\sqrt{x} - \sqrt{2})}{(\sqrt{x} + \sqrt{2})(\sqrt{x} - \sqrt{2})}$
   $= \dfrac{(x - 2)(\sqrt{x} - \sqrt{2})}{x - 2}$
   $= \sqrt{x} - \sqrt{2}$

10. a. $\left(\dfrac{1}{4}\right)^{1/2} = \sqrt{\dfrac{1}{4}} = \dfrac{1}{2}$

    b. $(125)^{1/3} = \sqrt[3]{125} = 5$

    c. $(-32)^{1/5} = \sqrt[5]{-32} = -2$

11. a. $(25)^{2/3} = \sqrt[3]{25^2} = \sqrt[3]{5^4} = 5\sqrt[3]{5}$

    b. $-36^{3/2} = -\left(\sqrt{36}\right)^3 = -(6)^3 = -216$

    c. $16^{-5/2} = \dfrac{1}{\left(\sqrt{16}\right)^5} = \dfrac{1}{4^5} = \dfrac{1}{1024}$

    d. $(-36)^{-1/2} = \dfrac{1}{\sqrt{-36}}$
       Not a real number

12. a. $4x^{1/2} \cdot 3x^{1/5} = 12x^{1/2 + 1/5} = 12x^{7/10}$

    b. $\dfrac{25x^{-1/4}}{5x^{1/3}} = 5x^{-1/4 - 1/3} = 5x^{-7/12} = \dfrac{5}{x^{7/12}}$

    c. $\left(x^{2/3}\right)^{-1/5} = x^{-2/15} = \dfrac{1}{x^{2/15}}$

13. a. $\sqrt[6]{x^4} = x^{4/6} = x^{2/3} = \sqrt[3]{x^2}$

    b. $\sqrt[4]{25}\sqrt{5} = \sqrt[4]{5^2}\sqrt{5} = 5^{2/4} \cdot 5^{1/2} = 5^{1/2} \cdot 5^{1/2} = 5$

    c. $\sqrt{\sqrt[3]{x^{12}}} = \sqrt{x^{12/3}} = \sqrt{x^4} = \left(x^4\right)^{1/2} = x^2$

14. $x(x+3)^{-1/2} + (x+3)^{1/2}$
    $= \dfrac{x}{(x+3)^{1/2}} + (x+3)^{1/2}$
    $= \dfrac{x}{(x+3)^{1/2}} + \dfrac{(x+3)^{1/2}(x+3)^{1/2}}{(x+3)^{1/2}}$
    $= \dfrac{x + (x+3)}{(x+3)^{1/2}} = \dfrac{2x+3}{(x+3)^{1/2}}$
    $= \dfrac{(2x+3)(x+3)^{1/2}}{(x+3)^{1/2}(x+3)^{1/2}} = \dfrac{(2x+3)(x+3)^{1/2}}{x+3}$

P.6 Concepts and Vocabulary

1. Any positive number has <u>two</u> square roots.

3. Radicals that have the same index and the same radicand are called <u>like radicals</u>.

5. False. For all real x, $\sqrt{x^2} = |x|$.

7. True

Section P.6 Rational Exponents and Radicals 23

P.6 Building Skills

9. $\sqrt{64} = \sqrt{8^2} = 8$

11. $\sqrt[3]{64} = \sqrt[3]{4^3} = 4$

13. $\sqrt[3]{-27} = \sqrt[3]{(-3)^3} = -3$

15. $\sqrt[3]{-\dfrac{1}{8}} = \sqrt[3]{\left(-\dfrac{1}{2}\right)^3} = -\dfrac{1}{2}$

17. $\sqrt{(-3)^2} = \sqrt{9} = 3$

19. There is no real number x such that $x^4 = -16$.

21. $-\sqrt[5]{-1} = -\sqrt[5]{(-1)^5} = -(-1) = 1$

23. $\sqrt[5]{(-7)^5} = -7$

25. $\sqrt{32} = \sqrt{16 \cdot 2} = \sqrt{16}\sqrt{2} = 4\sqrt{2}$

27. $\sqrt{18x^2} = \sqrt{9 \cdot 2x^2} = \sqrt{9}\sqrt{2}\sqrt{x^2} = 3x\sqrt{2}$

29. $\sqrt{9x^3} = \sqrt{9 \cdot x^2 \cdot x} = \sqrt{9}\sqrt{x^2}\sqrt{x} = 3x\sqrt{x}$

31. $\sqrt{6x}\sqrt{3x} = \sqrt{18x^2} = \sqrt{9 \cdot 2 \cdot x^2}$
$= \sqrt{9}\sqrt{2}\sqrt{x^2} = 3x\sqrt{2}$

33. $\sqrt{15x}\sqrt{3x^2} = \sqrt{45x^3} = \sqrt{9 \cdot 5 \cdot x^2 \cdot x}$
$= \sqrt{9}\sqrt{5}\sqrt{x^2}\sqrt{x} = 3x\sqrt{5x}$

35. $\sqrt[3]{-8x^3} = \sqrt[3]{-8}\sqrt[3]{x^3} = -2x$

37. $\sqrt[3]{-x^6} = -x^2$

38. $\sqrt[3]{-27x^6} = -3x^2$

39. $\sqrt{\dfrac{5}{32}} = \dfrac{\sqrt{5}}{\sqrt{32}} = \dfrac{\sqrt{5}}{\sqrt{16 \cdot 2}} = \dfrac{\sqrt{5}}{4\sqrt{2}}$
$= \dfrac{\sqrt{5}\sqrt{2}}{4\sqrt{2}\sqrt{2}} = \dfrac{\sqrt{10}}{8}$

41. $\sqrt[3]{\dfrac{8}{x^3}} = \dfrac{\sqrt[3]{8}}{\sqrt[3]{x^3}} = \dfrac{2}{x}$

43. $\sqrt[4]{\dfrac{2}{x^5}} = \dfrac{\sqrt[4]{2}}{\sqrt[4]{x^5}} = \dfrac{\sqrt[4]{2}}{\sqrt[4]{x^4 \cdot x}} = \dfrac{\sqrt[4]{2}}{\sqrt[4]{x^4}\sqrt[4]{x}} = \dfrac{\sqrt[4]{2}}{x\sqrt[4]{x}}$
$= \dfrac{\sqrt[4]{2}\sqrt[4]{x^3}}{x\sqrt[4]{x}\sqrt[4]{x^3}} = \dfrac{\sqrt[4]{2x^3}}{x\sqrt[4]{x^4}} = \dfrac{\sqrt[4]{2x^3}}{x^2}$

45. $\sqrt{\sqrt{x^8}} = \sqrt{x^4} = x^2$

47. $\sqrt{\dfrac{9x^6}{y^4}} = \dfrac{\sqrt{9x^6}}{\sqrt{y^4}} = \dfrac{3x^3}{y^2}$

49. $2\sqrt{3} + 5\sqrt{3} = 7\sqrt{3}$

51. $6\sqrt{5} - \sqrt{5} + 4\sqrt{5} = 9\sqrt{5}$

53. $\sqrt{98x} - \sqrt{32x} = 7\sqrt{2x} - 4\sqrt{2x} = 3\sqrt{2x}$

55. $\sqrt[3]{24} - \sqrt[3]{81} = \sqrt[3]{8 \cdot 3} - \sqrt[3]{27 \cdot 3}$
$= 2\sqrt[3]{3} - 3\sqrt[3]{3} = -\sqrt[3]{3}$

57. $\sqrt[3]{3x} - 2\sqrt[3]{24x} + \sqrt[3]{375x}$
$= \sqrt[3]{3x} - 2\sqrt[3]{8 \cdot 3x} + \sqrt[3]{125 \cdot 3x}$
$= \sqrt[3]{3x} - 2\sqrt[3]{8}\sqrt[3]{3x} + \sqrt[3]{125} \cdot \sqrt[3]{3x}$
$= \sqrt[3]{3x} - 2 \cdot 2\sqrt[3]{3x} + 5\sqrt[3]{3x}$
$= \sqrt[3]{3x} - 4\sqrt[3]{3x} + 5\sqrt[3]{3x} = 2\sqrt[3]{3x}$

59. $\sqrt{2x^5} - 5\sqrt{32x} + \sqrt{18x^3}$
$= \sqrt{2x^4 \cdot x} - 5\sqrt{16 \cdot 2x} + \sqrt{9 \cdot 2x^2 \cdot x}$
$= x^2\sqrt{2x} - 5 \cdot 4\sqrt{2x} + 3x\sqrt{2x}$
$= x^2\sqrt{2x} - 20\sqrt{2x} + 3x\sqrt{2x}$
$= \sqrt{2x}\left(x^2 + 3x - 20\right)$

61. $\sqrt{48x^5y} - 4y\sqrt{3x^3y} + y\sqrt{3xy^3}$
$= \sqrt{16 \cdot 3x^4xy} - 4y\sqrt{3x^2xy} + y\sqrt{3xy^2y}$
$= 4x^2\sqrt{3xy} - 4xy\sqrt{3xy} + y^2\sqrt{3xy}$
$= \sqrt{3xy}\left(4x^2 - 4xy + y^2\right) = \sqrt{3xy}(2x-y)^2$

63. $\dfrac{2}{\sqrt{3}} = \dfrac{2\sqrt{3}}{\sqrt{3}\sqrt{3}} = \dfrac{2\sqrt{3}}{3}$

65. $\dfrac{7}{\sqrt{15}} = \dfrac{7\sqrt{15}}{\sqrt{15}\sqrt{15}} = \dfrac{7\sqrt{15}}{15}$

67. $\dfrac{1}{\sqrt{2}+x} = \dfrac{1(\sqrt{2}-x)}{(\sqrt{2}+x)(\sqrt{2}-x)} = \dfrac{\sqrt{2}-x}{2-x^2}$

69. $\dfrac{3}{2-\sqrt{3}} = \dfrac{3(2+\sqrt{3})}{(2-\sqrt{3})(2+\sqrt{3})} = \dfrac{3(2+\sqrt{3})}{2^2-\sqrt{3}^2}$
$= \dfrac{3(2+\sqrt{3})}{4-3} = 3(2+\sqrt{3})$

Copyright © 2019 Pearson Education Inc.

71. $\dfrac{1}{\sqrt{3}+\sqrt{2}} = \dfrac{1(\sqrt{3}-\sqrt{2})}{(\sqrt{3}+\sqrt{2})(\sqrt{3}-\sqrt{2})} = \dfrac{\sqrt{3}-\sqrt{2}}{\sqrt{3}^2-\sqrt{2}^2} = \dfrac{\sqrt{3}-\sqrt{2}}{3-2} = \sqrt{3}-\sqrt{2}$

73. $\dfrac{\sqrt{5}-\sqrt{2}}{\sqrt{5}+\sqrt{2}} = \dfrac{(\sqrt{5}-\sqrt{2})(\sqrt{5}-\sqrt{2})}{(\sqrt{5}+\sqrt{2})(\sqrt{5}-\sqrt{2})} = \dfrac{\sqrt{5}^2-2\sqrt{10}+\sqrt{2}^2}{\sqrt{5}^2-\sqrt{2}^2} = \dfrac{7-2\sqrt{10}}{3}$

75. $\dfrac{\sqrt{x+h}-\sqrt{x}}{\sqrt{x+h}+\sqrt{x}} = \dfrac{(\sqrt{x+h}-\sqrt{x})(\sqrt{x+h}-\sqrt{x})}{(\sqrt{x+h}+\sqrt{x})(\sqrt{x+h}-\sqrt{x})} = \dfrac{(\sqrt{x+h})^2-2\sqrt{x}\sqrt{x+h}+\sqrt{x}^2}{(\sqrt{x+h})^2-\sqrt{x}^2}$

$= \dfrac{x+h-2\sqrt{x(x+h)}+x}{x+h-x} = \dfrac{2x+h-2\sqrt{x(x+h)}}{h}$

77. $\dfrac{\sqrt{4+h}-2}{h} = \dfrac{(\sqrt{4+h}-2)(\sqrt{4+h}+2)}{h(\sqrt{4+h}+2)} = \dfrac{4+h-4}{h(\sqrt{4+h}+2)} = \dfrac{h}{h(\sqrt{4+h}+2)} = \dfrac{1}{\sqrt{4+h}+2}$

79. $\dfrac{2-\sqrt{4-x}}{x} = \dfrac{(2-\sqrt{4-x})(2+\sqrt{4-x})}{x(2+\sqrt{4-x})} = \dfrac{4-(4-x)}{x(2+\sqrt{4-x})} = \dfrac{x}{x(2+\sqrt{4-x})} = \dfrac{1}{2+\sqrt{4-x}}$

81. $\dfrac{\sqrt{x}-2}{x-4} = \dfrac{(\sqrt{x}-2)(\sqrt{x}+2)}{(x-4)(\sqrt{x}+2)} = \dfrac{x-4}{(x-4)(\sqrt{x}+2)} = \dfrac{1}{\sqrt{x}+2}$

83. $\dfrac{\sqrt{x^2+4x}-x}{2} = \dfrac{(\sqrt{x^2+4x}-x)(\sqrt{x^2+4x}+x)}{2(\sqrt{x^2+4x}+x)} = \dfrac{x^2+4x-x^2}{2(\sqrt{x^2+4x}+x)} = \dfrac{4x}{2(\sqrt{x^2+4x}+x)} = \dfrac{2x}{\sqrt{x^2+4x}+x}$

85. $25^{1/2} = \sqrt{25} = 5$

87. $(-8)^{1/3} = \sqrt[3]{-8} = -2$

89. $8^{2/3} = \sqrt[3]{8^2} = \sqrt[3]{(2^3)^2} = \sqrt[3]{2^6} = 2^2 = 4$

91. $-25^{-3/2} = -\dfrac{1}{25^{3/2}} = -\dfrac{1}{\sqrt{25^3}} = -\dfrac{1}{\sqrt{(5^2)^3}}$

$= -\dfrac{1}{\sqrt{5^6}} = -\dfrac{1}{5^3} = -\dfrac{1}{125}$

93. $\left(\dfrac{9}{25}\right)^{-3/2} = \left(\dfrac{25}{9}\right)^{3/2} = \sqrt{\left(\dfrac{25}{9}\right)^3} = \dfrac{\sqrt{25^3}}{\sqrt{9^3}}$

$= \dfrac{\sqrt{(5^2)^3}}{\sqrt{(3^2)^3}} = \dfrac{\sqrt{5^6}}{\sqrt{3^6}} = \dfrac{5^3}{3^3} = \dfrac{125}{27}$

95. $x^{1/2} \cdot x^{2/5} = x^{1/2+2/5} = x^{9/10}$

97. $x^{3/5} \cdot x^{-1/2} = x^{3/5-1/2} = x^{1/10}$

99. $(8x^6)^{2/3} = 8^{2/3} x^{6(2/3)} = (2^3)^{2/3} x^{6(2/3)}$

$= 2^2 x^4 = 4x^4$

101. $(27x^6 y^3)^{-2/3} = 27^{-2/3} x^{6(-2/3)} y^{3(-2/3)}$

$= 3^{3(-2/3)} x^{6(-2/3)} y^{3(-2/3)}$

$= 3^{-2} x^{-4} y^{-2}$

$= \dfrac{1}{3^2 x^4 y^2} = \dfrac{1}{9x^4 y^2}$

103. $\dfrac{15x^{3/2}}{3x^{1/4}} = 5x^{3/2-1/4} = 5x^{5/4}$

105. $\left(\dfrac{x^{-1/4}}{y^{-2/3}}\right)^{-12} = \dfrac{x^{(-1/4)(-12)}}{y^{(-2/3)(-12)}} = \dfrac{x^3}{y^8}$

107. $\sqrt[4]{3^2} = 3^{2/4} = 3^{1/2}$

109. $\sqrt[3]{x^9} = x^{9/3} = x^3$

111. $\sqrt[3]{x^6 y^9} = x^{6/3} y^{9/3} = x^2 y^3$

113. $\sqrt[4]{9}\sqrt{3} = \sqrt[4]{3^2}\sqrt{3} = 3^{2/4} \cdot 3^{1/2}$

$= 3^{1/2} \cdot 3^{1/2} = 3^{1/2+1/2} = 3^1 = 3$

115. $\sqrt{\sqrt[3]{x^{10}}} = \sqrt{x^{10/3}} = \left(x^{10/3}\right)^{1/2} = x^{(10/3)(1/2)} = x^{5/3}$

117. $\sqrt[3]{2} \cdot \sqrt[5]{3} = \sqrt[3 \cdot 5]{2^5} \cdot \sqrt[3 \cdot 5]{3^3} = \sqrt[15]{32} \cdot \sqrt[15]{27} = \sqrt[15]{32 \cdot 27} = \sqrt[15]{864}$

119. $\sqrt[3]{x^2} \cdot \sqrt[4]{x^3} = \sqrt[3 \cdot 4]{\left(x^2\right)^4} \cdot \sqrt[3 \cdot 4]{\left(x^3\right)^3} = \sqrt[12]{x^8} \cdot \sqrt[12]{x^9} = \sqrt[12]{x^8 \cdot x^9} = \sqrt[12]{x^{17}} = \sqrt[12]{x^{12} \cdot x^5} = \sqrt[12]{x^{12}} \cdot \sqrt[12]{x^5} = x\sqrt[12]{x^5}$

121. $\sqrt[9]{2m^2n} \cdot \sqrt[3]{5m^5n^2} = \sqrt[9]{2m^2n} \cdot \sqrt[3 \cdot 3]{\left(5m^5n^2\right)^3} = \sqrt[9]{2m^2n} \cdot \sqrt[9]{5^3\left(m^5\right)^3\left(n^2\right)^3} = \sqrt[9]{2m^2n} \cdot \sqrt[9]{125m^{15}n^6}$
$= \sqrt[9]{2m^2n \cdot 125m^{15}n^6} = \sqrt[9]{250m^{2+15}n^{1+6}} = \sqrt[9]{250m^{17}n^7} = m\sqrt[9]{250m^8n^7}$

123. $\sqrt[5]{x^4y^3} \cdot \sqrt[6]{x^3y^5} = \sqrt[5 \cdot 6]{\left(x^4y^3\right)^6} \cdot \sqrt[5 \cdot 6]{\left(x^3y^5\right)^5} = \sqrt[30]{\left(x^4\right)^6\left(y^3\right)^6} \cdot \sqrt[30]{\left(x^3\right)^5\left(y^5\right)^5} = \sqrt[30]{x^{24}y^{18}} \cdot \sqrt[30]{x^{15}y^{25}}$
$= \sqrt[30]{x^{24}y^{18} \cdot x^{15}y^{25}} = \sqrt[30]{x^{24+15}y^{18+25}} = \sqrt[30]{x^{39}y^{43}} = \sqrt[30]{x^{30+9}y^{30+13}}$
$= \sqrt[30]{x^{30}x^9y^{30}y^{13}} = xy\sqrt[30]{x^9y^{13}}$

125. $\dfrac{4}{3}x^{1/3}(2x-3) + 2x^{4/3} = 2x^{1/3}\left(\dfrac{2}{3}(2x-3) + x\right) = 2x^{1/3}\left(\dfrac{4x}{3} - 2 + x\right) = 2x^{1/3}\left(\dfrac{7x}{3} - 2\right) = \dfrac{2}{3}x^{1/3}(7x-6)$

127. $4(3x+1)^{1/3}(2x-1) + 2(3x+1)^{4/3} = 2(3x+1)^{1/3}\left(2(2x-1) + (3x+1)\right) = 2(3x+1)^{1/3}(4x-2+3x+1)$
$= 2(3x+1)^{1/3}(7x-1)$

129. $3x\left(x^2+1\right)^{1/2}\left(2x^2-x\right) + 2\left(x^2+1\right)^{3/2}(4x-1) = \left(x^2+1\right)^{1/2}\left[3x\left(2x^2-x\right) + 2\left(x^2+1\right)(4x-1)\right]$
$= \left(x^2+1\right)^{1/2}\left(6x^3 - 3x^2 + 8x^3 - 2x^2 + 8x - 2\right)$
$= \left(x^2+1\right)^{1/2}\left(14x^3 - 5x^2 + 8x - 2\right)$

P.6 Applying the Concepts

131. Substituting 692 for A in $\sqrt{\dfrac{4A}{\sqrt{3}}}$ gives $\sqrt{\dfrac{4(692)}{1.73}} = \sqrt{\dfrac{2768}{1.73}} = \sqrt{1600} = 40$ cm.

133. Substituting 6 for w in $V = \sqrt{\dfrac{w}{1.5}}$ gives $V = \sqrt{\dfrac{6}{1.5}} = \sqrt{4} = 2$ cm/sec.

135. Substituting 1058 for W and 2 for R in $I = \sqrt{\dfrac{W}{R}}$ gives $I = \sqrt{\dfrac{1058}{2}} = \sqrt{529} = 23$ amp.

137. Substituting 16.25 for h in $P = 14.7(0.5)^{h/3.25}$ gives $P = 14.7(0.5)^{16.25/3.25} = 14.7(0.5)^5 = 14.7(0.03125) \approx 0.46$ lb/sq in.

P.6 Beyond the Basics

139. $\dfrac{1}{3-\sqrt{8}} - \dfrac{1}{\sqrt{8}-\sqrt{7}} + \dfrac{1}{\sqrt{7}-\sqrt{6}} - \dfrac{1}{\sqrt{6}-\sqrt{5}} + \dfrac{1}{\sqrt{5}-2}$

$= \dfrac{1(3+\sqrt{8})}{(3-\sqrt{8})(3+\sqrt{8})} - \dfrac{1(\sqrt{8}+\sqrt{7})}{(\sqrt{8}-\sqrt{7})(\sqrt{8}+\sqrt{7})} + \dfrac{1(\sqrt{7}+\sqrt{6})}{(\sqrt{7}-\sqrt{6})(\sqrt{7}+\sqrt{6})} - \dfrac{1(\sqrt{6}+\sqrt{5})}{(\sqrt{6}-\sqrt{5})(\sqrt{6}+\sqrt{5})} + \dfrac{1(\sqrt{5}+2)}{(\sqrt{5}-2)(\sqrt{5}+2)}$

$= \dfrac{3+\sqrt{8}}{9-8} - \dfrac{\sqrt{8}+\sqrt{7}}{8-7} + \dfrac{\sqrt{7}+\sqrt{6}}{7-6} - \dfrac{\sqrt{6}+\sqrt{5}}{6-5} + \dfrac{\sqrt{5}+2}{5-4}$

$= 3+\sqrt{8} - (\sqrt{8}+\sqrt{7}) + (\sqrt{7}+\sqrt{6}) - (\sqrt{6}+\sqrt{5}) + (\sqrt{5}+2)$

$= 3+\sqrt{8}-\sqrt{8}-\sqrt{7}+\sqrt{7}+\sqrt{6}-\sqrt{6}-\sqrt{5}+\sqrt{5}+2 = 3+2 = 5$

141. a. $(2+\sqrt{3}) + \dfrac{1}{2+\sqrt{3}} = 2+\sqrt{3} + \dfrac{2-\sqrt{3}}{(2+\sqrt{3})(2-\sqrt{3})} = 2+\sqrt{3} + \dfrac{2-\sqrt{3}}{4-3} = 2+\sqrt{3}+2-\sqrt{3} = 4$

b. $(2+\sqrt{3})^2 + \dfrac{1}{(2+\sqrt{3})^2} = 7+4\sqrt{3} + \dfrac{7-4\sqrt{3}}{(7+4\sqrt{3})(7-4\sqrt{3})} = 7+4\sqrt{3} + \dfrac{7-4\sqrt{3}}{49-48} = 7+4\sqrt{3}+7-4\sqrt{3} = 14$

143. Convert each radical to a radical with a common index. In this case, convert each radical to a radical with index 6.

$\sqrt{5} = \sqrt[6]{5^3} = \sqrt[6]{125}, \sqrt[3]{11} = \sqrt[6]{11^2} = \sqrt[6]{121}, \sqrt[6]{123}$

In increasing order of magnitude, the radicals are: $\sqrt[3]{11}, \sqrt[6]{123}, \sqrt{5}$

145. Convert each radical to a radical with a common index. In this case, convert each radical to a radical with index 12.

$\sqrt{3} = \sqrt[12]{3^6} = \sqrt[12]{729}, \sqrt[3]{7} = \sqrt[12]{7^4} = \sqrt[12]{2401},$
$\sqrt[4]{10} = \sqrt[12]{10^3} = \sqrt[12]{1000}$

In increasing order of magnitude, the radicals are: $\sqrt{3}, \sqrt[4]{10}, \sqrt[3]{7}$

147. $(3\sqrt{5}+4\sqrt{3})(3\sqrt{5}-4\sqrt{3}) = (3\sqrt{5})^2 - (4\sqrt{3})^2 = 45 - 48 = -3$

149. a. If $\sqrt{c}$ is an irrational number, $a+b\sqrt{c} = 0$ only if $a = b = 0$.

b. $a+b\sqrt{c} = x+y\sqrt{c} \Rightarrow a+b\sqrt{c} - (x+y\sqrt{c}) = 0 \Rightarrow (a-x)+(b-y)\sqrt{c} = 0$

From part (a), $(a-x)+(b-y)\sqrt{c} = 0 \Rightarrow a-x = 0$ and $b-y = 0 \Rightarrow a = x$ and $b = y$.

151. $\sqrt{16+2\sqrt{55}} = \sqrt{a+\sqrt{c}} = \sqrt{x}+\sqrt{y}$

From exercise 150, we have $x+y = a$ and $\sqrt{c} = 2\sqrt{xy}$. Then $a = 16 = x+y$ and $\sqrt{c} = \sqrt{55} = \sqrt{xy}$.

$\begin{cases} 16 = x+y \\ 55 = xy \end{cases} \Rightarrow \begin{cases} 16-x = y \\ 55 = xy \end{cases} \Rightarrow 55 = x(16-x) \Rightarrow -x^2+16x-55 = 0 \Rightarrow$

$x^2 - 16x + 55 = 0 \Rightarrow (x-11)(x-5) = 0 \Rightarrow x = 11$ or $x = 5$

If $x = 11$, $y = 5$, and $\sqrt{a+\sqrt{c}} = \sqrt{x}+\sqrt{y} \Rightarrow \sqrt{16+2\sqrt{55}} = \sqrt{11}+\sqrt{5}$.

P.6 Getting Ready for the Next Section

153. $\dfrac{7}{3} + \dfrac{3}{2} = \dfrac{7(2)}{3(2)} + \dfrac{3(3)}{3(2)} = \dfrac{14}{6} + \dfrac{9}{6} = \dfrac{23}{6}$

155. $\dfrac{3}{2} \cdot \dfrac{12}{7} = \dfrac{3 \cdot 6}{7} = \dfrac{18}{7}$

157. $-7x + 2x = -5x$

159. $(4x - 5) + (7x - 3) = 11x - 8$

Chapter P Review Exercises

Building Skills

1. a. $\{4\}$

   b. $\{0, 4\}$

   c. $\{-5, 0, 4\}$

   d. $\left\{-5, 0, 0.2, 0.\overline{31}, \dfrac{1}{2}, 4\right\}$

   e. $\{\sqrt{7}, \sqrt{12}\}$

   f. $\left\{-5, 0, 0.2, 0.\overline{31}, \dfrac{1}{2}, \sqrt{7}, \sqrt{12}, 4\right\}$

3. Distributive property

5. Multiplicative identity

7. $x \leq 1$

9. $(0, \infty)$

11. $|2 - |-3|| = |2 - 3| = |-1| = 1$

13. $|1 - \sqrt{15}| = \sqrt{15} - 1$

15. $(-4)^3 = -64$

17. $2^4 - 5 \cdot 3^2 = 16 - 5 \cdot 9 = 16 - 45 = -29$

19. $-25^{1/2} = -5$

21. $8^{4/3} = (2^3)^{4/3} = 2^{3(4/3)} = 2^4 = 16$

23. $\left(\dfrac{25}{36}\right)^{-3/2} = \left(\dfrac{36}{25}\right)^{3/2} = \dfrac{(6^2)^{3/2}}{(5^2)^{3/2}}$

   $= \dfrac{6^{2(3/2)}}{5^{2(3/2)}} = \dfrac{6^3}{5^3} = \dfrac{216}{125}$

25. $2 \cdot 5^3 = 2 \cdot 125 = 250$

27. $\dfrac{21 \times 10^6}{3 \times 10^7} = \dfrac{7}{10} = 0.7$

29. $\sqrt{5^2} = 5$

31. $\sqrt[3]{-125} = \sqrt[3]{(-5)^3} = -5$

33. $64^{-1/3} = \dfrac{1}{64^{1/3}} = \dfrac{1}{(4^3)^{1/3}} = \dfrac{1}{4}$

35. $(\sqrt{3} + 2)(\sqrt{3} - 2) = (\sqrt{3})^2 - 2^2 = 3 - 4 = -1$

37. $\dfrac{3^{-2} \cdot 7^0}{18^{-1}} = \dfrac{\frac{1}{3^2} \cdot 1}{\frac{1}{18}} = \dfrac{1}{9} \cdot 18 = 2$

39. $\dfrac{2 - 6\sqrt{4}}{4(2)} = \dfrac{2 - 6(2)}{8} = \dfrac{2 - 12}{8} = \dfrac{-10}{8} = -\dfrac{5}{4}$

41. $16^{-1/2} = \dfrac{1}{16^{1/2}} = \dfrac{1}{4}$

43. $\left(x^{-2}\right)^{-5} = x^{(-2)(-5)} = x^{10}$

45. $\left(\dfrac{x^{-3}}{y^{-1}}\right)^{-2} = \dfrac{x^{-3(-2)}}{y^{-1(-2)}} = \dfrac{x^6}{y^2}$

47. $\left(16 x^{-2/3} y^{-4/3}\right)^{3/2}$

   $= 16^{3/2} x^{(-2/3)(3/2)} y^{(-4/3)(3/2)}$

   $= 4^{2(3/2)} x^{-1} y^{-2} = \dfrac{4^3}{xy^2} = \dfrac{64}{xy^2}$

49. $\left(\dfrac{64 y^{-9/2}}{x^{-3}}\right)^{-2/3} = \dfrac{4^{3(-2/3)} y^{(-9/2)(-2/3)}}{x^{-3(-2/3)}}$

   $= \dfrac{4^{-2} y^3}{x^2} = \dfrac{y^3}{4^2 x^2} = \dfrac{y^3}{16 x^2}$

51. $\left(7 x^{1/4}\right)\left(3 x^{3/2}\right) = 21 x^{1/4 + 3/2} = 21 x^{7/4}$

53. $\dfrac{x^5(2x)^3}{4x^3} = \dfrac{x^5(2^3)x^3}{4x^3} = \dfrac{8x^{5+3}}{4x^3} = 2x^{8-3} = 2x^5$

55. $\left(\dfrac{x^2 y^{4/3}}{x^{1/3}}\right)^6 = \dfrac{x^{2(6)} y^{(4/3)(6)}}{x^{(1/3)(6)}} = \dfrac{x^{12} y^8}{x^2}$
$= x^{12-2} y^8 = x^{10} y^8$

57. $\dfrac{\sqrt{64}}{\sqrt{11}} = \dfrac{8}{\sqrt{11}} = \dfrac{8\sqrt{11}}{\sqrt{11} \cdot \sqrt{11}} = \dfrac{8\sqrt{11}}{11}$

59. $7\sqrt{6} - 3\sqrt{24} = 7\sqrt{6} - 3\sqrt{4 \cdot 6}$
$= 7\sqrt{6} - 3\sqrt{4}\sqrt{6} = 7\sqrt{6} - 3(2)\sqrt{6}$
$= 7\sqrt{6} - 6\sqrt{6} = \sqrt{6}$

61. $\sqrt{2x}\sqrt{6x} = \sqrt{12x^2} = \sqrt{4 \cdot 3x^2}$
$= \sqrt{4}\sqrt{3}\sqrt{x^2} = 2x\sqrt{3}$

63. $\dfrac{\sqrt{100x^3}}{\sqrt{4x}} = \sqrt{\dfrac{100x^3}{4x}} = \sqrt{25x^2} = 5x$

65. $4\sqrt[3]{135} + \sqrt[3]{40} = 4\sqrt[3]{27 \cdot 5} + \sqrt[3]{8 \cdot 5}$
$= 4\sqrt[3]{27}\sqrt[3]{5} + \sqrt[3]{8}\sqrt[3]{5}$
$= 4(3)\sqrt[3]{5} + 2\sqrt[3]{5}$
$= 12\sqrt[3]{5} + 2\sqrt[3]{5} = 14\sqrt[3]{5}$

67. $\dfrac{7}{\sqrt{3}} = \dfrac{7\sqrt{3}}{\sqrt{3}\sqrt{3}} = \dfrac{7\sqrt{3}}{3}$

69. $\dfrac{1-\sqrt{3}}{1+\sqrt{3}} = \dfrac{(1-\sqrt{3})(1-\sqrt{3})}{(1+\sqrt{3})(1-\sqrt{3})} = \dfrac{1 - 2\sqrt{3} + 3}{1^2 - (\sqrt{3})^2}$
$= \dfrac{4 - 2\sqrt{3}}{1 - 3} = \dfrac{4 - 2\sqrt{3}}{-2} = -2 + \sqrt{3}$

71. $3.7 \times (6.23 \times 10^{12}) = 23.051 \times 10^{12}$
$= 2.3051 \times 10^{13}$

73. $(x^3 - 6x^2 + 4x - 2) + (3x^3 - 6x^2 + 5x - 4)$
$= 4x^3 - 12x^2 + 9x - 6$

75. $(4x^4 + 3x^3 - 5x^2 + 9) + (5x^4 + 8x^3 - 7x^2 + 5)$
$= 9x^4 + 11x^3 - 12x^2 + 14$

77. $(x - 12)(x - 3) = x^2 - 3x - 12x + 36$
$= x^2 - 15x + 36$

79. $(x^5 - 2)(x^5 + 2) = (x^5)^2 - 2^2 = x^{10} - 4$

81. $(2x + 5)(3x - 11) = 6x^2 - 22x + 15x - 55$
$= 6x^2 - 7x - 55$

83. $x^2 - 3x - 10 = (x - 5)(x + 2)$

85. $24x^2 - 38x - 11 = (6x - 11)(4x + 1)$

87. $x(x + 11) + 5(x + 11) = (x + 11)(x + 5)$

89. $x^4 - x^3 + 7x - 7 = x^4 + 7x - x^3 - 7$
$= x(x^3 + 7) - 1(x^3 + 7)$
$= (x^3 + 7)(x - 1)$

91. $10x^2 + 23x + 12 = (5x + 4)(2x + 3)$

93. $12x^2 + 7x - 12 = (4x - 3)(3x + 4)$

95. $4x^2 - 49 = (2x - 7)(2x + 7)$

97. $x^2 + 12 + 36 = (x + 6)^2$

99. $64x^2 + 48x + 9 = (8x + 3)^2$

101. $8x^3 + 27 = (2x + 3)(4x^2 - 6x + 9)$

103. $x^3 + 5x^2 - 16x - 80 = x^3 - 16x + 5x^2 - 80$
$= x(x^2 - 16) + 5(x^2 - 16)$
$= (x^2 - 16)(x + 5)$
$= (x - 4)(x + 4)(x + 5)$

105. $\dfrac{4}{x-9} - \dfrac{10}{9-x} = \dfrac{4}{x-9} + \dfrac{10}{x-9} = \dfrac{14}{x-9}$

107. $\dfrac{3x+5}{x^2+14x+48} - \dfrac{3x-2}{x^2+10x+16}$
$= \dfrac{3x+5}{(x+6)(x+8)} - \dfrac{3x-2}{(x+2)(x+8)}$
$= \dfrac{(3x+5)(x+2)}{(x+6)(x+8)(x+2)} - \dfrac{(3x-2)(x+6)}{(x+2)(x+6)(x+8)}$
$= \dfrac{3x^2 + 11x + 10 - (3x^2 + 16x - 12)}{(x+2)(x+6)(x+8)}$
$= \dfrac{-5x + 22}{(x+2)(x+6)(x+8)}$

109. $\dfrac{x-1}{x+1} - \dfrac{x+1}{x-1} = \dfrac{(x-1)^2}{(x+1)(x-1)} - \dfrac{(x+1)^2}{(x+1)(x-1)}$
$= \dfrac{x^2 - 2x + 1 - (x^2 + 2x + 1)}{(x+1)(x-1)}$
$= -\dfrac{4x}{(x+1)(x-1)}$

111. $\dfrac{x-1}{2x-3} \cdot \dfrac{4x^2-9}{2x^2-x-1}$

$= \dfrac{\cancel{x-1}}{\cancel{2x-3}} \cdot \dfrac{(2x+3)\cancel{(2x-3)}}{(2x+1)\cancel{(x-1)}} = \dfrac{2x+3}{2x+1}$

113. $\dfrac{x^2-4}{4x^2-9} \cdot \dfrac{2x^2-3x}{2x+4}$

$= \dfrac{(x-2)\cancel{(x+2)}}{\cancel{(2x-3)}(2x+3)} \cdot \dfrac{x\cancel{(2x-3)}}{2\cancel{(x+2)}} = \dfrac{x(x-2)}{2(2x+3)}$

115. $\dfrac{\dfrac{1}{x^2}-x}{\dfrac{1}{x^2}+x} = \dfrac{\dfrac{1-x^3}{x^2}}{\dfrac{1+x^3}{x^2}} = \dfrac{1-x^3}{x^2} \cdot \dfrac{x^2}{1+x^3}$

$= \dfrac{1-x^3}{1+x^3} = \dfrac{(1-x)(1+x+x^2)}{(1+x)(1-x+x^2)}$

117. $\dfrac{\dfrac{x}{x-3}+x}{\dfrac{x}{3-x}-x} = \dfrac{\dfrac{x+x(x-3)}{x-3}}{\dfrac{x-x(3-x)}{3-x}} = \dfrac{x^2-2x}{x-3} \cdot \dfrac{3-x}{x^2-2x}$

$= \dfrac{x^2-2x}{x-3} \cdot \left(-\dfrac{x-3}{x^2-2x}\right) = -1$

Applying the Concepts

119. Using the Pythagorean Theorem, we have
$c^2 = 20^2 + 21^2 \Rightarrow c^2 = 841 \Rightarrow c = 29$

121. First, find the length of the missing leg:
$20^2 = 12^2 + x^2 \Rightarrow 400 = 144 + x^2 \Rightarrow$
$256 = x^2 \Rightarrow 16 = x$.
Area $= \dfrac{1}{2}bh \Rightarrow A = \dfrac{1}{2}(12)(16) = 96$ in.^{2}.
Perimeter $=$ the sum of the sides $\Rightarrow$
$P = 12 + 16 + 20 = 48$ in.

123. First, find the length of the missing side:
$10^2 = 6^2 + x^2 \Rightarrow 100 = 36 + x^2 \Rightarrow$
$64 = x^2 \Rightarrow 8 = x$. So the area $= (6)(8) = 48$
square feet.

125. $\sqrt{7920(0.7) + 0.7^2} = \sqrt{5544 + 0.49}$
≈ 74.46 miles

127. $\dfrac{36 - 3(2^2)}{8(2)} = \dfrac{36 - 3(4)}{16} = \dfrac{36 - 12}{16} = \dfrac{24}{16} = 1.5$ ft

Chapter P Practice Test

1. $|7 - |-3|| = |7 - 3| = |4| = 4$

2. $|\sqrt{2} - 100| = 100 - \sqrt{2}$

3. $\dfrac{5 - 3(-3)}{2} + (5)(-3) = \dfrac{5+9}{2} - 15 = -8$

4. $x \ne -9, x \ne 3 \Rightarrow (-\infty, -9) \cup (-9, 3) \cup (3, \infty)$

5. $\left(\dfrac{-3x^2 y}{x}\right)^3 = \dfrac{(-3)^3 x^{2(3)} y^3}{x^3} = \dfrac{-27 x^6 y^3}{x^3}$

$= -27 x^{6-3} y^3 = -27 x^3 y^3$

6. $\sqrt[3]{-8x^6} = \sqrt[3]{(-2)^3 x^6} = -2x^2$

7. $\sqrt{75x} - \sqrt{27x} = \sqrt{3 \cdot 25x} - \sqrt{3 \cdot 9x}$
$= 5\sqrt{3x} - 3\sqrt{3x} = 2\sqrt{3x}$

8. $-16^{-3/2} = -\dfrac{1}{16^{3/2}} = -\dfrac{1}{(4^2)^{3/2}}$

$= -\dfrac{1}{4^{2(3/2)}} = -\dfrac{1}{4^3} = -\dfrac{1}{64}$

9. $\left(\dfrac{x^{-2} y^4}{25 x^3 y^3}\right)^{-1/2}$

$= \left(\dfrac{25 x^3 y^3}{x^{-2} y^4}\right)^{1/2} = \dfrac{25^{1/2} \left(x^3\right)^{1/2} \left(y^3\right)^{1/2}}{\left(x^{-2}\right)^{1/2} \left(y^4\right)^{1/2}}$

$= \dfrac{5 x^{3/2} y^{3/2}}{x^{-1} y^2} = 5 x^{(3/2)-(-1)} y^{(3/2)-2}$

$= 5 x^{5/2} y^{-1/2} = \dfrac{5 x^{5/2}}{y^{1/2}} = \dfrac{5 x^{5/2} y^{1/2}}{y}$

10. $\dfrac{5}{1-\sqrt{3}} = \dfrac{5(1+\sqrt{3})}{(1-\sqrt{3})(1+\sqrt{3})}$

$= \dfrac{5 + 5\sqrt{3}}{1 - 3} = -\dfrac{5 + 5\sqrt{3}}{2}$

11. $4(x^2 - 3x + 2) + 3(5x^2 - 2x + 1)$
$= 4x^2 - 12x + 8 + 15x^2 - 6x + 3$
$= 19x^2 - 18x + 11$

12. $(x-2)(5x-1) = 5x^2 - 11x + 2$

13. $\left(x^2 + 3y^2\right)^2 = x^4 + 6x^2 y^2 + 9y^4$

14.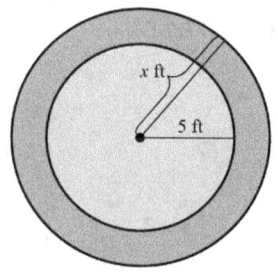

The total area is πx^2 square feet. The area of the rug without the border is 25π square feet. So, the area of the border = the total area – the area of the rug =
$\pi x^2 - 25\pi = \pi(x^2 - 25) = \pi(x-5)(x+5)$.

15. $x^2 - 5x + 6 = (x-3)(x-2)$

16. $9x^2 + 12x + 4 = (3x+2)^2$

17. $8x^3 - 27 = (2x-3)(4x^2 + 6x + 9)$

18. $\dfrac{6-3x}{x^2-4} - \dfrac{12}{2x+4}$

$= \dfrac{6-3x}{(x-2)(x+2)} - \dfrac{12}{2(x+2)}$

$= \dfrac{2(6-3x)}{2(x-2)(x+2)} - \dfrac{12(x-2)}{2(x-2)(x+2)}$

$= \dfrac{12-6x-(12x-24)}{2(x-2)(x+2)} = \dfrac{-18x+36}{2(x-2)(x+2)}$

$= \dfrac{-18(x-2)}{2(x-2)(x+2)} = -\dfrac{9}{x+2}$

19. $\dfrac{\frac{1}{x^2}}{9-\frac{1}{x^2}} = \dfrac{\frac{1}{x^2}}{\frac{9x^2-1}{x^2}} = \dfrac{1}{x^2} \cdot \dfrac{x^2}{(3x-1)(3x+1)}$

$= \dfrac{1}{(3x-1)(3x+1)}$

20. $c^2 = 35^2 + 12^2 \Rightarrow c^2 = 1369 \Rightarrow c = 37$

Chapter 1 Equations and Inequalities

1.1 Linear Equations in One Variable

1.1 Practice Problems

1. a. Both sides of the equation $\frac{x}{3} - 7 = 5$ are defined for all real numbers, so the domain is $(-\infty, \infty)$.

   b. The left side of the equation $\frac{2}{2-x} = 4$ is not defined if $x = 2$. The right side of the equation is defined for all real numbers, so the domain is $(-\infty, 2) \cup (2, \infty)$.

   c. The left side of the equation $\sqrt{x-1} = 0$ is not defined if $x < 1$. The right side of the equation is defined for all real numbers, so the domain is $[1, \infty)$.

2. $\frac{2}{3} - \frac{3}{2}x = \frac{1}{6} - \frac{7}{3}x$

   To clear the fractions, multiply both sides of the equation by the LCD, 6.
   $$4 - 9x = 1 - 14x$$
   $$4 - 9x + 14x = 1 - 14x + 14x$$
   $$4 + 5x = 1$$
   $$4 + 5x - 4 = 1 - 4$$
   $$5x = -3$$
   $$\frac{5x}{5} = \frac{-3}{5}$$
   $$x = -\frac{3}{5}$$
   Solution set: $\left\{-\frac{3}{5}\right\}$

3. $3x - [2x - 6(x+1)] = -1$
   $$3x - (2x - 6x - 6) = -1$$
   $$3x - (-4x - 6) = -1$$
   $$3x + 4x + 6 = -1$$
   $$7x + 6 = -1$$
   $$7x + 6 - 6 = -1 - 6$$
   $$7x = -7$$
   $$\frac{7x}{7} = \frac{-7}{7}$$
   $$x = -1$$
   Solution set: $\{-1\}$

4. $3x - [2x - 6(x+1)] = 7x - 1$
   $$3x - (2x - 6x - 6) = 7x - 1$$
   $$3x - (-4x - 6) = 7x - 1$$
   $$3x + 4x + 6 = 7x - 1$$
   $$7x + 6 = 7x - 1$$
   $$7x + 6 - 7x = 7x - 1 - 7x$$
   $$6 = -1$$
   Since $6 = -1$ is false, no number satisfies this equation. Thus, the equation is inconsistent, and the solution set is $\varnothing$.

5. $2(3x - 6) + 5 = 12 - (19 - 6x)$
   $$6x - 12 + 5 = 12 - 19 + 6x$$
   $$6x - 7 = -7 + 6x$$
   $$6x - 7 - 6x = -7 + 6x - 6x$$
   $$-7 = -7$$
   $$-7 + 7 = -7 + 7$$
   $$0 = 0$$
   The equation $0 = 0$ is always true. Therefore, the original equation is an identity, and the solution set is $(-\infty, \infty)$.

6. Following the reasoning in example 6, we have $x + 2x = 3x$ is the maximum extended length (in feet) of the cord.
   $$3x + 7 + 10 = 120$$
   $$3x + 17 = 120$$
   $$3x + 17 - 17 = 120 - 17$$
   $$3x = 103$$
   $$\frac{3x}{3} = \frac{103}{3} \Rightarrow x \approx 34.3$$
   The cord should be no longer than 34.3 feet.

7. $F = \frac{9}{5}C + 32$
   $$50 = \frac{9}{5}C + 32$$
   $$50 - 32 = \frac{9}{5}C + 32 - 32$$
   $$18 = \frac{9}{5}C$$
   $$18 \cdot \frac{5}{9} = \frac{5}{9} \cdot \frac{9}{5}C$$
   $$10 = C$$
   Thus, 50°F converts to 10°C.

8. $P = 2\ell + 2w$
   Subtract 2ℓ from both sides.
   $$P - 2\ell = 2w$$
   Now, divide both sides by 2.
   $$\frac{P - 2\ell}{2} = w$$

9. Use the formula $V = hwd$, where V is the volume, h is the height, w is the width, and d is the depth of the suitcase. So, $h = 21$, $w = 13$, and $d = 9$.
   $V = hwd = 21 \cdot 13 \cdot 9 = 2457$ in.^{3}

1.1 Concepts and Vocabulary

1. The domain of the variable in an equation is the set of all real number for which both sides of the equation are <u>defined</u>.

3. Two equations with the same solution set are called <u>equivalent</u>.

5. False. A linear equation that is equivalent to the equation $0 = 4$ is an inconsistent equation.

7. True

1.1 Building Skills

9. a. Substitute 0 for x in the equation $x - 2 = 5x + 6$:
   $0 - 2 = 5(0) + 6 \Rightarrow -2 \neq 6$
   So, 0 is not a solution of the equation.

   b. Substitute –2 for x in the equation $x - 2 = 5x + 6$:
   $-2 - 2 = 5(-2) + 6 \Rightarrow -4 = -10 + 6 \Rightarrow -4 = -4$
   So, –2 is a solution of the equation.

11. a. Substitute 4 for x in the equation $\dfrac{2}{x} = \dfrac{1}{3} + \dfrac{1}{x+2}$:
    $\dfrac{2}{4} = \dfrac{1}{3} + \dfrac{1}{4+2} \Rightarrow \dfrac{1}{2} = \dfrac{1}{3} + \dfrac{1}{6} \Rightarrow \dfrac{1}{2} = \dfrac{1}{2}$
    So, 4 is a solution of the equation.

    b. Substitute 1 for x in the equation $\dfrac{2}{x} = \dfrac{1}{3} + \dfrac{1}{x+2}$:
    $\dfrac{2}{1} = \dfrac{1}{3} + \dfrac{1}{1+2} \Rightarrow 2 = \dfrac{1}{3} + \dfrac{1}{3} \Rightarrow 2 \neq \dfrac{2}{3}$
    So, 1 is not a solution of the equation.

13. a. The equation $2x + 3x = 5x$ is an identity, so every real number is a solution of the equation. Thus 157 is a solution of the equation. This can be checked by substituting 157 for x in the equation:
    $2(157) + 3(157) = 5(157) \Rightarrow$
    $314 + 471 = 785 \Rightarrow 785 = 785$

   b. The equation $2x + 3x = 5x$ is an identity, so every real number is a solution of the equation. Thus –2046 is a solution of the equation. This can be checked by substituting –2046 for x in the equation:
   $2(-2046) + 3(-2046) = 5(-2046)$
   $-4092 - 6138 = -10,230$
   $-10,230 = -10,230$

15. The left side of the equation $\dfrac{y}{y-1} = \dfrac{3}{y+2}$ is not defined if $y = 1$, and the right side of the equation is not defined if $y = -2$. The domain is $(-\infty, -2) \cup (-2, 1) \cup (1, \infty)$.

17. The left side of the equation $\dfrac{3x}{(x-3)(x-4)} = 2x + 9$ is not defined if $x = 3$ or $x = 4$. The right side is defined for all real numbers. So, the domain is $(-\infty, 3) \cup (3, 4) \cup (4, \infty)$.

19. Substitute 0 for x in $2x + 3 = 5x + 1$. Because $3 \neq 1$, the equation is not an identity.

21. When the terms on the left side of the equation $\dfrac{1}{x} + \dfrac{1}{2} = \dfrac{2+x}{2x}$ are collected, the equation becomes $\dfrac{2+x}{2x} = \dfrac{2+x}{2x}$, which is an identity.

In exercises 23–48, solve the equations using the procedures listed on page 84 in your text: eliminate fractions, simplify, isolate the variable term, combine terms, isolate the variable term, and check the solution.

23. $3x + 5 = 14$
    $3x + 5 - 5 = 14 - 5$
    $3x = 9$
    $\dfrac{3x}{3} = \dfrac{9}{3}$
    $x = 3$
    Solution set: {3}

25. $-10x + 12 = 32$
    $-10x + 12 - 12 = 32 - 12$
    $-10x = 20$
    $\dfrac{-10x}{-10} = \dfrac{20}{-10}$
    $x = -2$
    Solution set: {–2}

27.
$$3 - y = -4$$
$$3 - y - 3 = -4 - 3$$
$$-y = -7 \Rightarrow y = 7$$
Solution set: $\{7\}$

29.
$$7x + 7 = 2(x + 1)$$
$$7x + 7 = 2x + 2$$
$$7x + 7 - 7 = 2x + 2 - 7$$
$$7x = 2x - 5$$
$$7x - 2x = 2x - 5 - 2x$$
$$5x = -5$$
$$\frac{5x}{5} = \frac{-5}{5} \Rightarrow x = -1$$
Solution set: $\{-1\}$

31.
$$3(2 - y) + 5y = 3y$$
$$6 - 3y + 5y = 3y$$
$$6 + 2y = 3y$$
$$6 + 2y - 2y = 3y - 2y \Rightarrow 6 = y$$
Solution set: $\{6\}$

33.
$$4y - 3y + 7 - y = 2 - (7 - y)$$
Distribute -1 to clear the parentheses.
$$7 = 2 - 7 + y$$
$$7 = -5 + y$$
$$7 + 5 = -5 + y + 5 \Rightarrow 12 = y$$
Solution set: $\{12\}$

35.
$$3(x - 2) + 2(3 - x) = 1$$
$$3x - 6 + 6 - 2x = 1 \Rightarrow x = 1$$
Solution set: $\{1\}$

37.
$$2x + 3(x - 4) = 7x + 10$$
$$2x + 3x - 12 = 7x + 10$$
$$5x - 12 = 7x + 10$$
$$5x - 12 + 12 = 7x + 10 + 12$$
$$5x = 7x + 22$$
$$5x - 7x = 7x + 22 - 7x$$
$$-2x = 22$$
$$\frac{-2x}{-2} = \frac{22}{-2} \Rightarrow x = -11$$
Solution set: $\{-11\}$

39.
$$4[x + 2(3 - x)] = 2x + 1$$
Distribute 2 to clear the inner parentheses.
$$4[x + 6 - 2x] = 2x + 1$$
Combine like terms within the brackets.
$$4[6 - x] = 2x + 1$$
Distribute 4 to clear the brackets.
$$24 - 4x = 2x + 1$$
$$24 - 4x - 24 = 2x + 1 - 24$$
$$-4x = 2x - 23$$
$$-4x - 2x = -23$$
$$-6x = -23$$
$$\frac{-6x}{-6} = \frac{-23}{-6} \Rightarrow x = \frac{23}{6}$$
Solution set: $\left\{\frac{23}{6}\right\}$

41.
$$3(4y - 3) = 4[y - (4y - 3)]$$
Distribute 3 on the left side and -1 on the right side to clear parentheses.
$$12y - 9 = 4[y - 4y + 3]$$
Combine like terms in the brackets.
$$12y - 9 = 4[-3y + 3]$$
Distribute 4 to clear the brackets.
$$12y - 9 = -12y + 12$$
$$12y - 9 + 9 = -12y + 12 + 9$$
$$12y = -12y + 21$$
$$12y + 12y = -12y + 21 + 12y$$
$$24y = 21$$
$$\frac{24y}{24} = \frac{21}{24} \Rightarrow y = \frac{21}{24} = \frac{7}{8}$$
Solution set: $\left\{\frac{7}{8}\right\}$

43.
$$2x - 3(2 - x) = (x - 3) + 2x + 1$$
Distribute -3 on the left to clear the parentheses.
$$2x - 6 + 3x = x - 3 + 2x + 1$$
$$5x - 6 = 3x - 2$$
$$5x - 6 + 6 = 3x - 2 + 6$$
$$5x = 3x + 4$$
$$5x - 3x = 3x + 4 - 3x$$
$$2x = 4$$
$$\frac{2x}{2} = \frac{4}{2} \Rightarrow x = 2$$
Solution set: $\{2\}$

45.
$$\frac{2x + 1}{9} - \frac{x + 4}{6} = 1$$
To clear the fractions, multiply both sides of the equation by the least common denominator, 36.
$$36\left(\frac{2x + 1}{9} - \frac{x + 4}{6}\right) = 36(1)$$
$$4(2x + 1) - 6(x + 4) = 36$$
$$8x + 4 - 6x - 24 = 36$$
$$2x - 20 = 36$$
$$2x - 20 + 20 = 36 + 20$$
$$2x = 56$$
$$\frac{2x}{2} = \frac{56}{2} \Rightarrow x = 28$$
Solution set: $\{28\}$

47. $\dfrac{1-x}{4} + \dfrac{5x+1}{2} = 3 - \dfrac{2(x+1)}{8}$

To clear the fractions, multiply both sides by the least common denominator, 8.

$8\left(\dfrac{1-x}{4} + \dfrac{5x+1}{2}\right) = 8\left(3 - \dfrac{2(x+1)}{8}\right)$

Distribute the 8 on both sides.

$8\left(\dfrac{1-x}{4}\right) + 8\left(\dfrac{5x+1}{2}\right) = 8(3) - 8\left(\dfrac{2(x+1)}{8}\right)$

$2(1-x) + 4(5x+1) = 8(3) - 2(x+1)$

Simplify by collecting like terms and combining constants.

$2 - 2x + 20x + 4 = 24 - 2x - 2$
$18x + 6 = 22 - 2x$
$18x + 6 + 2x = 22 - 2x + 2x$
$20x + 6 = 22$
$20x + 6 - 6 = 22 - 6$
$20x = 16$
$\dfrac{20x}{20} = \dfrac{16}{20}$
$x = \dfrac{16}{20} = \dfrac{4}{5}$

Solution set: $\left\{\dfrac{4}{5}\right\}$

49. To solve $d = rt$ for r, divide both sides of the equation by t. $r = \dfrac{d}{t}$.

51. To solve $C = 2\pi r$ for r, divide both sides of the equation by 2π. $r = \dfrac{C}{2\pi}$.

53. To solve $I = \dfrac{E}{R}$ for R, multiply both sides by R.

$RI = R\left(\dfrac{E}{R}\right) \Rightarrow RI = E$

Divide both sides by I.

$\dfrac{RI}{I} = \dfrac{E}{I} \Rightarrow R = \dfrac{E}{I}$

55. To solve $A = \dfrac{(a+b)h}{2}$ for h, multiply both sides by 2.

$2A = (a+b)h$

Divide both sides by $(a+b)$.

$\dfrac{2A}{a+b} = \dfrac{(a+b)h}{a+b} \Rightarrow \dfrac{2A}{a+b} = h$

57. To solve $\dfrac{1}{f} = \dfrac{1}{u} + \dfrac{1}{v}$ for u, clear the fractions by multiplying both sides by the least common denominator, fuv.

$fuv\left(\dfrac{1}{f}\right) = fuv\left(\dfrac{1}{u} + \dfrac{1}{v}\right)$

$fuv\left(\dfrac{1}{f}\right) = fuv\left(\dfrac{1}{u}\right) + fuv\left(\dfrac{1}{v}\right)$

Simplify.

$uv = fv + fu$

Subtract fu from both sides.

$uv - fu = fv + fu - fu$
$uv - fu = fv$

Factor the left side.

$u(v - f) = fv$

Divide both sides by $v - f$.

$\dfrac{u(v-f)}{v-f} = \dfrac{fv}{v-f} \Rightarrow u = \dfrac{fv}{v-f}$

59. To solve $y = mx + b$ for m, subtract b from both sides.

$y - b = mx + b - b \Rightarrow y - b = mx$

Divide both sides by x.

$\dfrac{y-b}{x} = \dfrac{mx}{x} \Rightarrow \dfrac{y-b}{x} = m$

1.1 Applying the Concepts

61. The formula for volume is $V = lwh$. Substitute 2808 for V, 18 for l, and 12 for h. Solve for w.

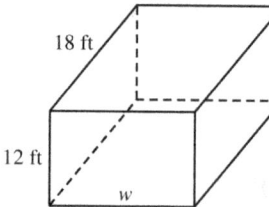

$2808 = 18 \cdot 12 \cdot w$
$2808 = 216w$
$\dfrac{2808}{216} = \dfrac{216w}{216}$
$13 = w$

The width of the pool is 13 ft.

63. Let $w =$ the width of the rectangle. Then $2w - 5 =$ the length of the rectangle.

$2w + 2(2w - 5) = 80$
$2w + 4w - 10 = 80$
$6w - 10 = 80$
$6w = 90$
$w = 15,\ 2w - 5 = 25$

The width of the rectangle is 15 ft and its length is 25 feet.

65. The formula for circumference of a circle is $C = 2\pi r$. Substitute 114π for C. Solve for r.
$$114\pi = 2\pi r \Rightarrow \frac{114\pi}{2\pi} = \frac{2\pi r}{2\pi} \Rightarrow 57 = r$$
The radius is 57 cm.

67. The formula for surface area of a cylinder is $S = 2\pi rh + 2\pi r^2$. Substitute 6π for S and 1 for r. Solve for h.

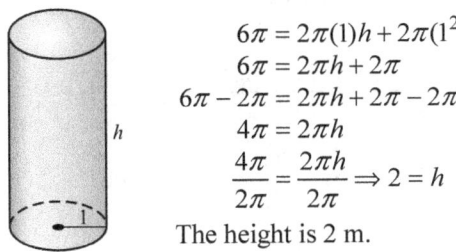

$$6\pi = 2\pi(1)h + 2\pi(1^2)$$
$$6\pi = 2\pi h + 2\pi$$
$$6\pi - 2\pi = 2\pi h + 2\pi - 2\pi$$
$$4\pi = 2\pi h$$
$$\frac{4\pi}{2\pi} = \frac{2\pi h}{2\pi} \Rightarrow 2 = h$$

The height is 2 m.

69. The formula for area of a trapezoid is $A = \frac{1}{2}h(b_1 + b_2)$. Substitute 66 for A, 6 for h, and 3 for b_1. Solve for b_2.

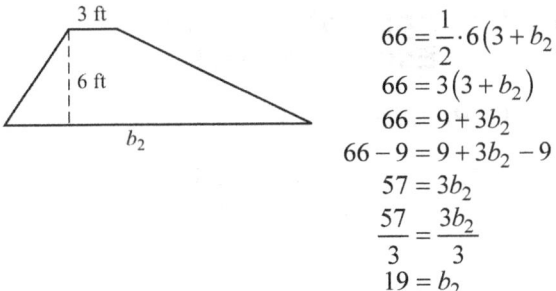

$$66 = \frac{1}{2} \cdot 6(3 + b_2)$$
$$66 = 3(3 + b_2)$$
$$66 = 9 + 3b_2$$
$$66 - 9 = 9 + 3b_2 - 9$$
$$57 = 3b_2$$
$$\frac{57}{3} = \frac{3b_2}{3}$$
$$19 = b_2$$

The length of the second base is 19 ft.

71. $F = \frac{9}{5}C + 32 = \frac{9}{5}(20) + 32 = 36 + 32 = 68°$

20°C is equivalent to 68°F.

73. $C = \frac{5}{9}(F - 32) = \frac{5}{9}(57 - 32) = \frac{5}{9}(25) \approx 13.9°$

57°F is equivalent to approximately 14°C.

1.1 Beyond the Basics

75. First, solve $7x + 2 = 16$. Subtracting 2 from both sides, we have $7x = 14$. Then divide both sides by 7; we obtain $x = 2$. Now substitute 2 for x in $3x - 1 = k$. This becomes $3(2) - 1 = k$, so $k = 5$.

77. If $k = -2$ then the equation becomes $\frac{3}{y-2} = \frac{4}{y-2}$, which is inconsistent. Note that two fractions with the same denominator are not equal if the numerators are not equal.

In exercises 79–84, assume $a \neq 0$, $b \neq 0$.

79. To solve $a(a + x) = b^2 - bx$ for x, distribute a on the left side of the equation.
$$a^2 + ax = b^2 - bx$$
Add bx to both sides.
$$a^2 + ax + bx = b^2 - bx + bx$$
$$a^2 + ax + bx = b^2$$
Subtract a^2 from both sides.
$$a^2 + ax + bx - a^2 = b^2 - a^2$$
$$ax + bx = b^2 - a^2$$
Factor both sides. The right side is the difference of two squares.
$$x(a + b) = (b - a)(b + a)$$
Divide both sides by $(a + b)$.
$$\frac{x(a+b)}{(a+b)} = \frac{(b-a)(b+a)}{(a+b)}$$
$$x = b - a$$

81. To solve $\frac{ax}{b} - \frac{bx}{a} = \frac{(a+b)^2}{ab}$ for x, multiply both sides by the common denominator, ab.
$$ab\left(\frac{ax}{b} - \frac{bx}{a}\right) = ab\left(\frac{(a+b)^2}{ab}\right)$$
$$a^2x - b^2x = (a+b)^2$$
Factor the left side.
$$x(a^2 - b^2) = (a+b)^2$$
Divide both sides by $(a^2 - b^2)$.
$$\frac{x(a^2-b^2)}{(a^2-b^2)} = \frac{(a+b)^2}{(a^2-b^2)} \Rightarrow x = \frac{(a+b)^2}{(a^2-b^2)}$$
Factor the numerator and denominator on the right side. Then simplify.
$$x = \frac{(a+b)(a+b)}{(a-b)(a+b)}$$
$$x = \frac{a+b}{a-b}$$

83. To solve $\dfrac{b(bx-1)}{a} - \dfrac{a(1+ax)}{b} = 1$ for x, multiply both sides by the common denominator, ab.

$$ab\left(\dfrac{b(bx-1)}{a} - \dfrac{a(1+ax)}{b}\right) = ab(1)$$
$$b(b(bx-1)) - a(a(1+ax)) = ab$$
$$b^2(bx-1) - a^2(1+ax) = ab$$
$$b^3x - b^2 - a^2 - a^3x = ab$$

Add $b^2 + a^2$ to both sides.
$$b^3x - b^2 - a^2 - a^3x + b^2 + a^2$$
$$= ab + b^2 + a^2$$
$$b^3x - a^3x = a^2 + ab + b^2$$

Factor both sides.
$$x(b-a)(b^2 + ab + a^2) = a^2 + ab + b^2$$

Divide both sides by $(b-a)(b^2+ab+a^2)$ and then simplify.

$$\dfrac{x(b-a)(b^2+ab+a^2)}{(b-a)(b^2+ab+a^2)} =$$
$$\dfrac{a^2+ab+b^2}{(b-a)(b^2+ab+a^2)}$$
$$x = \dfrac{1}{b-a}$$

1.1 Critical Thinking/Discussion/Writing

85. a. The solution set of $x^2 = x$ is $\{0,1\}$, while the solution set of $x = 1$ is $\{1\}$. Therefore, the equations are not equivalent.

b. The solution set of $x^2 = 9$ is $\{-3, 3\}$, while the solution set of $x = 3$ is $\{3\}$. Therefore, the equations are not equivalent.

c. The solution set of $x^2 - 1 = x - 1$ is $\{0, 1\}$, while the solution set of $x = 0$ is $\{0\}$. Therefore, the equations are not equivalent.

d. The equation $\dfrac{x}{x-2} = \dfrac{2}{x-2}$ is an inconsistent equation, so its solution set is $\varnothing$. The solution set of $x = 2$ is $\{2\}$. Therefore, the equations are not equivalent.

86. In the division step, there is division by zero because $x = 1$. The remainder of the argument is invalid due to the division by zero.

1.1 Getting Ready for the Next Section

87. $(3x+7) - x + 2(5-x) = 3x + 7 - x + 10 - 2x$
$= 17$

89. $-2x + 7(x-2) + 4 = -2x + 7x - 14 + 4$
$= 5x - 10$

91. $\dfrac{2x}{5} + \dfrac{x}{2} = \dfrac{4x}{10} + \dfrac{5x}{10} = \dfrac{9x}{10}$

93. $\dfrac{3x}{8} - \dfrac{2x}{3} = \dfrac{9x}{24} - \dfrac{16x}{24} = -\dfrac{7x}{24}$

95. 24 minutes = $\dfrac{24}{60} = \dfrac{2}{5} = 0.4$ hour

97. 0.25 hour = $\dfrac{1}{4} \times 60 = 15$ minutes

99. 5 yards = $5 \times 3 = 15$ feet

101. 3 inches = $3 \times 2.54 = 7.62$ centimeters

1.2 Applications of Linear Equations: Modeling

1.2 Practice Problems

1. Let x = the original price.
Then the discounted price is $x - 0.15x$.
So we have
$x - 0.15x = 29.75$
$0.85x = 29.75$
$x = 35$
The original price of the backpack was $35.

2. Let w = the width of the rectangle. Then $2w + 5$ = the length of the rectangle.
$P = 2l + 2w$, so we have
$28 = 2(2w + 5) + 2w$
$28 = 4w + 10 + 2w$
$28 = 6w + 10$
$18 = 6w$
$3 = w$
The width of the rectangle is 3 m and the length is $2(3) + 5 = 11$ m.

3. Let x = the amount invested in stocks. Then $15{,}000 - x$ = the amount invested in bonds.
$x = 3(15{,}000 - x)$
$x = 45{,}000 - 3x$
$4x = 45{,}000$
$x = 11{,}250$
Tyrick invested $11,250 in stocks and $15,000 − $11,250 = $3,750 in bonds.

4. Let x = the amount of capital. Then $\dfrac{x}{5}$ = the amount invested at 5%, $\dfrac{x}{6}$ = the amount invested at 8%, and $x - \left(\dfrac{x}{5} + \dfrac{x}{6}\right) = \dfrac{19x}{30}$ = the amount invested at 10%.

Principal	Rate	Time	Interest
$\dfrac{x}{5}$	0.05	1	$0.05\left(\dfrac{x}{5}\right)$
$\dfrac{x}{6}$	0.08	1	$0.08\left(\dfrac{x}{6}\right)$
$\dfrac{19x}{30}$	0.1	1	$0.1\left(\dfrac{19x}{30}\right)$

The total interest is $130, so
$$0.05\left(\dfrac{x}{5}\right) + 0.08\left(\dfrac{x}{6}\right) + 0.1\left(\dfrac{19x}{30}\right) = 130$$
Multiply by the LCD, 30.
$$0.3x + 0.4x + 1.9x = 3900$$
$$2.6x = 3900$$
$$x = 1500$$
The total capital is $1500.

5. Let x = the length of the bridge.
Then $x + 130$ = the distance the train travels.
$rt = d$, so
$25(21) = x + 130 \Rightarrow 525 = x + 130 \Rightarrow 395 = x$
The bridge is 395 m long.

6. The initial separation between the ship and the aircraft is 955 miles.
Let t = time elapsed when ship and aircraft meet. Then $32t$ = the distance the ship traveled and $350t$ = distance the aircraft traveled.
$$32t + 350t = 955$$
$$382t = 955$$
$$t = 2.5$$
The aircraft and ship meet after 2.5 hours.

7. Let x = the amount of time they worked together to complete the job. Then $\dfrac{1}{45}x$ = the portion of the job done by Jim and $\dfrac{1}{30}x$ = the portion of the job done by Anita.

$$\dfrac{1}{45}x + \dfrac{1}{30}x = 1$$
$$2x + 3x = 90 \quad \text{Multiply by the LCD, 90.}$$
$$5x = 90$$
$$x = 18$$
It took them 18 minutes to wash the car together.

8. Let x = the amount of 40% sulfuric acid.

Acid	Amount of acid	% acid	Amount of pure acid
40%	x	0.4	$0.4x$
20%	$50 - x$	0.2	$0.2(50 - x)$
25%	50	0.25	$0.25(50)$

Then, we have $0.4x + 0.2(50 - x) = 0.25(50)$.
$$0.4x + 0.2(50 - x) = 0.25(50)$$
$$0.4x + 10 - 0.2x = 12.5$$
$$0.2x + 10 = 12.5$$
$$0.2x = 2.5$$
$$x = 12.5$$
12.5 gallons of 40% sulfuric acid solution should be added.

1.2 Concepts and Vocabulary

1. If the sides of a rectangle are a and b units, the perimeter of the rectangle is <u>$2a + 2b$</u> units.

3. The distance d traveled by an object moving at rate r for time t is given by $d =$ <u>rt</u>.

5. False. The original price can be found by solving the equation $x - 0.3x = 238$ or $0.7x = 238$.

7. False. The interest $I = (100)(0.05)(3)$.

1.2 Building Skills

9. $0.065x$

11. $\$22,000 - x$

13. a. Natasha: $\dfrac{t}{4}$ b. Natasha's brother: $\dfrac{t}{5}$

15. a. $10 - x$ b. $4.60x$
    c. $7.40(10 - x)$

17. a. $8 + x$ b. $0.8 + x$

1.2 Applying the Concepts

19. Let x = the original price. Then,
$x - 0.2x = 1192 \Rightarrow 0.8x = 1192 \Rightarrow x = 1490$
The original price was $1490.

21. Let x = the fourth test score. Then,
$$\frac{81 + 75 + 77 + x}{4} = 80$$
$233 + x = 320 \Rightarrow x = 87$
Kylee needs an 87 on the next test to raise the average to 80.

23. Let w = the width of the rectangle.
Then $2w - 5$ = the length of the rectangle.
$$2w + 2(2w - 5) = 80$$
$$2w + 4w - 10 = 80$$
$$6w - 10 = 80$$
$$6w = 90$$
$$w = 15, \ 2w - 5 = 25$$
The width of the rectangle is 15 ft and its length is 25 feet.

25. Substitute 920 for A, 500 for P and 0.06 for r into the formula $A = P + Prt$. Solve for t.
$$920 = 500 + 500 \cdot 0.06t$$
$$920 = 500 + 30t$$
$$920 - 500 = 500 + 30t - 500$$
$$420 = 30t$$
$$\frac{420}{30} = \frac{30t}{30} \Rightarrow 14 = t$$
It will take 14 years until the amount of money available is $920.

27. Substitute 427 for M and 302 for b into the formula $M = \frac{m-b}{4}$. Solve for m.
$$427 = \frac{m - 302}{4}$$
$$4(427) = 4\left(\frac{m - 302}{4}\right)$$
$$1708 = m - 302$$
$$1708 + 302 = m - 302 + 302 \Rightarrow 2010 = m$$
The gross monthly income must be $2010.

29. Substitute 170 for P into the formula $P = 200 - 0.02q$. Solve for q.
$$170 = 200 - 0.02q$$
$$170 - 200 = 200 - 0.02q - 200$$
$$-30 = -0.02q$$
$$\frac{-30}{-0.02} = \frac{-0.02q}{-0.02} \Rightarrow 1500 = q$$
Note that the solution must fall between 100 and 2000 cameras. The retailer must order 1500 cameras.

31. Let x = the cost of the less expensive land. Then $x + 23,000$ = the cost of the more expensive land. Together they cost $147,000, so
$$x + (x + 23,000) = 147,000$$
$$2x + 23,000 = 147,000$$
$$2x = 124,000 \Rightarrow x = 62,000$$
The less expensive piece of land costs $62,000 and the more expensive piece of land costs $62,000 + $23,000 = $85,000.

33. Let x = the lottery ticket sales in July. Then $1.10x$ = the lottery ticket sales in August. A total of 1113 tickets were sold, so
$$x + 1.10x = 1113$$
$$2.10x = 1113 \Rightarrow x = 530$$
530 tickets were sold in July, and 1.10(530) = 583 tickets were sold in August.

35. Let x = the amount the younger son receives. Then $4x$ = the amount the older son receives. Together they receive $225,000, so
$$x + 4x = 225,000 \Rightarrow 5x = 225,000 \Rightarrow$$
$$x = 45,000$$
The younger son will received $45,000, and the older son will receive
$4(\$45,000) = \$180,000$.

37. a. Let x = the number of points needed to average 75.
$$\frac{87 + 59 + 73 + x}{4} = 75$$
$$219 + x = 300$$
$$x = 81$$
You need to score 81 in order to average 75.

b. $$\frac{87 + 59 + 73 + 2x}{5} = 75$$
$$219 + 2x = 375$$
$$2x = 156$$
$$x = 78$$
You need to score 78 in order to average 75 if the final carries double weight.

39. Let x = the amount invested in a tax shelter.
 Then $7000 - x$ = the amount invested in a bank.

Investment	Principal	Rate	Time	Interest
Tax shelter	x	0.09	1	$0.09x$
Bank	$7000 - x$	0.06	1	$0.06(7000 - x)$

The total interest was $540, so
$0.09x + 0.06(7000 - x) = 540$
$0.09x + 420 - 0.06x = 540$
$0.03x + 420 = 540$
$0.03x = 120 \Rightarrow x = 4000$
Mr. Mostafa invested $4000 in a tax shelter and $7000 - 4000 = \$3000$ in a bank.

41. Let x = the amount to be invested at 8%.

Principal	Rate	Time	Interest
5000	0.05	1	250
x	0.08	1	$0.08x$
$5000 + x$	0.06	1	$0.06(5000 + x)$

The amount of interest for the total investment is the sum of the interest earned on the individual investments, so
$0.06(5000 + x) = 250 + 0.08x$
$300 + 0.06x = 250 + 0.08x$
$50 + 0.06x = 0.08x$
$50 = 0.02x \Rightarrow 2500 = x$
So, $2500 must be invested at 8%.

43. There is a profit of $2 on each shaving set. They want to earn $40,000 + $30,000 = $70,000. Let x = the number of shaving sets to be sold. Then $2x$ = the amount of profit for x shaving sets. So, $2x = 70,000 \Rightarrow x = 35,000$
They must sell 35,000 shaving sets.

45. Let x = the time the second car travels.
 Then $1 + x$ = the time the first car travels. So,

	Rate	Time	Distance
First car	50	$1 + x$	$50(1 + x)$
Second car	70	x	$70x$

The distances are equal, so
$50(1 + x) = 70x$
$50 + 50x = 70x$
$50 = 20x \Rightarrow 2.5 = x$
So, it will take the second car 2.5 hours to overtake the first car.

47. At 20 miles per hour, it will take Lucas two minutes to bike the remaining 2/3 of a mile.
$\left(\dfrac{20\,\text{mi}}{1\,\text{hr}} = \dfrac{20\,\text{mi}}{60\,\text{min}} = \dfrac{1\,\text{mi}}{3\,\text{min}}\right)$ So his brother will have to bike 1 mile in 2 minutes:
$\dfrac{1\,\text{mi}}{2\,\text{min}} = \dfrac{30\,\text{mi}}{60\,\text{min}} = \dfrac{30\,\text{mi}}{1\,\text{hr}}$

49. Let x = the rate the slower car travels. Then $x + 7$ = the rate the faster car travels. So,

	Rate	Time	Distance
First car	x	3	$3x$
Second car	$x + 7$	3	$3(x + 7)$

The planes are 621 miles apart, so
$3x + 3(x + 7) = 621$
$3x + 3x + 21 = 621$
$6x + 21 = 621 \Rightarrow 6x = 600 \Rightarrow x = 100$
One car is traveling at 100 miles per hour, and the other car is traveling at 107 miles per hour.

51. Let x = the amount of time it takes both pumps to drain the pool together. The old pump drains 1/6 of the pool in one hour, so it will drain $x/6$ of the pool. The new pump drains 1/4 of the pool in one hour, so it will drain $x/4$ of the pool. So,
$\dfrac{x}{6} + \dfrac{x}{4} = 1$
$12\left(\dfrac{x}{6} + \dfrac{x}{4}\right) = 12(1)$
$2x + 3x = 12$
$5x = 12 \Rightarrow x = \dfrac{12}{5} = 2.4$
It will take the two pumps 2.4 hours or 2 hours and 24 minutes to drain the pool working together.

53. Let x = the amount of time it takes both blowers to fill the blimp together. The first blower fills 1/6 of the blimp in one hour, so it will fill $x/6$ of the blimp. The second blower fills 1/9 of the blimp in one hour, so it will fill $x/9$ of the blimp. So,
$\dfrac{x}{6} + \dfrac{x}{9} = 1$
$36\left(\dfrac{x}{6} + \dfrac{x}{9}\right) = 36(1)$
$6x + 4x = 36$
$10x = 36 \Rightarrow x = 3.6$

(continued on next page)

Chapter 1 Equations and Inequalities

(*continued*)

It will take the two blowers 3.6 hours or 3 hours and 36 minutes to fill the blimp working together.

55. Let $x =$ the time for the new sorter to complete the job alone. Then $2x =$ the time for the old sorter to complete the job alone.. So, $8/x$ is the portion of the job done by the new sorter and $8/(2x) = 4/x$ is the portion of the job done by the old sorter.

$$\frac{8}{x} + \frac{4}{x} = 1 \Rightarrow x\left(\frac{8}{x} + \frac{4}{x}\right) = x(1) \Rightarrow$$
$$8 + 4 = x \Rightarrow 12 = x$$

It will take 12 hours for the new sorter to complete the job working alone.

57. Let $x =$ the amount of time they worked together to complete the job. Then $x/9 =$ the portion of the job done by a professor and $x/6 =$ the portion of the job done by a student.

$$3\left(\frac{1}{9}x\right) + 2\left(\frac{1}{6}x\right) = 1$$
$$3\left(\frac{x}{3} + \frac{x}{3}\right) = 1(3)$$
$$2x = 3 \Rightarrow x = 1.5$$

The job will take 1.5 hours.

59. Let $x =$ the number of grams of pure gold to be added. Then $120 + x =$ the number of grams in the new alloy.

| Number of grams of gold in the new alloy | = | Number of grams of gold in the original alloy | + | Number of grams of pure gold added. |

$0.8(120 + x) = 0.75(120) + x \Rightarrow 96 + 0.8x = 90 + x \Rightarrow 6 + 0.8x = x \Rightarrow 6 = 0.2x \Rightarrow 30 = x$

So, 30 grams of pure gold must be added.

61. Let $x =$ the amount of 60-40 solder to be added.

Solder	Amount of solder	% tin	Amount of tin
60-40	x	0.6	$0.6x$
40-60	$600 - x$	0.4	$0.4(600 - x)$
55-45	600	0.55	$0.55(600)$

$$0.6x + 0.4(600 - x) = 0.55(600)$$
$$0.6x + 240 - 0.4x = 330$$
$$0.2x + 240 = 330$$
$$0.2x = 90$$
$$x = 450$$

450 g of 60-40 solder must be added to 150 g of 40-60 solder.

63. Let $x =$ the amount of 60% boric acid

Solution	Amount of solution	% boric acid	Amount of boric acid
60%	x	0.6	$0.6x$
8%	7.5	0.08	$0.08(7.5)$
20%	$x + 7.5$	0.2	$0.2(x + 7.5)$

$$0.6x + 0.08(7.5) = 0.2(x + 7.5)$$
$$0.6x + 0.6 = 0.2x + 1.5$$
$$0.4x = 0.9 \Rightarrow x = 2.25$$

2.25 liters of 60% boric acid solution are needed.

65. Let $x =$ the number of dimes. Then $3x =$ the number of nickels, and $4x =$ the number of quarters.

Coin	Number of coins	Value of each coin	Total
Nickels	$3x$	0.05	$0.15x$
Dimes	x	0.10	$0.10x$
Quarters	$4x$	0.25	x

$$0.15x + 0.10x + x = 96.25$$
$$1.25x = 96.25 \Rightarrow x = 77$$

So, there are 77 dimes, $3(77) = 231$ nickels, and $4(77) = 308$ quarters.

67. Let x = Eric's grandfather's age now. Then $x - 57$ = Eric's age now. $x + 5$ = Eric's grandfather's age five years from now, and $(x - 57) + 5 = x - 52$ = Eric's age five years from now. So,
$$x + 5 = 4(x - 52)$$
$$x + 5 = 4x - 208$$
$$x + 213 = 4x \Rightarrow 213 = 3x \Rightarrow 71 = x$$
Eric's grandfather is 71 years old now.

69. Let x = the length of the tin rectangle. Then $x - 2$ = the length of the box.

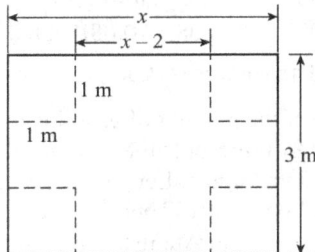

The formula for volume is $V = lwh$, so
$(x - 2)(1)(1) = 2 \Rightarrow x - 2 = 2 \Rightarrow x = 4$
The length of the tin rectangle is 4 m.

1.2 Beyond the Basics

71. Let x = the average speed for the second half of the trip.

	Rate	Distance	Time
1st half	75	D	$\dfrac{D}{75}$
2nd half	x	D	$\dfrac{D}{x}$
Whole trip	60	$2D$	$\dfrac{2D}{60}$

So,
$$\frac{D}{75} + \frac{D}{x} = \frac{2D}{60}$$
$$300x\left(\frac{D}{75} + \frac{D}{x}\right) = 300x\left(\frac{2D}{60}\right)$$
$$4Dx + 300D = 5x(2D)$$
$$4Dx + 300D = 10Dx$$
$$300D = 6Dx \Rightarrow 50 = x$$
The average speed for the second half of the drive is 50 mph.

73. Let x = the number of liters of water in the original mixture. Then $5x$ = the number of liters of alcohol in the original mixture, and $6x$ = the total number of liters in the original mixture.

$x + 5$ = the number of liters of water in the new mixture. Then $6x + 5$ = the total number of liters in the new mixture. Since the ratio of alcohol to water in the new mixture is 5:2, then the amount of alcohol in the new mixture is 5/7 of the total mixture or $\dfrac{5}{7}(6x + 5)$.

There was no alcohol added, so the amount of alcohol in the original mixture equals the amount of alcohol in the new mixture. This gives
$$\frac{5}{7}(6x + 5) = 5x$$
$$5(6x + 5) = 35x$$
$$30x + 25 = 35x \Rightarrow 25 = 5x \Rightarrow 5 = x$$
So, there were 5 liters of water in the original mixture and 25 liters of alcohol.

75. Let x = Democritus' age now. Then $x/6$ = the number of years as a boy, $x/8$ = the number of years as a youth, and $x/2$ = the number of years as a man. He has spent 15 years as a mature adult. So,
$$\frac{x}{6} + \frac{x}{8} + \frac{x}{2} + 15 = x$$
$$24\left(\frac{x}{6} + \frac{x}{8} + \frac{x}{2} + 15\right) = 24x$$
$$4x + 3x + 12x + 360 = 24x$$
$$19x + 360 = 24x$$
$$360 = 5x \Rightarrow 72 = x$$
Democritus is 72 years old.

77. There are 180 minutes from 3 p.m. to 6 p.m. So, the number of minutes before 6 p.m. plus 50 minutes plus 4 × the number of minutes before 6 p.m. equals 180 minutes. Let x = the number of minutes before 6 p.m. So,
$x + 50 + 4x = 180 \Rightarrow 5x + 50 = 180 \Rightarrow$
$5x = 130 \Rightarrow x = 26$
So it is 26 minutes before 6 p.m. or 5:34 p.m. Check this by verifying that $26 + 50 = 76$ minutes before 6 p.m. is the same time as $4(26) = 104$ minutes after 3 p.m. Seventy-six minutes before 6 p.m. is 4:44 p.m., while 104 minutes after 3 p.m. is also 4:44 p.m.

79. a. Because of the head wind, the plane flies at 140 mph from Atlanta to Washington and 160 mph from Washington to Atlanta. Let $x =$ the distance the plane flew before turning back. So,

	Rate	Distance	Time
to	140	x	$x/140$
from	160	x	$x/160$

$$\frac{x}{140} + \frac{x}{160} = 1.5$$
$$160x + 140x = 1.5(140)(160)$$
$$300x = 33{,}600 \Rightarrow x = 112$$

The plane flew 112 miles before turning back.

b. The plane traveled 224 miles in 1.5 hours, so the average speed is $\dfrac{224}{1.5} = 149.33$ mph.

81. The ends of the trains are 440 feet apart when the trains first meet. Because the speed is in miles per hour, but the distance is measured in feet, convert the speed to feet per hour. The distance is relatively short, so we need to convert the feet per hour to feet per second.

$$\frac{50 \text{ mi}}{\text{hr}} \cdot \frac{5280 \text{ ft}}{\text{mi}} \cdot \frac{1 \text{ hr}}{60 \text{ min}} \cdot \frac{1 \text{ min}}{60 \text{ sec}} = \frac{220}{3} \text{ feet}$$

per second.

$$\frac{60 \text{ mi}}{\text{hr}} \cdot \frac{5280 \text{ ft}}{\text{mi}} \cdot \frac{1 \text{ hr}}{60 \text{ min}} \cdot \frac{1 \text{ min}}{60 \text{ sec}} = 88 \text{ feet per second.}$$

	Rate	Time	Distance
Train A	$\frac{220}{3}$	t	$\frac{220t}{3}$
Train B	88	t	$88t$

230 ft, 50 mph, $\frac{220}{3}$ ft/sec ; 210 ft, 60 mph, 88 ft/sec

The total distance is 440 feet, so we have

$$\frac{220t}{3} + 88t = 440$$
$$220t + 264t = 1320$$
$$484t = 1320$$
$$t = 2.7$$

The ends of the trains will pass each other after 2.7 seconds.

1.2 Critical Thinking/Discussion/Writing

83. If x represents the amount the pawn shop owner paid for the first watch and the owner made a profit of 10%, then $1.1x = 499$, so $x = 453.64$. If y represents the amount the pawn shop owner paid for the second watch and the owner lost 10%, then $0.9y = 499$, so $y = 554.44$. Together the two watches cost $\$453.64 + \$554.44 = \$1008.08$. But the pawn shop owner sold the two watches for $998, so there was a loss. The amount of loss is $(1008.08 - 998)/1008.08 = 10.08/1008.08 \approx 0.01 = 1\%$. The answer is (C).

84. Let x represent the amount of gasoline used in July. Then $0.8x$ represents the amount of gasoline used in August. Let y represent the price of gasoline in July. Then $1.2y$ represents the cost of gasoline in August. The cost of gasoline used in July is xy (amount × price), and the cost of gasoline used in August is $0.8x \times 1.2y = 0.96xy$. So the cost of gasoline used in August is 96% of the cost of gasoline used in July, which is a decrease of 4%. The answer is (D).

1.2 Getting Ready for the Next Section

85. $\sqrt{8} = \sqrt{4 \cdot 2} = \sqrt{4}\sqrt{2} = 2\sqrt{2}$

87. $\sqrt{12} = \sqrt{4 \cdot 3} = \sqrt{4}\sqrt{3} = 2\sqrt{3}$

89. $\dfrac{2 + 3\sqrt{8}}{2} = \dfrac{2 + 3 \cdot 2\sqrt{2}}{2} = \dfrac{2}{2} + \dfrac{3 \cdot 2\sqrt{2}}{2} = 1 + 3\sqrt{2}$

91. $\dfrac{15 - \sqrt{75}}{5} = \dfrac{15 - 5\sqrt{3}}{5} = \dfrac{15}{5} - \dfrac{5\sqrt{3}}{5} = 3 - \sqrt{3}$

93. $x^2 + x = x(x + 1)$

95. $x^2 - 4 = (x - 2)(x + 2)$

97. $x^2 + 4x + 4 = (x + 2)^2$

99. $x^2 - 8x + 7 = (x - 1)(x - 7)$

101. $6x^2 - x - 1 = (3x + 1)(2x - 1)$

103. $-5x^2 + 3x + 2 = (5x + 2)(-x + 1)$
$ = (5x + 2)(1 - x)$

1.3 Quadratic Equations

1.3 Practice Problems

1. $x^2 + 25x = -84$
$x^2 + 25x + 84 = 0$
$(x+4)(x+21) = 0$
$x + 4 = 0 \mid x + 21 = 0$
$x = -4 \mid x = -21$
Solution set: $\{-21, -4\}$

2. $2m^2 = 5m$
$2m^2 - 5m = 0$
$m(2m - 5) = 0$
$m = 0 \mid 2m - 5 = 0$
$ 2m = 5$
$ m = \dfrac{5}{2}$
Solution set: $\left\{0, \dfrac{5}{2}\right\}$

3. $x^2 - 6x = -9$
$x^2 - 6x + 9 = 0$
$(x - 3)^2 = 0$
$x - 3 = 0$
$x = 3$
Solution set: $\{3\}$

4. $(x + 2)^2 = 5$
$x + 2 = \pm\sqrt{5}$
$x = -2 \pm \sqrt{5}$
Solution set: $\{-2 - \sqrt{5}, -2 + \sqrt{5}\}$

5. $x^2 - 6x + 7 = 0$
$x^2 - 6x = -7$
$x^2 - 6x + 9 = -7 + 9$
$(x - 3)^2 = 2$
$x - 3 = \pm\sqrt{2}$
$x = 3 \pm \sqrt{2}$
Solution set: $\{3 - \sqrt{2}, 3 + \sqrt{2}\}$

6. $4x^2 - 24x + 25 = 0$
$4x^2 - 24x = -25$
$x^2 - 6x = -\dfrac{25}{4}$
$x^2 - 6x + 9 = -\dfrac{25}{4} + 9$
Solution set: $\left\{3 - \dfrac{\sqrt{11}}{2}, 3 + \dfrac{\sqrt{11}}{2}\right\}$

7. $6x^2 - x - 2 = 0$
$a = 6, b = -1, c = -2$
$x = \dfrac{-b \pm \sqrt{b^2 - 4ac}}{2a}$
$= \dfrac{-(-1) \pm \sqrt{(-1)^2 - 4(6)(-2)}}{2(6)}$
$= \dfrac{1 \pm \sqrt{49}}{12} = \dfrac{1 \pm 7}{12} = \dfrac{-6}{12} = -\dfrac{1}{2}$ or $\dfrac{8}{12} = \dfrac{2}{3}$
Solution set: $\left\{-\dfrac{1}{2}, \dfrac{2}{3}\right\}$

8. a. $3x^2 - 2x - 5 = 0$
$a = 3, b = -2, c = -5$
$b^2 - 4ac = (-2)^2 - 4(3)(-5) = 64$
Because 64 is a perfect square, there are two distinct rational solutions.

   b. $x^2 + 4 = 4x$
Rewrite the equation in standard form.
$x^2 - 4x + 4 = 0$
$a = 1, b = -4, c = 4$
$b^2 - 4ac = (-4)^2 - 4(1)(4) = 0$
The discriminant is 0, so there is 1 rational solution (a double solution).

   c. $3x^2 + 6x = -6$
Rewrite the equation in standard form.
$3x^2 + 6x + 6 = 0$
$a = 3, b = 6, c = 6$
$b^2 - 4ac = 6^2 - 4(3)(6) = -36$
The discriminant is negative, so there are no real solutions.

44 Chapter 1 Equations and Inequalities

9. Let r = the radius of the battery. Then $2r$ = the width of the square, and $(2r)^2$ the area of the square. The area of the battery is $\pi r^2 = 64\pi$, so $r^2 = 64 \Rightarrow r = 8$ mm. (Note that the radius must be positive.) Therefore the area of the square is $(2r)^2 = (2 \cdot 8)^2 = 256$ sq mm. The area of the square not covered by the battery is $256 - 64\pi \approx 256 - 64(3.14) \approx 55$ sq mm.

10. Let x = the frontage of the building. Then $5x$ = the depth of the building and $5x - 45$ = the depth of the rear portion.
$$x(5x - 45) = 2100$$
$$5x^2 - 45x = 2100$$
$$5x^2 - 45x - 2100 = 0$$
$a = 5, b = -45, c = -2100$
$$x = \frac{-(-45) \pm \sqrt{(-45)^2 - 4(5)(-2100)}}{2(5)}$$
$$= \frac{45 \pm \sqrt{44,025}}{10} \approx -16.48 \text{ or } 25.48$$
Reject the negative solution.
$5x = 5 \cdot 25.482 = 127.41$
The building is approximately 25.48 ft by 127.41 ft.

11. $\Phi = \dfrac{\text{length}}{\text{width}} \Rightarrow \dfrac{1+\sqrt{5}}{2} = \dfrac{x}{36}$
$x = 36\left(\dfrac{1+\sqrt{5}}{2}\right) = 18 + 18\sqrt{5} \approx 58.25$ ft

1.3 Concepts and Vocabulary

1. Any equation of the form $ax^2 + bx + c = 0$ with $a \neq 0$, is called a quadratic equation.

3. From the Square Root Property, we know that if $u^2 = 5$, then $u = \pm\sqrt{5}$.

5. True

7. True

1.3 Building Skills

9. $(-6)^2 + 4(-6) - 12 = 36 - 24 - 12 = 0$
–6 is a solution of the equation.

11. $3\left(\dfrac{2}{3}\right)^2 + 7\left(\dfrac{2}{3}\right) - 6 = 3\left(\dfrac{4}{9}\right) + \dfrac{14}{3} - 6$
$= \dfrac{4}{3} + \dfrac{14}{3} - 6 = 0$
2/3 is a solution of the equation.

13. $\left(2-\sqrt{3}\right)^2 - 4\left(2-\sqrt{3}\right) + 1$
$= \left(4 - 4\sqrt{3} + 3\right) - 8 + 4\sqrt{3} + 1 = 0$
$2 - \sqrt{3}$ is a solution of the equation.

15. $4\left(2+\sqrt{3}\right)^2 - 8\left(2+\sqrt{3}\right) + 13$
$= 4\left(7 + 4\sqrt{3}\right) - 8\left(2+\sqrt{3}\right) + 13$
$= 28 + 16\sqrt{3} - 16 - 8\sqrt{3} + 13 = 25 + 8\sqrt{3} \neq 0$
$2 + \sqrt{3}$ is not a solution of the equation.

17. $k(1)^2 + 1 - 3 = 0 \Rightarrow k - 2 = 0 \Rightarrow k = 2$

19. $x^2 - 5x = 0 \Rightarrow x(x-5) = 0 \Rightarrow$
$x = 0$ or $x - 5 = 0 \Rightarrow x = 0$ or $x = 5$

21. $x^2 + 5x = 14$
$x^2 + 5x - 14 = 0$
$(x+7)(x-2) = 0$
$x + 7 = 0$ or $x - 2 = 0$
$x = -7$ or $x = 2$

23. $x^2 = 5x + 6$
$x^2 - 5x - 6 = 0$
$(x-6)(x+1) = 0$
$x - 6 = 0$ or $x + 1 = 0$
$x = 6$ or $x = -1$

25. $3x^2 = 48 \Rightarrow x^2 = 16 \Rightarrow x = \pm 4$

27. $x^2 + 1 = 5 \Rightarrow x^2 = 4 \Rightarrow x = \pm 2$

29. $(x-1)^2 = 16$
$x - 1 = -4$ or $x - 1 = 4 \Rightarrow x = -3$ or $x = 5$

31. To complete the square, find 1/2 of the coefficient of the x-term, $4/2 = 2$, and then square the answer. $2^2 = 4$.

33. To complete the square, find 1/2 of the coefficient of the x-term, $6/2 = 3$, and then square the answer. $3^2 = 9$.

35. To complete the square, find $1/2$ of the coefficient of the x-term and then square the answer. $\left(\dfrac{-7}{2}\right)^2 = \dfrac{49}{4}$.

37. To complete the square, find $1/2$ of the coefficient of the x-term, $\dfrac{1}{2} \cdot \dfrac{1}{3} = \dfrac{1}{6}$ and then square the answer. $\left(\dfrac{1}{6}\right)^2 = \dfrac{1}{36}$.

39. To complete the square, find $1/2$ of the coefficient of the x-term and then square the answer. $(a/2)^2 = a^2/4$.

41. $x^2 + 2x - 5 = 0 \Rightarrow x^2 + 2x = 5$
Now, complete the square.
$x^2 + 2x + 1 = 5 + 1 \Rightarrow (x+1)^2 = 6 \Rightarrow$
$x + 1 = \pm\sqrt{6} \Rightarrow x = -1 \pm \sqrt{6}$

43. $x^2 - 3x - 1 = 0$
$x^2 - 3x = 1$
Now, complete the square.
$x^2 - 3x + \dfrac{9}{4} = 1 + \dfrac{9}{4}$
$\left(x - \dfrac{3}{2}\right)^2 = \dfrac{13}{4}$
$x - \dfrac{3}{2} = \pm\dfrac{\sqrt{13}}{2}$
$x = \dfrac{3}{2} \pm \dfrac{\sqrt{13}}{2} = \dfrac{3 \pm \sqrt{13}}{2}$

45. $2r^2 + 3r = 9$
$r^2 + \dfrac{3}{2}r = \dfrac{9}{2}$
$r^2 + \dfrac{3}{2}r + \dfrac{9}{16} = \dfrac{9}{2} + \dfrac{9}{16}$
$\left(r + \dfrac{3}{4}\right)^2 = \dfrac{81}{16}$
$r + \dfrac{3}{4} = \pm\dfrac{9}{4} \Rightarrow r = \dfrac{-3 \pm 9}{4} = \dfrac{3}{2}$ or -3

In exercises 47–52, use the quadratic formula $x = \dfrac{-b \pm \sqrt{b^2 - 4ac}}{2a}$.

47. $x^2 + 2x - 4 = 0 \Rightarrow a = 1, b = 2, c = -4$
$x = \dfrac{-2 \pm \sqrt{2^2 - 4(1)(-4)}}{2(1)} = \dfrac{-2 \pm \sqrt{4 + 16}}{2}$
$= \dfrac{-2 \pm \sqrt{20}}{2} = \dfrac{-2 \pm 2\sqrt{5}}{2} = -1 \pm \sqrt{5}$

49. $6x^2 = 7x + 5 \Rightarrow 6x^2 - 7x - 5 = 0 \Rightarrow$
$a = 6, b = -7, c = -5$
$x = \dfrac{-(-7) \pm \sqrt{(-7)^2 - 4(6)(-5)}}{2(6)}$
$= \dfrac{7 \pm \sqrt{49 + 120}}{12} = \dfrac{7 \pm \sqrt{169}}{12} = \dfrac{7 \pm 13}{12}$
$x = \dfrac{7 + 13}{12} = \dfrac{20}{12} = \dfrac{5}{3}$ or
$x = \dfrac{7 - 13}{12} = \dfrac{-6}{12} = -\dfrac{1}{2}$

51. $3z^2 - 2z = 7 \Rightarrow 3z^2 - 2z - 7 = 0 \Rightarrow$
$a = 3, b = -2, c = -7$
$z = \dfrac{-(-2) \pm \sqrt{(-2)^2 - 4(3)(-7)}}{2(3)}$
$= \dfrac{2 \pm \sqrt{4 + 84}}{6} = \dfrac{2 \pm \sqrt{88}}{6} = \dfrac{2 \pm 2\sqrt{22}}{6}$
$= \dfrac{1 \pm \sqrt{22}}{3}$

53. $2x^2 + 5x - 3 = 0$
$(2x - 1)(x + 3) = 0$
$2x - 1 = 0$ or $x + 3 = 0$
$x = \dfrac{1}{2}$ or $x = -3$

55. $(3x - 2)^2 - 16 = 0$
$(3x - 2)^2 = 16$
$3x - 2 = \pm 4$
$3x - 2 = -4$ or $3x - 2 = 4$
$3x = -2$ $\qquad$ $3x = 6$
$x = -\dfrac{2}{3}$ or $\qquad x = 2$

57.
$$5x^2 - 6x = 4x^2 + 6x - 3$$
$$x^2 - 12x = -3$$
Now, complete the square.
$$x^2 - 12x + 36 = -3 + 36$$
$$(x-6)^2 = 33$$
$$x - 6 = \pm\sqrt{33} \Rightarrow x = 6 \pm \sqrt{33}$$

59. $3p^2 + 8p + 4 = 0 \Rightarrow a = 3, b = 8, c = 4$
$$p = \frac{-8 \pm \sqrt{8^2 - 4(3)(4)}}{2(3)}$$
$$= \frac{-8 \pm \sqrt{64 - 48}}{6} = \frac{-8 \pm \sqrt{16}}{6} = \frac{-8 \pm 4}{6}$$
$$p = \frac{-4}{6} = -\frac{2}{3} \text{ or } p = \frac{-12}{6} = -2$$

61. $3y^2 + 5y + 2 = 0$
$(3y + 2)(y + 1) = 0$
$3y + 2 = 0 \text{ or } y + 1 = 0$
$$y = -\frac{2}{3} \text{ or } y = -1$$

63. $5x^2 + 12x + 4 = 0$
$(5x + 2)(x + 2) = 0$
$5x + 2 = 0 \text{ or } x + 2 = 0$
$$x = -\frac{2}{5} \text{ or } x = -2$$

65. $5y^2 + 10y + 4 = 2y^2 + 3y + 1$
$3y^2 + 7y = -3$
$y^2 + \frac{7}{3}y = -1$
Now, complete the square.
$$y^2 + \frac{7}{3}y + \frac{49}{36} = -1 + \frac{49}{36}$$
$$\left(y + \frac{7}{6}\right)^2 = \frac{13}{36}$$
$$y + \frac{7}{6} = \pm\sqrt{\frac{13}{36}} = \pm\frac{\sqrt{13}}{6}$$
$$y = -\frac{7}{6} \pm \frac{\sqrt{13}}{6} = \frac{-7 \pm \sqrt{13}}{6}$$

67. $2x^2 + x = 15$
$2x^2 + x - 15 = 0$
$(2x - 5)(x + 3) = 0$
$2x - 5 = 0 \text{ or } x + 3 = 0 \Rightarrow x = \frac{5}{2} \text{ or } x = -3$

69.
$$12x^2 - 10x = 12$$
$$12x^2 - 10x - 12 = 0$$
$$2(6x^2 - 5x - 6) = 0$$
$$6x^2 - 5x - 6 = 0$$
$$(3x + 2)(2x - 3) = 0$$
$3x + 2 = 0 \text{ or } 2x - 3 = 0$
$$x = -\frac{2}{3} \text{ or } x = \frac{3}{2}$$

71. $(x + 13)(x + 5) = -2 \Rightarrow x^2 + 18x + 65 = -2 \Rightarrow$
$x^2 + 18x + 67 = 0 \Rightarrow a = 1, b = 18, c = 67$
$$x = \frac{-18 \pm \sqrt{18^2 - 4(1)(67)}}{2(1)}$$
$$= \frac{-18 \pm \sqrt{324 - 268}}{2} = \frac{-18 \pm \sqrt{56}}{2}$$
$$= \frac{-18 \pm 2\sqrt{14}}{2} = -9 \pm \sqrt{14}$$

73. $18x^2 - 45x = -7$
$18x^2 - 45x + 7 = 0$
$(3x - 7)(6x - 1) = 0$
$3x - 7 = 0 \text{ or } 6x - 1 = 0 \Rightarrow x = \frac{7}{3} \text{ or } x = \frac{1}{6}$

75. $2t^2 - 5 = 0 \Rightarrow a = 2, b = 0, c = -5$
$$t = \frac{0 \pm \sqrt{0^2 - 4(2)(-5)}}{2(2)} = \frac{\pm\sqrt{40}}{4}$$
$$= \pm\frac{2\sqrt{10}}{4} = \pm\frac{\sqrt{10}}{2}$$

77. $4x^2 - 10x - 750 = 0$
$2(2x^2 - 5x - 375) = 0$
$2x^2 - 5x - 375 = 0$
$(2x + 25)(x - 15) = 0$
$2x + 25 = 0 \text{ or } x - 15 = 0$
$$x = -\frac{25}{2} \text{ or } x = 15$$

79. $x^2 + 4x - 2 = 0$
$a = 1, b = 4, c = -2$
$b^2 - 4ac = 4^2 - 4(1)(-2) = 24$
Because the discriminant, 24, is not a perfect square, there are two distinct irrational solutions.

81. $-3x^2 = 4$
Write the equation in standard form.
Write the equation in standard form.
$-3x^2 - 4 = 0$
$a = -3, b = 0, c = -4$
$b^2 - 4ac = (0)^2 - 4(-3)(-4) = -48$
Because the discriminant is –48, there are no real solutions.

83. $6x^2 - 3x - 2 = 0$
$a = 6, b = -3, c = -2$
$b^2 - 4ac = (-3)^2 - 4(6)(-2) = 57$
Because the discriminant, 57, is not a perfect square, there are two distinct irrational solutions.

85. $2x^2 - 3x + 1 = 0$
$a = 2, b = -3, c = 1$
$b^2 - 4ac = (-3)^2 - 4(2)(1) = 1$
Because the discriminant, 1, is a perfect square, there are two distinct rational solutions.

87. $\Phi = \dfrac{\text{length}}{\text{width}} \Rightarrow \dfrac{1+\sqrt{5}}{2} = \dfrac{x}{14.72}$
$x = 14.72\left(\dfrac{1+\sqrt{5}}{2}\right) \approx 23.82$ in.

89. $\Phi = \dfrac{\text{length}}{\text{width}} \Rightarrow \dfrac{1+\sqrt{5}}{2} = \dfrac{8.46}{x}$
$x = 8.46\left(\dfrac{2}{1+\sqrt{5}}\right) \approx 5.23$ cm

1.3 Applying the Concepts

91. Let x = the width of the plot. Then $3x$ = the length of the plot. So, $3x^2 = 10{,}800 \Rightarrow$
$x^2 = 3600 \Rightarrow x = 60$.
The plot is 60 ft by 180 ft.

93. Let x = the first integer. Then $28 - x$ = the second integer. So, $x(28 - x) = 147 \Rightarrow$
$28x - x^2 = 147 \Rightarrow 0 = x^2 - 28x + 147 \Rightarrow$
$0 = (x - 7)(x - 21) \Rightarrow x = 7$ or $x = 21$.
The sum of the two numbers is 28. So, the numbers are 7 and 21.

95. $p = 0.003x^2 + 0.2573x + 83.975$

  a. The year 2014 corresponds to $x = 14$.
  $p = 0.003(14)^2 + 0.2573(14) + 83.975$
  $= 88.1652$
  In 2014, about 88% of people age 25 and older obtained a bachelor's degrees.

  b. The year 2020 corresponds to $x = 20$.
  $p = 0.003(20)^2 + 0.2573(20) + 83.975$
  $= 90.321$
  In 2020, about 90% of people age 25 and older will obtain a bachelor's degrees.

97. Let x = the width of the rectangle. Then $x + 5$ = the length of the rectangle. So,
$x(x + 5) = 500 \Rightarrow x^2 + 5x = 500 \Rightarrow$
$x^2 + 5x - 500 = 0 \Rightarrow (x + 25)(x - 20) = 0 \Rightarrow$
$x = -25$ or $x = 20$. The length cannot be negative, so reject that solution.
$x = 20 \Rightarrow x + 5 = 25$. The rectangle is 25 cm by 20 cm.

99. Let x = one piece of the wire. Then $16 - x$ = the other piece of the wire. Each piece is bent into a square, so the sides of the squares are $\dfrac{x}{4}$ and $\dfrac{16 - x}{4}$, respectively.

$\left(\dfrac{x}{4}\right)^2 + \left(\dfrac{16 - x}{4}\right)^2 = 10$
$\dfrac{x^2}{16} + \dfrac{256 - 32x + x^2}{16} = 10$
$x^2 + 256 - 32x + x^2 = 160$
$2x^2 - 32x + 96 = 0$
$x^2 - 16x + 48 = 0$
$(x - 12)(x - 4) = 0 \Rightarrow x = 12$ or $x = 4$

If one piece of the wire is 12, then the other piece is $16 - 12 = 4$. The pieces are 12 in. and 4 in.

101. Let r = the radius of the can. Then
$$32\pi = 2\pi r(6) + 2\pi r^2$$
$$32\pi = 12\pi r + 2\pi r^2$$
Divide both sides by 2π
$$16 = 6r + r^2$$
$$0 = r^2 + 6r - 16$$
$$0 = (r-2)(r+8)$$
$$r = 2 \text{ or } r = -8$$
The answer cannot be negative, so we reject -8. The radius of the can is 2 inches.

103. Let x = the length of the piece of tin. Then $x - 10$ = the length of the box.

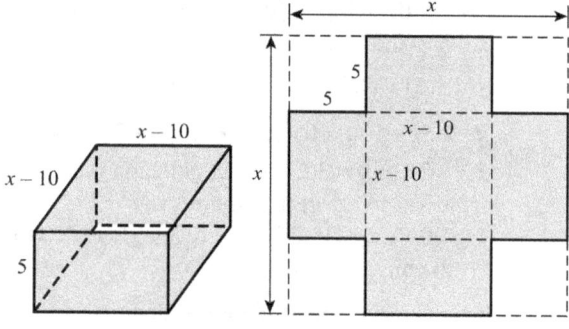

$$(x-10)(x-10)(5) = 480$$
$$5(x^2 - 20x + 100) = 480$$
$$x^2 - 20x + 100 = 96$$
$$x^2 - 20x + 4 = 0$$

Solve using the quadratic formula with $a = 1$, $b = -20$, and $c = 4$.
$$x = \frac{-(-20) \pm \sqrt{(-20)^2 - 4(1)(4)}}{2(1)}$$
$$= \frac{20 \pm \sqrt{400 - 16}}{2} = \frac{20 \pm \sqrt{384}}{2}$$
$$\approx \frac{20 + 19.6}{2} \text{ or } \frac{20 - 19.6}{2} \approx 19.8 \text{ or } 0.2$$

The length cannot be 0.2 inches, so we reject that solution. The tin is 19.8 in. by 19.8 in.

105. Let x = the time the buses travel. So the distance the first bus travels = $52x$ mi, and the distance the second bus travels = $39x$ mi.

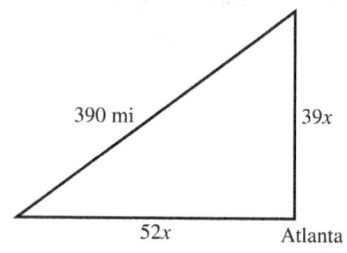

Using the Pythagorean theorem, we have
$$(52x)^2 + (39x)^2 = 390^2$$
$$2704x^2 + 1521x^2 = 152,100$$
$$4225x^2 = 152,100$$
$$x^2 = 36 \Rightarrow x = \pm 6$$
Time cannot be negative, so we reject -6. The buses will be 390 miles apart after 6 hours.

107. Let x = the width of the border. Then length of the garden with the border is $25 + 2x$, and the width of the garden with the border is $15 + 2x$. The area of the border = the area of the garden with the border – the area of the garden.

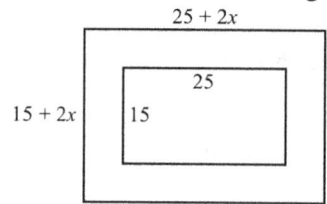

$$A_b = 624 = (25 + 2x)(15 + 2x) - (15)(25)$$
$$= 375 + 80x + 4x^2 - 375$$
$$= 80x + 4x^2$$
$$624 = 80x + 4x^2$$
$$0 = 4x^2 + 80x - 624$$
$$0 = x^2 + 20x - 156$$
$$0 = (x+26)(x-6)$$
$$x = -26 \text{ or } x = 6$$
We reject the negative solution. The width of the border is 6 feet.

109. a. $h = -16(2^2) + 112(2) = 160$ feet

b. $96 = -16t^2 + 112t \Rightarrow 16t^2 - 112t + 96 = 0$
$\Rightarrow t^2 - 7t + 6 = 0 \Rightarrow (t-1)(t-6) = 0 \Rightarrow$
$t = 1$ or $t = 6$
The ball will be at a height of 96 feet at 1 second and at 6 seconds.

c. $0 = -16t^2 + 112t \Rightarrow 16t^2 - 112t = 0 \Rightarrow$
$t^2 - 7t = 0 \Rightarrow t(t-7) = 0 \Rightarrow t = 0$ or $t = 7$
The ball will return to the ground at 7 seconds.

111. a. $-16t^2 + 96t + 480 = 592$
$-16t^2 + 96t - 112 = 0$
$a = -16, b = 96, c = -112$
$$x = \frac{-96 \pm \sqrt{96^2 - 4(-16)(-112)}}{2(-16)}$$
≈ 1.59 or 4.41
The projectile will be at a height of 592 ft at about 1.59 sec and 4.41 sec.

b. $-16t^2 + 96t + 480 = 0$
$a = -16, b = 96, c = 480$
$$x = \frac{-96 \pm \sqrt{96^2 - 4(-16)(480)}}{2(-16)}$$
≈ -3.24 (reject this) or 9.24
The projectile will crash on the ground after approximately 9.24 seconds.

113. a. $-16t^2 + 256t + 480 = 592$
$-16t^2 + 256t - 112 = 0$
$a = -16, b = 256, c = -112$
$$x = \frac{-256 \pm \sqrt{256^2 - 4(-16)(-112)}}{2(-16)}$$
≈ 0.45 or 15.55
The projectile will be at a height of 592 ft at about 0.45 sec and 15.55 sec.

b. $-16t^2 + 256t + 480 = 0$
$a = -16, b = 256, c = 480$
$$x = \frac{-256 \pm \sqrt{256^2 - 4(-16)(480)}}{2(-16)}$$
≈ -1.70 (reject this) or 17.70
The projectile will crash on the ground after 17.70 seconds.

1.3 Beyond the Basics

In exercises 115–120, the equation has equal roots if the discriminant equals 0.

115. $x^2 - kx + 3 = 0$
$a = 1, b = -k, c = 3$
$b^2 - 4ac = (-k)^2 - 4(1)(3) = k^2 - 12$
Now solve $k^2 - 12 = 0$.
$k^2 - 12 = 0 \Rightarrow k^2 = 12 \Rightarrow k = \pm\sqrt{12} = \pm 2\sqrt{3}$

117. $2x^2 + kx + k = 0$
$a = 2, b = k, c = k$
$b^2 - 4ac = k^2 - 4(2)k = k^2 - 8k$
Now solve $k^2 - 8k = 0$.
$k^2 - 8k = 0 \Rightarrow k(k-8) = 0 \Rightarrow k = 0$ or $k = 8$

119. Since r and s are roots of the equation, then
$ar^2 + br + c = 0 = as^2 + bs + c \Rightarrow$
$ar^2 + br + c - c = as^2 + bs + c - c$
$ar^2 + br = as^2 + bs$
$ar^2 - as^2 = bs - br$
$a(r^2 - s^2) = b(s - r)$
$a(r^2 - s^2) = -b(r - s)$
$\frac{r^2 - s^2}{r - s} = -\frac{b}{a} \Rightarrow r + s = -\frac{b}{a}$

To find $r \cdot s$, first divide both sides of $ar^2 + br + c = 0$ by a. We have
$r^2 + \frac{b}{a}r + \frac{c}{a} = 0$. Now use the results from the first part of the problem and substitute $-(r+s)$ for $\frac{b}{a}$. This gives
$r^2 - (r+s)r + \frac{c}{a} = 0$
$r^2 - r^2 - rs + \frac{c}{a} = 0 \Rightarrow \frac{c}{a} = rs$

Use the results from exercise 119 to solve exercises 121 and 122. The sum of the roots is $-\frac{b}{a}$, and the product of the roots is $\frac{c}{a}$.

121. $-\frac{k-3}{2} = \frac{3k-5}{2} \Rightarrow -k + 3 = 3k - 5 \Rightarrow$
$-4k = -8 \Rightarrow k = 2$

123. From exercise 119, we have
$r + s = -\frac{b}{a} \Rightarrow b = -a(r+s)$ and
$rs = \frac{c}{a} \Rightarrow c = ars$. Substitute these values into the equation:
$ax^2 + bx + c = ax^2 - a(r+s)x + ars$
$= a(x^2 - (r+s)x + rs)$
$= a(x^2 - rx - sx + rs)$
$= a[x(x-r) - s(x-r)]$
$= a(x-r)(x-s)$

125. a. $(x-(-3))(x-4) = 0 \Rightarrow (x+3)(x-4) = 0 \Rightarrow$
$x^2 - x - 12 = 0$

b. $(x-5)(x-5) = 0 \Rightarrow x^2 - 10x + 25 = 0$

c. $(x-(3+\sqrt{2}))(x-(3-\sqrt{2})) = 0 \Rightarrow$
$x^2 - 3x - x\sqrt{2} - 3x + x\sqrt{2} + 7 = 0 \Rightarrow$
$x^2 - 6x + 7 = 0$

1.3 Critical Thinking/Discussion/Writing

127. $(x-a)(x-b) = k^2 \Rightarrow$
$x^2 - (a+b)x + ab - k^2 = 0$
$[-(a+b)]^2 - 4(1)(ab - k^2)$
$= a^2 + 2ab + b^2 - 4ab + 4k^2$
$= a^2 - 2ab + b^2 + 4k^2$
$= (a-b)^2 + 4k^2 \geq 0$

Thus, the solutions of the equation are real.

128. $ax(1-x) = 1 \Rightarrow ax - ax^2 = 1 \Rightarrow$
$-ax^2 + ax - 1 = 0$

Examining the discriminant, we have
$a^2 - 4(-a)(-1) = a^2 - 4a$
$a^2 - 4a < 0 \Rightarrow ax(1-x) = 1$ has no real solutions for $0 < a < 4$.

1.3 Getting Ready for the Next Section

129. $(3+\sqrt{2})(3-\sqrt{2}) = 3^2 - (\sqrt{2})^2 = 9 - 2 = 7$

131. $(2+\sqrt{5})^2 = 2^2 + 2 \cdot 2 \cdot \sqrt{5} + (\sqrt{5})^2$
$= 4 + 4\sqrt{5} + 5 = 9 + 4\sqrt{5}$

133. $\dfrac{5 + 5\sqrt{20}}{5} = 1 + \sqrt{20} = 1 + \sqrt{4 \cdot 5} = 1 + 2\sqrt{5}$

135. $\dfrac{16 - \sqrt{100}}{2} = \dfrac{16 - 10}{2} = \dfrac{6}{2} = 3$

137. $(3x+2) + (x-7) = 4x - 5$

139. $(9x+4) - (2x+12) = 9x + 4 - 2x - 12$
$= 7x - 8$

141. $(3x+2)(x-9) = 3x^2 - 27x + 2x - 18$
$= 3x^2 - 25x - 18$

143. $(x-3)(x+3) = x^2 - 9$

Solve each equation in exercises 145–148 using the quadratic formula, $x = \dfrac{-b \pm \sqrt{b^2 - 4ac}}{2a}$.

145. $x^2 + 4x + 1 = 0$
$a = 1, b = 4, c = 1$
$x = \dfrac{-4 \pm \sqrt{4^2 - 4 \cdot 1 \cdot 1}}{2 \cdot 1} = \dfrac{-4 \pm \sqrt{12}}{2}$
$= \dfrac{-4 \pm 2\sqrt{3}}{2} = -2 \pm \sqrt{3}$

147. $5x^2 + 8x + 2 = 0$
$a = 5, b = 8, c = 2$
$x = \dfrac{-8 \pm \sqrt{8^2 - 4 \cdot 5 \cdot 2}}{2 \cdot 5} = \dfrac{-8 \pm \sqrt{24}}{10}$
$= \dfrac{-8 \pm 2\sqrt{6}}{10} = \dfrac{-4 \pm \sqrt{6}}{5}$

1.4 Complex Numbers: Quadratic Equations with Complex Solutions

1.4 Practice Problems

1. a. $-1 + 2i$
real part: -1, imaginary part: 2

b. $-\dfrac{1}{3} - 6i$
real part: $-\dfrac{1}{3}$, imaginary part: -6

c. $8 = 8 + 0i$
real part: 8, imaginary part: 0

2. a. $(1-4i) + (3+2i) = 4 - 2i$

b. $(4+3i) - (5-i) = -1 + 4i$

c. $(3-\sqrt{-9}) - (5-\sqrt{-64}) = (3-3i) - (5-8i)$
$= -2 + 5i$

3. a. $(2-6i)(1+4i) = 2 + 8i - 6i - 24i^2$
$= 2 + 2i + 24 = 26 + 2i$

b. $-3i(7-5i) = -21i + 15i^2 = -15 - 21i$

4. a. $(-3 + \sqrt{-4})^2 = (-3 + 2i)^2$
$= (-3)^2 + 2(-3)(2i) + (2i)^2$
$= 9 - 12i + 4i^2 = 9 - 12i - 4$
$= 5 - 12i$

Section 1.4 Complex Numbers: Quadratic Equations with Complex Solutions

b. $(5+\sqrt{-2})(4+\sqrt{-8}) = (5+i\sqrt{2})(4+2i\sqrt{2})$
$= 20 + 10\sqrt{2}i + 4\sqrt{2}i + 4i^2$
$= 20 + 14\sqrt{2}i - 4$
$= 16 + 14\sqrt{2}i$

5. a. $z = 1 + 6i \Rightarrow \bar{z} = 1 - 6i$
$z\bar{z} = (1+6i)(1-6i) = 1 - 36i^2 = 1 + 36 = 37$

b. $z = -2i \Rightarrow \bar{z} = 2i$
$z\bar{z} = (-2i)(2i) = -4i^2 = 4$

6. a. $\dfrac{2}{1-i} = \dfrac{2}{1-i} \cdot \dfrac{1+i}{1+i} = \dfrac{2+2i}{1-i^2} = \dfrac{2+2i}{1+1}$
$= \dfrac{2+2i}{2} = 1+i$

b. $\dfrac{-3i}{4+\sqrt{-25}} = \dfrac{-3i}{4+5i} = \dfrac{-3i}{4+5i} \cdot \dfrac{4-5i}{4-5i}$
$= \dfrac{-12i + 15i^2}{16 - 25i^2} = \dfrac{-15 - 12i}{16 + 25}$
$= \dfrac{-15 - 12i}{41} = -\dfrac{15}{41} - \dfrac{12i}{41}$

7. $Z_t = \dfrac{Z_1 Z_2}{Z_1 + Z_2} = \dfrac{(1+2i)(2-3i)}{(1+2i)+(2-3i)}$
$= \dfrac{2 - 3i + 4i - 6i^2}{3 - i} = \dfrac{2+i+6}{3-i} = \dfrac{8+i}{3-i}$
$= \dfrac{8+i}{3-i} \cdot \dfrac{3+i}{3+i} = \dfrac{24 + 8i + 3i + i^2}{9 - i^2}$
$= \dfrac{24 + 11i - 1}{9 + 1} = \dfrac{23 + 11i}{10} = \dfrac{23}{10} + \dfrac{11}{10}i$

8. a. $4x^2 + 9 = 0$
$4x^2 = -9$
$x^2 = -\dfrac{9}{4} \Rightarrow x = \pm\sqrt{-\dfrac{9}{4}} = \pm\dfrac{3}{2}i$
Solution set: $\left\{-\dfrac{3}{2}i, \dfrac{3}{2}i\right\}$

b. $x^2 = 4x - 13$
$x^2 - 4x + 13 = 0$
$x = \dfrac{-(-4) \pm \sqrt{(-4)^2 - 4(1)(13)}}{2(1)}$
$= \dfrac{4 \pm \sqrt{-36}}{2} = \dfrac{4 \pm 6i}{2} = 2 \pm 3i$
Solution set: $\{2 - 3i, 2 + 3i\}$

9. a. $9x^2 - 6x + 1 = 0 \Rightarrow a = 9, b = -6, c = 1$
So, $D = (-6)^2 - 4(9)(1) = 36 - 36 = 0$.
Therefore, there is one real root.

b. $x^2 - 5x + 3 = 0 \Rightarrow a = 1, b = -5, c = 3$
So, $D = (-5)^2 - 4(1)(3) = 25 - 12 = 13 > 0$
Therefore, there are two unequal real roots.

c. $2x^2 - 3x + 4 = 0 \Rightarrow a = 2, b = -3, c = 4$
So, $D = (-3)^2 - 4(2)(4) = 9 - 32 = -23 < 0$
Therefore, there are two nonreal complex solutions.

1.4 Concepts and Vocabulary

1. We define $i = \sqrt{-1}$, so that $i^2 = \underline{-1}$.

3. For $b > 0$, $\sqrt{-b} = \underline{i\sqrt{b}}$.

5. True

7. False. Every imaginary number has a conjugate.

9. $(5 + 2i) + (3 + i) = (5 + 3) + (2 + 1)i = 8 + 3i$

11. $(4 - 3i) - (5 + 3i) = (4 - 5) + (-3 - 3)i$
$= -1 - 6i$

13. $(-2 - 3i) + (-3 - 2i) = [-2 + (-3)] + (-3 - 2)i$
$= -5 - 5i$

15. $3(5 + 2i) = 3(5) + 3(2i) = 15 + 6i$

17. $-4(2 - 3i) = -4(2) - 4(-3i) = -8 + 12i$

19. $3i(5 + i) = 3i(5) + 3i(i) = 15i + 3i^2$
Because $i^2 = -1$, $3i^2 = -3$.
So, $15i + 3i^2 = 15i - 3 = -3 + 15i$.

21. $4i(2 - 5i) = 4i(2) + 4i(-5i) = 8i - 20i^2$.
Because $i^2 = -1$, $-20i^2 = (-20)(-1) = 20$.
So, $8i - 20i^2 = 8i + 20 = 20 + 8i$.

23. $(3 + i)(2 + 3i) = 3 \cdot 2 + 3 \cdot 3i + i \cdot 2 + i \cdot 3i$
$= 6 + 9i + 2i + 3i^2$
$= 6 + 11i + 3(-1)$
$= 3 + 11i$

25. $(2 - 3i)(2 + 3i) = 2 \cdot 2 + 2 \cdot 3i + (-3i) \cdot 2 + (-3i) \cdot 3i$
$= 4 + 6i - 6i - 9i^2$
$= 4 - 9(-1) = 4 + 9 = 13$

27. $(3 + 4i)(4 - 3i)$
$= 3 \cdot 4 + 3 \cdot (-3i) + 4i \cdot 4 + (4i) \cdot (-3i)$
$= 12 - 9i + 16i - 12i^2$
$= 12 + 7i - 12(-1)$
$= 12 + 7i + 12 = 24 + 7i$

29. $\left(\sqrt{3} - 12i\right)^2$
$= \left(\sqrt{3} - 12i\right)\left(\sqrt{3} - 12i\right)$
$= \sqrt{3} \cdot \sqrt{3} + \sqrt{3} \cdot (-12i)$
$\quad - 12i \cdot \sqrt{3} + (-12i) \cdot (-12i)$
$= 3 - 12\sqrt{3}i - 12\sqrt{3}i + 144i^2$
$= 3 - 24\sqrt{3}i + 144(-1) = 3 - 24\sqrt{3}i - 144$
$= -141 - 24\sqrt{3}i$

31. $\left(2 - \sqrt{-16}\right)(3 + 5i)$
$= (2 - 4i)(3 + 5i)$
$= 2 \cdot 3 + 2 \cdot 5i - 4i \cdot 3 - 4i \cdot 5i$
$= 6 + 10i - 12i - 20i^2$
$= 6 - 2i - 20(-1)$
$= 6 - 2i + 20$
$= 26 - 2i$

33. If $z = 2 - 3i$ then $\bar{z} = 2 + 3i$, and
$z\bar{z} = (2 - 3i)(2 + 3i) = 4 - 9i^2 = 4 + 9 = 13$.

35. If $z = \dfrac{1}{2} - 2i$ then $\bar{z} = \dfrac{1}{2} + 2i$, and
$z\bar{z} = \left(\dfrac{1}{2} - 2i\right)\left(\dfrac{1}{2} + 2i\right) = \dfrac{1}{4} - 4i^2 = \dfrac{1}{4} + 4 = \dfrac{17}{4}$

37. If $z = \sqrt{2} - 3i$ then $\bar{z} = \sqrt{2} + 3i$, and
$z\bar{z} = \left(\sqrt{2} - 3i\right)\left(\sqrt{2} + 3i\right) = 2 - 9i^2 = 2 + 9 = 11$

39. The denominator is $-i$, so its conjugate is i. Multiply the numerator and denominator by i.
$\dfrac{5}{-i} = \dfrac{5i}{-i \cdot i} = \dfrac{5i}{1} = 5i$

41. The denominator is $1 + i$, so its conjugate is $1 - i$. Multiply the numerator and denominator by $1 - i$.
$\dfrac{-1}{1+i} = \dfrac{-1(1-i)}{(1+i)(1-i)} = \dfrac{-1+i}{1+1}$
$= \dfrac{-1+i}{2} = -\dfrac{1}{2} + \dfrac{1}{2}i$

43. The denominator is $2 + i$, so its conjugate is $2 - i$. Multiply the numerator and denominator by $2 - i$.
$\dfrac{5i}{2+i} = \dfrac{5i(2-i)}{(2+i)(2-i)} = \dfrac{10i - 5i^2}{4+1}$
$= \dfrac{10i + 5}{5} = 1 + 2i$

45. The denominator is $1 + i$, so its conjugate is $1 - i$. Multiply the numerator and denominator by $1 - i$.
$\dfrac{2+3i}{1+i} = \dfrac{(2+3i)(1-i)}{(1+i)(1-i)} = \dfrac{2 - 2i + 3i - 3i^2}{1+1}$
$= \dfrac{2+i+3}{2} = \dfrac{5+i}{2} = \dfrac{5}{2} + \dfrac{1}{2}i$

47. The denominator is $4 - 7i$, so its conjugate is $4 + 7i$. Multiply the numerator and denominator by $4 + 7i$.
$\dfrac{2-5i}{4-7i} = \dfrac{(2-5i)(4+7i)}{(4-7i)(4+7i)} = \dfrac{8 + 14i - 20i - 35i^2}{16+49}$
$= \dfrac{8 - 6i + 35}{65} = \dfrac{43 - 6i}{65} = \dfrac{43}{65} - \dfrac{6}{65}i$

49. The denominator is $1 + i$, so its conjugate is $1 - i$. Multiply the numerator and denominator by $1 - i$.
$\dfrac{2+\sqrt{-4}}{1+i} = \dfrac{(2+2i)}{(1+i)} = \dfrac{(2+2i)(1-i)}{(1+i)(1-i)}$
$= \dfrac{2 - 2i + 2i - 2i^2}{1+1} = \dfrac{2 - 2i^2}{2} = \dfrac{2+2}{2} = 2$

51. The denominator is $2 - 3i$, so its conjugate is $2 + 3i$. Multiply the numerator and denominator by $2 + 3i$.
$\dfrac{-2+\sqrt{-25}}{2-3i} = \dfrac{-2+5i}{2-3i} = \dfrac{(-2+5i)(2+3i)}{(2-3i)(2+3i)}$
$= \dfrac{-4 - 6i + 10i + 15i^2}{4+9}$
$= \dfrac{-4 + 4i - 15}{13} = \dfrac{-19 + 4i}{13}$
$= -\dfrac{19}{13} + \dfrac{4}{13}i$

53. $x^2 + 5 = 1 \Rightarrow x^2 = -4 \Rightarrow x = \pm\sqrt{-4} = \pm 2i$

55. $z^2 - 2z + 2 = 0$
$z^2 - 2z = -2$
Now, complete the square.
$z^2 - 2z + 1 = -2 + 1$
$(z - 1)^2 = -1 \Rightarrow z - 1 = \pm i \Rightarrow z = 1 \pm i$

57. $2x^2 - 20x + 49 = -7$
$2x^2 - 20x = -56$
$x^2 - 10x = -28$
Now, complete the square.
$x^2 - 10x + 25 = -28 + 25 \Rightarrow (x-5)^2 = -3$
$x - 5 = \pm i\sqrt{3} \Rightarrow x = 5 \pm i\sqrt{3}$

59. $8(x^2 - x) = x^2 - 3 \Rightarrow 8x^2 - 8x = x^2 - 3 \Rightarrow$
$7x^2 - 8x + 3 = 0 \Rightarrow a = 7, b = -8, c = 3$

$x = \dfrac{-(-8) \pm \sqrt{(-8)^2 - 4(7)(3)}}{2(7)}$

$= \dfrac{8 \pm \sqrt{64 - 84}}{14} = \dfrac{8 \pm \sqrt{-20}}{14}$

$= \dfrac{8 \pm 2i\sqrt{5}}{14} = \dfrac{4 \pm i\sqrt{5}}{7}$

$x = \dfrac{4}{7} + \dfrac{\sqrt{5}}{7}i$ or $x = \dfrac{4}{7} - \dfrac{\sqrt{5}}{7}i$

61. $9k^2 + 25 = 0 \Rightarrow a = 9, b = 0, c = 25$

$t = \dfrac{0 \pm \sqrt{0^2 - 4(9)(25)}}{2(9)} = \dfrac{\pm\sqrt{-900}}{18}$

$= \pm\dfrac{30i}{18} = \pm\dfrac{5}{3}i$

1.4 Applying the Concepts

63. $Z_1 = 4 + 3i$ and $Z_2 = 5 - 2i$.
So, $Z_1 + Z_2 = (4 + 3i) + (5 - 2i) = 9 + i$.

65. $Z = \dfrac{V}{I}$, $I = 7 + 5i$, $V = 35 + 70i$. Then,

$Z = \dfrac{35 + 70i}{7 + 5i}$. Simplify the fraction by multiplying the numerator and denominator by $7 - 5i$.

$\dfrac{35 + 70i}{7 + 5i} = \dfrac{(35 + 70i)(7 - 5i)}{(7 + 5i)(7 - 5i)}$

$= \dfrac{245 - 175i + 490i - 350i^2}{49 + 25}$

$= \dfrac{245 + 315i + 350}{74} = \dfrac{595 + 315i}{74}$

$= \dfrac{595}{74} + \dfrac{315}{74}i$

67. $Z = \dfrac{V}{I}$, $Z = 5 - 7i$, $I = 2 + 5i$. Then,

$V = ZI = (5 - 7i)(2 + 5i)$
$= 10 + 25i - 14i - 35i^2$
$= 10 + 11i + 35 = 45 + 11i$

69. $Z = \dfrac{V}{I}$, $V = 12 + 10i$, $Z = 12 + 6i$. Then,

$I = \dfrac{V}{Z} = \dfrac{12 + 10i}{12 + 6i}$. Simplify the fraction by multiplying the numerator and denominator by $12 - 6i$.

$\dfrac{12 + 10i}{12 + 6i} = \dfrac{(12 + 10i)(12 - 6i)}{(12 + 6i)(12 - 6i)}$

$= \dfrac{144 - 72i + 120i - 60i^2}{144 + 36}$

$= \dfrac{144 + 48i + 60}{180}$

$= \dfrac{204 + 48i}{180} = \dfrac{17}{15} + \dfrac{4}{15}i$

1.4 Beyond the Basics

71. To find i^{17}, first divide 17 by 4. The remainder is 1, so $i^{17} = i^1 = i$.

73. To find i^{-7}, first rewrite it as $\dfrac{1}{i^7}$. Then divide 7 by 4. The remainder is 3, so $\dfrac{1}{i^7} = \dfrac{1}{i^3}$. Simplify the fraction by multiplying the numerator and denominator by i.

$\dfrac{1}{i^3} \cdot \dfrac{i}{i} = \dfrac{i}{i^4} = i$.

75. To find i^{10}, first divide 10 by 4. The remainder is 2, so $i^{10} = i^2 = -1$. So $i^{10} + 7 = -1 + 7 = 6$.

77. To find i^5, first divide 5 by 4. The remainder is 1, so $i^5 = i$. So $3i^5 = 3i$. $i^3 = -i$, so $-2i^3 = 2i$. Then, $3i^5 - 2i^3 = 3i + 2i = 5i$.

79. $i^3 = -i$, so $2i^3 = -2i$. $i^4 = 1$, so $1 + i^4$
$= 1 + 1 = 2$. Then $2i^3(1 + i^4) = -2i(2) = -4i$.

81. $\dfrac{1}{a + bi} = \dfrac{1(a - bi)}{(a + bi)(a - bi)} = \dfrac{a - bi}{a^2 + b^2}$

$= \dfrac{a}{a^2 + b^2} - \dfrac{b}{a^2 + b^2}i$

83. $\dfrac{z - \bar{z}}{2i} = \dfrac{(a + bi) - (a - bi)}{2i} = \dfrac{2bi}{2i} = b$

85. $\dfrac{1 - 2i}{5 - 5i} = \dfrac{1 - 2i}{5 - 5i} \cdot \dfrac{5 + 5i}{5 + 5i} = \dfrac{5 + 5i - 10i - 10i^2}{25 - 25i^2}$

$= \dfrac{5 - 5i + 10}{25 + 25} = \dfrac{15 - 5i}{50} = \dfrac{3}{10} - \dfrac{1}{10}i$

$\dfrac{2 - 3i}{9 - 7i} = \dfrac{2 - 3i}{9 - 7i} \cdot \dfrac{9 + 7i}{9 + 7i} = \dfrac{18 + 14i - 27i - 21i^2}{81 - 49i^2}$

$= \dfrac{18 - 13i + 21}{81 + 49} = \dfrac{39 - 13i}{130} = \dfrac{3}{10} - \dfrac{1}{10}i$

(continued on next page)

(continued)

$$\frac{1-2i}{5-5i} = \frac{3}{10} - \frac{1}{10}i = \frac{2-3i}{9-7i} \Rightarrow \frac{1-2i}{5-5i} = \frac{2-3i}{9-7i}$$

87. $(1+3i)z + (2+4i) = 7-3i$
$(1+3i)z = 7-3i-(2+4i) = 5-7i$
$z = \frac{5-7i}{1+3i} = \frac{5-7i}{1+3i} \cdot \frac{1-3i}{1-3i}$
$= \frac{5-15i-7i+21i^2}{1-9i^2}$
$= \frac{5-22i-21}{1+9} = \frac{-16-22i}{10}$
$= -\frac{8}{5} - \frac{11}{5}i$

1.4 Critical Thinking/Discussion/Writing

89. a. True. Every real number a can be written as a complex number $a + 0i$.

  b. False.

  c. False. A complex number with the form $a + 0i$ does not have an imaginary component.

  d. True

  e. True. $(a+bi)(a-bi) = a^2 + b^2$. There is no imaginary component.

  f. True

90. $\frac{1+i}{1-i} = \frac{1+i}{1-i} \cdot \frac{1+i}{1+i} = \frac{1+2i-1}{2} = \frac{2i}{2} = i$
$i^n = 1 \Rightarrow n = 4$

91. If $i = 0$, then $i^2 = 0^2 \Rightarrow -1 = 0$, which is a contradiction. If $i < 0$, then $i \cdot i > 0 \cdot i$ (since i is negative) $\Rightarrow i^2 > 0 \Rightarrow -1 > 0$, a contradiction.
If $i > 0$, then $i \cdot i > 0 \cdot i \Rightarrow i^2 > 0 \Rightarrow -1 > 0$, a contradiction. Thus, the set of complex numbers does not have the ordering properties of the set of real numbers.

92. Answers may vary. Sample answer:
$zw = 0 \Rightarrow (a+bi)(c+di) = 0 \Rightarrow$
$ac + (bd)i = 0 \Rightarrow ac = 0, bd = 0$
It is given that $z \neq 0$ and $w \neq 0$, so only one of a and b can be zero, and only one of c and d can be zero. If $a \neq 0$, then $b = 0$, $c = 0$, and $d \neq 0$. One example is $z = a + 0i$, $w = 0 + di$.

1.4 Getting Ready for the Next Section

93. $x^3 - x^2 - 9x + 9 = (x^3 - x^2) - (9x - 9)$
$= x^2(x-1) - 9(x-1)$
$= (x^2 - 9)(x-1)$
$= (x-3)(x+3)(x-1)$

95. $\frac{1}{x-1}, \frac{1}{x}$
LCD $= x(x-1)$

97. $\frac{15}{x+3}, \frac{2x+1}{x-3}$
LCD $= (x+3)(x-3)$

99. $\frac{x+1}{(x-3)(2-x)}, \frac{12}{(x-2)(x+1)}$
Note that $x - 2 = -(2-x)$.
The LCD is $(x-3)(x-2)(x+1)$.

101. $27^{2/3} = (\sqrt[3]{27})^2 = 3^2 = 9$

103. $4^{5/2} = (\sqrt{4})^5 = 2^5 = 32$

105. $(\sqrt{3+x^2})^2 = 3 + x^2$

107. $((3x^2+7)^{1/3})^3 = (3x^2+7)^{(1/3)(3)} = 3x^2+7$

1.5 Solving Other Types of Equations

1.5 Practice Problems

1. $x^3 + 2x^2 - x - 2 = 0$
$x^2(x+2) - (x+2) = 0$
$(x^2 - 1)(x+2) = 0$
$(x-1)(x+1)(x+2) = 0$
$x - 1 = 0 \Rightarrow x = 1$ or $x + 1 = 0 \Rightarrow x = -1$ or $x + 2 = 0 \Rightarrow x = -2$
Solution set: $\{-2, -1, 1\}$

2.
$$x^4 = 4x^2$$
$$x^4 - 4x^2 = 0$$
$$x^2(x^2 - 4) = 0$$
$$x^2(x-2)(x+2) = 0$$
$x^2 = 0 \Rightarrow x = 0$ or $x - 2 = 0 \Rightarrow x = 2$ or $x + 2 = 0 \Rightarrow x = -2$
Solution set: {−2, 0, 2}

3.
$$\frac{1}{x} - \frac{12}{5x+10} = \frac{1}{5}$$
$$\frac{1}{x} - \frac{12}{5(x+2)} = \frac{1}{5}$$
Multiply by the LCD, $5x(x+2)$.
$$5(x+2) - 12x = x(x+2)$$
$$5x + 10 - 12x = x^2 + 2x$$
$$0 = x^2 + 9x - 10$$
$$0 = (x+10)(x-1)$$
$x + 10 = 0 \Rightarrow x = -10$ or $x - 1 = 0 \Rightarrow x = 1$
Be sure to check the solutions in the original equation.
Solution set: {−10, 1}

4.
$$\frac{x}{x-2} - \frac{2}{x+2} = \frac{4x}{x^2-4}$$
$$\frac{x}{x-2} - \frac{2}{x+2} = \frac{4x}{(x-2)(x+2)}$$
Multiply by the common denominator $(x-2)(x+2)$.
$$x(x+2) - 2(x-2) = 4x$$
$$x^2 + 2x - 2x + 4 = 4x$$
$$x^2 - 4x + 4 = 0$$
$$(x-2)^2 = 0 \Rightarrow x = 2$$
Since neither $\frac{x}{x-2}$ nor $\frac{4x}{x^2-4}$ is defined for $x = 2$, this is an extraneous solution.
Solution set: ∅

5.
$$x = \sqrt{x^3 - 6x}$$
$$x^2 = \left(\sqrt{x^3-6x}\right)^2 = x^3 - 6x$$
$$0 = x^3 - x^2 - 6x$$
$$= x(x^2 - x - 6)$$
$$= x(x-3)(x+2)$$
$x = 0$ or $x - 3 = 0 \Rightarrow x = 3$ or $x + 2 = 0 \Rightarrow x = -2$

Now check to see if −2, 0, or 3 are extraneous solutions. If $x = -2$, then
$\sqrt{(-2)^3 - 6(-2)} = \sqrt{4} = 2 \neq -2$, so −2 is an extraneous solution. If $x = 0$, then
$\sqrt{0^3 - 6(0)} = \sqrt{0} = 0$, so 0 is a solution.
If $x = 3$, then $\sqrt{3^3 - 6(3)} = \sqrt{9} = 3$, so 3 is a solution.
Solution set: {0, 3}

6.
$$\sqrt{6x+4} + 2 = x$$
$$\sqrt{6x+4} = x - 2$$
$$\left(\sqrt{6x+4}\right)^2 = (x-2)^2$$
$$6x + 4 = x^2 - 4x + 4$$
$$0 = x^2 - 10x = x(x-10)$$
$x = 0$ or $x - 10 = 0 \Rightarrow x = 10$
Now check to see if 0 or 10 are extraneous solutions. If $x = 0$, then
$\sqrt{6(0)+4} + 2 = \sqrt{4} + 2 = 2 + 2 = 4 \neq 0$, so 0 is an extraneous solution. If $x = 10$, then
$\sqrt{6(10)+4} + 2 = \sqrt{64} + 2 = 8 + 2 = 10$, so 10 is a solution.
Solution set: {10}

7.
$$\sqrt{x-5} + \sqrt{x} = 5$$
$$\sqrt{x-5} = 5 - \sqrt{x}$$
$$\left(\sqrt{x-5}\right)^2 = \left(5 - \sqrt{x}\right)^2$$
$$x - 5 = 25 - 10\sqrt{x} + x$$
$$-30 = -10\sqrt{x}$$
$$3 = \sqrt{x} \Rightarrow 9 = x$$
Now check to see if 9 is an extraneous solution.
$\sqrt{9-5} + \sqrt{9} = \sqrt{4} + \sqrt{9} = 2 + 3 = 5$, so 9 is a solution.
Solution set: {9}

8. a.
$$3(x-2)^{3/5} + 4 = 7$$
$$3(x-2)^{3/5} = 3$$
$$(x-2)^{3/5} = 1$$
$$\left[(x-2)^{3/5}\right]^{5/3} = 1^{5/3}$$
$$x - 2 = 1$$
$$x = 3$$
Be sure to check that 3 satisfies the original equation.
Solution set: {3}

b. $(2x+1)^{4/3} - 7 = 9$

$(2x+1)^{4/3} = 16$

$\left[(2x+1)^{4/3}\right]^{3/4} = \pm\left(16^{3/4}\right)$

$2x+1 = \pm 8$

$2x+1 = -8$	$2x+1 = 8$
$2x = -9$	$2x = 7$
$x = -\dfrac{9}{2}$	$x = \dfrac{7}{2}$

Check:

$\left[2\left(-\dfrac{9}{2}\right)+1\right]^{4/3} - 7 \stackrel{?}{=} 9$

$(-9+1)^{4/3} - 7 \stackrel{?}{=} 9$

$(-8)^{4/3} - 7 \stackrel{?}{=} 9$

$16 - 7 = 9$

$\left[2\left(\dfrac{7}{2}\right)+1\right]^{4/3} - 7 \stackrel{?}{=} 9$

$(7+1)^{4/3} - 7 \stackrel{?}{=} 9$

$(8)^{4/3} - 7 \stackrel{?}{=} 9$

$16 - 7 = 9$

Solution set: $\left\{-\dfrac{9}{2}, \dfrac{7}{2}\right\}$

9. $x^{2/3} - 7x^{1/3} + 6 = 0$

Let $u = x^{1/3}$. Then the original equation becomes

$u^2 - 7u + 6 = 0 \Rightarrow (u-1)(u-6) = 0 \Rightarrow$

$u = 1, 6$

Now solve for x:

$1 = x^{1/3} \Rightarrow 1^3 = \left(x^{1/3}\right)^3 \Rightarrow 1 = x$

$6 = x^{1/3} \Rightarrow 6^3 = \left(x^{1/3}\right)^3 \Rightarrow 216 = x$

Be sure to check that both solutions satisfy the original equation.
Solution set: $\{1, 216\}$

10. $\left(1+\dfrac{1}{x}\right)^2 - 6\left(1+\dfrac{1}{x}\right) + 8 = 0$

Let $u = \left(1+\dfrac{1}{x}\right)$. Then the original equation becomes

$u^2 - 6u + 8 = 0 \Rightarrow (u-2)(u-4) = 0 \Rightarrow$

$u = 2, 4$

Now solve for x:

$1 + \dfrac{1}{x} = 2 \Rightarrow x+1 = 2x \Rightarrow x = 1$

$1 + \dfrac{1}{x} = 4 \Rightarrow x+1 = 4x \Rightarrow 1 = 3x \Rightarrow \dfrac{1}{3} = x$

Be sure to check that both solutions satisfy the original equation.

Solution set: $\left\{\dfrac{1}{3}, 1\right\}$

11. $t_0 = t\sqrt{1-\dfrac{v^2}{c^2}}$, $t_0 = 20$, $t = 25$

$20 = 25\sqrt{1-\dfrac{v^2}{c^2}} \Rightarrow \dfrac{4}{5} = \sqrt{1-\dfrac{v^2}{c^2}} \Rightarrow$

$\dfrac{16}{25} = 1 - \dfrac{v^2}{c^2} \Rightarrow \dfrac{v^2}{c^2} = 1 - \dfrac{16}{25} = \dfrac{9}{25} \Rightarrow$

$\dfrac{v}{c} = \dfrac{3}{5} \Rightarrow v = 0.6c$

The spacecraft must have been traveling at 60% of the speed of light.

1.5 Concepts and Vocabulary

1. If an apparent solution does not satisfy the equation, it is called an <u>extraneous</u> solution.

3. If $x^{3/4} = 8$, then $x = \underline{16}$.

5. False

7. False. The solution for the equation $x^2 = 9$ is $x = \pm 3$.

1.5 Building Skills

9. $x^3 = 2x^2 \Rightarrow x^3 - 2x^2 = 0 \Rightarrow x^2(x-2) = 0 \Rightarrow$
$x^2 = 0$ or $x - 2 = 0 \Rightarrow x = 0$ or $x = 2$

11. $\left(\sqrt{x}\right)^3 = \sqrt{x} \Rightarrow \left(\sqrt{x}\right)^2 \sqrt{x} = \sqrt{x} \Rightarrow$
$x\sqrt{x} = \sqrt{x} \Rightarrow x\sqrt{x} - \sqrt{x} = 0 \Rightarrow$
$\sqrt{x}(x-1) = 0 \Rightarrow \sqrt{x} = 0$ or $(x-1) = 0 \Rightarrow$
$x = 0$ or $x = 1$

13. $x^3 + x = 0 \Rightarrow x(x^2+1) = 0 \Rightarrow$
$x = 0$ or $x^2 + 1 = 0 \Rightarrow x = 0$ or $x^2 = -1 \Rightarrow$
$x = 0$ or $x = i$
We are looking for real roots only, so we reject $x = i$. Solution set: $\{0\}$

15.
$$x^4 - x^3 = x^2 - x$$
$$x^4 - x^3 - x^2 + x = 0$$
Factor by grouping.
$$x^3(x-1) - x(x-1) = 0$$
$$(x^3 - x)(x-1) = 0$$
$$x(x^2 - 1)(x-1) = 0$$
$$x(x+1)(x-1)(x-1) = 0$$
$$x = 0 \text{ or } x = -1 \text{ or } x = 1$$
Solution set: $\{-1, 0, 1\}$

17.
$$x^4 = 27x$$
$$x^4 - 27x = 0$$
$$x(x^3 - 27) = 0$$
$$x(x^3 - 3^3) = 0$$
$$x(x-3)(x^2 + 3x + 9) = 0$$
$$x = 0 \text{ or } x = 3 \text{ or } x^2 + 3x + 9 = 0$$
Solving $x^2 + 3x + 9 = 0$ gives
$x = -\dfrac{3}{2} \pm \dfrac{3i\sqrt{3}}{2}$. However, we are looking for the real roots only, so we reject these roots.
Solution set: $\{0, 3\}$

19.
$$\frac{x+1}{3x-2} = \frac{5x-4}{3x+2}$$
$$(x+1)(3x+2) = (5x-4)(3x-2)$$
$$3x^2 + 5x + 2 = 15x^2 - 22x + 8$$
$$-12x^2 + 27x - 6 = 0$$
$$-3(4x^2 - 9x + 2) = 0$$
$$4x^2 - 9x + 2 = 0 \Rightarrow (4x-1)(x-2) = 0$$
$$4x - 1 = 0 \text{ or } x - 2 = 0 \Rightarrow x = \frac{1}{4} \text{ or } x = 2$$

21.
$$\frac{1}{x} + \frac{2}{x+1} = 1$$
$$(x)(x+1)\left(\frac{1}{x} + \frac{2}{x+1}\right) = (x)(x+1)(1)$$
$$(x+1) + 2x = x^2 + x$$
$$3x + 1 = x^2 + x$$
$$0 = x^2 - 2x - 1$$
Solve using the quadratic formula.
$$x = \frac{-(-2) \pm \sqrt{(-2)^2 - 4(1)(-1)}}{2}$$
$$x = \frac{2 \pm \sqrt{4+4}}{2} = \frac{2 \pm \sqrt{8}}{2} = \frac{2 \pm 2\sqrt{2}}{2} = 1 \pm \sqrt{2}$$

23.
$$\frac{6x-7}{x} - \frac{1}{x^2} = 5$$
$$x^2\left(\frac{6x-7}{x} - \frac{1}{x^2}\right) = 5x^2$$
$$x(6x-7) - 1 = 5x^2$$
$$6x^2 - 7x - 1 = 5x^2$$
$$x^2 - 7x - 1 = 0$$
Solve using the quadratic formula.
$$x = \frac{-(-7) \pm \sqrt{(-7)^2 - 4(1)(-1)}}{2}$$
$$x = \frac{7 \pm \sqrt{49+4}}{2} = \frac{7 \pm \sqrt{53}}{2} = \frac{7}{2} \pm \frac{\sqrt{53}}{2}$$

25.
$$\frac{1}{x} + \frac{1}{x-3} = \frac{7}{3x-5}$$
$$x(x-3)(3x-5)\left(\frac{1}{x} + \frac{1}{x-3} = \frac{7}{3x-5}\right)$$
$$(x-3)(3x-5) + x(3x-5) = 7x(x-3)$$
$$3x^2 - 14x + 15 + 3x^2 - 5x = 7x^2 - 21x$$
$$6x^2 - 19x + 15 = 7x^2 - 21x$$
$$-x^2 + 2x + 15 = 0$$
$$-1(x^2 - 2x - 15) = 0$$
$$x^2 - 2x - 15 = 0$$
$$(x-5)(x+3) = 0 \Rightarrow$$
$$x - 5 = 0 \text{ or } x + 3 = 0 \Rightarrow x = 5 \text{ or } x = -3$$

27.
$$\frac{5}{x+1} - \frac{4}{2x+2} + \frac{2}{2x-1} = \frac{13}{18}$$
$$\frac{5}{x+1} - \frac{4}{2(x+1)} + \frac{2}{2x-1} = \frac{13}{18}$$
$$18(x+1)(2x-1)\left(\frac{5}{x+1} - \frac{4}{2(x+1)} + \frac{2}{2x-1}\right)$$
$$= 18(x+1)(2x-1)\left(\frac{13}{18}\right)$$
$$90(2x-1) - 36(2x-1) + 36(x+1)$$
$$= 13(x+1)(2x-1)$$
$$180x - 90 - 72x + 36 + 36x + 36$$
$$= 13(2x^2 + x - 1)$$
$$144x - 18 = 26x^2 + 13x - 13$$
$$0 = 26x^2 - 131x + 5$$
$$0 = (26x-1)(x-5)$$
$$26x - 1 = 0 \text{ or } x - 5 = 0$$
$$x = \frac{1}{26} \text{ or } x = 5$$

29. $\dfrac{x}{x-5} - \dfrac{5}{x+5} = \dfrac{10x}{x^2-25}$

$x(x+5) - 5(x-5) = 10x$

$x^2 + 5x - 5x + 25 = 10x$

$x^2 - 10x + 25 = 0$

$(x-5)^2 = 0 \Rightarrow x = 5$

Since $x = 5$ makes the denominator in the first fraction equal zero, there is no solution. Solution set: $\emptyset$

31. $\dfrac{x}{x-3} + \dfrac{3}{x+3} = \dfrac{6x}{x^2-9}$

$x(x+3) + 3(x-3) = 6x$

$x^2 + 3x + 3x - 9 = 6x$

$x^2 - 9 = 0$

$(x+3)(x-3) = 0 \Rightarrow x = -3, 3$

Since $x = -3$ makes the denominator in the second fraction equal zero and $x = 3$ makes the denominator in the first fraction equal zero, there is no solution. Solution set: $\emptyset$

33. $\dfrac{1}{x-1} + \dfrac{x}{x+3} = \dfrac{4}{x^2+2x-3}$

$\dfrac{1}{x-1} + \dfrac{x}{x+3} = \dfrac{4}{(x-1)(x+3)}$

$(x+3) + x(x-1) = 4$

$x + 3 + x^2 - x = 4 \Rightarrow x^2 = 1 \Rightarrow x = \pm 1$

If $x = 1$, then the denominator in the first fraction equals zero, so 1 is an extraneous solution. Solution set: $\{-1\}$

35. $\dfrac{2x}{x+3} - \dfrac{x}{x-1} = \dfrac{14}{x^2+2x-3}$

$\dfrac{2x}{x+3} - \dfrac{x}{x-1} = \dfrac{14}{(x+3)(x-1)}$

$2x(x-1) - x(x+3) = 14$

$2x^2 - 2x - x^2 - 3x = 14$

$x^2 - 5x - 14 = 0$

$(x-7)(x+2) = 0 \Rightarrow x = 7, -2$

Solution set: $\{-2, 7\}$

37. $\sqrt[3]{3x-1} = 2 \Rightarrow \left(\sqrt[3]{3x-1}\right)^3 = 2^3 \Rightarrow$

$3x - 1 = 8 \Rightarrow 3x = 9 \Rightarrow x = 3$

39. There is no solution for $\sqrt{x-1} = -2$ because the square root is not negative. The solution set is $\emptyset$.

41. $x + \sqrt{x+6} = 0 \Rightarrow x = -\sqrt{x+6} \Rightarrow$

$x^2 = \left(-\sqrt{x+6}\right)^2 \Rightarrow x^2 = x + 6 \Rightarrow$

$x^2 - x - 6 = 0 \Rightarrow (x-3)(x+2) = 0 \Rightarrow$

$x = 3$ or $x = -2$

Now check to see if 3 or -2 are extraneous roots. If $x = 3$, then $x + \sqrt{x+6} = 3 + \sqrt{3+6} = 3 + 3 \neq 0$. So 3 is extraneous. If $x = -2$, then $x + \sqrt{x+6} = -2 + \sqrt{-2+6} = -2 + \sqrt{4} = -2 + 2 = 0$. Solution set: $\{-2\}$

43. $\sqrt{y+6} = y \Rightarrow \left(\sqrt{y+6}\right)^2 = y^2 \Rightarrow y + 6 = y^2 \Rightarrow$

$0 = y^2 - y - 6 \Rightarrow 0 = (y-3)(y+2) \Rightarrow$

$y = 3$ or $y = -2$

Now check to see if 3 or -2 are extraneous roots. If $y = 3$, then $\sqrt{y+6} = y \Rightarrow$

$\sqrt{3+6} = 3 \Rightarrow \sqrt{9} = 3 \Rightarrow 3 = 3$. So 3 is a solution. If $y = -2$, then $\sqrt{y+6} = y \Rightarrow$

$\sqrt{-2+6} = -2 \Rightarrow \sqrt{4} = -2 \Rightarrow 2 \neq -2$. So -2 is extraneous. Solution set: $\{3\}$

45. $\sqrt{6y-11} = 2y - 7$

$\left(\sqrt{6y-11}\right)^2 = (2y-7)^2$

$6y - 11 = 4y^2 - 28y + 49$

$0 = 4y^2 - 34y + 60$

$0 = 2y^2 - 17y + 30$

$0 = (2y-5)(y-6) \Rightarrow y = \dfrac{5}{2}$ or $y = 6$

Now check to see if 5/2 or 6 are extraneous roots. If $y = \dfrac{5}{2}$, then $\sqrt{6y-11} = 2y - 7 \Rightarrow$

$\sqrt{6\left(\dfrac{5}{2}\right) - 11} = 2\left(\dfrac{5}{2}\right) - 7 \Rightarrow \sqrt{15-11} = 5 - 7 \Rightarrow$

$\sqrt{4} = -2 \Rightarrow 2 \neq -2$. So 5/2 is extraneous. If $y = 6$, then $\sqrt{6y-11} = 2y - 7 \Rightarrow$

$\sqrt{6(6) - 11} = 2(6) - 7 \Rightarrow \sqrt{36-11} = 12 - 7 \Rightarrow$

$\sqrt{25} = 5 \Rightarrow 5 = 5$. Solution set: $\{6\}$

47. $t - \sqrt{3t+6} = -2$

$t + 2 = \sqrt{3t+6}$

$(t+2)^2 = \left(\sqrt{3t+6}\right)^2$

$t^2 + 4t + 4 = 3t + 6$

$t^2 + t - 2 = 0$

$(t+2)(t-1) = 0 \Rightarrow t = -2$ or $t = 1$

(continued on next page)

(*continued*)

Now check to see if -2 or 1 are extraneous roots. If $t = -2$, then $t - \sqrt{3t + 6} = -2 \Rightarrow$
$-2 - \sqrt{3(-2) + 6} = -2 \Rightarrow -2 - \sqrt{-6 + 6} \Rightarrow$
$-2 - 0 = -2$. So -2 is a solution.
If $t = 1$, then $t - \sqrt{3t + 6} = -2 \Rightarrow$
$1 - \sqrt{3(1) + 6} = -2 \Rightarrow 1 - \sqrt{9} = -2 \Rightarrow$
$1 - 3 = -2 \Rightarrow -2 = -2$. So 1 is a solution.
Solution set: $\{-2, 1\}$

49. $\sqrt{x - 3} = \sqrt{2x - 5} - 1$
$\left(\sqrt{x - 3}\right)^2 = \left(\sqrt{2x - 5} - 1\right)^2$
$x - 3 = (2x - 5) - 2\sqrt{2x - 5} + 1$
$x - 3 = 2x - 4 - 2\sqrt{2x - 5}$
$2\sqrt{2x - 5} = x - 1 \Rightarrow \left(2\sqrt{2x - 5}\right)^2 = (x - 1)^2$
$4(2x - 5) = x^2 - 2x + 1$
$8x - 20 = x^2 - 2x + 1$
$0 = x^2 - 10x + 21$
$0 = (x - 7)(x - 3) \Rightarrow x = 7$ or $x = 3$
Now check to see if 3 or 7 are extraneous roots. If $x = 3$, then $\sqrt{x - 3} = \sqrt{2x - 5} - 1 \Rightarrow$
$\sqrt{3 - 3} = \sqrt{2(3) - 5} - 1 \Rightarrow 0 = 0$. So 3 is a solution.
If $x = 7$, then $\sqrt{x - 3} = \sqrt{2x - 5} - 1 \Rightarrow$
$\sqrt{7 - 3} = \sqrt{2(7) - 5} - 1 \Rightarrow \sqrt{4} = \sqrt{9} - 1 \Rightarrow$
$2 = 2$. So 7 is a solution.
Solution set: $\{3, 7\}$

51. $\sqrt{2y + 9} = 2 + \sqrt{y + 1}$
$\left(\sqrt{2y + 9}\right)^2 = \left(2 + \sqrt{y + 1}\right)^2$
$2y + 9 = 4 + 4\sqrt{y + 1} + y + 1$
$y + 4 = 4\sqrt{y + 1}$
$(y + 4)^2 = \left(4\sqrt{y + 1}\right)^2$
$y^2 + 8y + 16 = 16(y + 1)$
$y^2 + 8y + 16 = 16y + 16$
$y^2 - 8y = 0$
$y(y - 8) = 0 \Rightarrow y = 0$ or $y = 8$
Now check to see if 0 or 8 are extraneous roots. If $y = 0$, then $\sqrt{2y + 9} = 2 + \sqrt{y + 1} \Rightarrow$
$\sqrt{2(0) + 9} = 2 + \sqrt{0 + 1} \Rightarrow \sqrt{9} = 2 + 1 \Rightarrow 3 = 3$.

So 0 is a solution. If $y = 8$, then
$\sqrt{2y + 9} = 2 + \sqrt{y + 1} \Rightarrow \sqrt{2(8) + 9} =$
$2 + \sqrt{8 + 1} \Rightarrow \sqrt{25} = 2 + \sqrt{9} \Rightarrow 5 = 5$. So 8 is a solution. Solution set: $\{0, 8\}$

53. $\sqrt{7z + 1} - \sqrt{5z + 4} = 1$
$\sqrt{7z + 1} = \sqrt{5z + 4} + 1$
$\left(\sqrt{7z + 1}\right)^2 = \left(\sqrt{5z + 4} + 1\right)^2$
$7z + 1 = 5z + 4 + 2\sqrt{5z + 4} + 1$
$2z - 4 = 2\sqrt{5z + 4}$
$(2z - 4)^2 = \left(2\sqrt{5z + 4}\right)^2$
$4z^2 - 16z + 16 = 4(5z + 4)$
$4z^2 - 16z + 16 = 20z + 16$
$4z^2 - 36z = 0 \Rightarrow 4z(z - 9) = 0 \Rightarrow$
$z = 0$ or $z = 9$
Now check to see if 0 or 9 are extraneous roots. If $z = 0$, then $\sqrt{7z + 1} - \sqrt{5z + 4} = 1 \Rightarrow$
$\sqrt{7(0) + 1} - \sqrt{5(0) + 4} = 1 \Rightarrow \sqrt{1} - \sqrt{4} = 1 \Rightarrow$
$1 - 2 \neq 1$. So 0 is extraneous. If $z = 9$, then
$\sqrt{7z + 1} - \sqrt{5z + 4} = 1 \Rightarrow \sqrt{7(9) + 1} - \sqrt{5(9) + 4}$
$= 1 \Rightarrow \sqrt{64} - \sqrt{49} \Rightarrow 8 - 7 = 1 \Rightarrow 1 = 1$. So 9 is a solution. Solution set: $\{9\}$

55. $\sqrt{2x + 5} + \sqrt{x + 6} = 3$
$\sqrt{2x + 5} = 3 - \sqrt{x + 6}$
$\left(\sqrt{2x + 5}\right)^2 = \left(3 - \sqrt{x + 6}\right)^2$
$2x + 5 = 9 - 6\sqrt{x + 6} + x + 6$
$x - 10 = -6\sqrt{x + 6}$
$(x - 10)^2 = \left(-6\sqrt{x + 6}\right)^2$
$x^2 - 20x + 100 = 36x + 216$
$x^2 - 56x - 116 = 0$
$(x - 58)(x + 2) = 0 \Rightarrow x = 58$ or $x = -2$
Now check to see if 58 or -2 are extraneous roots. If $x = 58$, then $\sqrt{2x + 5} + \sqrt{x + 6} = 3 \Rightarrow$
$\sqrt{2(58) + 5} + \sqrt{58 + 6} = 3 \Rightarrow \sqrt{121} + \sqrt{64} = 3 \Rightarrow$
$11 + 8 \neq 3$. So 58 is extraneous. If $x = -2$, then $\sqrt{2x + 5} + \sqrt{x + 6} = 3 \Rightarrow \sqrt{2(-2) + 5} +$
$\sqrt{-2 + 6} = 3 \Rightarrow \sqrt{1} + \sqrt{4} \Rightarrow 1 + 2 = 3$.
So -2 is a solution. Solution set: $\{-2\}$

57. $\sqrt{2x-5} - \sqrt{x-3} = 1$
$$\sqrt{2x-5} = 1 + \sqrt{x-3}$$
$$\left(\sqrt{2x-5}\right)^2 = \left(1+\sqrt{x-3}\right)^2$$
$$2x - 5 = 1 + 2\sqrt{x-3} + x - 3$$
$$x - 3 = 2\sqrt{x-3}$$
$$(x-3)^2 = \left(2\sqrt{x-3}\right)^2$$
$$x^2 - 6x + 9 = 4x - 12$$
$$x^2 - 10x + 21 = 0$$
$$(x-7)(x-3) = 0 \Rightarrow x = 7 \text{ or } x = 3$$
Now check to see if 7 or 3 are extraneous roots. If $x = 7$, then $\sqrt{2x-5} - \sqrt{x-3} = 1 \Rightarrow \sqrt{2(7)-5} - \sqrt{7-3} = 1 \Rightarrow \sqrt{9} - \sqrt{4} = 1 \Rightarrow 3 - 2 = 1$. So 7 is a solution. If $x = 3$, then $\sqrt{2x-5} - \sqrt{x-3} = 1 \Rightarrow \sqrt{2(3)-5} - \sqrt{3-3} = 1 \Rightarrow \sqrt{1} - \sqrt{0} = 1 \Rightarrow 1 = 1$. So 3 is a solution. Solution set: $\{3, 7\}$

59. $(x-4)^{3/2} = 27$
$$\left[(x-4)^{3/2}\right]^{2/3} = 27^{2/3}$$
$$x - 4 = 9 \Rightarrow x = 13$$
Verify that 13 satisfies the original equation. Solution set: $\{13\}$

61. $(5x-3)^{2/3} - 5 = 4$
$$(5x-3)^{2/3} = 9$$
$$\left[(5x-3)^{2/3}\right]^{3/2} = \pm 9^{3/2}$$
$$5x - 3 = \pm 27$$

$5x - 3 = -27$	$5x - 3 = 27$
$5x = -24$	$5x = 30$
$x = -\dfrac{24}{5}$	$x = 6$

If $x = -24/5$, then
$$\left(5\left(-\frac{24}{5}\right) - 3\right)^{2/3} - 5 = (-24-3)^{2/3} - 5$$
$$= (-27)^{2/3} - 5$$
$$= 9(-1)^{2/3} - 5 = 4,$$
so $x = -\dfrac{24}{5}$ is a solution. Verify that 6 satisfies the original equation.
Solution set: $\left\{-\dfrac{24}{5}, 6\right\}$

63. Let $u = \sqrt{x}$. Then $u^2 = x$. So the equation becomes $u^2 - 5u + 6 = 0 \Rightarrow (u-3)(u-2) = 0 \Rightarrow u = 3$ or $u = 2$. Solve for x: $3 = \sqrt{x} \Rightarrow x = 9$ or $2 = \sqrt{x} \Rightarrow 4 = x$. Check to make sure that neither solution is extraneous. If $x = 9$, then $9 - 5\sqrt{9} + 6 = 0 \Rightarrow 9 - 15 + 6 = 0 \Rightarrow 0 = 0$. If $x = 4$, then $4 - 5\sqrt{4} + 6 = 0 \Rightarrow 4 - 10 + 6 = 0 \Rightarrow 0 = 0$.
Solution set: $\{4, 9\}$

65. Let $u = \sqrt{y}$. Then $u^2 = y$. So the equation becomes $2u^2 - 15u = -7 \Rightarrow 2u^2 - 15u + 7 = 0 \Rightarrow (2u-1)(u-7) = 0 \Rightarrow u = \dfrac{1}{2}$ or $u = 7$. Solve for y: $\dfrac{1}{2} = \sqrt{y} \Rightarrow \dfrac{1}{4} = y$ or $7 = \sqrt{y} \Rightarrow 49 = y$. Now check to make sure that neither solution is extraneous.
If $y = \dfrac{1}{4}$, then $2\left(\dfrac{1}{4}\right) - 15\sqrt{\dfrac{1}{4}} = -7 \Rightarrow$
$$\frac{1}{2} - \frac{15}{2} = -7 \Rightarrow -7 = -7.$$
If $y = 49$, then
$$2(49) - 15\sqrt{49} = -7 \Rightarrow 98 - 105 = -7 \Rightarrow$$
$-7 = -7$. Solution set: $\left\{\dfrac{1}{4}, 49\right\}$

67. $x^{-2} - x^{-1} - 42 = 0$
Let $u = x^{-1}$. Then the equation becomes $u^2 - u - 42 = 0 \Rightarrow (u-7)(u+6) = 0 \Rightarrow u = 7, -6$. Now solve for x:
$$7 = x^{-1} \Rightarrow x = \frac{1}{7}; \; -6 = x^{-1} \Rightarrow x = -\frac{1}{6}$$
Be sure to check that neither of the solutions are extraneous. Solution set: $\left\{-\dfrac{1}{6}, \dfrac{1}{7}\right\}$

69. $x^{2/3} - 6x^{1/3} + 8 = 0$
Let $u = x^{1/3}$. Then the equation becomes $u^2 - 6x + 8 = 0 \Rightarrow (u-4)(u-2) = 0 \Rightarrow u = 4, 2$. Now solve for x:
$x^{1/3} = 4 \Rightarrow x = 64$; $x^{1/3} = 2 \Rightarrow x = 8$
Be sure to check that neither of the solutions are extraneous. Solution set: $\{8, 64\}$

71. $2x^{1/2} + 3x^{1/4} - 2 = 0$

Let $u = x^{1/4}$. Then the equation becomes
$2u^2 + 3u - 2 = 0 \Rightarrow (2u - 1)(u + 2) = 0 \Rightarrow$
$u = \dfrac{1}{2}, -2$. Now solve for x:

$x^{1/4} = \dfrac{1}{2} \Rightarrow x = \dfrac{1}{16}$; $x^{1/4} = -2 \Rightarrow x$ is not a

real number. Be sure to check that $\dfrac{1}{16}$ is not

extraneous. Solution set: $\left\{\dfrac{1}{16}\right\}$

73. Let $u = x^2$. Then $x^4 = u^2$. So the equation
becomes $u^2 - 13u + 36 = 0 \Rightarrow$
$(u - 4)(u - 9) = 0 \Rightarrow u = 4$ or $u = 9$. Now
solve for x: $4 = x^2 \Rightarrow \pm 2 = x$ or
$9 = x^2 \Rightarrow \pm 3 = x$. Now check to make sure
that none of the solutions are extraneous.
If $x = \pm 2$, then $(\pm 2)^4 - 13(\pm 2)^2 + 36 = 0 \Rightarrow$
$16 - 52 + 36 = 0 \Rightarrow 0 = 0$. If $x = \pm 3$, then
$(\pm 3)^4 - 13(\pm 3)^2 + 36 = 0 \Rightarrow 81 - 117 + 36 = 0$
$\Rightarrow 0 = 0$. The solutions are ± 2 and ± 3.

75. Let $u = t^2$. Then $t^4 = u^2$. So the equation
becomes $2u^2 + u - 1 = 0 \Rightarrow (2u - 1)(u + 1) = 0$
$\Rightarrow u = \dfrac{1}{2}$ or $u = -1$. Now solve for t:

$\dfrac{1}{2} = t^2 \Rightarrow \pm\sqrt{\dfrac{1}{2}} = \pm\dfrac{\sqrt{2}}{2} = t$ or $-1 = t^2 \Rightarrow$

$\pm i = t$. Now check to make sure that none of
the solutions are extraneous.

If $t = \pm\dfrac{\sqrt{2}}{2}$, then $2\left(\pm\dfrac{\sqrt{2}}{2}\right)^4 + \left(\dfrac{\sqrt{2}}{2}\right)^2 - 1 = 0 \Rightarrow$

$\dfrac{1}{2} + \dfrac{1}{2} - 1 = 0 \Rightarrow 0 = 0$. If $x = \pm i$, then

$2(\pm i)^4 + (\pm i)^2 - 1 = 0 \Rightarrow 2 - 1 - 1 = 0 \Rightarrow 0 = 0$.

The solutions are $\pm i$ and $\pm\dfrac{\sqrt{2}}{2}$.

77. Let $u = \sqrt{p^2 - 3}$. Then $p^2 - 3 = u^2$. So the
equation becomes $u^2 + 4u - 5 = 0 \Rightarrow$
$(u + 5)(u - 1) = 0 \Rightarrow u = -5$ or $u = 1$.

Now solve for p: $-5 = \sqrt{p^2 - 3}$, which is not

possible, or $1 = \sqrt{p^2 - 3} \Rightarrow$

$1 = p^2 - 3 \Rightarrow 1 = p^2 - 3 \Rightarrow 4 = p^2 \Rightarrow \pm 2 = p$.
Now check to make sure that none of the
solutions are extraneous. If $p = \pm 2$, then
$(\pm 2)^2 - 3 + 4\sqrt{(\pm 2)^2 - 3} - 5 = 0 \Rightarrow$
$4 - 3 + 4\sqrt{1} - 5 = 0 \Rightarrow 0 = 0$.
Solution set: $\{\pm 2\}$.

79. Let $u = 3t + 1$. Then the equation becomes
$u^2 - 3u + 2 = 0 \Rightarrow (u - 2)(u - 1) = 0 \Rightarrow$
$u = 2$ or $u = 1$. Now solve for t: $2 = 3t + 1 \Rightarrow$
$\dfrac{1}{3} = t$ or $1 = 3t + 1 \Rightarrow 0 = t$. Now check to

make sure that neither solution is extraneous.
If $t = 0$,
$(3 \cdot 0 + 1)^2 - 3(3 \cdot 0 + 1) + 2 = 0 \Rightarrow$
$1 - 3 + 2 = 0 \Rightarrow 0 = 0$, so 0 is a solution. If

$t = \dfrac{1}{3}$, then $\left(3\left(\dfrac{1}{3}\right) + 1\right)^2 - 3\left(3\left(\dfrac{1}{3}\right) + 1\right) + 2 = 0 \Rightarrow$

$4 - 6 + 2 = 0 \Rightarrow 0 = 0$. So 1/3 is a solution.

Solution set: $\left\{0, \dfrac{1}{3}\right\}$.

81. Let $u = \sqrt{y} + 5$. Then the equation becomes
$u^2 - 9u + 20 = 0 \Rightarrow (u - 5)(u - 4) = 0 \Rightarrow$
$u = 5$ or $u = 4$. Now solve for y:
$5 = \sqrt{y} + 5 \Rightarrow y = 0$ or $4 = \sqrt{y} + 5 \Rightarrow$
$-1 = \sqrt{y}$, which is not defined. Now check to
see if $y = 0$ is extraneous.
$\left(\sqrt{0} + 5\right)^2 - 9\left(\sqrt{0} + 5\right) + 20 = 0 \Rightarrow$
$25 - 45 + 20 = 0 \Rightarrow 0 = 0$.
Solution set: $\{0\}$

83. Let $u = x^2 - 4$. So the equation becomes
$u^2 - 3u - 4 = 0 \Rightarrow (u-4)(u+1) = 0 \Rightarrow$
$u = 4$ or $u = -1$. Now solve for x:
$4 = x^2 - 4 \Rightarrow x = \pm 2\sqrt{2}$ or
$-1 = x^2 - 4 \Rightarrow x = \pm\sqrt{3}$. Now check to make sure that neither solution is extraneous. If $x = \pm 2\sqrt{2}$, then
$\left[(\pm 2\sqrt{2})^2 - 4\right]^2 - 3\left[(\pm 2\sqrt{2})^2 - 4\right] - 4 = 0 \Rightarrow$
$16 - 12 - 4 = 0 \Rightarrow 0 = 0$. So $\pm 2\sqrt{2}$ are solutions.
If $x = \pm\sqrt{3}$, then
$\left[(\pm\sqrt{3})^2 - 4\right]^2 - 3\left[(\pm\sqrt{3})^2 - 4\right] - 4 = 0 \Rightarrow$
$1 + 3 - 4 = 0 \Rightarrow 0 = 0$. So $\pm\sqrt{3}$ are solutions.
Solution set: $\{\pm\sqrt{3}, \pm 2\sqrt{2}\}$.

85. Let $u = x^2 - 3x$. So the equation becomes
$u^2 - 2u - 8 = 0 \Rightarrow (u+2)(u-4) = 0 \Rightarrow$
$u = -2$ or $u = 4$. Now solve for x:
$-2 = x^2 - 3x \Rightarrow x^2 - 3x + 2 = 0 \Rightarrow$
$(x-2)(x-1) = 0 \Rightarrow x = 2$ or $x = 1$ or
$4 = x^2 - 3x \Rightarrow x^2 - 3x - 4 = 0 \Rightarrow$
$(x-4)(x+1) = 0 \Rightarrow x = 4$ or $x = -1$.
Now check to make sure that none of the solutions are extraneous. If $x = 2$, then
$(2^2 - 3(2))^2 - 2(2^2 - 3(2)) - 8 = 0 \Rightarrow$
$4 - (-4) - 8 = 0 \Rightarrow 0 = 0$. So 2 is a solution.
If $x = 1$, then
$(1^2 - 3(1))^2 - 2(1^2 - 3(1)) - 8 = 0 \Rightarrow$
$4 + 4 - 8 = 0 \Rightarrow 0 = 0$. So 1 is a solution. If $x = 4$, then
$(4^2 - 3(4))^2 - 2(4^2 - 3(4)) - 8 = 0 \Rightarrow$
$16 - 8 - 8 = 0 \Rightarrow 0 = 0$. So 4 is a solution.
If $x = -1$, then
$\left[(-1)^2 - 3(-1)\right]^2 - 2\left[(-1)^2 - 3(-1)\right] - 8 = 0 \Rightarrow$
$16 - 8 - 8 = 0 \Rightarrow 0 = 0$. So -1 is a solution.
Solution set: $\{-1, 1, 2, 4\}$.

1.5 Applying the Concepts

87. Let $x =$ the amount of time it takes Todrick to do the job alone. Then, $x + 3 =$ the amount of time it takes his brother to do the job alone.

$\dfrac{2}{x} + \dfrac{2}{x+3} = 1$

$x(x+3)\left(\dfrac{2}{x} + \dfrac{2}{x+3}\right) = 1 \cdot x(x+3)$

$2(x+3) + 2x = x^2 + 3x$

$4x + 6 = x^2 + 3x$

$0 = x^2 - x - 6$

$0 = (x-3)(x+2)$

$x = 3$ or $x = -2$

Disregard the negative solution. Todrick can do the job in 3 days alone; his brother can complete the job in $3 + 3 = 6$ days alone.

89. $63 + \sqrt{t + 40} = 70$
$\sqrt{t + 40} = 7$
$t + 40 = 49 \Rightarrow t = 9$
The pressure will be 70 psi at time $t = 9$.

91. Let $x =$ the denominator. Then $x - 2 =$ the numerator.

$\dfrac{x-2}{x} + \dfrac{x}{x-2} = \dfrac{25}{12}$

$12x(x-2)\left(\dfrac{x-2}{x} + \dfrac{x}{x-2}\right) = 12x(x-2)\left(\dfrac{25}{12}\right)$

$12(x-2)^2 + 12x^2 = 25x(x-2)$

$24x^2 - 48x + 48 = 25x^2 - 50x$

$0 = x^2 - 2x - 48$

$0 = (x-8)(x+6)$

$x = 8$ or $x = -6$

The solution must be positive. So the fraction is $\dfrac{6}{8}$.

93. Let $x =$ the number of shares of stock. Then $1800/x =$ the price of each share of stock.

$\left(\dfrac{1800}{x} - 18\right)(x + 5) = 1800$

$\dfrac{1800(x+5)}{x} - 18(x+5) = 1800$

$\dfrac{1800x + 9000}{x} - 18x - 90 = 1800$

$1800x + 9000 - 18x^2 - 90x = 1800x$

$-18x^2 - 90x + 9000 = 0$

$-18(x^2 + 5x - 500) = 0$

$(x - 20)(x + 25) = 0$

$x = 20$ or $x = -25$

The answer must be positive, so we reject -25.

a. Latasha bought 20 shares of stock.

b. She paid $1800/20 = \$90$ for each share.

95. Let d = the depth of the well and let t_1 = the time it takes for the stone to hit the water. Using Galileo's formula, we have
$d = 16t_1^2 \Rightarrow \dfrac{\sqrt{d}}{4} = t_1$. Let t_2 = the time it takes for the sound of the splash to return to the top of the well. Then $t_2 = \dfrac{d}{1100}$, and
$$t_1 + t_2 = \dfrac{\sqrt{d}}{4} + \dfrac{d}{1100} = 4$$
$$1100\left(\dfrac{\sqrt{d}}{4} + \dfrac{d}{1100}\right) = 1100(4)$$
$$275\sqrt{d} + d = 4400$$
Now let $u = \sqrt{d}$ and $u^2 = d$. This gives $u^2 + 275u = 4400 \Rightarrow u^2 + 275u - 4400 = 0$.
Now solve for u using the quadratic formula with $a = 1$, $b = 275$, and $c = -4400$:
$$u = \dfrac{-275 \pm \sqrt{275^2 - 4(1)(-4400)}}{2(1)} \approx 15.2.$$
Note that only the positive answer has meaning, so we reject the negative answer. $u^2 = d \approx 230$. So the well is approximately 230 feet deep.

97. Let x = the speed of the current. Then $10 + x$ = the speed of the motorboat going downstream and $10 - x$ = the speed of the motorboat going upstream. $\dfrac{12}{10+x}$ = the time the boat took to go downstream and $\dfrac{12}{10-x}$ = the time the boat took to go upstream. So, we have
$$\dfrac{12}{10-x} = \dfrac{12}{10+x} + \dfrac{1}{2}$$
$$2(10+x)(10-x)\left(\dfrac{12}{10-x} = \dfrac{12}{10+x} + \dfrac{1}{2}\right)$$
$$240 + 24x = 240 - 24x + 100 - x^2$$
$$x^2 + 48x - 100 = 0$$
$$(x+50)(x-2) = 0 \Rightarrow x = -50 \text{ or } x = 2$$
Only the positive answer has meaning, so we reject –50. The rate of the current was 2 mph.

99. Let x = the time it took the wife to wash the car alone. Then $x + 20$ = the time it took the husband to wash the car alone. So we have
$$\dfrac{1}{x} + \dfrac{1}{x+20} = \dfrac{1}{24}$$
$$24x(x+20)\left(\dfrac{1}{x} + \dfrac{1}{x+20} = \dfrac{1}{24}\right)$$
$$24x + 480 + 24x = x^2 + 20x$$
$$0 = x^2 - 28x - 480$$
$$0 = (x-40)(x+12)$$
$$x = 40 \text{ or } x = -12$$
Only the positive answer has meaning, so we reject –12. The wife took 40 minutes to wash the car alone.

101. Let x = the length of the shorter side. Then $100 + x$ = the length of the longer side. Use the Pythagorean theorem to find the length of the diagonal $= \sqrt{x^2 + (x+100)^2}$. So we have
$$\sqrt{x^2 + (x+100)^2} + (x+100) = 3x \Rightarrow$$
$$\sqrt{x^2 + (x+100)^2} = 2x - 100 \Rightarrow$$
$$\left(\sqrt{x^2 + (x+100)^2}\right)^2 = (2x-100)^2 \Rightarrow$$
$$x^2 + (x+100)^2 = 4x^2 - 400x + 10,000 \Rightarrow$$
$$2x^2 + 200x + 10,000 = 4x^2 - 400x + 10,000 \Rightarrow$$
$$0 = 2x^2 - 600x \Rightarrow 0 = 2x(x-300) \Rightarrow$$
$$x = 0 \text{ or } x = 300$$
The shorter side is 300 feet and the longer side is 400 feet. So the area is 120,000 square feet.

103. Use the Pythagorean theorem to find the lengths of AE and BE. $AE = \sqrt{64 + x^2}$ and $BE = \sqrt{(18-x)^2 + 25}$.
So we have $AE + BE = 23$ or
$$\sqrt{64 + x^2} + \sqrt{(18-x)^2 + 25} = 23 \Rightarrow$$
$$\sqrt{(18-x)^2 + 25} = 23 - \sqrt{64 + x^2} \Rightarrow$$
$$\left(\sqrt{(18-x)^2 + 25}\right)^2 = \left(23 - \sqrt{64 + x^2}\right)^2 \Rightarrow$$
$$(18-x)^2 + 25 = 529 - 46\sqrt{64 + x^2} + 64 + x^2 \Rightarrow$$
$$324 - 36x + x^2 + 25 = 593 - 46\sqrt{64 + x^2} + x^2 \Rightarrow$$
$$-244 - 36x = -46\sqrt{64 + x^2} \Rightarrow$$
$$(244 + 36x)^2 = \left(46\sqrt{64 + x^2}\right)^2$$
$$1296x^2 + 17,568x + 59,536 = 135,424 + 2116x^2$$
$$-820x^2 + 17,568x - 75,888 = 0$$
$$-4(205x^2 - 4392x + 18,972) = 0$$
$$(x-6)(205x - 3162) = 0 \Rightarrow$$
$$x = 6 \text{ or } x = \dfrac{3162}{205} \approx 15.4$$
Check for extraneous solutions. Neither is extraneous, so $CE = 6$ mi or $CE \approx 15.4$ mi.

1.5 Beyond the Basics

105. $\left(\dfrac{x^2-3}{x}\right)^2 - 6\left(\dfrac{x^2-3}{x}\right) + 8 = 0$

 Let $u = \dfrac{x^2-3}{x}$, so we have
 $u^2 - 6u + 8 = 0 \Rightarrow (u-4)(u-2) = 0 \Rightarrow$
 $u = 4$ or $u = 2$.
 Now solve for x.
 $\dfrac{x^2-3}{x} = 4 \Rightarrow x^2 - 3 = 4x \Rightarrow$
 $x^2 - 4x - 3 = 0 \Rightarrow$
 $x = 2 \pm 2\sqrt{7}$ (using the quadratic formula) or
 $\dfrac{x^2-3}{x} = 2 \Rightarrow x^2 - 3 = 2x \Rightarrow$
 $x^2 - 2x - 3 = 0 \Rightarrow$
 $(x-3)(x+1) = 0 \Rightarrow x = 3$ or $x = -1$
 Be sure to check to make sure that none of the solutions are extraneous. None are extraneous.
 Solution set: $\{-1,\ 2 \pm 2\sqrt{7},\ 3\}$.

107. $\sqrt{\sqrt{3x-2} + 2\sqrt{4x+1}} = 2\sqrt{2}$
 $\sqrt{3x-2} + 2\sqrt{4x+1} = 8$
 $\sqrt{3x-2} = 8 - 2\sqrt{4x+1}$
 $3x - 2 = 64 - 32\sqrt{4x+1} + 4(4x+1)$
 $3x - 2 = 16x + 68 - 32\sqrt{4x+1}$
 $-13x - 70 = -32\sqrt{4x+1}$
 $169x^2 + 1820x + 4900 = 4096x + 1024$
 $169x^2 - 2276x + 3876 = 0$
 Solve for x using the quadratic formula.
 $x = \dfrac{2276 \pm \sqrt{2276^2 - 4(169)(3876)}}{2(169)}$
 $x = 2$ or $x \approx 11.47$
 Check to make sure that neither solution is extraneous. 11.47 is extraneous.
 Solution set: $\{2\}$.

109. $\dfrac{x+2}{x} + \dfrac{2x+6}{x^2+4x} = -\dfrac{1}{x+4}$
 $x(x+4)\left(\dfrac{x+2}{x} + \dfrac{2x+6}{x^2+4x}\right) = -\dfrac{1}{x+4}$
 $(x+4)(x+2) + 2x + 6 = -x$
 $x^2 + 9x + 14 = 0$
 $(x+7)(x+2) = 0$
 $x = -7$ or $x = -2$
 Solution set: $\{-7, -2\}$

111. $\left(\dfrac{t}{t+2}\right)^2 + \dfrac{2t}{t+2} - 15 = 0$

 Let $u = \dfrac{t}{t+2}$. This gives
 $u^2 + 2u - 15 = 0 \Rightarrow (u+5)(u-3) = 0 \Rightarrow$
 $u = -5$ or $u = 3$.
 Now solve for t.
 $-5 = \dfrac{t}{t+2} \Rightarrow -5t - 10 = t \Rightarrow t = -\dfrac{5}{3}$ or
 $3 = \dfrac{t}{t+2} \Rightarrow 3t + 6 = t \Rightarrow t = -3$.
 Be sure to check to make sure that neither solution is extraneous. Neither is extraneous.
 Solution set: $\left\{-3,\ -\dfrac{5}{3}\right\}$.

113. $x^{4/3} - 4x^{2/3} + 3 = 0$
 Let $u = x^{2/3}$. The equation becomes
 $u^2 - 4u + 3 = 0 \Rightarrow (u-1)(u-3) = 0 \Rightarrow$
 $u = 1, 3$

$x^{2/3} = 1$	$x^{2/3} = 3$
$\left(x^{2/3}\right)^{3/2} = \pm\left(1^{3/2}\right)$	$\left(x^{2/3}\right)^{3/2} = \pm\left(3^{3/2}\right)$
$x = \pm 1$	$x = \pm\sqrt{27} = \pm 3\sqrt{3}$

 Verify that none of the solutions are extraneous. Solution set: $\{\pm 1,\ \pm 3\sqrt{3}\}$.

115. $\dfrac{x+b}{x-b} = \dfrac{x-5b}{2x-5b}$
 $(2x-5b)(x+b) = (x-b)(x-5b)$
 $2x^2 - 3bx - 5b^2 = x^2 - 6bx + 5b^2$
 $x^2 + 3bx - 10b^2 = 0$
 $(x+5b)(x-2b) = 0 \Rightarrow x = -5b$ or $x = 2b$
 Solution set: $\{-5b, 2b\}$

1.5 Critical Thinking/Discussion/Writing

117. To find the value of x, we can let $x = \sqrt{1+x}$ because the square root contains a copy of itself. Now solve the equation:
 $x = \sqrt{1+x} \Rightarrow x^2 = 1 + x \Rightarrow x^2 - x - 1 = 0$.
 Now use the quadratic formula to solve for x:
 $x = \dfrac{1 \pm \sqrt{1+4}}{2} = \dfrac{1 \pm \sqrt{5}}{2}$.
 The original expression is positive, so
 $x = \dfrac{1-\sqrt{5}}{2}$ cannot be the solution.
 Solution set: $\left\{\dfrac{1+\sqrt{5}}{2}\right\}$.

Copyright © 2019 Pearson Education Inc.

118. To find the value of x, we can let $x = \sqrt{20 + x}$ because the square root contains a copy of itself. Now solve the equation:
$x = \sqrt{20 + x} \Rightarrow x^2 = 20 + x \Rightarrow$
$x^2 - x - 20 = 0$. Now use the quadratic formula. $x = \dfrac{1 \pm \sqrt{1 + 80}}{2} = \dfrac{1 \pm 9}{2} \Rightarrow x = 5$
or $x = -4$. The original expression is positive, so $x = -4$ cannot be a solution.
Solution set: $\{5\}$.

119. To find the value of x, we can let $x = \sqrt{n + x}$ because the square root contains a copy of itself. Now solve the equation:
$x = \sqrt{n + x} \Rightarrow x^2 = n + x \Rightarrow$
$x^2 - x - n = 0$. Now use the quadratic formula. $x = \dfrac{1 \pm \sqrt{1 + 4n}}{2} = \dfrac{1}{2} \pm \dfrac{\sqrt{1 + 4n}}{2}$.
The original expression is positive, so we reject the negative root.
Solution set: $\left\{\dfrac{1}{2} + \dfrac{\sqrt{1 + 4n}}{2}\right\}$.

120. $\sqrt{x + 2\sqrt{x - 1}} + \sqrt{x - 2\sqrt{x - 1}} = 4$
Note that $x + 2\sqrt{x - 1} = (x - 1) + 1 + 2\sqrt{x - 1}$
$= \left(\sqrt{x - 1} + 1\right)^2$. Similarly, $x - 2\sqrt{x - 1} =$
$(x - 1) - 2\sqrt{x - 1} + 1 = \left(\sqrt{x - 1} - 1\right)^2$.
Substitute these expressions into the original equation:
$\sqrt{\left(\sqrt{x-1}+1\right)^2} + \sqrt{\left(\sqrt{x-1}-1\right)^2} = 4 \Rightarrow$
$\sqrt{x - 1} + 1 + \sqrt{x - 1} - 1 = 4 \Rightarrow$
$2\sqrt{x - 1} = 4 \Rightarrow \sqrt{x - 1} = 2 \Rightarrow x - 1 = 4 \Rightarrow$
$x = 5$.
Solution set: $\{5\}$

1.5 Getting Ready for the Next Section

121. $5 > -1$

123. $7 \geq 7$

125. $3 > 3x + 4$

127. $x \geq 0$

129. $2(x - 3) + 17 = 4x + 1$
$2x - 6 + 17 = 4x + 1$
$2x + 11 = 4x + 1$
$-2x = -10$
$x = 5$
Solution set: $\{5\}$

131. $2x - 3 - (3x - 1) = 6$
$2x - 3 - 3x + 1 = 6$
$-x - 2 = 6$
$-x = 8 \Rightarrow x = -8$
Solution set: $\{-8\}$

133. $\dfrac{5}{2x} + 3 = \dfrac{5 + 3(2x)}{2x} = \dfrac{5 + 6x}{2x}$

135. $\dfrac{3}{x + 1} - \dfrac{x - 2}{x + 3} = \dfrac{3(x + 3) - (x + 1)(x - 2)}{(x + 1)(x + 3)}$
$= \dfrac{3x + 9 - \left(x^2 - x - 2\right)}{(x + 1)(x + 3)}$
$= \dfrac{-x^2 + 4x + 11}{(x + 1)(x + 3)}$

1.6 Inequalities

1.6 Practice Problems

1. a. $3x + 1 \leq 10 \Rightarrow 3x \leq 9 \Rightarrow x \leq 3$
Solution set: $(-\infty, 3]$

b. $4x + 9 > 2(x + 6) + 1 \Rightarrow 4x + 9 > 2x + 13 \Rightarrow$
$2x > 4 \Rightarrow x > 2$
Solution set: $(2, \infty)$

c. $7 - 2x \geq -3 \Rightarrow -2x \geq -10 \Rightarrow x \leq 5$
Solution set: $(-\infty, 5]$

2. Let t = time elapsed since the plane went on autopilot. Then $340t$ = distance the plane has flown in t hours, and $185 + 340t$ = the plane's distance from Miami after t hours.
$185 + 340t \geq 1035 \Rightarrow 340t \geq 850 \Rightarrow t \geq 2.5$
The tower will suspect trouble if the plane has not arrived in 2.5 hours.

3. a. $2(4 - x) + 6x < 4(x + 1) + 7$
$8 - 2x + 6x < 4x + 4 + 7$
$8 + 4x < 4x + 11$
$0 < 3$
The last inequality is always true, so the solution set is $(-\infty, \infty)$.

b. $3(x-2)+5 \geq 7(x-1)-4(x-2)$
$3x-6+5 \geq 7x-7-4x+8$
$3x-1 \geq 3x+1 \Rightarrow -1 \geq 1$
The last inequality is always false, so the solution set is $\emptyset$.

4. $3x-5 \geq 7$ or $5-2x \geq 1$
$3x \geq 12$ or $-2x \geq -4$
$x \geq 4$ or $x \leq 2$
Solution set: $(-\infty, 2] \cup [4, \infty)$

5. $2(3-x)-3 < 5$ and $2(x-5)+7 \leq 3$
$6-2x-3 < 5$ and $2x-10+7 \leq 3$
$3-2x < 5$ and $2x-3 \leq 3$
$-2x < 2$ and $2x \leq 6$
$x > -1$ and $x \leq 3$
Solution set: $(-1, 3]$

6. $-6 \leq 4x-2 < 4 \Rightarrow -4 \leq 4x < 6 \Rightarrow -1 \leq x < \dfrac{3}{2}$

Solution set: $\left[-1, \dfrac{3}{2}\right)$

7. Start with the interval for x.
$-3 \leq x \leq 2$
$3(-3) \leq 3x \leq 3(2)$
$-9 \leq 3x \leq 6$
$-9+5 \leq 3x+5 \leq 6+5$
$-4 \leq 3x+5 \leq 11$
Therefore, $a = -4$ and $b = 11$.

8. $15 < C < 25$
$\left(\dfrac{9}{5}\right)15 < \dfrac{9}{5}C < \left(\dfrac{9}{5}\right)25$
$27 < \dfrac{9}{5}C < 45$
$27 + 32 < \dfrac{9}{5}C + 32 < 45 + 32$
$59 < \dfrac{9}{5}C + 32 < 77$
The temperature range from 15°C to 25°C corresponds to a range from 59°F to 77°F.

9. $x^2 + 2 < 3x + 6$
$x^2 - 3x - 4 < 0 \Rightarrow (x+1)(x-4) < 0$
The factors equal 0 at $x = -1$ and $x = 4$.

Interval	Test point	Value of $x^2 - 3x - 4$	Result
$(-\infty, -1)$	-2	6	+
$(-1, 4)$	0	-4	$-$
$(4, \infty)$	5	6	+

The solution set is $(-1, 4)$.

10. $\dfrac{2x+5}{x-1} \leq 1$

$\dfrac{2x+5}{x-1} - 1 \leq 0$

$\dfrac{2x+5-x+1}{x-1} \leq 0 \Rightarrow \dfrac{x+6}{x-1} \leq 0$

$x+6 = 0 \Rightarrow x = -6$
$x-1 = 0 \Rightarrow x = 1$
The test points are -6 and 1.

Interval	Test point	Value of $\dfrac{x+6}{x-1}$	Result
$(-\infty, -6)$	-9	$\dfrac{3}{10}$	+
$(-6, 1)$	0	-6	$-$
$(1, \infty)$	2	8	+

Note that the expression is undefined for $x = 1$. The solution set is $[-6, 1)$.

1.6 Concepts and Vocabulary

1. < **3.** <

5. True **7.** True

1.6 Building Skills

9. $(-2, 5)$

11. $(0, 4]$

13. $[-1, \infty)$

15. $(-\infty, -2]$

17. $2x - 3 > -5 \Rightarrow 2x > -2 \Rightarrow x > -1$
$(-1, \infty)$

19. $x + 3 < 6 \Rightarrow x < 3$
$(-\infty, 3)$

21. $1 - x \leq 4 \Rightarrow -x \leq 3 \Rightarrow x \geq -3$
$[-3, \infty)$

23. $2x + 5 < 9 \Rightarrow 2x < 4 \Rightarrow x < 2$
$(-\infty, 2)$

25. $3 - 3x > 15 \Rightarrow -3x > 12 \Rightarrow x < -4$
$(-\infty, -4)$

26. $8 - 4x \geq 12 \Rightarrow -4x \geq 4 \Rightarrow x \leq -1$
$(-\infty, -1]$

29. $3(x - 3) \leq 3 - x \Rightarrow 3x - 9 \leq 3 - x \Rightarrow$
$4x \leq 12 \Rightarrow x \leq 3$
$(-\infty, 3]$

31. $6x + 4 > 3x + 10 \Rightarrow 3x > 6 \Rightarrow x > 2$
$(2, \infty)$

33. $8(x - 1) - x \leq 7x - 12$
$8x - 8 - x \leq 7x - 12$
$7x - 8 \leq 7x - 12 \Rightarrow -8 \leq -12$ False
The solution set is $\varnothing$.

35. $5(x + 2) \leq 3(x + 1) + 10 \Rightarrow$
$5x + 10 \leq 3x + 3 + 10 \Rightarrow 5x + 10 \leq 3x + 13 \Rightarrow$
$2x \leq 3 \Rightarrow x \leq \dfrac{3}{2}$
$\left(-\infty, \dfrac{3}{2}\right]$

37. $2(x + 1) + 3 \geq 2(x + 2) - 1$
$2x + 2 + 3 \geq 2x + 4 - 1$
$2x + 5 \geq 2x + 3 \Rightarrow 5 \geq 3$ True
$(-\infty, \infty)$

39. $2(x + 1) - 2 \leq 3(2 - x) + 9$
$2x + 2 - 2 \leq 6 - 3x + 9$
$2x \leq 15 - 3x$
$5x \leq 15 \Rightarrow x \leq 3$
$(\infty, 3]$

41. $9x - 6 \geq \dfrac{3}{2}x + 9$
$18x - 12 \geq 3x + 18$
$15x \geq 30 \Rightarrow x \geq 2$
Solution set: $[2, \infty)$

43. $\dfrac{x - 3}{3} \leq 2 + \dfrac{x}{2} \Rightarrow 2x - 6 \leq 12 + 3x \Rightarrow$
$-18 \leq x \Rightarrow x \geq -18$
Solution set: $[-18, \infty)$

45. $\dfrac{3x + 1}{2} < x - 1 + \dfrac{x}{2}$
$3x + 1 < 2x - 2 + x$
$3x + 1 < 3x - 2 \Rightarrow 1 < -2$ False
Solution set: $\varnothing$

47. $\dfrac{x - 3}{2} \geq \dfrac{x}{3} + 1$
$3x - 9 \geq 2x + 6 \Rightarrow x \geq 15$
Solution set: $[15, \infty)$

49. $\dfrac{3x + 1}{3} - \dfrac{x}{2} \leq \dfrac{x + 2}{2}$
$6x + 2 - 3x \leq 3x + 6$
$3x + 2 \leq 3x + 6 \Rightarrow 2 \leq 6$ True
Solution set: $(-\infty, \infty)$

51. $2x + 5 < 1 \quad$ or $\quad 2 + x > 4$
$2x < -4 \quad$ or $\quad x > 2$
$x < -2 \quad$ or
Solution set: $(-\infty, -2) \cup (2, \infty)$

53. $\dfrac{2x - 3}{4} \leq 2 \quad$ or $\quad \dfrac{4 - 3x}{2} \geq 2$
$2x - 3 \leq 8 \quad$ or $\quad 4 - 3x \geq 4$
$2x \leq 11 \quad$ or $\quad -3x \geq 0$
$x \leq \dfrac{11}{2} \quad$ or $\quad x \leq 0$
Solution set: $\left(-\infty, \dfrac{11}{2}\right]$

55. $\dfrac{2x + 1}{3} \geq x + 1 \quad$ or $\quad \dfrac{x}{2} - 1 > \dfrac{x}{3}$
$2x + 1 \geq 3x + 3 \quad$ or $\quad 3x - 6 > 2x$
$-2 \geq x \quad$ or $\quad -6 > -x$
$x \leq -2 \quad$ or $\quad 6 < x \Rightarrow x > 6$
Solution set: $(-\infty, -2] \cup (6, \infty)$

57. $\dfrac{x - 1}{2} > \dfrac{x}{3} - 1 \quad$ or $\quad \dfrac{2x + 5}{3} \leq \dfrac{x + 1}{6}$
$3x - 3 > 2x - 6 \quad$ or $\quad 4x + 10 \leq x + 1$
$x > -3 \quad$ or $\quad 3x \leq -9 \Rightarrow x \leq -3$
Solution set: $(-\infty, \infty)$

59. $3 - 2x \leq 7$ and $2x - 3 \leq 7$
$\quad\quad -2x \leq 4$ and $2x \leq 10$
$\quad\quad\quad x \geq -2$ and $x \leq 5$
Solution set: $[-2, 5]$

61. $2(x+1) + 3 \geq 1$ and $2(2-x) > -6$
$\quad\quad 2x + 2 + 3 \geq 1$ and $4 - 2x > -6$
$\quad\quad\quad 2x + 5 \geq 1$ and $-2x > -10$
$\quad\quad\quad\quad 2x \geq -4$ and $x < 5$
$\quad\quad\quad\quad\quad x \geq -2$
Solution set: $[-2, 5)$

63. $2(x+1) - 3 > 7$ and $3(2x+1) + 1 < 10$
$\quad\quad 2x + 2 - 3 > 7$ and $6x + 3 + 1 < 10$
$\quad\quad\quad 2x - 1 > 7$ and $6x + 4 < 10$
$\quad\quad\quad\quad 2x > 8$ and $6x < 6$
$\quad\quad\quad\quad\quad x > 4$ and $x < 1$
Since x cannot be less than 1 and greater than 4 at the same time, the solution set is $\varnothing$.

65. $5 + 3(x-1) < 3 + 3(x+1)$ and $3x - 7 \leq 8$
$\quad\quad 5 + 3x - 3 < 3 + 3x + 3$ and $3x \leq 15$
$\quad\quad 2 + 3x < 6 + 3x$ and $x \leq 5$
$\quad\quad 2 < 6$ True
Solution set: $(-\infty, 5]$

67. $3 < x + 5 < 4 \Rightarrow -2 < x < -1$ or $(-2, -1)$

69. $-4 \leq x - 2 < 2 \Rightarrow -2 \leq x < 4$ or $[-2, 4)$

71. $-9 \leq 2x + 3 \leq 5 \Rightarrow -12 \leq 2x \leq 2 \Rightarrow$
$-6 \leq x \leq 1$ or $[-6, 1]$

73. $0 \leq 1 - \dfrac{x}{3} < 2 \Rightarrow 0 \leq 3 - x < 6 \Rightarrow$
$-3 \leq -x < 3 \Rightarrow 3 \geq x > -3$ or $(-3, 3]$

75. $-1 < \dfrac{2x - 3}{5} \leq 0 \Rightarrow -5 < 2x - 3 \leq 0 \Rightarrow$
$-2 < 2x \leq 3 \Rightarrow -1 < x \leq \dfrac{3}{2}$ or $\left(-1, \dfrac{3}{2}\right]$

77. $5x \leq 3x + 1 < 4x - 2 \Rightarrow 2x \leq 1 < x - 2 \Rightarrow$
$x \leq \dfrac{1}{2}$ and $3 < x \Rightarrow x > 3$
x cannot be less than or equal to 1/2 at the same time that 3 is less than x, so the solution set is $\varnothing$.

79. $-2 < x < 1 \Rightarrow -2 + 7 < x + 7 < 1 + 7 \Rightarrow$
$5 < x + 7 < 8 \Rightarrow a = 5, b = 8$

81. $-1 < x < 1 \Rightarrow 1 > -x > -1 \Rightarrow$
$2 + 1 > 2 - x > 2 - 1 \Rightarrow 3 > 2 - x > 1 \Rightarrow$
$1 < 2 - x < 3 \Rightarrow a = 1, b = 3$

83. $0 < x < 4 \Rightarrow 0 < 5x < 20 \Rightarrow -1 < 5x - 1 < 19 \Rightarrow$
$a = -1, b = 19$

85. $x^2 + 4x - 12 \leq 0 \Rightarrow (x+6)(x-2) \leq 0$.
Now solve the associated equation:
$(x+6)(x-2) = 0 \Rightarrow x = -6$ or $x = 2$.
So, the intervals are
$(-\infty, -6], [-6, 2],$ and $[2, \infty)$.

Interval	Test point	Value of $x^2 + 4x - 12$	Result
$(-\infty, -6]$	-10	48	$+$
$[-6, 2]$	0	-12	$-$
$[2, \infty)$	3	9	$+$

The solution set is $[-6, 2]$.

87. $6x^2 + 7x - 3 \geq 0 \Rightarrow (3x - 1)(2x + 3) \geq 0$.
Now solve the associated equation:
$(3x - 1)(2x + 3) = 0 \Rightarrow x = 1/3$ or $x = -3/2$.
The intervals are $(-\infty, -3/2], [-3/2, 1/3],$
and $[1/3, \infty)$.

Interval	Test point	Value of $6x^2 + 7x - 3$	Result
$(-\infty, -3/2]$	-2	7	$+$
$[-3/2, 1/3]$	0	-3	$-$
$[1/3, \infty)$	1	10	$+$

The solution set is $(-\infty, -3/2] \cup [1/3, \infty)$.

89. $(x+3)(x+1)(x-1) \geq 0$.
Now solve the associated equation:
$(x+3)(x+1)(x-1) = 0 \Rightarrow$
$x = -3, x = -1$ or $x = 1$.
The intervals are
$(-\infty, -3], [-3, -1], [-1, 1],$ and $[1, \infty)$.

(continued on next page)

(continued)

Interval	Test point	Value of $(x+3)(x+1)(x-1)$	Result
$(-\infty, -3]$	-10	-693	$-$
$[-3, -1]$	-2	3	$+$
$[-1, 1]$	0	-3	$-$
$[1, \infty)$	10	1287	$+$

The solution set is $[-3, -1] \cup [1, \infty)$.

91. $x^3 - 4x^2 - 12x > 0$
Solve the associated equation
$x^3 - 4x^2 - 12x = 0$.
$x^3 - 4x^2 - 12x = 0 \Rightarrow x(x^2 - 4x - 12) = 0 \Rightarrow$
$x(x+2)(x-6) = 0 \Rightarrow x = 0, -2, 6$
The intervals are $(-\infty, -2)$, $(-2, 0)$, $(0, 6)$, and $(6, \infty)$.

Interval	Test point	Value of $x^3 - 4x^2 - 12x$	Result
$(-\infty, -2)$	-10	-1280	$-$
$(-2, 0)$	-1	7	$+$
$(0, 6)$	1	-15	$-$
$(6, \infty)$	10	480	$+$

The solution set is $(-2, 0) \cup (6, \infty)$.

93. $x^2 + 2x < -1 \Rightarrow x^2 + 2x + 1 < 0$.
Now solve the associated equation:
$x^2 + 2x + 1 = 0 \Rightarrow (x+1)^2 = 0 \Rightarrow x = -1$.
The intervals are $(-\infty, -1)$ and $(-1, \infty)$.

Interval	Test point	Value of $x^2 + 2x + 1$	Result
$(\infty, -1)$	-2	1	$+$
$(-1, \infty)$	0	1	$+$

Neither interval has a negative solution, so the solution set is $\varnothing$.

95. $x^3 - x^2 \geq 0$.
Now solve the associated equation:
$x^3 - x^2 = 0 \Rightarrow x^2(x-1) = 0 \Rightarrow$
$x = 0$ or $x = 1$.
The intervals are $(-\infty, 0], [0, 1]$ and $[1, \infty)$.

Interval	Test point	Value of $x^3 - x^2$	Result
$(-\infty, 0]$	-1	-2	$-$
$[0, 1]$	$\frac{1}{2}$	$-\frac{1}{8}$	$-$
$[1, \infty)$	2	4	$+$

Note that 0 is a solution. The solution set is $\{0\} \cup [1, \infty)$.

97. $x^2 \geq 1 \Rightarrow x^2 - 1 \geq 0$.
Now solve the associated equation:
$x^2 - 1 = 0 \Rightarrow (x-1)(x+1) = 0 \Rightarrow$
$x = -1$ or $x = 1$.
The intervals are $(-\infty, -1], [-1, 1]$, and $[1, \infty)$.

Interval	Test point	Value of $x^2 - 1$	Result
$(-\infty, -1]$	-2	3	$+$
$[-1, 1]$	0	-1	$-$
$[1, \infty)$	2	3	$+$

The solution set is $(-\infty, -1] \cup [1, \infty)$.

99. $x^3 < -8 \Rightarrow x^3 + 8 < 0$.
Now solve the associated equation:
$x^3 + 8 = 0 \Rightarrow (x+2)(x^2 - 2x + 4) = 0 \Rightarrow$
$x = -2$ or $x^2 - 2x + 4 = 0$.
Solve $x^2 - 2x + 4 = 0$ using the quadratic formula:
$x = \dfrac{-(-2) \pm \sqrt{(-2)^2 - 4(1)(4)}}{2(1)} = 1 \pm i\sqrt{3}$.
We cannot use complex intervals, so the only intervals we examine are $(-\infty, -2)$ and $(-2, \infty)$.

Interval	Test point	Value of $x^3 + 8$	Result
$(-\infty, -2)$	-10	-992	$-$
$(-2, \infty)$	10	1008	$+$

The solution set is $(-\infty, -2)$.

101. $\dfrac{x+2}{x-5} < 0$.

Now solve $x + 2 = 0 \Rightarrow x = -2$ and $x - 5 = 0 \Rightarrow x = 5$.
So the intervals are $(-\infty, -2), (-2, 5)$, and $(5, \infty)$.

Interval	Test point	Value of $\dfrac{x+2}{x-5}$	Result
$(-\infty, -2)$	-10	$8/15$	$+$
$(-2, 5)$	0	$-2/5$	$-$
$(5, \infty)$	10	$12/5$	$+$

Note that the fraction is undefined if $x = 5$.
The solution set is $(-2, 5)$.

103. $\dfrac{x+4}{x} < 0$.

Solve $x + 4 = 0 \Rightarrow x = -4$ and $x = 0$.
The fraction is undefined if $x = 0$. So the intervals are $(-\infty, -4), (-4, 0)$, and $(0, \infty)$.

Interval	Test point	Value of $\dfrac{x+4}{x}$	Result
$(-\infty, -4)$	-10	$3/5$	$+$
$(-4, 0)$	-2	-1	$-$
$(0, \infty)$	2	3	$+$

The solution set is $(-4, 0)$.

105. $\dfrac{x+1}{x+2} \le 3 \Rightarrow \dfrac{x+1}{x+2} - 3 \le 0 \Rightarrow$
$\dfrac{x+1-3(x+2)}{x+2} \le 0 \Rightarrow \dfrac{-2x-5}{x+2} \le 0$.

Now solve $-2x - 5 = 0 \Rightarrow x = -5/2$ and $x + 2 = 0 \Rightarrow x = -2$. So the intervals are $(-\infty, -5/2], [-5/2, -2)$, and $(-2, \infty)$. The original fraction is not defined if $x = -2$, so -2 is not included in the intervals.

Interval	Test point	Value of $\dfrac{-2x-5}{x+2}$	Result
$(-\infty, -5/2]$	-3	-1	$-$
$[-5/2, -2)$	$-9/4$	2	$+$
$(-2, \infty)$	3	$-11/5$	$-$

The solution set is $(-\infty, -5/2] \cup (-2, \infty)$.

107. $\dfrac{(x-2)(x+2)}{x} > 0$.

Now we have $x = 0$ and $(x-2)(x+2) = 0 \Rightarrow x = 2$ and $x = -2$.
So the intervals are $(-\infty, -2), (-2, 0), (0, 2)$, and $(2, \infty)$. Note that the original fraction is not defined if $x = 0$, so 0 is not included in the intervals.

Interval	Test point	Value of $\dfrac{(x-2)(x+2)}{x}$	Result
$(-\infty, -2)$	-3	$-5/3$	$-$
$(-2, 0)$	-1	3	$+$
$(0, 2)$	1	-3	$-$
$(2, \infty)$	3	$5/3$	$+$

The solution set is $(-2, 0) \cup (2, \infty)$.

109. $\dfrac{(x-2)(x+1)}{(x-3)(x+5)} \ge 0$.

Set the numerator and denominator equal to zero and solve for x.
$(x-2)(x+1) = 0 \Rightarrow x = 2, -1$
$(x-3)(x+5) = 0 \Rightarrow x = 3, -5$
The intervals are $(-\infty, -5), (-5, -1], [-1, 2], [2, 3)$, and $(3, \infty)$.

Interval	Test point	Value of $\dfrac{(x-2)(x+1)}{(x-3)(x+5)}$	Result
$(-\infty, -5)$	-6	$40/9$	$+$
$(-5, -1]$	-2	$-4/15$	$-$
$[-1, 2]$	0	$2/15$	$+$
$[2, 3)$	$5/2$	$-5/15$	$-$
$(3, \infty)$	5	$9/10$	$+$

The solution set is $(-\infty, -5) \cup [-1, 2] \cup (3, \infty)$.

111. $\dfrac{x^2-1}{x^2-4} \le 0$.

Set the numerator and denominator equal to zero and solve for x.

$x^2 - 1 = 0 \Rightarrow x = \pm 1$; $x^2 - 4 = 0 \Rightarrow x = \pm 2$

The intervals are $(-\infty, -2)$, $(-2, -1]$, $[-1, 1]$, $[1, 2)$, and $[2, \infty)$.

Interval	Test point	Value of $\dfrac{x^2-1}{x^2-4}$	Result
$(-\infty, -2)$	-3	$8/5$	$+$
$(-2, -1]$	$-3/2$	$-5/7$	$-$
$[-1, 1]$	0	$1/4$	$+$
$[1, 2)$	$3/2$	$-5/7$	$-$
$[2, \infty)$	3	$8/5$	$+$

The solution set is $(-2, -1] \cup [1, 2)$.

113. $\dfrac{x+4}{3x-2} \ge 1 \Rightarrow \dfrac{x+4}{3x-2} - 1 \ge 0 \Rightarrow$

$\dfrac{x+4-3x+2}{3x-2} \ge 0 \Rightarrow \dfrac{-2x+6}{3x-2} \ge 0$.

Now we have $-2x + 6 = 0 \Rightarrow x = 3$ and $3x - 2 = 0 \Rightarrow x = 2/3$. So the intervals are $(-\infty, 2/3)$, $(2/3, 3]$, and $[3, \infty)$. Note that the original fraction is not defined if $x = 2/3$, so $2/3$ is not included in the intervals.

Interval	Test point	Value of $\dfrac{x+4}{3x-2} - 1$	Result
$(-\infty, 2/3)$	0	-3	$-$
$(2/3, 3]$	1	4	$+$
$[3, \infty)$	4	$-1/5$	$-$

The solution set is $(2/3, 3]$.

115. $3 \le \dfrac{2x+6}{2x+1} \Rightarrow \dfrac{2x+6}{2x+1} - 3 \ge 0 \Rightarrow$

$\dfrac{2x+6-3(2x+1)}{2x+1} \ge 0 \Rightarrow \dfrac{-4x+3}{2x+1} \ge 0$.

This gives $-4x + 3 = 0 \Rightarrow x = 3/4$ and $2x + 1 = 0 \Rightarrow x = -1/2$. The intervals are $(-\infty, -1/2)$, $(-1/2, 3/4]$, and $[3/4, \infty)$. The original fraction is not defined if $x = -1/2$, so $-1/2$ is not included in the intervals.

Interval	Test point	Value of $\dfrac{2x+6}{2x+1} - 3$	Result
$(-\infty, -1/2)$	-1	-7	$-$
$(-1/2, 3/4]$	0	3	$+$
$[3/4, \infty)$	1	$-1/3$	$-$

The solution set is $(-1/2, 3/4]$.

117. $\dfrac{x+2}{x-3} \ge \dfrac{x-1}{x+3} \Rightarrow \dfrac{x+2}{x-3} - \dfrac{x-1}{x+3} \ge 0 \Rightarrow$

$\dfrac{(x+2)(x+3) - (x-1)(x-3)}{(x-3)(x+3)} \ge 0 \Rightarrow$

$\dfrac{(x^2+5x+6) - (x^2-4x+3)}{(x-3)(x+3)} \ge 0 \Rightarrow$

$\dfrac{9x+3}{(x-3)(x+3)} \ge 0$

Set the numerator and the denominator equal to zero and solve for x. $9x + 3 = 0 \Rightarrow x = -1/3$ and $(x-3)(x+3) = 0 \Rightarrow x = 3$ or $x = -3$. The intervals are $(-\infty, -3)$, $(-3, -1/3]$, $[-1/3, 3)$, and $(3, \infty)$. Note that the original fractions are not defined if $x = -3$ or $x = 3$, so -3 and 3 are not included in the intervals.

Interval	Test point	Value of $\dfrac{x+2}{x-3} - \dfrac{x-1}{x+3}$	Result
$(-\infty, -3)$	-6	$-17/9$	$-$
$(-3, -1/3]$	-1	$3/4$	$+$
$[-1/3, 3)$	1	$-3/2$	$-$
$(3, \infty)$	6	$19/9$	$+$

The solution set is $(-3, -1/3] \cup (3, \infty)$.

119. $\dfrac{x-1}{x+1} \leq \dfrac{x+2}{x-3} \Rightarrow \dfrac{x-1}{x+1} - \dfrac{x+2}{x-3} \leq 0 \Rightarrow$

$\dfrac{(x-1)(x-3)-(x+2)(x+1)}{(x+1)(x-3)} \leq 0 \Rightarrow$

$\dfrac{(x^2-4x+3)-(x^2+3x+2)}{(x+1)(x-3)} \leq 0 \Rightarrow$

$\dfrac{-7x+1}{(x+1)(x-3)} \leq 0$

Set the numerator and the denominator equal to zero and solve for x. $-7x+1=0 \Rightarrow x = 1/7$ and $(x+1)(x-3) = 0 \Rightarrow x = -1$ or $x = 3$. The intervals are $(-\infty, -1)$, $(-1, 1/7]$, $[1/7, 3)$, and $(3, \infty)$. Note that the original fractions are not defined if $x = -1$ or $x = 3$, so -1 and 3 are not included in the intervals.

Interval	Test point	Value of $\dfrac{x-1}{x+1} - \dfrac{x+2}{x-3}$	Result
$(-\infty, -1)$	-2	3	$+$
$(-1, 1/7]$	0	$-1/3$	$-$
$[1/7, 3)$	1	$3/2$	$+$
$(3, \infty)$	4	$-27/5$	$-$

The solution set is $(-1, 1/7] \cup (3, \infty)$.

1.6 Applying the Concepts

121. Let $x =$ the selling price of the refrigerator. Then
$1750 + 0.15(1750) \leq x \leq 1750 + 0.20(1750) \Rightarrow$
$2012.50 \leq x \leq 2100$
The refrigerator's selling price ranges from $2012.50 to $2100.

123. Let $x =$ the amount of gasoline in the car at the start of the trip. Then
$300 \leq 40x \leq 480 \Rightarrow 7.5 \leq x \leq 12$
The car had between 7.5 gallons and 12 gallons of gas at the start of the trip.

125. Let $x =$ the amount of cream.
Then $270 - x =$ the amount of milk. So,
$0.3x + 0.03(270-x) \geq 0.045(270) \Rightarrow$
$0.3x + 8.1 - 0.03x \geq 12.15 \Rightarrow$
$0.27x \geq 4.05 \Rightarrow x \geq 15$
At least 15 quarts of cream must be added.

127. Let $x =$ the number of 4-door sedans. Then $3x =$ the number of SUV's and $2x =$ the number of convertibles.
$x + 3x + 2x \geq 48 \Rightarrow 6x \geq 48 \Rightarrow x \geq 8$
There are at least 8 four-door sedans.

129. $132t - t^2 < 3200 \Rightarrow -t^2 + 132t - 3200 > 0 \Rightarrow$
$t^2 - 132t + 3200 < 0 \Rightarrow (t-100)(t-32) < 0$
Solving the associated equation gives $t = 100$ or $t = 32$. The intervals to be checked are $(-\infty, 32), (32, 100),$ and $(100, \infty)$. Checking a number in each interval shows that the temperature must fall in the range $(32°F, 100°F)$ for the number of mites to exceed 3200.

131. Probability must be less than 1, so we have
$0.5 < \dfrac{64 - 0.2x}{208 - x} \leq 1 \Rightarrow$
$0.5 < \dfrac{64 - 0.2x}{208 - x}$ and $\dfrac{64 - 0.2x}{208 - x} \leq 1$
Solving each inequality independently gives
$0.5 < \dfrac{64 - 0.2x}{208 - x} \Rightarrow 0 < \dfrac{64 - 0.2x}{208 - x} - 0.5 \Rightarrow$
$0 < \dfrac{64 - 0.2x - 0.5(208 - x)}{208 - x} \Rightarrow$
$0 < \dfrac{-40 + 0.3x}{208 - x}$
Now we have $-40 + 0.3x = 0 \Rightarrow x = 400/3$ and $208 - x = 0 \Rightarrow 208 = x$. The original fraction is not defined if $x = 208$, so 208 is not in the solution set. So there need to be more than 133 cards. (Note that the value of x is rounded up to account for the partial value.)
$\dfrac{64 - 0.2x}{208 - x} \leq 1 \Rightarrow \dfrac{64 - 0.2x}{208 - x} - 1 \leq 0 \Rightarrow$
$\dfrac{64 - 0.2x - 208 + x}{208 - x} \leq 0 \Rightarrow \dfrac{0.8x - 144}{208 - x} \leq 0 \Rightarrow$
$0.8x - 144 \leq 0 \Rightarrow 0.8x \leq 144 \Rightarrow x \leq 180$
So, the likelihood that the next card dealt would be a jack, queen, king, or ace is greater than 50% if more than 133 cards and less than 180 cards are dealt.

1.6 Beyond the Basics

133. $2x^2 + kx + 2 = 0$ has two real solutions if the discriminant is greater than zero. So $k^2 - 4(2)(2) > 0 \Rightarrow k^2 - 16 > 0$. Solving the associated equation gives $k = -4$ or $k = 4$. The intervals to be tested are $(-\infty, -4), (-4, 4),$ and $(4, \infty)$. $k^2 - 16 > 0$ for $(-\infty, -4) \cup (4, \infty)$.

135. $x^2 + kx + k = 0$ has two real solutions if the discriminant is greater than zero. So $k^2 - 4k > 0 \Rightarrow k(k-4) > 0$. Solving the associated equation gives $k = 0$ or $k = 4$. The intervals to be tested are $(-\infty, 0)$, $(0, 4)$, and $(4, \infty)$. $k^2 - 4k > 0$ for $(-\infty, 0) \cup (4, \infty)$.

137. $\dfrac{x}{2x+1} \geq \dfrac{1}{4}$ and $\dfrac{6x}{4x-1} < \dfrac{1}{2}$

We will solve each inequality independently and then determine where the solution sets intersect.

$\dfrac{x}{2x+1} \geq \dfrac{1}{4} \Rightarrow \dfrac{x}{2x+1} - \dfrac{1}{4} \geq 0 \Rightarrow$

$\dfrac{4x - (2x+1)}{4(2x+1)} \geq 0 \Rightarrow \dfrac{2x-1}{8x+4} \geq 0$

Now, we have $2x - 1 = 0 \Rightarrow x = 1/2$ and $8x + 4 = 0 \Rightarrow x = -1/2$. Note that the original fraction $\dfrac{x}{2x+1}$ is not defined if $x = -1/2$, so the intervals to be tested are $(-\infty, -1/2)$, $(-1/2, 1/2]$, and $[1/2, \infty)$.

Interval	Test point	Value of $\dfrac{x}{2x+1} - \dfrac{1}{4}$	Result
$(-\infty, -1/2)$	-1	$3/4$	$+$
$(-1/2, 1/2]$	0	$-1/4$	$-$
$[1/2, \infty)$	1	$1/12$	$+$

The solution set is $(-\infty, -1/2) \cup [1/2, \infty)$.

$\dfrac{6x}{4x-1} < \dfrac{1}{2} \Rightarrow \dfrac{6x}{4x-1} - \dfrac{1}{2} < 0 \Rightarrow$

$\dfrac{12x - (4x-1)}{2(4x-1)} < 0 \Rightarrow \dfrac{8x+1}{8x-2} < 0$

The original fractions $\dfrac{6x}{4x-1}$ and $\dfrac{8x+1}{8x-2}$ are not defined if $x = 1/4$. Also, $8x + 1 = 0 \Rightarrow x = -1/8$. The intervals to be tested are $(-\infty, -1/8)$, $(-1/8, 1/4)$, and $(1/4, \infty)$.

Interval	Test point	Value of $\dfrac{6x}{4x-1} - \dfrac{1}{2}$	Result
$(-\infty, -1/8)$	-1	$7/10$	$+$
$(-1/8, 1/4)$	0	$-1/2$	$-$
$(1/4, \infty)$	1	$3/2$	$+$

The solution set is $(-1/8, 1/4)$.

The two solution sets do not intersect, so the solution set of $\dfrac{x}{2x+1} \geq \dfrac{1}{4}$ and $\dfrac{6x}{4x-1} < \dfrac{1}{2}$ is $\varnothing$.

139. Let x = one number. Then $c - x$ = the other number. So we have

$x(c-x) = 36 \Rightarrow cx - x^2 = 36 \Rightarrow$

$x^2 - cx + 36 = 0$

$x^2 - cx + 36 = 0$ has one or two real solutions if the discriminant is greater than or equal to zero. So

$(-c)^2 - 4(1)(36) \geq 0 \Rightarrow c^2 - 144 \geq 0 \Rightarrow$

$c^2 \geq 144 \Rightarrow c \geq 12$ or $c \leq -12$

Solution set: $(-\infty, -12] \cup [12, \infty)$

141. Let x = the total number of radios imported. Then $x - 1000$ = the number of radios subject to the penalty tax. So we have

$10x + 0.05x(x - 1000) \leq 640{,}000$

$10x + 0.05x^2 - 50x \leq 640{,}000$

$0.05x^2 - 40x - 640{,}000 \leq 0$

$x^2 - 800x - 12{,}800{,}000 \leq 0$

$(x - 4000)(x + 3200) \leq 0$

Solving the associated equation gives $x = 4000$ or $x = -3200$. Checking the intervals $[0, 4000]$ and $[4000, \infty)$, we find that the solution is in the range $[0, 4000]$. So they can import no more than 4000 radios.

1.6 Critical Thinking/Discussion/Writing

In Exercises 143 and 144, answers may vary. Sample responses are given.

143. a. $(x+4)(x-5) < 0$

b. $(x+2)(x-6) \leq 0$

c. $x^2 \geq 0$ **d.** $x^2 < 0$

e. $(x-3)^2 \leq 0$ **f.** $(x-2)^2 > 0$

144. a. $\dfrac{x-4}{x+2} \le 0$ **b.** $\dfrac{x-3}{x-5} \le 0$

c. It is not possible to have a quadratic inequality with solution set (2, 5). If we try $(x-2)(x-5) < 0$, then the solution set will be (2, 5). If we try $(x-2)(x-5) \le 0$, then the solution set will be [2, 5].

1.6 Getting Ready for the Next Section

145. $|-3| = 3$ **147.** $|6-4| = |2| = 2$

149. $|0| = 0$

151. $d = |5-(-2)| = |7| = 7$

153. $d = |5.7 - 2.3| = |3.4| = 3.4$

155. $|x-(-2)| = 5$ or $|x+2| = 5$

157. $|x-4| \le 2$ **159.** $|x-5| \le 3$

1.7 Equations and Inequalities Involving Absolute Value

1.7 Practice Problems

1. a. $|x-2| = 0 \Rightarrow x-2 = 0 \Rightarrow x = 2$
Solution set: {2}

b. $|6x-3| - 8 = 1$
$|6x-3| = 9$
$6x-3 = 9$ or $6x-3 = -9$
$6x = 12$ $6x = -6$
$x = 2$ $x = -1$
Solution set: {−1, 2}

2. $|x+2| = |x-3|$
$x+2 = x-3$ or $x+2 = -(x-3)$
$0 = -5$ False $x+2 = -x+3$
$2x = 1 \Rightarrow x = \dfrac{1}{2}$

Solution set: $\left\{\dfrac{1}{2}\right\}$

3. $|3x-4| = 2(x-1)$
$3x-4 = 2(x-1)$ or $3x-4 = -2(x-1)$
$3x-4 = 2x-2$ $3x-4 = -2x+2$
$x = 2$
                  $5x = 6 \Rightarrow x = \dfrac{6}{5}$

Solution set: $\left\{\dfrac{6}{5}, 2\right\}$

4. $|3x+3| \le 6 \Rightarrow -6 \le 3x+3 \le 6 \Rightarrow$
$-9 \le 3x \le 3 \Rightarrow -3 \le x \le 1$
Solution set: $[-3, 1]$

5. Let x = the actual speed of the search plane, in miles per hour. Then,
$|x-115| \le 25 \Rightarrow -25 \le x-115 \le 25 \Rightarrow$
$90 \le x \le 140$

Thus, the actual speed of the plane is between 90 and 140 miles per hour. Since the plane uses 10 gallons of fuel per hour and has 30 gallons of fuel, it can fly for 3 hours. The actual number of miles the search plane can fly is $3x$:
$3(90) \le 3x \le 3(140) \Rightarrow 270 \le x \le 420$

The plane can fly between 270 miles and 420 miles, inclusive.

6. $|2x+3| \ge 6 \Rightarrow 2x+3 \le -6$ or $2x+3 \ge 6$
$\begin{array}{l|l} 2x+3 \le -6 & 2x+3 \ge 6 \\ 2x \le -9 & 2x \ge 3 \\ x \le -\dfrac{9}{2} & x \ge \dfrac{3}{2} \end{array}$

Solution set: $\left(-\infty, -\dfrac{9}{2}\right] \cup \left[\dfrac{3}{2}, \infty\right)$

7. a. $|5-9x| > -3$
Since absolute value is always nonnegative, $|5-9x| > -3$ is true for all real numbers.
Solution set: $(-\infty, \infty)$

b. $|7x-4| \le -1$
Since absolute value is always nonnegative, $|7x-4| \le -1$ is false for all real numbers.
Solution set: $\varnothing$

8. $|x-2| < 4|x+4| \Rightarrow \left|\dfrac{x-2}{x+4}\right| < 4 \Rightarrow$

$-4 < \dfrac{x-2}{x+4} < 4$

$0 < \dfrac{x-2}{x+4} + 4 \qquad\qquad \dfrac{x-2}{x+4} - 4 < 0$

$0 < \dfrac{x-2+4(x+4)}{x+4} \qquad \dfrac{x-2-4(x+4)}{x+4} < 0$

$0 < \dfrac{x-2+4x+16}{x+4} \qquad \dfrac{x-2-4x-16}{x+4} < 0$

$0 < \dfrac{5x+14}{x+4} \qquad\qquad \dfrac{-3x-18}{x+4} < 0$

$5x+14 = 0 \Rightarrow x = -\dfrac{14}{5};\ x+4 = 0 \Rightarrow x = -4$

$-3x-18 = 0 \Rightarrow x = -6$

Interval	Test point	Value of $\dfrac{5x+14}{x+4}$	Result
$(-\infty, -6)$	-10	6	$+$
$\left(-4, -\dfrac{14}{5}\right)$	-3	-1	$-$
$\left(-\dfrac{14}{5}, \infty\right)$	-1	3	$+$

Note that the expression is undefined for $x = -4$. The solution set is for this part of the original inequality is $(-\infty, -6) \cup \left(-\dfrac{14}{5}, \infty\right)$.

Interval	Test point	Value of $\dfrac{-3x-18}{x+4}$	Result
$(-\infty, -6)$	-7	-1	$-$
$(-6, -4)$	-5	3	$+$
$(-4, \infty)$	-1	-5	$-$

Note that the expression is undefined for $x = 4$. The solution set is for this part of the original inequality is $(-\infty, -6) \cup (-4, \infty)$.

Interval	Test point	Value of $\dfrac{-3x-18}{x+4}$	Result
$(-\infty, -6)$	-7	-1	$-$
$(-6, -4)$	-5	3	$+$
$(-4, \infty)$	-1	-5	$-$

Note that the expression is undefined for $x = 4$. The solution set is for this part of the original inequality is $(-\infty, -6) \cup (-4, \infty)$.

The figure shows that both inequalities are true on $(-\infty, -6) \cup \left(-\dfrac{14}{5}, \infty\right)$, so the solution set is $(-\infty, -6) \cup \left(-\dfrac{14}{5}, \infty\right)$.

1.7 Concepts and Vocabulary

1. The solution set of the equation $|x| = a$ is $\{-a, a\}$.

3. The solution set of the inequality $|x| \geq a$ is $(-\infty, -a] \cup [a, \infty)$.

5. True

7. True

1.7 Building Skills

9. $|3x| = 9 \Rightarrow 3x = 9$ or $3x = -9 \Rightarrow$ $x = 3$ or $x = -3$

11. $|-2x| = 6 \Rightarrow -2x = 6$ or $-2x = -6 \Rightarrow$ $x = -3$ or $x = 3$

13. $|x+3| = 2 \Rightarrow x+3 = 2$ or $x+3 = -2 \Rightarrow$ $x = -1$ or $x = -5$

15. $|6-2x| = 8 \Rightarrow 6-2x = 8$ or $6-2x = -8 \Rightarrow$ $x = -1$ or $x = 7$

17. $|6x-2| = 9 \Rightarrow 6x-2 = 9$ or $6x-2 = -9 \Rightarrow$ $x = \dfrac{11}{6}$ or $x = -\dfrac{7}{6}$

19. $|2x+3| - 1 = 0 \Rightarrow |2x+3| = 1 \Rightarrow 2x+3 = 1$ or $2x+3 = -1 \Rightarrow x = -1$ or $x = -2$

21. $\dfrac{1}{2}|x| = 3 \Rightarrow |x| = 6 \Rightarrow x = -6$ or $x = 6$

23. $\left|\dfrac{1}{4}x + 2\right| = 3 \Rightarrow \dfrac{1}{4}x + 2 = -3$ or $\dfrac{1}{4}x + 2 = 3 \Rightarrow$ $x = -20$ or $x = 4$

25. $6|1-2x|-8=10 \Rightarrow 6|1-2x|=18 \Rightarrow$
$|1-2x|=3 \Rightarrow 1-2x=3$ or $1-2x=-3 \Rightarrow$
$x=-1$ or $x=2$

27. $2|3x-4|+9=7 \Rightarrow 2|3x-4|=-2 \Rightarrow$
$|3x-4|=-1$
The solution set is $\varnothing$ because an absolute value cannot be negative.

29. $|2x+1|=-1$
The solution set is $\varnothing$ because an absolute value cannot be negative.

31. $|x^2-4|=0 \Rightarrow x^2-4=0 \Rightarrow x=\pm 2$
Solution set: $\{-2, 2\}$

33. $|1-2x|=3 \Rightarrow 1-2x=3$ or $1-2x=-3 \Rightarrow$
$x=-1$ or $x=2$
Solution set: $\{-1, 2\}$

35. $\left|\frac{1}{3}-x\right|=\frac{2}{3} \Rightarrow \frac{1}{3}-x=\frac{2}{3}$ or $\frac{1}{3}-x=-\frac{2}{3} \Rightarrow$
$x=-\frac{1}{3}$ or $x=1$
Solution set: $\left\{-\frac{1}{3}, 1\right\}$

In exercises 37–46, be sure to check answers to eliminate extraneous solutions.

37. $|x+3|=|x+5| \Rightarrow x+3=x+5$ (impossible)
or $x+3=-(x+5) \Rightarrow x+3=-x-5 \Rightarrow x=-4$
The solution set is $\{-4\}$.

39. $|3x-2|=|6x+7| \Rightarrow 3x-2=6x+7 \Rightarrow$
$x=-3$ or $3x-2=-(6x+7) \Rightarrow$
$3x-2=-6x-7 \Rightarrow x=-\frac{5}{9}$
The solution set is $\left\{-3, -\frac{5}{9}\right\}$.

41. $|2x-1|=x+1 \Rightarrow$
$2x-1=x+1$ or $2x-1=-(x+1)$
$x=2$ $\qquad 2x-1=-x-1$
$\qquad\qquad\qquad 3x=0 \Rightarrow x=0$
Solution set: $\{0, 2\}$

43. $|4-3x|=x-1 \Rightarrow$
$4-3x=x-1$ or $4-3x=-(x-1)$
$5=4x$ $\qquad 4-3x=-x+1$
$\frac{5}{4}=x$ $\qquad 3=2x \Rightarrow \frac{3}{2}=x$
Solution set: $\left\{\frac{5}{4}, \frac{3}{2}\right\}$

45. $|3x+2|=2(x-1) \Rightarrow$
$3x+2=2(x-1)$ or $3x+2=-2(x-1)$
$3x+2=2x-2$ $\qquad 3x+2=-2x+2$
$x=-4$ $\qquad 5x=0 \Rightarrow x=0$
If $x=-4$, then $|3x+2|=|3(-4)+2|=|-10|=10$, while $2(x-1)=2(-4-1)=2(-5)=-10$.
Therefore, -4 is not a solution.
If $x=0$, then $|3x+2|=|3(0)+2|=|2|=2$, while $2(x-1)=2(0-1)=2(-1)=-2$.
Therefore, 0 is not a solution.
Solution set: $\varnothing$

47. $|3x|<12 \Rightarrow -12<3x<12 \Rightarrow -4<x<4$
The solution set is $(-4, 4)$.

49. $|4x|>16 \Rightarrow 4x<-16$ or $4x>16 \Rightarrow$
$x<-4$ or $x>4$.
The solution set is $(-\infty, -4) \cup (4, \infty)$.

51. $|x+1|<3 \Rightarrow -3<x+1<3 \Rightarrow -4<x<2$.
The solution set is $(-4, 2)$.

53. $|x|+2 \geq -1 \Rightarrow |x| \geq -3$
Since absolute value is always nonnegative, the inequality is true for all real numbers.
Solution set: $(-\infty, \infty)$

55. $|2x-3|<4 \Rightarrow -4<2x-3<4 \Rightarrow$
$-1<2x<7 \Rightarrow -\frac{1}{2}<x<\frac{7}{2}$.
The solution set is $\left(-\frac{1}{2}, \frac{7}{2}\right)$.

57. $|5-2x|>3 \Rightarrow 5-2x<-3$ or $5-2x>3 \Rightarrow$
$x>4$ or $x<1$.
The solution set is $(-\infty, 1) \cup (4, \infty)$.

59. $|3x+4| \leq 19 \Rightarrow -19 \leq 3x+4 \leq 19 \Rightarrow$
$-23 \leq 3x \leq 15 \Rightarrow -\frac{23}{3} \leq x \leq 5$.
The solution set is $\left[-\frac{23}{3}, 5\right]$.

61. $|2x-15| < 0$. The solution set is $\varnothing$ because an absolute value cannot be negative.

63. $\left|\dfrac{x-2}{x+3}\right| < 1 \Rightarrow -1 < \dfrac{x-2}{x+3} < 1$

$0 < \dfrac{x-2}{x+3} + 1 \qquad \dfrac{x-2}{x+3} - 1 < 0$

$0 < \dfrac{x-2+(x+3)}{x+3} \qquad \dfrac{x-2-(x+3)}{x+3} < 0$

$0 < \dfrac{2x+1}{x+3} \qquad \dfrac{-5}{x+3} < 0$

$2x+1 = 0 \Rightarrow x = -\dfrac{1}{2}; \ x+3 = 0 \Rightarrow x = -3$

Interval	Test point	Value of $\dfrac{2x+1}{x+3}$	Result
$(-\infty, -3)$	-4	7	$+$
$\left(-3, -\tfrac{1}{2}\right)$	-2	-3	$-$
$\left(-\tfrac{1}{2}, \infty\right)$	2	1	$+$

Note that the expression is undefined for $x = -3$. The solution set is for this part of the original inequality is $(-\infty, -3) \cup \left(-\tfrac{1}{2}, \infty\right)$.

Interval	Test point	Value of $\dfrac{-5}{x+3}$	Result
$(-\infty, -3)$	-4	5	$+$
$\left(-3, -\tfrac{1}{2}\right)$	-2	-5	$-$
$\left(-\tfrac{1}{2}, \infty\right)$	2	-1	$-$

Note that the expression is undefined for $x = -3$. The solution set is for this part of the original inequality is $\left(-3, -\tfrac{1}{2}\right) \cup \left(-\tfrac{1}{2}, \infty\right)$.

Both inequalities are true on $\left(-\tfrac{1}{2}, \infty\right)$, so the solution set is $\left(-\tfrac{1}{2}, \infty\right)$.

65. $\left|\dfrac{2x-3}{x+1}\right| \le 3 \Rightarrow -3 \le \dfrac{2x-3}{x+1} \le 3$

$0 \le \dfrac{2x-3}{x+1} + 3 \qquad \dfrac{2x-3}{x+1} - 3 \le 0$

$0 \le \dfrac{2x-3+3(x+1)}{x+1} \qquad \dfrac{2x-3-3(x+1)}{x+1} \le 0$

$0 \le \dfrac{5x}{x+1} \qquad \dfrac{-x-6}{x+1} \le 0$

$5x = 0 \Rightarrow x = 0; \ x+1 = 0 \Rightarrow x = -1$
$-x-6 = 0 \Rightarrow x = -6$

Interval	Test point	Value of $\dfrac{5x}{x+1}$	Result
$(-\infty, -6]$	-11	$\tfrac{11}{2}$	$+$
$[-6, -1)$	-2	10	$+$
$(-1, 0]$	$-\tfrac{1}{2}$	-5	$-$
$[0, \infty)$	4	4	$+$

Note that the expression is undefined for $x = -1$. The solution set is for this part of the original inequality is $(-\infty, -6] \cup [-6, -1) \cup [0, \infty)$.

Interval	Test point	Value of $\dfrac{-x-6}{x+1}$	Result
$(-\infty, -6]$	-11	$-\tfrac{1}{2}$	$-$
$[-6, -1)$	-2	4	$+$
$(-1, 0]$	$-\tfrac{1}{2}$	-11	$-$
$[0, \infty)$	4	-2	$-$

Note that the expression is undefined for $x = -1$. The solution set is for this part of the original inequality is $(-\infty, -6] \cup (-1, 0] \cup [0, \infty)$.

Both inequalities are true on $(-\infty, -6] \cup [0, \infty)$, so the solution set is $(-\infty, -6] \cup [0, \infty)$.

67. $\left|\dfrac{x-1}{x+2}\right| \geq 2 \Rightarrow \dfrac{x-1}{x+2} \leq -2$ or $\dfrac{x-1}{x+2} \geq 2$

$\dfrac{x-1}{x+2} \leq -2$

$\dfrac{x-1}{x+2} + 2 \leq 0$

$\dfrac{x-1+2(x+2)}{x+2} \leq 0$

$\dfrac{3x+3}{x+2} \leq 0$

$\dfrac{x-1}{x+2} \geq 2$

$\dfrac{x-1}{x+2} - 2 \geq 0$

$\dfrac{x-1-2(x+2)}{x+2} \geq 0$

$\dfrac{-x-5}{x+2} \geq 0$

$3x+3 = 0 \Rightarrow x = -1;\ x+2 = 0 \Rightarrow x = -2$
$-x-5 = 0 \Rightarrow x = -5$

Interval	Test point	Value of $\dfrac{3x+3}{x+2}$	Result
$(-\infty, -5]$	-6	$\dfrac{15}{4}$	$+$
$[-5, -2)$	-3	6	$+$
$(-2, -1]$	$-\dfrac{3}{2}$	-3	$-$
$[-1, \infty)$	1	2	$+$

Note that the expression is undefined for $x = -2$. The solution set is for this part of the original inequality is $(-2, -1]$.

Interval	Test point	Value of $\dfrac{-x-5}{x+2}$	Result
$(-\infty, -5]$	-6	$-\dfrac{1}{4}$	$-$
$[-5, -2)$	-3	2	$+$
$(-2, -1]$	$-\dfrac{3}{2}$	-7	$-$
$[-1, \infty)$	1	-2	$-$

Note that the expression is undefined for $x = -2$. The solution set is for this part of the original inequality is $[-5, -2)$. Since the original inequality is an "or" inequality, the solution set of the original inequality is the union of the two solution sets.
The solution set is $(-2, -1] \cup [-5, -2)$.

69. $\left|\dfrac{2x+1}{x-1}\right| > 4 \Rightarrow \dfrac{2x+1}{x-1} < -4$ or $\dfrac{2x+1}{x-1} > 4$

$\dfrac{2x+1}{x-1} < -4$

$\dfrac{2x+1}{x-1} + 4 < 0$

$\dfrac{2x+1+4(x-1)}{x-1} < 0$

$\dfrac{6x-3}{x-1} < 0$

$\dfrac{2x+1}{x-1} > 4$

$\dfrac{2x+1}{x-1} - 4 > 0$

$\dfrac{2x+1-4(x-1)}{x-1} > 0$

$\dfrac{-2x+5}{x-1} > 0$

$6x - 3 = 0 \Rightarrow x = \dfrac{1}{2};\ x - 1 = 0 \Rightarrow x = 1$

$-2x + 5 = 0 \Rightarrow x = \dfrac{5}{2}$

Interval	Test point	Value of $\dfrac{6x-3}{x-1}$	Result
$\left(-\infty, \frac{1}{2}\right)$	0	3	$+$
$\left(\frac{1}{2}, 1\right)$	$\frac{3}{4}$	-6	$-$
$\left(1, \frac{5}{2}\right)$	2	9	$+$
$\left(\frac{5}{2}, \infty\right)$	4	7	$+$

Note that the expression is undefined for $x = -2$. The solution set is for this part of the original inequality is $\left(\frac{1}{2}, 1\right)$.

Interval	Test point	Value of $\dfrac{-2x+5}{x-1}$	Result
$\left(-\infty, \frac{1}{2}\right)$	0	-5	$-$
$\left(\frac{1}{2}, 1\right)$	$\frac{3}{4}$	-14	$-$
$\left(1, \frac{5}{2}\right)$	2	1	$+$
$\left(\frac{5}{2}, \infty\right)$	4	-1	$-$

Note that the expression is undefined for $x = -2$. The solution set is for this part of the original inequality is $\left(1, \frac{5}{2}\right)$. Since the original inequality is an "or" inequality, the solution set of the original inequality is the union of the two solution sets. The solution set is $\left(\frac{1}{2}, 1\right) \cup \left(1, \frac{5}{2}\right)$.

71. $|x-1| \le 2|2x-5| \Rightarrow \left|\dfrac{x-1}{2x-5}\right| \le 2 \Rightarrow$

$-2 \le \dfrac{x-1}{2x-5} \le 2$

$0 \le \dfrac{x-1}{2x-5} + 2$ | $\dfrac{x-1}{2x-5} - 2 \le 0$

$0 \le \dfrac{x-1+2(2x-5)}{2x-5}$ | $\dfrac{x-1-2(2x-5)}{2x-5} \le 0$

$0 \le \dfrac{x-1+4x-10}{2x-5}$ | $\dfrac{x-1-4x+10}{2x-5} \le 0$

$0 \le \dfrac{5x-11}{2x-5}$ | $\dfrac{-3x+9}{2x-5} \le 0$

$5x-11 = 0 \Rightarrow x = \dfrac{11}{5}$; $2x-5 = 0 \Rightarrow x = \dfrac{5}{2}$

$-3x+9 = 0 \Rightarrow x = 3$

Interval	Test point	Value of $\dfrac{5x-11}{2x-5}$	Result
$\left(-\infty, \dfrac{11}{5}\right]$	0	$\dfrac{11}{5}$	+
$\left[\dfrac{11}{5}, \dfrac{5}{2}\right)$	$\dfrac{23}{10}$	$-\dfrac{5}{4}$	−
$\left(\dfrac{5}{2}, \infty\right)$	3	4	+

Note that the expression is undefined for $x = \dfrac{5}{2}$. The solution set is for this part of the original inequality is $S_1 = \left(-\infty, \dfrac{11}{5}\right] \cup \left(\dfrac{5}{2}, \infty\right)$.

Interval	Test point	Value of $\dfrac{-3x+9}{2x-5}$	Result
$\left(-\infty, \dfrac{5}{2}\right)$	0	$-\dfrac{9}{5}$	−
$\left(\dfrac{5}{2}, 3\right]$	$\dfrac{11}{4}$	$\dfrac{3}{2}$	+
$[3, \infty)$	5	$-\dfrac{6}{5}$	−

Note that the expression is undefined for $x = \dfrac{5}{2}$. The solution set is for this part of the original inequality is $S_2 = \left(-\infty, \dfrac{5}{2}\right) \cup [3, \infty)$.

The figure shows that both inequalities are true on $\left(-\infty, \dfrac{11}{5}\right] \cup [3, \infty)$, so the solution set is $\left(-\infty, \dfrac{11}{5}\right] \cup [3, \infty)$.

1.7 Applying the Concepts

73. $|T - 75| = 20 \Rightarrow T - 75 = -20$ or
$T - 75 = 20 \Rightarrow T = 55$ or $T = 95$
The temperatures in Tampa during December are between 55°F and 95°F, inclusive.

75. $|x - 700| \le 50$

77. $|x - 120| \le 6.75$

79. Let x = the number of people at a party. Then
$|120 - x| \le 15 \Rightarrow -15 \le 120 - x \le 15 \Rightarrow$
$-135 \le -x \le -105 \Rightarrow 105 \le x \le 135$. So, between 105 and 135 people will be at the party. The total spent on food will be between $48(105) = \$5040$ and $48(135) = \$6480$, inclusive.

81. Let x = the actual weight of Sarah's catch.
Then $|32 - x| \le \dfrac{1}{2} \Rightarrow -\dfrac{1}{2} \le 32 - x \le \dfrac{1}{2} \Rightarrow$
$-1 \le 64 - 2x \le 1 \Rightarrow -65 \le -2x \le -63 \Rightarrow$.
$32.5 \ge x \ge 31.5$.
So, her catch is between 31.5 pounds and 32.5 pounds. She will be paid between $0.60(31.5) = \$18.90$ and $0.60(32.5) = \$19.50$, inclusive.

1.7 Beyond the Basics

83. a. $|x^2 - 9| = x - 3 \Rightarrow x^2 - 9 = -(x-3)$ or
$x^2 - 9 = x - 3$
$x^2 - 9 = -(x-3) \Rightarrow x^2 - 9 = -x + 3 \Rightarrow$
$x^2 + x - 12 = 0 \Rightarrow (x+4)(x-3) = 0 \Rightarrow$
$x = -4, 3$

$x^2 - 9 = x - 3 \Rightarrow x^2 - x - 6 = 0 \Rightarrow$
$(x+2)(x-3) = 0 \Rightarrow x = -2, 3$
Checking $x = -4$, $x = -2$, and $x = 3$ in the original equation shows that $x = -4$ and $x = -2$ are extraneous solutions.
The solution set is $\{3\}$.

b. $|x^2 - 8| = -2x \Rightarrow x^2 - 8 = -(-2x)$ or
$x^2 - 8 = -2x$
$x^2 - 8 = -(-2x) \Rightarrow x^2 - 8 = 2x \Rightarrow$
$x^2 - 2x + 8 = 0 \Rightarrow (x+2)(x-4) = 0 \Rightarrow$
$x = -2, 4$
$x^2 - 8 = -2x \Rightarrow x^2 + 2x - 8 = 0 \Rightarrow$
$(x+4)(x-2) = 0 \Rightarrow x = -4, 2$
Checking $x = -4$, $x = -2$, $x = 2$, and $x = 4$ in the original equation shows that $x = 2$, and $x = 4$ are extraneous solution.
The solution set is $\{-4, -2\}$.

c. $|x^2 - 5x| = 6 \Rightarrow x^2 - 5x = -6$ or
$x^2 - 5x = 6$
$x^2 - 5x = -6 \Rightarrow x^2 - 5x + 6 = 0 \Rightarrow$
$(x-2)(x-3) = 0 \Rightarrow x = 2, 3$
$x^2 - 5x = 6 \Rightarrow x^2 - 5x - 6 = 0 \Rightarrow$
$(x+1)(x-6) = 0 \Rightarrow x = -1, 6$
Checking $x = -1$, $x = 2$, $x = 3$, and $x = 6$ in the original equation shows that all values are solutions.
The solution set is $\{-1, 2, 3, 6\}$.

d. $|x^2 + 3x - 2| = 2 \Rightarrow x^2 + 3x - 2 = -2$ or
$x^2 + 3x - 2 = 2$. If $x^2 + 3x - 2 = -2$, then
$x^2 + 3x = 0 \Rightarrow x(x+3) = 0 \Rightarrow x = 0$ or $x = -3$.
If $x^2 + 3x - 2 = 2$, then $x^2 + 3x - 4 = 0 \Rightarrow$
$(x+4)(x-1) = 0 \Rightarrow x = 1$ or $x = -4$.
Checking $x = -4$, $x = -3$, $x = 0$, and $x = 1$ in the original equation shows that all values are solutions.
The solution set is $\{-4, -3, 0, 1\}$.

85. $|2x - 3| + |x - 2| = 4$

The points $x = \frac{3}{2}$ and $x = 2$ are the points where the absolute value expressions equal zero. Since these expressions must be negative or positive for other x-values, then these points divide the number line into intervals each of which should be considered separately. Thus, we will consider the intervals $\left(-\infty, \frac{3}{2}\right]$, $\left(\frac{3}{2}, 2\right]$, and $(2, \infty)$.

Interval	Test point	Sign of $2x - 3$	Sign of $x - 2$
$\left(-\infty, \frac{3}{2}\right]$	0	−	−
$\left(\frac{3}{2}, 2\right]$	$\frac{7}{4}$	+	−
$(2, \infty)$	3	+	+

On the first interval, both absolute-value expressions will have negative values, so change the signs on both of them when taking the bars off.
$-(2x - 3) - (x - 2) = 4$
$-2x + 3 - x + 2 = 4$
$-3x + 5 = 4$
$-3x = -1 \Rightarrow x = \dfrac{1}{3}$

Because $x = \frac{1}{3}$ lies in the interval $\left(-\infty, \frac{3}{2}\right]$, this is a valid solution.
On the second interval, $|2x - 3|$ is positive, so just take the bars off. But $|x - 2|$ is negative, change the sign when taking the bars off.
$(2x - 3) - (x - 2) = 4$
$2x - 3 - x + 2 = 4$
$x - 1 = 4 \Rightarrow x = 5$

Since $x = 5$ does not lie in the interval $\left(\frac{3}{2}, 2\right]$, it is not a valid solution of the original equation.
On the third interval, the arguments of both absolute values expressions are positive, so just remove the absolute value bars.
$(2x - 3) + (x - 2) = 4$
$2x - 3 + x - 2 = 4$
$3x - 5 = 4$
$3x = 9 \Rightarrow x = 3$

Because $x = 3$ lies in the interval $(2, \infty)$, the solution is valid. Thus, the solution set for the original equation is $\left\{\frac{1}{3}, 3\right\}$.

87. Let $u = |x|$. Then $|x|^2 - 4|x| - 7 = 5 \Rightarrow$
$u^2 - 4u - 7 = 5 \Rightarrow u^2 - 4u - 12 = 0 \Rightarrow$
$(u - 6)(u + 2) = 0 \Rightarrow u = -2, 6$.
Now solve for x: $-2 = |x|$ is not possible.
$6 = |x| \Rightarrow x = 6$ or $x = -6$
Solution set: $\{-6, 6\}$.

89. $0 < a < b \Rightarrow b - a > 0$ and
$0 < c < d \Rightarrow d - c > 0$. The product of two positive numbers is positive, so
$(b-a)c > 0 \Leftrightarrow bc > ac$ (1) and
$(d-c)a > 0 \Leftrightarrow ad > ac$ (2). We know that $bc > ac$ and $bd > ad$, so
$bd - bc > ad - ac \Rightarrow bd + ac > ad + bc$.
Substituting (1) and (2) into the inequality, we have $bd + ac > ad + bc > ac + ac \Rightarrow bd > ac$.

91. $x^2 < a \Rightarrow x^2 - a < 0 \Rightarrow \left(x - \sqrt{a}\right)\left(x + \sqrt{a}\right) < 0$
$\left(x - \sqrt{a}\right)\left(x + \sqrt{a}\right) < 0 \Rightarrow$
$\left(x - \sqrt{a}\right) > 0$ and $\left(x + \sqrt{a}\right) < 0 \Rightarrow$
$x > \sqrt{a}$ and $x < -\sqrt{a} \Rightarrow \sqrt{a} < x < -\sqrt{a} \Rightarrow$
$\sqrt{a} < -\sqrt{a}$, a contradiction
or
$\left(x - \sqrt{a}\right) < 0$ and $\left(x + \sqrt{a}\right) > 0 \Rightarrow$
$x > -\sqrt{a}$ and $x < \sqrt{a} \Rightarrow -\sqrt{a} < x < \sqrt{a}$
Thus, the solution set is $\left(-\sqrt{a}, \sqrt{a}\right)$.

In Exercises 93–104, answers may vary. Sample responses are given.

93. $|x - 4| < 3$ **95.** $|x - 4| \le 6$

97. $|x - 7| > 4$ **99.** $|2x - 5| \ge 15$

101. $\left|x - \dfrac{a+b}{2}\right| < \dfrac{b-a}{2} \Rightarrow |2x - a - b| < b - a$

103. $|2x - a - b| > b - a$

105. $|x - 39| \le 31$

107. $1 \le |x - 2| \le 3 \Rightarrow$
$1 \le x - 2 \le 3$ or $1 \le -(x - 2) \le 3$
$3 \le x \le 5$ $1 \le -x + 2 \le 3$
$ -1 \le -x \le 1$
$ 1 \ge x \ge -1 \Rightarrow -1 \le x \le 1$
The union of the two solution sets is the solution set of the original inequality.
Solution set: $[-1, 1] \cup [3, 5]$

109. $|x - 2| + |x - 4| \ge 8$.
Solve the equation $|x - 2| + |x - 4| = 8$ to find the critical values for the inequality.
The points $x = 2$ and $x = 4$ are the points where the absolute value expressions equal zero. Because these expressions must be negative or positive for other x-values, then these points divide the number line into intervals each of which should be considered separately. Thus, we will consider the intervals $(-\infty, 2]$, $(2, 4]$, and $(4, \infty)$.

Interval	Test point	Sign of $x - 2$	Sign of $x - 4$
$(-\infty, 2]$	0	−	−
$(2, 4]$	3	+	−
$(4, \infty)$	5	+	+

On the first interval, both absolute-value expressions will have negative values, so change the signs on both of them when taking the bars off.
$-(x - 2) + [-(x - 4)] = 8$
$-2x + 6 = 8$
$-2x = 2 \Rightarrow x = -1$
Since $x = -1$ lies in the interval $(-\infty, 2]$, $x = -1$ is a valid solution.
On the second interval, $x - 2$ is positive, so just take the bars off. But $x - 4$ is negative, change the sign when taking the bars off.
$(x - 2) + [-(x - 4)] = 8$
$2 = 8$
This is a false statement, so there are no values of x in the interval $(2, 4]$ that are valid solutions of the equation.
On the third interval, the arguments of both absolute values expressions are positive, so just remove the absolute value bars.
$x - 2 + x - 4 = 8$
$2x - 6 = 8$
$2x = 14 \Rightarrow x = 7$
Because 7 lies in the interval $(2, \infty)$, $x = 7$ is a valid solution. So $x = -1, 2, 4,$ and 7 are critical values. Test values in the intervals $(-\infty, -1], [-1, 2], [2, 4], [4, 7],$ and $[7, \infty)$ to see where the original inequality is true.

(continued on next page)

(*continued*)

Interval	Test point	Value of $\|x-2\|+\|x-4\|$	Value ≥ 8?
$(-\infty, -1]$	-2	10	Yes
$[-1, 2]$	0	6	No
$[2, 4]$	3	2	No
$[4, 7]$	5	4	No
$[7, \infty)$	10	14	Yes

Thus, the solution set is $(-\infty, -1] \cup [7, \infty)$.
Verify this by graphing $Y_1 = |x-2| + |x-4|$ and $Y_2 = 8$.

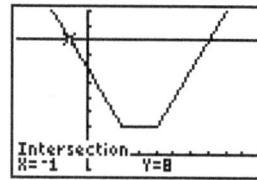
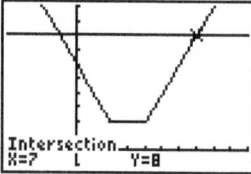

111. $\dfrac{|x-1|-(x+2)}{x+2} \geq 0$.

The value of $x - 1$ changes from negative to positive at $x = 1$, where it is 0. So, we will consider two cases,
$$\begin{cases} \dfrac{(x-1)-(x+2)}{x+2} \geq 0 & \text{if } x \geq 1 \\ \dfrac{-(x-1)-(x+2)}{x+2} \geq 0 & \text{if } x < 1 \end{cases}$$
Test the case for $x \geq 1$.
$\dfrac{(x-1)-(x+2)}{x+2} \geq 0 \Rightarrow \dfrac{-3}{x+2} \geq 0$
$x + 2 = 0 \Rightarrow x = -2$
Since this test case is for $x \geq 1$, it is only necessary to test the interval $[1, \infty)$.

Interval	Test point	Value of $\dfrac{-3}{x+2}$	Result
$[1, \infty)$	2	$-\tfrac{3}{4}$	$-$

$-\tfrac{3}{4}$ is not ≥ 1, so there is no solution in this interval.
Now test the case $x < 1$.
$\dfrac{-(x-1)-(x+2)}{x+2} \geq 0 \Rightarrow \dfrac{-2x-1}{x+2} \geq 0$
$-2x - 1 = 0 \Rightarrow x = -\tfrac{1}{2}$; $x + 2 = 0 \Rightarrow x = -2$

The intervals to be tested are $(-\infty, -2)$, $\left(-2, -\tfrac{1}{2}\right]$, and $\left[-\tfrac{1}{2}, 1\right]$.

Interval	Test point	Value of $\dfrac{-2x-1}{x+2}$	Result
$(-\infty, -2)$	-3	-5	$-$
$\left(-2, -\tfrac{1}{2}\right]$	-1	1	$+$
$\left[-\tfrac{1}{2}, 1\right]$	0	$-\tfrac{1}{2}$	$-$

The only interval where the value of $\dfrac{-2x-1}{x+2} \geq 0$ is $\left(-2, -\tfrac{1}{2}\right]$, so the solution set is $\left(-2, -\tfrac{1}{2}\right]$. Verify this by graphing
$Y_1 = \dfrac{|x-1|-(x+2)}{x+2}$.

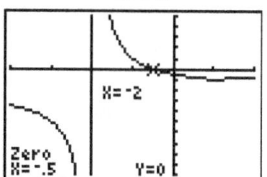

1.7 Critical Thinking/Discussion/Writing

112. To find what values make $\sqrt{(x-3)^2} = x-3$ true, solve $x - 3 = 0 \Rightarrow x = 3$, so we check the intervals $(-\infty, 3]$ and $[3, \infty)$ to see which makes the equation true. The equation is true for $[3, \infty)$.

113. To find what values make $\sqrt{\left(x^2-6x+8\right)^2} = x^2-6x+8$ true, solve $x^2 - 6x + 8 = 0 \Rightarrow (x-2)(x-4) = 0 \Rightarrow x = 2$ or $x = 4$. Then check the intervals $(-\infty, 2]$, $[2, 4]$, and $[4, \infty)$ to see which make the equation true. The equation is true for $(-\infty, 2] \cup [4, \infty)$.

114. To solve $|x-3|^2 - 7|x-3| + 10 = 0$, let $u = |x-3|$. So we have $u^2 - 7u + 10 = 0 \Rightarrow$ $(u-5)(u-2) = 0 \Rightarrow u = 5$ or $u = 2$. Now solve for x. $5 = |x-3| \Rightarrow -5 = x-3$ or $5 = x-3 \Rightarrow x = -2$ or $x = 8$. $2 = |x-3| \Rightarrow -2 = x-3$ or $2 = x-3 \Rightarrow x = 1$ or $x = 5$. So the solution set is $\{-2, 1, 5, 8\}$.

1.7 Getting Ready for the Next Section

115. $\dfrac{2+5}{2} = \dfrac{7}{2} = 3.5$

116. $\dfrac{-3+7}{2} = \dfrac{4}{2} = 2$

117. $\dfrac{-3-7}{2} = \dfrac{-10}{2} = -5$

118. $\dfrac{(a-b)-(a+b)}{2} = \dfrac{a-b-a-b}{2}$
$= \dfrac{-2b}{2} = -b$

119. $\sqrt{(5-2)^2 + (3-7)^2} = \sqrt{3^2 + (-4)^2}$
$= \sqrt{9+16} = \sqrt{25} = 5$

121. $\sqrt{(2-5)^2 + (8-6)^2} = \sqrt{(-3)^2 + 2^2} = \sqrt{9+4}$
$= \sqrt{13}$

123. $x^2 + 4x + \underline{4} = (x+2)^2$

125. $x^2 - 5x + \left(\dfrac{-5}{2}\right)^2 = x^2 - 5x + \underline{\dfrac{25}{4}} = \left(x - \dfrac{5}{2}\right)^2$

127. $x^2 + \dfrac{3}{2}x + \left(\dfrac{\frac{3}{2}}{2}\right)^2 = x^2 + \dfrac{3}{2}x + \left(\dfrac{3}{4}\right)^2$
$= x^2 + \dfrac{3}{2}x + \underline{\dfrac{9}{16}} = \left(x + \dfrac{3}{4}\right)^2$

Chapter 1 Review Exercises

Building Skills

1. $5x - 4 = 11 \Rightarrow 5x = 15 \Rightarrow x = 3$

3. $3(2x-4) = 9 - (x+7) \Rightarrow 6x - 12 = 2 - x \Rightarrow$
$7x = 14 \Rightarrow x = 2$

5. $3x + 8 = 3(x+2) + 2 \Rightarrow 3x + 8 = 3x + 8$
This is an identity, so the solution set is $(-\infty, \infty)$.

7. $x - (5x-2) = 7(x-1) - 2 \Rightarrow$
$-4x + 2 = 7x - 9 \Rightarrow -11x = -11 \Rightarrow x = 1$

9. $\dfrac{2}{x+3} = \dfrac{5}{11x-1} \Rightarrow 2(11x-1) = 5(x+3) \Rightarrow$
$22x - 2 = 5x + 15 \Rightarrow 17x = 17 \Rightarrow x = 1$

11. $\dfrac{y+5}{2} + \dfrac{y-1}{3} = \dfrac{7y+3}{8} + \dfrac{4}{3} \Rightarrow$
$12(y+5) + 8(y-1) = 3(7y+3) + 8(4) \Rightarrow$
$20y + 52 = 21y + 41 \Rightarrow 11 = y$

13. $|2x-3| = |4x+5| \Rightarrow 2x - 3 = 4x + 5 \Rightarrow$
$-2x = 8 \Rightarrow x = -4$
or $2x - 3 = -(4x+5) \Rightarrow$
$2x - 3 = -4x - 5 \Rightarrow 6x = -2 \Rightarrow x = -1/3$
The solution set is $\left\{-4, -\dfrac{1}{3}\right\}$.

15. $|2x-1| = |2x+7|$
$2x - 1 = 2x + 7$ or $2x - 1 = -(2x+7)$
$-2 = 7$ False $\quad 2x - 1 = -2x - 7$
$\quad\quad\quad\quad\quad\quad\quad 4x = -6 \Rightarrow x = -\dfrac{3}{2}$
Solution set: $\left\{-\dfrac{3}{2}\right\}$

17. $|3x-2| = 2x + 1$
$3x - 2 = 2x + 1$ or $3x - 2 = -(2x+1)$
$x = 3 \quad\quad\quad\quad\quad 3x - 2 = -2x - 1$
$\quad\quad\quad\quad\quad\quad\quad\quad 5x = 1$
$\quad\quad\quad\quad\quad\quad\quad\quad x = \dfrac{1}{5}$
Solution set: $\left\{\dfrac{1}{5}, 3\right\}$

19. $p = k + gt \Rightarrow p - k = gt \Rightarrow \dfrac{p-k}{t} = g$

21. $T = \dfrac{2B}{B-1} \Rightarrow TB - T = 2B \Rightarrow TB - 2B = T \Rightarrow$
$B(T-2) = T \Rightarrow B = \dfrac{T}{T-2}$

23. $x^2 - 7x = 0 \Rightarrow x(x-7) = 0 \Rightarrow$
$x = 0$ or $x = 7$

25. $x^2 - 3x - 10 = 0 \Rightarrow (x-5)(x+2) = 0 \Rightarrow$
$x = 5$ or $x = -2$

27. $(x-1)^2 = 2x^2 + 3x - 5 \Rightarrow$
    $x^2 - 2x + 1 = 2x^2 + 3x - 5 \Rightarrow$
    $0 = x^2 + 5x - 6 \Rightarrow 0 = (x+6)(x-1) \Rightarrow$
    $x = -6$ or $x = 1$

29. $\dfrac{x^2}{4} + x = \dfrac{5}{4} \Rightarrow x^2 + 4x = 5 \Rightarrow$
    $x^2 + 4x - 5 = 0 \Rightarrow (x+5)(x-1) = 0 \Rightarrow$
    $x = -5$ or $x = 1$

31. $3x(x+1) = 2x + 2 \Rightarrow 3x^2 + 3x = 2x + 2 \Rightarrow$
    $3x^2 + x - 2 = 0 \Rightarrow (3x-2)(x+1) = 0 \Rightarrow$
    $x = \dfrac{2}{3}$ or $x = -1$

33. Use the quadratic formula to solve
    $x^2 - 3x - 1 = 0$:
    $x = \dfrac{-(-3) \pm \sqrt{(-3)^2 - 4(1)(-1)}}{2(1)}$
    $= \dfrac{3 \pm \sqrt{13}}{2} = \dfrac{3}{2} \pm \dfrac{\sqrt{13}}{2}$

35. $2x^2 + x - 1 = 0 \Rightarrow (2x-1)(x+1) = 0 \Rightarrow$
    $x = \dfrac{1}{2}$ or $x = -1$

37. Divide $3x^2 - 12x - 24 = 0$ by 3 to obtain $x^2 - 4x - 8 = 0$. Now use the quadratic formula:
    $x = \dfrac{-(-4) \pm \sqrt{(-4)^2 - 4(1)(-8)}}{2(1)}$
    $= \dfrac{4 \pm \sqrt{48}}{2} = \dfrac{4 \pm 4\sqrt{3}}{2} = 2 \pm 2\sqrt{3}$

39. Use the quadratic formula to solve
    $2x^2 - x - 2 = 0$:
    $x = \dfrac{-(-1) \pm \sqrt{(-1)^2 - 4(2)(-2)}}{2(2)}$
    $= \dfrac{1 \pm \sqrt{17}}{4} = \dfrac{1}{4} \pm \dfrac{\sqrt{17}}{4}$

41. Use the quadratic formula to solve
    $x^2 - x + 1 = 0$:
    $x = \dfrac{-(-1) \pm \sqrt{(-1)^2 - 4(1)(1)}}{2(1)}$
    $= \dfrac{1 \pm \sqrt{-3}}{2} = \dfrac{1 \pm i\sqrt{3}}{2} = \dfrac{1}{2} \pm \dfrac{\sqrt{3}}{2}i$

43. Use the quadratic formula to solve
    $x^2 - 6x + 13 = 0$:
    $x = \dfrac{-(-6) \pm \sqrt{(-6)^2 - 4(1)(13)}}{2(1)}$
    $= \dfrac{6 \pm \sqrt{-16}}{2} = \dfrac{6 \pm 4i}{2} = 3 \pm 2i$

45. Use the quadratic formula to solve
    $4x^2 - 8x + 13 = 0$:
    $x = \dfrac{-(-8) \pm \sqrt{(-8)^2 - 4(4)(13)}}{2(4)}$
    $= \dfrac{8 \pm \sqrt{-144}}{8} = \dfrac{8 \pm 12i}{8} = 1 \pm \dfrac{3}{2}i$

47. The discriminant is $(-11)^2 - 4(3)(6) = 49 > 0$, so there are 2 real unequal roots.

49. The discriminant is $2^2 - 4(5)(1) = -16 < 0$, so there are 2 unequal complex roots.

51. $\sqrt{x^2 - 16} = 0 \Rightarrow x^2 - 16 = 0 \Rightarrow$
    $(x+4)(x-4) = 16 \Rightarrow x = \pm 4$

53. $\sqrt{4 - 7x} = \sqrt{2}x \Rightarrow 4 - 7x = 2x^2 \Rightarrow$
    $2x^2 + 7x - 4 = 0 \Rightarrow (2x-1)(x+4) = 0 \Rightarrow$
    $x = \dfrac{1}{2}$ or $x = -4$
    If $x = -4$, then the equation becomes
    $\sqrt{4 - 7(-4)} = \sqrt{2}(-4) \Rightarrow \sqrt{32} = -4\sqrt{2}$, which is not true, so we reject that root. The solution set is $\left\{\dfrac{1}{2}\right\}$.

55. $y - 2\sqrt{y} - 3 = 0$. Let $u = \sqrt{y}$, so the equation becomes $u^2 - 2u - 3 = 0 \Rightarrow$
    $(u-3)(u+1) = 0 \Rightarrow u = 3$ or $u = -1$.
    Now solve for y. $3 = \sqrt{y} \Rightarrow 9 = y$ or $-1 = \sqrt{y}$ (reject this). The solution set is $\{9\}$.

57. $\sqrt{x} - 1 = \sqrt{5 + \sqrt{x}}$. Let $u = \sqrt{x}$. The equation becomes $u - 1 = \sqrt{5 + u} \Rightarrow (u-1)^2 = 5 + u \Rightarrow$
    $u^2 - 2u + 1 = 5 + u \Rightarrow u^2 - 3u - 4 \Rightarrow$
    $(u-4)(u+1) = 0 \Rightarrow u = 4$ or $u = -1$.
    Now solve for x.
    $4 = \sqrt{x} \Rightarrow 16 = x$ or $-1 = \sqrt{x}$
    (not possible). The solution set is $\{16\}$.

59. $(7x+5)^2 + 2(7x+5) - 15 = 0$. Let $u = 7x + 5$.
Then the equation becomes
$u^2 + 2u - 15 = 0 \Rightarrow$
$(u+5)(u-3) = 0 \Rightarrow u = -5$ or $u = 3$.
Now solve for x: $-5 = 7x + 5 \Rightarrow x = -\frac{10}{7}$ or
$3 = 7x + 5 \Rightarrow x = -\frac{2}{7}$.
The solution set is $\{-10/7, -2/7\}$.

61. $x^{2/3} + 3x^{1/3} - 4 = 0$. Let $u = x^{1/3}$. So the
equation becomes $u^2 + 3u - 4 = 0 \Rightarrow$
$(u+4)(u-1) = 0 \Rightarrow u = -4$ or $u = 1$. Now
solve for u: $-4 = x^{1/3} \Rightarrow -64 = x$ or
$1 = x^{1/3} \Rightarrow 1 = x$. The solution set is $\{-64, 1\}$.

63. $(\sqrt{t}+5)^2 - 9(\sqrt{t}+5) + 20 = 0$.
Let $u = \sqrt{t} + 5$, so the equation becomes
$u^2 - 9u + 20 = 0 \Rightarrow (u-5)(u-4) = 0 \Rightarrow$
$u = 5$ or $u = 4$. Now solve for t. $5 = \sqrt{t} + 5 \Rightarrow$
$0 = t$ or $4 = \sqrt{t} + 5 \Rightarrow -1 = \sqrt{t}$. (reject this)
The solution set is $\{0\}$.

65. $4x^4 - 37x^2 + 9 = 0$. Let $u = x^2$. So the
equation becomes $4u^2 - 37u + 9 = 0 \Rightarrow$
$(4u-1)(u-9) = 0 \Rightarrow u = \frac{1}{4}$ or $u = 9$. Now
solve for x: $\frac{1}{4} = x^2 \Rightarrow x = \pm\frac{1}{2}$ or $9 = x^2 \Rightarrow$
$x = \pm 3$. The solution set is $\left\{-3, -\frac{1}{2}, \frac{1}{2}, 3\right\}$.

67. $\frac{2x+1}{2x-1} = \frac{x-1}{x+1}$
$(2x+1)(x+1) = (2x-1)(x-1)$
$2x^2 + 3x + 1 = 2x^2 - 3x + 1$
$6x = 0 \Rightarrow x = 0$
The solution set is $\{0\}$.

69. $\left(\frac{7x}{x+1}\right)^2 - 3\left(\frac{7x}{x+1}\right) = 18$. Let $u = \frac{7x}{x+1}$. So,
the equation becomes $u^2 - 3u = 18 \Rightarrow$
$u^2 - 3u - 18 = 0 \Rightarrow (u-6)(u+3) = 0 \Rightarrow$
$u = 6$ or $u = -3$. Now solve for x:

$-3 = \frac{7x}{x+1} \Rightarrow -3x - 3 = 7x \Rightarrow x = -\frac{3}{10}$ or
$6 = \frac{7x}{x+1} \Rightarrow 6x + 6 = 7x \Rightarrow x = 6$.
The solution set is $\left\{-\frac{3}{10}, 6\right\}$.

71. $x^2 + 2yx - 3y^2 = 0$
$(x+3y)(x-y) = 0$
$x + 3y = 0 \Rightarrow x = -3y$
$x - y = 0 \Rightarrow x = y$
Solution set: $\{-3y, y\}$

73. $x^2 + (3-2y)x + y^2 - 3y + 2 = 0$
Use the quadratic formula with $a = 1$,
$b = 3 - 2y$ and $c = y^2 - 3y + 2$.
$x = \frac{-(3-2y) \pm \sqrt{(3-2y)^2 - 4(1)(y^2 - 3y + 2)}}{2(1)}$
$x = \frac{2y - 3 \pm \sqrt{4y^2 - 12y + 9 - (4y^2 - 12y + 8)}}{2}$
$= \frac{(2y-3) \pm 1}{2} = \frac{2y-4}{2} = y - 2$ or
$\frac{2y-2}{2} = y - 1$
Solution set: $\{y-2, y-1\}$

75. $x + 5 < 3 \Rightarrow x < -2$
The solution set is $(-\infty, -2)$.

77. $3(x-3) \le 8 \Rightarrow 3x - 9 \le 8 \Rightarrow 3x \le 17 \Rightarrow x \le \frac{17}{3}$
The solution set is $\left(-\infty, \frac{17}{3}\right]$.

79. $x + 2 \ge \frac{2}{3}x - 2x \Rightarrow 3x + 6 \ge 2x - 6x \Rightarrow$
$3x + 6 \ge -4x \Rightarrow 6 \ge -7x \Rightarrow -\frac{6}{7} \le x$
The solution set is $\left[-\frac{6}{7}, \infty\right)$.

81. $\frac{1}{6} > \frac{4-3x}{3} \Rightarrow 1 > 2(4-3x) \Rightarrow 1 > 8 - 6x \Rightarrow$
$-7 > -6x \Rightarrow \frac{7}{6} < x$
Solution set: $\left(\frac{7}{6}, \infty\right)$

83. $\dfrac{x-3}{3} - 2 \leq \dfrac{x}{6} + \dfrac{1}{2} \Rightarrow 2(x-3) - 2(6) \leq x+3 \Rightarrow$
$2x - 6 - 12 \leq x + 3 \Rightarrow 2x - 18 \leq x + 3 \Rightarrow x \leq 21$
Solution set: $(-\infty, 21]$

85. $\dfrac{3-2x}{4} + 1 > \dfrac{x-5}{3}$
$3(3-2x) + 12 > 4(x-5)$
$9 - 6x + 12 > 4x - 20$
$-6x + 21 > 4x - 20$
$-10x > -41 \Rightarrow x < \dfrac{41}{10}$
Solution set: $\left(-\infty, \dfrac{41}{10}\right)$

87. $3x - 1 < 2$ or $11 - 2x < 5$
$\quad 3x < 3$ or $\quad -2x < -6$
$\quad x < 1$ or $\quad x > 3$
Solution set: $(-\infty, 1) \cup (3, \infty)$

89. $4x - 5 < 7$ and $7 - 3x < 1$
$\quad 4x < 12$ and $\quad -3x < -6$
$\quad x < 3$ and $\quad x > 2$
Solution set: $(2, 3)$

91. $-3 \leq 2x + 1 < 7 \Rightarrow -4 \leq 2x < 6 \Rightarrow -2 \leq x < 3$
Solution set: $[-2, 3)$

93. $-3 < 3 - 2x \leq 97 \Rightarrow -6 < -2x \leq 94 \Rightarrow$
$3 > x \geq -47 \Rightarrow -47 \leq x < 3$
Solution set: $[-47, 3)$

95. $x^2 + x - 6 \geq 0 \Rightarrow (x+3)(x-2) \geq 0$
Solve the associated equation:
$(x+3)(x-2) = 0 \Rightarrow x = -3$ or $x = 2$.
So, the intervals are $(-\infty, -3]$, $[-3, 2]$, and
and $[2, \infty)$.

Interval	Test point	Value of $x^2 + x - 6$	Result
$(-\infty, -3]$	-4	6	$+$
$[-3, 2]$	0	-6	$-$
$[2, \infty)$	3	6	$+$

The solution set is $(-\infty, -3] \cup [2, \infty)$.

97. $\dfrac{(x-1)(x+3)}{(x+2)(x+5)} \geq 0$
Set the numerator and denominator equal to zero and solve for x.
$(x-1)(x+3) = 0 \Rightarrow x = 1, -3$
$(x+2)(x+5) = 0 \Rightarrow x = -2, -5$
The intervals are $(-\infty, -5)$, $(-5, -3]$, $[-3, -2)$, $(-2, 1]$, and $(1, \infty)$.

Interval	Test point	Value of $\dfrac{(x-1)(x+3)}{(x+2)(x+5)}$	Result
$(-\infty, -5)$	-6	$\dfrac{21}{4}$	$+$
$(-5, -3]$	-4	$-\dfrac{5}{2}$	$-$
$[-3, -2)$	$-\dfrac{5}{2}$	$\dfrac{7}{5}$	$+$
$(-2, 1]$	0	$-\dfrac{3}{10}$	$-$
$(1, \infty)$	2	$\dfrac{5}{28}$	$+$

The solution set is
$(-\infty, -5) \cup [-3, -2) \cup (1, \infty)$.

99. $|3x + 2| \leq 7 \Rightarrow -7 \leq 3x + 2 \leq 7 \Rightarrow$
$-9 \leq 3x \leq 5 \Rightarrow -3 \leq x \leq \dfrac{5}{3}$
The solution set is $\left[-3, \dfrac{5}{3}\right]$.

101. $4|x-2| + 8 > 12 \Rightarrow 4|x-2| > 4 \Rightarrow$
$|x-2| > 1 \Rightarrow x - 2 < -1$ or $x - 2 > 1 \Rightarrow$
$x < 1$ or $x > 3$
The solution set is $(-\infty, 1) \cup (3, \infty)$.

103. $\left|\dfrac{4-x}{5}\right| \geq 1 \Rightarrow \dfrac{4-x}{5} \leq -1$ or $\dfrac{4-x}{5} \geq 1 \Rightarrow$
$4 - x \leq -5 \Rightarrow x \geq 9$ or $4 - x \geq 5 \Rightarrow x \leq -1$
The solution set is $(-\infty, -1] \cup [9, \infty)$.

105. $\left|\dfrac{x-1}{x+2}\right| \le 3 \Rightarrow -3 \le \dfrac{x-1}{x+2} \le 3$

$0 \le \dfrac{x-1}{x+2} + 3$ | $\dfrac{x-1}{x+2} - 3 \le 0$
$0 \le \dfrac{x-1+3(x+2)}{x+2}$ | $\dfrac{x-1-3(x+2)}{x+2} \le 0$
$0 \le \dfrac{4x+5}{x+2}$ | $\dfrac{-2x-7}{x+2} \le 0$

$4x+5 = 0 \Rightarrow x = -\dfrac{5}{4};\ x+2 = 0 \Rightarrow x = -2$

Interval	Test point	Value of $\dfrac{4x+5}{x+2}$	Result
$(-\infty, -2)$	-4	$\dfrac{11}{2}$	$+$
$\left(-2, -\dfrac{5}{4}\right]$	$-\dfrac{3}{2}$	-2	$-$
$\left[-\dfrac{5}{4}, \infty\right)$	0	$\dfrac{5}{2}$	$+$

Note that the expression is undefined for $x = -2$. The solution set for this part of the original inequality is $(-\infty, -2) \cup \left[-\dfrac{5}{4}, \infty\right)$.

$-2x - 7 = 0 \Rightarrow x = -\dfrac{7}{2};\ x+2 = 0 \Rightarrow x = -2$

Interval	Test point	Value of $\dfrac{-2x-7}{x+2}$	Result
$\left(-\infty, -\dfrac{7}{2}\right]$	-4	$-\dfrac{1}{2}$	$-$
$\left[-\dfrac{7}{2}, -2\right)$	-3	1	$+$
$(-2, \infty)$	0	$-\dfrac{7}{2}$	$-$

Note that the expression is undefined for $x = -2$. The solution set is for this part of the original inequality is $\left(-\infty, -\dfrac{7}{2}\right] \cup (-2, \infty)$.

Both inequalities are true on $\left(-\infty, -\dfrac{7}{2}\right]$ and $\left[-\dfrac{5}{4}, \infty\right)$, so the solution set is $\left(-\infty, -\dfrac{7}{2}\right] \cup \left[-\dfrac{5}{4}, \infty\right)$.

107. $\left|\dfrac{2x-3}{x+2}\right| \ge 2 \Rightarrow \dfrac{2x-3}{x+2} \le -2$ or $\dfrac{2x-3}{x+2} \ge 2$

$\dfrac{2x-3}{x+2} \le -2$ | $\dfrac{2x-3}{x+2} \ge 2$
$\dfrac{2x-3}{x+2} + 2 \le 0$ | $\dfrac{2x-3}{x+2} - 2 \ge 0$
$\dfrac{2x-3+2(x+2)}{x+2} \le 0$ | $\dfrac{2x-3-2(x+2)}{x+2} \ge 0$
$\dfrac{4x+1}{x+2} \le 0$ | $\dfrac{-7}{x+2} \ge 0$

$4x+1 = 0 \Rightarrow x = -\dfrac{1}{4};\ x+2 = 0 \Rightarrow x = -2$

Interval	Test point	Value of $\dfrac{4x+1}{x+2}$	Result
$(-\infty, -2)$	-3	11	$+$
$\left(-2, -\dfrac{1}{4}\right]$	-1	-3	$-$
$\left[-\dfrac{1}{4}, \infty\right)$	0	$\dfrac{1}{2}$	$+$

Note that the expression is undefined for $x = -2$. The solution set is for this part of the original inequality is $\left(-2, -\dfrac{1}{4}\right]$.

Interval	Test point	Value of $\dfrac{-7}{x+2}$	Result
$(-\infty, -2)$	-3	$\dfrac{7}{5}$	$+$
$(-2, \infty)$	0	$-\dfrac{7}{2}$	$-$

Note that the expression is undefined for $x = -2$. The solution set is for this part of the original inequality is $(-\infty, -2)$. Since the original inequality is an "or" inequality, the solution set of the original inequality is the union of the two solution sets.

The solution set is $(-\infty, -2) \cup \left(-2, -\dfrac{1}{4}\right]$.

Applying the Concepts

109. $C = 2\pi r \Rightarrow 22 = 2\pi r \Rightarrow \dfrac{11}{\pi} = r$

The radius is $11/\pi$ cm.

111. $A = \dfrac{1}{2}h(b_1 + b_2) \Rightarrow 32 = \dfrac{1}{2}(8)(5 + b_2) \Rightarrow$
$32 = 4(5 + b_2) \Rightarrow 8 = 5 + b_2 \Rightarrow 3 = b_2$
The other base is 3 meters.

113. $V = lwh \Rightarrow 4212 = 27(12)h \Rightarrow 13 = h$
     The box is 13 cm high.

115. $V = \pi r^2 h \Rightarrow 8750\pi = 5^2 \pi h \Rightarrow 350 = h$
     The height is 350 cm.

117. Let x = measure or the base angle.
     Then $40 + 2x$ = the measure of the third angle.
     So,
     $x + x + 40 + 2x = 180 \Rightarrow 4x + 40 = 180 \Rightarrow$
     $x = 35$.
     The three angles are $35°$, $35°$, and $110°$.

119. Let x = amount invested at 6%.
     Then $30{,}000 - x$ = amount invested at 8%. So,
     we have $0.06x + 0.08(30{,}000 - x) = 2160 \Rightarrow$
     $-0.02x + 2400 = 2160 \Rightarrow -0.02x = -240 \Rightarrow$
     $x = 12{,}000$. $12,000 was invested at 6%, and
     $18,000 was invested at 8%.

121. Let x = amount of 4 1/2% solution.
     Then $10 - x$ = the amount of 12% solution.
     So, we have
     $0.045x + 0.12(10 - x) = 0.06(10) \Rightarrow$
     $-0.075x + 1.2 = 0.6 \Rightarrow -0.075x = -0.6 \Rightarrow$
     $x = 8$. There are 8 liters of the 4 1/2%
     solution and 2 liters of the 12% solution.

123. Let x = the number of people at the party.
     Each person shook $x - 1$ hands. So, there are
     $x(x - 1)$ handshakes. However, each
     handshake is counted twice (if A shakes hands
     with B, that is the same as B shaking hands
     with A).
     So, we have $\frac{x(x-1)}{2} = 28 \Rightarrow x^2 - x = 56 \Rightarrow$
     $x^2 - x - 56 = 0 \Rightarrow (x - 8)(x + 7) = 0 \Rightarrow$
     $x = 8$ or $x = -7$. We reject the negative
     answer. There were 8 people at the party.

125. Let x = the number of shares that Lavina
     bought. She spent $18,040 for the stock, or
     $18{,}040/x$ per share.
     She sold $x - 20$ shares at $\frac{18{,}040}{x} + 18$ per
     share for a total of $20,000. Therefore,
     $(x - 20)\left(\frac{18{,}040}{x} + 18\right) = 20{,}000$
     $\frac{18{,}040 + 18x}{x} = \frac{20{,}000}{x - 20}$
     $(x - 20)(18{,}040 + 18x) = 20{,}000x$

$18x^2 + 17{,}680x - 360{,}800 = 20{,}000x$
$18x^2 - 2320x - 360{,}800 = 0$
$2(9x + 820)(x - 220) = 0$
$9x + 820 = 0 \Rightarrow x = -\frac{820}{9}$ (reject this)
$x - 220 = 0 \Rightarrow x = 220$
Lavina bought 220 shares of stock.

127. Let x = the number of horses in the herd. Then
     $x/4$ = the number of horses in the forest, and
     $2\sqrt{x}$ = the number of horses in the
     mountains.
     $\frac{x}{4} + 2\sqrt{x} + 15 = x \Rightarrow x + 8\sqrt{x} + 60 = 4x \Rightarrow$
     $8\sqrt{x} = 3x - 60 \Rightarrow 64x = (3x - 60)^2 \Rightarrow$
     $64x = 9x^2 - 360x + 3600 \Rightarrow$
     $9x^2 - 424x + 3600 = 0 \Rightarrow$
     $(x - 36)(9x - 100) = 0 \Rightarrow x = 36$ or $x = 100/9$
     Reject the fractional solution. There were 36
     horses in the herd.

129. Let x = the original number of members going
     on the trip. Then $x + 4$ = the number of members
     going on the trip. The cost per member for the
     original number going on the trip was $\frac{324}{x}$, and
     the cost for the final number going on the trip is
     $\frac{324}{x} - 0.9$. Therefore,
     $\frac{324}{x+4} = \frac{324}{x} - 0.9$
     $324x = 324(x+4) - 0.9x(x+4)$
     $324x = 324x + 1296 - 0.9x^2 - 3.6x$
     $0 = -0.9x^2 - 3.6x + 1296$
     $0 = x^2 + 4x - 1440$
     $0 = (x - 36)(x + 40) \Rightarrow x = 36$ or $x = -40$
     (Reject the negative solution). So, 36 people
     originally signed up for the trip, and 40 people
     went on the trip.

Chapter 1 Practice Test A

1. $5x - 9 = 3x - 5 \Rightarrow 2x = 4 \Rightarrow x = 2$

2. $\frac{7}{24} = \frac{x}{8} + \frac{1}{6} \Rightarrow 7 = 3x + 4 \Rightarrow x = 1$

Chapter 1 Practice Test A

3. $\dfrac{1}{x-2} - 5 = \dfrac{1}{x+2} \Rightarrow \dfrac{1-5x+10}{x-2} = \dfrac{1}{x+2} \Rightarrow$
$\dfrac{-5x+11}{x-2} = \dfrac{1}{x+2} \Rightarrow$
$(-5x+11)(x+2) = x-2$
$-5x^2 + x + 22 = x - 2$
$-5x^2 + 24 = 0 \Rightarrow -5x^2 = -24 \Rightarrow$
$x^2 = \dfrac{24}{5} \Rightarrow x = \pm\dfrac{2\sqrt{6}}{\sqrt{5}} = \pm\dfrac{2\sqrt{30}}{5}$

4. Let x = the length of the rectangle.
Then $x - 3$ = the width of the rectangle.
$x(x-3) = 54 \Rightarrow x^2 - 3x - 54 = 0 \Rightarrow$
$(x+6)(x-9) = 0 \Rightarrow x = -6$ or $x = 9$
We reject the negative answer. The rectangle is 9 cm by 6 cm.

5. $x^2 + 36 = -13x \Rightarrow x^2 + 13x + 36 = 0 \Rightarrow$
$(x+9)(x+4) = 0 \Rightarrow x = -9$ or $x = -4$

6. Let x = the amount to be invested at 8%.
Then the total amount invested is $x + 8200$.
$0.06(8200) + 0.08x = 0.07(8200 + x)$
$492 + 0.08x = 574 + 0.07x$
$0.01x = 82 \Rightarrow x = 8200$
Fran must invest $8200 at 7%.

7. Let x = the width of the border. Then the length of the border is $2x + 25$, and the width of the border is $2x + 11$.

$(2x+25)(2x+11) = 351$
$4x^2 + 72x + 275 = 351$
$4x^2 + 72x - 76 = 0$
$4(x+19)(x-1) = 0 \Rightarrow x = -19$ or $x = 1$
The border is 1 inch wide.

8. $\quad -6x - 15 = (2x+5)^2$
$-6x - 15 = 4x^2 + 20x + 25$
$4x^2 + 26x + 40 = 0$
$2(2x+5)(x+4) = 0 \Rightarrow x = -\dfrac{5}{2}$ or $x = -4$
The solution set is $\{-4, -5/2\}$.

9. To find the constant term, find 1/2 of 2/3 = 1/3 and then square the answer: $(1/3)^2 = 1/9$.
The trinomial is $x^2 + \dfrac{2}{3}x + \dfrac{1}{9}$, which factors into $\left(x + \dfrac{1}{3}\right)^2$.

10. $x = \dfrac{-(-5) \pm \sqrt{(-5)^2 - 4(3)(-1)}}{2(3)}$
$= \dfrac{5 \pm \sqrt{37}}{6} = \dfrac{5}{6} \pm \dfrac{\sqrt{37}}{6}$

11. Let x = the length of the side of the original piece of cardboard. Then $x - 4$ = the length of the side of the box.

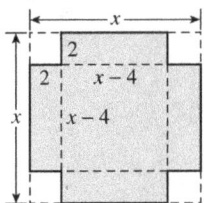

$2(x-4)^2 = 50 \Rightarrow (x-4)^2 = 25 \Rightarrow$
$x - 4 = 5 \Rightarrow x = 9$ (Reject the negative solution.) The side of the original cardboard square is 9 inches.

12. $x = \dfrac{-12 \pm \sqrt{12^2 - 4(1)(33)}}{2(1)}$
$= \dfrac{-12 \pm \sqrt{12}}{2} = \dfrac{-12 \pm 2\sqrt{3}}{2} = -6 \pm \sqrt{3}$

13. $3x^4 - 75x^2 = 0 \Rightarrow 3x^2(x^2 - 25) = 0 \Rightarrow$
$3x^2(x-5)(x+5) = 0 \Rightarrow 3x^2 = 0 \Rightarrow x = 0$
or $x - 5 = 0 \Rightarrow x = 5$
or $x + 5 = 0 \Rightarrow x = -5$
The solution set is $\{-5, 0, 5\}$.

14. Let $u = \sqrt{x}$. The equation becomes
$3u^2 - 2 - 5u = 0 \Rightarrow 3u^2 - 5u - 2 = 0 \Rightarrow$
$(3u+1)(u-2) = 0 \Rightarrow u = -\dfrac{1}{3}$ or $u = 2$
Now solve for x: $2 = \sqrt{x} \Rightarrow x = 4$ (We reject the negative solution). The solution set is $\{4\}$.

15. $\left|\frac{1}{3}x+5\right|=\left|\frac{2}{3}x+7\right| \Rightarrow \frac{1}{3}x+5=\frac{2}{3}x+7$

    or $\frac{1}{3}x+5=-\left(\frac{2}{3}x+7\right)$.

    $\frac{1}{3}x+5=\frac{2}{3}x+7 \Rightarrow x+15=2x+21 \Rightarrow x=-6$.

    $\frac{1}{3}x+5=-\left(\frac{2}{3}x+7\right) \Rightarrow \frac{1}{3}x+5=-\frac{2}{3}x-7 \Rightarrow$

    $x+15=-2x-21 \Rightarrow 3x=-36 \Rightarrow x=-12$

    The solution set is $\{-12, -6\}$.

16. $\frac{x}{2}-5 \geq \frac{4x}{9} \Rightarrow 9x-90 \geq 8x \Rightarrow x \geq 90$

    The solution set is $[90, \infty)$.

17. $-4 < 2x-3 < 4 \Rightarrow -1 < 2x < 7 \Rightarrow$

    $-\frac{1}{2} < x < \frac{7}{2}$

    The solution set is $\left(-\frac{1}{2}, \frac{7}{2}\right)$.

18. $\left|\frac{2}{3}x-1\right|-2 > \frac{1}{3} \Rightarrow \left|\frac{2}{3}x-1\right| > \frac{7}{3} \Rightarrow$

    $\frac{2}{3}x-1 < -\frac{7}{3}$ or $\frac{2}{3}x-1 > \frac{7}{3}$

    $\frac{2}{3}x-1 < -\frac{7}{3} \Rightarrow 2x-3 < -7 \Rightarrow 2x < -4 \Rightarrow$

    $x < -2$

    $\frac{2}{3}x-1 > \frac{7}{3} \Rightarrow 2x-3 > 7 \Rightarrow 2x > 10 \Rightarrow x > 5$

    The solution set is $(-\infty, -2) \cup (5, \infty)$.

19. $\frac{2}{5}y-13 \leq -\left(7+\frac{13}{5}y\right)$

    $\frac{2}{5}y-13 \leq -7-\frac{13}{5}y$

    $2y-65 \leq -35-13y \Rightarrow 15y \leq 30 \Rightarrow y \leq 2$

    The solution set is $(-\infty, 2]$.

20. $0 \leq 5x-2 \leq 8 \Rightarrow 2 \leq 5x \leq 10 \Rightarrow \frac{2}{5} \leq x \leq 2$

    The solution set is $\left[\frac{2}{5}, 2\right]$.

Chapter 1 Practice Test B

1. $2x-2 = 5x+34 \Rightarrow -3x = 36 \Rightarrow x = -12$.
   The answer is D.

2. $\frac{z}{2} = 2z+35 \Rightarrow z = 4z+70 \Rightarrow -3z = 70 \Rightarrow$

   $z = -\frac{70}{3}$. The answer is D.

3. $\frac{1}{t-2}-\frac{1}{2} = \frac{-2t}{4t-1} \Rightarrow$

   $2(4t-1)-(t-2)(4t-1) = -2t(2)(t-2) \Rightarrow$

   $8t-2-4t^2+9t-2 = -4t^2+8t \Rightarrow$

   $9t-4 = 0 \Rightarrow 9t = 4 \Rightarrow t = \frac{4}{9}$.

   The answer is A.

4. Let w = the width of the rectangle.
   Then $w+4$ = the length of the rectangle.
   $w(w+4) = 77 \Rightarrow w^2+4w-77 = 0 \Rightarrow$
   $(w+11)(w-7) = 0 \Rightarrow w = -11$ or $w = 7$
   We reject the negative solution. The rectangle is 7 cm by 11 cm. The answer is C.

5. $x^2+12 = -7x \Rightarrow x^2+7x+12 = 0 \Rightarrow$
   $(x+4)(x+3) = 0 \Rightarrow x = -4$ or $x = -3$
   The answer is B.

6. Let x = the amount to be invested at 12%.
   Then the total amount invested is $7500+x$.
   $0.07(7500)+0.12x = 0.10(7500+x)$
   $525+0.12x = 750+0.10x$
   $0.02x = 225 \Rightarrow x = 11{,}250$
   Rena must invest $11,250 at 12%.
   The answer is A.

7. Let x = the width of the border. Then the length of the border is $2x+20$, and the width of the border is $2x+13$.

   $(2x+20)(2x+13) = 368$

   $4x^2+66x+260 = 368$

   $4x^2+66x-108 = 0$

   $2(2x-3)(x+18) = 0 \Rightarrow x = \frac{3}{2}$ or $x = -18$

   The border is 1.5 inch wide. The answer is D.

8. $$-6x - 2 = (3x+1)^2$$
$$-6x - 2 = 9x^2 + 6x + 1$$
$$9x^2 + 12x + 3 = 0$$
$$3x^2 + 4x + 1 = 0$$
$(3x+1)(x+1) = 0 \Rightarrow x = -\dfrac{1}{3}$ or $x = -1$
The answer is D.

9. To find the constant term, find $1/2$ of $1/6 = 1/12$ and then square the answer: $(1/12)^2 = 1/144$.

The trinomial is $x^2 + \dfrac{1}{6}x + \dfrac{1}{144}$, which factors into $\left(x + \dfrac{1}{12}\right)^2$. The answer is B.

10. $x = \dfrac{-10 \pm \sqrt{10^2 - 4(7)(2)}}{2(7)}$
$= \dfrac{-10 \pm \sqrt{44}}{14} = \dfrac{-10 \pm 2\sqrt{11}}{14} = -\dfrac{5}{7} \pm \dfrac{\sqrt{11}}{7}$
The answer is A.

11. Let $x =$ the length of the side of the original piece of cardboard. Then $x - 6 =$ the length of the side of the box.

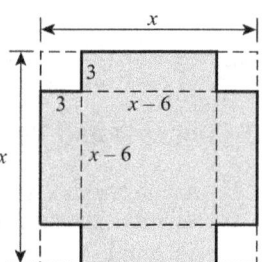

$3(x-6)^2 = 675 \Rightarrow (x-6)^2 = 225 \Rightarrow$
$x - 6 = 15 \Rightarrow x = 21$ (Note that we reject the negative solution.) The side of the original cardboard square is 21 inches.
The answer is B.

12. $x = \dfrac{-14 \pm \sqrt{14^2 - 4(1)(38)}}{2(1)}$
$= \dfrac{-14 \pm \sqrt{44}}{2} = \dfrac{-14 \pm 2\sqrt{11}}{2}$
$= -7 \pm \sqrt{11}$
The answer is D.

13. $5x^4 - 45x^2 = 0 \Rightarrow 5x^2(x^2 - 9) = 0 \Rightarrow$
$5x^2(x-3)(x+3) = 0 \Rightarrow 5x^2 = 0 \Rightarrow x = 0$
or $x - 3 = 0 \Rightarrow x = 3$
or $x + 3 = 0 \Rightarrow x = -3$
The solution set is $\{-3, 0, 3\}$.
The answer is A.

14. Let $u = \sqrt{x}$. The equation becomes
$u^2 - 2048 - 32u = 0 \Rightarrow$
$u^2 - 32u - 2048 = 0 \Rightarrow$
$(u - 64)(u + 32) = 0 \Rightarrow u = 64$ or $u = -32$
Now solve for x: $64 = \sqrt{x} \Rightarrow x = 4096$ (We reject the negative solution). The answer is A.

15. $\left|\dfrac{1}{2}x + 2\right| = \left|\dfrac{3}{4}x - 2\right| \Rightarrow \dfrac{1}{2}x + 2 = \dfrac{3}{4}x - 2$
or $\dfrac{1}{2}x + 2 = -\left(\dfrac{3}{4}x - 2\right)$.
$\dfrac{1}{2}x + 2 = \dfrac{3}{4}x - 2 \Rightarrow 2x + 8 = 3x - 8 \Rightarrow x = 16$
$\dfrac{1}{2}x + 2 = -\left(\dfrac{3}{4}x - 2\right) \Rightarrow \dfrac{1}{2}x + 2 = -\dfrac{3}{4}x + 2 \Rightarrow$
$2x + 8 = -3x + 8 \Rightarrow 5x = 0 \Rightarrow x = 0$
The answer is D.

16. $\dfrac{x}{6} - \dfrac{1}{3} \le \dfrac{x}{3} + 1 \Rightarrow x - 2 \le 2x + 6 \Rightarrow -8 \le x$
The answer is D.

17. $-13 \le -3x + 2 < -4 \Rightarrow -15 \le -3x < -6 \Rightarrow$
$5 \ge x > 2$
The answer is B.

18. $8 + \left|1 - \dfrac{x}{2}\right| \ge 10 \Rightarrow \left|1 - \dfrac{x}{2}\right| \ge 2 \Rightarrow$
$1 - \dfrac{x}{2} \le -2 \Rightarrow -\dfrac{x}{2} \le -3 \Rightarrow x \ge 6$ or
$1 - \dfrac{x}{2} \ge 2 \Rightarrow -\dfrac{x}{2} \ge 1 \Rightarrow x \le -2$
The solution set is $(-\infty, -2] \cup [6, \infty)$.
The answer is C.

19. $\dfrac{2}{3}x - 2 < \dfrac{5}{3}x \Rightarrow 2x - 6 < 5x \Rightarrow -6 < 3x \Rightarrow$
$-2 < x$
The solution set is $(-2, \infty)$. The answer is A.

20. $0 \le 7x - 1 \le 13 \Rightarrow 1 \le 7x \le 14 \Rightarrow \dfrac{1}{7} \le x \le 2$
The solution set is $\left[\dfrac{1}{7}, 2\right]$. The answer is A.

Chapter 2 Graphs and Functions

2.1 The Coordinate Plane

2.1 Practice Problems

1.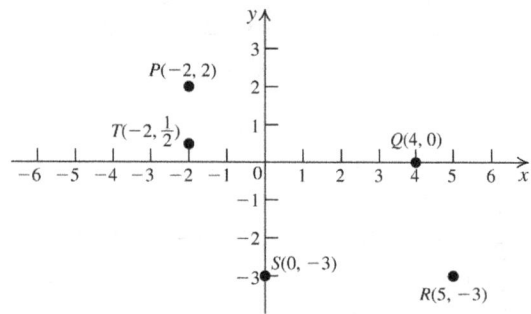

2.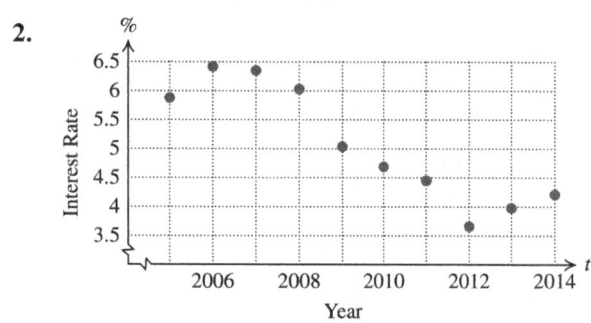

3. $(x_1, y_1) = (-5, 2);\ (x_2, y_2) = (-4, 1)$

$$d = \sqrt{(x_2 - x_1)^2 + (y_2 - y_1)^2}$$
$$= \sqrt{(-4 - (-5))^2 + (1 - 2)^2}$$
$$= \sqrt{1^2 + (-1)^2} = \sqrt{2} \approx 1.4$$

4. $(x_1, y_1) = (6, 2);\ (x_2, y_2) = (-2, 0)$
$(x_3, y_3) = (1, 5)$

$$d_1 = \sqrt{(x_2 - x_1)^2 + (y_2 - y_1)^2}$$
$$= \sqrt{(-2 - 6)^2 + (0 - 2)^2}$$
$$= \sqrt{(-8)^2 + (-2)^2} = \sqrt{68}$$

$$d_2 = \sqrt{(x_3 - x_1)^2 + (y_3 - y_1)^2}$$
$$= \sqrt{(1 - 6)^2 + (5 - 2)^2}$$
$$= \sqrt{(-5)^2 + (3)^2} = \sqrt{34}$$

$$d_3 = \sqrt{(x_3 - x_2)^2 + (y_3 - y_2)^2}$$
$$= \sqrt{(1 - (-2))^2 + (5 - 0)^2}$$
$$= \sqrt{(3)^2 + (5)^2} = \sqrt{34}$$

Yes, the triangle is an isosceles triangle.

5.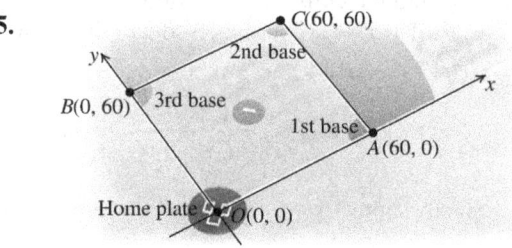

We are asked to find the distance between the points $A(60, 0)$ and $B(0, 60)$.

$$d(A, B) = \sqrt{(60 - 0)^2 + (0 - 60)^2}$$
$$= \sqrt{(60)^2 + (-60)^2} = \sqrt{2(60)^2}$$
$$= 60\sqrt{2} \approx 84.85 \text{ ft}$$

6. $M = \left(\dfrac{5+6}{2}, \dfrac{-2+(-1)}{2}\right) = \left(\dfrac{11}{2}, -\dfrac{3}{2}\right)$

2.1 Concepts and Vocabulary

1. A point with a negative first coordinate and a positive second coordinate lies in the <u>second</u> quadrant.

3. The distance between the points $P(x_1, y_1)$ and $Q(x_2, y_2)$ is given by the formula $d(P, Q) = \underline{\sqrt{(x_2 - x_1)^2 + (y_2 - y_1)^2}}$.

5. True

7. False. Every point in quadrant II has a negative x-coordinate.

2.1 Building Skills

9.

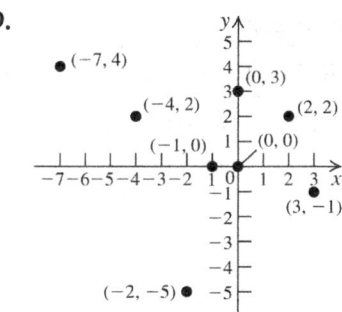

(2, 2): Q1; (3, −1): Q4; (−1, 0): x-axis
(−2, −5): Q3; (0, 0): origin; (−7, 4): Q2
(0, 3): y-axis; (−4, 2): Q2

11. a. If the x-coordinate of a point is 0, the point lies on the y-axis.

b.

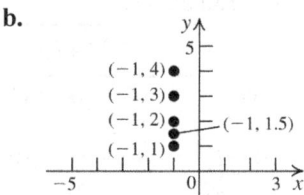

The set of all points of the form $(-1, y)$ is a vertical line that intersects the x-axis at -1.

13. a. $y > 0$ **b.** $y < 0$

c. $x < 0$ **d.** $x > 0$

In Exercises 15–24, use the distance formula, $d = \sqrt{(x_2 - x_1)^2 + (y_2 - y_1)^2}$ and the midpoint formula, $(x, y) = \left(\dfrac{x_1 + x_2}{2}, \dfrac{y_1 + y_2}{2}\right)$.

15. a. $d = \sqrt{(2-2)^2 + (5-1)^2} = \sqrt{4^2} = 4$

b. $M = \left(\dfrac{2+2}{2}, \dfrac{1+5}{2}\right) = (2, 3)$

17. a. $d = \sqrt{(2-(-1))^2 + (-3-(-5))^2}$
$= \sqrt{3^2 + 2^2} = \sqrt{13}$

b. $M = \left(\dfrac{-1+2}{2}, \dfrac{-5+(-3)}{2}\right) = (0.5, -4)$

19. a. $d = \sqrt{(3-(-1))^2 + (-6.5-1.5)^2}$
$= \sqrt{4^2 + (-8)^2} = \sqrt{80} = 4\sqrt{5}$

b. $M = \left(\dfrac{-1+3}{2}, \dfrac{1.5+(-6.5)}{2}\right) = (1, -2.5)$

21. a. $d = \sqrt{\left(\sqrt{2}-\sqrt{2}\right)^2 + (5-4)^2} = \sqrt{1^2} = 1$

b. $M = \left(\dfrac{\sqrt{2}+\sqrt{2}}{2}, \dfrac{4+5}{2}\right) = \left(\sqrt{2}, 4.5\right)$

23. a. $d = \sqrt{(k-t)^2 + (t-k)^2}$
$= \sqrt{(k^2 - 2tk + t^2) + (t^2 - 2kt + k^2)}$
$= \sqrt{2t^2 - 4tk + 2k^2} = \sqrt{2(t^2 - 2tk + k^2)}$
$= \sqrt{2(t-k)^2} = |t-k|\sqrt{2}$

b. $M = \left(\dfrac{t+k}{2}, \dfrac{k+t}{2}\right)$

25. $P = (-1, -2), Q = (0, 0), R = (1, 2)$
$d(P, Q) = \sqrt{(0-(-1))^2 + (0-(-2))^2} = \sqrt{5}$
$d(Q, R) = \sqrt{(1-0)^2 + (2-0)^2} = \sqrt{5}$
$d(P, R) = \sqrt{(1-(-1))^2 + (2-(-2))^2}$
$= \sqrt{2^2 + 4^2} = \sqrt{20} = 2\sqrt{5}$
Because $d(P, Q) + d(Q, R) = d(P, R)$, the points are collinear.

27. $P = (4, -2), Q = (1, 3), R = (-2, 8)$
$d(P, Q) = \sqrt{(1-4)^2 + (3-(-2))^2} = \sqrt{34}$
$d(Q, R) = \sqrt{(-2-1)^2 + (8-3)^2} = \sqrt{34}$
$d(P, R) = \sqrt{(-2-4)^2 + (8-(-2))^2}$
$= \sqrt{(-6)^2 + 10^2} = \sqrt{136} = 2\sqrt{34}$
Because $d(P, Q) + d(Q, R) = d(P, R)$, the points are collinear.

29. $P = (-1, 4), Q = (3, 0), R = (11, -8)$
$d(P, Q) = \sqrt{(3-(-1))^2 + (0-4)^2} = 4\sqrt{2}$
$d(Q, R) = \sqrt{(11-3)^2 + ((-8)-0)^2} = 8\sqrt{2}$
$d(P, R) = \sqrt{(11-(-1))^2 + (-8-4)^2}$
$= \sqrt{(12)^2 + (-12)^2} = \sqrt{288} = 12\sqrt{2}$
Because $d(P, Q) + d(Q, R) = d(P, R)$, the points are collinear.

31. It is not possible to arrange the points in such a way so that $d(P, Q) + d(Q, R) = d(P, R)$, so the points are not collinear.

33. First, find the midpoint M of PQ.
$$M = \left(\frac{-4+0}{2}, \frac{0+8}{2}\right) = (-2, 4)$$
Now find the midpoint R of PM.
$$R = \left(\frac{-4+(-2)}{2}, \frac{0+4}{2}\right) = (-3, 2)$$
Finally, find the midpoint S of MQ.
$$S = \left(\frac{-2+0}{2}, \frac{4+8}{2}\right) = (-1, 6)$$
Thus, the three points are $(-3, 2)$, $(-2, 4)$, and $(-1, 6)$.

35. $d(P,Q) = \sqrt{(-1-(-5))^2 + (4-5)^2} = \sqrt{17}$
$d(Q,R) = \sqrt{(-4-(-1))^2 + (1-4)^2} = 3\sqrt{2}$
$d(P,R) = \sqrt{(-4-(-5))^2 + (1-5)^2} = \sqrt{17}$
The triangle is isosceles.

37. $d(P,Q) = \sqrt{(0-(-4))^2 + (7-8)^2} = \sqrt{17}$
$d(Q,R) = \sqrt{(-3-0)^2 + (5-7)^2} = \sqrt{13}$
$d(P,R) = \sqrt{(-3-(-4))^2 + (5-8)^2} = \sqrt{10}$
The triangle is scalene.

39. $d(P,Q) = \sqrt{(9-0)^2 + (-9-(-1))^2} = \sqrt{145}$
$d(Q,R) = \sqrt{(5-9)^2 + (1-(-9))^2} = 2\sqrt{29}$
$d(P,R) = \sqrt{(5-0)^2 + (1-(-1))^2} = \sqrt{29}$
The triangle is scalene.

41. $d(P,Q) = \sqrt{(-1-1)^2 + (1-(-1))^2} = 2\sqrt{2}$
$d(Q,R) = \sqrt{\left(-\sqrt{3}-(-1)\right)^2 + \left(-\sqrt{3}-1\right)^2}$
$= \sqrt{(3-2\sqrt{3}+1)+(3+2\sqrt{3}+1)}$
$= \sqrt{8} = 2\sqrt{2}$
$d(P,R) = \sqrt{\left(-\sqrt{3}-1\right)^2 + \left(-\sqrt{3}-(-1)\right)^2}$
$= \sqrt{(3+2\sqrt{3}+1)+(3-2\sqrt{3}+1)}$
$= \sqrt{8} = 2\sqrt{2}$
The triangle is equilateral.

43. First find the lengths of the sides:
$d(P,Q) = \sqrt{(-1-7)^2 + (3-(-12))^2} = 17$
$d(Q,R) = \sqrt{(14-(-1))^2 + (11-3)^2} = 17$
$d(R,S) = \sqrt{(22-14)^2 + (-4-11)^2} = 17$
$d(S,P) = \sqrt{(22-7)^2 + (-4-(-12))^2} = 17$

All the sides are equal, so the quadrilateral is either a square or a rhombus. Now find the length of the diagonals:
$d(P,R) = \sqrt{(14-7)^2 + (11-(-12))^2} = 17\sqrt{2}$
$d(Q,S) = \sqrt{(22-(-1))^2 + (-4-3)^2} = 17\sqrt{2}$
The diagonals are equal, so the quadrilateral is a square.

45. $5 = \sqrt{(x-2)^2 + (2-(-1))^2}$
$= \sqrt{x^2 - 4x + 4 + 9} \Rightarrow$
$5 = \sqrt{x^2 - 4x + 13} \Rightarrow 25 = x^2 - 4x + 13 \Rightarrow$
$0 = x^2 - 4x - 12 \Rightarrow 0 = (x-6)(x+2) \Rightarrow$
$x = -2$ or $x = 6$

47. $P = (-5, 2)$, $Q = (2, 3)$, $R = (x, 0)$ (R is on the x-axis, so the y-coordinate is 0).
$$d(P,R) = \sqrt{(x-(-5))^2 + (0-2)^2}$$
$$d(Q,R) = \sqrt{(x-2)^2 + (0-3)^2}$$
$\sqrt{(x-(-5))^2 + (0-2)^2} = \sqrt{(x-2)^2 + (0-3)^2}$
$(x+5)^2 + (0-2)^2 = (x-2)^2 + (0-3)^2$
$x^2 + 10x + 25 + 4 = x^2 - 4x + 4 + 9$
$10x + 29 = -4x + 13$
$14x = -16$
$x = -\dfrac{8}{7}$
The coordinates of R are $\left(-\dfrac{8}{7}, 0\right)$.

2.1 Applying the Concepts

49.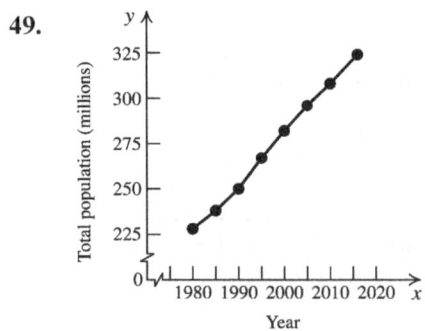

51.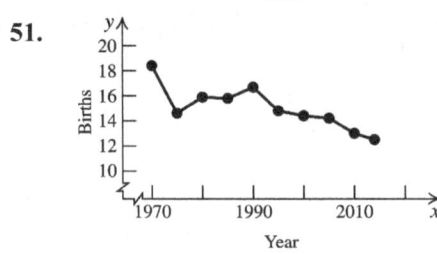

53.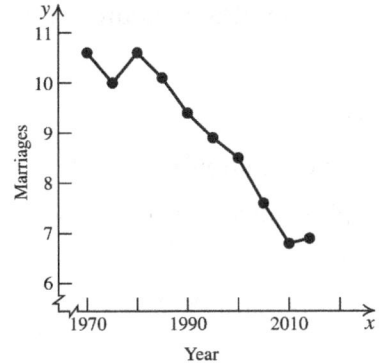

55. 2014 is the midpoint of the initial range, so
$$M = \left(\frac{2012+2016}{2}, \frac{326+425}{2}\right)$$
$$= (2014, 375.5)$$
Americans spent about $376 billion on prescription drugs in 2014.

57. Percentage of Android sales in June 2013: 51.5%

59. Android sales were at a maximum in June 2014.

61. Denote the diagonal connecting the endpoints of the edges a and b by d. Then a, b, and d form a right triangle. By the Pythagorean theorem, $a^2 + b^2 = d^2$. The edge c and the diagonals d and h also form a right triangle, so $c^2 + d^2 = h^2$. Substituting d^2 from the first equation, we obtain $a^2 + b^2 + c^2 = h^2$.

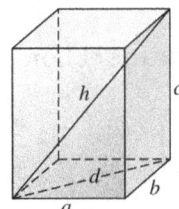

63. First, find the initial length of the rope using the Pythagorean theorem:
$$c = \sqrt{24^2 + 10^2} = 26.$$

After t seconds, the length of the rope is $26 - 3t$. Now find the distance from the boat to the dock, x, using the Pythagorean theorem again and solving for x:
$$(26-3t)^2 = x^2 + 10^2$$
$$676 - 156t + 9t^2 = x^2 + 100$$
$$576 - 156t + 9t^2 = x^2$$
$$\sqrt{576 - 156t + 9t^2} = x$$

2.1 Beyond the Basics

65. a. If AB is one of the diagonals, then DC is the other diagonal, and both diagonals have the same midpoint. The midpoint of AB is $\left(\frac{2+5}{2}, \frac{3+4}{2}\right) = (3.5, 3.5)$. The midpoint of $DC = (3.5, 3.5) = \left(\frac{x+3}{2}, \frac{y+8}{2}\right)$.

So we have $3.5 = \frac{x+3}{2} \Rightarrow x = 4$ and
$3.5 = \frac{y+8}{2} \Rightarrow y = -1$.
The coordinates of D are $(4, -1)$.

b. If AC is one of the diagonals, then DB is the other diagonal, and both diagonals have the same midpoint. The midpoint of AC is $\left(\frac{2+3}{2}, \frac{3+8}{2}\right) = (2.5, 5.5)$. The midpoint of $DB = (2.5, 5.5) = \left(\frac{x+5}{2}, \frac{y+4}{2}\right)$.

So we have $2.5 = \frac{x+5}{2} \Rightarrow x = 0$ and
$5.5 = \frac{y+4}{2} \Rightarrow y = 7$.
The coordinates of D are $(0, 7)$.

c. If BC is one of the diagonals, then DA is the other diagonal, and both diagonals have the same midpoint. The midpoint of BC is $\left(\frac{5+3}{2}, \frac{4+8}{2}\right) = (4, 6)$. The midpoint of DA is $(4,6) = \left(\frac{x+2}{2}, \frac{y+3}{2}\right)$. So we have
$4 = \frac{x+2}{2} \Rightarrow x = 6$ and $6 = \frac{y+3}{2} \Rightarrow y = 9$.
The coordinates of D are $(6, 9)$.

67. a. The midpoint of the diagonal connecting (1, 2) and (5, 8) is $\left(\frac{1+5}{2}, \frac{2+8}{2}\right) = (3, 5)$.
The midpoint of the diagonal connecting $(-2, 6)$ and $(8, 4)$ is $\left(\frac{-2+8}{2}, \frac{6+4}{2}\right) = (3, 5)$. Because the midpoints are the same, the figure is a parallelogram.

b. The midpoint of the diagonal connecting $(3, 2)$ and (x, y) is $\left(\dfrac{3+x}{2}, \dfrac{2+y}{2}\right)$. The midpoint of the diagonal connecting $(6, 3)$ and $(6, 5)$ is $(6, 4)$. So $\dfrac{3+x}{2} = 6 \Rightarrow x = 9$ and $\dfrac{2+y}{2} = 4 \Rightarrow y = 6$.

69. Let $P(0, 0)$, $Q(a, 0)$, and $R(0, b)$ be the vertices of the right triangle. The midpoint M of the hypotenuse is $\left(\dfrac{a}{2}, \dfrac{b}{2}\right)$.

$$d(Q, M) = \sqrt{\left(a - \dfrac{a}{2}\right)^2 + \left(0 - \dfrac{b}{2}\right)^2}$$
$$= \sqrt{\left(\dfrac{a}{2}\right)^2 + \left(-\dfrac{b}{2}\right)^2} = \dfrac{\sqrt{a^2 + b^2}}{2}$$

$$d(R, M) = \sqrt{\left(0 - \dfrac{a}{2}\right)^2 + \left(b - \dfrac{b}{2}\right)^2}$$
$$= \sqrt{\left(-\dfrac{a}{2}\right)^2 + \left(\dfrac{b}{2}\right)^2} = \dfrac{\sqrt{a^2 + b^2}}{2}$$

$$d(P, M) = \sqrt{\left(0 - \dfrac{a}{2}\right)^2 + \left(0 - \dfrac{b}{2}\right)^2}$$
$$= \sqrt{\left(-\dfrac{a}{2}\right)^2 + \left(-\dfrac{b}{2}\right)^2} = \dfrac{\sqrt{a^2 + b^2}}{2}$$

71. Since ABC is an equilateral triangle and O is the midpoint of AB, then the coordinates of A are $(-a, 0)$.

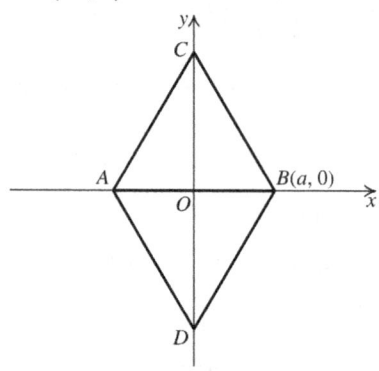

$AB = AC = AB = 2a$. Using triangle BOC and the Pythagorean theorem, we have
$BC^2 = OB^2 + OC^2 \Rightarrow (2a)^2 = a^2 + OC^2 \Rightarrow$
$4a^2 = a^2 + OC^2 \Rightarrow 3a^2 = OC^2 \Rightarrow OC = \sqrt{3}a$
Thus, the coordinates of C are $\left(0, \sqrt{3}a\right)$ and the coordinates of D are $\left(0, -\sqrt{3}a\right)$.

2.1 Critical Thinking/Discussion/Writing

73. a. y-axis

b. x-axis

74. a. The union of the x- and y-axes

b. The plane without the x- and y-axes

75. a. Quadrants I and III

b. Quadrants II and IV

76. a. The origin

b. The plane without the origin

77. a. Right half-plane

b. Upper half-plane

78. Let (x, y) be the point.

The point lies in	if
Quadrant I	$x > 0$ and $y > 0$
Quadrant II	$x < 0$ and $y > 0$
Quadrant III	$x < 0$ and $y < 0$
Quadrant IV	$x > 0$ and $y < 0$

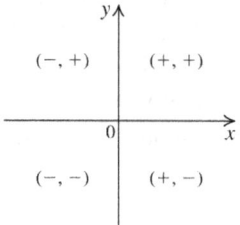

2.1 Getting Ready for the Next Section

79. a. $x^2 + y^2 = \left(\dfrac{1}{2}\right)^2 + \left(\dfrac{1}{2}\right)^2 = \dfrac{1}{4} + \dfrac{1}{4} = \dfrac{2}{4} = \dfrac{1}{2}$

b. $x^2 + y^2 = \left(\dfrac{\sqrt{2}}{2}\right)^2 + \left(\dfrac{\sqrt{2}}{2}\right)^2 = \dfrac{2}{4} + \dfrac{2}{4} = 1$

81. a. $\dfrac{x}{|x|} + \dfrac{|y|}{y} = \dfrac{2}{|2|} + \dfrac{|-3|}{-3} = \dfrac{2}{2} + \dfrac{3}{-3} = 1 - 1 = 0$

b. $\dfrac{x}{|x|} + \dfrac{|y|}{y} = \dfrac{-4}{|-4|} + \dfrac{|3|}{3} = \dfrac{-4}{4} + \dfrac{3}{3} = -1 + 1 = 0$

83. $x^2 - 6x + \left(\dfrac{-6}{2}\right)^2 = x^2 - 6x + 3^2$
$= x^2 - 6x + \underline{9}$

85. $y^2 + 3y = y^2 + 3y + \left(\dfrac{3}{2}\right)^2 = y^2 + 3y + \underline{\dfrac{9}{4}}$

87. $x^2 - ax + \left(\dfrac{-a}{2}\right)^2 = x^2 - ax + \underline{\dfrac{a^2}{4}}$

2.2 Graphs of Equations

2.2 Practice Problems

1. $y = -x^2 + 1$

x	$y = -x^2 + 1$	(x, y)
-2	$y = -(-2)^2 + 1$	$(-2, -3)$
-1	$y = -(-1)^2 + 1$	$(-1, 0)$
0	$y = -(0)^2 + 1$	$(0, 1)$
1	$y = -(1)^2 + 1$	$(1, 0)$
2	$y = -(2)^2 + 1$	$(2, -3)$

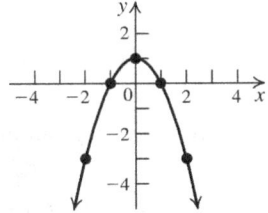

2. To find the x-intercept, let $y = 0$, and solve the equation for x: $0 = 2x^2 + 3x - 2 \Rightarrow$
$0 = (2x - 1)(x + 2) \Rightarrow x = \dfrac{1}{2}$ or $x = -2$. To find the y-intercept, let $x = 0$, and solve the equation for y:
$y = 2(0)^2 + 3(0) - 2 \Rightarrow y = -2$.
The x-intercepts are $\dfrac{1}{2}$ and -2; the y-intercept is -2.

3. To test for symmetry about the y-axis, replace x with $-x$ to determine if $(-x, y)$ satisfies the equation.
$(-x)^2 - y^2 = 1 \Rightarrow x^2 - y^2 = 1$, which is the same as the original equation. So the graph is symmetric about the y-axis.

4. x-axis: $x^2 = (-y)^3 \Rightarrow x^2 = -y^3$, which is not the same as the original equation, so the equation is not symmetric with respect to the x-axis.
y-axis: $(-x)^2 = y^3 \Rightarrow x^2 = y^3$, which is the same as the original equation, so the equation is symmetric with respect to the y-axis.

origin: $(-x)^2 = (-y)^3 \Rightarrow x^2 = -y^3$, which is not the same as the original equation, so the equation is not symmetric with respect to the origin.

5. $y = -t^4 + 77t^2 + 324$

a. First, find the intercepts. If $t = 0$, then $y = 324$, so the y-intercept is $(0, 324)$. If $y = 0$, then we have
$$0 = -t^4 + 77t^2 + 324$$
$$t^4 - 77t^2 - 324 = 0$$
$$(t^2 - 81)(t^2 + 4) = 0$$
$$(t + 9)(t - 9)(t^2 + 4) = 0 \Rightarrow t = -9, 9, \pm 2i$$
So, the t-intercepts are $(-9, 0)$ and $(9, 0)$.
Next, check for symmetry.
t-axis: $-y = -t^4 + 77t^2 + 324$ is not the same as the original equation, so the equation is not symmetric with respect to the t-axis.
y-axis: $y = -(-t)^4 + 77(-t)^2 + 324 \Rightarrow$
$y = -t^4 + 77t^2 + 324$, which is the same as the original equation. So the graph is symmetric with respect to the y-axis.
origin: $-y = -(-t)^4 + 77(-t)^2 + 324 \Rightarrow$
$-y = -t^4 + 77t^2 + 324$, which is not the same as the original equation. So the graph is not symmetric with respect to the origin.
Now, make a table of values. Since the graph is symmetric with respect to the y-axis, if (t, y) is on the graph, then so is $(-t, y)$. However, the graph pertaining to the physical aspects of the problem consists only of those values for $t \geq 0$.

t	$y = -t^4 + 77t^2 + 324$	(t, y)
0	324	(0, 324)
1	400	(1, 400)
2	616	(2, 616)
3	936	(3, 936)
4	1300	(4, 1300)
5	1624	(5, 1624)
6	1800	(6, 1800)
7	1696	(7, 1696)
8	1156	(8, 1156)
9	0	(9, 0)

(continued on next page)

(*continued*)

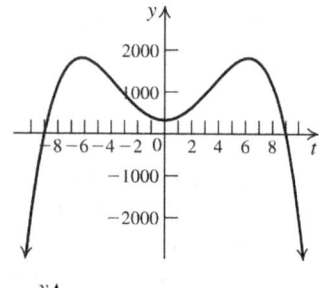

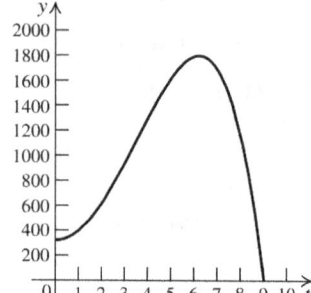

b.

c. The population becomes extinct after 9 years.

6. The standard form of the equation of a circle is $(x-h)^2 + (y-k)^2 = r^2$
$(h, k) = (3, -6)$ and $r = 10$
The equation of the circle is
$(x-3)^2 + (y+6)^2 = 100$.

7. $(x-2)^2 + (y+1)^2 = 36 \Rightarrow (h, k) = (2, -1), r = 6$
This is the equation of a circle with center $(2, -1)$ and radius 6.

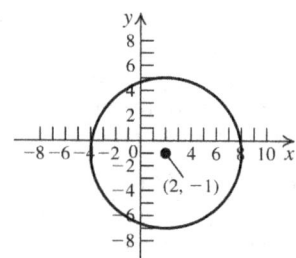

8. $x^2 + y^2 + 4x - 6y - 12 = 0 \Rightarrow$
$x^2 + 4x + y^2 - 6y = 12$
Now complete the square:
$x^2 + 4x + 4 + y^2 - 6y + 9 = 12 + 4 + 9 \Rightarrow$
$(x+2)^2 + (y-3)^2 = 25$
This is a circle with center $(-2, 3)$ and radius 5.

2.2 Concepts and Vocabulary

1. The graph of an equation in two variables, such as *x* and *y*, is the set of all ordered pairs (*a*, *b*) that satisfy the equation.

3. If $(0, -5)$ is a point of a graph, then -5 is a *y*- intercept of the graph.

5. False. The equation of a circle has both an x^2-term and a y^2-term. The given equation does not have a y^2-term.

7. False. The center of the circle with equation $(x+3)^2 + (y+4)^2 = 9$ is $(-3, -4)$.

2.2 Building Skills

In exercises 9–14, to determine if a point lies on the graph of the equation, substitute the point's coordinates into the equation to see if the resulting statement is true.

9. on the graph: $(-3, -4), (1, 0), (4, 3)$; not on the graph: $(2, 3)$

11. on the graph: $(3, 2), (0, 1), (8, 3)$; not on the graph: $(8, -3)$

13. on the graph: $(1, 0), (2, \sqrt{3}), (2, -\sqrt{3})$; not on the graph: $(0, -1)$

15.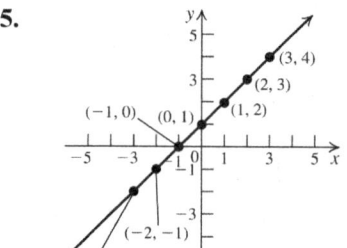

17.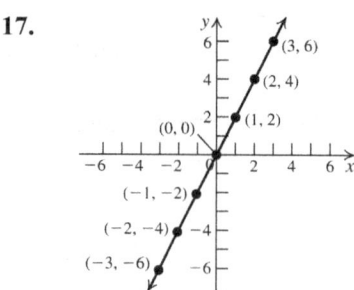

19.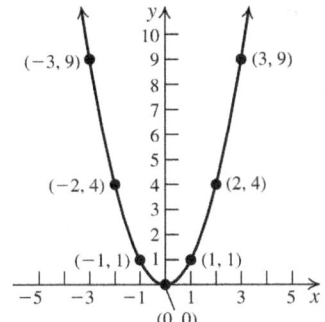

21.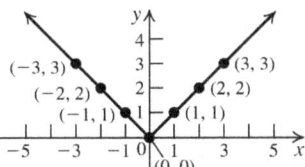

23.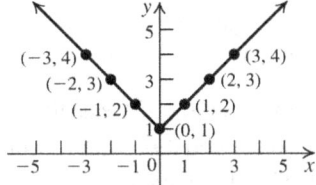

25.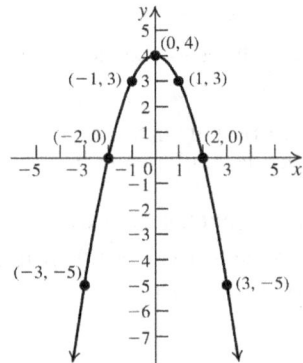

27.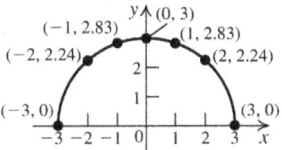

29.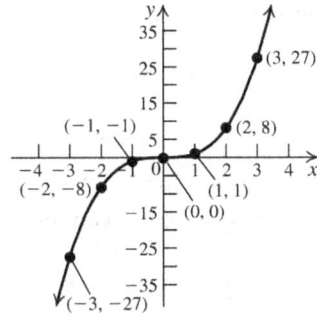

31.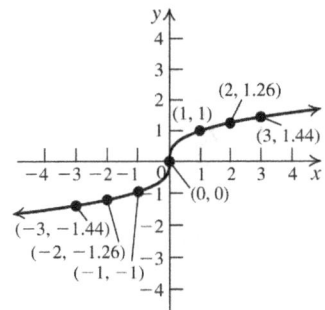

33.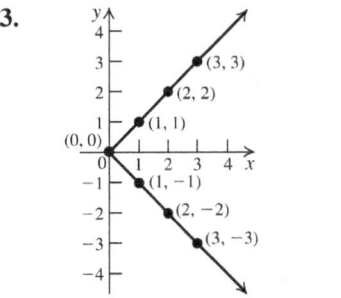

35.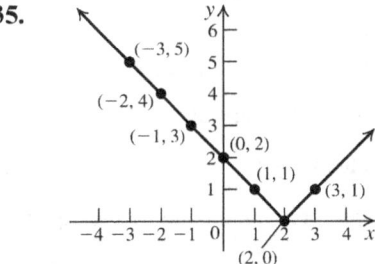

For exercises 37–46, read the answers directly from the given graphs.

37. x-intercepts: $-1, 1$
y-intercepts: none
symmetries: y-axis

39. x-intercepts: $-\pi, 0, \pi$
y-intercepts: 0
symmetries: origin

41. x-intercepts: $-3, 3$
y-intercepts: $-2, 2$
symmetries: x-axis, y-axis, origin

43. x-intercepts: $-2, 0, 2$
y-intercepts: 0
symmetries: origin

45. x-intercepts: $-2, 0, 2$
y-intercepts: 0, 3
symmetries: y-axis

47.

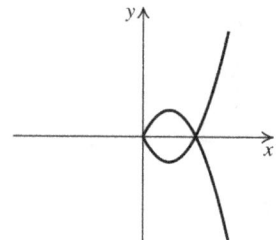

49.

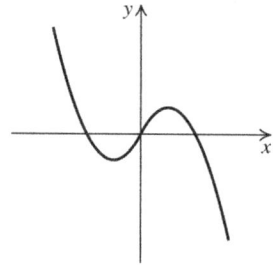

51. To find the x-intercept, let y = 0, and solve the equation for x: $3x + 4(0) = 12 \Rightarrow x = 4$. To find the y-intercept, let x = 0, and solve the equation for y: $3(0) + 4y = 12 \Rightarrow y = 3$. The x-intercept is 4; the y-intercept is 3.

53. To find the x-intercept, let y = 0, and solve the equation for x: $\frac{x}{5} + \frac{0}{3} = 1 \Rightarrow x = 5$. To find the y-intercept, let x = 0, and solve the equation for y: $\frac{0}{5} + \frac{y}{3} = 1 \Rightarrow y = 3$. The x-intercept is 5; the y-intercept is 3.

55. To find the x-intercept, let y = 0, and solve the equation for x: $0 = \frac{x+2}{x-1} \Rightarrow x = -2$. To find the y-intercept, let x = 0, and solve the equation for y: $y = \frac{0+2}{0-1} = -2$. The x-intercept is –2; the y-intercept is –2.

57. To find the x-intercept, let y = 0, and solve the equation for x: $0 = x^2 - 6x + 8 \Rightarrow x = 4$ or $x = 2$. To find the y-intercept, let x = 0, and solve the equation for y: $y = 0^2 - 6(0) + 8 \Rightarrow y = 8$. The x-intercepts are 2 and 4; the y-intercept is 8.

59. To find the x-intercept, let y = 0, and solve the equation for x: $x^2 + 0^2 = 4 \Rightarrow x = \pm 2$. To find the y-intercept, let x = 0, and solve the equation for y: $0^2 + y^2 = 4 \Rightarrow y = \pm 2$. The x-intercepts are –2 and 2; the y-intercepts are –2 and 2.

61. To find the x-intercept, let y = 0, and solve the equation for x: $0 = \sqrt{9 - x^2} \Rightarrow x = \pm 3$. To find the y-intercept, let x = 0, and solve the equation for y: $y = \sqrt{9 - 0^2} \Rightarrow y = 3$. The x-intercepts are –3 and 3; the y-intercept is 3.

63. To find the x-intercept, let y = 0, and solve the equation for x: $x(0) = 1 \Rightarrow$ no solution. To find the y-intercept, let x = 0, and solve the equation for y: $(0)y = 1 \Rightarrow$ no solution. There is no x-intercept; there is no y-intercept.

In exercises 65–74, to test for symmetry with respect to the x-axis, replace y with –y to determine if (x, –y) satisfies the equation. To test for symmetry with respect to the y-axis, replace x with –x to determine if (–x, y) satisfies the equation. To test for symmetry with respect to the origin, replace x with –x and y with –y to determine if (–x, –y) satisfies the equation.

65. $-y = x^2 + 1$ is not the same as the original equation, so the equation is not symmetric with respect to the x-axis.

$y = (-x)^2 + 1 \Rightarrow y = x^2 + 1$, so the equation is symmetric with respect to the y-axis.

$-y = (-x)^2 + 1 \Rightarrow -y = x^2 + 1$, is not the same as the original equation, so the equation is not symmetric with respect to the origin.

67. $-y = x^3 + x$ is not the same as the original equation, so the equation is not symmetric with respect to the x-axis.

$y = (-x)^3 - x \Rightarrow y = -x^3 - x \Rightarrow$

$y = -(x^3 + x)$ is not the same as the original equation, so the equation is not symmetric with respect to the y-axis.

$-y = (-x)^3 - x \Rightarrow -y = -x^3 - x \Rightarrow$

$-y = -(x^3 + x) \Rightarrow y = x^3 + x$, so the equation is symmetric with respect to the origin.

69. $-y = 5x^4 + 2x^2$ is not the same as the original equation, so the equation is not symmetric with respect to the x-axis.

$y = 5(-x)^4 + 2(-x)^2 \Rightarrow y = 5x^4 + 2x^2$, so the equation is symmetric with respect to the y-axis.

$-y = 5(-x)^4 + 2(-x) \Rightarrow -y = 5x^4 + 2x^2$ is not the same as the original equation, so the equation is not symmetric with respect to the origin.

71. $-y = -3x^5 + 2x^3$ is not the same as the original equation, so the equation is not symmetric with respect to the x-axis.
$y = -3(-x)^5 + 2(-x)^3 \Rightarrow y = 3x^5 - 2x^3$ is not the same as the original equation, so the equation is not symmetric with respect to the y-axis.
$-y = -3(-x)^5 + 2(-x)^3 \Rightarrow -y = 3x^5 - 2x^3 \Rightarrow$
$-y = -(-3x^5 + 2x^3) \Rightarrow y = -3x^5 + 2x^3$, so the equation is symmetric with respect to the origin.

73. $x^2(-y)^2 + 2x(-y) = 1 \Rightarrow x^2y^2 - 2xy = 1$ is not the same as the original equation, so the equation is not symmetric with respect to the x-axis.
$(-x)^2 y^2 + 2(-x)y = 1 \Rightarrow x^2 y^2 - 2xy = 1$ is not the same as the original equation, so the equation is not symmetric with respect to the y-axis.
$(-x)^2(-y)^2 + 2(-x)(-y) = 1 \Rightarrow$
$x^2 y^2 + 2xy = 1$, so the equation is symmetric with respect to the origin.

For exercises 75–78, use the standard form of the equation of a circle, $(x-h)^2 + (y-k)^2 = r^2$.

75. Center (2, 3); radius = 6

77. Center (–2, –3); radius = $\sqrt{11}$

79. $x^2 + (y-1)^2 = 4$

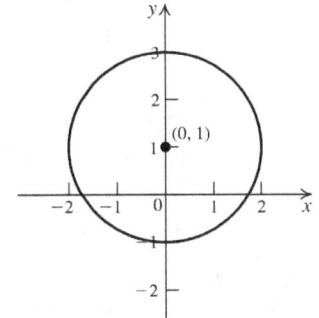

81. $(x+1)^2 + (y-2)^2 = 2$

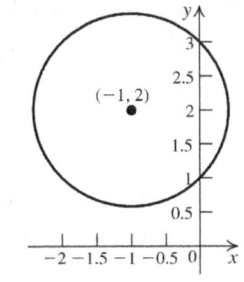

83. Find the radius by using the distance formula:
$d = \sqrt{(-1-3)^2 + (5-(-4))^2} = \sqrt{97}$.
The equation of the circle is
$(x-3)^2 + (y+4)^2 = 97$.

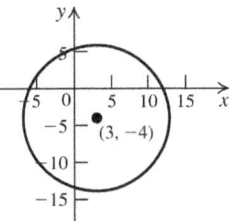

85. The circle touches the x-axis, so the radius is 2. The equation of the circle is
$(x-1)^2 + (y-2)^2 = 4$.

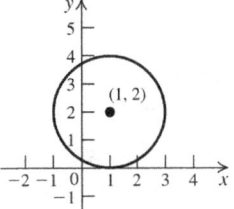

87. Find the diameter by using the distance formula:
$d = \sqrt{(-3-7)^2 + (6-4)^2} = \sqrt{104} = 2\sqrt{26}$.
So the radius is $\sqrt{26}$. Use the midpoint formula to find the center:
$M = \left(\dfrac{7+(-3)}{2}, \dfrac{4+6}{2}\right) = (2,5)$. The equation of the circle is $(x-2)^2 + (y-5)^2 = 26$.

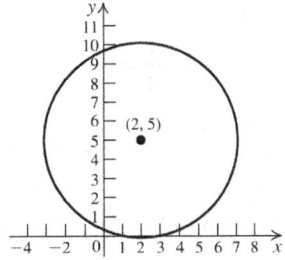

89. a. $x^2 + y^2 - 2x - 2y - 4 = 0 \Rightarrow$
$x^2 - 2x + y^2 - 2y = 4$
Now complete the square:
$x^2 - 2x + 1 + y^2 - 2y + 1 = 4 + 1 + 1 \Rightarrow$
$(x-1)^2 + (y-1)^2 = 6$. This is a circle with center (1, 1) and radius $\sqrt{6}$.

b. To find the x-intercepts, let $y = 0$ and solve for x:
$(x-1)^2 + (0-1)^2 = 6 \Rightarrow (x-1)^2 + 1 = 6 \Rightarrow$
$(x-1)^2 = 5 \Rightarrow x-1 = \pm\sqrt{5} \Rightarrow x = 1 \pm \sqrt{5}$
Thus, the x-intercepts are $\left(1+\sqrt{5}, 0\right)$ and $\left(1-\sqrt{5}, 0\right)$.
To find the y-intercepts, let $x = 0$ and solve for y:
$(0-1)^2 + (y-1)^2 = 6 \Rightarrow 1 + (y-1)^2 = 6 \Rightarrow$
$(y-1)^2 = 5 \Rightarrow y - 1 = \pm\sqrt{5} \Rightarrow y = 1 \pm \sqrt{5}$
Thus, the y-intercepts are $\left(0, 1+\sqrt{5}\right)$ and $\left(0, 1-\sqrt{5}\right)$.

91. a. $2x^2 + 2y^2 + 4y = 0 \Rightarrow$
$2(x^2 + y^2 + 2y) = 0 \Rightarrow x^2 + y^2 + 2y = 0$.
Now complete the square:
$x^2 + y^2 + 2y + 1 = 0 + 1 \Rightarrow x^2 + (y+1)^2 = 1$.
This is a circle with center $(0, -1)$ and radius 1.

b. To find the x-intercepts, let $y = 0$ and solve for x: $x^2 + (0+1)^2 = 1 \Rightarrow x^2 = 0 \Rightarrow x = 0$
Thus, the x-intercept is $(0, 0)$.
To find the y-intercepts, let $x = 0$ and solve for y:
$0^2 + (y+1)^2 = 1 \Rightarrow y + 1 = \pm 1 \Rightarrow y = 0, -2$
Thus, the y-intercepts are $(0, 0)$ and $(0, -2)$.

93. a. $x^2 + y^2 - x = 0 \Rightarrow x^2 - x + y^2 = 0$.
Now complete the square:
$x^2 - x + \frac{1}{4} + y^2 = 0 + \frac{1}{4} \Rightarrow$
$\left(x - \frac{1}{2}\right)^2 + y^2 = \frac{1}{4}$. This is a circle with center $\left(\frac{1}{2}, 0\right)$ and radius $\frac{1}{2}$.

b. To find the x-intercepts, let $y = 0$ and solve for x: $\left(x - \frac{1}{2}\right)^2 + 0^2 = \frac{1}{4} \Rightarrow x - \frac{1}{2} = \pm \frac{1}{2} \Rightarrow$
$x = 0, 1$. Thus, the x-intercepts are $(0, 0)$ and $(1, 0)$. To find the y-intercepts, let $x = 0$ and solve for y.

$\left(0 - \frac{1}{2}\right)^2 + y^2 = \frac{1}{4} \Rightarrow y^2 + \frac{1}{4} = \frac{1}{4} \Rightarrow$
$y^2 = 0 \Rightarrow y = 0$.
Thus, the y-intercept is $(0, 0)$.

2.2 Applying the Concepts

95. The distance from $P(x, y)$ to the x-axis is $|x|$ while the distance from P to the y-axis is $|y|$. So the equation of the graph is $|x| = |y|$.

97. If you save \$100 each month, it will take 24 months (or two years) to save \$2400. So, the graph starts at $(0, 0)$ and increases to $(2, 2400)$. It will take another 30 months (or 2.5) years to withdraw \$80 per month until the \$2400 is gone. Thus, the graph passes through $(4.5, 0)$.

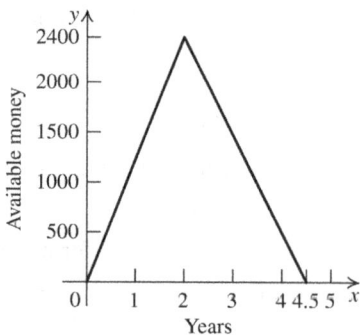

99. a. July 2018 is represented by $t = 0$, so March 2018 is represented by $t = -4$. The monthly profit for March is determined by
$P = -0.5(-4)^2 - 3(-4) + 8 = \$12$ million.

b. July 2018 is represented by $t = 0$, so October 2018 is represented by $t = 3$. So the monthly profit for October is determined by
$P = -0.5(3)^2 - 3(3) + 8 = -\$5.5$ million. This is a loss.

c.

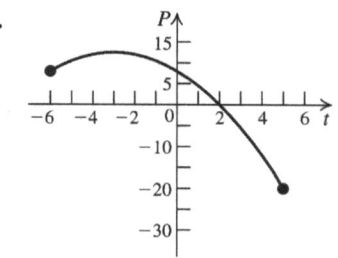

Because $t = 0$ represents July 2018, $t = -6$ represents January 2018, and $t = 5$ represents December 2018.

d. To find the *t*-intercept, set $P = 0$ and solve for *t*: $0 = -0.5t^2 - 3t + 8 \Rightarrow$
$$t = \frac{3 \pm \sqrt{(-3)^2 - 4(-0.5)(8)}}{2(-0.5)} = \frac{3 \pm \sqrt{25}}{-1}$$
$= 2$ or -8

The *t*-intercepts represent the months with no profit and no loss. In this case, $t = -8$ makes no sense in terms of the problem, so we disregard this solution. $t = 2$ represents Sept 2018.

e. To find the *P*-intercept, set $t = 0$ and solve for *P*: $P = -0.5(0)^2 - 3(0) + 8 \Rightarrow P = 8$.
The *P*-intercept represents the profit in July 2018.

101. a.

t	Height = $-16t^2 + 128t + 320$
0	320 feet
1	432 feet
2	512 feet
3	560 feet
4	576 feet
5	560 feet
6	512 feet

b.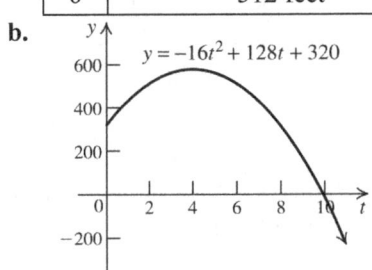

c. $0 \le t \le 10$

d. To find the *t*-intercept, set $y = 0$ and solve for *t*:
$0 = -16t^2 + 128t + 320 \Rightarrow$
$0 = -16(t^2 - 8t - 20) \Rightarrow$
$0 = (t - 10)(t + 2) \Rightarrow t = 10$ or $t = -2$.
The graph does not apply if $t < 0$, so the *t*-intercept is 10. This represents the time when the object hits the ground. To find the *y*-intercept, set $t = 0$ and solve for *y*:
$y = -16(0)^2 + 128(0) + 320 \Rightarrow y = 320$.
This represents the height of the building.

2.2 Beyond the Basics

103. $x^2 + y^2 - 4x + 2y - 20 = 0 \Rightarrow$
$x^2 - 4x + y^2 + 2y = 20 \Rightarrow$
$x^2 - 4x + 4 + y^2 + 2y + 1 = 20 + 4 + 1 \Rightarrow$
$(x-2)^2 + (y+1)^2 = 25$

So this is the graph of a circle with center $(2, -1)$ and radius 5. The area of this circle is 25π. $x^2 + y^2 - 4x + 2y - 31 = 0 \Rightarrow$
$x^2 - 4x + y^2 + 2y = 31 \Rightarrow$
$x^2 - 4x + 4 + y^2 + 2y + 1 = 31 + 4 + 1 \Rightarrow$
$(x-2)^2 + (y+1)^2 = 36$

So, this is the graph of a circle with center $(2, -1)$ and radius 6. The area of this circle is 36π. Both circles have the same center, so the area of the region bounded by the two circles equals $36\pi - 25\pi = 11\pi$.

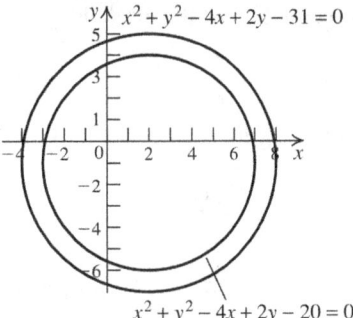

2.2 Critical Thinking/Discussion/Writing

105. The graph of $y^2 = 2x$ is the union of the graphs of $y = \sqrt{2x}$ and $y = -\sqrt{2x}$.

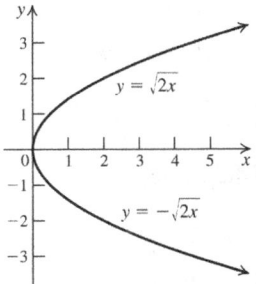

106. Let (x, y) be a point on the graph. The graph is symmetric with regard to the x-axis, so the point $(x, -y)$ is also on the graph. Because the graph is symmetric with regard to the y-axis, the point $(-x, y)$ is also on the graph. Therefore the point $(-x, -y)$ is on the graph, and the graph is symmetric with respect to the origin. The graph of $y = x^3$ is an example of a graph that is symmetric with respect to the origin but is not symmetric with respect to the x- and y-axes.

107. a. First find the radius of the circle:
$$d(A, B) = \sqrt{(6-0)^2 + (8-1)^2} = \sqrt{85} \Rightarrow$$
$$r = \frac{\sqrt{85}}{2}.$$
The center of the circle is
$$\left(\frac{6+0}{2}, \frac{1+8}{2}\right) = \left(3, \frac{9}{2}\right).$$
So the equation of the circle is
$$(x-3)^2 + \left(y - \frac{9}{2}\right)^2 = \frac{85}{4}.$$
To find the x-intercepts, set $y = 0$, and solve for x:
$$(x-3)^2 + \left(0 - \frac{9}{2}\right)^2 = \frac{85}{4} \Rightarrow$$
$$(x-3)^2 + \frac{81}{4} = \frac{85}{4} \Rightarrow x^2 - 6x + 9 = 1 \Rightarrow$$
$$x^2 - 6x + 8 = 0$$
The x-intercepts are the roots of this equation.

b. First find the radius of the circle:
$$d(A, B) = \sqrt{(a-0)^2 + (b-1)^2}$$
$$= \sqrt{a^2 + (b-1)^2} \Rightarrow$$
$$r = \frac{\sqrt{a^2 + (b-1)^2}}{2}.$$
The center of the circle is
$$\left(\frac{a+0}{2}, \frac{b+1}{2}\right) = \left(\frac{a}{2}, \frac{b+1}{2}\right)$$
So the equation of the circle is
$$\left(x - \frac{a}{2}\right)^2 + \left(y - \frac{b+1}{2}\right)^2 = \frac{a^2 + (b-1)^2}{4}.$$

To find the x-intercepts, set $y = 0$ and solve for x:
$$\left(x - \frac{a}{2}\right)^2 + \left(0 - \frac{b+1}{2}\right)^2 = \frac{a^2 + (b-1)^2}{4}$$
$$x^2 - ax + \frac{a^2}{4} + \frac{(b+1)^2}{4} = \frac{a^2 + (b-1)^2}{4}$$
$$4x^2 - 4ax + a^2 + b^2 + 2b + 1 = a^2 + b^2 - 2b + 1$$
$$4x^2 - 4ax + 4b = 0$$
$$x^2 - ax + b = 0$$
The x-intercepts are the roots of this equation.

c. $a = 3$ and $b = 1$. Approximate the roots of the equation by drawing a circle whose diameter has endpoints $A(0, 1)$ and $B(3, 1)$. The center of the circle is $\left(\frac{3}{2}, 1\right)$ and the radius is $\frac{3}{2}$. The roots are approximately $(0.4, 0)$ and $(2.6, 0)$.

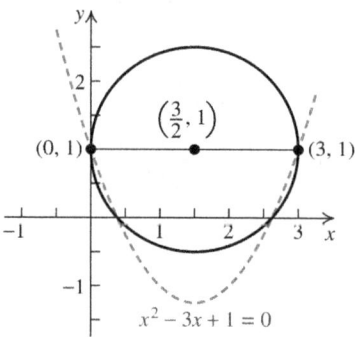

108. a. The coordinates of the center of each circle are (r, r) and $(3r, r)$.

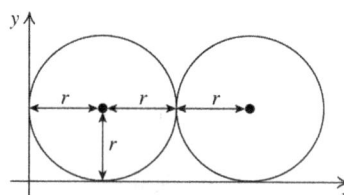

b. To find the area of the shaded region, first find the area of the rectangle shown in the figure below, and then subtract the sum of the areas of the two sectors, A and B.

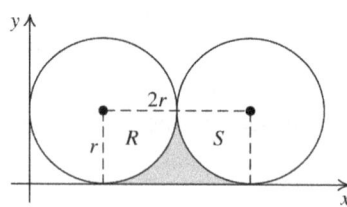

(continued on next page)

(continued)

$$A_{\text{rect}} = r(2r) = 2r^2$$
$$A_{\text{sector } R} = A_{\text{sector } S} = \frac{1}{4}\pi r^2$$
$$A_{\text{shaded region}} = 2r^2 - \left(\frac{1}{4}\pi r^2 + \frac{1}{4}\pi r^2\right)$$
$$= 2r^2 - \frac{1}{2}\pi r^2 = \left(2 - \frac{\pi}{2}\right)r^2$$

2.2 Getting Ready for the Next Section

109. $\dfrac{5-3}{6-2} = \dfrac{2}{4} = \dfrac{1}{2}$

111. $\dfrac{2-(-3)}{3-13} = \dfrac{5}{-10} = -\dfrac{1}{2}$

113. $\dfrac{\frac{1}{2}-\frac{1}{4}}{\frac{3}{8}-\left(-\frac{1}{4}\right)} = \dfrac{\frac{1}{4}}{\frac{5}{8}} = \dfrac{1}{4} \cdot \dfrac{8}{5} = \dfrac{2}{5}$

115. $-\dfrac{1}{2}$ 117. $\dfrac{3}{2}$

119. $-\dfrac{2+\frac{3}{4}}{1-\frac{1}{2}} = -\dfrac{\frac{11}{4}}{\frac{1}{2}} = -\dfrac{11}{4} \cdot 2 = -\dfrac{11}{2}$

121. $2x + 3y = 6 \Rightarrow 3y = -2x + 6 \Rightarrow y = -\dfrac{2}{3}x + 2$

123. $y - 2 - \dfrac{2}{3}(x+1) = 0 \Rightarrow y - 2 = \dfrac{2}{3}(x+1) \Rightarrow$
$y = \dfrac{2}{3}x + \dfrac{2}{3} + 2 = \dfrac{2}{3}x + \dfrac{8}{3}$

2.3 Lines

2.3 Practice Problems

1. $m = \dfrac{-3-5}{-6-(-7)} = -\dfrac{8}{13}$

    A slope of $-\dfrac{8}{13}$ means that the value of y decreases 8 units for every 13 units increase in x.

2. $P(-2, -3)$, $m = -\dfrac{2}{3}$
$$y - (-3) = -\dfrac{2}{3}\left[x - (-2)\right]$$
$$y + 3 = -\dfrac{2}{3}(x + 2)$$
$$y + 3 = -\dfrac{2}{3}x - \dfrac{4}{3} \Rightarrow y = -\dfrac{2}{3}x - \dfrac{13}{3}$$

3. $m = \dfrac{-4-6}{-3-(-1)} = \dfrac{-10}{-2} = 5$

    Use either point to determine the equation of the line. Using $(-3, -4)$, we have
    $y - (-4) = 5\left[x - (-3)\right] \Rightarrow y + 4 = 5(x+3) \Rightarrow$
    $y + 4 = 5x + 15 \Rightarrow y = 5x + 11$
    Using $(-1, 6)$, we have
    $y - 6 = 5\left[x - (-1)\right] \Rightarrow y - 6 = 5(x+1) \Rightarrow$
    $y - 6 = 5x + 5 \Rightarrow y = 5x + 11$

4. $y - y_1 = m(x - x_1) \Rightarrow y - (-3) = 2(x - 0)$
   point-slope form
   $y - (-3) = 2(x - 0) \Rightarrow$
   $y + 3 = 2x \Rightarrow y = 2x - 3$

5. The slope is $-\dfrac{2}{3}$ and the y-intercept is 4. The line goes through $(0, 4)$, so locate a second point by moving two units down and three units right. Thus, the line goes through $(3, 2)$.

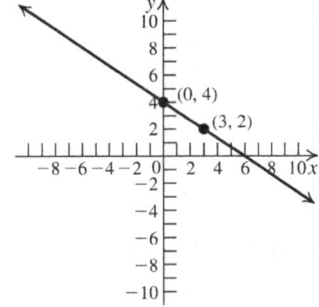

6. $x = -3$. The slope is undefined, and there is no y-intercept. The x-intercept is -3.
$y = 7$. The slope is 0, and the y-intercept is 7.

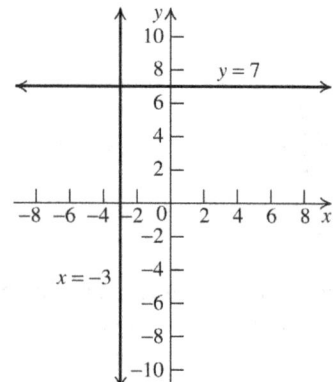

7. First, solve for y to write the equation in slope-intercept form:
$3x + 4y = 24 \Rightarrow 4y = -3x + 24 \Rightarrow$
$y = -\dfrac{3}{4}x + 6$. The slope is $-\dfrac{3}{4}$, and the y-intercept is 6. Find the x-intercept by setting $y = 0$ and solving the equation for x:
$0 = -\dfrac{3}{4}x + 6 \Rightarrow 6 = \dfrac{3}{4}x \Rightarrow 8 = x$. Thus, the graph passes through the points $(0, 6)$ and $(8, 0)$.

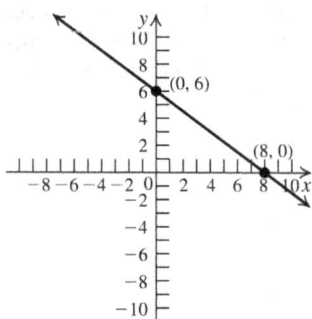

8. Use the equation $H = 2.6x + 65$.
$H_1 = 2.6(43) + 65 = 176.8$
$H_2 = 2.6(44) + 65 = 179.4$
The person is between 176.8 cm and 179.4 cm tall, or 1.768 m and 1.794 m.

9. a. Parallel lines have the same slope, so the slope of the line is $m = \dfrac{3-7}{2-5} = \dfrac{-4}{-3} = \dfrac{4}{3}$.
Using the point-slope form, we have
$y - 5 = \dfrac{4}{3}[x - (-2)] \Rightarrow 3y - 15 = 4(x + 2) \Rightarrow$
$3y - 15 = 4x + 8 \Rightarrow 4x - 3y + 23 = 0$

b. The slopes of perpendicular lines are negative reciprocals. Write the equation $4x + 5y + 1 = 0$ in slope-intercept form to find its slope: $4x + 5y + 1 = 0 \Rightarrow$
$5y = -4x - 1 \Rightarrow y = -\dfrac{4}{5}x - \dfrac{1}{5}$. The slope of a line perpendicular to this lines is $\dfrac{5}{4}$.
Using the point-slope form, we have
$y - (-4) = \dfrac{5}{4}(x - 3) \Rightarrow 4(y + 4) = 5(x - 3) \Rightarrow$
$4y + 16 = 5x - 15 \Rightarrow 5x - 4y - 31 = 0$

10. Because 2016 is 10 years after 2006, set $x = 10$. Then $y = 0.44(10) + 6.70 = 11.1$
There were 11.1 million registered motorcycles in the U.S. in 2016.

2.3 Concepts and Vocabulary

1. The slope of a horizontal line is $\underline{0}$; the slope of a vertical line is $\underline{\text{undefined}}$.

3. Every line parallel to the line $y = 3x - 2$ has slope, m, equal to $\underline{3}$.

5. False. The slope of the line $y = -\tfrac{1}{4}x + 5$ is equal to $-\tfrac{1}{4}$.

7. True

2.3 Building Skills

9. $m = \dfrac{7-3}{4-1} = \dfrac{4}{3}$; the graph is rising.

11. $m = \dfrac{-2-(-2)}{-2-6} = \dfrac{0}{-8} = 0$; the graph is horizontal.

13. $m = \dfrac{-3.5-2}{3-0.5} = \dfrac{-5.5}{2.5} = -2.2$; the graph is falling.

15. $m = \dfrac{5-1}{(1+\sqrt{2})-\sqrt{2}} = \dfrac{4}{1} = 4$; the graph is rising.

17. ℓ_3 **19.** ℓ_4

21. ℓ_1 passes through the points $(2, 3)$ and $(-5, 4)$. $m_{\ell_1} = \dfrac{-4-3}{-5-2} = \dfrac{-7}{-7} = 1$.

Section 2.3 Lines 107

23. ℓ_3 passes through the points (2, 3) and (0, −1). $m_{\ell_3} = \dfrac{-1-3}{0-2} = 2$.

25. $(0, 5); m = 3$
$y = 3x + 5$

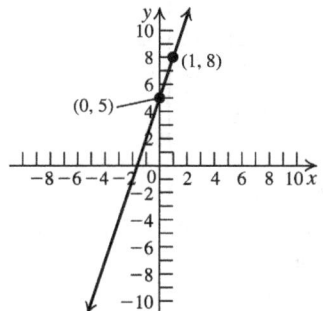

27. $y = \dfrac{1}{2}x + 4$

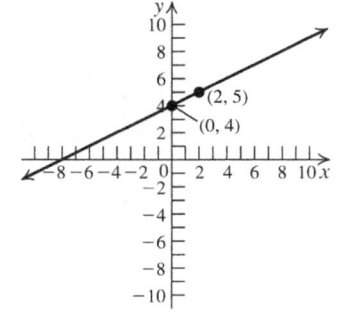

29. $y - 1 = -\dfrac{3}{2}(x - 2) \Rightarrow y - 1 = -\dfrac{3}{2}x + 3 \Rightarrow$
$y = -\dfrac{3}{2}x + 4$

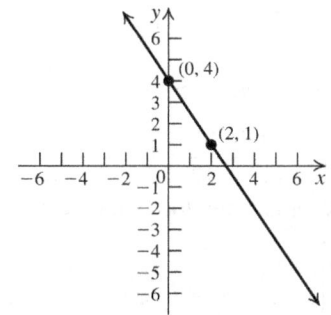

31. $y + 4 = 0(x - 5) \Rightarrow y + 4 = 0 \Rightarrow y = -4$

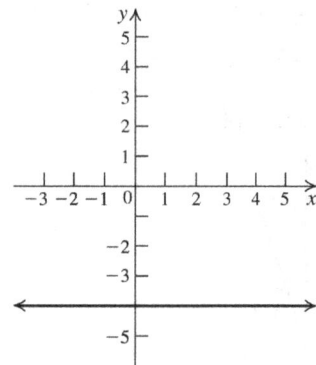

33. $m = \dfrac{0-1}{1-0} = -1$. The y-intercept is $(0, 1)$, so the equation is $y = -x + 1$.

35. $m = \dfrac{3-3}{3-(-1)} = 0$ Because the slope = 0, the line is horizontal. Its equation is $y = 3$.

37. $m = \dfrac{1-(-1)}{1-(-2)} = \dfrac{2}{3}$. Now write the equation in point-slope form, and then solve for y to write the equation in slope-intercept form.
$y + 1 = \dfrac{2}{3}(x + 2) \Rightarrow y + 1 = \dfrac{2}{3}x + \dfrac{4}{3} \Rightarrow$
$y = \dfrac{2}{3}x + \dfrac{1}{3}$

39. $m = \dfrac{2 - \dfrac{1}{4}}{0 - \dfrac{1}{2}} = \dfrac{\dfrac{7}{4}}{-\dfrac{1}{2}} = -\dfrac{7}{2}$. Now write the equation in point-slope form, and then solve for y to write the equation in slope-intercept form. $y - 2 = -\dfrac{7}{2}x \Rightarrow y = -\dfrac{7}{2}x + 2$

41. $x = 5$ 43. $y = 0$

45. $y = 14$ 47. $y = -\dfrac{2}{3}x - 4$

49. $m = \dfrac{4-0}{0-(-3)} = \dfrac{4}{3}; y = \dfrac{4}{3}x + 4$

51. $y = 7$ 53. $y = -5$

55. $y = 3x - 2$

The slope is 3 and the y-intercept is $(0, -2)$.

$0 = 3x - 2 \Rightarrow 3x = 2 \Rightarrow x = \dfrac{2}{3}$

The x-intercept is $\left(\dfrac{2}{3}, 0\right)$.

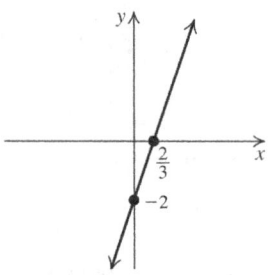

57. $x + 2y - 4 = 0 \Rightarrow 2y = -x + 4 \Rightarrow y = -\dfrac{1}{2}x + 2$.

The slope is $-1/2$, and the y-intercept is $(0, 2)$. To find the x-intercept, set $y = 0$ and solve for x: $x + 2(0) - 4 = 0 \Rightarrow x = 4$.

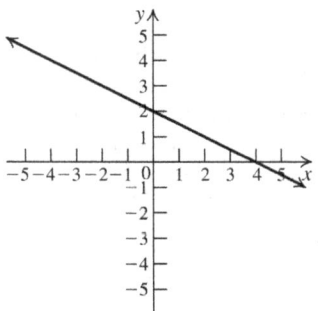

59. $3x - 2y + 6 = 0 \Rightarrow 3x + 6 = 2y \Rightarrow \dfrac{3}{2}x + 3 = y$.

The slope is $3/2$, and the y-intercept is $(0, 3)$. To find the x-intercept, set $y = 0$ and solve for x: $3x - 2(0) + 6 = 0 \Rightarrow 3x = -6 \Rightarrow x = -2$.

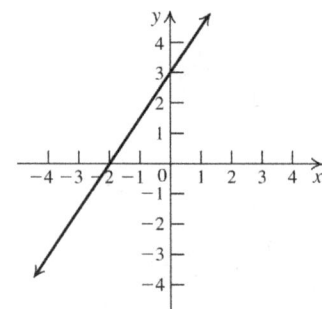

61. $x - 5 = 0 \Rightarrow x = 5$. The slope is undefined, and there is no y-intercept. The x-intercept is 5.

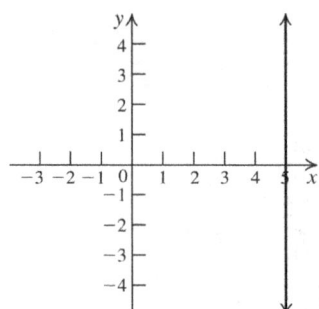

63. $x = 0$. The slope is undefined, and the y-intercepts are the y-axis. This is a vertical line whose x-intercept is 0.

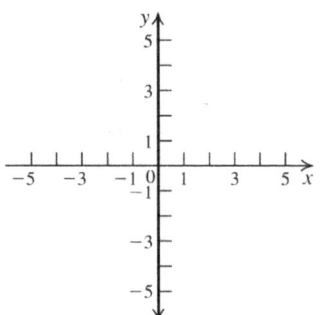

For exercises 65–68, the two-interecepts form of the equation of a line is $\dfrac{x}{a} + \dfrac{y}{b} = 1$.

65. $\dfrac{x}{4} + \dfrac{y}{3} = 1$

67. $2x + 3y = 6 \Rightarrow \dfrac{2x}{6} + \dfrac{3y}{6} = \dfrac{6}{6} \Rightarrow \dfrac{x}{3} + \dfrac{y}{2} = 1$;

x-intercept = 3; y-intercept = 2

69. $2x - 3y = 12 \Rightarrow \dfrac{2x}{12} - \dfrac{3y}{12} = 1 \Rightarrow \dfrac{x}{6} - \dfrac{y}{4} = 1$

The x-intercept is 6 and the y-intercept is –4.

71. $-5x+2y=10 \Rightarrow -\dfrac{5x}{10}+\dfrac{2y}{10}=1 \Rightarrow -\dfrac{x}{2}+\dfrac{y}{5}=1$

The x-intercept is –2 and the y-intercept is 5.

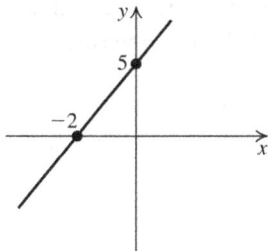

73. $m=\dfrac{9-4}{7-2}=\dfrac{5}{5}=1$. The equation of the line through (2, 4) and (7, 9) is $y-4=1(x-2) \Rightarrow y=x+2$. Check to see if (–1, 1) satisfies the equation by substituting $x=-1$ and $y=1$: $1=-1+2 \Rightarrow 1=1$. So (–1, 1) lies on the line.

75. The given line passes through the points (0, 3) and (4, 0), so its slope is $-\tfrac{3}{4}$. Any line parallel to this line will have the same slope. The line that passes through the origin and is parallel to the given line has equation $y=-\tfrac{3}{4}x$.

77. The red line passes through the points (–2, 0) and (0, 3), so its slope is $\tfrac{3}{2}$. The blue line passes through (4, 2) and has the same slope, so its equation is
$y-2=\tfrac{3}{2}(x-4) \Rightarrow 2y-4=3x-12 \Rightarrow 2y=3x-8 \Rightarrow y=\tfrac{3}{2}x-4$

79. The slope of $y=3x-1$ is 3. The slope of $y=3x+2$ is also 3. The lines are parallel.

81. The slope of $y=2x-4$ is 2. The slope of $y=-\tfrac{1}{2}x+4$ is $-\tfrac{1}{2}$. The lines are perpendicular.

83. The slope of $3x+8y=7$ is $-3/8$, while the slope of $5x-7y=0$ is $5/7$. The lines are neither parallel nor perpendicular.

85. The slope of $x=4y+8$ is $1/4$. The slope of $y=-4x+1$ is -4, so the lines are perpendicular.

87. Both lines are vertical lines. The lines are parallel.

89. The equation of the line through (2, –3) with slope 3 is
$y+3=3(x-2) \Rightarrow y+3=3x-6 \Rightarrow y=3x-9$.

91. A line perpendicular to a line with slope $-\tfrac{1}{2}$ has slope 2. The equation of the line through (–1, 2) with slope 2 is $y-2=2(x-(-1)) \Rightarrow$
$y-2=2(x+1) \Rightarrow y-2=2x+2 \Rightarrow$
$y=2x+4$.

93. The slope of the line joining (1, –2) and (–3, 2) is $\dfrac{2-(-2)}{-3-1}=-1$. The equation of the line through (–2, –5) with slope –1 is
$y-(-5)=-(x-(-2)) \Rightarrow y+5=-(x+2) \Rightarrow$
$y+5=-x-2 \Rightarrow y=-x-7$.

95. The slope of the line joining (–3, 2) and (–4, –1) is
$\dfrac{-1-2}{-4-(-3)}=3$.

A line perpendicular to this line has slope $-\tfrac{1}{3}$. The equation of the line through (1, –2) with slope $-\tfrac{1}{3}$ is
$y-(-2)=-\dfrac{1}{3}(x-1) \Rightarrow 3(y+2)=-(x-1) \Rightarrow$
$3y+6=-x+1 \Rightarrow 3y=-x-5 \Rightarrow$
$y=-\dfrac{1}{3}x-\dfrac{5}{3}$.

97. The slope of the line $y=6x+5$ is 6. The lines are parallel, so the slope of the new line is also 6. The equation of the line with slope 6 and y-intercept 4 is $y=6x+4$.

99. The slope of the line $y=6x+5$ is 6. The lines are perpendicular, so the slope of the new line is $-\tfrac{1}{6}$. The equation of the line with slope $-\tfrac{1}{6}$ and y-intercept 4 is $y=-\tfrac{1}{6}x+4$.

101. The slope of $x+y=1$ is –1. The lines are parallel, so they have the same slope. The equation of the line through (1, 1) with slope –1 is $y-1=-(x-1) \Rightarrow y-1=-x+1 \Rightarrow$
$y=-x+2$.

103. The slope of $3x - 9y = 18$ is $1/3$. The lines are perpendicular, so the slope of the new line is -3. The equation of the line through $(-2, 4)$ with slope -3 is $y - 4 = -3(x - (-2)) \Rightarrow$
$y - 4 = -3x - 6 \Rightarrow y = -3x - 2$.

2.3 Applying the Concepts

105. a. The y-intercept represents the initial expenses.

   b. The x-intercept represents the point at which the teacher breaks even, i.e., the expenses equal the income.

   c. The teacher's profit if there are 16 students in the class is $640.

   d. The slope of the line is
   $$\frac{640 - (-750)}{16 - 0} = \frac{1390}{16} = \frac{695}{8}$$
   The equation of the line is
   $P = \dfrac{695}{8} n - 750$.

107. slope $= \dfrac{\text{rise}}{\text{run}} \Rightarrow \dfrac{4}{40} = \dfrac{1}{10}$

109. 8 in. in two weeks $\Rightarrow$ the plant grows 4 in. per week. John wants to trim the hedge when it grows 6 in., so he should trim it every $\frac{6}{4} = 1.5$ weeks ≈ 10 days.

111. a. $x =$ the number of weeks; $y =$ the amount of money in the account after x weeks; $y = 7x + 130$

   b. The slope is the amount of money deposited each week; the y-intercept is the initial deposit.

113. a. $x =$ the number of months owed to pay off the refrigerator; $y =$ the amount owed; $y = -15x + 600$

   b. The slope is the amount that the balance due changes per month; the y-intercept is the initial amount owed.

115. a. $x =$ the number of years after 2010; $y =$ the life expectancy of a female born in the year $2010 + x$; $y = 0.17x + 80.8$

   b. The slope is the rate of increase in life expectancy; the y-intercept is the current life expectancy.

117. There are 30 days in June. For the first 13 days, you used data at a rate of $\dfrac{435}{13} \approx 33.5$ MB/day. At the same rate, you will use $33.5(17) = 569.5$ MB for the rest of the month.
$435 + 569.5 = 1004.5$
So, you don't need to buy extra data. You will have about 20 MB left.

119. $y = 5x + 40,000$

121. a. The two points are $(100, 212)$ and $(0, 32)$. So the slope is $\dfrac{212 - 32}{100 - 0} = \dfrac{180}{100} = \dfrac{9}{5}$.
The equation is
$F - 32 = \dfrac{9}{5}(C - 0) \Rightarrow F = \dfrac{9}{5} C + 32$

   b. One degree Celsius change in the temperature equals $9/5$ degrees change in degrees Fahrenheit.

   c.
C	$F = \dfrac{9}{5} C + 32$
40°C	104°F
25°C	77°F
−5°C	23°F
−10°C	14°F

   d. $100°F = \dfrac{9}{5} C + 32 \Rightarrow C = 37.78°C$
   $90°F = \dfrac{9}{5} C + 32 \Rightarrow C = 32.22°C$
   $75°F = \dfrac{9}{5} C + 32 \Rightarrow C = 23.89°C$
   $-10°F = \dfrac{9}{5} C + 32 \Rightarrow C = -23.33°C$
   $-20°F = \dfrac{9}{5} C + 32 \Rightarrow C = -28.89°C$

   e. $97.6°F = \dfrac{9}{5} C + 32 \Rightarrow C = 36.44°C$;
   $99.6°F = \dfrac{9}{5} C + 32 \Rightarrow C = 37.56°C$

   f. Let $x = °F = °C$. Then $x = \dfrac{9}{5} x + 32 \Rightarrow$
   $-\dfrac{4}{5} x = 32 \Rightarrow x = -40$. At $-40°$, °F = °C.

123. a. The year 2005 is represented by $t = 0$, and the year 2011 is represented by $t = 6$. The points are (0, 2425) and (6, 4026). So the slope is $\dfrac{4026 - 2425}{6} \approx 266.8$ The equation is $y - 2425 = 266.8(t - 0) \Rightarrow$
$y = 266.8t + 2425$

b.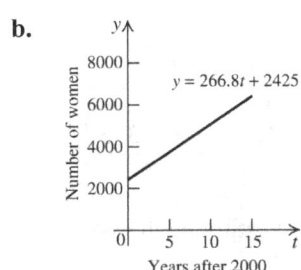

c. The year 2008 is represented by $t = 3$. So $y = 266.8(3) + 2425 \Rightarrow y = 3225.4$. Note that there cannot be a fraction of a person, so. there were 3225 women prisoners in 2008.

d. The year 2017 is represented by $t = 12$. So $y = 266.8(12) + 2425 \Rightarrow y = 5626.6$. There will be 5627 women prisoners in 2017.

125. The independent variable t represents the number of years after 2000, with $t = 0$ representing 2000. The two points are (0, 11.7) and (5, 12.7). So the slope is $\dfrac{12.7 - 11.7}{5} = 0.2$. The equation is
$p - 11.7 = 0.2(t - 0) \Rightarrow p = 0.2t + 11.7$. The year 2010 is represented by $t = 10$.
$p = 0.2(10) + 11.7 \Rightarrow p = 13.7\%$.

127. a.
```
LinReg
y=ax+b
a=-2
b=12.4
```
$y = -2x + 12.4$

b.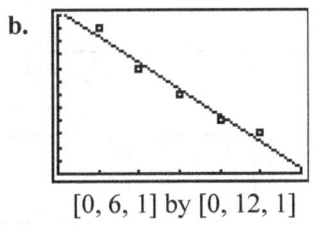

[0, 6, 1] by [0, 12, 1]

c. The price in the table is given as the number of nickels. 35¢ = 7 nickels, so let $x = 7$. $y = -2(7) + 12.4 = -1.6$
Thus, no newspapers will be sold if the price per copy is 35¢. Note that this is also clear from the graph, which appears to cross the x-axis at approximately $x = 6$.

2.3 Beyond the Basics

129. $3 = \dfrac{c - 3}{1 - (-2)} \Rightarrow 9 = c - 3 \Rightarrow c = 12$

131. a. Let $A = (0, 1)$, $B = (1, 3)$, $C = (-1, -1)$.
$m_{AB} = \dfrac{3-1}{1-0} = 2; m_{BC} = \dfrac{-1-3}{-1-1} = \dfrac{-4}{-2} = 2$
$m_{AC} = \dfrac{-1-1}{-1-0} = 2$
The slopes of the three segments are the same, so the points are collinear.

b. $d(A,B) = \sqrt{(1-0)^2 + (3-1)^2} = \sqrt{5}$
$d(B,C) = \sqrt{(-1-1)^2 + (-1-3)^2} = 2\sqrt{5}$
$d(A,C) = \sqrt{(-1-0)^2 + (-1-1)^2} = \sqrt{5}$
Because $d(B, C) = d(A, B) + d(A, C)$, the three points are collinear.

133. a. $m_{AB} = \dfrac{4-1}{-1-1} = -\dfrac{3}{2}; m_{BC} = \dfrac{8-4}{5-(-1)} = \dfrac{2}{3}$.
The product of the slopes $= -1$, so $AB \perp BC$.

b. $d(A,B) = \sqrt{(-1-1)^2 + (4-1)^2} = \sqrt{13}$
$d(B,C) = \sqrt{(5-(-1))^2 + (8-4)^2} = \sqrt{52}$
$d(A,C) = \sqrt{(5-1)^2 + (8-1)^2} = \sqrt{65}$
$(d(A,B))^2 + (d(B,C))^2 = (d(A,C))^2$, so the triangle is a right triangle.

112 Chapter 2 *Graphs and Functions*

For exercises 135 and 136, refer to the figures accompanying the exercises in your text.

135. $\overline{AD}$ and $\overline{BC}$ are parallel because they lie on parallel lines l_1 and l_2. $\overline{AB}$ and $\overline{CD}$ are parallel because they are parallel to the *x*-axis. Therefore, *ABCD* is a parallelogram. $\overline{AB} \cong \overline{CD}$ and $\overline{AD} \cong \overline{CB}$ because opposite sides of a parallelogram are congruent. $\triangle ABD \cong \triangle CDB$ by SSS. Then $m_1 = \dfrac{\text{rise}}{\text{run}} = \dfrac{BD}{CD}$ and $m_2 = \dfrac{\text{rise}}{\text{run}} = \dfrac{BD}{AB}$. Since $AB = CD$, $m_1 = \dfrac{BD}{CD} = \dfrac{BD}{AB} = m_2$.

137. Let the quadrilateral *ABCD* be such that $AB \cong CD$ and $AB \parallel CD$. Locate the points as shown in the figure.

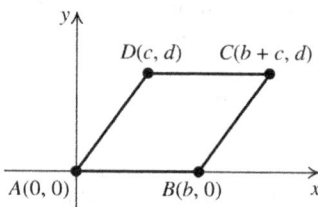

Because $AB \parallel CD$, the *y*-coordinates of *C* and *D* are equal. Because $AB \cong CD$, the *x*-coordinates of the points are as shown in the figure. The slope of *AD* is d/c. The slope of *BC* is $\dfrac{d-0}{b+c-b} = \dfrac{d}{c}$. So $AD \parallel BC$.

$d(A,D) = \sqrt{d^2 + c^2}$.

$d(B,C) = \sqrt{d^2 + ((b+c)-b)^2} = \sqrt{d^2 + c^2}$.

So $AD \cong BC$.

139. Let (x, y) be the coordinates of point *B*. Then
$d(A,B) = 12.5 = \sqrt{(x-2)^2 + (y-2)^2} \Rightarrow$
$(x-2)^2 + (y-2)^2 = 156.25$ and
$m_{AB} = \dfrac{4}{3} = \dfrac{y-2}{x-2} \Rightarrow 4(x-2) = 3(y-2) \Rightarrow$
$y = \dfrac{4}{3}x - \dfrac{2}{3}$. Substitute this into the first equation and solve for *x*:

$(x-2)^2 + \left(\left(\dfrac{4}{3}x - \dfrac{2}{3}\right) - 2\right)^2 = 156.25$

$(x-2)^2 + \left(\dfrac{4}{3}x - \dfrac{8}{3}\right)^2 = 156.25$

$x^2 - 4x + 4 + \dfrac{16}{9}x^2 - \dfrac{64}{9}x + \dfrac{64}{9} = 156.25$

$9x^2 - 36x + 36 + 16x^2 - 64x + 64 = 1406.25$

$25x^2 - 100x - 1306.25 = 0$

Solve this equation using the quadratic formula:

$x = \dfrac{100 \pm \sqrt{100^2 - 4(25)(-1306.25)}}{2(25)}$

$= \dfrac{100 \pm \sqrt{10,000 + 130,625}}{50}$

$= \dfrac{100 \pm \sqrt{140,625}}{50} = \dfrac{100 \pm 375}{50}$

$= 9.5$ or -5.5

Now find *y* by substituting the *x*-values into the slope formula: $\dfrac{4}{3} = \dfrac{y-2}{9.5-2} \Rightarrow y = 12$ or

$\dfrac{4}{3} = \dfrac{y-2}{-5.5-2} \Rightarrow y = -8$. So the coordinates of *B* are (9.5, 12) or (−5.5, −8).

141.

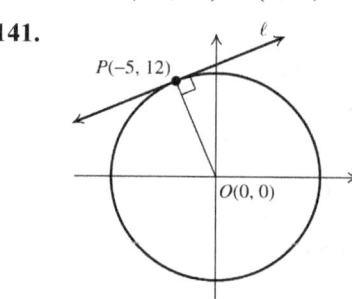

$m_{\overline{OP}} = \dfrac{12-0}{-5-0} = -\dfrac{12}{5}$

Since the tangent line ℓ is perpendicular to $\overline{OP}$, the slope of ℓ is the negative reciprocal of $-\dfrac{12}{5}$ or $\dfrac{5}{12}$. Using the point-slope form, we have

$y - 12 = \dfrac{5}{12}[x - (-5)] \Rightarrow y - 12 = \dfrac{5}{12}(x+5) \Rightarrow$

$y - 12 = \dfrac{5}{12}x + \dfrac{25}{12} \Rightarrow y = \dfrac{5}{12}x + \dfrac{169}{12}$

143.

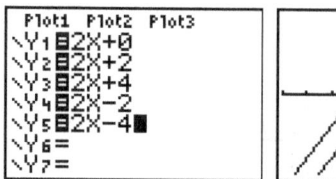 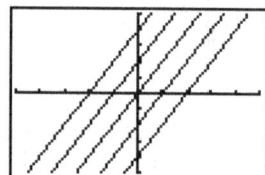

The family of lines has slope 2. The lines have different *y*-intercepts.

145.

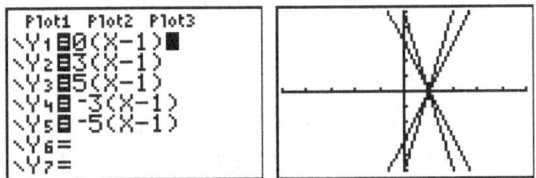

The lines pass through (1, 0). The lines have different slopes.

2.3 Critical Thinking/Discussion/Writing

147. a.

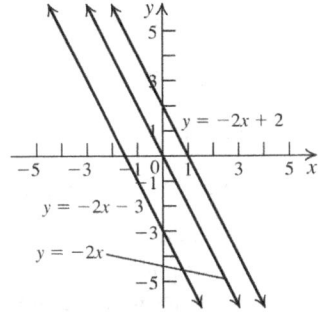

This is a family of lines parallel to the line $y = -2x$. They all have slope -2.

b.

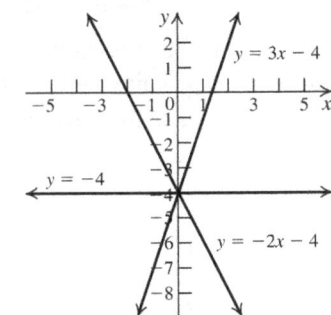

This is a family of lines that passes through the point $(0, -4)$. Their y-intercept is -4.

148. $\left. \begin{array}{l} y = m_1 x + b_1 \\ y = m_2 x + b_2 \end{array} \right\} \Rightarrow m_1 x + b_1 = m_2 x + b_2 \Rightarrow$

$m_1 x - m_2 x = b_2 - b_1 \Rightarrow x(m_1 - m_2) = b_2 - b_1 \Rightarrow$

$x = \dfrac{b_2 - b_1}{m_1 - m_2}$

a. If $m_1 > m_2 > 0$ and $b_1 > b_2$, then

$x = \dfrac{b_2 - b_1}{m_1 - m_2} = -\dfrac{b_1 - b_2}{m_1 - m_2}.$

b. If $m_1 > m_2 > 0$ and $b_1 < b_2$, then

$x = \dfrac{b_2 - b_1}{m_1 - m_2}.$

c. If $m_1 < m_2 < 0$ and $b_1 > b_2$, then

$x = \dfrac{b_2 - b_1}{m_1 - m_2} = \dfrac{b_1 - b_2}{m_2 - m_1}.$

d. If $m_1 < m_2 < 0$ and $b_1 < b_2$, then

$x = \dfrac{b_2 - b_1}{m_1 - m_2} = -\dfrac{b_2 - b_1}{m_2 - m_1}.$

2.3 Getting Ready for the Next Section

149. $x^2 - 4 = 0 \Rightarrow (x+2)(x-2) = 0 \Rightarrow$
$x + 2 = 0 \Rightarrow x = -2$ or
$x - 2 = 0 \Rightarrow x = 2$
Solution: $\{-2, 2\}$

151. $x^2 - x - 2 = 0$
$(x+1)(x-2) = 0$
$x + 1 = 0 \Rightarrow x = -1$ or $x - 2 = 0 \Rightarrow x = 2$
Solution: $\{-1, 2\}$

153. $(3(a+h) + 1) - (3a + 1) = (3a + 3h + 1) - (3a + 1)$
$= 3h$

155. $\dfrac{-(a+h)^2 + a^2}{h} = \dfrac{-(a^2 + 2ah + h^2) + a^2}{h}$

$= \dfrac{-2ah - h^2}{h} = \dfrac{h(-2a - h)}{h}$

$= -2a - h$

157. $(x-1)(x-3) < 0$
Solve the associated equation:
$(x-1)(x-3) = 0 \Rightarrow x = 1$ or $x = 3$.
So, the intervals are
$(-\infty, 1), (1, 3)$, and $(1, \infty)$.

Interval	Test point	Value of $(x-1)(x-3)$	Result
$(-\infty, 1)$	0	3	+
$(1, 3)$	2	-2	$-$
$(3, \infty)$	4	3	+

The solution set is $(1, 3)$.

2.4 Functions

2.4 Practice Problems

1. a. The domain of R is $\{2, -2, 3\}$ and its range is $\{1, 2\}$. The relation R is a function because no two ordered pairs in R have the same first component.

   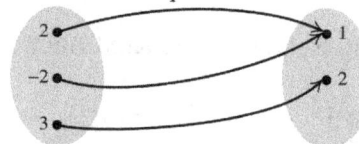

   b. The domain of S is $\{2, 3\}$ and its range is $\{5, -2\}$. The relation S is not a function because the ordered paired $(3, -2)$ and $(3, 5)$ have the same first component.

2. Solve each equation for y.

   a. $2x^2 - y^2 = 1 \Rightarrow 2x^2 - 1 = y^2 \Rightarrow$
   $\pm\sqrt{2x^2 - 1} = y$; not a function

   b. $x - 2y = 5 \Rightarrow x - 5 = 2y \Rightarrow \frac{1}{2}(x-5) = y$; a function

3. $g(x) = -2x^2 + 5x$

   a. $g(0) = -2(0)^2 + 5(0) = 0$

   b. $g(-1) = -2(-1)^2 + 5(-1) = -7$

   c. $g(x+h) = -2(x+h)^2 + 5(x+h)$
   $= -2(x^2 + 2xh + h^2) + 5x + 5h$
   $= -2x^2 - 4hx + 5x - 2h^2 + 5h$

4. $A_{TLMS} = (\text{length})(\text{height}) = (|3-1|)(22)$
   $= (2)(22) = 44$ sq. units

5. a. $f(x) = \dfrac{1}{\sqrt{1-x}}$ is not defined when
   $1 - x = 0 \Rightarrow x = 1$ or
   when $1 - x < 0 \Rightarrow 1 < x$. Thus, the domain of f is $(-\infty, 1)$.

   b. $g(x) = \sqrt{x^2 + 2x - 3}$ is not defined when $x^2 + 2x - 3 < 0$.
   Use the test point method to see that $x^2 + 2x - 3 < 0$ on the interval $(-3, 1)$.
   Thus, the domain of g is $(-\infty, -3] \cup [1, \infty)$.

   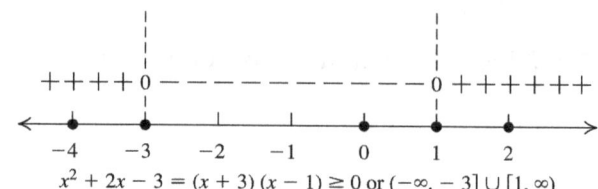

   $x^2 + 2x - 3 = (x+3)(x-1) \geq 0$ or $(-\infty, -3] \cup [1, \infty)$

6. $f(x) = x^2$, domain $X = [-3, 3]$

   a. $f(x) = 10 \Rightarrow x^2 = 10 \Rightarrow x = \pm\sqrt{10} \approx \pm 3.16$
   Since $\sqrt{10} > 3$ and $-\sqrt{10} < -3$, neither solution is in the interval $X = [-3, 3]$. Therefore, 10 is not in the range of f.

   b. $f(x) = 4 \Rightarrow x^2 = 4 \Rightarrow x = \pm 2$
   Since $-3 < -2 < 2 < 3$, 4 is in the range of f.

   c. The range of f is the interval $[0, 9]$ because for each number y in this interval, the number $x = \sqrt{y}$ is in the interval $[-3, 3]$.

7.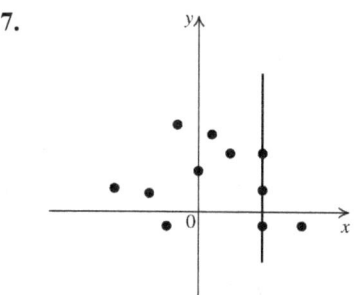

   The graph is not a function because a vertical line can be drawn through three points, as shown.

8.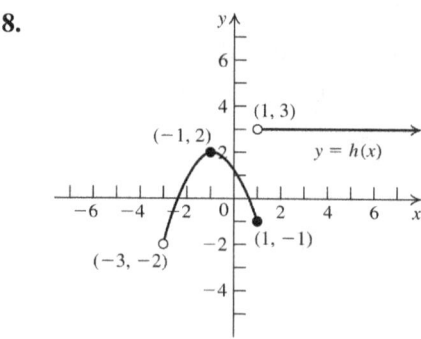

   Domain: $(-3, \infty)$; range: $(-2, 2] \cup \{3\}$

9. $y = f(x) = x^2 + 4x - 5$

   a. Check whether the ordered pair (2, 7) satisfies the equation:
   $$7 \stackrel{?}{=} 2^2 + 4(2) - 5$$
   $$7 = 7 \checkmark$$
   The point (2, 7) is on the graph.

   b. Let $y = -8$, then solve for x:
   $-8 = x^2 + 4x - 5 \Rightarrow 0 = x^2 + 4x + 3 \Rightarrow$
   $0 = (x+3)(x+1) \Rightarrow x = -3$ or $x = -1$
   The points $(-3, -8)$ and $(-1, -8)$ lie on the graph.

   c. Let $x = 0$, then solve for y:
   $y = 0^2 + 4(0) - 5 = -5$
   The y-intercept is -5.

   d. Let $y = 0$, then solve for x:
   $0 = x^2 + 4x - 5 \Rightarrow 0 = (x+5)(x-1) \Rightarrow$
   $x = -5$ or $x = 1$
   The x-intercepts are -5 and 1.

10. The range of $C(t)$ is $[6, 12)$.
    $C(11) = \frac{1}{2}C(10) + 6 = \frac{1}{2}(11.989) + 6 \approx 11.995$.

11. From Example 11, we have
    $AP = \sqrt{500^2 + x^2}$ and $\overline{PD} = 1200 - x$ feet. If $c =$ the cost on land, the total cost C is given by $C = 1.3c(PD) + c(AP)$
    $= 1.3c\sqrt{500^2 + x^2} + c(1200 - x)$
    $= c\left[1.3\sqrt{500^2 + x^2} + 1200 - x\right]$

12. a. $C(x) = 1200x + 100,000$

    b. $R(x) = 2500x$

    c. $P(x) = R(x) - C(x)$
       $= 2500x - (1200x + 100,000)$
       $= 1300x - 100,000$

    d. The break-even point occurs when $C(x) = R(x)$.
       $1200x + 100,000 = 2500x$
       $100,000 = 1300x \Rightarrow x \approx 77$
       Metro needs 77 shows to break even.

2.4 Basic Concepts and Skills

1. In the functional notation $y = f(x)$, x is the independent variable.

3. If the point $(9, -14)$ is on the graph of a function f, then $f(9) = -14$.

5. False.

7. True. $-x = 7$ and the square root function is defined for all positive numbers.

2.4 Building Skills

9. Domain: $\{a, b, c\}$; range: $\{d, e\}$; function

11. Domain: $\{a, b, c\}$; range: $\{1, 2\}$; function

13. Domain: $\{0, 3, 8\}$; range: $\{-3, -2, -1, 1, 2\}$; not a function

15. $x + y = 2 \Rightarrow y = -x + 2$; a function

17. $y = \frac{1}{x}$; a function

19. $y^2 = x^2 \Rightarrow y = \pm\sqrt{x^2} \Rightarrow y = \pm x$; not a function

21. $y = \frac{1}{\sqrt{2x - 5}}$; a function

23. $2 - y = 3x \Rightarrow y = 2 - 3x$; a function

25. $x + y^2 = 8 \Rightarrow y = \pm\sqrt{8 - x}$; not a function

27. $x^2 + y^3 = 5 \Rightarrow y = \sqrt[3]{5 - x^2}$; a function

In exercises 29–32, $f(x) = x^2 - 3x + 1$, $g(x) = \frac{2}{\sqrt{x}}$, and $h(x) = \sqrt{2 - x}$.

29. $f(0) = 0^2 - 3(0) + 1 = 1$
    $g(0) = \frac{2}{\sqrt{0}} \Rightarrow g(0)$ is undefined
    $h(0) = \sqrt{2 - 0} = \sqrt{2}$
    $f(a) = a^2 - 3a + 1$
    $f(-x) = (-x)^2 - 3(-x) + 1 = x^2 + 3x + 1$

31. $f(-1) = (-1)^2 - 3(-1) + 1 = 5$;
    $g(-1) = \frac{2}{\sqrt{-1}} \Rightarrow g(-1)$ is undefined;
    $h(-1) = \sqrt{2 - (-1)} = \sqrt{3}$; $h(c) = \sqrt{2 - c}$;
    $h(-x) = \sqrt{2 - (-x)} = \sqrt{2 + x}$

33. a. $f(0) = \dfrac{2(0)}{\sqrt{4-0^2}} = 0$

   b. $f(1) = \dfrac{2(1)}{\sqrt{4-1^2}} = \dfrac{2}{\sqrt{3}} = \dfrac{2\sqrt{3}}{3}$

   c. $f(2) = \dfrac{2(2)}{\sqrt{4-2^2}} = \dfrac{4}{0} \Rightarrow f(2)$ is undefined

   d. $f(-2) = \dfrac{2(-2)}{\sqrt{4-(-2)^2}} = \dfrac{-4}{0} \Rightarrow f(-2)$ is undefined

   e. $f(-x) = \dfrac{2(-x)}{\sqrt{4-(-x)^2}} = \dfrac{-2x}{\sqrt{4-x^2}}$

35. The width of each rectangle is 1. The height of the left rectangle is $f(1) = 1^2 + 2 = 3$. The height of the right rectangle is $f(2) = 2^2 + 2 = 6$.
$A = (1)(f(1)) + (1)(f(2))$
$= 1(3) + (1)(6) = 9$ sq. units

37. $(-\infty, \infty)$

39. The denominator is not defined for $x = 9$. The domain is $(-\infty, 9) \cup (9, \infty)$

41. The denominator is not defined for $x = -1$ or $x = 1$. The domain is $(-\infty, -1) \cup (-1, 1) \cup (1, \infty)$.

43. The numerator is not defined for $x < 3$, and the denominator is not defined for $x = -2$. The domain is $[3, \infty)$

45. The denominator equals 0 if $x = -1$ or $x = -2$. The domain is $(-\infty, -2) \cup (-2, -1) \cup (-1, \infty)$.

47. The denominator is not defined for $x = 0$. The domain is $(-\infty, 0) \cup (0, \infty)$

49. 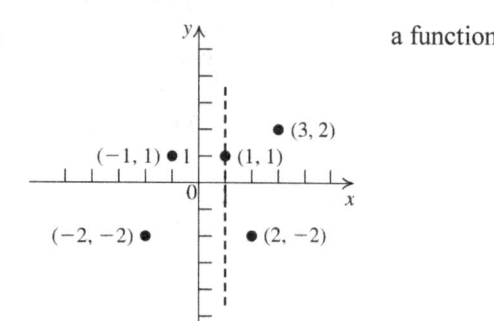 a function

51. 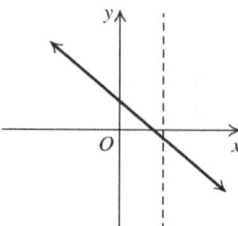 a function

53. 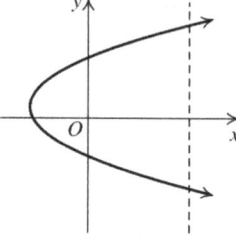 not a function

55. $f(-4) = -2; f(-1) = 1; f(3) = 5; f(5) = 7$

57. $h(-2) = -5; h(-1) = 4; h(0) = 3; h(1) = 4$

59. $h(x) = 7$, so solve the equation
$7 = x^2 - x + 1$.
$x^2 - x - 6 = 0 \Rightarrow (x-3)(x+2) = 0 \Rightarrow x = -2$ or $x = 3$.

61. a. $1 = -2(1+1)^2 + 7 \Rightarrow 1 = -1$, which is false. Therefore, $(1, 1)$ does not lie on the graph of f.

   b. $1 = -2(x+1)^2 + 7 \Rightarrow 2(x+1)^2 = 6 \Rightarrow (x+1)^2 = 3 \Rightarrow x+1 = \pm\sqrt{3} \Rightarrow x = -1 \pm \sqrt{3}$
   The points $\left(-1-\sqrt{3}, 1\right)$ and $\left(-1+\sqrt{3}, 1\right)$ lie on the graph of f.

   c. $y = -2(0+1)^2 + 7 \Rightarrow y = 5$
   The y-intercept is $(0, 5)$.

   d. $0 = -2(x+1)^2 + 7 \Rightarrow -7 = -2(x+1)^2 \Rightarrow \dfrac{7}{2} = (x+1)^2 \Rightarrow \pm\sqrt{\dfrac{7}{2}} = \pm\dfrac{\sqrt{14}}{2} = x+1 \Rightarrow x = -1 \pm \dfrac{\sqrt{14}}{2}$
   The x-intercepts are $\left(-1-\dfrac{\sqrt{14}}{2}, 0\right)$ and $\left(-1+\dfrac{\sqrt{14}}{2}, 0\right)$.

63. Domain: $[-3, 2]$; range: $[-3, 3]$

65. Domain: $[-4, \infty)$; range: $[-2, 3]$

67. Domain: $[-3, \infty)$; range: $[-1, 4] \cup \{-3\}$

69. Domain: $(-\infty, 4] \cup [-2, 2] \cup [4, \infty)$
    Range: $[-2, 2] \cup \{3\}$

71. $[-9, \infty)$

73. $-3, 4, 7, 9$

75. $f(-7) = 4$, $f(1) = 5$, $f(5) = 2$

77. $\{-3.75, -2.25, 3\} \cup [12, \infty)$

79. $[-9, \infty)$

81. $g(-4) = -1$, $g(1) = 3$, $g(3) = 4$

83. $[-9, -5)$

2.4 Applying the Concepts

85. A function because there is only one high temperature per day.

87. Not a function because there are several states that begin with N (i.e., New York, New Jersey, New Mexico, Nevada, North Carolina, North Dakota); there are also several states that begin with T and S.

89. $A(x) = x^2$; $A(4) = 16$; $A(4)$ represents the area of a tile with side 4.

91. It is a function. $S(x) = 6x^2$; $S(3) = 54$

93. a. The domain is $[0, 8]$.

    b. $h(2) = 128(2) - 16(2^2) = 192$
       $h(4) = 128(4) - 16(4^2) = 256$
       $h(6) = 128(6) - 16(6^2) = 192$

    c. $0 = 128t - 16t^2 \Rightarrow 0 = 16t(8-t) \Rightarrow$
       $t = 0$ or $t = 8$. It will take 8 seconds for the stone to hit the ground.

    d.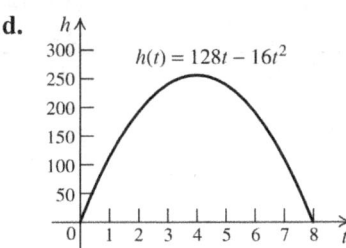

95. $x + y = 28 \Rightarrow y = 28 - x$
    $P = x(28 - x) = 28x - x^2$

97. Note that the length of the base = the width of the base = x.
    $V = lwh = x^2 h = 64 \Rightarrow h = \dfrac{64}{x^2}$
    $S = 2lw + 2lh + 2wh$
    $ = 2x^2 + 2x\left(\dfrac{64}{x^2}\right) + 2x\left(\dfrac{64}{x^2}\right)$
    $ = 2x^2 + \dfrac{128}{x} + \dfrac{128}{x} = 2x^2 + \dfrac{256}{x}$

99. The piece with length x is formed into a circle, so $C = x = 2\pi r \Rightarrow r = \dfrac{x}{2\pi}$. Thus, the area of the circle is $A = \pi r^2 = \pi\left(\dfrac{x}{2\pi}\right)^2 = \dfrac{x^2}{4\pi}$.
    The piece with length $20 - x$ is formed into a square, so $P = 20 - x = 4s \Rightarrow s = \dfrac{1}{4}(20 - x)$.
    Thus, the area of the square is
    $s^2 = \left[\dfrac{1}{4}(20-x)\right]^2 = \dfrac{1}{16}(20-x)^2$.
    The sum of the areas is $A = \dfrac{x^2}{4\pi} + \dfrac{1}{16}(20-x)^2$

101. The volume of the pool is
     $V = 288 = x^2 h \Rightarrow h = \dfrac{288}{x^2}$.

     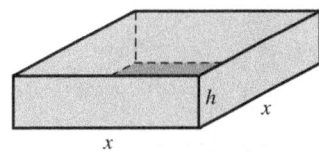

     The total area to be tiled is
     $4xh = 4x\left(\dfrac{288}{x^2}\right) = \dfrac{1152}{x}$
     The cost of the tile is $6\left(\dfrac{1152}{x}\right) = \dfrac{6912}{x}$.
     The area of the bottom of the pool is x^2, so the cost of the cement is $2x^2$. Therefore, the total cost is $C = 2x^2 + \dfrac{6912}{x}$.

103.

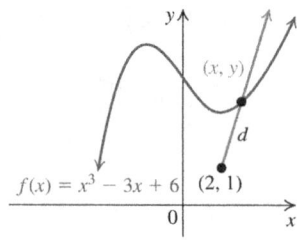

Using the distance formula we have
$$d = \sqrt{(x-2)^2 + (y-1)^2}$$
$$= \sqrt{(x-2)^2 + \left[\left(x^3 - 3x + 6\right) - 1\right]^2}$$
$$= \sqrt{(x-2)^2 + \left(x^3 - 3x + 5\right)^2}$$
$$= \left[(x-2)^2 + \left(x^3 - 3x + 5\right)^2\right]^{1/2}$$

105. a. $p(5) = 1275 - 25(5) = 1150$. If 5000 TVs can be sold, the price per TV is $1150.
$p(15) = 1275 - 25(15) = 900$. If 15,000 TVs can be sold, the price per TV is $900.
$p(30) = 1275 - 25(30) = 525$. If 30,000 TVs can be sold, the price per TV is $525.

b.

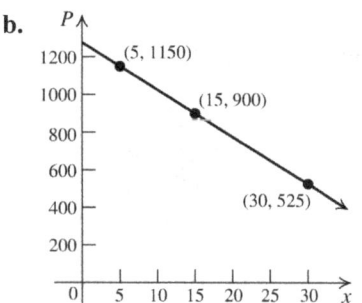

c. $650 = 1275 - 25x \Rightarrow -625 = -25x \Rightarrow x = 25$
25,000 TVs can be sold at $650 per TV.

107. a. $C(x) = 5.5x + 75,000$

b. $R(x) = 0.6(15)x = 9x$

c. $P(x) = R(x) - C(x) = 9x - (5.5x + 75,000)$
$= 3.5x - 75,000$

d. The break-even point is when the profit is zero: $3.5x - 75,000 = 0 \Rightarrow x = 21,429$

e. $P(46,000) = 3.5(46,000) - 75,000$
$= \$86,000$
The company's profit is $86,000 when 46,000 copies are sold.

2.4 Beyond the Basics

109. $x = \dfrac{2}{y-4} \Rightarrow xy - 4x = 2 \Rightarrow xy = 2 + 4x \Rightarrow$
$y = \dfrac{4x+2}{x} \Rightarrow f(x) = \dfrac{4x+2}{x};$
Domain: $(-\infty, 0) \cup (0, \infty)$. $f(4) = \dfrac{9}{2}$.

111. $\left(x^2 + 1\right)y + x = 2 \Rightarrow y = \dfrac{2-x}{x^2+1} \Rightarrow$
$f(x) = \dfrac{2-x}{x^2+1}$; Domain: $(-\infty, \infty); f(4) = -\dfrac{2}{17}$

113. $f(x) \ne g(x)$ because they have different domains.

115. $f(x) \ne g(x)$ because they have different domains. $g(x)$ is not defined for $x = -1$, while $f(x)$ is defined for all real numbers.

117. $f(x) = g(x)$ because $f(3) = 10 = g(3)$ and $f(5) = 26 = g(5)$.

119. $f(2) = 15 = a(2^2) + 2a - 3 \Rightarrow 15 = 6a - 3 \Rightarrow a = 3$.

121. $h(6) = 0 = \dfrac{3(6) + 2a}{2(6) - b} \Rightarrow 0 = 18 + 2a \Rightarrow a = -9$
$h(3)$ is undefined $\Rightarrow \dfrac{3(3) + 2(-9)}{2(3) - b}$ has a zero in the denominator. So $6 - b = 0 \Rightarrow b = 6$.

123. $g(x) = x^2 - \dfrac{1}{x^2} \Rightarrow g\!\left(\dfrac{1}{x}\right) = \dfrac{1}{x^2} - \dfrac{1}{\frac{1}{x^2}} = \dfrac{1}{x^2} - x^2$
$g(x) + g\!\left(\dfrac{1}{x}\right) = \left(x^2 - \dfrac{1}{x^2}\right) + \left(\dfrac{1}{x^2} - x^2\right) = 0$

125. $f(x) = \dfrac{x+3}{4x-5} \Rightarrow$
$f(t) = \dfrac{\frac{3+5x}{4x-1} + 3}{4\left(\frac{3+5x}{4x-1}\right) - 5} = \dfrac{\frac{(3+5x) + 3(4x-1)}{4x-1}}{\frac{(12+20x) - (5(4x-1))}{4x-1}}$
$= \dfrac{(3+5x) + (12x-3)}{(12+20x) - (20x-5)} = \dfrac{17x}{17} = x$

2.4 Critical Thinking/Discussion/Writing

126. Answers may vary. Sample answers are given

    a. $y = \sqrt{x-2}$ b. $y = \dfrac{1}{\sqrt{x-2}}$

    c. $y = \sqrt{2-x}$ d. $y = \dfrac{1}{\sqrt{2-x}}$

127. a. $ax^2 + bx + c = 0$

    b. $y = c$

    c. The equation will have no x-intercepts if $b^2 - 4ac < 0$.

    d. It is not possible for the equation to have no y-intercepts because $y = f(x)$.

128. a. $f(x) = |x|$

    b. $f(x) = 0$

    c. $f(x) = x$

    d. $f(x) = \sqrt{-x^2}$ (Note: the point is the origin.)

    e. $f(x) = 1$

    f. A vertical line is not a function.

129. a. $\{(a, 1), (b, 1)\}$ $\{(a, 2), (b, 1)\}$
       $\{(a, 1), (b, 2)\}$ $\{(a, 2), (b, 2)\}$
       $\{(a, 1), (b, 3)\}$ $\{(a, 2), (b, 3)\}$

       $\{(a, 3), (b, 1)\}$
       $\{(a, 3), (b, 2)\}$
       $\{(a, 3), (b, 3)\}$
       There are nine functions from X to Y.

    b. $\{(1, a)\}, \{(2, a)\}, \{(3, a)\}$
       $\{(1, a)\}, \{(2, a)\}, \{(3, b)\}$
       $\{(1, a)\}, \{(2, b)\}, \{(3, a)\}$
       $\{(1, b)\}, \{(2, a)\}, \{(3, a)\}$
       $\{(1, b)\}, \{(2, a)\}, \{(3, a)\}$
       $\{(1, b)\}, \{(2, b)\}, \{(3, a)\}$
       $\{(1, b)\}, \{(2, a)\}, \{(3, b)\}$
       $\{(1, b)\}, \{(2, b)\}, \{(3, b)\}$
       There are eight functions from Y to X.

130. If a set X has m elements and a set of Y has n elements, there are n^m functions that can be defined from X to Y. This is true since a function assigns each element of X to an element of Y. There are m possibilities for each element of X, so there are

    $\underbrace{n \cdot n \cdot n \cdots n}_{m} = n^m$ possible functions.

2.4 Getting Ready for the Next Section

131. $2x - 4 < 12 \Rightarrow 2x < 16 \Rightarrow x < 8$

    The solution set is $(-\infty, 8)$.

133. $x^2 > 0$

    Solve the associated equation:

    $x^2 = 0 \Rightarrow x = 0$.

    So, the intervals are $(-\infty, 0)$ and $(0, \infty)$.

Interval	Test point	Value of x^2	Result
$(-\infty, 0)$	-1	1	$+$
$(0, \infty)$	1	1	$+$

    The solution set is $(-\infty, 0) \cup (0, \infty)$

For exercises 135–140, $f(x) = 3 - 2x^2$, $g(x) = \sqrt{x+3}$, and $h(x) = \dfrac{2}{x^2 + 1}$.

135. $f(0) = 3 - 2(0)^2 = 3$

    $g(0) = \sqrt{0+3} = \sqrt{3}$

    $h(0) = \dfrac{2}{0^2 + 1} = 2$

137. $f(-2) = 3 - 2(-2)^2 = 3 - 8 = -5$

    $g(-2) = \sqrt{-2+3} = \sqrt{1} = 1$

    $h(-2) = \dfrac{2}{(-2)^2 + 1} = \dfrac{2}{5}$

139. $f(-x) = 3 - 2(-x)^2 = 3 - x^2$

    $g(-x) = \sqrt{-x+3}$

    $h(-x) = \dfrac{2}{(-x)^2 + 1} = \dfrac{2}{x^2 + 1}$

2.5 Properties of Functions

2.5 Practice Problems

1. The function is decreasing on $(0, 3)$, $(12, 13)$, and $(15, 24)$; increasing on $(3, 12)$ and $(13, 15)$

2. Relative maxima of 3640 at $x = 12$ and 4070 at $x = 15$; relative minima of 40 at $x = 3$ and 3490 at $x = 13$.

3.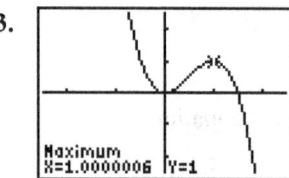
   Relative minimum of 0 at $x = 0$
   Relative maximum of 1 at $x = 1$

4. $v = (11 - r)r^2$
   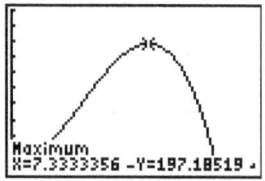
   $[0, 13, 1]$ by $[0, 250, 25]$

   Mrs. Osborn's windpipe should be contracted to a radius of 7.33 mm for maximizing the airflow velocity.

5. $f(x) = -x^2$
   Replace x with $-x$:
   $f(-x) = -(-x)^2 = -x^2 = f(x)$
   Thus, the function is even.
   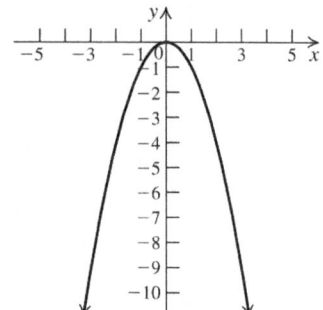

6. $f(x) = -x^3$
   Replace x with $-x$:
   $f(-x) = -(-x)^3 = x^3 = -f(x)$
   Thus, the function is odd.
   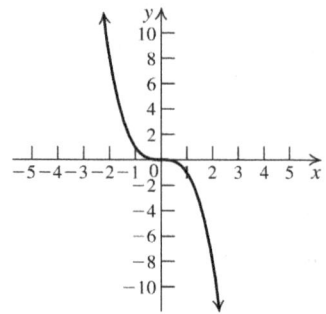

7. a. $g(-x) = 3(-x)^4 - 5(-x)^2$
      $= 3x^4 - 5x^2 = f(x) \Rightarrow$
      $g(x)$ is even.

   b. $f(-x) = 4(-x)^5 + 2(-x)^3 = -4x^5 - 2x^3$
      $= -(4x^5 + 2x^3) = -f(x) \Rightarrow$
      $f(x)$ is odd.

   c. $h(-x) = 2(-x) + 1 = -2x + 1$
      $\neq h(x)$
      $\neq h(-x) \Rightarrow h$ is neither even nor odd.

8. $f(x) = 1 - x^2$; $a = 2, b = 4$
   $f(2) = 1 - 2^2 = -3$; $f(4) = 1 - 4^2 = -15$
   $\dfrac{f(b) - f(a)}{b - a} = \dfrac{-15 - (-3)}{4 - 2} = \dfrac{-12}{2} = -6$
   The average rate of change is -6.

9. $f(t) = 1 - t$; $a = 2, b = x, x \neq 2$
   $f(a) = f(2) = 1 - 2 = -1$
   $f(b) = f(x) = 1 - x$
   $\dfrac{f(b) - f(a)}{b - a} = \dfrac{(1 - x) - (-1)}{x - 2} = \dfrac{2 - x}{x - 2}$
   $= \dfrac{-1(x - 2)}{x - 2} = -1$
   The average rate of change is -1.

10. $f(x) = 100x^2 - 800x + 2000$
    $f(0) = 100(0)^2 - 800(0) + 2000 = 2000$
    $f(3) = 100(3)^2 - 800(3) + 2000 = 500$
    $\dfrac{f(3) - f(0)}{3 - 0} = \dfrac{500 - 2000}{3} = \dfrac{-1500}{3} = -500$

(continued on next page)

(*continued*)

The average rate of change is −500, so the number of bacteria per cubic centimeter decreases by 500 each day after adding the chlorine.

11. $f(x) = -x^2 + x - 3$
$f(x+h) = -(x+h)^2 + (x+h) - 3$
$= -x^2 - 2xh - h^2 + x + h - 3$

$\dfrac{f(x+h) - f(x)}{h}$

$= \dfrac{\left(-x^2 - 2xh - h^2 + x + h - 3\right) - \left(-x^2 + x - 3\right)}{h}$

$= \dfrac{-2xh - h^2 + h}{h} = -2x - h + 1$

2.5 Concepts and Vocabulary

1. A function f is decreasing if $x_1 < x_2$ implies that $\underline{f(x_1) > f(x_2)}$.

3. A function f is even if $\underline{f(-x) = f(x) \text{ for all } x \text{ in the domain of } f}$.

5. True

7. True

2.5 Building Skills

9. Increasing on $(-\infty, \infty)$

11. Increasing on $(-\infty, 2)$, decreasing on $(2, \infty)$

13. Increasing on $(-\infty, -2)$, constant on $(-2, 2)$, increasing on $(2, \infty)$

15. Increasing on $(-\infty, -3)$ and $\left(-\frac{1}{2}, 2\right)$, decreasing on $\left(-3, -\frac{1}{2}\right)$ and $(2, \infty)$

17. No relative extrema

19. (2, 10) is a relative maximum point and a turning point.

21. Any point on $(x, 2)$ is a relative maximum and a relative minimum point on the interval $(-2, 2)$. Relative maximum at $(-2, 2)$; relative minimum at $(2, 2)$. None of these points are turning points.

23. $(-3, 4)$ and $(2, 5)$ are relative maxima points and turning points. $\left(-\frac{1}{2}, -2\right)$ is a relative minimum and a turning point.

For exercises 25–34, recall that the graph of an even function is symmetric about the y-axis, and the graph of an odd function is symmetric about the origin.

25. The graph is symmetric with respect to the origin. The function is odd.

27. The graph has no symmetries, so the function is neither odd nor even.

29. The graph is symmetric with respect to the origin. The function is odd.

31. The graph is symmetric with respect to the y-axis. The function is even.

33. The graph is symmetric with respect to the origin. The function is odd.

For exercises 35–48, $f(-x) = f(x) \Rightarrow f(x)$ is even and $f(-x) = -f(x) \Rightarrow f(x)$ is odd.

35. $f(-x) = 2(-x)^4 + 4 = 2x^4 + 4 = f(x) \Rightarrow$
$f(x)$ is even.

37. $f(-x) = 5(-x)^3 - 3(-x) = -5x^3 + 3x$
$= -(5x^3 - 3x) = -f(x) \Rightarrow$
$f(x)$ is odd.

39. $f(-x) = 2(-x) + 4 = -2x + 4$
$\neq -f(x) \neq f(x) \Rightarrow$
$f(x)$ is neither even nor odd.

41. $f(-x) = \dfrac{1}{(-x)^2 + 4} = \dfrac{1}{x^2 + 4} = f(x) \Rightarrow$
$f(x)$ is even.

43. $f(-x) = \dfrac{(-x)^3}{(-x)^2 + 1} = -\dfrac{x^3}{x^2 + 1} = -f(x) \Rightarrow$
$f(x)$ is odd.

45. $f(-x) = \dfrac{-x}{(-x)^5 - 3(-x)^3} = \dfrac{-x}{-x^5 + 3x^3}$
$= \dfrac{(-1)(-x)}{(-1)\left(-x^5 + 3x^3\right)} = \dfrac{x}{x^5 - 3x^3} = f(x)$

Thus, $f(x)$ is even.

47. $f(-x) = \dfrac{(-x)^2 - 2(-x)}{5(-x)^4 + 4(-x)^2 + 7} = \dfrac{x^2 + 2x}{5x^4 + 4x^2 + 7}$
$\neq -f(x) \neq f(x)$

Thus, $f(x)$ is neither even nor odd.

49. a. domain: $(-\infty, \infty)$; range: $(-\infty, 3]$

   b. x-intercepts: $(-3, 0), (3, 0)$
   y-intercept: $(0, 3)$

c. increasing on $(-\infty, 0)$, decreasing on $(0, \infty)$

d. relative maximum at $(0, 3)$

e. even

51. a. domain: $(-3, 4)$; range: $[-2, 2]$

   b. x-intercept: $(1, 0)$; y-intercept: $(0, -1)$

   c. constant on $(-3, -1)$ and $(3, 4)$
   increasing on $(-1, 3)$

   d. Since the function is constant on $(-3, -1)$, any point $(x, -2)$ is both a relative maximum and a relative minimum on that interval. Since the function is constant on $(3, 4)$, any point $(x, 2)$ is both a relative maximum and a relative minimum on that interval.

   e. neither even not odd

53. a. domain: $(-2, 4)$; range: $[-2, 3]$

   b. x-intercept: $(0, 0)$; y-intercept: $(0, 0)$

   c. decreasing on $(-2, -1)$ and $(3, 4)$
   increasing on $(-1, 3)$

   d. relative maximum: $(3, 3)$
   relative minimum: $(-1, -2)$

   e. neither even nor odd

55. a. domain: $(-\infty, \infty)$; range: $(0, \infty)$

   b. no x-intercept; y-intercept: $(0, 1)$

   c. increasing on $(-\infty, \infty)$

   d. no relative minimum or relative maximum

   e. neither even nor odd

57. $f(x) = -2x + 7$; $a = -1$, $b = 3$
$f(3) = -2(3) + 7 = 1$; $f(-1) = -2(-1) + 7 = 9$
$$\text{average rate of change} = \frac{f(3) - f(-1)}{3 - (-1)}$$
$$= \frac{1 - 9}{4} = -2$$

59. $f(x) = 3x + c$; $a = 1$, $b = 5$
$f(5) = 3 \cdot 5 + c = 15 + c$; $f(1) = 3 \cdot 1 + c = 3 + c$
$$\text{average rate of change} = \frac{f(5) - f(1)}{5 - 1}$$
$$= \frac{15 + c - (3 + c)}{4}$$
$$= \frac{12}{4} = 3$$

61. $h(x) = x^2 - 1$; $a = -2$, $b = 0$
$h(0) = 0^2 - 1 = -1$; $h(-2) = (-2)^2 - 1 = 3$
$$\text{average rate of change} = \frac{h(0) - h(-2)}{0 - (-2)}$$
$$= \frac{-1 - 3}{2} = -2$$

63. $f(x) = (3 - x)^2$; $a = 1$, $b = 3$
$f(4) = (3 - 3)^2 = 0$; $f(1) = (3 - 1)^2 = 4$
$$\text{average rate of change} = \frac{f(3) - f(1)}{3 - 1}$$
$$= \frac{0 - 4}{2} = -2$$

65. $g(x) = x^3$; $a = -1$, $b = 3$
$g(3) = 3^3 = 27$; $g(-1) = (-1)^3 = -1$
$$\text{average rate of change} = \frac{g(3) - g(-1)}{3 - (-1)}$$
$$= \frac{27 - (-1)}{4} = 7$$

67. $h(x) = \frac{1}{x}$; $a = 2$, $b = 6$
$h(2) = \frac{1}{2}$; $h(6) = \frac{1}{6}$
$$\text{average rate of change} = \frac{h(6) - h(2)}{6 - 2}$$
$$= \frac{\frac{1}{6} - \frac{1}{2}}{4} = -\frac{1}{12}$$

69. $f(x + h) = x + h$
$f(x + h) - f(x) = x + h - x = h$
$$\frac{f(x + h) - f(x)}{h} = \frac{h}{h} = 1$$

71. $f(x + h) = -2(x + h) + 3 = -2x - 2h + 3$
$f(x + h) - f(x) = -2x - 2h + 3 - (-2x + 3)$
$= -2h$
$$\frac{f(x + h) - f(x)}{h} = \frac{-2h}{h} = -2$$

73. $f(x + h) = m(x + h) + b = mx + mh + b$
$f(x + h) - f(x) = mx + mh + b - (mx + b)$
$= mh$
$$\frac{f(x + h) - f(x)}{h} = \frac{mh}{h} = m$$

75. $f(x+h) = (x+h)^2 = x^2 + 2xh + h^2$
$f(x+h) - f(x) = x^2 + 2xh + h^2 - x^2$
$= 2xh + h^2$
$\dfrac{f(x+h) - f(x)}{h} = \dfrac{2xh + h^2}{h} = 2x + h$

77. $f(x+h) = 2(x+h)^2 + 3(x+h)$
$= 2x^2 + 4xh + 2h^2 + 3x + 3h$
$= 2x^2 + 4xh + 3x + 2h^2 + 3h$
$f(x+h) - f(x)$
$= 2x^2 + 4xh + 3x + 2h^2 + 3h - (2x^2 + 3x)$
$= 4xh + 2h^2 + 3h$
$\dfrac{f(x+h) - f(x)}{h} = \dfrac{4xh + 2h^2 + 3h}{h}$
$= 4x + 2h + 3$

79. $f(x+h) = 4$
$f(x+h) - f(x) = 4 - 4 = 0$
$\dfrac{f(x+h) - f(x)}{h} = \dfrac{0}{h} = 0$

81. $f(x+h) = \dfrac{1}{x+h}$
$f(x+h) - f(x) = \dfrac{1}{x+h} - \dfrac{1}{x}$
$= \dfrac{x}{x(x+h)} - \dfrac{x+h}{x(x+h)}$
$= -\dfrac{h}{x(x+h)}$
$\dfrac{f(x+h) - f(x)}{h} = \dfrac{-\dfrac{h}{x(x+h)}}{h} = -\dfrac{1}{x(x+h)}$

2.5 Applying the Concepts

83. a. Increasing: (2006, 2009), (2011, 2012), (2013, 2014)
Decreasing: (2009, 2011), (2012, 2013)

b. Relative maxima: 251.1 at $x = 2009$, 293.2 at $x = 2012$
Relative minima: 21.5 at $x = 2011$, 187.0 at $x = 2013$

85. domain: $[0, \infty)$
The particle's motion is tracked indefinitely from time $t = 0$.

87. The graph is above the t-axis on the intervals (0, 9) and (21, 24). This means that the particle was moving forward between 0 and 9 seconds and between 21 and 24 seconds.

89. The function is increasing on (0, 3), (5, 6), (16, 19), and (21, 23). However, the speed $|v|$ of the particle is increasing on (0, 3), (5, 6), (11, 15), and (21, 23). Note that the particle is moving forward on (0, 3), (5, 6), and (21, 23), and moving backward on (11, 15).

91. The maximum speed is between times $t = 15$ and $t = 16$.

93. The particle is moving forward with increasing velocity.

95.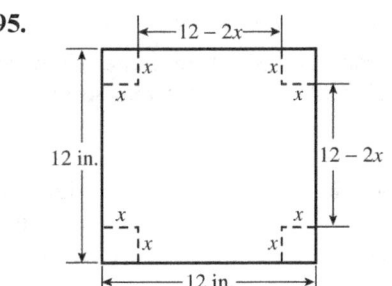

a. $V = lwh = (12 - 2x)(12 - 2x)x$
$= (144 - 48x + 4x^2)x$
$= 4x^3 - 48x^2 + 144x$

b. The length of the squares in the corners must be greater than 0 and less than 6, so the domain of V is (0, 6).

c.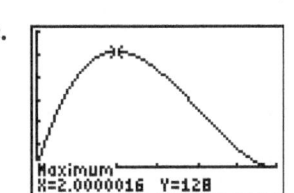
[0, 6, 1] by [−25, 150, 25]
range: [0, 128]

d. V is at its maximum when $x = 2$.

97. a. $C(x) = 210x + 10,500$

b. $C(50) = 210(50) + 10,500 = \$21,000$
It costs $21,000 to produce 50 notebooks per day.

c. average cost $= \dfrac{\$21,000}{50} = \$420$

d. $\dfrac{210x + 10,500}{x} = 315$
$210x + 10,500 = 315x$
$10,500 = 105x \Rightarrow x = 100$
The average cost per notebook will be $315 when 100 notebooks are produced.

99. average rate of increase $= \dfrac{f(2014) - f(2000)}{2014 - 2000}$
$= \dfrac{20.2 - 15.3}{14} = 0.35$

The average rate of increase was 0.35 million, or 350,000, students per year.

101. a. $f(0) = 0^2 + 3(0) + 4 = 4$

The particle is 4 ft to the right from the origin.

b. $f(4) = 4^2 + 3(4) + 4 = 32$

The particle started 4 ft from the origin, so it traveled $32 - 4 = 28$ ft in four seconds.

c. $f(3) = 3^2 + 3(3) + 4 = 22$

The particle started 4 ft from the origin, so it traveled $22 - 4 = 18$ ft in three seconds. The average velocity is $18/3 = 6$ ft/sec

d. $f(2) = 2^2 + 3(2) + 4 = 14$
$f(5) = 5^2 + 3(5) + 4 = 44$

The particle traveled $44 - 14 = 30$ ft between the second and fifth seconds. The average velocity is $30/(5-2) = 10$ ft/sec

2.5 Beyond the Basics

103. $f(x) = \dfrac{x-1}{x+1}$

$f(2x) = \dfrac{2x-1}{2x+1}$

$\dfrac{3f(x)+1}{f(x)+3} = \dfrac{3\left(\dfrac{x-1}{x+1}\right)+1}{\dfrac{x-1}{x+1}+3} = \dfrac{\dfrac{3x-3}{x+1}+1}{\dfrac{x-1+3(x+1)}{x+1}}$

$= \dfrac{\dfrac{3x-3+x+1}{x+1}}{\dfrac{x-1+3(x+1)}{x+1}} = \dfrac{4x-2}{4x+2}$

$= \dfrac{2(2x-1)}{2(2x+1)} = \dfrac{2x-1}{2x+1} = f(2x)$

105. In order to find the relative maximum, first observe that the relative maximum of $-(x+1)^2 \leq 0$. Then $-(x+1)^2 \leq 0 \Rightarrow$
$(x+1)^2 \geq 0 \Rightarrow x \geq -1$.

Thus, the x-coordinate of the relative maximum is -1. $f(-1) = -(-1+1)^2 + 5 = 5$
The relative maximum is $(-1, 5)$.
There is no relative minimum.

107. $f(x) = \sqrt{x}$

$\dfrac{f(x+h) - f(x)}{h} = \dfrac{\sqrt{x+h} - \sqrt{x}}{h} \cdot \dfrac{\sqrt{x+h}+\sqrt{x}}{\sqrt{x+h}+\sqrt{x}}$

$= \dfrac{x+h-x}{h\left(\sqrt{x+h}+\sqrt{x}\right)}$

$= \dfrac{h}{h\left(\sqrt{x+h}+\sqrt{x}\right)}$

$= \dfrac{1}{\sqrt{x+h}+\sqrt{x}}$

109. $f(x) = -\dfrac{1}{\sqrt{x}}$

$\dfrac{f(x+h) - f(x)}{h}$

$= \dfrac{-\dfrac{1}{\sqrt{x+h}} + \dfrac{1}{\sqrt{x}}}{h} = \dfrac{\dfrac{\sqrt{x+h}-\sqrt{x}}{\sqrt{x}\sqrt{x+h}}}{h}$

$= \dfrac{\dfrac{\sqrt{x+h}-\sqrt{x}}{\sqrt{x(x+h)}}}{h} = \dfrac{\sqrt{x+h}-\sqrt{x}}{h\sqrt{x(x+h)}}$

$= \dfrac{\sqrt{x+h}-\sqrt{x}}{h\sqrt{x(x+h)}} \cdot \dfrac{\sqrt{x+h}+\sqrt{x}}{\sqrt{x+h}+\sqrt{x}}$

$= \dfrac{(x+h)-x}{h\sqrt{x(x+h)}\left(\sqrt{x+h}+\sqrt{x}\right)}$

$= \dfrac{1}{\sqrt{x(x+h)}\left(\sqrt{x}+\sqrt{x+h}\right)}$

2.5 Critical Thinking/Discussion/Writing

111. f has a relative maximum at $x = a$ if there is an interval $[a, x_1)$ with $a < x_1 < b$ such that $f(a) \geq f(x)$, or $f(x) \leq f(a)$, for every x in the interval $(x_1, b]$.

112. f has a relative minimum at $x = b$ if there is x_1 in $[a, b]$ such that $f(x) \geq f(b)$ for every x in the interval $(x_1, b]$.

113. Answers will vary. Sample answers are given.

a. $f(x) = x$ on the interval $[-1, 1]$

b. $f(x) = \begin{cases} x & \text{if } 0 \le x < 1 \\ 0 & \text{if } x = 1 \end{cases}$

c. $f(x) = \begin{cases} x & \text{if } 0 < x \le 1 \\ 1 & \text{if } x = 0 \end{cases}$

d. $f(x) = \begin{cases} 0 & \text{if } x = 0 \text{ or } x = 1 \\ 1 & \text{if } 0 < x < 1 \text{ and } x \text{ is rational} \\ -1 & \text{if } 0 < x < 1 \text{ and } x \text{ is irrational} \end{cases}$

2.5 Getting Ready for the Next Section

115. $m = \dfrac{-1-2}{7-6} = \dfrac{-3}{1} = -3$
$y - 2 = -3(x - 6) \Rightarrow y - 2 = -3x + 18 \Rightarrow$
$y = -3x + 20$

117.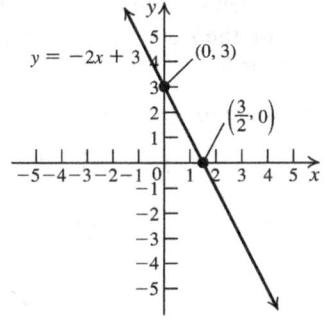

119. If $x = -3$, then $|x + 1| = \underline{2}$.

121. $f(x) = x^{3/2}$

   (i) $f(2) = 2^{3/2} = (\sqrt{2})^3 = \sqrt{8} = 2\sqrt{2}$

   (ii) $f(4) = 4^{3/2} = (\sqrt{4})^3 = 2^3 = 8$

   (iii) $f(-4) = (-4)^{3/2} = (\sqrt{-4})^3 = (2i)^3 = -8i$

2.6 A Library of Functions

2.6 Practice Problems

1. Because $g(-2) = 2$ and $g(1) = 8$, the line passes through the points $(-2, 2)$ and $(1, 8)$.
$m = \dfrac{8-2}{1-(-2)} = \dfrac{6}{3} = 2$
Use the point-slope form:
$y - 8 = 2(x - 1) \Rightarrow y - 8 = 2x - 2 \Rightarrow$
$y = 2x + 6 \Rightarrow g(x) = 2x + 6$

2. Using the formula
Shark length = (0.96)(tooth height) − 0.22,
gives:
Shark length = (0.96)(16.4) − 0.22 = 15.524 m

3.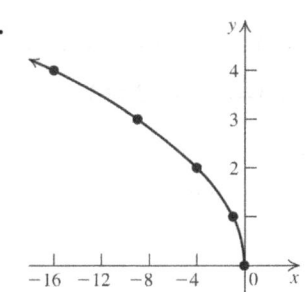

Domain: $(-\infty, 0]$; range: $[0, \infty)$

4.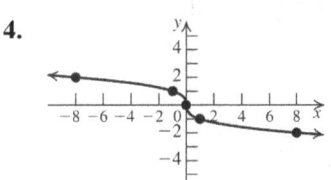

Domain: $(-\infty, \infty)$; range: $(-\infty, \infty)$

5. $f(x) = \begin{cases} x^2 & \text{if } x \le -1 \\ 2x & \text{if } x > -1 \end{cases}$

$f(-2) = (-2)^2 = 4$; $f(3) = 2(3) = 6$

6. a. $f(x) = \begin{cases} 50 + 4(x - 55) & 56 \le x < 75 \\ 200 + 5(x - 75) & x \ge 75 \end{cases}$

   b. The fine for driving 60 mph is
   $50 + 4(60 - 55) = \$70$.

   c. The fine for driving 90 mph is
   $200 + 5(90 - 75) = \$275$.

7. The graph of f is made up of two parts: a line segment passing through $(1, 5)$ and $(3, 2)$ on the interval $[1, 3]$, and a line segment passing through $(3, 2)$ and $(5, 4)$ on the interval $[3, 5]$.

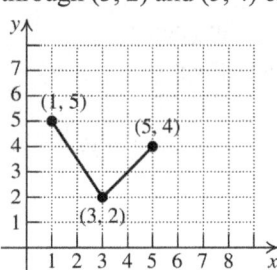

(continued on next page)

(*continued*)

For the first line segment:
$$m = \frac{2-5}{3-1} = -\frac{3}{2}$$
$$y - 5 = -\frac{3}{2}(x-1) \Rightarrow 2y - 10 = -3(x-1) \Rightarrow$$
$$2y - 10 = -3x + 3 \Rightarrow 2y = -3x + 13 \Rightarrow$$
$$y = -\frac{3}{2}x + \frac{13}{2}$$

For the second line segment:
$$m = \frac{4-2}{5-3} = 1$$
$$y - 4 = x - 5 \Rightarrow y = x - 1$$

The piecewise function is
$$g(x) = \begin{cases} -\frac{3}{2}x + \frac{13}{2} & \text{if } 1 \le x \le 3 \\ x - 1 & \text{if } 3 < x \le 5 \end{cases}$$

8. $f(x) = \begin{cases} -3x & \text{if } x \le -1 \\ 2x & \text{if } x > -1 \end{cases}$

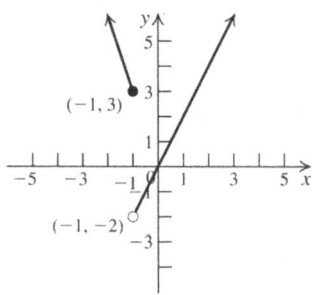

Graph $f(x) = -3x$ on the interval $(-\infty, -1]$, and graph $f(x) = 2x$ on the interval $(-1, \infty)$.

9. $f(x) = [\![x]\!]$
   $f(-3.4) = -4$; $f(4.7) = 4$

2.6 Concepts and Vocabulary

1. The graph of the linear function $f(x) = b$ is a <u>horizontal</u> line.

3. The graph of the function
$$f(x) = \begin{cases} x^2 + 2 & \text{if } x \le 1 \\ ax & \text{if } x > 1 \end{cases}$$
will have a break at $x = 1$ unless $a = \underline{3}$.

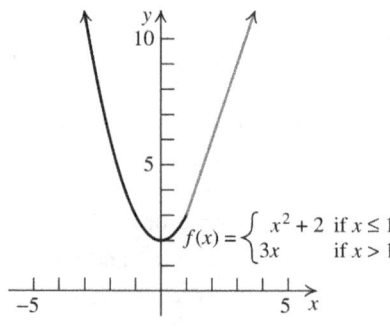

5. True. The equation of the graph of a vertical line has the format $x = a$.

7. True

2.6 Building Skills

In exercises 9–18, first find the slope of the line using the two points given. Then substitute the coordinates of one of the points into the slope-intercept form of the equation to solve for b.

9. The two points are $(0, 1)$ and $(-1, 0)$.
$$m = \frac{0-1}{-1-0} = 1. \quad 1 = 1(0) + b \Rightarrow b = 1.$$
$$f(x) = x + 1$$

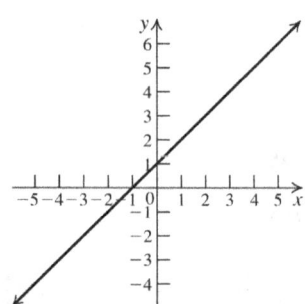

11. The two points are $(-1, 1)$ and $(2, 7)$.
$$m = \frac{7-1}{2-(-1)} = 2. \quad 1 = 2(-1) + b \Rightarrow 3 = b.$$
$$f(x) = 2x + 3$$

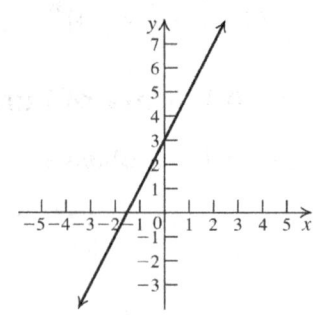

13. The two points are (1, 1) and (2, –2).
$m = \dfrac{-2-1}{2-1} = -3$. $1 = -3(1) + b \Rightarrow b = 4$.
$f(x) = -3x + 4$.

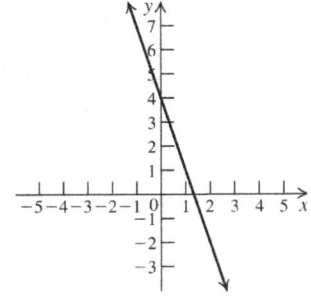

15. The two points are (–2, 2) and (2, 4).
$m = \dfrac{4-2}{2-(-2)} = \dfrac{1}{2}$. $4 = \dfrac{1}{2}(2) + b \Rightarrow b = 3$.
$f(x) = \dfrac{1}{2}x + 3$.

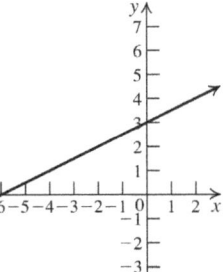

17. The two points are (0, –1) and (3, –3).
$m = \dfrac{-3-(-1)}{3-0} = -\dfrac{2}{3}$.
$-1 = -\dfrac{2}{3}(0) + b \Rightarrow b = -1$.
$f(x) = -\dfrac{2}{3}x - 1$.

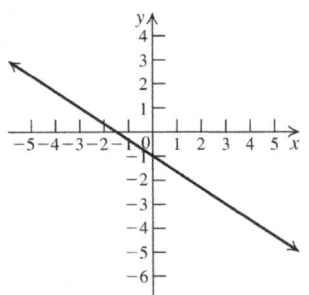

19. $f(x) = \begin{cases} x & \text{if } x \geq 2 \\ 2 & \text{if } x < 2 \end{cases}$

  a. $f(1) = 2;\ f(2) = 2;\ f(3) = 3$

  b.

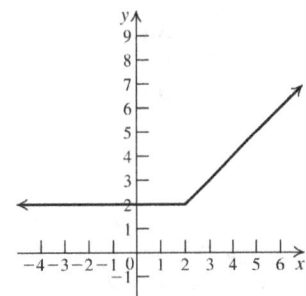

21. $g(x) = \begin{cases} 1 & \text{if } x > 0 \\ -1 & \text{if } x < 0 \end{cases}$

  a. $f(-15) = -1;\ f(12) = 1$

  b.

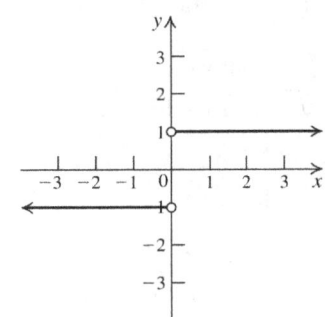

  c. domain: $(-\infty, 0) \cup (0, \infty)$
  range: $\{-1, 1\}$

23. $f(x) = \begin{cases} 2x & \text{if } x < 0 \\ x^2 & \text{if } x \geq 0 \end{cases}$

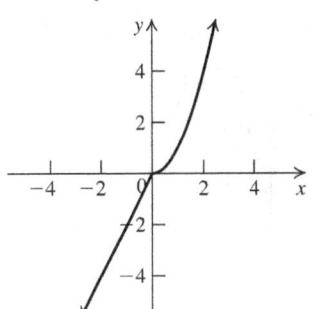

Domain: $(-\infty, \infty)$; range: $(-\infty, \infty)$

25. $g(x) = \begin{cases} \dfrac{1}{x} & \text{if } x < 0 \\ \sqrt{x} & \text{if } x \geq 0 \end{cases}$

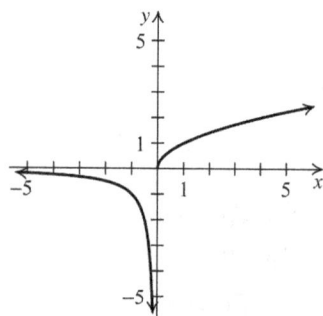

Domain: $(-\infty, \infty)$; range: $(-\infty, \infty)$

27. $f(x) = \begin{cases} [\![x]\!] & \text{if } x < 1 \\ \sqrt[3]{x} & \text{if } x \geq 1 \end{cases}$

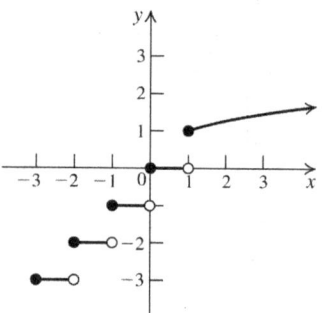

Domain: $(-\infty, \infty)$;
range: $\{\ldots, -3, -2, -1, 0\} \cup [1, \infty)$

29. $g(x) = \begin{cases} |x| & \text{if } x < 1 \\ x^3 & \text{if } x \geq 1 \end{cases}$

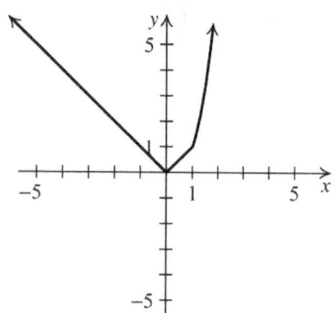

Domain: $(-\infty, \infty)$; range: $[0, \infty)$

31. $f(x) = \begin{cases} |x| & \text{if } -2 \leq x < 1 \\ \sqrt{x} & \text{if } x \geq 1 \end{cases}$

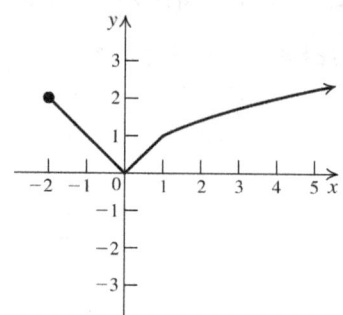

Domain: $[-2, \infty)$; range: $[0, \infty)$

33. $f(x) = \begin{cases} [\![x]\!] & \text{if } x < 1 \\ -2x & \text{if } 1 \leq x \leq 4 \end{cases}$

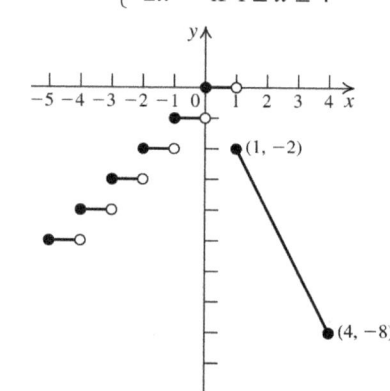

Domain: $(-\infty, 4]$;
range: $\{\ldots, -3, -2, -1, 0\} \cup [-8, -2]$

35. $f(x) = \begin{cases} 2x+3 & \text{if } x < -2 \\ x+1 & -2 \leq x < 1 \\ -x+3 & \text{if } x \geq 1 \end{cases}$

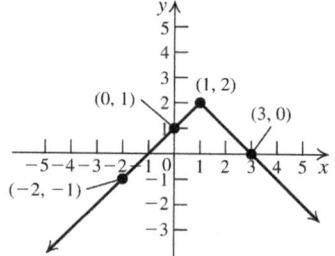

Domain: $(-\infty, \infty)$; range: $(-\infty, 2]$

37. The graph of f is made up of two parts.
For $x < 2$, the graph is made up of the half-line passing through the points $(-1, 0)$ and $(2, 3)$.
$$m = \frac{3-0}{2-(-1)} = \frac{3}{3} = 1$$
$$y - 0 = x - (-1) \Rightarrow y = x + 1$$
For $x \geq 2$, the graph is a half-line passing through the points $(2, 3)$ and $(3, 0)$.
$$m = \frac{0-3}{3-2} = -3$$
$$y - 0 = -3(x-3) \Rightarrow y = -3x + 9$$
Combining the two parts, we have
$$f(x) = \begin{cases} x+1 & \text{if } x < -2 \\ -3x+9 & \text{if } x \geq 2 \end{cases}$$

39. The graph of f is made up of three parts.
For $x < -2$, the graph is the half-line passing through the points $(-2, 2)$ and $(-3, 0)$.
$$m = \frac{0-2}{-3-(-2)} = \frac{-2}{-1} = 2$$
$$y - 0 = 2(x-(-3)) \Rightarrow y = 2(x+3) \Rightarrow$$
$$y = 2x + 6$$
For $-2 \leq x < 2$, the graph is a horizontal line segment passing through the points $(-2, 4)$ and $(2, 4)$, so the equation is $y = 4$.
For $x \geq 2$, the graph is the half-line passing through the points $(2, 1)$ and $(3, 0)$.
$$m = \frac{0-1}{3-2} = -1$$
$$y - 0 = -(x-3) \Rightarrow y = -x + 3$$
Combining the three parts, we have
$$f(x) = \begin{cases} 2x+6 & \text{if } x < -2 \\ 4 & \text{if } -2 \leq x < 2 \\ -x+3 & \text{if } x \geq 2 \end{cases}$$

2.6 Applying the Concepts

41. a. $f(x) = \dfrac{x}{33.81}$
Domain: $[0, \infty)$; range: $[0, \infty)$.

b. $f(3) = \dfrac{3}{33.81} \approx 0.0887$
This means that 3 oz ≈ 0.0887 liter.

c. $f(12) = \dfrac{12}{33.81} \approx 0.3549$ liter.

43. a. $P(0) = \dfrac{1}{33}(0) + 1 = 1$. The y-intercept is 1.
This means that the pressure at sea level $(d = 0)$ is 1 atm.
$$0 = \frac{1}{33}d + 1 \Rightarrow d = -33.$$
d can't be negative, so there is no d-intercept.

b. $P(0) = 1$ atm; $P(10) = \dfrac{1}{33}(10) + 1 \approx 1.3$ atm;
$P(33) = \dfrac{1}{33}(33) + 1 = 2$ atm;
$P(100) = \dfrac{1}{33}(100) + 1 \approx 4.03$ atm.

c. $5 = \dfrac{1}{33}d + 1 \Rightarrow d = 132$ feet
The pressure is 5 atm at 132 feet.

45. a. $C(x) = 50x + 6000$

b. The y-intercept is the fixed overhead cost.

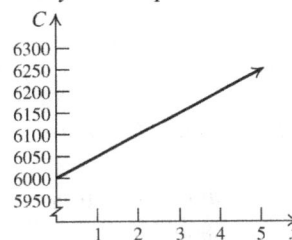

c. $11{,}500 = 50x + 6000 \Rightarrow 110$
110 printers were manufactures on a day when the total cost was $11,500.

47. a. $R = 900 - 30x$

b. $R(6) = 900 - 30(6) = 720$
If you move in 6 days after the first of the month, the rent is $720.

c. $600 = 900 - 30x \Rightarrow x = 10$
You moved in ten days after first of the month.

49. The rate of change (slope) is $\dfrac{100-40}{20-80} = -1$.
Use the point $(20, 100)$ to find the equation of the line: $100 = -20 + b \Rightarrow b = 120$. The equation of the line is $y = -x + 120$. Now solve $50 = -x + 120 \Rightarrow x = 70$.
Age 70 corresponds to 50% capacity.

51. a. The rate of change (slope) is
$$\frac{50-30}{420-150} = \frac{2}{27}.$$

The equation of the line is

$y - 30 = \dfrac{2}{27}(x - 150) \Rightarrow$

$y = \dfrac{2}{27}(x - 150) + 30.$

b. $y = \dfrac{2}{27}(350 - 150) + 30 \Rightarrow y = \dfrac{1210}{27} \approx 44.8$

There can't be a fractional number of deaths, so round up. There will be about 45 deaths when $x = 350$ milligrams per cubic meter.

c. $70 = \dfrac{2}{27}(x - 150) + 30 \Rightarrow x = 690$

If the number of deaths per month is 70, the concentration of sulfur dioxide in the air is 690 mg/m^3.

53. a.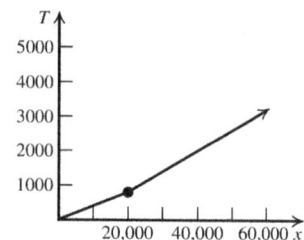

b. (i) $T(12,000) = 0.04(12,000) = \$480$

(ii) $T(20,000) = 800 + 0.06(20,000 - 20,000)$
$= \$800$

(iii) $T(50,000) = 800 + 0.06(50,000 - 20,000)$
$= \$2600$

c. (i) $600 = 0.04x \Rightarrow x = \$15,000$

(ii) $1200 = 0.04x \Rightarrow x = \$30,000$, which is outside of the domain. Try
$1200 = 800 + 0.06(x - 20,000) \Rightarrow$
$x \approx \$26,667$

(iii) $2300 = 800 + 0.06(x - 20,000) \Rightarrow$
$x = \$45,000$

2.6 Beyond the Basics

55. $2(3) - 1 = a - 3(3) \Rightarrow 5 = a - 9 \Rightarrow a = 14$

57. a. Domain: $(-\infty, \infty)$; range: $[0, 1)$

b. The function is increasing on $(n, n+1)$ for every integer n.

c. $f(-x) = -x - [\![-x]\!] \neq -f(x) \neq f(x)$, so the function is neither even nor odd.

59. a. (i) $WCI(2) = 40$

(ii)
$WCI(16)$
$= 91.4 + (91.4 - 40)$
$\cdot \left(0.0203(16) - 0.304\sqrt{16} - 0.474\right)$
≈ 21

(iii) $WCI(50) = 1.6(40) - 55 = 9$

b. (i) $-58 = 91.4 + (91.4 - T) \cdot$
$(0.0203(36) - 0.304\sqrt{36} - 0.474)$
$-58 = 91.4 + (91.4 - T)(-1.5672)$
$-58 = 91.4 - 143.24 + 1.5672T$
$-58 = -51.84 + 1.5672T \Rightarrow T \approx -4°F$

(ii) $-10 = 1.6T - 55 \Rightarrow T \approx 28°F$

61. $C(x) = 2[\![x]\!] + 4$

2.6 Critical Thinking/Discussion/Writing

63. D   64. C

65. a. If f is even, then f is increasing on $(-\infty, -1)$ and decreasing on $(-1, 0)$.

b. If f is odd, then f is decreasing on $(-\infty, -1)$ and increasing on $(-1, 0)$.

66. a. If f is even, then f has a relative maximum at $x = -1$ and a relative minimum at $x = -3$.

b. If f is odd, then f has a relative minimum at $x = -1$ and a relative maximum at $x = -3$.

2.6 Getting Ready for the Next Section

67. If we add 3 to each y-coordinate of the graph of f, we will obtain the graph of $y = \underline{f(x) + 3}$.

69. If we replace each x-coordinate with its opposite in the graph of f, we will obtain the graph of $y = \underline{f(-x)}$.

71.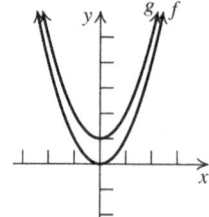

73.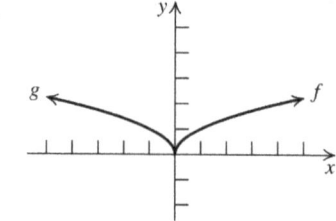

For exercises 75–78, refer to section 1.3 in your text for help on completing the square.

75. $x^2 + 8x + \underline{16}$; $(x+4)^2$

77. $x^2 - \dfrac{2}{3}x + \dfrac{1}{\underline{9}}$; $\left(x - \dfrac{1}{3}\right)^2$

2.7 Transformations of Functions

2.7 Practice Problems

1.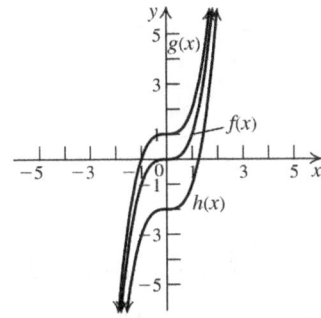

The graph of g is the graph of f shifted one unit up. The graph of h is the graph of f shifted two units down.

2.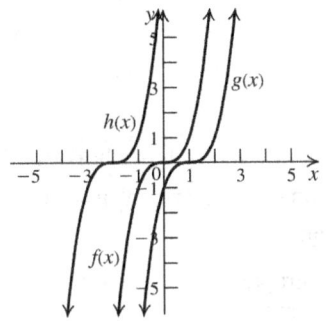

The graph of g is the graph of f shifted one unit to the right. The graph of h is the graph of f shifted two units to the left.

3. The graph of $f(x) = \sqrt{x-2} + 3$ is the graph of $g(x) = \sqrt{x}$ shifted two units to the right and three units up.

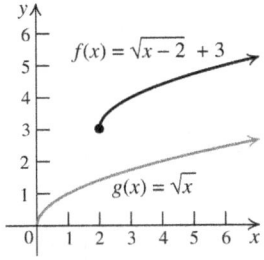

4. The graph of $y = -(x-1)^2 + 2$ can be obtained from the graph of $y = x^2$ by first shifting the graph of $y = x^2$ one unit to the right. Reflect the resulting graph about the x-axis, and then shift the graph two units up

5. The graph of $y = 2x - 4$ is obtained from the graph of $y = 2x$ by shifting it down by four units. We know that
$$|y| = \begin{cases} y & \text{if } y \geq 0 \\ -y & \text{if } y < 0. \end{cases}$$
This means that the portion of the graph on or above the x-axis $(y \geq 0)$ is unchanged while the portion of the graph below the x-axis $(y < 0)$ is reflected above the x-axis. The graph of $y = |f(x)| = |2x - 4|$ is given on the right.

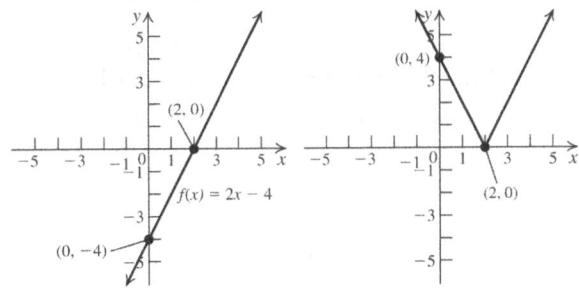

6.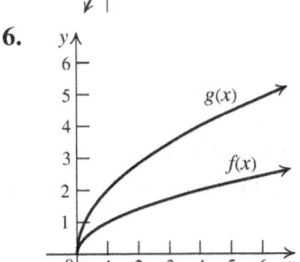

The graph of g is the graph of f stretched vertically by multiplying each of its y-coordinates by 2.

7. a.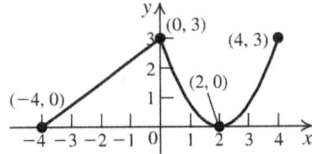

b.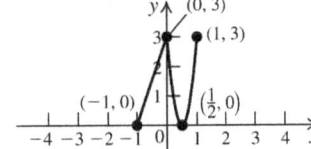

8. Start with the graph of $y = \sqrt{x}$. Shift the graph one unit to the left, then stretch the graph vertically by a factor of three. Shift the resulting graph down two units.

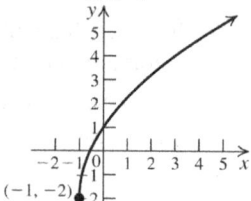

9.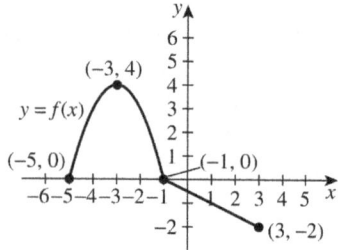

Shift the graph one unit right to graph $y = f(x-1)$.

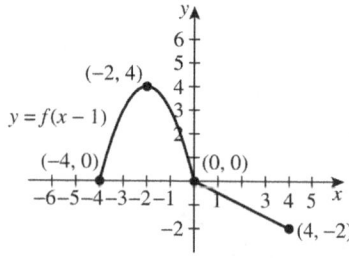

Compress horizontally by a factor of 2. Multiply each x-coordinate by $\frac{1}{2}$ to graph $y = f(2x-1)$.

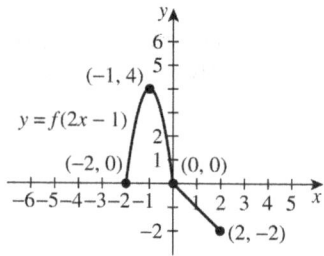

Compress vertically by a factor of $\frac{1}{2}$. Multiply each y-coordinate by $\frac{1}{2}$ to graph $y = \frac{1}{2}f(2x-1)$.

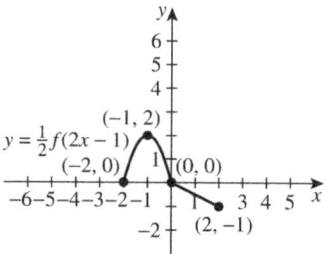

Shift the graph up three units to graph $y = \frac{1}{2}f(2x-1) + 3$.

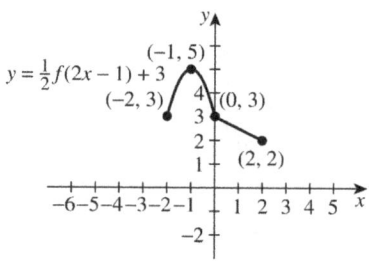

$y = f(x) \to y = f(x-1) \to y = f(2x-1) \to$
$y = \frac{1}{2}f(2x-1) \Rightarrow y = \frac{1}{2}f(2x-1) + 3$

2.7 Concepts and Vocabulary

1. The graph of $y = f(x) - 3$ is found by vertically shifting the graph of $y = f(x)$ three units down.

3. The graph of $y = f(bx)$ is a horizontal compression of the graph of $y = f(x)$ is b is greater than 1.

5. False. The graphs are the same if the function is an even function.

7. False. The graph on the left shows $y = x^2$ first shifted up two units and then reflected about the x-axis, while the graph on the right shows $y = x^2$ reflected about the x-axis and then shifted up two units.

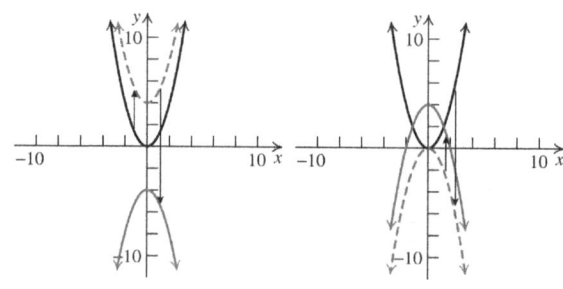

2.7 Building Skills

9. a. The graph of g is the graph of f shifted two units up.

b. The graph of h is the graph of f shifted one unit down.

11. a. The graph of g is the graph of f shifted one unit to the left.

b. The graph of h is the graph of f shifted two units to the right.

13. a. The graph of g is the graph of f shifted one unit left, then two units down.

b. The graph of h is the graph of f shifted one unit right, then three units up.

15. a. The graph of g is the graph of f reflected about the x-axis.

b. The graph of h is the graph of f reflected about the y-axis.

17. a. The graph of g is the graph of f vertically stretched by a factor of 2.

b. The graph of h is the graph of f horizontally compressed by a factor of 2.

19. a. The graph of g is the graph of f reflected about the x-axis and then shifted one unit up.

b. The graph of h is the graph of f reflected about the y-axis and then shifted one unit up.

21. a. The graph of g is the graph of f shifted one unit up.

b. The graph of h is the graph of f shifted one unit to the left.

23. e **25.** g

27. i **29.** b

31. l **33.** d

35.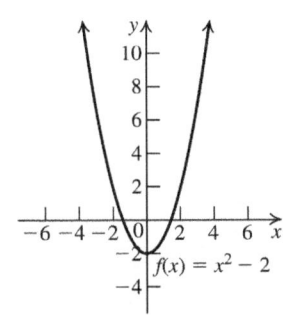

37.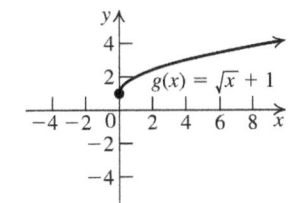

39. $f(x) = |x| + 2$

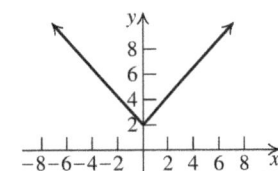

41. $f(x) = x^3 + 2$

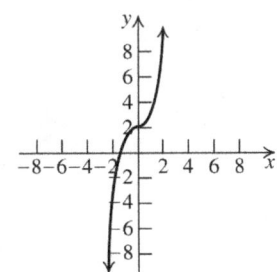

43. $f(x) = \dfrac{1}{x} + 1$

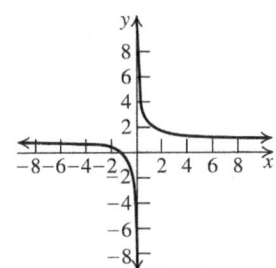

45.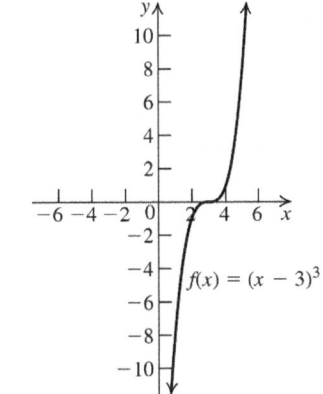

47. $f(x) = \sqrt{x-1}$

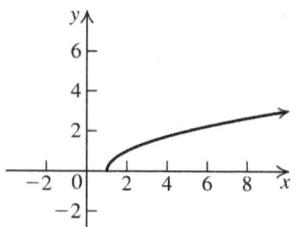

49.

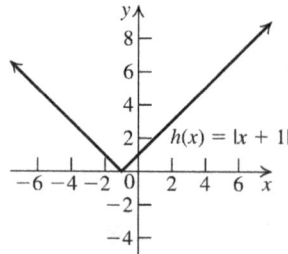

51. $f(x) = (x+1)^3$

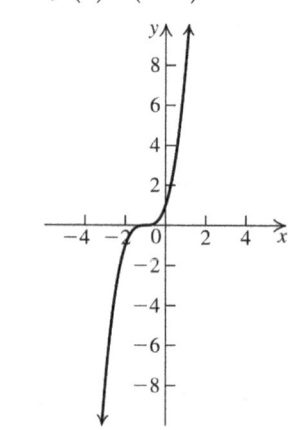

53. $f(x) = \dfrac{1}{x-3}$

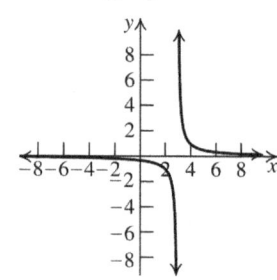

55. $f(x) = \sqrt{-x}$

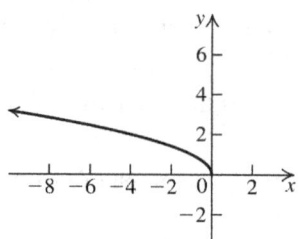

57. $f(x) = -x^2$

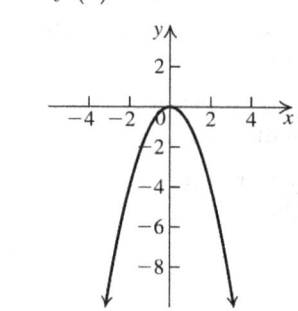

59. $f(x) = 2x^2$

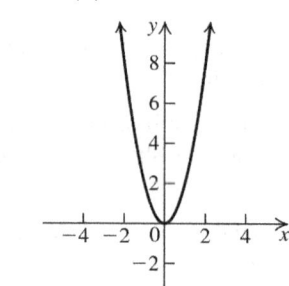

61. $f(x) = 2|x|$

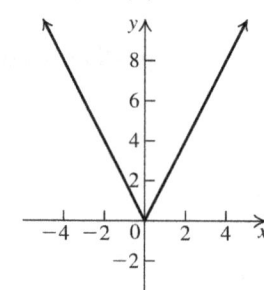

63. $f(x) = (x-2)^2 + 1$

Start with the graph of $f(x) = x^2$, then shift it two units right and one unit up.

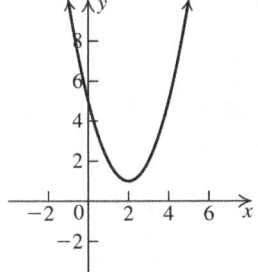

65. $f(x) = 5 - (x-3)^2$

Start with the graph of $f(x) = x^2$, then shift it three units right. Reflect the graph across the
x-axis. Shift it five units up.

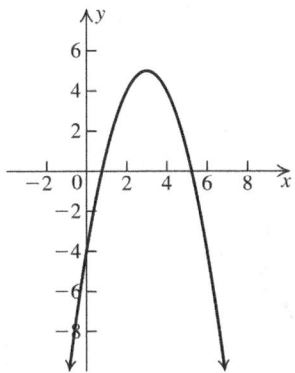

67. $f(x) = \sqrt{x+1} - 3$

Start with the graph of $f(x) = \sqrt{x}$, then shift it one unit left and three units down.

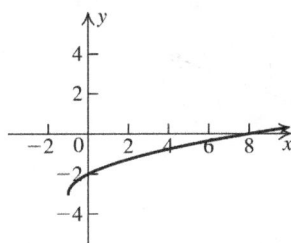

69. $f(x) = \sqrt{1-x} + 2$

Start with the graph of $f(x) = \sqrt{x}$, then shift it one unit left. Reflect the graph across the y-axis, and then shift it two units up.

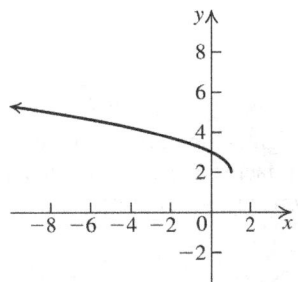

71. $f(x) = |x-1| - 2$

Start with the graph of $f(x) = |x|$, then shift it one unit right and two units down.

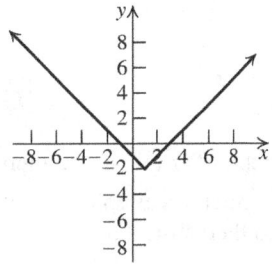

73. $f(x) = \dfrac{1}{x-1} + 3$

Start with the graph of $f(x) = \frac{1}{x}$, then shift it one unit right and three units up.

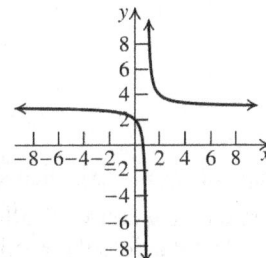

75. $f(x) = 2(x+1)^2 - 1$

Start with the graph of $f(x) = x^2$, then shift it one unit left. Stretch the graph vertically by a factor of 2, then shift it one unit down.

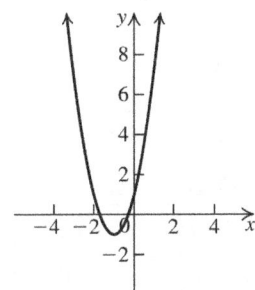

77. $f(x) = 2 - \frac{1}{2}(x-3)^2$

Start with the graph of $f(x) = x^2$, then shift it three units right. Compress the graph vertically by a factor of 1/2, reflect it across the x-axis, then shift it two units up.

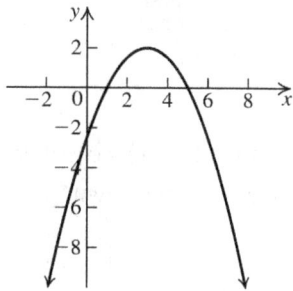

79. $f(x) = 2\sqrt{x+1} - 3$

Start with the graph of $f(x) = \sqrt{x}$, then shift it one unit left. Stretch the graph vertically by a factor of 2, and then shift it three units down.

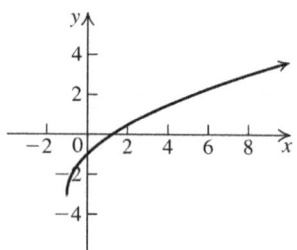

81. $f(x) = -2|x-1| + 2$

Start with the graph of $f(x) = |x|$, then shift it one unit right. Stretch the graph vertically by a factor of 2, then reflect it across the x-axis. Shift the graph up two units.

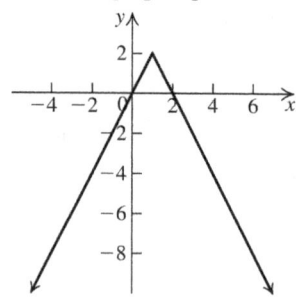

83. $y = x^3 + 2$

85. $y = -|x|$

87. $y = (x-3)^2 + 2$

89. $y = -\sqrt{x+3} - 2$

91. $y = 3(-x+4)^3 + 2$

92. $y = -(-x+1)^3 + 1$

93. $y = -2|x-4| - 3$

95.

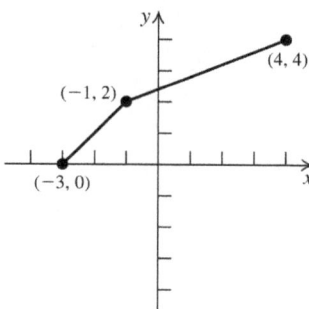

97.

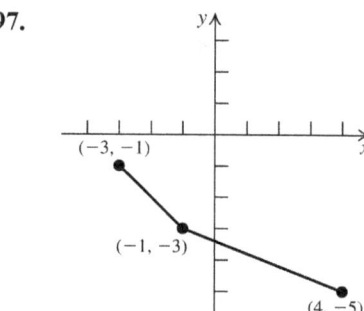

99.

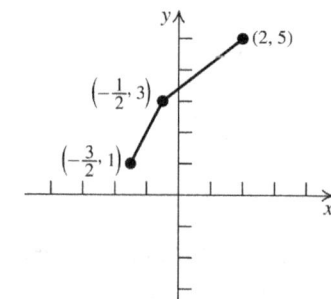

101.

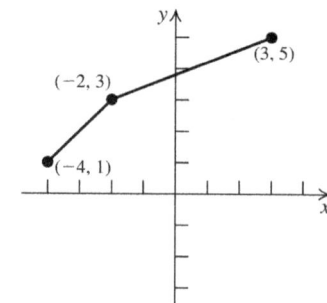

103.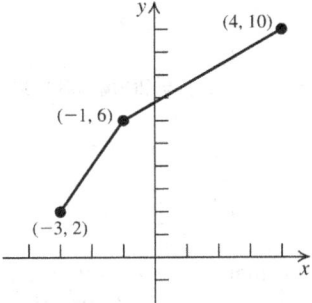

105.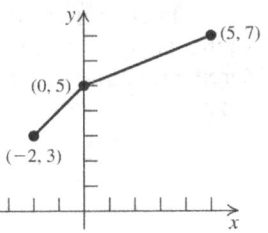

107.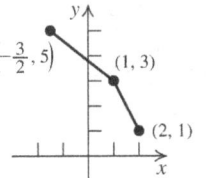

109. a.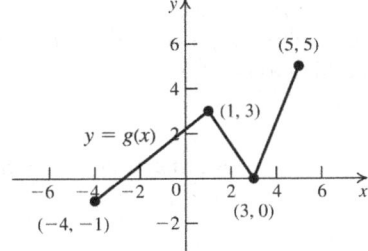

b.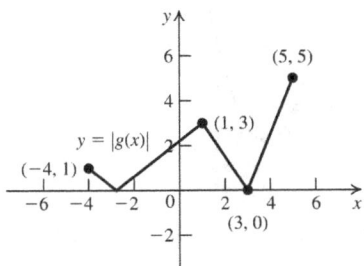

111. a.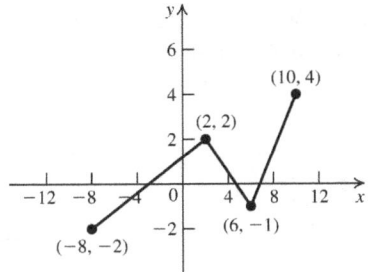

b.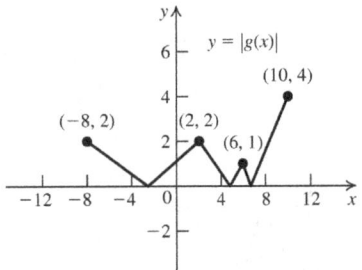

113. a.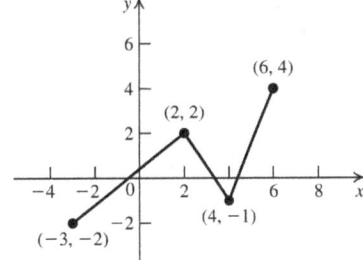

b.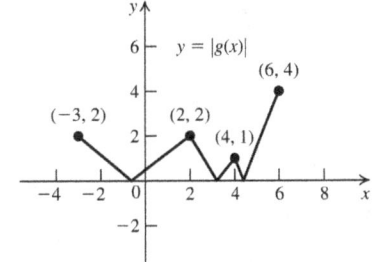

115. a.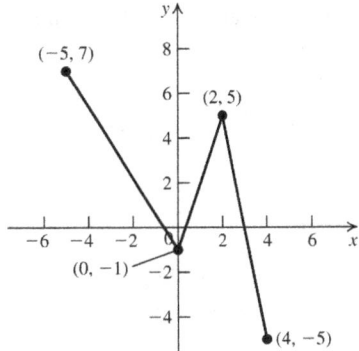

b.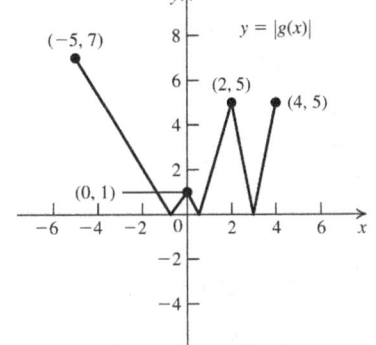

117. a.

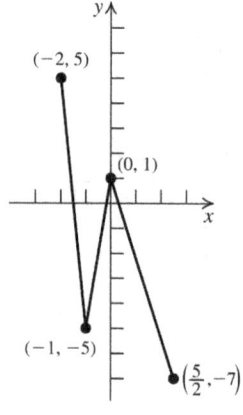

b.

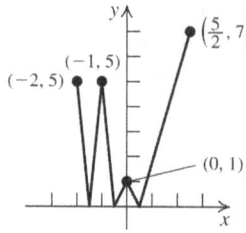

2.7 Applying the Concepts

119. $g(x) = f(x) + 800$

121. $p(x) = 1.02(f(x) + 500)$

123. a. Shift one unit right, stretch vertically by a factor of 10, and shift 5000 units up.

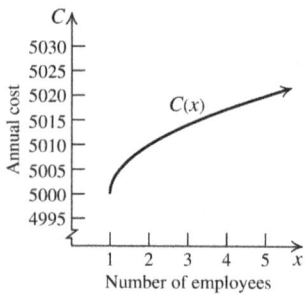

b. $C(400) = 5000 + 10\sqrt{400-1} = \$5199.75$

125. a. Shift one unit left, reflect across the x-axis, and shift up 109,561 units.

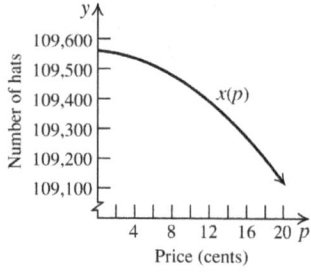

b.
$$69,160 = 109,561 - (p+1)^2$$
$$40,401 = (p+1)^2$$
$$201 = p+1 \Rightarrow p = 200¢ = \$2.00$$

c.
$$0 = 109,561 - (p+1)^2$$
$$109,561 = (p+1)^2$$
$$331 = p+1 \Rightarrow p = 330¢ = \$3.30$$

127. The first coordinate gives the month; the second coordinate gives the hours of daylight. From March to September, there is daylight more than half of the day each day. From September to March, more than half of the day is dark each day.

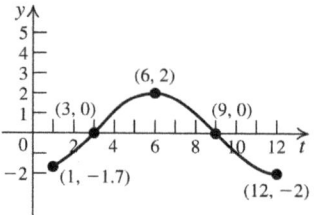

2.7 Beyond the Basics

129. The graph is shifted one unit right, then reflected about the x-axis, and finally reflected about the y-axis. The equation is $g(x) = -\sqrt{1-x}$.

131. Shift two units left, then 4 units down.

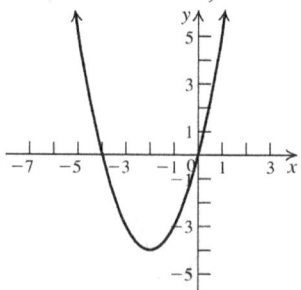

133. $f(x) = -x^2 + 2x = -(x^2 - 2x + 1) + 1$
$= -(x-1)^2 + 1$

Shift one unit right, reflect about the x-axis, then shift one unit up.

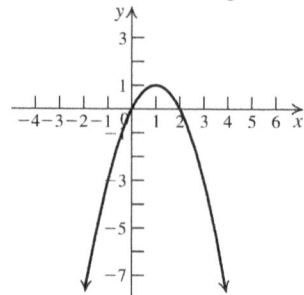

135. $f(x) = 2x^2 - 4x = 2(x^2 - 2x + 1) - 2$
$= 2(x-1)^2 - 2$
Shift one unit right, stretch vertically by a factor of 2, then shift two units down.

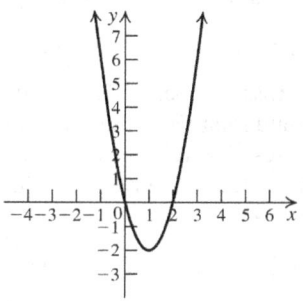

137. $f(x) = -2x^2 - 8x + 3 = -2(x^2 + 4x) + 3$
$= -2(x^2 + 4x + 4) + 3 + 2(4)$
$= -2(x+2)^2 + 11$
Shift two units left, stretch vertically by a factor of 2, reflect across the x-axis, then shift eleven units up.

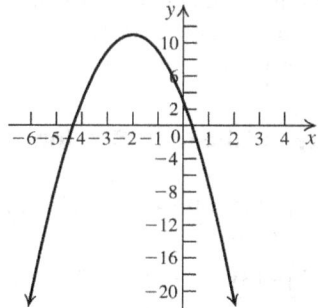

139.

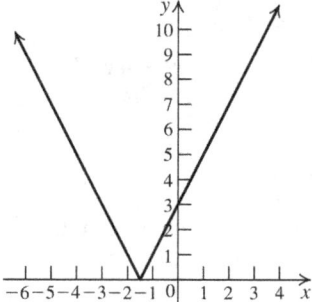

141.

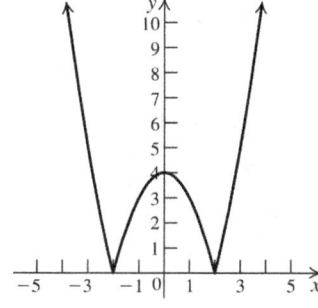

143.

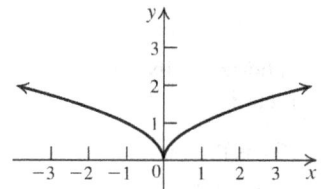

2.7 Critical Thinking/Discussion/Writing

145. a. $y = f(x+2)$ is the graph of $y = f(x)$ shifted two units left. So the x-intercepts are $-1-2 = -3$, $0-2 = -2$, and $2-2 = 0$.

b. $y = f(x-2)$ is the graph of $y = f(x)$ shifted two units right. So the x-intercepts are $-1+2 = 1$, $0+2 = 2$, and $2+2 = 4$.

c. $y = -f(x)$ is the graph of $y = f(x)$ reflected across the x-axis. The x-intercepts are the same, $-1, 0, 2$.

d. $y = f(-x)$ is the graph of $y = f(x)$ reflected across the y-axis. The x-intercepts are the opposites, $1, 0, -2$.

e. $y = f(2x)$ is the graph of $y = f(x)$ compressed horizontally by a factor of 1/2. The x-intercepts are $-\frac{1}{2}, 0, 1$.

f. $y = f\left(\frac{1}{2}x\right)$ is the graph of $y = f(x)$ stretched horizontally by a factor of 2. The x-intercepts are $-2, 0, 4$.

146. a. $y = f(x) + 2$ is the graph of $y = f(x)$ shifted two units up. The y-intercept is $2 + 2 = 4$.

b. $y = f(x) - 2$ is the graph of $y = f(x)$ shifted two units down. The y-intercept is $2 - 2 = 0$.

c. $y = -f(x)$ is the graph of $y = f(x)$ reflected across the x-axis. The y-intercept is the opposite, -2.

d. $y = f(-x)$ is the graph of $y = f(x)$ reflected across the y-axis. The y-intercept is the same, 2.

e. $y = 2f(x)$ is the graph of $y = f(x)$ stretched vertically by a factor of 2. The y-intercept is 4.

f. $y = \frac{1}{2}f(x)$ is the graph of $y = f(x)$ compressed horizontally by a factor of 1/2. The y-intercept is 1.

147. a. $y = f(x+2)$ is the graph of $y = f(x)$ shifted two units left. The domain is $[-1-2, 3-2] = [-3, 1]$. The range is the same, $[-2, 1]$.

b. $y = f(x) - 2$ is the graph of $y = f(x)$ shifted two units down. The domain is the same, $[-1, 3]$. The range is $[-2-2, 1-2] = [-4, -1]$.

c. $y = -f(x)$ is the graph of $y = f(x)$ reflected across the x-axis. The domain is the same, $[-1, 3]$. The range is the opposite, $[-1, 2]$.

d. $y = f(-x)$ is the graph of $y = f(x)$ reflected across the y-axis. The domain is the opposite, $[-3, 1]$. The range is the same, $[-2, 1]$.

e. $y = 2f(x)$ is the graph of $y = f(x)$ stretched vertically by a factor of 2. The domain is the same, $[-1, 3]$. The range is $[2(-2), 2(1)] = [-4, 2]$.

f. $y = \frac{1}{2}f(x)$ is the graph of $y = f(x)$ compressed horizontally by a factor of 1/2. The domain is the same, $[-1, 3]$. The range is $\left[\frac{1}{2}(-2), \frac{1}{2}(1)\right] = \left[-1, \frac{1}{2}\right]$.

148. a. $y = f(x+2)$ is the graph of $y = f(x)$ shifted two units left. So the relative maximum is at $x = 1 - 2 = -1$, and the relative minimum is at $x = 2 - 2 = 0$.

b. $y = f(x) - 2$ is the graph of $y = f(x)$ shifted two units down. The locations of the relative maximum and minimum do not change. Relative maximum at $x = 1$, relative minimum at $x = 2$.

c. $y = -f(x)$ is the graph of $y = f(x)$ reflected across the x-axis. The relative maximum and relative minimum switch. The relative maximum occurs at $x = 2$, and the relative minimum occurs at $x = 1$.

d. $y = f(-x)$ is the graph of $y = f(x)$ reflected across the y-axis. The relative maximum and relative minimum occur at their opposites. The relative maximum occurs at $x = -1$, and the relative minimum occurs at $x = -2$.

e. $y = 2f(x)$ is the graph of $y = f(x)$ stretched vertically by a factor of 2. The locations of the relative maximum and minimum do not change. Relative maximum at $x = 1$, relative minimum at $x = 2$.

f. $y = \frac{1}{2}f(x)$ is the graph of $y = f(x)$ compressed horizontally by a factor of 1/2. The locations of the relative maximum and minimum do not change. Relative maximum at $x = 1$, relative minimum at $x = 2$.

2.7 Getting Ready for the Next Section

149. $(5x^2 + 5x + 7) + (x^2 + 9x - 4) = 6x^2 + 14x + 3$

151. $(5x^2 + 6x - 2) - (3x^2 - 9x + 1) = 2x^2 + 15x - 3$

153. $(x-2)(x^2 + 2x + 4)$
$= x^3 + 2x^2 + 4x - 2x^2 - 4x - 8$
$= x^3 - 8$

155. $f(x) = \dfrac{2x - 3}{x^2 - 5x + 6}$
The function is not defined when the denominator is zero.
$x^2 - 5x + 6 = 0 \Rightarrow (x-2)(x-3) = 0 \Rightarrow x = 2, 3$
The domain is $(-\infty, 2) \cup (2, 3) \cup (3, \infty)$.

157. $f(x) = \sqrt{2x - 3}$
The function is defined only if $2x - 3 \geq 0$.
$2x - 3 \geq 0 \Rightarrow 2x \geq 3 \Rightarrow x \geq \dfrac{3}{2}$
The domain is $\left[\dfrac{3}{2}, \infty\right)$.

159. $\dfrac{x-1}{x-10} < 0$

First, solve $x - 1 = 0 \Rightarrow x = 1$ and $x - 10 = 0 \Rightarrow x = 10$.
So the intervals are $(-\infty, 1), (1, 10)$, and $(10, \infty)$.

Interval	Test point	Value of $\dfrac{x-1}{x-10}$	Result
$(-\infty, 1)$	0	$\frac{1}{10}$	+
$(1, 10)$	5	$-\frac{4}{5}$	−
$(10, \infty)$	15	$\frac{14}{5}$	+

Note that the fraction is undefined if $x = 10$.
The solution set is $(1, 10)$.

161. $\dfrac{-2x+8}{x^2+1} \le 0$

Set the numerator and denominator equal to zero and solve for x.
$-2x + 8 = 0 \Rightarrow -2x = -8 \Rightarrow x = 4$
$x^2 + 1 = 0$ has no real solution.
The intervals are $(-\infty, 4]$ and $[4, \infty)$.

Interval	Test point	Value of $\dfrac{-2x+8}{x^2+1}$	Result
$(-\infty, 4]$	0	8	+
$[4, \infty)$	5	$-\frac{1}{13}$	−

The solution set is $[4, \infty)$.

Interval	Test point	Value of $\dfrac{(x-3)(x-1)}{(x-5)(x+1)}$	Result
$(-\infty, -1)$	−2	$\frac{15}{7}$	+
$(-1, 1]$	0	$-\frac{3}{5}$	−
$[1, 3]$	2	$\frac{1}{9}$	+
$[3, 5)$	4	$-\frac{3}{5}$	−
$(5, \infty)$	6	$\frac{15}{7}$	+

The solution set is $(-\infty, -1) \cup [1, 3] \cup (5, \infty)$.

2.8 Combining Functions; Composite Functions

2.8 Practice Problems

1. $f(x) = 3x - 1,\ g(x) = x^2 + 2$
$(f+g)(x) = f(x) + g(x)$
$\qquad = 3x - 1 + x^2 + 2 = x^2 + 3x + 1$
$(f-g)(x) = f(x) - g(x)$
$\qquad = (3x - 1) - (x^2 + 2) = -x^2 + 3x - 3$

$(fg)(x) = f(x) \cdot g(x)$
$\qquad = (3x - 1)(x^2 + 2) = 3x^3 - x^2 + 6x - 2$

$\left(\dfrac{f}{g}\right)(x) = \dfrac{f(x)}{g(x)} = \dfrac{3x-1}{x^2+2}$

2. $f(x) = \sqrt{x-1},\ g(x) = \sqrt{3-x}$
The domain of f is $[1, \infty)$ and the domain of g is $(-\infty, 3]$. The intersection of D_f and D_g, $D_f \cap D_g = [1, 3]$.
The domain of fg is $[1, 3]$.
The domain of $\dfrac{f}{g}$ is $[1, 3)$.
The domain of $\dfrac{g}{f}$ is $(1, 3]$.

3. $f(x) = -5x,\ g(x) = x^2 + 1$
   a. $(f \circ g)(0) = f(g(0))$
   $\qquad = f(0^2 + 1) = f(1) = -5$
   b. $(g \circ f)(0) = g(f(0)) = g(-5 \cdot 0) = g(0) = 1$

4. $f(x) = 2 - x,\ g(x) = 2x^2 + 1$
   a. $(g \circ f)(x) = g(f(x)) = g(2 - x)$
   $\qquad = 2(2-x)^2 + 1$
   $\qquad = 2(4 - 4x + x^2) + 1$
   $\qquad = 8 - 8x + 2x^2 + 1 = 2x^2 - 8x + 9$
   b. $(f \circ g)(x) = f(g(x)) = f(2x^2 + 1)$
   $\qquad = 2 - (2x^2 + 1) = 1 - 2x^2$

c. $(g \circ g)(x) = g(g(x)) = g(2x^2 + 1)$
$= 2(2x^2 + 1)^2 + 1$
$= 2(4x^4 + 4x^2 + 1) + 1$
$= 8x^4 + 8x^2 + 3$

5. $f(x) = \sqrt{x+1},\ g(x) = \dfrac{2}{x-3}$

Let $A = \{x \mid g(x) \text{ is defined}\}$.

$g(x)$ is not defined if $x = 3$, so
$A = (-\infty, 3) \cup (3, \infty)$.

Let $B = \{x \mid f(g(x)) \text{ is defined}\}$.

$f(g(x)) = \sqrt{\dfrac{2}{x-3} + 1} = \sqrt{\dfrac{2 + x - 3}{x - 3}} = \sqrt{\dfrac{x-1}{x-3}}$

$f(g(x))$ is not defined if $x = 3$ or if
$\dfrac{x-1}{x-3} < 0$.

$x - 1 = 0 \Rightarrow x = 1$

Interval	Test point	Value of $\dfrac{x-1}{x-3}$	Result
$(-\infty, 1]$	0	$\dfrac{1}{3}$	+
$[1, 3)$	2	-1	−
$(3, \infty)$	4	3	+

$f(g(x))$ is not defined for $[1, 3)$, so
$B = (-\infty, 1] \cup (3, \infty)$.

The domain of $f \circ g$ is
$A \cap B = (-\infty, 1] \cup (3, \infty)$.

6. $f(x) = \sqrt{x};\ g(x) = 3 - x$

$f \circ g = f(3-x) = \sqrt{3-x}$

$f \circ g$ is not defined if $3 - x < 0 \Rightarrow 3 < x$ or $x > 3$. $f \circ g$ is defined for $x < 3$. Thus, the domain of $f \circ g$ is $(-\infty, 3]$.

7. $f(x) = \sqrt{x-1},\ g(x) = \sqrt{4-x^2}$

a. $(f \circ g)(x) = f(g(x)) = f\left(\sqrt{4-x^2}\right)$
$= \sqrt{\sqrt{4-x^2} - 1}$

The function $g(x) = \sqrt{4-x^2}$ is defined for $4 - x^2 \geq 0 \Rightarrow x^2 \leq 4 \Rightarrow -2 \leq x \leq 2$. So, $A = [-2, 2]$.

The function $f(g(x))$ is defined for
$\sqrt{4-x^2} - 1 \geq 0 \Rightarrow \sqrt{4-x^2} \geq 1 \Rightarrow$
$4 - x^2 \geq 1 \Rightarrow -x^2 \geq -3 \Rightarrow x^2 \leq 3 \Rightarrow$
$-\sqrt{3} \leq x \leq \sqrt{3}$

So, $B = \left[-\sqrt{3},\ \sqrt{3}\right]$.

The domain of $f \circ g$ is
$A \cap B = \left[-\sqrt{3},\ \sqrt{3}\right]$.

b. $(g \circ f)(x) = g(f(x)) = \sqrt{4 - \left(\sqrt{x-1}\right)^2}$
$= \sqrt{4 - (x-1)} = \sqrt{5-x}$

The function $f(x) = \sqrt{x-1}$ is defined for
$x - 1 \geq 0 \Rightarrow x \geq 1$. So, $A = [1, \infty)$.

The function $g(f(x))$ is defined for
$5 - x \geq 0 \Rightarrow 5 \geq x,$ or $x \leq 5$. So,
$B = (-\infty, 5]$.

The domain of $g \circ f$ is $A \cap B = [1, 5]$.

8. $H(x) = \dfrac{1}{\sqrt{2x^2 + 1}},\ g(x) = \sqrt{2x^2 + 1}$

If $f(x) = \dfrac{1}{x}$, then

$H(x) = (f \circ g)(x) = f\left(\sqrt{2x^2 + 1}\right) = \dfrac{1}{\sqrt{2x^2 + 1}}$.

9. a. $A = f(g(t)) = f(g(3)) = f(3t)$
$= \pi(3t)^2 = 9\pi t^2$

  b. $A = 9\pi t^2 = 9\pi(6)^2 = 324\pi$
  The area covered by the oil slick is $324\pi \approx 1018$ square miles.

10. a. $r(x) = x - 4500$

  b. $d(x) = x - 0.06x = 0.94x$

  c. i. $(r \circ d)(x) = r(0.94x) = 0.94x - 4500$

  ii. $(d \circ r)(x) = d(x - 4500)$
  $= 0.94(x - 4500)$
  $= 0.94x - 4230$

  d. $(d \circ r)(x) - (r \circ d)(x)$
  $= (0.94x - 4230) - (0.94x - 4500)$
  $= 270$

2.8 Concepts and Vocabulary

1. $(f \cdot g)(x) = \underline{f(x) \cdot g(x)}$.

3. The composition of the function f with the function g is written as $f \circ g$ and is defined by $f \circ g(x) = \underline{f(g(x))}$.

5. False. For example, if $f(x) = 2x$ and $g(x) = x^2$, then
$(f \circ g)(x) = f(g(x)) = f(x^2) = 2x^2$, while
$(g \circ f)(x) = g(f(x)) = g(2x) = 4x^2$.

7. False. The domain of $f \cdot g$ may include $g(x) = 0$, but the domain of $\frac{f}{g}$ cannot include $g(x) = 0$.

2.8 Building Skills

9. $(f + g)(-2) = f(-2) + g(-2) = 1 + 2 = 3$

11. $(f - g)(4) = f(4) - g(4) = -2 - 1 = -3$

13. $(f \cdot g)(-1) = f(-1) \cdot g(-1) = 1 \cdot (-4) = -4$

15. $\left(\frac{f}{g}\right)(-2) = \frac{f(-2)}{g(-2)} = \frac{1}{2}$

17. $(f \circ g)(1) = f(g(1)) = f(-2) = 1$

19. $(f \circ g)(-3) = f(g(-3)) = f(0) = 0$

21. a. $(f + g)(-1) = f(-1) + g(-1)$
$= 2(-1) + -(-1) = -2 + 1 = -1$

  b. $(f - g)(0) = f(0) - g(0) = 2(0) - (-0) = 0$

  c. $(f \cdot g)(2) = f(2) \cdot g(2) = 2(2) \cdot (-2) = -8$

  d. $\left(\frac{f}{g}\right)(1) = \frac{f(1)}{g(1)} = \frac{2(1)}{-1} = -2$

23. a. $(f + g)(-1) = f(-1) + g(-1)$
$= \frac{1}{\sqrt{-1+2}} + (2(-1) + 1) = 0$

  b. $(f - g)(0) = f(0) - g(0)$
  $= \frac{1}{\sqrt{0+2}} - (2(0) + 1) = \frac{\sqrt{2}}{2} - 1$

  c. $(f \cdot g)(2) = f(2) \cdot g(2)$
  $= \frac{1}{\sqrt{2+2}} \cdot (2(2) + 1) = \frac{5}{2}$

  d. $\left(\frac{f}{g}\right)(1) = \frac{f(1)}{g(1)} = \frac{\frac{1}{\sqrt{1+2}}}{2(1)+1} = \frac{1}{3\sqrt{3}} = \frac{\sqrt{3}}{9}$

25. a. $f + g = x^2 + x - 3$; domain: $(-\infty, \infty)$

  b. $f - g = x - 3 - x^2 = -x^2 + x - 3$; domain: $(-\infty, \infty)$

  c. $f \cdot g = (x - 3)x^2 = x^3 - 3x^2$; domain: $(-\infty, \infty)$

  d. $\frac{f}{g} = \frac{x-3}{x^2}$; domain: $(-\infty, 0) \cup (0, \infty)$

  e. $\frac{g}{f} = \frac{x^2}{x-3}$; domain: $(-\infty, 3) \cup (3, \infty)$

27. a. $f + g = (x^3 - 1) + (2x^2 + 5) = x^3 + 2x^2 + 4$; domain: $(-\infty, \infty)$

  b. $f - g = (x^3 - 1) - (2x^2 + 5) = x^3 - 2x^2 - 6$; domain: $(-\infty, \infty)$

  c. $f \cdot g = (x^3 - 1)(2x^2 + 5)$
  $= 2x^5 + 5x^3 - 2x^2 - 5$; domain: $(-\infty, \infty)$

d. $\dfrac{f}{g} = \dfrac{x^3-1}{2x^2+5}$; domain: $(-\infty, \infty)$

e. $\dfrac{g}{f} = \dfrac{2x^2+5}{x^3-1}$; domain: $(-\infty, 1) \cup (1, \infty)$

29. a. $f+g = 2x + \sqrt{x} - 1$; domain: $[0, \infty)$

b. $f-g = 2x - \sqrt{x} - 1$; domain: $[0, \infty)$

c. $f \cdot g = (2x-1)\sqrt{x} = 2x\sqrt{x} - \sqrt{x}$; domain: $[0, \infty)$

d. $\dfrac{f}{g} = \dfrac{2x-1}{\sqrt{x}}$; domain: $(0, \infty)$

e. $\dfrac{g}{f} = \dfrac{\sqrt{x}}{2x-1}$; the numerator is defined only for $x \geq 0$, while the denominator $= 0$ when $x = \dfrac{1}{2}$, so the domain is $\left[0, \dfrac{1}{2}\right) \cup \left(\dfrac{1}{2}, \infty\right)$.

31. a. $f+g = x - 6 + \sqrt{x-3}$; domain: $[3, \infty)$

b. $f-g = x - 6 - \sqrt{x-3}$; domain: $[3, \infty)$

c. $f \cdot g = (x-6)\sqrt{x-3}$; domain: $[3, \infty)$

d. $\dfrac{f}{g} = \dfrac{x-6}{\sqrt{x-3}}$; domain: $(3, \infty)$

e. $\dfrac{g}{f} = \dfrac{\sqrt{x-3}}{x-6}$; the numerator is defined only for $x \geq 3$, while the denominator $= 0$ when $x = 6$, so the domain is $[3, 6) \cup (6, \infty)$.

33. a. $f+g = 1 - \dfrac{2}{x+1} + \dfrac{1}{x}$
$= \dfrac{x(x+1) - 2x + (x+1)}{x(x+1)}$
$= \dfrac{x^2 + x - 2x + x + 1}{x(x+1)} = \dfrac{x^2+1}{x(x+1)}$
domain: $(-\infty, -1) \cup (-1, 0) \cup (0, \infty)$

b. $f-g = 1 - \dfrac{2}{x+1} - \dfrac{1}{x}$
$= \dfrac{x(x+1) - 2x - (x+1)}{x(x+1)}$
$= \dfrac{x^2 + x - 2x - x - 1}{x(x+1)} = \dfrac{x^2 - 2x - 1}{x(x+1)}$
domain: $(-\infty, -1) \cup (-1, 0) \cup (0, \infty)$

c. $f \cdot g = \left(1 - \dfrac{2}{x+1}\right)\dfrac{1}{x} = \left(\dfrac{x+1-2}{x+1}\right)\dfrac{1}{x}$
$= \left(\dfrac{x-1}{x+1}\right)\dfrac{1}{x} = \dfrac{x-1}{x(x+1)}$
domain: $(-\infty, -1) \cup (-1, 0) \cup (0, \infty)$

d. $\dfrac{f}{g} = \dfrac{1 - \frac{2}{x+1}}{\frac{1}{x}} = \left(1 - \dfrac{2}{x+1}\right)(x)$
$= \left(\dfrac{x+1-2}{x+1}\right)x = \dfrac{x(x-1)}{x+1}$
domain: $(-\infty, -1) \cup (-1, 0) \cup (0, \infty)$

e. $\dfrac{g}{f} = \dfrac{\frac{1}{x}}{1 - \frac{2}{x+1}} = \dfrac{\frac{1}{x}(x+1)}{\left(1 - \frac{2}{x+1}\right)(x+1)} = \dfrac{\frac{x+1}{x}}{x+1-2}$
$= \dfrac{\frac{x+1}{x}}{x-1} = \dfrac{x+1}{x(x-1)}$

The denominator equals zero when $x = 0$ or $x = 1$, so the domain is $(-\infty, -1) \cup (-1, 0) \cup (0, 1) \cup (1, \infty)$.

35. a. $f+g = \dfrac{2}{x+1} + \dfrac{x}{x+1} = \dfrac{2+x}{x+1}$
Neither f nor g is defined for $x = -1$, so the domain is $(-\infty, -1) \cup (-1, \infty)$.

b. $f-g = \dfrac{2}{x+1} - \dfrac{x}{x+1} = \dfrac{2-x}{x+1}$.
domain: $(-\infty, -1) \cup (-1, \infty)$.

c. $f \cdot g = \left(\dfrac{2}{x+1}\right)\left(\dfrac{x}{x+1}\right) = \dfrac{2x}{(x+1)^2}$.
domain: $(-\infty, -1) \cup (-1, \infty)$.

d. $\dfrac{f}{g} = \dfrac{\frac{2}{x+1}}{\frac{x}{x+1}} = \dfrac{2}{x}$. Neither f nor g is defined for $x = -1$, and f/g is not defined for $x = 0$, so the domain is $(-\infty, -1) \cup (-1, 0) \cup (0, \infty)$.

e. $\dfrac{f}{g} = \dfrac{\frac{x}{x+1}}{\frac{2}{x+1}} = \dfrac{x}{2}$. Neither f nor g is defined for $x = -1$, so the domain is $(-\infty, -1) \cup (-1, \infty)$.

37. $f(x) = \dfrac{x^2}{x+1}$; $g(x) = \dfrac{2x}{x^2-1}$

   a. $f+g = \dfrac{x^2}{x+1} + \dfrac{2x}{x^2-1} = \dfrac{x^2(x-1)}{x^2-1} + \dfrac{2x}{x^2-1}$
   $= \dfrac{x^3 - x^2 + 2x}{x^2-1}$

   f is not defined for $x = -1$, g is not defined for $x = \pm 1$, and $f+g$ is not defined for either -1 or 1, so the domain is $(-\infty, -1) \cup (-1, 1) \cup (1, \infty)$.

   b. $f-g = \dfrac{x^2}{x+1} - \dfrac{2x}{x^2-1} = \dfrac{x^2(x-1)}{x^2-1} - \dfrac{2x}{x^2-1}$
   $= \dfrac{x^3 - x^2 - 2x}{x^2-1} = \dfrac{x(x^2 - x - 2)}{x^2-1}$
   $= \dfrac{x(x-2)(x+1)}{(x-1)(x+1)} = \dfrac{x^2 - 2x}{x-1}$

   f is not defined for $x = -1$, g is not defined for $x = \pm 1$, and $f-g$ is not defined for 1, so the domain is $(-\infty, -1) \cup (-1, 1) \cup (1, \infty)$.

   c. $f \cdot g = \dfrac{x^2}{x+1} \cdot \dfrac{2x}{x^2-1} = \dfrac{2x^3}{x^3 + x^2 - x - 1}$

   f is not defined for $x = -1$, g is not defined for $x = \pm 1$, and fg is not defined for either -1 or 1, so the domain is $(-\infty, -1) \cup (-1, 1) \cup (1, \infty)$.

   d. $\dfrac{f}{g} = \dfrac{\dfrac{x^2}{x+1}}{\dfrac{2x}{x^2-1}} = \dfrac{x^2}{x+1} \cdot \dfrac{x^2-1}{2x} = \dfrac{x(x-1)}{2}$

   f is not defined for $x = -1$, g is not defined for $x = \pm 1$, and f/g is not defined for either -1, 0, or 1, so the domain is $(-\infty, -1) \cup (-1, 0) \cup (0, 1) \cup (1, \infty)$.

   e. $\dfrac{g}{f} = \dfrac{\dfrac{2x}{x^2-1}}{\dfrac{x^2}{x+1}} = \dfrac{2x}{x^2-1} \cdot \dfrac{x+1}{x^2} = \dfrac{2}{x(x-1)}$

   Neither f nor g is defined for $x = -1$ and g/f is not defined for $x = 0$ or $x = 1$, so the domain is $(-\infty, -1) \cup (-1, 0) \cup (0, 1) \cup (1, \infty)$.

39. $f(x) = \sqrt{x-1}$; $g(x) = \sqrt{5-x}$

   a. $f \cdot g = \sqrt{x-1} \cdot \sqrt{5-x}$
   f is not defined for $x < 1$, g is not defined for $x > 5$. The domain is $[1, 5]$.

   b. $\dfrac{f}{g} = \dfrac{\sqrt{x-1}}{\sqrt{5-x}}$
   f is not defined for $x < 1$, g is not defined for $x > 5$. The denominator is zero when $x = 5$. The domain is $[1, 5)$.

41. $f(x) = \sqrt{x+2}$; $g(x) = \sqrt{9-x^2}$

   a. $f \cdot g = \sqrt{x+2} \cdot \sqrt{9-x^2}$
   f is not defined for $x < -2$, g is defined for $[-3, 3]$. The domain is $[-2, 3]$.

   b. $\dfrac{f}{g} = \dfrac{\sqrt{x+2}}{\sqrt{9-x^2}}$
   f is not defined for $x < -2$, g is defined for $[-3, 3]$. The denominator is zero when $x = -3$ or $x = 3$. The domain is $[-2, 3)$.

43. $(g \circ f)(x) = 2(x^2 - 1) + 3 = 2x^2 + 1$;
   $(g \circ f)(2) = 2(2^2 - 1) + 3 = 9$;
   $(g \circ f)(-3) = 2((-3)^2 - 1) + 3 = 19$

45. $(f \circ g)(2) = 2(2(2^2) - 3) + 1 = 11$

47. $(f \circ g)(-3) = 2(2(-3)^2 - 3) + 1 = 31$

49. $(f \circ g)(0) = 2(2(0^2) - 3) + 1 = -5$

51. $(f \circ g)(-c) = 2(2(-c)^2 - 3) + 1 = 4c^2 - 5$

53. $(g \circ f)(a) = 2(2a+1)^2 - 3$
   $= 2(4a^2 + 4a + 1) - 3$
   $= 8a^2 + 8a - 1$

55. $(f \circ f)(1) = 2(2(1) + 1) + 1 = 7$

57. $f(x) = \dfrac{1}{x}$; $g(x) = 10 - 5x$

   $(f \circ g)(x) = \dfrac{1}{10 - 5x}$

   The domain of $f \circ g$ is the set of all real numbers such that $10 - 5x \neq 0$, or $x \neq 2$. The domain of $f \circ g$ is $(-\infty, 2) \cup (2, \infty)$.

59. $f(x) = \sqrt{x}$; $g(x) = 2x - 8$
$(f \circ g)(x) = \sqrt{2x - 8}$
The domain of $f \circ g$ is the set of all real numbers such that $2x - 8 \geq 0$, or $x \geq 4$. The domain of $f \circ g$ is $[4, \infty)$.

61. $(f \circ g)(x) = \dfrac{2}{\dfrac{1}{x}+1} = \dfrac{2}{\dfrac{x+1}{x}} = \dfrac{2x}{x+1}$

The domain of g is $(-\infty, 0) \cup (0, \infty)$. Since -1 is not in the domain of f, we must exclude those values of x that make $g(x) = -1$.
$\dfrac{1}{x} = -1 \Rightarrow x = -1$
Thus, the domain of $f \circ g$ is
$(-\infty, -1) \cup (-1, 0) \cup (0, \infty)$.

63. $(f \circ g)(x) = \sqrt{(2 - 3x) - 3} = \sqrt{-1 - 3x}$.
The domain of g is $(-\infty, \infty)$. Since f is not defined for $(-\infty, 3)$, we must exclude those values of x that make $g(x) < 3$.
$2 - 3x < 3 \Rightarrow -3x < 1 \Rightarrow x > -\dfrac{1}{3}$
Thus, the domain of $f \circ g$ is $\left(-\infty, -\dfrac{1}{3}\right]$.

65. $(f \circ g)(x) = |x^2 - 1|$; domain: $(-\infty, \infty)$

67. a. $(f \circ g)(x) = 2(x + 4) - 3 = 2x + 5$;
domain: $(-\infty, \infty)$

b. $(g \circ f)(x) = (2x - 3) + 4 = 2x + 1$;
domain: $(-\infty, \infty)$

c. $(f \circ f)(x) = 2(2x - 3) - 3 = 4x - 9$;
domain: $(-\infty, \infty)$

d. $(g \circ g)(x) = (x + 4) + 4 = x + 8$;
domain: $(-\infty, \infty)$

69. a. $(f \circ g)(x) = 1 - 2(1 + x^2) = -2x^2 - 1$;
domain: $(-\infty, \infty)$

b. $(g \circ f)(x) = 1 + (1 - 2x)^2 = 4x^2 - 4x + 2$;
domain: $(-\infty, \infty)$

c. $(f \circ f)(x) = 1 - 2(1 - 2x) = 4x - 1$;
domain: $(-\infty, \infty)$

d. $(g \circ g)(x) = 1 + (1 + x^2)^2 = x^4 + 2x^2 + 2$;
domain: $(-\infty, \infty)$

71. a. $(f \circ g)(x) = 2(2x - 1)^2 + 3(2x - 1)$
$= 2(4x^2 - 4x + 1) + 6x - 3$
$= 8x^2 - 2x - 1$; domain: $(-\infty, \infty)$

b. $(g \circ f)(x) = 2(2x^2 + 3x) - 1 = 4x^2 + 6x - 1$;
domain: $(-\infty, \infty)$

c. $(f \circ f)(x) = 2(2x^2 + 3x)^2 + 3(2x^2 + 3x)$
$= 2(4x^4 + 12x^3 + 9x^2) + 6x^2 + 9x$
$= 8x^4 + 24x^3 + 24x^2 + 9x$;
domain: $(-\infty, \infty)$

d. $(g \circ g)(x) = 2(2x - 1) - 1 = 4x - 3$;
domain: $(-\infty, \infty)$

73. a. $(f \circ g)(x) = \left(\sqrt{x}\right)^2 = x$; domain: $[0, \infty)$

b. $(g \circ f)(x) = \sqrt{x^2} = |x|$; domain: $(-\infty, \infty)$

c. $(f \circ f)(x) = \left(x^2\right)^2 = x^4$; domain: $(-\infty, \infty)$

d. $(g \circ g)(x) = \sqrt{\sqrt{x}} = \sqrt[4]{x}$; domain: $[0, \infty)$

75. a. $(f \circ g)(x) = \dfrac{1}{2\left(\dfrac{1}{x^2}\right) - 1} = \dfrac{1}{\dfrac{2 - x^2}{x^2}}$
$= \dfrac{x^2}{2 - x^2} = -\dfrac{x^2}{x^2 - 2}$.

The domain of g is $(-\infty, 0) \cup (0, \infty)$. Since $\dfrac{1}{2}$ is not in the domain of f, we must find those values of x that make $g(x) = \dfrac{1}{2}$.
$\dfrac{1}{x^2} = \dfrac{1}{2} \Rightarrow x^2 = 2 \Rightarrow x = \pm\sqrt{2}$
Thus, the domain of $f \circ g$ is
$\left(-\infty, -\sqrt{2}\right) \cup \left(-\sqrt{2}, 0\right) \cup \left(0, \sqrt{2}\right) \cup \left(\sqrt{2}, \infty\right)$.

b. $(g \circ f) = \dfrac{1}{\left(\dfrac{1}{2x-1}\right)^2} = (2x - 1)^2$

The domain of f is $\left(-\infty, \dfrac{1}{2}\right) \cup \left(\dfrac{1}{2}, \infty\right)$. Since 0 is not in the domain of g, we must find those values of x that make $f(x) = 0$. However, there are no such values, so the domain of $g \circ f$ is $\left(-\infty, \dfrac{1}{2}\right) \cup \left(\dfrac{1}{2}, \infty\right)$.

c. $(f \circ f)(x) = \dfrac{1}{2\left(\dfrac{1}{2x-1}\right)-1} = \dfrac{1}{\dfrac{2-2x+1}{2x-1}}$

$= \dfrac{2x-1}{3-2x} = -\dfrac{2x-1}{2x-3}$

The domain of f is $\left(-\infty, \dfrac{1}{2}\right) \cup \left(\dfrac{1}{2}, \infty\right)$.

$-\dfrac{2x-1}{2x-3}$ is defined for $\left(-\infty, \dfrac{3}{2}\right) \cup \left(\dfrac{3}{2}, \infty\right)$,

so the domain of $f \circ f$ is

$\left(-\infty, \dfrac{1}{2}\right) \cup \left(\dfrac{1}{2}, \dfrac{3}{2}\right) \cup \left(\dfrac{3}{2}, \infty\right)$.

d. $(g \circ g)(x) = \dfrac{1}{\left(\dfrac{1}{x^2}\right)} = x^4$.

The domain of g is $(-\infty, 0) \cup (0, \infty)$, while $g \circ g = x^4$ is defined for all real numbers. Thus, the domain of $g \circ g$ is $(-\infty, 0) \cup (0, \infty)$.

77. a. $(f \circ g)(x) = \sqrt{\sqrt{4-x}-1}$; domain: $(-\infty, 3]$

b. $(g \circ f)(x) = \sqrt{4-\sqrt{x-1}}$; domain: $[1, 17]$

c. $(f \circ f)(x) = \sqrt{\sqrt{x-1}-1}$; domain: $[2, \infty)$

d. $(g \circ g)(x) = \sqrt{4-\sqrt{4-x}}$; domain: $[-12, 4]$

79. a. $(f \circ g)(x) = \dfrac{1-\dfrac{x+3}{x-4}}{\dfrac{x+3}{x-4}+2} = \dfrac{\left(1-\dfrac{x+3}{x-4}\right)(x-4)}{\left(\dfrac{x+3}{x-4}+2\right)(x-4)}$

$= \dfrac{(x-4)-(x+3)}{(x+3)+2(x-4)} = -\dfrac{7}{3x-5}$

The domain of g is $(-\infty, 4) \cup (4, \infty)$. The denominator of $f \circ g$ is 0 when $x = \dfrac{5}{3}$, so the domain of $f \circ g$ is

$\left(-\infty, \dfrac{5}{3}\right) \cup \left(\dfrac{5}{3}, 4\right) \cup (4, \infty)$.

b. $(g \circ f)(x) = \dfrac{\dfrac{1-x}{x+2}+3}{\dfrac{1-x}{x+2}-4} = \dfrac{\left(\dfrac{1-x}{x+2}+3\right)(x+2)}{\left(\dfrac{1-x}{x+2}-4\right)(x+2)}$

$= \dfrac{(1-x)+3(x+2)}{(1-x)-4(x+2)} = \dfrac{2x+7}{-5x-7}$

$= -\dfrac{2x+7}{5x+7}$

The domain of f is $(-\infty, -2) \cup (-2, \infty)$. The denominator of $g \circ f$ is 0 when $x = -\dfrac{7}{5}$, so, the domain of $g \circ f$ is

$(-\infty, -2) \cup \left(-2, -\dfrac{7}{5}\right) \cup \left(-\dfrac{7}{5}, \infty\right)$.

c. $(f \circ f)(x) = \dfrac{1-\dfrac{1-x}{x+2}}{\dfrac{1-x}{x+2}+2} = \dfrac{\left(1-\dfrac{1-x}{x+2}\right)(x+2)}{\left(\dfrac{1-x}{x+2}+2\right)(x+2)}$

$= \dfrac{(x+2)-(1-x)}{(1-x)+2(x+2)} = \dfrac{2x+1}{x+5}$

The domain of f is $(-\infty, -2) \cup (-2, \infty)$. The denominator of $f \circ f$ is 0 when $x = -5$, so, the domain of $f \circ f$ is

$(-\infty, -5) \cup (-5, -2) \cup (-2, \infty)$.

d. $(g \circ g)(x) = \dfrac{\dfrac{x+3}{x-4}+3}{\dfrac{x+3}{x-4}-4} = \dfrac{\left(\dfrac{x+3}{x-4}+3\right)(x-4)}{\left(\dfrac{x+3}{x-4}-4\right)(x-4)}$

$= \dfrac{(x+3)+3(x-4)}{(x+3)-4(x-4)} = \dfrac{4x-9}{-3x+19}$

$= -\dfrac{4x-9}{3x-19}$

The domain of g is $(-\infty, 4) \cup (4, \infty)$. The denominator of $g \circ g$ is 0 when $x = \dfrac{19}{3}$, so the domain of $g \circ g$ is

$(-\infty, 4) \cup \left(4, \dfrac{19}{3}\right) \cup \left(\dfrac{19}{3}, \infty\right)$.

In exercises 81–90, sample answers are given. Other answers are possible.

81. $H(x) = \sqrt{x+2} \Rightarrow f(x) = \sqrt{x}, g(x) = x+2$

83. $H(x) = \left(x^2-3\right)^{10} \Rightarrow f(x) = x^{10}, g(x) = x^2-3$

85. $H(x) = \dfrac{1}{3x-5} \Rightarrow f(x) = \dfrac{1}{x}, g(x) = 3x-5$

87. $H(x) = \sqrt[3]{x^2-7} \Rightarrow f(x) = \sqrt[3]{x}, g(x) = x^2-7$

89. $H(x) = \dfrac{1}{\left|x^3-1\right|} \Rightarrow f(x) = \dfrac{1}{|x|}, g(x) = x^3-1$

2.8 Applying the Concepts

91. a. $f(x)$ is the cost function.

b. $g(x)$ is the revenue function.

c. $h(x)$ is the selling price of x shirts including sales tax.

d. $P(x)$ is the profit function.

93. a. $P(x) = R(x) - C(x) = 25x - (350 + 5x)$
$= 20x - 350$

b. $P(20) = 20(20) - 350 = 50$. This represents the profit when 20 radios are sold.

c. $P(x) = 20x - 350; 500 = 20x - 350 \Rightarrow x = 43$

d. $C = 350 + 5x \Rightarrow x = \dfrac{C - 350}{5} = x(C)$.

$(R \circ x)(C) = 25\left(\dfrac{C - 350}{5}\right) = 5C - 1750$.

This function represents the revenue in terms of the cost C.

95. a. $f(x) = 0.7x$

b. $g(x) = x - 5$

c. $(g \circ f)(x) = 0.7x - 5$

d. $(f \circ g)(x) = 0.7(x - 5)$

e. $(f \circ g) - (g \circ f) = 0.7(x - 5) - (0.7x - 5)$
$= 0.7x - 3.5 - 0.7x + 5$
$= \$1.50$

97. a. $f(x) = 1.1x; g(x) = x + 8$

b. $(f \circ g)(x) = 1.1(x + 8) = 1.1x + 8.8$
. This represents a final test score computed by first adding 8 points to the original score and then increasing the total by 10%.

c. $(g \circ f)(x) = 1.1x + 8$
This represents a final test score computed by first increasing the original score by 10% and then adding 8 points.

d. $(f \circ g)(70) = 1.1(70 + 8) = 85.8;$
$(g \circ f)(70) = 1.1(70) + 8 = 85.0;$

e. $(f \circ g)(x) \neq (g \circ f)(x)$

f. (i) $(f \circ g)(x) = 1.1x + 8.8 \geq 90 \Rightarrow x \geq 73.82$

(ii) $(g \circ f)(x) = 1.1x + 8 \geq 90 \Rightarrow x \geq 74.55$

99. a. $f(x) = \pi x^2$

b. $g(x) = \pi(x + 30)^2$

c. $g(x) - f(x)$ represents the area between the fountain and the fence.

d. The circumference of the fence is $2\pi(x + 30)$.
$10.5(2\pi(x + 30)) = 4200 \Rightarrow$
$\pi(x + 30) = 200 \Rightarrow$
$\pi x + 30\pi = 200 \Rightarrow \pi x = 200 - 30\pi$.

$g(x) - f(x) = \pi(x + 30)^2 - \pi x^2$
$= \pi(x^2 + 60x + 900) - \pi x^2$
$= 60\pi x + 900\pi$. Now substitute $200 - 30\pi$ for πx to compute the estimate:
$1.75[60(200 - 30\pi) + 900\pi]$
$= 1.75(12,000 - 900\pi) \approx \$16,052$.

101. a. $(f \circ g)(t) = \pi(2t + 1)^2$

b. $A(t) = f(2t + 1) = \pi(2t + 1)^2$

c. They are the same.

2.8 Beyond the Basics

103. a. When you are looking for the domain of the sum of two functions that are given as sets, you are looking for the intersection of their domains. Since the x-values that f and g have in common are -2, 1, and 3, the domain of $f + g$ is $\{-2, 1, 3\}$. Now add the y-values.
$(f + g)(-2) = 3 + 0 = 3$
$(f + g)(1) = 2 + (-2) = 0$
$(f + g)(3) = 0 + 2 = 2$
Thus, $f + g = \{(-2, 3), (1, 0), (3, 2)\}$.

b. When you are looking for the domain of the product of two functions that are given as sets, you are looking for the intersection of their domains. Since the x-values that f and g have in common are -2, 1, and 3, the domain of $f + g$ is $\{-2, 1, 3\}$. Now multiply the y-values.
$(fg)(-2) = 3 \cdot 0 = 0$
$(fg)(1) = 2 \cdot (-2) = -4$
$(fg)(3) = 0 \cdot 2 = 0$
Thus, $fg = \{(-2, 0), (1, -4), (3, 0)\}$.

c. When you are looking for the domain of the quotient of two functions that are given as sets, you are looking for the intersection of their domains and values of x that do not cause the denominator to equal zero. The x-values that f and g have in common are $-2, 1,$ and 3; however, $g(-2) = 0$, so the domain is $\{1, 3\}$. Now divide the y-values.

$\left(\dfrac{f}{g}\right)(1) = \dfrac{2}{-2} = -1$

$\left(\dfrac{f}{g}\right)(3) = \dfrac{0}{2} = 0$

Thus, $\dfrac{f}{g} = \{(1, -1), (3, 0)\}$.

d. When you are looking for the domain of the composition of two functions that are given as sets, you are looking for values that come from the domain of the inside function and when you plug those values of x into the inside function, the output is in the domain of the outside function.

$f(g(-2)) = f(0)$, which is undefined

$f(g(0)) = f(2) = 1$,

$f(g(1)) = f(-2) = 3$,

$f(g(3)) = f(2) = 1$

Thus, $f \circ g = \{(0, 1), (1, 3), (3, 1)\}$.

105. a. $f(-x) = h(-x) + h(-(-x)) = h(-x) + h(x)$
$= f(x) \Rightarrow f(x)$ is an even function.

b. $g(-x) = h(-x) - h(-(-x)) = h(-x) - h(x)$
$= -g(x) \Rightarrow g(x)$ is an odd function.

c. $\begin{cases} f(x) = h(x) + h(-x) \\ g(x) = h(x) - h(-x) \end{cases} \Rightarrow$
$f(x) + g(x) = 2h(x) \Rightarrow$
$h(x) = \dfrac{f(x) + g(x)}{2} = \dfrac{f(x)}{2} + \dfrac{g(x)}{2} \Rightarrow$
$h(x)$ is the sum of an even function and an odd function.

107. $f(x) = \sqrt{\dfrac{1-|x|}{2-|x|}}$

$f(x)$ is defined if $\dfrac{1-|x|}{2-|x|} \geq 0$ and $2-|x| \neq 0$.

$2 - |x| = 0 \Rightarrow 2 = |x| \Rightarrow x = \pm 2$

Thus, the values -2 and 2 are not in the domain of f.

$\dfrac{1-|x|}{2-|x|} \geq 0$ if $1-|x| \geq 0$ and $2-|x| > 0$, or if $1-|x| \leq 0$ and $2-|x| < 0$.

Case 1: $1-|x| \geq 0$ and $2-|x| > 0$.

$1 - |x| \geq 0 \Rightarrow 1 \geq |x| \Rightarrow -1 \leq x \leq 1$

$2 - |x| > 0 \Rightarrow 2 > |x| \Rightarrow -2 < x < 2$

Thus, $1 - |x| \geq 0$ and $2 - |x| > 0 \Rightarrow -1 \leq x \leq 1$.

Case 2: $1-|x| \leq 0$ and $2-|x| < 0$.

$1 - |x| \leq 0 \Rightarrow 1 \leq |x| \Rightarrow (-\infty, -1] \cup [1, \infty)$

$2 - |x| < 0 \Rightarrow 3 \leq |x| \Rightarrow (-\infty, -2) \cup (2, \infty)$

Thus, $1 - |x| \leq 0$ and $2 - |x| < 0 \Rightarrow$
$(-\infty, -2) \cup (2, \infty)$.

The domain of f is
$(-\infty, -2) \cup [-1, 1] \cup (2, \infty)$.

2.8 Critical Thinking/Discussion/Writing

109. a. The domain of $f(x)$ is $(-\infty, 0) \cup [1, \infty)$.

b. The domain of $g(x)$ is $[0, 2]$.

c. The domain of $f(x) + g(x)$ is $[1, 2]$.

d. The domain of $\dfrac{f(x)}{g(x)}$ is $[1, 2)$.

110. a. The domain of f is $(-\infty, 0)$. The domain of $f \circ f$ is $\varnothing$ because $f \circ f = \dfrac{1}{\sqrt{-\dfrac{1}{\sqrt{-x}}}}$ and the denominator is the square root of a negative number.

b. The domain of f is $(-\infty, 1)$. The domain of $f \circ f$ is $(-\infty, 0)$ because
$f \circ f = \dfrac{1}{\sqrt{1 - \dfrac{1}{\sqrt{1-x}}}}$ and the denominator must be greater than 0. If $x = 0$, then the denominator $= 0$.

111. a. The sum of two even functions is an even function.
$f(x) = f(-x)$ and $g(x) = g(-x) \Rightarrow$
$(f + g)(x) = f(x) + g(x) = f(-x) + g(-x)$
$= (f + g)(-x)$.

b. The sum of two odd functions is an odd function.
$f(-x) = -f(x)$ and $g(-x) = -g(x) \Rightarrow$
$(f+g)(-x) = f(-x) + g(-x) = -f(x) - g(x)$
$= -(f+g)(x).$

c. The sum of an even function and an odd function is neither even nor odd.
$f(x)$ even $\Rightarrow f(x) = f(-x)$ and $g(x)$ odd $\Rightarrow$
$g(-x) = -g(x) \Rightarrow f(-x) + g(-x) =$
$f(x) + (-g(x)),$ which is neither even nor odd.

d. The product of two even functions is an even function.
$f(x) = f(-x)$ and $g(x) = g(-x) \Rightarrow$
$(f \cdot g)(x) = f(x) \cdot g(x) = f(-x) \cdot (g(-x))$
$= (f \cdot g)(-x).$

e. The product of two odd functions is an even function.
$f(-x) = -f(x)$ and $g(-x) = -g(x) \Rightarrow$
$(f \cdot g)(-x) = f(-x) \cdot g(-x) = -f(x) \cdot (-g(x))$
$= (f \cdot g)(x).$

f. The product of an even function and an odd function is an odd function.
$f(x)$ even $\Rightarrow f(x) = f(-x)$ and $g(x)$ odd $\Rightarrow$
$g(-x) = -g(x) \Rightarrow$
$f(-x) \cdot g(-x) = f(x) \cdot (-g(x)) = -(f \cdot g)(x)$

112. a. $f(-x) = -f(x)$ and $g(-x) = -g(x) \Rightarrow$
$(f \circ g)(-x) = f(g(-x)) = f(-g(x)) =$
$-f(g(x)) \Rightarrow (f \circ g)(x)$ is odd.

b. $f(x) = f(-x)$ and $g(x) = g(-x) \Rightarrow$
$(f \circ g)(-x) = f(g(-x)) = f(g(x)) \Rightarrow$
$(f \circ g)(x)$ is even.

c. $f(x)$ odd $\Rightarrow f(-x) = -f(x)$ and
$g(x)$ even $\Rightarrow g(x) = g(-x) \Rightarrow (f \circ g)(-x)$
$f(g(x)) = f(g(-x)) \Rightarrow (f \circ g)(x)$ is even.

d. $f(x)$ even $\Rightarrow f(x) = f(-x)$ and $g(x)$ odd $\Rightarrow$
$g(-x) = -g(x) \Rightarrow (f \circ g)(-x) = f(-g(x))$
$= f(g(x)) = (f \circ g)(-x) \Rightarrow (f \circ g)(x)$ is even.

2.8 Getting Ready for the Next Section

113. a. Yes, R defines a function.

b. $S = \{(2, -3), (1, -1), (3, 1), (1, 2)\}$
No, S does not define a function since the first value 1 maps to two different second values, -1 and 2.

115. $x = 2y + 3 \Rightarrow x - 3 = 2y \Rightarrow \dfrac{x-3}{2} = y$

117. $x^2 + y^2 = 4, x \leq 0 \Rightarrow x^2 = 4 - y^2 \Rightarrow$
$x = -\sqrt{4 - y^2}$

2.9 Inverse Functions

2.9 Practice Problems

1. $f(x) = (x-1)^2$ is not one-to-one because the horizontal line $y = 1$ intersects the graph at two different points.

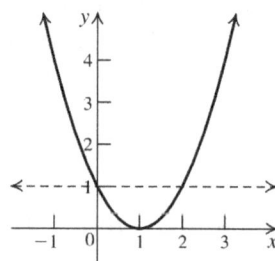

2. a. $f^{-1}(12) = -3$

b. $f(9) = 4$

3. $f(x) = 3x - 1,\ g(x) = \dfrac{x+1}{3}$

$(f \circ g)(x) = f\left(\dfrac{x+1}{3}\right) = 3\left(\dfrac{x+1}{3}\right) - 1 = x$

$(g \circ f)(x) = g(3x - 1) = \dfrac{3x - 1 + 1}{3} = x$

Because $f(g(x)) = g(f(x)) = x,$ the two functions are inverses.

4. The graph of f^{-1} is the reflection of the graph of f about the line $y = x$.

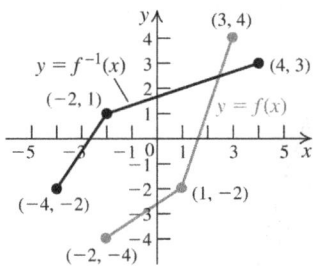

5. $f(x) = -2x + 3$ is a one-to-one function, so the function has an inverse. Interchange the variables and solve for y:
$f(x) = y = -2x + 3 \Rightarrow x = -2y + 3 \Rightarrow$
$\dfrac{x-3}{-2} = y \Rightarrow y = f^{-1}(x) = \dfrac{3-x}{2}$.

6. Interchange the variables and solve for y:
$f(x) = y = \dfrac{x}{x+3}, x \neq -3$
$x = \dfrac{y}{y+3} \Rightarrow xy + 3x = y \Rightarrow 3x = y - xy \Rightarrow$
$3x = y(1-x) \Rightarrow \dfrac{3x}{1-x} = y \Rightarrow$
$f^{-1}(x) = \dfrac{3x}{1-x}, x \neq 1$

7. $f(x) = \dfrac{x}{x+3}$
The function is not defined if the denominator is zero, so the domain is $(-\infty, -3) \cup (-3, \infty)$.
The range of the function is the same as the domain of the inverse, thus the range is $(-\infty, 1) \cup (1, \infty)$.

8. G is one-to-one because the domain is restricted, so an inverse exists.
$G(x) = y = x^2 - 1, x \leq 0$. Interchange the variables and solve for y:
$x = y^2 - 1, y \leq 0 \Rightarrow y = G^{-1}(x) = -\sqrt{x+1}$.

9. From the text, we have $d = \dfrac{11p}{5} - 33$.
$d = \dfrac{11 \cdot 1650}{5} - 33 = 3597$
The bell was 3597 feet below the surface when the gauge failed.

2.9 Concepts and Vocabulary

1. If no horizontal line intersects the graph of a function f in more than one point, the f is a one-to-one function.

3. If $f(x) = 3x$, then $f^{-1}(x) = \dfrac{1}{3}x$.

5. True

7. False. $f^{-1}(x)$ means the inverse of f.

2.9 Building Skills

9. One-to-one   11. Not one-to-one

13. Not one-to-one   15. One-to-one

17. $f(2) = 7 \Rightarrow f^{-1}(7) = 2$

19. $f(-1) = 2 \Rightarrow f^{-1}(2) = -1$

21. $f(a) = b \Rightarrow f^{-1}(b) = a$

23. $(f^{-1} \circ f)(337) = f^{-1}(f(337)) = 337$

25. a. $f(3) = 2(3) - 3 = 3$

    b. Using the result from part (a), $f^{-1}(3) = 3$.

    c. $(f \circ f^{-1})(19) = f(f^{-1}(19)) = 19$

    d. $(f \circ f^{-1})(5) = f(f^{-1}(5)) = 5$

27. a. $f(1) = 1^3 + 1 = 2$

    b. Using the result from part (a), $f^{-1}(2) = 1$.

    c. $(f \circ f^{-1})(269) = f(f^{-1}(269)) = 269$

29. $f(g(x)) = 3\left(\dfrac{x-1}{3}\right) + 1 = x - 1 + 1 = x$
$g(f(x)) = \dfrac{(3x+1)-1}{3} = \dfrac{3x}{3} = x$

31. $f(g(x)) = \left(\sqrt[3]{x}\right)^3 = x$
$g(f(x)) = \sqrt[3]{x^3} = x$

33. $f(g(x)) = \dfrac{\dfrac{1+2x}{1-x} - 1}{\dfrac{1+2x}{1-x} + 2} = \dfrac{\dfrac{1+2x-(1-x)}{1-x}}{\dfrac{1+2x+2(1-x)}{1-x}}$

$= \dfrac{3x}{3} = x$

$g(f(x)) = \dfrac{1 + 2\left(\dfrac{x-1}{x+2}\right)}{1 - \dfrac{x-1}{x+2}} = \dfrac{1 + \dfrac{2x-2}{x+2}}{1 - \dfrac{x-1}{x+2}}$

$= \dfrac{\dfrac{x+2+2x-2}{x+2}}{\dfrac{x+2-(x-1)}{x+2}} = \dfrac{3x}{3} = x$

35.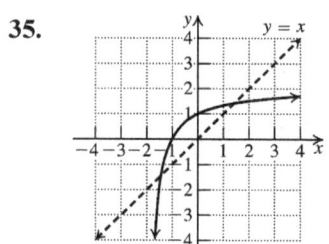

37.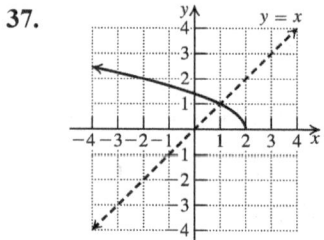

39.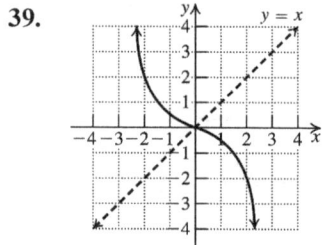

41.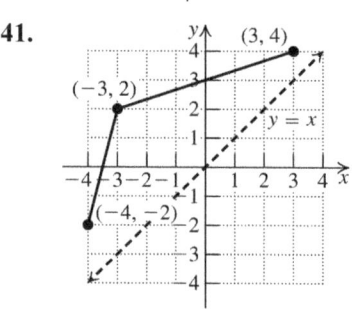

43. a. One-to-one

 b. $f(x) = y = 15 - 3x$. Interchange the variables and solve for y: $x = 15 - 3y \Rightarrow$
 $y = f^{-1}(x) = \dfrac{15-x}{3} = 5 - \dfrac{1}{3}x$.

 c.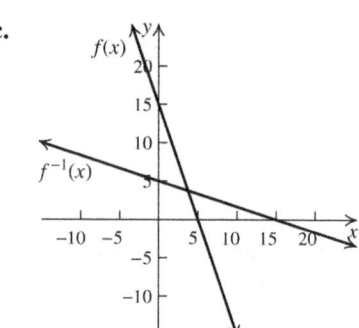

 d. Domain of f: $(-\infty, \infty)$; x-intercept of f: 5; y-intercept of f: 15
 domain of f^{-1}: $(-\infty, \infty)$; x-intercept of f^{-1}: 15; y-intercept of f^{-1}: 5

45. a. Not one-to-one

47. a. One-to-one

 b. $f(x) = y = \sqrt{x} + 3$ Interchange the variables and solve for y: $x = \sqrt{y} + 3 \Rightarrow$
 $x - 3 = \sqrt{y} \Rightarrow y = f^{-1}(x) = (x-3)^2$.

 c.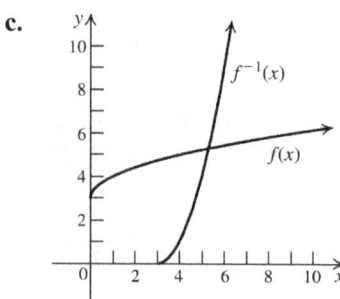

 d. Domain of f: $[0, \infty)$; x-intercept of f: none; y-intercept of f: 3
 domain of f^{-1}: $[3, \infty)$; x-intercept of f^{-1}: 3; y-intercept of f^{-1}: none

49. a. One-to-one

 b. $g(x) = y = \sqrt[3]{x+1}$. Interchange the variables and solve for y: $x = \sqrt[3]{y+1} \Rightarrow$
 $x^3 = y + 1 \Rightarrow y = g^{-1}(x) = x^3 - 1$

c.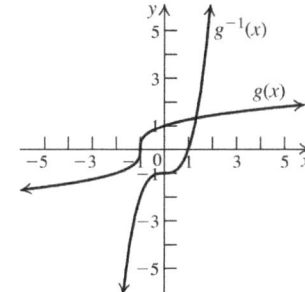

d. Domain of g: $(-\infty, \infty)$; x-intercept of g: -1;
y-intercept of g: 1
domain of g^{-1}: $(-\infty, \infty)$; x-intercept of g^{-1}: 1; y-intercept of g^{-1}: -1

51. a. One-to-one

b. $f(x) = y = \dfrac{1}{x-1}$. Interchange the variables and solve for y: $x = \dfrac{1}{y-1} \Rightarrow x(y-1) = 1 \Rightarrow$
$\dfrac{1}{x} = y - 1 \Rightarrow y = f^{-1}(x) = \dfrac{1}{x} + 1 = \dfrac{1+x}{x}$.

c.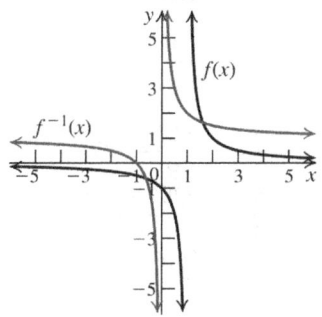

d. Domain of f: $(-\infty, 1) \cup (1, \infty)$
x-intercept of f: none; y-intercept of f: -1
domain of f^{-1}: $(-\infty, 0) \cup (0, \infty)$
x-intercept of f^{-1}: -1
y-intercept of f^{-1}: none

53. a. One-to-one

b. $f(x) = y = 2 + \sqrt{x+1}$. Interchange the variables and solve for y: $x = 2 + \sqrt{y+1} \Rightarrow$
$x - 2 = \sqrt{y+1} \Rightarrow (x-2)^2 = y + 1 \Rightarrow$
$y = f^{-1}(x) = (x-2)^2 - 1 = x^2 - 4x + 3$

c.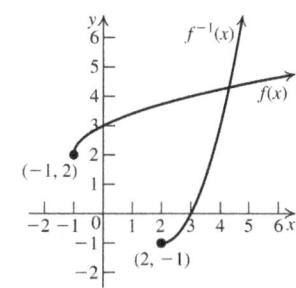

d. Domain of f: $[-1, \infty)$; x-intercept of f: none; y-intercept of f: 3
Domain of f^{-1}: $[2, \infty)$;
x-intercept of f^{-1}: 3
y-intercept of f^{-1}: none

55. The range of f is the same as the domain of f^{-1}.
Domain: $(-\infty, -2) \cup (-2, \infty)$
Range: $(-\infty, 1) \cup (1, \infty)$

57. $f(x) = y = \dfrac{x+1}{x-2}$. Interchange the variables
and solve for y: $x = \dfrac{y+1}{y-2} \Rightarrow xy - 2x = y + 1 \Rightarrow$
$xy - y = 2x + 1 \Rightarrow y(x-1) = 2x + 1 \Rightarrow$
$y = f^{-1}(x) = \dfrac{2x+1}{x-1}$.
Domain of f: $(-\infty, 2) \cup (2, \infty)$
Range of f: $(-\infty, 1) \cup (1, \infty)$.

59. $f(x) = y = \dfrac{1-2x}{1+x}$. Interchange the variables
and solve for y: $x = \dfrac{1-2y}{1+y} \Rightarrow$
$x + xy = 1 - 2y \Rightarrow xy + 2y = 1 - x \Rightarrow$
$y(x+2) = 1 - x \Rightarrow y = f^{-1}(x) = \dfrac{1-x}{x+2}$.
Domain of f: $(-\infty, -1) \cup (-1, \infty)$
Range of f: $(-\infty, -2) \cup (-2, \infty)$.

61. f is one-to-one since the domain is restricted, so an inverse exists.
$f(x) = y = -x^2, x \geq 0$. Interchange the variables and solve for y:
$x = -y^2 \Rightarrow y = \sqrt{-x}, x \leq 0$.

(*continued on next page*)

(*continued*)

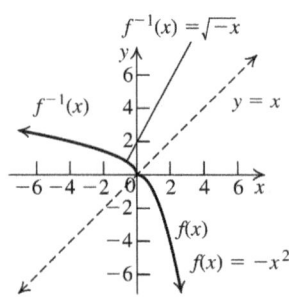

63. f is one-to-one since the domain is restricted, so an inverse exists.
$f(x) = y = |x| = x,\ x \geq 0$. Interchange the variables and solve for y: $y = x,\ x \geq 0$.

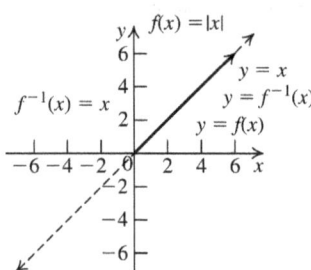

65. f is one-to-one since the domain is restricted, so an inverse exists.
$f(x) = y = x^2 + 1,\ x \leq 0$. Interchange the variables and solve for y:
$x = y^2 + 1 \Rightarrow y = -\sqrt{x-1},\ x \geq 1$.

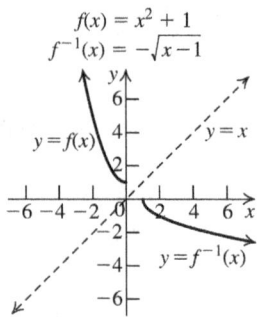

67. f is one-to-one since the domain is restricted, so an inverse exists.
$f(x) = y = -x^2 + 2,\ x \leq 0$. Interchange the variables and solve for y:
$x = -y^2 + 2 \Rightarrow y = -\sqrt{2-x},\ x \leq 2$.

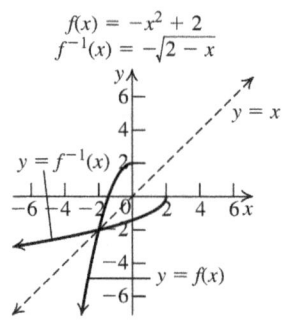

2.9 Applying the Concepts

69. a. $K(C) = C + 273 \Rightarrow$
$C(K) = K - 273 = K^{-1}(C)$.
This represents the Celsius temperature corresponding to a given Kelvin temperature.

   b. $C(300) = 300 - 273 = 27°C$

   c. $K(22) = 22 + 273 = 295°K$

71. a. $F(K(C)) = \dfrac{9}{5}(C + 273) - \dfrac{2297}{5}$
$= \dfrac{9}{5}C + \dfrac{9(273)}{5} - \dfrac{2297}{5}$
$= \dfrac{9}{5}C + \dfrac{160}{5} = \dfrac{9}{5}C + 32$

   b. $C(K(F)) = \dfrac{5}{9}F + \dfrac{2297}{9} - 273$
$= \dfrac{5}{9}F + \dfrac{2297 - 2457}{9}$
$= \dfrac{5}{9}F - \dfrac{160}{9}$

73. a. $E(x) = 0.75x$, where x represents the number of dollars
$D(x) = 1.25x$, where x represents the number of euros.

   b. $E(D(x)) = 0.75(1.25x) = 0.9375x \neq x$.
Therefore, the two functions are not inverses.

   c. She loses money either way.

75. a. $7 = 4 + 0.05x \Rightarrow x = \$60$. This means that if food sales $\leq \$60$, he will receive the minimum hourly wage. If food sales $> \$60$, his wages will be based on food sales.
$w = \begin{cases} 4 + 0.05x & \text{if } x > 60 \\ 7 & \text{if } x \leq 60 \end{cases}$

b. The function does not have an inverse because it is constant on (0, 60), and it is not one-to-one.

c. If the domain is restricted to [60, ∞), the function has an inverse.

77. a. $V = 8\sqrt{x} \Rightarrow \dfrac{V}{8} = \sqrt{x} \Rightarrow \dfrac{1}{64}V^2 = x = V^{-1}(x)$

This represents the height of the water in terms of the velocity.

b. (i) $x = \dfrac{1}{64}(30^2) = 14.0625$ ft

(ii) $x = \dfrac{1}{64}(20^2) = 6.25$ ft

79. a. The function represents the amount she still owes after x months.

b. $y = 36{,}000 - 600x$. Interchange the variables and solve for y: $x = 36{,}000 - 600y \Rightarrow$

$600y = 36{,}000 - x \Rightarrow y = 60 - \dfrac{x}{600} \Rightarrow$

$f^{-1}(x) = 60 - \dfrac{1}{600}x.$

This represents the number of months that have passed from the first payment until the balance due is $x.

c. $y = 60 - \dfrac{1}{600}(22{,}000) = 23.33 \approx 24$ months

There are 24 months remaining.

2.9 Beyond the Basics

81. $f(g(3)) = f(1) = 3, f(g(5)) = f(3) = 5$, and $f(g(2)) = f(4) = 2 \Rightarrow f(g(x)) = x$ for each x.
$g(f(1)) = g(3) = 1, g(f(3)) = g(5) = 3$, and $g(f(4)) = g(2) = 4 \Rightarrow g(f(x)) = x$ for each x.
So, f and g are inverses.

83. a.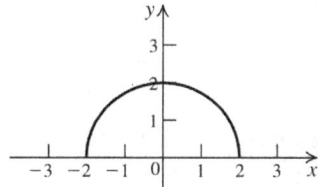

b. f is not one-to-one

c. Domain: [–2, 2]; range: [0, 2]

85. a. f satisfies the horizontal line test.

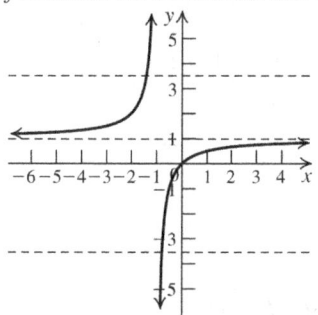

b. $y = 1 - \dfrac{1}{x+1}$. Interchange the variables and solve for y: $x = 1 - \dfrac{1}{y+1} \Rightarrow$

$\dfrac{1}{y+1} = 1 - x \Rightarrow 1 = y + 1 - xy - x \Rightarrow$

$xy - y = -x \Rightarrow y(x-1) = -x \Rightarrow$

$y = f^{-1}(x) = -\dfrac{x}{x-1} = \dfrac{x}{1-x}$

c. Domain of f: $(-\infty, -1) \cup (-1, \infty)$; range of f: $(-\infty, 1) \cup (1, \infty)$.

87. a. (i) $f(x) = y = 2x - 1$. Interchange the variables and solve for y: $x = 2y - 1 \Rightarrow$

$y = f^{-1}(x) = \dfrac{1}{2}x + \dfrac{1}{2}$

(ii) $g(x) = y = 3x + 4$. Interchange the variables and solve for y: $x = 3y + 4 \Rightarrow$

$y = g^{-1}(x) = \dfrac{1}{3}x - \dfrac{4}{3}$

(iii) $(f \circ g)(x) = 2(3x+4) - 1 = 6x + 7$

(iv) $(g \circ f)(x) = 3(2x-1) + 4 = 6x + 1$

(v) $(f \circ g)(x) = y = 6x + 7$. Interchange the variables and solve for y:

$x = 6y + 7 \Rightarrow (f \circ g)^{-1}(x) = \dfrac{1}{6}x - \dfrac{7}{6}$

(vi) $(g \circ f)(x) = y = 6x + 1$. Interchange the variables and solve for y:

$x = 6y + 1 \Rightarrow (f \circ g)^{-1}(x) = \dfrac{1}{6}x - \dfrac{1}{6}$

(vii) $(f^{-1} \circ g^{-1})(x) = \dfrac{1}{2}\left(\dfrac{1}{3}x - \dfrac{4}{3}\right) + \dfrac{1}{2}$

$= \dfrac{1}{6}x - \dfrac{2}{3} + \dfrac{1}{2} = \dfrac{1}{6}x - \dfrac{1}{6}$

156 Chapter 2 Graphs and Functions

(viii) $\left(g^{-1}\circ f^{-1}\right)(x)=\dfrac{1}{3}\left(\dfrac{1}{2}x+\dfrac{1}{2}\right)-\dfrac{4}{3}$

$=\dfrac{1}{6}x+\dfrac{1}{6}-\dfrac{4}{3}=\dfrac{1}{6}x-\dfrac{7}{6}$

b. (i) $(f\circ g)^{-1}(x)=\dfrac{1}{6}x-\dfrac{7}{6}$

$=\left(g^{-1}\circ f^{-1}\right)(x)$

(ii) $(g\circ f)^{-1}(x)=\dfrac{1}{6}x-\dfrac{1}{6}$

$=\left(f^{-1}\circ g^{-1}\right)(x)$

2.9 Critical Thinking/Discussion/Writing

89. No. For example, $f(x)=x^3-x$ is odd, but it does not have an inverse, because $f(0)=f(1)$, so it is not one-to-one.

90. Yes. The function $f=\{(0,1)\}$ is even, and it has an inverse: $f^{-1}=\{(1,0)\}$.

91. Yes, because increasing and decreasing functions are one-to-one.

92. a. $R=\{(-1,1),(0,0),(1,1)\}$

b. $R=\{(-1,1),(0,0),(1,2)\}$

2.9 Getting Ready for the Next Section

93. $x^2-x-12=(x+3)(x-4)$

95. $x^2+2x-8=(x+4)(x-2)$

97. $x^2-7x+12=0$
$(x-3)(x-4)=0$
$x-3=0\ |\ x-4=0$
$x=3\ \ \ |\ \ x=4$
Solution: $\{3,4\}$

99. $3x^2+7x+2=0$
$(3x+1)(x+2)=0$
$3x+1=0\ |\ x+2=0$
$x=-\dfrac{1}{3}\ \ |\ \ x=-2$
Solution: $\left\{-2,-\dfrac{1}{3}\right\}$

101. $y=(x+2)^2-3$

Start with the graph of $f(x)=x^2$, then shift it two units left and three units down.

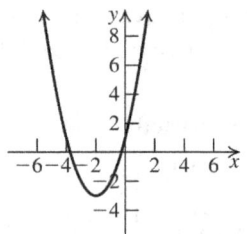

103. $y=-(x+1)^2+2$

Start with the graph of $f(x)=x^2$, then shift it one unit left. Reflect the graph across the x-axis and then shift it two units up

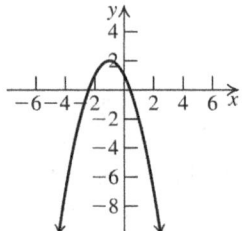

Chapter 2 Review Exercises
Building Skills

1. False. The midpoint is
$\left(\dfrac{-3+3}{2},\dfrac{1+11}{2}\right)=(0,6)$.

3. True

5. False.
The slope is 4/3 and the y-intercept is 3.

7. True

9. a. $d(P,Q)=\sqrt{(-1-3)^2+(3-5)^2}=2\sqrt{5}$

b. $M=\left(\dfrac{3+(-1)}{2},\dfrac{5+3}{2}\right)=(1,4)$

c. $m=\dfrac{3-5}{-1-3}=\dfrac{1}{2}$

11. a. $d(P,Q)=\sqrt{(9-4)^2+(-8-(-3))^2}=5\sqrt{2}$

b. $M=\left(\dfrac{4+9}{2},\dfrac{-3+(-8)}{2}\right)=\left(\dfrac{13}{2},-\dfrac{11}{2}\right)$

c. $m=\dfrac{-8-(-3)}{9-4}=-1$

13. a. $D(P,Q)=\sqrt{(5-2)^2+(-2-(-7))^2}=\sqrt{34}$

b. $M=\left(\dfrac{2+5}{2},\dfrac{-7+(-2)}{2}\right)=\left(\dfrac{7}{2},-\dfrac{9}{2}\right)$

c. $m = \dfrac{-2-(-7)}{5-2} = \dfrac{5}{3}$

15. $d(A,B) = \sqrt{(-2-0)^2 + (-3-5)^2} = \sqrt{68}$
$d(A,C) = \sqrt{(3-0)^2 + (0-5)^2} = \sqrt{34}$
$d(B,C) = \sqrt{(3-(-2))^2 + (0-(-3))^2} = \sqrt{34}$
Using the Pythagorean theorem, we have
$AC^2 + BC^2 = \left(\sqrt{34}\right)^2 + \left(\sqrt{34}\right)^2$
$= 68 = \left(\sqrt{68}\right)^2 = AB^2$
Alternatively, we can show that AC and CB are perpendicular using their slopes.
$m_{AC} = \dfrac{0-5}{3-0} = -\dfrac{5}{3}; m_{CB} = \dfrac{0-(-3)}{3-(-2)} = \dfrac{3}{5}$
$m_{AC} \cdot m_{CB} = -1 \Rightarrow AC \perp CB$, so $\triangle ABC$ is a right triangle.

17. $A = (-6,3), B = (4,5)$
$d(A,O) = \sqrt{(-6-0)^2 + (3-0)^2} = \sqrt{45}$
$d(B,O) = \sqrt{(4-0)^2 + (5-0)^2} = \sqrt{41}$
$(4, 5)$ is closer to the origin.

19. $A = (-5,3), B = (4,7), C = (x,0)$
$d(A,C) = \sqrt{(x-(-5))^2 + (0-3)^2}$
$= \sqrt{(x+5)^2 + 9}$
$d(B,C) = \sqrt{(x-4)^2 + (0-7)^2}$
$= \sqrt{(x-4)^2 + 49}$
$d(A,C) = d(B,C) \Rightarrow$
$\sqrt{(x+5)^2 + 9} = \sqrt{(x-4)^2 + 49}$
$(x+5)^2 + 9 = (x-4)^2 + 49$
$x^2 + 10x + 34 = x^2 - 8x + 65$
$x = \dfrac{31}{18} \Rightarrow$ The point is $\left(\dfrac{31}{18}, 0\right)$.

21. Not symmetric with respect to the x-axis; symmetric with respect to the y-axis; not symmetric with respect to the origin.

23. Symmetric with respect to the x-axis; not symmetric with respect to the y-axis; not symmetric with respect to the origin.

25. x-intercept: 4; y-intercept: 2; not symmetric with respect to the x-axis; not symmetric with respect to the y-axis; not symmetric with respect to the origin.

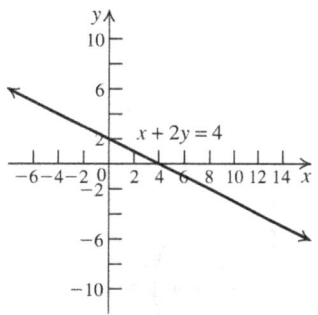

27. x-intercept: 0; y-intercept: 0; not symmetric with respect to the x-axis; symmetric with respect to the y-axis; not symmetric with respect to the origin.

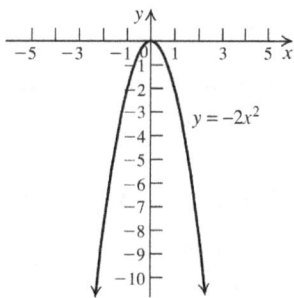

29. x-intercept: 0; y-intercept: 0; not symmetric with respect to the x-axis; not symmetric with respect to the y-axis; symmetric with respect to the origin.

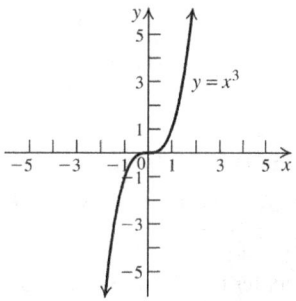

31. No x-intercept; y-intercept: 2; not symmetric with respect to the x-axis; symmetric with respect to the y-axis; not symmetric with respect to the origin.

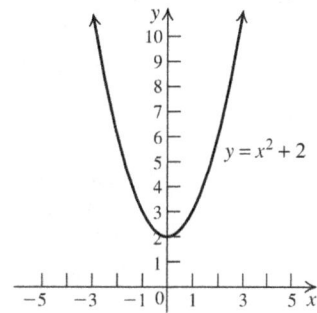

33. x-intercepts: –4, 4; y-intercepts: –4, 4; symmetric with respect to the x-axis; symmetric with respect to the y-axis; symmetric with respect to the origin.

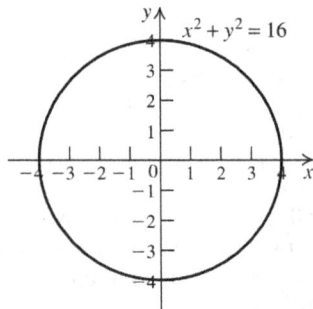

35. $(x-2)^2 + (y+3)^2 = 25$

37. The radius is 2, so the equation of the circle is $(x+2)^2 + (y+5)^2 = 4$.

39. $\dfrac{x}{2} - \dfrac{y}{5} = 1 \Rightarrow 5x - 2y = 10 \Rightarrow \dfrac{5}{2}x - 5 = y$. Line with slope 5/2 and y-intercept –5.

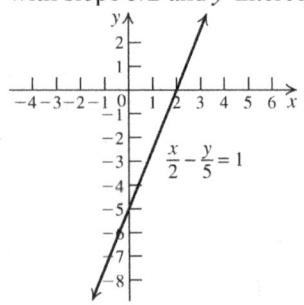

41. $x^2 + y^2 - 2x + 4y - 4 = 0 \Rightarrow$
$x^2 - 2x + 1 + y^2 + 4y + 4 = 4 + 1 + 4 \Rightarrow$
$(x-1)^2 + (y+2)^2 = 9$.
Circle with center (1, –2) and radius 3.

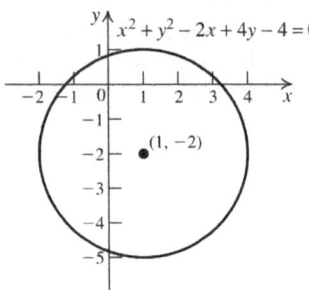

43. $y - 2 = -2(x - 1) \Rightarrow y = -2x + 4$

45. $m = \dfrac{7-3}{-1-1} = -2; 3 = -2(1) + b \Rightarrow 5 = b \Rightarrow$
$y = -2x + 5$

47. a. $y = 3x - 2 \Rightarrow m = 3; y = 3x + 2 \Rightarrow m = 3$
The slopes are equal, so the lines are parallel.

b. $3x - 5y + 7 \Rightarrow m = 3/5;$
$5x - 3y + 2 = 0 \Rightarrow m = 5/3$
The slopes are neither equal nor negative reciprocals, so the lines are neither parallel nor perpendicular.

c. $ax + by + c = 0 \Rightarrow m = -a/b;$
$bx - ay + d = 0 \Rightarrow m = b/a$
The slopes are negative reciprocals, so the lines are perpendicular.

d. $y + 2 = \dfrac{1}{3}(x - 3) \Rightarrow m = \dfrac{1}{3};$
$y - 5 = 3(x - 3) \Rightarrow m = 3$
The slopes are neither equal nor negative reciprocals, so the lines are neither parallel nor perpendicular.

49. Domain: {–1, 0, 1, 2}; range: {–1, 0, 1, 2}. This is a function.

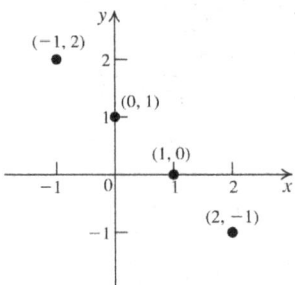

51. Domain: $(-\infty, \infty)$; range: $(-\infty, \infty)$. This is a function.

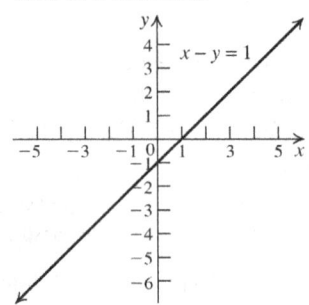

53. Domain: $[-0.2, 0.2]$; range: $[-0.2, 0.2]$.
This is not a function.

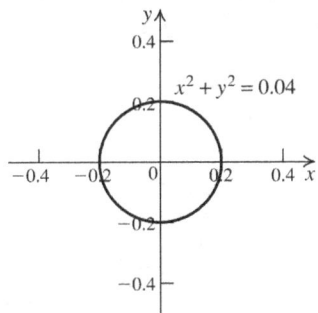

55. Domain: $\{1\}$; range: $(-\infty, \infty)$.
This is not a function.

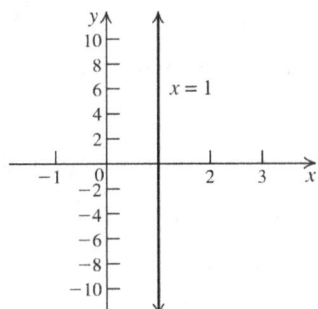

57. Domain: $(-\infty, \infty)$; range: $[0, \infty)$.
This is a function.

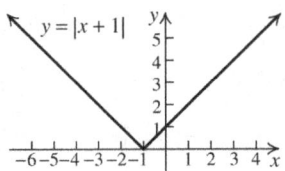

59. $f(-2) = 3(-2) + 1 = -5$

61. $f(x) = 4 \Rightarrow 3x + 1 = 4 \Rightarrow x = 1$

63. $(f+g)(1) = f(1) + g(1)$
$= (3(1)+1) + (1^2 - 2) = 3$

65. $(f \cdot g)(-2) = f(-2) \cdot g(-2)$
$= (3(-2)+1) \cdot ((-2)^2 - 2) = -10$

67. $(f \circ g)(3) = 3(3^2 - 2) + 1 = 22$

69. $(f \circ g)(x) = 3(x^2 - 2) + 1 = 3x^2 - 5$

71. $(f \circ f)(x) = 3(3x+1) + 1 = 9x + 4$

73. $f(a+h) = 3(a+h) + 1 = 3a + 3h + 1$

75. $\dfrac{f(x+h) - f(x)}{h} = \dfrac{(3(x+h)+1) - (3x+1)}{h}$
$= \dfrac{3x + 3h + 1 - 3x - 1}{h} = \dfrac{3h}{h} = 3$

77. Domain: $(-\infty, \infty)$; range: $\{-3\}$.
Constant on $(-\infty, \infty)$.

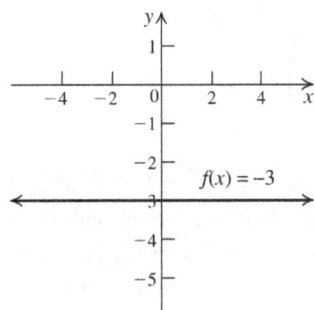

79. Domain: $\left[\dfrac{2}{3}, \infty\right)$; range: $[0, \infty)$

Increasing on $\left(\dfrac{2}{3}, \infty\right)$.

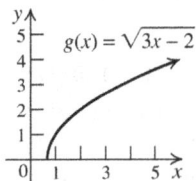

81. Domain: $(-\infty, \infty)$; range: $[1, \infty)$. Decreasing on $(-\infty, 0)$; increasing on $(0, \infty)$.

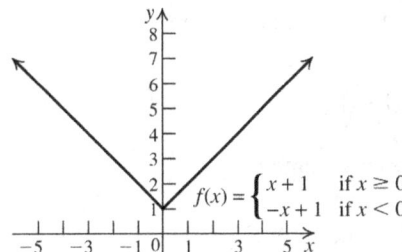

83. The graph of g is the graph of f shifted one unit left.

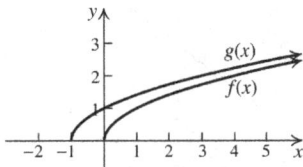

85. The graph of g is the graph of f shifted two units right, and then reflected in the x-axis.

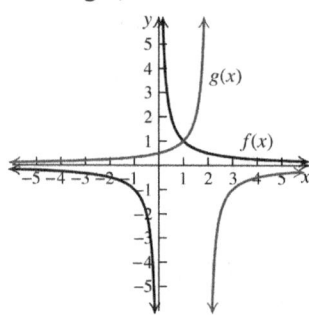

87. $f(-x) = (-x)^2 - (-x)^4 = x^2 - x^4 = f(x) \Rightarrow$
$f(x)$ is even. Not symmetric with respect to the x-axis; symmetric with respect to the y-axis; not symmetric with respect to the origin.

89. $f(-x) = |-x| + 3 = |x| + 3 = f(x) \Rightarrow$
$f(x)$ is even. Not symmetric with respect to the x-axis; symmetric with respect to the y-axis; not symmetric with respect to the origin.

91. $f(-x) = \sqrt{-x} \neq f(x)$ or $f(-x) \Rightarrow f(x)$ is neither even nor odd. Not symmetric with respect to the x-axis; not symmetric with respect to the y-axis; not symmetric with respect to the origin.

93. $f(x) = \sqrt{x^2 - 4} \Rightarrow f(x) = (g \circ h)(x)$ where $g(x) = \sqrt{x}$ and $h(x) = x^2 - 4$.

95. $h(x) = \sqrt{\dfrac{x-3}{2x+5}} \Rightarrow h(x) = (f \circ g)(x)$ where $f(x) = \sqrt{x}$ and $g(x) = \dfrac{x-3}{2x+5}$.

97. $f(x)$ is one-to-one. $f(x) = y = x + 2$.
Interchange the variables and solve for y:
$x = y + 2 \Rightarrow y = x - 2 = f^{-1}(x)$.

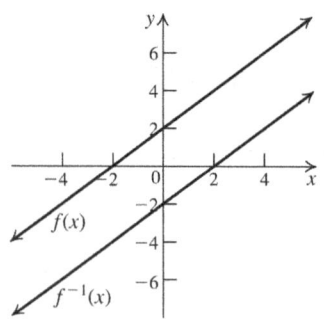

99. $f(x)$ is one-to-one. $f(x) = y = \sqrt[3]{x-2}$.
Interchange the variables and solve for y:
$x = \sqrt[3]{y-2} \Rightarrow y = x^3 + 2 = f^{-1}(x)$.

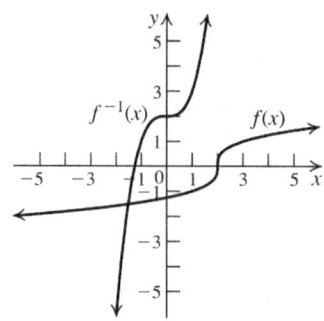

101. $f(x) = y = \dfrac{x-1}{x+2}, x \neq -2$.
Interchange the variables and solve for y.
$x = \dfrac{y-1}{y+2} \Rightarrow xy + 2x = y - 1 \Rightarrow$
$xy - y = -2x - 1 \Rightarrow y(x-1) = -2x - 1 \Rightarrow$
$y = \dfrac{-2x-1}{x-1} \Rightarrow y = f^{-1}(x) = \dfrac{2x+1}{1-x}$
Domain of f: $(-\infty, -2) \cup (-2, \infty)$
Range of f: $(-\infty, 1) \cup (1, \infty)$

103. a. $A = (-3, -3), B = (-2, 0), C = (0, 1), D = (3, 4)$.
Find the equation of each segment:
$m_{AB} = \dfrac{0-(-3)}{-2-(-3)} = 3.0 = 3(-2) + b \Rightarrow b = 6$.
The equation of AB is $y = 3x + 6$.
$m_{BC} = \dfrac{1-0}{0-(-2)} = \dfrac{1}{2}; b = 1$.
The equation of BC is $y = \dfrac{1}{2}x + 1$.
$m_{CD} = \dfrac{4-1}{3-0} = 1; b = 1$.
The equation of CD is $y = x + 1$.
So,
$f(x) = \begin{cases} 3x + 6 & \text{if } -3 \leq x \leq -2 \\ \dfrac{1}{2}x + 1 & \text{if } -2 < x < 0 \\ x + 1 & \text{if } 0 \leq x \leq 3 \end{cases}$

b. Domain: $[-3, 3]$; range: $[-3, 4]$

c. x-intercept: -2; y-intercept: 1

d.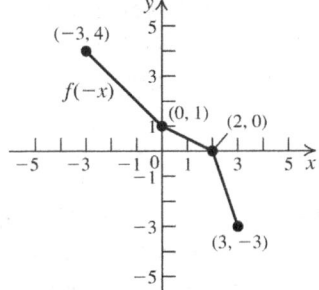

e.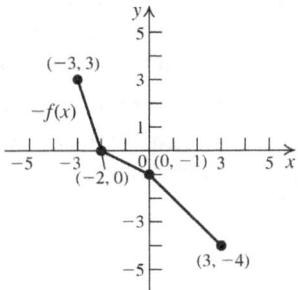

f.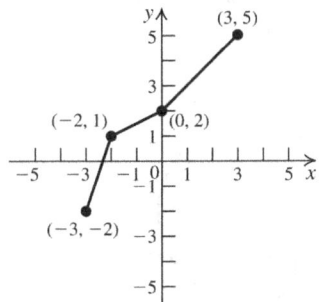

g.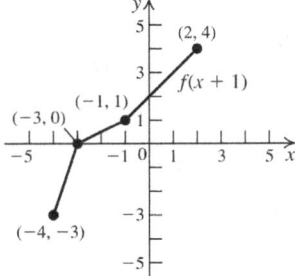

h.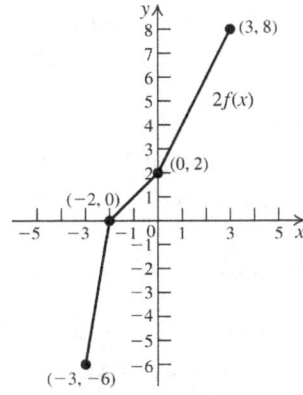

i.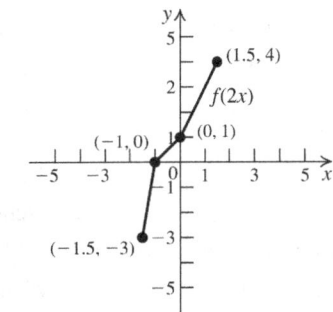

j.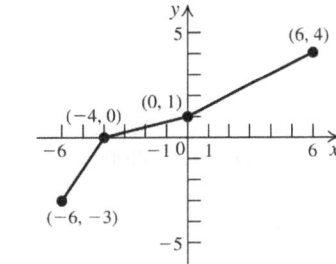

k. f is one-to-one because it satisfies the horizontal line test.

l.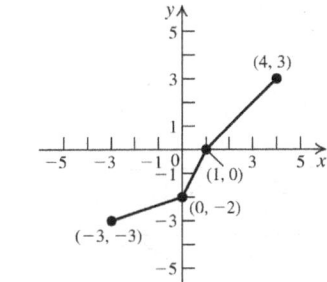

Applying the Concepts

105. a. rate of change (slope) $= \dfrac{173{,}000 - 54{,}000}{223{,}000 - 87{,}000}$
$= 0.875$
$54{,}000 = 0.875(87{,}000) + b \Rightarrow$
$b = -22{,}125.$
The equation is $C = 0.875w - 22{,}125.$

b. The slope represents the cost to dispose of one pound of waste. The x-intercept represents the amount of waste that can be disposed with no cost. The y-intercept represents the fixed cost.

c. $C = 0.875(609{,}000) - 22{,}125 = \$510{,}750$

d. $1{,}000{,}000 = 0.875w - 22{,}125 \Rightarrow$
$w = 1{,}168{,}142.86$ pounds

107.
a. $f(2) = 100 + 55(2) - 3(2)^2 = \$198$.
   She started with $100, so she won $98.
b. She was winning at a rate of $49/hour.
c. $0 = 100 + 55t - 3t^2 \Rightarrow (-t+20)(3t+5) \Rightarrow$
   $t = 20$, $t = -5/3$. Since t represents the amount of time, we reject $t = -5/3$.
   Chloe will lose all her money after playing for 20 hours.
d. $\$100/20 = \$5$/hour.

109.
a. $(L \circ x)(t) = 0.5\sqrt{(1+0.002t^2)^2 + 4}$
   $= 0.5\sqrt{0.000004t^4 + 0.004t^2 + 5}$
b. $(L \circ x)(5) = 0.5\sqrt{(1+0.002(5^2))^2 + 4}$
   $= 0.5\sqrt{(1.05)^2 + 4} = 0.5\sqrt{5.1025}$
   ≈ 1.13

111.
a.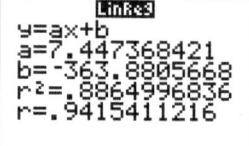

   $y \approx 7.4474x - 363.88$

b.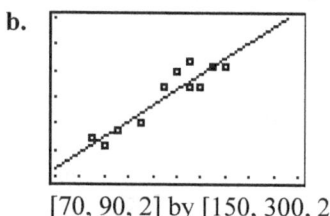

   [70, 90, 2] by [150, 300, 25]

c. $y \approx 7.4474(76) - 363.88 \approx 202$
   A player whose height is 76 inches weighs about 202 pounds.

Chapter 2 Practice Test A

1. The endpoints of the diameter are $(-2, 3)$ and $(-4, 5)$, so the center of the circle is
$C = \left(\dfrac{-2+(-4)}{2}, \dfrac{3+5}{2}\right) = (-3, 4)$.
The length of the diameter is
$\sqrt{(-4-(-2))^2 + (5-3)^2} = \sqrt{8} = 2\sqrt{2}$.
Therefore, the length of the radius is $\sqrt{2}$.
The equation of the circle is
$(x+3)^2 + (y-4)^2 = 2$.

2. To test if the graph is symmetric with respect to the y-axis, replace x with $-x$:
$3(-x) + 2(-x)y^2 = 1 \Rightarrow -3x - 2xy^2 = 1$, which is not the same as the original equation, so the graph is not symmetric with respect to the y-axis. To test if the graph is symmetric with respect to the x-axis, replace y with $-y$:
$3x + 2x(-y)^2 = 1 \Rightarrow 3x + 2xy^2 = 1$, which is the same as the original equation, so the graph is symmetric with respect to the x-axis. To test if the graph is symmetric with respect to the origin, replace x with $-x$ and y with $-y$:
$3(-x) + 2(-x)(-y)^2 = 1 \Rightarrow -3x - 2xy^2 = 1$,
which is not the same as the original equation, so the graph is not symmetric with respect to the origin.

3. $0 = x^2(x-3)(x+1) \Rightarrow x = 0$ or $x = 3$ or $x = -1$
$y = 0^2(0-3)(0+1) \Rightarrow y = 0$. The x-intercepts are 0, 3, and -1; the y-intercept is 0.

4.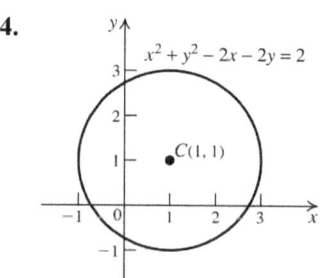

Intercepts:
$y^2 - 2y = 2 \Rightarrow y = 1 \pm \sqrt{3}$
$x^2 - 2x = 2 \Rightarrow x = 1 \pm \sqrt{3}$

5. $7 = -1(2) + b \Rightarrow 9 = b$
The equation is $y = -x + 9$.

6. $8x - 2y = 7 \Rightarrow y = 4x - \dfrac{7}{2} \Rightarrow$ the slope of the line is 4. $-1 = 4(2) + b \Rightarrow b = -9$. So the equation is $y = 4x - 9$.

7. $(fg)(2) = f(2) \cdot g(2)$
$= (-2(2)+1)(2^2 + 3(2) + 2)$
$= (-3)(12) = -36$

8. $g(f(2)) = g(2(2)-3) = g(1) = 1 - 2(1)^2 = -1$

9. $(f \circ f)(x) = (x^2 - 2x)^2 - 2(x^2 - 2x)$
$= x^4 - 4x^3 + 4x^2 - 2x^2 + 4x$
$= x^4 - 4x^3 + 2x^2 + 4x$

10. a. $f(-1) = (-1)^3 - 2 = -3$

    b. $f(0) = 0^3 - 2 = -2$

    c. $f(1) = 1 - 2(1)^2 = -1$

11. $1 - x > 0 \Rightarrow x < 1$; x must also be greater than or equal to 0, so the domain is [0, 1).

12. $\dfrac{f(4) - f(1)}{4 - 1} = \dfrac{(2(4) + 7) - (2(1) + 7)}{3} = 2$

13. $f(-x) = 2(-x)^4 - \dfrac{3}{(-x)^2} = 2x^4 - \dfrac{3}{x^2} = f(x) \Rightarrow$

    $f(x)$ is even.

14. Increasing on $(-\infty, 0)$ and $(2, \infty)$; decreasing on $(0, 2)$.

15. Shift the graph of $y = \sqrt{x}$ three units to the right, then stretch the graph vertically by a factor of 2, and then shift the resulting graph four units up.

16. $25 = 25 - (2t - 5)^2 \Rightarrow 0 = -(2t - 5)^2 \Rightarrow$
    $0 = 2t - 5 \Rightarrow t = 5/2 = 2.5$ seconds

17. $f(2) = 7 \Rightarrow f^{-1}(7) = 2$

18. $f(x) = y = \dfrac{2x}{x - 1}$. Interchange the variables and solve for y: $x = \dfrac{2y}{y - 1} \Rightarrow$
    $xy - x = 2y \Rightarrow xy - 2y = x \Rightarrow$
    $y(x - 2) = x \Rightarrow y = f^{-1}(x) = \dfrac{x}{x - 2}$

19. $A(x) = 100x + 1000$

20. a. $C(230) = 0.25(230) + 30 = \$87.50$

    b. $57.50 = 0.25m + 30 \Rightarrow m = 110$ miles

Chapter 2 Practice Test B

1. To test if the graph is symmetric with respect to the y-axis, replace x with $-x$:
   $|-x| + 2|y| = 2 \Rightarrow |x| + 2|y| = 2$, which is the same as the original equation, so the graph is symmetric with respect to the y-axis. To test if the graph is symmetric with respect to the x-axis, replace y with $-y$:
   $|x| + 2|-y| = 2 \Rightarrow |x| + 2|y| = 2$, which is the same as the original equation, so the graph is symmetric with respect to the x-axis. To test if the graph is symmetric with respect to the origin, replace x with $-x$, and y with $-y$:
   $|-x| + 2|-y| = 2 \Rightarrow |x| + 2|y| = 2$, which is the same as the original equation, so the graph is symmetric with respect to the origin. The answer is D.

2. $0 = x^2 - 9 \Rightarrow x = \pm 3$; $y = 0^2 - 9 \Rightarrow y = -9$.
   The x-intercepts are ± 3; the y-intercept is -9. The answer is B.

3. D   4. D   5. C

6. Suppose the coordinates of the second point are (a, b). Then $-\dfrac{1}{2} = \dfrac{b - 2}{a - 3}$. Substitute each of the points given into this equation to see which makes it true. The answer is C.

7. Find the slope of the original line:
   $6x - 3y = 5 \Rightarrow y = 2x - \dfrac{5}{3}$. The slope is 2.
   The equation of the line with slope 2, passing through $(-1, 2)$ is $y - 2 = 2(x + 1)$.
   The answer is D.

8. $(f \circ g)(x) = 3(2 - x^2) - 5 = 1 - 3x^2$.
   The answer is B.

9. $(f \circ f)(x) = 2(2x^2 - x)^2 - (2x^2 - x)$
    $= 8x^4 - 8x^3 + x$
   The answer is A.

10. $g(a - 1) = \dfrac{1 - (a - 1)}{1 + (a - 1)} = \dfrac{2 - a}{a}$.
    The answer is C.

11. $1 - x \geq 0 \Rightarrow x \leq 1$; x must also be greater than or equal to 0, so the domain is [0, 1].
    The answer is A.

12. $x^2 + 3x - 4 = 6 \Rightarrow x^2 + 3x - 10 = 0 \Rightarrow$
    $(x + 5)(x - 2) = 0 \Rightarrow x = -5, 2$
    The answer is D.

13. A   14. A   15. B

16. D   17. C

18. $f(x) = y = \dfrac{x}{3x+2}$. Interchange the variables and solve for y: $x = \dfrac{y}{3y+2} \Rightarrow$
$3xy + 2x = y \Rightarrow 3xy - y = -2x \Rightarrow$
$y(3x-1) = -2x \Rightarrow y = f^{-1}(x) = -\dfrac{2x}{3x-1} \Rightarrow$
$f^{-1}(x) = \dfrac{2x}{1-3x}$
The answer is C.

19. $w = 5x - 190; w = 5(70) - 190 = 160.$
The answer is B.

20. $50 = 0.2m + 25 \Rightarrow m = 125$. The answer is A.

Cumulative Review Exercises (Chapters P–2)

1. a. $\left(\dfrac{x^3}{y^2}\right)^2 \left(\dfrac{y^2}{x^3}\right)^3 = \left(\dfrac{x^6}{y^4}\right)\left(\dfrac{y^6}{x^9}\right) = \dfrac{y^2}{x^3}$

   b. $\dfrac{x^{-1}y^{-1}}{x^{-1}+y^{-1}} = \dfrac{\dfrac{1}{x}\cdot\dfrac{1}{y}}{\dfrac{1}{x}+\dfrac{1}{y}} = \dfrac{\dfrac{xy}{}}{\dfrac{y+x}{xy}} = \dfrac{1}{x+y}$

2. a. $2x^2 + x - 15 = (2x - 5)(x + 3)$

   b. $x^3 - 2x^2 + 4x - 8 = x^2(x-2) + 4(x-2)$
   $= (x^2 + 4)(x - 2)$

3. a. $\sqrt{75} + \sqrt{108} - \sqrt{192} = 5\sqrt{3} + 6\sqrt{3} - 8\sqrt{3}$
   $= 3\sqrt{3}$

   b. $\dfrac{x-1}{x+1} - \dfrac{x-2}{x+2} = \dfrac{(x-1)(x+2)-(x-2)(x+1)}{(x+1)(x+2)}$
   $= \dfrac{(x^2+x-2)-(x^2-x-2)}{(x+1)(x+2)}$
   $= \dfrac{2x}{(x+1)(x+2)}$

4. a. $\dfrac{1}{2+\sqrt{3}} = \dfrac{1}{2+\sqrt{3}} \cdot \dfrac{2-\sqrt{3}}{2-\sqrt{3}} = \dfrac{2-\sqrt{3}}{4-3} = 2-\sqrt{3}$

   b. $\dfrac{1}{\sqrt{5}-2} = \dfrac{1}{\sqrt{5}-2} \cdot \dfrac{\sqrt{5}+2}{\sqrt{5}+2} = \dfrac{\sqrt{5}+2}{5-4} = \sqrt{5}+2$

5. a. $3x - 7 = 5 \Rightarrow 3x = 12 \Rightarrow x = 4$

   b. $\dfrac{1}{x-1} = \dfrac{3}{x-1} \Rightarrow$ There is no solution.

6. a. $x^2 - 3x = 0 \Rightarrow x(x-3) = 0 \Rightarrow$
   $x = 0$ or $x = 3$

   b. $x^2 + 3x - 10 = 0 \Rightarrow (x+5)(x-2) = 0 \Rightarrow$
   $x = -5$ or $x = 2$

7. a. $2x^2 - x + 3 = 0 \Rightarrow x = \dfrac{1 \pm \sqrt{1-4(2)(3)}}{2(2)} \Rightarrow$
   $x = \dfrac{1 \pm \sqrt{-23}}{4} \Rightarrow x = \dfrac{1 \pm i\sqrt{23}}{4}$

   b. $4x^2 - 12x + 9 = 0 \Rightarrow (2x-3)^2 = 0 \Rightarrow x = \dfrac{3}{2}$

8. a. $x - 6\sqrt{x} + 8 = 0 \Rightarrow (\sqrt{x}-4)(\sqrt{x}-2) = 0 \Rightarrow$
   $\sqrt{x} = 4 \Rightarrow x = 16$ or $\sqrt{x} = 2 \Rightarrow x = 4$

   b. $\left(x-\dfrac{1}{x}\right)^2 - 10\left(x-\dfrac{1}{x}\right) + 21 = 0.$
   Let $u = x - \dfrac{1}{x}$.
   $u^2 - 10u + 21 = 0 \Rightarrow$
   $(u-7)(u-3) = 0 \Rightarrow u = 7$ or $u = 3$;
   $x - \dfrac{1}{x} = 7 \Rightarrow x^2 - 1 = 7x \Rightarrow$
   $x^2 - 7x - 1 = 0 \Rightarrow x = \dfrac{7 \pm \sqrt{7^2-(4)(-1)}}{2} \Rightarrow$
   $x = \dfrac{7 \pm \sqrt{53}}{2}; x - \dfrac{1}{x} = 3 \Rightarrow x^2 - 1 = 3x \Rightarrow$
   $x^2 - 3x - 1 = 0 \Rightarrow x = \dfrac{3 \pm \sqrt{3^2-4(-1)}}{2} \Rightarrow$
   $x = \dfrac{3 \pm \sqrt{13}}{2}$
   The solution set is
   $\left\{\dfrac{7-\sqrt{53}}{2}, \dfrac{7+\sqrt{53}}{2}, \dfrac{3-\sqrt{13}}{2}, \dfrac{3+\sqrt{13}}{2}\right\}.$

9. a. $\sqrt{3x-1} = 2x-1 \Rightarrow 3x-1 = (2x-1)^2 \Rightarrow$
$3x-1 = 4x^2 - 4x + 1 \Rightarrow 4x^2 - 7x + 2 = 0 \Rightarrow$
$x = \dfrac{7 \pm \sqrt{(-7)^2 - 4(4)(2)}}{2(4)} = \dfrac{7 \pm \sqrt{17}}{8}$. If
$x = \dfrac{7-\sqrt{17}}{8}, \sqrt{3\left(\dfrac{7-\sqrt{17}}{8}\right)-1} \approx 0.281$
while $2\left(\dfrac{7-\sqrt{17}}{8}\right)-1 \approx -0.281$, so the
solution set is $\left\{\dfrac{7+\sqrt{17}}{8}\right\}$.

b. $\sqrt{1-x} = 2 - \sqrt{2x+1}$
$\left(\sqrt{1-x}\right)^2 = \left(2 - \sqrt{2x+1}\right)^2$
$1 - x = 4 - 4\sqrt{2x+1} + 2x + 1$
$-4 - 3x = -4\sqrt{2x+1}$
$(-4-3x)^2 = \left(-4\sqrt{2x+1}\right)^2$
$16 + 24x + 9x^2 = 16(2x+1)$
$16 + 24x + 9x^2 = 32x + 16$
$9x^2 - 8x = 0 \Rightarrow x(9x-8) = 0$
$x = 0$ or $x = \dfrac{8}{9}$.
Check to make sure that neither solution is extraneous. The solution set is $\{0, 8/9\}$.

10. a. $2x - 5 < 11 \Rightarrow x < 8 \Rightarrow (-\infty, 8)$

b. $-3x + 4 > -5 \Rightarrow x < 3 \Rightarrow (-\infty, 3)$

11. a. $-3 < 2x - 3 < 5 \Rightarrow 0 < 2x < 8 \Rightarrow 0 < x < 4$.
The solution set is $(0, 4)$.

b. $5 \leq 1 - 2x \leq 7 \Rightarrow 4 \leq -2x \leq 6 \Rightarrow -2 \geq x \geq -3$.
The solution set is $[-3, -2]$.

12. a. $|2x - 1| \leq 7 \Rightarrow -7 \leq 2x - 1 \leq 7 \Rightarrow$
$-6 \leq 2x \leq 8 \Rightarrow -3 \leq x \leq 4$
The solution set is $[-3, 4]$.

b. $|2x - 3| \geq 5 \Rightarrow 2x - 3 \geq 5 \Rightarrow x \geq 4$ or
$2x - 3 \leq -5 \Rightarrow x \leq -1$.
The solution set is $(-\infty, -1] \cup [4, \infty)$.

13. $d(A,C) = \sqrt{(2-5)^2 + (2-(-2))^2} = 5$
$d(B,C) = \sqrt{(2-6)^2 + (2-5)^2} = 5$
Since the lengths of the two sides are equal, the triangle is isosceles.

14.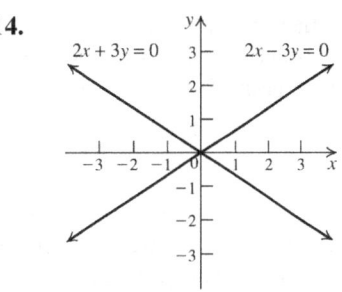

15. First, find the equation of the circle with center $(2, -1)$ and radius determined by $(2, -1)$ and $(-3, -1)$:
$r = \sqrt{(2-(-3))^2 + (-1-(-1))^2} = 5$.
The equation is $(x-2)^2 + (y+1)^2 = 5^2$. Now check to see if the other three points satisfy the equation:
$(2-2)^2 + (4+1)^2 = 5^2 \Rightarrow 5^2 = 5^2$,
$(5-2)^2 + (3+1)^2 = 5^2 \Rightarrow 3^2 + 4^2 = 5^2$ (true because 3, 4, 5 is a Pythagorean triple), and
$(6-2)^2 + (2+1)^2 = 5^2 \Rightarrow 4^2 + 3^2 = 5^2$.
Since all the points satisfy the equation, they lie on the circle.

16. $x^2 + y^2 - 6x + 4y + 9 = 0 \Rightarrow$
$x^2 - 6x + y^2 + 4y = -9$.
Now complete both squares:
$x^2 - 6x + 9 + y^2 + 4y + 4 = -9 + 9 + 4 \Rightarrow$
$(x-3)^2 + (y+2)^2 = 4$.
The center is $(3, -2)$ and the radius is 2.

17. $y = -3x + 5$

18. The x-intercept is 4, so $(4, 0)$ satisfies the equation. To write the equation in slope-intercept form, find the y-intercept:
$0 = 2(4) + b \Rightarrow -8 = b$
The equation is $y = 2x - 8$.

19. The slope of the perpendicular line is the negative reciprocal of the slope of the original line. The slope of the original line is 2, so the slope of the perpendicular is $-1/2$. Now find the y-intercept of the perpendicular:
$-1 = -\dfrac{1}{2}(2) + b \Rightarrow b = 0$. The equation of the perpendicular is $y = -\dfrac{1}{2}x$.

20. The slope of the parallel line is the same as the slope of the original line, 2. Now find the y-intercept of the parallel line: $-1 = 2(2) + b \Rightarrow b = -5$. The equation of the parallel line is $y = 2x - 5$.

21. The slope of the perpendicular line is the negative reciprocal of the slope of the original line. The slope of the original line is $\frac{7-(-1)}{5-3} = 4$, so the slope of the perpendicular is $-1/4$. The perpendicular bisector passes through the midpoint of the original segment. The midpoint is $\left(\frac{3+5}{2}, \frac{-1+7}{2}\right) = (4, 3)$. Use this point and the slope to find the y-intercept: $3 = -\frac{1}{4}(4) + b \Rightarrow b = 4$. The equation of the perpendicular bisector is $y = -\frac{1}{4}x + 4$.

22. The slope is undefined because the line is vertical. Because it passes through (5, 7), the equation of the line is $x = 5$.

23. Use the slope formula to solve for x:
$2 = \frac{5-11}{x-5} \Rightarrow 2(x-5) = -6 \Rightarrow 2x - 10 = -6 \Rightarrow x = 2$

24. The line through $(x, 3)$ and $(3, 7)$ has slope -2 because it is perpendicular to a line with slope $1/2$. Use the slope formula to solve for x:
$-2 = \frac{3-7}{x-3} \Rightarrow -2(x-3) = -4 \Rightarrow x - 3 = 2 \Rightarrow x = 5$

25.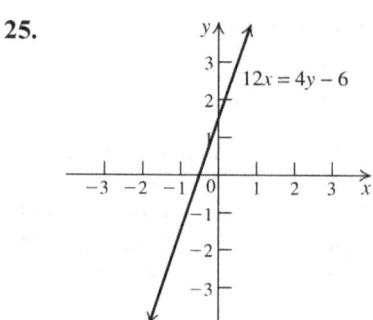

26.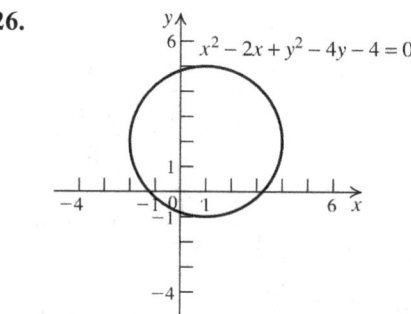

27.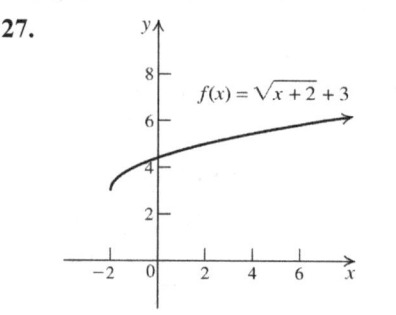

28.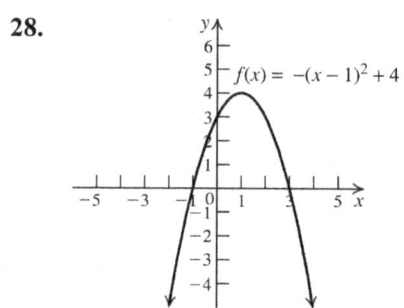

29. Let x = the number of books initially purchased, and $\frac{1650}{x}$ = the cost of each book. Then $x - 16$ = the number of books sold, and $\frac{1650}{x-16}$ = the selling price of each book. The profit = the selling price – the cost, so
$\frac{1650}{x-16} - \frac{1650}{x} = 10 \Rightarrow$
$1650x - 1650(x-16) = 10x(x-16) \Rightarrow$
$1650x - 1650x + 26,400 = 10x^2 - 160x \Rightarrow$
$10x^2 - 160x - 26,400 = 0 \Rightarrow$
$x^2 - 16x - 2640 = 0 \Rightarrow (x-60)(x+44) = 0 \Rightarrow$
$x = 60, x = -44$. Reject -44 because there cannot be a negative number of books. So she bought 60 books.

30. Let x = the monthly note on the 1.5 year lease, and $1.5(12)x = 18x$ = the total expense for the 1.5 year lease. Then $x - 250$ = the monthly note on the 2 year lease, and $2(12)(x - 250) = 24x - 6000$ the total expenses for the 2 year lease. Then
$18x + 24x - 6000 = 21,000 \Rightarrow$
$42x = 27,000 \Rightarrow x = 642.86$. So the monthly note for the 1.5 year lease is $642.86, and the monthly note for the 2 year lease is $642.86 - 250 = $392.86.

31. a. The domain of f is the set of all values of x which make $x + 1 \geq 0$ (because the square root of a negative number is not a real value.) So $x \geq -1$ or $[-1, \infty)$ in interval notation is the domain.

b. $y = \sqrt{0+1} - 3 \Rightarrow y = -2; 0 = \sqrt{x+1} - 3 \Rightarrow$
$3 = \sqrt{x+1} \Rightarrow 9 = x + 1 \Rightarrow 8 = x$. The x-intercept is 8, and the y-intercept is -2.

c. $f(-1) = \sqrt{-1+1} - 3 = -3$

d. $f(x) > 0 \Rightarrow \sqrt{x+1} - 3 > 0 \Rightarrow \sqrt{x+1} > 3 \Rightarrow$
$x + 1 > 9 \Rightarrow x > 8$. In interval notation, this is $(8, \infty)$.

32. a. $f(-2) = -(-2) = 2; f(0) = 0^2 = 0;$
$f(2) = 2^2 = 4$

b. f decreases on $(-\infty, 0)$ and increases on $(0, \infty)$.

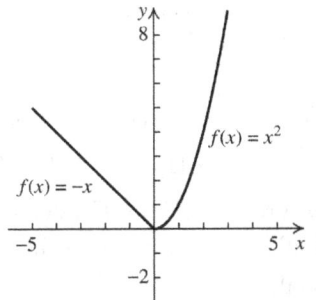

33. a. $(f \circ g)(x) = \dfrac{1}{\dfrac{2}{x} - 2} = \dfrac{1}{\dfrac{2-2x}{x}} = \dfrac{x}{2-2x}$.

Because 0 is not in the domain of g, it must be excluded from the domain of $(f \circ g)$.
Because 2 is not in the domain of f, any values of x for which $g(x) = 2$ must also be excluded from the domain of
$(f \circ g): \dfrac{2}{x} = 2 \Rightarrow x = 1$, so 1 is excluded also. The domain of $(f \circ g)$ is
$(-\infty, 0) \cup (0, 1) \cup (1, \infty)$.

b. $(g \circ f)(x) = \dfrac{2}{\dfrac{1}{x-2}} = 2(x-2) = 2x - 4$.

Because 2 is not in the domain of f, it must be excluded from the domain of $(g \circ f)$.
Because 0 is not in the domain of g, any values of x for which $f(x) = 0$ must also be excluded from the domain of $(g \circ f)$.
However, there is no value for x which makes $f(x) = 0$. So the domain of $(g \circ f)$ is
$(-\infty, 2) \cup (2, \infty)$.

Chapter 3 Polynomial and Rational Functions

3.1 Quadratic Functions

3.1 Practice Problems

1. Substitute 1 for h, -5 for k, 3 for x, and 7 for y in the standard form for a quadratic equation to solve for a:
$$7 = a(3-1)^2 - 5 \Rightarrow 7 = 4a - 5 \Rightarrow a = 3$$
The equation is $y = 3(x-1)^2 - 5$.
Since $a = 3 > 0$, f has a minimum value of -5 at $x = 1$.

2. The graph of $f(x) = -2(x+1)^2 + 3$ is a parabola with $a = -2$, $h = -1$ and $k = 3$. Thus, the vertex is $(-1, 3)$, and the maximum value of the function is 3. The parabola opens down because $a < 0$. Now, find the x-intercepts:
$$0 = -2(x+1)^2 + 3 \Rightarrow 2(x+1)^2 = 3 \Rightarrow$$
$$(x+1)^2 = \frac{3}{2} \Rightarrow x+1 = \pm\sqrt{\frac{3}{2}} \Rightarrow x = \pm\sqrt{\frac{3}{2}} - 1 \Rightarrow$$
$x \approx 0.22$ or $x \approx -2.22$. Next, find the y-intercept: $f(0) = -2(0+1)^2 + 3 = 1$
Plot the vertex, the x-intercepts, and the y-intercept, and join them with a parabola.

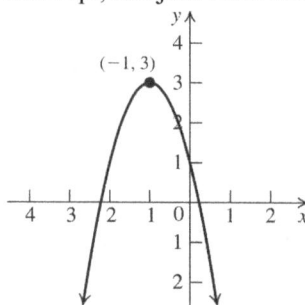

3. The graph of $f(x) = 3x^2 - 3x - 6$ is a parabola with $a = 3$, $b = -3$ and $c = -6$. The parabola opens up because $a > 0$. Now, find the vertex:
$$h = -\frac{b}{2a} = -\frac{-3}{2(3)} = \frac{1}{2}$$
$$k = f(h) = f\left(\frac{1}{2}\right) = 3\left(\frac{1}{2}\right)^2 - 3\left(\frac{1}{2}\right) - 6 = -\frac{27}{4}$$
Thus, the vertex (h, k) is $\left(\frac{1}{2}, -\frac{27}{4}\right)$, and the minimum value of the function is $-\frac{27}{4}$.

Next, find the x-intercepts:
$$3x^2 - 3x - 6 = 0 \Rightarrow 3(x^2 - x - 2) = 0 \Rightarrow$$
$$(x-2)(x+1) = 0 \Rightarrow x = 2 \text{ or } x = -1$$
Now, find the y-intercept:
$$f(0) = 3(0)^2 - 3(0) - 6 = -6.$$
Thus, the intercepts are $(-1, 0)$, $(2, 0)$ and $(0, -6)$. Use the fact that the parabola is symmetric with respect to its axis, $x = \frac{1}{2}$, to locate additional points. Plot the vertex, the x-intercepts, the y-intercept, and any additional points, and join them with a parabola.

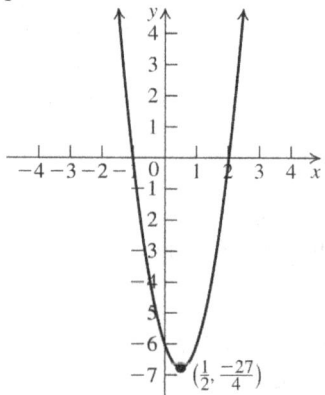

4. The graph of $f(x) = 3x^2 - 6x - 1$ is a parabola with $a = 3$, $b = -6$ and $c = -1$. The parabola opens up because $a > 0$. Complete the square to write the equation in standard form:
$$f(x) = 3x^2 - 6x - 1 = 3(x^2 - 2x) - 1$$
$$= 3(x^2 - 2x + 1) - 1 - 3 = 3(x-1)^2 - 4$$
Thus, the vertex is $(1, -4)$. The domain of f is $(-\infty, \infty)$ and the range is $[-4, \infty)$.
Next, find the x-intercepts:
$$0 = 3(x-1)^2 - 4 \Rightarrow \frac{4}{3} = (x-1)^2 \Rightarrow$$
$$\pm\frac{2\sqrt{3}}{3} = x - 1 \Rightarrow 1 \pm \frac{2\sqrt{3}}{3} = x \Rightarrow x \approx 2.15 \text{ or }$$
$x \approx -0.15$. Now, find the y-intercept: $f(0) = 3(0)^2 - 6(0) - 1 = -1$.

(continued on next page)

(continued)

Use the fact that the parabola is symmetric with respect to its axis, $x = 1$, to locate additional points. Plot the vertex, the x-intercepts, the y-intercept, and any additional points, and join them with a parabola.

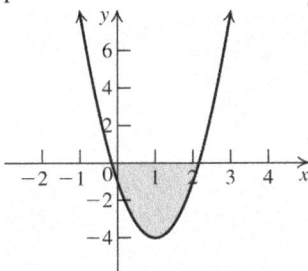

The graph of f is below the x-axis between the x-intercepts, so the solution set for $f(x) = 3x^2 - 6x - 1 \leq 0$ is
$$\left[1 - \frac{2\sqrt{3}}{3}, 1 + \frac{2\sqrt{3}}{3}\right] \text{ or } \left[\frac{3 - 2\sqrt{3}}{3}, \frac{3 + 2\sqrt{3}}{3}\right].$$

5. $h(t) = -\dfrac{g_E}{2}t^2 + v_0 t + h_0$

   $g_E = 32 \text{ ft/s}^2$, $h_0 = 100$ ft, max height = 244 ft

   a. Using the given values, we have
   $$h(t) = -\frac{32}{2}t^2 + v_0 t + 100$$
   $$= -16t^2 + v_0 t + 100$$
   Using the formula for the vertex of a parabola gives $t = -\dfrac{v_0}{2(-16)} = \dfrac{v_0}{32}$. This is the time at which the maximum height $h(t) = 244$ ft is attained. Thus,
   $$h(t) = 244 = h\left(\frac{v_0}{32}\right)$$
   $$= -16\left(\frac{v_0}{32}\right)^2 + v_0\left(\frac{v_0}{32}\right) + 100$$
   $$= -\frac{v_0^2}{64} + \frac{v_0^2}{32} + 100 = \frac{v_0^2}{64} + 100$$
   Solving for v_0 yields
   $$244 = \frac{v_0^2}{64} + 100 \Rightarrow 144 = \frac{v_0^2}{64} \Rightarrow$$
   $$v_0^2 = 144 \cdot 64 \Rightarrow v_0 = \sqrt{144 \cdot 64} = 12 \cdot 8 = 96$$
   Thus, $h(t) = -16t^2 + 96t + 100$ feet.

   b. Using the formula for the vertex of a parabola gives $t = -\dfrac{96}{2(-16)} = \dfrac{96}{32} = 3$.
   The ball reached its highest point 3 seconds after it was released.

6. Let x = the length of the playground and y = the width of the playground.
   Then $2(x + y) = 1000 \Rightarrow x + y = 500 \Rightarrow y = 500 - x$.
   The area of the playground is
   $A(x) = xy = x(500 - x) = 500x - x^2$.
   The vertex for the parabola is (h, k) where
   $h = -\dfrac{500}{2(-1)} = 250$ and
   $k = 500(250) - 250^2 = 62,500$.
   Thus, the maximum area that can be enclosed is 62,500 ft². The playground is a square with side length 250 ft.

3.1 Concepts and Vocabulary

1. A point where the axis meets the parabola is called the vertex.

3. The vertical line passing through the vertex of a parabola is called the axis or axis of symmetry.

5. False. $a = 1 > 0$.

7. True. The x-coordinate of the vertex of the parabola is $h = -\dfrac{b}{2a}$, so the y-coordinate of the vertex is $f\left(-\dfrac{b}{2a}\right)$.

3.1 Building Skills

9. f

11. a

13. h

15. g

17. Substitute -8 for y and 2 for x to solve for a:
    $-8 = a(2)^2 \Rightarrow -2 = a$.
    The equation is $y = -2x^2$.

19. Substitute 20 for y and 2 for x to solve for a:
    $20 = a(2)^2 \Rightarrow 5 = a$.
    The equation is $y = 5x^2$.

21. Substitute 0 for h, 0 for k, 8 for y, and -2 for x in the standard form for a quadratic equation to solve for a: $8 = a(-2-0)^2 + 0 \Rightarrow 8 = 4a \Rightarrow 2 = a$. The equation is $y = 2x^2$.

23. Substitute -3 for h, 0 for k, -4 for y, and -5 for x in the standard form for a quadratic equation to solve for a:
$-4 = a(-5-(-3))^2 + 0 \Rightarrow -1 = a$.
The equation is $y = -(x+3)^2$.

25. Substitute 2 for h, 5 for k, 7 for y, and 3 for x in the standard form for a quadratic equation to solve for a: $7 = a(3-2)^2 + 5 \Rightarrow 2 = a$.
The equation is $y = 2(x-2)^2 + 5$.

27. Substitute 2 for h, -3 for k, 8 for y, and -5 for x in the standard form for a quadratic equation to solve for a: $8 = a(-5-2)^2 - 3 \Rightarrow \dfrac{11}{49} = a$.
The equation is $y = \dfrac{11}{49}(x-2)^2 - 3$.

29. Substitute $\dfrac{1}{2}$ for h, $\dfrac{1}{2}$ for k, $-\dfrac{1}{4}$ for y, and $\dfrac{3}{4}$ for x in the standard form for a quadratic equation to solve for a:
$-\dfrac{1}{4} = a\left(\dfrac{3}{4} - \dfrac{1}{2}\right)^2 + \dfrac{1}{2} \Rightarrow -12 = a$.
The equation is $y = -12\left(x - \dfrac{1}{2}\right)^2 + \dfrac{1}{2}$.

31. The vertex is $(-2, 0)$, and the graph passes through $(0, 3)$. Substitute -2 for h, 0 for k, 3 for y, and 0 for x in the standard form for a quadratic equation to solve for a:
$3 = a(0-(-2))^2 + 0 \Rightarrow \dfrac{3}{4} = a$.
The equation is $y = \dfrac{3}{4}(x+2)^2$.

33. The vertex is $(3, -1)$, and the graph passes through $(5, 2)$. Substitute 3 for h, -1 for k, 2 for y, and 5 for x in the standard form for a quadratic equation to solve for a:
$2 = a(5-3)^2 - 1 \Rightarrow \dfrac{3}{4} = a$.
The equation is $y = \dfrac{3}{4}(x-3)^2 - 1$.

35. $f(x) = 3x^2$

Stretch the graph of $y = x^2$ vertically by a factor of 3.

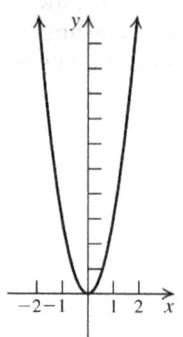

37. $g(x) = (x-4)^2$

Shift the graph of $y = x^2$ right four units.

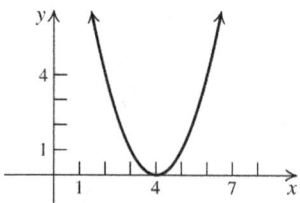

39. $f(x) = -2x^2 - 4$

Stretch the graph of $y = x^2$ vertically by a factor of 2, reflect the resulting graph in the x-axis, then shift the resulting graph down 4 units.

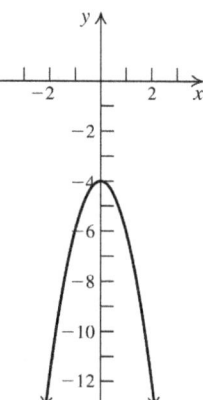

41. $g(x) = (x-3)^2 + 2$

Shift the graph of $y = x^2$ right three units, then shift the resulting graph up two units.

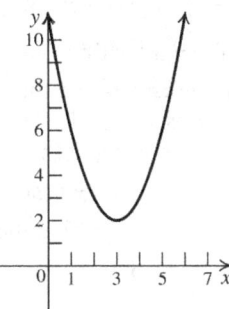

43. $f(x) = -3(x-2)^2 + 4$

Shift the graph of $y = x^2$ right two units, stretch the resulting graph vertically by a factor of 3, reflect the resulting graph about the x-axis, and then shift the resulting graph up four units.

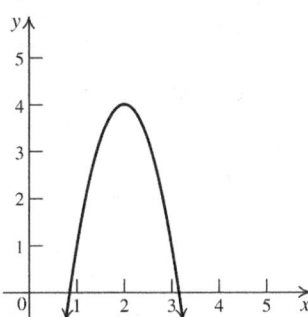

45. Complete the square to write the equation in standard form: $y = x^2 + 4x \Rightarrow$

$y + 4 = x^2 + 4x + 4 \Rightarrow y = (x+2)^2 - 4$. This is the graph of $y = x^2$ shifted two units left and four units down. The vertex is $(-2, -4)$. The axis of symmetry is $x = -2$. To find the x-intercepts, let $y = 0$ and solve
$0 = (x+2)^2 - 4 \Rightarrow (x+2)^2 = 4 \Rightarrow$
$x + 2 = \pm 2 \Rightarrow x = -4$ or $x = 0$
To find the y-intercept, let $x = 0$ and solve
$y = (0+2)^2 - 4 \Rightarrow y = 0$.

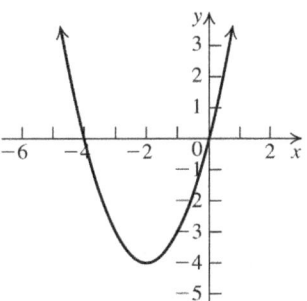

47. Complete the square to write the equation in standard form:
$y = 6x - 10 - x^2 \Rightarrow y = -(x^2 - 6x + 10) \Rightarrow$
$y - 9 = -(x^2 - 6x + 9) - 10 \Rightarrow y = -(x-3)^2 - 1$.

This is the graph of $y = x^2$ shifted three units right, reflected about the x-axis, and then shifted one unit down. The vertex is $(3, -1)$. The axis of symmetry is $x = 3$. To find the x-intercepts, let $y = 0$ and solve
$0 = -(x-3)^2 - 1 \Rightarrow -1 = (x-3)^2 \Rightarrow$ there is no x-intercept. To find the y-intercept, let $x = 0$ and solve $y = -(0-3)^2 - 1 \Rightarrow y = -10$.

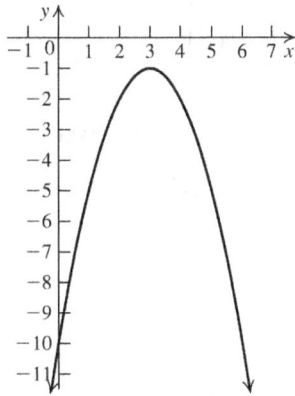

49. Complete the square to write the equation in standard form:
$y = 2x^2 - 8x + 9 \Rightarrow y = 2(x^2 - 4x) + 9 \Rightarrow$
$y + 8 = 2(x^2 - 4x + 4) + 9 \Rightarrow y = 2(x-2)^2 + 1$.

This is the graph of $y = x^2$ shifted 2 units right, stretched vertically by a factor of 2, and then shifted one unit up. The vertex is $(2, 1)$. The axis of symmetry is $x = 2$. To find the x-intercepts, let $y = 0$ and solve
$0 = 2(x-2)^2 + 1 \Rightarrow -\frac{1}{2} = (x-2)^2 \Rightarrow$ there is no x-intercept. To find the y-intercept, let $x = 0$ and solve $y = 2(0-2)^2 + 1 \Rightarrow y = 9$.

(continued on next page)

(*continued*)

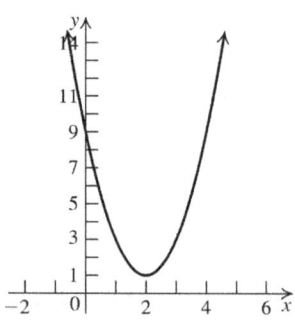

51. Complete the square to write the equation in standard form:

$y = -3x^2 + 18x - 11 \Rightarrow y = -3(x^2 - 6x) - 11 \Rightarrow$

$y - 27 = -3(x^2 - 6x + 9) - 11 \Rightarrow$

$y = -3(x - 3)^2 + 16$. This is the graph of

$y = x^2$ shifted three units right, stretched vertically by a factor of three, reflected about the *x*-axis, and then shifted 16 units up. The vertex is (3, 16). The axis of symmetry is $x = 3$. To find the *x*-intercepts, let $y = 0$ and solve $y = 0 = -3(x - 3)^2 + 16 \Rightarrow$

$\frac{16}{3} = (x - 3)^2 \Rightarrow \pm \frac{4\sqrt{3}}{3} = x - 3 \Rightarrow$

$3 \pm \frac{4}{3}\sqrt{3} = x$. To find the *y*-intercept, let $x = 0$

and solve $y = -3(0 - 3)^2 + 16 \Rightarrow y = -11$.

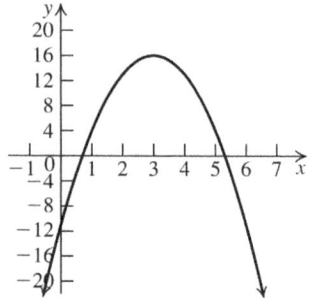

53. Complete the square to write the equation in standard form:

$y = -x^2 - 4x \Rightarrow y = -(x^2 + 4x) \Rightarrow$

$y - 4 = -(x^2 + 4x + 4) \Rightarrow y = -(x + 2)^2 + 4$

This is the graph of $y = x^2$ shifted two units left, reflected about the *x*-axis, and then shifted four units up. The vertex is (–2, 4). The axis is $x = -2$. To find the *x*-intercepts, let $y = 0$ and solve $0 = -(x + 2)^2 + 4 \Rightarrow 4 = (x + 2)^2 \Rightarrow$

$\pm 2 = x + 2 \Rightarrow x = -4, x = 0$. To find the *y*-intercept, let $x = 0$ and solve

$y = -(0 + 2)^2 + 4 \Rightarrow y = 0$.

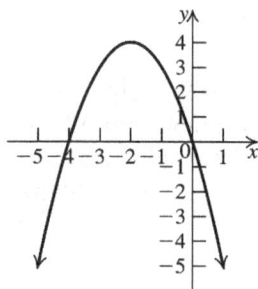

55. a. $a = 1 > 0$, so the graph opens up.

b. The vertex is

$\left(-\frac{-8}{2(1)}, f\left(-\frac{-8}{2(1)}\right)\right) = (4, -1)$.

c. The axis of symmetry is $x = 4$.

d. To find the *x*-intercepts, let $y = 0$ and solve

$0 = x^2 - 8x + 15 \Rightarrow 0 = (x - 3)(x - 5) \Rightarrow$

$x = 3$ or $x = 5$. To find the *y*-intercept, let $x = 0$ and solve $y = 0^2 - 8(0) + 15 \Rightarrow$

$y = 15$.

e.

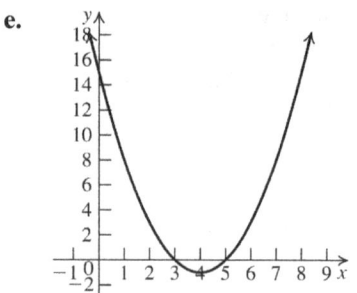

57. a. $a = 1 > 0$, so the graph opens up.

b. The vertex is

$\left(-\frac{-1}{2(1)}, f\left(-\frac{-1}{2(1)}\right)\right) = \left(\frac{1}{2}, -\frac{25}{4}\right)$.

c. The axis of symmetry is $x = \dfrac{1}{2}$.

d. To find the x-intercepts, let $y = 0$ and solve
$0 = x^2 - x - 6 \Rightarrow 0 = (x-3)(x+2) \Rightarrow x = 3$ or $x = -2$. To find the y-intercept, let $x = 0$ and solve $y = 0^2 - (0) - 6 \Rightarrow y = -6$.

e.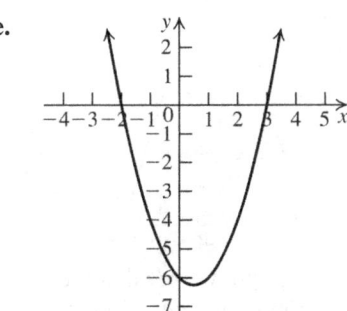

59. a. $a = 1 > 0$, so the graph opens up.

b. The vertex is $\left(-\dfrac{-2}{2(1)}, f\left(-\dfrac{-2}{2(1)}\right)\right) = (1, 3)$.

c. The axis of symmetry is $x = 1$.

d. To find the x-intercepts, let $y = 0$ and solve
$0 = x^2 - 2x + 4 \Rightarrow$
$x = \dfrac{-(-2) \pm \sqrt{(-2)^2 - 4(1)(4)}}{2(1)} \Rightarrow$
$x = \dfrac{2 \pm \sqrt{-12}}{2} \Rightarrow$ there are no x-intercepts.
To find the y-intercept, let $x = 0$ and solve $y = 0^2 - 2(0) + 4 \Rightarrow y = 4$.

e.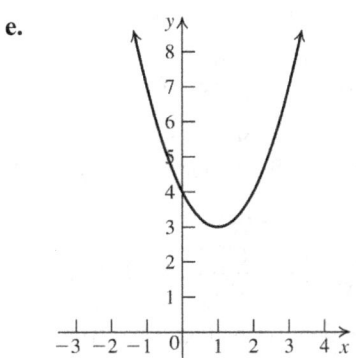

61. a. $a = -1 < 0$, so the graph opens down.

b. The vertex is
$\left(-\dfrac{-2}{2(-1)}, f\left(-\dfrac{-2}{2(-1)}\right)\right) = (-1, 7)$.

c. The axis of symmetry is $x = -1$.

d. To find the x-intercepts, let $y = 0$ and solve
$0 = 6 - 2x - x^2 \Rightarrow$
$x = \dfrac{-(-2) \pm \sqrt{(-2)^2 - 4(-1)(6)}}{2(-1)} \Rightarrow x = \dfrac{2 \pm \sqrt{28}}{-2} \Rightarrow$
$x = -1 \pm \sqrt{7}$.
To find the y-intercept, let $x = 0$ and solve
$y = 6 - 2(0) - 0^2 \Rightarrow y = 6$.

e.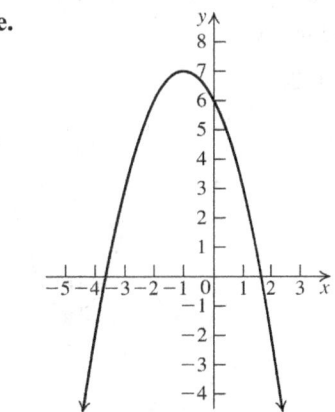

63. a. $a = 1 > 0$, so the graph opens up and has a minimum value. Find the minimum value by finding the vertex:
$\left(-\dfrac{-4}{2(1)}, f\left(-\dfrac{-4}{2(1)}\right)\right) = (2, -1)$
The minimum value is -1.

b. The range of f is $[-1, \infty)$.

65. a. $a = -1 < 0$, so the graph opens down and has a maximum value. Find the maximum value by finding the vertex:
$\left(-\dfrac{4}{2(-1)}, f\left(-\dfrac{4}{2(-1)}\right)\right) = (2, 0)$
The maximum value is 0.

b. The range of f is $(-\infty, 0]$.

67. a. $a = 2 > 0$, so the graph opens up and has a minimum value. Find the minimum value by finding the vertex:
$\left(-\dfrac{-8}{2(2)}, f\left(-\dfrac{-8}{2(2)}\right)\right) = (2, -5)$
The minimum value is -5.

b. The range of f is $[-5, \infty)$.

69. a. $a = -4 < 0$, so the graph opens down and has a maximum value. Find the maximum value by finding the vertex:

$$\left(-\frac{12}{2(-4)}, f\left(-\frac{12}{2(-4)}\right)\right) = \left(\frac{3}{2}, 16\right)$$

The maximum value is 16.

b. The range of f is $(-\infty, 16]$.

In exercises 71–76, the x-intercepts are the boundaries of the intervals.

71. Solution: $[-2, 2]$

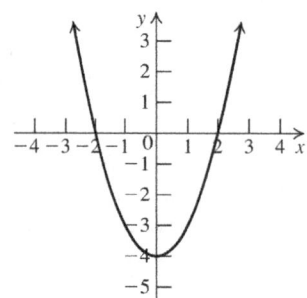

73. Solution: $(-\infty, 1) \cup (3, \infty)$

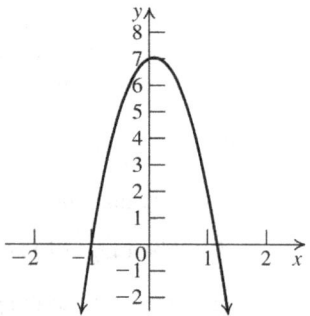

75. Solution: $\left(-1, \frac{7}{6}\right)$

3.1 Applying the Concepts

77. $h(x) = -\dfrac{32}{132^2}x^2 + x + 3$

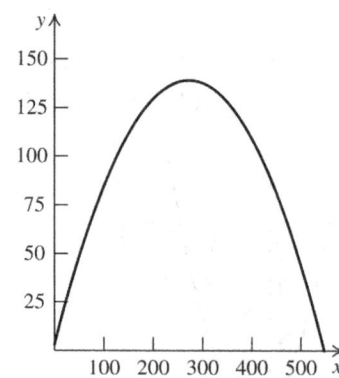

a. Using the graph, we see that the ball traveled approximately 550 ft horizontally.

b. Using the graph, we see that the ball went approximately 140 high.

c. $0 = -\dfrac{32}{132^2}x^2 + x + 3$

$$x = \frac{-1 \pm \sqrt{1^2 - 4\left(-\dfrac{32}{132^2}\right)(3)}}{2\left(-\dfrac{32}{132^2}\right)} \approx -3 \text{ or } 547$$

Thus, the ball traveled approximately 547 ft horizontally. The ball reached its maximum height at the vertex of the function,

$$\left(-\frac{b}{2a}, h\left(-\frac{b}{2a}\right)\right).$$

$$-\frac{b}{2a} = -\frac{1}{2\left(-32/(132^2)\right)} \approx 272.25$$

$$h(272.25) = -\frac{32}{132^2}(272.25)^2 + 272.25 + 3$$

$$\approx 139$$

The ball reached approximately 139 ft

79. The vertex of the function is

$$\left(-\frac{114}{2(-3)}, f\left(-\frac{114}{2(-3)}\right)\right) = (19, 1098).$$

The revenue is at its maximum when $x = 19$.

81. $\left(-\dfrac{-50}{2(1)}, f\left(-\dfrac{-50}{2(1)}\right)\right) = (25, -425).$

The total cost is minimum when $x = 25$.

83. Let $x=$ the length of the rectangle. Then $\frac{80-2x}{2} = 40-x =$ the width of the rectangle.

The area of the rectangle $= x(40-x)$
$= 40x - x^2$. Find the vertex to find the maximum value:

$$\left(-\frac{40}{2(-1)}, f\left(-\frac{40}{2(-1)}\right)\right) = (20, 400).$$

The rectangle with the maximum area is a square with sides of length 20 units. Its area is 400 square units.

85. Let $x =$ the width of the fields. Then, $\frac{600-3x}{2} = 300 - \frac{3}{2}x =$ the length of the two fields together. (Note that there is fencing between the two fields, so there are three "widths.")

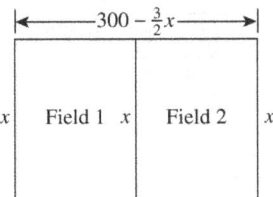

The total area $= x\left(300 - \frac{3}{2}x\right) = 300x - \frac{3}{2}x^2$.

Find the vertex to find the dimensions and maximum value:

$$\left(-\frac{300}{2(-3/2)}, f\left(-\frac{300}{2(-3/2)}\right)\right) = (100, 15,000).$$

So the width of each field is 100 meters. The length of the two fields together is $300 - 1.5(100) = 150$ meters, so the length of each field is $150/2 = 75$ meters. The area of each field is $100(75) = 7500$ square meters.

87. The yield per tree is modeled by the equation of a line passing through (26, 500) where the x-coordinate represents the number of trees planted, and the y-coordinate represents the number of apples per tree. The rate of change is -10; that is, for each tree planted the yield decreases by 10. So, the yield per tree is $y - 500 = -10(x-26) \Rightarrow y = -10x + 760$.

Since there are x trees, the total yield $=$
$x(-10x + 760) = -10x^2 + 760x.$

Use the vertex to find the number of trees that will maximize the yield:

$$\left(-\frac{760}{2(-10)}, f\left(-\frac{760}{2(-10)}\right)\right) = (38, 14440).$$

So the maximum yield occurs when 38 trees are planted per acre.

89. If 20 students or less go on the trip, the cost is $72 per students. If more than 20 students go on the trip, the cost is reduced by $2 per the number of students over 20. So the cost per student is a piecewise function based on the number of students, n, going on the trip:

$$f(n) = \begin{cases} 72 & \text{if } n \leq 20 \\ 72 - 2(n-20) = 112 - 2n & \text{if } n > 20 \end{cases}$$

The total revenue is

$$nf(n) = \begin{cases} 72n & \text{if } n \leq 20 \\ n(112-2n) = 112n - 2n^2 & \text{if } n > 20 \end{cases}$$

The maximum revenue is either 1440 (the revenue if 20 students go on the trip) or the maximum of $112n - 2n^2$. Find this by using the vertex:

$$\left(-\frac{112}{(2)(-2)}, f\left(-\frac{112}{(2)(-2)}\right)\right) = (28, 1568)$$

The maximum revenue is $1568 when 28 students go on the trip.

91. $h(t) = -\frac{g_M}{2}t^2 + v_0 t + h_0$

$g_M = 1.6\,\text{m/s}^2$, $h_0 = 5$ m, max height $= 25$ m

a. Using the given values, we have

$$h(t) = -\frac{1.6}{2}t^2 + v_0 t + 5 = -0.8t^2 + v_0 t + 5$$

Using the formula for the vertex of a parabola gives $t = -\frac{v_0}{2(-0.8)} = \frac{v_0}{1.6}$. This is the time at which the height $h(t) = 25$ m is attained. Thus,

$$h(t) = 25 = h\left(\frac{v_0}{1.6}\right)$$
$$= -0.8\left(\frac{v_0}{1.6}\right)^2 + v_0\left(\frac{v_0}{1.6}\right) + 5$$
$$= -\frac{v_0^2}{3.2} + \frac{v_0^2}{1.6} + 5 = \frac{v_0^2}{3.2} + 5$$

Solving for v_0 yields

$$25 = \frac{v_0^2}{3.2} + 5 \Rightarrow 20 = \frac{v_0^2}{3.2} \Rightarrow$$
$$v_0^2 = 64 \Rightarrow v_0 = 8$$

Thus, $h(t) = -0.8t^2 + 8t + 5$.

b. Using the formula for the vertex of a parabola gives $t = -\dfrac{8}{2(-0.8)} = \dfrac{8}{1.6} = 5.$
The stone reached its highest point 5 seconds after it was released.

b. Using the formula for the vertex of a parabola gives $t = -\dfrac{12}{2(-0.8)} = \dfrac{12}{1.6} = 7.5.$
The stone reached its highest point 7.5 seconds after it was released.

93. a. The maximum height occurs at the vertex:
$\left(-\dfrac{64}{2(-16)}, f\left(-\dfrac{64}{2(-16)}\right)\right) = (2, 64),$ so the maximum height is 64 feet.

b. When the projectile hits the ground, $h = 0$, so solve
$0 = -16t^2 + 64t \Rightarrow -16t(t - 4) = 0 \Rightarrow$
$t = 0$ or $t = 4.$
The projectile hits the ground at 4 seconds.

95. Let x = the radius of the semicircle. Then the length of the rectangle is $2x$. The circumference of the semicircle is πx, so the perimeter of the rectangular portion of the window is $18 - \pi x$.

The width of the rectangle = $\dfrac{18 - \pi x - 2x}{2}.$

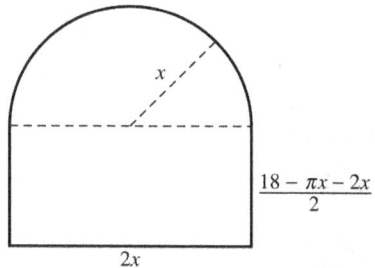

The area of the semicircle is $\pi x^2/2$, and the area of the rectangle is $2x\left(\dfrac{18 - \pi x - 2x}{2}\right) = 18x - \pi x^2 - 2x^2.$ So the total area is
$18x - \pi x^2 - 2x^2 + \dfrac{\pi x^2}{2} = 18x - 2x^2 - \dfrac{\pi x^2}{2} \Rightarrow$
$18x - \left(2 + \dfrac{\pi}{2}\right)x^2.$

The maximum area occurs at the x-coordinate of the vertex:
$-\dfrac{18}{2\left(-2 - \dfrac{\pi}{2}\right)} = -\dfrac{18}{-4 - \pi} = \dfrac{18}{4 + \pi}.$

This is the radius of the semicircle. The length of the rectangle is $2\left(\dfrac{18}{4+\pi}\right) = \dfrac{36}{4+\pi}$ ft.
The width of the rectangle is
$\dfrac{18 - \pi\left(\dfrac{18}{4+\pi}\right) - 2\left(\dfrac{18}{4+\pi}\right)}{2} = \dfrac{18}{4+\pi}$ ft.

97. a. $V = \left(-\dfrac{1.18}{2(-0.01)}, f\left(-\dfrac{1.18}{2(-0.01)}\right)\right)$
$= (59, 36.81)$

b. $0 = -0.01x^2 + 1.18x + 2 \Rightarrow$
$x = \dfrac{-1.18 \pm \sqrt{1.18^2 - 4(-0.01)(2)}}{2(-0.01)}$
$= \dfrac{-1.18 \pm \sqrt{1.4724}}{-0.02} \approx \dfrac{-1.18 \pm 1.21}{-0.02}$
≈ -1.5 or $119.5.$
Reject the negative answer, and round the positive answer to the nearest whole number. The ball hits the ground approximately 120 feet from the punter.

c. The maximum height occurs at the vertex. (See part (a).) The maximum height is approximately 37 ft.

d. The player is at $x = 6$ feet.
$f(6) = -0.01(6)^2 + 1.18(6) + 2 = 8.72.$
The player must reach approximately 9 feet to block the ball.

e. $7 = -0.01x^2 + 1.18x + 2$
$0 = -0.01x^2 + 1.18x - 5$
$x = \dfrac{-1.18 \pm \sqrt{1.18^2 - 4(-0.01)(-5)}}{2(-0.01)}$
$= \dfrac{-1.18 \pm \sqrt{1.1924}}{-0.02} \Rightarrow$
$x \approx 4.4$ feet or $x \approx 113.6$ feet

99. $f(x) = 1.51 - 0.22x + 0.029x^2$

a. We must complete the square to write the function in standard form.
$$0.029x^2 - 0.22x + 1.51$$
$$= 0.029\left(x^2 - \frac{0.22}{0.029}x\right) + 1.51$$
$$= 0.029\left(x^2 - \frac{0.22}{0.029}x + \frac{0.22^2}{4 \cdot 0.029^2}\right)$$
$$+ 1.51 - \left(\frac{0.22^2}{4 \cdot 0.029^2}\right)(0.029)$$
$$= 0.029\left(x - \frac{0.22}{2 \cdot 0.029}\right)^2 + 1.09$$
$$= 0.029(x - 3.79)^2 + 1.09$$

b. Sales were at a minimum at $x \approx 3.79$, or during year 4 (2010). This fits the original data.

c. The year 2011 is represented by $x = 5$.
$f(5) = 1.51 - 0.22(5) + 0.029(5)^2 = 1.135$
In 2011, there were about 1.14 million vehicles sold.

3.1 Beyond the Basics

101. $y = 3(x + 2)^2 + 3 \Rightarrow y = 3(x^2 + 4x + 4) + 3 \Rightarrow$
$y = 3x^2 + 12x + 15$

103. The y-intercept is 4, so the graph passes through (0, 4). Substitute the coordinates (0, 4) for x and y and (1, –2) for h and k into the standard form $y = a(x - h)^2 + k$ to solve for a: $4 = a(0 - 1)^2 - 2 \Rightarrow 6 = a$.
The x-coordinate of the vertex is
$1 = -\frac{b}{2a} = -\frac{b}{2(6)} \Rightarrow b = -12$. The y-intercept $= c$, so the equation is $f(x) = 6x^2 - 12x + 4$.

105. The x-coordinate of the vertex is 3. Substitute the coordinates of the x-intercept and the x-coordinate of the vertex into the standard form $y = a(x - h)^2 + k$ to find an expression relating a and k:
$0 = a(7 - 3)^2 + k \Rightarrow -16a = k$.

Now substitute the coordinates of the y-intercept, the x-coordinate of the vertex, and the expression for k into the standard form $y = a(x - h)^2 + k$ to solve for a:
$14 = a(0 - 3)^2 - 16a \Rightarrow -2 = a$. Use this value to find k: $k = -16(-2) = 32$. The equation is
$y = -2(x - 3)^2 + 32 \Rightarrow y = -2x^2 + 12x + 14$.

107. The x-coordinate of the vertex is $\frac{-2 + 6}{2} = 2$.
Substitute the coordinates of one of the x-intercepts and the x-coordinate of the vertex into the standard form $y = a(x - h)^2 + k$ to find an expression relating a and k:
$0 = a(-2 - 2)^2 + k \Rightarrow -16a = k$.
Now substitute the coordinates of the other x-intercept, the x-coordinate of the vertex, and the expression for k into the standard form $y = a(x - h)^2 + k$ to solve for a:
$0 = a(6 - 2)^2 - 16a \Rightarrow 1 = a$. Use this value to find k: $-16(1) = -16 = k$.
So, one equation is
$y = (x - 2)^2 - 16 = x^2 - 4x - 12$. The graph of this equation opens upward. To find the equation of the graph that opens downward, multiply the equation by –1:
$y = -x^2 + 4x + 12$.

109. The x-coordinate of the vertex is $\frac{-7 - 1}{2} = -4$.
Substitute the coordinates of one of the x-intercepts and the x-coordinate of the vertex into the standard form $y = a(x - h)^2 + k$ to find an expression relating a and k:
$0 = a(-7 + 4)^2 + k \Rightarrow -9a = k$.
Now substitute the coordinates of the other x-intercept, the x-coordinate of the vertex, and the expression for k into the standard form $y = a(x - h)^2 + k$ to solve for a:
$0 = a(-1 + 4)^2 - 9a \Rightarrow 1 = a$. Use this value to find k: $-9(1) = -9 = k$. So one equation is
$y = (x + 4)^2 - 9 = x^2 + 8x + 7$. The graph of this equation opens upward. To find the equation of the graph that opens downward, multiply the equation by –1:
$y = -x^2 - 8x - 7$.

178 Chapter 3 Polynomial and Rational Functions

3.1 Critical Thinking/Discussion/Writing

111. $f(h+p) = a(h+p)^2 + b(h+p) + c$
$f(h-p) = a(h-p)^2 + b(h-p) + c$
$f(h+p) = f(h-p) \Rightarrow a(h+p)^2 + b(h+p) + c = a(h-p)^2 + b(h-p) + c \Leftrightarrow$
$ah^2 + 2ahp + ap^2 + bh + bp + c = ah^2 - 2ahp + ap^2 + bh - bp + c \Rightarrow 4ahp = -2bp \Rightarrow 2ah = -b$.
(We can divide by p because $p \neq 0$.) Since $a \neq 0$, $2ah = -b \Rightarrow h = -\dfrac{b}{2a}$.

112. Write $y = 2x^2 - 8x + 9$ in standard form to find the axis of symmetry:
$y = 2x^2 - 8x + 9 \Rightarrow y + 8 = 2(x^2 - 4x + 4) + 9 \Rightarrow y = 2(x-2)^2 + 1$.
The axis of symmetry is $x = 2$.
Using the results of exercise 111, we know that $f(2+p) = f(-1) = f(2-p)$.
$2 + p = -1 \Rightarrow p = -3$, so $2 - p = 2 - (-3) = 5$.
The point symmetric to the point $(-1, 19)$ across the axis of symmetry is $(5, 19)$.

113. Answers will vary. If $a > 0$, then the parabola $f(x) = ax^2 + bx + c$ opens upward and the solution of the inequality is the portion of the graph that is above the x-axis. If the x-intercepts are represented by x_0 and x_1 with $x_0 \leq x_1$, then the solution of the inequality is $(-\infty, x_0) \cup (x_1, \infty)$. If $a < 0$, then the parabola $f(x) = ax^2 + bx + c$ opens downward and the solution of the inequality is the portion of the graph that is below the x-axis. If the x-intercepts are represented by x_0 and x_1 with $x_0 \leq x_1$, then the solution of the inequality is (x_0, x_1).

114. a. $(f \circ g)(x) = f(mx+b) = a\left[(mx+b) - h\right]^2 + k$
$= a\left[(mx+b)^2 - 2h(mx+b) + h^2\right] + k$
$= a\left[m^2x^2 + 2bmx + b^2 - 2hmx - 2hb + h^2\right] + k$
$= am^2x^2 + 2abmx + ab^2 - 2ahmx - 2ahb + ah^2 + k$
$= am^2x^2 + (2abm - 2ahm)x + (ab^2 - 2ahb + ah^2 + k)$

This is the equation of a parabola. The x-coordinate of the vertex is
$-\dfrac{2abm - 2ahm}{2am^2} = -\dfrac{2am(b-h)}{2am^2} = -\dfrac{b-h}{m}$ or $\dfrac{h-b}{m}$.

The y-coordinate of the vertex is
$am^2\left(\dfrac{h-b}{m}\right)^2 + (2abm - 2ahm)\left(\dfrac{h-b}{m}\right) + (ab^2 - 2ahb + k)$
$= a(h-b)^2 + 2a(b-h)(h-b) + (ab^2 - 2ahb + ah^2 + k)$
$= ah^2 - 2ahb + ab^2 - 2ab^2 + 4abh - 2ah^2 + ab^2 - 2ahb + ah^2 + k$
$= k$

The vertex is $\left(\dfrac{h-b}{m}, k\right)$.

b. $(g \circ f)(x) = g\left[a(x-h)^2 + k\right] = m\left[a(x-h)^2 + k\right] + b$
$= m\left(ax^2 - 2ahx + ah^2 + k\right) + b = max^2 - 2mahx + mah^2 + mk + b$

This is the equation of a parabola. The x-coordinate of the vertex is $-\dfrac{-2mah}{2ma} = h$.

The y-coordinate of the vertex is $mah^2 - 2mah^2 + mah^2 + mk + b = mk + b$.
Thus, the vertex is $(h, mk + b)$.

115. If the discriminant equals zero, there is exactly one real solution. Thus, the vertex of $y = f(x)$ lies on the x-axis at $x = -\dfrac{b}{2a}$.
If the discriminant > 0, there are two unequal real solutions. This means that the graph of $y = f(x)$ crosses the x-axis in two places. If $a > 0$, then the vertex lies below the x-axis and the parabola crosses the x-axis; if $a < 0$, then the vertex lies above the x-axis and the parabola crosses the x-axis. If the discriminant < 0, there are two nonreal complex solutions. If $a > 0$, then the vertex lies above the x-axis and the parabola does not cross the x-axis (it opens upward); if $a < 0$, then the vertex lies below the x-axis and the parabola does not cross the x-axis (it opens downward).

116. Let $g(x) = x^2$ and $k(x) = x - h$. Then $f(x) = g(k(x)) = g(x - h) = (x - h)^2$, which is a horizontal translation of g.

3.1 Getting Ready for the Next Section

117. $-5^0 = -(5^0) = -1$

119. $-4^{-2} = -(4)^{-2} = -\dfrac{1}{(4)^2} = -\dfrac{1}{16}$

121. $x^7 \cdot x^{-7} = x^{7+(-7)} = x^0 = 1$

123. $x^2\left(3 - \dfrac{3}{4}\right) = x^2\left(\dfrac{12}{4} - \dfrac{3}{4}\right) = \dfrac{9}{4}x^2$

125. $4x^2 - 9 = (2x + 3)(2x - 3)$

127. $15x^2 + 11x - 12 = (3x + 4)(5x - 3)$

129. $x^2(x - 1) - 4(x - 1) = (x^2 - 4)(x - 1)$
$= (x + 2)(x - 2)(x - 1)$

131. $x^3 + 4x^2 + 3x + 12 = (x^3 + 4x^2) + (3x + 12)$
$= x^2(x + 4) + 3(x + 4)$
$= (x^2 + 3)(x + 4)$

133. $2x + 9 = 5x + 3 \Rightarrow -3x = -6 \Rightarrow x = 2$
Solution set: $\{2\}$

135. $6x^2 - x - 2 = 0 \Rightarrow (3x - 2)(2x + 1) = 0 \Rightarrow$
$x = \dfrac{2}{3}, x = -\dfrac{1}{2}$
Solution set: $\left\{-\dfrac{1}{2}, \dfrac{2}{3}\right\}$

3.2 Polynomial Functions

3.2 Practice Problems

1. a. $f(x) = \dfrac{x^2 + 1}{x - 1}$ is not a polynomial function because its domain is not $(-\infty, \infty)$.

b. $g(x) = 2x^7 + 5x^2 - 17$ is a polynomial function. Its degree is 7, the leading term is $2x^7$, and the leading coefficient is 2.

2. $P(x) = 4x^3 + 2x^2 + 5x - 17$
$= x^3\left[4 + \dfrac{2}{x} + \dfrac{5}{x^2} - \dfrac{17}{x^3}\right]$

When $|x|$ is large, the terms $\dfrac{2}{x}, \dfrac{5}{x^2}$, and $-\dfrac{17}{x^3}$ are close to 0. Therefore,
$P(x) = x^3(4 + 0 + 0 - 0) \approx 4x^3$.

3. Use the leading-term test to determine the end behavior of $y = f(x) = -2x^4 + 5x^2 + 3$. Here $n = 4$ and $a_n = -2 < 0$. Thus, Case 2 applies. The end behavior is described as
$y \to -\infty$ as $x \to -\infty$ and $y \to -\infty$ as $x \to \infty$.

4. First group the terms, then factor and solve $f(x) = 0$:

$$f(x) = 2x^3 - 3x^2 + 4x - 6$$
$$= 2x^3 + 4x - 3x^2 - 6$$
$$= 2x(x^2 + 2) - 3(x^2 + 2)$$
$$= (2x - 3)(x^2 + 2)$$

$0 = (2x - 3)(x^2 + 2)$

$0 = 2x - 3 \Rightarrow x = \dfrac{3}{2}$ or

$0 = x^2 + 2$ (no real solution)

The only real zero is $\dfrac{3}{2}$.

5. $f(x) = 2x^3 - 3x - 6$

$f(1) = -7$ and $f(2) = 4$. Since $f(1)$ and $f(2)$ have opposite signs, by the Intermediate Value Theorem, f has a real zero between 1 and 2.

6. $f(x) = (x + 1)^2(x - 3)(x + 5) = 0 \Rightarrow$
$(x + 1)^2 = 0$ or $x - 3 = 0$ or $x + 5 = 0 \Rightarrow$
$x = -1$ or $x = 3$ or $x = -5$
$f(x)$ has three distinct zeros.

7. $f(x) = (x - 1)^2(x + 3)(x + 5) = 0 \Rightarrow$
$(x - 1)^2 = 0$ or $x + 3 = 0$ or $x + 5 = 0 \Rightarrow$
$x = 1$ (multiplicity 2) or $x = -3$ (multiplicity 1) or $x = -5$ (multiplicity 1)

8. $f(x) = -x^4 + 3x^2 - 2$ has at most three turning points. Using a graphing calculator, we see that there are indeed, three turning points.

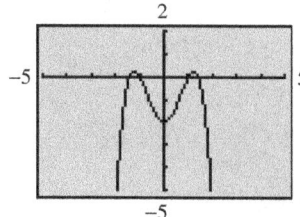

9. $f(x) = -x^4 + 5x^2 - 4$

Since the degree, 4, is even and the leading coefficient is -1, the end behavior is as shown:

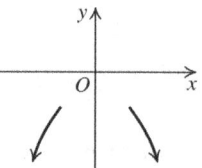

$y \to -\infty$ as $x \to -\infty$ and $y \to -\infty$ as $x \to \infty$.

Now find the zeros of the function:
$0 = -x^4 + 5x^2 - 4 \Rightarrow 0 = -(x^4 - 5x^2 + 4)$
$= -(x^2 - 4)(x^2 - 1) \Rightarrow$
$0 = x^2 - 4$ or $0 = x^2 - 1 \Rightarrow$
$x = \pm 2$ or $x = \pm 1$

There are four zeros, each of multiplicity 1, so the graph crosses the x-axis at each zero. Next, find the y-intercept:

$f(0) = -x^4 + 5x^2 - 4 = -4$

Now find the intervals on which the graph lies above or below the x-axis. The four zeros divide the x-axis into five intervals, $(-\infty, -2)$, $(-2, -1), (-1, 1), (1, 2)$, and $(2, \infty)$. Determine the sign of a test value in each interval

Interval	Test point	Value of $f(x)$	Above/below x-axis
$(-\infty, -2)$	-3	-40	below
$(-2, -1)$	-1.5	2.1875	above
$(-1, 1)$	0	-4	below
$(1, 2)$	1.5	2.1875	above
$(2, \infty)$	3	-40	below

Plot the zeros, y-intercepts, and test points, and then join the points with a smooth curve.

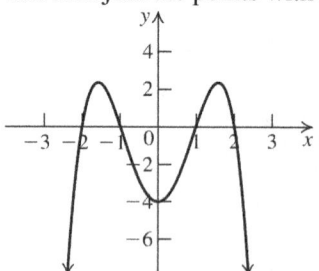

10. $f(x) = x(x+2)^3(x-1)^2$

First, find the zeros of the function.
$x(x+2)^3(x-1)^2 = 0 \Rightarrow$
$x = 0 \mid x+2 = 0 \mid x-1 = 0$
$ x = -2 \mid x = 1$

The function has three distinct zeros, -2, 1, and 0. The zero $x = -2$ has multiplicity 3, so the graph crosses the x-axis at -2, and near -2 the function looks like
$f(x) = -2(x+2)^3(-2-1)^2 = -18(x+2)^3$.

The zero $x = 0$ has multiplicity 1, so the graph crosses the x-axis at 0, and near 0 the function looks like $f(x) = x(0+2)^3(0-1)^2 = 8x$.

The zero $x = 1$ has multiplicity 2, so the graph touches the x-axis at 1, and near 1 the function looks like
$f(x) = 1(1+2)^3(x-1)^2 = 27(x-1)^2$.

We can determine the end behavior of the polynomial by finding its leading term as a product of the leading terms of each factor.
$f(x) = x(x+2)^3(x-1)^2 \approx x(x^3)(x^2) = x^6$

Therefore, the end behavior of $f(x)$ is

Case 1:

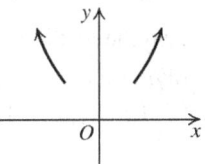

Now find the intervals on which the graph lies above or below the x-axis. The three zeros divide the x-axis into four intervals, $(-\infty, -2)$, $(-2, 0), (0, 1)$, and $(1, \infty)$. Determine the sign of a test value in each interval.

Interval	Test point	Value of $f(x)$	Above/below x-axis
$(-\infty, -2)$	-3	48	above
$(-2, 0)$	-1	-4	below
$(0, 1)$	0.5	1.95	above
$(1, \infty)$	2	128	above

Plot the zeros, y-intercepts, and test points, and then join the points with a smooth curve.

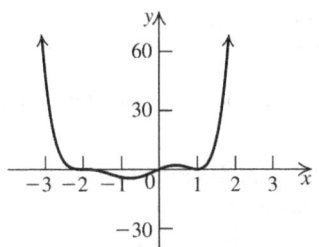

11. $V = \dfrac{\pi}{3\sqrt{3}} x^3 = \dfrac{\pi}{3\sqrt{3}} \cdot 7^3 \approx 207.378 \text{ dm}^3$
    $\approx 207.378 \text{ L}$

3.2 Concepts and Vocabulary

1. Consider the polynomial $2x^5 - 3x^4 + x - 6$. The degree of this polynomial is $\underline{5}$, its leading term is $\underline{2x^5}$, its leading coefficient is $\underline{2}$, and its constant term is $\underline{-6}$.

3. If c is a zero of even multiplicity for a polynomial function f, then the graph of f $\underline{\text{touches}}$ the x-axis at c.

5. False. The graph of a polynomial function is a smooth curve. That means it has no corners or cusps.

7. False. The polynomial function of degree n has, at most, n zeros.

3.2 Building Skills

9. Polynomial function; degree: 5; leading term: $2x^5$; leading coefficient: 2

11. Polynomial function; degree: 3; leading term: $\dfrac{2}{3}x^3$; leading coefficient: $\dfrac{2}{3}$

13. Polynomial function; degree: 4; leading term: πx^4; leading coefficient: π

15. Not a polynomial function because the graph has sharp corners. It is not a smooth curve. There is a presence of $|x|$.

17. Not a polynomial function because the domain is not $(-\infty, \infty)$.

19. Not a polynomial function because of the presence of $\sqrt{x}$.

21. Not a polynomial function because the domain is not $(-\infty, \infty)$.

23. Not a polynomial function because the graph is not continuous.

25. Not a polynomial function because the graph is not continuous.

27. Not a polynomial function because it is not the graph of a function.

29. c **31.** a **33.** d

35. $f(x) = x - x^3$

The leading term is $-x^3$, so $f(x) \to \infty$ as $x \to -\infty$ and $f(x) \to -\infty$ as $x \to \infty$.

37. $f(x) = 4x^4 + 2x^3 + 1$

The leading term is $4x^4$, so $f(x) \to \infty$ as $x \to -\infty$ and $f(x) \to \infty$ as $x \to \infty$.

39. $f(x) = (x+2)^2(2x-1)$

The leading term is $x^2(2x) = 2x^3$, so $f(x) \to -\infty$ as $x \to -\infty$ and $f(x) \to \infty$ as $x \to \infty$.

41. $f(x) = (x+2)^2(4-x)$

The leading term is $x^2(-x) = -x^3$, so $f(x) \to \infty$ as $x \to -\infty$ and $f(x) \to -\infty$ as $x \to \infty$.

43. $f(x) = 3(x-1)(x+2)(x-3)$

Zeros: $-2, 1, 3$
$x = -2$: multiplicity: 1, crosses the x-axis
$x = 1$: multiplicity: 1, crosses the x-axis
$x = 3$: multiplicity: 1, crosses the x-axis

45. $f(x) = (x+2)^2(2x-1)$

Zeros: $-2, \frac{1}{2}$
$x = -2$: multiplicity 2, touches but does not cross the x-axis
$x = \frac{1}{2}$: multiplicity 1, crosses the x-axis

47. $f(x) = x^2(x^2-9)(3x+2)^3$

Zeros: $-3, -\frac{2}{3}, 0, 3$
$x = -3$: multiplicity 1, crosses the x-axis
$x = -\frac{2}{3}$: multiplicity 3, crosses the x-axis
$x = 0$: multiplicity 2, touches but does not cross the x-axis
$x = 3$: multiplicity 1, crosses the x-axis

49. $f(x) = (x^2+1)(3x-2)^2$

Zero: $\frac{2}{3}$, multiplicity 2, touches but does not cross the x-axis

51. $f(2) = 2^4 - 2^3 - 10 = -2$;
$f(3) = 3^4 - 3^3 - 10 = 44$.

Because the sign changes, there is a real zero between 2 and 3. The zero is approximately 2.09.

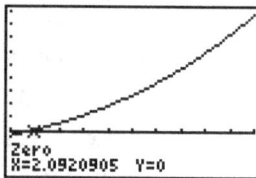

53. $f(2) = 2^5 - 9(2)^2 - 15 = -19$;
$f(3) = 3^5 - 9(3)^2 - 15 = 147$.

Because the sign changes, there is a real zero between 2 and 3. The zero is approximately 2.28.

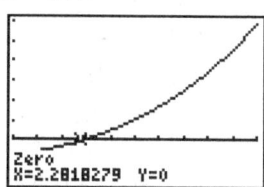

55. The least possible degree is 3. The zeros are -2, 1, and 3, each with multiplicity 1. The end behavior is $f(x) \to -\infty$ as $x \to -\infty$ and $f(x) \to \infty$ as $x \to \infty$, so the leading coefficient is 1. The polynomial with smallest possible degree is $(x+2)(x-1)(x-3)$.

57. The least possible degree is 3. The zeros are -3 (multiplicity 1) and 2 (multiplicity 2). The end behavior is $f(x) \to -\infty$ as $x \to -\infty$ and $f(x) \to \infty$ as $x \to \infty$, so the leading coefficient is 1. The polynomial with smallest possible degree is $(x+3)(x-2)^2$.

59. The least possible degree is 4. The zeros are -1 (multiplicity 2) and 2 (multiplicity 2). The end behavior is $f(x) \to \infty$ as $x \to -\infty$ and $f(x) \to \infty$ as $x \to \infty$, so the leading coefficient is 1. The polynomial with smallest possible degree is $(x+1)^2(x-2)^2$.

61. The zeros are –2 (multiplicity 2), 2 (multiplicity 1), and 3 (multiplicity 1), so the least possible degree is 4. The end behavior is $f(x) \to \infty$ as $x \to -\infty$ and $f(x) \to \infty$ as $x \to \infty$, so the leading coefficient is 1. The polynomial with smallest possible degree is $(x+2)^2(x-2)(x-3)$.

For exercises 63–74, use the procedures shown in Examples 9 and 10 in the text to graph the function.

63. $f(x) = 2(x+1)(x-2)(x+4)$

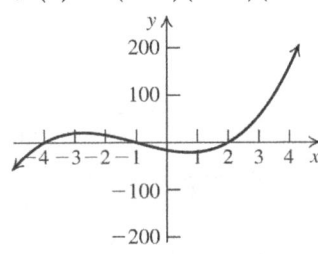

65. $f(x) = (x-1)^2(x+3)(x-4)$

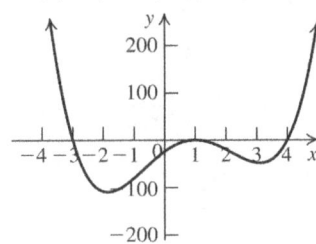

67. $f(x) = -x^2(x-3)^2$

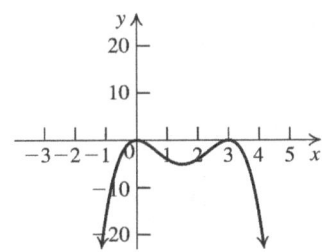

69. $f(x) = (x-1)^2(x+2)^3(x-3)$

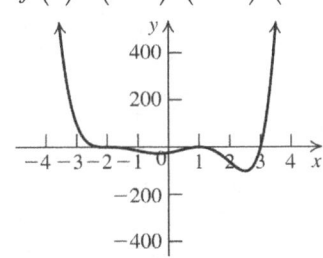

71. $f(x) = -x^2(x^2-1)(x+1)$

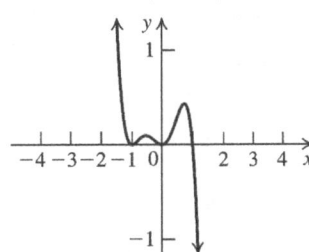

73. $f(x) = x^2(x^2+1)(x-2)$

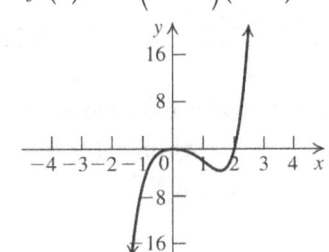

3.2 Applying the Concepts

75. a. $P = \dfrac{0.04825}{746}(40)^3 \approx 4.14$ hp

b. $P = \dfrac{0.04825}{746}(65)^3 \approx 17.76$ hp

c. $P = \dfrac{0.04825}{746}(80)^3 \approx 33.12$ hp

d. $\dfrac{P(2v)}{P(v)} = \dfrac{\frac{0.04825}{746}(2v)^3}{\frac{0.04825}{746}v^3} = \dfrac{8v^3}{v^3} = 8$

77. a. $3x^2(4-x) = 0 \Rightarrow x = 0$ or $x = 4$. $x = 0$, multiplicity 2; $x = 4$, multiplicity 1.

b.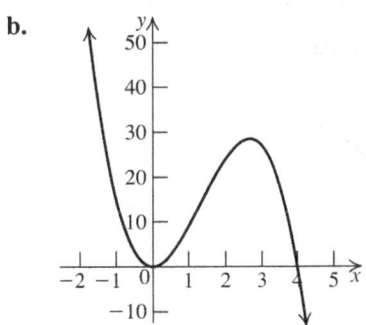

c. There are 2 turning points.

d. Domain: [0, 4]. The portion between the x-intercepts is the graph of $R(x)$.

79. a. Domain [0, 1]

b.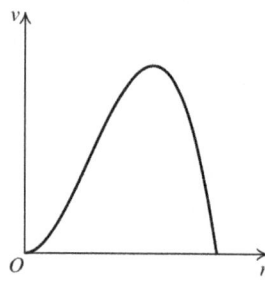

81. a. $R(x) = x\left(27 - \left(\dfrac{x}{300}\right)^2\right) = 27x - \dfrac{x^3}{90,000}$

b. The domain of $R(x)$ is the same as the domain of p. $p \geq 0$ when $p \leq 900\sqrt{3}$. The domain is $\left[0, 900\sqrt{3}\right]$.

c.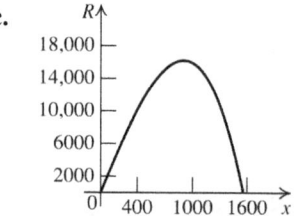

83. a. $N(x) = (x+12)\left(400 - 2x^2\right)$

b. The low end of the domain is 0. (There cannot be fewer than 0 workers.) The upper end of the domain is the value where productivity is 0, so solve $N(x) = 0$ to find the upper end of the domain.

$(x+12)(400 - 2x^2) = 0 \Rightarrow x = -12$ (reject this) or $400 - 2x^2 = 0 \Rightarrow 200 = x^2 \Rightarrow x = \pm 10\sqrt{2}$ (reject the negative solution).

The domain is $[0, 10\sqrt{2}]$.

c.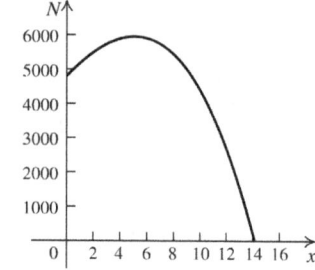

85. a. $V(x) = x(8 - 2x)(15 - 2x)$

b.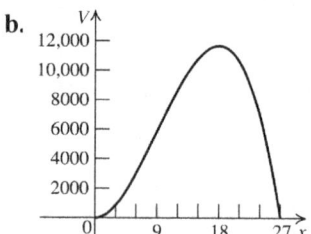

87. a. $V(x) = x^2(108 - 4x)$

b.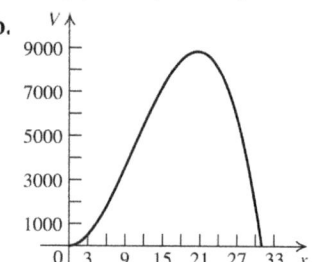

89. a. $V(x) = x^2(62 - 2x)$

b.

91. a. $V(x) = 2x^2(45 - 3x)$

b.

93.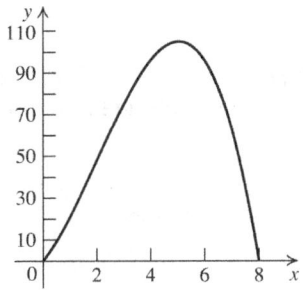

95. $f(x) = 0.963 + 0.88x - 0.192x^2 + 0.0117x^3$

   a. The year 2015 is represented by $x = 7$.
   $f(7) \approx 1.73$ trillion

   b.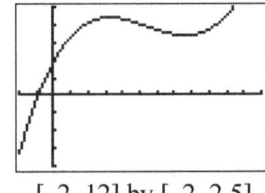

   [–2, 12] by [–2, 2.5]
   It appears that there is one zero and two turning points.

   c.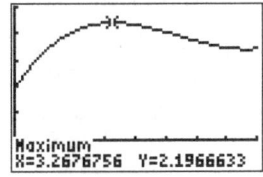

   [0, 8] by [–0.5, 2.5]
   The maximum occurs at $x \approx 3.27$, which is during 2011. This fits the original data.

3.2 Beyond the Basics

97. The graph of $f(x) = (x-1)^4$ is the graph of $y = x^4$ shifted one unit right.
$(x-1)^4 = 0 \Rightarrow x = 1$. The zero is $x = 1$ with multiplicity 4.

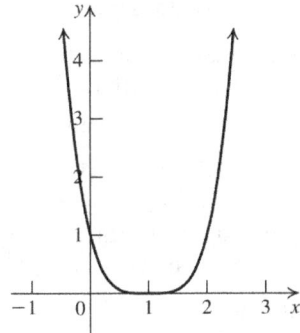

99. The graph of $f(x) = x^4 + 2$ is the graph of $y = x^4$ shifted two units up.
$x^4 + 2 = 0 \Rightarrow x = \sqrt[4]{-2}$. There are no zeros.

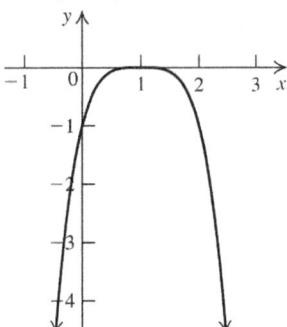

101. The graph of $f(x) = -(x-1)^4$ is the graph of $y = x^4$ shifted one unit right and then reflected across the x-axis.
$-(x-1)^4 = 0 \Rightarrow x = 1$. The zero is $x = 1$ with multiplicity 4.

103. The graph of $f(x) = x^5 + 1$ is the graph of $y = x^5$ shifted one unit up. $x^5 + 1 = 0 \Rightarrow$
$x = \sqrt[5]{-1} \Rightarrow x = -1$. The zero is –1 with multiplicity 1.

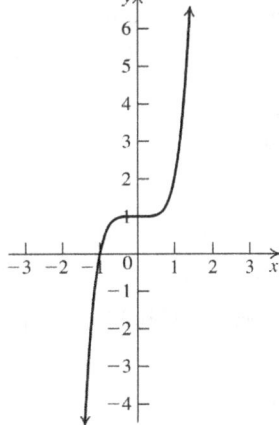

105. The graph of $f(x) = 8 - \dfrac{(x+1)^5}{4}$ is the graph of $y = x^5$ shifted one unit left, compressed vertically by one-fourth, reflected in the x-axis, and then shifted up eight units.

$8 - \dfrac{(x+1)^5}{4} = 0 \Rightarrow -\dfrac{(x+1)^5}{4} = -8 \Rightarrow$

$(x+1)^5 = 32 \Rightarrow x+1 = 2 \Rightarrow x = 1$. The zero is 1 with multiplicity 1.

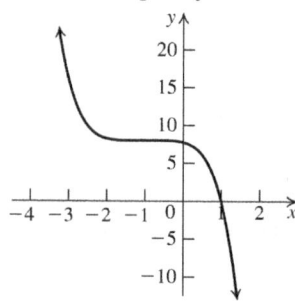

107. If $0 < x < 1$, then $x^2 < 1$. Multiplying both sides of the inequality by x^2, we obtain $x^4 < x^2$. Thus, $0 < x^4 < x^2$.

109.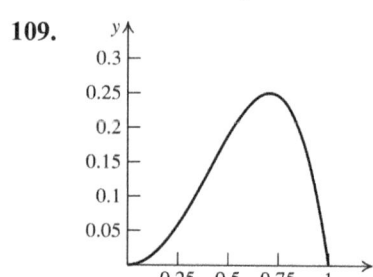

The graph of $y = x^2 - x^4$ represents the difference between the two functions $y = x^2$ and $y = x^4$, so the maximum distance between the graphs occurs at the local maximum of $y = x^2 - x^4$. The maximum vertical distance is 0.25. It occurs at $x \approx 0.71$.

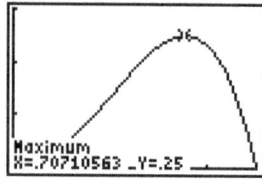

Answers will vary in exercises 111–114. Sample answers are given.

111. $f(x) = x^2(x+1)$

113. $f(x) = 1 - x^4$

115. The smallest possible degree is 5, because the graph has five x-intercepts and four turning points.

117. The smallest possible degree is 6, because the graph has five turning points.

119. $f(x) = 10x^4 + x + 1$

For large values of x, $f(x) \approx 10x^4$. For small values of x, $f(x) \approx x + 1$.

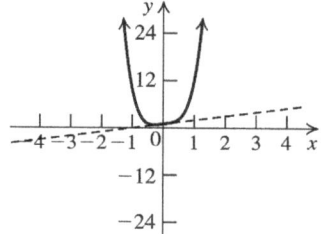

121. $f(x) = x^4 - x^2 + 1$

For large values of x, $f(x) \approx x^4$. For small values of x, $f(x) \approx -x^2 + 1$.

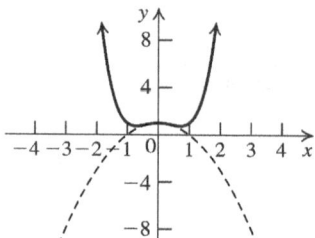

3.2 Critical Thinking/Discussion/Writing

123. If a human is scaled up by a factor of 20, then the total weight will increase $20^3 = 8000$ times but the bone strength will increase only $20^2 = 400$ times. The bones would probably break under the increased weight.

124. If a human is scaled down by a factor of 70, then the body surface area will decrease $70^2 = 4900$ times while the volume will decrease $70^3 = 343,000$ times. The body will rapidly lose heat and will have a very hard time maintaining body temperature unless its metabolic rate increases drastically.

125. It is not possible for a polynomial function to have no y-intercepts because the domain of any polynomial function is $(-\infty, \infty)$, which includes the point $x = 0$.

126. It is possible for a polynomial function to have no x-intercepts because the function can be shifted above the x-axis. An example is the function $y = x^2 + 1$.

127. It is not possible for the graph of a polynomial function of degree 3 to have exactly one local maximum and no local minimum because the graph of a function of degree 3 rises in one direction and falls in the other. This requires an even number of turning points. Since the degree is 3, there can be only zero or two turning points. Therefore, if there is a local maximum, there must also be another turning point, which will be a local minimum.

128. It is not possible for the graph of a polynomial function of degree 4 to have exactly one local maximum and exactly one local minimum because the graph rises in both directions or falls in both directions. This requires an odd number of turning points. Since the degree is 4, there can be only one or three turning points. If there were exactly one local maximum and exactly one local minimum, there would be two turning points.

129. If $P(x) = a_m x^m + a_{m-1} x^{m-1} + \cdots + a_1 x + a_0$ and $Q(x) = b_n x^n + b_{n-1} x^{n-1} + \cdots + b_1 x + b_0$, then

$$(P \circ Q)(x) = a_m \left(b_n x^n + b_{n-1} x^{n-1} + \cdots + b_1 x + b_0 \right)^m + a_{m-1} \left(b_n x^n + b_{n-1} x^{n-1} + \cdots + b_1 x + b_0 \right)^{m-1} + \cdots$$
$$+ a_1 \left(b_n x^n + b_{n-1} x^{n-1} + \cdots + b_1 x + b_0 \right) + a_0,$$

which is a polynomial of degree mn.

130. If $P(x) = a_m x^m + a_{m-1} x^{m-1} + \cdots + a_1 x + a_0$ and $Q(x) = b_n x^n + b_{n-1} x^{n-1} + \cdots + b_1 x + b_0$, then

$$(P \circ Q)(x) = a_m \left(b_n x^n + b_{n-1} x^{n-1} + \cdots + b_1 x + b_0 \right)^m + a_{m-1} \left(b_n x^n + b_{n-1} x^{n-1} + \cdots + b_1 x + b_0 \right)^{m-1} + \cdots$$
$$+ a_1 \left(b_n x^n + b_{n-1} x^{n-1} + \cdots + b_1 x + b_0 \right) + a_0,$$

and the leading coefficient is $a_m (b_n)^m$.

$$(Q \circ P)(x) = b_n \left(a_m x^m + a_{m-1} x^{m-1} + \cdots + a_1 x + a_0 \right)^n + b_{n-1} \left(a_m x^m + a_{m-1} x^{m-1} + \cdots + a_1 x + a_0 \right)^{n-1} + \cdots$$
$$+ b_1 \left(a_m x^m + a_{m-1} x^{m-1} + \cdots + a_1 x + a_0 \right) + b_0,$$

and the leading coefficient is $b_n (a_m)^n$.

3.2 Getting Ready for the Next Section

131. $f(x) = x^3 - 3x^2 + 2x - 9$
$f(-2) = (-2)^3 - 3(-2)^2 + 2(-2) - 9 = -33$

133. $g(x) = \left(-7x^5 + 2x \right)(x - 13)$
$g(13) = \left[-7(13)^5 + 2(13) \right] \left[(13) - 13 \right] = 0$

135. $x^2 - 3x - 10 = (x - 5)(x + 2)$

137. $(2x + 3)\left(x^2 + 6x - 7 \right) = (2x + 3)(x + 7)(x - 1)$

139. $3x^3 - 6x^2 + 5x - 10 = \left(3x^3 - 6x^2 \right) + (5x - 10)$
$= 3x^2 (x - 2) + 5(x - 2)$
$= \left(3x^2 + 5 \right)(x - 2)$

3.3 Dividing Polynomials

3.3 Practice Problems

1.
$$\begin{array}{r} 3x + 4 \\ x^2 + 0x + 1 \overline{\smash{\big)}\, 3x^3 + 4x^2 + x + 7} \\ (-)\underline{3x^3 + 0x^2 + 3x} \\ 4x^2 - 2x + 7 \\ (-)\underline{4x^2 + 0x + 4} \\ -2x + 3 \end{array}$$

Quotient: $3x + 4$; remainder: $-2x + 3$

2. $x^2 - x + 3 \overline{\smash{\big)}\, x^4 + 0x^3 + 5x^2 + 2x + 6}$

$\underline{x^4 - x^3 + 3x^2}$
$x^3 + 2x^2 + 2x$
$\underline{x^3 - x^2 + 3x}$
$3x^2 - x + 6$
$\underline{3x^2 - 3x + 9}$
$2x - 3$

Quotient: $x^2 + x + 3$; remainder: $2x - 3$ or

$x^2 + x + 3 + \dfrac{2x - 3}{x^2 - x + 3}$

3. $(2x^3 - 7x^2 + 5) \div (x - 3)$

$\begin{array}{r|rrrr} 3 & 2 & -7 & 0 & 5 \\ & & 6 & -3 & -9 \\ \hline & 2 & -1 & -3 & -4 \end{array}$

The quotient is $2x^2 - x - 3$ remainder -4 or

$2x^2 - x - 3 - \dfrac{4}{x - 3}$.

4. $(2x^3 + x^2 - 18x - 7) \div (x + 3)$

$\begin{array}{r|rrrr} -3 & 2 & 1 & -18 & -7 \\ & & -6 & 15 & 9 \\ \hline & 2 & -5 & -3 & 2 \end{array}$

The quotient is $2x^2 - 5x - 3$ remainder 2 or

$2x^2 - 5x - 3 + \dfrac{2}{x + 3}$.

5. $F(1) = 1^{110} - 2 \cdot 1^{57} + 5 = 1 - 2 + 5 = 4$, so the remainder when $F(x) = x^{110} - 2x^{57} + 5$ is divided by $x - 1$ is 4.

6. $\begin{array}{r|rrrrr} -2 & 1 & 0 & 10 & 2 & -20 \\ & & -2 & 4 & -28 & 52 \\ \hline & 1 & -2 & 14 & -26 & 32 \end{array}$

The remainder is 32, so $f(-2) = 32$.

7. Because -2 is a zero of the function $3x^3 - x^2 - 20x - 12$, $x + 2$ is a factor. Use synthetic division to find the depressed equation.

$\begin{array}{r|rrrr} -2 & 3 & -1 & -20 & -12 \\ & & -6 & 14 & 12 \\ \hline & 3 & -7 & -6 & 0 \end{array}$

Thus,
$3x^3 - x^2 - 20x - 12 = (x + 2)(3x^2 - 7x - 6)$.

Now solve $3x^2 - 7x - 6 = 0$ to find the two remaining zeros:

$3x^2 - 7x - 6 = 0 \Rightarrow (x - 3)(3x + 2) = 0 \Rightarrow$

$x - 3 = 0$ or $3x + 2 = 0 \Rightarrow x = 3$ or $x = -\dfrac{2}{3}$

The solution set is $\left\{-2, -\dfrac{2}{3}, 3\right\}$.

8. $C(x) = 0.23x^3 - 4.255x^2 + 0.345x + 41.05$
$C(3) = 10 \Rightarrow C(x) = (x - 3)Q(x) + 10 \Rightarrow$
$C(x) - 10 = (x - 3)Q(x)$
So, 3 is zero of $C(x) - 10 =$
$0.23x^3 - 4.255x^2 + 0.345x + 31.05$.
We need to find another positive zero of $C(x) - 10$. Use synthetic division to find the depressed equation.

$\begin{array}{r|rrrr} 3 & 0.23 & -4.255 & 0.345 & 31.05 \\ & & 0.69 & -10.695 & -31.05 \\ \hline & 0.23 & -3.565 & -10.35 & 0 \end{array}$

Solve the depressed equation
$0.23x^2 - 3.565x - 10.35 = 0$ using the quadratic formula:

$x = \dfrac{3.565 \pm \sqrt{(-3.565)^2 - 4(0.23)(-10.35)}}{2(0.23)}$

≈ 18 or -2.5

The positive zero is 18.
Check by verifying that $C(18) = 10$.

3.3 Concepts and Vocabulary

1. In the division
$\dfrac{x^4 - 2x^3 + 5x^2 - 2x + 1}{x^2 - 2x + 3} = x^2 + 2 + \dfrac{2x - 5}{x^2 - 2x + 3}$,
the dividend is $\underline{x^4 - 2x^3 + 5x^2 - 2x + 1}$, the divisor is $\underline{x^2 - 2x + 3}$, the quotient is $\underline{x^2 + 2}$, and the remainder is $\underline{2x - 5}$.

3. The Remainder Theorem states that if a polynomial $F(x)$ is divided by $(x - a)$, then the remainder $R = \underline{F(a)}$.

5. True

7. False. If the polynomial $F(x)$ has $(x + 2)$ as a factor, then $F(-2) = 0$.

3.3 Building Skills

9.
$$\begin{array}{r} 3x-2 \\ 2x+1\overline{)6x^2 - x - 2} \\ \underline{-(6x^2+3x)} \\ -4x-2 \\ \underline{-(-4x-2)} \\ 0 \end{array}$$

In exercises 11–16, insert zero coefficients for missing terms.

11.
$$\begin{array}{r} 3x^3 - 3x^2 - 3x + 6 \\ x+1\overline{)3x^4 + 0x^3 - 6x^2 + 3x - 7} \\ \underline{-(3x^4 + 3x^3)} \\ -3x^3 - 6x^2 \\ \underline{-(-3x^3 - 3x^2)} \\ -3x^2 + 3x \\ \underline{-(-3x^2 - 3x)} \\ 6x - 7 \\ \underline{-(6x+6)} \\ -13 \end{array}$$

13.
$$\begin{array}{r} 2x-1 \\ 2x^2 - x - 5\overline{)4x^3 - 4x^2 - 9x + 5} \\ \underline{-(4x^3 - 2x^2 - 10x)} \\ -2x^2 + x + 5 \\ \underline{-(-2x^2 + x + 5)} \\ 0 \end{array}$$

15.
$$\begin{array}{r} z^2 + 2z + 1 \\ z^2 - 2z + 1\overline{)z^4 + 0z^3 - 2z^2 + 0z + 1} \\ \underline{-(z^4 - 2z^3 + z^2)} \\ 2z^3 - 3z^2 + 0z \\ \underline{-(2z^3 - 4z^2 + 2z)} \\ z^2 - 2z + 1 \\ \underline{-(z^2 - 2z + 1)} \\ 0 \end{array}$$

17.
$$\begin{array}{r|rrrr} 1 & 1 & -1 & -7 & 2 \\ & & 1 & 0 & -7 \\ \hline & 1 & 0 & -7 & -5 \end{array}$$

The quotient is $x^2 - 7$ and the remainder is -5.

19.
$$\begin{array}{r|rrrr} -2 & 1 & 4 & -7 & -10 \\ & & -2 & -4 & 22 \\ \hline & 1 & 2 & -11 & 12 \end{array}$$

The quotient is $x^2 + 2x - 11$ and the remainder is 12.

21.
$$\begin{array}{r|rrrrr} 2 & 1 & -3 & 2 & 4 & 5 \\ & & 2 & -2 & 0 & 8 \\ \hline & 1 & -1 & 0 & 4 & 13 \end{array}$$

The quotient is $x^3 - x^2 + 4$ and the remainder is 13.

23.
$$\begin{array}{r|rrrr} \frac{1}{2} & 2 & 4 & -3 & 1 \\ & & 1 & \frac{5}{2} & -\frac{1}{4} \\ \hline & 2 & 5 & -\frac{1}{2} & \frac{3}{4} \end{array}$$

The quotient is $2x^2 + 5x - \frac{1}{2}$ and the remainder is $\frac{3}{4}$.

25.
$$\begin{array}{r|rrrr} -\frac{1}{2} & 2 & -5 & 3 & 2 \\ & & -1 & 3 & -3 \\ \hline & 2 & -6 & 6 & -1 \end{array}$$

The quotient is $2x^2 - 6x + 6$ and the remainder is -1.

27.
$$\begin{array}{r|rrrrrr} 1 & 1 & 1 & -7 & 2 & 1 & -1 \\ & & 1 & 2 & -5 & -3 & -2 \\ \hline & 1 & 2 & -5 & -3 & -2 & -3 \end{array}$$

The quotient is $x^4 + 2x^3 - 5x^2 - 3x - 2$ and the remainder is -3.

29.
$$\begin{array}{r|rrrrrr} -1 & 1 & 0 & 0 & 0 & 0 & 1 \\ & & -1 & 1 & -1 & 1 & -1 \\ \hline & 1 & -1 & 1 & -1 & 1 & 0 \end{array}$$

The quotient is $x^4 - x^3 + x^2 - x + 1$ remainder 0.

31.
$$\begin{array}{r|rrrrrr} -1 & 1 & 0 & 2 & -1 & 0 & 0 & 5 \\ & & -1 & 1 & -3 & 4 & -4 & 4 \\ \hline & 1 & -1 & 3 & -4 & 4 & -4 & 9 \end{array}$$

The quotient is $x^5 - x^4 + 3x^3 - 4x^2 + 4x - 4$ remainder 9.

33. a.
$$\begin{array}{r|rrrr} 1 & 1 & 3 & 0 & 1 \\ & & 1 & 4 & 4 \\ \hline & 1 & 4 & 4 & 5 \end{array}$$

The remainder is 5, so $f(1) = 5$.

b.
```
-1| 1   3   0   1
      -1  -2   2
   ─────────────
    1   2  -2   3
```
The remainder is 3, so $f(-1) = 3$.

c.
```
1/2| 1   3    0    1
        1/2  7/4  7/8
    ──────────────────
     1   7/2  7/4  15/8
```
The remainder is $\frac{15}{8}$, so $f\left(\frac{1}{2}\right) = \frac{15}{8}$.

d.
```
10| 1   3    0     1
       10  130  1300
   ──────────────────
    1  13  130  1301
```
The remainder is 1301, so $f(10) = 1301$.

35. a.
```
1| 1   5  -3   0  -20
       1   6   3    3
  ──────────────────────
   1   6   3   3  -17
```
The remainder is -17, so $f(1) = -17$.

b.
```
-1| 1   5  -3   0  -20
       -1  -4   7   -7
   ──────────────────────
    1   4  -7   7  -27
```
The remainder is -27, so $f(-1) = -27$.

c.
```
-2| 1   5  -3   0   -20
       -2  -6  18  -36
   ──────────────────────
    1   3  -9  18  -56
```
The remainder is -56, so $f(-2) = -56$.

d.
```
2| 1   5  -3   0  -20
       2  14  22   44
  ──────────────────────
   1   7  11  22   24
```
The remainder is 24, so $f(2) = 24$.

37. $f(1) = 2(1)^3 + 3(1)^2 - 6(1) + 1 = 0 \Rightarrow x - 1$ is a factor of $2x^3 + 3x^2 - 6x + 1$.
Check as follows:
```
1| 2   3  -6   1
       2   5  -1
  ────────────────
   2   5  -1   0
```

39. $f(-1) = 5(-1)^4 + 8(-1)^3 + (-1)^2 + 2(-1) + 4 = 0 \Rightarrow x + 1$ is a factor of $5x^4 + 8x^3 + x^2 + 2x + 4$.
Check as follows:
```
-1| 5   8   1   2   4
       -5  -3   2  -4
   ──────────────────
    5   3  -2   4   0
```

41. $f(2) = 2^4 + 2^3 - 2^2 - 2 - 18 = 0 \Rightarrow x - 2$ is a factor of $x^4 + x^3 - x^2 - x - 18$.
Check as follows:
```
2| 1   1  -1  -1  -18
       2   6  10   18
  ──────────────────────
   1   3   5   9    0
```

43. $f(-2) = (-2)^6 - (-2)^5 - 7(-2)^4 + (-2)^3 + 8(-2)^2 + 5(-2) + 2 = 0 \Rightarrow x + 2$ is a factor of
$x^6 - x^5 - 7x^4 + x^3 + 8x^2 + 5x + 2$.
Check as follows:
```
-2| 1  -1  -7   1   8   5   2
       -2   6   2  -6  -4  -2
   ──────────────────────────────
    1  -3  -1   3   2   1   0
```

45. $f(-1) = 0 = (-1)^3 + 3(-1)^2 + (-1) + k \Rightarrow 0 = 1 + k \Rightarrow k = -1$

47. $f(2) = 0 = 2(2)^3 + (2^2)k - 2k - 2 \Rightarrow 14 + 2k = 0 \Rightarrow k = -7$

In exercises 49–52, use synthetic division to find the remainder.

49.
```
2| -2   4  -4   9
       -4   0  -8
  ──────────────────
   -2   0  -4   1
```
The remainder is 1, so $x - 2$ is not a factor of $-2x^3 + 4x^2 - 4x + 9$.

51.
```
-2| 4   9   3   1   4
       -8  -2  -2   2
   ──────────────────────
    4   1   1  -1   6
```
The remainder is 6, so $x + 2$ is not a factor of $4x^4 + 9x^3 + 3x^2 + x + 4$.

3.3 Applying the Concepts

53. $A = lw \Rightarrow l = \dfrac{A}{w} \Rightarrow$

$$\begin{array}{r}
2x^2 + 1 \\
x^2 - x + 2 \overline{\smash{\big)}\, 2x^4 - 2x^3 + 5x^2 - x + 2}\\
-(2x^4 - 2x^3 + 4x^2)\\
\hline
x^2 - x + 2\\
-(x^2 - x + 2)\\
\hline
0
\end{array}$$

The width is $2x^2 + 1$.

55. a. $R(40) = 3000 \Rightarrow R(40) - 3000 = 0$ and $R(60) = 3000 \Rightarrow R(60) - 3000 = 0$. Thus, 40 and 60 are zeros of $R(x) - 3000$. Therefore, $R(x) - 3000 = a(x - 40)(x - 60) = a(x^2 - 100x + 2400)$

Since (30, 2400) lies on $R(x)$, we have
$2400 - 3000 = a(30 - 40)(30 - 60) \Rightarrow -600 = a(-10)(-30) \Rightarrow a = -2$.

Thus, $R(x) - 3000 = -2(x - 40)(x - 60) = -2x^2 + 200x - 4800$.

b. $R(x) - 3000 = -2x^2 + 200x - 4800 \Rightarrow R(x) = -2x^2 + 200x - 1800$

c. The maximum weekly revenue occurs at the vertex of the function, $\left(-\dfrac{b}{2a}, R\left(-\dfrac{b}{2a}\right)\right)$.

$-\dfrac{b}{2a} = -\dfrac{200}{2(-2)} = 50$

$R(50) = -2(50)^2 + 200(50) - 1800 = 3200$

The maximum revenue is $3200 if the phone is priced at $50.

57. Since $t = 11$ represents 2002, we have $C(11) = 97.6 \Rightarrow C(x) = (x - 11)Q(x) + 97.6 \Rightarrow$
$C(t) - 97.6 = (t - 11)Q(t)$. So
$-0.0006t^3 - 0.0613t^2 + 2.0829t + 82.904 - 97.6 = -0.0006t^3 - 0.0613t^2 + 2.0829t - 14.6960$
$= (x - 11)Q(x)$.

Use synthetic division to find $Q(x)$:

$$\begin{array}{r|rrrr}
11 & -0.0006 & -0.0613 & 2.0829 & -14.6960\\
 & & -0.0066 & -0.7469 & 14.6960\\
\hline
 & -0.0006 & -0.0679 & 1.3360 & 0
\end{array}$$

Now solve the depressed equation to find another zero:

$-0.0006^2 - 0.0679t + 1.3360 = 0 \Rightarrow t = \dfrac{0.0679 \pm \sqrt{0.0679^2 - 4(-0.0006)(1.3360)}}{2(-0.0006)} \Rightarrow$

$t = \dfrac{0.0679 \pm \sqrt{0.00781681}}{-0.0012} = \dfrac{0.0679 \pm 0.0884}{-0.0012} \Rightarrow t \approx -130.26 \text{ or } t \approx 17.0939$.

Since we must find t greater than 0, $t = 17$, and the year is $1991 + 17 = 2008$.

59. $M(t) = -0.0027t^3 + 0.3681t^2 - 5.8645t + 195.2782$

Since $M(2) = 191.736$, $M(t) = (t-2)Q(t) + 195.2782 \Rightarrow M(t) - 195.2782 = (t-2)Q(t)$

We must find two other zeros of

$F(t) = M(t) - 185 = -0.0027t^3 + 0.3681t^2 - 5.8645t + 195.2782 - 185$

$ = -0.0027t^3 + 0.3681t^2 - 5.8645t + 10.2782$

Because 2 is a zero of $F(t)$, use synthetic division to find $Q(t)$.

```
2| -0.0027    0.3681   -5.8645    10.2782
            -0.0054    0.7254   -10.2782
   -0.0027   0.3627   -5.1391         0
```

Now use the quadratic formula to solve the depressed equation.

$t = \dfrac{-0.3627 \pm \sqrt{0.3627^2 - 4(-0.0027)(-5.1391)}}{2(-0.0027)} = \dfrac{-0.3627 \pm \sqrt{0.0760}}{-0.0054} \Rightarrow t \approx 16.1 \text{ or } t \approx 118$

Thus, the model shows that the Marine Corps had about 186,000 when $t \approx 16$, or in the year $1990 + 16 = 2006$.

3.3 Beyond the Basics

61. Divide $4x^3 + 8x^2 - 11x + 3$ by $\left(x - \dfrac{1}{2}\right)$:

```
1/2|  4    8   -11    3
           2    5   -3
      4   10   -6    0
```

Now divide $4x^2 + 10x - 6$ by $\left(x - \dfrac{1}{2}\right)$:

```
1/2|  4   10   -6
           2    6
      4   12    0
```

Since $\left(x - \dfrac{1}{2}\right)$ does not divide $4x + 12$, $\left(x - \dfrac{1}{2}\right)$ is a root of multiplicity 2 of $4x^3 + 8x^2 - 11x + 3$.

63. a. $x + a$ is a factor of $x^n + a^n$ if n is an odd integer. The possible rational zeros of $x^n + a^n$ are $\{\pm a, \pm a^2, \pm a^3, \ldots, \pm a^n\}$. Since $x + a$ is a factor means that $-a$ is a root, then $(-a)^n + a^n = 0 \Rightarrow (-a)^n = -a^n$ only for odd values of n.

b. $x + a$ is a factor of $x^n - a^n$ if n is an even integer. The possible rational zeros of $x^n - a^n$ are $\{\pm a, \pm a^2, \pm a^3, \ldots, \pm a^n\}$. Since $x + a$ is a factor means that $-a$ is a root, then $(-a)^n - a^n = 0 \Rightarrow (-a)^n = a^n$ only for even values of n.

c. There is no value of n for which $x - a$ is a factor of $x^n + a^n$. The possible rational zeros of $x^n + a^n$ are $\{\pm a, \pm a^2, \pm a^3, \ldots, \pm a^n\}$. If $x - a$ is a factor, then a is a root, and $a^n + a^n = 0 \Rightarrow a^n = -a^n$, which is not possible.

d. $x - a$ is a factor of $x^n - a^n$ for all positive integers n. The possible rational zeros of $x^n - a^n$ are $\{\pm a, \pm a^2, \pm a^3, \ldots, \pm a^n\}$. If $x - a$ is a factor, then a is a root, and $a^n - a^n = 0$, which is true for all values of n. However, if n is negative, then $x^n - a^n = \dfrac{1}{x^{-n}} - \dfrac{1}{a^{-n}}$ and $x - a$ is not a factor.

65. a. Divide the divisor and the dividend by 2 so that the leading coefficient of the divisor is 1:

$$\frac{2x^3+3x^2+6x-2}{2x-1} = \frac{2x^3+3x^2+6x-2}{2\left(x-\frac{1}{2}\right)}$$

$$= \frac{x^3+\frac{3}{2}x^2+3x-1}{\left(x-\frac{1}{2}\right)}$$

$$\begin{array}{r|rrrr} \frac{1}{2} & 1 & \frac{3}{2} & 3 & -1 \\ & & \frac{1}{2} & 1 & 2 \\ \hline & 1 & 2 & 4 & 1 \end{array}$$

Because the original polynomials were divided by 2 to use synthetic division, multiply the remainder by 2 to find the remainder for the original division, 2.

b. Divide the divisor and the dividend by 2 so that the leading coefficient of the divisor is 1:

$$\frac{2x^3-x^2-4x+1}{2x+3} = \frac{2x^3-x^2-4x+1}{2\left(x+\frac{3}{2}\right)}$$

$$= \frac{x^3-\frac{1}{2}x^2-2x+\frac{1}{2}}{\left(x+\frac{3}{2}\right)}$$

$$\begin{array}{r|rrrr} -\frac{3}{2} & 1 & -\frac{1}{2} & -2 & \frac{1}{2} \\ & & -\frac{3}{2} & 3 & -\frac{3}{2} \\ \hline & 1 & -2 & 1 & -1 \end{array}$$

Because the original polynomials were divided by 2 to use synthetic division, multiply the remainder by 2 to find the remainder for the original division, –2.

67. $f(x) = ax^3 + bx^2 + cx + d$, $a \neq 0$

There are two possibilities for the end behavior of f:

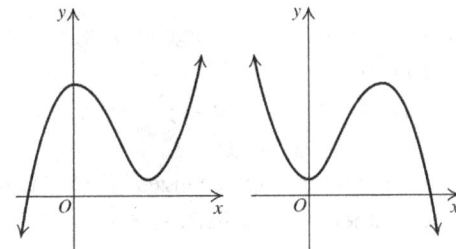

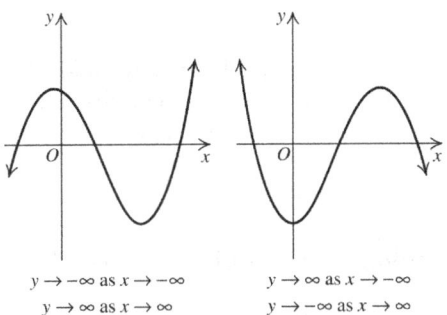

$y \to -\infty$ as $x \to -\infty$
$y \to \infty$ as $x \to \infty$

$y \to \infty$ as $x \to -\infty$
$y \to -\infty$ as $x \to \infty$

Therefore, there is one sign change from the left end of f to the right side of f or there are three sign changes. By the Intermediate Value Theorem, then, there is one zero (if there is one sign change) or there are three zeros (if there are three sign changes.)

69. $dp < 0$ means that d and p have opposite signs, and, thus, are on opposite sides of the x-axis. If $d > 0$ then $p < 0$. Since $kp < 0$, $k > 0$ also. The end behavior is $g(x) \to \infty$ as $x \to -\infty$ and $g(x) \to \infty$ as $x \to \infty$. However, f has end behavior shown in exercise 65. Therefore, the two graphs must intersect. Similar reasoning applies if $d < 0$.

Algebraic proof: Assume $d > 0$ and $p < 0$. Consider the fourth degree polynomial $g - f$. Then $(g-f)(0) = p - d < 0$. Since $k > 0$, $(g-f)(x)$ has upward end behavior on both ends, so $(g-f)(x)$ is eventually > 0. By the Intermediate Value Theorem, there is some c such that $(g-f)(c) = 0 \Rightarrow g(c) = f(c)$. Therefore, the graphs intersect.

71. Answers will vary. One example is $f(x) = x^4 + 2$

3.3 Critical Thinking/Discussion/Writing

72. a. The quotient when $F(x)$ is divided by $2x - 6$ is one-half the quotient when $F(x)$ is divided by $x - 3$, but the remainders are identical.

b. If $\dfrac{F(x)}{x+\frac{b}{a}} = Q(x) + \dfrac{R(x)}{x+\frac{b}{a}}$, then

$$\frac{F(x)}{ax+b} = \frac{1}{a}\left[\frac{F(x)}{x+\frac{b}{a}}\right] = \frac{1}{a}\left[Q(x) + \frac{R(x)}{x+\frac{b}{a}}\right]$$

$$= \frac{1}{a}Q(x) + \frac{R(x)}{ax+b}$$

(continued on next page)

(*continued*)

So, the remainders are again identical and the quotient when $F(x)$ is divided by $ax + b$ is $1/a$ times the quotient when $F(x)$ is divided by $x + \dfrac{b}{a}$.

3.3 Getting Ready for the Next Section

73.

$$\begin{array}{r|rrrr} 1 & 1 & -1 & 2 & 1 & -3 \\ & & 1 & 0 & 2 & 3 \\ \hline & 1 & 0 & 2 & 3 & 0 \end{array}$$

The quotient is $x^3 + 2x + 3$ and the remainder is 0.

75.

$$\begin{array}{r|rrrrrrr} 5 & -1 & 5 & 0 & 0 & 2 & -10 & 3 \\ & & -5 & 0 & 0 & 0 & 10 & 0 \\ \hline & -1 & 0 & 0 & 0 & 2 & 0 & 3 \end{array}$$

The quotient is $-x^5 + 2x$ and the remainder is 3.

77. The factors of 11 are $\pm 1, \pm 11$.

79. The factors of 51 are $\pm 1, \pm 3, \pm 17, \pm 51$.

81. $x^2 + 2x - 1$
Degree: 2
Leading coefficient: 1
Constant term: -1

83. $-7x^{10} + 3x^3 - 2x + 4$
Degree: 10
Leading coefficient: -7
Constant term: 4

85. $f(x) = (x-1)(x+2)^2$
The zeros are 1, with multiplicity 1, and -2, with multiplicity 2.

3.4 The Real Zeros of a Polynomial Function

3.4 Practice Problems

1. $F(x) = 2x^3 + 3x^2 - 6x - 8$

The factors of the constant term, -8, are $\{\pm 1, \pm 2, \pm 4, \pm 8\}$, and the factors of the leading coefficient, 2, are $\{\pm 1, \pm 2\}$. The possible rational zeros are
$\left\{\pm \dfrac{1}{2}, \pm 1, \pm 2, \pm 4, \pm 8\right\}$.

Use synthetic division to find one rational root:

$$\begin{array}{r|rrrr} -2 & 2 & 3 & -6 & -8 \\ & & -4 & 2 & 8 \\ \hline & 2 & -1 & -4 & 0 \end{array}$$

The remainder is 0, so -2 is a zero of the function.
$2x^3 + 3x^2 - 6x - 8 = (x+2)(2x^2 - x - 4)$

Now find the zeros of $2x^2 - x - 4$ using the quadratic formula:

$$x = \dfrac{1 \pm \sqrt{(-1)^2 - 4(2)(-4)}}{2(1)} = \dfrac{1 \pm \sqrt{33}}{2}, \text{ which}$$

are not rational roots.
The only rational zero is $\{-2\}$.

2. $2x^3 - 9x^2 + 6x - 1 = 0$

The factors of the constant term, -1, are $\{\pm 1\}$, and the factors of the leading coefficient, 2, are $\{\pm 1, \pm 2\}$. The possible rational zeros are $\left\{\pm \dfrac{1}{2}, \pm 1\right\}$. Use synthetic division to find one rational zero:

$$\begin{array}{r|rrrr} \tfrac{1}{2} & 2 & -9 & 6 & -1 \\ & & 1 & -4 & 1 \\ \hline & 2 & -8 & 2 & 0 \end{array}$$

Thus, $2x^3 - 9x^2 + 6x - 1 = 0 \Rightarrow$

$\left(x - \dfrac{1}{2}\right)(2x^2 - 8x + 2) = 0 \Rightarrow$

$x - \dfrac{1}{2} = 0 \Rightarrow x = \dfrac{1}{2}$ or

$2x^2 - 8x + 2 = 0$

$x = \dfrac{-(-8) \pm \sqrt{(-8)^2 - 4(2)(2)}}{2(2)} = \dfrac{8 \pm \sqrt{48}}{4}$

$= 2 \pm \sqrt{3}$

Solution set: $\left\{\dfrac{1}{2}, 2 - \sqrt{3}, 2 + \sqrt{3}\right\}$

3. $f(x) = 2x^5 + 3x^2 + 5x - 1$
There is one sign change in $f(x)$, so there is one positive zero.
$f(-x) = 2(-x)^5 + 3(-x)^2 + 5(-x) - 1$
$= -2x^5 + 3x^2 - 5x - 1$
There are two sign changes in $f(-x)$, so there are either 2 or 0 negative zeros.

Section 3.4 The Real Zeros of a Polynomial Function   195

4. $f(x) = 2x^3 + 5x^2 + x - 2$

   The possible rational zeros are $\left\{\pm\dfrac{1}{2}, \pm 1, \pm 2\right\}$.

   There is one sign change in $f(x)$, so there is one positive zero.
   $$f(-x) = 2(-x)^3 + 5(-x)^2 + (-x) - 2$$
   $$= -2x^3 + 5x^2 - x - 2$$
   There are two sign changes in $f(-x)$, so there are either 2 or 0 negative zeros. Try synthetic division by $x - k$ with $k = 1, 2, 3, \ldots$. The first integer that makes each number in the last row a 0 or a positive number is an upper bound on the zeros of $F(x)$. Then use synthetic division by $x - k$ with $k = -1, -2, -3, \ldots$.
   The first negative integer for which the numbers in the last row alternate in sign is a lower bound on the zeros of $F(x)$. In this case, 1 is an upper bound and -3 is a lower bound.

   ```
   1| 2   5   1  -2        -3| 2   5   1   -2
         2   7   8                -6   3  -12
      ─────────────            ─────────────────
         2   7   8   6              2  -1   4  -14
   ```

5. $f(x) = 3x^4 - 11x^3 + 22x - 12$

   Step 1: f has at most 4 real zeros.
   Step 2: $f(x) = 3x^4 - 11x^3 + 0x^2 + 22x - 12$
   There are three sign changes in f, so f has 1 or 3 positive zeros.
   $$f(-x) = 3(-x)^4 - 11(-x)^3 + 0(-x)^2$$
   $$+ 22(-x) - 12$$
   $$= 3x^4 + 11x^3 + 0x^2 - 22x - 12$$
   There is one sign change in $f(-x)$, so f has 1 negative zero.
   Step 3: The factors of the constant term, -12, are $\{\pm 1, \pm 2, \pm 3, \pm 4, \pm 6, \pm 12\}$, and the factors of the leading coefficient, 3, are $\{\pm 1, \pm 3\}$. The possible rational zeros are
   $$\left\{\pm 1, \pm 2, \pm 3, \pm 4, \pm 6, \pm 12, \pm\dfrac{1}{3}, \pm\dfrac{2}{3}, \pm\dfrac{4}{3}\right\}.$$
   Steps 4–7: Test for zeros until a zero or an upper bound is found. Try a positive possibility:

   ```
   3| 3  -11    0   22  -12
           9   -6  -18   12
       ───────────────────────
       3   -2   -6    4    0
   ```
   Since 3 is a zero, we have
   $f(x) = (x - 3)(3x^3 - 2x^2 - 6x + 4)$.
   Now find the a zero of
   $Q_1(x) = 3x^3 - 2x^2 - 6x + 4$. The factors of the constant term, 4, are $\{\pm 1, \pm 2, \pm 4\}$, and the factors of the leading coefficient, 3, are $\{\pm 1, \pm 3\}$. The possible rational zeros are
   $\left\{\pm 1, \pm 2, \pm 4, \pm\dfrac{1}{3}, \pm\dfrac{2}{3}, \pm\dfrac{4}{3}\right\}$. Try the positive possibilities first.

   ```
   2/3| 3  -2  -6   4
              2   0  -4
        ─────────────────
        3   0  -6   0
   ```

   Thus, $f(x) = (x - 3)\left(x - \dfrac{2}{3}\right)(3x^2 - 6)$.

   Now solve $3x^2 - 6 = 0$.
   $3x^2 - 6 \Rightarrow x^2 = 2 \Rightarrow x = \pm\sqrt{2}$

   Solution set: $\left\{\dfrac{2}{3}, 3, \sqrt{2}, -\sqrt{2}\right\}$

6. $f(x) = 3x^3 - x^2 - 9x + 3$

   Step 1: Since the degree, 3, is odd, and the leading coefficient, 3, is positive, the end behavior is similar to that of $y = x^3$.

   Step 2: Solve $3x^3 - x^2 - 9x + 3 = 0$ to find the real zeros. There are two sign changes in f, so f has either 2 or 0 positive zeros.
   $$f(-x) = 3(-x)^3 - (-x)^2 - 9(-x) + 3$$
   $$= -3x^3 - x^2 + x + 3$$
   There is one sign change in $f(-x)$, so there is one negative zero. By the Rational Root Theorem, the possible rational roots are
   $\left\{\pm 1, \pm 3, \pm\dfrac{1}{3}\right\}$. Trying each value, we find that $\dfrac{1}{3}$ is a rational zero.

   ```
   1/3| 3  -1  -9   3
             1   0  -3
        ─────────────────
        3   0  -9   0
   ```

   (continued on next page)

(continued)

Solve the depressed equation $3x^2 - 9 = 0$ to find the remaining zeros.

$3x^2 - 9 = 0 \Rightarrow 3x^2 = 9 \Rightarrow x^2 = 3 \Rightarrow x = \pm\sqrt{3}$

Step 3: The three zeros divide the x-axis into four intervals, $(-\infty, -\sqrt{3}), \left(-\sqrt{3}, \frac{1}{3}\right),$ $\left(\frac{1}{3}, \sqrt{3}\right),$ and $(\sqrt{3}, \infty)$.

Step 4: Determine the sign of a test value in each interval

Interval	Test point	Value of $f(x)$	Above/below x-axis
$(-\infty, -\sqrt{3})$	-2	-7	below
$\left(-\sqrt{3}, \frac{1}{3}\right)$	0	3	above
$\left(\frac{1}{3}, \sqrt{3}\right)$	1	-4	below
$(\sqrt{3}, \infty)$	2	5	above

Plot the zeros, y-intercepts, and test points, and then join the points with a smooth curve.

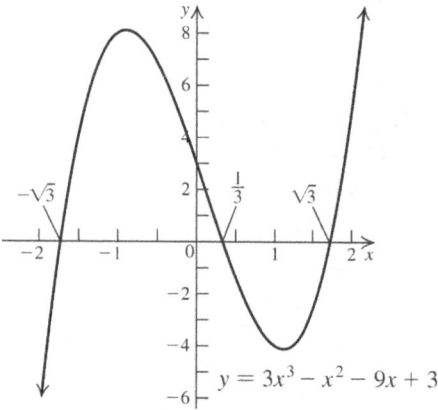

$y = 3x^3 - x^2 - 9x + 3$

3.4 Concepts and Vocabulary

1. Let $P(x) = a_n x^n + \cdots + a_0$ with integer coefficients. If $\frac{p}{q}$ is a rational zero of $P(x)$, then $\frac{p}{q} = \frac{\text{possible factors of } \underline{a_0}}{\text{possible factors of } \underline{a_n}}$.

3. The number of positive zeros of a polynomial function $P(x)$, is equal to the number of <u>variations</u> of sign of $P(x)$ or less than that number by <u>an even integer</u>.

5. False. Possible rational zeros of $P(x) = 9x^3 - 9x^2 - x + 1$ are $\left\{\pm\frac{1}{9}, \pm\frac{1}{3}, \pm 1\right\}$.

7. True

3.4 Building Skills

9. The factors of the constant term, 5, are $\{\pm 1, \pm 5\}$, and the factors of the leading coefficient, 3, are $\{\pm 1, \pm 3\}$. The possible rational zeros are $\left\{\pm\frac{1}{3}, \pm 1, \pm\frac{5}{3}, \pm 5\right\}$.

11. The factors of the constant term, 6, are $\{\pm 1, \pm 2, \pm 3, \pm 6\}$, and the factors of the leading coefficient, 4, are $\{\pm 1, \pm 2, \pm 4\}$. The possible rational zeros are $\left\{\pm\frac{1}{4}, \pm\frac{1}{2}, \pm\frac{3}{4}, \pm 1, \pm\frac{3}{2},\right.$ $\left.\pm 2, \pm 3, \pm 6\right\}$.

13. The factors of the constant term, 4, are $\{\pm 1, \pm 2, \pm 4\}$, and the factors of the leading coefficient, 1, are $\{\pm 1\}$. The possible rational zeros are $\{\pm 1, \pm 2, \pm 4\}$. Use synthetic division to find one rational root:

$$\begin{array}{r|rrrr} \underline{1} & 1 & -1 & -4 & 4 \\ & & 1 & 0 & -4 \\ \hline & 1 & 0 & -4 & 0 \end{array}$$

So,
$x^3 - x^2 - 4x + 4 = (x-1)(x^2 - 4)$
$= (x-1)(x-2)(x+2) \Rightarrow$
the rational zeros are $\{-2, 1, 2\}$.

15. The factors of the constant term, 2, are $\{\pm 1, \pm 2\}$, and the factors of the leading coefficient, 1, are $\{\pm 1\}$. The possible rational zeros are $\{\pm 1, \pm 2\}$. Use synthetic division to find one rational root:

$$\begin{array}{r|rrrr} \underline{-1} & 1 & 1 & 2 & 2 \\ & & -1 & 0 & -2 \\ \hline & 1 & 0 & 2 & 0 \end{array}$$

So, $x^3 + x^2 + 2x + 2 = (x+1)(x^2 + 2) \Rightarrow$ the rational zero is $\{-1\}$.

17. The factors of the constant term, 6, are $\{\pm 1, \pm 2, \pm 3, \pm 6\}$, and the factors of the leading coefficient, 2, are $\{\pm 1, \pm 2\}$. The possible rational zeros are $\left\{\pm \dfrac{1}{2}, \pm 1, \pm \dfrac{3}{2}, \pm 2, \pm 3, \pm 6\right\}$.

Use synthetic division to find one rational root:

$$\underline{2|} \begin{array}{cccc} 2 & 1 & -13 & 6 \\ & 4 & 10 & -6 \\ \hline 2 & 5 & -3 & 0 \end{array}$$

$2x^3 + x^2 - 13x + 6 = (x-2)(2x^2 + 5x - 3)$
$\qquad = (x-2)(x+3)(2x-1) \Rightarrow$

the rational zeros are $\left\{-3, \dfrac{1}{2}, 2\right\}$.

19. The factors of the constant term, -2, are $\{\pm 1, \pm 2\}$, and the factors of the leading coefficient, 3, are $\{\pm 1, \pm 3\}$. The possible rational zeros are $\left\{\pm \dfrac{1}{3}, \pm \dfrac{2}{3}, \pm 1, \pm 2\right\}$. Use synthetic division to find one rational root:

$$\underline{\tfrac{2}{3}|} \begin{array}{cccc} 3 & -2 & 3 & -2 \\ & 2 & 0 & 2 \\ \hline 3 & 0 & 3 & 0 \end{array}$$

$3x^3 - 2x^2 + 3x - 2 = \left(x - \dfrac{2}{3}\right)(3x^2 + 3) \Rightarrow$ the rational zero is $\left\{\dfrac{2}{3}\right\}$.

21. The factors of the constant term, 2, are $\{\pm 1, \pm 2\}$, and the factors of the leading coefficient, 3, are $\{\pm 1, \pm 3\}$. The possible rational zeros are $\left\{\pm \dfrac{1}{3}, \pm \dfrac{2}{3}, \pm 1, \pm 2\right\}$.

Use synthetic division to find one rational root:

$$\underline{-\tfrac{1}{3}|} \begin{array}{cccc} 3 & 7 & 8 & 2 \\ & -1 & -2 & -2 \\ \hline 3 & 6 & 6 & 0 \end{array}$$

$3x^3 + 7x^2 + 8x + 2 = \left(x + \dfrac{1}{3}\right)(3x^2 + 6x + 6) \Rightarrow$

the rational zero is $\left\{-\dfrac{1}{3}\right\}$.

23. The factors of the constant term, -2, are $\{\pm 1, \pm 2\}$, and the factors of the leading coefficient, 1, are $\{\pm 1\}$. The possible rational zeros are $\{\pm 1, \pm 2\}$. Use synthetic division to find one rational root:

$$\underline{-1|} \begin{array}{ccccc} 1 & -1 & -1 & -1 & -2 \\ & -1 & 2 & -1 & 2 \\ \hline 1 & -2 & 1 & -2 & 0 \end{array}$$

So, -1 is a rational zero. Use synthetic division again to find another rational zero:

$x^4 - x^3 - x^2 - x - 2 = (x+1)(x^3 - 2x^2 + x - 2)$

$$\underline{2|} \begin{array}{cccc} 1 & -2 & 1 & -2 \\ & 2 & 0 & 2 \\ \hline 1 & 0 & 1 & 0 \end{array}$$

$x^4 - x^3 - x^2 - x - 2 = (x+1)(x-2)(x^2+1) \Rightarrow$
the rational zeros are $\{-1, 2\}$.

25. The factors of the constant term, 12, are $\{\pm 1, \pm 2, \pm 3, \pm 4, \pm 6, \pm 12\}$, and the factors of the leading coefficient, 1, are $\{\pm 1\}$. The possible rational zeros are $\{\pm 1, \pm 2, \pm 3, \pm 4, \pm 6, \pm 12\}$. Use synthetic division to find one rational root:

$$\underline{-3|} \begin{array}{ccccc} 1 & -1 & -13 & 1 & 12 \\ & -3 & 12 & 3 & -12 \\ \hline 1 & -4 & -1 & 4 & 0 \end{array}$$

So, -3 is a rational zero. Use synthetic division again to find another rational zero:
$x^4 - x^3 - 13x^2 + x + 12$
$\qquad = (x+3)(x^3 - 4x^2 - x + 4)$

$$\underline{-1|} \begin{array}{cccc} 1 & -4 & -1 & 4 \\ & -1 & 5 & -4 \\ \hline 1 & -5 & 4 & 0 \end{array}$$

So, -1 is also a rational zero.
$x^4 - x^3 - 13x^2 + x + 12$
$\qquad = (x+3)(x+1)(x^2 - 5x + 4)$
$\qquad = (x+3)(x+1)(x-4)(x-1) \Rightarrow$
the rational zeros are $\{-3, -1, 1, 4\}$.

27. The factors of the constant term, 1, are $\{\pm 1\}$, and the factors of the leading coefficient, 1, are $\{\pm 1\}$. The possible rational zeros are $\{\pm 1\}$. Use synthetic division to find one rational root:

$$\underline{1|} \begin{array}{ccccc} 1 & -2 & 10 & -1 & 1 \\ & 1 & -1 & 9 & 8 \\ \hline 1 & -1 & 9 & 8 & 9 \end{array}$$

(continued on next page)

(continued)

The remainder is 9 so, 1 is not a rational zero.
Try −1:

```
−1 | 1   −2   10   −1    1
            −1    3  −13   14
      1   −3   13  −14   15
```

The remainder is 15 so, −1 is not a rational zero. Therefore, there are no rational zeros.

29. $f(x) = 6x^3 + 13x^2 + x - 2$

Zeros: $\left\{-2, -\dfrac{1}{2}, \dfrac{1}{3}\right\}$

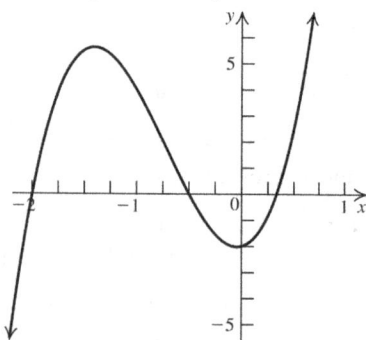

31. $f(x) = 2x^3 - 3x^2 - 14x + 21$

Zeros: $\left\{\pm\sqrt{7}, \dfrac{3}{2}\right\}$

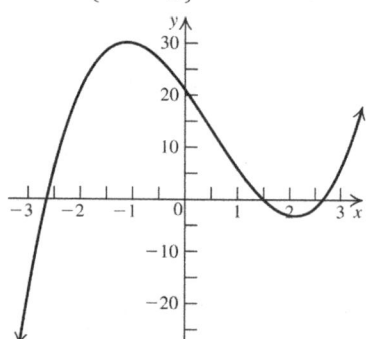

33. $f(x) = 4x^4 + 4x^3 - 3x^2 - 2x + 1$

Zeros: $\left\{-1, \dfrac{1}{2}\right\}$

35. $f(x) = 5x^3 - 2x^2 - 3x + 4;$
$f(-x) = 5(-x)^3 - 2(-x)^2 - 3(-x) + 4$
$= -5x^3 - 2x^2 + 3x + 4$

There are two sign changes in $f(x)$, so there are either 2 or 0 positive zeros. There is one sign change in $f(-x)$, so there is one negative zero.

37. $f(x) = 2x^3 + 5x^2 - x + 2;$
$f(-x) = 2(-x)^3 + 5(-x)^2 - (-x) + 2$
$= -2x^3 + 5x^2 + x + 2$

There are two sign changes in $f(x)$, so there are either 2 or 0 positive zeros. There is one sign change in $f(-x)$, so there is one negative zero.

39. $h(x) = 2x^5 - 5x^3 + 3x^2 + 2x - 1$
$h(-x) = 2(-x)^5 - 5(-x)^3 + 3(-x)^2 + 2(-x) - 1$
$= -2x^5 + 5x^3 + 3x^2 - 2x - 1$

There are three sign changes in $h(x)$, so there are either 1 or 3 positive zeros. There are two sign changes in $h(-x)$, so there are either 2 or 0 negative zeros.

41. $G(x) = -3x^4 - 4x^3 + 5x^2 - 3x + 7;$
$G(-x) = -3(-x)^4 - 4(-x)^3 + 5(-x)^2 - 3(-x) + 7$
$= -3x^4 + 4x^3 + 5x^2 + 3x + 7$

There are three sign changes in $G(x)$, so there are either 3 or 1 positive zeros. There is one sign change in $G(-x)$, so there is 1 negative zero.

43. $f(x) = x^4 + 2x^2 + 4$
$f(-x) = (-x)^4 + 2(-x)^2 + 4 = x^4 + 2x^2 + 4$

There are no sign changes in $f(x)$, nor are there sign changes in $f(-x)$. Therefore, there are no positive zeros and no negative zeros.

45. $g(x) = 2x^5 + x^3 + 3x$
$g(-x) = 2(-x)^5 + (-x)^3 + 3(-x)$
$= -2x^5 - x^3 - 3x$

There are no sign changes in $g(x)$, nor are there sign changes in $g(-x)$. Therefore, there are no positive zeros and no negative zeros.

47. $h(x) = -x^5 - 2x^3 + 4$

    $h(-x) = -(-x)^5 - 2(-x)^3 + 4$
    $= x^5 + x^3 + 4$

    There is one sign change in $h(x)$, and no sign changes in $g(-x)$. Therefore, there is one positive zero and there are no negative zeros.

49. The possible rational zeros are $\left\{\pm\frac{1}{3}, \pm 1, \pm 3\right\}$.

    There are three sign changes in $f(x)$, so there are either 3 or 1 positive zeros. There are no sign changes in $f(-x)$, so there are no negative zeros. Using synthetic division with $k = 1, 2, 3, \ldots$ and $k = -1, -2, -3, \ldots$ gives an upper bound of 1 and a lower bound of -1.

    ```
    1|  3  -1   9  -3        -1|  3  -1   9   -3
           3   2  11                -3   4  -13
        ─────────────             ──────────────
        3   2  11   8             3  -4  13  -16
    ```

51. The possible rational zeros are $\left\{\pm\frac{1}{3}, \pm 1, \pm\frac{7}{3}, \pm 7\right\}$. There are no sign changes in $F(x)$, so there are no positive zeros. There are three sign changes in $F(-x)$, so there are either 3 or 1 negative zeros. Using synthetic division with $k = 1, 2, 3, \ldots$ and $k = -1, -2, -3, \ldots$ gives an upper bound of 1 and a lower bound of -3.

    ```
    1|  3   2   5   7       -3|  3   2    5    7
            3   5  10              -9  21  -75
        ─────────────             ─────────────────
        3   5  10  17             3  -7   25  -68
    ```

53. The possible rational zeros are $\{\pm 1, \pm 31\}$.

    There are two sign changes in $h(x)$, so there are either 2 or 0 positive zeros. There are two sign changes in $h(-x)$, so there are either 2 or 0 negative zeros. Using synthetic division with $k = 1, 2, 3, \ldots$ and $k = -1, -2, -3, \ldots$ gives an upper bound of 31 and a lower bound of -31.

    ```
    31|  1    3    -15       -9         31
              31  1054   32,209    998,200
         ──────────────────────────────────
         1   34   1039   32,200    998,231

    -31|  1    3    -15       -9          31
              -31    868   -26,443    820,012
         ────────────────────────────────────
          1  -28    853   -26,452    820,043
    ```

55. The possible rational zeros are $\left\{\pm\frac{1}{6}, \pm\frac{1}{3}, \pm\frac{1}{2}, \pm 1, \pm\frac{7}{6}, \pm\frac{7}{3}, \pm\frac{7}{2}, \pm 7\right\}$.

    There are two sign changes in $f(x)$, so there are either 2 or 0 positive zeros. There are two sign changes in $f(-x)$, so there are either 2 or 0 negative zeros. Using synthetic division with $k = 1, 2, 3, \ldots$ and $k = -1, -2, -3, \ldots$ gives an upper bound of 4 and a lower bound of -4.

    ```
    4|  6   1  -43   -7     7
           24  100  228   884
        ───────────────────────
        6  25   57  221   891

    -4|  6   1   -43    -7      7
           -24   92  -196    812
        ──────────────────────────
        6  -23   49  -203    819
    ```

In exercises 57–76, first check to see the number of possible positive zeros and the number of possible negative zeros.

57. $f(x) = x^3 + 5x^2 - 8x + 2$

    The function has degree 3, so there are three zeros, zero or two possible positive zeros; one possible negative zero.

    The possible rational zeros are $\{\pm 1, \pm 2\}$.

    ```
    1|  1   5  -8   2
            1   6  -2
        ──────────────
        1   6  -2   0
    ```

    So, 1 is a zero. Now solve the depressed equation $x^2 + 6x - 2 = 0$.

    $$x = \frac{-6 \pm \sqrt{36 - 4(1)(-2)}}{2(1)} = \frac{-6 \pm \sqrt{44}}{2}$$
    $= -3 \pm \sqrt{11}$

    The solution set is $\left\{1, -3 \pm \sqrt{11}\right\}$.

59. $f(x) = 2x^3 - x^2 - 6x + 3$

    The function has degree 3, so there are three zeros, zero or two possible positive zeros; one possible negative zero. The possible rational zeros are $\left\{\pm 1, \pm 3, \pm\frac{1}{2}, \pm\frac{3}{2}\right\}$.

    ```
    1/2|  2  -1  -6   3
              1   0  -3
          ──────────────
          2   0  -6   0
    ```

    So, 1/2 is a zero.

    (continued on next page)

(continued)

Now solve the depressed equation
$2x^2 - 6 = 0$.

$2x^2 - 6 = 0 \Rightarrow x^2 = 3 \Rightarrow x = \pm\sqrt{3}$

Solution set: $\left\{-\sqrt{3}, \dfrac{1}{2}, \sqrt{3}\right\}$

61. $f(x) = 2x^3 - 9x^2 + 6x - 1$

The function has degree 3, so there are three zeros, one or three possible positive zeros; no possible negative zeros.

The possible rational zeros are $\left\{\pm 1, \pm\dfrac{1}{2}\right\}$.

$$\begin{array}{r|rrrr} \frac{1}{2} & 2 & -9 & 6 & -1 \\ & & 1 & -4 & 1 \\ \hline & 2 & -8 & 2 & 0 \end{array}$$

So, 1/2 is a zero. Now solve the depressed equation $2x^2 - 8x + 2 = 0$

$x = \dfrac{-(-8) \pm \sqrt{(-8)^2 - 4(2)(2)}}{2(2)} = \dfrac{8 \pm \sqrt{48}}{4}$

$= 2 \pm \sqrt{3}$

Solution set: $\left\{2 - \sqrt{3}, \dfrac{1}{2}, 2 + \sqrt{3}\right\}$.

63. $f(x) = x^4 + x^3 - 5x^2 - 3x + 6$

The function has degree 4, so there are four zeros, zero or two possible positive zeros; zero or two possible negative zeros.
The possible rational zeros are
$\{\pm 1, \pm 2, \pm 3, \pm 6\}$.

$$\begin{array}{r|rrrrr} 1 & 1 & 1 & -5 & -3 & 6 \\ & & 1 & 2 & -3 & -6 \\ \hline & 1 & 2 & -3 & -6 & 0 \end{array}$$

So, 1 is a zero. Now find a zero of the depressed equation $x^3 + 2x^2 - 3x - 6 = 0$.

$$\begin{array}{r|rrrr} -2 & 1 & 2 & -3 & -6 \\ & & -2 & 0 & 6 \\ \hline & 1 & 0 & -3 & 0 \end{array}$$

So, −2 is a zero. Now solve the depressed equation $x^2 - 3 = 0$.

$x^2 - 3 = 0 \Rightarrow x^2 = 3 \Rightarrow x = \pm\sqrt{3}$

Solution set: $\left\{-2, -\sqrt{3}, 1, \sqrt{3}\right\}$

65. $f(x) = x^4 - 3x^3 + 3x - 1$

The function has degree 4, so there are four zeros, one or three possible positive real zeros; one possible negative real zero. The possible rational zeros are $\{\pm 1\}$.

$$\begin{array}{r|rrrrr} 1 & 1 & -3 & 0 & 3 & -1 \\ & & 1 & -2 & -2 & 1 \\ \hline & 1 & -2 & -2 & 1 & 0 \end{array}$$

So, 1 is a zero. Now find a zero of the depressed equation $g(x) = x^3 - 2x^2 - 2x + 1$.

$$\begin{array}{r|rrrr} -1 & 1 & -2 & -2 & 1 \\ & & -1 & 3 & -1 \\ \hline & 1 & -3 & 1 & 0 \end{array}$$

So, −1 is a zero. Now solve the depressed equation $x^2 - 3x + 1 = 0$

$x = \dfrac{3 \pm \sqrt{9 - 4(1)(1)}}{2(1)} = \dfrac{3 \pm \sqrt{5}}{2}$

The solution set is $\left\{\pm 1, \dfrac{3 \pm \sqrt{5}}{2}\right\}$.

67. $f(x) = 2x^4 - 5x^3 - 4x^2 + 15x - 6$

The function has degree 4, so there are four zeros, one or three possible positive real zeros; one possible negative real zero. The possible rational zeros are

$\left\{\pm 1, \pm 2, \pm 3, \pm 6, \pm\dfrac{1}{2}, \pm\dfrac{3}{2}\right\}$.

$$\begin{array}{r|rrrrr} 2 & 2 & -5 & -4 & 15 & -6 \\ & & 4 & -2 & -12 & 6 \\ \hline & 2 & -1 & -6 & 3 & 0 \end{array}$$

So, 2 is a zero. Now find a zero of the depressed equation $g(x) = 2x^3 - x^2 - 6x + 3$.

$$\begin{array}{r|rrrr} \frac{1}{2} & 2 & -1 & -6 & 3 \\ & & 1 & 0 & -3 \\ \hline & 2 & 0 & -6 & 0 \end{array}$$

So, 1/2 is a zero. Now solve the depressed equation $2x^2 - 6 = 0$.

$2x^2 - 6 = 0 \Rightarrow x^2 = 3 \Rightarrow x = \pm\sqrt{3}$

The solution set is $\left\{-\sqrt{3}, \dfrac{1}{2}, \sqrt{3}, 2\right\}$.

69. $f(x) = 6x^4 - x^3 - 13x^2 + 2x + 2$

The function has degree 4, so there are four zeros, zero or two possible positive real zeros; zero or two possible negative real zeros. The possible rational zeros are
$\left\{\pm 1, \pm 2, \pm \dfrac{1}{2}, \pm \dfrac{2}{3}, \pm \dfrac{1}{6}, \pm \dfrac{1}{3}\right\}$.

$$\begin{array}{r|rrrrr} \frac{1}{2} & 6 & -1 & -13 & 2 & 2 \\ & & 3 & 1 & -6 & -2 \\ \hline & 6 & 2 & -12 & -4 & 0 \end{array}$$

So, $1/2$ is a zero. Now find a zero of the depressed equation
$g(x) = 6x^3 + 2x^2 - 12x - 4$.

$$\begin{array}{r|rrrr} -\frac{1}{3} & 6 & 2 & -12 & -4 \\ & & -2 & 0 & 4 \\ \hline & 6 & 0 & -12 & 0 \end{array}$$

So, $-1/3$ is a zero. Now solve the depressed equation $6x^2 - 12 = 0$.
$6x^2 - 12 = 0 \Rightarrow x^2 = 2 \Rightarrow x = \pm\sqrt{2}$

The solution set is $\left\{-\sqrt{2}, -\dfrac{1}{3}, \dfrac{1}{2}, \sqrt{2}\right\}$.

71. $f(x) = x^5 - 2x^4 - 4x^3 + 8x^2 + 3x - 6$

The function has degree 5, so there are five zeros. There are three sign changes, so there are either 1, 3, or 5 positive real zeros. There are two sign changes in $f(-x)$, so there are zero or two negative real zeros. The possible rational zeros are $\{\pm 1, \pm 3, \pm 6\}$. Using synthetic division to test the positive values, we find that one zero is 1:

$$\begin{array}{r|rrrrrr} 1 & 1 & -2 & -4 & 8 & 3 & -6 \\ & & 1 & -1 & -5 & 3 & 6 \\ \hline & 1 & -1 & -5 & 3 & 6 & 0 \end{array}$$

The zeros of the depressed function $x^4 - x^3 - 5x^2 + 3x + 6$ are also zeros of f. Use synthetic division again to find the next zero, 2:

$$\begin{array}{r|rrrrr} 2 & 1 & -1 & -5 & 3 & 6 \\ & & 2 & 2 & -6 & -6 \\ \hline & 1 & 1 & -3 & -3 & 0 \end{array}$$

$x^5 - 2x^4 - 4x^3 + 8x^2 + 3x - 6$
$= (x - 1)(x - 2)(x^3 + x^2 - 3x - 3)$

Now find a zero of the depressed function $x^3 + x^2 - 3x - 3$. Use synthetic division again to find the next zero, -1:

$$\begin{array}{r|rrrr} -1 & 1 & 1 & -3 & -3 \\ & & -1 & 0 & 3 \\ \hline & 1 & 0 & -3 & 0 \end{array}$$

$x^5 - 2x^4 - 4x^3 + 8x^2 + 3x - 6$
$= (x - 1)(x - 2)(x + 1)(x^2 - 3)$

Now solve the depressed equation: $x^2 - 3 = 0$.
$x^2 - 3 = 0 \Rightarrow x^2 = 3 \Rightarrow x = \pm\sqrt{3}$

Solution set: $\left\{-\sqrt{3}, -1, 1, \sqrt{3}, 2\right\}$

73. $f(x) = 2x^5 + x^4 - 11x^3 - x^2 + 15x - 6$

The function has degree 5, so there are five zeros. There are three sign changes, so there are either 1 or 3 positive real zeros. There are two sign changes in $f(-x)$, so there are zero or two negative real zeros. The possible rational zeros are $\left\{\pm 1, \pm 2, \pm 3, \pm 6, \pm \dfrac{1}{2}, \pm \dfrac{3}{2}\right\}$. Using synthetic division to test the positive values, we find that one zero is 1:

$$\begin{array}{r|rrrrrr} 1 & 2 & 1 & -11 & -1 & 15 & -6 \\ & & 2 & 3 & -8 & -9 & 6 \\ \hline & 2 & 3 & -8 & -9 & 6 & 0 \end{array}$$

The zeros of the depressed function $2x^4 + 3x^3 - 8x^2 - 9x + 6$ are also zeros of f. Use synthetic division again to find the next zero, -2:

$$\begin{array}{r|rrrrr} -2 & 2 & 3 & -8 & -9 & 6 \\ & & -4 & 2 & 12 & -6 \\ \hline & 2 & -1 & -6 & 3 & 0 \end{array}$$

$2x^5 + x^4 - 11x^3 - x^2 + 15x - 6$
$= (x - 1)(x + 2)(2x^3 - x^2 - 6x + 3)$

Now find a zero of the depressed function $2x^3 - x^2 - 6x + 3$. Use synthetic division again to find the next zero, $1/2$:

$$\begin{array}{r|rrrr} \frac{1}{2} & 2 & -1 & -6 & 3 \\ & & 1 & 0 & -3 \\ \hline & 2 & 0 & -6 & 0 \end{array}$$

$2x^5 + x^4 - 11x^3 - x^2 + 15x - 6$
$= (x - 1)(x + 2)\left(x - \dfrac{1}{2}\right)(2x^2 - 6)$

(continued on next page)

(*continued*)

Now solve the depressed equation:
$2x^2 - 6 = 0$.
$2x^2 - 6 = 0 \Rightarrow x^2 = 3 \Rightarrow x = \pm\sqrt{3}$
Solution set: $\left\{-2, -\sqrt{3}, \dfrac{1}{2}, 1, \sqrt{3}\right\}$

75. $f(x) = 2x^5 - 13x^4 + 27x^3 - 17x^2 - 5x + 6$

The function has degree 5, so there are five zeros. There are four sign changes, so there are either 0, 2, or 4 positive real zeros. There is one sign change in $f(-x)$, so there is one possible negative real zero. The possible rational zeros are
$\left\{\pm 1, \pm 2, \pm 3, \pm 6, \pm\dfrac{1}{2}, \pm\dfrac{3}{2}\right\}$.

Using synthetic division to test the positive values, we find that one zero is 1:

$\underline{1|}$ 2 −13 27 −17 −5 6
           2 −11 16 −1 −6
     2 −11 16 −1 −6   0

The zeros of the depressed function $2x^4 - 11x^3 + 16x^2 - x - 6$ are also zeros of f. Use synthetic division again to find the next zero, 1:

$\underline{1|}$ 2 −11 16 −1 −6
            2 −9 7   6
     2 −9   7   6   0

$2x^5 - 13x^4 + 27x^3 - 17x^2 - 5x + 6$
$= (x-1)(x-1)(2x^3 - 9x^2 + 7x + 6)$

The zeros of the depressed function $2x^3 - 9x^2 + 7x + 6$ are also zeros of P. Use synthetic division again to find the next zero, 2.

$\underline{2|}$ 2 −9 7   6
            4 −10 −6
     2 −5 −3   0

$2x^5 - 13x^4 + 27x^3 - 17x^2 - 5x + 6$
$= (x-1)(x-1)(x-2)(2x^2 - 5x - 3)$

Now solve the depressed equation:
$2x^2 - 5x - 3 = 0$.
$2x^2 - 5x - 3 = 0 \Rightarrow (2x+1)(x-3) = 0 \Rightarrow$
$x = -\dfrac{1}{2}, 3$

Solution set: $\left\{-\dfrac{1}{2}, 1 \text{ (multiplicity 2)}, 2, 3\right\}$

3.4 Applying the Concepts

77. The length of the rectangle is $x^2 - 2x + 3$ and its width is $x - 2$. Its area is 306 square units, so we have $(x^2 - 2x + 3)(x - 2) = 306 \Rightarrow$
$x^3 - 4x^2 + 7x - 6 = 306 \Rightarrow$
$x^3 - 4x^2 + 7x - 312 = 0$.
There are 3 sign changes in $f(x)$ and no sign changes in $f(-x)$, so there are 1 or 3 possible positive real zeros and no possible negative real zeros. The possible rational zeros are
$\{\pm 1, \pm 2, \pm 3, \pm 4, \pm 6, \pm 8, \pm 12, \pm 13, \pm 24,$
$\pm 26, \pm 39, \pm 52, \pm 78, \pm 104, \pm 156, \pm 312\}$

Using synthetic division, we find that 8 is a zero:

$\underline{8|}$ 1 −4 7   −312
           8 32 312
     1   4 39   0

Note that the discriminant of the depressed equation $x^2 + 4x + 39 = 0$, $4^2 - 4(1)(39) < 0$, so there are 2 complex solutions to the depressed equation. Therefore, $x = 8$. The width of the rectangle is $8 - 2 = 6$ units and the length of the rectangle is $8^2 - 2 \cdot 8 + 3 = 51$ units.

79. Use synthetic division to solve the equation $628 = 3x^3 - 6x^2 + 108x + 100 \Rightarrow$
$3x^3 - 6x^2 + 108x - 528 = 0$. The factors of the constant term, −528, are
$\{\pm 1, \pm 2, \pm 3, \pm 4, \pm 6, \pm 8, \pm 11, \pm 48, \pm 66, \pm 88,$
$\pm 132, \pm 176, \pm 264, \pm 528\}$. The factors of the leading coefficient, 3, are $\{\pm 1, \pm 3\}$. Only the positive, whole number possibilities make sense for the problem, so the possible rational zeros are $\{1, 2, 3, 4, 6, 8, 11, 16, 22, 44, 48, 66, 88, 132, 176, 264, 528\}$.

$\underline{4|}$ 3 −6 108 −528
           12 24   528
     3   6 132   0

Thus, $x = 4$.

81. The cost function gives the result as a number of thousands, so set it equal to 125:
$x^3 - 15x^2 + 5x + 50 = 125 \Rightarrow$
$x^3 - 15x^2 + 5x - 75 = 0$. The factors of the constant term, -75, are $\{\pm 1, \pm 3, \pm 5, \pm 15, \pm 25, \pm 75\}$. The factors of the leading coefficient, 1, are $\{\pm 1\}$. Only the positive solutions make sense for the problem, so the possible rational zeros are $\{1, 3, 5, 15, 25, 75\}$. Use synthetic division to find the zero:

$$\begin{array}{r|rrrr} 15 & 1 & -15 & 5 & -75 \\ & & 15 & 0 & 75 \\ \hline & 1 & 0 & 5 & 0 \end{array}$$

The total monthly cost is $125,000 when 1500 units are produced.

83. a. The country exported oil for 5 years, from 2010 to 2015.

   b. There are three zeros, so the minimum degree of the polynomial is 3. The zeros are 0, 3, and 8, so the polynomial is of the form $p(x) = a(x-0)(x-3)(x-8)$. The graph goes through the point $(9, -3)$, so we have
   $-3 = a(9-0)(9-3)(9-8) \Rightarrow -3 = 54a \Rightarrow$
   $a = -\dfrac{1}{18}$
   Thus, the equation is
   $p(x) = y = -\dfrac{1}{18}x(x-3)(x-8)$.

   c. The year 2012 is represented by $x = 5$.
   $p(5) = -\dfrac{1}{18}(5)(5-3)(5-8) \approx 1.7$
   In 2012, the country exported about 1.7 million barrels of oil.

85. a. There are three zeros, so the minimum degree of the polynomial is 3. The zeros are 0, 6, and 8, so the polynomial is of the form $p(x) = a(x-0)(x-6)(x-8)$. The graph goes through the point $(9, 0.5)$, so we have
   $0.5 = a(9-0)(9-6)(9-8) \Rightarrow$
   $0.5 = 27a \Rightarrow a = \dfrac{1}{54}$
   Thus, the equation is
   $p(x) = y = \dfrac{1}{54}x(x-6)(x-8)$.

   b. $p(4) = \dfrac{1}{54}(4)(4-6)(4-8) \approx 0.59259$
   The profit in 2008 was about $592.59.

   c. Set $Y_1 = \dfrac{1}{54}x(x-6)(x-8)$ and $Y_2 = 0.6$.
   Find the intersection of the two graphs.

   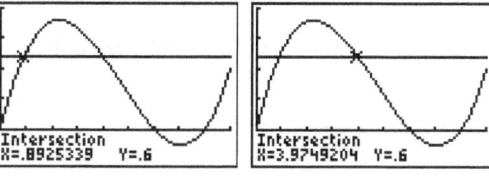

   [0, 9, 1] by [−0.25, 1, 0.25]
   Ms. Sharpy realized a profit of $600 in 2004 and 2007.

   d.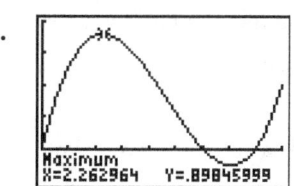

   The maximum profit was approximately $898.46 in 2006.

3.4 Beyond the Basics

87. Let $x = \sqrt{3} \Rightarrow x^2 = 3 \Rightarrow x^2 - 3 = 0$. The only possible rational zeros of this equation are ± 1 and ± 3. Because $\sqrt{3}$ is neither of these, it must be irrational.

89. $x = 3 - \sqrt{2} \Rightarrow 3 - x = \sqrt{2} \Rightarrow 9 - 6x + x^2 = 2 \Rightarrow$
   $x^2 - 6x + 7 = 0$. The possible rational zeros of this equation are $\{\pm 1, \pm 7\}$. Since $3 - \sqrt{2}$ is not in the set, it must be irrational.

91. $y = (x-1)(x+1)(x-0) = x^3 - x$

93. $y = \left(x + \dfrac{1}{2}\right)(x-2)\left(x - \dfrac{7}{3}\right) \Rightarrow$
   $y = (2x+1)(x-2)(3x-7)$
   $= 6x^3 - 23x^2 + 15x + 14$

95. $y = \left(x - (1+\sqrt{2})\right)\left(x - (1-\sqrt{2})\right)(x-3)$
   $= (x^2 - 2x - 1)(x-3) = x^3 - 5x^2 + 5x + 3$

204 Chapter 3 Polynomial and Rational Functions

97. $y = \left(x-\left(1+\sqrt{3}\right)\right)\left(x-\left(1-\sqrt{3}\right)\right) \cdot \left(x-\left(3-\sqrt{2}\right)\right)\left(x-\left(3+\sqrt{2}\right)\right) = \left(x^2 - 2x - 2\right)\left(x^2 - 6x + 7\right)$
$= x^4 - 8x^3 + 17x^2 - 2x - 14$

99. $y = \left(x-\left(\sqrt{2}+\sqrt{3}\right)\right)\left(x-\left(\sqrt{2}-\sqrt{3}\right)\right) = x^2 - 2x\sqrt{2} - 1$

Since we must have integer coefficients, there must be additional zeros. Try $-\left(\sqrt{2}+\sqrt{3}\right)$:

$y = \left(x-\left(\sqrt{2}+\sqrt{3}\right)\right)\left(x-\left(\sqrt{2}-\sqrt{3}\right)\right) \cdot \left(x+\left(\sqrt{2}+\sqrt{3}\right)\right)\left(x+\left(\sqrt{2}-\sqrt{3}\right)\right) = x^4 - 10x^2 + 1$

101. $x = 3 - \sqrt{2} + \sqrt{5} \Rightarrow x - 3 = -\sqrt{2} + \sqrt{5} \Rightarrow (x-3)^2 = \left(-\sqrt{2}+\sqrt{5}\right)^2 \Rightarrow x^2 - 6x + 9 = 7 - 2\sqrt{10} \Rightarrow$
$x^2 - 6x + 2 = -2\sqrt{10} \Rightarrow \left(x^2 - 6x + 2\right)^2 = \left(-2\sqrt{10}\right)^2 \Rightarrow x^4 - 12x^3 + 40x^2 - 24x + 4 = 40 \Rightarrow$
$y = x^4 - 12x^3 + 40x^2 - 24x - 36$

103.
i. Simplifying the fraction if necessary, we can assume that $\dfrac{p}{q}$ is in lowest terms.

 Since $\dfrac{p}{q}$ is a zero of F, we have

 $F\left(\dfrac{p}{q}\right) = 0$.

ii. Substitute $\dfrac{p}{q}$ for x in the equation $F(x) = 0$.

iii. Multiply the equation in (ii) by q^n.

iv. Subtract $a_0 q^n$ from both sides of the equation.

v. The left side of the equation in (iv) is
 $a_n p^n + a_{n-1} p^{n-1} q + \cdots + a_1 p q^{n-1} =$
 $p\left(a_n p^{n-1} + a_{n-1} p^{n-2} q + \cdots + a_1 q^{n-1}\right)$.
 Therefore p is a factor.

vi. $a = b \Leftrightarrow \dfrac{a}{p} = \dfrac{b}{p}$.

vii. Since p and q have no common prime factors, p must be a factor of a_0.

viii. Rearrange the terms of the equation in (iii).

ix. The left side of the equation in (viii) is
 $a_{n-1} p^{n-1} q + \cdots + a_1 p q^{n-1} + a_0 q^n =$
 $q\left(a_{n-1} p^{n-1} + \cdots + a_1 p q^{n-2} + a_0 q^{n-1}\right)$.
 Therefore q is a factor.

x. $a = b \Leftrightarrow \dfrac{a}{q} = \dfrac{b}{q}$.

xi. Since p and q have no common prime factors, q must be a factor of a_n.

3.4 Critical Thinking/Discussion/Writing

105. a. False. The factors of the constant term, 3, are $\{\pm 1, \pm 3\}$ and the factors of the leading coefficient, 1, are $\{\pm 1\}$. The possible rational zeros are $\{\pm 1, \pm 3\}$.

b. False. The factors of the constant term, 25, are $\{\pm 1, \pm 5, \pm 25\}$ and the factors of the leading coefficient, 2, are $\{\pm 1, \pm 2\}$. The possible rational zeros are
$\left\{\pm\dfrac{1}{2}, \pm 1, \pm\dfrac{5}{2}, \pm 5, \pm\dfrac{25}{2}, \pm 25\right\}$.

106. Since $f(x)$ has integer coefficients, if $x = \sqrt{2}$ is a zero, then so is $x = -\sqrt{2}$. Then, two of the factors of $f(x)$ are $\left(x+\sqrt{2}\right)$ and $\left(x-\sqrt{2}\right)$.
$\left(x+\sqrt{2}\right)\left(x-\sqrt{2}\right) = x^2 - 2$, so divide $f(x)$ by $x^2 - 2$ to find another factor.

$$\begin{array}{r}
x^2 - 2x - 2 \\
x^2 + 0x - 2 \overline{\smash{\big)}\, x^4 - 2x^3 - 4x^2 + 4x + 4} \\
\underline{x^4 + 0x^3 - 2x^2} \\
-2x^3 - 2x^2 + 4x \\
\underline{-2x^3 - 0x^2 + 4x} \\
-2x^2 + 0x + 4 \\
\underline{-2x^2 + 0x + 4}
\end{array}$$

(*continued on next page*)

(*continued*)

Use the quadratic formula to solve $x^2 - 2x - 2 = 0$.

$$x = \frac{-(-2) \pm \sqrt{(-2)^2 - 4(1)(-2)}}{2(1)} = \frac{2 \pm \sqrt{4+8}}{2} = \frac{2 \pm 2\sqrt{3}}{2} = 1 \pm \sqrt{3}$$

The zeros of the equation are $x = \pm\sqrt{2}$ and $x = 1 \pm \sqrt{3}$.

3.4 Getting Ready for the Next Section

107. $(-2i)^5 = (-2)^5 (i)^5 = (-2)^5 (i)^4 i = -32i$

109. $(2x+i)(-2x+i) = -4x^2 + i^2 = -4x^2 - 1$

111. $(5x+2i)(5x-2i) = 25x^2 - 4i^2 = 25x^2 + 4$

113. $(x-1+2i)(x-1-2i) = ((x-1)+2i)((x-1)-2i) = (x-1)^2 - 4i^2 = x^2 - 2x + 1 + 4 = x^2 - 2x + 5$

115. $[x-(2+3i)][x-(2-3i)] = x^2 - (2-3i)x - (2+3i)x + (2+3i)(2-3i)$
$= x^2 - 2x + 3ix - 2x - 3ix + 4 - 9i^2 = x^2 - 4x + 4 + 9 = x^2 - 4x + 13$

3.5 The Complex Zeros of a Polynomial Function

3.5 Practice Problems

1. a. $P(x) = 3(x+2)(x-1)[x-(1+i)][x-(1-i)] = 3(x+2)(x-1)(x-1-i)(x-1+i)$

   b. $P(x) = 3(x+2)(x-1)[x-(1+i)][x-(1-i)] = 3(x+2)(x-1)(x^2 - 2x + 2)$
   $= 3(x+2)(x^3 - 3x^2 + 4x - 2) = 3(x^4 - x^3 - 2x^2 + 6x - 4) = 3x^4 - 3x^3 - 6x^2 + 18x - 12$

2. Since $2 - 3i$ is a zero of multiplicity 2, so is $2 + 3i$. Since i is a zero, so is $-i$. The eight zeros are $-3, -3, 2 - 3i, 2 - 3i, 2 + 3i, 2 + 3i, i$, and $-i$.

3. The function has degree four, so there are four zeros. Since one zero is $2i$, another zero is $-2i$. So $(x - 2i)(x + 2i) = x^2 + 4$ is a factor of $P(x)$. Now divide to find the other factors.

$$\begin{array}{r} x^2 - 3x + 2 \\ x^2 + 4 \overline{) x^4 - 3x^3 + 6x^2 - 12x + 8} \\ \underline{x^4 + 4x^2} \\ -3x^3 + 2x^2 - 12x \\ \underline{-3x^3 - 12x} \\ 2x^2 + 8 \\ \underline{2x^2 + 8} \\ 0 \end{array}$$

So,
$P(x) = x^4 - 3x^3 + 6x^2 - 12x + 8$
$= (x - 2i)(x + 2i)(x^2 - 3x + 2)$
$= (x - 2i)(x + 2i)(x - 2)(x - 1)$

The zeros of $P(x)$ are $1, 2, 2i$, and $-2i$.

4. $f(x) = x^4 - 8x^3 + 22x^2 - 28x + 16$

The function has degree 4, so there are four zeros. There are four sign changes, so there are either 0, 2, or 4 positive zeros.

$f(-x) = (-x)^4 - 8(-x)^3 + 22(-x)^2 - 28(-x) + 16$
$= x^4 + 8x^3 + 22x + 28x + 16$

There are no sign changes in $f(-x)$, so there are no negative zeros. The possible rational zeros are $\{\pm 1, \pm 2, \pm 4, \pm 8, \pm 16\}$. Using synthetic division to test the positive values, we find that one zero is 2:

$$\begin{array}{r|rrrrr} 2 & 1 & -8 & 22 & -28 & 16 \\ & & 2 & -12 & 20 & -16 \\ \hline & 1 & -6 & 10 & -8 & 0 \end{array}$$

(*continued on next page*)

(*continued*)

The zeros of the depressed function $x^3 - 6x^2 + 10x - 8$ are also zeros of P. The possible rational zeros of the depressed function are $\{\pm 1, \pm 2, \pm 4, \pm 8\}$. Examine only the positive possibilities and find that 4 is a zero:

$$\underline{4|}\;\; 1 \;\; -6 \;\; 10 \;\; -8$$
$$\;\; 4 \;\; -8 \;\;\;\; 8$$
$$\overline{\;\; 1 \;\; -2 \;\;\;\; 2 \;\;\;\; 0}$$

So,
$x^4 - 8x^3 + 22x^2 - 28x + 16$
$= (x-2)(x-4)(x^2 - 2x + 2)$

Now find the zeros of $x^2 - 2x + 2$ using the quadratic formula:

$$x = \frac{2 \pm \sqrt{(-2)^2 - 4(1)(2)}}{2(1)} = \frac{2 \pm \sqrt{-4}}{2} = 1 \pm i$$

The zeros are 2, 4, $1+i$, and $1-i$.

3.5 Concepts and Vocabulary

1. The Fundamental Theorem of Algebra states that a polynomial function of degree $n \geq 1$ has at least one <u>complex</u> zero.

3. If P is a polynomial function with real coefficients and if $z = a + bi$ is a zero of P, then $\overline{z} = a - bi$ is also a zero of $P(x)$.

5. False. A polynomial function of degree $n \geq 1$ has at least one complex zero.

7. True

3.5 Building Skills

9. $x^2 + 25 = 0 \Rightarrow x^2 = -25 \Rightarrow x = \pm 5i$

11. $x^2 + 4x + 4 = -9 \Rightarrow (x+2)^2 = -9 \Rightarrow$
    $x + 2 = \pm 3i \Rightarrow x = -2 \pm 3i$

13. $(x-2)(x-3i)(x+3i) = 0 \Rightarrow x = 2$ or $x = \pm 3i$

15. The remaining zero is $3 - i$.

17. The remaining zero is $-5 - i$.

19. The remaining zeros are $-i$ and $-3i$.

21. $P(x) = 2(x - (5-i))(x - (5+i))(x - 3i)(x + 3i)$
    $= 2x^4 - 20x^3 + 70x^2 - 180x + 468$

23. $P(x) = 7(x-5)^2(x-1)(x-(3-i))(x-(3+i))$
    $= 7x^5 - 119x^4 + 777x^3 - 2415x^2$
    $+ 3500x - 1750$

25. The function has degree four, so there are four zeros. Since one zero is $3i$, another zero is $-3i$. So $(x - 3i)(x + 3i) = x^2 + 9$ is a factor of $P(x)$. Now divide to find the other factor:

$$\begin{array}{r} x^2 + x \\ x^2 + 9 \overline{) x^4 + x^3 + 9x^2 + 9x} \\ \underline{x^4 + 9x^2 } \\ x^3 + 9x \\ \underline{x^3 + 9x} \\ 0 \end{array}$$

So,
$P(x) = x^4 + x^3 + 9x^2 + 9x$
$= (x^2 + 9)(x^2 + x) = x(x+1)(x^2 + 9)$

27. The function has degree five, so there are five zeros. Since one zero is $3 - i$, another zero is $3 + i$. So
$(x - (3-i))(x - (3+i)) = x^2 - 6x + 10$ is a factor of $P(x)$. Now divide to find the other factor:

$$\begin{array}{r} x^3 + x^2 - 2x \\ x^2 - 6x + 10 \overline{) x^5 - 5x^4 + 2x^3 + 22x^2 - 20x} \\ \underline{x^5 - 6x^4 + 10x^3 } \\ x^4 - 8x^3 + 22x^2 \\ \underline{x^4 - 6x^3 + 10x^2 } \\ -2x^3 + 12x^2 - 20x \\ \underline{-2x^3 + 12x^2 - 20x} \\ 0 \end{array}$$

$P(x) = x^5 - 5x^4 + 2x^3 + 22x^2 - 20x$
$= (x^2 - 6x + 10)(x^3 + x^2 - 2x)$
$= (x^2 - 6x + 10)(x^2 + x - 2)x$
$= x(x+2)(x-1)(x^2 - 6x + 10)$

29. The function has degree 3, so there are three zeros. There are three sign changes, so there are 1 or 3 positive zeros. The possible rational zeros are $\{\pm 1, \pm 17\}$. Using synthetic division we find that one zero is 1:

$$\underline{1|}\ \ \begin{array}{cccc} 1 & -9 & 25 & -17 \\ & 1 & -8 & 17 \\ \hline 1 & -8 & 17 & 0 \end{array}$$

$x^3 - 9x^2 + 25x - 17 = (x-1)(x^2 - 8x + 17)$.
Now solve the depressed equation
$x^2 - 8x + 17 = 0 \Rightarrow$

$$x = \frac{8 \pm \sqrt{(-8)^2 - 4(1)(17)}}{2(1)} \Rightarrow x = \frac{8 \pm \sqrt{-4}}{2} \Rightarrow$$

$x = 4 \pm i$. The zeros are $1, 4 \pm i$.

31. The function has degree 3, so there are three zeros. There are two sign changes, so there are either 2 or 0 positive zeros. There is one sign change in $f(-x)$, so there is one negative zero. The possible rational zeros are
$\left\{\pm\frac{1}{3}, \pm\frac{2}{3}, \pm 1, \pm\frac{4}{3}, \pm\frac{5}{3}, \pm 2, \pm\frac{8}{3}, \pm\frac{10}{3}, \pm 4, \right.$
$\left. \pm 5, \pm\frac{20}{3}, \pm 8, \pm 10, \pm\frac{40}{3}, \pm 20, \pm 40\right\}$. Using synthetic division to test the negative values we find that one zero is $-4/3$.

$$\underline{-\frac{4}{3}|}\ \ \begin{array}{cccc} 3 & -2 & 22 & 40 \\ & -4 & 8 & -40 \\ \hline 3 & -6 & 30 & 0 \end{array}$$

$3x^3 - 2x^2 + 22x + 40 = \left(x + \frac{4}{3}\right)(3x^2 - 6x + 30)$
$= 3\left(x + \frac{4}{3}\right)(x^2 - 2x + 10)$

Now solve the depressed equation
$x^2 - 2x + 10 = 0 \Rightarrow$

$$x = \frac{2 \pm \sqrt{(-2)^2 - 4(1)(10)}}{2(1)} \Rightarrow x = \frac{2 \pm \sqrt{-36}}{2} \Rightarrow$$

$x = 1 \pm 3i$. The zeros are $-\frac{4}{3}, 1 \pm 3i$.

33. The function has degree 4, so there are four zeros. There are four sign changes, so there are 4, 2, or 0 positive zeros. There are no sign changes in $f(-x)$, so there are no negative zeros. The possible rational zeros are
$\left\{\pm\frac{1}{2}, \pm 1, \pm\frac{3}{2}, \pm 3, \pm\frac{9}{2}, \pm 9\right\}$.

Using synthetic division to test the positive values we find that one zero is 1:

$$\underline{1|}\ \ \begin{array}{ccccc} 2 & -10 & 23 & -24 & 9 \\ & 2 & -8 & 15 & -9 \\ \hline 2 & -8 & 15 & -9 & 0 \end{array}$$

The zeros of the depressed function
$2x^3 - 8x^2 + 15x - 9$ are also zeros of P.
The possible rational zeros are
$\left\{\pm\frac{1}{2}, \pm 1, \pm\frac{3}{2}, \pm 3, \pm\frac{9}{2}, \pm 9\right\}$.

Using synthetic division to test the positive values we find that one zero is 1:

$$\underline{1|}\ \ \begin{array}{cccc} 2 & -8 & 15 & -9 \\ & 2 & -6 & 9 \\ \hline 2 & -6 & 9 & 0 \end{array}$$

Thus,
$2x^4 - 10x^3 + 23x^2 - 24x + 9$
$= (x-1)^2(2x^2 - 6x + 9)$.

Now solve the depressed equation
$2x^2 - 6x + 9 = 0$.

$$x = \frac{6 \pm \sqrt{(-6)^2 - 4(2)(9)}}{2(2)} \Rightarrow x = \frac{6 \pm \sqrt{-36}}{4} \Rightarrow$$

$$x = \frac{6 \pm 6i}{4} = \frac{3}{2} \pm \frac{3i}{2}.$$

The zeros of P are 1 (multiplicity 2), $\frac{3}{2} \pm \frac{3i}{2}$.

35. The function has degree 4, so there are four zeros. There are three sign changes, so there are either 3 or 1 positive zeros. There is one sign change in $f(-x)$, so there is one negative zero. The possible rational zeros are $\{\pm 1, \pm 2, \pm 3, \pm 5, \pm 6, \pm 10, \pm 15, \pm 30\}$. Using synthetic division to test the negative values, we find that one zero is -3:

$$\underline{-3|}\ \ \begin{array}{ccccc} 1 & -4 & -5 & 38 & -30 \\ & -3 & 21 & -48 & 30 \\ \hline 1 & -7 & 16 & -10 & 0 \end{array}$$

The zeros of the depressed function
$x^3 - 7x^2 + 16x - 10$ are also zeros of P.
The possible rational zeros of the depressed function are $\{\pm 1, \pm 2, \pm 5, \pm 10\}$. Since we have already found the negative zero, we examine only the positive possibilities and find that 1 is a zero:

(continued on next page)

(continued)

$$\begin{array}{r|rrrr} 1 & 1 & -7 & 16 & -10 \\ & & 1 & -6 & 10 \\ \hline & 1 & -6 & 10 & 0 \end{array}$$

$x^4 - 4x^3 - 5x^2 + 38x - 30$
$= (x+3)(x-1)(x^2 - 6x + 10)$

Now solve the depressed equation:

$x^2 - 6x + 10 = 0 \Rightarrow x = \dfrac{6 \pm \sqrt{(-6)^2 - 4(1)(10)}}{2(1)} \Rightarrow$

$x = \dfrac{6 \pm \sqrt{-4}}{2} \Rightarrow x = 3 \pm i$.

The zeros are $-3, 1, 3 \pm i$.

37. The function has degree 5, so there are five zeros. There are five sign changes, so there are either 5, 3, or 1 positive zeros. There are no sign changes in $f(-x)$, so there are no negative zeros. The possible rational zeros are $\left\{\dfrac{1}{2}, 1, \dfrac{3}{2}, 2, 3, 6\right\}$. Using synthetic division to test the positive values, we find that one zero is 1/2.

$$\begin{array}{r|rrrrrr} \frac{1}{2} & 2 & -11 & 19 & -17 & 17 & -6 \\ & & 1 & -5 & 7 & -5 & 6 \\ \hline & 2 & -10 & 14 & -10 & 12 & 0 \end{array}$$

The zeros of the depressed function $2x^4 - 10x^3 + 14x^2 - 10x + 12$ are also zeros of P. Use synthetic division again to find the next zero, 2:

$$\begin{array}{r|rrrrr} 2 & 2 & -10 & 14 & -10 & 12 \\ & & 4 & -12 & 4 & -12 \\ \hline & 2 & -6 & 2 & -6 & 0 \end{array}$$

$2x^5 - 11x^4 + 19x^3 - 17x^2 + 17x - 6$

$= \left(x - \dfrac{1}{2}\right)(x-2)(2x^3 - 6x^2 + 2x - 6)$

$= \left(x - \dfrac{1}{2}\right)(x-2)(2x^3 - 6x^2 + 2x - 6)$

Use factoring by grouping to factor $x^3 - 3x^2 + x - 3$:

$x^3 - 3x^2 + x - 3 = x^2(x-3) + 1(x-3)$
$= (x^2 + 1)(x-3)$. So the remaining zeros are 3 and $\pm i$. The zeros are $\dfrac{1}{2}, 2, 3, \pm i$.

39. Since one zero is $3i$, $-3i$ is another zero. There are two zeros, so the degree of the equation is at least 2. Thus, the equation is of the form
$f(x) = a(x - 3i)(x + 3i) = a(x^2 + 9)$.
The graph passes through (0, 3), so we have
$3 = a(0^2 + 9) \Rightarrow a = \dfrac{1}{3}$. Thus, the equation of the function is $f(x) = \dfrac{1}{3}(x^2 + 9)$.

41. Since i and $2i$ are zeros, so are $-i$, and $-2i$. From the graph, we see that 2 is also a zero. There are five zeros, so the degree of the equation is at least 5. Thus, the equation is of the form
$f(x) = a(x-i)(x+i)(x-2i)(x+2i)(x-2)$
$= a(x^2 + 1)(x^2 + 4)(x-2)$

The graph passes through (0, 4), so we have
$4 = a(0^2 + 1)(0^2 + 4)(0 - 2) \Rightarrow a = -\dfrac{1}{2}$.

Thus, the equation is

$f(x) = -\dfrac{1}{2}(x^2 + 1)(x^2 + 4)(x - 2)$ or

$f(x) = \dfrac{1}{2}(x^2 + 1)(x^2 + 4)(2 - x)$.

3.5 Beyond the Basics

43. There are three cube roots. We know that one root is 1. Using synthetic division, we find
$0 = x^3 - 1 = (x-1)(x^2 + x + 1)$:

$$\begin{array}{r|rrrr} 1 & 1 & 0 & 0 & -1 \\ & & 1 & 1 & 1 \\ \hline & 1 & 1 & 1 & 0 \end{array}$$

Solve the depressed equation

$x^2 + x + 1 = 0 \Rightarrow x = \dfrac{-1 \pm \sqrt{1^2 - 4(1)(1)}}{2(1)}$

$= \dfrac{-1 \pm \sqrt{-3}}{2} = -\dfrac{1}{2} \pm \dfrac{i\sqrt{3}}{2}$.

So, the cube roots of 1 are 1 and $-\dfrac{1}{2} \pm \dfrac{i\sqrt{3}}{2}$.

45. $P(x) = x^2 + (i - 2)x - 2i, \ x = -i$

$$\begin{array}{r|rrr} -i & 1 & i-2 & -2i \\ & & -i & 2i \\ \hline & 1 & -2 & 0 \end{array}$$

$x^2 + (i-2)x - 2i = (x+i)(x-2)$

47. $P(x) = x^3 - (3+i)x^2 - (4-3i)x + 4i,\ x = i$

$$\begin{array}{r|rrrr} i & 1 & -(3+i) & -(4-3i) & 4i \\ & & i & -3i & -4i \\ \hline & 1 & -3 & -4 & 0 \end{array}$$

$$\begin{aligned} &x^3 - (3+i)x^2 - (4-3i)x + 4i \\ &\quad = (x^2 - 3x - 4)(x - i) = (x-4)(x+1)(x-i) \end{aligned}$$

49. Since $1 + 2i$ is a zero, so is $1 - 2i$. The equation is of the form

$$\begin{aligned} f(x) &= a(x-2)(x-(1+2i))(x-(1-2i)) \\ &= a(x-2)(x^2 - 2x + 5) \end{aligned}$$

The y-intercept is 40, so we have

$$40 = a(0-2)(0^2 - 2(0) + 5) \Rightarrow a = -4$$

Thus, the equation is

$$f(x) = -4(x-2)(x^2 - 2x + 5).$$

Because $a < 0$, $y \to \infty$ as $x \to -\infty$ and $y \to -\infty$ as $x \to \infty$.

51. Since $3 + i$ is a zero, so is $3 - i$. The equation is of the form

$$\begin{aligned} f(x) &= a(x-1)(x+1)(x-(3+i))(x+(3+i)) \\ &= a(x^2 - 1)(x^2 - 6x + 10) \end{aligned}$$

The y-intercept is 20, so we have

$$20 = a(0^2 - 1)(0^2 - 6(0) + 10) \Rightarrow a = -2$$

Thus, the equation is

$$f(x) = -2(x^2 - 1)(x^2 - 6x + 10).$$

Because $a < 0$, $y \to -\infty$ as $x \to -\infty$ and $y \to -\infty$ as $x \to \infty$.

3.5 Critical Thinking/Discussion/Writing

53. Factoring the polynomial we have

$$\begin{aligned} &a_n x^n + a_{n-1}x^{n-1} + \cdots + a_1 x + a_0 \\ &= a_n(x - r_1)(x - r_2)\cdots(x - r_n). \end{aligned}$$

Expanding the right side, we find that the coefficient of x^{n-1} is $-a_n r_1 - a_n r_2 - \cdots - a_n r_n$ and the constant term is $(-1)^n a_n r_1 r_2 \cdots r_n$. Comparing the coefficients with those on the left side, we obtain

$$a_{n-1} = -a_n(r_1 + r_2 + \cdots + r_n).$$

Because $a_n \neq 0$, $-\dfrac{a_{n-1}}{a_n} = r_1 + r_2 + \cdots + r_n$ and $(-1)^n \dfrac{a_0}{a_n} = r_1 r_2 \cdots r_n$.

54. a. $x^3 + 6x = 20 \Rightarrow x^3 + 6x - 20 = 0$. There is one sign change, so there is one positive root. There are no sign changes in $f(-x)$, so there are no negative roots. Therefore, there is only one real solution.

b. Substituting $v - u$ for x and remembering that $v^3 - u^3 = 20$ and $uv = 2$, we have

$$\begin{aligned} &(v-u)^3 + 6(v-u) \\ &= v^3 - 3v^2u + 3vu^2 - u^3 + 6(v-u) \\ &= (v^3 - u^3) - (3v^2u - 3vu^2) + 6(v-u) \\ &= (v^3 - u^3) - 3vu(v-u) + 6(v-u) \\ &= (v^3 - u^3) - (6 - 3uv)(v-u) \\ &= 20 - (6 - 3(2))(v-u) = 20. \end{aligned}$$

Therefore, $x = v - u$ is the solution.

c. To solve the system $v^3 - u^3 = 20, vu = 2$, solve the second equation for v, and substitute that value into the first equation, keeping in mind that u cannot be zero, so division by u is permitted:

$$\begin{aligned} &v^3 - u^3 = 20 \Rightarrow \left(\frac{2}{u}\right)^3 - u^3 \\ &\quad = \frac{8}{u^3} - u^3 = 20 \Rightarrow \\ &-u^3 + \frac{8}{u^3} - 20 = 0. \end{aligned}$$

Let $-u^3 = a$, so

$$a - \frac{8}{a} - 20 = 0 \Rightarrow a^2 - 20a - 8 = 0 \Rightarrow$$

$$a = \frac{20 \pm \sqrt{400 + 32}}{2} = 10 \pm 6\sqrt{3} = -u^3 \Rightarrow$$

$$\sqrt[3]{-10 \pm 6\sqrt{3}} = u \Rightarrow v = \frac{2}{\sqrt[3]{-10 \pm 6\sqrt{3}}}$$

$$= \frac{2}{\sqrt[3]{-10 \pm 6\sqrt{3}}} \cdot \frac{\sqrt[3]{10 \pm 6\sqrt{3}}}{\sqrt[3]{10 \pm 6\sqrt{3}}} = \sqrt[3]{10 \pm 6\sqrt{3}}$$

d. If q is positive, then, according to Descartes's Rule of Signs, the polynomial $x^3 + px - q$ has one positive zero and no negative zeros. If q is negative, then it has no positive zeros and one negative zero. Either way, there is exactly one real solution. From (b), we have $v^3 - u^3 = q$, $vu = \dfrac{p}{3}$, and $x = v - u$. Substituting, we have

$x^3 + px = (v-u)^3 + p(v-u)$
$= v^3 - 3v^2u + 3vu^2 - u^3 + p(v-u)$
$= (v^3 - u^3) - (3v^2u - 3vu^2) + p(v-u)$
$= (v^3 - u^3) - 3vu(v-u) + p(v-u)$
$= (v^3 - u^3) + (p - 3uv)(v-u)$
$= q + \left(p - 3 \cdot \dfrac{p}{3}\right)x = q \Rightarrow x = v - u$ is the solution.

Solving the system $v^3 - u^3 = q$, $vu = \dfrac{p}{3}$, we obtain

$v^3 - u^3 = q \Rightarrow \left(\dfrac{p}{3u}\right)^3 - u^3 = \dfrac{p^3}{3^3 u^3} - u^3 = q$

$\Rightarrow -u^3 + \dfrac{p^3}{3^3 u^3} - q = 0$. Let $-u^3 = a$, so

we have $a - \dfrac{p^3}{3^3 a} - q = 0 \Rightarrow$

$a^2 - qa - \dfrac{p^3}{3^3} = 0 \Rightarrow$

$a = \dfrac{q \pm \sqrt{q^2 + 4\left(\dfrac{p^3}{3^3}\right)}}{2} = \dfrac{q}{2} \pm \sqrt{\dfrac{q^2 + 4\left(\dfrac{p^3}{3^3}\right)}{4}}$

$= \dfrac{q}{2} \pm \sqrt{\left(\dfrac{q}{2}\right)^2 + \left(\dfrac{p}{3}\right)^3} = -u^3 \Rightarrow$

$u = \sqrt[3]{-\dfrac{q}{2} + \sqrt{\left(\dfrac{q}{2}\right)^2 + \left(\dfrac{p}{3}\right)^3}}$ and

$v = \sqrt[3]{\dfrac{q}{2} + \sqrt{\left(\dfrac{q}{2}\right)^2 + \left(\dfrac{p}{3}\right)^3}}$ or

$u = \sqrt[3]{-\dfrac{q}{2} - \sqrt{\left(\dfrac{q}{2}\right)^2 + \left(\dfrac{p}{3}\right)^3}}$ and

$v = \sqrt[3]{\dfrac{q}{2} - \sqrt{\left(\dfrac{q}{2}\right)^2 + \left(\dfrac{p}{3}\right)^3}}$.

The difference $v - u$ is the same in both cases, so
$x = v - u$

$= \sqrt[3]{-\dfrac{q}{2} + \sqrt{\left(\dfrac{q}{2}\right)^2 + \left(\dfrac{p}{3}\right)^3}}$

$- \sqrt[3]{\dfrac{q}{2} + \sqrt{\left(\dfrac{q}{2}\right)^2 + \left(\dfrac{p}{3}\right)^3}}$.

e. Substituting $x = y - \dfrac{a}{3}$, we have

$x^3 + ax^2 + bx + c = 0 \Rightarrow$

$\left(y - \dfrac{a}{3}\right)^3 + a\left(y - \dfrac{a}{3}\right)^2 + b\left(y - \dfrac{a}{3}\right) + c = 0 \Rightarrow$

$\left(y^3 - ay^2 + \dfrac{a^2}{3}y - \dfrac{a^3}{27}\right) + \left(ay^2 - \dfrac{2a^2y}{3} + \dfrac{a^3}{9}\right)$

$+ by - \dfrac{ab}{3} + c = 0 \Rightarrow$

$y^3 + py = q$, where $p = b - \dfrac{a^2}{3}$ and

$q = -\dfrac{2a^3}{27} + \dfrac{ab}{3} - c$

$y^3 - \dfrac{a^2 y}{3} + by + \dfrac{2a^3}{27} - \dfrac{ab}{3} + c = 0 \Rightarrow$

$y^3 + \left(b - \dfrac{a^2}{3}\right)y = -\dfrac{2a^3}{27} + \dfrac{ab}{3} - c$

55. $x^3 + 6x^2 + 10x + 8 = 0 \Rightarrow a = 6, b = 10, c = 8$.

Substituting $x = y - \dfrac{6}{3} = y - 2$ as in (1e), we have

$(y-2)^3 + 6(y-2)^2 + 10(y-2) + 8 = 0 \Rightarrow$
$(y^3 - 6y^2 + 12y - 8) + (6y^2 - 24y + 24)$
$\qquad + 10y - 20 + 8 = 0 \Rightarrow$

$y^3 - 2y + 4 = 0 \Rightarrow y^3 - 2y = -4$.

Then, using the results of (1d), we have

$y = \sqrt[3]{\dfrac{-4}{2} + \sqrt{\left(\dfrac{-4}{2}\right)^2 + \left(\dfrac{-2}{3}\right)^3}}$

$\qquad - \sqrt[3]{-\dfrac{-4}{2} + \sqrt{\left(\dfrac{-4}{2}\right)^2 + \left(\dfrac{-2}{3}\right)^3}}$.

Using a calculator, we find that $y = -2$. So $x = -2 - 2 = -4$. Now use synthetic division to find the depressed equation:

$\begin{array}{r|rrrr} -4 & 1 & 6 & 10 & 8 \\ & & -4 & -8 & -8 \\ \hline & 1 & 2 & 2 & 0 \end{array}$

$x^3 + 6x^2 + 10x + 8 = (x+4)(x^2 + 2x + 2) = 0$.

$x^2 + 2x + 2 = 0 \Rightarrow$

$x = \dfrac{-2 \pm \sqrt{2^2 - 4(1)(2)}}{2(1)} = \dfrac{-2 \pm \sqrt{-4}}{2}$

$= \dfrac{-2 \pm 2i}{2} = -1 \pm i$.

The solution set is $\{-4, -1 \pm i\}$.

3.5 Getting Ready for the Next Section

57. $2x^2 + x - 3 = 0 \Rightarrow (2x+3)(x-1) = 0 \Rightarrow$
$x = -\dfrac{3}{2},\ x = 1$
Solution set: $\left\{-\dfrac{3}{2},\ 1\right\}$

59. $x^4 - x^3 - 12x^2 = 0 \Rightarrow x^2(x^2 - x - 12) = 0 \Rightarrow$
$x^2(x-4)(x+3) = 0 \Rightarrow x = 0,\ x = 4,\ x = -3$
Solution set: $\{-3, 0, 4\}$

61. $\dfrac{x^2 - x - 4}{x - 2}$

$\underline{2\rfloor}\ \ 1\ \ -1\ \ -4$
$\ \ \ \ 2\ \ \ \ 2$
$\overline{\ 1\ \ \ \ 1\ \ -2}$

$Q(x) = x + 1;\ R(x) = -2$

63. $2x^3 + 0x^2 + x \overline{\smash{\big)}\,8x^4 + 6x^3 - 0x^2 + 0x - 5}$ with quotient $4x + 3$

$\underline{8x^4 + 0x^3 + 4x^2}$
$ 6x^3 - 4x^2 + 0x$
$ \underline{6x^3 + 0x^2 + 3x}$
$ -4x^2 - 3x - 5$

$Q(x) = 4x + 3;\ R(x) = -4x^2 - 3x - 5$

65. $\dfrac{7 - (-1)}{2(-1)^2 + 3(-1)} = \dfrac{8}{-1} = -8$

67. $\dfrac{(2)^2 + 4(2) - 1}{9 - (2)^3} = 11$

3.6 Rational Functions

3.6 Practice Problems

1. $f(x) = \dfrac{x - 3}{x^2 - 4x - 5}$

The domain of f consists of all real numbers for which $x^2 - 4x - 5 \neq 0$.
$x^2 - 4x - 5 = 0 \Rightarrow (x-5)(x+1) = 0 \Rightarrow$
$x = 5$ or $x = -1$
Thus, the domain is
$(-\infty, -1) \cup (-1, 5) \cup (5, \infty)$.

2. a. $g(x) = \dfrac{3}{x - 2}$

Let $f(x) = \dfrac{1}{x}$. Then
$g(x) = \dfrac{3}{x-2} = 3\left(\dfrac{1}{x-2}\right) = 3f(x-2)$.

The graph of $y = f(x - 2)$ is the graph of $y = f(x)$ shifted two units to the right. This moves the vertical asymptote two units to the right. The graph of $y = 3f(x - 2)$ is the graph of $y = f(x-2)$ stretched vertically three units.
The domain of g is $(-\infty, 2) \cup (2, \infty)$.
The range of g is $(-\infty, 0) \cup (0, \infty)$.
The vertical asymptote is $x = 2$.
The horizontal asymptote is $y = 0$.

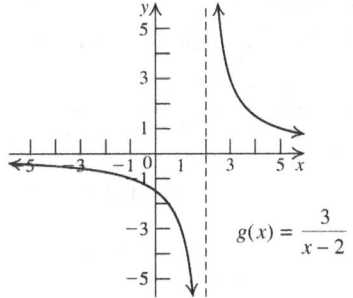

$g(x) = \dfrac{3}{x - 2}$

b. $h(x) = \dfrac{2x + 5}{x + 1}$

$x+1 \overline{\smash{\big)}\,2x + 5}$ with quotient 2
$\underline{2x + 2}$
3

$h(x) = \dfrac{2x + 5}{x + 1} = 2 + \dfrac{3}{x + 1}$

Let $f(x) = \dfrac{1}{x}$. Then
$h(x) = \dfrac{2x + 5}{x + 1} = 2 + \dfrac{3}{x + 1} = 2 + 3f(x + 1)$.

The graph of $y = h(x)$ is the graph of $y = f(x)$ shifted one units to the left and then stretched vertically three units. The graph is then shifted two units up. This moves the vertical asymptote one unit to the left. The horizontal asymptote is shifted two units up. The domain of h is
$(-\infty, -1) \cup (-1, \infty)$.

(continued on next page)

(continued)

The range of h is $(-\infty, 2) \cup (2, \infty)$.
The vertical asymptote is $x = -1$.
The horizontal asymptote is $y = 2$.

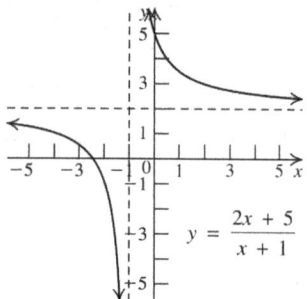

3. $f(x) = \dfrac{x+1}{x^2 + 3x - 10}$

   The vertical asymptotes are located at the zeros of the denominator.
   $x^2 + 3x - 10 = 0 \Rightarrow (x+5)(x-2) = 0 \Rightarrow$
   $x = -5$ or $x = 2$
   The vertical asymptotes are $x = -5$ and $x = 2$.

4. $f(x) = \dfrac{3-x}{x^2 - 9} = \dfrac{-(x-3)}{(x-3)(x+3)} = -\dfrac{1}{x+3}$
   $x + 3 = 0 \Rightarrow x = -3$
   The vertical asymptote is $x = -3$.

5. a. $f(x) = \dfrac{2x - 5}{3x + 4}$

   Since the numerator and denominator both have degree 1, the horizontal asymptote is $y = \dfrac{2}{3}$.

   b. $g(x) = \dfrac{x^2 + 3}{x - 1}$

   Since the degree of the numerator is greater than the degree of the denominator, there are no horizontal asymptotes.

   c. $h(x) = \dfrac{100x + 57}{0.01x^3 + 8x - 9}$

   Since the degree of the numerator is less than the degree of the denominator, the horizontal asymptote is the line $y = 0$.

6. $f(x) = \dfrac{2x}{x^2 - 1}$

   There are no common factors of the form $x - a$ between $2x$ and $x^2 - 1$.
   First find the intercepts:
   $\dfrac{2x}{x^2 - 1} = 0 \Rightarrow x = 0$
   The graph passes through the origin.
   Find the vertical asymptotes:
   $x^2 - 1 = 0 \Rightarrow (x-1)(x+1) = 0 \Rightarrow$
   $x = 1$ or $x = -1$
   Find the horizontal asymptote: The degree of the numerator is less than the degree of the denominator, so the horizontal asymptote is the x-axis.

   By long division, $f(x) = \dfrac{2x}{x^2 - 1} = 0 + \dfrac{2x}{x^2 - 1}$.

   $R(x) = 2x$ has zero 0 and $D(x) = x^2 - 1$ has zeros -1 and 1. These zeros divide the x-axis into four intervals, $(-\infty, -1)$, $(-1, 0)$, $(0, 1)$, and $(1, \infty)$. Use test values to determine where the graph of f is above and below the x-axis.

Interval	Test point	Value of $f(x)$	Above/below x-axis
$(-\infty, -1)$	-3	$-\dfrac{3}{4}$	below
$(-1, 0)$	$-\dfrac{1}{2}$	$\dfrac{4}{3}$	above
$(0, 1)$	$\dfrac{1}{2}$	$-\dfrac{4}{3}$	below
$(1, \infty)$	2	$\dfrac{4}{3}$	above

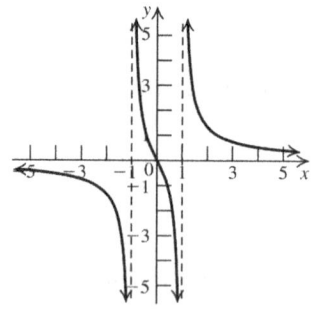

7. $f(x) = \dfrac{2x^2 - 1}{2x^2 + x - 3}$

   $f(x)$ is in lowest terms.
   Find the intercepts:
   $\dfrac{2x^2 - 1}{2x^2 + x - 3} = 0 \Rightarrow x = \pm\dfrac{\sqrt{2}}{2}$; $f(0) = \dfrac{1}{3}$
   Find the vertical asymptotes:
   $2x^2 + x - 3 = 0 \Rightarrow (2x+3)(x-1) = 0 \Rightarrow$
   $x = -\dfrac{3}{2}$ or $x = 1$
   Find the horizontal asymptote: The degree of the numerator is the same as the degree of the denominator, so the horizontal asymptote is
   $y = \dfrac{2}{2} = 1$.

   $f(x) = \dfrac{2x^2 - 1}{2x^2 + x - 3} = 1 + \dfrac{2 - x}{2x^2 + x - 3}$

   The zero of $R(x) = 2 - x$ is 2 and the zeros of $D(x) = 2x^2 + x - 3 = (2x+3)(x-1)$ are
   $x = -\dfrac{3}{2}$ and $x = 1$.
   Use test values to determine where the graph of f is above and below the horizontal asymptote $y = 1$.

Interval	Test point	Value of $f(x)$	Above/below $y = 1$
$\left(-\infty, -\dfrac{3}{2}\right)$	-2	$\dfrac{7}{3}$	above
$\left(-\dfrac{3}{2}, 1\right)$	0	$\dfrac{1}{3}$	below
$(1, 2)$	$\dfrac{3}{2}$	$\dfrac{7}{6}$	above
$(2, \infty)$	3	$\dfrac{17}{18}$	above

   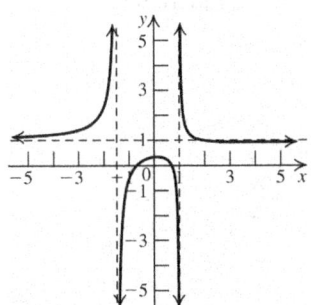

   Notice that the graph crosses the horizontal asymptote at $(2, 1)$.

8. $f(x) = \dfrac{x^2 + 1}{x^2 + 2}$

   $f(x)$ is in lowest terms.
   Find the intercepts:
   $\dfrac{x^2 + 1}{x^2 + 2} = 0 \Rightarrow x = \pm i \Rightarrow$ there is no x-intercept.
   $f(0) = \dfrac{1}{2} \Rightarrow \left(0, \dfrac{1}{2}\right)$ is the y-intercept.
   Find the vertical asymptotes:
   $x^2 + 2 = 0 \Rightarrow x = i\sqrt{2} \Rightarrow$ there is no vertical asymptote.
   Find the horizontal asymptote: The degree of the numerator is the same as the degree of the denominator, so the horizontal asymptote is
   $y = \dfrac{1}{1} = 1$.

   $f(x) = \dfrac{x^2 + 1}{x^2 + 2} = 1 + \dfrac{-1}{x^2 + 2}$

   Neither $R(x) = -1$ nor $D(x) = x^2 + 2$ have real zeros. Since $\dfrac{-1}{x^2 + 2}$ is negative for all values of x, the graph of $f(x)$ is always below the line
   $y = 1$.

   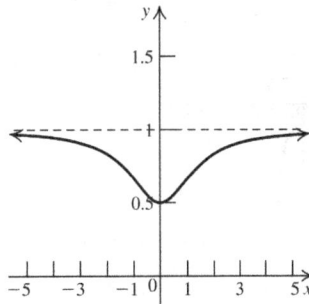

9. $f(x) = \dfrac{x^2 + 2}{x - 1}$

   $f(x)$ is in lowest terms.
   Find the intercepts:
   $\dfrac{x^2 + 2}{x - 1} = 0 \Rightarrow x = \pm i\sqrt{2} \Rightarrow$ there is no x-intercept. $f(0) = -2 \Rightarrow (0, -2)$ is the y-intercept.
   Find the vertical asymptotes:
   $x - 1 = 0 \Rightarrow x = 1$

   (continued on next page)

(*continued*)

Find the horizontal asymptote: The degree of the numerator is greater than the degree of the denominator, so there is no horizontal asymptote. However, there is an oblique asymptote.

$\dfrac{x^2+2}{x-1} = x+1+\dfrac{3}{x-1} \Rightarrow y = x+1$ is the oblique asymptote.

The graph is above the line $y = x + 1$ on $(1, \infty)$ and below the line on $(-\infty, 1)$.

The intervals determined by the zeros of the numerator and of the denominator of

$f(x) - (x+1) = \dfrac{x^2+2}{x-1} - (x+1) = \dfrac{3}{x-1}$ divide

the *x*-axis into two intervals, $(-\infty, 1)$ and $(1, \infty)$. Use test numbers to determine where the graph of *f* is above and below the *x*-axis.

Interval	Test point	Value of $f(x)$	Above/below $y = x+1$
$(-\infty, 1)$	-1	$-\dfrac{3}{2}$	below
$(1, \infty)$	2	6	above

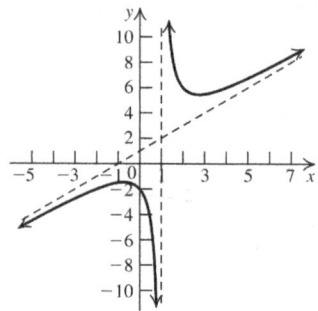

10. $R(x) = \dfrac{x(100-x)}{x+20}$

   a. $R(10) = \dfrac{10(100-10)}{10+20} = 30$ billion dollars.
   This means that if income is taxed at a rate of 10%, then the total revenue for the government will be 30 billion dollars. Similarly, $R(20) = \$40$ billion, $R(30) = \$42$ billion, $R(40) = \$40$ billion, $R(50) \approx \$35.7$ billion, $R(60) = \$30$ billion.

   b.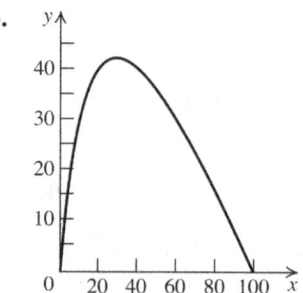

   c.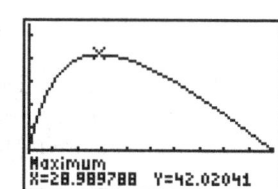

   From the graphing calculator screen, we see that a tax rate of about 29% generates the maximum tax revenue of about $42.02 billion.

3.6 Concepts and Vocabulary

1. A rational function can be expressed in the form $\dfrac{N(x)}{D(x)}$, where *N(x)* and *D(x)* are <u>polynomials and *D(x)* is not the zero polynomial</u>.

3. The line $y = k$ is a horizontal asymptote of *f* if $f(x) \to k$ as $x \to \underline{\infty}$ or as $x \to \underline{-\infty}$.

5. False. A rational function has a vertical asymptote only if the denominator has a real zero and the numerator and denominator have no common factors.

7. False. The graph of a rational function may cross its *horizontal* asymptote.

3.6 Building Skills

9. $x + 4 = 0 \Rightarrow x = -4$
   The domain of the function is $(-\infty, -4) \cup (-4, \infty)$

11. $(-\infty, \infty)$

13. $x^2 - x - 6 = 0 \Rightarrow (x-3)(x+2) = 0 \Rightarrow x = -2, 3$
    The domain of the function is $(-\infty, -2) \cup (-2, 3) \cup (3, \infty)$.

15. $x^2 - 6x + 8 = 0 \Rightarrow (x-4)(x-2) = 0 \Rightarrow x = 2, 4$
    The domain of the function is $(-\infty, 2) \cup (2, 4) \cup (4, \infty)$.

17. As $x \to 1^+$, $f(x) \to \underline{\infty}$.

19. As $x \to -2^+$, $f(x) \to \underline{\infty}$.

21. As $x \to \infty$, $f(x) \to \underline{1}$.

23. The domain of f is $\underline{(-\infty, -2) \cup (-2, 1) \cup (1, \infty)}$.

25. The equations of the vertical asymptotes of the graph are $x = \underline{-2}$ and $x = \underline{1}$.

27.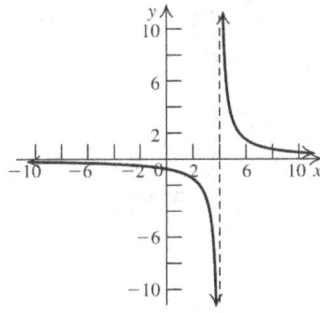

Domain: $(-\infty, 4) \cup (4, \infty)$
Range: $(-\infty, 0) \cup (0, \infty)$
Vertical asymptote: $x = 4$
Horizontal asymptote: $y = 0$

29.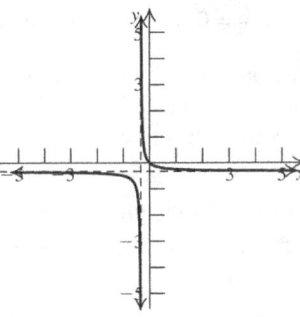

Domain: $\left(-\infty, -\frac{1}{3}\right) \cup \left(-\frac{1}{3}, \infty\right)$
Range: $\left(-\infty, -\frac{1}{3}\right) \cup \left(-\frac{1}{3}, \infty\right)$
Vertical asymptote: $x = -\frac{1}{3}$
Horizontal asymptote: $y = -\frac{1}{3}$

31.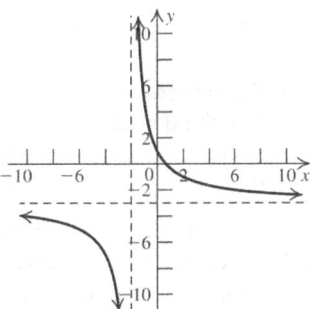

Domain: $(-\infty, -2) \cup (-2, \infty)$
Range: $(-\infty, -3) \cup (-3, \infty)$
Vertical asymptote: $x = -2$
Horizontal asymptote: $y = -3$

33.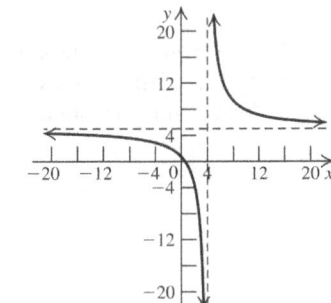

Domain: $(-\infty, 4) \cup (4, \infty)$
Range: $(-\infty, 5) \cup (5, \infty)$
Vertical asymptote: $x = 4$
Horizontal asymptote: $y = 5$

In exercises 35–44, to find the vertical asymptotes, first eliminate any common factors in the numerator and denominator, and then set the denominator equal to zero and solve for x.

35. $x = 1$

37. $x = -4, x = 3$

39. $h(x) = \dfrac{x^2 - 1}{x^2 + x - 6} = \dfrac{(x-1)(x+1)}{(x-2)(x+3)}$.
The equations of the vertical asymptotes are $x = -3$ and $x = 2$.

41. $f(x) = \dfrac{x^2 - 6x + 8}{x^2 - x - 12} = \dfrac{(x-4)(x-2)}{(x-4)(x+3)}$.
Disregard the common factor. The vertical asymptote is $x = -3$.

43. There is no vertical asymptote.

For exercises 45–52, locate the horizontal asymptote as follows:

- If the degree of the numerator of a rational function is less than the degree of the denominator, then the x-axis ($y = 0$) if the horizontal asymptote.

- If the degree of the numerator of a rational function equals the degree of the denominator, the horizontal asymptote is the line with the equation $y = \dfrac{a_n}{b_m}$, where a_n is the coefficient of the leading term of the numerator and b_m is the coefficient of the leading term of the denominator.

- If the degree of the numerator of a rational function is greater than the degree of the denominator, then there is no horizontal asymptote.

45. $y = 0$

47. $y = \dfrac{2}{3}$

49. There is no horizontal asymptote.

51. $y = 0$

53. d **55.** e **57.** a

59. $0 = \dfrac{2x}{x-3} \Rightarrow x = 0$ is the x-intercept.

$\dfrac{2(0)}{0-3} = 0 \Rightarrow y = 0$ is the y-intercept. The vertical asymptote is $x = 3$. The horizontal asymptote is $y = 2$. The intervals to be tested are $(-\infty, 3)$ and $(3, \infty)$. The graph is above the horizontal asymptote on $(3, \infty)$ and below the horizontal asymptote on $(-\infty, 3)$.

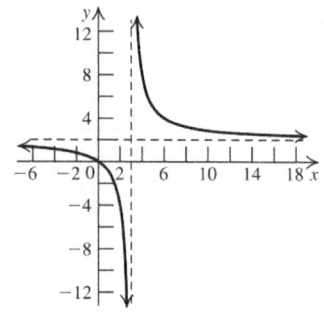

61. $0 = \dfrac{x}{x^2 - 4} \Rightarrow x = 0$ is the x-intercept.

$\dfrac{0}{0^2 - 4} = 0 \Rightarrow y = 0$ is the y-intercept. The vertical asymptotes are $x = -2$ and $x = 2$. The horizontal asymptote is the x-axis. The intervals to be tested are $(-\infty, -2), (-2, 0), (0, 2)$ and $(2, \infty)$. The graph is above the x-axis on $(-2, 0) \cup (2, \infty)$ and below the x-axis on $(-\infty, -2) \cup (0, 2)$.

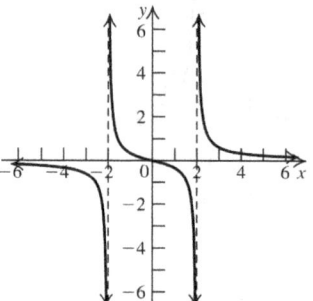

63. $0 = \dfrac{-2x^2}{x^2 - 9} \Rightarrow x = 0$ is the x-intercept.

$\dfrac{0^2}{0^2 - 4} = 0 \Rightarrow y = 0$ is the y-intercept. The vertical asymptotes are $x = -3$ and $x = 3$. The horizontal asymptote is $y = -2$. The intervals to be tested are $(-\infty, -3), (-3, 3)$, and $(3, \infty)$. The graph is above the horizontal asymptote on $(-3, 3)$ and below the horizontal asymptote on $(-\infty, -3) \cup (3, \infty)$.

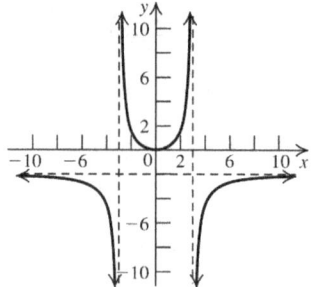

65. $0 = \dfrac{2}{x^2 - 2} \Rightarrow$ there is no x-intercept.

$\dfrac{2}{0^2 - 2} = 0 \Rightarrow y = -1$ is the y-intercept. The vertical asymptotes are $x = \pm\sqrt{2}$. The horizontal asymptote is the x-axis.

(continued on next page)

(*continued*)

The intervals to be tested are
$(-\infty, -\sqrt{2}), (-\sqrt{2}, \sqrt{2})$ and $(\sqrt{2}, \infty)$.
he graph is above the *x*-axis on
$(-\infty, -\sqrt{2}) \cup (\sqrt{2}, \infty)$ and below the *x*-axis on
$(-\sqrt{2}, \sqrt{2})$.

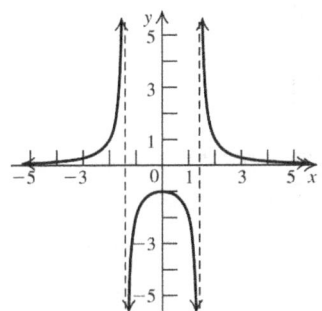

67. $0 = \dfrac{x+1}{(x-2)(x+3)} \Rightarrow x = -1$ is the *x*-intercept.

$\dfrac{0+1}{(0-2)(0+3)} = -\dfrac{1}{6} \Rightarrow y = -\dfrac{1}{6}$ is the

y-intercept. The vertical asymptotes are $x = -3$ and $x = 2$. The horizontal asymptote is the *x*-axis. The intervals to be tested are $(-\infty, -3), (-3, -1), (-1, 2)$, and $(2, \infty)$. The graph is above the *x*-axis on $(-3, -1) \cup (2, \infty)$ and below the *x*-axis on $(-\infty, -3) \cup (-1, 2)$.

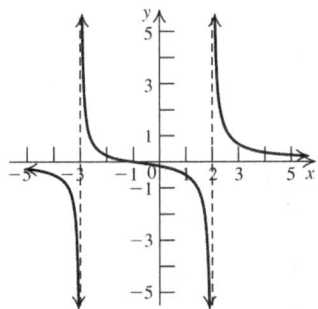

69. $0 = \dfrac{x^2}{x^2+1} \Rightarrow x = 0$ is the *x*-intercept.

$\dfrac{0^2}{0^2+1} = 0 \Rightarrow y = 0$ is the *y*-intercept. There is no vertical asymptote. The horizontal asymptote is $y = 1$. The intervals to be tested are $(-\infty, 0)$ and $(0, \infty)$. The graph is above the *x*-axis on $(-\infty, 0) \cup (0, \infty)$ and below the horizontal asymptote on $(-\infty, \infty)$.

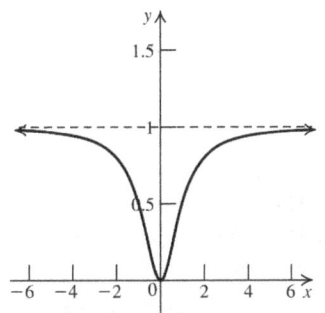

71. $f(x) = \dfrac{x^3 - 4x}{x^3 - 9x} = \dfrac{x(x-2)(x+2)}{x(x-3)(x+3)}$
$= \dfrac{(x-2)(x+2)}{(x-3)(x+3)}$

$f(x) = 0 \Rightarrow x = \pm 2$ are the *x*-intercepts.

$\dfrac{x^3 - 4x}{x^3 - 9x} = \dfrac{x(x^2-4)}{x(x^2-9)} = \dfrac{x^2-4}{x^2-9} \Rightarrow$

$\dfrac{0^2 - 4}{0^2 - 9} = \dfrac{4}{9} \Rightarrow \dfrac{4}{9}$ is the *y*-intercept. However,

there is a hole at $\left(0, \dfrac{4}{9}\right)$ since $0^3 - 9(0) = 0$

$x^2 - 9 = 0 \Rightarrow x(x+3)(x-3) = 0 \Rightarrow x = -3$
and $x = 3$ are the vertical asymptotes. The degree of the numerator is the same as the degree of the denominator, so the horizontal asymptote is $y = 1$. The intervals to be tested are $(-\infty, -3), (-3, 0), (0, 3)$, and $(3, \infty)$. The graph is above the horizontal asymptote on $(-\infty, -3) \cup (3, \infty)$ and below the horizontal asymptote on $(-3, 0) \cup (0, 3)$.

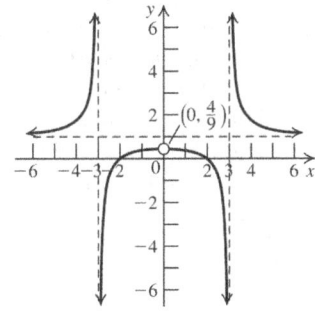

73. $0 = \dfrac{(x-2)^2}{x-2}$ ⇒ there is no x-intercept. There is a hole at $(2, 0)$. $\dfrac{(0-2)^2}{0-2} = -2 \Rightarrow y = -2$ is the y-intercept. There are no vertical asymptotes. There are no horizontal asymptotes. The intervals to be tested are $(-\infty, 2)$ and $(2, \infty)$.

The graph is above the x-axis on $(2, \infty)$ and below the x-axis on $(-\infty, 2)$.

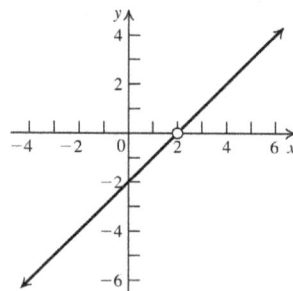

75. The x-intercept is 1 and the vertical asymptote is $x = 2$. The horizontal asymptote is $y = -2$, so the degree of the numerator equals the degree of the denominator, and the ratio of the leading terms of the numerator and the denominator is -2. Thus, the equation is of the form $y = a\left(\dfrac{x-1}{x-2}\right)$. The y-intercept is -1, so we have $-1 = a\left(\dfrac{0-1}{0-2}\right) \Rightarrow a = -2$.

Thus, the equation is $f(x) = \dfrac{-2(x-1)}{x-2}$.

77. The x-intercepts are 1 and 3, and the vertical asymptotes are $x = 0$ and $x = 2$. The horizontal asymptote is $y = 1$, so the degree of the numerator equals the degree of the denominator, and the ratio of the leading terms of the numerator and the denominator is 1. Thus, the equation is of the form $y = a\dfrac{(x-1)(x-3)}{x(x-2)}$. There is no y-intercept, so the equation is $f(x) = \dfrac{(x-1)(x-3)}{x(x-2)}$.

79. $\dfrac{2x^2+1}{x} = 2x + \dfrac{1}{x}$.

The oblique asymptote is $y = 2x$.

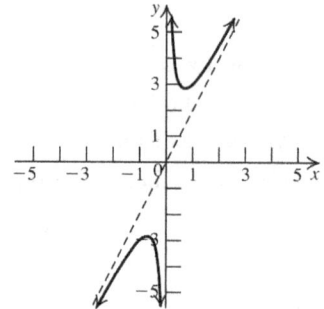

81. $\dfrac{x^3-1}{x^2} = x - \dfrac{1}{x^2}$.

The oblique asymptote is $y = x$.

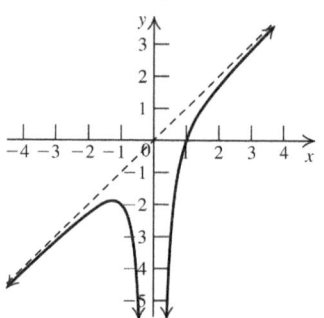

83. $\begin{array}{r} x-2 \\ x+1 \overline{\smash{)}\, x^2 - x + 1} \\ \underline{x^2 + x} \\ -2x + 1 \\ \underline{-2x - 2} \\ 3 \end{array}$

The oblique asymptote is $y = x - 2$.

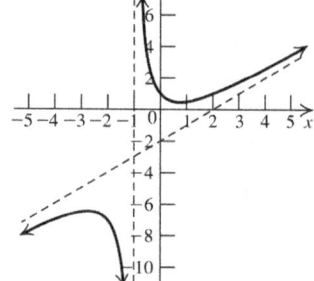

85. $\begin{array}{r} x-2 \\ x^2-1 \overline{\smash{)}\, x^3 - 2x^2 + 0x + 1} \\ \underline{x^3 - x} \\ -2x^2 + x + 1 \\ \underline{-2x^2 + 2} \\ x - 1 \end{array}$

The oblique asymptote is $y = x - 2$. Note that there is a hole in the graph at $x = 1$.

(continued on next page)

(continued)

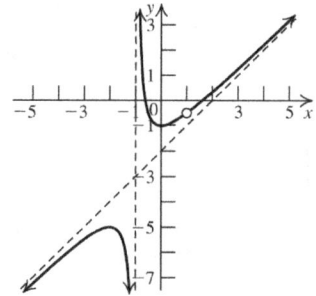

3.6 Applying the Concepts

87. a. $C(x) = 0.5x + 2000$

b. $\overline{C}(x) = \dfrac{C(x)}{x} = \dfrac{0.5x + 2000}{x} = 0.5 + \dfrac{2000}{x}$

c. $\overline{C}(100) = 0.5 + \dfrac{2000}{100} = 20.5$

$\overline{C}(500) = 0.5 + \dfrac{2000}{500} = 4.5$

$\overline{C}(1000) = 0.5 + \dfrac{2000}{1000} = 2.5$

These show the average cost of producing 100, 500, and 1000 trinkets, respectively.

d. The horizontal asymptote of $\overline{C}(x)$ is $y = 0.5$. It means that the average cost approaches the daily fixed cost of producing each trinket as the number of trinkets approaches ∞.

89. a.

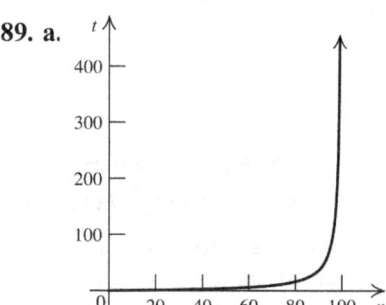

b. $f(50) = \dfrac{4(50)+1}{100-50} \approx 4$ min

$f(75) = \dfrac{4(75)+1}{100-75} \approx 12$ min

$f(95) = \dfrac{4(95)+1}{100-95} \approx 76$ min

$f(99) = \dfrac{4(99)+1}{100-99} \approx 397$ min

c. **(i)** As $x \to 100^-$, $f(x) \to \infty$.

(ii) The statement is not applicable because the domain is $x < 100$.

d. No, the bird doesn't ever collect all the seed from the field.

91. a. $C(50) = \dfrac{3(50^2)+50}{50(100-50)} = \$3.02$ billion

b.

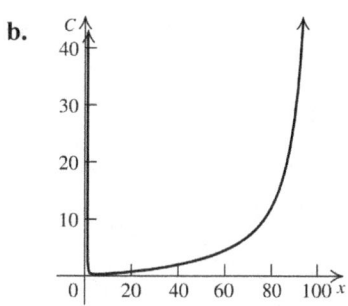

c. $30 = \dfrac{3x^2+50}{x(100-x)} = \dfrac{3x^2+50}{100x-x^2} \Rightarrow$

$3000x - 30x^2 = 3x^2 + 50 \Rightarrow$

$-33x^2 + 3000x - 50 = 0 \Rightarrow$

$x = \dfrac{-3000 \pm \sqrt{3000^2 - 4(-33)(-50)}}{2(-33)}$

$= \dfrac{-3000 \pm \sqrt{8,993,400}}{-66} \approx \dfrac{-3000 \pm 2998.9}{-66}$

≈ 90.89 or 0.017.

From the graph, we see that approximately 90.89% of the impurities can be removed at a cost of $30 billion.

93. a. $P(0) = \dfrac{8(0)+16}{2(0)+1} = 16$ thousand $= 16{,}000$

b. The horizontal asymptote is $y = 4$. This means that the population will stabilize at 4000.

95. a. $f(x) = \dfrac{10x + 200{,}000}{x - 2500}$

b. $C(10{,}000) = \dfrac{10(10{,}000) + 200{,}000}{10{,}000 - 2500} = \$40$

c. $20 > \dfrac{10x + 200{,}000}{x - 2500} \Rightarrow$

$20x - 50{,}000 > 10x + 200{,}000 \Rightarrow$

$10x > 250{,}000 \Rightarrow x > 25{,}000$

More than 25,000 books must be sold to bring the average cost under $20.

d. The vertical asymptote is $x = 2500$. This represents the number of free samples. The horizontal asymptote is $y = 10$. This represents the cost of printing and binding one book.

3.6 Beyond the Basics

97. Stretch the graph of $y = \dfrac{1}{x}$ vertically by a factor of 2, and then reflect the graph about the x-axis.

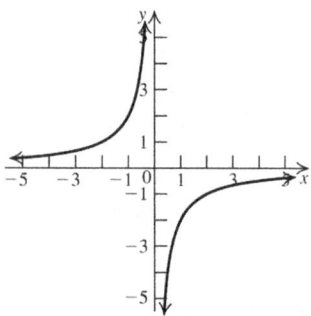

99. Shift the graph of $y = \dfrac{1}{x^2}$ two units right.

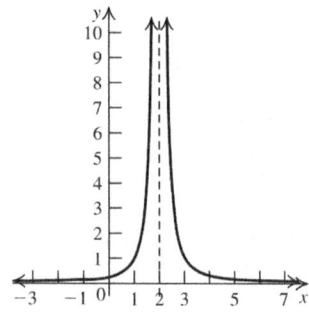

101. Shift the graph of $y = \dfrac{1}{x^2}$ one unit right and two units down.

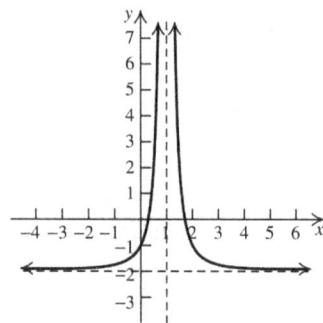

103. Shift the graph of $y = \dfrac{1}{x^2}$ six units left.

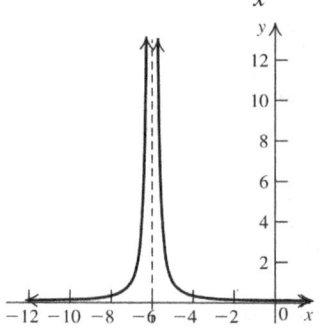

105. $x^2 - 2x + 1 \overline{)x^2 - 2x + 2} \Rightarrow$
$\underline{x^2 - 2x + 1}$
1

$\dfrac{x^2 - 2x + 2}{x^2 - 2x + 1} = \dfrac{(x^2 - 2x + 1) + 1}{x^2 - 2x + 1} =$
$\dfrac{x^2 - 2x + 1}{x^2 - 2x + 1} + \dfrac{1}{x^2 - 2x + 1} = 1 + \dfrac{1}{(x-1)^2}$

Shift the graph of $y = \dfrac{1}{x^2}$ one unit right and one unit up.

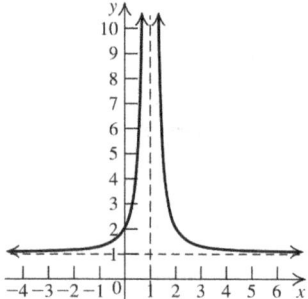

107. a. $f(x) \to 0$ as $x \to -\infty$; $f(x) \to 0$ as $x \to \infty$; $f(x) \to -\infty$ as $x \to 0^-$; $f(x) \to \infty$ as $x \to 0^+$. There are no x- or y-intercepts. The horizontal asymptote is the x-axis. The vertical asymptote is the y-axis. The graph is above the x-axis on $(0, \infty)$ and below the x-axis on $(-\infty, 0)$.

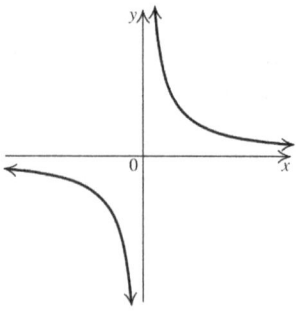

b. $f(x) \to 0$ as $x \to -\infty$; $f(x) \to 0$ as $x \to \infty$; $f(x) \to \infty$ as $x \to 0^-$; $f(x) \to -\infty$ as $x \to 0^+$. There are no x- or y-intercepts. The horizontal asymptote is the x-axis. The vertical asymptote is the y-axis. The graph is above the x-axis on $(-\infty, 0)$ and below the x-axis on $(0, \infty)$.

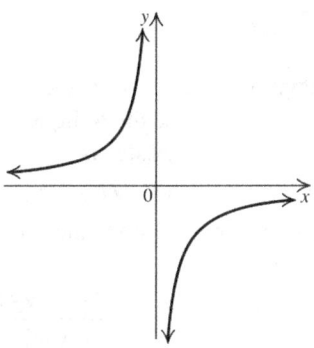

c. $f(x) \to 0$ as $x \to -\infty$; $f(x) \to 0$ as $x \to \infty$; $f(x) \to \infty$ as $x \to 0^-$; $f(x) \to \infty$ as $x \to 0^+$. There are no x- or y-intercepts. The horizontal asymptote is the x-axis. The vertical asymptote is the y-axis. The graph is above the x-axis on $(-\infty, 0) \cup (0, \infty)$.

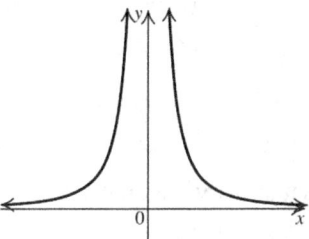

d. $f(x) \to 0$ as $x \to -\infty$; $f(x) \to 0$ as $x \to \infty$; $f(x) \to -\infty$ as $x \to 0^-$; $f(x) \to -\infty$ as $x \to 0^+$. There are no x- or y-intercepts. The horizontal asymptote is the x-axis. The vertical asymptote is the y-axis. The graph is below the x-axis on $(-\infty, 0) \cup (0, \infty)$.

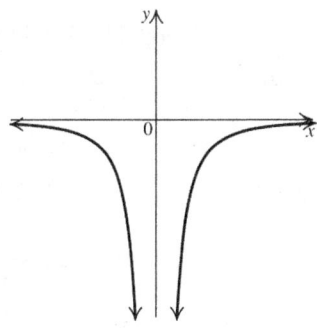

109.

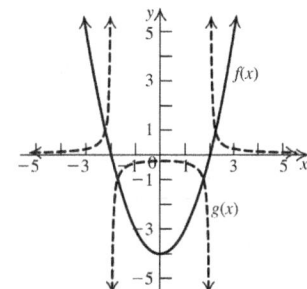

111. $f(x) = 2x + 3 \Rightarrow [f(x)]^{-1} = \dfrac{1}{2x+3}$ and $y = 2x + 3$ becomes $x = 2y + 3 \Rightarrow \dfrac{x-3}{2} = \dfrac{x}{2} - \dfrac{3}{2} = y = f^{-1}(x)$.

Thus, the functions are different.

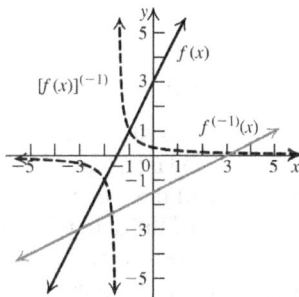

113. $g(x) = \dfrac{2x^3 + 3x^2 + 2x - 4}{x^2 - 1} = 2x + 3 + \dfrac{-1}{x^2 - 1}$

$g(x)$ has the oblique asymptote $y = 2x + 3$.

$g(x) \to -\infty$ as $x \to -\infty$, and $g(x) \to \infty$ as $x \to \infty$.

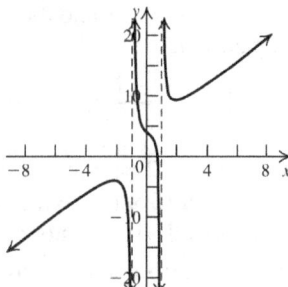

115. The horizontal asymptote is $y = 1$.

$\dfrac{x^2 + x - 2}{x^2 - 2x - 3} = 1 \Rightarrow x^2 + x - 2 = x^2 - 2x - 3 \Rightarrow$

$3x = -1 \Rightarrow x = -\dfrac{1}{3}$.

The point of intersection is $\left(-\dfrac{1}{3}, 1\right)$.

(continued on next page)

(*continued*)

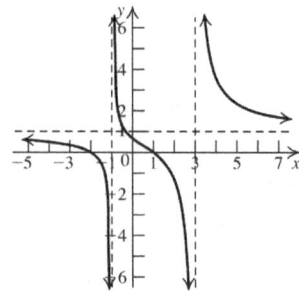

117. Vertical asymptote at $x = 3 \Rightarrow$ the denominator is $x - 3$. Horizontal asymptote at $y = -1 \Rightarrow$ the ratio of the leading coefficients of the numerator and denominator $= -1$. The x–intercept $= 2 \Rightarrow$ the numerator is $x - 2$. So the equation is of the form
$$f(x) = -\frac{x-2}{x-3} \text{ or } \frac{2-x}{x-3}.$$

119. Vertical asymptote at $x = 0 \Rightarrow$ the denominator is x. The slant asymptote $y = -x \Rightarrow$ the degree of the numerator is one more than the degree of the denominator and the quotient of the numerator and the denominator is $-x$. The x-intercepts -1 and $1 \Rightarrow$ the numerator is $(x-1)(x+1)$. So the equation is of the form
$$f(x) = a\frac{(x-1)(x+1)}{x} = a\left(\frac{x^2-1}{x}\right).$$

$$\begin{array}{r} x \\ x\overline{)x^2 - 1} \\ \underline{x^2} \\ -1 \end{array}$$

However, the slant asymptote is $y = -x$, so $a = -1$ and the equation is
$$f(x) = -\frac{x^2-1}{x} = \frac{-x^2+1}{x}.$$

121. Vertical asymptote at $x = 2 \Rightarrow$ the denominator is $x - 2$. Horizontal asymptote at $y = 1 \Rightarrow$ the degrees of the numerator and denominator are the same and the leading coefficients of the numerator and denominator are the same. So the numerator is $x + a$, where a is chosen so that $f(0) = -2 \Rightarrow \frac{x+4}{x-2} = f(x)$.

123. $f(x) \to -\infty$ as $x \to 1^-$ and $f(x) \to \infty$ as $x \to 1^+ \Rightarrow$ the denominator is zero if $x = 1$. Since $x \to 1$ from both directions, the denominator is $(x-1)^2$. $f(x) \to 4$ as $x \to \pm\infty \Rightarrow$ the horizontal asymptote is 4.

So the leading coefficient of the numerator is 4 and the degree of the numerator is the same as the degree of the denominator. The numerator is $4x^2 + a$, where a is chosen so that
$$f\left(\frac{1}{2}\right) = 0 \Rightarrow \frac{4(1/2)^2 + a}{(1/2 - 1)^2} \Rightarrow a = -1. \text{ So,}$$
$$f(x) = \frac{4x^2 - 1}{(x-1)^2}.$$

125. Vertical asymptote at $x = 1 \Rightarrow$ the denominator could be $x - 1$. Because the oblique asymptote is $y = 3x + 2$, the numerator is $(3x+2)(x-1) + a$, where a can be any number (no x-intercepts or function values are given). Let $a = 1$. So,
$$f(x) = \frac{(3x+2)(x-1)+1}{x-1} = \frac{3x^2 - x + 1}{x-1}.$$

3.6 Critical Thinking/Discussion/Writing

127. Answers may vary. Sample answers are given:
   a. $f(x) = \dfrac{1}{x}$
   b. $f(x) = \dfrac{x^2 + 2}{x^2 + x - 2}$
   c. $f(x) = \dfrac{x^3 + 2x^2 - 7x - 1}{x^3 + x^2 - 6x + 5}$

128. Answers may vary. Sample answers are given:
   a. $f(x) = \dfrac{x^2 + 1}{x}$
   b. $f(x) = \dfrac{x^3 + x - 2}{x^2}$
   c. $f(x) = \dfrac{x^5 + x^4 + x^2 - 4}{x^4}$

129. Since $y = x + 1$ is the oblique asymptote, we know that $R(x) = x + 1 + \dfrac{r(x)}{D(x)}$. We are told that the graph of $R(x)$ intersects the asymptote at the points $(2, R(2))$ and $(5, R(5))$, so $R(x) = x + 1$ when $\dfrac{r(x)}{D(x)} = 0$.
   Thus, $r(x) = (x-2)(x-5)$.

(*continued on next page*)

(*continued*)

Recall that the numerator of a rational function must have degree greater than that of the denominator in order for there to be an oblique asymptote, so the denominator, $D(x)$, can be any polynomial whose degree is greater than or equal to 3. Thus, a possible rational function is

$$R(x) = x + 1 + \frac{(x-2)(x-5)}{x^5}$$

$$= \frac{(x+1)x^5 + (x-2)(x-5)}{x^5}.$$

130. See exercise 129 for explanation.

$$R(x) = (ax+b) + \frac{K(x-c_1)(x-c_2)\cdots(x-c_n)}{D(x)}$$

$$= \frac{(ax+b) + D(x) + K(x-c_1)(x-c_2)\cdots(x-c_n)}{D(x)},$$

where $K \neq 0$, $D(x)$ is a polynomial of degree $n+1$ and none of its zeros are at $c_1, c_2, \ldots, c_n$.

3.6 Getting Ready for the Next Section

131. a. $y - 3 = -\frac{2}{3}(x-(-5)) \Rightarrow y - 3 = -\frac{2}{3}(x+5) \Rightarrow$

$y = -\frac{2}{3}x - \frac{2}{3}(5) + 3 \Rightarrow y = -\frac{2}{3}x - \frac{1}{3}$

b. $y = -\frac{2}{3}(1) - \frac{1}{3} = -1$

133. $3(1) + 2c = 11 \Rightarrow 2c = 8 \Rightarrow c = 4$

135. $12 = \frac{k}{\left(\frac{1}{2}\right)^2} \Rightarrow k = 3$

$y = \frac{3}{2^2} \Rightarrow y = \frac{3}{4}$

3.7 Variation

3.7 Practice Problems

1. $y = kx \Rightarrow 6 = 30k \Rightarrow \frac{1}{5} = k$

$y = \left(\frac{1}{5}\right)120 = 24$

2. $I = kV \Rightarrow 60 = 220k \Rightarrow \frac{3}{11} = k$

$75 = \left(\frac{3}{11}\right)V \Rightarrow 275 = V$

A battery of 275 volts is needed to produce 60 amperes of current.

3. $y = kx^2 \Rightarrow 48 = k(2)^2 \Rightarrow 12 = k$

$y = 12(5)^2 = 300$

4. $A = \frac{k}{B} \Rightarrow 12 = \frac{k}{5} \Rightarrow 60 = k$

$A = \frac{60}{3} = 20$

5. $y = \frac{k}{\sqrt{x}} \Rightarrow \frac{3}{4} = \frac{k}{\sqrt{16}} \Rightarrow k = 3$

$2 = \frac{3}{\sqrt{x}} \Rightarrow \sqrt{x} = \frac{3}{2} \Rightarrow x = \frac{9}{4}$

6. $F = G \cdot \frac{m_1 m_2}{r^2} \Rightarrow m \cdot g = G \cdot \frac{mM_{Mars}}{R_{Mars}^2} \Rightarrow$

$g = G \cdot \frac{M_{Mars}}{R_{Mars}^2}$

Note that the radius of Mars is given in kilometers, which must be converted to meters.

$g = \frac{(6.67 \times 10^{-11} \text{ m}^3/\text{kg}/\text{sec}^2)(6.42 \times 10^{23} \text{ kg})}{(3397 \text{ km})^2}$

$= \frac{(6.67 \times 10^{-11} \text{ m}^3/\text{kg}/\text{sec}^2)(6.42 \times 10^{23} \text{ kg})}{(3.397 \times 10^6 \text{ m})^2}$

$\approx 3.7 \text{ m/sec}^2$

3.7 Concepts and Vocabulary

1. y varies directly as x if $y = kx$.

3. y varies directly as the *n*th power of x if $y = kx^n$.

5. True

7. False.

3.7 Building Skills

9. $x = ky; 15 = 30k \Rightarrow k = \frac{1}{2}; x = \frac{1}{2}(28) = 14$

11. $s = kt^2; 64 = 2^2 k \Rightarrow k = 16; s = 5^2(16) = 400$

13. $r = \frac{k}{u}; 3 = \frac{k}{11} \Rightarrow k = 33; r = \frac{33}{1/3} = 99$

15. $B = \dfrac{k}{A^3}; 1 = \dfrac{k}{2^3} \Rightarrow k = 8; B = \dfrac{8}{4^3} = \dfrac{1}{8}$

17. $z = kxy; 42 = (2)(3)k \Rightarrow k = 7; 56 = 2(7)y \Rightarrow y = 4$

19. $z = kx^2; 32 = 4^2 k \Rightarrow k = 2; z = 2(5^2) = 50$

21. $P = kTQ^2; 36 = (17)(6^2)k \Rightarrow k = \dfrac{1}{17};$
    $P = \dfrac{1}{17}(4)(9^2) = \dfrac{324}{17}$

23. $z = \dfrac{k\sqrt{x}}{y^2}; 24 = \dfrac{\sqrt{16}k}{3^2} \Rightarrow k = 54$
    $27 = \dfrac{54\sqrt{x}}{2^2} \Rightarrow 2 = \sqrt{x} \Rightarrow x = 4$

25. $\dfrac{16}{12} = \dfrac{8}{y} \Rightarrow y = 6$

27. $\dfrac{100}{x_0} = \dfrac{y}{2x_0} \Rightarrow y = 200$

3.7 Applying the Concepts

29. $y = Hx$, where y is the speed of the galaxies and x is the distance between them.

31. a. $y = 30.5x$, where y is the length in centimeters and x is the length in feet.

    b. (i) $\quad y = 30.5(8) = 244$ cm

    (ii) $\quad y = 30.5\left(5\dfrac{1}{3}\right) \approx 162.67$ cm

    c. (i) $\quad 57 = 30.5x \Rightarrow x \approx 1.87$ ft
    (ii) $\quad 124 = 30.5x \Rightarrow x \approx 4.07$ ft

33. $P = kQ; 7 = 20k \Rightarrow \dfrac{7}{20} = k; P = \dfrac{7}{20}(100) = 35$ g

35. $d = kt^2; 64 = 2^2 k \Rightarrow k = 16; 9 = 16t^2 \Rightarrow$
    $t = \dfrac{3}{4}$ sec

37. $P = \dfrac{k}{V}; 20 = \dfrac{k}{300} \Rightarrow k = 6000;$
    $P = \dfrac{6000}{100} = 60$ lb/in.^{2}

39. a. The astronaut is $6000 + 3960$ miles from the Earth's center.
    $W = \dfrac{k}{d^2}; 120 = \dfrac{k}{3960^2} \Rightarrow k = 120(3960^2)$
    $W = \dfrac{120(3960^2)}{(6000+3960)^2} \approx 18.97$ lb

    b. $200 = \dfrac{k}{3960^2} \Rightarrow k = 200(3960^2)$
    $W = \dfrac{200(3960^2)}{3950^2} \approx 201.01$ lb

41. 1740 km $= 1{,}740{,}000$ m $= 1.740 \times 10^6$ m
    $g = G \cdot \dfrac{M}{R^2} = 6.67 \times 10^{-11} \times \dfrac{7.4 \times 10^{22}}{\left(1.740 \times 10^6\right)^2}$
    $= \dfrac{(6.67)(7.4)(10^{11})}{1.740^2 \times 10^{12}} \approx 1.63$ m/sec^2

43. a. $I = \dfrac{k}{d^2}; 320 = \dfrac{k}{10^2} \Rightarrow k = 32{,}000$
    $I = \dfrac{32{,}000}{5^2} = 1280$ candlepower

    b. $400 = \dfrac{32{,}000}{d^2} \Rightarrow d^2 = 80 \Rightarrow d \approx 8.94$ ft from the source.

45. $p = k\sqrt{l} \Rightarrow k = \dfrac{p}{\sqrt{l}} = \dfrac{2p}{2\sqrt{l}} = \dfrac{2p}{\sqrt{4l}} \Rightarrow$ the length is multiplied by 4 if the period is doubled.

47. a. $H = kR^2 N$, where k is a constant

    b. $k(2R)^2 N = 4kR^2 N = 4H \Rightarrow$ the horsepower is multiplied by 4 if the radius is doubled.

    c. $kR^2(2N) = 2kR^2 N = 2H \Rightarrow$ the horsepower is doubled if the number of pistons is doubled.

    d. $k\left(\dfrac{R}{2}\right)^2 (2N) = \dfrac{2kR^2 N}{4} = \dfrac{kR^2 N}{2} = \dfrac{H}{2} \Rightarrow$
    the horsepower is halved if the radius is halved and the number of pistons is doubled.

3.7 Beyond the Basics

49. a. $E = kl^2 v^3$

b. $1920 = k(10^2)(8^3) \Rightarrow k = \dfrac{3}{80} = 0.0375$

c. $E = 0.0375(8^2)(25^3) = 37,500$ watts

d. $kl^2(2v)^3 = 8kl^2v^3 = 8E$

e. $k(2l)^2 v^3 = 4kl^2v^3 = 4E$

f. $k(2l)^2(2v)^3 = 4(8)kl^2v^3 = 32E$

51. a. $m = kw^{3/4}; 75 = k(75^{3/4}) \Rightarrow$
$k = \sqrt[4]{3}\sqrt{5} \approx 2.94$

b. $m = 2.94(450^{3/4}) \approx 287.25$ watts

c. $k(4w)^{3/4} = 4^{3/4}kw^{3/4} \approx 2.83kw^{3/4} \approx 2.83$ m

d. $250 \approx 2.94w^{3/4} \Rightarrow w^{3/4} \approx 85.034 \Rightarrow$
$w \approx 373.93$ kg

53. a. $T^2 = \dfrac{4\pi^2 r^3}{G(M_1 + M_2)}$

b. Because gravity is in terms of cubic meters per kilogram per second squared, convert the distance from kilometers to meters:
$1.5 \times 10^8 \text{ km} = 1.5 \times 10^{11} \text{ m}$.
$(3.15 \times 10^7)^2 \approx \dfrac{4\pi^2(1.5 \times 10^{11})^3}{6.67 \times 10^{-11} M_{sun}} \Rightarrow$
$9.9225 \times 10^{14} \approx \dfrac{4\pi^2(3.375) \times 10^{33}}{6.67 \times 10^{-11} M_{sun}} \Rightarrow$
$M_{sun} \approx \dfrac{133.24 \times 10^{24}}{6.67 \times 10^{-11} \times 9.9225 \times 10^{14}}$
$\approx 2.01 \times 10^{30}$ kg

55. a. $R = kN(P - N)$, where k is the constant of proportionality.

b. $45 = 1000(9000)k \Rightarrow k = 5 \times 10^{-6}$

c. $R = 5 \times 10^{-6}(5000)(5000) = 125$ people per day.

d. $100 = 5 \times 10^{-6} N(10,000 - N) \Rightarrow$
$20,000,000 = 10,000N - N^2 \Rightarrow$

$N^2 - 10,000N + 20,000,000 = 0 \Rightarrow$
$N = \dfrac{10,000 \pm \sqrt{10,000^2 - 4(1)(2 \times 10^7)}}{2(1)}$
$= \dfrac{10,000 \pm \sqrt{20,000,000}}{2} \approx 2764$ or 7236

3.7 Critical Thinking/Discussion/Writing

56. $I = \dfrac{kV}{R} \Rightarrow IR = kV$
$1.3I = \dfrac{kV}{1.2R} \Rightarrow 1.56IR = kV \Rightarrow$ the voltage must increase by 56%.

57. a. $v = kw^2$. The diamond is cut into two pieces whose weights are $\dfrac{2w}{5}$ and $\dfrac{3w}{5}$.
The value of the first piece is $\left(\dfrac{4}{25}\right)(1000) = \$160$, and the value of the second piece is $\left(\dfrac{9}{25}\right)(1000) = \$360$.
The two pieces together are valued at $520, a loss of $480.

b. The stone is broken into three pieces whose weights are $\dfrac{5w}{25} = \dfrac{w}{5}, \dfrac{9w}{25}$, and $\dfrac{11w}{25}$. So, the values of the three pieces are
$\left(\dfrac{1}{5}\right)^2(25,000) = \$1000,$
$\left(\dfrac{9}{25}\right)^2(25,000) = \$3240,$ and
$\left(\dfrac{11}{25}\right)^2(25,000) = \$4840,$ respectively.
The total value is $9,080, a loss of $15,920.

c. The weights of the pieces are
$\dfrac{w}{15}, \dfrac{2w}{15}, \dfrac{3w}{15} = \dfrac{w}{5}, \dfrac{4w}{15},$ and $\dfrac{5w}{15} = \dfrac{w}{3}$,
respectively. If the original value is x, then
$x - 85,000 = \left(\dfrac{1}{15}\right)^2 x + \left(\dfrac{2}{15}\right)^2 x + \left(\dfrac{1}{5}\right)^2 x$
$+ \left(\dfrac{4}{15}\right)^2 x + \left(\dfrac{1}{3}\right)^2 x \Rightarrow$
$x - 85,000 = \dfrac{11}{45}x \Rightarrow x = \$112,500 =$ the original value of the diamond.

(*continued on next page*)

(*continued*)

A diamond whose weight is twice that of the original diamond is worth 4 times the value of the original diamond = $450,000.

58. $p = kd(n - f)$
$80 = k(40)(30 - f)$ and
$180 = k(60)(35 - f)$. Solving the first equation for k we have $k = -\dfrac{2}{f - 30}$.

Substitute that value into the second equation and solve for f:
$180 = \left(\dfrac{2}{30 - f}\right)(60)(35 - f) \Rightarrow$
$180 = \dfrac{4200 - 120f}{30 - f} \Rightarrow$
$5400 - 180f = 4200 - 120f \Rightarrow$
$1200 = 60f \Rightarrow f = 20$

59. $w_{\text{solid}} = kr_o^3; w_{\text{hollow}} = kr_o^3 - kr_i^3 = \dfrac{7}{8}kr_o^3$
$kr_o^3 - kr_i^3 = \dfrac{7}{8}kr_o^3 \Rightarrow r_o^3 - r_i^3 = \dfrac{7}{8}r_o^3 \Rightarrow$
$\dfrac{1}{8}r_o^3 = r_i^3 \Rightarrow \dfrac{1}{8} = \dfrac{r_i^3}{r_o^3} \Rightarrow \dfrac{1}{2} = \dfrac{r_i}{r_o}$

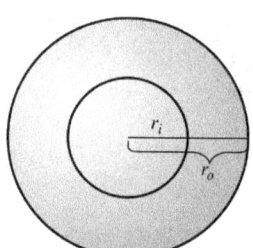

60. $s = 24 - k\sqrt{w}; 20 = 24 - k\sqrt{4} \Rightarrow k = 2$
$0 \le 24 - 2\sqrt{w} \Rightarrow w \le 144$
The greatest number of wagons the engine can move is 144.

3.7 Getting Ready for the Next Sections

61. $5^0 = 1$

63. $3^{-2} = \dfrac{1}{3^2} = \dfrac{1}{9}$

65. $\left(\dfrac{1}{2}\right)^{-4} = 2^4 = 16$

67. $5^{2x-3} \cdot 5^{3-x} = 5^{(2x-3)+(3-x)} = 5^x$

69. $\left(2^{x-1}\right)^x = 2^{(x-1)(x)} = 2^{x^2 - x}$

71. $y = mx + b \Rightarrow y - b = mx \Rightarrow m = \dfrac{y - b}{x}$

73. $A = B(1 + C) \Rightarrow \dfrac{A}{B} = 1 + C \Rightarrow \dfrac{A}{B} - 1 = C$

75. $A = B \cdot 10^{-n} \Rightarrow \dfrac{A}{10^{-n}} = B \Rightarrow A \cdot 10^n = B$

Chapter 3 Review Exercises

Building Skills

1. (i) Opens up
(ii) Vertex: (1, 2)
(iii) Axis: $x = 1$
(iv) $0 = (x-1)^2 + 2 \Rightarrow -2 = (x-1)^2 \Rightarrow$ there are no x-intercepts.
(v) $y = (0-1)^2 + 2 \Rightarrow y = 3$ is the y-intercept.
(vi) The function is decreasing on $(-\infty, 1)$ and increasing on $(1, \infty)$.

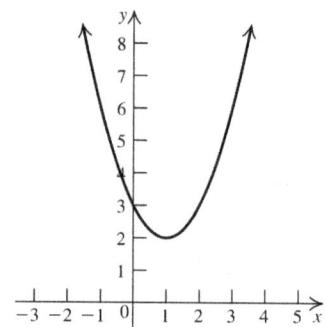

3. (i) Opens down
(ii) vertex: (3, 4)
(iii) Axis: $x = 3$
(iv) $0 = -2(x-3)^2 + 4 \Rightarrow 2 = (x-3)^2 \Rightarrow x = 3 \pm \sqrt{2}$ are the x-intercepts.
(v) $y = -2(0-3)^2 + 4 \Rightarrow y = -14$ is the y-intercept.
(vi) The function is increasing on $(-\infty, 3)$ and decreasing on $(3, \infty)$.

(*continued on next page*)

(*continued*)

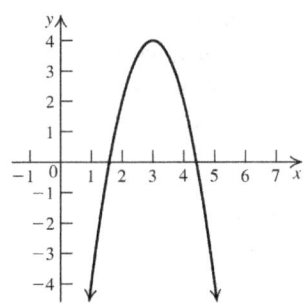

5. (i) Opens down
   (ii) Vertex: (0, 3)
   (iii) Axis: $x = 0$ (y-axis)
   (iv) $0 = -2x^2 + 3 \Rightarrow \dfrac{3}{2} = x^2 \Rightarrow$

   $x = \pm \dfrac{\sqrt{6}}{2}$ are the x-intercepts.

   (v) $y = -2(0)^2 + 3 \Rightarrow y = 3$ is the y-intercept.
   (vi) The function is increasing on $(-\infty, 0)$ and decreasing on $(0, \infty)$.

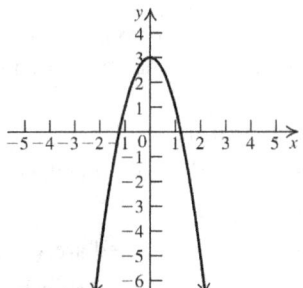

7. (i) Opens up
   (ii) To find the vertex, write the equation in standard form by completing the square:

   $$y = 2x^2 - 4x + 3$$
   $$y - 3 + 2 = 2(x^2 - 2x + 1)$$
   $$y = 2(x - 1)^2 + 1$$

   The vertex is (1, 1).
   (iii) Axis: $x = 1$
   (iv) $0 = 2x^2 - 4x + 3 \Rightarrow$

   $x = \dfrac{4 \pm \sqrt{(-4)^2 - 4(2)(3)}}{2(2)} = \dfrac{4 \pm \sqrt{-8}}{4} \Rightarrow$

   there are no x-intercepts.
   (v) $y = 2(0)^2 - 4(0) + 3 \Rightarrow y = 3$ is the y-intercept.
   (vi) The function is decreasing on $(-\infty, 1)$ and increasing on $(1, \infty)$.

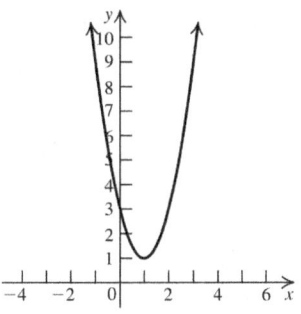

9. (i) Opens up
   (ii) To find the vertex, write the equation in standard form by completing the square.

   $$y = 3x^2 - 2x + 1$$
   $$y - 1 + \dfrac{1}{3} = 3\left(x^2 - \dfrac{2}{3}x + \dfrac{1}{9}\right)$$
   $$y = 3\left(x - \dfrac{1}{3}\right)^2 + \dfrac{2}{3}$$

   The vertex is $\left(\dfrac{1}{3}, \dfrac{2}{3}\right)$.

   (iii) Axis: $x = \dfrac{1}{3}$
   (iv) $0 = 3x^2 - 2x + 1 \Rightarrow$

   $x = \dfrac{2 \pm \sqrt{(-2)^2 - 4(3)(1)}}{2(3)} = \dfrac{2 \pm \sqrt{-8}}{6} \Rightarrow$

   there are no x-intercepts.
   (v) $y = 3(0)^2 - 3(0) + 1 \Rightarrow y = 1$ is the y-intercept.
   (vi) The function is decreasing on $\left(-\infty, \dfrac{1}{3}\right)$ and increasing on $\left(\dfrac{1}{3}, \infty\right)$.

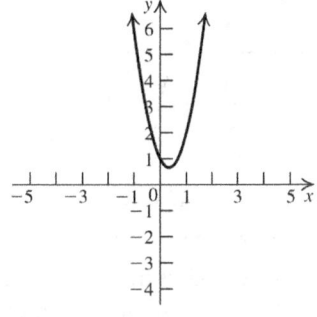

228 Chapter 3 Polynomial and Rational Functions

In exercises 11–14, find the vertex using the formula $\left(-\dfrac{b}{2a}, f\left(-\dfrac{b}{2a}\right)\right)$.

11. $a > 0 \Rightarrow$ the graph opens up, so f has a minimum value at its vertex. The vertex is
$$\left(-\dfrac{-4}{2(1)}, f\left(-\dfrac{-4}{2(1)}\right)\right) = (2, -1).$$
The minimum value is -1.

13. $a < 0 \Rightarrow$ the graph opens down, so f has a maximum value at its vertex. The vertex is
$$\left(-\dfrac{-3}{2(-2)}, f\left(-\dfrac{-3}{2(-2)}\right)\right) = \left(-\dfrac{3}{4}, \dfrac{25}{8}\right).$$
The maximum value is $\dfrac{25}{8}$.

15. Shift the graph of $y = x^3$ one unit left and two units down.

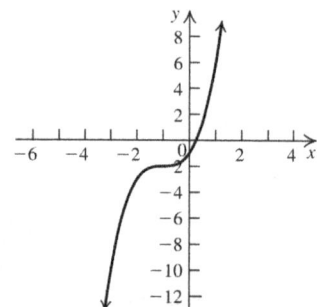

17. Shift the graph of $y = x^3$ one unit right, reflect the resulting graph in the y-axis, and shift it one unit up.

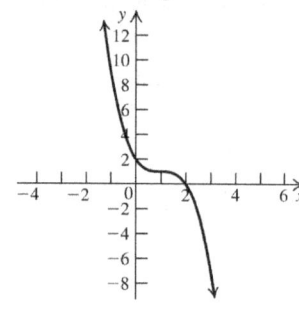

19. (i) $f(x) \to -\infty$ as $x \to -\infty$
$f(x) \to \infty$ as $x \to \infty$

(ii) Zeros: $x = -2$, multiplicity 1, crosses the x-axis; $x = 0$, multiplicity 1, crosses the x-axis; $x = 1$, multiplicity 1, crosses the x-axis.

(iii) x-intercepts: $-2, 0, 1$;
$y = 0(0-1)(0+2) \Rightarrow y = 0$ is the y-intercept.

(iv) The intervals to be tested are $(-\infty, -2)$, $(-2, 0)$, $(0, 1)$, and $(1, \infty)$.
The graph is above the x-axis on $(-2, 0) \cup (1, \infty)$ and below the x-axis on $(-\infty, -2) \cup (0, 1)$.

(v) $f(-x) = -x(-x-1)(-x+2) \neq f(x) \Rightarrow$ f is not even.
$-f(x) = -(x(x-1)(x+2)) \neq f(-x) \Rightarrow$ f is not odd. There are no symmetries.

(vi)

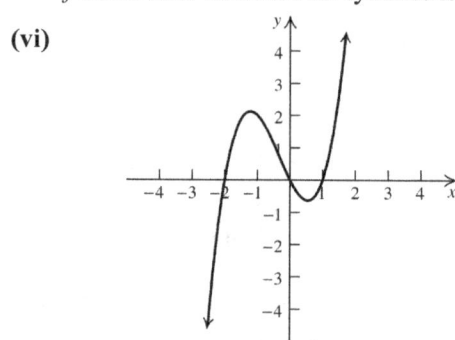

21. (i) $f(x) \to -\infty$ as $x \to -\infty$
$f(x) \to -\infty$ as $x \to \infty$

(ii) Zeros: $x = 0$, multiplicity 2, touches but does not cross the x-axis; $x = 1$, multiplicity 2, touches but does not cross the x-axis.

(iii) x-intercepts: 0, 1;
$y = -0^2(0-1)^2 \Rightarrow y = 0$ is the y-intercept.

(iv) The intervals to be tested are $(-\infty, 0)$, $(0, 1)$, and $(1, \infty)$. The graph is below the x-axis on $(-\infty, 0) \cup (0, 1) \cup (1, \infty)$.

(v) $f(-x) = -(-x)^2(-x-1)$
$= -x^2(-x-1) \neq -f(x)$ or $f(x) \Rightarrow$
$= -f(x) \Rightarrow f$ is neither even nor odd.
There are no symmetries.

(vi)

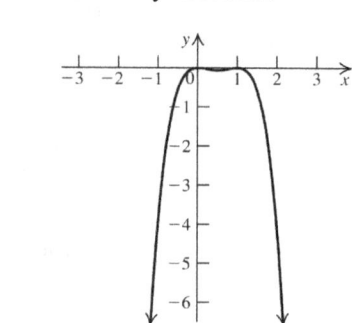

23. (i) $f(x) \to -\infty$ as $x \to -\infty$
$f(x) \to -\infty$ as $x \to \infty$

(ii) Zeros: $x = -1$, multiplicity 1, crosses the x-axis; $x = 0$, multiplicity 2, touches but does not cross; $x = 1$, multiplicity 1, crosses the x-axis.

(iii) x-intercepts: $-1, 0, 1$;
$y = -0^2(0^2 - 1) \Rightarrow y = 0$ is the y-intercept.

(iv) The intervals to be tested are $(-\infty, -1)$, $(-1, 0), (0, 1)$, and $(1, \infty)$. The graph is above the x-axis on $(-1, 0) \cup (0, 1)$ and below the x-axis on $(-\infty, -1) \cup (1, \infty)$.

(v) $f(-x) = -(-x)^2((-x)^2 - 1)$
$= -x^2(x^2 - 1) = f(x) \Rightarrow f$ is even, and f is symmetric with respect to the y-axis.

(vi)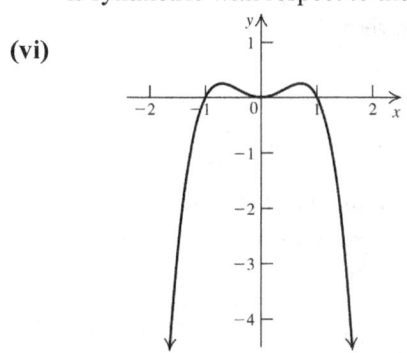

25.
$$\begin{array}{r} 2x + 3 \\ 3x - 2 \overline{\smash{)}6x^2 + 5x - 13} \\ \underline{6x^2 - 4x} \\ 9x - 13 \\ \underline{9x - 6} \\ -7 \end{array}$$

27.
$$\begin{array}{r} 8x^3 - 12x^2 + 14x - 21 \\ x + 1 \overline{\smash{)}8x^4 - 4x^3 + 2x^2 - 7x + 165} \\ \underline{8x^4 + 8x^3} \\ -12x^3 + 2x^2 \\ \underline{-12x^3 - 12x^2} \\ 14x^2 - 7x \\ \underline{14x^2 + 14x} \\ -21x + 165 \\ \underline{-21x - 21} \\ 186 \end{array}$$

29.
$$\begin{array}{r|rrrr} 3 & 1 & 0 & -12 & 3 \\ & & 3 & 9 & -9 \\ \hline & 1 & 3 & -3 & -6 \end{array}$$
Quotient: $x^2 + 3x - 3$ remainder -6.

31.
$$\begin{array}{r|rrrrr} -1 & 2 & -3 & 5 & -7 & 165 \\ & & -2 & 5 & -10 & 17 \\ \hline & 2 & -5 & 10 & -17 & 182 \end{array}$$
Quotient: $2x^3 - 5x^2 + 10x - 17$ remainder 182.

33. (i) $f(2) = 2^3 - 3(2^2) + 11(2) - 29 = -11$

(ii)
$$\begin{array}{r|rrrr} 2 & 1 & -3 & 11 & -29 \\ & & 2 & -2 & 18 \\ \hline & 1 & -1 & 9 & -11 \end{array}$$
The remainder is -11, so $f(2) = -11$.

35. (i) $f(2) = (-3)^4 - 2(-3)^2 - 5(-3) + 10 = 88$

(ii)
$$\begin{array}{r|rrrrr} -3 & 1 & 0 & -2 & -5 & 10 \\ & & -3 & 9 & -21 & 78 \\ \hline & 1 & -3 & 7 & -26 & 88 \end{array}$$
The remainder is 88, so $f(-3) = 88$.

37.
$$\begin{array}{r|rrrr} 2 & 1 & -7 & 14 & -8 \\ & & 2 & -10 & 8 \\ \hline & 1 & -5 & 4 & 0 \end{array}$$
The remainder is 0, so 2 is a zero. Now find the zeros of the depressed function $x^2 - 5x + 4 = (x - 4)(x - 1) \Rightarrow 4$ and 1 are also zeros. So the zeros of $x^3 - 7x^2 + 14x - 8$ are 1, 2, and 4.

39.
$$\begin{array}{r|rrrr} \frac{1}{3} & 3 & 14 & 13 & -6 \\ & & 1 & 5 & 6 \\ \hline & 3 & 15 & 18 & 0 \end{array}$$
The remainder is 0, so $1/3$ is a zero. Now find the zeros of the depressed function $3x^2 + 15x + 18 = 3(x + 2)(x + 3) \Rightarrow -2$ and -3 are also zeros. So the zeros of $3x^3 + 14x^2 + 13x - 6$ are $-3, -2$, and $1/3$.

41. The factors of the constant term are $\{\pm 1, \pm 2, \pm 3, \pm 6\}$, and the factors of the leading coefficient are $\{\pm 1\}$. So the possible rational zeros are $\{\pm 1, \pm 2, \pm 3, \pm 6\}$.

43. $f(x) = 5x^3 + 11x^2 + 2x;$
$f(-x) = 5(-x)^3 + 11(-x)^2 + 2(-x)$
$= -5x^3 + 11x^2 - 2x$
There are no sign changes in $f(x)$, so there are no positive zeros. There are two sign changes in $f(-x)$, so there are 2 or 0 negative zeros.
$5x^3 + 11x^2 + 2x = x(5x^2 + 11x + 2)$
$= x(5x+1)(x+2) \Rightarrow$ the zeros are $-2, -\frac{1}{5}, 0$.

45. $f(x) = x^3 + 3x^2 - 4x - 12$
$f(-x) = (-x)^3 + 3(-x)^2 - 4(-x) - 12$
$= -x^3 + 3x^2 + 4x - 12$
There is one sign change in $f(x)$, so there is one positive zero. There are two sign changes in $f(-x)$, so there are 2 or 0 negative zeros. The possible rational zeros are $\{\pm 1, \pm 2, \pm 3, \pm 4, \pm 6, \pm 12\}$. Use synthetic division to find one zero:

$\underline{-3|}\ 1\quad 3\quad -4\quad -12$
$\phantom{\underline{-3|}\ 1}\ \ -3\quad 0\quad\ \ 12$
$\phantom{\underline{-3|}\ }\ 1\quad 0\quad -4\quad\ \ 0$

The remainder is 0, so –3 is a zero. Now find the zeros of the depressed function
$x^2 - 4 = (x-2)(x+2) \Rightarrow -2$ and 2 are also zeros. So the zeros of $x^3 + 3x^2 - 4x - 12$ are –3, –2, and 2.

47. $f(x) = x^3 - 4x^2 - 5x + 14$
$f(-x) = (-x)^3 - 4(-x)^2 - 5(-x) + 14$
$= -x^3 - 4x^2 + 5x + 14$
There are two sign changes in $f(x)$, so there are 0 or 2 real positive zeros. There is one sign change in $f(-x)$, so there is one real negative zero. The possible rational zeros are $\{\pm 1, \pm 2, \pm 7, \pm 14\}$. Use synthetic division to find one zero:

$\underline{-2|}\ 1\quad -4\quad -5\quad\ \ 14$
$\phantom{\underline{-2|}\ 1}\ \ -2\quad\ \ 12\quad -14$
$\phantom{\underline{-2|}\ }\ 1\quad -6\quad\ \ 7\quad\ \ 0$

The remainder is 0, so –2 is a zero. Now find the zeros of the depressed function
$x^2 - 6x + 7$.

$x = \dfrac{-(-6) \pm \sqrt{(-6)^2 - 4(1)(7)}}{2(1)}$
$= \dfrac{6 \pm \sqrt{8}}{2} = \dfrac{6 \pm 2\sqrt{2}}{2} = 3 \pm \sqrt{2}$

So, the zeros of $x^3 - 4x^2 - 5x + 14$ are –2, $3 - \sqrt{2}$, and $3 + \sqrt{2}$.

49. $f(x) = 2x^3 - 5x^2 - 2x + 2$
$f(-x) = 2(-x)^3 - 5(-x)^2 - 2(-x) + 2$
$= -2x^3 - 5x^2 + 2x + 2$
There are two sign changes in $f(x)$, so there are 0 or 2 real positive zeros. There is one sign change in $f(-x)$, so there is one real negative zero. The possible rational zeros are $\left\{\pm 1, \pm 2, \pm\dfrac{1}{2}\right\}$. Use synthetic division to find one zero:

$\underline{\tfrac{1}{2}|}\ 2\quad -5\quad -2\quad\ \ 2$
$\phantom{\underline{\tfrac{1}{2}|}\ 2}\ \ \ 1\quad -2\quad -2$
$\phantom{\underline{\tfrac{1}{2}|}\ }\ 2\quad -4\quad -4\quad\ \ 0$

The remainder is 0, so 1/2 is a zero. Now find the zeros of the depressed function
$2x^2 - 4x - 4$.

$x = \dfrac{-(-4) \pm \sqrt{(-4)^2 - 4(2)(-4)}}{2(2)}$
$= \dfrac{4 \pm \sqrt{48}}{4} = \dfrac{4 \pm 4\sqrt{3}}{4} = 1 \pm \sqrt{3}$

So, the zeros of $2x^3 - 4x^2 - 2x + 2$ are 1/2, $1 - \sqrt{3}$, and $1 + \sqrt{3}$.

51. The function has degree three, so there are three zeros. Use synthetic division to find the depressed function:

$\underline{2|}\ 1\quad 0\quad -7\quad\ \ 6$
$\phantom{\underline{2|}\ 1}\ \ 2\quad\ \ 4\quad -6$
$\phantom{\underline{2|}\ }\ 1\quad 2\quad -3\quad\ \ 0$

Now find the zeros of the depressed function
$x^2 + 2x - 3 = (x+3)(x-1) \Rightarrow -3$ and 1 are zeros. The zeros of the original function are –3, 1, 2.

53. The function has degree four, so there are four zeros. Use synthetic division twice to find the depressed function:

```
-1| 1  -2   6  -18  -27      3| 1  -3   9  -27
      -1   3   -9   27           3   0   27
   ─────────────────────       ─────────────────
      1  -3   9  -27   0          1   0   9   0
```

Alternatively, divide $x^4 - 2x^3 + 6x^2 - 18x - 27$ by $(x+1)(x-3) = x^2 - 2x - 3$.

Now find the zeros of the depressed function $x^2 + 9 \Rightarrow \pm 3i$ are zeros. The zeros of the original function are $-1, 3$, and $\pm 3i$.

55. The function has degree four, so there are four zeros. Since one zero is $-1 + 2i$, another zero is $-1 - 2i$. Divide $x^4 + 2x^3 + 9x^2 + 8x + 20$ by $(x-(-1+2i))(x-(-1-2i)) = x^2 + 2x + 5$ to find the depressed function:

$$\begin{array}{r} x^2 + 4 \\ x^2 + 2x + 5 \overline{) x^4 + 2x^3 + 9x^2 + 8x + 20} \\ \underline{x^4 + 2x^3 + 5x^2} \\ 4x^2 + 8x + 20 \\ \underline{4x^2 + 8x + 20} \\ 0 \end{array}$$

Now find the zeros of the depressed function: $x^2 + 4 = 0 \Rightarrow x = \pm 2i$. The zeros of the original function are $\pm 2i$ and $-1 \pm 2i$.

57. The function has degree three, so there are three zeros. The possible rational zeros are $\{\pm 1, \pm 2, \pm 4\}$. Using synthetic division, we find that one zero is 1:

```
1| 1  -1  -4   4
       1   0  -4
   ────────────────
   1   0  -4   0
```

Now find the zeros of the depressed function: $x^2 - 4 = 0 \Rightarrow x = \pm 2$. The solution set is $\{-2, 1, 2\}$.

59. The function has degree three, so there are three zeros. The possible rational zeros are $\left\{\pm\frac{1}{4}, \pm\frac{1}{2}, \pm\frac{3}{4}, \pm 1, \pm\frac{3}{2}, \pm 3\right\}$. Using synthetic division, we find that one zero is -1:

```
-1| 4   0  -7  -3
       -4   4   3
   ────────────────
    4  -4  -3   0
```

Now find the zeros of the depressed function:
$4x^2 - 4x - 3 = 0 \Rightarrow (2x-3)(2x+1) = 0 \Rightarrow$
$x = \frac{3}{2}$ or $x = -\frac{1}{2}$.
The solution set is $\{-1/2, -1, 3/2\}$.

61. $x^3 - 8x^2 + 23x - 22 = 0$

The function has degree three, so there are three zeros. The possible rational zeros are $\{\pm 1, \pm 2, \pm 11, \pm 22\}$. Using synthetic division, we find that one zero is 2:

```
2| 1  -8   23  -22
       2  -12   22
   ──────────────────
    1  -6   11    0
```

Now find the zeros of the depressed function: $x^2 - 6x + 11 = 0$.

$$x = \frac{-(-6) \pm \sqrt{(-6)^2 - 4(1)(11)}}{2(1)}$$

$$= \frac{6 \pm \sqrt{-8}}{2} = \frac{6 \pm 2i\sqrt{2}}{2} = 3 \pm i\sqrt{2}$$

Solution set: $\{2, 3 \pm i\sqrt{2}\}$

63. $3x^3 - 5x^2 + 16x + 6 = 0$

The function has degree three, so there are three zeros. The possible rational zeros are $\left\{\pm 1, \pm 2, \pm 3, \pm 6, \pm\frac{1}{3}, \pm\frac{2}{3}\right\}$. Using synthetic division, we find that one zero is $-1/3$:

```
-1/3| 3  -5   16   6
          -1    2  -6
     ──────────────────
      3  -6   18   0
```

Now find the zeros of the depressed function:
$3x^2 - 6x + 18 = 0 \Rightarrow 3(x^2 - 2x + 6) = 0$.

$$x = \frac{-(-2) \pm \sqrt{(-2)^2 - 4(1)(6)}}{2(1)}$$

$$= \frac{2 \pm \sqrt{-20}}{2} = \frac{2 \pm 2i\sqrt{5}}{2} = 1 \pm i\sqrt{5}$$

Solution set: $\{-1/3, 1 \pm i\sqrt{5}\}$

232 Chapter 3 Polynomial and Rational Functions

65. $x^4 - x^3 - x^2 - x - 2 = 0$
The function has degree four, so there are four zeros. The possible rational zeros are $\{\pm 1, \pm 2\}$. Using synthetic division, we find that one zero is 2:

```
2| 1  -1  -1  -1  -2
       2   2   2   2
   1   1   1   1   0
```

Now find the zeros of the depressed function: $x^3 + x^2 + x + 1 = 0$. We can use synthetic division or factor to find another zero:
$$x^3 + x^2 + x + 1 = 0$$
$$x^2(x+1) + (x+1) = 0$$
$$(x^2 + 1)(x + 1) = 0$$
$$x^2 + 1 = 0 \Rightarrow x^2 = -1 \Rightarrow x = \pm i \text{ or}$$
$$x + 1 = 0 \Rightarrow x = -1$$
Solution set: $\{-1, 2, \pm i\}$

67. $2x^4 - x^3 - 2x^2 + 13x - 6 = 0$
The function has degree four, so there are four zeros. The possible rational zeros are $\left\{\pm 1, \pm 2, \pm 3, \pm 6, \pm \dfrac{1}{2}, \pm \dfrac{3}{2}\right\}$. Using synthetic division, we find that one zero is -2:

```
-2| 2  -1  -2   13  -6
       -4  10  -16   6
    2  -5   8   -3   0
```

Now find the zeros of the depressed function: $2x^3 - 5x^2 + 8x - 3 = 0$. Again using synthetic division, we find that one zero is $1/2$:

```
1/2| 2  -5   8  -3
         1  -2   3
     2  -4   6   0
```

Now find the zeros of the depressed function: $2x^2 - 4x + 6 = 0 \Rightarrow 2(x^2 - 2x + 3) = 0$.
$$x = \dfrac{-(-2) \pm \sqrt{(-2)^2 - 4(1)(3)}}{2(1)}$$
$$= \dfrac{2 \pm \sqrt{-8}}{2} = \dfrac{2 \pm 2i\sqrt{2}}{2} = 1 \pm i\sqrt{2}$$
Solution set: $\left\{-2, \dfrac{1}{2}, 1 \pm i\sqrt{2}\right\}$.

69. The only possible rational roots are $\{\pm 1, \pm 2\}$. None of these satisfies the equation.

71. $f(1) = 1^3 + 6(1)^2 - 28 = -21$;
$f(2) = 2^3 + 6(2)^2 - 28 = 4$. Because the sign changes, there is a real zero between 1 and 2. The zero is approximately 1.88.

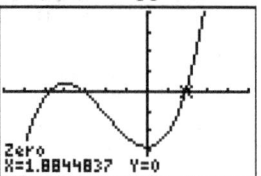

73. $1 + \dfrac{1}{x} = 0 \Rightarrow x = -1$ is the x-intercept. There is no y-intercept. The vertical asymptote is the y-axis ($x = 0$). The horizontal asymptote is $y = 1$. Testing the intervals $(-\infty, -1), (-1, 0),$ and $(0, \infty)$, we find that the graph is above the x-axis on $(-\infty, -1) \cup (0, \infty)$ and below the x-axis on $(-1, 0)$.

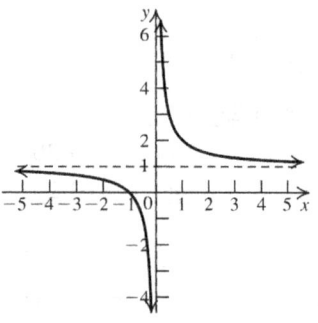

75. $\dfrac{x}{x^2 - 1} = 0 \Rightarrow x = 0$ is the x-intercept.
$\dfrac{0}{0^2 - 1} = 0 \Rightarrow y = 0$ is the y-intercept. The vertical asymptotes are $x = 1$ and $x = -1$. The horizontal asymptote is the x-axis. Testing the intervals $(-\infty, -1), (-1, 0), (0, 1),$ and $(1, \infty)$, we find that the graph is above the x-axis on $(-\infty, -1) \cup (0, \infty)$ and below the x-axis on $(-1, 0) \cup (0, 1)$.

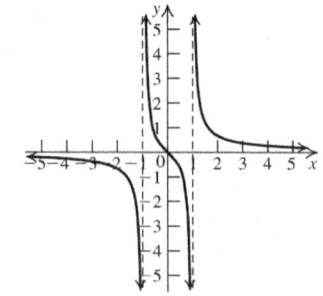

Copyright © 2019 Pearson Education Inc.

77. $\dfrac{x^3}{x^2-9} = 0 \Rightarrow x = 0$ is the x-intercept.

$\dfrac{0^3}{0^2-9} = 0 \Rightarrow y = 0$ is the y-intercept. The vertical asymptotes are $x = 3$ and $x = -3$. There is no horizontal asymptote. The oblique asymptote is $y = x$. Testing the intervals $(-\infty, -3), (-3, 0), (0, 3)$, and $(3, \infty)$, we find that the graph is above the x-axis on $(-3, 0) \cup (3, \infty)$, and below the x-axis on $(-\infty, -3) \cup (0, 3)$.

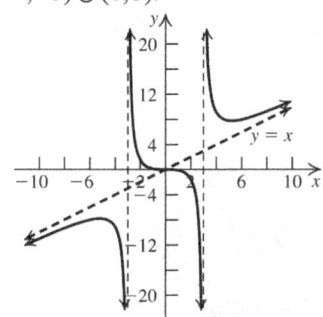

79. $\dfrac{x^4}{x^2-4} = 0 \Rightarrow x = 0$ is the x-intercept.

$\dfrac{0^4}{0^2-4} = 0 \Rightarrow y = 0$ is the y-intercept. The vertical asymptotes are $x = 2$ and $x = -2$. There is no horizontal asymptote. Testing the intervals $(-\infty, -2), (-2, 0), (0, 2)$, and $(2, \infty)$, we find that the graph is above the x-axis on $(-\infty, -2) \cup (2, \infty)$ and below the x-axis on $(-2, 0) \cup (0, 2)$.

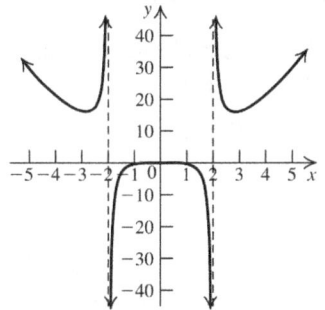

Applying the Concepts

81. $y = kx; 12 = 4k \Rightarrow k = 3; y = 3(5) = 15$

83. $s = kt^2; 20 = 2^2 k \Rightarrow 5 = k; s = 3^2(5) = 45$

85. The maximum height occurs at the vertex, $\left(-\dfrac{20}{2(-1/10)}, f\left(-\dfrac{20}{2(-1/10)}\right)\right) = (100, 1000).$

The maximum height is 1000. To find where the missile hits the ground, solve $-\dfrac{1}{10}x^2 + 20x = 0 \Rightarrow x = 0$ or $x = 200$. The missile hits the ground at $x = 200$.

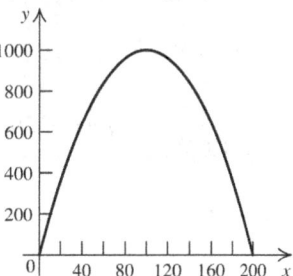

87. The area of each section is 400 square feet. Since the width is x, $y = \dfrac{400}{x}$. The total amount of fencing needed is $4x + \dfrac{2400}{x}$.

Using a graphing calculator, we find that this is a minimum at $x \approx 24.5$ feet. So $y \approx \dfrac{400}{24.5} \approx 16.3$ feet.

The dimensions of each pen should be approximately 24.5 ft by 16.3 ft.

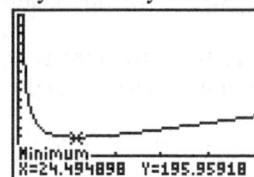

89. a.

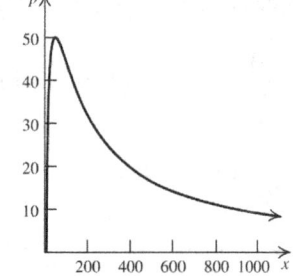

b.

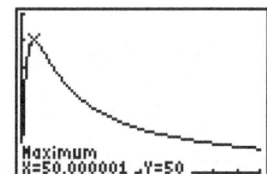

The maximum occurs at $x = 50$.

234 Chapter 3 Polynomial and Rational Functions

91. a. The revenue is $24x$. Profit = revenue − cost, so
$$P(x) = 24x - \left(150 + 3.9x + \frac{3}{1000}x^2\right)$$
$$= -\frac{3}{1000}x^2 + 20.1x - 150$$

b. The maximum occurs at the x-coordinate of the vertex: $-\dfrac{20.1}{2\left(-\dfrac{3}{1000}\right)} = 3350$

c. $\bar{C}(x) = \dfrac{C(x)}{x} = \dfrac{\frac{3}{1000}x^2 + 3.9x + 150}{x}$
$= \dfrac{3}{1000}x + 3.9 + \dfrac{150}{x}$

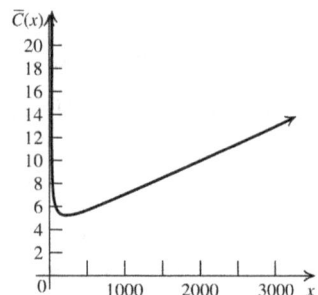

93. $280 = 40k \Rightarrow k = 7; s = 7(35) = \$245$

95. $I = \dfrac{ki}{d^2}$, where I = illumination, i = intensity, and d = distance from the source. The illumination 6 inches from the source is $I_6 = \dfrac{300k}{6^2}$, while the illumination x inches from the source is $I_x = \dfrac{300k}{x^2}$. $I_6 = 2I_x \Rightarrow$
$\dfrac{300k}{36} = 2\left(\dfrac{300k}{x^2}\right) \Rightarrow x^2 = 72 \Rightarrow x = 6\sqrt{2}$ in.

97. $I = \dfrac{k}{R}; 30 = \dfrac{k}{300} \Rightarrow k = 9000$

a. $I = \dfrac{9000}{250} = 36$ amp

b. $60 = \dfrac{9000}{R} \Rightarrow R = 150$ ohms

99. $R = ki(p-i)$

a. $255 = k(0.15)(20,000)(0.85)(20,000)$
$= \dfrac{1}{200,000}$

b. $R = \dfrac{1}{200,000}(10,000)(10,000) = 500$ people per day.

c. $95 = \dfrac{1}{200,000}x(20,000 - x) \Rightarrow$
$x^2 - 20,000x + 19,000,000 = 0 \Rightarrow$
$(x - 1000)(x - 19,000) = 0 \Rightarrow x = 1000$ or $x = 19,000$

Chapter 3 Practice Test A

1. $x^2 - 6x + 2 = 0 \Rightarrow x = \dfrac{6 \pm \sqrt{(-6)^2 - 4(1)(2)}}{2(1)} \Rightarrow$
$x = \dfrac{6 \pm \sqrt{28}}{2} = 3 \pm \sqrt{7}$ are the x-intercepts

2.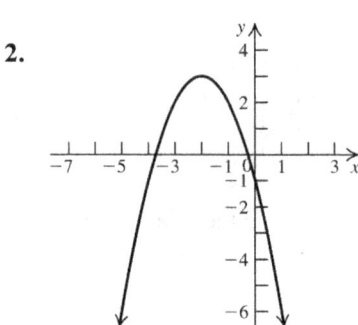

3. The vertex is at $\left(-\dfrac{b}{2a}, f\left(-\dfrac{b}{2a}\right)\right)$
$= \left(-\dfrac{14}{2(-7)}, f\left(-\dfrac{14}{2(-7)}\right)\right) = (1, 10).$

4. The denominator is 0 when $x = -4$ or $x = 1$. The domain is $(-\infty, -4) \cup (-4, 1) \cup (1, \infty)$.

5. Using either long division or synthetic division, we find that the quotient is $x^2 - 4x + 3$ and the remainder is 0.

$$\begin{array}{r|rrrr} -2 & 1 & -2 & -5 & 6 \\ & & -2 & 8 & -6 \\ \hline & 1 & -4 & 3 & 0 \end{array}$$

6.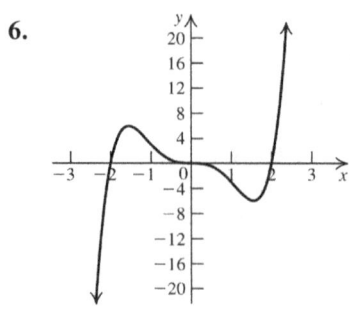

Copyright © 2019 Pearson Education Inc.

7. The function has degree three, so there are three zeros. Since 2 is a zero, use synthetic division to find the depressed function:

   $\underline{2|}\ \ 2\ \ -2\ \ -8\ \ \ \ 8$
   $\ \ \ \ \ \ \ \ \ \ \ \ \ \ \ \ 4\ \ \ \ 4\ \ -8$
   $\ \ \ \ \ \ \overline{\ \ 2\ \ \ \ 2\ \ -4\ \ \ \ 0\ }$

   Now find the zeros of $2x^2 + 2x - 4$:
   $2x^2 + 2x - 4 = 2(x^2 + x - 2) = 2(x+2)(x-1) \Rightarrow$ the zeros are $x = -2$ and $x = 1$. The zeros of the original function are $-2, 1, 2$.

8. $2x+3\overline{)\,-6x^3 + x^2 + 17x + 3\,}$ with quotient $-3x^2 + 5x + 1$

   $\ \ \ \ \ \ \ \ \ \ -6x^3 - 9x^2$
   $\ 10x^2 + 17x$
   $\ 10x^2 + 15x$
   $\ 2x + 3$
   $\ 2x + 3$
   $\ 0$

9. Using synthetic division to find the remainder, we have $P(-2) = -53$.

   $\underline{-2|}\ \ 1\ \ \ \ 5\ \ \ -7\ \ \ \ \ 9\ \ \ \ \ 17$
   $\ \ \ \ \ \ \ \ \ \ \ \ \ \ \ \ -2\ \ \ -6\ \ \ 26\ \ -70$
   $\ \ \ \ \ \ \ \overline{\ \ 1\ \ \ \ 3\ \ -13\ \ \ 35\ \ -53\ }$

10. The function has degree three, so there are three zeros. There are two sign changes in $f(x)$, so there are either 2 or 0 positive zeros. There is one sign change in $f(-x)$, so there is one negative zero. The possible rational zeros are $\{\pm 1, \pm 2, \pm 4, \pm 5, \pm 10, \pm 20\}$.

    Using synthetic division, we find that -2 is a zero:

    $\underline{-2|}\ \ 1\ \ -5\ \ -4\ \ \ \ 20$
    $\ \ \ \ \ \ \ \ \ \ \ \ \ \ \ -2\ \ \ 14\ -20$
    $\ \ \ \ \ \ \overline{\ \ 1\ \ -7\ \ \ 10\ \ \ \ 0\ }$

    Now find the zeros of the depressed function $x^2 - 7x + 10$: $x^2 - 7x + 10 = (x-5)(x-2) \Rightarrow$ the zeros are $x = 2$ and $x = 5$. The zeros of the original function are $-2, 2, 5$.

11. The function has degree four, so there are four zeros. Factoring, we have $x^4 + x^3 - 15x^2 = x^2(x^2 + x - 15) \Rightarrow 0$ is a zero of multiplicity 2. Now find the zeros of $x^2 + x - 15$:
    $$x = \frac{-1 \pm \sqrt{1^2 - 4(1)(-15)}}{2(1)} = \frac{-1 \pm \sqrt{61}}{2}.$$

    The zeros of the original function are 0 and $\frac{-1 \pm \sqrt{61}}{2}$.

12. The factors of the constant term, 9, are $\{\pm 1, \pm 3, \pm 9\}$, while the factors of the leading coefficient are $\{\pm 1, \pm 2\}$. The possible rational zeros are $\left\{\pm \frac{1}{2}, \pm 1, \pm \frac{3}{2}, \pm 3, \pm \frac{9}{2}, \pm 9\right\}$.

13. $f(x) \to -\infty$ as $x \to -\infty$; $f(x) \to \infty$ as $x \to \infty$.

14. $f(x) = (x^2 - 4)(x+2)^2$
    $= (x-2)(x+2)(x+2)^2 \Rightarrow$
    the zeros are -2 (multiplicity 3) and 2 (multiplicity 1).

15. There are two sign changes in $f(x)$, so there are either two or zero positive zeros. There is one sign change in $f(-x)$, so there is one negative zero.

16. The zeros of the denominator are $x = 5$ and $x = -4$, so those are the vertical asymptotes. The degree of the numerator equals the degree of the denominator, so the horizontal asymptote is $y = 2$.

17. $y = \dfrac{kx}{t^2}$

18. $6 = \dfrac{8k}{2^2} \Rightarrow k = 3$; $y = \dfrac{3(12)}{3^2} = 4$

19. The minimum occurs at the x-coordinate of the vertex: $-\dfrac{-30}{2(1)} = 15$ (which represents 15 thousand units.)

20. $V = x(17 - 2x)(8 - 2x)$

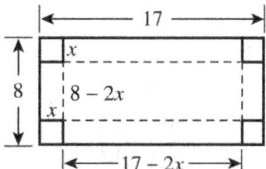

Chapter 3 Practice Test B

1. $x^2 + 5x + 3 = 0 \Rightarrow x = \dfrac{-5 \pm \sqrt{5^2 - 4(1)(3)}}{2(1)}$
   $= \dfrac{-5 \pm \sqrt{13}}{2}$. The answer is B.

2. The graph of $f(x) = 4 - (x-2)^2$ is the graph of $f(x) = x^2$ shifted two units right, reflected across the x-axis, and then shifted 4 units up. The answer is D.

3. $\left(-\dfrac{b}{2a}, f\left(-\dfrac{b}{2a}\right)\right) = \left(-\dfrac{12}{2(6)}, f\left(-\dfrac{12}{2(6)}\right)\right)$
   $= (-1, -11)$. The answer is A.

4. The denominator is 0 when $x = -3$ or $x = 2$. The answer is B.

5. $\begin{array}{r|rrrr} -3 & 1 & 0 & -8 & 6 \\ & & -3 & 9 & -3 \\ \hline & 1 & -3 & 1 & 3 \end{array}$
   The answer is D.

6. $P(x) = x^4 + 2x^3 = x^3(x+2)$. So the zeros are 0 (multiplicity 3) and –2 (multiplicity 1). The only graph with those zeros is C.

7. $\begin{array}{r|rrrr} 3 & 3 & -26 & 61 & -30 \\ & & 9 & -51 & 30 \\ \hline & 3 & -17 & 10 & 0 \end{array}$
   The zeros of the depressed function $3x^2 - 17x + 10$ are $x = \dfrac{2}{3}$ and $x = 5$. The answer is C.

8. $2x - 3 \overline{\smash{\big)}\,\begin{array}{l} -5x^2 + 3x - 4 \\ -10x^3 + 21x^2 - 17x + 12 \end{array}}$
   $\underline{-10x^3 + 15x^2}$
   $6x^2 - 17x$
   $\underline{6x^2 - 9x}$
   $-8x + 12$
   $\underline{-8x + 12}$
   0
   The answer is B.

9. $\begin{array}{r|rrrrr} -3 & 1 & 4 & 7 & 10 & 15 \\ & & -3 & -3 & -12 & 6 \\ \hline & 1 & 1 & 4 & -2 & 21 \end{array}$
   $P(-3) = 21$. The answer is C.

10. The polynomial has degree three, so there are three zeros. The possible rational zeros are $\{\pm 1, \pm 2, \pm 3, \pm 4, \pm 6, \pm 12\}$. Using synthetic division, we find that one zero is –3:
    $\begin{array}{r|rrrr} -3 & -1 & 1 & 8 & -12 \\ & & 3 & -12 & 12 \\ \hline & -1 & 4 & -4 & 0 \end{array}$
    The zeros of the depressed function $-x^2 + 4x - 4$ are $x = 2$ (multiplicity 2). The answer is A.

11. The polynomial has degree three, so there are three zeros. Factoring, we have
    $x^3 + x^2 - 30x = x(x^2 + x - 30) = x(x+6)(x-5)$
    The answer is A.

12. The factors of the constant term, 60, are $\{\pm 1, \pm 2, \pm 3, \pm 4, \pm 5, \pm 6, \pm 10, \pm 12, \pm 15, \pm 20, \pm 30, \pm 60\}$. Since the leading coefficient is 1, these are also the possible rational zeros. The answer is D.

13. The answer is C.

14. $(x^2 - 1)(x+1)^2 = (x-1)(x+1)(x+1)^2$. The answer is B.

15. There are three sign changes in $P(x)$, so there are 3 or 1 positive zeros. There are two sign changes in $P(-x)$, so there are 2 or 0 negative zeros. The answer is C.

16. The zeros of the denominator are $x = 3$ and $x = -4$, so those are the vertical asymptotes. The degree of the numerator equals the degree of the denominator, so the horizontal asymptote is $y = 1$. The answer is C.

17. The answer is D.

18. $27 = \dfrac{3^2 k}{1^3} \Rightarrow k = 3; S = \dfrac{3(6^2)}{3^3} = 4$. The answer is B.

19. The minimum occurs at the x-coordinate of the vertex: $-\dfrac{-24}{2(1)} = 12$. The answer is B.

20. $V = x(10 - 2x)(12 - 2x)$. The answer is D.

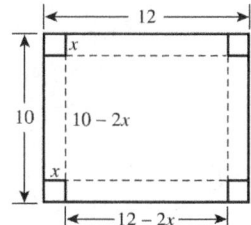

Cumulative Review Exercises
Chapters P–3

1. $d = \sqrt{(x_2 - x_1)^2 + (y_2 - y_1)^2}$
   $= \sqrt{(-1-2)^2 + (3-5)^2} = \sqrt{13}$

2. $M = \left(\dfrac{x_1 + x_2}{2}, \dfrac{y_1 + y_2}{2}\right)$
   $= \left(\dfrac{2 + (-8)}{2}, \dfrac{-5 + (-3)}{2}\right) = (-3, -4)$

3. $0 = x^2 - 2x - 8 \Rightarrow 0 = (x-4)(x+2) \Rightarrow$
   $x - 4 = 0 \Rightarrow x = 4$ or $x + 2 = 0 \Rightarrow x = -2$
   $f(0) = 0^2 - 2(0) - 8 = -8$
   The x-intercepts are $(-2, 0)$ and $(4, 0)$, and the y-intercept is $(0, -8)$.

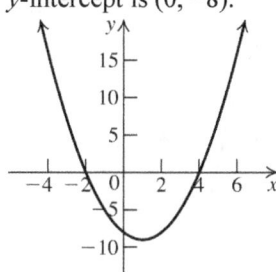

4. Write the equation in slope-intercept form:
   $x + 3y - 6 = 0 \Rightarrow 3y = -x + 6 \Rightarrow y = -\dfrac{1}{3}x + 2$
   The slope is $-\dfrac{1}{3}$, and the y-intercept is $(0, 2)$.
   $x + 3(0) - 6 = 0 \Rightarrow x - 6 = 0 \Rightarrow x = 6$, so the x-intercept is $(6, 0)$.

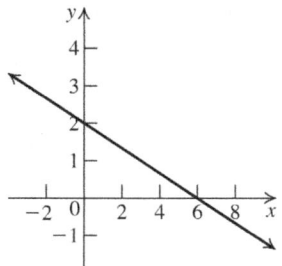

5. Write the equation in slope-intercept form:
   $x = 2y - 6 \Rightarrow x + 6 = 2y \Rightarrow \dfrac{1}{2}x + 3 = y$
   The slope is $\dfrac{1}{2}$, and the y-intercept is $(0, 3)$. The x-intercept is $x = 2(0) - 6 = -6$.

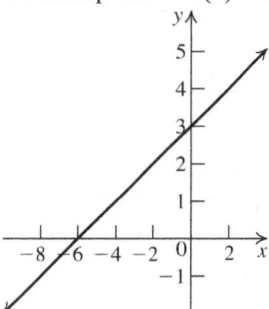

6. $(x - 2)^2 + (y + 3)^2 = 16$

7. $x^2 + y^2 + 2x - 4y - 4 = 0 \Rightarrow$
   $(x^2 + 2x) + (y^2 - 4y) = 4 \Rightarrow$
   $(x^2 + 2x + 1) + (y^2 - 4y + 4) = 4 + 1 + 4 \Rightarrow$
   $(x + 1)^2 + (y - 2)^2 = 9$. The center is $(-1, 2)$. The radius is 3.

8. $y + 2 = 3(x - 1) \Rightarrow y = 3x - 5$

9. $2x + 3y = 5 \Rightarrow y = -\dfrac{2}{3}x + \dfrac{5}{3} \Rightarrow$ the slope is $-\dfrac{2}{3}$.
   $y - 3 = -\dfrac{2}{3}(x - 1) \Rightarrow y = -\dfrac{2}{3}x + \dfrac{11}{3}$.

10. $2x + 3 = 0 \Rightarrow x = -\dfrac{3}{2}$.
    The domain is $\left(-\infty, -\dfrac{3}{2}\right) \cup \left(-\dfrac{3}{2}, \infty\right)$.

11. $4 - 2x > 0 \Rightarrow x < 2$. The domain is $(-\infty, 2)$.

12. $f(-2) = (-2)^2 - 2(-2) + 3 = 11$;
    $f(3) = 3^2 - 2(3) + 3 = 6$;
    $f(x + h) = (x + h)^2 - 2(x + h) + 3$
    $= x^2 + 2xh + h^2 - 2x - 2h + 3$
    $= x^2 + 2(h - 1)x + h^2 - 2h + 3$

(continued on next page)

(*continued*)

$$\frac{f(x+h)-f(x)}{h}$$
$$=\frac{(x^2+2(h-1)x+h^2-2h+3)-(x^2-2x+3)}{h}$$
$$=\frac{2(h-1)x+2x+h^2-2h}{h}$$
$$=\frac{2hx-2x+2x+h^2-2h}{h}=\frac{h^2+2hx-2h}{h}$$
$$=h+2x-2$$

13. a. $f(g(x))=\sqrt{x^2+1}$

 b. $g(f(x))=\left(\sqrt{x}\right)^2+1=x+1$

 c. $f(f(x))=\sqrt{\sqrt{x}}=\sqrt[4]{x}$

 d. $g(g(x))=\left(x^2+1\right)^2+1=x^4+2x^2+2$

14. a. $f(1)=3(1)+2=5; f(3)=4(3)-1=11;$
 $f(4)=6$

 b.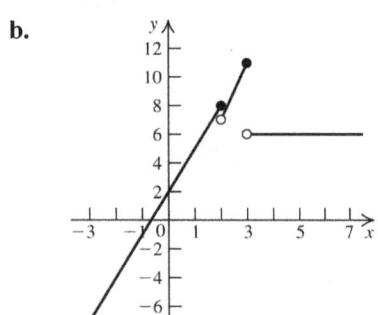

15. $y=2x-3$. Interchange x and y, and then solve for y.
$$x=2y-3 \Rightarrow y=\frac{x+3}{2}=\frac{1}{2}x+\frac{3}{2}=f^{-1}(x).$$

16. a. Shift the graph of $y=\sqrt{x}$ two units left.

 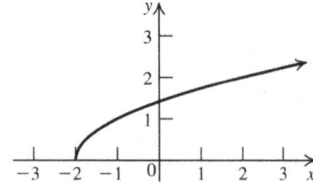

 b. Shift the graph of $y=\sqrt{x}$ one unit left, stretch vertically by a factor of two, reflect about the *x*-axis, and then shift up three units.

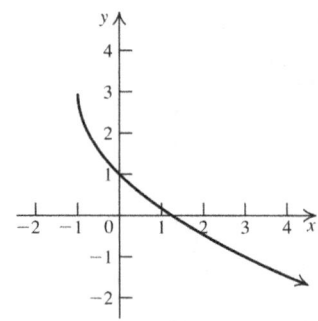

17. The factors of the constant term, –6, are $\{\pm 1,\pm 2,\pm 3,\pm 6\}$, and the factors of the leading coefficient, 2, are $\{\pm 1,\pm 2\}$. The possible rational zeros are $\left\{\pm\frac{1}{2},\pm 1,\pm\frac{3}{2},\pm 2,\pm 3,\pm 6\right\}$.

18.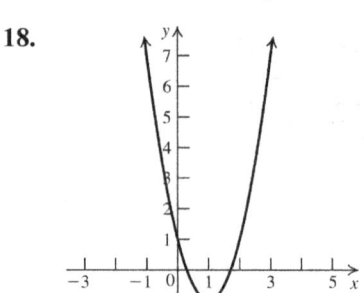

19.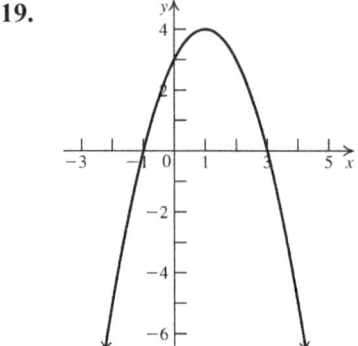

20.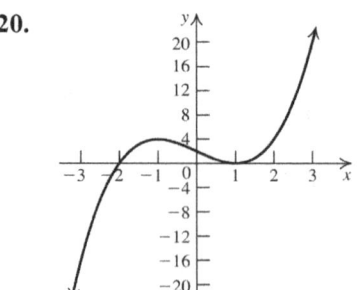

21.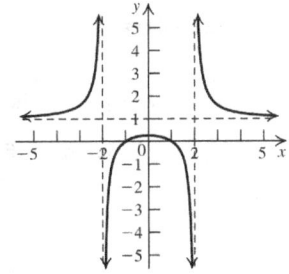

22. Since one zero is $1 + i$, another zero is $1 - i$. So $(x-(1-i))(x-(1+i)) = x^2 - 2x + 2$ is a factor of $f(x)$. Now divide to find the other factor:

$$\begin{array}{r} x^2 - x - 2 \\ x^2 - 2x + 2 \overline{\smash{\big)}\, x^4 - 3x^3 + 2x^2 + 2x - 4} \\ \underline{x^4 - 2x^3 + 2x^2} \\ -x^3 + 0x^2 + 2x \\ \underline{-x^3 + 2x^2 - 2x} \\ -2x^2 + 4x - 4 \\ \underline{-2x^2 + 4x - 4} \\ 0 \end{array}$$

$f(x) = x^4 - 3x^3 + 2x^2 + 2x - 4$
$= (x-(1-i))(x-(1+i))(x^2 - x - 2)$
$= (x-(1-i))(x-(1+i))(x-2)(x+1) \Rightarrow$
the zeros of $f(x)$ are $-1, 2, 1-i, 1+i$.

23. $y = k\sqrt{x}; 6 = k\sqrt{4} \Rightarrow k = 3; y = 3\sqrt{9} = 9$

24. Profit = revenue − cost
$150x - (0.02x^2 + 100x + 3000)$
$= -0.02x^2 + 50x - 3000.$
The maximum occurs at the vertex:
$$\left(-\frac{50}{2(-0.02)}, f\left(-\frac{50}{2(-0.02)}\right)\right)$$
$= (1250, \$28,250).$

25. $\underline{10|}\ \ 0.02\ \ 48.8\ \ -2990\ \ 25{,}000$
$\ 0.2\ \ \ \ 490\ \ -25{,}000$
$\ 0.02\ \ 49\ \ -2500\ \ \phantom{-25{,}00}0$

Now solve the depressed equation
$0.02x^2 + 49x - 2500 = 0.$

$x = \dfrac{-49 \pm \sqrt{49^2 - 4(0.02)(-2500)}}{2(0.02)}$

$= \dfrac{-49 \pm \sqrt{2601}}{0.04} = \dfrac{-49 \pm 51}{0.04} \Rightarrow x = 50$ or

$x = -2500$. There cannot be a negative amount of units sold, so another break-even point is 50.

Chapter 4 Exponential and Logarithmic Functions

4.1 Exponential Functions

4.1 Practice Problems

1. $f(x) = \left(\dfrac{1}{4}\right)^x$

   $f(2) = \left(\dfrac{1}{4}\right)^2 = \dfrac{1}{16}$

   $f(0) = \left(\dfrac{1}{4}\right)^0 = 1$

   $f(-1) = \left(\dfrac{1}{4}\right)^{-1} = 4$

   $f\left(\dfrac{5}{2}\right) = \left(\dfrac{1}{4}\right)^{5/2} = \left(\sqrt{\dfrac{1}{4}}\right)^5 = \left(\dfrac{1}{2}\right)^5 = \dfrac{1}{32}$

   $f\left(-\dfrac{3}{2}\right) = \left(\dfrac{1}{4}\right)^{-3/2} = 4^{3/2} = \left(\sqrt{4}\right)^3 = 2^3 = 8$

2. a. $3^{\sqrt{8}} \cdot 3^{\sqrt{2}} = 3^{\sqrt{8}+\sqrt{2}} = 3^{2\sqrt{2}+\sqrt{2}} = 3^{3\sqrt{2}} = 27^{\sqrt{2}}$

   b. $\left(a^{\sqrt{8}}\right)^{\sqrt{2}} = a^{\sqrt{8}\cdot\sqrt{2}} = a^{\sqrt{16}} = a^4$

3. a.

x	2^x
-3	1/8
-2	1/4
-1	1/2
0	1
1	2
2	4
3	8

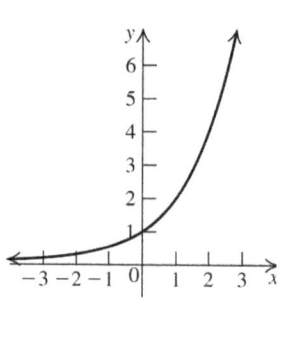

   b.

x	$(1/3)^x$
-3	27
-2	9
-1	3
0	1
1	1/3
2	1/9
3	1/27

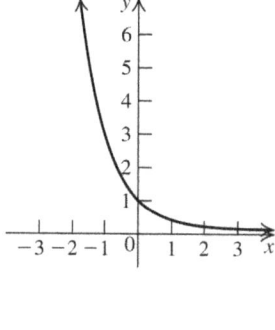

4. First reflect the graph of $f(x) = 2^x$ about the x-axis to obtain $f(x) = -2^x$. Then, shift the graph up three units.

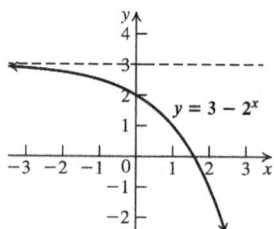

5. $f(x) = 2^{x-1} + 3$

   Shift the graph of $y = 2^x$ one unit to the right, then three units up.

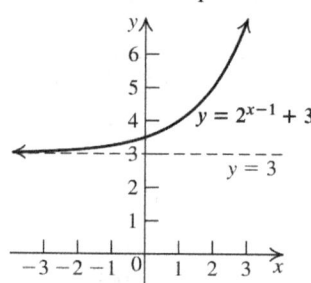

   Domain: $(-\infty, \infty)$; range: $(3, \infty)$; horizontal asymptote: $y = 3$

6. $(-2, 16)$ and $\left(3, \dfrac{1}{2}\right)$

   $16 = ca^{-2}$

   $\dfrac{1}{2} = ca^3$

   $\dfrac{16}{1/2} = \dfrac{ca^{-2}}{ca^3}$

   $32 = a^{-5} \Rightarrow a = \dfrac{1}{2}$

   $16 = c\left(\dfrac{1}{2}\right)^{-2} \Rightarrow 16 = 4c \Rightarrow 4 = c$

   So, $f(x) = 4\left(\dfrac{1}{2}\right)^x$.

7. $A = P + I = P + Prt$
   $= 10{,}000 + 10{,}000(0.075)(2)$
   $= 10{,}000 + 1500 = 11{,}500$
   There will be $11,500 in the account.

8. a. $A = P(1+r)^t = 8000(1+0.075)^5$
$\approx 11,485.03$
There will be $11,485.03 in the account.

 b. Interest $= A - P = 11,485.03 - 8000$
$= 3485.03$
She will receive $3485.03.

9. The formula for compound interest is
$A = P\left(1+\dfrac{r}{n}\right)^{nt}$.
For each of the following, $t = 1$.
 (i) Annual compounding ($n = 1$)
$A = \$5000\left(1+\dfrac{0.065}{1}\right)^1 = \$5325.00$

 (ii) Semiannual compounding ($n = 2$)
$A = \$5000\left(1+\dfrac{0.065}{2}\right)^2 \approx \$5330.28$

 (iii) Quarterly compounding ($n = 4$)
$A = \$5000\left(1+\dfrac{0.065}{4}\right)^4 \approx \$5333.01$

 (iv) Monthly compounding ($n = 12$)
$A = \$5000\left(1+\dfrac{0.065}{12}\right)^{12} \approx \$5334.86$

 (v) Daily compounding ($n = 365$)
$A = \$5000\left(1+\dfrac{0.065}{365}\right)^{365} \approx \$5335.76$

10. $20,000 = 9000\left(1+\dfrac{r}{12}\right)^{12 \cdot 8}$

$\left(1+\dfrac{r}{12}\right)^{96} = \dfrac{20,000}{9000} = \dfrac{20}{9}$

$1+\dfrac{r}{12} = \left(\dfrac{20}{9}\right)^{1/96}$

$\dfrac{r}{12} = \left(\dfrac{20}{9}\right)^{1/96} - 1$

$r = 12\left[\left(\dfrac{20}{9}\right)^{1/96} - 1\right] \approx 0.10023$

Carmen needs an interest rate of about 10.023%.

11. $A = Pe^{rt} = \$9000e^{(0.06)(8.25)} \approx \$14,764.48$

12. Use the exponential growth/decay formula
$A(t) = A_0 e^{kt}$

 a. $A(30) = 6.08e^{0.016(30)} \approx 9.8257$
If the rate of growth is 1.6% per year, there will be about 9.83 billion people in the world in the year 2030.

 b. $A(-10) = 6.08e^{0.016(-10)} \approx 5.1810$
If the rate of growth is 1.6% per year, there were about 5.18 billion people in the world in the year 1990.

13. $A(t) = A_0 e^{-kt}$
$A(6) = 22,000e^{(-0.18)(6)} \approx \$7471.10$

4.1 Concepts and Vocabulary

1. For the exponential function
$f(x) = ca^x, a > 0, \ a \neq 1$, the domain is
$(-\infty, \infty)$, and for $c > 0$, the range is $(0, \infty)$.

3. The horizontal asymptote of the graph of
$y = \left(\dfrac{1}{3}\right)^x$ is the x-axis.

5. False. The graphs are symmetric with respect to the y-axis.

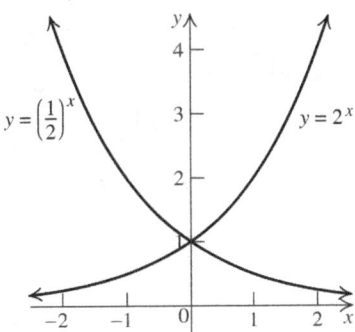

7. True

4.1 Building Skills

9. Not an exponential function. The base is not a constant.

11. Exponential function. The base is a constant, 1/2.

13. Not an exponential function. The base is not a constant.

15. Not an exponential function. The base is not a positive constant.

17. $f(0) = 5^{0-1} = 5^{-1} = \dfrac{1}{5}$

19. $g(3.2) = 3^{1-3.2} = 3^{-2.2} = 0.0892$
$g(-1.2) = 3^{1-(-1.2)} = 3^{2.2} = 11.2116$

21. $h(1.5) = \left(\dfrac{2}{3}\right)^{2(1.5)-1} = \left(\dfrac{2}{3}\right)^2 = \dfrac{4}{9}$
$h(-2.5) = \left(\dfrac{2}{3}\right)^{2(-2.5)-1} = \left(\dfrac{2}{3}\right)^{-6} = \dfrac{729}{64}$

23. $3^{\sqrt{2}} \cdot 3^{\sqrt{2}} = 3^{\sqrt{2}+\sqrt{2}} = 3^{2\sqrt{2}}$

25. $8^\pi \div 4^\pi = (2^3)^\pi \div (2^2)^\pi = 2^{3\pi} \div 2^{2\pi}$
$= 2^{3\pi - 2\pi} = 2^\pi$

27. $\left(3^{\sqrt{2}}\right)^{\sqrt{3}} = 3^{\sqrt{2}\cdot\sqrt{3}} = 3^{\sqrt{6}}$

29. $\left(a^{\sqrt{3}}\right)^{\sqrt{12}} = a^{\sqrt{3}\cdot\sqrt{12}} = a^{\sqrt{36}} = a^6$

31. a. $(0, 1)$ and $(2, 16)$
$f(0) = 1 \Rightarrow 1 = ca^0 \Rightarrow 1 = c$
$f(2) = 16 \Rightarrow 16 = 1 \cdot a^2 \Rightarrow a = 4$
$f(x) = 4^x$

b. $(0, 1)$ and $\left(-2, \dfrac{1}{9}\right)$
$f(0) = 1 \Rightarrow 1 = ca^0 \Rightarrow 1 = c$
$f(-2) = \dfrac{1}{9} \Rightarrow \dfrac{1}{9} = 1 \cdot a^{-2} \Rightarrow a = 3$
$f(x) = 3^x$

33. a. $(1, 1)$ and $(2, 5)$
$f(1) = 1 \Rightarrow 1 = ca^1$
$f(2) = 5 \Rightarrow 5 = ca^2$
$\dfrac{1}{5} = \dfrac{ca^1}{ca^2} \Rightarrow \dfrac{1}{5} = a^{-1} \Rightarrow 5 = a$
$f(1) = 1 \Rightarrow 1 = c \cdot 5^1 \Rightarrow c = \dfrac{1}{5}$
$f(x) = \dfrac{1}{5} \cdot (5)^x$

b. $(1, 1)$ and $\left(2, \dfrac{1}{5}\right)$
$f(1) = 1 \Rightarrow 1 = ca^1$
$f(2) = \dfrac{1}{5} \Rightarrow \dfrac{1}{5} = ca^2$
$\dfrac{1}{1/5} = \dfrac{ca^1}{ca^2} \Rightarrow 5 = a^{-1} \Rightarrow \dfrac{1}{5} = a$
$f(1) = 1 \Rightarrow 1 = c\left(\dfrac{1}{5}\right)^1 \Rightarrow c = 5$
$f(x) = 5 \cdot \left(\dfrac{1}{5}\right)^x$

35.

x	4^x
-2	$1/16$
-1	$1/4$
0	1
1	4
2	16

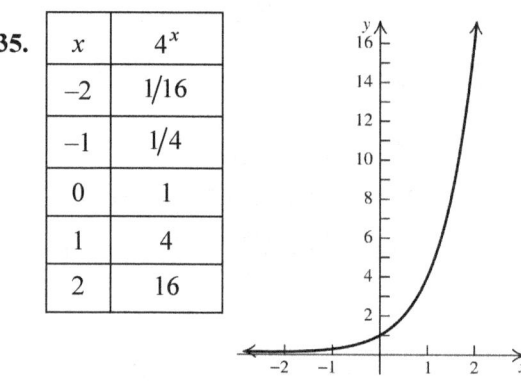

37.

x	$(3/2)^{-x}$
-4	$81/16$
-2	$9/4$
-1	$3/2$
0	1
1	$2/3$
2	$4/9$
4	$16/81$

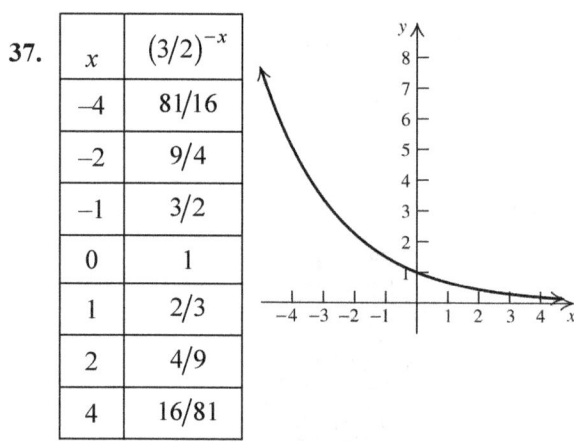

39.

x	$(1/4)^x$
-2	16
-1	4
0	1
1	$1/4$
2	$1/16$

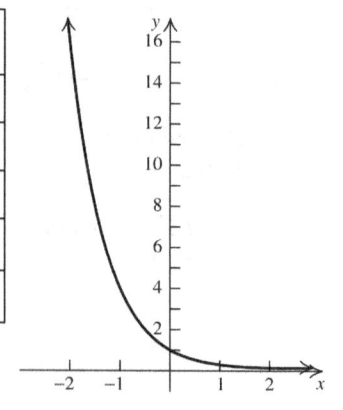

41.

x	1.3^{-x}
−4	≈ 2.86
−3	≈ 2.2
0	1
2	≈ 0.59
4	≈ 0.35

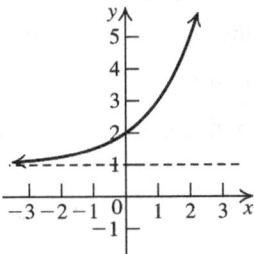

43. c   **45.** a

47. $g(x) = 2^x + 1$

Shift the graph of $f(x) = 2^x$ one unit up.
Domain: $(-\infty, \infty)$; Range: $(1, \infty)$
Horizontal asymptote: $y = 1$

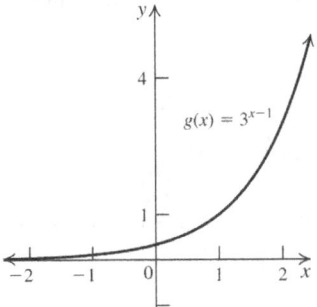

49. $g(x) = 3^{x-1}$

Shift the graph of $f(x) = 3^x$ one unit right.
Domain: $(-\infty, \infty)$; Range: $(0, \infty)$
Horizontal asymptote: $y = 0$

51. $g(x) = 4^{-x}$

Reflect the graph of $f(x) = 4^x$ about the y-axis.
Domain: $(-\infty, \infty)$; Range: $(0, \infty)$
Horizontal asymptote: $y = 0$

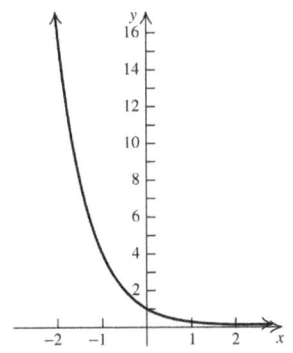

53. $g(x) = 2^{x-1} + 3$

Shift the graph of $f(x) = 2^x$ one unit right, then up three units.
Domain: $(-\infty, \infty)$; Range: $(3, \infty)$
Horizontal asymptote: $y = 3$

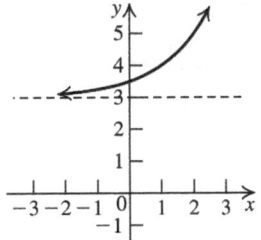

55. $g(x) = -3^{x-2} + 1$

Shift the graph of $f(x) = 3^x$ two units right, reflect across the x-axis, then shift the graph up one unit.
Domain: $(-\infty, \infty)$; Range: $(-\infty, 1)$
Horizontal asymptote: $y = 1$

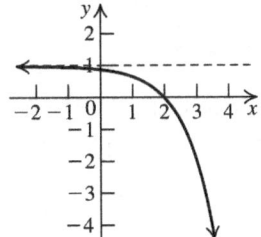

57. $g(x) = \left(\frac{1}{2}\right)^{x+1} - 1$

Shift the graph of $f(x) = \left(\frac{1}{2}\right)^x$ one unit left, then shift the graph down one unit.
Domain: $(-\infty, \infty)$; Range: $(-1, \infty)$
Horizontal asymptote: $y = -1$

(continued on next page)

(continued)

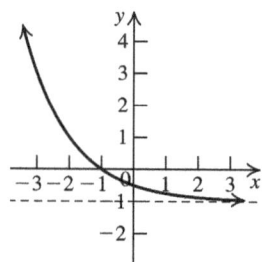

59. $g(x) = -2 \cdot 5^{x-1} + 4$

Shift the graph of $f(x) = 5^x$ one unit to the right, reflect it about the x-axis, stretch vertically by a factor of 2, then shift four units up. Domain: $(-\infty, \infty)$; Range: $(-\infty, 4)$. Horizontal asymptote: $y = 4$.

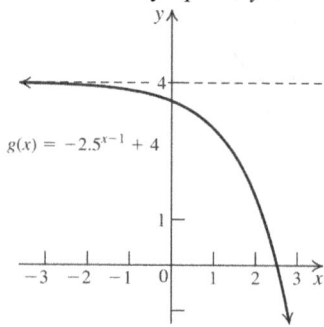

61. The graph passes through the points $(0, 3)$ and $(1, 3.5)$.

$f(0) = 3 \Rightarrow 3 = a^0 + b \Rightarrow 3 = 1 + b \Rightarrow b = 2$
$f(1) = 3.5 \Rightarrow 3.5 = a^1 + 2 \Rightarrow 1.5 = a$
$f(x) = 1.5^x + 2$
$f(2) = 1.5^2 + 2 = 4.25$

63. The graph passes through $(-2, 7)$ and $(-1, 1)$.

$f(-2) = 7 \Rightarrow 7 = a^{-2} + b$ (1)
$f(-1) = 1 \Rightarrow 1 = a^{-1} + b$ (2)
Subtract (2) from (1).
$6 = a^{-2} - a^{-1} \Rightarrow 6 = \dfrac{1}{a^2} - \dfrac{1}{a} \Rightarrow$
$6a^2 = 1 - a \Rightarrow 6a^2 + a - 1 = 0 \Rightarrow$
$(2a+1)(3a-1) = 0 \Rightarrow a = -\dfrac{1}{2}, \dfrac{1}{3}$

If $a = -\dfrac{1}{2}$, then
$1 = \left(-\dfrac{1}{2}\right)^{-1} + b \Rightarrow 1 = -2 + b \Rightarrow b = 3$.

If $a = \dfrac{1}{3}$, then $1 = \left(\dfrac{1}{3}\right)^{-1} + b \Rightarrow 1 = 3 + b \Rightarrow$

$b = -2$. The horizontal asymptote of the given graph is $y = -2$, so the equation of the graph is

$f(x) = \left(\dfrac{1}{3}\right)^x - 2$.

$f(2) = \left(\dfrac{1}{3}\right)^2 - 2 = \dfrac{1}{9} - 2 = -\dfrac{17}{9}$

65. $y = 2^{x+2} + 5$

67. $y = 2\left(\dfrac{1}{2}\right)^x - 5$

69. The graph of $f(x) = 2^x$ was reflected about the x-axis, and then shifted up two units.

71. $I = Prt = \$5000 \cdot 0.1 \cdot 5 = \$2500$

73. $I = Prt = \$7800 \cdot 0.06875 \cdot 10.75 = \$5764.69$

75. a. $A = 3500\left(1 + \dfrac{0.065}{1}\right)^{13} = \$7936.21$

b. interest $= \$7936.21 - \$3500 = \$4436.21$

77. a. $A = 7500e^{0.05(10)} = \$12,365.41$

b. interest $= \$12,365.41 - \$7500 = \$4865.41$

79. $10,000 = P\left(1 + \dfrac{0.08}{1}\right)^{10} = P(1.08)^{10} \Rightarrow$
$P = \$4631.93$

81. $10,000 = P\left(1 + \dfrac{0.08}{365}\right)^{365(10)} \Rightarrow P = \$4493.68$

83. Reflect the graph of $y = e^x$ about the y-axis.

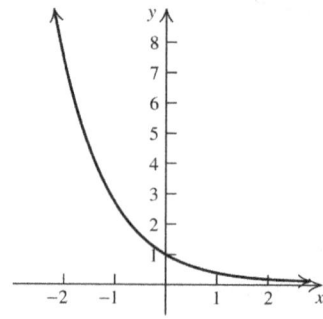

Horizontal asymptote: $y = 0$

85. Shift the graph of $y = e^x$ two units right.

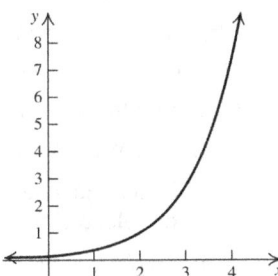

Horizontal asymptote: $y = 0$

87. Shift the graph of $y = e^x$ one unit up.

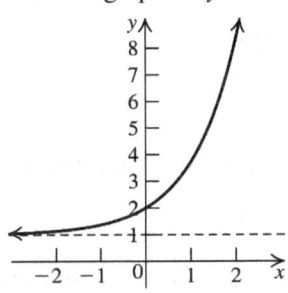

Horizontal asymptote: $y = 1$

89. Shift the graph of $y = e^x$ two units right, reflect the graph about the x-axis, and then shift it three units up.

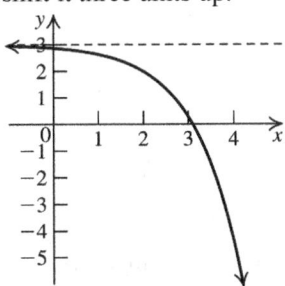

Horizontal asymptote: $y = 3$

4.1 Applying the Concepts

91. a. Exponential growth
    b. Linear growth
    c. Exponential growth
    d. Linear growth

93. $f(t) = 7.89e^{-0.42t}$

   a. The function represents exponential decay.

   b. $f(3) = 7.89e^{-0.42(3)} \approx 2.24$
   The average hard drive cost per gigabyte in 2003 was about $2.24.

   c. The price decreases quickly in the beginning, but then decreasing the price becomes more difficult as time goes on.

95. a. (i) $T = 200 \cdot 4^{-0.1(2)} + 25 = 176.6°C$

    (ii) $T = 200 \cdot 4^{-0.1(3.5)} + 25 = 148.1°C$

   b. $125 = 200 \cdot 4^{-0.1t} + 25 \Rightarrow 100 = 200 \cdot 4^{-0.1t} \Rightarrow$
   $\frac{1}{2} = 4^{-0.1t} \Rightarrow 2^{-1} = 2^{-0.2t} \Rightarrow -1 = -0.2t \Rightarrow$
   $t = 5$ hours

   c. As $t \to \infty$, $T \to 25$. Verify graphically.

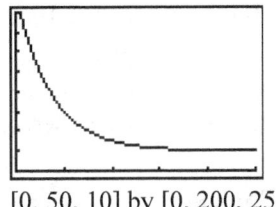

[0, 50, 10] by [0, 200, 25]

97. $A = 190,000\left(1 + \frac{0.03}{1}\right)^{1(5)} \approx \$220,262$

99. $A = 80,000\left(1 - \frac{0.15}{1}\right)^{1(5)} \approx \$35,496.43$

101. Assume that interest is compounded annually.
$22,000,000 = 5000\left(1 + \frac{r}{1}\right)^{1(54)} \Rightarrow$
$4400 = (1+r)^{54} \Rightarrow 4400^{1/54} = 1+r \Rightarrow$
$1.1681 \approx 1 + r \Rightarrow r \approx 0.1681 \approx 16.81\%$

103. $A = 10e^{-0.43(10)} \approx 0.1357 \text{ mm}^2$

105. $95,600 = 60,000\left(1 + \frac{r}{4}\right)^{4(12)} \Rightarrow$
$\frac{239}{150} = \left(1 + \frac{r}{4}\right)^{48} \Rightarrow \sqrt[48]{\frac{239}{150}} = 1 + \frac{r}{4} \Rightarrow$
$\sqrt[48]{\frac{239}{150}} - 1 = \frac{r}{4} \Rightarrow 4\left(\sqrt[48]{\frac{239}{150}} - 1\right) \approx 0.039 = r$

The interest rate on the bond was about 3.9%.

107. The amount Javier pays interest on is the principal plus the initial transfer fee

Bank A:

Transfer fee $= 0.02(1000) = 20$

$$I = (1000 + 20)\left(1 + \frac{0.08}{12}\right)^{12(1)} \approx 1104.66$$

Bank B:

Transfer fee $= 0.06(1000) = 60$

$$I = (1000 + 60)\left(1 + \frac{0.06}{12}\right)^{12(1)} \approx 1125.38$$

Javier will save $1125.38 - 1104.66 = \$20.72$ by choosing Bank A over Bank B.

109. Because the jar is full in 60 minutes, and the number of bacteria doubles every minute, it will be half full one minute earlier, or in 59 minutes.

4.1 Beyond the Basics

111. a. $f(x) + g(x) = \left(3^x + 3^{-x}\right) + \left(3^x - 3^{-x}\right)$
$= 2 \cdot 3^x$

b. $f(x) - g(x) = \left(3^x + 3^{-x}\right) - \left(3^x - 3^{-x}\right)$
$= 2 \cdot 3^{-x}$

c. $[f(x)]^2 - [g(x)]^2$
$= \left[\left(3^x + 3^{-x}\right)\right]^2 - \left[\left(3^x - 3^{-x}\right)\right]^2$
$= \left(3^{2x} + 3^{-2x} + 2\right) - \left(3^{2x} + 3^{-2x} - 2\right) = 4$

d. $[f(x)]^2 + [g(x)]^2$
$= \left[\left(3^x + 3^{-x}\right)\right]^2 + \left[\left(3^x - 3^{-x}\right)\right]^2$
$= \left(3^{2x} + 3^{-2x} + 2\right) + \left(3^{2x} + 3^{-2x} - 2\right)$
$= 2\left(3^{2x} + 3^{-2x}\right)$

113. $A = P\left(1 + \frac{r}{m}\right)^{m(1)} = P(1+y) \Rightarrow$
$\left(1 + \frac{r}{m}\right)^m = 1 + y \Rightarrow y = \left(1 + \frac{r}{m}\right)^m - 1$

115. $y = e^x \to y = e^{x-1} \to y = e^{2x-1} \to y = 3e^{2x-1}$

117. $y = e^x \to y = e^{2+x} \to y = e^{2+3x} \to$
$y = 5e^{2+3x} \to y = 5e^{2-3x} \to y = 5e^{2-3x} + 4$

4.1 Critical Thinking/Discussion/Writing

119. The base a cannot be 1 because this makes the function a constant, $f(x) = 1$. Similarly, the base a cannot be 0 because this becomes $f(x) = 0^x = 0$, a constant. We rule out negative bases so that the domain can include all real numbers. For example, a cannot be -3 because $f\left(\frac{1}{2}\right) = (-3)^{1/2} = \sqrt{-3}$ is not a real number.

120. $g(x)$ is defined for the set of integers. For example, $g(-2) = (-3)^{-2} = \frac{1}{9}$ and $g(3) = (-3)^3 = -27$.

121. If $0 < a < 1$, then, for $x < 0$, the denominator becomes very large, so $y = \frac{b}{1 + ca^x}$ approaches 0. For $x \geq 0$, ca^x approaches 0, so the denominator $1 + ca^x$ approaches 1 and $y = \frac{b}{1 + ca^x}$ approaches b.

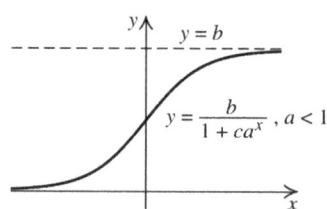

If $a > 1$, then, for $x < 0$, ca^x approaches 0, so the denominator $1 + ca^x$ approaches 1 and $y = \frac{b}{1 + ca^x}$ approaches b.
For $x \geq 0$, the denominator becomes very large, so $y = \frac{b}{1 + ca^x}$ approaches 0.

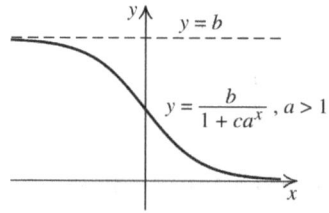

122. There are four possibilities, $0 < a < 1$ with $c < 0$, $0 < a < 1$ with $c > 0$, $a > 1$ with $c < 0$, and $a > 1$ with $c > 0$. They are illustrated below

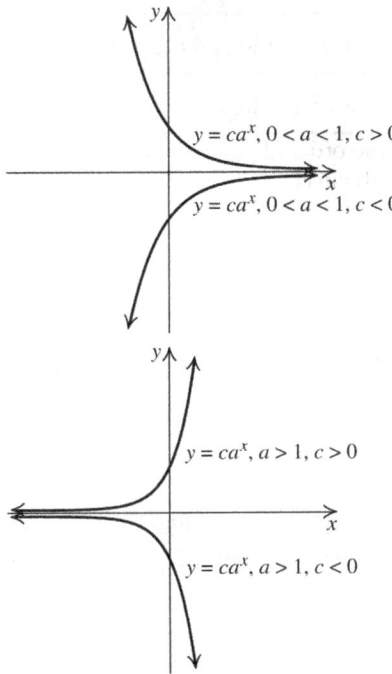

123. The function $y = 2^x$ is an increasing function, so it is one-to-one. The horizontal line $y = k$ for $k > 0$ intersects the graph of $y = 2^x$ in exactly one point, so there is exactly one solution for each value of k.

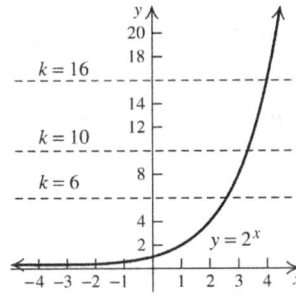

124. The graphs of $y = 2^x$ and $y = x$ intersect in exactly two points: $(1, 2)$ and $(2, 4)$. So $2^x = 2x \Rightarrow x = 1$ or 2.

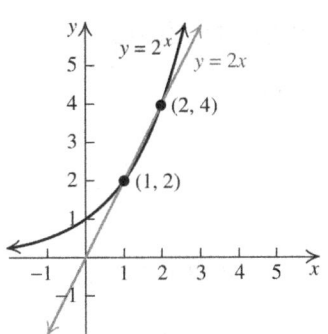

4.1 Getting Ready for the Next Section

125. $10^0 = 1$

127. $(-8)^{1/3} = \sqrt[3]{-8} = -2$

129. $\left(\dfrac{1}{7}\right)^{-2} = 7^2 = 49$

131. $18^{1/2} = \sqrt{18} = \sqrt{9 \cdot 2} = 3\sqrt{2}$

133. $f(x) = y = 3x + 4$
Interchange x and y, then solve for y.
$x = 3y + 4 \Rightarrow x - 4 = 3y \Rightarrow$
$\dfrac{x-4}{3} = y = f^{-1}(x)$

The domain and range of f^{-1} are $(-\infty, \infty)$.

135. $f(x) = y = \sqrt{x},\ x \geq 0$
Interchange x and y, then solve for y.
$x = \sqrt{y} \Rightarrow x^2 = y = f^{-1}(x),\ x \geq 0$

The domain and range of f^{-1} are $[0, \infty)$.

137. $f(x) = y = \dfrac{1}{x-1},\ x \neq 1$
Interchange x and y, then solve for y.
$x = \dfrac{1}{y-1} \Rightarrow x(y-1) = 1 \Rightarrow y - 1 = \dfrac{1}{x} \Rightarrow$
$y = f^{-1}(x) = 1 + \dfrac{1}{x},\ x \neq 0$

The domain of f^{-1} is $(-\infty, 0) \cup (0, \infty)$. The range of f^{-1} is $(-\infty, 1) \cup (1, \infty)$.

4.2 Logarithmic Functions

4.2 Practice Problems

1. a. $2^{10} = 1024$ is equivalent to $\log_2 1024 = 10$.

   b. $9^{-1/2} = \dfrac{1}{3}$ is equivalent to $\log_9\left(\dfrac{1}{3}\right) = -\dfrac{1}{2}$.

   c. $p = a^q$ is equivalent to $\log_a p = q$

2. a. $\log_2 64 = 6$ is equivalent to $2^6 = 64$.

   b. $\log_v u = w$ is equivalent to $v^w = u$.

3. a. $\log_3 9 = y \Rightarrow 3^y = 9 \Rightarrow 3^y = 3^2 \Rightarrow y = 2$
   Thus, $\log_3 9 = 2$.

   b. $\log_9 \dfrac{1}{3} = y \Rightarrow 9^y = \dfrac{1}{3} \Rightarrow 3^{2y} = 3^{-1} \Rightarrow y = -\dfrac{1}{2}$
   Thus, $\log_9 \dfrac{1}{3} = -\dfrac{1}{2}$.

   c. $\log_{1/2} 32 = y \Rightarrow \left(\dfrac{1}{2}\right)^y = 32 \Rightarrow 2^{-y} = 2^5 \Rightarrow$
   $y = -5$
   Thus, $\log_{1/2} 32 = -5$.

4. a. $\log_5 1 = 0$ b. $\log_3 3^5 = 5$

   c. $7^{\log_7 5} = 5$

5. Since the domain of the logarithmic function is $(0, \infty)$, the expression $\sqrt{1-x}$ must be positive. $\sqrt{1-x}$ is defined for $(-\infty, 1)$, so the domain of $\log_{10}\sqrt{1-x}$ is $(-\infty, 1)$.

6. Create a table of values to find ordered pairs on the graph of $y = \log_2 x$.

x	$y = \log_2 x$	(x, y)
$\dfrac{1}{8}$	$2^{-3} = \dfrac{1}{8} \Rightarrow y = \log_2 \dfrac{1}{8} = -3$	$\left(\dfrac{1}{8}, -3\right)$
$\dfrac{1}{4}$	$2^{-2} = \dfrac{1}{4} \Rightarrow y = \log_2 \dfrac{1}{4} = -2$	$\left(\dfrac{1}{4}, -2\right)$
$\dfrac{1}{2}$	$2^{-1} = \dfrac{1}{2} \Rightarrow y = \log_2 \dfrac{1}{2} = -1$	$\left(\dfrac{1}{2}, -1\right)$
1	$2^0 = 1 \Rightarrow y = \log_2 1 = 0$	$(1, 0)$
2	$2^1 = 2 \Rightarrow y = \log_2 2 = 1$	$(2, 1)$
4	$2^2 = 4 \Rightarrow y = \log_2 4 = 2$	$(4, 2)$
8	$2^3 = 8 \Rightarrow y = \log_2 8 = 3$	$(8, 3)$

Plot the ordered pairs and connect them with a smooth curve.

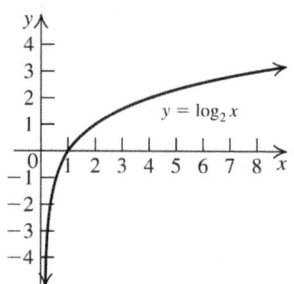

7. Shift the graph of $y = \log_2 x$ three units right, then reflect the resulting graph about the x-axis.

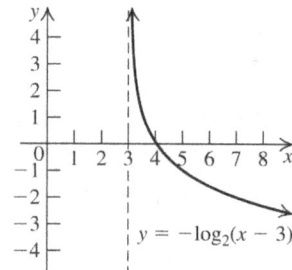

8. To sketch the graph of $f(x) = 3 - \log_2 x$, reflect the graph of $y = \log_2 x$ about the x-axis, and then shift three units up.

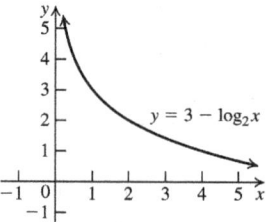

9. $P_2 = \log 3 - \log 2 \approx 0.176$
   This means that about 17.6% of the data is expected to have 2 as the first digit.

10. a. $y = \ln\dfrac{1}{e} \Rightarrow e^y = \dfrac{1}{e} \Rightarrow e^y = e^{-1} \Rightarrow y = -1$
    Thus, $\ln\dfrac{1}{e} = -1$.

    b. Using a calculator, $\ln 2 \approx 0.693$.

11. a. If P dollars are invested, then the amount $A = 3P$.
$A = Pe^{rt} \Rightarrow 3P = Pe^{0.065t} \Rightarrow 3 = e^{0.065t} \Rightarrow$
$\ln 3 = 0.065t \Rightarrow t = \dfrac{\ln 3}{0.065} \approx 16.9$
It will take approximately 17 years to triple your money.

b. $A = Pe^{rt} \Rightarrow 3P = Pe^{5r} \Rightarrow 3 = e^{5r} \Rightarrow$
$\ln 3 = 5r \Rightarrow t = \dfrac{\ln 3}{5} \approx 0.2197$
The investment will triple in 5 years at the rate of 21.97%.

12. From Example 12, we have $k = -\ln\left(\dfrac{93}{108}\right)$.
With this value of k, find t when $T = 120°$.
$$\ln\left(\dfrac{T(t) - T_s}{T_0 - T_s}\right) = -kt$$
$$\ln\left(\dfrac{120 - 72}{180 - 72}\right) = -\left(-\ln\left(\dfrac{93}{108}\right)\right) \cdot t$$
$$\ln\left(\dfrac{48}{108}\right) = \ln\left(\dfrac{93}{108}\right) t$$
$$\left[\dfrac{1}{\ln\left(\frac{93}{108}\right)}\right] \ln\left(\dfrac{48}{108}\right) = t \Rightarrow t \approx 5.423$$
The employees should wait about 5.4 minutes (realistically 5.5 minutes) to deliver the coffee at 120°F.

4.2 Concepts and Vocabulary

1. The domain of the function $y = \log_a x$ is $(0, \infty)$ and it range is $(-\infty, \infty)$.

3. The logarithm with base 10 is called the common logarithm, and the logarithm with base e is called the natural logarithm.

5. False

7. False. The domain of $f(x) = \log_a x$, $a > 0$ and $a \neq 1$ is $(0, \infty)$.

4.2 Building Skills

9. $\log_5 25 = 2$

11. $\log_{1/16} 4 = -\dfrac{1}{2}$

13. $\log_{10} 1 = 0$

15. $\log_{10} 0.1 = -1$

17. $a^2 + 2 = 7 \Rightarrow a^2 = 5 \Rightarrow \log_a 5 = 2$

19. $2a^3 - 3 = 10 \Rightarrow 2a^3 = 13 \Rightarrow a^3 = \dfrac{13}{2} \Rightarrow$
$\log_a\left(\dfrac{13}{2}\right) = 3$

21. $2^5 = 32$

23. $10^2 = 100$

25. $10^0 = 1$

27. $10^{-2} = 0.01$

29. $3 \log_8 2 = 1 \Rightarrow \log_8 2 = \dfrac{1}{3} \Rightarrow 8^{1/3} = 2$

31. $e^x = 2$

33. $\log_5 125 = 3$ because $5^3 = 125$

35. $\log 10,000 = 4$ because $10^4 = 10,000$

37. $\log_2 \dfrac{1}{8} = -3$ because $2^{-3} = \dfrac{1}{8}$

39. $\log_3 \sqrt{27} = \dfrac{3}{2}$ because $3^{3/2} = \sqrt{27}$

41. $\log_{16} 2 = \dfrac{1}{4}$ because $16^{1/4} = 2$

43. $\log_3 1 = 0$

45. $\log_7 7 = 1$

47. $\log_6 6^7 = 7$

49. $3^{\log_3 5} = 5$

51. $2^{\log_2 7} + \log_5 5^{-3} = 7 + (-3) = 4$

53. $4^{\log_4 6} - \log_4 4^{-2} = 6 - (-2) = 8$

55. Since the domain of the logarithmic function is $(0, \infty)$, the expression $x + 1$ must be positive. This occurs in the interval $(-1, \infty)$, so the domain of $\log_2 (x + 1)$ is $(-1, \infty)$.

57. Since the domain of the logarithmic function is $(0, \infty)$, the expression $\sqrt{x - 1}$ must be positive. This occurs in the interval $(1, \infty)$, so the domain of $\log_3 \sqrt{x - 1}$ is $(1, \infty)$.

59. Since the domain of the logarithmic function is $(0, \infty)$, the expressions $(x - 2)$ and $(2x - 1)$ must be positive. This occurs in the interval $(2, \infty)$ for $x - 2$ and in the interval $\left(\frac{1}{2}, \infty\right)$ for $(2x - 1)$. The intersection of the two intervals is $(2, \infty)$, so the domain of f is $(2, \infty)$.

61. Since the domain of the logarithmic function is $(0,\infty)$, the expressions $(x-1)$ and $(2-x)$ must be positive. This occurs in the interval $(1,\infty)$ for $x-1$ and in the interval $(-\infty, 2)$ for $(2-x)$. The intersection of the two intervals is $(1, 2)$, so the domain of h is $(1, 2)$.

63. $f(x) = \ln|x|$

    The domain of the logarithmic function is $(0,\infty)$, and the domain of $g(x) = |x|$ is $(-\infty, \infty)$. Thus, the domain of $f(x)$ is $(-\infty, 0) \cup (0, \infty)$.

65. $f(x) = \log_2\left(\dfrac{x}{3} - 4\right)$

    The domain of the logarithmic function is $(0,\infty)$, so $\dfrac{x}{3} - 4 > 0 \Rightarrow \dfrac{x}{3} > 4 \Rightarrow x > 12$.
    The domain of $f(x)$ is $(12, \infty)$.

67. $f(x) = \ln\left(\dfrac{x}{x+1}\right)$

    The domain of the logarithmic function is $(0,\infty)$, so $\dfrac{x}{x+1} > 0$
    Using the test interval method from section 2.5, we have
    numerator: $x = 0$
    denominator: $x = -1$
    The boundary points are -1 and 0. The intervals are $(-\infty, -1)$, $(-1, 0)$, and $(0, \infty)$.

Interval	Test point	Value of $R(x) = \dfrac{x}{x+1}$	Result
$(-\infty, -1)$	-2	2	$+$
$(-1, 0)$	$-\dfrac{1}{2}$	-1	$-$
$(0, \infty)$	1	$\dfrac{1}{2}$	$+$

    The domain of $f(x)$ is $(-\infty, -1) \cup (0, \infty)$

69. $f(x) = \log_3\left(\dfrac{x-2}{x+1}\right)$

    The domain of the logarithmic function is $(0,\infty)$, so $\dfrac{x-2}{x+1} > 0$
    Using the test interval method from section 2.5, we have
    numerator: $x = 2$
    denominator: $x = -1$
    The boundary points are -1 and 2. The intervals are $(-\infty, -1)$, $(-1, 2)$, and $(2, \infty)$.

Interval	Test point	Value of $R(x) = \dfrac{x-2}{x+1}$	Result
$(-\infty, -1)$	-2	4	$+$
$(-1, 2)$	0	-2	$-$
$(2, \infty)$	3	$\dfrac{1}{4}$	$+$

    The domain of $f(x)$ is $(-\infty, -1) \cup (2, \infty)$

71. a. F b. A
    c. D d. B
    e. E f. C

73. $f(x) = \log_4(x+3)$

    Shift the graph of $y = \log_4 x$ three units left.
    Domain: $(-3, \infty)$; range: $(-\infty, \infty)$; asymptote: $x = -3$

75. $f(x) = \log_{1/2}(x-1)$

Shift the graph of $y = \log_{1/2} x$ one unit right.

Domain: $(1, \infty)$; range: $(-\infty, \infty)$; asymptote: $x = 1$

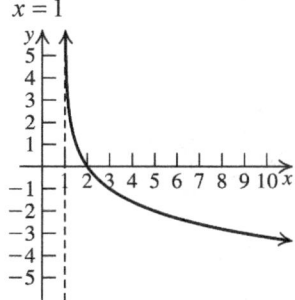

77. $f(x) = -\log_5 x$

Reflect the graph of $y = \log_5 x$ about the x-axis. Domain: $(0, \infty)$; range: $(-\infty, \infty)$; asymptote: $x = 0$

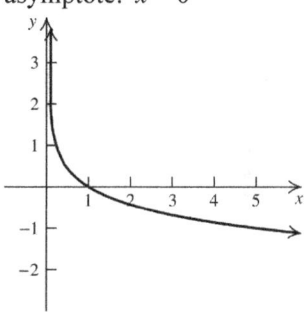

79. $f(x) = 2 - \log_4(x-1)$

Shift the graph of $y = \log_4 x$ one unit right, reflect across the x-axis, then shift two units up. Domain: $(1, \infty)$; range: $(-\infty, \infty)$; asymptote: $x = 1$

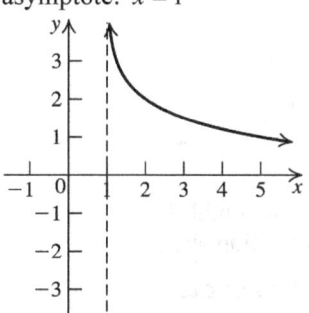

81. $f(x) = 1 + \log_{1/5}(-x)$

Reflect the graph of $y = \log_{1/5} x$ in the y-axis, and then shift it up one unit. Domain: $(-\infty, 0)$; range: $(-\infty, \infty)$; asymptote: $x = 0$

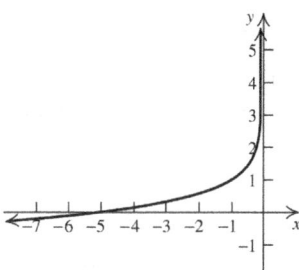

83. $f(x) = |\log_3 x|$

On $(0,1)$, reflect the graph of $y = \log_3 x$ about the x-axis. Domain: $(0, \infty)$; range: $[0, \infty)$; asymptote: $x = 0$ (the y-axis)

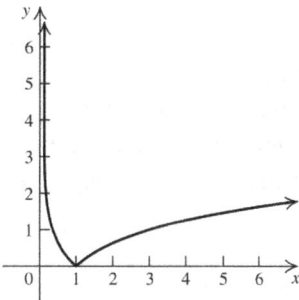

85. $f(x) = \log_2(x-1)$

Shift the graph of $y = \log_2 x$ one unit right.

Domain: $(1, \infty)$; range: $(-\infty, \infty)$; asymptote: $x = 1$

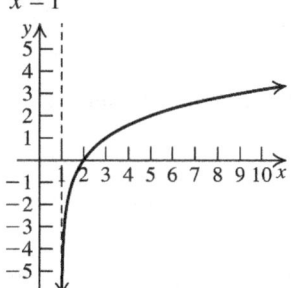

87. $f(x) = \log_2(3-x)$

Shift the graph of $y = \log_2 x$ three units right and then reflect it about the y-axis.

Domain: $(-\infty, 3)$; range: $(-\infty, \infty)$; asymptote: $x = 3$

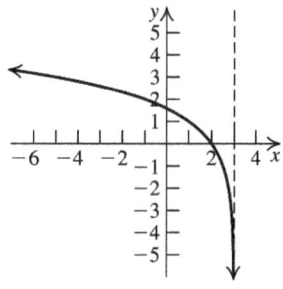

89. $f(x) = -\log_2 x - 2$

Reflect the graph of $y = \log_2 x$ about the x-axis and then shift it down two units. Domain: $(0, \infty)$; range: $(-\infty, \infty)$; asymptote: $x = 0$ (the y-axis)

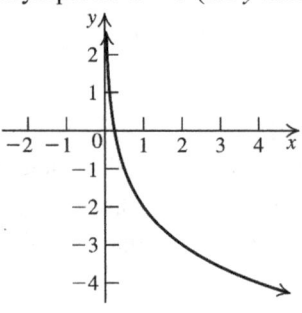

91. $f(x) = 2 + \log_2(3 - x)$

Shift the graph of $y = \log_2 x$ three units right, then reflect it about the y-axis, and then shift it 2 units up. Domain: $(-\infty, 3)$; range: $(-\infty, \infty)$; asymptote: $x = 3$

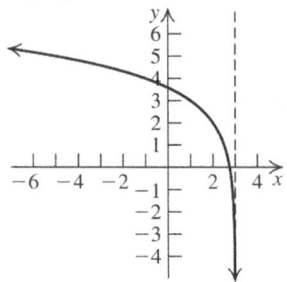

93. $\log_4(\log_3 81) = \log_4 4 = 1$ because $3^4 = 81$ and $4^1 = 4$

95. $\log_{\sqrt{2}} 2 = 2$ because $\sqrt{2}^2 = 2$.

97. $\log_{\sqrt{2}} 4 = 4$ because $(\sqrt{2})^4 = 4$.

99. $f(x) = \ln(x + 2)$

Shift the graph of $y = \ln x$ two units left.

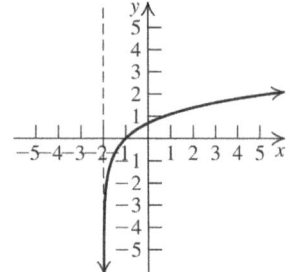

101. $f(x) = -\ln(x - 1)$

Shift the graph of $y = \ln x$ graph one unit right, then reflect the graph about the x-axis.

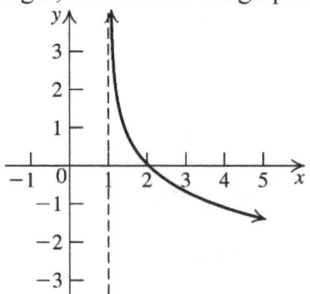

103. $f(x) = 2 + \ln(-x)$

Reflect the graph of $y = \ln x$ graph about the y-axis and then shift it up two units.

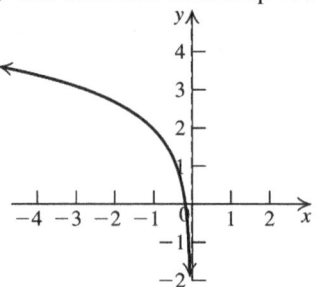

105. $f(x) = 3 - 2\ln x$

Stretch the graph of $y = \ln x$ vertically by a factor of 2, reflect the resulting graph about the x-axis, and shift it three units up.

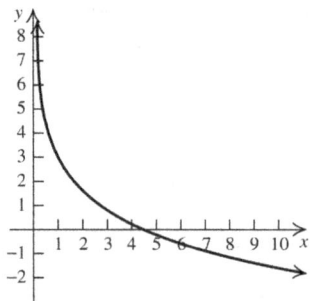

107. There was a horizontal shift of one unit left and then a reflection about the x-axis.

4.2 Applying the Concepts

109. $f(t) = 221.69 - 66.9 \ln t$

a. Because $\ln 0$ is not defined, the year 2003 is represented by $t = 1$ and the year 2008 is represented by $t = 6$.

$f(6) = 221.69 - 66.9 \ln 6 \approx 102$

In 2008, the consumer price index for personal computers was about 102.

b. $221.69 - 66.9 \ln t = 30$
$-66.9 \ln t = -191.69$
$\ln t = \dfrac{191.69}{66.9}$
$t = e^{191.69/66.9} \approx 17.6$
The consumer price index will reach 30 sometime during $2003 + 17.6 = 2020$.

c. The price decreases faster in the beginning, but then the decreases in price become slower as time goes on.

111. $2P = Pe^{0.08t} \Rightarrow 2 = e^{0.08t} \Rightarrow \ln 2 = 0.08t \Rightarrow$
$t \approx 8.66$ years

113. $2P = Pe^{6k} \Rightarrow 2 = e^{6k} \Rightarrow \ln 2 = 6k \Rightarrow$
$k \approx 0.1155 = 11.55\%$

115. a. The population in 2010 was 109.7% of the population in 2000.
$308 = 1.097 A_0 \Rightarrow A_0 \approx 280.7657247$
The population in 2000 was about 280.8 million.

b. $308 = 280.7657247 e^{10r} \Rightarrow$
$\dfrac{308}{280.7657247} = e^{10r} \Rightarrow$
$\ln\left(\dfrac{308}{280.7657247}\right) = 10r \Rightarrow$
$r = \dfrac{\ln\left(\dfrac{308}{280.7657247}\right)}{10} \approx 0.009257 \approx 0.93\%$

c. $400 = 308 e^{0.0092t} \Rightarrow \dfrac{400}{308} = e^{0.0092t} \Rightarrow$
$\ln\left(\dfrac{400}{308}\right) = 0.0092t \Rightarrow t = \dfrac{\ln\left(\dfrac{400}{308}\right)}{0.0092} \approx 28.4$
years after 2010, or in the year 2039.

117. Find k using $T = 50, T_0 = 75, T_s = 20,$ and $t = 1$ minute.
$50 = 20 + (75 - 20)e^{-k(1)} \Rightarrow \dfrac{30}{55} = e^{-k} \Rightarrow$
$\ln \dfrac{30}{55} = -k \Rightarrow 0.6061 \approx k$

a. (i) $T = 20 + (75 - 20)e^{-0.6061(5)} \Rightarrow$
$T \approx 22.66°F$

(ii) $T = 20 + (75 - 20)e^{-0.6061(10)} \Rightarrow$
$T \approx 20.13°F$

(iii) $T = 20 + (75 - 20)e^{-0.6061(60)} \Rightarrow$
$T \approx 20°F$

b. $22 = 20 + (75 - 20)e^{-0.6061t} \Rightarrow$
$\dfrac{2}{55} = e^{-0.6061t} \Rightarrow \ln \dfrac{2}{55} = -0.6061t \Rightarrow$
$t \approx 5.5$ minutes

119. Find k using $T = 160, T_0 = 50, T_s = 400,$ and $t = 10$ minutes.
$160 = 400 + (50 - 400)e^{-10k} \Rightarrow \dfrac{240}{350} = e^{-10k} \Rightarrow$
$\ln \dfrac{24}{35} = -10k \Rightarrow k \approx 0.03773$
Use this value of k to find the time given $T = 220, T_0 = 50,$ and $T_s = 400$.
$220 = 400 + (50 - 400)e^{-0.03773t} \Rightarrow$
$\dfrac{18}{35} = e^{-0.03773t} \Rightarrow \ln \dfrac{18}{35} = -0.03773t \Rightarrow$
$t \approx 17.6$ minutes

121. a. $A(1) = 0.8(0.04) = 0.032$
Using the model $A(t) = A_0 e^{rt}$, we have
$0.032 = 0.04 e^r \Rightarrow 0.8 = e^r \Rightarrow r = \ln 0.8$
So, in one year (12 months), the concentration of the contaminant will be
$A(12) = 0.04 e^{12 \ln 0.8} = 0.0027 = 0.27\%$

b. $0.0001 \geq 0.04 e^{\ln 0.8 t} \Rightarrow 0.0025 \geq e^{\ln 0.8 t} \Rightarrow$
$\ln 0.0025 \geq \ln 0.8 t \Rightarrow t \geq 26.85$ months

123. a. $3000 = 1500 e^{3r} \Rightarrow \ln 2 = 3r \Rightarrow$
$r = \dfrac{\ln 2}{3} \approx 0.231$
Thus, the function is $P \approx 1500 e^{0.231t}$.
Alternatively, since the population doubles every three years, we have the geometric progression given by $P = 1500 \cdot 2^{t/3}$.

b. $P = 1500 e^{0.231(7)} \approx 7557$
There will be about 7557 sheep in the herd seven years from now.

c. $15,000 = 1500 e^{0.231t} \Rightarrow 10 = e^{0.231t} \Rightarrow$
$\ln 10 = 0.231t \Rightarrow t = \dfrac{\ln 10}{0.231} \approx 10$
The herd will have 15,000 sheep about 10 years from now.

125. a. $P_3 = \log 4 - \log 3 = \log\left(\dfrac{4}{3}\right) \approx 0.125$
About 12.5% of the data can be expected to have 3 as the first digit.

b. $P_1 = \log 2 - \log 1 \approx 0.3010$
$P_2 = \log 3 - \log 2 \approx 0.1761$
$P_3 = \log 4 - \log 3 \approx 0.1249$
$P_4 = \log 5 - \log 4 \approx 0.0969$
$P_5 = \log 6 - \log 5 \approx 0.0792$
$P_6 = \log 7 - \log 6 \approx 0.0669$
$P_7 = \log 8 - \log 7 \approx 0.0580$
$P_8 = \log 9 - \log 8 \approx 0.0512$
$P_9 = \log 10 - \log 9 \approx 0.0458$
$P_1 + P_2 + P_3 + \cdots + P_8 + P_9 = 1$
This means that one of the digits $1, \ldots, 9$ will appear as the first digit.

4.2 Beyond the Basics

127. a. $h(x) = \log_3 x$ and $g(x) = \log_2 x$. So, $f(x) = g(h(x))$. The domain of $h(x)$ is $(0, \infty) \Rightarrow$ the domain of $f(x)$ is $(1, \infty)$.

b. $h(x) = \ln(x-1)$ and $g(x) = \log x$. So, $f(x) = g(h(x))$. The domain of $h(x)$ is $(1, \infty) \Rightarrow$ the domain of $f(x)$ is $(2, \infty)$.

c. $h(x) = \log(x-1)$ and $g(x) = \ln x$. So, $f(x) = g(h(x))$. The domain of $h(x)$ is $(1, \infty) \Rightarrow$ the domain of $f(x)$ is $(2, \infty)$.

d. $h(x) = \log(x-1)$ and $g(x) = \log x$. So, $f(x) = g(g(h(x)))$. The domain of $h(x)$ is $(1, \infty)$. The domain of $g(h(x))$ is $(2, \infty)$. (Note that $\log(x-1) = 0 \Rightarrow x = 2$.) So, the domain of $g(g(h(x)))$ is $(11, \infty)$.

129. $y = \log x \to y = \log(x-2) \to$
$y = \log\left(\frac{1}{2}x - 2\right) \to y = 3\log\left(\frac{1}{2}x - 2\right) \to$
$y = -3\log\left(\frac{1}{2}x - 2\right) \to y = -3\log\left(\frac{1}{2}x - 2\right) + 4$

131. a. $P = 100{,}000 e^{-0.07(20)} \approx \$24{,}659.70$

b. $50{,}000 = 75{,}000 e^{-10r} \Rightarrow \frac{2}{3} = e^{-10r} \Rightarrow$
$\ln(2/3) = -10r \Rightarrow r \approx 0.0405 = 4.05\%$

4.2 Critical Thinking/Discussion/Writing

133. $2^{\log_2 3} - 3^{\log_3 2} = 3 - 2 = 1$

134. $\log_3 4 = \log_3\left(2^2\right) = 2\log_3 2$
$\log_2 9 = \log_2\left(3^2\right) = 2\log_2 3$
Let $x = \log_3 2$ and let $y = \log_2 3$.
$\left(\log_3 4 + \log_2 9\right)^2 - \left(\log_3 4 - \log_2 9\right)^2$
$= (2x + 2y)^2 - (2x - 2y)^2$
$= \left[(2x + 2y) + (2x - 2y)\right] \cdot$
$\quad \left[(2x + 2y) - (2x - 2y)\right]$
$= (4x)(4y) = 16xy$
Note that
$x = \log_3 2 \Rightarrow 3^x = 2$ and $y = \log_2 3 \Rightarrow 2^y = 3$
$3^{xy} = \left(3^x\right)^y = 2^y = 3 \Rightarrow 3^{xy} = 3 \Rightarrow xy = 1$
Alternatively,
$xy = \log_3 2 \cdot \log_2 3 = \frac{\log 2}{\log 3} \cdot \frac{\log 3}{\log 2} = 1$.
Therefore, $16xy = 16 \cdot 1 = 16$, so
$\left(\log_3 4 + \log_2 9\right)^2 - \left(\log_3 4 - \log_2 9\right)^2 = 16$.

135. $\log_3\left[\log_4\left(\log_2 x\right)\right] = 0 \Rightarrow \log_4\left(\log_2 x\right) = 3^0 \Rightarrow$
$\log_4\left(\log_2 x\right) = 1 \Rightarrow \log_2 x = 4 \Rightarrow x = 2^4 = 16$

136. a. $f(x) = |\log x| = \begin{cases} -\log x & \text{if } 0 < x < 1 \\ \log x & \text{if } x \geq 1 \end{cases}$

137. a. Yes, the statement is always true.

b. The increasing property is used.

138. There are four possibilities, $0 < a < 1$ with $c < 0$, $0 < a < 1$ with $c > 0$, $a > 1$ with $c < 0$, and $a > 1$ with $c > 0$. They are illustrated below.

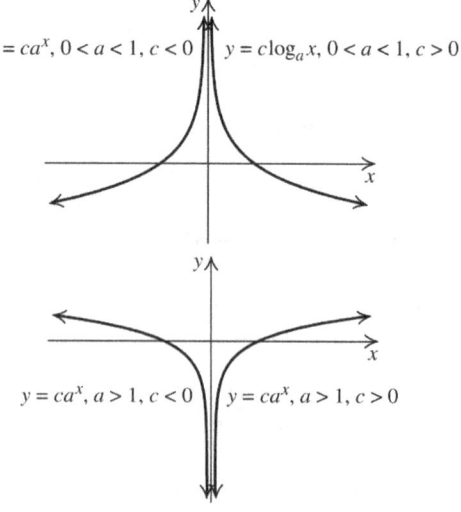

4.2 Getting Ready for the Next Sections

139. $a^2 \cdot a^7 = a^9$

141. $\sqrt{a^8} = \left(a^8\right)^{1/2} = a^{8 \cdot (1/2)} = a^4$

143. $\left(\dfrac{243}{32}\right)^{4/5} = \left[\left(\dfrac{243}{32}\right)^{1/5}\right]^4 = \left(\dfrac{3}{2}\right)^4$

145. $\left(4.7 \times 10^7\right)\left(8.1 \times 10^5\right) = (4.7 \times 8.1)\left(10^7 \times 10^5\right)$
$= 38.07 \times 10^{12}$
$= 3.807 \times 10^{13}$

147. $\log_3 81 = 4$ because $3^4 = 81$.
$\log_3 3 + \log_3 27 = 1 + 3 = 4$ because $\log_3 3 = 1$ and $3^3 = 27$. Therefore,
$\log_3 81 = \log_3 3 + \log_3 27$.

149. $\log_2 16 = 4$ because $2^4 = 16$.
$2\log_2 4 = 2 \cdot 2 = 4$ because $2^2 = 4$.
Therefore, $\log_2 16 = 2\log_2 4$.

4.3 Rules of Logarithms

4.3 Practice Problems

1. Given $\log_5 z = 3$ and $\log_5 y = 2$

a. $\log_5 (y/z) = \log_5 y - \log_5 z = 2 - 3 = -1$

b. $\log_5 \left(y^2 z^3\right) = \log_5 \left(y^2\right) + \log_5 \left(z^3\right)$
$= 2\log_5 y + 3\log_5 z$
$= 2 \cdot 2 + 3 \cdot 3 = 13$

2. a. $\ln \dfrac{2x-1}{x+4} = \ln(2x-1) - \ln(x+4)$

b. $\log\sqrt{\dfrac{4xy}{z}} = \log\left(\dfrac{4xy}{z}\right)^{1/2} = \dfrac{1}{2}\log\left(\dfrac{4xy}{z}\right)$
$= \dfrac{1}{2}(\log 4xy - \log z)$
$= \dfrac{1}{2}(\log 4 + \log x + \log y - \log z)$
$= \dfrac{1}{2}\left(\log\left(2^2\right) + \log x + \log y - \log z\right)$
$= \log 2 + \dfrac{1}{2}\log x + \dfrac{1}{2}\log y - \dfrac{1}{2}\log z$

3. $\dfrac{1}{2}\left[\log(x+1) + \log(x-1)\right]$
$= \dfrac{1}{2}\left[\log\left((x+1)(x-1)\right)\right]$
$= \dfrac{1}{2}\log\left(x^2 - 1\right) = \log\left(x^2 - 1\right)^{1/2}$
$= \log\sqrt{x^2 - 1}$

4. $K = 234^{567}$
$\log K = \log\left(234^{567}\right)$
$\log K = 567\log 234 \approx 1343.345391$
Since $\log K$ lies between the integers 1343 and 1344, the number K requires 1344 digits to the left of the decimal point. By definition of the common logarithm, we have
$K = 10^{1343.345391} = 10^{0.345391} \times 10^{1343}$
$\approx 2.215088 \times 10^{1343}$

5. $\log_3 15 = \dfrac{\log 15}{\log 3} \approx 2.46497$

6. a., b., Substitute $(3, 3)$ and $(9, 1)$ in the equation $y = c + b\log x$ to obtain
$3 = c + b\log 3$ (1)
$1 = c + b\log 9$ (2)
Subtract equation (1) from equation (2) and solve the resulting equation for b.
$-2 = b\log 9 - b\log 3 = b(\log 9 - \log 3)$
$= b\log\dfrac{9}{3} = b\log 3$
$b = -\dfrac{2}{\log 3}$
Substitute the value for b into equation (1) and solve for c.
$3 = c + \left(-\dfrac{2}{\log 3}\right)\log 3 \Rightarrow 3 = c - 2 \Rightarrow c = 5$
Substituting the values for b and c into $y = c + b\log x$ gives
$y = 5 - \dfrac{2}{\log 3}(\log x) = 5 - 2\left(\dfrac{\log x}{\log 3}\right)$
$= 5 - 2\log_3 x$

7. $\ln\left(\dfrac{A(t)}{A_0}\right) = kt \Rightarrow \ln\left(\dfrac{66}{100}\right) = 15k \Rightarrow$

   $\ln(0.66) = 15k \Rightarrow k = \dfrac{\ln(0.66)}{15}$

   To find the half-life, we use the formula
   $h = -\dfrac{\ln 2}{k} = -\dfrac{\ln 2}{\ln(0.66)/15} \approx 25.$

   The half-life of strontium-90 is about 25 years.

8. King Tut died in 1346 B.C., so the object was made in 1540 B.C., so the time elapsed between when the object was made and 1960 is 1540 + 1960 = 3500 years. The decay function for carbon-14 in exponential form is $A(t) = A_0 e^{-0.0001216t}$. (See example 8 in the text.) Let x = the percent of the original amount of carbon-14 in the object remaining after t years. Then

   $xA_0 = A_0 e^{-0.0001216t} \Rightarrow x = e^{-0.0001216t}$.

   $t = 3500$, so

   $x = e^{-0.0001216(3500)} \approx 0.6534 \approx 65.34\%$.

4.3 Concepts and Vocabulary

1. $\log_a MN = \underline{\log_a M} + \underline{\log_a N}$.

3. $\log_a M^r = \underline{r \log_a M}$.

5. False. There is no rule for the logarithm of a sum. $\log_a u + \log_a v = \log_a(uv)$.

7. False. $\ln(2\sqrt{x}) = \ln 2 + \ln \sqrt{x} = \ln 2 + \dfrac{1}{2}\ln x$

4.3 Building Skills

9. $\log 6 = \log(2 \cdot 3) = \log 2 + \log 3$
   $= 0.3 + 0.48 = 0.78$

11. $\log 5 = \log\left(\dfrac{10}{2}\right) = \log 10 - \log 2 = 1 - 0.3 = 0.7$

13. $\log\left(\dfrac{2}{x}\right) = \log 2 - \log x = 0.3 - 2 = -1.7$

15. $\log(2x^2 y) = \log 2 + 2\log x + \log y$
    $= 0.3 + 2(2) + 3 = 7.3$

17. $\log \sqrt[3]{x^2 y^4} = \log(x^{2/3} y^{4/3}) = \dfrac{2}{3}\log x + \dfrac{4}{3}\log y$
    $= \dfrac{2}{3}(2) + \dfrac{4}{3}(3) = \dfrac{16}{3}$

19. $\log \sqrt[3]{48} = \log(2^4 \cdot 3)^{1/3} = \dfrac{4}{3}\log 2 + \dfrac{1}{3}\log 3$
    $= \dfrac{4}{3}(0.3) + \dfrac{1}{3}(0.48) = 0.56$

21. $\ln[x(x-1)] = \ln x + \ln(x-1)$

23. $\ln\dfrac{x+1}{(x-2)^2} = \ln(x+1) - \ln(x-2)^2$
    $= \ln(x+1) - 2\ln(x-2)$

25. $\log_a \sqrt{x} y^3 = \log_a \sqrt{x} + \log_a y^3$
    $= \log_a x^{1/2} + \log_a y^3$
    $= \dfrac{1}{2}\log_a x + 3\log_a y$

27. $\log_a \sqrt[3]{\dfrac{x}{y}} = \log_a \left(\dfrac{x}{y}\right)^{1/3} = \dfrac{1}{3}\log_a \left(\dfrac{x}{y}\right)$
    $= \dfrac{1}{3}\log_a x - \dfrac{1}{3}\log_a y$

29. $\log_2 \sqrt[4]{\dfrac{xy^2}{8}} = \log_2 \left(\dfrac{xy^2}{8}\right)^{1/4} = \dfrac{1}{4}\log_2 \left(\dfrac{xy^2}{8}\right)$
    $= \dfrac{1}{4}\log_2(xy^2) - \dfrac{1}{4}\log_2 8$
    $= \dfrac{1}{4}\log_2 x + \dfrac{1}{4}\log_2 y^2 - \dfrac{1}{4}(3)$
    $= \dfrac{1}{4}\log_2 x + 2 \cdot \dfrac{1}{4}\log_2 y - \dfrac{3}{4}$
    $= \dfrac{1}{4}\log_2 x + \dfrac{1}{2}\log_2 y - \dfrac{3}{4}$

31. $\log \dfrac{\sqrt{x^2+1}}{x+3} = \log \sqrt{x^2+1} - \log(x+3)$
    $= \dfrac{1}{2}\log(x^2+1) - \log(x+3)$

33. $\log_3 \dfrac{(x-1)(x+1)}{x^2-4} = \log_3\left[(x-1)(x+1)\right] - \log_3\left(x^2-4\right)$
$= \log_3(x-1) + \log_3(x+1) - \log_3\left[(x-2)(x+2)\right]$
$= \log_3(x-1) + \log_3(x+1) - \left(\log_3(x-2) + \log_3(x+2)\right)$
$= \log_3(x-1) + \log_3(x+1) - \log_3(x-2) - \log_3(x+2)$

35. $\log_b x^2 y^3 z = 2\log_b x + 3\log_b y + \log_b z$

37. $\ln\left(\dfrac{x\sqrt{x-1}}{x^2+2}\right) = \ln\left(x\sqrt{x-1}\right) - \ln\left(x^2+2\right) = \ln x + \ln\left((x-1)^{1/2}\right) - \ln\left(x^2+2\right) = \ln x + \dfrac{1}{2}\ln(x-1) - \ln\left(x^2+2\right)$

39. $\ln\left(\dfrac{(x+1)^2}{(x-3)\sqrt{x+4}}\right) = \ln\left((x+1)^2\right) - \ln\left((x-3)\sqrt{x+4}\right) = 2\ln(x+1) - \left(\ln(x-3) + \ln(x+4)^{1/2}\right)$
$= 2\ln(x+1) - \ln(x-3) - \dfrac{1}{2}\ln(x+4)$

41. $\ln\left((x+1)\sqrt{\dfrac{x^2+2}{x^2+5}}\right) = \ln(x+1) + \ln\left(\dfrac{x^2+2}{x^2+5}\right)^{1/2} = \ln(x+1) + \dfrac{1}{2}\left(\ln\left(x^2+2\right) - \ln\left(x^2+5\right)\right)$
$= \ln(x+1) + \dfrac{1}{2}\ln\left(x^2+2\right) - \dfrac{1}{2}\ln\left(x^2+5\right)$

43. $\ln\left(\dfrac{x^3(3x+1)^4}{\sqrt{x^2+1}(x+2)^{-5}(x-3)^2}\right) = \ln\left(\dfrac{x^3(3x+1)^4(x+2)^5}{(x^2+1)^{1/2}(x-3)^2}\right)$
$= \ln\left[x^3(3x+1)^4(x+2)^5\right] - \ln\left[(x^2+1)^{1/2}(x-3)^2\right]$
$= 3\ln x + 4\ln(3x+1) + 5\ln(x+2) - \dfrac{1}{2}\ln\left(x^2+1\right) - 2\ln(x-3)$

45. $\log_2 x + \log_2 7 = \log_2(7x)$

47. $\log(3x+2) - \log x = \log\dfrac{3x+2}{x}$

49. $\dfrac{1}{2}\ln x + 2\ln y = \ln x^{1/2} + \ln y^2 = \ln\sqrt{x} + \ln y^2 = \ln y^2\sqrt{x}$

51. $\dfrac{1}{2}\log x - \log y + \log z = \log\left(\dfrac{z\sqrt{x}}{y}\right)$

53. $\dfrac{1}{5}(\log_2 z + 2\log_2 y) = \log_2\left(y^2 z\right)^{1/5} = \log_2 \sqrt[5]{y^2 z}$

55. $\ln x + 2\ln y + 3\ln z = \ln(xy^2 z^3)$

57. $3\log x - \log y + \dfrac{1}{2}\log\left(x^2+4\right) = \log x^3 - \log y + \log\left(x^2+4\right)^{1/2} = \log x^3 - \log y + \log\sqrt{x^2+4}$
$= \log\dfrac{x^3\sqrt{x^2+4}}{y}$

59. $2\ln x - \dfrac{1}{2}\ln(x^2+1) = \ln x^2 - \ln\sqrt{x^2+1}$

$= \ln\left(\dfrac{x^2}{\sqrt{x^2+1}}\right)$

61. $K = e^{500}$

$\log K = \log\left(e^{500}\right) = 500\log e = 217.147241 \Rightarrow$

$K = 10^{217.147241} = 10^{0.147241} \times 10^{217}$

$\approx 1.4036 \times 10^{217}$

63. $K = 324^{756}$

$\log K = \log\left(324^{756}\right) = 756\log 324$

$= 1897.972028 \Rightarrow$

$K = 10^{1897.972028} = 10^{0.972028} \times 10^{1897}$

$\approx 9.3762 \times 10^{1897}$

65. $K = 234^{567}$

$\log K = \log\left(234^{567}\right) = 567\log 234$

$= 1343.345$

$M = 567^{234}$

$\log M = \log\left(567^{234}\right) = 234\log 567$

$= 664.338$

Since $\log K > \log M$, $K > M$. Thus, $234^{567} > 567^{234}$.

67. $K = 17^{200} \cdot 53^{67}$

$\log K = \log\left(17^{200} \cdot 53^{67}\right)$

$= \log\left(17^{200}\right) + \log\left(53^{67}\right)$

$= 200\log 17 + 67\log 53$

$= 361.6163$

There are 362 digits in the given product.

69. $\log_2 5 = \dfrac{\log 5}{\log 2} \approx 2.322$

71. $\log_{1/2} 3 = \dfrac{\log 3}{\log\frac{1}{2}} \approx -1.585$

73. $\log_{\sqrt{5}}\sqrt{17} = \dfrac{\log\sqrt{17}}{\log\sqrt{5}} \approx 1.760$

75. $\log_2 7 + \log_4 3 = \dfrac{\log 7}{\log 2} + \dfrac{\log 3}{\log 4} \approx 3.6$

77. $\log_3 \sqrt{3} = \log_3 3^{1/2} = \dfrac{1}{2}$

79. $\log_3(\log_2 8) = \log_3(\log_2 2^3) = \log_3 3 = 1$

81. $5^{2\log_5 3 + \log_5 2} = 5^{\log_5 3^2 + \log_5 2} = 5^{\log_5(9\cdot 2)}$

$= 5^{\log_5 18} = 18$

83. $\log 4 + 2\log 5 = \log(4 \cdot 5^2) = \log 100$

$= \log 10^2 = 2$

85. Substitute (10, 1) and (1, 2) in the equation $y = c + b\log x$ to obtain

$1 = c + b\log 10$ (1)
$2 = c + b\log 1$ (2)

Subtract equation (1) from equation (2) and solve the resulting equation for b.

$1 = b\log 1 - b\log 10 = b\cdot 0 - b\cdot 1 = -b \Rightarrow$
$b = -1$

Substitute the value for b into equation (1) and solve for c.

$1 = c - \log 10 \Rightarrow 1 = c - 1 \Rightarrow c = 2$

Substituting the values for b and c into $y = c + b\log x$ gives $y = 2 - \log x$.

87. Substitute $(e, 1)$ and $(1, 2)$ in the equation $y = c + b\log x$ to obtain

$1 = c + b\log e$ (1)
$2 = c + b\log 1$ (2)

Since $\log 1 = 0$, equation (2) becomes $c = 2$. Substitute the value for c into equation (1) and solve for b.

$1 = 2 + b\log e \Rightarrow -1 = b\log e \Rightarrow b = -\dfrac{1}{\log e}$

Substituting the values for b and c into $y = c + b\log x$ gives

$y = 2 - \left(\dfrac{1}{\log e}\right)\log x = 2 - \left(\dfrac{\log x}{\log e}\right) = 2 - \ln x.$

89. Substitute (5, 4) and (25, 7) in the equation $y = c + b\log x$ to obtain

$4 = c + b\log 5$ (1)
$7 = c + b\log 25$ (2)

Subtract equation (1) from equation (2) and solve the resulting equation for b.

$3 = b\log 25 - b\log 5 = b(\log 25 - \log 5)$

$= b\log\left(\dfrac{25}{5}\right) = b\log 5 \Rightarrow$

$b = \dfrac{3}{\log 5}$

Substitute the value for b into equation (1) and solve for c.

$4 = c + \left(\dfrac{3}{\log 5}\right)\log 5 \Rightarrow 4 = c + 3 \Rightarrow c = 1$

(*continued on next page*)

Section 4.3 Rules of Logarithms 259

(continued)

Substituting the values for b and c into
$y = c + b \log x$ gives

$$y = 1 + \left(\frac{3}{\log 5}\right)\log x = 1 + 3\left(\frac{\log x}{\log 5}\right)$$
$$= 1 + 3\log_5 x.$$

91. Substitute (2, 4) and (4, 9) in the equation
$y = c + b \log x$ to obtain
$4 = c + b \log 2$ (1)
$9 = c + b \log 4$ (2)
Subtract equation (1) from equation (2) and solve the resulting equation for b.
$5 = b \log 4 - b \log 2 = b(\log 4 - \log 2)$
$= b \log\left(\frac{4}{2}\right) \Rightarrow 5 = b \log 2 \Rightarrow b = \frac{5}{\log 2}$

Substitute the value for b into equation (1) and solve for c.
$4 = c + \left(\frac{5}{\log 2}\right)\log 2 = c + 5 \Rightarrow -1 = c$

Substituting the values for b and c into $y = c + b \log x$ gives
$$y = -1 + \left(\frac{5}{\log 2}\right)\log x = -1 + 5\left(\frac{\log x}{\log 2}\right)$$
$$= -1 + 5 \log_2 x.$$

93. $A(t) = A_0 e^{kt} \Rightarrow 23 = 50 e^{12k} \Rightarrow$
$0.46 = e^{12k} \Rightarrow \ln(0.46) = 12k \Rightarrow$
$k = \frac{\ln(0.46)}{12}$
To find the half-life, we use the formula
$h = -\frac{\ln 2}{k} = -\frac{\ln 2}{\ln(0.46)/12} \approx 10.7.$
The half-life is about 10.7 years.

95. $A(t) = A_0 e^{kt} \Rightarrow 3.8 = 10.3 e^{15k} \Rightarrow$
$\frac{3.8}{10.3} = e^{15k} \Rightarrow \ln\left(\frac{3.8}{10.3}\right) = 15k \Rightarrow$
$k = \frac{\ln\left(\frac{3.8}{10.3}\right)}{15}$
To find the half-life, we use the formula
$h = -\frac{\ln 2}{k} = -\frac{\ln 2}{\ln(3.8/10.3)/15} \approx 10.4.$
The half-life is about 10.4 hours.

4.3 Applying the Concepts

97. $7.09 = 7e^{1k} \Rightarrow$
$\ln\frac{7.09}{7} = k \approx 0.012775 = 1.2775\%$

99. $7e^{100k} < 20 \Rightarrow 100k < \ln\left(\frac{20}{7}\right) \Rightarrow$
$0.010498 = 1.0498\% < k$
The maximum rate of growth is 1.0498%.

101. $3500 = 1000 e^{0.1t} \Rightarrow \ln 3.5 = 0.1t \Rightarrow$
$t \approx 12.53$ years

103. Because the half-life is 8 days, there will be 10 grams left after 8 days. Use this to find k:
$10 = 20 e^{8k} \Rightarrow \frac{1}{2} = e^{8k} \Rightarrow \ln\left(\frac{1}{2}\right) = 8k \Rightarrow$
$k \approx -0.08664$
$A = 20 e^{-0.08664(5)} \approx 12.969$ grams

105. $\frac{1}{2} = e^{-0.055t} \Rightarrow \ln\left(\frac{1}{2}\right) = -0.055t \Rightarrow t \approx 12.6$ yr.

107. $1.5 = 5e^{-10k} \Rightarrow 0.3 = e^{-10k} \Rightarrow \ln 0.3 = -10k \Rightarrow$
$k \approx 0.1204$

109. Find k using $A = 8, A_0 = 16,$ and $t = 36$:
$8 = 16 e^{-36k} \Rightarrow \frac{1}{2} = e^{-36k} \Rightarrow$
$\ln\left(\frac{1}{2}\right) = 36k \Rightarrow k \approx 0.0193$
$A = 16 e^{-0.0193(8)} \Rightarrow A \approx 13.7$ grams.

In exercises 111–114, we use the formula
$$P = \frac{r \cdot M}{1 - \left(1 + \frac{r}{n}\right)^{-nt}} \div n,$$

where P = the payment, r = the annual interest rate, M = the mortgage amount, t = the number of years, and n = the number of payments per year.

111. $P = \dfrac{0.06 \cdot 120{,}000}{1 - \left(1 + \frac{0.06}{12}\right)^{-12 \cdot 20}} \div 12 = 859.72$

The monthly payment is $859.72.
There are 240 payments so the total amount paid is $240 \cdot \$859.72 = \$206{,}332.80$.
The amount of interest paid is
$206{,}332.80 - 120{,}000 = \$86{,}332.80$.

113. $850 = \dfrac{0.085 \cdot M}{1 - \left(1 + \frac{0.085}{12}\right)^{-12 \cdot 30}} \div 12$

$10{,}200 = \dfrac{0.085 \cdot M}{1 - \left(1 + \frac{0.085}{12}\right)^{-12 \cdot 30}}$

(continued on next page)

(continued)

$$M = \frac{10,200\left[1-\left(1+\frac{0.085}{12}\right)^{-12\cdot 30}\right]}{0.085}$$
$$\approx 110,545.60$$

Andy can afford a mortgage of about $110,545.60.

4.3 Beyond the Basics

115. $\log_b\left(\sqrt{x^2+1}-x\right)+\log_b\left(\sqrt{x^2+1}+x\right)$
$$= \log_b\left(\left(\sqrt{x^2+1}-x\right)\left(\sqrt{x^2+1}+x\right)\right)$$
$$= \log_b\left(x^2+1-x^2\right) = \log_b 1 = 0$$

117. $(\log_b a)(\log_a b) = \dfrac{\log a}{\log b} \cdot \dfrac{\log b}{\log a} = 1$

119.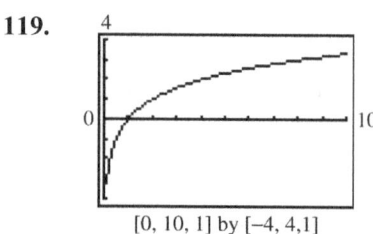

[0, 10, 1] by [−4, 4, 1]

121. a. $\log\left(\dfrac{a}{b}\right)+\log\left(\dfrac{b}{a}\right)+\log\left(\dfrac{c}{a}\right)+\log\left(\dfrac{a}{c}\right)=0$
(See exercise 118.)

b. $\log\left(\dfrac{a^2}{bc}\right)+\log\left(\dfrac{b^2}{ca}\right)+\log\left(\dfrac{c^2}{ab}\right)$
$= 2\log a - (\log b + \log c)$
$\quad + 2\log b - (\log c + \log a)$
$\quad + 2\log c - (\log a + \log b) = 0$

c. $\log_2 3 \cdot \log_3 4 = \dfrac{\log 3}{\log 2} \cdot \dfrac{\log 2^2}{\log 3}$
$= \dfrac{\log 3}{\log 2} \cdot \dfrac{2\log 2}{\log 3} = 2$

d. $\log_a b \cdot \log_b c \cdot \log_c a$
$= \dfrac{\log b}{\log a} \cdot \dfrac{\log c}{\log b} \cdot \dfrac{\log a}{\log c} = 1$

123. $f(x) = \log_4\left(\log_5\left(\log_3\left(18x-x^2-77\right)\right)\right)$
$18x - x^2 - 77 = -\left(x^2 - 18x + 77\right)$
$\quad = -(x-11)(x-7)$

$\log_3\left(18x-x^2-77\right)$ is defined only for those values of x which make $18x-x^2-77 > 0$.
Thus, the domain of $\log_3\left(18x-x^2-77\right)$ is $(7, 11)$.

$\log_5\left[\log_3\left(18x-x^2-77\right)\right]$ is defined only for those values of x in the interval $(7, 11)$ which make $\log_3\left(18x-x^2-77\right) > 0$. We use a graphing calculator to solve this. Note that we use the change of base formula to define the function.

$$Y_1 = \log_3\left(18x-x^2-77\right) = \frac{\log\left(18x-x^2-77\right)}{\log 3}.$$

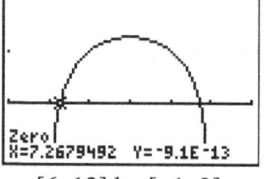 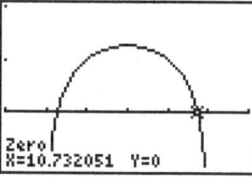

[6, 12] by [−1, 2]

Thus, the domain of $\log_5\left[\log_3\left(18x-x^2-77\right)\right]$ is approximately $(7.27, 10.73)$.

$\log_4\left(\log_5\left(\log_3\left(18x-x^2-77\right)\right)\right)$ is defined only for those values of x in the interval $(7.27, 10.73)$ which make
$\log_5\left[\log_3\left(18x-x^2-77\right)\right] > 0$.

We use a graphing calculator to solve this. Note that we use the change of base formula to define the function.

$$Y_2 = \log_5\left(\log_3\left(18x-x^2-77\right)\right)$$
$$= \log_5(Y_1) = \frac{Y_1}{\log 5}$$

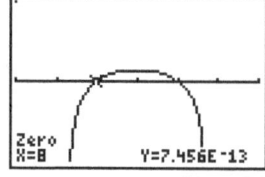 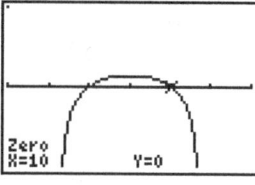

[6, 12] by [−1, 1]

Thus, the domain of
$\log_4\left(\log_5\left(\log_3\left(18x-x^2-77\right)\right)\right)$ is $(8, 10)$.

125.
a. False b. True
c. False d. False
e. True   f. True
g. False h. True
i. False j. True

127.
$$\log\left(\frac{a+b}{3}\right) = \frac{1}{2}(\log a + \log b)$$
$$2\log\left(\frac{a+b}{3}\right) = \log(ab)$$
$$2\log(a+b) - 2\log 3 = \log(ab)$$
$$\log(a+b)^2 - \log 9 = \log(ab)$$
$$\log(a+b)^2 = \log(ab) + \log 9$$
$$\log(a+b)^2 = \log(9ab)$$
$$(a+b)^2 = 9ab$$
$$a^2 + 2ab + b^2 = 9ab \Rightarrow a^2 + b^2 - 7ab = 0$$

4.3 Critical Thinking/Discussion/Writing

129. In step 2, $\log\left(\frac{1}{2}\right)$ is negative, so $3 < 4 \Rightarrow$
$3\log\left(\frac{1}{2}\right) > 4\log\left(\frac{1}{2}\right)$.

130. The domain of $2\log x$ is $(0, \infty)$, while the domain of $\log(x^2)$ is $(-\infty, 0) \cup (0, \infty)$.

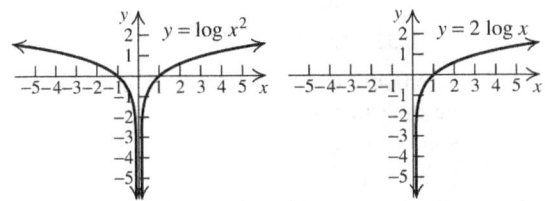

131. If $p = 2^m - 1$ is a prime number, then the only way p and 2^m have a different number of digits is if $2^m = 10^k$ or $2^{n-k} = 5^k$. However, this is impossible because 2^{n-k} is even and 5^k is odd. Thus, p and 2^m have the same number of digits.
$$K = 2^{43112609}$$
$$\log K = \log\left(2^{43112609}\right) = 43112609 \log 2$$
$$\approx 12978188.5$$
There are 12,978,189 digits in this prime number.

132. $\dfrac{1}{\log(1-x)}$ is defined only for those values of $\log(1-x) \neq 0$ and for those values of x such that $\log(1-x)$ is defined.
$\log(1-x) = 0 \Rightarrow 1 - x = 1 \Rightarrow x = 0$
$\log(1-x)$ is defined for
$1 - x > 0 \Rightarrow 1 > x$ or $x < 1$.
Thus, the domain of $\dfrac{1}{\log(1-x)}$ is
$(-\infty, 0) \cup (0, 1)$.

$\dfrac{1}{\ln(x+2)}$ is defined only for those values of $\ln(x+2) \neq 0$ and for those values of x such that $\ln(x+2)$ is defined.
$\ln(x+2) = 0 \Rightarrow x + 2 = 1 \Rightarrow x = -1$
$\ln(x+2)$ is defined for
$x + 2 > 0 \Rightarrow x > -2$ or $-2 < x$.
Thus, the domain of $\dfrac{1}{\ln(x+2)}$ is
$(-2, -1) \cup (-1, \infty)$.
The intersection of the two domains is the domain of $\dfrac{1}{\log(1-x)} + \dfrac{1}{\ln(x+2)}$. Thus, the domain is $(-2, -1) \cup (-1, 0) \cup (0, 1)$.

4.3 Getting Ready for the Next Section

133. $11 \cdot 3^0 = 11 \cdot 1 = 11$

135. $4^x \cdot 2^{-2x+1} = \left(2^2\right)^x \cdot 2^{-2x+1} = 2^{2x} \cdot 2^{-2x+1}$
$= 2^{2x+(-2x+1)} = 2^1 = 2$

For exercises 137–140, let $t = 5^x$. Then $5^{2x} = t^2$.

137. $5^{2x} - 5^x = -1 \Rightarrow t^2 - t = -1 \Rightarrow t^2 - t + 1 = 0$

139. $\dfrac{5^x + 3 \cdot 5^{-x}}{5^x} = \dfrac{1}{4} \Rightarrow \dfrac{t + \frac{3}{t}}{t} = \dfrac{1}{4} \Rightarrow \dfrac{\frac{t^2+3}{t}}{t} = \dfrac{1}{4} \Rightarrow$
$\dfrac{t^2 + 3}{t^2} = \dfrac{1}{4} \Rightarrow 4t^2 + 12 = t^2 \Rightarrow 3t^2 + 12 = 0 \Rightarrow$
$t^2 + 4 = 0$

141. $2x - (11 + x) = 8x + (7 + 2x)$
$x - 11 = 10x + 7$
$-18 = 9x \Rightarrow x = -2$
Solution set: $\{-2\}$

143. $x^2 + 3x - 1 = 3$
$x^2 + 3x - 4 = 0$
$(x-1)(x+4) = 0$
$x - 1 = 0 \mid x + 4 = 0$
$x = 1 \mid x = -4$
Solution set: $\{-4, 1\}$

145. $2x < 7 + x \Rightarrow x < 7$
Solution set: $(-\infty, 7)$

147. $12x > 30 - 3x$
$15x > 30 \Rightarrow x > 2$
Solution set: $(2, \infty)$

4.4 Exponential and Logarithmic Equations and Inequalities

4.4 Practice Problems

1. a. $3^x = 243 \Rightarrow 3^x = 3^5 \Rightarrow x = 5$

b. $8^x = 4 \Rightarrow (2^3)^x = 2^2 \Rightarrow 2^{3x} = 2^2 \Rightarrow$
$3x = 2 \Rightarrow x = \dfrac{2}{3}$

2. $7 \cdot 3^{x+1} = 11 \Rightarrow 3^{x+1} = \dfrac{11}{7} \Rightarrow$
$\ln(3^{x+1}) = \ln\left(\dfrac{11}{7}\right) \Rightarrow (x+1)\ln 3 = \ln\left(\dfrac{11}{7}\right) \Rightarrow$
$x + 1 = \dfrac{\ln(11/7)}{\ln 3} \Rightarrow x = \dfrac{\ln(11/7)}{\ln 3} - 1 \approx -0.589$

3. $3^{x+1} = 2^{2x}$
$\ln(3^{x+1}) = \ln(2^{2x})$
$(x+1)\ln 3 = 2x \ln 2$
$x \ln 3 + \ln 3 = 2x \ln 2$
$x \ln 3 - 2x \ln 2 = -\ln 3$
$x(\ln 3 - 2 \ln 2) = -\ln 3$
$x = -\dfrac{\ln 3}{\ln 3 - 2\ln 2} \approx 3.819$

4. $e^{2x} - 4e^x - 5 = 0$
$(e^x - 5)(e^x + 1) = 0$
$e^x - 5 = 0$ or $e^x + 1 = 0$
$e^x - 5 = 0 \Rightarrow e^x = 5 \Rightarrow \ln(e^x) = \ln 5 \Rightarrow$
$x \ln e = \ln 5 \Rightarrow x = \ln 5 \approx 1.609$
$e^x + 1 = 0 \Rightarrow e^x = -1$, which is not possible.

5. The given model is $P(t) = P_0(1+r)^t$.

a. In 2020, ten years after the base year, the population of the United States will be
$P(10) = 308(1 + 0.011)^{10} \approx 343.61$ million.
The population of Pakistan will be
$P(10) = 185(1 + 0.033)^{10} \approx 255.96$ million.

b. To find when the population of the U.S. will be 350 million, solve for t:
$350 = 308(1 + 0.011)^t \Rightarrow \dfrac{350}{308} = 1.011^t \Rightarrow$
$\ln\left(\dfrac{350}{308}\right) = t \ln 1.011 \Rightarrow$
$t = \dfrac{\ln(350/308)}{\ln 1.011} \approx 11.69$
The population of the U.S. will be 350 million approximately 11.69 years after 2010, sometime in the year 2022.

c. To find when the population of the two countries will be the same, solve for t:
$308(1+.011)^t = 185(1+.033)^t$
$308(1.011)^t = 185(1.033)^t$
$\dfrac{308}{185} = \dfrac{1.033^t}{1.011^t} = \left(\dfrac{1.033}{1.011}\right)^t$
$\ln\left(\dfrac{308}{185}\right) = t \ln\left(\dfrac{1.033}{1.011}\right)$
$\dfrac{\ln(308/185)}{\ln(1.033/1.011)} \approx 23.68 = t$
The populations of the two countries will be the same about 23.68 years after 2010, sometime in the year 2033.

6. $1 + 2\ln x = 4 \Rightarrow 2\ln x = 3 \Rightarrow \ln x = \dfrac{3}{2} \Rightarrow x = e^{3/2}$

7. $\log_3(x-8) + \log_3 x = 2$
$\log_3[x(x-8)] = 2$
$x^2 - 8x = 3^2$
$x^2 - 8x - 9 = 0$
$(x-9)(x+1) = 0$
$x - 9 = 0$ or $x + 1 = 0$
$x = 9 \qquad x = -1$
Now check each possible solution in the original equation.
$\log_3(9-8) + \log_3 9 \stackrel{?}{=} 2$
$\log_3 1 + \log_3 9 \stackrel{?}{=} 2$
$0 + 2 = 2$ ✓

(continued on next page)

(*continued*)

$\log_3(-1-8) + \log_3(-1) \stackrel{?}{=} 2$

This is not possible because logarithms are not defined for negative values.
The solution set is {9}.

8. $\ln(x+5) + \ln(x+1) = \ln(x-1)$
    $\ln[(x+5)(x+1)] = \ln(x-1)$
    $\ln(x^2 + 6x + 5) = \ln(x-1)$
    $x^2 + 6x + 5 = x - 1$
    $x^2 + 5x + 6 = 0$
    $(x+2)(x+3) = 0$
    $x + 2 = 0 \Rightarrow x = -2$ or $x + 3 = 0 \Rightarrow x = -3$
    Now check each possible solution in the original equation.
    $\ln(-2+5) + \ln(-2+1) \stackrel{?}{=} \ln(-2-1)$
    $\ln 3 + \ln(-1) = \ln(-3)$
    This is not possible because logarithms are not defined for negative values.
    $\ln(-3+5) + \ln(-3+1) \stackrel{?}{=} \ln(-3-1)$
    $\ln 2 + \ln(-2) = \ln(-4)$
    This is not possible because logarithms are not defined for negative values.
    The solution set is $\emptyset$.

9. The year 1987 represents $t = 0$, so the year 2005 represents $t = 18$. Using equation (19) in example 9 in the text, we have
    $6.5 = \dfrac{35}{1 + 6e^{-18k}} \Rightarrow 6.5 + 39e^{-18k} = 35 \Rightarrow$
    $e^{-18k} = \dfrac{28.5}{39} \Rightarrow -18k = \ln\left(\dfrac{28.5}{39}\right) \Rightarrow$
    $k = -\dfrac{\ln(28.5/39)}{18} \approx 0.0174 = 1.74\%$
    The growth rate was about 1.74%.

10. $3(0.5)^x + 7 > 19 \Rightarrow 3(0.5)^x > 12 \Rightarrow$
    $(0.5)^x > 4 \Rightarrow x \ln 0.5 > \ln 4 \Rightarrow x < \dfrac{\ln 4}{\ln 0.5} \Rightarrow$
    $x < -2$

11. Since $1 - 3x$ must be positive, $x < \dfrac{1}{3}$.
    $\ln(1 - 3x) > 2 \Rightarrow 1 - 3x > e^2 \Rightarrow$
    $-3x > e^2 - 1 \Rightarrow x < \dfrac{e^2 - 1}{-3} \Rightarrow$
    $x < \dfrac{1 - e^2}{3}$ or $x < -2.13$

4.4 Concepts and Vocabulary

1. An equation that contains terms of the form a^x is called a(n) <u>exponential</u> equation.

3. The equation $y = \dfrac{m}{1 + ae^{-bx}}$ represents a(n) <u>logistic</u> model.

5. True

7. True.
    $8^{2x} = 4^{3x} \Rightarrow (2^3)^{2x} = (2^2)^{3x} \Rightarrow 2^{6x} = 2^{6x}$

4.4 Building Skills

9. $2^x = 16 \Rightarrow 2^x = 2^4 \Rightarrow x = 4$

11. $8^x = 32 \Rightarrow 2^{3x} = 2^5 \Rightarrow 3x = 5 \Rightarrow x = \dfrac{5}{3}$

13. $4^{|x|} = 128 \Rightarrow 2^{2|x|} = 2^7 \Rightarrow 2|x| = 7 \Rightarrow x = \pm\dfrac{7}{2}$

15. $5^{-|x|} = 625 \Rightarrow 5^{-|x|} = 5^4 \Rightarrow -|x| = 4 \Rightarrow$ there is no solution.

17. $\ln x = 0 \Rightarrow x = 1$

19. $\log_2 x = -1 \Rightarrow 2^{-1} = x \Rightarrow x = \dfrac{1}{2}$

21. $\log_3 |x| = 2 \Rightarrow 3^2 = |x| \Rightarrow x = \pm 9$

23. $\dfrac{1}{2}\log x - 2 = 0 \Rightarrow \dfrac{1}{2}\log x = 2 \Rightarrow \log x = 4 \Rightarrow$
    $10^4 = 10{,}000 = x$

In exercises 25–60, the equations can be solved using either the common logarithm or the natural logarithm.

25. $2^x = 3 \Rightarrow x \ln 2 = \ln 3 \Rightarrow x = \dfrac{\ln 3}{\ln 2} \approx 1.585$

27. $2^{2x+3} = 15 \Rightarrow (2x+3) \ln 2 = \ln 15 \Rightarrow$
    $2x + 3 = \dfrac{\ln 15}{\ln 2} \Rightarrow$
    $x = \dfrac{\dfrac{\ln 15}{\ln 2} - 3}{2} = \dfrac{\ln 15 - 3 \ln 2}{2 \ln 2} \approx 0.453$

29. $e^{x+1} = 3 \Rightarrow (x+1) \ln e = \ln 3 \Rightarrow x + 1 = \ln 3 \Rightarrow$
    $x = \ln 3 - 1 \approx 0.099$

31. $5 \cdot 2^x - 7 = 10 \Rightarrow 2^x = \dfrac{17}{5} \Rightarrow$
$x \ln 2 = \ln\left(\dfrac{17}{5}\right) \Rightarrow x = \dfrac{\ln 17 - \ln 5}{\ln 2} \approx 1.766$

33. $3 \cdot 4^{2x-1} + 4 = 14 \Rightarrow 4^{2x-1} = \dfrac{10}{3} \Rightarrow$
$(2x - 1) \ln 4 = \ln\left(\dfrac{10}{3}\right) \Rightarrow$
$x = \dfrac{\dfrac{\ln 10 - \ln 3}{\ln 4} + 1}{2} = \dfrac{\ln 10 - \ln 3 + \ln 4}{2 \ln 4} \approx 0.934$

35. $2e^{x-2} + 3 = 7 \Rightarrow 2e^{x-2} = 4 \Rightarrow e^{x-2} = 2 \Rightarrow$
$(x - 2) \ln e = \ln 2 \Rightarrow x - 2 = \ln 2 \Rightarrow$
$x = \ln 2 + 2 \approx 2.693$

37. $5^{1-x} = 2^x \Rightarrow (1 - x) \ln 5 = x \ln 2 \Rightarrow$
$\ln 5 - x \ln 5 = x \ln 2 \Rightarrow \ln 5 = x(\ln 5 + \ln 2) \Rightarrow$
$x = \dfrac{\ln 5}{\ln 5 + \ln 2} \approx 0.699$

39. $2^{1-x} = 3^{4x+6} \Rightarrow (1 - x) \ln 2 = (4x + 6) \ln 3 \Rightarrow$
$\ln 2 - x \ln 2 = 4x \ln 3 + 6 \ln 3 \Rightarrow$
$\ln 2 - 6 \ln 3 = 4x \ln 3 + x \ln 2 \Rightarrow$
$\ln 2 - 6 \ln 3 = x(4 \ln 3 + \ln 2) \Rightarrow$
$\dfrac{\ln 2 - 6 \ln 3}{4 \ln 3 + \ln 2} = x \Rightarrow x \approx -1.159$

41. $\qquad 2 \cdot 3^{x-1} = 5^{x+1}$
$\ln\left(2 \cdot 3^{x-1}\right) = \ln\left(5^{x+1}\right)$
$\ln 2 + (x - 1) \ln 3 = (x + 1) \ln 5$
$\ln 2 + x \ln 3 - \ln 3 = x \ln 5 + \ln 5$
$x \ln 3 - x \ln 5 = \ln 5 - \ln 2 + \ln 3$
$x(\ln 3 - \ln 5) = \ln 5 - \ln 2 + \ln 3$
$x = \dfrac{\ln 5 - \ln 2 + \ln 3}{\ln 3 - \ln 5}$
$= \dfrac{\ln 2 - \ln 3 - \ln 5}{\ln 5 - \ln 3} \approx -3.944$

43. $1.065^t = 2 \Rightarrow t \ln 1.065 = \ln 2 \Rightarrow$
$t = \dfrac{\ln 2}{\ln 1.065} \approx 11.007$

45. Let $y = 2^x$. Then $2^{2x} - 4 \cdot 2^x = 21 \Rightarrow$
$y^2 - 4y - 21 = 0 \Rightarrow (y - 7)(y + 3) = 0 \Rightarrow$
$y = 7$ or $y = -3$. Reject the negative solution.
$2^x = 7 \Rightarrow x \ln 2 = \ln 7 \Rightarrow x = \dfrac{\ln 7}{\ln 2} \approx 2.807$

47. $9^x - 6 \cdot 3^x + 8 = 0 \Rightarrow 3^{2x} - 6 \cdot 3^x + 8 = 0$.
Let $y = 3^x$. Then $3^{2x} - 6 \cdot 3^x + 8 = 0 \Rightarrow$
$y^2 - 6y + 8 = 0 \Rightarrow (y - 4)(y - 2) = 0 \Rightarrow y = 2$
or $y = 4$. Substituting, we have $3^x = 2 \Rightarrow$
$x \ln 3 = \ln 2 \Rightarrow x = \dfrac{\ln 2}{\ln 3} \approx 0.631$ or $3^x = 4 \Rightarrow$
$x \ln 3 = \ln 4 \Rightarrow x = \dfrac{\ln 4}{\ln 3} \approx 1.262$.
The solution set is $\{0.631, 1.262\}$.

49. $\qquad 3^{3x} - 4 \cdot 3^{2x} + 2 \cdot 3^x = 8$
$3^{3x} - 4 \cdot 3^{2x} + 2 \cdot 3^x - 8 = 0$
$3^{2x}\left(3^x - 4\right) + 2\left(3^x - 4\right) = 0$
$\left(3^{2x} + 2\right)\left(3^x - 4\right) = 0$
$3^{2x} + 2 = 0 \Rightarrow 3^{2x} = -2 \Rightarrow$ there is no solution.
$3^x - 4 = 0 \Rightarrow 3^x = 4 \Rightarrow x \ln 3 = \ln 4 \Rightarrow$
$x = \dfrac{\ln 4}{\ln 3} \approx 1.262$
Solution set: $\left\{\dfrac{\ln 4}{\ln 3} \approx 1.262\right\}$

51. $e^{2x} - 2e^x - 3 = 0$
$\left(e^x - 3\right)\left(e^x + 1\right) = 0$
$e^x - 3 = 0 \Rightarrow e^x = 3 \Rightarrow x \ln e = \ln 3 \Rightarrow x = \ln 3$
$e^x + 1 = 0 \Rightarrow e^x = -1 \Rightarrow$ there is no solution.
Solution set: $\{\ln 3 \approx 1.099\}$

53. $\dfrac{3^x - 3^{-x}}{3^x + 3^{-x}} = \dfrac{1}{4} \Rightarrow 4 \cdot 3^x - 4 \cdot 3^{-x} = 3^x + 3^{-x} \Rightarrow$
$3 \cdot 3^x - 5 \cdot 3^{-x} = 0 \Rightarrow 3^x\left(3 \cdot 3^x - 5 \cdot 3^{-x}\right) = 0 \Rightarrow$
$3 \cdot 3^{2x} - 5 \cdot 3^0 = 0 \Rightarrow 3 \cdot 3^{2x} = 5 \Rightarrow 3^{2x} = \dfrac{5}{3} \Rightarrow$
$2x \ln 3 = \ln\left(\dfrac{5}{3}\right) \Rightarrow$
$x = \dfrac{\ln(5/3)}{2 \ln 3} = \dfrac{\ln 5 - \ln 3}{2 \ln 3} \approx 0.232$

55. $\dfrac{4}{2 + 3^x} = 1 \Rightarrow 4 = 2 + 3^x \Rightarrow 2 = 3^x \Rightarrow$
$\ln 2 = x \ln 3 \Rightarrow x = \dfrac{\ln 2}{\ln 3} \approx 0.631$

57. $\dfrac{17}{5 - 3^x} = 7 \Rightarrow 17 = 35 - 7 \cdot 3^x \Rightarrow$
$7 \cdot 3^x = 18 \Rightarrow \ln 7 + x \ln 3 = \ln 18 \Rightarrow$
$x = \dfrac{\ln 18 - \ln 7}{\ln 3} \approx 0.860$

59. $\dfrac{5}{2+3^x} = 4 \Rightarrow 5 = 4(2+3^x) \Rightarrow$

$5 = 8 + 4 \cdot 3^x \Rightarrow -\dfrac{3}{4} = 3^x \Rightarrow$ there is no solution.

61. $3 + \log(2x+5) = 2 \Rightarrow \log(2x+5) = -1 \Rightarrow$
$10^{-1} = 2x+5 \Rightarrow -\dfrac{49}{10} = 2x \Rightarrow x = -\dfrac{49}{20}$

63. $\log\left(x^2 - x - 5\right) = 0 \Rightarrow x^2 - x - 5 = 10^0 \Rightarrow$
$x^2 - x - 6 = 0 \Rightarrow (x-3)(x+2) = 0 \Rightarrow x = -2$ or $x = 3$

65. $\log_4\left(x^2 - 7x + 14\right) = 1 \Rightarrow x^2 - 7x + 14 = 4^1 \Rightarrow$
$x^2 - 7x + 10 = (x-2)(x-5) \Rightarrow x = 2$ or $x = 5$

67. $\ln(2x-3) - \ln(x+5) = 0 \Rightarrow$
$\ln(2x-3) = \ln(x+5) \Rightarrow 2x - 3 = x + 5 \Rightarrow x = 8$

69. $\log x + \log(x+9) = 1 \Rightarrow \log(x(x+9)) = 1 \Rightarrow$
$x^2 + 9x = 10 \Rightarrow x^2 + 9x - 10 = 0 \Rightarrow$
$(x+10)(x-1) = 0 \Rightarrow x = -10$ or $x = 1$.
Reject the negative solution because the logarithm of a negative number is undefined. The solution is $\{1\}$.

71. $\log_a(5x-2) - \log_a(3x+4) = 0 \Rightarrow$
$\log_a \dfrac{5x-2}{3x+4} = 0 \Rightarrow \dfrac{5x-2}{3x+4} = a^0 = 1 \Rightarrow$
$5x - 2 = 3x + 4 \Rightarrow x = 3$

73. $\log_6(x+2) + \log_6(x-3) = 1 \Rightarrow$
$\log_6\left((x+2)(x-3)\right) = 1 \Rightarrow x^2 - x - 6 = 6 \Rightarrow$
$x^2 - x - 12 = 0 \Rightarrow (x-4)(x+3) = 0 \Rightarrow$
$x = 4$ or $x = -3$.
Reject the negative solution because the logarithm of a negative number is undefined. The solution is $\{4\}$.

75. $\log_3(2x-7) - \log_3(4x-1) = 2 \Rightarrow$
$\log_3 \dfrac{2x-7}{4x-1} = 2 \Rightarrow \dfrac{2x-7}{4x-1} = 3^2 = 9 \Rightarrow$
$2x - 7 = 36x - 9 \Rightarrow 2 = 34x \Rightarrow x = \dfrac{1}{17}$.
$2\left(\dfrac{1}{17}\right) - 7 = -\dfrac{117}{17}$, so there is no solution.

77. $\log_7 3x + \log_7(2x-1) = \log_7(16x-10) \Rightarrow$
$\log_7\left(3x(2x-1)\right) = \log_7(16x-10) \Rightarrow$
$6x^2 - 3x = 16x - 10 \Rightarrow 6x^2 - 19x + 10 = 0 \Rightarrow$
$(2x-5)(3x-2) = 0 \Rightarrow x = \dfrac{5}{2}$ or $x = \dfrac{2}{3}$
The solution is $\{1\}$.

79. $f(x) = 20 + a \cdot 2^{kx};\ f(0) = 50;\ f(1) = 140$
$50 = 20 + a \cdot 2^{k \cdot 0} \Rightarrow 50 = 20 + a \Rightarrow a = 30$
$140 = 20 + 30 \cdot 2^{k \cdot 1} \Rightarrow 4 = 2^k \Rightarrow k = 2$
$f(2) = 20 + 30 \cdot 2^{2 \cdot 2} = 20 + 480 = 500$

81. $f(x) = 16 + a \cdot 3^{kx};\ f(0) = 21;\ f(4) = 61$
$21 = 16 + a \cdot 3^{k \cdot 0} \Rightarrow a = 5$
$61 = 16 + 5 \cdot 3^{k \cdot 4} \Rightarrow 9 = 3^{4k} \Rightarrow 3^2 = 3^{4k} \Rightarrow$
$2 = 4k \Rightarrow k = \dfrac{1}{2}$
$f(2) = 16 + 5 \cdot 3^{\frac{1}{2} \cdot 2} = 16 + 15 = 31$

83. $f(x) = \dfrac{10}{3 + ae^{kx}};\ f(0) = 2;\ f(1) = \dfrac{1}{2}$
$2 = \dfrac{10}{3 + ae^{k \cdot 0}} \Rightarrow 2 = \dfrac{10}{3+a} \Rightarrow 6 + 2a = 10 \Rightarrow$
$2a = 4 \Rightarrow a = 2$
$\dfrac{1}{2} = \dfrac{10}{3 + 2e^{k \cdot 1}} \Rightarrow \dfrac{1}{2} = \dfrac{10}{3 + 2e^k} \Rightarrow$
$3 + 2e^k = 20 \Rightarrow e^k = \dfrac{17}{2} \Rightarrow k = \ln \dfrac{17}{2}$
$f(2) = \dfrac{10}{3 + 2e^{2\left(\ln \frac{17}{2}\right)}} \approx 0.068 = \dfrac{4}{59}$

85. $f(x) = \dfrac{4}{a + 4e^{kx}};\ f(0) = 2;\ f(1) = 9$
$2 = \dfrac{4}{a + 4e^{k \cdot 0}} = \dfrac{4}{a+4} \Rightarrow 2a + 8 = 4 \Rightarrow a = -2$
$9 = \dfrac{4}{-2 + 4e^{k \cdot 1}} = \dfrac{4}{-2 + 4e^k} \Rightarrow$
$-18 + 36e^k = 4 \Rightarrow e^k = \dfrac{22}{36} = \dfrac{11}{18} \Rightarrow k = \ln \dfrac{11}{18}$
$f(2) = \dfrac{4}{-2 + 4e^{2\ln(11/18)}} \approx -7.902$

87. $5(0.3)^x + 1 \leq 11 \Rightarrow (0.3)^x \leq 2 \Rightarrow$
$x \ln 0.3 \leq \ln 2 \Rightarrow x \geq \dfrac{\ln 2}{\ln 0.3}$ (Note that $\ln 0.3 < 0$)

89. $-3(1.2)^x + 11 \geq 8 \Rightarrow -3(1.2)^x \geq -3 \Rightarrow$
$1.2^x \leq 1 \Rightarrow x \ln 1.2 \leq \ln 1 \Rightarrow x \leq 0$

91. Note that the domain of $\log(5x+15)$ is $(-3, \infty)$ since $5x + 15$ must be greater than 0.
$\log(5x+15) < 2 \Rightarrow 5x + 15 < 10^2 \Rightarrow$
$5x < 85 \Rightarrow x < 17$
Solution set: $(-3, 17)$

93. Note that the domain of $\ln(x-5)$ is $(5, \infty)$ since $x-5$ must be greater than 0.
 $\ln(x-5) \geq 1 \Rightarrow x-5 \geq e \Rightarrow x \geq e+5$
 Solution set: $[e+5, \infty)$

95. Note that the domain of $\log_2(3x-7)$ is $\left(\frac{7}{3}, \infty\right)$ since $3x-7$ must be greater than 0.
 $\log_2(3x-7) < 3 \Rightarrow 3x-7 < 2^3 \Rightarrow 3x < 15 \Rightarrow x < 5$
 Solution set: $\left(\frac{7}{3}, 5\right)$

97. $1 + 4e^{2x} > 9 \Rightarrow 4e^{2x} > 8 \Rightarrow e^{2x} > 2 \Rightarrow$
 $2x \ln e > \ln 2 \Rightarrow 2x > \ln 2 \Rightarrow x > \frac{\ln 2}{2}$

99. $\frac{50}{1 + 2e^{-x}} \geq 10 \Rightarrow 50 \geq 10(1 + 2e^{-x}) \Rightarrow$
 $50 \geq 10 + 20e^{-x} \Rightarrow 40 \geq 20e^{-x} \Rightarrow$
 $2 \geq e^{-x} \Rightarrow \ln 2 \geq -x \ln e \Rightarrow \ln 2 \geq -x \Rightarrow$
 $x \geq -\ln 2$

4.4 Applying the Concepts

101. $f(x) = 0.885 - 0.14 \ln x$

 a. $f(1) = 0.885 - 0.14 \ln 1 \approx 0.885 \approx 89\%$
 The relative risk of mortality associated with exercising one hour per week is about 89%, so the reduced risk is $100\% - 89\% = 11\%$.

 b. If the risk of mortality is reduced by 20%, then it is 80% or 0.80.
 $0.885 - 0.14 \ln x = 0.80$
 $-0.14 \ln x = -0.085$
 $\ln x = \frac{0.085}{0.14}$
 $x = e^{0.085/0.14} \approx 1.835$
 To reduce the risk of mortality by 20%, a person should exercise about 1.835 hours or 1 hour, 50 minutes per week.

103. a. $18{,}000 = 10{,}000\left(1 + \frac{0.06}{1}\right)^{(1)t} \Rightarrow$
 $1.8 = 1.06^t \Rightarrow t = \frac{\ln 1.8}{\ln 1.06} \approx 10.087$ years

 b. $18{,}000 = 10{,}000\left(1 + \frac{0.06}{4}\right)^{4t} \Rightarrow$
 $1.8 = 1.015^{4t} \Rightarrow 4t = \frac{\ln 1.8}{\ln 1.015} \Rightarrow t \approx 9.870$ yr

 c. $18{,}000 = 10{,}000\left(1 + \frac{0.06}{12}\right)^{12t} \Rightarrow$
 $1.8 = 1.005^{12t} \Rightarrow 12t = \frac{\ln 1.8}{\ln 1.005} \Rightarrow$
 $t \approx 9.821$ years

 d. $18{,}000 = 10{,}000\left(1 + \frac{0.06}{365}\right)^{365t} \Rightarrow$
 $1.8 = \left(1 + \frac{0.06}{365}\right)^{365t} \Rightarrow$
 $365t = \frac{\ln 1.8}{\ln\left(1 + \frac{0.06}{365}\right)} \Rightarrow t \approx 9.797$ years

 e. $18{,}000 = 10{,}000 e^{0.06t} \Rightarrow 1.8 = e^{0.06t} \Rightarrow$
 $\ln 1.8 = 0.06t \Rightarrow t \approx 9.796$ years

105. $40{,}000 = 20{,}000 e^{8r} \Rightarrow 2 = e^{8r} \Rightarrow$
 $\ln 2 = 8r \Rightarrow r = \frac{\ln 2}{8} \approx 0.0866 = 8.66\%$
 $c_5 = 20{,}000 e^{5(0.0866)} \approx \$30{,}837.52$
 The car will cost \$30,837.52 in five years. The annual rate of inflation is about 8.66%.

107. a. $\log\left(\frac{I}{12}\right) = -0.025(30) \Rightarrow$
 $\frac{I}{12} = 10^{-0.025(30)} \Rightarrow I \approx 2.134$ lumens

 b. $\log\left(\frac{4}{12}\right) = -0.025x \Rightarrow x \approx 19.08$ feet

109. a. First, find a by substituting $t = 0$ and
 $f(0) = 1000$: $1000 = \frac{20{,}000}{1 + ae^{-k(0)}} \Rightarrow$
 $1000(1 + a) = 20{,}000 \Rightarrow 1 + a = 20 \Rightarrow a = 19$.
 Now find k using $t = 4$ and $f(t) = 8999$.
 $8999 = \frac{20{,}000}{1 + 19e^{-k(4)}} \Rightarrow$
 $8999 + 170{,}981 e^{-4k} = 20{,}000 \Rightarrow$
 $e^{-4k} = \frac{11{,}001}{170{,}981} \Rightarrow -4k = \ln\left(\frac{11{,}001}{170{,}981}\right) \Rightarrow$
 $k \approx 0.68589$
 Use this value of k to find the number of people infected:
 $f(8) = \frac{20{,}000}{1 + 19e^{-0.68589(8)}} \approx 18{,}542$

b. $12,400 = \dfrac{20,000}{1+19e^{-0.68589t}} \Rightarrow$
$12,400 + 12,400 \cdot 19e^{-0.68589t} = 20,000 \Rightarrow$
$12,400 \cdot 19e^{-0.68589t} = 7600 \Rightarrow$
$e^{-0.68589t} = \dfrac{7600}{12,400 \cdot 19} \Rightarrow$
$-0.68589t = \ln\left(\dfrac{7600}{12,400 \cdot 19}\right) \Rightarrow$
$t = \dfrac{\ln\left(\dfrac{7600}{12,400 \cdot 19}\right)}{-0.68589} \approx 5$

12,400 people will be infected after 5 weeks.

111. a. $\dfrac{4490}{1+e^{5.4094-1.0255(0)}} \approx 20$

b. The carrying capacity is the numerator, 4490.

c.
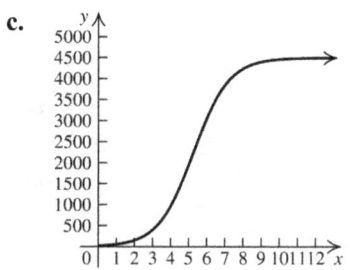

113. $14.7e^{-0.21h} \leq 8 \Rightarrow e^{-0.21h} \leq \dfrac{8}{14.7} \Rightarrow$
$-0.21h \ln e \leq \ln\left(\dfrac{8}{14.7}\right) \Rightarrow$
$-0.21h \leq \ln 8 - \ln 14.7 \Rightarrow h \geq \dfrac{\ln 8 - \ln 14.7}{-0.21} \Rightarrow$
$h \geq 2.9$
P is at most 8 psi for $h \geq 2.9$ miles.

4.4 Beyond the Basics

115. $P = \dfrac{M}{1+e^{-kt}} \Rightarrow P\left(1+e^{-kt}\right) = M \Rightarrow$
$P + Pe^{-kt} = M \Rightarrow \dfrac{M-P}{P} = e^{-kt} \Rightarrow$
$\dfrac{M-P}{P} = \dfrac{1}{e^{kt}} \Rightarrow \dfrac{P}{M-P} = e^{kt} \Rightarrow$
$\ln\left(\dfrac{P}{M-P}\right) = kt \Rightarrow t = \dfrac{1}{k}\ln\left(\dfrac{P}{M-P}\right)$

117. a. $\dfrac{\log x}{2} = \dfrac{\log y}{3} = \dfrac{\log z}{5} = k \Rightarrow$
$\log x = 2k \Rightarrow x = 10^{2k}$.
$\log y = 3k \Rightarrow y = 10^{3k}$.
$\log z = 5k \Rightarrow z = 10^{5k}$.
So, $xy = 10^{2k} \cdot 10^{3k} = 10^{5k} = z$.

b. $\dfrac{\log x}{2} = \dfrac{\log y}{3} = \dfrac{\log z}{5} \Rightarrow$
$3\log x = 2\log y \Rightarrow \log x^3 = \log y^2 \Rightarrow$
$x^3 = y^2$.
$5\log x = 2\log z \Rightarrow \log x^5 = \log z^2 \Rightarrow$
$x^5 = z^2$.
So, $y^2 z^2 = x^3 x^5 = x^8$

119. $(\log x)^2 = \log x \Rightarrow (\log x)^2 - \log x = 0 \Rightarrow$
$\log x(\log x - 1) = 0 \Rightarrow \log x = 0 \Rightarrow x = 1$ or
$\log x - 1 = 0 \Rightarrow \log x = 1 \Rightarrow x = 10$

121. $(\log_3 x)(\log_3 3x) = 2 \Rightarrow$
$(\log_3 x)(\log_3 3 + \log_3 x) = 2 \Rightarrow$
$(\log_3 x)(1 + \log_3 x) = 2$
Let $u = \log_3 x$. Then we have
$u(1+u) = 2 \Rightarrow u^2 + u - 2 = 0 \Rightarrow$
$(u+2)(u-1) = 0 \Rightarrow u = -2, 1 \Rightarrow$
$\log_3 x = -2 \Rightarrow x = 3^{-2} = \dfrac{1}{9}$ or
$\log_3 x = 1 \Rightarrow x = 3^1 = 3$

123. $\log_4 x^2(x-1)^2 - \log_2(x-1) = 1 \Rightarrow$
$\dfrac{\log_2 x^2(x-1)^2}{\log_2 4} - \log_2(x-1) = 1 \Rightarrow$
$\dfrac{\log_2 [x(x-1)]^2}{2} - \log_2(x-1) = 1 \Rightarrow$
$\dfrac{2\log_2 [x(x-1)]}{2} - \log_2(x-1) = 1 \Rightarrow$
$\log_2 [x(x-1)] - \log_2(x-1) = 1 \Rightarrow$
$\log_2 \dfrac{x(x-1)}{x-1} = 1 \Rightarrow \log_2 x = 1 \Rightarrow x = 2$.

125. $\dfrac{\log(3x-5)}{2} = \log x \Rightarrow \log(3x-5) = 2\log x \Rightarrow \log(3x-5) = \log x^2 \Rightarrow 3x-5 = x^2 \Rightarrow$
$x^2 - 3x + 5 = 0$
$x = \dfrac{-(-3) \pm \sqrt{(-3)^2 - 4(1)(5)}}{2(1)} = \dfrac{3 \pm \sqrt{9-20}}{2} = \dfrac{3 \pm \sqrt{-11}}{2}$
Since logarithms are not defined for complex numbers, there is no solution. Solution set: $\varnothing$

127. $f(x) = y = 3^x + 5$
Interchange x and y, then solve for y.
$x = 3^y + 5 \Rightarrow x - 5 = 3^y \Rightarrow \log_3(x-5) = y = f^{-1}(x)$

129. $f(x) = y = 3 \cdot 4^x + 7$
Interchange x and y, then solve for y.
$x = 3 \cdot 4^y + 7 \Rightarrow \dfrac{x-7}{3} = 4^y \Rightarrow \log_4\left(\dfrac{x-7}{3}\right) = y = f^{-1}(x)$

131. $f(x) = y = 1 + \log_2(x-1)$
Interchange x and y, then solve for y.
$x = 1 + \log_2(y-1) \Rightarrow x - 1 = \log_2(y-1) \Rightarrow 2^{x-1} = y - 1 \Rightarrow 2^{x-1} + 1 = y = f^{-1}(x)$

133. $f(x) = y = \dfrac{1}{2}\ln\left(\dfrac{x-1}{x+1}\right)$
Interchange x and y, then solve for y.
$x = \dfrac{1}{2}\ln\left(\dfrac{y-1}{y+1}\right) \Rightarrow 2x = \ln\left(\dfrac{y-1}{y+1}\right) \Rightarrow e^{2x} = \dfrac{y-1}{y+1} \Rightarrow (y+1)e^{2x} = y - 1 \Rightarrow$
$ye^{2x} + e^{2x} = y - 1 \Rightarrow e^{2x} + 1 = y - ye^{2x} \Rightarrow e^{2x} + 1 = y(1 - e^{2x}) \Rightarrow \dfrac{1 + e^{2x}}{1 - e^{2x}} = y = f^{-1}(x)$

135. $7^n > 43^{67} \Rightarrow n\ln 7 > 67\ln 43 \Rightarrow n > \dfrac{67\ln 43}{\ln 7} \Rightarrow n > 129.5$
Thus, the smallest integer for which $7^n > 43^{67}$ is 130.

137. $8^{1/n} < 1.01 \Rightarrow \dfrac{1}{n}\ln 8 < \ln 1.01 \Rightarrow \dfrac{\ln 8}{\ln 1.01} < n \Rightarrow 208.98 < n$
Thus, the smallest integer for which $8^{1/n} < 1.01$ is 209.

139. If 31^n has 567 digits, then $566 \le \log 31^n < 567$. (See Section 4.3, Example 4.)
$566 \le \log 31^n < 567 \Rightarrow 566 \le n\log 31 < 567 \Rightarrow \dfrac{566}{\log 31} \le n < \dfrac{567}{\log 31} \Rightarrow 379.5 \le n < 380.2$
Thus, $n = 380$.

141. $\dfrac{x^{\log y}}{x^{\log z}} \cdot \dfrac{y^{\log z}}{y^{\log x}} \cdot \dfrac{z^{\log x}}{z^{\log y}} = \dfrac{\left(10^{\log x}\right)^{\log y}}{\left(10^{\log x}\right)^{\log z}} \cdot \dfrac{\left(10^{\log y}\right)^{\log z}}{\left(10^{\log y}\right)^{\log x}} \cdot \dfrac{\left(10^{\log z}\right)^{\log x}}{\left(10^{\log z}\right)^{\log y}} = \dfrac{10^{\log x \log y}}{10^{\log x \log z}} \cdot \dfrac{10^{\log y \log z}}{10^{\log x \log y}} \cdot \dfrac{10^{\log x \log z}}{10^{\log y \log z}}$
$= \dfrac{10^{\log x \log y + \log y \log z + \log x \log z}}{10^{\log x \log z + \log x \log y + \log y \log z}} = 1$

4.4 Critical Thinking/Discussion/Writing

143. $\dfrac{P}{2} = \dfrac{P}{1+ae^{-kt}} \Rightarrow \dfrac{1}{2} = \dfrac{1}{1+ae^{-kt}} \Rightarrow$
$1 + ae^{-kt} = 2 \Rightarrow ae^{-kt} = 1 \Rightarrow e^{-kt} = \dfrac{1}{a} = a^{-1} \Rightarrow$
$-kt = \ln\left(a^{-1}\right) = -\ln a \Rightarrow t = \dfrac{\ln a}{k}$

144. a. $\log_4(x-1)^2 = 3 \Rightarrow (x-1)^2 = 4^3 = 64 \Rightarrow$
$x^2 - 2x - 63 = 0 \Rightarrow (x+7)(x-9) = 0 \Rightarrow$
$x = -7, 9$

   b. $2\log_4(x-1) = 3 \Rightarrow \log_4(x-1) = \dfrac{3}{2} \Rightarrow$
$x - 1 = 4^{3/2} = 8 \Rightarrow x = 9$

   c. $2\log_4|x-1| = 3 \Rightarrow \log_4|x-1| = \dfrac{3}{2} \Rightarrow$
$|x-1| = 4^{3/2} = 8 \Rightarrow x - 1 = 8 \Rightarrow x = 9$ or
$x - 1 = -8 \Rightarrow x = -7$
   The three equations do not have identical solutions because the equation in (a) has a quadratic term, the equation in (b) has a linear term, and the equation in (c) has an absolute value.

4.4 Getting Ready for the Next Section

145. $9^2 = 81 \Rightarrow \log_9 81 = 2$

147. $2 \cdot 10^x + 1 = 7 \Rightarrow 2 \cdot 10^x = 6 \Rightarrow 10^x = 3 \Rightarrow$
$\log 3 = x$

149. $\log_2 64 = 6 \Rightarrow 2^6 = 64$

151. $\log\left(\dfrac{A}{2}\right) = 3 \Rightarrow 10^3 = \dfrac{A}{2} \Rightarrow A = 2 \cdot 10^3$

153. $\log_5 x = -3 \Rightarrow x = 5^{-3} = \dfrac{1}{125}$

155. $\log_x 1000 = 3 \Rightarrow 1000 = x^3 \Rightarrow 10^3 = x^3 \Rightarrow$
$x = 10$

157. $2^{x+1} = 32 \Rightarrow 2^{x+1} = 2^5 \Rightarrow x + 1 = 5 \Rightarrow x = 4$

159. $3^{x+1} = 5^{2x-3}$
$(x+1)\ln 3 = (2x-3)\ln 5$
$x \ln 3 + \ln 3 = 2x \ln 5 - 3 \ln 5$
$\ln 3 + 3 \ln 5 = 2x \ln 5 - x \ln 3$
$\ln 3 + 3 \ln 5 = x(2 \ln 5 - \ln 3)$
$x = \dfrac{\ln 3 + 3 \ln 5}{2 \ln 5 - \ln 3}$

4.5 Logarithmic Scales; Modeling

4.5 Practice Problems

1. a. $\text{pH} = -\log\left[\text{H}^+\right] = -\log\left(2.68 \times 10^{-6}\right)$
$= -\left(\log 2.68 + \log 10^{-6}\right)$
$= -\log 2.68 + 6 \approx 5.57$

   b. $8.47 = -\log\left[\text{H}^+\right]$
$-8.47 = \log\left[\text{H}^+\right]$
$\left[\text{H}^+\right] = 10^{-8.47} = 10^{0.53} \times 10^{-9} \approx 3.39 \times 10^{-9}$

2. $2.8 = \text{pH}_{\text{acid rain}} = -\log\left[\text{H}^+\right]_{\text{acid rain}} \Rightarrow$
$\left[\text{H}^+\right]_{\text{acid rain}} = 10^{-2.8}$
$6.2 = \text{pH}_{\text{ordinary rain}} = -\log\left[\text{H}^+\right]_{\text{ordinary rain}} \Rightarrow$
$\left[\text{H}^+\right]_{\text{ordinary rain}} = 10^{-6.2}$
$\dfrac{\left[\text{H}^+\right]_{\text{acid rain}}}{\left[\text{H}^+\right]_{\text{ordinary rain}}} = \dfrac{10^{-2.8}}{10^{-6.2}} = 10^{-2.8-(-6.2)}$
$= 10^{3.4} \approx 2512$
This acid rain is about 2512 times more acidic than the ordinary rain.

3. $M = \log\left(\dfrac{I}{I_0}\right)$
$6.5 = \log\left(\dfrac{I}{I_0}\right) \Rightarrow \left(\dfrac{I}{I_0}\right) = 10^{6.5} \Rightarrow$
$I = 10^{6.5} I_0 \approx 3{,}162{,}278 I_0$

4. Let I_M denote the intensity of the Mozambique earthquake and let I_C denote the intensity of the southern California earthquake. Then we have
$7.0 = \log\left(\dfrac{I_M}{I_0}\right) \Rightarrow \dfrac{I_M}{I_0} = 10^{7.0} \Rightarrow I_M = 10^{7.0} I_0$
$5.2 = \log\left(\dfrac{I_C}{I_0}\right) \Rightarrow \dfrac{I_C}{I_0} = 10^{5.2} \Rightarrow I_C = 10^{5.2} I_0$
$\dfrac{I_M}{I_C} = \dfrac{10^{7.0} I_0}{10^{5.2} I_0} = \dfrac{10^{7.0}}{10^{5.2}} = 10^{7.0-5.2} = 10^{1.8} \approx 63.1$
The intensity of the Mozambique earthquake was about 63 times that of the southern California earthquake.

5. Let M_{2011} and E_{2011} represent the magnitude and energy of the 2011 Japan earthquake. Let M_{1997} and E_{1997} represent the magnitude and energy of the 1977 Iran earthquake. From Table 4.7, we have $M_{2011} = 9.0$ and $M_{1997} = 7.5$.

$$\frac{E_{2011}}{E_{1997}} = 10^{1.5(M_{2011}-M_{1997})} = 10^{1.5(9.0-7.5)} = 10^{2.25}$$
$$\approx 177.8$$

The energy released by the 2011 Japan earthquake was about 178 times that released by the 1997 Iran earthquake.

6. $I = 200 \times 10^{-7}$ W/m² and $I_0 = 10^{-12}$ W/m².

$$L = 10\log\left(\frac{I}{I_0}\right) = 10\log\left(\frac{200 \times 10^{-7}}{10^{-12}}\right)$$
$$= 10\log(200 \times 10^5) = 10[\log 200 + \log 10^5]$$
$$= 10[\log 200 + 5\log 10] = 10[\log 200 + 5]$$
$$= 10\log 200 + 50 \approx 73.01$$

The decibel level is approximately 73 dB.

7. $I = I_0 \times 10^{L/10}$
$I_{75} = I_0 \times 10^{75/10}$
$I_{55} = I_0 \times 10^{55/10}$
$$\frac{I_{75}}{I_{55}} = \frac{I_0 \times 10^{75/10}}{I_0 \times 10^{55/10}} = 10^{20/10} = 10^2 = 100$$

A 75 dB sound is 100 times more intense than a 55 dB sound.

8. $I = I_0 \times 10^{L/10}$
$I_{48} = 10^{-12} \times 10^{48/10} = 10^{-12} \times 10^{4.8} = 10^{-7.2}$
$= 10^{0.8} \times 10^{-8} \approx 6.3 \times 10^{-8}$ W/m²

The intensity of a 48 dB sound is about 6.3×10^{-8} W/m².

9. B is two semitones above A, so $P(f) = 200$.
$f_0 = 440$ Hz
$$P(f) = 1200\log_2 \frac{f}{f_0}$$
$$200 = 1200\log_2 \frac{f}{440} \Rightarrow \frac{1}{6} = \log_2 \frac{f}{440} \Rightarrow$$
$$2^{1/6} = \frac{f}{440} \Rightarrow f = 2^{1/6} \cdot 440 \approx 493.9$$

The frequency of B is about 494 Hz.

10. If the reference frequency is f_0 and it increases by 25%, then $f = f_0 + 0.25f_0 = 1.25f_0$. Then,

$$P(f) = 1200\log_2 \frac{1.25 f_0}{f_0}$$
$$= 1200\log_2 1.25 = 1200\frac{\log 1.25}{\log 2}$$
$$\approx 386.3$$

Since an equal tempered major third is 400 cents, the difference is about $400 - 386.3 = 13.7$ cents, or a little less than 14 cents.

11. a. $m_2 - m_1 = 2.5\log\left(\frac{b_1}{b_2}\right)$

$$2 - 0 = \frac{5}{2}\log\left(\frac{b_1}{b_2}\right) \Rightarrow \frac{4}{5} = \log\left(\frac{b_1}{b_2}\right) \Rightarrow$$
$$\frac{b_1}{b_2} = 10^{4/5} \approx 6.3$$

A magnitude 0 star is approximately 6.3 times brighter than a magnitude 2 star.

b. Let $m_1 = 4.6$ and $b_2 = 150b_1$.

$$m_2 - m_1 = 2.5\log\left(\frac{b_1}{b_2}\right)$$
$$m_2 - 4.6 = 2.5\log\left(\frac{b_1}{1.50b_1}\right) = 2.5\log\left(\frac{1}{1.50}\right)$$
$$= 2.5\log 1.50^{-1} = -2.5\log 1.50$$
$$m_2 = -2.5\log 1.50 + 4.6 \approx 4.1598$$

The magnitude of the star that is 50% brighter than a star of magnitude 4.6 is about 4.1598.

12. Set x as the elapsed time and y as the temperature inside the car. Using a graphing calculator, the logarithmic regression function is $y = 72.88 + 11.66\ln x$.

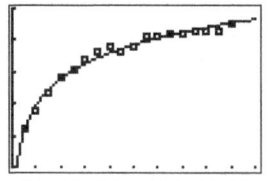 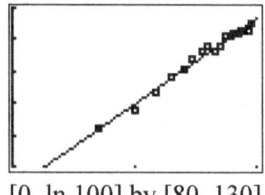

[0, 100] by [80, 130] [0, ln 100] by [80, 130]

The graph on the left shows a scatterplot of the data along with the logarithmic regression function. The graph on the right shows a scatter-plot of the transformed data along with the linear regression, $Y = 72.88 + 11.66X$ for the transformed data $Y = y$ and $X = \ln x$. From the scatterplot of the transformed data, we see that the logarithmic model is a good choice.

13. Set x as the body weight and y as the metabolism rate. Using a graphing calculator, the power regression function is $y = 2.00x^{0.75}$.

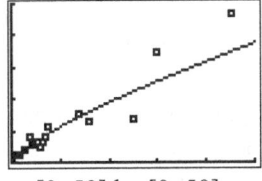
[0, 50] by [0, 50]

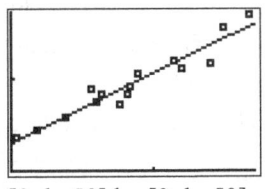
[0, ln 50] by [0, ln 50]

The graph on the left shows a scatterplot of the data along with the power regression function. The graph on the right shows a scatter-plot of the transformed data along with the linear regression, $Y = 0.69 + 0.75X$ for the transformed data $Y = \ln y$ and $X = \ln x$. From the scatterplot of the transformed data, we see that the power model is a good choice. Convert the linear function to the power function: $y = e^{0.69}x^{0.75} \approx 1.99x^{0.75}$, which is close to the power regression function determined with the calculator.

4.5 Concepts and Vocabulary

1. On the Richter scale, the magnitude of an earthquake $M = \underline{\log \frac{I}{I_0}}$. The energy E released by an earthquake of magnitude M is given by $\log E = \underline{4.4 + 1.5M}$.

3. If f_0 is a reference frequency and f is the frequency of a note, then the change in pitch $P(f) = 1200 \log_2 \left(\frac{f}{f_0}\right)$.

5. False. If the pH value of a solution is more than 7, the solution is alkaline.

7. True

4.5 Building Skills

9. $\text{pH} = -\log\left[\text{H}^+\right] = -\log\left(10^{-8}\right) = 8\log 10 = 8$
   The substance is a base.

11. $\text{pH} = -\log\left[\text{H}^+\right] = -\log\left(2.3 \times 10^{-5}\right)$
    $= -\left(\log 2.3 + \log 10^{-5}\right)$
    $= -\left(\log 2.3 - 5\log 10\right) = -\left(\log 2.3 - 5\right)$
    $= -\log 2.3 + 5 \approx 4.64$
    The substance is an acid.

13. $\text{pH} = 6 = -\log\left[\text{H}^+\right] \Rightarrow \left[\text{H}^+\right] = 10^{-6}$
    $\left[\text{H}^+\right]\left[\text{OH}^-\right] = 10^{-14}$
    $10^{-6}\left[\text{OH}^-\right] = 10^{-14}$
    $\left[\text{OH}^-\right] = \frac{10^{-14}}{10^{-6}} = 10^{-8}$

15. $\text{pH} = 9.5 = -\log\left[\text{H}^+\right] \Rightarrow \left[\text{H}^+\right] = 10^{-9.5} \Rightarrow$
    $\left[\text{H}^+\right] = 10^{0.5} \times 10^{-10} \approx 3.16 \times 10^{-10}$
    $\left[\text{H}^+\right]\left[\text{OH}^-\right] = 10^{-14}$
    $\left(3.16 \times 10^{-10}\right)\left[\text{OH}^-\right] = 10^{-14}$
    $\left[\text{OH}^-\right] = \frac{10^{-14}}{3.16 \times 10^{-10}} = \frac{1}{3.16} \times 10^{-4}$
    $\approx 0.316 \times 10^{-4} = 3.16 \times 10^{-5}$

17. a. $M = \log\left(\frac{I}{I_0}\right)$
    $5 = \log\left(\frac{I}{I_0}\right) \Rightarrow \left(\frac{I}{I_0}\right) = 10^5 \Rightarrow I = 10^5 I_0$

    b. $\log E = 4.4 + 1.5M$
    $\log E = 4.4 + 1.5(5) = 11.9 \Rightarrow$
    $E = 10^{11.9} = 10^{0.9} \times 10^{11}$
    $\approx 7.94 \times 10^{11}$ joules

19. a. $M = \log\left(\frac{I}{I_0}\right)$
    $7.8 = \log\left(\frac{I}{I_0}\right) \Rightarrow \left(\frac{I}{I_0}\right) = 10^{7.8} \Rightarrow$
    $I = 10^{0.8} \times 10^7 I_0 \approx 6.3 \times 10^7 I_0$

    b. $\log E = 4.4 + 1.5M$
    $\log E = 4.4 + 1.5(7.8) = 16.1 \Rightarrow$
    $E = 10^{16.1} = 10^{0.1} \times 10^{16}$
    $\approx 1.26 \times 10^{16}$ joules

21. a. $\log E = 4.4 + 1.5M$
    $\log 10^{13.4} = 4.4 + 1.5M \Rightarrow 13.4 = 4.4 + 1.5M$
    $M = \frac{13.4 - 4.4}{1.5} = \frac{9}{1.5} = 6$

    b. $M = \log\left(\frac{I}{I_0}\right)$
    $6 = \log\left(\frac{I}{I_0}\right) \Rightarrow \left(\frac{I}{I_0}\right) = 10^6 \Rightarrow I = 10^6 I_0$

23. a. $\log E = 4.4 + 1.5M$
$\log 10^{12} = 4.4 + 1.5M \Rightarrow 12 = 4.4 + 1.5M$
$M = \dfrac{12 - 4.4}{1.5} = \dfrac{7.6}{1.5} \approx 5.1$

b. $M = \log\left(\dfrac{I}{I_0}\right)$

$5.1 = \log\left(\dfrac{I}{I_0}\right) \Rightarrow \left(\dfrac{I}{I_0}\right) = 10^{5.1}$

$I = 10^{5.1} I_0 = 10^{0.1} \times 10^5 I_0 \approx 1.26 \times 10^5 I_0$

25. $L = 10\log\left(\dfrac{I}{I_0}\right) = 10\log\left(\dfrac{10^{-8}}{10^{-12}}\right) = 10\log 10^4$
$= 40$

27. $L = 10\log\left(\dfrac{I}{I_0}\right) = 10\log\left(\dfrac{3.5 \times 10^{-7}}{10^{-12}}\right)$
$= 10\log(3.5 \times 10^5) = 10(\log 3.5 + \log 10^5)$
$= 10\log 3.5 + 50 \approx 55.4$

29. $I = I_0 \times 10^{L/10} = 10^{-12} \times 10^{80/10}$
$= 10^{-12} \times 10^8 = 10^{-4}$

31. $I = I_0 \times 10^{L/10} = 10^{-12} \times 10^{64.7/10}$
$= 10^{-12} \times 10^{6.47} = 10^{-12} \times 10^6 \times 10^{0.47}$
$\approx 2.95 \times 10^{-6}$

33. A# is one semitone above A, so $P(f) = 100$.
$f_0 = 440$ Hz
$P(f) = 1200 \log_2 \dfrac{f}{f_0}$
$100 = 1200 \log_2 \dfrac{f}{440} \Rightarrow \dfrac{1}{12} = \log_2 \dfrac{f}{440} \Rightarrow$
$2^{1/12} = \dfrac{f}{440} \Rightarrow f = 2^{1/12} \cdot 440 \approx 466$ Hz
The frequency of A# is about 466 Hz.
C is three semitones above A, so $P(f) = 300$.
$f_0 = 440$ Hz
$P(f) = 1200 \log_2 \dfrac{f}{f_0}$
$300 = 1200 \log_2 \dfrac{f}{440} \Rightarrow \dfrac{1}{4} = \log_2 \dfrac{f}{440} \Rightarrow$
$2^{1/4} = \dfrac{f}{440} \Rightarrow f = 2^{1/4} \cdot 440 \approx 523$ Hz
The frequency of C is about 523 Hz.

35. $P(f) = 1200\log_2 \dfrac{f}{f_0} = 1200\log_2\left(\dfrac{10}{9}\right)$
$= 1200\left(\dfrac{\log\left(\dfrac{10}{9}\right)}{\log 2}\right) \approx 182$

The difference, $200 - 182 = 18$ cents, is noticeable.

37. Let the magnitude and brightness of star A be m_1 and b_1, and let the magnitude and brightness of star B be m_2 and b_2.

$m_2 - m_1 = 2.5\log\left(\dfrac{b_1}{b_2}\right)$

$20 - 4 = 2.5\log\left(\dfrac{b_1}{b_2}\right) \Rightarrow 16 = 2.5\log\left(\dfrac{b_1}{b_2}\right) \Rightarrow$

$6.4 = \log\left(\dfrac{b_1}{b_2}\right) \Rightarrow \dfrac{b_1}{b_2} = 10^{6.4} \Rightarrow$

$b_1 = 10^{0.4} \times 10^6 b_2 \approx 2.5 \times 10^6 b_2$

Star A is 2.5×10^6 times as bright as star B.

39. Let the magnitude and brightness of star A be m_1 and b_1, and let the magnitude of star B be $m_2 = m_1 - 2$ and b_2.

$m_2 - m_1 = 2.5\log\left(\dfrac{b_1}{b_2}\right)$

$(m_1 - 2) - m_1 = 2.5\log\left(\dfrac{b_1}{b_2}\right) \Rightarrow$

$-2 = 2.5\log\left(\dfrac{b_1}{b_2}\right) \Rightarrow -0.8 = \log\left(\dfrac{b_1}{b_2}\right) \Rightarrow$

$\dfrac{b_1}{b_2} = 10^{-0.8} \approx \dfrac{1}{6.3} \Rightarrow 6.3 b_1 = b_2$

Star B is about 6.3 times brighter than star A.

4.5 Applying the Concepts

41. $\text{pH} = -\log[H^+] = -\log(3.98 \times 10^{-8})$
$= -(\log 3.98 + \log 10^{-8}) = -(\log 3.98 - 8)$
$= -\log 3.98 + 8 \approx 7.4$

The pH of human blood is about 7.4; it is basic.

43. a. $\text{pH} = 3.15 = -\log[H^+] \Rightarrow [H^+] = 10^{-3.15} \Rightarrow$
$[H^+] = 10^{0.85} \times 10^{-4} \approx 7.1 \times 10^{-4}$

b. $\text{pH} = 7.2 = -\log[H^+] \Rightarrow [H^+] = 10^{-7.2} \Rightarrow$
$[H^+] = 10^{0.8} \times 10^{-8} \approx 6.3 \times 10^{-8}$

c. $pH = 7.78 = -\log[H^+] \Rightarrow [H^+] = 10^{-7.78} \Rightarrow$
$[H^+] = 10^{0.22} \times 10^{-8} \approx 1.7 \times 10^{-8}$

d. $pH = 3 = -\log[H^+] \Rightarrow [H^+] = 10^{-3} \Rightarrow$
$[H^+] = 10^{-3}$

45. $pH_{\text{acid rain}} = 3.8 = -\log[H^+] \Rightarrow$
$[H^+] = 10^{-3.8} \Rightarrow$
$[H^+] = 10^{0.2} \times 10^{-4} \approx 1.6 \times 10^{-4}$

The average concentration of hydrogen ions in the acid rain is 1.6×10^{-4} moles per liter.

$pH_{\text{ordinary rain}} = 6 = -\log[H^+] \Rightarrow$
$[H^+] = 10^{-6}$

$\dfrac{[H^+]_{\text{acid rain}}}{[H^+]_{\text{ordinary rain}}} = \dfrac{1.6 \times 10^{-4}}{10^{-6}} = 1.6 \times 10^2 = 160$

The acid rain is 160 times more acidic than ordinary rain.

47. $[H^+_A] = 100[H^+_B]$
$pH_A = -\log[H^+_A] = -\log(100[H^+_B])$
$= -(\log 100 + \log[H^+_B])$
$= -(2 + \log[H^+_B]) = -\log[H^+_B] - 2$
$= pH_B - 2$

49. $[H^+_{\text{new}}] = 50[H^+_{\text{original}}]$
$pH_{\text{new}} = -\log(50[H^+_{\text{original}}])$
$= -\log 50 - \log[H^+_{\text{original}}]$
$= -\log 50 - pH_{\text{original}}$

The pH decreases by $\log 50 \approx 1.7$.
The solution becomes more acidic.

51. a. $M = \log\left(\dfrac{I}{I_0}\right)$
$7.8 = \log\left(\dfrac{I}{I_0}\right) \Rightarrow \left(\dfrac{I}{I_0}\right) = 10^{7.8} \Rightarrow$
$I = 10^{7.8} I_0 = 10^{0.8} \times 10^7 I_0 \approx 6.31 \times 10^7 I_0$

b. $\log E = 4.4 + 1.5(7.8) = 16.1 \Rightarrow$
$E = 10^{16.1} = 10^{0.1} \times 10^{16} \approx 1.259 \times 10^{16}$ j

53. a. $M_A = M_B + 1$. Then we have
$M_A = M_B + 1 = \log\left(\dfrac{I_A}{I_0}\right) \Rightarrow$
$10^{(M_B+1)} = \dfrac{I_A}{I_0} \Rightarrow 10^{M_B} \times 10 = \dfrac{I_A}{I_0} \Rightarrow$
$10^{M_B} \times 10 I_0 = I_A$ (1)

$M_B = \log\left(\dfrac{I_B}{I_0}\right) \Rightarrow 10^{M_B} = \dfrac{I_B}{I_0} \Rightarrow$
$10^{M_B} \times I_0 = I_B$

Using equation (1), we have
$10^{M_B} \times 10 I_0 = I_A \Rightarrow 10^{M_B} \times I_0 \times 10 = I_A \Rightarrow$
$I_B \times 10 = I_A$

The intensity of the earthquake A was 10 times that of earthquake B.

b. $\dfrac{E_A}{E_B} = 10^{1.5(M_A - M_B)} = 10^{1.5(M_B + 1 - M_B)}$
$= 10^{1.5} \approx 31.6 \Rightarrow E_A \approx 31.6 E_B$

The energy released by the earthquake A was about 31.6 times that released by earthquake B.

55. $M = \log\left(\dfrac{I}{I_0}\right)$

$I_A = 150 I_B \Rightarrow$
$\log\left(\dfrac{I_A}{I_0}\right) = \log\left(\dfrac{150 I_B}{I_0}\right) = \log\left(150 \cdot \dfrac{I_B}{I_0}\right)$
$= \log 150 + \log\left(\dfrac{I_B}{I_0}\right) = \log 150 + M_B$

The difference in Richter scale readings is $\log 150 \approx 2.18$.

57. $L = 10\log\left(\dfrac{I}{I_0}\right) = 10\log\left(\dfrac{5.2 \times 10^{-5}}{10^{-12}}\right)$
$= 10\log(5.2 \times 10^7) = 10[\log 5.2 + \log 10^7]$
$= 10[\log 5.2 + 7] = 10\log 5.2 + 70 \approx 77.2$ dB

59. $I = I_0 \times 10^{L/10}$
$I_{130} = I_0 \times 10^{130/10}$
$I_{65} = I_0 \times 10^{65/10}$
$\dfrac{I_{130}}{I_{65}} = \dfrac{I_0 \times 10^{130/10}}{I_0 \times 10^{65/10}} = 10^{65/10} = 10^{6.5}$
$= 10^{0.5} \times 10^6 \approx 3.16 \times 10^6$

The 130 dB sound is about 3.16×10^6 times as intense as the 65 dB sound.

61. A sound at the threshold of pain has intensity 10 W/m^2. A sound that is 1000 times as intense has intensity 10^4 W/m^2.
$$L = 10\log\left(\frac{I}{I_0}\right) = 10\log\left(\frac{10^4}{10^{-12}}\right) = 10\log 10^{16}$$
$$= 160 \text{ dB}$$

63. $L_2 = L_1 + 1$
$$I_{L_2} = I_0 \times 10^{L_2/10} = I_0 \times 10^{0.1(L_1+1)}$$
$$I_{L_1} = I_0 \times 10^{L_1/10} = I_0 \times 10^{0.1L_1}$$
$$\frac{I_{L_2}}{I_{L_1}} = \frac{I_0 \times 10^{0.1(L_1+1)}}{I_0 \times 10^{0.1L_1}} = 10^{0.1(L_1+1-L_1)} \text{ The}$$
$$= 10^{0.1} \approx 1.26 \Rightarrow I_{L_2} \approx 1.26 \times I_{L_1}$$
intensity of the louder sound is about 1.26 times the intensity of the softer sound.

65. $P(f) = 1200\log_2 \frac{f}{f_0}$
$$P(441) = 1200\log_2\left(\frac{441}{440}\right) = 1200\left(\frac{\log\left(\frac{441}{440}\right)}{\log 2}\right)$$
$$\approx 3.93 \approx 4$$
The difference in pitch is about 4 cents higher.

67. Let m_1 represent the magnitude of the sun and let m_2 represent the magnitude of the full Moon.
$$m_2 - m_1 = 2.5\log\left(\frac{b_1}{b_2}\right)$$
$$-13 - (-27) = 2.5\log\left(\frac{b_1}{b_2}\right) \Rightarrow$$
$$\frac{14}{2.5} = 5.6 = \log\left(\frac{b_1}{b_2}\right) \Rightarrow$$
$$\frac{b_1}{b_2} = 10^{5.6} = 10^{0.6} \times 10^5 \approx 3.98 \times 10^5 \Rightarrow$$
$$b_1 = 3.98 \times 10^5 b_2$$
The sun is about 3.98×10^5 times brighter than the full Moon.

69. Let m_1 represent the magnitude of the Sun and let m_2 represent the magnitude of Venus.
$$m_2 - m_1 = 2.5\log\left(\frac{b_1}{b_2}\right)$$
$$-4 - (-27) = 2.5\log\left(\frac{b_1}{b_2}\right)$$

$$\frac{23}{2.5} = 9.2 = \log\left(\frac{b_1}{b_2}\right)$$
$$\frac{b_1}{b_2} = 10^{9.2} = 10^{0.2} \times 10^9 \approx 1.58 \times 10^9 \Rightarrow$$
$$b_1 = 1.58 \times 10^9 b_2$$
The Sun is about 1.58×10^9 times brighter than Venus.

71. Let m_1 and b_1 represent the magnitude and brightness of the unknown star and let m_2 and b_2 represent the magnitude and brightness of Saturn.
$$m_2 - m_1 = 2.5\log\left(\frac{b_1}{b_2}\right); \ b_1 = 560 b_2$$
$$1.47 - m_1 = 2.5\log\left(\frac{560 b_2}{b_2}\right) \Rightarrow$$
$$1.47 - m_1 = 2.5\log 560 \Rightarrow$$
$$1.47 - 2.5\log 560 = m_1 \Rightarrow -5.4 \approx m_1$$
The magnitude of the unknown star is about -5.4.

73. $y = 25.72 \cdot 1.6667^x = 25.72 e^{x\ln 1.6667}$
$$= 25.72 e^{0.511x}$$

75. $y = 66.654 - 10.367\ln x$

4.5 Beyond the Basics

77. $L = 10\log\left(\dfrac{4 \times 10^4}{r^2}\right)$

a. $r = 10$ ft
$$L = 10\log\left(\frac{4 \times 10^4}{10^2}\right) \approx 26 \text{ dB}$$
$r = 25$ ft
$$L = 10\log\left(\frac{4 \times 10^4}{25^2}\right) \approx 18 \text{ dB}$$
$r = 50$ ft
$$L = 10\log\left(\frac{4 \times 10^4}{50^2}\right) \approx 12 \text{ dB}$$
$r = 100$ ft
$$L = 10\log\left(\frac{4 \times 10^4}{100^2}\right) \approx 6 \text{ dB}$$

b. $0 = 10\log\left(\dfrac{4\times 10^4}{r^2}\right) \Rightarrow 0 = \log\left(\dfrac{4\times 10^4}{r^2}\right) \Rightarrow$

$10^0 = 1 = \dfrac{4\times 10^4}{r^2} \Rightarrow r^2 = 40{,}000 \Rightarrow$

$r = 200$ ft

The decibel level drops to 0 when a listener is 200 ft away.

c. $L = 10\log\left(\dfrac{4\times 10^4}{r^2}\right)$

$= 10\left(\log 4 + \log\left(10^4\right) - \log\left(r^2\right)\right)$

$= 10\left((4 + \log 4) - 2\log r\right)$

$= 40 + 10\log 4 - 20\log r$

$a = 40 + 10\log 4 = 10(4 + \log 4)$

$b = -20$

d. $L = 10\log\left(\dfrac{4\times 10^4}{r^2}\right) = 10\log\left(\dfrac{2\times 10^2}{r}\right)^2 \Rightarrow$

$L = 20\log\left(\dfrac{200}{r}\right) \Rightarrow \dfrac{L}{20} = \log\left(\dfrac{200}{r}\right) \Rightarrow$

$10^{L/20} = \dfrac{200}{r} \Rightarrow r = \dfrac{200}{10^{L/20}}$

79. The energy released by the earthquake is given by

$E = \left(2.5\times 10^4\right)\times 10^{1.5(8)} = 2.5\times 10^{16}$ joules.

$\dfrac{E_{\text{earthquake}}}{E_{\text{bomb}}} = \dfrac{2.5\times 10^{16}}{5\times 10^{15}} = 0.5\times 10 = 5$

The earthquake released five times as much energy as the nuclear bomb did.

81. $10 = 1200\log_2 \dfrac{f}{880} \Rightarrow \dfrac{1}{120} = \log_2 \dfrac{f}{880} \Rightarrow$

$2^{1/120} = \dfrac{f}{880} \Rightarrow f = 880\times 2^{1/120} \approx 885$

$-10 = 1200\log_2 \dfrac{f}{880} \Rightarrow -\dfrac{1}{120} = \log_2 \dfrac{f}{880} \Rightarrow$

$2^{-1/120} = \dfrac{f}{880} \Rightarrow f = 880\times 2^{-1/120} \approx 875$

The frequency range is [875, 885].

83. Let m_1 and b_1 represent the magnitude and brightness of the unknown star and let m_2 and b_2 represent the magnitude and brightness of the given star.

$m_2 - m_1 = 2.5\log\left(\dfrac{b_1}{b_2}\right);\ b_2 = 176 b_1$

$m_2 - 3.42 = 2.5\log\left(\dfrac{b_1}{176 b_1}\right) \Rightarrow$

$m_2 - 3.42 = 2.5\log\left(\dfrac{1}{176}\right) \Rightarrow$

$m_2 = 2.5\log\left(\dfrac{1}{176}\right) + 3.42 \approx -2.2$

The magnitude of the unknown star is about -2.2.

4.5 Getting Ready for the Next Section

85. $\pi d = 180\left(\dfrac{\pi}{4}\right) \Rightarrow \pi d = 45\pi \Rightarrow d = 45$

87. $30\pi = 180x \Rightarrow \dfrac{\pi}{6} = x$

89. $\left(-\dfrac{1}{2}, \dfrac{\sqrt{3}}{2}\right)$ is located in Quadrant II.

91. $\left(\dfrac{1}{2}, \dfrac{\sqrt{3}}{2}\right)$ is located in Quadrant I.

Chapter 4 Review Exercises

Concepts and Vocabulary

1. False. $f(x) = a^x$ is an exponential function if $a > 0$ and $a \neq 1$.

3. False. The domain of $f(x) = \log(2 - x)$ is $(-\infty, 2)$.

5. True

7. False. $\ln M + \ln N = \ln(MN)$

9. True

Building Skills

11. h   13. f

15. d   17. a

19. Domain: $(-\infty, \infty)$
range: $(0, \infty)$
asymptote: $y = 0$

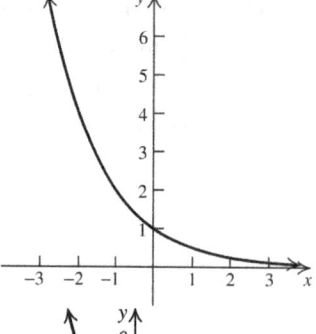

21. Domain: $(-\infty, \infty)$
range: $(3, \infty)$
asymptote: $y = 3$

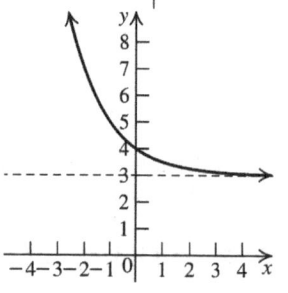

23. Domain: $(-\infty, \infty)$
range: $(0, 1]$
asymptote: $y = 0$

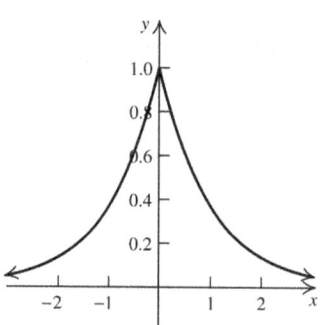

25. Domain: $(-\infty, 0)$
range: $(-\infty, \infty)$
asymptote: $x = 0$

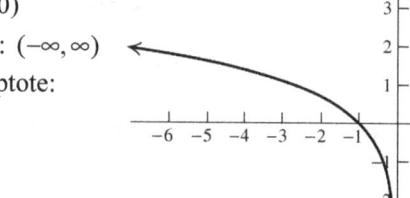

27. Domain: $(1, \infty)$
range: $(-\infty, \infty)$
asymptote: $x = 1$

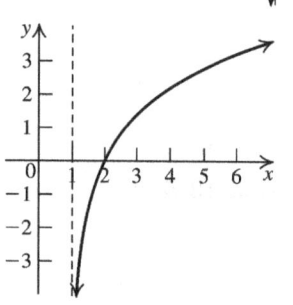

29. Domain: $(-\infty, 0)$
range: $(-\infty, \infty)$
asymptote: $x = 0$

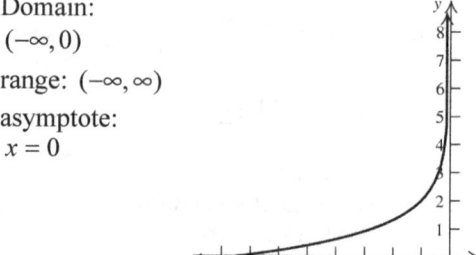

31. a. y-intercept: $y = 3 - 2e^0 = 1$

x-intercept: $0 = 3 - 2e^{-x} \Rightarrow 2e^{-x} = 3 \Rightarrow$
$e^{-x} = \dfrac{3}{2} \Rightarrow e^x = \dfrac{2}{3} \Rightarrow x = \ln\left(\dfrac{2}{3}\right)$

b. As $x \to \infty$, $f(x) \to 3$.
As $x \to -\infty$, $f(x) \to -\infty$.

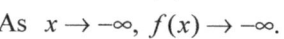
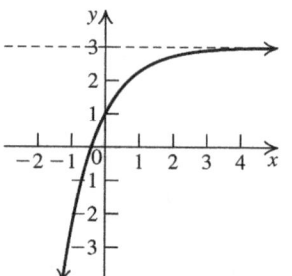

33. a. y-intercept: $y = e^{0^2} = 1$

x-intercept: $0 = e^{-x^2} \Rightarrow \ln 0 = -x^2 \Rightarrow$
there is no x-intercept.

b. As $x \to \infty$, $f(x) \to 0$.
As $x \to -\infty$, $f(x) \to 0$.

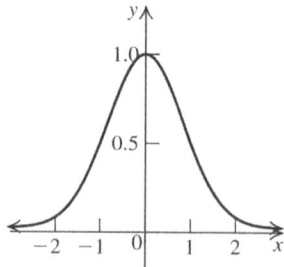

35. $10 = a(2^{0k}) \Rightarrow a = 10$
$640 = 10(2^{3k}) \Rightarrow 64 = 8^k \Rightarrow k = 2$
$f(2) = 10(2^{2(2)}) = 160$

37. $1 = \dfrac{4}{1+ae^{-k(0)}} = \dfrac{4}{1+a} \Rightarrow 1+a=4 \Rightarrow a=3$

$\dfrac{1}{2} = \dfrac{4}{1+3e^{-3k}} \Rightarrow 1+3e^{-3k}=8 \Rightarrow e^{-3k}=\dfrac{7}{3} \Rightarrow$

$-3k = \ln\left(\dfrac{7}{3}\right) \Rightarrow k \approx -0.2824$

$f(4) = \dfrac{4}{1+3e^{-(-0.2824)(4)}} \approx 0.38898$

39. The graph passes through (0, 3), so we have $3 = ca^0 \Rightarrow c = 3$. The graph passes through (2, 12), so we have $12 = 3a^2 \Rightarrow 4 = a^2 \Rightarrow a = 2$. The equation is $f(x) = 3 \cdot 2^x$.

41. The graph has been shifted right one unit, so the equation is of the form $y = \log_a(x-1)$. The graph passes through the point (6, 1), so we have $1 = \log_a(6-1) \Rightarrow 1 = \log_a 5 \Rightarrow a = 5$. Thus, the equation is $y = \log_5(x-1)$.

43. a. $y = -\left(2^{x-1}+3\right) = -2^{x-1}-3$

   b. $y = -2^{x-1}+3$

45. $\ln(xy^2z^3) = \ln x + 2\ln y + 3\ln z$

47. $\ln\left[\dfrac{x\sqrt{x^2+1}}{(x^2+3)^2}\right] = \ln x + \dfrac{1}{2}\ln(x^2+1) - 2\ln(x^2+3)$

49. $\ln y = \ln x + \ln 3 = \ln(3x) \Rightarrow y = 3x$

51. $\ln y = \ln(x-3) - \ln(y+2) \Rightarrow$
$\ln y + \ln(y+2) = \ln(x-3) \Rightarrow$
$\ln(y(y+2)) = \ln(x-3) \Rightarrow$
$\ln(y^2+2y) = \ln(x-3) \Rightarrow y^2+2y = x-3 \Rightarrow$
$y^2+2y+1 = x-3+1 \Rightarrow (y+1)^2 = x-2 \Rightarrow$
$y+1 = \pm\sqrt{x-2} \Rightarrow y = -1 \pm \sqrt{x-2}$
We disregard the negative solution, so $y = -1+\sqrt{x-2}$.

53. $\ln y = \dfrac{1}{2}\ln(x-1) + \dfrac{1}{2}\ln(x+1) - \ln(x^2+1) \Rightarrow$

$\ln y = \ln\left(\dfrac{\sqrt{x-1}\sqrt{x+1}}{x^2+1}\right) = \ln\left(\dfrac{\sqrt{x^2-1}}{x^2+1}\right) \Rightarrow$

$y = \dfrac{\sqrt{x^2-1}}{x^2+1}$

55. $3^x = 81 \Rightarrow 3^x = 3^4 \Rightarrow x = 4$

57. $2^{x^2+2x} = 16 \Rightarrow 2^{x^2+2x} = 2^4 \Rightarrow x^2+2x=4 \Rightarrow$
$x^2+2x-4 = 0 \Rightarrow x = \dfrac{-2\pm\sqrt{4+16}}{2} = -1\pm\sqrt{5}$

59. $3^x = 23 \Rightarrow x\ln 3 = \ln 23 \Rightarrow x = \dfrac{\ln 23}{\ln 3} \approx 2.854$

61. $273^x = 19 \Rightarrow x\ln 273 = \ln 19 \Rightarrow$
$x = \dfrac{\ln 19}{\ln 273} \approx 0.525$

63. $3^{2x} = 7^x \Rightarrow 2x\ln 3 = x\ln 7 \Rightarrow$
$2x\ln 3 - x\ln 7 = 0 \Rightarrow x(2\ln 3 - \ln 7) = 0 \Rightarrow$
$x = 0$

65. $1.7^{3x} = 3^{2x-1} \Rightarrow 3x\ln 1.7 = (2x-1)\ln 3 \Rightarrow$
$3x\ln 1.7 = 2x\ln 3 - \ln 3 \Rightarrow$
$3x\ln 1.7 - 2x\ln 3 = -\ln 3 \Rightarrow$
$x(3\ln 1.7 - 2\ln 3) = -\ln 3 \Rightarrow$
$x = \dfrac{-\ln 3}{3\ln 1.7 - 2\ln 3} \approx 1.815$

67. $\log_3(x+2) - \log_3(x-1) = 1 \Rightarrow$
$\log_3\left(\dfrac{x+2}{x-1}\right) = 1 \Rightarrow \dfrac{x+2}{x-1} = 3 \Rightarrow$
$x+2 = 3x-3 \Rightarrow 2x = 5 \Rightarrow x = \dfrac{5}{2}$

69. $\log_2(x+2) + \log_2(x+4) = 3 \Rightarrow$
$\log_2[(x+2)(x+4)] = 3 \Rightarrow$
$x^2+6x+8 = 2^3 \Rightarrow x^2+6x = 0 \Rightarrow$
$x(x+6) = 0 \Rightarrow x = -6, 0$
Reject -6 since $\log_2(-6+2) = \log_2(-4)$ is not defined. Solution set: $\{0\}$

71. $\log_5(x^2-5x+6) - \log_5(x-2) = 1 \Rightarrow$
$\log_5\left(\dfrac{x^2-5x+6}{x-2}\right) = 1 \Rightarrow \dfrac{x^2-5x+6}{x-2} = 5 \Rightarrow$
$x^2-5x+6 = 5x-10 \Rightarrow x^2-10x+16 = 0 \Rightarrow$
$(x-2)(x-8) = 0 \Rightarrow x = 2, 8$
If $x = 2$, then $\log_5(x-2) = \log_5 0$, which is undefined, so reject $x = 2$. Solution set: $\{8\}$

73. $\log_6(x-2) + \log_6(x+1)$
$= \log_6(x+4) + \log_6(x-3) \Rightarrow$
$\log_6[(x-2)(x+1)] = \log_6[(x+4)(x-3)] \Rightarrow$
$x^2-x-2 = x^2+x-12 \Rightarrow -2x = -10 \Rightarrow x = 5$

75. $2\ln 3x = 3\ln x \Rightarrow \ln(3x)^2 = \ln x^3 \Rightarrow$
$9x^2 = x^3 \Rightarrow x = 9$

77. Multipling both sides of $2^x - 8 \cdot 2^{-x} - 7 = 0$ by 2^x we have $2^{2x} - 8 \cdot 2^{-x+x} - 7 \cdot 2^x = 0 \Rightarrow$ $2^{2x} - 7 \cdot 2^x - 8 = 0$. Letting $u = 2^x$, we have $u^2 - 7u - 8 = 0 \Rightarrow (u-8)(u+1) = 0 \Rightarrow u = 8$ or $u = -1$. (Reject this solution) $2^x = 8 \Rightarrow x = 3$

79. $3(0.2^x) + 5 \le 20 \Rightarrow 3(0.2^x) \le 15 \Rightarrow$
$(0.2^x) \le 5 \Rightarrow x \ln 0.2 \le \ln 5 \Rightarrow x \ge \dfrac{\ln 5}{\ln 0.2} \Rightarrow$
$x \ge -1$

81. Note that the domain of $\log(2x+7)$ is $\left(-\dfrac{7}{2}, \infty\right)$ since $2x + 7$ must be greater than 0.
$\log(2x+7) < 2 \Rightarrow 2x + 7 < 10^2 \Rightarrow 2x < 93 \Rightarrow$
$x < \dfrac{93}{2}$
Solution set: $\left(-\dfrac{7}{2}, \dfrac{93}{2}\right)$

Applying the Concepts

83. $2P = P(1+0.0625)^t \Rightarrow 2 = 1.0625^t \Rightarrow$
$\ln 2 = t \ln 1.0625 \Rightarrow t = \dfrac{\ln 2}{\ln 1.0625} \approx 11.4$
It will take about 11.4 years to double the investment.

85. Find k: $\dfrac{1}{2} = e^{20k} \Rightarrow \ln\left(\dfrac{1}{2}\right) = 20k \Rightarrow$
$k = \dfrac{\ln(1/2)}{20}$
$0.25 = e^{\frac{\ln(1/2)}{20}t} \Rightarrow \ln 0.25 = \dfrac{\ln(1/2)}{20} t \Rightarrow$
$t = \dfrac{\ln 0.25}{\frac{\ln(1/2)}{20}} = 40$
It will take 40 hours to decay to 25% of its original value.

87. At 5% compounded yearly,
$A = 7000\left(1 + \dfrac{0.05}{1}\right)^{7(1)} = 9849.70$.
At 4.75% compounded monthly,
$A = 7000\left(1 + \dfrac{0.0475}{12}\right)^{7(12)} = 9754.74$.
The 5% investment provides the greater return.

89. a. $t = 10; P(10) = 33e^{0.003(10)} \approx 34$ million

b. $60 = 33e^{0.003t} \Rightarrow \dfrac{60}{33} = e^{0.003t} \Rightarrow$
$\ln\left(\dfrac{60}{33}\right) = 0.003t \Rightarrow t = \dfrac{\ln\left(\dfrac{60}{33}\right)}{0.003} \approx 199.3$ yr
The population will be 60 million sometime during the year 2206 (after 199.3 years).

91. a.

b. $0.029 = 0.3e^{-0.47t} \Rightarrow \dfrac{0.029}{0.3} = e^{-0.47t} \Rightarrow$
$\ln\left(\dfrac{0.029}{0.3}\right) = -0.47t \Rightarrow t = \dfrac{\ln\left(\dfrac{0.029}{0.3}\right)}{-0.47} \Rightarrow$
$t \approx 4.97 \approx 5$ hours

93. First find a and k:
$f(0) = 212 = 75 + ae^{-k(0)} \Rightarrow 137 = a$
$f(1) = 200 = 75 + 137e^{-k(1)} \Rightarrow \dfrac{125}{137} = e^{-k} \Rightarrow$
$\ln(125/137) = -k \Rightarrow k \approx 0.0917$
$150 = 75 + 137e^{-0.0917t} \Rightarrow \dfrac{75}{137} = e^{-0.0917t} \Rightarrow$
$t = \dfrac{\ln(75/137)}{-0.0917} \approx 6.57$ minutes

95. a. $D(0) = 5e^{0.08(0)} \approx 5$ thousand = 5000 people per square mile

b. $D(5) = 5e^{0.08(5)} \approx 7.459$ thousand = 7459 people per square mile

c. $15 = 5e^{0.08t} \Rightarrow 3 = e^{0.08t} \Rightarrow$
$t = \dfrac{\ln 3}{0.08} \approx 13.73$ miles

97. $I = I_0 e^{-0.73(2)} \approx 0.2322 I_0$

99. a. $s(470,000) = 0.04 + 0.86\ln(470,000)$
≈ 11.27 feet per second

b. $s(450) = 0.04 + 0.86\ln(450)$
≈ 5.29 feet per second

c. $4.6 = 0.04 + 0.86 \ln p \Rightarrow \dfrac{4.56}{0.86} = \ln p \Rightarrow$
$p = e^{4.56/0.86} \approx 200.8 \approx 201$ people

101. a. $m(0) = \dfrac{6}{1+5e^{-0.7(0)}} = \dfrac{6}{1+5} = 1$ gram

b. The mass approaches 6 grams

c. $5 = \dfrac{6}{1+5e^{-0.7t}} \Rightarrow 5 + 25e^{-0.7t} = 6 \Rightarrow$
$e^{-0.7t} = \dfrac{1}{25} \Rightarrow t = \dfrac{\ln\left(\dfrac{1}{25}\right)}{-0.7} \approx 4.6$ days

103. $P(f) = 1200 \log_2 \dfrac{f}{f_0}$
$P(f) = 1200 \log_2 \left(\dfrac{1.2 f_0}{f_0}\right) = 1200 \log_2 (1.2)$
$= 1200 \left(\dfrac{\log 1.2}{\log 2}\right) \approx 315.6 \approx 316$

The difference, $316 - 300 = 16$ cents, is noticeable.

105. $\text{pH}_{\text{very acid rain}} = 2.4 = -\log\left[\text{H}^+\right] \Rightarrow$
$\left[\text{H}^+\right] = 10^{-2.4} \Rightarrow$
$\left[\text{H}^+\right] = 10^{0.6} \times 10^{-3} \approx 4.0 \times 10^{-3}$

The average concentration of hydrogen ions in the very acid rain is 4.0×10^{-3} moles per liter.

$\text{pH}_{\text{acid rain}} = 5.6 = -\log\left[\text{H}^+\right] \Rightarrow$
$\left[\text{H}^+\right] = 10^{-5.6} = 10^{0.4} \times 10^{-6} \approx 2.5 \times 10^{-6}$

$\dfrac{\left[\text{H}^+\right]_{\text{very acid rain}}}{\left[\text{H}^+\right]_{\text{acid rain}}} = \dfrac{10^{-2.4}}{10^{-5.6}} = 10^{3.2} = 10^{0.2} \times 10^3$
$\approx 1.585 \times 10^3 = 1585$

The very acid rain is 1585 times more acidic than acid rain.

107. $I = I_0 \times 10^{L/10}$
$I_{115} = I_0 \times 10^{115/10}$
$I_{95} = I_0 \times 10^{95/10}$
$\dfrac{I_{115}}{I_{95}} = \dfrac{I_0 \times 10^{115/10}}{I_0 \times 10^{95/10}} = 10^{20/10} = 10^2 = 100$

The 115 dB sound is 100 times as intense as the 95 dB sound.

109. Let m_1 represent the magnitude of Sirius and let m_2 represent the magnitude of Saturn.
$m_2 - m_1 = 2.5 \log\left(\dfrac{b_1}{b_2}\right)$
$1.47 - (-1) = 2.5 \log\left(\dfrac{b_1}{b_2}\right) \Rightarrow$
$\dfrac{2.47}{2.5} = 0.988 = \log\left(\dfrac{b_1}{b_2}\right) \Rightarrow$
$\dfrac{b_1}{b_2} = 10^{0.988} \approx 9.7 \Rightarrow b_1 = 9.7 b_2$

Sirius is about 9.7 times brighter than Saturn.

Chapter 4 Practice Test A

1. $5^{-x} = 125 \Rightarrow 5^{-x} = 5^3 \Rightarrow x = -3$

2. $\log_2 x = 5 \Rightarrow x = 2^5 = 32$

3. Range: $(-\infty, 1)$; asymptote: $y = 1$

4. $\log_2 \dfrac{1}{8} \Rightarrow 2^x = \dfrac{1}{8} = 2^{-3} \Rightarrow x = -3$

5. $\left(\dfrac{1}{4}\right)^{2-x} = 4 \Rightarrow 4^{x-2} = 4 \Rightarrow x - 2 = 1 \Rightarrow x = 3$

6. $\log 0.001 \Rightarrow 10^x = 0.001 = 10^{-3} \Rightarrow x = -3$

7. $\ln 3 + 5 \ln x = \ln 3 + \ln x^5 = \ln(3x^5)$

8. $2^{x+1} = 5 \Rightarrow (x+1)\ln 2 = \ln 5 \Rightarrow$
$x \ln 2 + \ln 2 = \ln 5 \Rightarrow x \ln 2 = \ln 5 - \ln 2 \Rightarrow$
$x \ln 2 = \ln\left(\dfrac{5}{2}\right) \Rightarrow x = \dfrac{\ln(5/2)}{\ln 2}$

9. $e^{2x} + e^x - 6 = 0 \Rightarrow (e^x + 3)(e^x - 2) = 0 \Rightarrow$
$e^x = -3$ (reject this) or $e^x - 2 = 0 \Rightarrow$
$e^x = 2 \Rightarrow x = \ln 2$

10. $\ln \dfrac{2x^3}{(x+1)^5} = \ln(2x^3) - \ln(x+1)^5$
$= \ln 2 + 3 \ln x - 5 \ln(x+1)$

11. $\ln e^{-5} = -5$

12. $y = \ln(x-1) + 3$

13.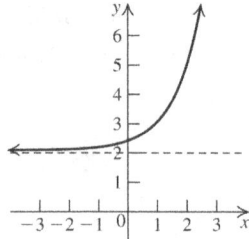

14. Domain: $(-\infty, 0)$

15. $3\ln x + \ln(x^3 + 2) - \dfrac{1}{2}\ln(3x^2 + 2)$
    $= \ln x^3 + \ln(x^3 + 2) - \ln\sqrt{3x^2 + 2}$
    $= \ln\dfrac{x^3(x^3 + 2)}{\sqrt{3x^2 + 2}}$

16. $\log x = \log 6 - \log(x-1) \Rightarrow \log x = \log\dfrac{6}{x-1} \Rightarrow$
    $x = \dfrac{6}{x-1} \Rightarrow x^2 - x = 6 \Rightarrow x^2 - x - 6 = 0 \Rightarrow$
    $(x-3)(x+2) = 0 \Rightarrow x = -2$ (reject this) or $x = 3$

17. $\log_x 9 = 2 \Rightarrow x^2 = 9 \Rightarrow x = 3$

18. $A = 15{,}000\left(1 + \dfrac{0.07}{4}\right)^{4t} = 15{,}000(1.0175)^{4t}$

19. $1{,}500{,}000 = 15{,}000 e^{0.2t} \Rightarrow 100 = e^{0.2t} \Rightarrow$
    $\ln 100 = 0.2t \Rightarrow t = \dfrac{\ln 100}{0.2} \approx 23.03$
    The Hispanic population reached 1.5 million in 1983, 23.03 years after 1960.

20. Let I_{1960} denote the intensity of the 1960 Morocco earthquake and let I_{1972} denote the intensity of the 1972 Nicaragua earthquake. Then,
    $5.8 = \log\left(\dfrac{I_{1960}}{I_0}\right) \Rightarrow \dfrac{I_{1960}}{I_0} = 10^{5.8} \Rightarrow$
    $I_{1960} = 10^{5.8} I_0$
    $6.2 = \log\left(\dfrac{I_{1972}}{I_0}\right) \Rightarrow \dfrac{I_{1972}}{I_0} = 10^{6.2} \Rightarrow$
    $I_{1972} = 10^{6.2} I_0$
    $\dfrac{I_{1972}}{I_{1960}} = \dfrac{10^{6.2} I_0}{10^{5.8} I_0} = 10^{0.4} \approx 2.5 \Rightarrow I_{1972} \approx 2.5 I_{1960}$
    The intensity of the 1972 Nicarague earthquake was about 2.5 times that of the 1960 Morocco earthquake.

Chapter 4 Practice Test B

1. $3^{-x} = 9 \Rightarrow 3^{-x} = 3^2 \Rightarrow x = -2$.
   The answer is B.

2. $\log_5 x = 2 \Rightarrow 5^2 = 25 = x$. The answer is B.

3. The answer is B.

4. $\log_4 64 \Rightarrow 4^x = 64 = 4^3 \Rightarrow x = 3$.
   The answer is D.

5. $\left(\dfrac{1}{3}\right)^{1-x} = 3 \Rightarrow 3^{x-1} = 3 \Rightarrow x - 1 = 1 \Rightarrow x = 2$.
   The answer is D.

6. $\log 0.01 \Rightarrow 10^x = 10^{-2} \Rightarrow x = -2$.
   The answer is B.

7. $\ln 7 + 2\ln x = \ln(7x^2)$. The answer is B.

8. $2^{x^2} = 3^x \Rightarrow x^2 \ln 2 = x \ln 3 \Rightarrow$
   $x^2 \ln 2 - x \ln 3 = 0 \Rightarrow x(x \ln 2 - \ln 3) = 0 \Rightarrow$
   $x = 0$ or $x\ln 2 - \ln 3 = 0 \Rightarrow x = \dfrac{\ln 3}{\ln 2}$
   The answer is C.

9. $e^{2x} - e^x - 6 = 0 \Rightarrow (e^x - 3)(e^x + 2) = 0 \Rightarrow$
   $e^x = -2$ (reject this) or $e^x - 3 = 0 \Rightarrow$
   $e^x = 3 \Rightarrow x = \ln 3$.
   The answer is D.

10. $\ln\dfrac{3x^2}{(x+1)^{10}} = \ln(3x^2) - \ln(x+1)^{10}$
    $= \ln 3 + 2\ln x - 10\ln(x+1)$.
    The answer is D.

11. $\ln e^{3x} = 3x$. The answer is B.

12. The answer is D.

13. The answer is B.

14. The answer is A.

15. $\ln x - 2\ln(x^2 + 1) + \dfrac{1}{2}\ln(x^4 + 1)$
    $= \ln x - \ln(x^2 + 1)^2 + \ln\sqrt{x^4 + 1}$
    $= \ln\dfrac{x\sqrt{x^4 + 1}}{(x^2 + 1)^2}$.
    The answer is A.

16. $\log x = \log 12 - \log(x+1) \Rightarrow \log x = \log \dfrac{12}{x+1} \Rightarrow$
$x = \dfrac{12}{x+1} \Rightarrow x^2 + x = 12 \Rightarrow x^2 + x - 12 = 0 \Rightarrow$
$(x+4)(x-3) = 0 \Rightarrow x = -4$ (reject this) or $x = 3$
The answer is D.

17. $\log_x 16 = 4 \Rightarrow x^4 = 16 = 2^4 \Rightarrow x = 2$
The answer is B.

18. $A = 12{,}000\left(1 + \dfrac{0.105}{12}\right)^{12t} = 12{,}000(1.00875)^{12t}$
The answer is D.

19. $t = 2020 - 2000 = 20$
$P(20) = 10{,}000 \log_5(20+5) = 10{,}000 \log_5 25$
$= 20{,}000$. The answer is B.

20. Let I_{1994} denote the intensity of the 1994 Northridge, CA earthquake and let I_{1988} denote the intensity of the 1988 Armenia earthquake. Then,
$6.7 = \log\left(\dfrac{I_{1994}}{I_0}\right) \Rightarrow \dfrac{I_{1994}}{I_0} = 10^{6.7} \Rightarrow$
$I_{1994} = 10^{6.7} I_0$
$7.0 = \log\left(\dfrac{I_{1988}}{I_0}\right) \Rightarrow \dfrac{I_{1988}}{I_0} = 10^{7.0} \Rightarrow$
$I_{1988} = 10^{7.0} I_0$
$\dfrac{I_{1988}}{I_{1994}} = \dfrac{10^{7.0} I_0}{10^{6.7} I_0} = 10^{0.3} \approx 2.0 \Rightarrow$
$I_{1988} \approx 2.0 I_{1994}$
The intensity of the 1988 Armenia earthquake was about twice that of the 1994 Northridge, CA earthquake. The answer is B.

Cumulative Review Exercises (Chapters P–4)

1. $x^2 + (0-1)^2 = 1 \Rightarrow x^2 = 0 \Rightarrow x = 0$.
$0^2 + (y-1)^2 = 1 \Rightarrow y^2 - 2y + 1 = 1 \Rightarrow$
$y^2 - 2y = 0 \Rightarrow y(y-2) = 0 \Rightarrow y = 0$ or $y = 2$.
The x-intercept is 0.
The y-intercepts are 0 and 2.
$x^2 + (y-1)^2 = 1 \Rightarrow (y-1)^2 = 1 - x^2 \Rightarrow$
$y - 1 = \pm\sqrt{1-x^2} \Rightarrow y = 1 \pm \sqrt{1-x^2} = f(x)$
$f(-x) = 1 \pm \sqrt{1-(-x)^2} = 1 \pm \sqrt{1-x^2} = f(x) \Rightarrow$
$f(x)$ is even $\Rightarrow f$ is symmetric about the y-axis.

2.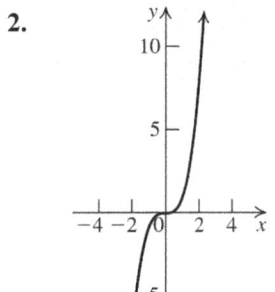

3. $x^2 + 2x + y^2 - 4y - 20 = 0$
Group the terms and complete the squares.
$(x^2 + 2x + 1) + (y^2 - 4y + 4) = 20 + 1 + 4$
$(x+1)^2 + (y-2)^2 = 25$
The center of the circle is $(-1, 2)$ and its radius is 5.

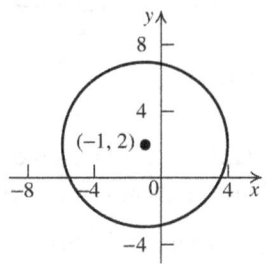

4. $2x + 3y = 17 \Rightarrow y = -\dfrac{2}{3}x + \dfrac{17}{3} \Rightarrow$ the slope of the perpendicular line is $\dfrac{3}{2}$.
$4 = \dfrac{3}{2}(-2) + b \Rightarrow 7 = b$.
The equation is $y = \dfrac{3}{2}x + 7$.

5. Logarithms are positive, so the domain is $(-\infty, -3) \cup (2, \infty)$.

6. a. $f(-2) = 2(-2) - 3 = -7$

  b. $f(0) = 2(0) - 3 = -3$

  c. $f(3) = 2(3^2) + 1 = 19$

7. Shift the graph of $f(x) = \sqrt{x}$ one unit left, stretch vertically by a factor of 3, then shift the graph two units down.

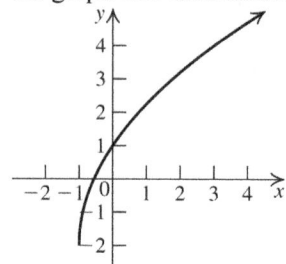

8. $(f \circ g)(x) = f\left(-\dfrac{1}{x}\right) = 3\sqrt{-\dfrac{1}{x}} \Rightarrow$ the domain is $(-\infty, 0)$.

9. a. The graph of $f(x)$ passes the horizontal line test, so there is an inverse. $y = 2x^3 + 1$ becomes $x = 2y^3 + 1 \Rightarrow \dfrac{x-1}{2} = y^3 \Rightarrow$
$y = \sqrt[3]{\dfrac{x-1}{2}} = f^{-1}(x)$.

   b. The graph of $f(x)$ does not pass the horizontal line test, so there is no inverse.

   c. The graph of $f(x)$ passes the horizontal line test, so there is an inverse. $y = \ln x$ becomes $x = \ln y \Rightarrow e^x = y = f^{-1}(x)$.

10. a. $x^2 + 1 \overline{\smash{\big)}\, 2x^3 + 0x^2 + 3x + 1}$
$\underline{2x^3 + 0x^2 + 2x}$
$x + 1$

The quotient is $2x + \dfrac{x+1}{x^2+1}$.

   b. $\begin{array}{r|rrrrr} -3 & 2 & 7 & 0 & -2 & 3 \\ & & -6 & -3 & 9 & -21 \\ \hline & 2 & 1 & -3 & 7 & -18 \end{array}$

$\dfrac{2x^4 + 7x^3 - 2x + 3}{x+3} = 2x^3 + x^2 - 3x + 7 - \dfrac{18}{x+3}$

11. a. $(x-1)(x-2)(x+3) = x^3 - 7x + 6$

   b. $(x-1)(x+1)(x-i)(x+i) = x^4 - 1$

12. a. $\log_2 5x^3 = \log_2 5 + 3\log_2 x$

   b. $\log_a \sqrt[3]{\dfrac{xy^2}{z}} = \dfrac{1}{3}\log_a \dfrac{xy^2}{z}$
   $= \dfrac{1}{3}(\log_a x + 2\log_a y - \log_a z)$

   c. $\ln \dfrac{3\sqrt{x}}{5y} = \ln 3\sqrt{x} - \ln 5y$
   $= \ln 3 + \dfrac{1}{2}\ln x - (\ln 5 + \ln y)$
   $= \ln 3 + \dfrac{1}{2}\ln x - \ln 5 - \ln y$

13. a. $\dfrac{1}{2}(\log x + \log y) = \dfrac{1}{2}\log(xy) = \log\sqrt{xy}$

   b. $3\ln x - 2\ln y = \ln x^3 - \ln y^2 = \ln \dfrac{x^3}{y^2}$

14. a. $\log_9 17 = \dfrac{\log 17}{\log 9} \approx 1.289$

   b. $\log_3 25 = \dfrac{\log 25}{\log 3} \approx 2.930$

   c. $\log_{1/2} 0.3 = \dfrac{\log 0.3}{\log (1/2)} \approx 1.737$

15. a. $\log_6(x+3) + \log_6(x-2) = 1 \Rightarrow$
   $\log_6((x+3)(x-2)) = 1 \Rightarrow$
   $(x+3)(x-2) = 6 \Rightarrow x^2 + x - 6 = 6 \Rightarrow$
   $x^2 + x - 12 = 0 \Rightarrow (x+4)(x-3) = 0 \Rightarrow$
   $x = -4$ (reject this) or $x = 3$

   b. $5^{x^2 - 4x + 5} = 25 = 5^2 \Rightarrow x^2 - 4x + 5 = 2 \Rightarrow$
   $x^2 - 4x + 3 = 0 \Rightarrow (x-3)(x-1) = 0 \Rightarrow$
   $x = 1$ or $x = 3$

   c. $3.1^{x-1} = 23 \Rightarrow (x-1)\ln 3.1 = \ln 23 \Rightarrow$
   $x - 1 = \dfrac{\ln 23}{\ln 3.1} \Rightarrow x = 1 + \dfrac{\ln 23}{\ln 3.1} \approx 3.771$

16. The possible rational zeros are $\{\pm 1, \pm 2, \pm 4\}$.
Using synthetic division, we find that one zero is 1:

$$\underline{1|}\begin{array}{rrrrr} 1 & -1 & -2 & -2 & 4 \\ & 1 & 0 & -2 & -4 \\ \hline 1 & 0 & -2 & -4 & 0 \end{array}$$

Using synthetic division again, we find that 2 is a root of the depressed equation $f(x) = x^3 - 2x - 4$:

$$\underline{2|}\begin{array}{rrrr} 1 & 0 & -2 & -4 \\ & 2 & 4 & 4 \\ \hline 1 & 2 & 2 & 0 \end{array}$$

There are no rational zeros of the depressed equation $f(x) = x^2 + 2x + 2$. So the only rational zeros are $\{1, 2\}$.
Find the upper bound for the zeros by testing 1, 2, 3,... using synthetic division. The smallest value that makes all the terms in the bottom row nonnegative is the upper bound. The upper bound is 3.

$$\underline{3|}\begin{array}{rrrrr} 1 & -1 & -2 & -2 & 4 \\ & 3 & 6 & 12 & 30 \\ \hline 1 & 2 & 4 & 10 & 34 \end{array}$$

Find the lower bound for the zeros by testing −1, −2, −3,...using synthetic division. The largest value that makes the terms in the bottom row alternate signs is the lower bound. The lower bound is −1.

$$\underline{-1|}\begin{array}{rrrrr} 1 & -1 & -2 & -2 & 4 \\ & -1 & 2 & 0 & 2 \\ \hline 1 & -2 & 0 & -2 & 6 \end{array}$$

17. a. The zeros are 1 (multiplicity 3), −2 (multiplicity 2), and 3 (multiplicity 1).

b. At $x = 1$ and $x = 3$, the graph crosses the x-axis. At $x = -2$, the graph touches, but does not cross, the x-axis.

c. As $x \to -\infty, f(x) \to \infty$.
As $x \to \infty, f(x) \to \infty$.

d.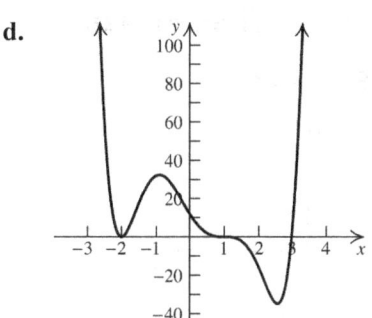

18. a. The vertical asymptotes are $x = 3$ and $x = -4$.

b. $f(x) = \dfrac{(x-1)(x+2)}{(x-3)(x+4)} = \dfrac{x^2 + x - 2}{x^2 + x - 12} \Rightarrow$ the horizontal asymptote is $y = 1$.

c. $f(x)$ lies above the horizontal asymptote on $(-\infty, -4) \cup (3, \infty)$. $f(x)$ lies below the horizontal asymptote on $(-4, 3)$.

d.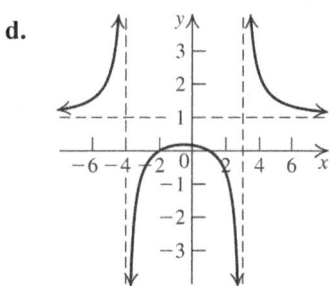

Chapter 5 Trigonometric Functions

5.1 Angles and Their Measure

5.1 Practice Problems

1.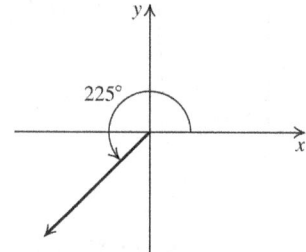

2. a. $13°9'22'' = \left(13 + \dfrac{9}{60} + \dfrac{22}{3600}\right)° \approx 13.16°$

   b. $41.275° = 41° + 0.275(60') = 41° + 16.5'$
   $= 41° + 16' + 0.5(60'') = 41°16'30''$

3. $-45° = -45 \cdot \dfrac{\pi}{180} = -\dfrac{\pi}{4}$ radian

4. $\dfrac{3\pi}{2} = \dfrac{3\pi}{2} \cdot \dfrac{180}{\pi} = 270°$

5. a. $1230° = 3(360°) + 150°$, so a $1230°$ angle is coterminal with an angle of $150°$.

   b. $-1560° + 5(360°) = -1560° + 1800°$
   $= 240°$
   An angle of $-1560°$ is coterminal with an angle of $240°$.

6. complement: $90° - 67° = 23°$
   supplement: $180° - 67° = 113°$

7. First convert the angle measurement from degrees to radians:
   $225° = 225°\left(\dfrac{\pi}{180°}\right) = \dfrac{5\pi}{4}$ radians.
   Then, $s = r\theta = 2\left(\dfrac{5\pi}{4}\right) = \dfrac{5\pi}{2} \approx 7.85$ m

8. The difference in the latitudes is
   $41°51' - 30°25' = 11°26'$
   $= 11° + \left(\dfrac{26}{60}\right)° \approx 11.43°$
   $= 11.43°\left(\dfrac{\pi}{180°}\right)$
   ≈ 0.1995 radian
   $s = 0.1995 \cdot 3960 \approx 790$ miles.

9. First convert the angle measurement from degrees to radians:
   $\theta = 60° \cdot \dfrac{\pi}{180°} = \dfrac{\pi}{3}$ radians.
   $A = \dfrac{1}{2}r^2\theta = \dfrac{1}{2} \cdot 10^2 \cdot \dfrac{\pi}{3} = \dfrac{50\pi}{3} \approx 52.36$ in.^{2}

10. First convert revolutions per minute into radians per minute:
    18 revolutions per minute $= 18 \cdot 2\pi = 36\pi$ radians per minute. Thus, the angular speed $\omega = 36\pi$ radians per minute.
    To find the linear speed, use the formula $v = r\omega$: $v = 10 \cdot 36\pi = 360\pi$ radians per minute or about 1131 feet per minute.

5.1 Concepts and Vocabulary

1. A negative angle is formed by rotating the initial side in the clockwise direction.

3. Two angles with the same initial and terminal sides are coterminal angles.

5. False. One radian $= \dfrac{180}{\pi}° \approx 57.3°$.

7. True. Acute angles measure more than $0°$ and less than $90°$.

5.1 Building Skills

9.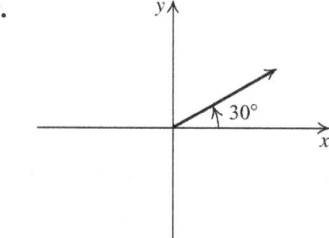

11.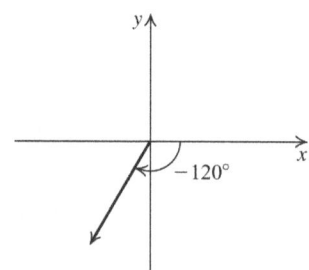

13.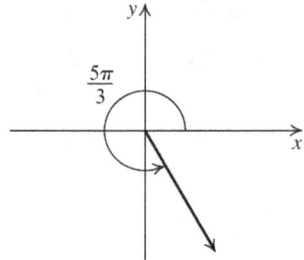

15.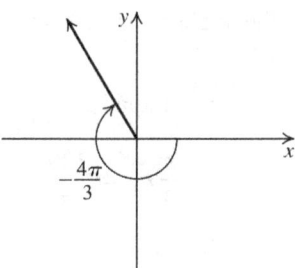

17. $70°45' = \left(70 + \dfrac{45}{60}\right)° = 70.75°$

19. $23°42'30'' = \left(23 + \dfrac{42}{60} + \dfrac{30}{3600}\right)° \approx 23.71°$

21. $-15°42'57'' = -\left(15 + \dfrac{42}{60} + \dfrac{57}{3600}\right)° \approx -15.72°$

23. $27.32° = 27° + 0.32(60') = 27° + 19.2'$
    $= 27° + 19' + 0.2(60'') = 27°19'12''$

25. $13.347° = 13° + 0.347(60') = 13° + 20.82'$
    $= 13° + 20' + 0.82(60'') = 13°20'49''$

27. $19.0511° = 19° + 0.0511(60') = 19° + 3.066'$
    $= 19° + 3' + 0.066(60'') = 19°3'4''$

29. $20° = 20 \cdot \dfrac{\pi}{180} = \dfrac{\pi}{9}$ radian

31. $-180° = -180 \cdot \dfrac{\pi}{180} = -\pi$ radians

33. $315° = 315 \cdot \dfrac{\pi}{180} = \dfrac{7\pi}{4}$ radians

35. $-510° = -510 \cdot \dfrac{\pi}{180} = -\dfrac{17\pi}{6}$ radians

37. $\dfrac{\pi}{12} = \dfrac{\pi}{12} \cdot \dfrac{180}{\pi} = 15°$

39. $-\dfrac{5\pi}{9} = -\dfrac{5\pi}{9} \cdot \dfrac{180}{\pi} = -100°$

41. $\dfrac{5\pi}{3} = \dfrac{5\pi}{3} \cdot \dfrac{180}{\pi} = 300°$

43. $-\dfrac{11\pi}{4} = -\dfrac{11\pi}{4} \cdot \dfrac{180}{\pi} = -495°$

For exercises 45–48, make sure your calculator is in Radian mode.

45. $12° = 12°\left(\dfrac{\pi}{180°}\right) \approx 0.21$ radian

    ```
    12°
        .2094395102
    ```

47. $-84° = -84°\left(\dfrac{\pi}{180°}\right) \approx -1.47$ radians

    ```
    -84°
        -1.466076572
    ```

For exercises 49–52, make sure your calculator is in Degree mode.

49. 0.94 radians $= 0.94\left(\dfrac{180°}{\pi}\right) \approx 53.86°$

    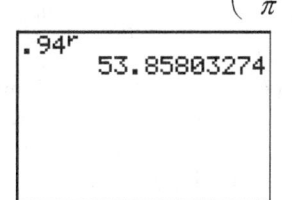

51. -8.21 radians $= -8.21\left(\dfrac{180°}{\pi}\right) \approx -470.40°$

    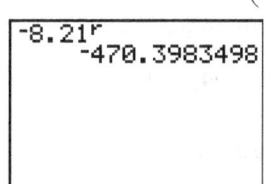

53. $-\dfrac{\pi}{4} + 2\pi = \dfrac{7\pi}{4}$

    An angle of $-\dfrac{\pi}{4}$ is coterminal with an angle of $\dfrac{7\pi}{4}$.

55. $-\dfrac{7\pi}{4} + 2\pi = \dfrac{\pi}{4}$

    An angle of $-\dfrac{7\pi}{4}$ is coterminal with an angle of $\dfrac{\pi}{4}$.

57. $\frac{16\pi}{3} - 4\pi = \frac{4\pi}{3}$

An angle of $\frac{16\pi}{3}$ is coterminal with an angle of $\frac{4\pi}{3}$.

59. $-65° + 360° = 295°$

An angle of $-65°$ is coterminal with an angle of $295°$.

61. $-200° + 360° = 160°$

An angle of $-200°$ is coterminal with an angle of $160°$.

63. $700° - 360° = 340°$

An angle of $700°$ is coterminal with an angle of $340°$.

65. complement: $43°$; supplement: $133°$

67. complement: none because the measure of the angle is greater than $90°$; supplement: $60°$

69. complement: none because the measure of the angle is greater than $90°$; supplement: none because the measure of the angle is greater than $180°$

In exercises 71–90, use the formulas
$$s = r\theta, v = \frac{s}{t}, \omega = \frac{\theta}{t}, v = r\omega, \text{ and } A = \frac{1}{2}r^2\theta,$$
where θ is the radian measure of the central angle that intercepts an arc of length s in a circle of radius r, v is the linear velocity, ω is the angular velocity, A is the area of a sector of the circle, and t is the time.

71. $s = r\theta \Rightarrow 7 = 25\theta \Rightarrow \theta = \frac{7}{25} = 0.28$ radian

73. $s = r\theta \Rightarrow 22 = 10.5\theta \Rightarrow$
$\theta = \frac{22}{10.5} = \frac{44}{21} \approx 2.095$ radians

75. First convert the angle measurement from degrees to radians: $25° = 25°\left(\frac{\pi}{180°}\right) = \frac{5\pi}{36}$.

Then $s = r\theta \Rightarrow s = 3 \cdot \frac{5\pi}{36} = \frac{5\pi}{12} \approx 1.309$ m.

77. $s = r\theta \Rightarrow s = 6.5 \cdot 12 = 78$ m

79. $s = r\theta \Rightarrow 3 = r \cdot \frac{1}{2} \Rightarrow r = 6$ cm

81. First convert the angle measurement from degrees to radians: $30° = 30°\left(\frac{\pi}{180°}\right) = \frac{\pi}{6}$

Then $s = r\theta \Rightarrow \frac{3\pi}{2} = r \cdot \frac{\pi}{6} \Rightarrow 9$ m

83. $A = \frac{1}{2}r^2\theta \Rightarrow \frac{1}{2} \cdot 10^2 \cdot 4 = 200$ in.^{2}

85. First convert the angle measurement from degrees to radians:
$60° = 60°\left(\frac{\pi}{180°}\right) = \frac{\pi}{3}$ radians

$A = \frac{1}{2}r^2\theta \Rightarrow 20 = \frac{1}{2} \cdot \frac{\pi}{3}r^2 \Rightarrow \frac{120}{\pi} = r^2 \Rightarrow$
$r = \sqrt{\frac{120}{\pi}} \approx 6.180$ ft

87. $v = r\omega \Rightarrow v = 6 \cdot 10 = 60$ m/min

89. $v = r\omega \Rightarrow 20 = 10\omega \Rightarrow \omega = 2$ radians/sec

5.1 Applying the Concepts

91. Three-quarters of a revolution is $3\pi/2$ radians. So the arc length (the distance the car moves) is $s = r\theta = 15\left(\frac{3\pi}{2}\right) \approx 70.69$ inches

93. The hands of the clock form an angle of
$\frac{4}{12} \cdot 360° = 120° \cdot \frac{\pi}{180°} = \frac{2\pi}{3}$ radians.

95. First convert the angle measurement from degrees to radians:
$\theta = 25° \cdot \frac{\pi}{180°} = \frac{5\pi}{36}$ radians. Using
$s = r\theta \Rightarrow \frac{5\pi}{36} = \frac{4}{r} \Rightarrow r \approx 9$ in. $\Rightarrow d \approx 18$ in.

97. The arc length is 6 inches and the radius is 4 inches, so $\theta = \frac{6}{4} = 1.5$ radians.

1 radian $= \frac{180°}{\pi}$, so $\theta = 1.5\left(\frac{180°}{\pi}\right) \approx 86°$.

99. a. The linear speed, v, of the larger pulley is 600 inches per minute. To find the angular speed, use the formula
$v = r\omega \Rightarrow 600 = 5\omega \Rightarrow$
$\omega = 120$ radians per minute.

b. The linear speed, v, of the smaller pulley is 600 inches per minute. To find the angular speed, use the formula $v = r\omega$.
$600 = 2\omega \Rightarrow \omega = 300$ radians per minute.

101. The radius of each wheel is 1 foot, while the linear velocity, v, is 25 miles per hour = $5280 \times 25 = 132{,}000$ feet per hour. $v = r\omega \Rightarrow 132{,}000 = 1 \cdot \omega$, so the angular speed is $132{,}000$ radians per hour.

103. The radius of the disk is 1.875 inches. Each revolution is 2π radians, so the disk rotating at 7200 revolutions per minute gives
$\omega = 7200 \dfrac{\text{revolutions}}{\text{minute}} \cdot \dfrac{2\pi \text{ radians}}{\text{revolution}} = 14{,}400\pi$
radians per minute.
Then $v = 1.875 \cdot 14{,}400\pi \approx 84{,}823$ inches per minute.

105. The difference in the latitudes is
$39°44' - 32°23' = 7°21' = 7 + \left(\dfrac{21}{60}\right)° = 7.35°$
$= 7.35°\left(\dfrac{\pi}{180°}\right) \approx 0.1283$ radian.
$s = 0.1283 \cdot 3960 \approx 508$ miles.

107. The difference in the latitudes is
$52°23' - 45°42' = 6°41' = 6 + \left(\dfrac{41}{60}\right)° = 6.683°$.
$s = 6.683°\left(\dfrac{\pi}{180°}\right) \cdot 3960 \approx 462$ miles.

109. Let the central angle θ = the difference in latitude. Then
$\theta = \dfrac{s}{r} = \dfrac{440}{3960} \approx 0.1111$ radian $\Rightarrow$
$\theta \approx 0.1111\left(\dfrac{180°}{\pi}\right) \approx 6.366°$.

111.

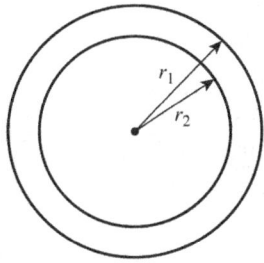

We are seeking $r_1 - r_2$. Use the formula $C = 2\pi r$ to find the radius of each circle.
$396 = 2\pi r_1 \Rightarrow \dfrac{396}{2\left(\frac{22}{7}\right)} = 63 = r_1$

$352 = 2\pi r_2 \Rightarrow \dfrac{352}{2\left(\frac{22}{7}\right)} = 56 = r_2$
$r_1 - r_2 = 63 - 56 = 7$
The width of the track is 7 m.

113. 66 km/hr = 66 km/60 min = 1.1 km/min.
 = 110,000 cm/min.
$C = \pi d = \dfrac{22}{7} \cdot 140 = 440$ cm
$\dfrac{110{,}000 \text{ cm/min}}{440 \text{ cm}} = 250$ rev/min

115. We are seeking the area of a sector with radius $\dfrac{3}{2} = 1.5$ ft and central angle $\dfrac{2\pi}{8} = \dfrac{\pi}{4}$.
$A = \dfrac{1}{2}r^2\theta = \dfrac{1}{2} \cdot 1.5^2 \cdot \dfrac{\pi}{4} = \dfrac{1}{2} \cdot 1.5^2 \cdot \dfrac{1}{4} \cdot \dfrac{22}{7}$
$= \dfrac{99}{112} \approx 0.884$ ft^2

5.1 Critical Thinking/Discussion/Writing

117. $3, \pi, 4, \dfrac{3\pi}{2}$

118. The line of latitude through $40°40'13''$ is further north, so its radius is smaller.

119. The length of the arc $\dfrac{\pi r}{2} - s = \dfrac{\pi r}{2} - r\theta$
$= r\left(\dfrac{\pi}{2} - \theta\right)$. So $\dfrac{\pi}{2} - \theta$ is the complement of θ.

120. Eratosthenes used the fact that the sun's rays are effectively parallel to determine that the central angle that subtends the arc from Syene to Alexandria is the same as the angle formed by an upright rod and its shadow at Alexandria.

Let x = the circumference of the earth, and then solve the proportion $\dfrac{7.2°}{360°} = \dfrac{500 \text{ miles}}{x \text{ miles}}$.
This gives $x = 25{,}000$ miles

5.1 Getting Ready for the Next Section

121. $5^2 + 12^2 = c^2 \Rightarrow 169 = c^2 \Rightarrow 13 = c$

123. $a^2 + 6^2 = 10^2 \Rightarrow a^2 + 36 = 100 \Rightarrow a^2 = 64 \Rightarrow a = 8$

125. $a^2 + 3^2 = 6^2 \Rightarrow a^2 + 9 = 36 \Rightarrow a^2 = 27 \Rightarrow a = \sqrt{27} = 3\sqrt{3}$

127. $\dfrac{A}{B} = \dfrac{\frac{1}{2}}{\frac{\sqrt{3}}{2}} = \dfrac{\frac{1}{2} \cdot 2}{\frac{\sqrt{3}}{2} \cdot 2} = \dfrac{1}{\sqrt{3}} \cdot \dfrac{\sqrt{3}}{\sqrt{3}} = \dfrac{\sqrt{3}}{3}$

5.2 Right-Triangle Trigonometry

5.2 Practice Problems

1.

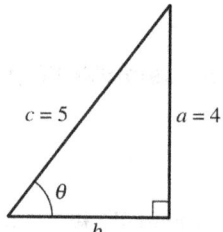

Use the Pythagorean theorem to find the length of the missing leg:

$25^2 = 16^2 + b^2 \Rightarrow b = 3$

$\sin\theta = \dfrac{\text{opposite}}{\text{hypotenuse}} = \dfrac{4}{5}$ \quad $\csc\theta = \dfrac{\text{hypotenuse}}{\text{opposite}} = \dfrac{5}{4}$

$\cos\theta = \dfrac{\text{adjacent}}{\text{hypotenuse}} = \dfrac{3}{5}$ \quad $\sec\theta = \dfrac{\text{hypotenuse}}{\text{adjacent}} = \dfrac{5}{3}$

$\tan\theta = \dfrac{\text{opposite}}{\text{adjacent}} = \dfrac{4}{3}$ \quad $\cot\theta = \dfrac{\text{adjacent}}{\text{opposite}} = \dfrac{3}{4}$

2. $\sin\theta = \dfrac{3}{5}$, $\cos\theta = \dfrac{4}{5}$

$\tan\theta = \dfrac{\sin\theta}{\cos\theta} = \dfrac{\frac{3}{5}}{\frac{4}{5}} = \dfrac{3}{4}$ \quad $\csc\theta = \dfrac{1}{\sin\theta} = \dfrac{1}{\frac{3}{5}} = \dfrac{5}{3}$

$\sec\theta = \dfrac{1}{\cos\theta} = \dfrac{1}{\frac{4}{5}} = \dfrac{5}{4}$ \quad $\cot\theta = \dfrac{1}{\tan\theta} = \dfrac{1}{\frac{3}{4}} = \dfrac{4}{3}$

3. $\cos\theta = \dfrac{1}{3} \Rightarrow b = \text{adjacent} = 1$ and $c = \text{hypotenuse} = 3$. Using the Pythagorean theorem, we have $a^2 + 1^2 = 3^3 \Rightarrow a = 2\sqrt{2}$.

$\sin\theta = \dfrac{\text{opposite}}{\text{hypotenuse}} = \dfrac{2\sqrt{2}}{3}$

$\csc\theta = \dfrac{\text{hypotenuse}}{\text{opposite}} = \dfrac{3}{2\sqrt{2}} = \dfrac{3\sqrt{2}}{4}$

$\sec\theta = \dfrac{\text{hypotenuse}}{\text{adjacent}} = 3$

$\tan\theta = \dfrac{\text{opposite}}{\text{adjacent}} = 2\sqrt{2}$

$\cot\theta = \dfrac{\text{adjacent}}{\text{opposite}} = \dfrac{1}{2\sqrt{2}} = \dfrac{\sqrt{2}}{4}$

4.

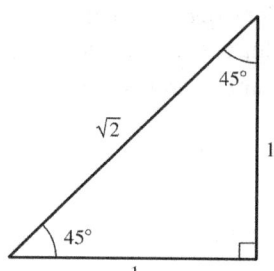

$\csc 45° = \dfrac{\text{hypotenuse}}{\text{opposite}} = \dfrac{\sqrt{2}}{1} = \sqrt{2}$

$\sec 45° = \dfrac{\text{hypotenuse}}{\text{adjacent}} = \dfrac{\sqrt{2}}{1} = \sqrt{2}$

$\cot 45° = \dfrac{\text{adjacent}}{\text{opposite}} = \dfrac{1}{1} = 1$

5.

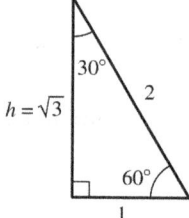

$\sin 60° = \dfrac{\text{opposite}}{\text{hypotenuse}} = \dfrac{\sqrt{3}}{2}$

$\cos 60° = \dfrac{\text{adjacent}}{\text{hypotenuse}} = \dfrac{1}{2}$

$\tan 60° = \dfrac{\text{opposite}}{\text{adjacent}} = \dfrac{\sqrt{3}}{1} = \sqrt{3}$

$\cot 60° = \dfrac{\text{adjacent}}{\text{opposite}} = \dfrac{1}{\sqrt{3}} = \dfrac{\sqrt{3}}{3}$

$\sec 60° = \dfrac{\text{hypotenuse}}{\text{adjacent}} = \dfrac{2}{1} = 2$

$\csc 60° = \dfrac{\text{hypotenuse}}{\text{opposite}} = \dfrac{2}{\sqrt{3}} = \dfrac{2\sqrt{3}}{3}$

6. a. $\csc 21° \approx 2.7904 \Rightarrow$
$\sec(90° - 21°) = \sec 69° \approx 2.7904$

b. $\dfrac{\cot 15°}{\tan 75°} = \dfrac{\cot 15°}{\cot(90°-75°)} = \dfrac{\cot 15°}{\cot 15°} = 1$

7.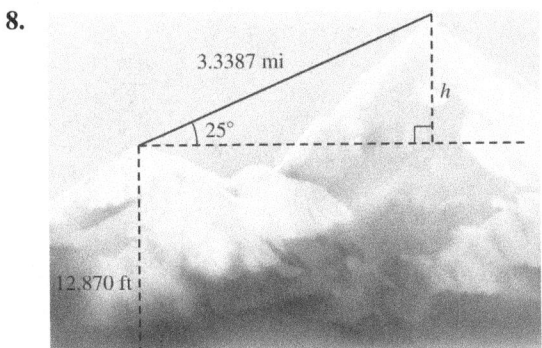

Not drawn to scale

From geometry, we know that $\theta = 4.2°$.

Thus, $\tan\theta = \dfrac{425}{d} \Rightarrow \tan 4.2° = \dfrac{425}{d} \Rightarrow$

$d = \dfrac{425}{\tan 4.2°} \approx 5787.3988$ ft ≈ 1.096 mi

8.

The sum of the side length h and the location height of 12,870 feet give the approximate height of Mount McKinley.

$\sin 25° = \dfrac{h}{3.3387}$
$h = 3.3387 \sin 25° \approx 1.4110$ mi
$\approx 1.4110(5280\text{ ft}) \approx 7450$ ft

The height of Mount McKinley is approximately $7450 + 12{,}870 = 20{,}320$ ft.

9.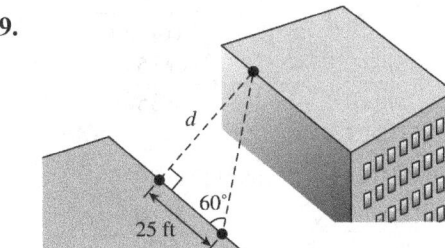

$\tan 60° = \dfrac{d}{25} \Rightarrow d = 25 \tan 60° \approx 43$ ft

5.2 Concepts and Vocabulary

1. If θ is an acute angle and $\sin\theta = \dfrac{2\sqrt{2}}{3}$, then $\cos\theta = \underline{\dfrac{1}{3}}$.

3. If θ is an acute angle (measured in degrees) in a right triangle and $\cos\theta = 0.7$, then $\sin(90°-\theta) = \underline{0.7}$.

5. False. The length of the leg is a multiple of 2.

7. True. The hypotenuse of a right triangle is the longest side in the triangle.

5.2 Building Skills

9. $\sin\theta = \dfrac{2\sqrt{5}}{25}$ $\csc\theta = \dfrac{5\sqrt{5}}{2}$

   $\cos\theta = \dfrac{11\sqrt{5}}{25}$ $\sec\theta = \dfrac{5\sqrt{5}}{11}$

   $\tan\theta = \dfrac{2}{11}$ $\cot\theta = \dfrac{11}{2}$

11. Use the Pythagorean theorem to find the length of the hypotenuse: $c = \sqrt{6^2 + 8^2} = 10$.

    $\sin\theta = \dfrac{6}{10} = \dfrac{3}{5}$ $\csc\theta = \dfrac{10}{6} = \dfrac{5}{3}$

    $\cos\theta = \dfrac{8}{10} = \dfrac{4}{5}$ $\sec\theta = \dfrac{10}{8} = \dfrac{5}{4}$

    $\tan\theta = \dfrac{6}{8} = \dfrac{3}{4}$ $\cot\theta = \dfrac{8}{6} = \dfrac{4}{3}$

13. Use the Pythagorean theorem to find the length of the hypotenuse:

    $c = \sqrt{7^2 + 1^2} = 5\sqrt{2}$.

    $\sin\theta = \dfrac{1}{5\sqrt{2}} = \dfrac{\sqrt{2}}{10}$ $\csc\theta = 5\sqrt{2}$

    $\cos\theta = \dfrac{7}{5\sqrt{2}} = \dfrac{7\sqrt{2}}{10}$ $\sec\theta = \dfrac{5\sqrt{2}}{7}$

    $\tan\theta = \dfrac{1}{7}$ $\cot\theta = 7$

15. $\sin\theta = \dfrac{12}{13}$, $\cos\theta = \dfrac{5}{13}$

    $\tan\theta = \dfrac{\sin\theta}{\cos\theta} = \dfrac{\tfrac{12}{13}}{\tfrac{5}{13}} = \dfrac{12}{5}$

    $\csc\theta = \dfrac{1}{\sin\theta} = \dfrac{1}{\tfrac{12}{13}} = \dfrac{13}{12}$

    $\sec\theta = \dfrac{1}{\cos\theta} = \dfrac{1}{\tfrac{5}{13}} = \dfrac{13}{5}$

    $\cot\theta = \dfrac{1}{\tan\theta} = \dfrac{1}{\tfrac{12}{5}} = \dfrac{5}{12}$

17. $\sin\theta = \dfrac{21}{29}$, $\cos\theta = \dfrac{20}{29}$

$\tan\theta = \dfrac{\sin\theta}{\cos\theta} = \dfrac{\frac{21}{29}}{\frac{20}{29}} = \dfrac{21}{20}$

$\csc\theta = \dfrac{1}{\sin\theta} = \dfrac{1}{\frac{21}{29}} = \dfrac{29}{21}$

$\sec\theta = \dfrac{1}{\cos\theta} = \dfrac{1}{\frac{20}{29}} = \dfrac{29}{20}$

$\cot\theta = \dfrac{1}{\tan\theta} = \dfrac{1}{\frac{21}{20}} = \dfrac{20}{21}$

19. $\sin\theta = \dfrac{2}{7}$, $\cos\theta = \dfrac{3\sqrt{5}}{7}$

$\tan\theta = \dfrac{\sin\theta}{\cos\theta} = \dfrac{\frac{2}{7}}{\frac{3\sqrt{5}}{7}} = \dfrac{2}{3\sqrt{5}} = \dfrac{2\sqrt{5}}{15}$

$\csc\theta = \dfrac{1}{\sin\theta} = \dfrac{1}{\frac{2}{7}} = \dfrac{7}{2}$

$\sec\theta = \dfrac{1}{\cos\theta} = \dfrac{1}{\frac{3\sqrt{5}}{7}} = \dfrac{7}{3\sqrt{5}} = \dfrac{7\sqrt{5}}{15}$

$\cot\theta = \dfrac{1}{\tan\theta} = \dfrac{1}{\frac{2}{3\sqrt{5}}} = \dfrac{3\sqrt{5}}{2}$

For exercises 21–26, use this triangle to help identify the opposite and adjacent legs.

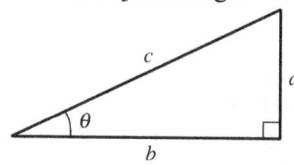

21. $\cos\theta = \dfrac{2}{3} \Rightarrow b = 2, c = 3$.

$3^2 = a^2 + 2^2 \Rightarrow a = \sqrt{5}$

$\sin\theta = \dfrac{\sqrt{5}}{3}$ $\csc\theta = \dfrac{3\sqrt{5}}{5}$

$\cos\theta = \dfrac{2}{3}$ $\sec\theta = \dfrac{3}{2}$

$\tan\theta = \dfrac{\sqrt{5}}{2}$ $\cot\theta = \dfrac{2\sqrt{5}}{5}$

23. $\tan\theta = \dfrac{5}{3} \Rightarrow a = 5, b = 3$.

$c^2 = 5^2 + 3^2 \Rightarrow c = \sqrt{34}$

$\sin\theta = \dfrac{5\sqrt{34}}{34}$ $\csc\theta = \dfrac{\sqrt{34}}{5}$

$\cos\theta = \dfrac{3\sqrt{34}}{34}$ $\sec\theta = \dfrac{\sqrt{34}}{3}$

$\tan\theta = \dfrac{5}{3}$ $\cot\theta = \dfrac{3}{5}$

25. $\sec\theta = \dfrac{13}{12} \Rightarrow b = 12, c = 13$.

$13^2 = a^2 + 12^2 \Rightarrow a = 5$

$\sin\theta = \dfrac{5}{13}$ $\csc\theta = \dfrac{13}{5}$

$\cos\theta = \dfrac{12}{13}$ $\sec\theta = \dfrac{13}{12}$

$\tan\theta = \dfrac{5}{12}$ $\cot\theta = \dfrac{12}{5}$

For exercises 27–32, use the fact that the value of any trigonometric function of an acute angle is equal to the co-function of the angle's complement.

27. $\sin 58° \approx 0.8480 \Rightarrow \cos 32° \approx 0.8480$

29. $\dfrac{\cos 42°}{\sin 48°} = \dfrac{\cos 42°}{\cos(90° - 48°)} = \dfrac{\cos 42°}{\cos 42°} = 1$

31. $\dfrac{\sin 10°}{\cos 80°} + \dfrac{\sec 20°}{\csc 70°} + \dfrac{\cot 35°}{\tan 55°}$

$= \dfrac{\sin 10°}{\sin(90° - 80°)} + \dfrac{\sec 20°}{\sec(90° - 70°)}$

$\qquad + \dfrac{\cot 35°}{\cot(90° - 55°)}$

$= \dfrac{\sin 10°}{\sin 10°} + \dfrac{\sec 20°}{\sec 20°} + \dfrac{\cot 35°}{\cot 35°}$

$= 1 + 1 + 1 = 3$

33. $\sin 60° \cos 30° + \cos 60° \sin 30°$

$= \dfrac{\sqrt{3}}{2} \cdot \dfrac{\sqrt{3}}{2} + \dfrac{1}{2} \cdot \dfrac{1}{2} = \dfrac{3}{4} + \dfrac{1}{4} = 1$

35. $\cos 60° \cos 30° + \sin 60° \sin 30°$

$= \dfrac{1}{2} \cdot \dfrac{\sqrt{3}}{2} + \dfrac{\sqrt{3}}{2} \cdot \dfrac{1}{2} = \dfrac{\sqrt{3}}{4} + \dfrac{\sqrt{3}}{4} = \dfrac{\sqrt{3}}{2}$

37. $\cot 45° + \csc 30° = 1 + 2 = 3$

39. $\cos 45° \cos 30° - \sin 45° \sin 30°$

$= \dfrac{\sqrt{2}}{2} \cdot \dfrac{\sqrt{3}}{2} - \dfrac{\sqrt{2}}{2} \cdot \dfrac{1}{2} = \dfrac{\sqrt{6} - \sqrt{2}}{4}$

Use this figure for exercises 41–50.

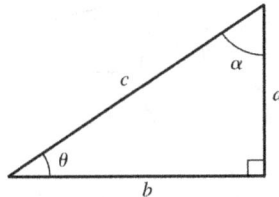

41. $8^2 + 10^2 = c^2 \Rightarrow c^2 = 164 \Rightarrow c \approx 12.806$
$\sin\theta = \dfrac{a}{c} = \dfrac{8}{12.806} \approx 0.625$
$\tan\theta = \dfrac{a}{b} = \dfrac{8}{10} = 0.8$

43. $c^2 = 23^2 + 7^2 \Rightarrow c \approx 24.042$.
$\cos\theta = \dfrac{7}{24.042} \approx 0.291,\ \tan\theta = \dfrac{23}{7} \approx 3.286$

45. $\theta = 30° \Rightarrow \sin\theta = 0.5, \sin 30° = \dfrac{9}{c} \Rightarrow c = 18$.
Use the Pythagorean theorem to find b:
$18^2 = 9^2 + b^2 \Rightarrow b \approx 15.588$

47. $c^2 = 12.5^2 + 6.2^2 \Rightarrow c \approx 13.953$
$\sin\theta = \dfrac{a}{c} = \dfrac{12.5}{13.953} \approx 0.896$
$\cos\theta = \dfrac{b}{c} = \dfrac{6.2}{13.953} \approx 0.444$

49. $14.5^2 = a^2 + 9.4^2 \Rightarrow a \approx 11.040$
$\sin\theta = \dfrac{a}{c} \approx \dfrac{11.040}{14.5} \approx 0.761$
$\tan\theta = \dfrac{a}{b} = \dfrac{11.040}{9.4} \approx 1.174$

51. $\sin\theta = \dfrac{1}{3}$

a. $\csc\theta = \dfrac{1}{\sin\theta} = 3$

b. $\sin^2\theta + \cos^2\theta = 1 \Rightarrow \left(\dfrac{1}{3}\right)^2 + \cos^2\theta = 1 \Rightarrow$
$\cos^2\theta = \dfrac{8}{9} \Rightarrow \cos\theta = \dfrac{2\sqrt{2}}{3}$

c. $\tan\theta = \dfrac{\sin\theta}{\cos\theta} = \dfrac{1/3}{2\sqrt{2}/3} = \dfrac{1}{2\sqrt{2}} = \dfrac{\sqrt{2}}{4}$

d. $\cot(90° - \theta) = \tan\theta = \dfrac{\sqrt{2}}{4}$

53. $\sec\theta = 5$

a. $\cos\theta = \dfrac{1}{\sec\theta} = \dfrac{1}{5}$

b. $\sin^2\theta + \cos^2\theta = 1 \Rightarrow \sin^2\theta + \left(\dfrac{1}{5}\right)^2 = 1 \Rightarrow$
$\sin^2\theta = \dfrac{24}{25} \Rightarrow \sin\theta = \dfrac{2\sqrt{6}}{5}$

c. $\tan\theta = \dfrac{\sin\theta}{\cos\theta} = \dfrac{2\sqrt{6}/5}{1/5} = 2\sqrt{6}$

d. $\tan(90° - \theta) = \cot\theta = \dfrac{1}{\tan\theta} = \dfrac{1}{2\sqrt{6}} = \dfrac{\sqrt{6}}{12}$

55. $\tan\theta = \dfrac{5}{2}$

a. $\cot\theta = \dfrac{1}{\tan\theta} = \dfrac{2}{5}$

b. $1 + \tan^2\theta = \sec^2\theta \Rightarrow 1 + \left(\dfrac{5}{2}\right)^2 = \sec^2\theta \Rightarrow$
$\sec^2\theta = \dfrac{29}{4} \Rightarrow \sec\theta = \dfrac{\sqrt{29}}{2}$

c. $\cos\theta = \dfrac{1}{\sec\theta} = \dfrac{1}{\sqrt{29}/2} = \dfrac{2}{\sqrt{29}} = \dfrac{2\sqrt{29}}{29}$

d. $\cos(90° - \theta) = \sin\theta$
$\sin^2\theta + \cos^2\theta = 1 \Rightarrow \sin^2\theta + \left(\dfrac{2\sqrt{29}}{29}\right)^2 = 1$
$\sin^2\theta + \dfrac{4}{29} = 1 \Rightarrow \sin^2\theta = \dfrac{25}{29} \Rightarrow$
$\sin\theta = \sqrt{\dfrac{25}{29}} = \dfrac{5}{\sqrt{29}} = \dfrac{5\sqrt{29}}{29}$

5.2 Applying the Concepts

57. $18^2 = 14^2 + b^2 \Rightarrow b^2 = 128 \Rightarrow b \approx 11.3$ ft

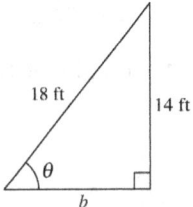

59. $1070^2 = 295^2 + b^2 \Rightarrow b \approx 1029$ m

61. The ski lift is the hypotenuse and $\theta = 30°$. The vertical rise is the side opposite θ, so

$$\sin 30° = \frac{a}{5000} \Rightarrow 0.5 = \frac{a}{5000} \Rightarrow a = 2500 \text{ ft}$$

63.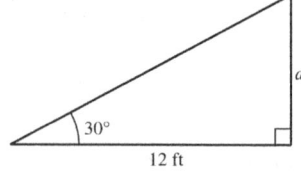

$$\tan 30° = \frac{a}{12} \Rightarrow \frac{\sqrt{3}}{3} = \frac{a}{12} \Rightarrow a = 4\sqrt{3} \approx 6.9 \text{ ft}$$

65.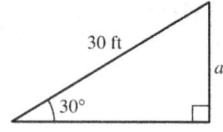

$$\sin 30° = \frac{a}{30} \Rightarrow \frac{1}{2} = \frac{a}{30} \Rightarrow a = 15 \text{ ft}$$

67.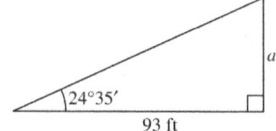

$$\tan 24°35' = \frac{a}{93} \Rightarrow 0.4575 \approx \frac{a}{93} \Rightarrow c \approx 43 \text{ ft}$$

69.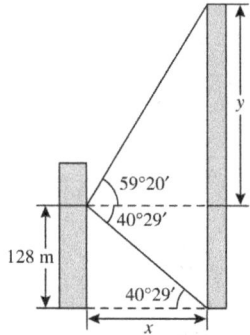

To find how far the second building is from the Empire State Building, we have

$$\tan 40°29' = \frac{128}{x} \Rightarrow 0.8536 \approx \frac{128}{x} \Rightarrow x \approx 150 \text{ m}$$

. To find the height of the Empire State Building, we have height $= y + 128$.

$$\tan 59°20' = \frac{y}{150} \Rightarrow$$

$1.6864 \approx \frac{y}{150} \Rightarrow y \approx 253$ m, so the height of the Empire State Building is approximately $128 + 253 = 381$ m.

71.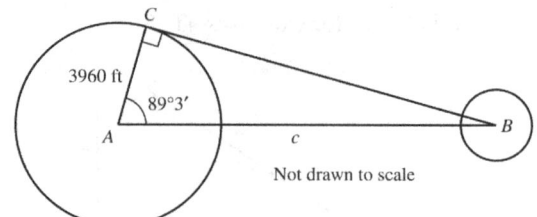

Note that BC is perpendicular to AC because the tangent drawn at the point of intersection of a radius and a circle is perpendicular to the radius.

The latitude of a point on the Earth is the measure of the angle formed by the radius connecting the point with the radius at the equator, so $m\angle A = 89°3'$. Then

$$\cos 89°3' = \frac{3960}{c} \Rightarrow 0.0166 \approx \frac{3960}{c} \Rightarrow$$

$c \approx 238,844$ miles.

73.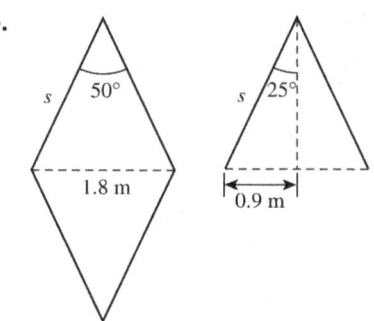

Drop a perpendicular from the top of the sign to form a right triangle as shown.

$$\sin 25° = \frac{0.9}{s} \Rightarrow s = \frac{0.9}{\sin 25°} \approx 2.1 \text{ m}$$

75.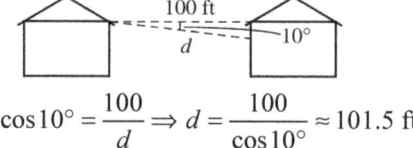

$$\cos 10° = \frac{100}{d} \Rightarrow d = \frac{100}{\cos 10°} \approx 101.5 \text{ ft}$$

77. If B is at the 53th parallel, then $m\angle CBA = 53°$. $AB = 3960$, so

$$\cos 53° = \frac{r}{3960} \Rightarrow$$

$$0.6018 \approx \frac{r}{3960} \Rightarrow r \approx 2383 \text{ miles}.$$

5.2 Beyond the Basics

For exercises 78–82, refer to Table 5.1 on page 534 in your text.

79. $2\sin^2\dfrac{\pi}{6} + \csc^2\dfrac{\pi}{6}\cos^2\dfrac{\pi}{3} = 2\left(\dfrac{1}{2}\right)^2 + 2^2\left(\dfrac{1}{2}\right)^2$

$= 2\cdot\dfrac{1}{4} + 4\cdot\dfrac{1}{4}$

$= \dfrac{1}{2} + 1 = \dfrac{3}{2}$

81. $8\sin^2\dfrac{\pi}{4} - 4\cos^2\dfrac{\pi}{4} + 2\sec^2\dfrac{\pi}{3}$

$= 8\left(\dfrac{\sqrt{2}}{2}\right)^2 - 4\left(\dfrac{\sqrt{2}}{2}\right)^2 + 2(2)^2$

$= 8\left(\dfrac{2}{4}\right) - 4\left(\dfrac{2}{4}\right) + 8 = 4 - 2 + 8 = 10$

83.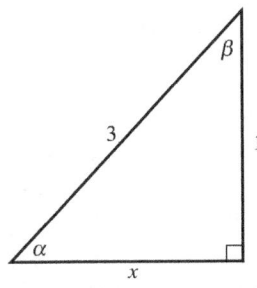

$\sin\alpha = \dfrac{1}{3}$

$3^2 = 1^2 + x^2 \Rightarrow x^2 = 8 \Rightarrow x = 2\sqrt{2}$

$\cos\alpha = \dfrac{2\sqrt{2}}{3}$ $\qquad\sec\alpha = \dfrac{3}{2\sqrt{2}} = \dfrac{3\sqrt{2}}{4}$

$\tan\alpha = \dfrac{1}{2\sqrt{2}} = \dfrac{\sqrt{2}}{4}$ $\qquad\csc\alpha = 3$

$\cos\alpha\csc\alpha + \tan\alpha\sec\alpha = \dfrac{2\sqrt{2}}{3}\cdot 3 + \dfrac{\sqrt{2}}{4}\cdot\dfrac{3\sqrt{2}}{4}$

$= 2\sqrt{2} + \dfrac{3}{8} = \dfrac{16\sqrt{2}+3}{8}$

85. $\tan\theta = \dfrac{4}{5} = \dfrac{\text{opposite}}{\text{adjacent}}$

$\text{opposite}^2 + \text{adjacent}^2 = \text{hypotenuse}^2 \Rightarrow$

$4^2 + 5^2 = \text{hypotenuse}^2 \Rightarrow \sqrt{41} = \text{hypotenuse}$

$\cos\theta = \dfrac{5}{\sqrt{41}} = \dfrac{5\sqrt{41}}{41}$ $\qquad\sin\theta = \dfrac{4}{\sqrt{41}} = \dfrac{4\sqrt{41}}{41}$

$\dfrac{\cos\theta - \sin\theta}{\cos\theta + \sin\theta} = \dfrac{\dfrac{5\sqrt{41}}{41} - \dfrac{4\sqrt{41}}{41}}{\dfrac{5\sqrt{41}}{41} + \dfrac{4\sqrt{41}}{41}} = \dfrac{\sqrt{41}}{9\sqrt{41}} = \dfrac{1}{9}$

87.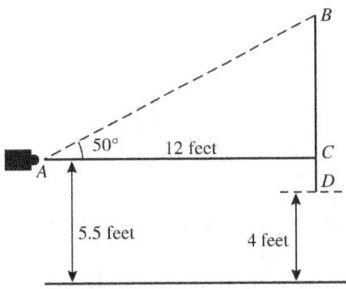

The height of the painting $= BC + CD$.
$CD = 5.5 - 4 = 1.5$ ft. Find BC as follows:
$\tan 50° = BC/12 \Rightarrow BC \approx 14.3$. The height of the painting is $14.3 + 1.5 = 15.8$ feet.

89.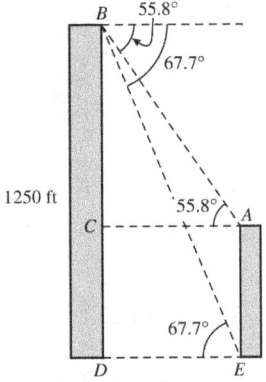

The height of the building $AE = 1250 - BC$.

In $\triangle BDE$, we have $\tan 67.7° = \dfrac{1250}{DE} \Rightarrow$

$DE = \dfrac{1250}{\tan 67.7°} = AC$.

Then in $\triangle BCA$, we have

$\tan 55.8° = \dfrac{BC}{\dfrac{1250}{\tan 67.7°}} \Rightarrow$

$BC = \dfrac{1250\tan 55.8°}{\tan 67.7°} \approx 754$ ft. The height of the building is $AE \approx 1250 - 754 \approx 496$ ft.

91.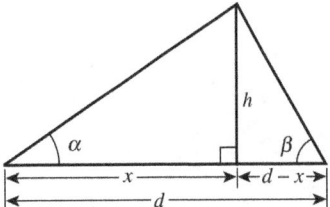

$\cot\alpha = \dfrac{x}{h} \Rightarrow h\cot\alpha = x$, and $\cot\beta = \dfrac{d-x}{h} \Rightarrow$
$d - h\cot\beta = x$.
$h\cot\alpha = d - h\cot\beta \Rightarrow h\cot\alpha + h\cot\beta = d \Rightarrow$
$h(\cot\alpha + \cot\beta) = d \Rightarrow h = \dfrac{d}{\cot\alpha + \cot\beta}$

93.

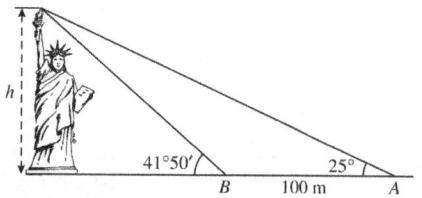

Using the result from exercise 90, we have
$$h = \frac{100}{\cot 25° - \cot 41°50'} \approx 97 \text{ m}$$

5.2 Critical Thinking/Discussion/Writing

95.

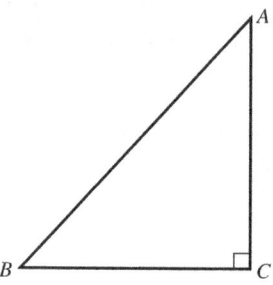

a. By the Pythagorean theorem, we have
$AC^2 + BC^2 = AB^2$.
$\sin A = \dfrac{BC}{AB}$ and $\cos A = \dfrac{AC}{AB}$.

$$\sin^2 A + \cos^2 A = \left(\frac{BC}{AB}\right)^2 + \left(\frac{AC}{AB}\right)^2$$
$$= \frac{BC^2 + AC^2}{AB^2} = \frac{AB^2}{AB^2} = 1$$

b. From part (a), we have $\sin A = \dfrac{BC}{AB}$ and $\cos A = \dfrac{AC}{AB}$. The cofunction identities give us
$\cos B = \sin A = \dfrac{BC}{AB}$ and $\sin B = \cos A = \dfrac{AC}{AB}$.
$$\sin A \cos B + \cos A \sin B = \frac{BC}{AB} \cdot \frac{BC}{AB} + \frac{AC}{AB} \cdot \frac{AC}{AB}$$
$$= \frac{BC^2 + AC^2}{AB^2}$$
$$= \frac{AB^2}{AB^2} = 1$$

96.

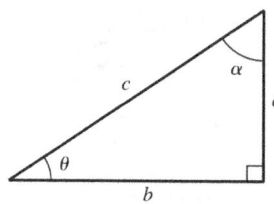

In the triangle, $\tan \theta = a$ if and only if $b = 1$.

97.

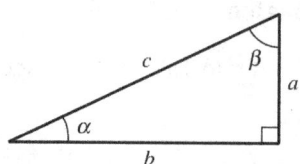

From geometry, we know that in any triangle, the longest side is opposite the largest angle and the shortest side is opposite the smallest angle.

So, if $\alpha < \beta$, then $a < b$. Using the result from exercise 96, if $\tan \alpha = a$, then $b = 1$, and $\tan \beta = \dfrac{1}{a}$. Therefore $0 < a < 1 \Rightarrow \dfrac{1}{a} > 1$ and $\tan \beta > \tan \alpha$.

98.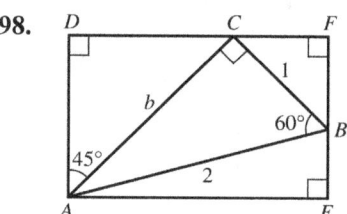

(i) $\sin 60° = \dfrac{b}{2} \Rightarrow b = 2\sin 60° = 2\left(\dfrac{\sqrt{3}}{2}\right) = \sqrt{3}$

$\angle CAB = 180° - 90° - 60° = 30°$

(ii) $\cos 45° = \dfrac{AD}{b} \Rightarrow AD = b\cos 45° \Rightarrow$
$AD = \sqrt{3}\left(\dfrac{\sqrt{2}}{2}\right) = \dfrac{\sqrt{6}}{2}$

Triangle ADC is an isosceles right triangle, so $DC = AD = \dfrac{\sqrt{6}}{2}$ and $m\angle ACD = 45°$.

(iii) $m\angle ACD + m\angle ACB + m\angle BCF = 180° \Rightarrow$
$45° + 90° + m\angle BCF = 180° \Rightarrow m\angle BCF = 45°$
$m\angle CBF = 180° - 45° - 90° = 45°$

(iv) $\cos 45° = \dfrac{CF}{1} \Rightarrow CF = \dfrac{\sqrt{2}}{2}$

Triangle FCB is an isosceles right triangle, so $FB = CF = \dfrac{\sqrt{2}}{2}$.

(v) $AE = DF = DC + CF = \dfrac{\sqrt{6}}{2} + \dfrac{\sqrt{2}}{2} = \dfrac{\sqrt{6}+\sqrt{2}}{2}$

$BE = FE - FB = DA - FB = \dfrac{\sqrt{6}}{2} - \dfrac{\sqrt{2}}{2}$
$= \dfrac{\sqrt{6}-\sqrt{2}}{2}$

(vi) $m\angle BAE = 90° - (45° + m\angle BAC)$
$= 90° - (45° + 30°) = 15°$
$m\angle ABE + m\angle BAE = 90° \Rightarrow$
$m\angle ABE + 15° = 90° \Rightarrow m\angle ABE = 75°$

(vii) $\sin \angle BAE = \sin 15° = \dfrac{BE}{BA} = \dfrac{\frac{\sqrt{6}-\sqrt{2}}{2}}{2}$
$= \dfrac{\sqrt{6}-\sqrt{2}}{4}$

$\cos \angle BAE = \cos 15° = \dfrac{AE}{BA} = \dfrac{\frac{\sqrt{6}+\sqrt{2}}{2}}{2}$
$= \dfrac{\sqrt{6}+\sqrt{2}}{4}$

(viii) $\sin 75° = \cos(90° - 75°) = \cos 15° = \dfrac{\sqrt{6}+\sqrt{2}}{4}$
$\cos 75° = \sin(90° - 75°) = \sin 15° = \dfrac{\sqrt{6}-\sqrt{2}}{4}$

5.2 Getting Ready for the Next Section

99. $\dfrac{\sqrt{3}-1}{\sqrt{3}+1} = \dfrac{\sqrt{3}-1}{\sqrt{3}+1} \cdot \dfrac{\sqrt{3}-1}{\sqrt{3}-1} = \dfrac{3-2\sqrt{3}+1}{3-1}$
$= \dfrac{4-2\sqrt{3}}{2} = 2 - \sqrt{3}$

101. $\dfrac{5+2\sqrt{3}}{7+4\sqrt{3}} = \dfrac{5+2\sqrt{3}}{7+4\sqrt{3}} \cdot \dfrac{7-4\sqrt{3}}{7-4\sqrt{3}} = \dfrac{35-6\sqrt{3}-24}{49-48}$
$= 11 - 6\sqrt{3}$

103. $\dfrac{1220}{360} = 3\,R\,140 \Rightarrow 1220 = 360 \cdot 3 + 140$

105. $\dfrac{-480}{360} = -2\,R\,240 \Rightarrow -480 = 360 \cdot (-2) + 240$

5.3 Trigonometric Functions of Any Angle; The Unit Circle

5.3 Practice Problems

1. $x = 2, y = -5 \Rightarrow r = \sqrt{2^2 + (-5)^2} = \sqrt{29}$

$\sin\theta = \dfrac{y}{r} = \dfrac{-5}{\sqrt{29}} = -\dfrac{5\sqrt{29}}{29}$ $\csc\theta = \dfrac{r}{y} = -\dfrac{\sqrt{29}}{5}$

$\cos\theta = \dfrac{x}{r} = \dfrac{2}{\sqrt{29}} = \dfrac{2\sqrt{29}}{29}$ $\sec\theta = \dfrac{r}{x} = \dfrac{\sqrt{29}}{2}$

$\tan\theta = \dfrac{y}{x} = \dfrac{-5}{2} = -\dfrac{5}{2}$ $\cot\theta = \dfrac{x}{y} = -\dfrac{2}{5}$

2. a. $\theta = 180° = \pi \Rightarrow$
$x = -1, y = 0, r = \sqrt{(-1)^2 + 0^2} = 1$
$\sin\theta = \dfrac{y}{r} = \dfrac{0}{1} = 0$
$\cos\theta = \dfrac{x}{r} = \dfrac{-1}{1} = -1$
$\tan\theta = \dfrac{y}{x} = \dfrac{0}{-1} = 0$
$\cot\theta = \dfrac{x}{y} = \dfrac{-1}{0}$, undefined
$\sec\theta = \dfrac{r}{x} = \dfrac{1}{-1} = -1$
$\csc\theta = \dfrac{r}{y} = \dfrac{1}{0}$, undefined

b. $\theta = 270° = \dfrac{3\pi}{2} \Rightarrow$
$x = 0, y = -1, r = \sqrt{0^2 + (-1)^2} = 1$
$\sin\theta = \dfrac{y}{r} = \dfrac{-1}{1} = -1$
$\cos\theta = \dfrac{x}{r} = \dfrac{0}{1} = 0$
$\tan\theta = \dfrac{y}{x} = \dfrac{-1}{0}$, undefined
$\cot\theta = \dfrac{x}{y} = \dfrac{0}{-1} = 0$
$\sec\theta = \dfrac{r}{x} = \dfrac{1}{0}$, undefined
$\csc\theta = \dfrac{r}{y} = \dfrac{1}{-1} = -1$

3. a. $360\overline{)1830}$
$\quad\;\;\underline{1800}$
$\quad\quad\;\;30$

$1830° = 30° + 5 \cdot 360°$, so $1830°$ is coterminal with $30°$. Thus,
$\cos 1830° = \cos 30° = \dfrac{\sqrt{3}}{2}$.

b. $\dfrac{31\pi}{3} = \dfrac{30\pi}{3} + \dfrac{\pi}{3} = 5(2\pi) + \dfrac{\pi}{3}$, so $\dfrac{31\pi}{3}$ is coterminal with $\dfrac{\pi}{3}$. Thus,
$\sin\dfrac{\pi}{3} = \dfrac{\sqrt{3}}{2} \Rightarrow \sin\dfrac{31\pi}{3} = \dfrac{\sqrt{3}}{2}$.

4. Because $\sin\theta > 0$ and $\cos\theta < 0$, θ lies in quadrant II.

5. Because $\tan\theta < 0$ and $\cos\theta > 0$, θ lies in quadrant IV. $\tan\theta = -\dfrac{4}{5}$, so let $x = 5$ and $y = -4$. $r = \sqrt{5^2 + (-4)^2} = \sqrt{41}$

   $\sin\theta = \dfrac{y}{r} = -\dfrac{4}{\sqrt{41}} = -\dfrac{4\sqrt{41}}{41}$;

   $\sec\theta = \dfrac{r}{x} = \dfrac{\sqrt{41}}{5}$

6. a. $\theta = 175° \Rightarrow \theta' = 180° - 175° = 5°$

   b. $\theta = \dfrac{5\pi}{3} \Rightarrow \theta' = 2\pi - \dfrac{5\pi}{3} = \dfrac{\pi}{3}$

   c. $\theta = 8.22$ lies in quadrant II and is coterminal with $8.22 - 2\pi \approx 1.94$. The reference angle $\theta' \approx \pi - 1.94 \approx 1.20$.

7. $1020° = 300° + 2 \cdot 360°$, so $1020°$ is coterminal with $300°$. Because $300°$ lies in quadrant IV, $\theta' = 360° - 300° = 60°$. In quadrant IV, $\cos\theta$ is positive, so $\cos 1020° = \cos 60° = 0.5$.

8. a. $1035° = 315° + 2 \cdot 360°$, so $1035°$ is coterminal with $315°$. Because $315°$ lies in quadrant IV, $\theta' = 360° - 315° = 45°$. In quadrant IV, $\csc\theta$ is negative, so $\csc 1035° = -\csc 45° = -\sqrt{2}$.

   b. $\dfrac{17\pi}{6} = \dfrac{5\pi}{6} + 2\pi$, so $\dfrac{17\pi}{6}$ is coterminal with $5\pi/6$, which lies in quadrant II. Thus, $\theta' = \pi - \dfrac{5\pi}{6} = \dfrac{\pi}{6}$. In quadrant II, $\cot\theta$ is negative, so $\cot\dfrac{17\pi}{6} = -\cot\dfrac{\pi}{6} = -\sqrt{3}$.

9. $\theta = 600° = 240° + 360°$.
   $240°$ lies in quadrant III, thus, $\theta' = 60°$, and $\cos\theta < 0$.

   $\cos 60° = \dfrac{1}{2} \Rightarrow \cos 600° = \cos 240° = -\dfrac{1}{2}$.

   $h = 67.8 - 67.2\left(-\dfrac{1}{2}\right) = 101.4$ m

10. a. From figure 5.46, we see that the endpoint of the arc of length $t = \dfrac{7\pi}{6}$ is $\left(-\dfrac{\sqrt{3}}{2}, -\dfrac{1}{2}\right)$. So, $\cos\dfrac{7\pi}{6} = -\dfrac{\sqrt{3}}{2}$,

    $\sin\dfrac{7\pi}{6} = -\dfrac{1}{2}$, and

    $\tan\dfrac{7\pi}{6} = \dfrac{\sin(7\pi/6)}{\cos(7\pi/6)}$

    $= \dfrac{-1/2}{-\sqrt{3}/2} = \dfrac{1}{\sqrt{3}} = \dfrac{\sqrt{3}}{3}$.

    b. $\theta = \dfrac{7\pi}{6}$ lies in quadrant III, so its sine and cosine are both negative and its tangent is positive. The reference angle is

    $\dfrac{7\pi}{6} - \pi = \dfrac{\pi}{6}$.

    Then $\sin\dfrac{7\pi}{6} = -\sin\dfrac{\pi}{6} = -\dfrac{1}{2}$,

    $\cos\dfrac{7\pi}{6} = -\cos\dfrac{\pi}{6} = -\dfrac{\sqrt{3}}{2}$, and

    $\tan\dfrac{7\pi}{6} = \tan\dfrac{\pi}{6} = \dfrac{\sqrt{3}}{3}$.

    c. $\dfrac{7\pi}{6} = \dfrac{7\pi}{6} \cdot \dfrac{180}{\pi} = 210°$

    Because $210°$ is in quadrant III, its reference angle is $210° - 180° = 30°$. Then,

    $\cos\dfrac{7\pi}{6} = \cos 210° = -\cos 30° = -\dfrac{\sqrt{3}}{2}$,

    $\sin\dfrac{7\pi}{6} = \sin 210° = -\sin 30° = -\dfrac{1}{2}$, and

    $\tan\dfrac{7\pi}{6} = \tan 210° = \tan 30° = \dfrac{\sqrt{3}}{3}$.

5.3 Concepts and Vocabulary

1. For a point $P(x, y)$ on the terminal side of an angle θ in standard position, we let $r = \sqrt{x^2 + y^2}$. Then, $\sin\theta = \dfrac{y}{r}$, $\cos\theta = \dfrac{x}{r}$, and $\tan\theta = \dfrac{y}{x}$.

3. The reference angle θ' for a nonquadrantal angle θ in standard position is the acute angle formed by the terminal side of θ and the x-axis.

5. False. The value of a trigonometric function for any angle is the same for any point on the terminal side of θ.

7. True.

5.3 Building Skills

9. $x = -4, y = 3 \Rightarrow r = \sqrt{(-4)^2 + 3^2} = 5$
$\sin\theta = \frac{3}{5}, \cos\theta = -\frac{4}{5}, \tan\theta = -\frac{3}{4}$

11. $x = -\sqrt{3}, y = -1 \Rightarrow r = \sqrt{\left(-\sqrt{3}\right)^2 + (-1)^2} = 2$
$\sin\theta = -\frac{1}{2}, \cos\theta = -\frac{\sqrt{3}}{2}, \tan\theta = \frac{\sqrt{3}}{3}$

13. $x = 3, y = 3 \Rightarrow r = \sqrt{3^2 + 3^2} = 3\sqrt{2}$
$\sin\theta = \frac{\sqrt{2}}{2}, \cos\theta = \frac{\sqrt{2}}{2}, \tan\theta = 1$

15. $x = -7, y = -24 \Rightarrow$
$r = \sqrt{(-7)^2 + (-24)^2} = \sqrt{625} = 25$
$\sin\theta = -\frac{24}{25}, \cos\theta = -\frac{7}{25}, \tan\theta = \frac{24}{7}$

17. $x = 12, y = -5 \Rightarrow r = \sqrt{12^2 + (-5)^2} = 13$
$\cot\theta = -\frac{12}{5}, \sec\theta = \frac{13}{12}, \csc\theta = -\frac{13}{5}$

19. $x = 7, y = -24 \Rightarrow$
$r = \sqrt{7^2 + (-24)^2} = \sqrt{625} = 25$
$\csc\theta = -\frac{25}{24}, \sec\theta = \frac{25}{7}, \cot\theta = -\frac{7}{24}$

21. $x = -\sqrt{2}, y = \sqrt{6} \Rightarrow$
$r = \sqrt{\left(-\sqrt{2}\right)^2 + \left(\sqrt{6}\right)^2} = \sqrt{8}$
$\csc\theta = \frac{\sqrt{8}}{\sqrt{6}} = \frac{\sqrt{48}}{6} = \frac{4\sqrt{3}}{6} = \frac{2\sqrt{3}}{3}$,
$\sec\theta = \frac{\sqrt{8}}{-\sqrt{2}} = -\sqrt{4} = -2$,
$\cot\theta = \frac{-\sqrt{2}}{\sqrt{6}} = -\frac{1}{\sqrt{3}} = -\frac{\sqrt{3}}{3}$

23. $x = 0, y = 4$
This point lies on the positive y-axis, and therefore, $\theta = 90°$
$\csc\theta = 1, \sec\theta$ is undefined, $\cot\theta = 0$

25. $450°$ is coterminal with $450° - 360° = 90°$. Thus, $\sin 450° = \sin 90° = 1$.

27. $-90°$ is coterminal with $360° - 90° = 270°$. Thus, $\cos(-90°) = \cos 270° = 0$.

29. $450°$ is coterminal with $450° - 360° = 90°$. Thus, $\tan 450° = \tan 90°$, which is undefined.

31. $-540°$ is coterminal with $540° + 2(360°) = 180°$. Thus, $\tan 540° = \tan 180° = 0$.

33. $900°$ is coterminal with $900° - 2(360°) = 180°$. Thus, $\csc 900° = \csc 180°$, which is undefined.

35. $-1530°$ is coterminal with $5(360°) - 1530° = 270°$. Thus, $\sin(-1530°) = \sin 270° = -1$.

37. $\cos(5\pi) = \cos(\pi + 2 \cdot 2\pi) = \cos\pi = -1$

39. $\tan(4\pi) = \tan(0 + 2 \cdot 2\pi) = \tan 0 = 0$

41. $\csc\left(\frac{5\pi}{2}\right) = \csc\left(\frac{\pi}{2} + 2\pi\right) = \csc\left(\frac{\pi}{2}\right) = 1$

43. $\sin(-2\pi) = \sin(0 + (-1)2\pi) = \sin 0 = 0$

45. $\cos\left(-\frac{3\pi}{2}\right) = \cos\left(\frac{\pi}{2} + (-1)2\pi\right) = \cos\left(\frac{\pi}{2}\right) = 0$

47. $\sin\theta < 0$ and $\cos\theta < 0 \Rightarrow \theta$ is in quadrant III.

49. $\sin\theta > 0$ and $\cos\theta < 0 \Rightarrow \theta$ is in quadrant II.

51. $\cos\theta > 0$ and $\csc\theta < 0 \Rightarrow \theta$ is in quadrant IV.

53. $\sec\theta < 0$ and $\csc\theta > 0 \Rightarrow \theta$ is in quadrant II.

55. $\sin\theta = -\frac{5}{13} \Rightarrow y = -5, r = 13$
$\sqrt{x^2 + (-5)^2} = 13 \Rightarrow x^2 + 25 = 169 \Rightarrow$
$x^2 = 144 \Rightarrow x = \pm 12$
Since θ is in quadrant III, x is negative, so $x = -12$.

57. $\cos\theta = \frac{7}{25} \Rightarrow x = 7, r = 25$
$\sqrt{7^2 + y^2} = 25 \Rightarrow 49 + y^2 = 625 \Rightarrow$
$y^2 = 576 \Rightarrow y = \pm 24$
Since θ is in quadrant I, y is positive, so $y = 24$.

59. $\cos\theta = -\frac{5}{13}, \theta$ in Quadrant III $\Rightarrow x < 0, y < 0$
$x = -5, r = 13 \Rightarrow 13^2 = (-5)^2 + y^2 \Rightarrow y = -12$
$\tan\theta = \frac{12}{5}$

61. $\cot\theta = -\dfrac{3}{4}, \theta$ in Quadrant II $\Rightarrow x < 0, y > 0$
$x = -3, y = 4 \Rightarrow r = \sqrt{(-3)^2 + 4^2} = 5$
$\cos\theta = -\dfrac{3}{5}$

63. $\sin\theta = \dfrac{3}{5}, \tan\theta < 0 \Rightarrow \theta$ is in Quadrant II and
$x < 0.\ y = 3, r = 5 \Rightarrow 5^2 = x^2 + 3^2 \Rightarrow x = -4$
$\sec\theta = -\dfrac{5}{4}$

65. $\sec\theta = 3, \sin\theta < 0 \Rightarrow \theta$ is in Quadrant IV and $y < 0$.
$r = 3, x = 1 \Rightarrow 3^2 = 1^2 + y^2 \Rightarrow y = -2\sqrt{2}$
$\cot\theta = \dfrac{1}{-2\sqrt{2}} = -\dfrac{\sqrt{2}}{4}$

67. Because $120°$ lies in Quadrant II, the reference angle is
$180° - \theta = 180° - 120° = 60°$.

69. Because $50°$ lies in Quadrant I, the reference angle is $50°$.

71. $420°$ is co-terminal with $60°$, which lies in Quadrant I. The reference angle is $60°$.

73. $\dfrac{19\pi}{4} = 4\pi + \dfrac{3\pi}{4} \Rightarrow \dfrac{19\pi}{4}$ is coterminal with $\dfrac{3\pi}{4}$ and lies in Quadrant II. The reference angle is $\pi - \dfrac{3\pi}{4} = \dfrac{\pi}{4}$.

75. $\dfrac{3\pi}{4}$ lies in Quadrant II. The reference angle is $\pi - \dfrac{3\pi}{4} = \dfrac{\pi}{4}$.

77. $\dfrac{5\pi}{6}$ lies in Quadrant II. The reference angle is $\pi - \dfrac{5\pi}{6} = \dfrac{\pi}{6}$.

79. $1470°$ is coterminal with $1470° - 4 \cdot 360° = 30°$.
Thus, $\sin 1470° = \sin 30° = \dfrac{1}{2}$.

81. $1125°$ is coterminal with $1125° - 3 \cdot 360° = 45°$. Thus, $\tan 1125° = \tan 45° = 1$.

83. $2130°$ is coterminal with $2130° - 5 \cdot 360° = 330°$.
$330°$ lies in quadrant IV with reference angle $\theta' = 360° - 330° = 30°$. In quadrant IV, $\csc\theta < 0$, so $\csc 2130° = -\csc 30° = -2$.

85. $690°$ is coterminal with $690° - 360° = 330°$.
$330°$ lies in quadrant IV with reference angle $\theta' = 360° - 330° = 30°$. In quadrant IV, $\tan\theta < 0$, so $\tan 690° = -\tan 30° = -\dfrac{\sqrt{3}}{3}$.

87. $\dfrac{13\pi}{3} = 4\pi + \dfrac{\pi}{3} \Rightarrow \dfrac{13\pi}{3}$ is coterminal with $\dfrac{\pi}{3}$ and lies in quadrant I. In quadrant I, $\sin\theta > 0$, so $\sin\left(\dfrac{13\pi}{3}\right) = \sin\dfrac{\pi}{3} = \dfrac{\sqrt{3}}{2}$.

89. $\dfrac{29\pi}{6} = 4\pi + \dfrac{5\pi}{6} \Rightarrow \dfrac{29\pi}{6}$ is coterminal with $\dfrac{5\pi}{6}$ and lies in quadrant II. The reference angle is $\pi - \dfrac{5\pi}{6} = \dfrac{\pi}{6}$. In quadrant II, $\cos\theta < 0$, so $\cos\dfrac{29\pi}{6} = -\cos\dfrac{\pi}{6} = -\dfrac{\sqrt{3}}{2}$.

91. $\dfrac{37\pi}{4} = 8\pi + \dfrac{5\pi}{4} \Rightarrow \dfrac{37\pi}{4}$ is coterminal with $\dfrac{5\pi}{4}$ and lies in quadrant III. The reference angle is $\dfrac{5\pi}{4} - \pi = \dfrac{\pi}{4}$. In quadrant III, $\tan\theta > 0$, so $\tan\dfrac{37\pi}{4} = \tan\dfrac{\pi}{4} = 1$.

93. $\dfrac{29\pi}{3} = 8\pi + \dfrac{5\pi}{3} \Rightarrow \dfrac{29\pi}{3}$ is coterminal with $\dfrac{5\pi}{3}$ and lies in quadrant IV. In quadrant IV, $\sec\theta > 0$, so $\sec\left(\dfrac{29\pi}{3}\right) = \sec\dfrac{5\pi}{3} = 2$.

95. a. From figure 5.46, we see that the endpoint of the arc of length $s = \dfrac{2\pi}{3}$ is $\left(-\dfrac{1}{2}, \dfrac{\sqrt{3}}{2}\right)$.

So, $\sin\dfrac{2\pi}{3} = \dfrac{\sqrt{3}}{2}$, $\cos\dfrac{2\pi}{3} = -\dfrac{1}{2}$, and
$\tan\dfrac{2\pi}{3} = \dfrac{\sin(2\pi/3)}{\cos(2\pi/3)} = \dfrac{\sqrt{3}/2}{-1/2} = -\sqrt{3}$.

b. $\theta = \dfrac{2\pi}{3}$ lies in quadrant II, so its sine is positive and its cosine and tangent are both negative. The reference angle is

$\pi - \dfrac{2\pi}{3} = \dfrac{\pi}{3}$. Then $\sin\dfrac{2\pi}{3} = \sin\dfrac{\pi}{3} = \dfrac{\sqrt{3}}{2}$,

$\cos\dfrac{2\pi}{3} = -\cos\dfrac{\pi}{3} = -\dfrac{1}{2}$, and

$\tan\dfrac{2\pi}{3} = -\tan\dfrac{\pi}{3} = -\sqrt{3}$.

c. $\dfrac{2\pi}{3} = \dfrac{2\pi}{3} \cdot \dfrac{180}{\pi} = 120°$

Because 120° is in quadrant II, its reference angle is 180° − 120° = 60°. Then,

$\sin\dfrac{2\pi}{3} = \sin 60° = \dfrac{\sqrt{3}}{2}$,

$\cos\dfrac{2\pi}{3} = -\cos 60° = -\dfrac{1}{2}$, and

$\tan\dfrac{2\pi}{3} = \tan 120° = -\tan 60° = -\sqrt{3}$.

97. a. From figure 5.46, we see that the endpoint of the arc of length $s = \dfrac{11\pi}{6}$ is $\left(\dfrac{\sqrt{3}}{2}, -\dfrac{1}{2}\right)$.

So, $\sin\dfrac{11\pi}{6} = -\dfrac{1}{2}$, $\cos\dfrac{11\pi}{6} = \dfrac{\sqrt{3}}{2}$, and

$\tan\dfrac{11\pi}{6} = \dfrac{\sin(2\pi/3)}{\cos(2\pi/3)} = \dfrac{-1/2}{\sqrt{3}/2}$

$= -\dfrac{1}{\sqrt{3}} = -\dfrac{\sqrt{3}}{3}$.

b. $\theta = \dfrac{11\pi}{6}$ lies in quadrant IV, so its sine and tangent are both negative and its cosine is positive. The reference angle is

$2\pi - \dfrac{11\pi}{6} = \dfrac{\pi}{6}$. Then

$\sin\dfrac{11\pi}{6} = -\sin\dfrac{\pi}{6} = -\dfrac{1}{2}$,

$\cos\dfrac{11\pi}{6} = \cos\dfrac{\pi}{6} = \dfrac{\sqrt{3}}{2}$, and

$\tan\dfrac{11\pi}{6} = -\tan\dfrac{\pi}{6} = -\dfrac{\sqrt{3}}{3}$.

c. $\dfrac{11\pi}{6} = \dfrac{11\pi}{6} \cdot \dfrac{180}{\pi} = 330°$

Because 330° is in quadrant IV, its reference angle is 360° − 330° = 30°. Then,

$\sin\dfrac{11\pi}{6} = -\sin 30° = -\dfrac{1}{2}$,

$\cos\dfrac{11\pi}{6} = \cos 30° = \dfrac{\sqrt{3}}{2}$, and

$\tan\dfrac{11\pi}{6} = -\tan 30° = -\dfrac{\sqrt{3}}{3}$.

99. a. From figure 5.46, we see that the endpoint of the arc of length $s = \dfrac{5\pi}{4}$ is

$\left(-\dfrac{\sqrt{2}}{2}, -\dfrac{\sqrt{2}}{2}\right)$. So,

$\sin\dfrac{5\pi}{4} = -\dfrac{\sqrt{2}}{2}$, $\cos\dfrac{5\pi}{4} = -\dfrac{\sqrt{2}}{2}$, and

$\tan\dfrac{5\pi}{4} = \dfrac{\sin(5\pi/4)}{\cos(5\pi/4)} = \dfrac{-\sqrt{2}/2}{-\sqrt{2}/2} = 1$.

b. $\theta = \dfrac{5\pi}{4}$ lies in quadrant III, so its sine and cosine are both negative and its tangent is positive. The reference angle is

$\dfrac{5\pi}{4} - \pi = \dfrac{\pi}{4}$.

Then $\sin\dfrac{5\pi}{4} = -\sin\dfrac{\pi}{4} = -\dfrac{\sqrt{2}}{2}$,

$\cos\dfrac{5\pi}{4} = -\cos\dfrac{\pi}{4} = -\dfrac{\sqrt{2}}{2}$, and

$\tan\dfrac{5\pi}{4} = \tan\dfrac{\pi}{4} = 1$.

c. $\dfrac{5\pi}{4} = \dfrac{5\pi}{4} \cdot \dfrac{180}{\pi} = 225°$

Because 225° is in quadrant III, its reference angle is 225° − 180° = 45°. Then,

$\sin\dfrac{5\pi}{4} = -\sin 45° = -\dfrac{\sqrt{2}}{2}$,

$\cos\dfrac{5\pi}{4} = -\cos 45° = -\dfrac{\sqrt{2}}{2}$, and

$\tan\dfrac{5\pi}{4} = \tan 45° = 1$.

5.3 Applying the Concepts

101.

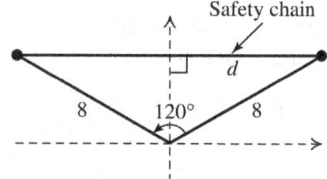

From geometry, we know that an altitude drawn from the vertex of an isosceles triangle bisects the vertex angle and also bisects the base of the triangle. Therefore,

$$\sin 60° = \frac{d}{8} \Rightarrow d = 8 \sin 60° = 8\left(\frac{\sqrt{3}}{2}\right) = 4\sqrt{3}.$$

The length of the chain is
$2 \cdot 4\sqrt{3} = 8\sqrt{3} \approx 13.86$ ft.

Use the figure below for exercises 103–110.

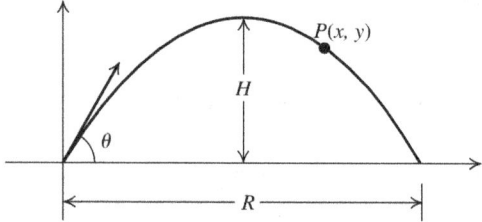

103. $H = \frac{1}{64}(44 \sin 30°)^2 = 7.5625$ ft

$t = \frac{44 \sin 30°}{16} = 1.375$ sec

$R = \frac{44^2 \sin 30° \cos 30°}{16} \approx 52.39$ ft

105. $H = \frac{1}{64}(44 \sin 60°)^2 = 22.6875$ ft

$t = \frac{44 \sin 60°}{16} \approx 2.38$ sec

$R = \frac{44^2 \sin 60° \cos 60°}{16} \approx 52.39$ ft

107. a. $y = x \tan 45° - \frac{16 \sec^2 45°}{80^2} x^2$

$= x(1) - \frac{16(\sqrt{2})^2}{80^2} x^2 = x - \frac{x^2}{200}$

b. $y = 100 - \frac{1}{200}(100^2) = 50$ ft

109. $t = \frac{80 \sin 45°}{16} \approx 3.54$ sec

111. $A = \frac{3(2)\cos 60°}{\sqrt{7 + 9\cos^2 60°}} \approx 0.99$ ft

113. a. $h = 100 \sin 30° = 100 \cdot \frac{1}{2} = 50$ ft

b. $h = 100 \sin 60° = 100 \cdot \frac{\sqrt{3}}{2} = 50\sqrt{3} \approx 86.6$ ft

5.3 Beyond the Basics

For exercises 115–118, refer to the following figure.

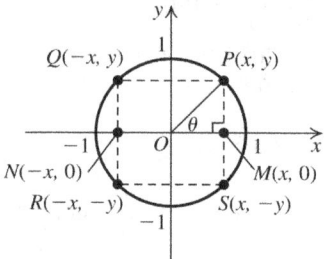

115. Since $\sin(-\theta) = -\sin \theta$, the y-value of the point on the unit circle on the terminal side of $-\theta$ equals the opposite of y-value of the point on the unit circle on the terminal side of θ. Since $\cos(-\theta) = \cos \theta$, the x-value of the point on the unit circle on the terminal side of $-\theta$ equals the x-value of the point on the unit circle on the terminal side of θ. Thus, the coordinates of the point on the unit circle are $(x, -y)$, and the point on the unit circle is S. The triangle we are seeking is $\triangle SOM$.

117. Since $\sin \theta = -\sin(180° + \theta)$, the y-value of the point on the unit circle on the terminal side of θ is the opposite the y-value of the point on the unit circle on the terminal side of $180° + \theta$. Since $\cos \theta = -\cos(180° + \theta)$, the x-value of the point on the unit circle on the terminal side of θ is the opposite of the x-value of the point on the unit circle on the terminal side of $180° + \theta$. Thus, the coordinates of the point on the unit circle are $(-x, -y)$, and the point on the unit circle is R. The triangle we are seeking is $\triangle RON$.

119. $\sin(-45°) = -\sin 45° = -\frac{\sqrt{2}}{2}$

$\tan(-60°) = -\tan 60° = -\sqrt{3}$

121. $\sin 225° = \sin(180° + 45°) = -\sin 45° = -\dfrac{\sqrt{2}}{2}$

$\cos 240° = \cos(180° + 60°) = -\cos 60° = -\dfrac{1}{2}$

$\tan 210° = \tan(180° + 30°) = \tan 30° = \dfrac{\sqrt{3}}{3}$

123.

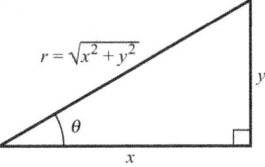

For acute angle θ,

$\sin \theta = \dfrac{y}{r} = \dfrac{y}{\sqrt{x^2+y^2}} \Rightarrow \sin^2 \theta = \dfrac{y^2}{x^2+y^2}$.

For any real numbers x and y, $x^2 + y^2 \geq y^2$,

so $\dfrac{y^2}{x^2+y^2} \leq 1 \Rightarrow \sin^2 \theta \leq 1 \Rightarrow$

$|\sin \theta| \leq 1 \Rightarrow -1 \leq \sin \theta \leq 1$.

Similarly, we can show that

$\cos \theta = \dfrac{x}{r} = \dfrac{x}{\sqrt{x^2+y^2}} \Rightarrow$

$\cos^2 \theta = \dfrac{x^2}{x^2+y^2} \Rightarrow$

$\cos^2 \theta \leq 1 \Rightarrow |\cos \theta| \leq 1 \Rightarrow -1 \leq \cos \theta \leq 1$.

125. In triangle ABC, $A + B + C = 180° \Rightarrow$
$A + B = 180° - C$. Substituting we have
$\tan(A+B) + \tan C = \tan(180° - C) + \tan C$
$= -\tan C + \tan C = 0$

5.3 Critical Thinking/Discussion/Writing

127. $\cos 105° = \cos 60° + \cos 45°$ is false because $\cos 105° < 0$, while $\cos 60° > 0$ and $\cos 45° > 0$.
The sum of two positive numbers is positive.

128. $\sin 260° = 2 \sin 130°$ is false because $\sin 260° < 0$ while $\sin 130° > 0$.

129. $\tan 123° = \tan 61° + \tan 62°$ is false because $\tan 123° < 0$, while $\tan 61° > 0$ and $\tan 62° > 0$.
The sum of two positive numbers is positive.

130. $\sec 380° = \sec 185° + \sec 195°$ is false because $\sec 380° > 0$ while $\sec 185° < 0$ and $\sec 195° < 0$. The sum of two negative numbers is negative.

131. False. For example, if $\theta = 90°$, then we have
$\cos(\sin 90°)° = \cos 1° \approx 0.9998$ while
$\sin(\cos 90°)° = \sin 0° = 0$

132. False. For example, $\sec 135° = -\sqrt{2}$ while $\tan 135° = -1$, so $\tan 135° > \sec 135°$.

5.3 Getting Ready for the Next Section

133. $3x - 4 = 0 \Rightarrow x = \dfrac{4}{3}$

135. $x^2 + 3x - 10 = 0 \Rightarrow (x+5)(x-2) = 0 \Rightarrow$
$x = -5$, $x = 2$

137. $f(x) = \sqrt{-x}$
Reflect the graph of $y = \sqrt{x}$ across the y-axis.

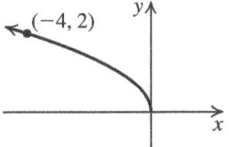

139. $f(x) = \sqrt{x+1}$
Shift the graph of $y = \sqrt{x}$ one unit left.

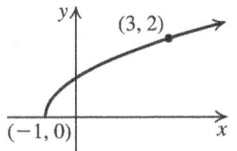

141. $f(x) = -3\sqrt{2x+1}$
Shift the graph of $y = \sqrt{x}$ one unit left, then compress it horizontally by a factor of 2. Stretch the graph vertically by a factor of three. Reflect the graph about the x-axis.

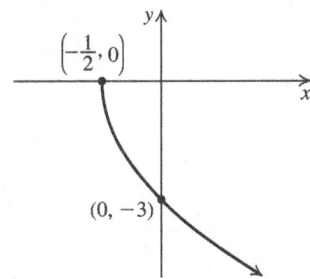

143. $-f(x)$

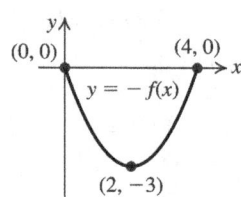

145. $2f(x)$

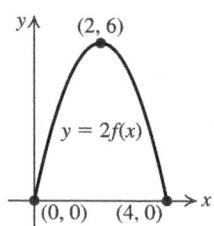

147. $-2f\left(\frac{1}{2}x\right)$

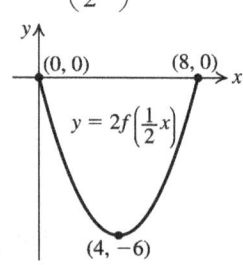

5.4 Graphs of the Sine and Cosine Functions

5.4 Practice Exercises

1.

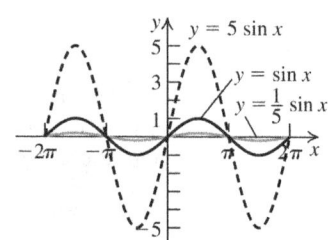

2.

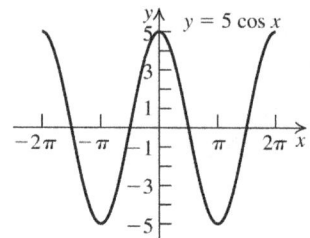

Amplitude = 5, range = $[-5, 5]$

3.

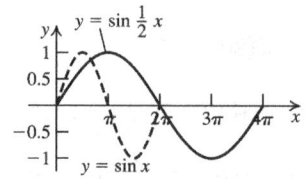

4.

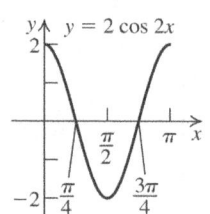

5.

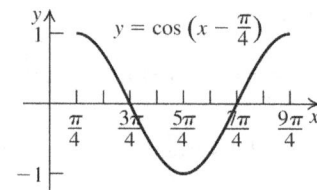

6. $y = \frac{1}{3}\cos\left[\frac{2}{5}\left(x + \frac{\pi}{6}\right)\right]$

$a = \frac{1}{3} \Rightarrow$ the amplitude is $\frac{1}{3}$.

$b = \frac{2}{5} \Rightarrow$ the period is $\frac{2\pi}{\frac{2}{5}} = 5\pi$.

$c = -\frac{\pi}{6} \Rightarrow$ the phase shift is $\frac{\pi}{6}$ unit left

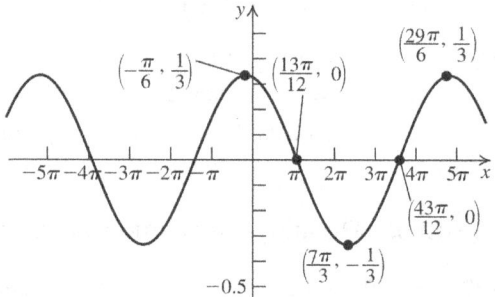

7. $\cos x \to \cos\left(\frac{\pi}{2} + x\right) \to \cos\left(\frac{\pi}{2} + \frac{x}{2}\right) \to$

$\cos\left(\frac{\pi}{2} - \frac{x}{2}\right) \to -3\cos\left(\frac{\pi}{2} - \frac{x}{2}\right)$

8. a. $y = 3\sin\left(2x + \dfrac{\pi}{4}\right) = 3\sin\left[2\left(x + \dfrac{\pi}{8}\right)\right]$

$= 3\sin\left[2\left(x - \left(-\dfrac{\pi}{8}\right)\right)\right]$

$b = 2,$ so the period is $\dfrac{2\pi}{2} = \pi.$

$c = -\dfrac{\pi}{8},$ so the phase shift is $\dfrac{\pi}{8}$ unit left.

b. $y = \cos(\pi x - 1) = \cos\pi\left(x - \dfrac{1}{\pi}\right)$

$b = \pi,$ so the period is $\dfrac{2\pi}{\pi} = 2.$

$c = \dfrac{1}{\pi},$ so the phase shift is $\dfrac{1}{\pi}$ unit right.

9.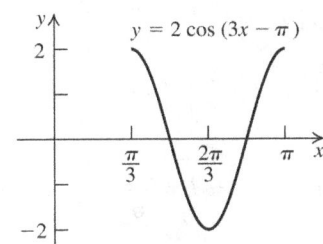

10.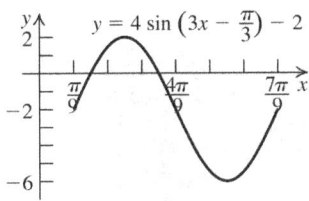

11. Answers will vary.

12. Plot the values given in the table by months using January = 1, ..., December = 12. Then sketch a function of the form $y = a\sin b(x+c) + d$ that models the points just graphed.

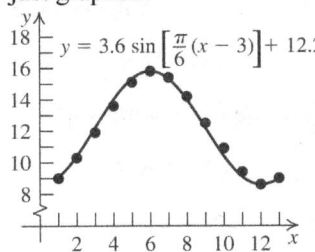

amplitude $= a = \dfrac{\text{highest value} - \text{lowest value}}{2}$

$= \dfrac{15.8 - 8.6}{2} = 3.6$

$d = \dfrac{1}{2}(15.8 + 8.6) = 12.2$

The weather repeats every 12 months, so the period is 12. Then, $12 = \dfrac{2\pi}{b} \Rightarrow b = \dfrac{\pi}{6}.$ The highest point on the graph, 15.8, occurs at $x = 6,$ so the phase shift is 3. The equation is

$y = 3.6\sin\dfrac{\pi}{6}(x - 3) + 12.2$

13. Using the reasoning presented in the example in the text, the form of the equation is $y = -4\cos\omega t.$ The period is

$\dfrac{2\pi}{\omega} = 3 \Rightarrow \dfrac{2\pi}{3} = \omega.$ So the equation is

$y = -4\cos\dfrac{2\pi}{3}t.$

5.4 Concepts and Vocabulary

1. The lowest point on the graph of $y = \cos x, 0 \le x \le 2\pi,$ occurs when $x = \underline{\pi}.$

3. The range of the cosine function is $\underline{[-1, 1]}.$

5. False. The amplitude is 3.

7. True

9.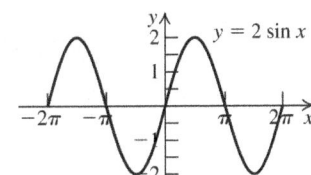

11.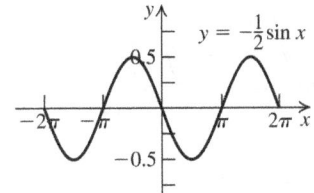

13.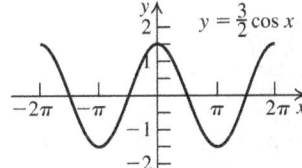

15.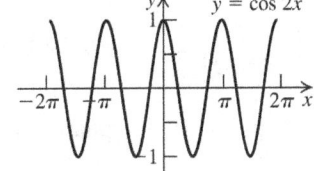

17.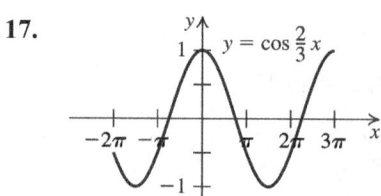

19.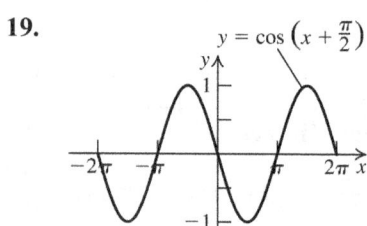

21.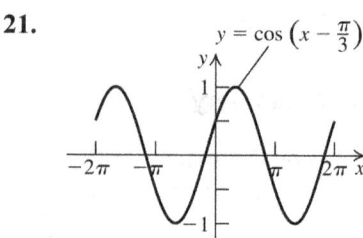

23.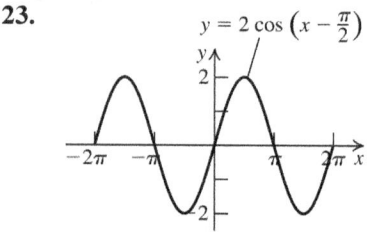

25.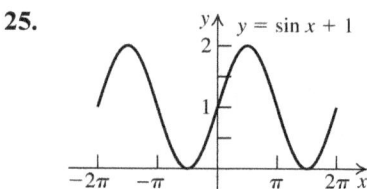

27.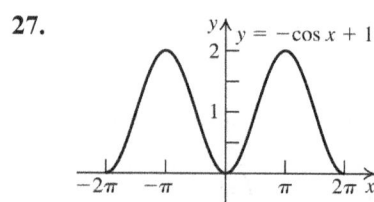

29. $y = 5\cos(x - \pi) \Rightarrow a = 5, b = 1, c = \pi \Rightarrow$ amplitude = 5, period = 2π, phase shift = π

31. $y = 7\cos\left[9\left(x + \dfrac{\pi}{6}\right)\right] \Rightarrow a = 7, b = 9, c = -\dfrac{\pi}{6} \Rightarrow$ amplitude = 7, period = $\dfrac{2\pi}{9}$, phase shift = $-\dfrac{\pi}{6}$

33. $y = -6\cos\left[\dfrac{1}{2}(x+2)\right] \Rightarrow a = -6, b = \dfrac{1}{2}, c = -2 \Rightarrow$ amplitude = 6, period = 4π, phase shift = -2

35. $y = 0.9\sin\left[0.25\left(x - \dfrac{\pi}{4}\right)\right] \Rightarrow a = 0.9, b = 0.25,$ $c = \dfrac{\pi}{4} \Rightarrow$ amplitude = 0.9, period = 8π, phase shift = $\dfrac{\pi}{4}$

37. $a = \dfrac{1}{4}$, period = $\dfrac{\pi}{8} \Rightarrow b = \dfrac{2\pi}{\pi/8} = 16$, $c = \dfrac{\pi}{16}$
$y = \dfrac{1}{4}\sin\left[16\left(x - \dfrac{\pi}{16}\right)\right]$

39. $a = 6$, period = $3\pi \Rightarrow b = \dfrac{2\pi}{3\pi} = \dfrac{2}{3}$, $c = -\dfrac{\pi}{4}$
$y = 6\sin\left[\dfrac{2}{3}\left(x - \left(-\dfrac{\pi}{4}\right)\right)\right] = 6\sin\left[\dfrac{2}{3}\left(x + \dfrac{\pi}{4}\right)\right]$

41. $a = 2.4$, period = $6 \Rightarrow b = \dfrac{2\pi}{6} = \dfrac{\pi}{3}$, $c = 3$
$y = 2.4\sin\left[\dfrac{\pi}{3}(x - 3)\right]$

43. $a = \dfrac{\pi}{2}$, period = $\dfrac{5}{8} \Rightarrow b = \dfrac{2\pi}{5/8} = \dfrac{16\pi}{5}$, $c = -\dfrac{1}{4}$
$y = \dfrac{\pi}{2}\sin\left[\dfrac{16\pi}{5}\left(x - \left(-\dfrac{1}{4}\right)\right)\right]$
$= \dfrac{\pi}{2}\sin\left[\dfrac{16\pi}{5}\left(x + \dfrac{1}{4}\right)\right]$

Answers may vary in exercises 45–59. Sample answers are given.

45. The graph is a cosine curve with amplitude 3 and period π, so $a = 3$ and $b = \dfrac{2\pi}{\pi} = 2$.
An equation is $y = 3\cos 2x$.

47. The graph is a sine curve with amplitude 2 and period $\dfrac{\pi}{2}$, so $a = 2$ and $b = \dfrac{2\pi}{\pi/2} = 4$.
An equation is $y = 2\sin 4x$.

Section 5.4 Graphs of the Sine and Cosine Functions   305

49. The graph is a sine curve with amplitude 3 and period π, so $a = 3$ and $b = \dfrac{2\pi}{\pi} = 2$. The graph is shifted $\dfrac{\pi}{4}$ units right, so $c = \dfrac{\pi}{4}$.
An equation is $y = 3\sin\left[2\left(x - \dfrac{\pi}{4}\right)\right]$.

51. The graph is a reflection of the sine curve about the x-axis. The amplitude 2 and the period is $\dfrac{7\pi}{8} - \left(-\dfrac{\pi}{8}\right) = \pi$, so $a = -2$ and $b = \dfrac{2\pi}{\pi} = 2$. The graph is shifted $\dfrac{\pi}{8}$ units left, so $c = \dfrac{\pi}{8}$.
An equation is
$y = -2\sin\left[2\left(x - \left(-\dfrac{\pi}{8}\right)\right)\right]$
$= -2\sin\left[2\left(x + \dfrac{\pi}{8}\right)\right]$.

53.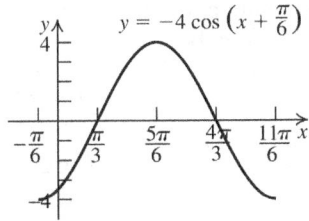
$y = -4\cos\left(x + \dfrac{\pi}{6}\right)$

55. $y = \dfrac{5}{2}\sin\left[2\left(x - \dfrac{\pi}{4}\right)\right]$

57. $y = -5\cos\left[4\left(x - \dfrac{\pi}{6}\right)\right]$

59. $y = \dfrac{1}{2}\sin\left[4\left(x + \dfrac{\pi}{4}\right)\right] + 2$

61. $y = 4\cos\left(2x + \dfrac{\pi}{3}\right) = 4\cos\left[2\left(x + \dfrac{\pi}{6}\right)\right]$
period = π; phase shift = $-\pi/6$

63. $y = -\dfrac{3}{2}\sin(2x - \pi) = -\dfrac{3}{2}\sin\left[2\left(x - \dfrac{\pi}{2}\right)\right]$
period = π; phase shift = $\pi/2$

65. $y = 3\cos \pi x$; period = 2; phase shift = 0

67. $y = \dfrac{1}{2}\cos\left(\dfrac{\pi x}{4} + \dfrac{\pi}{4}\right) = \dfrac{1}{2}\cos\left[\dfrac{\pi}{4}(x+1)\right]$
period = 8; phase shift = -1

69. $y = 2\sin(\pi x + 3) = 2\sin\left[\pi\left(x + \dfrac{3}{\pi}\right)\right]$
period = 2; phase shift = $-3/\pi$

71. $y = 3\sin(4t)$
period = $\dfrac{2\pi}{4} = \dfrac{\pi}{2}$; frequency = $\dfrac{2}{\pi}$

73. $y = 2\cos\left(\dfrac{\pi}{3}t\right)$
period = $\dfrac{2\pi}{\pi/3} = 6$; frequency = $\dfrac{1}{6}$

75. $y = -2\cos\left(\dfrac{1}{2}t\right)$
period = $\dfrac{2\pi}{1/2} = 4\pi$; frequency = $\dfrac{1}{4\pi}$

5.4 Applying the Concepts

77. $P = \sin\left(\dfrac{2\pi}{23}t\right)$, $E = \sin\left(\dfrac{2\pi}{28}t\right)$, $I = \sin\left(\dfrac{2\pi}{33}t\right)$,
where t represents the number of days since birth.
First, calculate t. Number of years from July 22, 1994 to July 22, 2015 = 2015 − 1994 = 21.
Number of leap years = 5 (1996, 2000, 2004, 2008, 2012)
Number of days from April 1, 20051 to July 22, 2005 = 29 (rest of April) + 31 (May) + 30 (June) + 22 (July) = 112
Number of days from July 22, 1994 to July 22, 2015 = (21)(365) + 5 − 112 = 7558
$P = \sin\left(\dfrac{2\pi(7558)}{23}\right) \approx -0.63$

(continued on next page)

(*continued*)

$$E = \sin\left(\frac{2\pi(7558)}{28}\right) \approx -0.43$$

$$I = \sin\left(\frac{2\pi(7558)}{33}\right) \approx 0.19$$

79. a. The pulse shows how many times the heart beats in one minute. Therefore, the pulse rate is the frequency of the function. The period is $\frac{2\pi}{140\pi} = \frac{1}{70}$. So the frequency is 70 pulses per minute.

b. 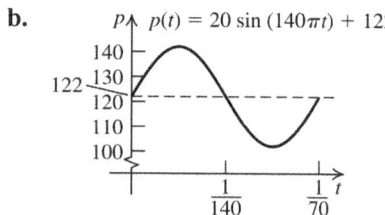 $p(t) = 20\sin(140\pi t) + 122$

c. The systolic reading is the maximum value of the graph, and the diastolic reading in the minimum value of the graph. The maximum value is $122 + 20 = 142$. The minimum value is $122 - 20 = 102$. So, Desmond's blood pressure is $\frac{142}{102}$ mm of mercury.

81. a. amplitude = 156; period $= \frac{2\pi}{120\pi} = \frac{1}{60}$

b. frequency $= \frac{1}{\text{period}} = 60$ cycles per second

c. 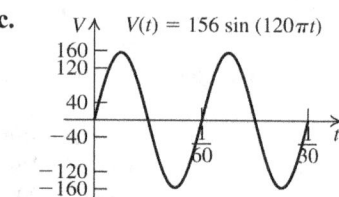 $V(t) = 156\sin(120\pi t)$

83. a. The largest number of kangaroos is when $\sin 2t = 1 \Rightarrow t = \frac{\pi}{4} \Rightarrow 650 + 150 = 800$.

b. The smallest number of kangaroos is when $\sin 2t = -1 \Rightarrow t = \frac{3\pi}{4} \Rightarrow 650 - 150 = 500$.

c. The time between occurrences of the largest and the smallest kangaroo population is 1/2 of the period. The period of the function is $\frac{2\pi}{2} = \pi$, so the time between the maximum and the minimum population is $\frac{\pi}{2} \approx 1.57$ years.

85. Plot the values given in the table by months using January = 1, …, December = 12. Then sketch a function of the form $y = a\sin b(x + c) + d$ that models the points just graphed.

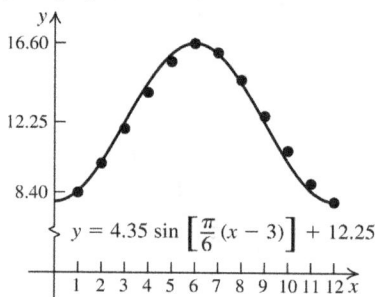

The highest number of daylight hours is 16.6 and the lowest number of daylight hours is 7.9, so $a = \frac{16.6 - 7.9}{2} = 4.35$. The vertical shift $d = \frac{16.6 + 7.9}{2} = 12.25$. The period is $12 = \frac{2\pi}{b} \Rightarrow b = \frac{\pi}{6}$. The highest point on the graph, 16.6, occurs at $x = 6$, so the phase shift is 3. The equation is

$$y = 4.35\sin\left[\frac{\pi}{6}(x - 3)\right] + 12.25.$$

87. a. The highest temperature is 70°F, and the lowest temperature is 19°F, so

$$a = \frac{70 - 19}{2} = 25.5.$$

The vertical shift is $d = \frac{70 + 19}{2} = 44.5$

The period is $12 = \frac{2\pi}{b} \Rightarrow b = \frac{\pi}{6}$.

The highest temperature occurs at $x = 7$, so $c = 7 - 3 = 4$. The equation is

$$y = 25.5\sin\left[\frac{\pi}{6}(x - 4)\right] + 44.5.$$

b.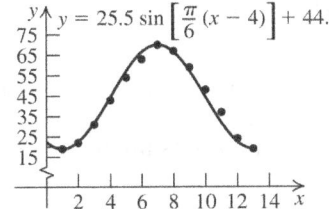

c. January = $f(1) = 19$; April = $f(4) = 44.5$; July = $f(7) = 70$; October = $f(10) = 44.5$. The computed values are very close to the measured values.

5.4 Beyond the Basics

89.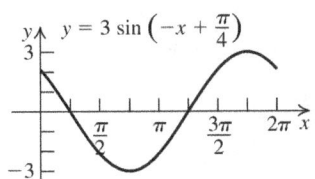

91.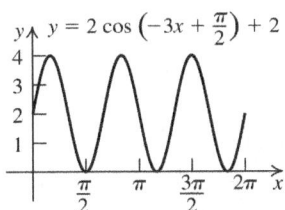

93.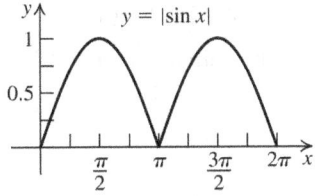

95.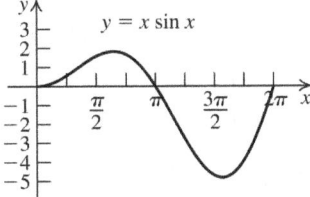

97.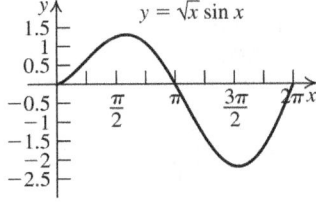

99.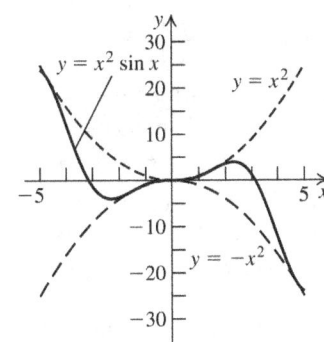

Exercises 101–103 can be verified using a graphing calculator.

101. $T = 25\sin\left[\dfrac{2\pi}{365}(t-110)\right] + 75$

a. The maximum value of $y = \sin x$ is 1. This function has amplitude is 25, meaning it is stretched vertically by a factor of 25, and the graph has been shifted vertically 75, so the highest average daily temperature is $25(1) + 75 = 100°$. The maximum value of $y = \sin x$ occurs at $x = \dfrac{\pi}{2}$.

This function stretched horizontally by a factor of $\dfrac{365}{2\pi}$ and then shifted right by 110, so the maximum value of this function occurs at $x = \dfrac{365}{2\pi}\left(\dfrac{\pi}{2}\right) + 110 = 201.25$, which is day 202 or July 21.

b. The minimum value of $y = \sin x$ is –1. This function has amplitude is 25, meaning it is stretched vertically by a factor of 25, and the graph has been shifted vertically 75, so the lowest average daily temperature is $25(-1) + 75 = 50°$. The minimum value of $y = \sin x$ occurs at $x = \dfrac{3\pi}{2}$. This function stretched horizontally by a factor of $\dfrac{365}{2\pi}$ and then shifted right by 110, so the minimum value of this function occurs at $x = \dfrac{365}{2\pi}\left(\dfrac{3\pi}{2}\right) + 110 \Rightarrow x \approx 384 - 365 = 19$, or January 19.

c. August 15 is day 227.

$T = 25\sin\left[\dfrac{2\pi}{365}(227-110)\right] + 75 \approx 97.6$

The average daily temperature on August 15 is about 97.6° F.

103. $f(x) = -6\cos\left(3x - \dfrac{\pi}{4}\right) - 5$

The minimum value of $y = \cos x$ is -1. This function has amplitude 6, so the minimum value of this function is $6(-1) - 5 = -11$. The maximum value of $y = \cos x$ is 1, so the maximum value of this function is $6(1) - 5 = 1$. The range is $[-11, 1]$.

105. a. If $\sin(x + p) = \sin x$ for some p with $0 < p < 2\pi$ and x a real number, then for $x = 0$ we have $\sin(0 + p) = \sin 0 = 0$. But, if $\sin p = 0$ and $0 < p < 2\pi$, then p must equal π, and then $\sin(x + \pi) = \sin x$ for all real numbers x.

b. $\sin\left(\dfrac{\pi}{2} + \pi\right) = \sin\dfrac{3\pi}{2} = -1 \ne \sin\dfrac{\pi}{2}$. So $p \ne \pi$. Therefore, $p = 2\pi$.

5.4 Critical Thinking/Discussion/Writing

106. a. The period is 20. So period $= \dfrac{2\pi}{b} \Rightarrow$

$20 = \dfrac{2\pi}{b} \Rightarrow b = \dfrac{\pi}{10}$. Graphing the function $y = \sin\dfrac{\pi}{10}x$, we find that there are 10 units between the value of x at which the maximum and minimum values are attained.

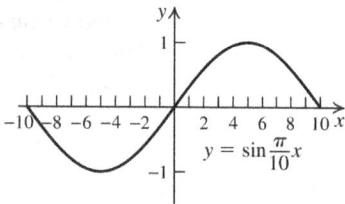

b. For $y = \sin\dfrac{\pi}{10}x$, $d = 0$. So, from $y = 0$ to the maximum of y is 5 units.

107. Since the minimum value occurs at a value of x that is five units from the value of x at which the function has the maximum value, the period of the function is 10 units.

5.4 Getting Ready for the Next Section

109. $y - (-3) = 5(x - 2) \Rightarrow y + 3 = 5x - 10 \Rightarrow$
$y = 5x - 13$

111. $f(x) = \dfrac{x^2 + 6x + 8}{x^2 + 1} = \dfrac{(x+2)(x+4)}{x^2 + 1}$

There are no common factors in the numerator and denominator, so find the x-intercepts and vertical asymptotes as follows.

x-intercepts:
$x^2 + 6x + 8 = 0 \Rightarrow (x+2)(x+4) = 0 \Rightarrow$
$x = -2, -4$

vertical asymptotes: $x^2 + 1 = 0 \Rightarrow$ there is no real solution. There are no vertical asymptotes.

For exercises 113–120, recall that
$f(-x) = f(x) \Rightarrow f(x)$ is even and
$f(-x) = -f(x) \Rightarrow f(x)$ is odd.

113. $f(x) = x^3 - 2x$
$f(-x) = (-x)^3 - 2(-x) = -x^3 + 2x = -f(x)$
$f(x)$ is odd.

115. $f(x) + g(x) = x^3 - 2x + 3x^2 + 1$
$\qquad = x^3 + 3x^2 - 2x + 1$
$f(-x) + g(-x) = (-x)^3 + 3(-x)^2 - 2(-x) + 1$
$\qquad = -x^3 + 3x^2 + 2x + 1$
$\qquad \ne f(-x) + g(-x)$
$\qquad \ne f(x) + g(x)$
$f(x) + g(x)$ is neither even nor odd.

117. $\dfrac{f(x)}{g(x)} = \dfrac{x^3 - 2x}{3x^2 + 1}$

$\dfrac{f(-x)}{g(-x)} = \dfrac{(-x)^3 - 2(-x)}{3(-x)^2 + 1} = \dfrac{-x^3 + 2x}{3x^2 + 1}$
$\qquad = \dfrac{f(-x)}{g(-x)}$

$\dfrac{f(x)}{g(x)}$ is odd.

119. $\dfrac{1}{f(x)} = \dfrac{1}{x^3 - 2x}$

$\dfrac{1}{f(-x)} = \dfrac{1}{(-x)^3 - 2(-x)} = \dfrac{1}{-x^3 + 2x}$
$\qquad = \dfrac{1}{f(-x)}$

$\dfrac{1}{f(x)}$ is odd.

5.5 Graphs of the Other Trigonometric Functions

5.5 Practice Problems

1. We have $m = \tan 30° = \dfrac{\sqrt{3}}{3}$, so the slope of the line is $\dfrac{\sqrt{3}}{3}$. Using the point-slope form, we have $y - 3 = \dfrac{\sqrt{3}}{3}(x - 2)$ as the equation of the line. In slope-intercept form, the equation is $y = \dfrac{\sqrt{3}}{3}x + 3 - \dfrac{2\sqrt{3}}{3}$.

2. $y = -3\tan\left(x + \dfrac{\pi}{4}\right)$

   $a = -3 \Rightarrow$ the vertical stretch factor is 3.
   $b = 1 \Rightarrow$ the period is π.
   $c = -\dfrac{\pi}{4} \Rightarrow$ the phase shift is $\dfrac{\pi}{4}$ units left.

   Find asymptotes: $x + \dfrac{\pi}{4} = -\dfrac{\pi}{2} \Rightarrow x = -\dfrac{3\pi}{4}$

   and $x + \dfrac{\pi}{4} = \dfrac{\pi}{2} \Rightarrow x = \dfrac{\pi}{4}$. Divide the interval $\left(-\dfrac{3\pi}{4}, \dfrac{\pi}{4}\right)$ into four equal parts: $-\dfrac{\pi}{2}, -\dfrac{\pi}{4}$, and 0.

x	$y = -3\tan\left(x + \dfrac{\pi}{4}\right)$
$-\dfrac{\pi}{2}$	3
$-\dfrac{\pi}{4}$	0
0	-3

   Sketch the vertical asymptotes and the points $\left(-\dfrac{\pi}{2}, 3\right), \left(-\dfrac{\pi}{4}, 0\right)$, and $(0, -3)$. Connect the points with a smooth curve. Repeat the graph to the left and right over intervals of length π.

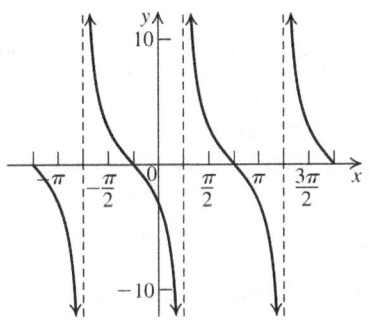

3. $y = -3\cot\left(x + \dfrac{\pi}{4}\right)$

   $a = -3 \Rightarrow$ the vertical stretch factor is 3.
   $b = 1 \Rightarrow$ the period is π.
   $c = -\dfrac{\pi}{4} \Rightarrow$ the phase shift is $\dfrac{\pi}{4}$ units left.

   Find asymptotes: $x + \dfrac{\pi}{4} = 0 \Rightarrow x = -\dfrac{\pi}{4}$ and

   $x + \dfrac{\pi}{4} = \pi \Rightarrow x = \dfrac{3\pi}{4}$. Divide the interval $\left(-\dfrac{\pi}{4}, \dfrac{3\pi}{4}\right)$ into four equal parts: $0, \dfrac{\pi}{4}$, and $\dfrac{\pi}{2}$.

x	$y = -3\cot\left(x + \dfrac{\pi}{4}\right)$
0	-3
$\dfrac{\pi}{4}$	0
$\dfrac{\pi}{2}$	3

   Sketch the vertical asymptotes and the points $(0, -3), \left(\dfrac{\pi}{4}, 0\right)$, and $\left(\dfrac{\pi}{2}, 3\right)$.

   Connect the points with a smooth curve. Repeat the graph to the left and right over intervals of length π to graph over the interval $\left(-\dfrac{3\pi}{4}, \dfrac{5\pi}{4}\right)$.

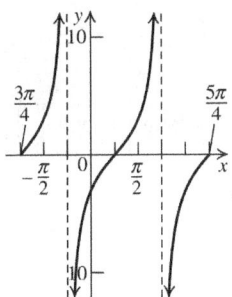

4. For $x = \dfrac{\pi}{8}, y = \csc\dfrac{x}{2} = \csc\dfrac{\pi}{16} \approx 5.1$ and for
$x = \dfrac{\pi}{4}, y = \csc\dfrac{x}{2} = \csc\dfrac{\pi}{8} \approx 2.6$. The range of Mach numbers associated with the interval $\left[\dfrac{\pi}{8}, \dfrac{\pi}{4}\right]$ is about [2.6, 5.1].

5.5 Concepts and Vocabulary

1. The tangent function has period $\underline{\pi}$.

3. If $0 \le x \le \pi$ and $\cot x = 0$, then $x = \dfrac{\pi}{2}$.

5. False. The slope of the line $y = -x + 5$ is -1, so $-1 = \tan\theta \Rightarrow \theta = 135°$. The line makes an angle of $135°$ with the positive x-axis.

7. False. The period of $y = \cot(2x)$ is $\dfrac{\pi}{2}$.

5.5 Building Skills

9. $\theta = 45° \Rightarrow \tan 45° = 1 =$ slope of the line
$y - 3 = 1(x - (-2)) \Rightarrow y - 3 = x + 2 \Rightarrow y = x + 5$

11. $\theta = 120° \Rightarrow \tan 60° = -\sqrt{3} =$ slope of the line
$y - (-2) = -\sqrt{3}(x - (-3)) \Rightarrow$
$y + 2 = -\sqrt{3}(x + 3) \Rightarrow y + 2 = -\sqrt{3}x - 3\sqrt{3} \Rightarrow$
$y = -\sqrt{3}x - 3\sqrt{3} - 2 = -\sqrt{3}x - (3\sqrt{3} + 2)$

13. $y = 3\tan\left(x - \dfrac{\pi}{4}\right)$

period $= \dfrac{\pi}{b} = \dfrac{\pi}{1} = \pi$

phase shift $= c = \dfrac{\pi}{4}$

15. $y = -2\cot\left(x + \dfrac{\pi}{3}\right) = -2\cot\left(x - \left(-\dfrac{\pi}{3}\right)\right)$

period $= \dfrac{\pi}{b} = \dfrac{\pi}{1} = \pi$

phase shift $= c = -\dfrac{\pi}{3}$

17. $y = 2\tan\left(3x + \dfrac{\pi}{2}\right) = 2\tan\left[3\left(x + \dfrac{\pi}{6}\right)\right]$
$= 2\tan\left[3\left(x - \left(-\dfrac{\pi}{6}\right)\right)\right]$

period $= \dfrac{\pi}{b} = \dfrac{\pi}{3}$

phase shift $= c = -\dfrac{\pi}{6}$

19. $y = -3\cot\left(2x + \dfrac{\pi}{3}\right) = -3\cot\left[2\left(x + \dfrac{\pi}{6}\right)\right]$
$= -3\cot\left[2\left(x - \left(-\dfrac{\pi}{6}\right)\right)\right]$

period $= \dfrac{\pi}{b} = \dfrac{\pi}{2}$

phase shift $= c = -\dfrac{\pi}{6}$

21. $y = 3\csc\left(2x - \dfrac{\pi}{8}\right) = 3\csc\left[2\left(x - \dfrac{\pi}{16}\right)\right]$

period $= \dfrac{2\pi}{b} = \dfrac{2\pi}{2} = \pi$

phase shift $= c = \dfrac{\pi}{16}$

23. $y = 2\sec\left(4x + \dfrac{\pi}{2}\right) = 2\sec\left[4\left(x + \dfrac{\pi}{8}\right)\right]$
$= 2\sec\left[4\left(x - \left(-\dfrac{\pi}{8}\right)\right)\right]$

period $= \dfrac{2\pi}{b} = \dfrac{2\pi}{4} = \dfrac{\pi}{2}$

phase shift $= c = -\dfrac{\pi}{8}$

25.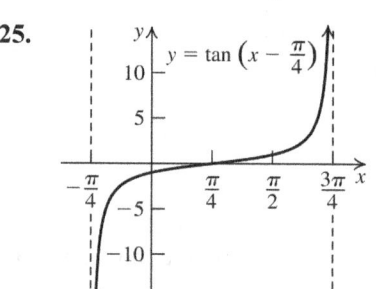

27.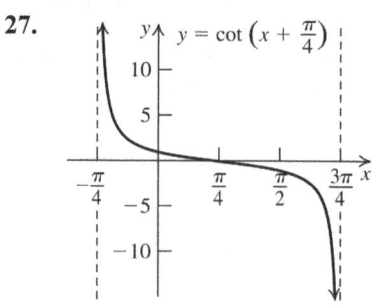

29.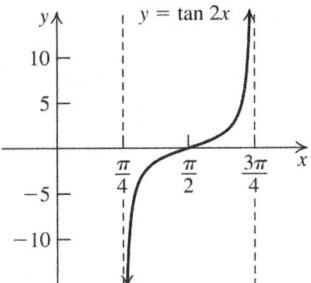

31.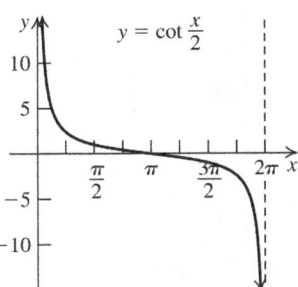

33.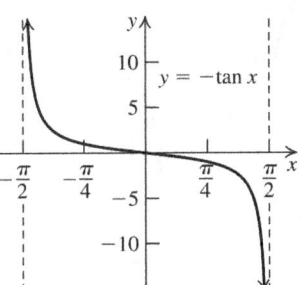

35.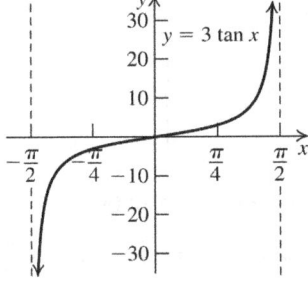

37.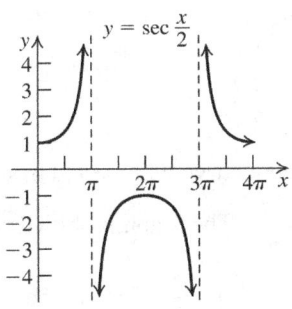

39.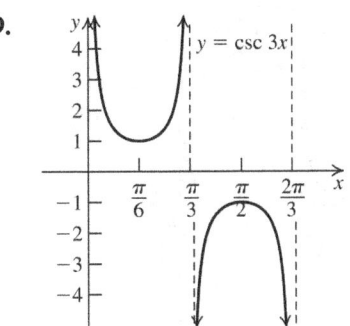

41.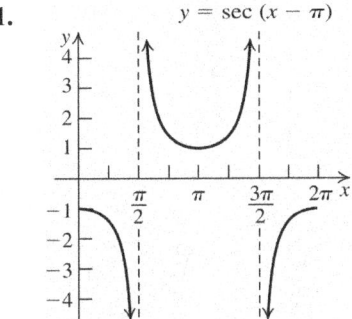

43.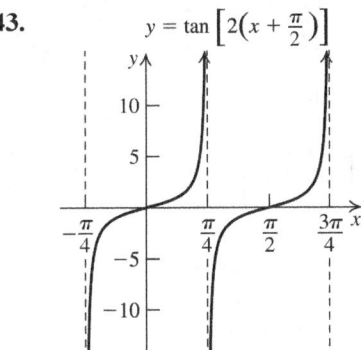

45.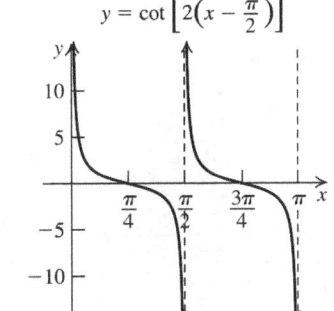

47.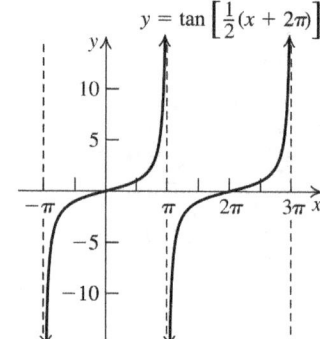

49.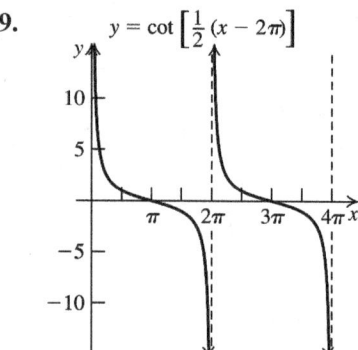

51.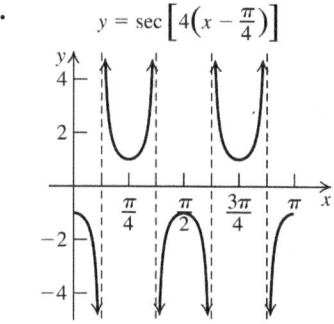

53.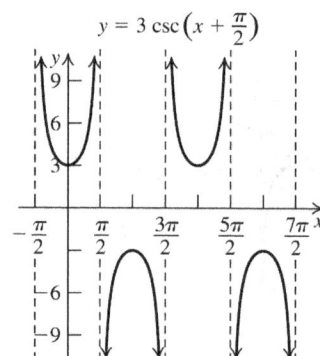

55.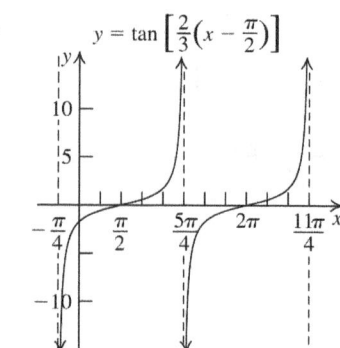

57.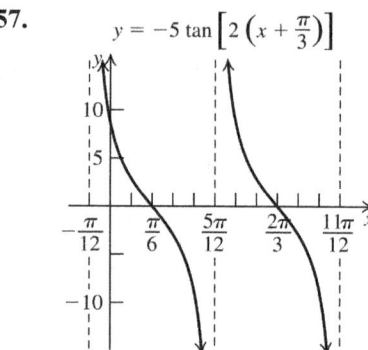

59.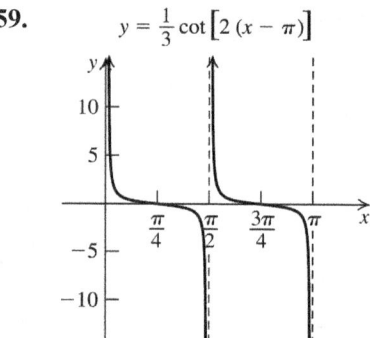

61. This is a tangent curve with period π, so $\pi = \dfrac{\pi}{b} \Rightarrow b = 1$. The curve is shifted $\dfrac{\pi}{2}$ units to the right and stretched vertically by a factor of 2, so $a = 2$. An equation is
$$y = 2\tan\left(x - \dfrac{\pi}{2}\right).$$

63. This is a tangent curve with period 6, so $6 = \dfrac{\pi}{b} \Rightarrow b = \dfrac{\pi}{6}$. The graph is shifted right 2 units, so $c = 2\left(\dfrac{\pi}{6}\right) = \dfrac{\pi}{3}$. The graphed is reflected across the vertical axis. So, we have
$$y = -a\tan\left(\dfrac{\pi}{6}x - \dfrac{\pi}{3}\right).$$

(continued on next page)

(*continued*)

Substitute $x = \frac{7}{2}$ and $y = -1$ to solve for a.

$-1 = -a \tan\left(\frac{\pi}{6}\left(\frac{7}{2}\right) - \frac{\pi}{3}\right) \Rightarrow$

$-1 = -a \tan\left(\frac{\pi}{4}\right) \Rightarrow 1 = a(1) \Rightarrow a = 1$

Thus, the equation is $y = -\tan\left(\frac{\pi}{6}x - \frac{\pi}{3}\right)$.

65. This is a cosecant curve with period π, so $\pi = \frac{2\pi}{b} \Rightarrow b = 2$. The maximum and minimum values of the associated sine curve are 3 and -3, so $a = 3$. There is no horizontal or vertical shift. The equation is $y = 3\csc(2x)$.

67. This is a cosecant curve with period 4, so $4 = \frac{2\pi}{b} \Rightarrow b = \frac{\pi}{2}$. The maximum and minimum values of the associated sine curve are 2 and -2, so $a = 2$. The graph is shifted $1\left(\frac{\pi}{2}\right) = \frac{\pi}{2}$ units to the left. The equation is $y = 2\csc\left(\frac{\pi}{2}x + \frac{\pi}{2}\right)$.

5.5 Applying the Concepts

69. a.

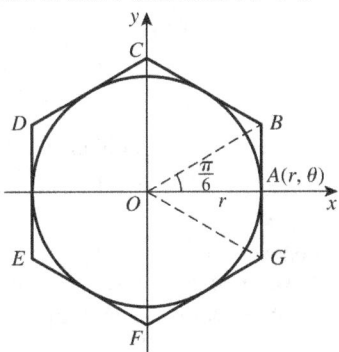

b. When $t = 2.5$, the light is pointing parallel to the wall.

Use the figure to solve exercises 71–74.

71. a. From the figure, it is clear that the x-coordinate of B is r. Find the y-coordinate of B as follows:

$\tan\frac{\pi}{6} = \frac{AB}{r} \Rightarrow AB = r\tan\frac{\pi}{6}$. From geometry, we know that A is the midpoint of the side. Therefore, the length of each side of the hexagon is $2r\tan\frac{\pi}{6}$.

b. $P = 6\left(2r\tan\frac{\pi}{6}\right) = 12r\tan\frac{\pi}{6}$

$= 12r\left(\frac{\sqrt{3}}{3}\right) = 4r\sqrt{3}$

73. a. Using the same reasoning as in exercise 71, we know that the length of each side of the circumscribed n-gon is $2r\tan\frac{\pi}{n}$. Thus, the perimeter of the n-gon is $P = 2nr\tan\frac{\pi}{n}$.

b. $C = 2\pi r = 20\pi \approx 62.8319$

c. $P = 2(4)(10)\tan\frac{\pi}{4} \approx 80.0$

$P = 2(10)(10)\tan\frac{\pi}{10} \approx 64.9839$

$P = 2(50)(10)\tan\frac{\pi}{50} \approx 62.9147$

$P = 2(100)(10)\tan\frac{\pi}{100} \approx 62.8525$

The perimeters approach the perimeter of the circle.

75. Solve the system of equations.
$$-7 = a\tan\left[b(-3-c)\right] \quad (1)$$
$$0 = a\tan\left[b(-1-c)\right] \quad (2)$$
$$7 = a\tan\left[b(1-c)\right] \quad (3)$$

Examining equation (2), either $a = 0$ or $\tan\left[b(-1-c)\right] = 0 \Rightarrow b(-1-c) = 0 \Rightarrow$ $b = 0$ or $c = -1$
If $a = 0$ or $b = 0$, then equations (1) and (3) are false, so we disregard these solutions. Thus, $c = -1$. Substitute this value into equations (1) and (3).
$$-7 = a\tan\left[b(-3-(-1))\right] = a\tan(-2b) \quad (4)$$
$$7 = a\tan\left[b(1-(-1))\right] = a\tan(2b) \quad (5)$$

If we let $a = 7$, then
$$\tan(2b) = 1 \Rightarrow 2b = \frac{\pi}{4} \Rightarrow b = \frac{\pi}{8}.$$

Verify that $a = 7$, $b = \frac{\pi}{8}$, $c = -1$ makes equations (1)–(3) true.

The model is $y = 7\tan\left[\frac{\pi}{8}(x+1)\right]$.

$$y(-1.5) = 7\tan\left[\frac{\pi}{8}(-1.5+1)\right] \approx -1.39$$

$$y(1.5) = 7\tan\left[\frac{\pi}{8}(1.5+1)\right] \approx 10.48$$

5.5 Beyond the Basics

77. For any x,
$$\frac{1}{f}(x+p) = \frac{1}{f(x+p)} = \frac{1}{f(x)} = \frac{1}{f}(x)$$

79. a. True
If f is an odd function, then $f(-x) = -f(x)$, and if g is an even function, $g(-x) = g(x)$.
$$\frac{f(-x)}{g(-x)} = \frac{-f(x)}{g(x)} = -\frac{f(x)}{g(x)}, \text{ an odd}$$
function. Similarly, we can show that $\frac{g}{f}$ is an odd function.

b. True
If f is an odd function, then $f(-x) = -f(x)$, and if g is an even function, $g(-x) = g(x)$.
$$f \circ g = f(g(-x)) = f(g(x)) \Rightarrow f \circ g \text{ is an even function.}$$
$$g \circ f = g(f(-x)) = g(-f(x))$$
$$= g(f(x)) \Rightarrow g \circ f \text{ is an even function.}$$

81.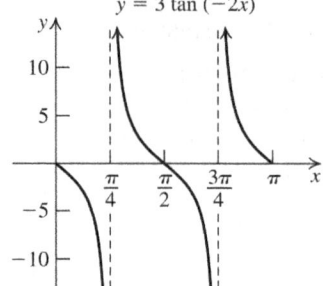

83.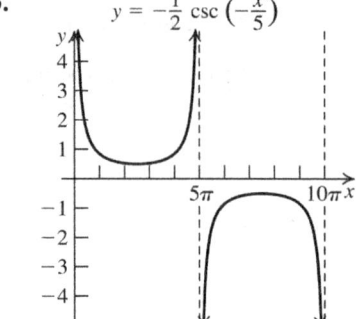

85.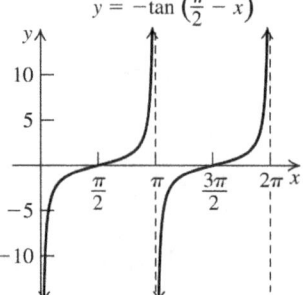

87. $y = 2\sec(\pi - 2x)$

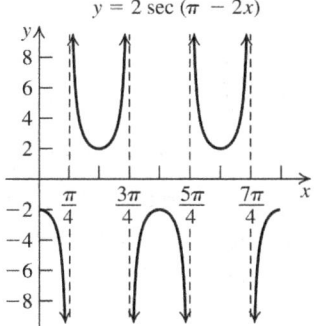

89. $y = 4\cot(\pi - 4x) + 3$

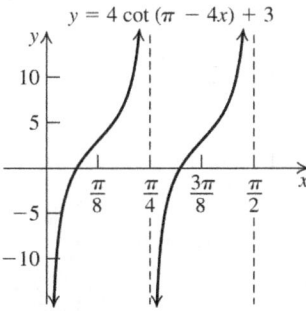

5.5 Critical Thinking/Discussion/Writing

91. The equation represents a cotangent curve with $b = \dfrac{1}{2} \Rightarrow \text{period} = \dfrac{\pi}{1/2} = 2\pi$. Since the period of a cotangent curve is π, this curve has been stretched horizontally by a factor of 2. The curve is then shifted $\dfrac{\pi}{8}(2) = \dfrac{\pi}{4}$ unit to the right. The cotangent curve has an asymptote at $x = -2\pi$, so the asymptote for this curve is also shifted $\pi/4$ unit to the right. Therefore, $k = -2\pi + \dfrac{\pi}{4} = -\dfrac{7\pi}{4}$.

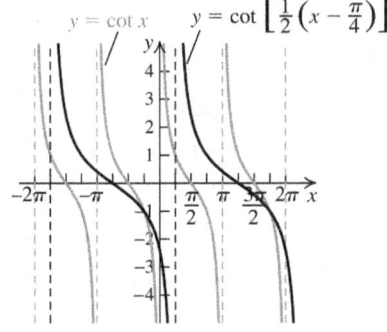

92. $\text{period} = \dfrac{\pi}{3} = \dfrac{2\pi}{b} \Rightarrow b = 6$

93. $y = -2\sec x = -2\csc\left(x + \dfrac{\pi}{2}\right)$

94. $\cot x = -\tan\left(x - \dfrac{\pi}{2}\right)$

95. $y = 5\cot\left(\dfrac{1}{2}x - \dfrac{\pi}{8}\right) = 5\cot\dfrac{1}{2}\left(x - \dfrac{\pi}{4}\right)$

The graph is the graph of $y = \cot x$ shifted $\dfrac{\pi}{4}$ units to the right, with period $= 2\pi$. So, an asymptote is located at $x = \dfrac{\pi}{4} - 2\pi = -\dfrac{7\pi}{4}$.

96. $y = \cot 3x$ has period $\dfrac{\pi}{3}$. $y = \sec bx$ has period $\dfrac{2\pi}{b}$. So, $\dfrac{\pi}{3} = \dfrac{2\pi}{b} \Rightarrow b = 6$.

5.5 Getting Ready for the Next Section

97. $f(x) = 3x + 7$ is a linear function that is not horizontal, so it's a one-to-one function. Find the inverse by interchanging x and y, and then solving for y.
$f(x) = y = 3x + 7$
$x = 3y + 7 \Rightarrow x - 7 = 3y \Rightarrow y = f^{-1}(x) = \dfrac{x - 7}{3}$

99. True

101. True

103. Using the horizontal line test (see Section 2.9), we see that any horizontal line $y = n$, where $n > 0$, intersects the graph of f in more than one point.

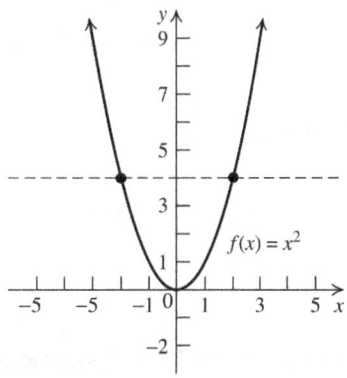

If the domain is restricted to $[0, \infty)$, then f is one-to-one.

105. Using the horizontal line test (see Section 2.9), we see that any horizontal line $y = n$, where $-1 < n < 1$, intersects the graph of f in more than one point.

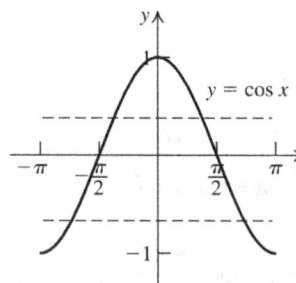

If the domain is restricted to $[0, \pi]$, then f is one-to-one.

5.6 Inverse Trigonometric Functions

5.6 Practice Problems

1. a. $y = \sin^{-1}\left(-\dfrac{\sqrt{3}}{2}\right) \Rightarrow \sin y = -\dfrac{\sqrt{3}}{2}$ and

$-\dfrac{\pi}{2} \leq y \leq \dfrac{\pi}{2} \Rightarrow y = -\dfrac{\pi}{3}$

b. $y = \sin^{-1}(-1) \Rightarrow \sin y = -1$ and

$-\dfrac{\pi}{2} \leq y \leq \dfrac{\pi}{2} \Rightarrow y = -\dfrac{\pi}{2}$

c. $y = \sin^{-1}(2)$ is undefined because 2 is not in the domain of $\sin^{-1} x$.

2. a. $y = \cos^{-1}\left(-\dfrac{\sqrt{2}}{2}\right) \Rightarrow \cos y = -\dfrac{\sqrt{2}}{2}$ and

$0 \leq y \leq \pi \Rightarrow y = \dfrac{3\pi}{4}$

b. $y = \cos^{-1}\dfrac{1}{2} \Rightarrow \cos y = \dfrac{1}{2}$ and

$0 \leq y \leq \pi \Rightarrow y = \dfrac{\pi}{3}$

c. $y = \arccos(\pi)$ is undefined because π is not in the domain of $\arccos x$.

3. $y = \tan^{-1}\dfrac{\sqrt{3}}{3} \Rightarrow \tan y = \dfrac{\sqrt{3}}{3}$ and

$-\dfrac{\pi}{2} \leq y \leq \dfrac{\pi}{2} \Rightarrow y = \dfrac{\pi}{6}$

4. $y = \sec^{-1} 2 \Rightarrow \sec y = 2$ and

$0 \leq y \leq \pi \Rightarrow y = \dfrac{\pi}{3}$

5. The function f with domain $[1, 1+\pi]$ is one-to-one, so it has an inverse. Replace $f(x)$ with y, interchange x and y, solve for y.

$f(x) = y = 3\cos(x-1) + 2$

$x = 3\cos(y-1) + 2$

$\dfrac{x-2}{3} = \cos(y-1)$

$y - 1 = \cos^{-1}\left(\dfrac{x-2}{3}\right)$

$y = \cos^{-1}\left(\dfrac{x-2}{3}\right) + 1$

$\cos^{-1}\left(\dfrac{x-2}{3}\right)$ is defined for

$-1 \leq \dfrac{x-2}{3} \leq 1 \Rightarrow$

$-3 \leq x - 2 \leq 3 \Rightarrow -1 \leq x \leq 5$.

The domain is $[-1, 5]$ and the range is $[1, 1+\pi]$.

6. a. For $|x| \leq 1$, let $\theta_1 = \sin^{-1}(-x)$ with θ_1 in the interval $\left[-\dfrac{\pi}{2}, \dfrac{\pi}{2}\right]$. Then,

$\sin \theta_1 = -x$, $|x| \leq 1$, or $x = -\sin \theta_1$.

Similarly, let $\theta_2 = -\sin^{-1} x \Rightarrow$

$-\theta_2 = \sin^{-1} x \Rightarrow x = \sin^{-1}(-\theta_2)$

So, $x = -\sin \theta_1 = \sin^{-1}(-\theta_2) \Rightarrow \theta_1 = \theta_2$

because $\sin t$ is an even function. Therefore,

$-\sin^{-1} x = \sin^{-1}(-x)$.

b. $\sin\left[\dfrac{\pi}{3} - \sin^{-1}\left(-\dfrac{1}{2}\right)\right] = \sin\left[\dfrac{\pi}{3} + \sin^{-1}\left(\dfrac{1}{2}\right)\right]$

$= \sin\left(\dfrac{\pi}{3} + \dfrac{\pi}{6}\right)$

$= \sin\left(\dfrac{\pi}{2}\right) = 1$

7. a. $y = \cot^{-1}(0.75) = \tan^{-1}\left(\dfrac{1}{0.75}\right) \approx 53.1301°$

b. $y = \csc^{-1}(13) = \sin^{-1}\left(\dfrac{1}{13}\right) \approx 4.4117°$

c. $y = \tan^{-1}(-12) \approx -85.2364°$

8. a. We cannot use the formula
   $\sin^{-1}(\sin x) = x$ for $x = \dfrac{3\pi}{2}$ because $\dfrac{3\pi}{2}$
   is not in the interval $-\dfrac{\pi}{2} \le x \le \dfrac{\pi}{2}$ for
   which $\sin^{-1} x$ is defined. However,
   $\sin \dfrac{3\pi}{2} = \sin\left(\dfrac{3\pi}{2} - 2\pi\right) = \sin\left(-\dfrac{\pi}{2}\right)$, so
   $\sin^{-1}\left(\sin \dfrac{3\pi}{2}\right) = -\dfrac{\pi}{2}$.

   b. $\cos\left(\cos^{-1}(-2)\right)$ is undefined because
   $\cos^{-1}(-2)$ is undefined.

9. $\cos\left[\sin^{-1}\left(-\dfrac{1}{3}\right)\right]$

   Let $\theta = \sin^{-1}\left(-\dfrac{1}{3}\right)$. Then $\sin\theta = -\dfrac{1}{3}$. Using
   the Pythagorean theorem with $y = -1$ and $r = 3$, we have $x = \sqrt{3^2 - (-1)^2} = \sqrt{8} = 2\sqrt{2}$. So,
   $\cos\left[\sin^{-1}\left(-\dfrac{1}{3}\right)\right] = \cos\theta = \dfrac{2\sqrt{2}}{3}$.

10. $\theta = \tan^{-1}\dfrac{30}{x} + \tan^{-1}\dfrac{10}{x}$
    $\theta = \tan^{-1}\dfrac{30}{25} + \tan^{-1}\dfrac{10}{25} \approx 72°$
    Set the camera to rotate through 72° to scan the entire counter.

5.6 Concepts and Vocabulary

1. The domain of $f(x) = \sin^{-1} x$ is [−1, 1].

3. The exact value of $y = \cos^{-1}\dfrac{1}{2}$ is $\dfrac{\pi}{3}$.

5. True

7. False. The domain of $\cos^{-1} x$ is [−1, 1].

5.6 Building Skills

9. $y = \sin^{-1} 0 = 0$

11. $y = \sin^{-1}\left(-\dfrac{1}{2}\right) = -\dfrac{\pi}{6}$

13. $y = \arccos(-1) = \pi$

15. $y = \arccos\left(\dfrac{\pi}{2}\right)$ is not defined

17. $y = \tan^{-1}\sqrt{3} = \dfrac{\pi}{3}$

19. $y = \arctan(-1) = -\dfrac{\pi}{4}$

21. $y = \cot^{-1}(-1) = \dfrac{3\pi}{4}$

23. $y = \cos^{-1}(-2)$ is not defined.

25. $y = \sec^{-1}(-2) = \dfrac{2\pi}{3}$

27. $y = \arcsin 1 = \dfrac{\pi}{2}$

29. $y = \text{arccot}(-\sqrt{3}) = \dfrac{5\pi}{6}$

31. Replace $f(x)$ with y. Then interchange x and y, and solve for y.
    $f(x) = y = 2\sin x + 1$, $-\dfrac{\pi}{2} \le x \le \dfrac{\pi}{2}$
    $x = 2\sin y + 1 \Rightarrow \sin y = \dfrac{x-1}{2} \Rightarrow$
    $y = f^{-1}(x) = \sin^{-1}\left(\dfrac{x-1}{2}\right)$

33. Replace $f(x)$ with y. Then interchange x and y, and solve for y.
    $f(x) = y = 3\cos(2x-1)$, $\dfrac{1}{2} \le x \le \dfrac{\pi}{2} + \dfrac{1}{2}$
    $x = 3\cos(2y-1) \Rightarrow \cos(2y-1) = \dfrac{x}{3} \Rightarrow$
    $2y - 1 = \cos^{-1}\left(\dfrac{x}{3}\right) \Rightarrow 2y = \cos^{-1}\left(\dfrac{x}{3}\right) + 1 \Rightarrow$
    $y = f^{-1}(x) = \dfrac{1}{2}\cos^{-1}\left(\dfrac{x}{3}\right) + \dfrac{1}{2}$

35. Replace $f(x)$ with y. Then interchange x and y, and solve for y.
    $f(x) = y = \tan(x-1) + 2$, $1 - \dfrac{\pi}{2} < x < 1 + \dfrac{\pi}{2}$
    $x = \tan(y-1) + 2 \Rightarrow \tan(y-1) = x - 2 \Rightarrow$
    $y - 1 = \tan^{-1}(x-2) \Rightarrow$
    $y = f^{-1}(x) = \tan^{-1}(x-2) + 1$

37. $y = \sin\left(\sin^{-1}\dfrac{1}{8}\right) = \dfrac{1}{8}$

39. $y = \tan^{-1}\left(\tan\dfrac{\pi}{7}\right) = \dfrac{\pi}{7}$

41. $y = \tan\left(\tan^{-1} 247\right) = 247$

43. $y = \sin^{-1}\left(\sin\dfrac{4\pi}{3}\right) = \sin^{-1}\left(-\dfrac{\sqrt{3}}{2}\right) = -\dfrac{\pi}{3}$

45. $y = \tan^{-1}\left(\tan\dfrac{2\pi}{3}\right) = \tan^{-1}(-\sqrt{3}) = -\dfrac{\pi}{3}$

47. $y = \sin^{-1}\left(\sin\dfrac{3\pi}{4}\right)$

$\dfrac{3\pi}{4}$ is not in the interval $\left[-\dfrac{\pi}{2}, \dfrac{\pi}{2}\right]$ so we must find θ in the interval $\left[-\dfrac{\pi}{2}, \dfrac{\pi}{2}\right]$ where $\sin\dfrac{3\pi}{4} = \sin\theta$. $\sin\dfrac{3\pi}{4} = \sin\dfrac{\pi}{4}$ so
$y = \sin^{-1}\left(\sin\dfrac{3\pi}{4}\right) = \sin^{-1}\left(\sin\dfrac{\pi}{4}\right) = \dfrac{\pi}{4}$.

49. $y = \sin\left(\sin^{-1}\sqrt{2}\right)$ is undefined because $x = \sqrt{2}$ is not in the interval $[-1, 1]$.

51. $y = \cos^{-1}(\cos(-\pi))$

$-\pi$ is not in the interval $[0, \pi]$ so we must find θ in the interval $[0, \pi]$ where $\cos(-\pi) = \cos\theta$. $\cos(-\pi) = \cos\pi$ so $y = \cos^{-1}(\cos(-\pi)) = \cos^{-1}(\cos\pi) = \pi$.

53. $\cot^{-1}\left(\dfrac{1}{\sqrt{3}}\right) = \tan^{-1}(\sqrt{3}) = \dfrac{\pi}{3}$

55. $\csc^{-1}(2) = \sin^{-1}\left(\dfrac{1}{2}\right) = \dfrac{\pi}{6}$

57. $\cot^{-1}\left(-\dfrac{1}{\sqrt{3}}\right) = \tan^{-1}(-\sqrt{3}) = \dfrac{2\pi}{3}$

59. $\sin^{-1}\left(-\dfrac{\sqrt{3}}{2}\right) = -\sin^{-1}\left(\dfrac{\sqrt{3}}{2}\right) = -\dfrac{\pi}{3}$

61. $\cos^{-1}\left(-\dfrac{1}{2}\right) = \pi - \cos^{-1}\left(\dfrac{1}{2}\right) = \pi - \dfrac{\pi}{3} = \dfrac{2\pi}{3}$

63. $\sin\left[\dfrac{\pi}{3} - \sin^{-1}\left(-\dfrac{1}{2}\right)\right] = \sin\left[\dfrac{\pi}{3} + \sin^{-1}\left(\dfrac{1}{2}\right)\right]$
$= \sin\left(\dfrac{\pi}{3} + \dfrac{\pi}{6}\right) = \sin\dfrac{\pi}{2}$
$= 1$

65. $\sin\left[\dfrac{\pi}{2} - \cos^{-1}(-1)\right] = \sin\left(\dfrac{\pi}{2} - \pi\right)$
$= \sin\left(-\dfrac{\pi}{2}\right) = -\sin\dfrac{\pi}{2}$
$= -1$

67. $\sin\left[\tan^{-1}(-\sqrt{3}) + \cos^{-1}\left(-\dfrac{\sqrt{3}}{2}\right)\right]$
$= \sin\left(-\dfrac{\pi}{3} + \dfrac{5\pi}{6}\right) = \sin\dfrac{\pi}{2} = 1$

69. $y = \cos^{-1} 0.6 \approx 53.13°$

71. $y = \sin^{-1}(-0.69) \approx -43.63°$

73. $y = \sec^{-1}(3.5) \approx 73.40°$

75. $y = \tan^{-1} 14 \approx 85.91°$

77. $y = \tan^{-1}(-42.147) \approx -88.64°$

In exercises 63–86, use the Pythagorean theorem to find the length of the third side of the triangle.

79. $y = \cos\left(\sin^{-1}\dfrac{2}{3}\right) = \dfrac{\sqrt{5}}{3}$

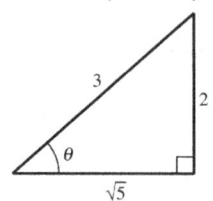

81. $y = \sin\left(\cos^{-1}\left(-\dfrac{4}{5}\right)\right) = \dfrac{3}{5}$

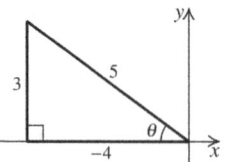

83. $y = \cos\left(\tan^{-1}\dfrac{5}{2}\right) = \dfrac{2}{\sqrt{29}} = \dfrac{2\sqrt{29}}{29}$

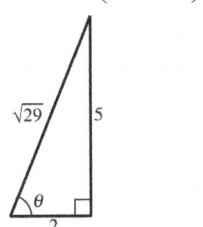

85. $y = \tan\left(\cos^{-1}\dfrac{4}{5}\right) = \dfrac{3}{4}$

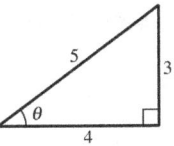

87. $y = \sin\left(\tan^{-1} 4\right) = \dfrac{4\sqrt{17}}{17}$

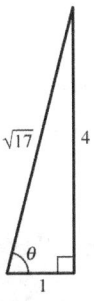

89. $y = \tan\left(\sec^{-1} 2\right) = \sqrt{3}$

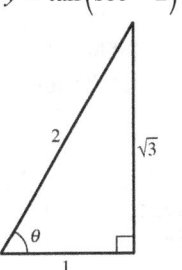

91. $y = \sin\left(\cos^{-1} x\right),\ |x| < 1$

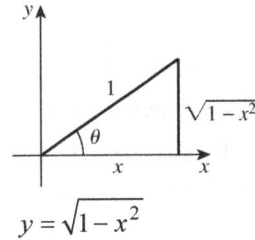

$y = \sqrt{1 - x^2}$

Applying the Concepts

93. $x = 110$ ft

$\theta = \tan^{-1}\left(\dfrac{440}{110}\right) + \tan^{-1}\left(\dfrac{880}{110}\right) \approx 159°$

95. a. Area $= \tan^{-1}\dfrac{1}{\sqrt{3}} - \tan^{-1}(-1) = \dfrac{\pi}{6} - \left(-\dfrac{\pi}{4}\right) = \dfrac{5\pi}{12}$ units^2

  b. Area $= \tan^{-1}\sqrt{3} - \tan^{-1}\left(-\sqrt{3}\right) = \dfrac{\pi}{3} - \left(-\dfrac{\pi}{3}\right) = \dfrac{2\pi}{3}$ units^2

97. $2\left[\cos\left(\sin^{-1}\dfrac{1}{2}\right) - \cos\left(\sin^{-1}\dfrac{\sqrt{3}}{2}\right)\right] = 2\left[\cos\dfrac{\pi}{6} - \cos\dfrac{\pi}{3}\right] = 2\left(\dfrac{\sqrt{3}}{2} - \dfrac{1}{2}\right) = \left(\sqrt{3} - 1\right)$ units^2

99. $T_1 = 19°26'$ N, $T_2 = 28°35'$ N
$N_1 = -99°7'$, $N_2 = 77°12'$
Make sure your calculator is in degree mode.
$x = \sin T_1 \sin T_2 + \cos T_1 \cos T_2 \cos(N_1 - N_2)$
$= \sin 19°26' \sin 28°35' + \cos 19°26' \cos 28°35' \cos(-99°7' - 77°12')$
≈ -0.6672038499
Now change your calculator to radian mode.
$d = 3960 \cos^{-1}(-0.6672038499) \approx 9113$ miles

5.6 Beyond the Basics

101. Examining the graph of $y = \sin^{-1} x$ in the text (Figure 5.77), we see that the function is increasing on its domain.

103. Examining the graph of $y = \tan^{-1} x$ in the text (Figure 5.80), we see that the function is increasing on its domain.

105. $\cot 0$ is undefined so $\cot^{-1}(\cot x) = x$ is true for $(0, \pi)$

107. $\cos^{-1}(\cos 10)$

10 radians is not in the interval $[0, \pi]$ so we must find θ in the interval $[0, \pi]$ where $\cos 10 = \cos \theta$. An angle of 10 radians lies in quadrant III, so its cosine is negative, and the reference angle is $4\pi - 10$. Therefore,
$$\cos^{-1}(\cos 10) = \cos^{-1}(\cos(4\pi - 10))$$
$$= 4\pi - 10$$

109. $y = \sin^{-1}(2x), -\dfrac{1}{2} \le x \le \dfrac{1}{2}$

Compress the graph of $y = \sin^{-1} x$ horizontally by a factor of $\tfrac{1}{2}$.

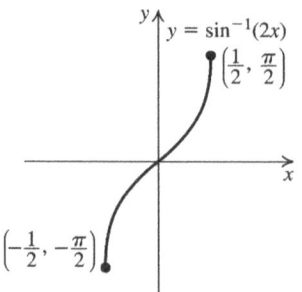

111. $y = \sin^{-1}(2x - 1), 0 \le x \le 1$

Shift the graph of $y = \sin^{-1} x$ one unit right and then compress the graph horizontally by a factor of $\tfrac{1}{2}$.

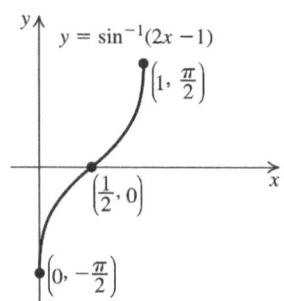

113. $y = \cos^{-1}\left(\dfrac{x}{3} - 1\right), 0 \le x \le 6$

Shift the graph of $y = \cos^{-1} x$ one unit right and then stretch the graph horizontally by a factor of 3.

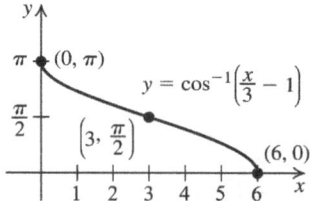

5.6 Critical Thinking/Discussion/Writing

115.

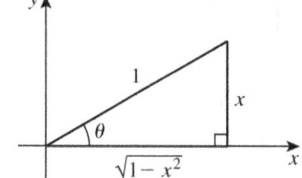

From the figure, we see that
$\sin^{-1} x = \cos^{-1} \sqrt{1 - x^2}$ if $0 \le x \le 1$.
For $-1 \le x < 0$, $\sin^{-1} x = -\cos^{-1} \sqrt{1 - x^2}$
since $\sin^{-1}(-x) = -\sin^{-1} x$.

116. Let $\sin^{-1} x = \alpha$ and let $\cos^{-1} x = \beta$.

By definition $\sin \alpha = x, -\dfrac{\pi}{2} < \alpha \le \dfrac{\pi}{2}$, and $\cos \beta = x, 0 \le \beta < \pi$. Therefore $\sin \alpha = \cos \beta$ for $0 \le \alpha, \beta < \dfrac{\pi}{2}$.

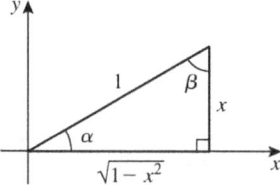

From the diagram, we see that $\beta = \dfrac{\pi}{2} - \alpha$, or $\alpha + \beta = \dfrac{\pi}{2}$. Substituting gives us
$\sin^{-1} x + \cos^{-1} x = \dfrac{\pi}{2}$.

117. We can find all value of x in the interval $[0, 2\pi]$ for which $\sin x \leq \csc x$ by examining the graphs.

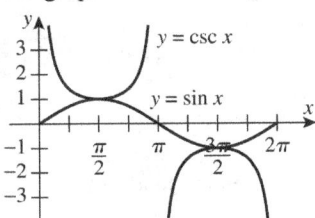

$\sin x \leq \csc x$ for $(0, \pi) \cup \left\{\dfrac{3\pi}{2}\right\}$. Note that 0 and π are not included in the solution set because cosecant is not defined for those values.

118. $\sin^{-1} x = \cos^{-1} x \Rightarrow$
$\sin(\sin^{-1} x) = \sin(\cos^{-1} x) \Rightarrow x = \sin(\cos^{-1} x)$
From the figure below, we see that
$\sin(\cos^{-1} x) = \sqrt{1-x^2}$

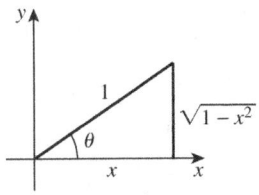

$x = \sqrt{1-x^2} \Rightarrow x^2 = 1 - x^2 \Rightarrow 2x^2 = 1 \Rightarrow$
$x = \pm \dfrac{\sqrt{2}}{2}$

Now substitute each value in the original equation to verify.

$\sin^{-1} \dfrac{\sqrt{2}}{2} = \dfrac{\pi}{4} = \cos^{-1} \dfrac{\sqrt{2}}{2}$ ✓

$\sin^{-1}\left(-\dfrac{\sqrt{2}}{2}\right) = -\dfrac{\pi}{4} \neq \cos^{-1} \dfrac{\sqrt{2}}{2} = \dfrac{3\pi}{4}$

Solution set: $\{\sqrt{2}/2\}$

5.6 Getting Ready for the Next Section

119. $\dfrac{1}{1-x} + \dfrac{1}{1+x} = \dfrac{1+x+(1-x)}{(1-x)(1+x)} = \dfrac{2}{1-x^2}$

121. $\dfrac{1}{\sqrt{2}+1} + \dfrac{1}{\sqrt{3}+\sqrt{2}}$
$= \dfrac{1}{\sqrt{2}+1} \cdot \dfrac{\sqrt{2}-1}{\sqrt{2}-1} + \dfrac{1}{\sqrt{3}+\sqrt{2}} \cdot \dfrac{\sqrt{3}-\sqrt{2}}{\sqrt{3}-\sqrt{2}}$
$= \dfrac{\sqrt{2}-1}{2-1} + \dfrac{\sqrt{3}-\sqrt{2}}{3-2} = \sqrt{3} - 1$

123. $x^2 + y^2 = 1$

$\dfrac{x}{1+y} - \dfrac{1-y}{x} = \dfrac{x^2 - (1-y)(1+y)}{x(1+y)}$
$= \dfrac{x^2 - (1-y^2)}{x(1+y)} = \dfrac{x^2 + y^2 - 1}{x(1+y)}$
$= \dfrac{1-1}{x(1+y)} = 0$

125. $\tan x \cos x \csc x = \dfrac{\sin x}{\cos x} \cdot \cos x \cdot \dfrac{1}{\sin x} = 1$

Chapter 5 Review Exercises

1.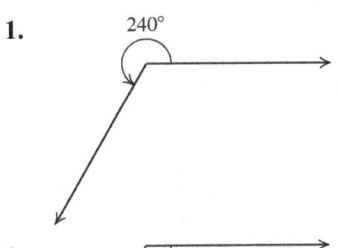

3.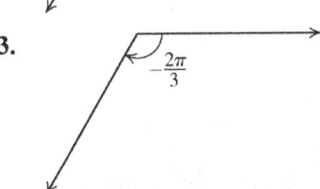

5. $20° = 20 \cdot \dfrac{\pi}{180} = \dfrac{\pi}{9}$ radian

7. $-60° = -60 \cdot \dfrac{\pi}{180} = -\dfrac{\pi}{3}$ radians

9. $\dfrac{5\pi}{18} = \dfrac{5\pi}{18} \cdot \dfrac{180}{\pi} = 50°$

11. $\theta = \dfrac{s}{r} = \dfrac{40}{15} = 2.667$ radians

13. Convert the angle measurement from degrees to radians:
$36° = 36°\left(\dfrac{\pi}{180°}\right) = \dfrac{s}{2} \Rightarrow s = 1.257$ m.

15. $A = \dfrac{1}{2}r^2\theta = \dfrac{1}{2} \cdot 3^2 \cdot 32° \cdot \dfrac{\pi}{180°} \approx 2.513$ ft^2

For exercises 17–20, use this triangle to help identify the opposite and adjacent legs:

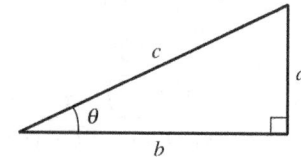

17. $\cos\theta = \dfrac{2}{9} \Rightarrow b = 2, c = 9$
$9^2 = a^2 + 2^2 \Rightarrow a = \sqrt{77}$
$\sin\theta = \dfrac{\sqrt{77}}{9}, \tan\theta = \dfrac{\sqrt{77}}{2}, \cot\theta = \dfrac{2\sqrt{77}}{77},$
$\sec\theta = \dfrac{9}{2}, \csc\theta = \dfrac{9\sqrt{77}}{77}$

19. $\tan\theta = \dfrac{5}{3} \Rightarrow a = 5, b = 3$
$c^2 = 5^2 + 3^2 \Rightarrow c = \sqrt{34}$
$\sin\theta = \dfrac{5\sqrt{34}}{34}, \cos\theta = \dfrac{3\sqrt{34}}{34}, \cot\theta = \dfrac{3}{5},$
$\sec\theta = \dfrac{\sqrt{34}}{3}, \csc\theta = \dfrac{\sqrt{34}}{5}$

21. $x = 2, y = 8 \Rightarrow r^2 = 8^2 + 2^2 \Rightarrow r = 2\sqrt{17}$
$\sin\theta = \dfrac{4\sqrt{17}}{17}, \cos\theta = \dfrac{\sqrt{17}}{17}, \tan\theta = 4,$
$\cot\theta = \dfrac{1}{4}, \sec\theta = \sqrt{17}, \csc\theta = \dfrac{\sqrt{17}}{4}$

23. $x = -\sqrt{5}, y = 2 \Rightarrow r^2 = 2^2 + \left(-\sqrt{5}\right)^2 \Rightarrow r = 3$
$\sin\theta = \dfrac{2}{3}, \cos\theta = -\dfrac{\sqrt{5}}{3}, \tan\theta = -\dfrac{2\sqrt{5}}{5},$
$\cot\theta = -\dfrac{\sqrt{5}}{2}, \sec\theta = -\dfrac{3\sqrt{5}}{5}, \csc\theta = \dfrac{3}{2}$

25. Quadrant IV

27. Quadrant III

29. $\cos\theta = -\dfrac{4}{5}, \theta$ in Quadrant III $\Rightarrow x < 0, y < 0$
$x = -4, r = 5 \Rightarrow 5^2 = (-4)^2 + y^2 \Rightarrow y = -3$
$\sin\theta = -\dfrac{3}{5}, \tan\theta = \dfrac{3}{4}, \cot\theta = \dfrac{4}{3}, \sec\theta = -\dfrac{5}{4},$
$\csc\theta = -\dfrac{5}{3}$

31. $\sin\theta = \dfrac{3}{5}, \theta$ in Quadrant II $\Rightarrow x < 0, y > 0$
$y = 3, r = 5 \Rightarrow 5^2 = x^2 + 3^2 \Rightarrow x = -4$
$\cos\theta = -\dfrac{4}{5}, \tan\theta = -\dfrac{3}{4}, \cot\theta = -\dfrac{4}{3},$
$\sec\theta = -\dfrac{5}{4}, \csc\theta = \dfrac{5}{3}$

33. Because $150°$ lies in Quadrant II, the reference angle is
$180° - \theta = 180° - 150° = 30°$. In Quadrant II, the cosine is negative, so
$\cos 150° = -\cos 30° = -\dfrac{\sqrt{3}}{2}$.

35. $390°$ is co-terminal with $30°$. Because $30°$ lies in Quadrant I, the reference angle is $30°$. The tangent is positive in Quadrant I, so
$\tan 390° = \tan 30° = \dfrac{\sqrt{3}}{3}$.

37.

39.

41. $y = 14\sin\left(2x + \dfrac{\pi}{7}\right) = 14\sin\left[2\left(x + \dfrac{\pi}{14}\right)\right] \Rightarrow$
$a = 14, b = 2, c = -\dfrac{\pi}{14} \Rightarrow$ amplitude $= 14$,
period $= \dfrac{2\pi}{2} = \pi$, phase shift $= -\dfrac{\pi}{14}$

43. $y = 6\tan\left(2x + \dfrac{\pi}{5}\right) = 6\tan\left[2\left(x + \dfrac{\pi}{10}\right)\right] \Rightarrow$
$a = 6, b = 2, c = -\dfrac{\pi}{10} \Rightarrow$ vertical stretch $= 6$,
period $= \dfrac{\pi}{2}$, phase shift $= -\dfrac{\pi}{10}$

45.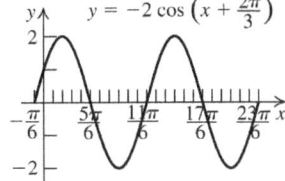

47.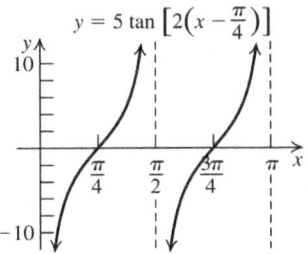

49.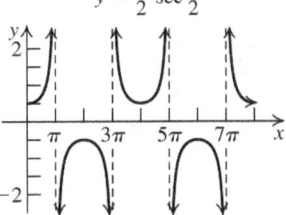

51. $y = \cos^{-1}\left(\cos\dfrac{5\pi}{8}\right) = \dfrac{5\pi}{8}$

53. $y = \tan^{-1}\left(\tan\left(-\dfrac{2\pi}{3}\right)\right) = \dfrac{\pi}{3}$

In exercises 54–58, use the Pythagorean theorem and the triangle below to find the length of the third side of the triangle.

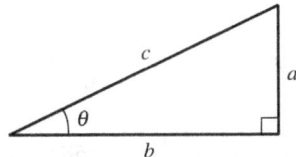

55. $\sin^{-1}\dfrac{\sqrt{2}}{2} \Rightarrow a = \sqrt{2}, c = 2 \Rightarrow$
$2^2 = \left(\sqrt{2}\right)^2 + b^2 \Rightarrow b = \sqrt{2}$
$y = \cos\left(\sin^{-1}\dfrac{\sqrt{2}}{2}\right) = \dfrac{\sqrt{2}}{2}$

57. $\cos^{-1}\left(-\dfrac{1}{2}\right) \Rightarrow b = -1, c = 2 \Rightarrow$
$2^2 = a^2 + (-1)^2 \Rightarrow a = \sqrt{3}$
$y = \tan\left(\cos^{-1}\left(-\dfrac{1}{2}\right)\right) = -\sqrt{3}$

59. The wheels turn a quarter of a revolution, so
$\theta = \dfrac{\pi}{2}$. $\theta = \dfrac{s}{r} \Rightarrow \dfrac{\pi}{2} = \dfrac{s}{28} \Rightarrow s \approx 44$ inches

61. The difference in the latitudes is
$42°31' - 29°59' = 12°32' = 12 + \dfrac{32}{60}° \approx 12.533°$
$= 12.533°\left(\dfrac{\pi}{180°}\right) \approx 0.2187$ radian.
$s = 0.2187 \cdot 3960 \approx 866$ miles.

63. $d = 28$ in. $\Rightarrow r = 14$ in.
$v = r\omega \Rightarrow 21 = 14\omega \Rightarrow \omega = 1.5$ radians per second

65. The distance of the satellite from the center of Earth is about $36,000 + 6400 = 42,400$ km. In one orbit, $\theta = 2\pi$ and
$s = r\theta \Rightarrow s = 42,400 \cdot 2\pi \Rightarrow s \approx 266,407$ km.
$t = 24$ hr, so $v = \dfrac{s}{t} = \dfrac{266,407 \text{ km}}{24 \text{ hr}} \approx 11,100$
km per hour.

67. The highest number of daylight hours is 14.3 and the lowest number of daylight hours is 9.8, so $a = \dfrac{14.3 - 9.8}{2} = 2.25$.
The vertical shift $d = \dfrac{14.3 + 9.8}{2} = 12.05$.
The period is $12 = \dfrac{2\pi}{b} \Rightarrow b = \dfrac{\pi}{6}$.
The highest point on the graph, 14.3, occurs at $x = 7$, so the phase shift $c = 7 - 3 = 4$. The equation is $y = 2.25\sin\left[\dfrac{\pi}{6}(x - 4)\right] + 12.05$.

Chapter 5 Practice Test A

1. $140° = 140 \cdot \dfrac{\pi}{180} = \dfrac{7\pi}{9} \approx 2.4435$ radians

2. $\dfrac{7\pi}{5} = \dfrac{7\pi}{5} \cdot \dfrac{180}{\pi} = 252°$

3. $x = 2, y = -5 \Rightarrow r^2 = 2^2 + (-5)^2 \Rightarrow r = \sqrt{29}$
$\cos\theta = \dfrac{x}{r} = \dfrac{2\sqrt{29}}{29}$

4. $\theta = \dfrac{s}{r} = \dfrac{15}{10} = 1.5$ radians

5. From exercise 4, we have $\theta = 1.5$.
$A = \dfrac{1}{2}r^2\theta = \dfrac{1}{2} \cdot 10^2 \cdot 1.5 = 75$ cm^2

6. $\cos\theta = \dfrac{2}{7} \Rightarrow x = 2, r = 7 \Rightarrow 7^2 = 2^2 + y^2 \Rightarrow$

   $y = 3\sqrt{5} \Rightarrow \sin\theta = \dfrac{3\sqrt{5}}{7}$

7. Quadrant II

8. $\cot\theta = -\dfrac{5}{12}, \theta$ in Quadrant IV $\Rightarrow x > 0, y < 0$

   $x = 5, y = -12 \Rightarrow r = \sqrt{5^2 + (-12)^2} = 13 \Rightarrow$

   $\sec\theta = \dfrac{r}{x} = \dfrac{13}{5}$

9. Because 217° lies in Quadrant III, the reference angle is
   $\theta - 180° = 217° - 180° = 37°$.

10. $0.223 = \cos\dfrac{3\pi}{7} = \sin\left(\dfrac{\pi}{2} - \dfrac{3\pi}{7}\right)$

    $= \sin\dfrac{\pi}{14} = 0.223$

11.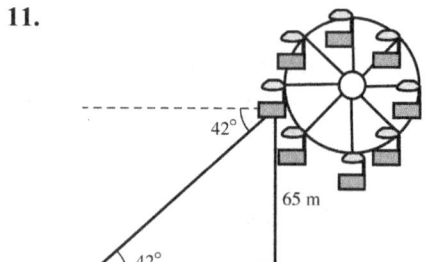

    $\cot 42° = \dfrac{x}{65} \Rightarrow x = 65\cot 42° \Rightarrow x \approx 72$ m

12. Amplitude = 7, range = [−7, 7]

13. $y = 19\sin[12(x + 3\pi)] \Rightarrow a = 19, b = 12,$

    $c = -3\pi \Rightarrow$ amplitude = 19, period = $\dfrac{2\pi}{12} = \dfrac{\pi}{6}$,

    phase shift = -3π

14. Because the ball was pulled down 7 inches
    $a = -7$. The period is 4, so $\omega = \dfrac{2\pi}{4} = \dfrac{\pi}{2}$. The
    equation is $y = -7\cos\left(\dfrac{\pi}{2}t\right)$.

15.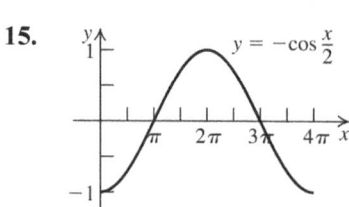

16.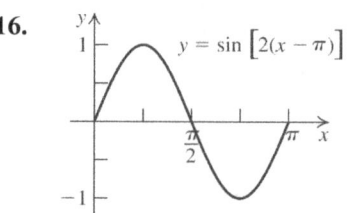

17.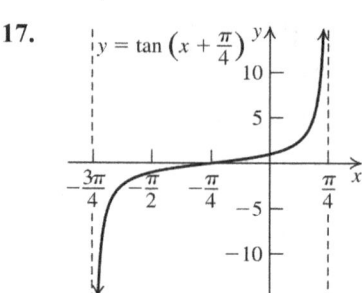

18. $y = \sin^{-1}\left(-\dfrac{\sqrt{3}}{2}\right) = -\dfrac{\pi}{3}$

19. $y = \cos^{-1}\left[\cos\dfrac{4\pi}{3}\right] = \dfrac{2\pi}{3}$

    Note that $\dfrac{4\pi}{3}$ is not in the range of $\cos^{-1}\theta$.

20. $\sin^{-1}\left(-\dfrac{1}{3}\right) \Rightarrow a = -1, c = 3 \Rightarrow 3^2 = (-1)^2 + b^2$

    $\Rightarrow b = 2\sqrt{2}$

    $y = \tan\left(\sin^{-1}\left(-\dfrac{1}{3}\right)\right) = \dfrac{a}{b} = -\dfrac{1}{2\sqrt{2}} = -\dfrac{\sqrt{2}}{4}$

Chapter 5 Practice Test B

1. $160° = 160 \cdot \dfrac{\pi}{180} = \dfrac{8\pi}{9}$. The answer is B.

2. $\dfrac{13\pi}{5} = \dfrac{13\pi}{5} \cdot \dfrac{180}{\pi} = 468°$. The answer is D.

3. $x = -3, y = 1 \Rightarrow r^2 = (-3)^2 + 1^2 \Rightarrow r = \sqrt{10}$

   $\cos\theta = \dfrac{x}{r} = -\dfrac{3\sqrt{10}}{10}$. The answer is A.

4. $\theta = \dfrac{s}{r} = \dfrac{4}{8} = \dfrac{1}{2}$ radian. The answer is A.

5. $A = \dfrac{1}{2}r^2\theta = \dfrac{1}{2} \cdot 6^2 \cdot \dfrac{3}{6} = 9$ cm^2.
   The answer is D.

6. $\cos\theta = \dfrac{3}{4} \Rightarrow x = 3, r = 4 \Rightarrow 4^2 = 3^2 + y^2 \Rightarrow$

   $y = \sqrt{7} \Rightarrow \sin\theta = \sqrt{7}/4$. The answer is B.

7. The answer is D.

8. $\tan\theta = 12/5$, θ in Quadrant III $\Rightarrow x < 0, y < 0$
   $x = -5, y = -12 \Rightarrow$
   $r = \sqrt{(-5)^2 + (-12)^2} = 13 \Rightarrow$
   $\csc\theta = -13/12$. The answer is B.

9. $640°$ is co-terminal with $280°$. Because $280°$ lies in Quadrant IV, the reference angle is $360° - \theta = 360° - 280° = 80°$.
   The answer is C.

10. The answer is A.

11. $v = r\omega = 50(0.8) = 40$ cm/sec.
    The answer is C.

12. $a = \dfrac{1}{2}, 4\pi = \dfrac{2\pi}{b} \Rightarrow b = \dfrac{1}{2}$. The answer is A.

13. The answer is B.

14. $y = -5\cos\left(\dfrac{\pi}{4}t\right) \Rightarrow$ period $= \dfrac{2\pi}{\pi/4} = 8$.
    The answer is D.

15. $y = -7\cos\left[\dfrac{1}{6}\left(x - \dfrac{\pi}{12}\right)\right] \Rightarrow a = -7, b = \dfrac{1}{6}$,
    $c = \dfrac{\pi}{12} \Rightarrow$ amplitude = 7, period $= \dfrac{2\pi}{1/6} = 12\pi$,
    phase shift $= \pi/12$.
    The answer is D.

16. The answer is C.

17. $\cos^{-1}\left(\cos\dfrac{11\pi}{3}\right) = \dfrac{\pi}{3}$. The answer is C.

18. $\sin^{-1}\left(-\dfrac{4}{5}\right) \Rightarrow y = -4, r = 5 \Rightarrow$
    $5^2 = x^2 + (-4)^2 \Rightarrow x = 3$
    $\cos\left(\sin^{-1}\left(-\dfrac{4}{5}\right)\right) = \dfrac{3}{5}$. The answer is A.

19. $y = \cos^{-1}\left(\cos\dfrac{7\pi}{4}\right) = \cos^{-1}\left(\dfrac{\sqrt{2}}{2}\right) = \dfrac{\pi}{4}$.
    The answer is D.

20. $\tan^{-1}\left(\dfrac{5}{3}\right) \Rightarrow x = 3, y = 5 \Rightarrow r^2 = 3^2 + 5^2 \Rightarrow$
    $r = \sqrt{34}$
    $\cos\left(\tan^{-1}\left(\dfrac{5}{3}\right)\right) = \dfrac{3\sqrt{34}}{34}$.
    The answer is D.

Cumulative Review Exercises (Chapters P–5)

1. $3x^2 - 30x + 50 = -24 \Rightarrow 3x^2 - 30x + 74 = 0 \Rightarrow$
   $x = \dfrac{30 \pm \sqrt{(-30)^2 - 4(3)(74)}}{2(3)} = 5 \pm \dfrac{\sqrt{3}}{3}$

2. Let $u = \sqrt{t} + 1$. Then
   $(\sqrt{t} + 1)^2 + 2(\sqrt{t} + 1) = 3 \Rightarrow$
   $u^2 + 2u - 3 = 0 \Rightarrow (u + 3)(u - 1) = 0 \Rightarrow$
   $u = -3$ or $u = 1$. Reject the negative root.
   $1 = \sqrt{t} + 1 \Rightarrow t = 0$

3. $\dfrac{4 - x}{2x - 4} > 0$.
   $4 - x = 0 \Rightarrow x = 4$ and $2x - 4 = 0 \Rightarrow x = 2$.
   The intervals to be tested are
   $(-\infty, 2), (2, 4)$, and $(4, \infty)$.

Interval	Test point	Value of $\dfrac{4-x}{2x-4}$	Result
$(-\infty, 2)$	0	-1	$-$
$(2, 4)$	3	$1/2$	$+$
$(4, \infty)$	5	$-1/6$	$-$

   The solution set is $(2, 4)$.

4. $m = \dfrac{-1 - 5}{3 - (-6)} = -\dfrac{2}{3}; -1 = -\dfrac{2}{3}(3) + b \Rightarrow b = 1$
   $y = -\dfrac{2}{3}x + 1$

5. Logarithms are defined only for positive numbers, so $\dfrac{x}{2 - x} > 0 \Rightarrow 0 < x < 2$.
   The domain is $(0, 2)$.

6. This is a parabola. Shift the graph of $y = x^2$ three units to the right, stretch vertically by a factor of 2, then shift the graph down five units.

(continued on next page)

(*continued*)

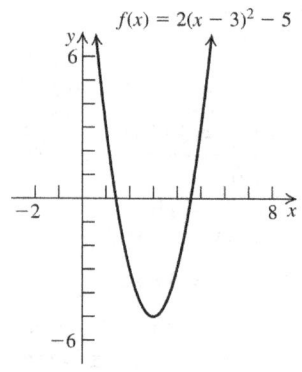

7. Stretch the graph of $y = e^x$ by a factor of 4, then shift the graph one unit up.

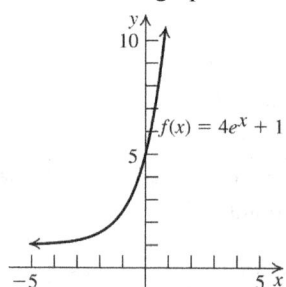

8. $(f \circ g)(x) = \ln\left(-\dfrac{1}{x}\right)$. Since the natural logarithm is defined only for positive numbers, $-\dfrac{1}{x} > 0 \Rightarrow$ the domain is $(-\infty, 0)$.

9. a. $f(x) = 3x + 7$ is a one-to-one function, so an inverse exists. Find the inverse by interchanging the variables and solving for y:
$y = 3x + 7 : x = 3y + 7 \Rightarrow y = \dfrac{x-7}{3} = f^{-1}(x)$

   b. $f(x) = |x|$ is not one-to-one, so no inverse exists.

10. $\begin{array}{r|rrrrr} -2 & 2 & 4 & 1 & 1 & -2 \\ & & -4 & 0 & -2 & 2 \\ \hline & 2 & 0 & 1 & -1 & 0 \end{array}$

    $\dfrac{2x^4 + 4x^3 + x^2 + x - 2}{x+2} = 2x^3 + x - 1$

11. Solve $x^2 + 4x - 5 = 0$ to find the vertical asymptotes: $x^2 + 4x - 5 = 0 \Rightarrow$ $(x+5)(x-1) = 0 \Rightarrow x = -5$ or $x = 1$. These are the vertical asymptotes. The numerator and denominator both have degree 2, so the horizontal asymptote is $y = -3$.

12. a. $\log \sqrt[3]{\dfrac{x^2 y}{z}} = \dfrac{1}{3} \log\left(\dfrac{x^2 y}{z}\right)$
    $= \dfrac{1}{3}\left(\log(x^2) + \log y - \log z\right)$
    $= \dfrac{2}{3}\log x + \dfrac{1}{3}\log y - \dfrac{1}{3}\log z$

    b. $\ln\left(\dfrac{7x^4}{5\sqrt{y}}\right) = \ln\left(\dfrac{7}{5}\right) + \ln x^4 - \ln \sqrt{y}$
    $= \ln 1.4 + 4\ln x - \dfrac{1}{2}\ln y$

13. $f(x) = \begin{cases} -\ln x & \text{if } x > 0 \\ \ln |x| & \text{if } x < 0 \end{cases}$

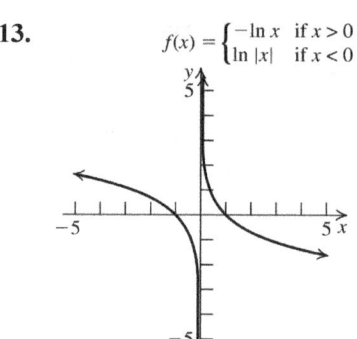

14. $\log(x^2 + 9x) = 1 \Rightarrow x^2 + 9x = 10^1 \Rightarrow$
    $x^2 + 9x - 10 = 0 \Rightarrow (x+10)(x-1) = 0 \Rightarrow$
    $x = -10$ or $x = 1$

15. The possible rational zeros are $\left\{\pm\dfrac{1}{2}, \pm 1, \pm\dfrac{3}{2},\right.$
    $\left.\pm 2, \pm 3, \pm 6\right\}$.

16. a. $\cos\theta = \dfrac{3}{7} \Rightarrow x = 3, r = 7 \Rightarrow 7^2 = 3^2 + y^2 \Rightarrow$
    $y = 2\sqrt{10}$
    $\sin\theta = \dfrac{2\sqrt{10}}{7}, \tan\theta = \dfrac{2\sqrt{10}}{3}, \cot\theta = \dfrac{3\sqrt{10}}{20},$
    $\sec\theta = \dfrac{7}{3}, \csc\theta = \dfrac{7\sqrt{10}}{20}$

    b. $\sin\theta = \dfrac{2}{11} \Rightarrow y = 2, r = 11 \Rightarrow$
    $11^2 = x^2 + 2^2 \Rightarrow x = \sqrt{117} = 3\sqrt{13}$
    $\cos\theta = \dfrac{3\sqrt{13}}{11}, \tan\theta = \dfrac{2\sqrt{13}}{39}, \cot\theta = \dfrac{3\sqrt{13}}{2},$
    $\sec\theta = \dfrac{11\sqrt{13}}{39}, \csc\theta = \dfrac{11}{2}$

17. a. $\cos\theta = -\dfrac{4}{5}, \theta$ in Quadrant II $\Rightarrow x = -4, r = 5$
$5^2 = (-4)^2 + y^2 \Rightarrow y = 3$
$\sin\theta = \dfrac{3}{5}, \tan\theta = -\dfrac{3}{4}, \cot\theta = -\dfrac{4}{3},$
$\sec\theta = -\dfrac{5}{4}, \csc\theta = \dfrac{5}{3}$

b. $\cot\theta = \dfrac{5}{12}, \theta$ is Quadrant III $\Rightarrow$
$x = -5, y = -12$
$r^2 = (-5)^2 + (-12)^2 \Rightarrow r = 13$
$\sin\theta = -\dfrac{12}{13}, \cos\theta = -\dfrac{5}{13}, \tan\theta = \dfrac{12}{5},$
$\sec\theta = -\dfrac{13}{5}, \csc\theta = -\dfrac{13}{12}$

18. a.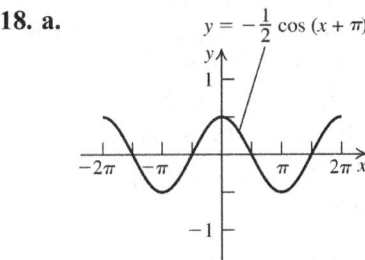

b.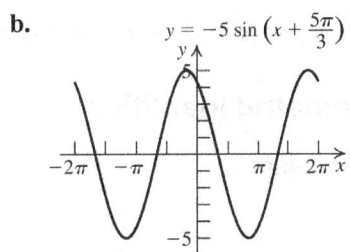

19. a. $\sin^{-1}\left(-\dfrac{1}{5}\right) \Rightarrow y = -1, r = 5 \Rightarrow$
$5^2 = x^2 + (-1)^2 \Rightarrow x = 2\sqrt{6}$
$\cos\left(\sin^{-1}\left(-\dfrac{1}{5}\right)\right) = \dfrac{2\sqrt{6}}{5}$

b. $\tan^{-1}\dfrac{7}{2} \Rightarrow y = 7, x = 2 \Rightarrow r^2 = 2^2 + 7^2 \Rightarrow$
$r = \sqrt{53}$
$\sin\left(\tan^{-1}\dfrac{7}{2}\right) = \dfrac{7}{\sqrt{53}} = \dfrac{7\sqrt{53}}{53}$

20. $y = \cos^{-1}\left(\cos\dfrac{5\pi}{3}\right) = \cos^{-1}\left(\dfrac{1}{2}\right) = \dfrac{\pi}{3}$

Chapter 6 Trigonometric Identities and Equations

6.1 Trigonometric Identities

6.1 Practice Problems

1. $\sin\theta = \dfrac{4}{5} \Rightarrow \csc\theta = \dfrac{5}{4}$

   $1 + \cot^2\theta = \left(\dfrac{5}{4}\right)^2 \Rightarrow \cot\theta = \pm\dfrac{3}{4}$

   $\dfrac{\pi}{2} < \theta < \pi \Rightarrow \cot\theta = -\dfrac{3}{4}$

2. $\dfrac{\tan x}{\sec x + 1} + \dfrac{\tan x}{\sec x - 1}$

   $= \dfrac{\tan x(\sec x - 1) + \tan x(\sec x + 1)}{(\sec x + 1)(\sec x - 1)}$

   $= \dfrac{\dfrac{\sin x}{\cos x}\left(\dfrac{1}{\cos x} - 1\right) + \dfrac{\sin x}{\cos x}\left(\dfrac{1}{\cos x} + 1\right)}{\sec^2 x - 1}$

   $= \dfrac{\dfrac{\sin x - \sin x\cos x + \sin x + \sin x\cos x}{\cos^2 x}}{\tan^2 x}$

   $= \dfrac{\dfrac{2\sin x}{\cos^2 x}}{\dfrac{\sin^2 x}{\cos^2 x}} = \dfrac{2\sin x}{\cos^2 x}\left(\dfrac{\cos^2 x}{\sin^2 x}\right) = \dfrac{2}{\sin x}$

3. $\cos x = 1 - \sin x$ is not an identity because for $x = \dfrac{\pi}{3}, \sin x = \dfrac{\sqrt{3}}{2}$, while $\cos x = \dfrac{1}{2} \ne 1 - \dfrac{\sqrt{3}}{2}$.

4. $\dfrac{1 - \sin^2 x}{\cos x} = \dfrac{\cos^2 x}{\cos x} = \cos x$

5. $\dfrac{\tan\theta}{\sec\theta - \cos\theta} = \dfrac{\tan\theta}{\dfrac{1}{\cos\theta} - \cos\theta} = \dfrac{\dfrac{\sin\theta}{\cos\theta}}{\dfrac{1 - \cos^2\theta}{\cos\theta}}$

   $= \dfrac{\sin\theta}{1 - \cos^2\theta} = \dfrac{\sin\theta}{\sin^2\theta}$

   $= \dfrac{1}{\sin\theta} = \dfrac{\csc\theta}{1}$

6. $\tan^4 x + \tan^2 x = \dfrac{\sin^4 x}{\cos^4 x} + \dfrac{\sin^2 x}{\cos^2 x}$

   $= \dfrac{\sin^4 x + \sin^2 x\cos^2 x}{\cos^4 x}$

   $= \dfrac{\sin^2 x(\sin^2 x + \cos^2 x)}{\cos^4 x}$

   $= \dfrac{\sin^2 x(1)}{\cos^4 x} = \dfrac{\sin^2 x}{\cos^2 x}\cdot\dfrac{1}{\cos^2 x}$

   $= \tan^2 x \sec^2 x$

7. $\tan\theta + \sec\theta = \dfrac{\sin\theta}{\cos\theta} + \dfrac{1}{\cos\theta} = \dfrac{\sin\theta + 1}{\cos\theta}$

   $\dfrac{\csc\theta + 1}{\cot\theta} = \dfrac{\dfrac{1}{\sin\theta} + 1}{\dfrac{\cos\theta}{\sin\theta}} = \dfrac{\dfrac{1 + \sin\theta}{\sin\theta}}{\dfrac{\cos\theta}{\sin\theta}} = \dfrac{1 + \sin\theta}{\cos\theta}$

   Because both sides of the original identity are equal to $\dfrac{1 + \sin\theta}{\cos\theta}$, the identity is verified.

8. $\dfrac{\sin\theta}{1 + \cos\theta}\cdot\dfrac{1 - \cos\theta}{1 - \cos\theta} = \dfrac{\sin\theta(1 - \cos\theta)}{1 - \cos^2\theta}$

   $= \dfrac{\sin\theta(1 - \cos\theta)}{\sin^2\theta}$

   $= \dfrac{1 - \cos\theta}{\sin\theta}$

9. $\sqrt{9 - (3\sin\theta)^2} = \sqrt{9 - 9\sin^2\theta} = \sqrt{9(1 - \sin^2\theta)}$

   $= \sqrt{9\cos^2\theta} = 3\cos\theta$

10. **a.** $\sin^3 x\cos^2 x = \sin x\sin^2 x\cos^2 x$

    $= \sin x(1 - \cos^2 x)\cos^2 x$

    $= \sin x(\cos^2 x - \cos^4 x)$

    $= (\cos^2 x - \cos^4 x)\sin x$

    b. Start with the right side.

    $\tan^2 x\sec^2 x - \sec^2 x + 1$

    $= \tan^2 x(\tan^2 x + 1) - (\sec^2 x - 1)$

    $= \tan^4 x + \tan^2 x - \tan^2 x = \tan^4 x$

6.1 Concepts and Vocabulary

1. An equation that is true for all values of the variable in its domain is called an identity.

3. $\sin^2 x + \underline{\cos^2 x} = 1$; $1 + \underline{\tan^2 x} = \sec^2 x$;
   $\csc^2 x - \cot^2 x = \underline{1}$

5. False. The statement is an identity.
   $\tan^2 x \cot x = \tan^2 x \cdot \dfrac{1}{\tan x} = \tan x$

7. False. θ lies in quadrant III, so $\tan \theta > 0$.

6.1 Building Skills

9. $\sin \theta = -\dfrac{12}{13} \Rightarrow \left(-\dfrac{12}{13}\right)^2 + \cos^2 \theta = 1 \Rightarrow$
   $\cos \theta = \pm \dfrac{5}{13}$. $\pi < \theta < \dfrac{3\pi}{2} \Rightarrow \cos \theta = -\dfrac{5}{13}$

11. $\cot \theta = \dfrac{1}{2} \Rightarrow 1 + \left(\dfrac{1}{2}\right)^2 = \csc^2 \theta \Rightarrow \csc \theta = \pm \dfrac{\sqrt{5}}{2}$
    $\pi < \theta < \dfrac{3\pi}{2} \Rightarrow \csc \theta = -\dfrac{\sqrt{5}}{2}$

13. $\sin x = -\dfrac{2}{3} \Rightarrow \csc x = -\dfrac{3}{2} \Rightarrow$
    $1 + \cot^2 x = \left(-\dfrac{3}{2}\right)^2 \Rightarrow \cot x = \pm \dfrac{\sqrt{5}}{2}$
    $\pi < x < \dfrac{3\pi}{2} \Rightarrow \cot x = \dfrac{\sqrt{5}}{2}$

15. $(1 + \tan x)(1 - \tan x) + \sec^2 x$
    $= 1 - \tan^2 x + \sec^2 x$
    $= 1 - \tan^2 x + 1 + \tan^2 x = 2$

17. $(\sec x + \tan x)(\sec x - \tan x)$
    $= \sec^2 x - \tan^2 x = 1 + \tan^2 x - \tan^2 x = 1$

19. $(\csc^4 x - \cot^4 x)$
    $= (\csc^2 x - \cot^2 x)(\csc^2 x + \cot^2 x)$
    $= \csc^2 x + \cot^2 x$

21. $\dfrac{\sec x \csc x (\sin x + \cos x)}{\sec x + \csc x}$
    $= \dfrac{\dfrac{1}{\cos x \sin x}(\sin x + \cos x)}{\dfrac{1}{\cos x} + \dfrac{1}{\sin x}}$
    $= \dfrac{\dfrac{\sin x + \cos x}{\cos x \sin x}}{\dfrac{\cos x + \sin x}{\sin x \cos x}} = 1$

23. $\dfrac{\tan^2 x - 2\tan x - 3}{\tan x + 1} = \dfrac{(\tan x - 3)(\tan x + 1)}{\tan x + 1}$
    $= \tan x - 3$

Answers may vary for exercises 25–30.

25. $\sin x = 1 - \cos x$ is not an identity because for $x = \pi$, $\sin x = 0$, while $1 - \cos x = 2$.

27. $\cos x = \sqrt{1 - \sin^2 x}$ is not an identity because for $x = \pi$, $\cos x = -1$, while $\sqrt{1 - \sin^2 x} = 1$.

29. $\sin^2 x = (1 - \cos x)^2$ is not an identity because for $x = \pi$, $\sin^2 x = 0$, while $(1 - \cos x)^2 = 4$.

31. $\sin x \tan x + \cos x$
    $= \sin x \cdot \dfrac{\sin x}{\cos x} + \cos x = \dfrac{\sin^2 x}{\cos x} + \cos x$
    $= \dfrac{\sin^2 x + \cos^2 x}{\cos x} = \dfrac{1}{\cos x} = \sec x$

33. $\tan x + \cot x = \dfrac{\sin x}{\cos x} + \dfrac{\cos x}{\sin x} = \dfrac{\sin^2 x + \cos^2 x}{\cos x \sin x}$
    $= \dfrac{1}{\cos x \sin x} = \dfrac{1}{\cos x} \cdot \dfrac{1}{\sin x}$
    $= \sec x \csc x$

35. $\dfrac{1 - 4\cos^2 x}{1 - 2\cos x} = \dfrac{(1 + 2\cos x)(1 - 2\cos x)}{1 - 2\cos x}$
    $= 1 + 2\cos x$

37. $(\cos x - \sin x)(\cos x + \sin x) = \cos^2 x - \sin^2 x$
    $= (1 - \sin^2 x) - \sin^2 x = 1 - 2\sin^2 x$

39. $\sin^2 x \cot^2 x + \sin^2 x = \sin^2 x \cdot \dfrac{\cos^2 x}{\sin^2 x} + \sin^2 x$
    $= \cos^2 x + \sin^2 x = 1$

41. $\tan^2 x - \sin^2 x = \dfrac{\sin^2 x}{\cos^2 x} - \sin^2 x$
    $= \dfrac{\sin^2 x - \sin^2 x \cos^2 x}{\cos^2 x} = \dfrac{\sin^2 x(1 - \cos^2 x)}{\cos^2 x}$
    $= \dfrac{\sin^2 x (\sin^2 x)}{\cos^2 x} = \dfrac{\sin^4 x}{\cos^2 x} = \sin^4 x \cdot \dfrac{1}{\cos^2 x}$
    $= \sin^4 x \sec^2 x$

43. $\sin^3 x - \cos^3 x = (\sin x - \cos x)(\sin^2 x + \sin x \cos x + \cos^2 x) = (\sin x - \cos x)(1 + \sin x \cos x)$

45. $\cos^4 x - \sin^4 x = (\cos^2 x + \sin^2 x)(\cos^2 x - \sin^2 x) = 1\left((1-\sin^2 x) - \sin^2 x\right) = 1 - 2\sin^2 x$

47. $\dfrac{1}{1-\sin x} + \dfrac{1}{1+\sin x} = \dfrac{(1+\sin x)+(1-\sin x)}{1-\sin^2 x} = \dfrac{2}{\cos^2 x} = 2\sec^2 x$

49. $\dfrac{\sin x}{1-\sin x} - \dfrac{\sin x}{1+\sin x} = \dfrac{\sin x(1+\sin x) - \sin x(1-\sin x)}{(1-\sin x)(1+\sin x)} = \dfrac{\sin x + \sin^2 x - \sin x + \sin^2 x}{1-\sin^2 x} = \dfrac{2\sin^2 x}{\cos^2 x} = 2\tan^2 x$

51. $\dfrac{1}{\sec x - \tan x} + \dfrac{1}{\sec x + \tan x} = \dfrac{\sec x + \tan x + \sec x - \tan x}{\sec^2 x - \tan^2 x} = \dfrac{2\sec x}{\sec^2 x - (\sec^2 x - 1)} = 2\sec x = \dfrac{2}{\cos x}$

53. $(\sin x + \cos x)^2 = \sin^2 x + 2\sin x \cos x + \cos^2 x = 1 + 2\sin x \cos x$

55. $(1 + \tan x)^2 = 1 + 2\tan x + \tan^2 x = \sec^2 x + 2\tan x$

57. $\sec^2 x + \csc^2 x = \dfrac{1}{\cos^2 x} + \dfrac{1}{\sin^2 x} = \dfrac{\sin^2 x + \cos^2 x}{\sin^2 x \cos^2 x} = \dfrac{1}{\sin^2 x \cos^2 x} = \sec^2 x \csc^2 x$

59. $\dfrac{\sin x}{1+\cos x} = \dfrac{\sin x}{1+\cos x} \cdot \dfrac{1-\cos x}{1-\cos x} = \dfrac{\sin x(1-\cos x)}{1-\cos^2 x} = \dfrac{\sin x(1-\cos x)}{\sin^2 x} = \dfrac{1-\cos x}{\sin x} = \dfrac{1}{\sin x} - \dfrac{\cos x}{\sin x} = \csc x - \cot x$

61. $\dfrac{\sin x}{1-\cos x} = \dfrac{\sin x}{1-\cos x} \cdot \dfrac{1+\cos x}{1+\cos x} = \dfrac{\sin x(1+\cos x)}{1-\cos^2 x} = \dfrac{\sin x(1+\cos x)}{\sin^2 x} = \dfrac{1+\cos x}{\sin x} = \dfrac{1}{\sin x} + \dfrac{\cos x}{\sin x} = \csc x + \cot x$

63. $\dfrac{\tan x \sin x}{\tan x + \sin x} = \dfrac{\tan x \sin x}{\tan x + \sin x} \cdot \dfrac{\tan x - \sin x}{\tan x - \sin x} = \dfrac{\tan x \sin x(\tan x - \sin x)}{\tan^2 x - \sin^2 x} = \dfrac{\tan x \sin x(\tan x - \sin x)}{\dfrac{\sin^2 x}{\cos^2 x} - \sin^2 x}$

$= \dfrac{\tan x \sin x(\tan x - \sin x)}{\dfrac{\sin^2 x - \sin^2 x \cos^2 x}{\cos^2 x}} = \dfrac{\tan x \sin x(\tan x - \sin x)}{\dfrac{\sin^2 x(1-\cos^2 x)}{\cos^2 x}} = \dfrac{\tan x \sin x(\tan x - \sin x)}{\dfrac{\sin^2 x \sin^2 x}{\cos^2 x}}$

$= \dfrac{\tan x \sin x(\tan x - \sin x)}{\tan^2 x \sin^2 x} = \dfrac{\tan x - \sin x}{\tan x \sin x}$

65. $(\tan x + \cot x)^2 = \tan^2 x + 2\tan x \cot x + \cot^2 x = \tan^2 x + \cot^2 x + 2\tan x\left(\dfrac{1}{\tan x}\right) = \tan^2 x + \cot^2 x + 2$
$= (\tan^2 x + 1) + (\cot^2 x + 1) = \sec^2 x + \csc^2 x$

67. $(1 + \cot^2 x)(1 + \tan^2 x) = \left(1 + \dfrac{\cos^2 x}{\sin^2 x}\right)\left(1 + \dfrac{\sin^2 x}{\cos^2 x}\right) = \left(\dfrac{\sin^2 x + \cos^2 x}{\sin^2 x}\right)\left(\dfrac{\cos^2 x + \sin^2 x}{\cos^2 x}\right) = \dfrac{1}{\sin^2 x \cos^2 x}$

69. $\dfrac{\sin^2 x - \cos^2 x}{\sec^2 x - \csc^2 x} = \dfrac{\sin^2 x - \cos^2 x}{\dfrac{1}{\cos^2 x} - \dfrac{1}{\sin^2 x}} = \dfrac{\sin^2 x - \cos^2 x}{\dfrac{\sin^2 x - \cos^2 x}{\sin^2 x \cos^2 x}} = (\sin^2 x - \cos^2 x) \cdot \dfrac{\sin^2 x \cos^2 x}{\sin^2 x - \cos^2 x} = \sin^2 x \cos^2 x$

71. $\dfrac{\tan x}{1+\sec x} + \dfrac{1+\sec x}{\tan x} = \dfrac{\tan^2 x + (1+\sec x)^2}{\tan x(1+\sec x)} = \dfrac{\tan^2 x + 1 + 2\sec x + \sec^2 x}{\tan x(1+\sec x)} = \dfrac{\sec^2 x + 2\sec x + \sec^2 x}{\tan x(1+\sec x)}$

$= \dfrac{2\sec^2 x + 2\sec x}{\tan x(1+\sec x)} = \dfrac{2\sec x(1+\sec x)}{\tan x(1+\sec x)} = \dfrac{2\sec x}{\tan x} = \dfrac{\dfrac{2}{\cos x}}{\dfrac{\sin x}{\cos x}} = \dfrac{2}{\sin x} = 2\csc x$

73. $\dfrac{\sin x + \tan x}{\cos x + 1} = \dfrac{\sin x + \dfrac{\sin x}{\cos x}}{\cos x + 1} = \dfrac{\dfrac{\sin x \cos x + \sin x}{\cos x}}{\cos x + 1} = \dfrac{\dfrac{\sin x(\cos x + 1)}{\cos x}}{\cos x + 1} = \dfrac{\sin x}{\cos x} = \tan x$

75. $\dfrac{\sec x - 1}{\tan x} \cdot \dfrac{\sec x + 1}{\sec x + 1} = \dfrac{\sec^2 x - 1}{\tan x(\sec x + 1)} = \dfrac{\tan^2 x}{\tan x(\sec x + 1)} = \dfrac{\tan x}{\sec x + 1}$

77. $\sqrt{1 + \tan^2 \theta} = \sec \theta$

79. $\sqrt{\sec^2 \theta - 1} = \tan \theta$

81. $\dfrac{\cos \theta}{\sqrt{1 - \cos^2 \theta}} = \dfrac{\cos \theta}{\sin \theta} = \cot \theta$

83. $\dfrac{\sec \theta}{\sqrt{\sec^2 \theta - 1}} = \dfrac{\sec \theta}{\tan \theta} = \dfrac{\dfrac{1}{\cos \theta}}{\dfrac{\sin \theta}{\cos \theta}} = \dfrac{1}{\sin \theta} = \csc \theta$

85. $\dfrac{\sqrt{3}\tan \theta}{\sqrt{(\sqrt{3}\tan \theta)^2 + 3}} = \dfrac{\sqrt{3}\tan \theta}{\sqrt{3\tan^2 \theta + 3}} = \dfrac{\sqrt{3}\tan \theta}{\sqrt{3(\tan^2 \theta + 1)}} = \dfrac{\sqrt{3}\tan \theta}{\sqrt{3\sec^2 \theta}} = \dfrac{\sqrt{3}\tan \theta}{\sqrt{3}\sec \theta} = \dfrac{\dfrac{\sin \theta}{\cos \theta}}{\dfrac{1}{\cos \theta}} = \sin \theta$

87. $\sin^3 x = (\sin^2 x)\sin x = (1 - \cos^2 x)\sin x$

89. $\cos^5 x = \cos^4 x \cos x = (\cos^2 x)^2 \cos x = (1 - \sin^2 x)^2 \cos x = (1 - 2\sin^2 x + \sin^4 x)\cos x$

91. $\cos^2 x \sin^5 x = \cos^2 x \sin^4 x \sin x = \cos^2 x (\sin^2 x)^2 \sin x = \cos^2 x (1 - \cos^2 x)^2 \sin x$
$= \cos^2 x (1 - 2\cos^2 x + \cos^4 x)\sin x = (\cos^2 x - 2\cos^4 x + \cos^6 x)\sin x$

93. $\cot^4 x = (\cot^2 x)(\cot^2 x) = (\cot^2 x)(\csc^2 x - 1) = \cot^2 x \csc^2 x - \cot^2 x = \cot^2 x \csc^2 x - (\csc^2 x - 1)$
$= \cot^2 x \csc^2 x - \csc^2 x + 1$

95. $\tan^3 x \sec^4 x = \tan^3 x \sec^2 x \sec^2 x = \tan^3 x \sec^2 x (1 + \tan^2 x) = \tan^3 x \sec^2 x + \tan^5 x \sec^2 x$

6.1 Applying the Concepts

97. $x = 20 \csc \theta$

99. $m_1 \cos \theta - \sin \theta = m_2 \cos \theta + m_1 m_2 \sin \theta$
$m_1 \cos \theta - m_2 \cos \theta = \sin \theta + m_1 m_2 \sin \theta$
$\cos \theta (m_1 - m_2) = \sin \theta (1 + m_1 m_2)$
$\dfrac{m_1 - m_2}{1 + m_1 m_2} = \dfrac{\sin \theta}{\cos \theta}$ or $\dfrac{\cos \theta}{\sin \theta} = \dfrac{1 + m_1 m_2}{m_1 - m_2}$
$\dfrac{m_1 - m_2}{1 + m_1 m_2} = \tan \theta$ or $\cot \theta = \dfrac{1 + m_1 m_2}{m_1 - m_2}$
$m_1 - m_2 = \tan \theta (1 + m_1 m_2) \Rightarrow m_1 - m_2 = \tan \theta + m_1 m_2 \tan \theta \Rightarrow m_1 - \tan \theta = m_2 + m_1 m_2 \tan \theta$
or $1 + m_1 m_2 = \cot \theta (m_1 - m_2) \Rightarrow 1 + m_1 m_2 = \cot \theta m_1 - \cot \theta m_2$

6.1 Beyond the Basics

101. $\dfrac{1-\sin x}{1-\sec x} - \dfrac{1+\sin x}{1+\sec x} = \dfrac{(1-\sin x)(1+\sec x)-(1+\sin x)(1-\sec x)}{1-\sec^2 x}$

$= \dfrac{1+\sec x - \sin x - \sin x \sec x - 1 + \sec x - \sin x + \sin x \sec x}{1-\sec^2 x}$

$= \dfrac{2(\sec x - \sin x)}{-\tan^2 x} = \dfrac{2\sin x - 2\sec x}{\dfrac{\sin^2 x}{\cos^2 x}} = \dfrac{2\sin x \cos^2 x - 2\sec x \cos^2 x}{\sin^2 x}$

$= \dfrac{2\sin x \cos^2 x}{\sin^2 x} - \dfrac{2\sec x \cos^2 x}{\sin^2 x} = \dfrac{2\cos^2 x}{\sin x} - \dfrac{\dfrac{2}{\cos x}\cos^2 x}{\sin^2 x}$

$= 2\cos x \cot x - \dfrac{2}{\sin x}\cot x = 2\cos x \cot x - 2\csc x \cot x = 2\cot x(\cos x - \csc x)$

103. $(1-\tan x)^2 + (1-\cot x)^2 = 1 - 2\tan x + \tan^2 x + 1 - 2\cot x + \cot^2 x = \sec^2 x - 2(\tan x + \cot x) + \csc^2 x$

$= \sec^2 x - 2\left(\dfrac{\sin x}{\cos x} + \dfrac{\cos x}{\sin x}\right) + \csc^2 x = \sec^2 x - 2\left(\dfrac{\sin^2 x + \cos^2 x}{\sin x \cos x}\right) + \csc^2 x$

$= \sec^2 x - \dfrac{2}{\sin x \cos x} + \csc^2 x = \sec^2 x - 2\sec x \csc x + \csc^2 x = (\sec x - \csc x)^2$

105. $\dfrac{\cot x + \csc x - 1}{\cot x + \csc x + 1} = \dfrac{\dfrac{\cos x}{\sin x} + \dfrac{1}{\sin x} - 1}{\dfrac{\cos x}{\sin x} + \dfrac{1}{\sin x} + 1} \cdot \dfrac{\sin x}{\sin x} = \dfrac{\cos x + 1 - \sin x}{\cos x + 1 + \sin x} = \dfrac{(\cos x + 1) - \sin x}{(\cos x + 1) + \sin x} \cdot \dfrac{(\cos x + 1) - \sin x}{(\cos x + 1) - \sin x}$

$= \dfrac{(\cos x + 1)^2 - 2\sin x(\cos x + 1) + \sin^2 x}{(\cos x + 1)^2 - \sin^2 x} = \dfrac{\cos^2 x + 2\cos x + 1 - 2\sin x \cos x - 2\sin x + \sin^2 x}{\cos^2 x + 2\cos x + 1 - \sin^2 x}$

$= \dfrac{2\cos x + 2 - 2\sin x \cos x - 2\sin x}{\cos^2 x + 2\cos x + 1 - (1-\cos^2 x)} = \dfrac{2(\cos x + 1) - 2\sin x(\cos x + 1)}{2\cos^2 x + 2\cos x}$

$= \dfrac{2(1-\sin x)(\cos x + 1)}{2\cos x(\cos x + 1)} = \dfrac{1-\sin x}{\cos x}$

107. $\sqrt{\dfrac{1-\cos x}{1+\cos x}} = \sqrt{\dfrac{1-\cos x}{1+\cos x} \cdot \dfrac{1-\cos x}{1-\cos x}} = \sqrt{\dfrac{(1-\cos x)^2}{1-\cos^2 x}} = \dfrac{1-\cos x}{\sqrt{\sin^2 x}} = \dfrac{1-\cos x}{\sin x} = \dfrac{1}{\sin x} - \dfrac{\cos x}{\sin x} = \csc x - \cot x$

109. $\sqrt{\dfrac{\sec x + 1}{\sec x - 1}} = \sqrt{\dfrac{\sec x + 1}{\sec x - 1} \cdot \dfrac{\sec x + 1}{\sec x + 1}} = \sqrt{\dfrac{(\sec x + 1)^2}{\sec^2 x - 1}} = \dfrac{\sec x + 1}{\sqrt{\tan^2 x}} = \dfrac{\sec x + 1}{\tan x} = \dfrac{\dfrac{1}{\cos x} + 1}{\dfrac{\sin x}{\cos x}}$

$= \dfrac{1}{\sin x} + \dfrac{\cos x}{\sin x} = \csc x + \cot x$

Now manipulate the right side.

$\dfrac{1}{\csc x - \cot x} = \dfrac{1}{\csc x - \cot x} \cdot \dfrac{\csc x + \cot x}{\csc x + \cot x} = \dfrac{\csc x + \cot x}{\csc^2 x - \cot^2 x} = \dfrac{\csc x + \cot x}{1 + \cot^2 x - \cot^2 x} = \csc x + \cot x$

Both sides of the original identity are equal to $\csc x + \cot x$, so the identity is verified.

111. $(\sin u \cos v + \cos u \sin v)^2 + (\cos u \cos v - \sin u \sin v)^2$

$= (\sin^2 u \cos^2 v + 2\sin u \cos v \cos u \sin v + \cos^2 u \sin^2 v) + (\cos^2 u \cos^2 v - 2\sin u \cos v \cos u \sin v + \sin^2 u \sin^2 v)$

$= \sin^2 u \cos^2 v + \cos^2 u \sin^2 v + \cos^2 u \cos^2 v + \sin^2 u \sin^2 v$

$= \sin^2 u(\cos^2 v + \sin^2 v) + \cos^2 u(\sin^2 v + \cos^2 v) = (\sin^2 u + \cos^2 u)(\sin^2 v + \cos^2 v) = 1$

113. $3(\sin^4 x + \cos^4 x) - 2(\sin^6 x + \cos^6 x) = 3(\sin^4 x + \cos^4 x) - 2\left((\sin^2 x)^3 + (\cos^2 x)^3\right)$

$= 3\sin^4 x + 3\cos^4 x - 2(\sin^2 x + \cos^2 x)(\sin^4 x - \sin^2 x \cos^2 x + \cos^4 x)$

$= 3\sin^4 x + 3\cos^4 x - 2\sin^4 x + 2\sin^2 x \cos^2 x - 2\cos^4 x$

$= \sin^4 x + 2\sin^2 x \cos^2 x + \cos^4 x = (\sin^2 x + \cos^2 x)^2 = 1$

Remember that
$u^3 + v^3$
$= (u+v)(u^2 - uv + v^2)$

115. $\dfrac{1-\cos\theta}{1+\cos\theta} + \dfrac{1+\cos\theta}{1-\cos\theta} - 4\cot^2\theta = \dfrac{1 - 2\cos\theta + \cos^2\theta + 1 + 2\cos\theta + \cos^2\theta}{1 - \cos^2\theta} - 4\cot^2\theta = \dfrac{2 + 2\cos^2\theta}{\sin^2\theta} - 4\cot^2\theta$

$= \dfrac{2}{\sin^2\theta} + \dfrac{2\cos^2\theta}{\sin^2\theta} - 4\cot^2\theta = 2\csc^2\theta - 2\cot^2\theta = 2(\csc^2\theta - \cot^2\theta) = 2$

117. $\dfrac{\sec^2\theta + 2\tan^2\theta}{1 + 3\tan^2\theta} = \dfrac{1 + \tan^2\theta + 2\tan^2\theta}{1 + 3\tan^2\theta} = \dfrac{1 + 3\tan^2\theta}{1 + 3\tan^2\theta} = 1$

119. $\cos^2 x(3 - 4\cos^2 x)^2 + \sin^2 x(3 - 4\sin^2 x)^2 = \cos^2 x(9 - 24\cos^2 x + 16\cos^4 x) + \sin^2 x(9 - 24\sin^2 x + 16\sin^4 x)$

$= 16\cos^6 x - 24\cos^4 x + 9\cos^2 x + 16\sin^6 x - 24\sin^4 x + 9\sin^2 x$

$= 16(\sin^6 x + \cos^6 x) - 24(\sin^4 x + \cos^4 x) + 9(\sin^2 x + \cos^2 x)$

$= 16\left((\sin^2 x)^3 + (\cos^2 x)^3\right) - 24(\sin^4 x + \cos^4 x) + 9$

$= 16(\sin^2 x + \cos^2 x)(\sin^4 x - \sin^2 x \cos^2 x + \cos^4 x) - 24(\sin^4 x + \cos^4 x) + 9$

$= 16\sin^4 x - 16\sin^2 x \cos^2 x + 16\cos^4 x - 24\sin^4 x - 24\cos^4 x + 9$

$= -8\sin^4 x - 16\sin^2 x \cos^2 x - 8\cos^4 x + 9 = -8(\sin^2 x + \cos^2 x)^2 + 9 = 1$

121. $x^2 + y^2 + z^2 = (r\cos u \cos v)^2 + (r\cos u \sin v)^2 + (r\sin u)^2$

$= r^2 \cos^2 u \cos^2 v + r^2 \cos^2 u \sin^2 v + r^2 \sin^2 u$

$= r^2(\cos^2 u \cos^2 v + \cos^2 u \sin^2 v + \sin^2 u)$

$= r^2(\cos^2 u(\cos^2 v + \sin^2 v) + \sin^2 u) = r^2(\cos^2 u + \sin^2 u) = r^2$

123. a. $\sec x + \cos x = 2 \Rightarrow (\sec x + \cos x)^2 = 4 \Rightarrow \sec^2 x + 2\sec x \cos x + \cos^2 x = 4 \Rightarrow$
$\sec^2 x + 2(1) + \cos^2 x = 4 \Rightarrow \sec^2 x + \cos^2 x = 2$

$\sec x + \cos x = 2 \Rightarrow (\sec x + \cos x)^4 = 16 \Rightarrow$
$\sec^4 x + 4\sec^3 x \cos x + 6\sec^2 x \cos^2 x + 4\sec x \cos^3 x + \cos^4 x = 16 \Rightarrow$
$\sec^4 x + \cos^4 x + 4\sec^2 x + 6\sec^2 x \cos^2 x + 4\cos^2 x = 16 \Rightarrow$
$\sec^4 x + \cos^4 x + 4(\sec^2 x + \cos^2 x) + 6 = 16 \Rightarrow \sec^4 x + \cos^4 x + 4(2) + 6 = 16 \Rightarrow$
$\sec^4 x + \cos^4 x = 2$

b. $\sec x + \cos x = 2 \Rightarrow (\sec x + \cos x)^3 = 8 \Rightarrow \sec^3 x + 3\sec^2 x \cos x + 3\sec x \cos^2 x + \cos^3 x = 8 \Rightarrow$
$\sec^3 x + \cos^3 x + 3\sec x + 3\cos x = 8 \Rightarrow \sec^3 x + \cos^3 x + 3(\sec x + \cos x) = 8 \Rightarrow$
$\sec^3 x + \cos^3 x + 3(2) = 8 \Rightarrow \sec^3 x + \cos^3 x = 2$

6.1 Critical Thinking/Discussion/Writing

125. a. Answers may vary. Sample answer:

If $x = \dfrac{3\pi}{2}$, then $\sqrt{\sin^2\left(\dfrac{3\pi}{2}\right)} = 1$, while $\sin\dfrac{3\pi}{2} = -1$.

b. $\left(\sqrt{\sin^2 x}\right)^2 = \sin^2 x \Rightarrow \sin^2 x = \sin^2 x$ and $\sin\dfrac{\pi}{2} = 1$.

126. If $\sec\theta = \dfrac{xy}{x^2+y^2}$, we have a right triangle with hypotenuse xy and one leg x^2+y^2. Using the Pythagorean theorem to find the length of the other leg, we have

$b = \sqrt{(xy)^2-(x^2+y^2)^2} = \sqrt{x^2y^2-(x^4+2x^2y^2+y^4)} = \sqrt{-x^4-x^2y^2-y^4} = \sqrt{-(x^4+x^2y^2+y^4)}$, which is the root of a negative number. Therefore $\sec\theta \neq \dfrac{xy}{x^2+y^2}$.

127. We want to solve $-1 \leq \sin\theta \leq 1 \Rightarrow -1 \leq \dfrac{1+t^2}{1-t^2} \leq 1$. Because $1+t^2$ is positive, $1-t^2$ has a similar sign to $\sin\theta$, which is sometimes negative and sometimes positive. So, we must use two compound inequalities.

$-1 \leq \dfrac{1+t^2}{1-t^2} < 0$ and $0 < \dfrac{1+t^2}{1-t^2} \leq 1$.

$-1 \leq \dfrac{1+t^2}{1-t^2} \leq 0 \Rightarrow$

$-1 \leq \dfrac{1+t^2}{1-t^2}$ | $\dfrac{1+t^2}{1-t^2} \leq 0$

$0 \leq \dfrac{1+t^2}{1-t^2}+1$

$0 \leq \dfrac{1+t^2+1-t^2}{1-t^2}$

$0 \leq \dfrac{2}{1-t^2}$

$1-t^2 = 0 \Rightarrow t^2 = 1 \Rightarrow t = \pm 1$

Interval	Test point	$\dfrac{2}{1-t^2}$	Result
$(-\infty, -1)$	-2	$-\tfrac{2}{3}$	$-$
$(-1, 1)$	0	2	$+$
$(1, \infty)$	2	$-\tfrac{2}{3}$	$-$

$0 \leq \dfrac{1+t^2}{1-t^2} \leq 1 \Rightarrow$

$0 \leq \dfrac{1+t^2}{1-t^2}$ | $\dfrac{1+t^2}{1-t^2} \leq 1$

$\dfrac{1+t^2}{1-t^2} - 1 \leq 0$

$\dfrac{1+t^2-(1-t^2)}{1-t^2} \leq 0$

$\dfrac{2t^2}{1-t^2} \leq 0$

$1-t^2 = 0 \Rightarrow t^2 = 1 \Rightarrow t = \pm 1$
$2t^2 = 0 \Rightarrow t = 0$

Interval	Test point	$\dfrac{2t^2}{1-t^2}$	Result
$(-\infty, -1)$	-2	$-\tfrac{8}{3}$	$-$
$(-1, 0)$	$-\tfrac{1}{2}$	$\tfrac{2}{3}$	$+$
$(0, 1)$	$\tfrac{1}{2}$	$\tfrac{2}{3}$	$+$
$(1, \infty)$	2	$-\tfrac{8}{3}$	$-$

Because the intervals for both sets of inequalities overlap, we must then look at test points for t within those intervals for the original inequality to see which intervals (if any) work. All test points, except for zero, will give a value that is out of the range for $\sin\theta$. This is easy to verify by graphing $Y_1 = \dfrac{1+t^2}{1-t^2}$ in the window $[-5, 5]$ by $[-5, 5]$.

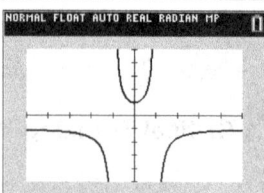

128. $0 \leq \cos^2\theta \leq 1 \Rightarrow 0 \leq \dfrac{a^2+b^2}{2ab} \leq 1 \Rightarrow 0 \leq a^2+b^2 \leq 2ab \Rightarrow a^2-2ab+b^2 \leq 0 \Rightarrow (a-b)^2 \leq 0$

This is true only if $a = b$ for $a, b \neq 0$.

129. $\csc^2\theta = \dfrac{2ab}{a^2+b^2} \Rightarrow \dfrac{2ab}{a^2+b^2} \geq 1 \Rightarrow 2ab \geq a^2+b^2 \Rightarrow 0 \geq a^2-2ab+b^2 \Rightarrow 0 \geq (a-b)^2$.

This is true only if $a=b$ for $a,b \neq 0$.

130. a. We know that $(\sin\theta - \csc\theta)^2 \geq 0$ because $a^2 \geq 0$ for any real number a. So

$\sin^2\theta - 2\sin\theta\csc\theta + \csc^2\theta \geq 0$.

$-2\sin\theta\csc\theta = -2\sin\theta\left(\dfrac{1}{\sin\theta}\right) = -2$, so $\sin^2\theta + \csc^2\theta - 2 \geq 0 \Rightarrow \sin^2\theta + \csc^2\theta \geq 2$

b. $(\cos\theta - \sec\theta)^2 \geq 0 \Rightarrow \cos^2\theta - 2\cos\theta\sec\theta + \sec^2\theta \geq 0$.

$2\cos\theta\sec\theta = 2\cos\theta\left(\dfrac{1}{\cos\theta}\right) = 2$, so $\cos^2\theta + \sec^2\theta \geq 2$

c. $(\sin\theta - \cos\theta)^2 \geq 0 \Rightarrow \sin^2\theta - 2\sin\theta\cos\theta + \cos^2\theta \geq 0 \Rightarrow 1 \geq 2\sin\theta\cos\theta \Rightarrow \dfrac{1}{2} \geq \sin\theta\cos\theta \Rightarrow$

$\dfrac{1}{2} \geq \dfrac{1}{\sec\theta\csc\theta} \Rightarrow \sec\theta\csc\theta \geq 2$

$(\sec\theta - \csc\theta)^2 \geq 0 \Rightarrow \sec^2\theta - 2\sec\theta\csc\theta + \csc^2\theta \geq 0 \Rightarrow \sec^2\theta + \csc^2\theta \geq 2\sec\theta\csc\theta$

Since $\sec\theta\csc\theta \geq 2$, $\sec^2\theta + \csc^2\theta \geq 2\sec\theta\csc\theta \Rightarrow \sec^2\theta + \csc^2\theta \geq 4$

6.1 Getting Ready for the Next Section

131. $d = \sqrt{(x_2-x_1)^2+(y_2-y_1)^2} = \sqrt{(-2-3)^2+(5-4)^2} = \sqrt{26}$

133. $d = \sqrt{(x_2-x_1)^2+(y_2-y_1)^2} = \sqrt{(\cos\theta-1)^2+(\sin\theta-0)^2} = \sqrt{\cos^2\theta-2\cos\theta+1+\sin^2\theta}$
$= \sqrt{\cos^2\theta+\sin^2\theta-2\cos\theta+1} = \sqrt{2-2\cos\theta} = \sqrt{2(1-\cos\theta)}$

135. False. $\cos 30° + \cos 60° = \dfrac{\sqrt{3}}{2} + \dfrac{1}{2} \neq 0 = \cos 90°$

137. If $\dfrac{\pi}{2} < \theta < \pi$, then θ lies in quadrant II and $\cos\theta < 0$, $\tan\theta < 0$.

$\sin\theta = \dfrac{y}{r} = \dfrac{1}{3} \Rightarrow 3^2 = x^2 + 1^2 \Rightarrow x^2 = 8 \Rightarrow$

$x = -2\sqrt{2}$ (because θ lies in quadrant II)

$\cos\theta = -\dfrac{2\sqrt{2}}{3}$, $\tan\theta = -\dfrac{1}{2\sqrt{2}} = -\dfrac{\sqrt{2}}{4}$

139. $\cos\theta = \dfrac{4}{5} \Rightarrow \cos(-\theta) = \cos\theta = \dfrac{4}{5}$

$\cos\theta - \cos(-\theta) = \dfrac{4}{5} - \dfrac{4}{5} = 0$

141. $u = \sin^{-1}\left(-\dfrac{3}{5}\right) \Rightarrow \sin u = -\dfrac{3}{5}$

The domain of $\sin^{-1}x$ is $\left[-\dfrac{\pi}{2}, \dfrac{\pi}{2}\right]$, so u lies in quadrant IV and $\cos u > 0$.

$\sin u = -\dfrac{3}{5} = \dfrac{y}{r} \Rightarrow y = -3$, $r = 5$

$5^2 = x^2 + (-3)^2 \Rightarrow x^2 = 16 \Rightarrow x = 4$

$\cos u = \dfrac{4}{5}$

6.2 Sum and Difference Formulas

6.2 Practice Problems

1. $\cos 15° = \cos(45°-30°)$
$= \cos 45°\cos 30° + \sin 45°\sin 30°$
$= \left(\dfrac{\sqrt{2}}{2}\right)\left(\dfrac{\sqrt{3}}{2}\right) + \left(\dfrac{\sqrt{2}}{2}\right)\left(\dfrac{1}{2}\right)$
$= \dfrac{\sqrt{6}+\sqrt{2}}{4}$

2. $\cos\dfrac{7\pi}{12} = \cos\left(\dfrac{\pi}{3}+\dfrac{\pi}{4}\right)$
$= \cos\dfrac{\pi}{3}\cos\dfrac{\pi}{4} - \sin\dfrac{\pi}{3}\sin\dfrac{\pi}{4}$
$= \dfrac{1}{2}\left(\dfrac{\sqrt{2}}{2}\right) - \dfrac{\sqrt{3}}{2}\left(\dfrac{\sqrt{2}}{2}\right)$
$= \dfrac{\sqrt{2}-\sqrt{6}}{4}$

3. $\sec\left(\dfrac{\pi}{2}-v\right) = \dfrac{1}{\cos\left(\dfrac{\pi}{2}-v\right)} = \dfrac{1}{\sin v} = \csc v$

4. $\sin(\pi + x) = \sin\pi\cos x + \cos\pi\sin x$
$= 0\cdot\cos x + (-1)\sin x$
$= -\sin x$

5. $\sin 43°\cos 13° - \cos 43°\sin 13° = \sin(43°-13°)$
$= \sin 30° = \dfrac{1}{2}$

6. $\cos(u+v) = \cos u\cos v - \sin u\sin v$
$= \left(-\dfrac{4}{5}\right)\left(\dfrac{12}{13}\right) - \left(-\dfrac{3}{5}\right)\left(-\dfrac{5}{13}\right) = -\dfrac{63}{65}$

7. $\dfrac{\cos(x+y)}{\sin x\sin y} = \dfrac{\cos x\cos y - \sin x\sin y}{\sin x\sin y}$
$= \dfrac{\cos x\cos y}{\sin x\sin y} - \dfrac{\sin x\sin y}{\sin x\sin y}$
$= \cot x\cot y - 1$

8. $\tan(\pi + x) = \dfrac{\tan\pi + \tan x}{1 - \tan\pi\tan x} = \dfrac{0 + \tan x}{1 - 0} = \tan x$

9. $\tan\left[\cos^{-1}\dfrac{3}{5} - \tan^{-1}\left(-\dfrac{1}{2}\right)\right]$

$u = \cos^{-1}\dfrac{3}{5} \Rightarrow \cos u = \dfrac{3}{5} = \dfrac{x}{r},\ 0\le u\le\pi$
u lies in quadrant I, so $y\ge 0$.
$5^2 = 3^2 + y^2 \Rightarrow y^2 = 16 \Rightarrow y = 4$
Thus, $\tan u = \dfrac{4}{3}$.

$v = \tan^{-1}\left(-\dfrac{1}{2}\right) \Rightarrow \tan v = -\dfrac{1}{2},\ -\dfrac{\pi}{2}<v<\dfrac{\pi}{2}$
v lies in quadrant IV, so $x\ge 0,\ y\le 0$.

$\tan(u-v) = \dfrac{\tan u - \tan v}{1 + \tan u\tan v}$

$\tan\left[\cos^{-1}\dfrac{3}{5} - \tan^{-1}\left(-\dfrac{1}{2}\right)\right]$
$= \dfrac{\dfrac{4}{3} - \left(-\dfrac{1}{2}\right)}{1 + \dfrac{4}{3}\left(-\dfrac{1}{2}\right)} = \dfrac{8-(-3)}{6+(-4)} = \dfrac{11}{2}$

10. $y = \sin x + \sqrt{3}\cos x = a\sin x + b\cos x \Rightarrow$
$a = 1, b = \sqrt{3} \Rightarrow \sqrt{a^2+b^2} = \sqrt{1+3} = 2.$
So, θ is an angle in standard position that has $(1,\sqrt{3})$ on its terminal side $\Rightarrow \theta = \dfrac{\pi}{3}$. Thus,
$\sin x + \sqrt{3}\cos x = 2\sin\left(x+\dfrac{\pi}{3}\right).$

11. From practice problem 10,
$\sin x + \sqrt{3}\cos x = 2\sin\left(x+\dfrac{\pi}{3}\right).$ This is the graph of $y = \sin x$, shifted $\dfrac{\pi}{3}$ units to the left and stretched vertically by a factor of 2.

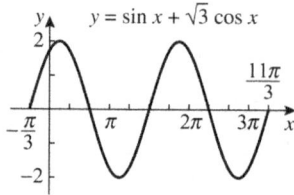

12. $y = y_1 + y_2 = 0.1\sin(400t) + 0.2\cos(400t) \Rightarrow$
$a = 0.1,\ b = 0.2 \Rightarrow$
$\sqrt{0.1^2 + 0.2^2} = \sqrt{0.05} = 0.1\sqrt{5} = A$
$\theta = \tan^{-1}\left(\dfrac{0.2}{0.1}\right) = \tan^{-1}2$
$y = 0.1\sin(400t) + 0.2\cos(400t) \Rightarrow$
$y = 0.1\sqrt{5}\sin(400t + \theta)$
$= 0.1\sqrt{5}\sin 400\left(t + \dfrac{\theta}{400}\right)$

Amplitude: $0.1\sqrt{5}$;

phase shift: $-\dfrac{\theta}{400} = -\dfrac{\tan^{-1}2}{400} \approx -0.0028$

period: $\dfrac{2\pi}{400} = \dfrac{\pi}{200}$; frequency: $\dfrac{400}{2\pi} = \dfrac{200}{\pi}$

6.2 Concepts and Vocabulary

1. $\sin(A+B) = \underline{\sin A\cos B + \cos A\sin B}$

3. $\tan(A+B) = \dfrac{\tan A + \tan B}{1 - \tan A\tan B}$

5. False.
$$\cos\left(\frac{\pi}{2}+x\right) = \cos\frac{\pi}{2}\cos x - \sin\frac{\pi}{2}\sin x$$
$$= 0\cdot\cos x - 1\cdot\sin x$$
$$= -\sin x$$

7. False. $\tan(x+y) = \dfrac{\tan x + \tan y}{1-\tan x \tan y}$

6.2 Building Skills

9. $\sin(45°+30°) = \sin 45°\cos 30° + \cos 45°\sin 30°$
$$= \left(\frac{\sqrt{2}}{2}\right)\left(\frac{\sqrt{3}}{2}\right) + \left(\frac{\sqrt{2}}{2}\right)\left(\frac{1}{2}\right)$$
$$= \frac{\sqrt{6}+\sqrt{2}}{4}$$

11. $\sin(60°-45°) = \sin 60°\cos 45° - \cos 60°\sin 45°$
$$= \left(\frac{\sqrt{3}}{2}\right)\left(\frac{\sqrt{2}}{2}\right) - \left(\frac{1}{2}\right)\left(\frac{\sqrt{2}}{2}\right)$$
$$= \frac{\sqrt{6}-\sqrt{2}}{4}$$

13. $\sin(-105°) = -\sin(105°) = -\sin(60°+45°)$
$$= -(\sin 60°\cos 45° + \cos 60°\sin 45°)$$
$$= -\left[\left(\frac{\sqrt{3}}{2}\right)\left(\frac{\sqrt{2}}{2}\right) + \left(\frac{1}{2}\right)\left(\frac{\sqrt{2}}{2}\right)\right]$$
$$= -\frac{\sqrt{6}+\sqrt{2}}{4}$$

15. $\tan 225° = \tan(180°+45°)$
$$= \frac{\tan 180° + \tan 45°}{1-\tan 180°\tan 45°} = 1$$

17. $\sin\left(\dfrac{\pi}{6}+\dfrac{\pi}{4}\right) = \sin\dfrac{\pi}{6}\cos\dfrac{\pi}{4} + \cos\dfrac{\pi}{6}\sin\dfrac{\pi}{4}$
$$= \left(\frac{1}{2}\right)\left(\frac{\sqrt{2}}{2}\right) + \left(\frac{\sqrt{3}}{2}\right)\left(\frac{\sqrt{2}}{2}\right)$$
$$= \frac{\sqrt{6}+\sqrt{2}}{4}$$

19. $\tan\left(\dfrac{\pi}{4}-\dfrac{\pi}{6}\right) = \dfrac{\tan(\pi/4)-\tan(\pi/6)}{1+\tan(\pi/4)\tan(\pi/6)}$
$$= \frac{1-\sqrt{3}/3}{1+1(\sqrt{3}/3)} = \frac{3-\sqrt{3}}{3+\sqrt{3}}$$
$$= 2-\sqrt{3}$$

21. $\sec\left(\dfrac{\pi}{3}+\dfrac{\pi}{4}\right)$
$$= \frac{1}{\cos\left(\dfrac{\pi}{3}+\dfrac{\pi}{4}\right)} = \frac{1}{\cos\dfrac{\pi}{3}\cos\dfrac{\pi}{4} - \sin\dfrac{\pi}{3}\sin\dfrac{\pi}{4}}$$
$$= \frac{1}{\left(\dfrac{1}{2}\right)\left(\dfrac{\sqrt{2}}{2}\right)-\left(\dfrac{\sqrt{3}}{2}\right)\left(\dfrac{\sqrt{2}}{2}\right)} = \frac{1}{\dfrac{\sqrt{2}-\sqrt{6}}{4}}$$
$$= \frac{4}{\sqrt{2}-\sqrt{6}}\cdot\frac{\sqrt{2}+\sqrt{6}}{\sqrt{2}+\sqrt{6}} = \frac{4(\sqrt{2}+\sqrt{6})}{-4}$$
$$= -\sqrt{2}-\sqrt{6}$$

23. $\cos\left(-\dfrac{5\pi}{12}\right) = \cos\left(\dfrac{5\pi}{12}\right) = \cos\left(\dfrac{\pi}{6}+\dfrac{\pi}{4}\right)$
$$= \cos\frac{\pi}{6}\cos\frac{\pi}{4} - \sin\frac{\pi}{6}\sin\frac{\pi}{4}$$
$$= \left(\frac{\sqrt{3}}{2}\right)\left(\frac{\sqrt{2}}{2}\right) - \frac{1}{2}\left(\frac{\sqrt{2}}{2}\right)$$
$$= \frac{\sqrt{6}-\sqrt{2}}{4}$$

25. $\tan\dfrac{19\pi}{12} = \tan\left(\dfrac{3\pi}{4}+\dfrac{5\pi}{6}\right) = \dfrac{\tan\dfrac{3\pi}{4}+\tan\dfrac{5\pi}{6}}{1-\tan\dfrac{3\pi}{4}\tan\dfrac{5\pi}{6}}$
$$= \frac{-1-\sqrt{3}/3}{1-(-1)(-\sqrt{3}/3)} = \frac{\sqrt{3}+3}{\sqrt{3}-3}$$
$$= -2-\sqrt{3}$$

27. $\tan\left(\dfrac{17\pi}{12}\right) = \tan\left(\dfrac{\pi}{4}+\dfrac{7\pi}{6}\right) = \dfrac{\tan\dfrac{\pi}{4}+\tan\dfrac{7\pi}{6}}{1-\tan\dfrac{\pi}{4}\tan\dfrac{7\pi}{6}}$
$$= \frac{1+\sqrt{3}/3}{1-(1)(\sqrt{3}/3)} = \frac{3+\sqrt{3}}{3-\sqrt{3}} = 2+\sqrt{3}$$

29. $\sin\left(x+\dfrac{\pi}{2}\right) = \sin x\cos\dfrac{\pi}{2} + \cos x\sin\dfrac{\pi}{2}$
$$= 0 + \cos x(1) = \cos x$$

31. $\sin\left(x-\dfrac{\pi}{2}\right) = \sin x\cos\dfrac{\pi}{2} - \cos x\sin\dfrac{\pi}{2}$
$$= 0 - \cos x(1) = -\cos x$$

33. $\tan\left(x+\dfrac{\pi}{2}\right) = \dfrac{\sin\left(x+\dfrac{\pi}{2}\right)}{\cos\left(x+\dfrac{\pi}{2}\right)} = \dfrac{\cos x}{-\sin x} = -\cot x$

35. $\csc(x+\pi) = \dfrac{1}{\sin(x+\pi)}$

$= \dfrac{1}{\sin x \cos \pi + \cos x \sin \pi}$

$= \dfrac{1}{-\sin x} = -\csc x$

37. $\cos\left(x + \dfrac{3\pi}{2}\right) = \cos x \cos \dfrac{3\pi}{2} - \sin x \sin \dfrac{3\pi}{2}$

$= 0 - (-1)\sin x = \sin x$

39. $\tan\left(x - \dfrac{3\pi}{2}\right) = \dfrac{\sin\left(x - \dfrac{3\pi}{2}\right)}{\cos\left(x - \dfrac{3\pi}{2}\right)}$

$= \dfrac{\sin x \cos \dfrac{3\pi}{2} - \sin \dfrac{3\pi}{2} \cos x}{\cos x \cos \dfrac{3\pi}{2} + \sin x \sin \dfrac{3\pi}{2}}$

$= \dfrac{\cos x}{-\sin x} = -\cot x$

41. $\cot(3\pi - x) = \dfrac{\cos(3\pi - x)}{\sin(3\pi - x)}$

$= \dfrac{\cos 3\pi \cos x + \sin 3\pi \sin x}{\sin 3\pi \cos x - \sin x \cos 3\pi}$

$= \dfrac{-\cos x}{\sin x} = -\cot x$

43. $\sin 56° \cos 34° + \cos 56° \sin 34° = \sin(56° + 34°)$
$= \sin 90° = 1$

45. $\cos 331° \cos 61° + \sin 331° \sin 61°$
$= \cos(331° - 61°) = \cos 270° = 0$

47. $\dfrac{\tan 129° - \tan 84°}{1 + \tan 129° \tan 84°} = \tan(129° - 84°)$
$= \tan 45° = 1$

49. $\sin \dfrac{7\pi}{12} \cos \dfrac{3\pi}{12} - \cos \dfrac{7\pi}{12} \sin \dfrac{3\pi}{12}$

$= \sin\left(\dfrac{7\pi}{12} - \dfrac{3\pi}{12}\right) = \sin\left(\dfrac{\pi}{3}\right) = \dfrac{\sqrt{3}}{2}$

51. $\dfrac{\tan \dfrac{5\pi}{12} - \tan \dfrac{2\pi}{12}}{1 + \tan \dfrac{5\pi}{12} \tan \dfrac{2\pi}{12}} = \tan\left(\dfrac{5\pi}{12} - \dfrac{2\pi}{12}\right) = \tan \dfrac{\pi}{4} = 1$

In exercises 53–58, $\tan u = \dfrac{3}{4}$ with u in Quadrant III $\Rightarrow \sin u = -\dfrac{3}{5}$ and $\cos u = -\dfrac{4}{5}$

$\sin v = \dfrac{5}{13}$ with v in Quadrant II $\Rightarrow \cos v = -\dfrac{12}{13}$

and $\tan v = -\dfrac{5}{12}$.

53. $\sin(u - v) = \sin u \cos v - \cos u \sin v$

$= \left(-\dfrac{3}{5}\right)\left(-\dfrac{12}{13}\right) - \left(-\dfrac{4}{5}\right)\left(\dfrac{5}{13}\right) = \dfrac{56}{65}$

55. $\cos(u + v) = \cos u \cos v - \sin u \sin v$

$= \left(-\dfrac{4}{5}\right)\left(-\dfrac{12}{13}\right) - \left(-\dfrac{3}{5}\right)\left(\dfrac{5}{13}\right) = \dfrac{63}{65}$

57. $\tan(u + v) = \dfrac{\tan u + \tan v}{1 - \tan u \tan v}$

$= \dfrac{\dfrac{3}{4} + \left(-\dfrac{5}{12}\right)}{1 - \left(\dfrac{3}{4}\left(-\dfrac{5}{12}\right)\right)} = \dfrac{16}{63}$

In exercises 59–64, $\cos \alpha = -\dfrac{2}{5}$ with α in Quadrant II $\Rightarrow \sin \alpha = \dfrac{\sqrt{21}}{5}$ and $\tan \alpha = -\dfrac{\sqrt{21}}{2}$

$\sin \beta = -\dfrac{3}{7}$ with β in Quadrant IV $\Rightarrow \cos \beta = \dfrac{2\sqrt{10}}{7}$

and $\tan \beta = -\dfrac{3\sqrt{10}}{20}$.

59. $\sin(\alpha - \beta) = \sin \alpha \cos \beta - \cos \alpha \sin \beta$

$= \left(\dfrac{\sqrt{21}}{5}\right)\left(\dfrac{2\sqrt{10}}{7}\right) - \left(-\dfrac{2}{5}\right)\left(-\dfrac{3}{7}\right)$

$= \dfrac{2\sqrt{210} - 6}{35}$

61. $\csc(\alpha+\beta) = \dfrac{1}{\sin(\alpha+\beta)}$

$\sin(\alpha+\beta) = \sin\alpha\cos\beta + \cos\alpha\sin\beta = \left(\dfrac{\sqrt{21}}{5}\right)\left(\dfrac{2\sqrt{10}}{7}\right) + \left(-\dfrac{2}{5}\right)\left(-\dfrac{3}{7}\right) = \dfrac{2\sqrt{210}+6}{35} \Rightarrow$

$\csc(\alpha+\beta) = \dfrac{35}{2\sqrt{210}+6} = \dfrac{35\sqrt{210}-105}{402}$

63. Use the results from exercises 59 and 60: $\cot(\alpha-\beta) = \dfrac{\cos(\alpha-\beta)}{\sin(\alpha-\beta)} = \dfrac{\frac{-4\sqrt{10}-3\sqrt{21}}{35}}{\frac{2\sqrt{210}-6}{35}} = \dfrac{-4\sqrt{10}-3\sqrt{21}}{2\sqrt{210}-6}$

65. $\dfrac{\sin(x+y)}{\cos x \cos y} = \dfrac{\sin x \cos y + \sin y \cos x}{\cos x \cos y} = \dfrac{\sin x \cos y}{\cos x \cos y} + \dfrac{\sin y \cos x}{\cos x \cos y} = \dfrac{\sin x}{\cos x} + \dfrac{\sin y}{\cos y} = \tan x + \tan y$

67. $\dfrac{\cos(x+y)}{\cos x \cos y} = \dfrac{\cos x \cos y - \sin x \sin y}{\cos x \cos y} = \dfrac{\cos x \cos y}{\cos x \cos y} - \dfrac{\sin x \sin y}{\cos x \cos y} = 1 - \tan x \tan y$

69. We use the results from exercises 66 and 68 in the first step.

$\dfrac{1+\cot x \cot y}{\cot x + \cot y} = \dfrac{\cos(x-y)}{\sin x \sin y} \div \dfrac{\sin(x+y)}{\sin x \sin y} = \dfrac{\cos(x-y)}{\sin x \sin y} \cdot \dfrac{\sin x \sin y}{\sin(x+y)} = \dfrac{\cos(x-y)}{\sin(x+y)}$

71. $\dfrac{\sin(x-y)}{\sin(x+y)} = \dfrac{\frac{\sin(x-y)}{\sin x \sin y}}{\frac{\sin(x+y)}{\sin x \sin y}} = \dfrac{\frac{\sin x \cos y - \sin y \cos x}{\sin x \sin y}}{\frac{\sin x \cos y + \sin y \cos x}{\sin x \sin y}} = \dfrac{\frac{\sin x \cos y}{\sin x \sin y} - \frac{\sin y \cos x}{\sin x \sin y}}{\frac{\sin x \cos y}{\sin x \sin y} + \frac{\sin y \cos x}{\sin x \sin y}} = \dfrac{\cot y - \cot x}{\cot y + \cot x}$

73. $\sin\left[\tan^{-1}\left(-\dfrac{3}{4}\right) + \cos^{-1}\left(\dfrac{4}{5}\right)\right] = \sin\left(\tan^{-1}\left(-\dfrac{3}{4}\right)\right)\cos\left(\cos^{-1}\left(\dfrac{4}{5}\right)\right) + \cos\left(\tan^{-1}\left(-\dfrac{3}{4}\right)\right)\sin\left(\cos^{-1}\left(\dfrac{4}{5}\right)\right)$

$= \left(-\dfrac{3}{5}\right)\left(\dfrac{4}{5}\right) + \left(\dfrac{4}{5}\right)\left(\dfrac{3}{5}\right) = 0$

75. $\sin\left[\sin^{-1}\left(\dfrac{3}{5}\right) - \cos^{-1}\left(\dfrac{4}{5}\right)\right] = \sin\left(\sin^{-1}\left(\dfrac{3}{5}\right)\right)\cos\left(\cos^{-1}\left(\dfrac{4}{5}\right)\right) - \cos\left(\sin^{-1}\left(\dfrac{3}{5}\right)\right)\sin\left(\cos^{-1}\left(\dfrac{4}{5}\right)\right)$

$= \left(\dfrac{3}{5}\right)\left(\dfrac{4}{5}\right) - \left(\dfrac{4}{5}\right)\left(\dfrac{3}{5}\right) = 0$

77. $\tan\left[\cos^{-1}\left(\dfrac{4}{5}\right) + \tan^{-1}\left(\dfrac{2}{3}\right)\right] = \dfrac{\tan\left(\cos^{-1}\left(\dfrac{4}{5}\right)\right) + \tan\left(\tan^{-1}\left(\dfrac{2}{3}\right)\right)}{1 - \tan\left(\cos^{-1}\left(\dfrac{4}{5}\right)\right)\tan\left(\tan^{-1}\left(\dfrac{2}{3}\right)\right)} = \dfrac{\frac{3}{4}+\frac{2}{3}}{1-\left(\frac{3}{4}\right)\left(\frac{2}{3}\right)} = \dfrac{17}{6}$

79. $y = \sin x + \cos x = a\sin x + b\cos x \Rightarrow a=1, b=1 \Rightarrow \sqrt{a^2+b^2} = \sqrt{2} = A$.
So θ is any angle in standard position that has $(1, 1)$ on its terminal side
$\Rightarrow \theta = \dfrac{\pi}{4}$. $y = \sin x + \cos x = \sqrt{2}\sin\left(x+\dfrac{\pi}{4}\right)$.

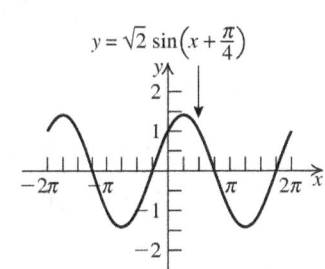

81. $y = 3\sin x + 4\cos x = a\sin x + b\cos x \Rightarrow a = 3, b = 4 \Rightarrow \sqrt{a^2 + b^2} = 5 = A$.
So θ is any angle in standard position that has $(3, 4)$ on its terminal side
$\Rightarrow \tan\theta = \dfrac{4}{3} \Rightarrow \theta \approx 0.927$.
$y = 3\sin x + 4\cos x = 5\sin(x + 0.927)$

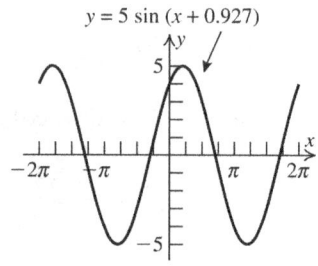

83. $y = 5\sin x - 12\cos x = a\sin x + b\cos x \Rightarrow a = 5, b = -12 \Rightarrow$
$\sqrt{a^2 + b^2} = 13 = A$. So θ is any angle in standard position that has
$(5, -12)$ on its terminal side $\Rightarrow \tan\theta = -\dfrac{12}{5} \Rightarrow \theta \approx -1.176$.
$y = 5\sin x - 12\cos x = 5\sin(x - 1.176)$.

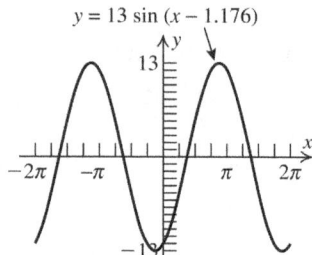

85. $y = \cos x - 3\sin x = a\sin x + b\cos x \Rightarrow a = -3, b = 1 \Rightarrow$
$\sqrt{a^2 + b^2} = \sqrt{10} = A$. So θ is any angle in standard position that has
$(-3, 1)$ on its terminal side $\Rightarrow \tan\theta = -\dfrac{1}{3} \Rightarrow \theta \approx -0.322 + \pi = 2.820$.
$y = \cos x - 3\sin x = \sqrt{10}\sin(x + 2.820)$.

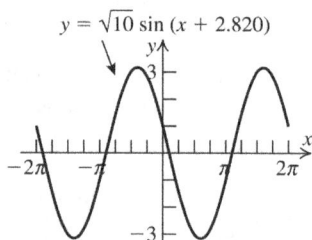

6.2 Applying the Concepts

87. $\alpha + \theta_1 + (180° - \theta_2) = 180° \Rightarrow \alpha = -(\theta_1 + \theta_2)$
If $\alpha = \theta_2 - \theta_1$, $\tan\alpha = \dfrac{\tan\theta_2 - \tan\theta_1}{1 + \tan\theta_1\tan\theta_2} = \dfrac{m_2 - m_1}{1 + m_1 m_2}$.
If $\alpha = \theta_2 - \theta_1 = \pi$, $\tan\alpha = -\dfrac{\tan\theta_2 - \tan\theta_1}{1 + \tan\theta_1\tan\theta_2} = -\dfrac{m_2 - m_1}{1 + m_1 m_2}$. So, $\tan\alpha = \left|\dfrac{m_2 - m_1}{1 + m_1 m_2}\right|$.

89. $\ell_1 : y = 2x + 5 \Rightarrow m_1 = 2$
$\ell_2 : 6x - 3y = 21 \Rightarrow y = 2x - 7 \Rightarrow m_2 = 2$
$\tan\alpha = \left|\dfrac{2 - 2}{1 + (2)(2)}\right| = 0 \Rightarrow \alpha = \tan^{-1}(0) \approx 0$

91. $y = 0.4\sin(400t) + 0.3\cos(400t) \Rightarrow a = 0.4, b = 0.3 \Rightarrow \sqrt{0.4^2 + 0.3^2} = 0.5 = A$
$\theta = \tan^{-1}\left(\dfrac{0.3}{0.4}\right)$
$y = 0.4\sin(400t) + 0.3\cos(400t) \Rightarrow y = 0.5\sin(400t + \theta) = 0.5\sin 400\left(t + \dfrac{\theta}{400}\right) \Rightarrow$ amplitude $= 0.5$
phase shift $= -\dfrac{\theta}{400} = -\dfrac{\tan^{-1}(0.3/0.4)}{400} \approx -0.00161$

93. a. $x = 0.12\sin 2t + 0.5\cos 2t \Rightarrow a = 0.12$,
$b = 0.5 \Rightarrow A = \sqrt{0.12^2 + 0.5^2} = 0.5142$

b. $\theta = \tan^{-1}\left(\dfrac{0.5}{0.12}\right) = \tan^{-1}\left(\dfrac{25}{6}\right)$

$y = 0.12\sin 2t + 0.5\cos 2t \approx 0.5142\sin(2t+\theta) \approx 0.5142\sin 2\left(t+\dfrac{\theta}{2}\right) \approx 0.5142\sin 2\left(t+\dfrac{\tan^{-1}(25/6)}{2}\right)$

$\approx 0.5142\sin 2(t+0.6676) \Rightarrow$ phase shift $= -0.6676.$

Period $= 2\pi/2 = \pi \Rightarrow$ frequency $= 1/\pi$.

6.2 Beyond the Basics

95. $\dfrac{\sin(x+h)-\sin x}{h} = \dfrac{\sin x\cos h + \sin h\cos x - \sin x}{h} = \dfrac{\sin x(\cos h - 1) + \sin h\cos x}{h} = \sin x\dfrac{(\cos h - 1)}{h} + \cos x\dfrac{\sin h}{h}$

97. $\cos u\cos(u+v) + \sin u\sin(u+v) = \cos u(\cos u\cos v - \sin u\sin v) + \sin u(\sin u\cos v + \sin v\cos u)$
$= \cos^2 u\cos v - \cos u\sin u\sin v + \sin^2 u\cos v + \sin u\sin v\cos u$
$= \cos v(\cos^2 u + \sin^2 u) = \cos v$

99. $\sin 5x\cos 3x - \cos 5x\sin 3x = \sin(3x+2x)\cos 3x - \cos(3x+2x)\sin 3x$
$= (\cos 3x)(\sin 3x\cos 2x + \sin 2x\cos 3x) - \sin 3x(\cos 3x\cos 2x - \sin 3x\sin 2x)$
$= \sin 3x\cos 3x\cos 2x + \sin 2x\cos^2 3x - (\sin 3x\cos 3x\cos 2x - \sin^2 3x\sin 2x)$
$= \sin 2x(\cos^2 3x + \sin^2 3x) = \sin 2x$

101. $\sin\left(\dfrac{\pi}{2}+x-y\right) = \sin\left(\dfrac{\pi}{2}-(-x+y)\right) = \cos(-x+y) = \cos(y-x) = \cos x\cos y + \sin x\sin y$

103. $\sin(x+y)\sin(x-y) = (\sin x\cos y + \sin y\cos x)(\sin x\cos y - \sin y\cos x) = (\sin^2 x\cos^2 y - \sin^2 y\cos^2 x)$
$= \sin^2 x(1-\sin^2 y) - \sin^2 y(1-\sin^2 x) = \sin^2 x - \sin^2 x\sin^2 y - \sin^2 y + \sin^2 x\sin^2 y$
$= \sin^2 x - \sin^2 y$

105. $\sin^2\left(\dfrac{\pi}{4}+\dfrac{x}{2}\right) - \sin^2\left(\dfrac{\pi}{4}-\dfrac{x}{2}\right) = \left(\sin\dfrac{\pi}{4}\cos\dfrac{x}{2} + \sin\dfrac{x}{2}\cos\dfrac{\pi}{4}\right)^2 - \left(\sin\dfrac{\pi}{4}\cos\dfrac{x}{2} - \sin\dfrac{x}{2}\cos\dfrac{\pi}{4}\right)^2$

$= \left(\dfrac{\sqrt{2}}{2}\cos\dfrac{x}{2} + \dfrac{\sqrt{2}}{2}\sin\dfrac{x}{2}\right)^2 - \left(\dfrac{\sqrt{2}}{2}\cos\dfrac{x}{2} - \dfrac{\sqrt{2}}{2}\sin\dfrac{x}{2}\right)^2$

$= \left(\dfrac{1}{2}\cos^2\dfrac{x}{2} + \cos\dfrac{x}{2}\sin\dfrac{x}{2} + \dfrac{1}{2}\sin^2\dfrac{x}{2}\right) - \left(\dfrac{1}{2}\cos^2\dfrac{x}{2} - \cos\dfrac{x}{2}\sin\dfrac{x}{2} + \dfrac{1}{2}\sin^2\dfrac{x}{2}\right)$

$= 2\cos\dfrac{x}{2}\sin\dfrac{x}{2} = \sin x$

107. $\dfrac{\sin(\alpha-\beta)}{\sin\alpha\sin\beta} + \dfrac{\sin(\beta-\gamma)}{\sin\beta\sin\gamma} + \dfrac{\sin(\gamma-\alpha)}{\sin\gamma\sin\alpha}$
$= \dfrac{\sin\gamma(\sin\alpha\cos\beta - \sin\beta\cos\alpha) + \sin\alpha(\sin\beta\cos\gamma - \cos\beta\sin\gamma) + \sin\beta(\sin\gamma\cos\alpha - \sin\alpha\cos\gamma)}{\sin\alpha\sin\beta\sin\gamma}$
$= \dfrac{\sin\alpha\sin\gamma\cos\beta - \sin\beta\sin\gamma\cos\alpha + \sin\alpha\sin\beta\cos\gamma - \sin\alpha\sin\gamma\cos\beta + \sin\beta\sin\gamma\cos\alpha - \sin\alpha\sin\beta\cos\gamma}{\sin\alpha\sin\beta\sin\gamma}$
$= 0$

109. $\dfrac{\sin(x-y)}{\sin(x+y)} = \dfrac{\sin x\cos y - \cos x\sin y}{\sin x\cos y + \cos x\sin y} \cdot \dfrac{\dfrac{1}{\cos x\cos y}}{\dfrac{1}{\cos x\cos y}} = \dfrac{\dfrac{\sin x}{\cos x} - \dfrac{\sin y}{\cos y}}{\dfrac{\sin x}{\cos x} + \dfrac{\sin y}{\cos y}} = \dfrac{\tan x - \tan y}{\tan x + \tan y}$

111. Start with the right side.

$$\frac{\sin 2x}{\sin 8x} = \frac{\sin(5x-3x)}{\sin(5x+3x)} = \frac{\sin 5x \cos 3x - \cos 5x \sin 3x}{\sin 5x \cos 3x + \cos 5x \sin 3x} \cdot \frac{\frac{1}{\cos 5x \cos 3x}}{\frac{1}{\cos 5x \cos 3x}} = \frac{\frac{\sin 5x}{\cos 5x} - \frac{\sin 3x}{\cos x}}{\frac{\sin 5x}{\cos 5x} + \frac{\sin 3x}{\cos 3x}} = \frac{\tan 5x - \tan 3x}{\tan 5x + \tan 3x}$$

113. Start with the right side.

$$\frac{\cos 3x}{\cos 5x} = \frac{\cos(4x-x)}{\cos(4x+x)} = \frac{\cos 4x \cos x + \sin 4x \sin x}{\cos 4x \cos x - \sin 4x \sin x} \cdot \frac{\frac{1}{\sin 4x \sin x}}{\frac{1}{\sin 4x \sin x}} = \frac{\frac{\cos 4x \cos x}{\sin 4x \sin x} + 1}{\frac{\cos 4x \cos x}{\sin 4x \sin x} - 1} = \frac{\cot 4x \cot x + 1}{\cot 4x \cot x + 1}$$

115. $\cos^2 15° = \cos^2(45° - 30°) = (\cos 45° \cos 30° + \sin 45° \sin 30°)^2$

$$= \left(\left(\frac{\sqrt{2}}{2}\right)\left(\frac{\sqrt{3}}{2}\right) + \left(\frac{\sqrt{2}}{2}\right)\left(\frac{1}{2}\right)\right)^2 = \left(\frac{\sqrt{6}+\sqrt{2}}{4}\right)^2 = \frac{2+\sqrt{3}}{4}$$

$\cos^2 75° = \cos^2(45° + 30°) = (\cos 45° \cos 30° - \sin 45° \sin 30°)^2$

$$= \left(\left(\frac{\sqrt{2}}{2}\right)\left(\frac{\sqrt{3}}{2}\right) - \left(\frac{\sqrt{2}}{2}\right)\left(\frac{1}{2}\right)\right)^2 = \left(\frac{\sqrt{6}-\sqrt{2}}{4}\right)^2 = \frac{2-\sqrt{3}}{4}$$

$$\cos^2 15° - \cos^2 30° + \cos^2 45° - \cos^2 60° + \cos^2 75° = \frac{2+\sqrt{3}}{4} - \left(\frac{\sqrt{3}}{2}\right)^2 + \left(\frac{\sqrt{2}}{2}\right)^2 - \left(\frac{1}{2}\right)^2 + \frac{2-\sqrt{3}}{4} = \frac{1}{2}$$

117. $\cos 60° + \cos 80° + \cos 100° = \cos 60° + \cos(90° - 10°) + \cos(90° + 10°)$

$$= \cos 60° + (\cos 90° \cos 10° + \sin 90° \sin 10°) + (\cos 90° \cos 10° - \sin 90° \sin 10°) = \frac{1}{2}$$

118. $\sin 30° - \sin 70° + \sin 110° = \sin 30° - \sin(90° - 20°) + \sin(90° + 20°)$

$$= \sin 30° - (\sin 90° \cos 20° - \cos 90° 20°) + (\sin 90° \cos 20° + \cos 90° 20°) = \frac{1}{2}$$

119. $\sin(\theta + n\pi) = \sin\theta \cos n\pi + \sin n\pi \cos\theta = \sin\theta \cos n\pi$ (because $\sin n\pi = 0$)

$\cos n\pi = 1$ for n even; $\cos n\pi = -1$ for n odd. So, $\sin\theta \cos n\pi = (-1)^n \sin\theta$

Thus, $\sin(\theta + n\pi) = (-1)^n \sin\theta$.

121. Pair the factors as follows:

$\tan 1° \tan 2° \tan 3° \cdots \tan 87° \tan 88° \tan 89° = (\tan 1° \tan 89°)(\tan 2° \tan 88°)(\tan 3° \tan 87°) \cdots \tan 89° \tan 1°$

$$= \tan(90° - 89°) \tan 89° = \cot 89° \tan 89° = \frac{1}{\tan 89°} \cdot \tan 89° = 1$$

Similarly, we can show that $(\tan 2° \tan 88°) = 1$, etc. Therefore, the product of the tangents is 1.

6.2 Critical Thinking/Discussion/Writing

123. Let $u = \sin^{-1} x$ and $v = \sin^{-1} y$. Then $\sin(u+v) = \sin u \cos v + \sin v \cos u$.

$\cos u = \sqrt{1-x^2}$ and $\cos v = \sqrt{1-y^2}$, so $\sin(u+v) = x\sqrt{1-y^2} + y\sqrt{1-x^2} \Rightarrow$

$\sin^{-1}(\sin(u+v)) = \sin^{-1}\left(x\sqrt{1-y^2} + y\sqrt{1-x^2}\right) \Rightarrow u+v = \sin^{-1}\left(x\sqrt{1-y^2} + y\sqrt{1-x^2}\right) \Rightarrow$

$\sin^{-1} x + \sin^{-1} y = \sin^{-1}\left(x\sqrt{1-y^2} + y\sqrt{1-x^2}\right)$.

Similarly, we can show that $\sin^{-1} x - \sin^{-1} y = \sin^{-1}\left(x\sqrt{1-y^2} - y\sqrt{1-x^2}\right)$.

Section 6.2 Sum and Difference Formulas   343

124. Let $u = \cos^{-1} x$ and let $v = \cos^{-1} y$. Then $\cos(u+v) = \cos u \cos v - \sin u \sin v$.
$\sin u = \sqrt{1-x^2}$ and $\sin v = \sqrt{1-y^2}$, so $\cos(u+v) = \cos u \cos v - \sqrt{1-x^2}\sqrt{1-y^2} \Rightarrow$
$\cos^{-1}(\cos(u+v)) = \cos^{-1}\left(\cos u \cos v - \sqrt{1-x^2}\sqrt{1-y^2}\right) \Rightarrow u+v = \cos^{-1}\left(\cos u \cos v - \sqrt{1-x^2}\sqrt{1-y^2}\right) \Rightarrow$
$\cos^{-1} x + \cos^{-1} y = \cos^{-1}\left(\cos u \cos v - \sqrt{1-x^2}\sqrt{1-y^2}\right) = \cos^{-1}\left(\cos(\cos^{-1} x)\cos(\cos^{-1} y) - \sqrt{1-x^2}\sqrt{1-y^2}\right)$
$= \cos^{-1}\left(xy - \sqrt{1-x^2}\sqrt{1-y^2}\right)$

125. Let $u = \tan^{-1} x$ and let $v = \tan^{-1} y$, so $\tan u = x$ and $\tan v = y$.
$\tan(u+v) = \dfrac{\tan u + \tan v}{1 - \tan u \tan v} = \dfrac{x+y}{1-xy} \Rightarrow \tan^{-1}(\tan(u+v)) = \tan^{-1}\left(\dfrac{x+y}{1-xy}\right) \Rightarrow u+v = \tan^{-1}\left(\dfrac{x+y}{1-xy}\right)$

If x and y are positive numbers in the domain of $\tan^{-1}\theta$ and $xy < 1$, then $\dfrac{x+y}{1-xy}$ is positive and therefore

$u + v = \tan^{-1}\left(\dfrac{x+y}{1-xy}\right)$.

126. Let $u = \tan^{-1} x$ and let $v = \tan^{-1} y$, so $\tan u = x$ and $\tan v = y$.
$\tan(u+v) = \dfrac{\tan u + \tan v}{1 - \tan u \tan v} = \dfrac{x+y}{1-xy} \Rightarrow \tan^{-1}(\tan(u+v)) = \tan^{-1}\left(\dfrac{x+y}{1-xy}\right) \Rightarrow u+v = \tan^{-1}\left(\dfrac{x+y}{1-xy}\right)$

If x and y are positive numbers in the domain of $\tan^{-1}\theta$ and $xy > 1$, then $\dfrac{x+y}{1-xy}$ is negative and therefore

$u + v = \pi + \tan^{-1}\left(\dfrac{x+y}{1-xy}\right)$.

127. $\tan\left(\dfrac{\pi}{2} - x\right) = \dfrac{\tan\dfrac{\pi}{2} - \tan x}{1 + \tan\dfrac{\pi}{2}\tan x}$.

However, $\tan\dfrac{\pi}{2}$ is undefined, so the fraction is also undefined.

128. a. $\tan(A+B+C) = \tan((A+B)+C) = \dfrac{\tan(A+B) + \tan C}{1 - \tan(A+B)\tan C} = \dfrac{\dfrac{\tan A + \tan B}{1 - \tan A \tan B} + \tan C}{1 - \dfrac{\tan A + \tan B}{1 - \tan A \tan B}\tan C}$

$= \dfrac{\dfrac{\tan A + \tan B + \tan C(1 - \tan A \tan B)}{1 - \tan A \tan B}}{\dfrac{1 - \tan A \tan B - (\tan A + \tan B)\tan C}{1 - \tan A \tan B}} = \dfrac{\tan A + \tan B + \tan C - \tan A \tan B \tan C}{1 - \tan A \tan B - \tan A \tan C - \tan B \tan C}$

b. Let $A = \tan^{-1} 1$, $B = \tan^{-1} 2$, $C = \tan^{-1} 3$. Then $\tan A = 1$, $\tan B = 2$, $\tan C = 3$.
$\tan\left(\tan^{-1} 1 + \tan^{-1} 2 + \tan^{-1} 3\right) = \tan(A+B+C) = \dfrac{1+2+3-1(2)(3)}{1-1(2)-1(3)-2(3)} = 0$

c. If $\tan\left(\tan^{-1} 1 + \tan^{-1} 2 + \tan^{-1} 3\right) = 0$, then $\tan^{-1}\left(\tan\left(\tan^{-1} 1 + \tan^{-1} 2 + \tan^{-1} 3\right)\right) = \tan^{-1} 0 \Rightarrow$
$\tan^{-1} 1 + \tan^{-1} 2 + \tan^{-1} 3 = \pi$.

6.2 Getting Ready for the Next Section

129. False. $\cos 120° = -\dfrac{1}{2}$ while
$2\cos 60° = 2\left(\dfrac{1}{2}\right) = 1.$

131. i. $y = \cos^2 x - \sin^2 x = \cos^2 x - (1 - \cos^2 x)$
$= 2\cos^2 x - 1$

ii. $y = 2\cos^2 x - 1 \Rightarrow 2\cos^2 x = 1 + y \Rightarrow$
$\cos^2 x = \dfrac{1+y}{2}$

133. Using the results from exercises 131 and 132, we
have $\tan x = \dfrac{\sin x}{\cos x} = \pm \dfrac{\sqrt{\dfrac{1-y}{2}}}{\sqrt{\dfrac{1+y}{2}}} = \pm\sqrt{\dfrac{1-y}{1+y}}$

6.3 Double-Angle and Half-Angle Formulas

6.3 Practice Problems

1. $\sin x = \dfrac{12}{13}, \dfrac{\pi}{2} < x < \pi \Rightarrow x$ in Quadrant II $\Rightarrow$
$\cos x = -\dfrac{5}{13}, \tan x = -\dfrac{12}{5}$

a. $\sin 2x = 2\sin x\cos x = 2\left(\dfrac{12}{13}\right)\left(-\dfrac{5}{13}\right) = -\dfrac{120}{169}$

b. $\cos 2x = \cos^2 x - \sin^2 x = \left(-\dfrac{5}{13}\right)^2 - \left(\dfrac{12}{13}\right)^2$
$= -\dfrac{119}{169}$

c. $\tan 2x = \dfrac{2\tan x}{1-\tan^2 x} = \dfrac{2\left(-\dfrac{12}{5}\right)}{1-\left(-\dfrac{12}{5}\right)^2} = \dfrac{-\dfrac{24}{5}}{-\dfrac{119}{25}}$
$= \dfrac{120}{119}$

2. a. $2\cos^2\left(\dfrac{\pi}{12}\right) - 1 = \cos\left(2\cdot\dfrac{\pi}{12}\right) = \cos\dfrac{\pi}{6} = \dfrac{\sqrt{3}}{2}$

b. $\cos^2 22.5° - \sin^2 22.5° = \cos(2\cdot 22.5°)$
$= \cos 45° = \dfrac{\sqrt{2}}{2}$

3. a. $\cos 3x = \cos(2x + x)$
$= \cos 2x\cos x - \sin 2x\sin x$
$= (2\cos^2 x - 1)\cos x$
$\quad - (2\sin x\cos x)\sin x$
$= 2\cos^3 x - \cos x - 2\sin^2 x\cos x$
$= 2\cos^3 x - \cos x - 2\cos x(1 - \cos^2 x)$
$= 2\cos^3 x - \cos x - 2\cos x + 2\cos^3 x$
$= 4\cos^3 x - 3\cos x$

b. $\tan 3x = \tan(2x + x) = \dfrac{\tan 2x + \tan x}{1 - \tan 2x\tan x}$
$= \dfrac{\dfrac{2\tan x}{1-\tan^2 x} + \tan x}{1 - \left(\dfrac{2\tan x}{1-\tan^2 x}\right)\tan x}$
$= \dfrac{\dfrac{2\tan x + \tan x - \tan^3 x}{1-\tan^2 x}}{1 - \dfrac{2\tan^2 x}{1-\tan^2 x}}$
$= \dfrac{\dfrac{2\tan x + \tan x - \tan^3 x}{1-\tan^2 x}}{\dfrac{1-\tan^2 x - 2\tan^2 x}{1-\tan^2 x}}$
$= \dfrac{3\tan x - \tan^3 x}{1 - 3\tan^2 x}$

4. $\sin^4 x = (\sin^2 x)^2 = \left(\dfrac{1-\cos 2x}{2}\right)^2$
$= \dfrac{1}{4}(1 - 2\cos 2x + \cos^2 2x)$
$= \dfrac{1}{4}\left(1 - 2\cos 2x + \dfrac{1+\cos 4x}{2}\right)$
$= \dfrac{1}{4} - \dfrac{1}{2}\cos 2x + \dfrac{1}{8} + \dfrac{1}{8}\cos 4x$
$= \dfrac{3}{8} - \dfrac{1}{2}\cos 2x + \dfrac{1}{8}\cos 4x$

5. $P = VI = 170\sin(120\pi t)\cdot 0.83\sin(120\pi t)$
$= 141.1\sin^2(120\pi t)$

The maximum value of $\sin^2(120\pi t)$ is 1, so
$P_{\max} = 141.1$ watts $\Rightarrow$
wattage rating $= \dfrac{141.1}{\sqrt{2}} \approx 99.8$ watts

6. $\sin(112.5°) = \sin\left(\dfrac{225°}{2}\right) = \sqrt{\dfrac{1-\cos(225°)}{2}}$
$= \sqrt{\dfrac{1-(-\sqrt{2}/2)}{2}} = \dfrac{\sqrt{2+\sqrt{2}}}{2}$

7. $\cos\dfrac{\theta}{2} = \sqrt{\dfrac{1+\cos\theta}{2}}$

   From Example 7, $\cos\theta = -\dfrac{12}{13}$, and $\dfrac{\theta}{2}$ lies in quadrant II. Therefore, $\cos\dfrac{\theta}{2}$ is negative.

   $\cos\dfrac{\theta}{2} = -\sqrt{\dfrac{1+\left(-\dfrac{12}{13}\right)}{2}} = -\dfrac{\sqrt{26}}{26}$

6.3 Concepts and Vocabulary

1. The double-angle formula for $\sin 2x = \underline{2\sin x \cos x}$.

3. The formula for $\cos 2x$ in terms of $\cos^2 x$ is $\cos 2x = \underline{2\cos^2 x - 1}$. Solve this formula for $\cos^2 x$ to obtain the power-reducing formula $\cos^2 x = \dfrac{1+\cos 2x}{2}$.

5. False. $\dfrac{1-\cos 2x}{1+\cos 2x} = \dfrac{1-(1-2\sin^2 x)}{1+(2\cos^2 x - 1)} = \dfrac{2\sin^2 x}{2\cos^2 x}$
   $= \tan^2 x \ne \tan 2x$

7. True

9. $\sin\theta = \dfrac{3}{5}, \theta$ in Quadrant II $\Rightarrow \cos\theta = -\dfrac{4}{5}$,
   $\tan\theta = -\dfrac{3}{4}$

   a. $\sin 2\theta = 2\sin\theta\cos\theta = 2\left(\dfrac{3}{5}\right)\left(-\dfrac{4}{5}\right) = -\dfrac{24}{25}$

   b. $\cos 2\theta = \cos^2\theta - \sin^2\theta$
   $= \left(-\dfrac{4}{5}\right)^2 - \left(\dfrac{3}{5}\right)^2 = \dfrac{7}{25}$

   c. $\tan 2\theta = \dfrac{2\tan\theta}{1-\tan^2\theta} = \dfrac{2(-3/4)}{1-(-3/4)^2} = -\dfrac{24}{7}$

11. $\tan\theta = 4, \sin\theta < 0 \Rightarrow \theta$ is in Quadrant III $\Rightarrow$
    $\sin\theta = -\dfrac{4}{\sqrt{17}}, \cos\theta = -\dfrac{1}{\sqrt{17}}$

    a. $\sin 2\theta = 2\sin\theta\cos\theta$
    $= 2\left(-\dfrac{4}{\sqrt{17}}\right)\left(-\dfrac{1}{\sqrt{17}}\right) = \dfrac{8}{17}$

    b. $\cos 2\theta = \cos^2\theta - \sin^2\theta$
    $= \left(-\dfrac{1}{\sqrt{17}}\right)^2 - \left(-\dfrac{4}{\sqrt{17}}\right)^2 = -\dfrac{15}{17}$

    c. $\tan 2\theta = \dfrac{2\tan\theta}{1-\tan^2\theta} = \dfrac{2(4)}{1-(4)^2} = -\dfrac{8}{15}$

13. $\tan\theta = -2, \dfrac{\pi}{2} < \theta < \pi \Rightarrow$
    θ is in Quadrant II $\Rightarrow \sin\theta = \dfrac{2}{\sqrt{5}}, \cos\theta = -\dfrac{1}{\sqrt{5}}$

    a. $\sin 2\theta = 2\sin\theta\cos\theta$
    $= 2\left(\dfrac{2}{\sqrt{5}}\right)\left(-\dfrac{1}{\sqrt{5}}\right) = -\dfrac{4}{5}$

    b. $\cos 2\theta = \cos^2\theta - \sin^2\theta$
    $= \left(-\dfrac{1}{\sqrt{5}}\right)^2 - \left(\dfrac{2}{\sqrt{5}}\right)^2 = -\dfrac{3}{5}$

    c. $\tan 2\theta = \dfrac{2\tan\theta}{1-\tan^2\theta} = \dfrac{2(-2)}{1-(-2)^2} = \dfrac{4}{3}$

15. $1 - 2\sin^2 75° = \cos(2\cdot 75°) = \cos 150° = -\dfrac{\sqrt{3}}{2}$

17. $2\cos^2 105° - 1 = \cos 210° = -\dfrac{\sqrt{3}}{2}$

19. $\dfrac{2\tan 165°}{1-\tan^2 165°} = \tan(2\cdot 165°) = \tan 330° = -\dfrac{\sqrt{3}}{3}$

21. $1 - 2\sin^2\dfrac{\pi}{8} = \cos\dfrac{\pi}{4} = \dfrac{\sqrt{2}}{2}$

23. $\dfrac{2\tan\left(-\dfrac{5\pi}{12}\right)}{1-\tan^2\left(-\dfrac{5\pi}{12}\right)} = \tan\left(-\dfrac{5\pi}{6}\right) = \tan\dfrac{\pi}{6} = \dfrac{\sqrt{3}}{3}$

25. $\sin 4\theta = \sin[2(2\theta)] = 2\sin 2\theta\cos 2\theta$
    $= 2(2\sin\theta\cos\theta)(1-2\sin^2\theta)$
    $= \cos\theta(4\sin\theta - 8\sin^3\theta)$

27. $\cos^4 x - \sin^4 x$
    $= (\cos^2 x - \sin^2 x)(\cos^2 x + \sin^2 x)$
    $= \cos^2 x - \sin^2 x = \cos 2x$

29. $(\sin x - \cos x)^2 = \sin^2 x - 2\sin x \cos x + \cos^2 x = 1 - 2\sin x \cos x = 1 - \sin 2x$

31. $\sin 4x = \sin 2(2x) = 2\sin 2x \cos 2x = 2(2\sin x \cos x)\cos 2x = 4\sin x \cos x \cos 2x$

33. $\dfrac{\sin 3x}{\sin x} - \dfrac{\cos 3x}{\cos x} = \dfrac{3\sin x - 4\sin^3 x}{\sin x} - \dfrac{4\cos^3 x - 3\cos x}{\cos x} = \dfrac{\sin x(3 - 4\sin^2 x)}{\sin x} - \dfrac{\cos x(4\cos^2 x - 3)}{\cos x}$
$= (3 - 4\sin^2 x) - (4\cos^2 x - 3) = 6 - 4(\sin^2 x + \cos^2 x) = 6 - 4 = 2$

35. $\dfrac{1 - \cos 2x}{\sin 2x} = \dfrac{1 - (1 - 2\sin^2 x)}{2\sin x \cos x}$
$= \dfrac{2\sin^2 x}{2\sin x \cos x} = \dfrac{\sin x}{\cos x} = \tan x$

37. Start with the right side.
$\dfrac{2\tan x}{1 + \tan^2 x} = \dfrac{2\tan x}{\sec^2 x} = \dfrac{\dfrac{2\sin x}{\cos x}}{\dfrac{1}{\cos^2 x}}$
$= 2\sin x \cos x = \sin 2x$

39. Start with the right side.
$\dfrac{\cos x + \sin x}{\cos x - \sin x} = \dfrac{\cos x + \sin x}{\cos x - \sin x} \cdot \dfrac{\cos x + \sin x}{\cos x + \sin x}$
$= \dfrac{\cos^2 x + 2\sin x \cos x + \sin^2 x}{\cos^2 x - \sin^2 x}$
$= \dfrac{1 + 2\sin x \cos x}{\cos 2x} = \dfrac{1 + \sin 2x}{\cos 2x}$

41. $4\sin^2 x \cos^2 x = 4\left(\dfrac{1 - \cos 2x}{2}\right)\left(\dfrac{1 + \cos 2x}{2}\right)$
$= 1 - \cos^2 2x = 1 - \dfrac{1 + \cos 4x}{2}$
$= \dfrac{1 - \cos 4x}{2}$

43. $4\sin x \cos x(1 - 2\sin^2 x) = 4\sin x \cos x(\cos 2x)$
$= 2\sin 2x \cos 2x$
$= \sin 4x$

45. $2\sin 3x \cos 3x(2\cos^2 3x - 1) = \sin 6x \cos 6x$
$= \dfrac{\sin 12x}{2}$

47. $\sin\dfrac{x}{2}\cos\dfrac{x}{2}\left(1 - 2\sin^2\dfrac{x}{2}\right)$
$= \sin\dfrac{x}{2}\cos\dfrac{x}{2}\cos x$
$= \sqrt{\dfrac{1 - \cos x}{2}}\sqrt{\dfrac{1 + \cos x}{2}}\cos x$
$= \dfrac{\sqrt{1 - \cos^2 x}}{2}\cos x = \dfrac{\sin x \cos x}{2} = \dfrac{\sin 2x}{4}$

49. $8\sin^4\dfrac{x}{2} = 8\left(\sin^2\dfrac{x}{2}\right)^2 = 8\left(\dfrac{1 - \cos x}{2}\right)^2$
$= 8\left(\dfrac{1 - 2\cos x + \cos^2 x}{4}\right)$
$= 2 - 4\cos x + 2\cos^2 x$
$= 2 - 4\cos x + 2\left(\dfrac{1 + \cos 2x}{2}\right)$
$= \cos 2x - 4\cos x + 3$

51. $\sin\left(\dfrac{\pi}{12}\right) = \sin\left(\dfrac{\pi/6}{2}\right) = \sqrt{\dfrac{1 - \cos(\pi/6)}{2}}$
$= \sqrt{\dfrac{1 - \sqrt{3}/2}{2}} = \dfrac{\sqrt{2 - \sqrt{3}}}{2}$

53. $\cos\dfrac{\pi}{8} = \cos\left(\dfrac{\pi/4}{2}\right) = \sqrt{\dfrac{1 + \cos(\pi/4)}{2}}$
$= \sqrt{\dfrac{1 + \sqrt{2}/2}{2}} = \dfrac{\sqrt{2 + \sqrt{2}}}{2}$

55. $\sin\left(-\dfrac{3\pi}{8}\right) = -\sin\left(\dfrac{3\pi/4}{2}\right) = -\sqrt{\dfrac{1 - \cos(3\pi/4)}{2}}$
$= -\sqrt{\dfrac{1 - (-\sqrt{2}/2)}{2}} = -\dfrac{\sqrt{2 + \sqrt{2}}}{2}$

57. $\tan\left(\dfrac{7\pi}{8}\right) = \tan\left(\dfrac{7\pi/4}{2}\right) = -\sqrt{\dfrac{1 - \cos(7\pi/4)}{1 + \cos(7\pi/4)}}$
$= -\sqrt{\dfrac{1 - \sqrt{2}/2}{1 + \sqrt{2}/2}} = -\sqrt{\dfrac{2 - \sqrt{2}}{2 + \sqrt{2}}}$

59. $\tan 112.5° = \tan\left(\dfrac{225°}{2}\right) = -\sqrt{\dfrac{1 - \cos 225°}{1 + \cos 225°}}$
$= -\sqrt{\dfrac{1 - (-\sqrt{2}/2)}{1 + (-\sqrt{2}/2)}} = -\sqrt{\dfrac{2 + \sqrt{2}}{2 - \sqrt{2}}}$
$= -1 - \sqrt{2}$

61. $\sin(-75°) = -\sin\left(\dfrac{150°}{2}\right) = -\sqrt{\dfrac{1-\cos(150°)}{2}}$
$= -\sqrt{\dfrac{1-\left(-\sqrt{3}/2\right)}{2}} = -\dfrac{\sqrt{2+\sqrt{3}}}{2}$

63. $\sin\theta = \dfrac{4}{5}, \dfrac{\pi}{2} < \theta < \pi \Rightarrow \cos\theta = -\dfrac{3}{5}$
$\dfrac{\pi}{4} < \dfrac{\theta}{2} < \dfrac{\pi}{2} \Rightarrow \sin\dfrac{\theta}{2} > 0, \cos\dfrac{\theta}{2} > 0, \tan\dfrac{\theta}{2} > 0$

   a. $\sin\dfrac{\theta}{2} = \sqrt{\dfrac{1-(-3/5)}{2}} = \dfrac{2\sqrt{5}}{5}$

   b. $\cos\dfrac{\theta}{2} = \sqrt{\dfrac{1+(-3/5)}{2}} = \dfrac{\sqrt{5}}{5}$

   c. $\tan\dfrac{\theta}{2} = \sqrt{\dfrac{1-(-3/5)}{1+(-3/5)}} = 2$

65. $\tan\theta = -\dfrac{2}{3}, \dfrac{\pi}{2} < \theta < \pi \Rightarrow \cos\theta = -\dfrac{3\sqrt{13}}{13}$
$\dfrac{\pi}{4} < \dfrac{\theta}{2} < \dfrac{\pi}{2} \Rightarrow \sin\dfrac{\theta}{2} > 0, \cos\dfrac{\theta}{2} > 0, \tan\dfrac{\theta}{2} > 0$

   a. $\sin\dfrac{\theta}{2} = \sqrt{\dfrac{1-\left(-3\sqrt{13}/13\right)}{2}} = \sqrt{\dfrac{13+3\sqrt{13}}{26}}$

   b. $\cos\dfrac{\theta}{2} = \sqrt{\dfrac{1+\left(-3\sqrt{13}/13\right)}{2}} = \sqrt{\dfrac{13-3\sqrt{13}}{26}}$

   c. $\tan\dfrac{\theta}{2} = -\sqrt{\dfrac{1-\left(-3\sqrt{13}/13\right)}{1+\left(-3\sqrt{13}/13\right)}} = \sqrt{\dfrac{13+3\sqrt{13}}{13-3\sqrt{13}}}$

67. $\sin\theta = \dfrac{1}{5}, \cos\theta < 0 \Rightarrow \cos\theta = -\dfrac{2\sqrt{6}}{5}$ and
$\dfrac{\pi}{2} < \theta < \pi \Rightarrow \dfrac{\pi}{4} < \dfrac{\theta}{2} < \dfrac{\pi}{2} \Rightarrow \sin\dfrac{\theta}{2} > 0,$
$\cos\dfrac{\theta}{2} > 0, \tan\dfrac{\theta}{2} > 0$

   a. $\sin\dfrac{\theta}{2} = \sqrt{\dfrac{1-\left(-2\sqrt{6}/5\right)}{2}} = \sqrt{\dfrac{5+2\sqrt{6}}{10}}$

   b. $\cos\dfrac{\theta}{2} = \sqrt{\dfrac{1+\left(-2\sqrt{6}/5\right)}{2}} = \sqrt{\dfrac{5-2\sqrt{6}}{10}}$

   c. $\tan\dfrac{\theta}{2} = \sqrt{\dfrac{1-\left(-2\sqrt{6}/5\right)}{1+\left(-2\sqrt{6}/5\right)}} = \sqrt{\dfrac{5+2\sqrt{6}}{5-2\sqrt{6}}}$

69. $\sec\theta = \sqrt{5} \Rightarrow$
$\cos\theta = \dfrac{\sqrt{5}}{5}, \sin\theta > 0 \Rightarrow 0 < \theta < \dfrac{\pi}{2}$
$0 < \dfrac{\theta}{2} < \dfrac{\pi}{4} \Rightarrow \sin\dfrac{\theta}{2} > 0, \cos\dfrac{\theta}{2} > 0, \tan\dfrac{\theta}{2} > 0$

   a. $\sin\dfrac{\theta}{2} = \sqrt{\dfrac{1-\left(\sqrt{5}/5\right)}{2}} = \sqrt{\dfrac{5-\sqrt{5}}{10}}$

   b. $\cos\dfrac{\theta}{2} = \sqrt{\dfrac{1+\left(\sqrt{5}/5\right)}{2}} = \sqrt{\dfrac{5+\sqrt{5}}{10}}$

   c. $\tan\dfrac{\theta}{2} = \sqrt{\dfrac{1-\left(\sqrt{5}/5\right)}{1+\left(\sqrt{5}/5\right)}} = \sqrt{\dfrac{5-\sqrt{5}}{5+\sqrt{5}}}$

71. $\left(\sin\dfrac{t}{2} + \cos\dfrac{t}{2}\right)^2$
$= \sin^2\dfrac{t}{2} + 2\sin\dfrac{t}{2}\cos\dfrac{t}{2} + \cos^2\dfrac{t}{2} = 1 + \sin t$

73. $2\cos^2\dfrac{x}{2} = 2\left(\dfrac{1+\cos 2\left(\dfrac{x}{2}\right)}{2}\right)$
$= (1+\cos x)\cdot\dfrac{1-\cos x}{1-\cos x}$
$= \dfrac{1-\cos^2 x}{1-\cos x} = \dfrac{\sin^2 x}{1-\cos x}$

75. $\tan\dfrac{x}{2} = \sqrt{\dfrac{1-\cos x}{1+\cos x}} = \dfrac{\sqrt{1-\cos x}}{\sqrt{1+\cos x}}\cdot\dfrac{\sqrt{1+\cos x}}{\sqrt{1+\cos x}}$
$= \dfrac{\sqrt{1-\cos^2 x}}{1+\cos x} = \dfrac{\sin x}{1+\cos x}$

77. $\sin^2\dfrac{x}{2} + \cos x = \dfrac{1-\cos 2\left(\dfrac{x}{2}\right)}{2} + \cos x$
$= \dfrac{1-\cos x}{2} + \cos x = \dfrac{1+\cos x}{2}$
$= \dfrac{1+\cos 2\left(\dfrac{x}{2}\right)}{2} = \cos^2\dfrac{x}{2}$

79. $\dfrac{2\tan\dfrac{x}{2}}{1+\tan^2\dfrac{x}{2}} = \dfrac{\dfrac{2\sin(x/2)}{\cos(x/2)}}{\sec^2\dfrac{x}{2}} = \dfrac{2\sin\dfrac{x}{2}}{\cos\dfrac{x}{2}}\cdot\cos^2\dfrac{x}{2} = 2\sin\left(\dfrac{x}{2}\right)\cdot\cos\left(\dfrac{x}{2}\right) = \sin x$

6.3 Applying the Concepts

81. $P = VI = 170\sin(120\pi t)\cdot 0.832\sin(120\pi t) = 141.44\sin^2(120\pi t)$

The maximum value of $\sin^2(120\pi t)$ is 1, so $P_{\max} = 141.44$ watts $\Rightarrow$ wattage rating $= \dfrac{141.44}{\sqrt{2}} \approx 100$ watts

83. $P = VI = 170\sin(120\pi t)\cdot 9.983\sin(120\pi t) = 1697.11\sin^2(120\pi t)$

The maximum value of $\sin^2(120\pi t)$ is 1, so $P_{\max} = 1697.11$ watts $\Rightarrow$ wattage rating $= \dfrac{1697.11}{\sqrt{2}} \approx 1200$ watts

85. $P = VI = 170\sin(120\pi t)\cdot 4.991\sin(120\pi t) = 848.47\sin^2(120\pi t)$

The maximum value of $\sin^2(120\pi t)$ is 1, so $P_{\max} = 848.47$ watts $\Rightarrow$ wattage rating $= \dfrac{848.47}{\sqrt{2}} \approx 600$ watts

87. $x = \dfrac{v_0^2}{16}\sin\theta\cos\theta$. Let $v_0 = \sqrt{32}$. Then $x = \left(\dfrac{(\sqrt{32})^2}{16}\right)\sin\theta\cos\theta = 2\sin\theta\cos\theta = \sin 2\theta$.

$\sin 2\theta$ is at its maximum, 1, when $2\theta = 90°$, so $\theta = 45°$.

   b. The maximum of $R^2\sin 2\theta$ occurs when $\sin 2\theta = 1$, or when $2\theta = 90° \Rightarrow \theta = 45°$.

89. $y = \sin^2 x = \dfrac{1-\cos 2x}{2} = \dfrac{1}{2} - \dfrac{1}{2}\cos 2x \Rightarrow a = -\dfrac{1}{2},\ b = 2$

Period $= \dfrac{2\pi}{2} = \pi$

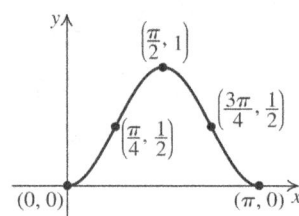

6.3 Beyond the Basics

91. $\tan 3x = \tan(x+2x) = \dfrac{\tan x + \tan 2x}{1-\tan x\tan 2x} = \dfrac{\tan x + \dfrac{2\tan x}{1-\tan^2 x}}{1-\tan x\left(\dfrac{2\tan x}{1-\tan^2 x}\right)} = \dfrac{\dfrac{\tan x - \tan^3 x + 2\tan x}{1-\tan^2 x}}{\dfrac{1-\tan^2 x - 2\tan^2 x}{1-\tan^2 x}} = \dfrac{3\tan x - \tan^3 x}{1-3\tan^2 x}$

93. $\dfrac{\sin x - \cos x}{\sin x + \cos x} - \dfrac{\sin x + \cos x}{\sin x - \cos x} = \dfrac{\sin^2 x - 2\sin x\cos x + \cos^2 x - (\sin^2 x + 2\sin x\cos x + \cos^2 x)}{\sin^2 x - \cos^2 x}$

$= \dfrac{-2\sin 2x}{\sin^2 x - \cos^2 x} = \dfrac{2\sin 2x}{\cos^2 x - \sin^2 x} = \dfrac{2\sin 2x}{\cos 2x} = 2\tan 2x$

95. $\tan 3x - \tan x = \dfrac{3\tan x - \tan^3 x}{1-3\tan^2 x} - \tan x = \dfrac{3\tan x - \tan^3 x - \tan x + 3\tan^3 x}{1-3\tan^2 x} = \dfrac{2\tan x + 2\tan^3 x}{1-3\tan^2 x}$

$= \dfrac{2\tan x(1+\tan^2 x)}{1-3\tan^2 x} = \dfrac{2\tan x\sec^2 x}{1-3\tan^2 x} = \dfrac{2\tan x}{\cos^2 x(1-3\tan^2 x)} = \dfrac{\dfrac{2\sin x}{\cos x}}{\cos^2 x - 3\sin^2 x}$

$= \dfrac{2\sin x}{\cos^3 x - 3\cos x(1-\cos^2 x)} = \dfrac{2\sin x}{4\cos^3 x - 3\cos x} = \dfrac{2\sin x}{\cos 3x}$

97. Let $u = \dfrac{\pi}{4} - x$.

$$\dfrac{1-\tan^2\left(\dfrac{\pi}{4}-x\right)}{1+\tan^2\left(\dfrac{\pi}{4}-x\right)} \Rightarrow \dfrac{1-\tan^2 u}{1+\tan^2 u} = \dfrac{\dfrac{2\tan u}{\tan 2u}}{\sec^2 u} = \dfrac{\dfrac{2\sin u}{\cos u}}{\dfrac{\sin 2u}{\cos 2u}} \cdot \cos^2 u = \dfrac{2\sin u}{\cos u} \cdot \dfrac{\cos 2u}{\sin 2u} \cdot \cos^2 u = \dfrac{2\sin u \cos u \cos 2u}{\sin 2u}$$

$$= \cos 2u = \cos\left[2\left(\dfrac{\pi}{4}-x\right)\right] = \cos\left(\dfrac{\pi}{2}-2x\right) = \sin 2x$$

99. $\sin^2\dfrac{\pi}{8} + \sin^2\dfrac{3\pi}{8} + \sin^2\dfrac{5\pi}{8} + \sin^2\dfrac{7\pi}{8} = \dfrac{1-\cos\dfrac{\pi}{4}}{2} + \dfrac{1-\cos\dfrac{3\pi}{4}}{2} + \dfrac{1-\cos\dfrac{5\pi}{4}}{2} + \dfrac{1-\cos\dfrac{7\pi}{4}}{2}$

$$= \dfrac{1-\dfrac{\sqrt{2}}{2}}{2} + \dfrac{1+\dfrac{\sqrt{2}}{2}}{2} + \dfrac{1+\dfrac{\sqrt{2}}{2}}{2} + \dfrac{1-\dfrac{\sqrt{2}}{2}}{2} = 2$$

101. $\sqrt{2+\sqrt{2+2\cos 4x}} = \sqrt{2+\sqrt{2+2\cos 2(2x)}} = \sqrt{2+\sqrt{2+2(2\cos^2 2x - 1)}} = \sqrt{2+\sqrt{4\cos^2 2x}} = \sqrt{2+2\cos 2x}$

$$= \sqrt{2+2(2\cos^2 x - 1)} = \sqrt{4\cos^2 x} = 2\cos x$$

103. $\sin\left(2\sin^{-1}\dfrac{1}{2}\right) = \sin\left(2\left(\dfrac{\pi}{6}\right)\right) = \sin\dfrac{\pi}{3} = \dfrac{\sqrt{3}}{2}$

Alternatively, let $u = \sin^{-1}\dfrac{1}{2}$. Then, $\sin u = \dfrac{1}{2} = \dfrac{y}{r} \Rightarrow 2^2 = x^2 + 1^2 \Rightarrow x = \sqrt{3}$, and $\cos u = \dfrac{\sqrt{3}}{2}$.

$\sin\left(2\sin^{-1}\dfrac{1}{2}\right) = \sin 2u = 2\sin u \cos u = 2\left(\dfrac{1}{2}\right)\left(\dfrac{\sqrt{3}}{2}\right) = \dfrac{\sqrt{3}}{2}$

105. $\cos\left(2\tan^{-1}\dfrac{1}{2}\right)$

Let $u = \tan^{-1}\dfrac{1}{2}$. Then $\tan u = \dfrac{1}{2} = \dfrac{y}{x} \Rightarrow r^2 = 2^2 + 1^2 \Rightarrow r = \sqrt{5}$, and $\cos u = \dfrac{2}{\sqrt{5}}$.

$\cos\left(2\tan^{-1}\dfrac{1}{2}\right) = \cos 2u = 2\cos^2 u - 1 = 2\left(\dfrac{2}{\sqrt{5}}\right)^2 - 1 = \dfrac{8}{5} - 1 = \dfrac{3}{5}$

107. $\tan\left(2\sin^{-1}\dfrac{3}{4}\right)$

Let $u = \sin^{-1}\dfrac{3}{4}$. Then, $\sin u = \dfrac{3}{4} = \dfrac{y}{r} \Rightarrow$

$4^2 = x^2 + 3^2 \Rightarrow x = \sqrt{7}$, and $\tan u = \dfrac{3}{\sqrt{7}}$.

$\tan\left(2\sin^{-1}\dfrac{3}{4}\right) = \tan 2u = \dfrac{2\tan u}{1-\tan^2 u}$

$= \dfrac{2\left(\dfrac{3}{\sqrt{7}}\right)}{1-\left(\dfrac{3}{\sqrt{7}}\right)^2} = \dfrac{\dfrac{6\sqrt{7}}{7}}{-\dfrac{2}{7}} = -3\sqrt{7}$

109. $\sin\left[2\cos^{-1}\left(-\dfrac{3}{5}\right)\right]$

Let $u = \cos^{-1}\left(-\dfrac{3}{5}\right)$. Then, u lies in quadrant II and $\cos u = -\dfrac{3}{5} = \dfrac{x}{r} \Rightarrow$

$5^2 = (-3)^2 + y^2 \Rightarrow y = 4$, and $\sin u = \dfrac{4}{5}$.

$\sin\left[2\cos^{-1}\left(-\dfrac{3}{5}\right)\right] = \sin 2u = 2\sin u \cos u$

$= 2\left(\dfrac{4}{5}\right)\left(-\dfrac{3}{5}\right) = -\dfrac{24}{25}$

111. $\cos\left[2\sin^{-1}\left(-\dfrac{1}{2}\right)\right]$

Let $u = \sin^{-1}\left(-\dfrac{1}{2}\right)$. Then u lies in quadrant IV and $\sin u = -\dfrac{1}{2} = \dfrac{y}{r} \Rightarrow 2^2 = x^2 + (-1)^2 \Rightarrow x = \sqrt{3} \Rightarrow \cos u = \dfrac{\sqrt{3}}{2}$.

$\cos\left[2\sin^{-1}\left(-\dfrac{1}{2}\right)\right] = \cos 2u = 2\cos^2 u - 1$

$= 2\left(\dfrac{\sqrt{3}}{2}\right)^2 - 1 = \dfrac{1}{2}$

113. $\tan\left[2\sin^{-1}\left(-\dfrac{2}{3}\right)\right]$

Let $u = \sin^{-1}\left(-\dfrac{2}{3}\right)$. Then u lies in quadrant IV and $\sin u = -\dfrac{2}{3} = \dfrac{y}{r} \Rightarrow 3^2 = x^2 + (-2)^2 \Rightarrow x = \sqrt{5} \Rightarrow \tan u = -\dfrac{2}{\sqrt{5}}$.

$\tan\left[2\sin^{-1}\left(-\dfrac{2}{3}\right)\right] = \tan 2u = \dfrac{2\tan u}{1 - \tan^2 u}$

$= \dfrac{2\left(-\dfrac{2}{\sqrt{5}}\right)}{1 - \left(-\dfrac{2}{\sqrt{5}}\right)^2} = \dfrac{-\dfrac{4\sqrt{5}}{5}}{\dfrac{1}{5}}$

$= -4\sqrt{5}$

115. $\sin\left[\dfrac{1}{2}\sin^{-1}\left(-\dfrac{3}{5}\right)\right]$

Let $u = \sin^{-1}\left(-\dfrac{3}{5}\right)$. Then, u lies in quadrant IV and $\dfrac{u}{2}$ lies in quadrant II.

$\sin u = -\dfrac{3}{5} = \dfrac{y}{r} \Rightarrow 5^2 = x^2 + (-3)^2 \Rightarrow x = 4$, and $\cos u = \dfrac{4}{5}$.

$\sin\left[\dfrac{1}{2}\sin^{-1}\left(-\dfrac{3}{5}\right)\right] = \sin\dfrac{u}{2} = \sqrt{\dfrac{1-\cos u}{2}}$

$= \sqrt{\dfrac{1 - \dfrac{4}{5}}{2}} = \sqrt{\dfrac{1}{10}} = \dfrac{\sqrt{10}}{10}$

117. $\tan\left[\dfrac{1}{2}\cos^{-1}\left(-\dfrac{5}{13}\right)\right]$

Let $u = \cos^{-1}\left(-\dfrac{5}{13}\right)$. Then u lies in quadrant II and $\dfrac{u}{2}$ lies in quadrant I.

$\cos u = -\dfrac{5}{13} = \dfrac{x}{r} \Rightarrow 13^2 = (-5)^2 + y^2 \Rightarrow y = 12$ and $\sin u = \dfrac{12}{13}$.

$\tan\left[\dfrac{1}{2}\cos^{-1}\left(-\dfrac{5}{13}\right)\right] = \tan\dfrac{u}{2} = \dfrac{\sin u}{1 + \cos u}$

$= \dfrac{\dfrac{12}{13}}{1 + \left(-\dfrac{5}{13}\right)} = \dfrac{\dfrac{12}{13}}{\dfrac{8}{13}} = \dfrac{3}{2}$

119. $\sin^2\left[\dfrac{1}{2}\sin^{-1}\left(-\dfrac{2}{3}\right)\right]$

Let $u = \sin^{-1}\left(-\dfrac{2}{3}\right)$. Then u lies in quadrant IV and $\dfrac{u}{2}$ lies in quadrant II.

$\sin u = -\dfrac{2}{3} = \dfrac{y}{r} \Rightarrow 3^2 = x^2 + (-2)^2 \Rightarrow x = \sqrt{5} \Rightarrow \cos u = \dfrac{\sqrt{5}}{3}$.

$\sin^2\left[\dfrac{1}{2}\sin^{-1}\left(-\dfrac{2}{3}\right)\right] = \sin^2\dfrac{u}{2} = \dfrac{1 - \cos 2\left(\dfrac{u}{2}\right)}{2}$

$= \dfrac{1 - \cos u}{2} = \dfrac{1 - \dfrac{\sqrt{5}}{3}}{2}$

$= \dfrac{\dfrac{3 - \sqrt{5}}{3}}{2} = \dfrac{3 - \sqrt{5}}{6}$

121. Let $u = \sin^{-1} x$, so $2\sin^{-1} x = 2u$.

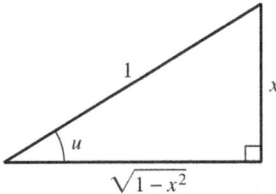

From the diagram, $\cos u = \sqrt{1-x^2}$.

$\sin 2u = 2\sin u \cos u = 2x\sqrt{1-x^2}$

$\sin^{-1}(\sin 2u) = \sin^{-1}\left(2x\sqrt{1-x^2}\right)$

$2u = \sin^{-1}\left(2x\sqrt{1-x^2}\right)$

$2\sin^{-1} x = \sin^{-1}\left(2x\sqrt{1-x^2}\right)$

$\sin^{-1}\theta$ is defined for $[-1, 1]$, so

$2x\sqrt{1-x^2} = \pm 1 \Rightarrow x\sqrt{1-x^2} = \pm\dfrac{1}{2} \Rightarrow$

$x^2(1-x^2) = \dfrac{1}{4} \Rightarrow x^2 - x^4 - \dfrac{1}{4} = 0 \Rightarrow$

$x^4 - x^2 + \dfrac{1}{4} = 0 \Rightarrow \left(x^2 - \dfrac{1}{2}\right)^2 = 0 \Rightarrow$

$x^2 - \dfrac{1}{2} = 0 \Rightarrow x^2 = \dfrac{1}{2} \Rightarrow x = \pm\dfrac{\sqrt{2}}{2}$

Thus, the restriction is $-\dfrac{\sqrt{2}}{2} \le x \le \dfrac{\sqrt{2}}{2}$.

123. Let $u = \tan^{-1} x$, or $x = \tan u$.

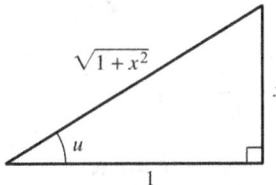

Then, starting with the right side, we have

$\tan^{-1}\left(\dfrac{2x}{1-x^2}\right) = \tan^{-1}\left(\dfrac{2\tan u}{1-\tan^2 u}\right)$

$= \tan^{-1}(\tan 2u)$

$= 2u = 2\tan^{-1} x$

125. a. $\sin\dfrac{\pi}{5} = \sin\left(2 \cdot \dfrac{\pi}{10}\right) = 2\sin\dfrac{\pi}{10}\cos\dfrac{\pi}{10}$

$= 2\left(\dfrac{\sqrt{5}-1}{4}\right)\left(\dfrac{\sqrt{10+2\sqrt{5}}}{4}\right)$

$= \dfrac{(\sqrt{5}-1)\left(\sqrt{10+2\sqrt{5}}\right)}{8}$

b. $\cos\dfrac{\pi}{5} = \cos\left(2 \cdot \dfrac{\pi}{10}\right) = 2\cos^2\dfrac{\pi}{10} - 1$

$= 2\left(\dfrac{\sqrt{10+2\sqrt{5}}}{4}\right)^2 - 1 = \dfrac{\sqrt{5}+1}{4}$

c. $\sin\dfrac{3\pi}{10} = \sin\left(\dfrac{\pi}{10} + \dfrac{\pi}{5}\right)$

$= \sin\dfrac{\pi}{10}\cos\dfrac{\pi}{5} + \sin\dfrac{\pi}{5}\cos\dfrac{\pi}{10}$

(using the results from parts a and b)

$= \left(\dfrac{\sqrt{5}-1}{4}\right)\left(\dfrac{\sqrt{5}+1}{4}\right)$

$+ \left(\dfrac{(\sqrt{5}-1)\left(\sqrt{10+2\sqrt{5}}\right)}{8}\right)\left(\dfrac{\sqrt{10+2\sqrt{5}}}{4}\right)$

$= \dfrac{\sqrt{5}+1}{4}$

d. $\cos\dfrac{7\pi}{10} = \cos\left(\dfrac{\pi}{2} + \dfrac{\pi}{5}\right)$

$= \cos\dfrac{\pi}{2}\cos\dfrac{\pi}{5} - \sin\dfrac{\pi}{2}\sin\dfrac{\pi}{5}$

$= -\sin\dfrac{\pi}{5} = -\dfrac{(\sqrt{5}-1)\left(\sqrt{10+2\sqrt{5}}\right)}{8}$

127. a. $4\sin\dfrac{\theta}{4}\sin\left(\dfrac{\pi}{2} - \dfrac{\theta}{4}\right)\sin\left(\dfrac{\pi}{2} - \dfrac{\theta}{2}\right)$

$= 4\sin\dfrac{\theta}{4}\cos\dfrac{\theta}{4}\cos\dfrac{\theta}{2}$

$= 2\left(2\sin\dfrac{\theta}{4}\cos\dfrac{\theta}{4}\right)\cos\dfrac{\theta}{2}$

$= 2\sin\dfrac{\theta}{2}\cos\dfrac{\theta}{2} = \sin\theta$

b. $\theta = \dfrac{\pi}{2} - \dfrac{\theta}{2} \Rightarrow 2\theta = \pi - \theta \Rightarrow \theta = \dfrac{\pi}{3}$

Using this result and the identity in (a), we have

$\sin\theta = 4\sin\dfrac{\theta}{4}\sin\left(\dfrac{\pi}{2} - \dfrac{\theta}{4}\right)\sin\left(\dfrac{\pi}{2} - \dfrac{\theta}{2}\right)$

$\sin\dfrac{\pi}{3}$

$= 4\sin\dfrac{\pi/3}{4}\sin\left(\dfrac{\pi}{2} - \dfrac{\pi/3}{4}\right)\sin\left(\dfrac{\pi}{2} - \dfrac{\pi/3}{2}\right)$

$\sin\dfrac{\pi}{3} = 4\sin\dfrac{\pi}{12}\sin\dfrac{5\pi}{12}\sin\dfrac{\pi}{3}$

$\sin\dfrac{\pi}{12}\sin\dfrac{5\pi}{12} = \dfrac{1}{4}$

(*continued on next page*)

(*continued*)

An alternative solution is

$$\sin\frac{\pi}{12}\sin\frac{5\pi}{12} = \sin\frac{\pi}{12}\sin\left(\frac{\pi}{2}-\frac{\pi}{12}\right)$$
$$= \sin\frac{\pi}{12}\cos\frac{\pi}{12} = \frac{1}{2}\sin\frac{\pi}{6} = \frac{1}{4}$$

c. $\theta = \frac{\pi}{2} - \frac{\theta}{4} \Rightarrow 4\theta = 2\pi - \theta \Rightarrow \theta = \frac{2\pi}{5}$

Using this result and the identity in (a), we have

$$\sin\theta = 4\sin\frac{\theta}{4}\sin\left(\frac{\pi}{2}-\frac{\theta}{4}\right)\sin\left(\frac{\pi}{2}-\frac{\theta}{2}\right)$$

$$\sin\frac{2\pi}{5}$$
$$= 4\sin\frac{2\pi/5}{4}\sin\left(\frac{\pi}{2}-\frac{2\pi/5}{4}\right)\sin\left(\frac{\pi}{2}-\frac{2\pi/5}{2}\right)$$

$$\sin\frac{2\pi}{5} = 4\sin\frac{\pi}{10}\sin\frac{2\pi}{5}\sin\frac{3\pi}{10}$$

$$\sin\frac{\pi}{10}\sin\frac{3\pi}{10} = \frac{1}{4}$$

6.3 Critical Thinking/Discussion/Writing

128. $\sin 2x = 2\sin x \Rightarrow 2\sin x \cos x = 2\sin x \Rightarrow$
$2\sin x \cos x - 2\sin x = 0 \Rightarrow$
$2\sin x(\cos x - 1) = 0 \Rightarrow$
$\sin x = 0$ or $\cos x - 1 = 0$
If $\sin x = 0$, then $x = 0$ or $x = \pi$. If
$\cos x - 1 = 0$, then $\cos x = 1 \Rightarrow x = 0$.
Substituting the values for x into the original equation gives $\sin(2\pi) = 0 = 2\sin\pi$. So, both $x = 0$ and $x = \pi$ are possible solutions.

129. Exercise 128 does not show that $\sin 2x = 2\sin x$ is an identity, because it is true only for some values of x. An identity is true for all domain values of x.

130. The range of $\cos 2x$ is $[-1, 1]$, so $0 \leq \frac{1-\cos 2x}{2}$.

131. a. $\sin\theta = \frac{4}{5}$ **b.** $\cos\theta = \frac{3}{5}$

c. $\sin 2\theta = 2\sin\theta\cos\theta = 2\left(\frac{4}{5}\right)\left(\frac{3}{5}\right) = \frac{24}{25}$

d. $\cos 2\theta = \cos^2\theta - \sin^2\theta$
$= \left(\frac{3}{5}\right)^2 - \left(\frac{4}{5}\right)^2 = -\frac{7}{25}$

e. $\tan 2\theta = \frac{2\tan\theta}{1-\tan^2\theta} = \frac{2\left(\frac{4}{3}\right)}{1-\left(\frac{4}{3}\right)^2} = \frac{\frac{8}{3}}{-\frac{7}{9}} = -\frac{24}{7}$

f. $\sin\frac{\theta}{2} = \sqrt{\frac{1-\cos\theta}{2}} = \sqrt{\frac{1-\left(\frac{3}{5}\right)}{2}} = \frac{\sqrt{5}}{5}$

Note that the answer is positive because $\frac{\theta}{2}$ is an acute angle.

g. $\cos\frac{\theta}{2} = \sqrt{\frac{1+\cos\theta}{2}} = \sqrt{\frac{1+\left(\frac{3}{5}\right)}{2}} = \frac{2\sqrt{5}}{5}$

Note that the answer is positive because $\frac{\theta}{2}$ is an acute angle.

h. $\tan\frac{\theta}{2} = \frac{\sin\frac{\theta}{2}}{\cos\frac{\theta}{2}} = \frac{\frac{\sqrt{5}}{5}}{\frac{2\sqrt{5}}{5}} = \frac{1}{2}$

132. a. $\cos 30° = \cos\frac{60°}{2} = \sqrt{\frac{1+\cos 60°}{2}} = \sqrt{\frac{1+\frac{1}{2}}{2}}$
$= \sqrt{\frac{3}{4}} = \frac{\sqrt{3}}{2}$

b. $\cos 15° = \cos\frac{30°}{2} = \sqrt{\frac{1+\cos 30°}{2}}$
$= \sqrt{\frac{1+\frac{\sqrt{3}}{2}}{2}} = \sqrt{\frac{2+\sqrt{3}}{4}} = \frac{\sqrt{2+\sqrt{3}}}{2}$

c. $\cos 7.5° = \cos\frac{15°}{2} = \sqrt{\frac{1+\cos 15°}{2}}$
$= \sqrt{\frac{1+\frac{\sqrt{2+\sqrt{3}}}{2}}{2}} = \sqrt{\frac{2+\sqrt{2+\sqrt{3}}}{4}}$
$= \frac{\sqrt{2+\sqrt{2+\sqrt{3}}}}{2}$

6.3 Getting Ready for the Next Section

133. $f(u+v) + f(u-v) = 2f(u)g(v)$

a. $f(u)g(v) = \frac{1}{2}\left[f(u+v) + f(u-v)\right]$

b. $\left.\begin{array}{l}x=u+v\\y=u-v\end{array}\right\}x+y=2u\Rightarrow u=\dfrac{x+y}{2}$

$\left.\begin{array}{l}x=u+v\\y=u-v\end{array}\right\}x-y=2v\Rightarrow v=\dfrac{x-y}{2}$

$f(x)+f(y)=2f\left(\dfrac{x+y}{2}\right)g\left(\dfrac{x-y}{2}\right)$

6.4 Product-to-Sum and Sum-to-Product Formulas

6.4 Practice Problems

1. $\cos 3x \cos x = \dfrac{1}{2}\left[\cos(3x+x)+\cos(3x-x)\right]$
   $= \dfrac{1}{2}\cos 4x + \dfrac{1}{2}\cos 2x$

2. $\sin 15° \cos 75°$
   $= \dfrac{1}{2}\left[\sin(15°+75°)+\sin(15°-75°)\right]$
   $= \dfrac{1}{2}\sin 90° + \dfrac{1}{2}\sin(-60°)$
   $= \dfrac{1}{2}(1)+\dfrac{1}{2}\left(-\dfrac{\sqrt{3}}{2}\right)=\dfrac{2-\sqrt{3}}{4}$

3. **a.** $\cos 2x - \cos 4x$
   $= -2\sin\left(\dfrac{2x+4x}{2}\right)\sin\left(\dfrac{2x-4x}{2}\right)$
   $= -2\sin 3x \sin(-x)$
   $= 2\sin 3x \sin x$

   b. $\sin 43° + \sin 17°$
   $= 2\sin\left(\dfrac{43°+17°}{2}\right)\cos\left(\dfrac{43°-17°}{2}\right)$
   $= 2\sin 30° \cos 13°$
   $= 2\left(\dfrac{1}{2}\right)\cos 13° = \cos 13°$

4. $\sin 3x - \cos x$
   $= \sin 3x - \sin\left(\dfrac{\pi}{2}-x\right)$
   $= 2\cos\left(\dfrac{3x+\left(\dfrac{\pi}{2}-x\right)}{2}\right)\sin\left(\dfrac{3x-\left(\dfrac{\pi}{2}-x\right)}{2}\right)$
   $= 2\cos\left(x+\dfrac{\pi}{4}\right)\sin\left(2x-\dfrac{\pi}{4}\right)$

5. $\dfrac{\cos 5x - \cos x}{\sin x - \sin 5x}$
   $= \dfrac{-2\sin\left(\dfrac{5x+x}{2}\right)\sin\left(\dfrac{5x-x}{2}\right)}{2\cos\left(\dfrac{x+5x}{2}\right)\sin\left(\dfrac{x-5x}{2}\right)}$
   $= \dfrac{-2\sin 3x \sin 2x}{2\cos 3x \sin(-2x)} = \dfrac{-2\sin 3x \sin 2x}{-2\cos 3x \sin 2x}$
   $= \dfrac{\sin 3x}{\cos 3x} = \tan 3x$

6. **a.** When you press the "1" key, the sound produced is given by
   $y = \sin[2\pi \cdot 697t] + \sin[2\pi \cdot 1209t]$.

   b. $y = \sin[2\pi(697)t] + \sin[2\pi(1209)t]$
   $= 2\sin\left[2\pi\left(\dfrac{697+1209}{2}\right)t\right] \cdot$
   $\quad \cos\left[2\pi\left(\dfrac{697-1209}{2}\right)t\right]$
   $= 2\sin[2\pi(953)t]\cos[2\pi(-256)t]$
   $= 2\sin(1906\pi t)\cos(512\pi t)$

   c. The frequency is 953 Hz. The variable amplitude is $2\cos(512\pi t)$.

6.4 Concepts and Vocabulary

1. We can rewrite the product of two sines as a difference of two cosines by using the formula
   $\sin x \sin y = \underline{\dfrac{1}{2}\left[\cos(x-y)-\cos(x+y)\right]}$.

3. We can rewrite the sum of two cosines as a product of two cosines by using the formula
   $\cos x + \cos y = \underline{2\cos\left(\dfrac{x+y}{2}\right)\cos\left(\dfrac{x-y}{2}\right)}$.

5. True

7. False.
   $\sin(x+y) = \sin x \cos y + \cos x \sin y$
   $\neq \sin x + \sin y$

6.4 Building Skills

9. $\sin x \cos x = \dfrac{1}{2}(\sin 2x + \sin 0) = \dfrac{1}{2}\sin 2x$

11. $\sin x \sin x = \dfrac{1}{2}(\cos 0 - \cos 2x) = \dfrac{1}{2} - \dfrac{1}{2}\cos 2x$

13. $\sin 25° \cos 5° = \dfrac{1}{2}(\sin 30° + \sin 20°)$
    $= \dfrac{1}{4} + \dfrac{1}{2}\sin 20°$

15. $\cos 140° \cos 20° = \dfrac{1}{2}(\cos 120° + \cos 160°)$
    $= -\dfrac{1}{4} + \dfrac{1}{2}\cos 160°$
    $= -\dfrac{1}{4} - \dfrac{1}{2}\cos 20°$

17. $\sin \dfrac{7\pi}{12} \sin \dfrac{\pi}{12}$
    $= \dfrac{1}{2}\left[\cos\left(\dfrac{7\pi}{12} - \dfrac{\pi}{12}\right) - \cos\left(\dfrac{7\pi}{12} + \dfrac{\pi}{12}\right)\right]$
    $= \dfrac{1}{2}\left(\cos \dfrac{\pi}{2} - \cos \dfrac{2\pi}{3}\right) = \dfrac{1}{4}$

19. $\cos \dfrac{5\pi}{8} \sin \dfrac{\pi}{8}$
    $= \dfrac{1}{2}\left[\sin\left(\dfrac{5\pi}{8} + \dfrac{\pi}{8}\right) - \sin\left(\dfrac{5\pi}{8} - \dfrac{\pi}{8}\right)\right]$
    $= \dfrac{1}{2}\left(\sin \dfrac{3\pi}{4} - \sin \dfrac{\pi}{2}\right) = \dfrac{\sqrt{2}}{4} - \dfrac{1}{2}$

21. $\sin 5\theta \cos \theta = \dfrac{1}{2}(\sin 6\theta + \sin 4\theta)$
    $= \dfrac{1}{2}\sin 6\theta + \dfrac{1}{2}\sin 4\theta$

23. $\cos 4x \cos 3x = \dfrac{1}{2}(\cos x + \cos 7x)$
    $= \dfrac{1}{2}\cos x + \dfrac{1}{2}\cos 7x$

25. $\sin 37.5° \sin 7.5° = \dfrac{1}{2}(\cos 30° - \cos 45°)$
    $= \dfrac{\sqrt{3} - \sqrt{2}}{4}$

27. $\sin 67.5° \cos 22.5° = \dfrac{1}{2}(\sin 90° + \sin 45°)$
    $= \dfrac{1}{2} + \dfrac{\sqrt{2}}{4}$

29. $\sin \dfrac{5\pi}{24} \cos \dfrac{\pi}{24} = \dfrac{1}{2}\left(\sin \dfrac{\pi}{6} + \sin \dfrac{\pi}{4}\right) = \dfrac{1+\sqrt{2}}{4}$

31. $\cos \dfrac{13\pi}{24} \cos \dfrac{5\pi}{24} = \dfrac{1}{2}\left(\cos \dfrac{\pi}{3} + \cos \dfrac{3\pi}{4}\right) = \dfrac{1-\sqrt{2}}{4}$

33. $\cos 40° - \cos 20°$
    $= -2\sin\left(\dfrac{40° + 20°}{2}\right)\sin\left(\dfrac{40° - 20°}{2}\right)$
    $= -2\sin 30° \sin 10° = -\sin 10°$

35. $\sin 32° - \sin 16°$
    $= 2\sin\left(\dfrac{32° - 16°}{2}\right)\cos\left(\dfrac{32° + 16°}{2}\right)$
    $= 2\sin 8° \cos 24°$

37. $\sin \dfrac{\pi}{5} + \sin \dfrac{2\pi}{5} = 2\sin \dfrac{3\pi}{10}\cos\left(-\dfrac{\pi}{10}\right)$
    $= 2\sin \dfrac{3\pi}{10}\cos \dfrac{\pi}{10}$

39. $\cos \dfrac{1}{2} + \cos \dfrac{1}{3} = 2\cos \dfrac{\frac{1}{2}+\frac{1}{3}}{2}\cos \dfrac{\frac{1}{2}-\frac{1}{3}}{2}$
    $= 2\cos \dfrac{5}{12}\cos \dfrac{1}{12}$

41. $\cos 3x + \cos 5x = 2\cos 4x \cos(-x)$
    $= 2\cos 4x \cos x$

43. $\sin 7x + \sin(-x) = 2\sin 3x \cos 4x$

45. $\sin x + \cos x = \sin x + \sin\left(\dfrac{\pi}{2} - x\right)$
    $= 2\sin \dfrac{\pi}{4}\cos\left(x - \dfrac{\pi}{4}\right)$
    $= \sqrt{2}\cos\left(x - \dfrac{\pi}{4}\right)$

47. $\sin 2x - \cos 2x = \sin 2x - \sin\left(\dfrac{\pi}{2} - 2x\right)$
    $= 2\sin\left(2x - \dfrac{\pi}{4}\right)\cos \dfrac{\pi}{4}$
    $= \sqrt{2}\sin\left(2x - \dfrac{\pi}{4}\right)$

49. $\sin 3x + \cos 5x = \sin 3x + \sin\left(\dfrac{\pi}{2} - 5x\right)$
    $= 2\sin\left(\dfrac{\pi}{4} - x\right)\cos\left(4x - \dfrac{\pi}{4}\right)$

51. $a(\sin x + \cos x) = a\left(\sin x + \sin\left(\dfrac{\pi}{2} - x\right)\right)$
    $= 2a\sin \dfrac{\pi}{4}\cos\left(x - \dfrac{\pi}{4}\right)$
    $= a\sqrt{2}\cos\left(x - \dfrac{\pi}{4}\right)$

53. $\dfrac{\sin x + \sin 3x}{\cos x + \cos 3x} = \dfrac{2\sin 2x \cos(-x)}{2\cos 2x \cos(-x)} = \tan 2x$

Section 6.4 Product-to-Sum and Sum-to-Product Formulas   355

55. $\dfrac{\cos 3x - \cos 7x}{\sin 7x + \sin 3x} = \dfrac{-2\sin 5x \sin(-2x)}{2\sin 5x \cos 2x} = \dfrac{\sin 2x}{\cos 2x} = \tan 2x$

57. $\dfrac{\cos 2x + \cos 2y}{\cos 2x - \cos 2y} = \dfrac{2\cos\left(\dfrac{2x+2y}{2}\right)\cos\left(\dfrac{2x-2y}{2}\right)}{-2\sin\left(\dfrac{2x+2y}{2}\right)\sin\left(\dfrac{2x-2y}{2}\right)} = \dfrac{2\cos(x+y)\cos(x-y)}{-2\sin(x+y)\sin(x-y)}$

$= -\cot(x+y)\cot(x-y) = \cot(y+x)\cot(y-x)$

59. $\sin x + \sin 2x + \sin 3x = \sin 2x + (\sin x + \sin 3x) = \sin 2x + 2\sin 2x\cos(-x) = \sin 2x(1 + 2\cos x)$

61. $\sin 2x + \sin 4x + \sin 6x = \sin 0x + \sin 2x + \sin 4x + \sin 6x = (\sin 2x + \sin 4x) + (\sin 0x + \sin 6x)$
$= 2\sin 3x\cos x + 2\sin 3x\cos 3x = 2\sin 3x(\cos x + \cos 3x)$
$= 2\sin 3x(2\cos 2x\cos(-x)) = 4\cos x\cos 2x\sin 3x$

6.4 Applying the Concepts

63. a. $y = \sin(2\pi(852)t) + \sin(2\pi(1209)t)$

  b. $y = \sin(2\pi(852)t) + \sin(2\pi(1209)t) = 2\sin\left[2\pi\left(\dfrac{852+1209}{2}\right)t\right]\cdot\cos\left[2\pi\left(\dfrac{852-1209}{2}\right)t\right]$
  $= 2\sin(2\pi(1030.5)t)\cos(2\pi(-357)t) = 2\sin(2\pi(1030.5)t)\cos(357\pi t)$

  c. frequency = 1030.5 Hz, amplitude = $2\cos(357\pi t)$.

65. a. $y = 0.05\cos(112\pi t) + 0.05\cos(120\pi t) = 0.05(\cos(112\pi t) + \cos(120\pi t))$
  $= 0.05(2\cos(116\pi t)\cos(4\pi t)) = 0.1\cos(116\pi t)\cos(4\pi t)$

  b. The frequency of $y_1 = \dfrac{112\pi}{2\pi} = 56$. The frequency of $y_2 = \dfrac{120\pi}{2\pi} = 60$.
  The beat frequency is $|56 - 60| = 4$.

6.4 Beyond the Basics

67. $y = \cos(2x + 1) + \cos(2x - 1)$

    $\cos(2x+1) + \cos(2x-1) = 2\cos 2x\cos 1 = 2\cos 1\cos 2x$

    Amplitude: $2\cos 1 \approx 1.0806$

    Period: $\dfrac{2\pi}{2} = \pi$

69. $\cos 40° + \cos 50° + \cos 70° + \cos 80° = (\cos 40° + \cos 80°) + (\cos 50° + \cos 70°)$
    $= 2\cos 60°\cos(-20°) + 2\cos 60°\cos(-10°)$
    $= 2\cos 60°(\cos(-20°) + \cos(-10°)) = \cos 10° + \cos 20°$

71. $\cos 4\theta\cos\theta - \cos 6\theta\cos 9\theta = \dfrac{1}{2}(\cos 3\theta + \cos 5\theta) - \dfrac{1}{2}(\cos 3\theta + \cos 15\theta)$
    $= \dfrac{1}{2}\cos 3\theta + \dfrac{1}{2}\cos 5\theta - \dfrac{1}{2}\cos 3\theta - \dfrac{1}{2}\cos 15\theta = \dfrac{1}{2}\cos 5\theta - \dfrac{1}{2}\cos 15\theta$
    $= \dfrac{1}{2}(\cos 5\theta - \cos 15\theta) = \dfrac{1}{2}(-2\sin 10\theta\sin(-5\theta)) = \sin 5\theta\sin 10\theta$

73. $\sin 25° \sin 35° - \sin 25° \sin 85° - \sin 35° \sin 85°$

$= \dfrac{1}{2}(\cos 10° - \cos 60°) - \dfrac{1}{2}(\cos 60° - \cos 110°) - \dfrac{1}{2}(\cos 50° - \cos 120°)$

$= \dfrac{1}{2}\cos 10° - \dfrac{1}{4} - \dfrac{1}{4} + \dfrac{1}{2}\cos 110° - \dfrac{1}{2}\cos 50° - \dfrac{1}{4} = -\dfrac{3}{4} + \dfrac{1}{2}(\cos 10° + \cos 110° - \cos 50°)$

$= -\dfrac{3}{4} + \dfrac{1}{2}(2\cos 60° \cos(-50°) - \cos 50°) = -\dfrac{3}{4} + \dfrac{1}{2}\cos 50° - \dfrac{1}{2}\cos 50° = -\dfrac{3}{4}$

75. $\dfrac{\sin 3x \cos 5x - \sin x \cos 7x}{\sin x \sin 7x + \cos 3x \cos 5x} = \dfrac{\dfrac{1}{2}(\sin 8x + \sin(-2x)) - \dfrac{1}{2}(\sin 8x + \sin(-6x))}{\dfrac{1}{2}(\cos(-6x) - \cos 8x) + \dfrac{1}{2}(\cos(-2x) + \cos 8x)}$

$= \dfrac{-\sin 2x + \sin 6x}{\cos 6x + \cos 2x} = \dfrac{2 \sin 2x \cos 4x}{2 \cos 4x \cos 2x} = \dfrac{2 \sin 2x \cos 4x}{2 \cos 4x \cos 2x} = \dfrac{\sin 2x}{\cos 2x} = \tan 2x$

77. $\dfrac{\cos x + \cos 3x + \cos 5x + \cos 7x}{\sin x + \sin 3x + \sin 5x + \sin 7x} = \dfrac{4 \cos x \cos 2x \cos 4x}{2 \sin 2x \cos x + 2 \sin 6x \cos x}$ (Use the result from exercise 60.)

$= \dfrac{4 \cos x \cos 2x \cos 4x}{2 \cos x(\sin 2x + \sin 6x)} = \dfrac{4 \cos x \cos 2x \cos 4x}{2 \cos x(2 \sin 4x \cos(-2x))}$

$= \dfrac{4 \cos x \cos 2x \cos 4x}{4 \cos x \cos 2x \sin 4x} = \cot 4x$

79. $2 \cos 6x \cos x - 2 \cos 4x \cos x + 2 \cos 2x \cos x - \cos x$

$= 2\left[\dfrac{1}{2}(\cos 5x + \cos 7x)\right] - 2\left[\dfrac{1}{2}(\cos 3x + \cos 5x)\right] + 2\left[\dfrac{1}{2}(\cos x + \cos 3x)\right] - \cos x$

$= \cos 5x + \cos 7x - \cos 3x - \cos 5x + \cos x + \cos 3x - \cos x = \cos 7x$

81. $\dfrac{\sin(x + 3y) + \sin(3x + y)}{\sin 2x + \sin 2y} = \dfrac{2 \sin(2x + 2y) \cos(-x + y)}{\sin 2x + \sin 2y} = \dfrac{2 \sin(2x + 2y) \cos(-(x - y))}{2 \sin(x + y) \cos(x - y)}$

$= \dfrac{2 \sin 2(x + y) \cos(x - y)}{2 \sin(x + y) \cos(x - y)} = \dfrac{4 \sin(x + y) \cos(x + y) \cos(x - y)}{2 \sin(x + y) \cos(x - y)} = 2 \cos(x + y)$

83. a. $-2 \sin\dfrac{x + y}{2} \sin\dfrac{x - y}{2} = -2 \cdot \dfrac{1}{2}\left[\cos\left(\dfrac{x + y}{2} - \dfrac{x - y}{2}\right) - \cos\left(\dfrac{x + y}{2} + \dfrac{x - y}{2}\right)\right]$

$= -1[\cos y - \cos x] = \cos x - \cos y$

b. $2 \sin\dfrac{x + y}{2} \cos\dfrac{x - y}{2} = 2 \cdot \dfrac{1}{2}\left[\sin\left(\dfrac{x + y}{2} + \dfrac{x - y}{2}\right) + \sin\left(\dfrac{x + y}{2} - \dfrac{x - y}{2}\right)\right] = \sin x + \sin y$

c. $2 \sin\dfrac{x - y}{2} \cos\dfrac{x + y}{2} = 2 \cdot \dfrac{1}{2}\left[\sin\left(\dfrac{x - y}{2} + \dfrac{x + y}{2}\right) + \sin\left(\dfrac{x - y}{2} - \dfrac{x + y}{2}\right)\right] = \sin x + \sin(-y) = \sin x - \sin y$

6.4 Critical Thinking/Discussion/Writing

84.

Step	Reason
$\sin\dfrac{\pi}{14}\sin\dfrac{3\pi}{14}\sin\dfrac{5\pi}{14} = \cos\dfrac{3\pi}{7}\cos\dfrac{2\pi}{7}\cos\dfrac{\pi}{7}$	$\sin x = \cos\left(\dfrac{\pi}{2}-x\right)$
$= \dfrac{8\sin(\pi/7)\cos(\pi/7)\cos(2\pi/7)\cos(3\pi/7)}{8\sin(\pi/7)}$	Multiply the numerator and denominator by $8\sin\dfrac{\pi}{7}$
$= \dfrac{2\sin(4\pi/7)\cos(3\pi/7)}{8\sin(\pi/7)}$	$\sin 2\theta = 2\sin\theta\cos\theta$ (applied twice)
$= \dfrac{\sin\pi + \sin(\pi/7)}{8\sin(\pi/7)}$	Product-to-sum formula: $\sin x\cos y = \dfrac{1}{2}[\sin(x+y)+\sin(x-y)]$
$= 1/8$	$\sin\pi = 0$; simplify

85. $\sin 2A + \sin 2B + \sin 2C = (\sin 2A + \sin 2B) + \sin 2C = 2\sin(A+B)\cos(A-B) + 2\sin C\cos C$
$A+B+C = 180° \Rightarrow C = 180° - (A+B)$, so
$2\sin(A+B)\cos(A-B) + 2\sin C\cos C = 2\sin(180° - (A+B))\cos(A-B) + 2\sin C\cos C$
$= 2\sin C\cos(A-B) + 2\sin C\cos C = 2\sin C(\cos(A-B) + \cos C)$
$= 2\sin C[\cos(A-B) + \cos(180° - (A+B))]$
$= 2\sin C(\cos(A-B) - \cos(A+B))$
$= 2\sin C(-2\sin A\sin(-2B)) = 4\sin A\sin B\sin C$

86. $\cos 2A + \cos 2B + \cos 2C = (\cos 2A + \cos 2B) + \cos 2C = 2\cos(A+B)\cos(A-B) + \cos 2C$
$A+B+C = 180° \Rightarrow A+B = 180° - C$, so
$2\cos(A+B)\cos(A-B) + \cos 2C = 2\cos(180° - C)\cos(A-B) + \cos 2C = -2\cos C\cos(A-B) + 2\cos^2 C - 1$
$= -1 - 2\cos C[\cos(A-B) - \cos C]$
$= -1 - 2\cos C[\cos(A-B) - \cos(180° - (A+B))]$
$= -1 - 2\cos C[\cos(A-B) + \cos(A+B)]$
$= -1 - 2\cos C[2\cos A\cos B] = -1 - 4\cos A\cos B\cos C$

87. $\cos^2 x = \cos x\cos x = \dfrac{1}{2}[\cos(x-x) + \cos(x+x)] = \dfrac{1}{2}[\cos 0 + \cos 2x] = \dfrac{1}{2}(1+\cos 2x) = \dfrac{1+\cos 2x}{2}$

88. $\sin^2 x = \sin x\sin x = \dfrac{1}{2}[\cos(x-x) - \cos(x+x)] = \dfrac{1}{2}[\cos 0 - \cos 2x] = \dfrac{1}{2}(1 - \cos 2x) = \dfrac{1-\cos 2x}{2}$

6.4 Getting Ready for the Next Section

89. $2x - 3 = 9$, conditional equation

91. $x^2 + 4 = 0$, inconsistent equation

93. $x^2 - 4 = (x+2)(x-2)$, identity

95. $\sin x = \dfrac{1}{2}$, conditional equation

97. $\sin x = -2$, inconsistent equation

99. $\sin 2x = 2\sin x\cos x$, identity

101. $(2x-1) - 3 = 6$
$2x - 4 = 6$
$2x = 10 \Rightarrow x = 5$
Solution: $\{5\}$

103. $(x+2)(x-3) = 0$
$x+2=0 \ | \ x-3=0$
$x = -2 \ | \ x = 3$
Solution: $\{-2, 3\}$

105. $\sqrt{x+1} = x-5$
$x+1 = (x-5)^2$
$x+1 = x^2 - 10x + 25$
$0 = x^2 - 11x + 24$
$0 = (x-3)(x-8)$

$x - 3 = 0 \mid x - 8 = 0$
$x = 3 \mid x = 8$

Verify each potential solution.
$\sqrt{3+1} \stackrel{?}{=} 3 - 5 \quad \sqrt{8+1} \stackrel{?}{=} 8 - 5$
$2 \neq -2 \qquad 3 = 3 \checkmark$

Solution: $\{8\}$

107. $\sin x = \frac{1}{2} \Rightarrow x = \frac{\pi}{6}, \frac{5\pi}{6}$

109. $\tan x = -1 \Rightarrow x = \frac{3\pi}{4}, \frac{7\pi}{4}$

111. $\csc x = -\frac{2\sqrt{3}}{3} \Rightarrow x = \frac{4\pi}{3}, \frac{5\pi}{3}$

113. $\cos x = -\frac{1}{2} \Rightarrow x = \frac{2\pi}{3}, \frac{4\pi}{3}$

6.5 Trigonometric Equations I

6.5 Practice Problems

1. a. $\sin x = 1 \Rightarrow x = \frac{\pi}{2} + 2n\pi$

b. $\cos x = 1 \Rightarrow x = 0 + 2n\pi = 2n\pi$

c. $\tan x = 1 \Rightarrow x = \frac{\pi}{4} + n\pi$

2. a. $\sec x = 1.5 \Rightarrow x$ is in Quadrant I or Quadrant IV.

$x = \sec^{-1}(1.5) \Rightarrow \sec x = 1.5 \Rightarrow \cos x = \frac{2}{3}$

$x = \cos^{-1} \frac{2}{3} \approx 48.2°$ or
$x \approx 360° - 48.2° = 311.8°$
Solution set:
$\{48.2° + n \cdot 360°, 311.8° + n \cdot 360°\}$

b. $\tan x = -2 \Rightarrow x$ is in Quadrant II or Quadrant IV.

$x = \tan^{-1}(-2) \approx \pi - 1.1071 \approx 2.0344$ or
$x \approx 2\pi - 1.1071 \approx 5.1760$
Solution set: $\{2.0345, 5.1761\}$

3. $\tan x = -\sqrt{3}$

$\tan^{-1}(-\sqrt{3}) = \frac{2\pi}{3}$, so

$\tan x = \tan \frac{2\pi}{3} \Rightarrow x = \frac{2\pi}{3} + n\pi$, n an integer.

4. $3\sec(x-30)° - 1 = \sec(x-30)° + 3 \Rightarrow$
$2\sec(x-30)° = 4 \Rightarrow \sec(x-30)° = 2 \Rightarrow$
$\cos(x-30)° = \frac{1}{2} \Rightarrow x - 30° = \cos^{-1}\left(\frac{1}{2}\right) \Rightarrow$
$x = \cos^{-1}\left(\frac{1}{2}\right) + 30°$

$\cos x$ is positive in quadrants I and IV, so

$\cos^{-1}\left(\frac{1}{2}\right) = 60°, 300°,$ and

$x = \cos^{-1}\left(\frac{1}{2}\right) + 30° = 60° + 30° = 90°$

$x = \cos^{-1}\left(\frac{1}{2}\right) + 30° = 300° + 30° = 330°$

Solution set: $\{90°, 330°\}$.

5. $d = tv_0 \cos\theta \Rightarrow 2500 = (1)(3280)\cos\theta \Rightarrow$
$\cos\theta = \frac{2500}{3280} \Rightarrow \theta = \cos^{-1}\left(\frac{2500}{3280}\right) \approx 40°$

6. $(\sin x - 1)(\sqrt{3}\tan x + 1) = 0 \Rightarrow$

$\sin x = 1 \Rightarrow x = \frac{\pi}{2}$ or

$\sqrt{3}\tan x + 1 = 0 \Rightarrow \tan x = -\frac{1}{\sqrt{3}} = -\frac{\sqrt{3}}{3} \Rightarrow$

$x = \frac{5\pi}{6}$ or $x = \frac{11\pi}{6}$

Solution set: $\left\{\frac{\pi}{2}, \frac{5\pi}{6}, \frac{11\pi}{6}\right\}$

7. $2\cos^2\theta - \cos\theta - 1 = 0 \Rightarrow$
$(2\cos\theta + 1)(\cos\theta - 1) = 0 \Rightarrow$

$2\cos\theta + 1 = 0 \Rightarrow \cos\theta = -\frac{1}{2}$ or

$\cos\theta - 1 = 0 \Rightarrow \cos\theta = 1$

$\cos\theta = -\frac{1}{2} \Rightarrow \theta = \frac{2\pi}{3}$ or $\theta = \frac{4\pi}{3}$

$\cos\theta = 1 \Rightarrow \theta = 0$

Solution set: $\left\{2n\pi, \frac{2\pi}{3} + 2n\pi, \frac{4\pi}{3} + 2n\pi\right\}$

8. $2\cos^2\theta + 3\sin\theta - 3 = 0$
    $2(1-\sin^2\theta) + 3\sin\theta - 3 = 0$
    $2 - 2\sin^2\theta + 3\sin\theta - 3 = 0$
    $-2\sin^2\theta + 3\sin\theta - 1 = 0$
    $2\sin^2\theta - 3\sin\theta + 1 = 0$
    $(2\sin\theta - 1)(\sin\theta - 1) = 0$
    $2\sin\theta - 1 = 0$ or $\sin\theta - 1 = 0$
    $\sin\theta = \dfrac{1}{2}$ $\sin\theta = 1$
    $\theta = \dfrac{\pi}{6}$ or $\theta = \dfrac{5\pi}{6}$ $\theta = \dfrac{\pi}{2}$
    Solution set: $\left\{\dfrac{\pi}{6}, \dfrac{\pi}{2}, \dfrac{5\pi}{6}\right\}$

9. $\sqrt{3}\cot\theta + 1 = \sqrt{3}\csc\theta$
    $(\sqrt{3}\cot\theta + 1)^2 = (\sqrt{3}\csc\theta)^2$
    $3\cot^2\theta + 2\sqrt{3}\cot\theta + 1 = 3\csc^2\theta$
    $3\cot^2\theta + 2\sqrt{3}\cot\theta + 1 = 3(1 + \cot^2\theta)$
    $3\cot^2\theta + 2\sqrt{3}\cot\theta + 1 = 3 + 3\cot^2\theta$
    $2\sqrt{3}\cot\theta = 2 \Rightarrow \cot\theta = \dfrac{\sqrt{3}}{3} \Rightarrow$
    $\theta = \dfrac{\pi}{3}$ or $\theta = \dfrac{4\pi}{3}$
    Check each answer:
    $\sqrt{3}\cot\dfrac{\pi}{3} + 1 \stackrel{?}{=} \sqrt{3}\csc\dfrac{\pi}{3}$
    $\sqrt{3} \cdot \dfrac{\sqrt{3}}{3} + 1 \stackrel{?}{=} \sqrt{3} \cdot \dfrac{2}{\sqrt{3}}$
    $2 = 2$ ✓
    $\sqrt{3}\cot\dfrac{4\pi}{3} + 1 \stackrel{?}{=} \sqrt{3}\csc\dfrac{4\pi}{3}$
    $\sqrt{3} \cdot \dfrac{\sqrt{3}}{3} + 1 \stackrel{?}{=} \sqrt{3}\left(-\dfrac{2}{\sqrt{3}}\right)$
    $2 \ne -2$
    Thus, $\dfrac{4\pi}{3}$ is extraneous and the solution is $\left\{\dfrac{\pi}{3}\right\}$.

6.5 Concepts and Vocabulary

1. The equation $\sin x = \dfrac{1}{2}$ has <u>two</u> solutions in $[0, 2\pi)$.

3. The equation $\cos x = 1$ has <u>one</u> solution in $[0, 2\pi)$.

5. False. The equation $\sec x = \dfrac{1}{2}$ has no solutions because $\dfrac{1}{2}$ is not the in range of $\sec x$.

7. False. All solutions of $\cos x = -\dfrac{1}{2}$ are given by $x = \dfrac{2\pi}{3} + 2n\pi$ or
    $x = \left(2\pi - \dfrac{2\pi}{3}\right) + 2n\pi = \dfrac{4\pi}{3} + 2n\pi$.

6.5 Building Skills

9. $\cos x = 0 \Rightarrow x = \dfrac{\pi}{2} + 2n\pi$ or $x = \dfrac{3\pi}{2} + 2n\pi$

11. $\tan x = -1 \Rightarrow x = \dfrac{3\pi}{4} + n\pi$

13. $\cos x = \dfrac{\sqrt{2}}{2} \Rightarrow x = \dfrac{\pi}{4} + 2n\pi$ or $x = \dfrac{7\pi}{4} + 2n\pi$

15. $\cot x = \sqrt{3} \Rightarrow x = \dfrac{\pi}{6} + n\pi$

17. $\cos x = -\dfrac{1}{2} \Rightarrow x = \dfrac{2\pi}{3} + 2n\pi$ or $x = \dfrac{4\pi}{3} + 2n\pi$

19. $\tan x = \dfrac{\sqrt{3}}{3} \Rightarrow x = 30° + 180n$

21. $\sin x = -\dfrac{1}{2} \Rightarrow x = 210° + 360°n$ or $x = 330° + 360°n$

23. $\csc x = 1 \Rightarrow x = 90° + 360°n$

25. $\sqrt{3}\csc x - 2 = 0 \Rightarrow \csc x = \dfrac{2\sqrt{3}}{3} \Rightarrow$
    $x = 60° + 360°n$ or $120° + 360°n$

27. $2\sec x - 4 = 0 \Rightarrow \sec x = 2 \Rightarrow$
    $x = 60° + 360°n$ or $300° + 360°n$

29. $\sin\theta = 0.4 \Rightarrow \theta$ is in Quadrant I or Quadrant II. $\theta = \sin^{-1}(0.4) = 23.6°$ or
    $\theta = 180° - 23.6° = 156.4°$

31. $\sec\theta = 7.2 \Rightarrow \theta$ is in Quadrant I or Quadrant IV. $\theta = \sec^{-1}(7.2) = 82.0°$ or
    $\theta = 360° - 82.0° = 278.0°$

33. $\tan(\theta - 30°) = -5 \Rightarrow \theta - 30°$ is in Quadrant II or Quadrant IV. Let $x = \theta - 30°$. Then,
$x = \tan^{-1}(-5) \approx -78.7° \Rightarrow$
$x \approx 180° - 78.7° \approx 101.3°$ or
$x \approx 360° - 78.7° \approx 281.3° \Rightarrow \theta \approx 131.3°$ or
$\theta \approx 311.3°$

35. $\csc x = -3 \Rightarrow x$ is in Quadrant III or Quadrant IV.
$x = \csc^{-1}(-3) \approx -0.3398 \Rightarrow$
$x \approx \pi - (-0.3398) \approx 3.4814$ or
$x \approx 2\pi + (-0.3398) \approx 5.9433$

37. $3 \tan x + 4 = 0 \Rightarrow \tan x = -4/3 \Rightarrow x$ is in Quadrant II or Quadrant IV.
$x = \tan^{-1}(-4/3) \approx -0.9273$
$\approx -0.9273 + \pi \approx 2.2143$ or
$x \approx -0.9273 + 2\pi \approx 5.3559$

39. $2 \csc x + 5 = 0 \Rightarrow \csc x = -5/2 \Rightarrow x$ is in Quadrant III or Quadrant IV.
$x = \csc^{-1}(-5/2)$
$\approx -0.4115 \Rightarrow x \approx \pi + 0.4115 \approx 3.5531$ or
$x \approx 2\pi - 0.4115 \approx 5.8717$

41. $\sin\left(x + \dfrac{\pi}{4}\right) = \dfrac{1}{2} \Rightarrow x + \dfrac{\pi}{4} = \sin^{-1}\left(\dfrac{1}{2}\right) \Rightarrow$
$x + \dfrac{\pi}{4} = \dfrac{\pi}{6} \Rightarrow -\dfrac{\pi}{12} = \dfrac{23\pi}{12}$ or
$x + \dfrac{\pi}{4} = \dfrac{5\pi}{6} \Rightarrow x = \dfrac{7\pi}{12}$
The solution is $\left\{\dfrac{7\pi}{12}, \dfrac{23\pi}{12}\right\}$.

43. $\sec\left(x - \dfrac{\pi}{8}\right) + 2 = 0 \Rightarrow x - \dfrac{\pi}{8} = \sec^{-1}(-2) \Rightarrow$
$x - \dfrac{\pi}{8} = \dfrac{2\pi}{3} \Rightarrow x = \dfrac{19\pi}{24}$ or $x - \dfrac{\pi}{8} = \dfrac{4\pi}{3} \Rightarrow$
$x = \dfrac{35\pi}{24}$
The solution is $\left\{\dfrac{19\pi}{24}, \dfrac{35\pi}{24}\right\}$.

45. $\sqrt{3} \tan\left(x - \dfrac{\pi}{6}\right) - 1 = 0 \Rightarrow \tan\left(x - \dfrac{\pi}{6}\right) = \dfrac{\sqrt{3}}{3} \Rightarrow$
$x - \dfrac{\pi}{6} = \tan^{-1}\left(\dfrac{\sqrt{3}}{3}\right) \Rightarrow x - \dfrac{\pi}{6} = \dfrac{\pi}{6} \Rightarrow x = \dfrac{\pi}{3}$
or $x - \dfrac{\pi}{6} = \dfrac{7\pi}{6} \Rightarrow x = \dfrac{4\pi}{3}$.
The solution is $\left\{\dfrac{\pi}{3}, \dfrac{4\pi}{3}\right\}$.

47. $2 \sin\left(x - \dfrac{\pi}{3}\right) + 1 = 0 \Rightarrow \sin\left(x - \dfrac{\pi}{3}\right) = -\dfrac{1}{2} \Rightarrow$
$x - \dfrac{\pi}{3} = \sin^{-1}\left(-\dfrac{1}{2}\right) \Rightarrow x - \dfrac{\pi}{3} = \dfrac{7\pi}{6} \Rightarrow x = \dfrac{3\pi}{2}$
or $x - \dfrac{\pi}{3} = \dfrac{11\pi}{6} \Rightarrow x = \dfrac{13\pi}{6} = \dfrac{\pi}{6}$.
The solution is $\left\{\dfrac{\pi}{6}, \dfrac{3\pi}{2}\right\}$.

49. $(\sin x + 1)(\tan x - 1) = 0 \Rightarrow$
$\sin x = -1$ or $\tan x = 1$
$\sin x = -1 \Rightarrow x = \dfrac{3\pi}{2}$
However, $\tan \dfrac{3\pi}{2}$ is undefined, so $x = \dfrac{3\pi}{2}$ is not a solution.
$\tan x = 1 \Rightarrow x = \dfrac{\pi}{4}$ or $x = \dfrac{5\pi}{4}$.
The solution is $\left\{\dfrac{\pi}{4}, \dfrac{5\pi}{4}\right\}$.

51. $(\csc x - 2)(\cot x + 1) = 0 \Rightarrow \csc x = 2$ or
$\cot x = -1$; $\csc x = 2 \Rightarrow x = \dfrac{\pi}{6}$ or $x = \dfrac{5\pi}{6}$
$\cot x = -1 \Rightarrow x = \dfrac{3\pi}{4}$ or $x = \dfrac{7\pi}{4}$.
The solution is $\left\{\dfrac{\pi}{6}, \dfrac{3\pi}{4}, \dfrac{5\pi}{6}, \dfrac{7\pi}{4}\right\}$.

53. $(\tan x + 1)(2 \sin x - 1) = 0 \Rightarrow \tan x = -1$ or
$\sin x = \dfrac{1}{2}$; $\tan x = -1 \Rightarrow x = \dfrac{3\pi}{4}$ or $\dfrac{7\pi}{4}$
$\sin x = \dfrac{1}{2} \Rightarrow x = \dfrac{\pi}{6}$ or $x = \dfrac{5\pi}{6}$
The solution is $\left\{\dfrac{\pi}{6}, \dfrac{3\pi}{4}, \dfrac{5\pi}{6}, \dfrac{7\pi}{4}\right\}$.

55. $\left(\sqrt{2} \sec x - 2\right)(2 \sin x + 1) = 0 \Rightarrow \sec x = \sqrt{2}$ or
$\sin x = -\dfrac{1}{2}$; $\sec x = \sqrt{2} \Rightarrow x = \dfrac{\pi}{4}$ or $x = \dfrac{7\pi}{4}$
$\sin x = -\dfrac{1}{2} \Rightarrow x = \dfrac{7\pi}{6}$ or $x = \dfrac{11\pi}{6}$.
The solution is $\left\{\dfrac{\pi}{4}, \dfrac{7\pi}{6}, \dfrac{7\pi}{4}, \dfrac{11\pi}{6}\right\}$.

57. $4\sin^2 x = 1 \Rightarrow \sin x = \pm\dfrac{1}{2} \Rightarrow x = \dfrac{\pi}{6}$ or $x = \dfrac{5\pi}{6}$ or $x = \dfrac{7\pi}{6}$ or $x = \dfrac{11\pi}{6}$.
The solution is $\left\{\dfrac{\pi}{6}, \dfrac{5\pi}{6}, \dfrac{7\pi}{6}, \dfrac{11\pi}{6}\right\}$.

59. $\tan^2 x = 1 \Rightarrow \tan x = \pm 1 \Rightarrow x = \dfrac{\pi}{4}$ or $x = \dfrac{3\pi}{4}$ or $x = \dfrac{5\pi}{4}$ or $x = \dfrac{7\pi}{4}$.
The solution is $\left\{\dfrac{\pi}{4}, \dfrac{3\pi}{4}, \dfrac{5\pi}{4}, \dfrac{7\pi}{4}\right\}$.

61. $3\csc^2 x = 4 \Rightarrow \csc x = \pm\dfrac{2\sqrt{3}}{3} \Rightarrow x = \dfrac{\pi}{3}$ or $x = \dfrac{2\pi}{3}$ or $x = \dfrac{4\pi}{3}$ or $x = \dfrac{5\pi}{3}$.
The solution is $\left\{\dfrac{\pi}{3}, \dfrac{2\pi}{3}, \dfrac{4\pi}{3}, \dfrac{5\pi}{3}\right\}$.

63. $2\sin^2\theta - \sin\theta - 1 = 0 \Rightarrow$
$(2\sin\theta + 1)(\sin\theta - 1) = 0 \Rightarrow \sin\theta = -\dfrac{1}{2}$
or $\sin\theta = 1$; $\sin\theta = -\dfrac{1}{2} \Rightarrow \theta = \dfrac{7\pi}{6}$ or $\theta = \dfrac{11\pi}{6}$
$\sin\theta = 1 \Rightarrow \theta = \dfrac{\pi}{2}$
The solution is $\left\{\dfrac{\pi}{2}, \dfrac{7\pi}{6}, \dfrac{11\pi}{6}\right\}$.

65. $2\cos^2\theta + 3\cos\theta = 2 \Rightarrow$
$2\cos^2\theta + 3\cos\theta - 2 = 0 \Rightarrow$
$(2\cos\theta - 1)(\cos\theta + 2) = 0 \Rightarrow \cos\theta = \dfrac{1}{2}$ or
$\cos\theta = -2$. (Reject this). $\cos\theta = \dfrac{1}{2} \Rightarrow$
$\theta = \dfrac{\pi}{3}$ or $\theta = \dfrac{5\pi}{3}$. The solution is $\left\{\dfrac{\pi}{3}, \dfrac{5\pi}{3}\right\}$.

67. $6\sin^2\theta - 11\sin\theta = -4$
$6\sin^2\theta - 11\sin\theta + 4 = 0 \Rightarrow$
$(2\sin\theta - 1)(3\sin\theta - 4) = 0 \Rightarrow \sin\theta = \dfrac{1}{2}$ or
$\sin\theta = \dfrac{4}{3}$. (Reject this). $\sin\theta = \dfrac{1}{2} \Rightarrow \theta = \dfrac{\pi}{6}$ or
$\theta = \dfrac{5\pi}{6}$. The solution is $\left\{\dfrac{\pi}{6}, \dfrac{5\pi}{6}\right\}$.

69. $3\tan^2\theta - 4\sqrt{3}\tan\theta = -3 \Rightarrow$
$3\tan^2\theta - 4\sqrt{3}\tan\theta + 3 = 0 \Rightarrow$
$(3\tan\theta - \sqrt{3})(\tan\theta - \sqrt{3}) = 0 \Rightarrow$
$\tan\theta = \dfrac{\sqrt{3}}{3}$ or $\tan\theta = \sqrt{3}$
$\tan\theta = \dfrac{\sqrt{3}}{3} \Rightarrow \theta = \dfrac{\pi}{6}, \theta = \dfrac{\pi}{6} + \pi = \dfrac{7\pi}{6}$
$\tan\theta = \sqrt{3} \Rightarrow \theta = \dfrac{\pi}{3}, \theta = \dfrac{\pi}{3} + \pi = \dfrac{4\pi}{3}$
The solution is $\left\{\dfrac{\pi}{6}, \dfrac{\pi}{3}, \dfrac{7\pi}{6}, \dfrac{4\pi}{3}\right\}$.

71. $\sin x = \cos x \Rightarrow \dfrac{\sin x}{\cos x} = 1 \Rightarrow \tan x = 1 \Rightarrow x = \dfrac{\pi}{4}$
or $x = \dfrac{5\pi}{4}$. The solution is $\left\{\dfrac{\pi}{4}, \dfrac{5\pi}{4}\right\}$.

73. $3\sin^2 x = \cos^2 x \Rightarrow 3 = \cot^2 x \Rightarrow$
$\pm\sqrt{3} = \cot x \Rightarrow x = \dfrac{\pi}{6}$ or $x = \dfrac{5\pi}{6}$ or
$x = \dfrac{7\pi}{6}$ or $x = \dfrac{11\pi}{6}$
The solution is $\left\{\dfrac{\pi}{6}, \dfrac{5\pi}{6}, \dfrac{7\pi}{6}, \dfrac{11\pi}{6}\right\}$.

75. $\cos^2 x - \sin^2 x = 1 \Rightarrow 1 - \sin^2 x - \sin^2 x = 1 \Rightarrow$
$-2\sin^2 x = 0 \Rightarrow \sin x = 0 \Rightarrow x = 0$ or $x = \pi$
The solution is $\{0, \pi\}$.

77. $2\cos^2 x - 3\sin x - 3 = 0 \Rightarrow$
$2(1 - \sin^2 x) - 3\sin x - 3 = 0 \Rightarrow$
$2\sin^2 x + 3\sin x + 1 = 0 \Rightarrow$
$(2\sin x + 1)(\sin x + 1) = 0 \Rightarrow \sin x = -1/2$ or
$\sin x = -1$; $\sin x = -\dfrac{1}{2} \Rightarrow x = \dfrac{7\pi}{6}$ or $x = \dfrac{11\pi}{6}$
$\sin x = -1 \Rightarrow x = \dfrac{3\pi}{2}$.
The solution is $\left\{\dfrac{7\pi}{6}, \dfrac{3\pi}{2}, \dfrac{11\pi}{6}\right\}$.

79. $\sqrt{3}\sec^2 x - 2\tan x - 2\sqrt{3} = 0 \Rightarrow$
$\sqrt{3}(1+\tan^2 x) - 2\tan x - 2\sqrt{3} = 0 \Rightarrow$
$\sqrt{3}\tan^2 x - 2\tan x - \sqrt{3} = 0 \Rightarrow$
$(\sqrt{3}\tan x + 1)(\tan x - \sqrt{3}) = 0 \Rightarrow \tan x = -\frac{\sqrt{3}}{3}$
or $\tan x = \sqrt{3}$; $\tan x = -\frac{\sqrt{3}}{3} \Rightarrow x = \frac{5\pi}{6}$ or
$x = \frac{11\pi}{6}$; $\tan x = \sqrt{3} \Rightarrow x = \frac{\pi}{3}$ or $x = \frac{4\pi}{3}$.
The solution is $\left\{\frac{\pi}{3}, \frac{5\pi}{6}, \frac{4\pi}{3}, \frac{11\pi}{6}\right\}$.

81. $\sqrt{3}\sin x = 1 + \cos x \Rightarrow$
$(\sqrt{3}\sin x)^2 = (1+\cos x)^2 \Rightarrow$
$3\sin^2 x = \cos^2 x + 2\cos x + 1 \Rightarrow$
$3(1-\cos^2 x) - \cos^2 x - 2\cos x - 1 = 0 \Rightarrow$
$-4\cos^2 x - 2\cos x + 2 = 0 \Rightarrow$
$-2(2\cos x - 1)(\cos x + 1) = 0 \Rightarrow \cos x = \frac{1}{2} \Rightarrow$
$x = \frac{\pi}{3}, \frac{5\pi}{3}$ or $\cos x = -1 \Rightarrow x = \pi$
Checking each answer, we find that $x = \frac{5\pi}{3}$
is extraneous. The solution is $\left\{\frac{\pi}{3}, \pi\right\}$.

83. $\tan x + 1 = \sec x \Rightarrow (\tan x + 1)^2 = \sec^2 x \Rightarrow$
$\tan^2 x + 2\tan x + 1 = \tan^2 x + 1 \Rightarrow$
$2\tan x = 0 \Rightarrow x = 0$ or $x = \pi$
Checking each answer, we find that $x = \pi$ is
extraneous. The solution is $\{0\}$.

85. $\cot x + 1 = \csc x$
$(\cot x + 1)^2 = \csc^2 x$
$\cot^2 x + 2\cot x + 1 = 1 + \cot^2 x$
$2\cot x = 0 \Rightarrow x = \frac{\pi}{2}, \frac{3\pi}{2}$
Checking each answer, we find that $\theta = \frac{3\pi}{2}$
is extraneous. The solution is $\left\{\frac{\pi}{2}\right\}$.

6.5 Applying the Concepts

87. $\tan\theta = \frac{24}{8\sqrt{3}} = \sqrt{3} \Rightarrow \theta = 60°$

89. $\sin\theta = \frac{60}{605} \Rightarrow \theta \approx 5.7°$

91. From the section opener, we know that the height traveled by a projectile is
$h = tv_0\sin\theta - 16t^2$, where h = the height, t = the time in seconds, θ = the initial angle the projectile makes with the ground, and v_0 is the projectile's initial velocity.
$h = tv_0\sin\theta - 16t^2 \Rightarrow$
$25 = (4)(150)\sin\theta - 16(4^2) \Rightarrow$
$\sin\theta = \frac{281}{600} \Rightarrow \theta = \sin^{-1}\left(\frac{281}{600}\right) \approx 28°$

6.5 Beyond the Basics

93. Using factoring by grouping, we have
$\sin x\cos x - \frac{\sqrt{3}}{2}\sin x + \frac{1}{2}\cos x - \frac{\sqrt{3}}{4} = 0 \Rightarrow$
$\sin x\left(\cos x - \frac{\sqrt{3}}{2}\right) + \frac{1}{2}\left(\cos x - \frac{\sqrt{3}}{2}\right) = 0 \Rightarrow$
$\left(\sin x + \frac{1}{2}\right)\left(\cos x - \frac{\sqrt{3}}{2}\right) = 0 \Rightarrow \sin x = -\frac{1}{2}$
or $\cos x = \frac{\sqrt{3}}{2}$. If $\sin x = -\frac{1}{2} \Rightarrow$
$x = \frac{7\pi}{6} + 2n\pi \Rightarrow x = \frac{7\pi}{6}$ or
$x = \frac{11\pi}{6} + 2n\pi \Rightarrow x = \frac{11\pi}{6}$
If $\cos x = \frac{\sqrt{3}}{2} \Rightarrow x = \frac{\pi}{6} + 2n\pi \Rightarrow x = \frac{\pi}{6}$ or
$x = \frac{11\pi}{6} + 2n\pi \Rightarrow x = \frac{11\pi}{6}$.
The solution is $\left\{\frac{\pi}{6}, \frac{7\pi}{6}, \frac{11\pi}{6}\right\}$.

95. $5\sin^2\theta + \cos^2\theta - \sec^2\theta = 0$
$5(1-\cos^2\theta) + \cos^2\theta - \frac{1}{\cos^2\theta} = 0$
$5 - 4\cos^2\theta - \frac{1}{\cos^2\theta} = 0$
$-4\cos^4\theta + 5\cos^2\theta - 1 = 0$
$4\cos^4\theta - 5\cos^2\theta + 1 = 0$
$(4\cos^2\theta - 1)(\cos^2\theta - 1) = 0 \Rightarrow$
$(2\cos\theta - 1)(2\cos\theta + 1)(\cos\theta - 1)(\cos\theta + 1) = 0$
$2\cos\theta - 1 = 0 \Rightarrow \cos\theta = \frac{1}{2} \Rightarrow \theta = \frac{\pi}{3}, \frac{5\pi}{3}$
$2\cos\theta + 1 = 0 \Rightarrow \cos\theta = -\frac{1}{2} \Rightarrow \theta = \frac{2\pi}{3}, \frac{4\pi}{3}$

(continued on next page)

(continued)

$\cos\theta - 1 = 0 \Rightarrow \cos\theta = 1 \Rightarrow \theta = 0$
$\cos\theta + 1 = 0 \Rightarrow \cos\theta = -1 \Rightarrow \theta = \pi$
Solution set: $\left\{0, \dfrac{\pi}{3}, \dfrac{2\pi}{3}, \pi, \dfrac{4\pi}{3}, \dfrac{5\pi}{3}\right\}$

97. $2\cos x = 1 - \sin x$
$(2\cos x)^2 = (1 - \sin x)^2$
$4\cos^2 x = \sin^2 x - 2\sin x + 1$
$4(1 - \sin^2 x) = \sin^2 x - 2\sin x + 1$
$4 - 4\sin^2 x = \sin^2 x - 2\sin x + 1$
$5\sin^2 x - 2\sin x - 3 = 0$
$(5\sin x + 3)(\sin x - 1) = 0 \Rightarrow$
$5\sin x + 3 = 0 \Rightarrow \sin x = -\dfrac{3}{5} \Rightarrow x \approx 5.6397$ or
$\sin x - 1 = 0 \Rightarrow \sin x = 1 \Rightarrow x = \dfrac{\pi}{2}$
Solution set: $\left\{\dfrac{\pi}{2}, 5.6397\right\}$

99. $\cot^2\theta + 3\csc\theta + 3 = 0$
$(\csc^2\theta - 1) + 3\csc\theta + 3 = 0$
$\csc^2\theta + 3\csc\theta + 2 = 0$
$(\csc\theta + 2)(\csc\theta + 1) = 0 \Rightarrow$
$\csc\theta = -2$ or $\csc\theta = -1$
$\csc\theta = -2 \Rightarrow \sin\theta = -\dfrac{1}{2} \Rightarrow \theta = \dfrac{7\pi}{6}, \dfrac{11\pi}{6}$
$\csc\theta = -1 \Rightarrow \sin\theta = -1 \Rightarrow \theta = \dfrac{3\pi}{2}$
Solution set: $\left\{\dfrac{7\pi}{6}, \dfrac{3\pi}{2}, \dfrac{11\pi}{6}\right\}$

6.5 Critical Thinking/Discussion/Writing

101. $\tan^2 x = 4 \Rightarrow \tan x = \pm 2$
If $\tan x = 2$, then $x \approx 1.107$ or
$x \approx 1.107 + \pi \approx 4.249$. If $\tan x = -2$, then
$x \approx -1.107$, which is not in $[0, 2\pi)$. So,
$x \approx -1.107 + \pi \approx 2.034$, and
$x \approx -1.107 + 2\pi \approx 5.176$.
Solution: $\{1.107, 2.034, 4.249, 5.176\}$

102. $3\cos^2 x + \cos x = 0 \Rightarrow \cos x(3\cos x + 1) = 0 \Rightarrow$
$\cos x = 0$ or $\cos x = -\dfrac{1}{3}$
If $\cos x = 0$, then $x = \dfrac{\pi}{2} \approx 1.5708$ or
$x = \dfrac{3\pi}{2} \approx 4.7124$. If $\cos x = -\dfrac{1}{3}$, then
$x \approx 1.9106$ (which is in quadrant II) or
$x \approx 4.3726$ (which is in quadrant III).
Solution: $\{1.571, 1.911, 4.373, 4.712\}$

103. $\cos x - \sec x + 1 = 0 \Rightarrow \cos x - \dfrac{1}{\cos x} + 1 = 0 \Rightarrow$
$\cos^2 x - 1 + \cos x = 0 \Rightarrow$
$\cos^2 x + \cos x + \dfrac{1}{4} = 1 + \dfrac{1}{4} \Rightarrow \left(\cos x + \dfrac{1}{2}\right)^2 = \dfrac{5}{4} \Rightarrow$
$\cos x + \dfrac{1}{2} = \pm\dfrac{\sqrt{5}}{2} \Rightarrow \cos x = -\dfrac{1}{2} \pm \dfrac{\sqrt{5}}{2}$
If $\cos x = -\dfrac{1}{2} + \dfrac{\sqrt{5}}{2}$, then $x \approx 0.9046$ or
$x \approx 2\pi - 0.9046 \approx 5.3786$.
$\cos x \neq -\dfrac{1}{2} - \dfrac{\sqrt{5}}{2}$, since $-\dfrac{1}{2} - \dfrac{\sqrt{5}}{2} \approx -1.6$,
which is not in the range of $\cos x$.
Solution: $\{0.905, 5.379\}$

104. $3\csc x + 2\cot^2 x = 5$
$3\csc x + 2(\csc^2 x - 1) = 5$
$2\csc^2 x + 3\csc x - 7 = 0$
Use the quadratic formula to solve for $\csc x$:
$\csc x = \dfrac{-3 \pm \sqrt{3^2 - 4(2)(-7)}}{2(2)} = \dfrac{-3 \pm \sqrt{65}}{4}$
If $\csc x = \dfrac{-3 + \sqrt{65}}{4}$, then $x \approx 0.9111$ or
$x \approx \pi - 0.9111 \approx 2.2305$. If
$\csc x = \dfrac{-3 - \sqrt{65}}{4}$, then $x \approx 3.5116$ or
$x \approx 5.9132$.
Solution: $\{0.911, 2.231, 3.512, 5.913\}$

105. $\sin x \cos x = \dfrac{1}{2} \Rightarrow 2\sin x \cos x = 1 \Rightarrow$
$\sin 2x = 1 \Rightarrow 2x = \dfrac{\pi}{2}$ or $2x = \dfrac{\pi}{2} + 2\pi \Rightarrow$
$x = \dfrac{\pi}{4}$ or $x = \dfrac{\pi}{4} + \pi = \dfrac{5\pi}{4}$
The solution is $\left\{\dfrac{\pi}{4}, \dfrac{5\pi}{4}\right\}$.

106.
$$7\cos^2 x + 3\sin^2 x = 4$$
$$7\cos^2 x + 3(1-\cos^2 x) - 4 = 0$$
$$7\cos^2 x + 3 - 3\cos^2 x - 4 = 0$$
$$4\cos^2 x - 1 = 0$$
$$\cos^2 x = \frac{1}{4} \Rightarrow \cos x = \pm\frac{1}{2}$$

$\cos x = \frac{1}{2} \Rightarrow x = \frac{\pi}{3}, x = \frac{5\pi}{3}$

$\cos x = -\frac{1}{2} \Rightarrow x = \frac{2\pi}{3}, x = \frac{4\pi}{3}$

Solution set: $\left\{\dfrac{\pi}{3}, \dfrac{2\pi}{3}, \dfrac{4\pi}{3}, \dfrac{5\pi}{3}\right\}$.

6.5 Getting Ready for the Next Section

107. $\sin x = -\dfrac{1}{2} \Rightarrow x = \dfrac{7\pi}{6}, \dfrac{11\pi}{6}$

109. $2\cos x = \sqrt{3} \Rightarrow \cos x = \dfrac{\sqrt{3}}{2} \Rightarrow x = \dfrac{\pi}{6}, \dfrac{11\pi}{6}$

111. Note that the range of $\cos^{-1} y$ is $[0, \pi]$.

$x = \cos^{-1}\left(-\dfrac{\sqrt{2}}{2}\right) \Rightarrow \cos x = -\dfrac{\sqrt{2}}{2} \Rightarrow x = \dfrac{3\pi}{4}$

113. Note that the range of $\tan^{-1} y$ is $\left(-\dfrac{\pi}{2}, \dfrac{\pi}{2}\right)$.

$x = \tan^{-1}(1) \Rightarrow \tan x = 1 \Rightarrow x = \dfrac{\pi}{4}$

115. $\cos 3x + \cos 5x = 2\cos\left(\dfrac{3x+5x}{2}\right)\cos\left(\dfrac{3x-5x}{2}\right)$
$$= 2\cos 4x \cos(-x)$$
$$= 2\cos 4x \cos x$$

6.6 Trigonometric Equations II

6.6 Practice Problems

1. $\sin 2x = \dfrac{1}{2} \Rightarrow 2x = \sin^{-1}\left(\dfrac{1}{2}\right) \Rightarrow$

$2x = \dfrac{\pi}{6} + 2n\pi$ or $\dfrac{5\pi}{6} + 2n\pi \Rightarrow$

$x = \dfrac{\pi}{12} + n\pi$ or $x = \dfrac{5\pi}{12} + n\pi$

Now find the values of n that result in solutions in the interval $[0, 2\pi)$: $n = 0, 1$.

$x = \dfrac{\pi}{12}$ or $x = \dfrac{5\pi}{12}$ or $x = \dfrac{13\pi}{12}$ or $x = \dfrac{17\pi}{12}$

Solution set: $\left\{\dfrac{\pi}{12}, \dfrac{5\pi}{12}, \dfrac{13\pi}{12}, \dfrac{17\pi}{12}\right\}$

2. $\tan\dfrac{x}{2} = \dfrac{1}{\sqrt{3}} \Rightarrow \dfrac{x}{2} = \tan^{-1}\left(\dfrac{1}{\sqrt{3}}\right) \Rightarrow$

$\dfrac{x}{2} = \dfrac{\pi}{6} + n\pi \Rightarrow x = \dfrac{\pi}{3} + 2n\pi$

Now find the values of n that result in solutions in the interval $[0, 2\pi)$:

$n = 0$: $x = \dfrac{\pi}{3}$

Solution set: $\left\{\dfrac{\pi}{3}\right\}$

3. $\sin 2x = \sin\left(\dfrac{3\pi}{4} - x\right)$

$2x = \dfrac{3\pi}{4} - x + 2n\pi$	$2x = \pi - \left(\dfrac{3\pi}{4} - x\right) + 2n\pi$
$3x = \dfrac{3\pi}{4} + 2n\pi$	$2x = \pi - \dfrac{3\pi}{4} + x + 2n\pi$
$x = \dfrac{\pi}{4} + \dfrac{2}{3}n\pi$	$x = \dfrac{\pi}{4} + 2n\pi$
$(n = 0)$	
$x = \dfrac{\pi}{4}$	$x = \dfrac{\pi}{4}$ $(n = 0)$
$(n = 1)$	
$x = \dfrac{\pi}{4} + \dfrac{2\pi}{3} = \dfrac{11\pi}{12}$	
$(n = 2)$	
$x = \dfrac{\pi}{4} + \dfrac{4\pi}{3} = \dfrac{19\pi}{12}$	

All other values of n result in solutions outside the interval $[0, 2\pi)$.

Solution set: $\left\{\dfrac{\pi}{4}, \dfrac{11\pi}{12}, \dfrac{19\pi}{12}\right\}$

4. $0.3 = 0.5\sin\left[\dfrac{\pi}{14.75}\left(x - \dfrac{14.75}{2}\right)\right] + 0.5$

$-0.2 = 0.5\sin\left[\dfrac{\pi}{14.75}\left(x - \dfrac{14.75}{2}\right)\right]$

$-0.4 = \sin\left[\dfrac{\pi}{14.75}\left(x - \dfrac{14.75}{2}\right)\right]$

Thus, $\theta = \sin^{-1}(-0.4) \approx -0.4115$ or
$\theta \approx \pi - 0.4115 \approx 3.5561$

(continued on next page)

(*continued*)

$$-0.4115 = \frac{\pi}{14.75}\left(x - \frac{14.75}{2}\right)$$

$$\frac{14.75(-0.4115)}{\pi} = x - \frac{14.75}{2}$$

$$x = -\frac{6.0699}{\pi} + \frac{14.75}{2} \approx 5.44$$

$$3.5561 = \frac{\pi}{14.75}\left(x - \frac{14.75}{2}\right)$$

$$\frac{14.75(3.5561)}{\pi} = x - \frac{14.75}{2}$$

$$x = 16.6821 + \frac{14.75}{2} \approx 24.06$$

So, 30% of the Moon is visible about 5 days and about 24 days after the new moon.

5. $\cot 3\theta = 1 \Rightarrow 3\theta = 45° + 180°(n) \Rightarrow$
$\theta = 15° + 60°(n)$

n	θ
0	$\theta = 15° + 60°(0) = 15°$
1	$\theta = 15° + 60°(1) = 75°$
2	$\theta = 15° + 60°(2) = 135°$
3	$\theta = 15° + 60°(3) = 195°$
4	$\theta = 15° + 60°(4) = 255°$
5	$\theta = 15° + 60°(5) = 315°$

Solution set: $\{15°, 75°, 135°, 195°, 255°, 315°\}$

6. $\sin 3x + \cos 3x = \sqrt{\frac{3}{2}} \Rightarrow$

$(\sin 3x + \cos 3x)^2 = \left(\sqrt{\frac{3}{2}}\right)^2 \Rightarrow$

$\sin^2 3x + 2\sin 3x \cos 3x + \cos^2 3x = \frac{3}{2} \Rightarrow$

$1 + \sin 6x = \frac{3}{2} \Rightarrow \sin 6x = \frac{1}{2}$

$\sin 6x = \frac{1}{2} \Rightarrow 6x = \frac{\pi}{6} + 2n\pi \Rightarrow x = \frac{\pi}{36} + \frac{n\pi}{3}$

or $6x = \frac{5\pi}{6} + 2n\pi \Rightarrow x = \frac{5\pi}{36} + \frac{n\pi}{3}$

n	x
0	$x = \frac{\pi}{36} + \frac{(0)\pi}{3} = \frac{\pi}{36}$
0	$x = \frac{5\pi}{36} + \frac{(0)\pi}{3} = \frac{5\pi}{36}$
1	$x = \frac{\pi}{36} + \frac{(1)\pi}{3} = \frac{13\pi}{36}$
1	$x = \frac{5\pi}{36} + \frac{(1)\pi}{3} = \frac{17\pi}{36}$
2	$x = \frac{\pi}{36} + \frac{(2)\pi}{3} = \frac{25\pi}{36}$
2	$x = \frac{5\pi}{36} + \frac{(2)\pi}{3} = \frac{29\pi}{36}$
3	$x = \frac{\pi}{36} + \frac{(3)\pi}{3} = \frac{37\pi}{36}$
3	$x = \frac{5\pi}{36} + \frac{(3)\pi}{3} = \frac{41\pi}{36}$
4	$x = \frac{\pi}{36} + \frac{(4)\pi}{3} = \frac{49\pi}{36}$
4	$x = \frac{5\pi}{36} + \frac{(4)\pi}{3} = \frac{53\pi}{36}$

Since squaring both sides of an equation may result in extraneous solutions, check all possible solutions.

Check:

x	$\sin 3x + \cos 3x \stackrel{?}{=} \sqrt{\frac{3}{2}}$
$\frac{\pi}{36}$	$\sin\left[3\left(\frac{\pi}{36}\right)\right] + \cos\left[3\left(\frac{\pi}{36}\right)\right] = \sqrt{\frac{3}{2}}$ ✓
$\frac{5\pi}{36}$	$\sin\left[3\left(\frac{5\pi}{36}\right)\right] + \cos\left[3\left(\frac{5\pi}{36}\right)\right] = \sqrt{\frac{3}{2}}$ ✓
$\frac{13\pi}{36}$	$\sin\left[3\left(\frac{13\pi}{36}\right)\right] + \cos\left[3\left(\frac{13\pi}{36}\right)\right] = -\sqrt{\frac{3}{2}}$ ✗
$\frac{17\pi}{36}$	$\sin\left[3\left(\frac{17\pi}{36}\right)\right] + \cos\left[3\left(\frac{17\pi}{36}\right)\right] = -\sqrt{\frac{3}{2}}$ ✗
$\frac{25\pi}{36}$	$\sin\left[3\left(\frac{25\pi}{36}\right)\right] + \cos\left[3\left(\frac{25\pi}{36}\right)\right] = \sqrt{\frac{3}{2}}$ ✓

(*continued on next page*)

(*continued*)

x	$\sin 3x + \cos 3x \stackrel{?}{=} \sqrt{\dfrac{3}{2}}$
$\dfrac{29\pi}{36}$	$\sin\left[3\left(\dfrac{29\pi}{36}\right)\right] + \cos\left[3\left(\dfrac{29\pi}{36}\right)\right] = \sqrt{\dfrac{3}{2}}$ ✓
$\dfrac{37\pi}{36}$	$\sin\left[3\left(\dfrac{37\pi}{36}\right)\right] + \cos\left[3\left(\dfrac{37\pi}{36}\right)\right] = -\sqrt{\dfrac{3}{2}}$ ✗
$\dfrac{41\pi}{36}$	$\sin\left[3\left(\dfrac{41\pi}{36}\right)\right] + \cos\left[3\left(\dfrac{41\pi}{36}\right)\right] = -\sqrt{\dfrac{3}{2}}$ ✗
$\dfrac{49\pi}{36}$	$\sin\left[3\left(\dfrac{49\pi}{36}\right)\right] + \cos\left[3\left(\dfrac{49\pi}{36}\right)\right] = \sqrt{\dfrac{3}{2}}$ ✓
$\dfrac{53\pi}{36}$	$\sin\left[3\left(\dfrac{53\pi}{36}\right)\right] + \cos\left[3\left(\dfrac{53\pi}{36}\right)\right] = \sqrt{\dfrac{3}{2}}$ ✓

Solution set:
$$\left\{\dfrac{\pi}{36}, \dfrac{5\pi}{36}, \dfrac{25\pi}{36}, \dfrac{29\pi}{36}, \dfrac{49\pi}{36}, \dfrac{53\pi}{36}\right\}$$

7. $\sin 2x + \sin 3x = 0$

$2\sin\dfrac{5x}{2}\cos\left(-\dfrac{x}{2}\right) = 0$

$2\sin\dfrac{5x}{2}\cos\dfrac{x}{2} = 0 \Rightarrow \sin\dfrac{5x}{2}\cos\dfrac{x}{2} = 0 \Rightarrow$

$\sin\dfrac{5x}{2} = 0$ or $\cos\dfrac{x}{2} = 0$.

$\sin\theta = 0 \Rightarrow \theta = 0 + 2n\pi = 2n\pi$ or

$\theta = \pi + 2n\pi \Rightarrow \dfrac{5x}{2} = 2n\pi \Rightarrow x = \dfrac{4n\pi}{5}$ or

$\dfrac{5x}{2} = \pi + 2n\pi \Rightarrow x = \dfrac{2\pi + 4n\pi}{5}$

$\cos\theta = 0 \Rightarrow \theta = \dfrac{\pi}{2} + 2n\pi$ or $\theta = \dfrac{3\pi}{2} + 2n\pi$.

$\dfrac{x}{2} = \dfrac{\pi}{2} + 2n\pi \Rightarrow x = \pi + 4n\pi$ or

$\dfrac{x}{2} = \dfrac{3\pi}{2} + 2n\pi \Rightarrow x = \dfrac{3\pi}{4} + 4n\pi$

n	x
0	$x = \dfrac{4(0)\pi}{5} = 0$
0	$x = \dfrac{2\pi + 4(0)\pi}{5} = \dfrac{2\pi}{5}$
0	$x = \pi + 4(0)\pi = \pi$
0	$x = \dfrac{3\pi}{4} + 4(0)\pi = \dfrac{3\pi}{4}$

n	x
1	$x = \dfrac{4(1)\pi}{5} = \dfrac{4\pi}{5}$
1	$x = \dfrac{2\pi + 4(1)\pi}{5} = \dfrac{6\pi}{5}$
1	$x = \pi + 4(1)\pi = 5\pi$ out of range
1	$x = \dfrac{3\pi}{4} + 4(1)\pi = \dfrac{19\pi}{4}$ out of range
2	$x = \dfrac{4(2)\pi}{5} = \dfrac{8\pi}{5}$
2	$x = \dfrac{2\pi + 4(2)\pi}{5} = \dfrac{10\pi}{5} = 2\pi$ (out of range)

Check each possible solution.

x	$\sin 2x + \sin 3x \stackrel{?}{=} 0$
0	$\sin(2 \cdot 0) + \sin(3 \cdot 0) = 0$ ✓
$\dfrac{2\pi}{5}$	$\sin\left(2 \cdot \dfrac{2\pi}{5}\right) + \sin\left(3 \cdot \dfrac{2\pi}{5}\right) = 0$ ✓
π	$\sin(2\pi) + \sin(3\pi) = 0$ ✓
$\dfrac{3\pi}{4}$	$\sin\left(2 \cdot \dfrac{3\pi}{4}\right) + \sin\left(3 \cdot \dfrac{3\pi}{4}\right) = 0$ ✗
$\dfrac{4\pi}{5}$	$\sin\left(2 \cdot \dfrac{4\pi}{5}\right) + \sin\left(3 \cdot \dfrac{4\pi}{5}\right) = 0$ ✓
$\dfrac{6\pi}{5}$	$\sin\left(2 \cdot \dfrac{6\pi}{5}\right) + \sin\left(3 \cdot \dfrac{6\pi}{5}\right) = 0$ ✓
$\dfrac{8\pi}{5}$	$\sin\left(2 \cdot \dfrac{8\pi}{5}\right) + \sin\left(3 \cdot \dfrac{8\pi}{5}\right) = 0$ ✓

Solution set: $\left\{0, \dfrac{2\pi}{5}, \dfrac{4\pi}{5}, \pi, \dfrac{6\pi}{5}, \dfrac{8\pi}{5}\right\}$

8. $\dfrac{\pi}{4} + 3\sin^{-1}(x+1) = \dfrac{5\pi}{4}$

$3\sin^{-1}(x+1) = \pi$

$\sin^{-1}(x+1) = \dfrac{\pi}{3} \Rightarrow x + 1 = \sin\dfrac{\pi}{3}$

$x = \sin\dfrac{\pi}{3} - 1 = \dfrac{\sqrt{3}}{2} - 1$

$= \dfrac{\sqrt{3} - 2}{2}$

Section 6.6 Trigonometric Equations II   **367**

9. a. $\tan^{-1} x + \cot^{-1} x = \dfrac{\pi}{2}$, x in $(-\infty, \infty)$.

   Let $\theta = \tan^{-1} x$. Then $-\dfrac{\pi}{2} < \theta < \dfrac{\pi}{2}$, so

   $\dfrac{\pi}{2} > -\theta > -\dfrac{\pi}{2}$ and $\pi > \dfrac{\pi}{2} - \theta > 0$ or

   $0 < \dfrac{\pi}{2} - \theta < \pi$. Then, we have

   $x = \tan \theta \Rightarrow x = \cot\left(\dfrac{\pi}{2} - \theta\right)$ (using a cofunction identity).

   $x = \cot\left(\dfrac{\pi}{2} - \theta\right) \Rightarrow \cot^{-1} x = \dfrac{\pi}{2} - \theta \Rightarrow$

   $\theta + \cot^{-1} x = \dfrac{\pi}{2} \Rightarrow \tan^{-1} x + \cot^{-1} x = \dfrac{\pi}{2}$

   b. $\tan^{-1} x - \cot^{-1} x = \dfrac{\pi}{4}$

   Using the result from part (a), we have

   $\tan^{-1} x - \left(\dfrac{\pi}{2} - \tan^{-1} x\right) = \dfrac{\pi}{4}$

   $2\tan^{-1} x - \dfrac{\pi}{2} = \dfrac{\pi}{4}$

   $2\tan^{-1} x = \dfrac{3\pi}{4} \Rightarrow$

   $\tan^{-1} x = \dfrac{3\pi}{8} \Rightarrow x = \tan\dfrac{3\pi}{8}$

   Now use a half-angle formula for $\tan\dfrac{3\pi}{4}$.

   $x = \tan\left(\dfrac{\frac{3\pi}{4}}{2}\right) = \dfrac{\sin\frac{3\pi}{4}}{1 + \cos\frac{3\pi}{4}} = \dfrac{\frac{\sqrt{2}}{2}}{1 - \frac{\sqrt{2}}{2}}$

   $= \dfrac{\sqrt{2}}{2 - \sqrt{2}} = \dfrac{\sqrt{2}}{2 - \sqrt{2}} \cdot \dfrac{2 + \sqrt{2}}{2 + \sqrt{2}} = \dfrac{2\sqrt{2} + 2}{2}$

   $= \sqrt{2} + 1$

   Alternatively, we have

   $x = \tan\left(\dfrac{\frac{3\pi}{4}}{2}\right) = \sqrt{\dfrac{1 - \cos\frac{3\pi}{4}}{1 + \cos\frac{3\pi}{4}}} = \sqrt{\dfrac{1 + \frac{\sqrt{2}}{2}}{1 - \frac{\sqrt{2}}{2}}}$

   $= \sqrt{\dfrac{2 + \sqrt{2}}{2 - \sqrt{2}}}$

6.6 Concepts and Vocabulary

1. If $\sin 2x_1 = \sin 2x_2$ and $0 < x_1 < x_2 < \dfrac{\pi}{2}$,

   then $x_2 = \underline{\dfrac{\pi}{2} - x_1}$.

3. If $\tan 2x_1 = \tan 2x_2$ and $0 < x_1 < x_2 < \pi$,

   then $x_2 = \underline{\dfrac{\pi}{2} + x_1}$.

5. True

7. False. For example, if $x = \pi$, then

   $\dfrac{2\tan\dfrac{\pi}{2}}{\pi}$ is undefined $\neq 1$.

6.6 Building Skills

9. $\cos 2x = \dfrac{\sqrt{3}}{2} \Rightarrow 2x = \dfrac{\pi}{6} + 2n\pi \Rightarrow x = \dfrac{\pi}{12} + n\pi$

   or $2x = \left(2\pi - \dfrac{\pi}{6}\right) + 2n\pi \Rightarrow$

   $2x = \dfrac{11\pi}{6} + 2n\pi \Rightarrow x = \dfrac{11\pi}{12} + n\pi$

   Now find the values of n that result in solutions in the interval $[0, 2\pi)$:

   $n = 0 \Rightarrow x = \dfrac{\pi}{12}$ or $x = \dfrac{11\pi}{12}$

   $n = 1 \Rightarrow x = \dfrac{13\pi}{12}$ or $x = \dfrac{23\pi}{12}$

   If $n > 1$, the solutions are out of the domain, so the solutions are $\left\{\dfrac{\pi}{12}, \dfrac{11\pi}{12}, \dfrac{13\pi}{12}, \dfrac{23\pi}{12}\right\}$.

11. $\sin 2x = -\dfrac{1}{2} \Rightarrow 2x = -\dfrac{\pi}{6} + 2n\pi \Rightarrow$

   $x = -\dfrac{\pi}{12} + n\pi$ or $2x = \left(\pi - \left(-\dfrac{\pi}{6}\right)\right) + 2n\pi \Rightarrow$

   $2x = \dfrac{7\pi}{6} + 2n\pi \Rightarrow x = \dfrac{7\pi}{12} + n\pi$

   Now find the values of n that result in solutions in the interval $[0, 2\pi)$: $n = 0 \Rightarrow x = \dfrac{7\pi}{12}$;

   $n = 1 \Rightarrow x = \dfrac{11\pi}{12}$ or $x = \dfrac{19\pi}{12}$;

   $n = 2 \Rightarrow x = \dfrac{23\pi}{12}$.

   (continued on next page)

368 Chapter 6 *Trigonometric Identities and Equations*

(*continued*)

If $n > 2$, the solutions are out of the domain, so the solutions are $\left\{\dfrac{7\pi}{12}, \dfrac{11\pi}{12}, \dfrac{19\pi}{12}, \dfrac{23\pi}{12}\right\}$.

13. $\sec 2x = \dfrac{1}{2}$ has no solution because the range of $\sec x$ is $-\infty < x < -1$ or $1 < x < \infty$

15. $\tan 2x = \dfrac{\sqrt{3}}{3} \Rightarrow 2x = \dfrac{\pi}{6} + n\pi \Rightarrow x = \dfrac{\pi}{12} + \dfrac{\pi}{2}n$.

Now find the values of n that result in solutions in the interval $[0, 2\pi)$:

$n = 0 \Rightarrow x = \dfrac{\pi}{12}$,

$n = 1 \Rightarrow x = \dfrac{7\pi}{12}, n = 2 \Rightarrow x = \dfrac{13\pi}{12}$.

$n = 3 \Rightarrow \dfrac{19\pi}{12}$

If $n > 3$, the solutions are out of the domain, so the solutions are $\left\{\dfrac{\pi}{12}, \dfrac{7\pi}{12}, \dfrac{13\pi}{12}, \dfrac{19\pi}{12}\right\}$.

17. $\sin\left(2x - \dfrac{\pi}{3}\right) = \dfrac{1}{2} \Rightarrow 2x - \dfrac{\pi}{3} = \sin^{-1}\left(\dfrac{1}{2}\right) \Rightarrow$

$2x - \dfrac{\pi}{3} = \dfrac{\pi}{6} + 2n\pi \Rightarrow 2x = \dfrac{\pi}{2} + 2n\pi \Rightarrow$

$x = \dfrac{\pi}{4} + n\pi$ or

$2x - \dfrac{\pi}{3} = \pi - \dfrac{\pi}{6} + 2n\pi = \dfrac{5\pi}{6} + 2n\pi \Rightarrow$

$2x = \dfrac{7\pi}{6} + 2n\pi \Rightarrow x = \dfrac{7\pi}{12} + n\pi$

Now find the values of n that result in solutions in the interval $[0, 2\pi)$:

$n = 0: x = \dfrac{\pi}{4}$ or $x = \dfrac{7\pi}{12}$

$n = 1: x = \dfrac{5\pi}{4}$ or $x = \dfrac{19\pi}{12}$

If $n > 1$, the solutions are out of the domain, so the solutions are $\left\{\dfrac{\pi}{4}, \dfrac{7\pi}{12}, \dfrac{5\pi}{4}, \dfrac{19\pi}{12}\right\}$.

19. $\sin 3x = \dfrac{1}{2} \Rightarrow 3x = \dfrac{\pi}{6} + 2n\pi$ or

$3x = \dfrac{5\pi}{6} + 2n\pi \Rightarrow x = \dfrac{\pi}{18} + \dfrac{2n\pi}{3}$ or

$x = \dfrac{5\pi}{18} + \dfrac{2n\pi}{3}$

Now find the values of n that result in solutions in the interval $[0, 2\pi)$:

$n = 0: \dfrac{\pi}{18}$ or $\dfrac{5\pi}{18}$

$n = 1: \dfrac{13\pi}{18}$ or $\dfrac{17\pi}{18}$

$n = 2: \dfrac{25\pi}{18}$ or $\dfrac{29\pi}{18}$

If $n > 3$, the solutions are out of the domain, so the solutions are $\left\{\dfrac{\pi}{18}, \dfrac{5\pi}{18}, \dfrac{13\pi}{18}, \dfrac{17\pi}{18}, \dfrac{25\pi}{18}, \dfrac{29\pi}{18}\right\}$.

21. $\cos 3x = \dfrac{1}{2} \Rightarrow 3x = \dfrac{\pi}{3} + 2n\pi$ or

$3x = \left(2\pi - \dfrac{\pi}{3}\right) + 2n\pi = \dfrac{5\pi}{3} + 2n\pi$

$3x = \dfrac{\pi}{3} + 2n\pi \Rightarrow x = \dfrac{\pi}{9} + \dfrac{2}{3}n\pi$

$3x = \dfrac{5\pi}{3} + 2n\pi \Rightarrow x = \dfrac{5\pi}{9} + \dfrac{2}{3}n\pi$

Now find the values of n that result in solutions in the interval $[0, 2\pi)$.

$n = 0: \dfrac{\pi}{9}$ or $\dfrac{5\pi}{9}$

$n = 1: \dfrac{7\pi}{9}$ or $\dfrac{11\pi}{9}$

$n = 2: \dfrac{13\pi}{9}$ or $\dfrac{17\pi}{9}$

If $n > 3$, the solutions are out of the domain, so the solutions are $\left\{\dfrac{\pi}{9}, \dfrac{5\pi}{9}, \dfrac{7\pi}{9}, \dfrac{11\pi}{9}, \dfrac{13\pi}{9}, \dfrac{17\pi}{9}\right\}$.

23. $\cos \dfrac{x}{2} = \dfrac{1}{2} \Rightarrow \dfrac{x}{2} = \dfrac{\pi}{3} + 2n\pi$ or

$\dfrac{x}{2} = \dfrac{5\pi}{3} + 2n\pi \Rightarrow x = \dfrac{2\pi}{3} + 4n\pi$ or

$x = \dfrac{10\pi}{3} + 4n\pi$

(Note that the second value is not in the domain.). The only value of n that results in a solution in the interval $[0, 2\pi)$ is $n = 0$. The solution is $\left\{\dfrac{2\pi}{3}\right\}$.

25. $\sin \dfrac{x}{2} = -\dfrac{\sqrt{3}}{2}$ has no solution because $-\dfrac{\sqrt{3}}{2}$ is not in the range of $\sin x$.

27. $\tan \dfrac{x}{3} = 1 \Rightarrow \dfrac{x}{3} = \dfrac{\pi}{4} + n\pi$ or $\dfrac{x}{3} = \dfrac{5\pi}{4} + n\pi \Rightarrow$
$x = \dfrac{3\pi}{4} + 3n\pi$. The only value of n that results in a solution in the interval $[0, 2\pi)$ is $n = 0$.
The solution is $\left\{\dfrac{3\pi}{4}\right\}$.

29. $\sin 2x = \sin x \Rightarrow \sin 2x - \sin x = 0 \Rightarrow$
$2 \sin x \cos x - \sin x = 0 \Rightarrow$
$\sin x (2 \cos x - 1) = 0 \Rightarrow \sin x = 0$ or
$2 \cos x - 1 = 0 \Rightarrow \cos x = \dfrac{1}{2}$
If $\sin x = 0 \Rightarrow x = 0$ or $x = \pi$.
If $\cos x = \dfrac{1}{2} \Rightarrow x = \dfrac{\pi}{3}$ or $x = \dfrac{5\pi}{3}$.
The solution is $\left\{0, \dfrac{\pi}{3}, \pi, \dfrac{5\pi}{3}\right\}$.

31. $\sin 2x = \cos x \Rightarrow \sin 2x - \cos x = 0 \Rightarrow$
$2 \sin x \cos x - \cos x = 0 \Rightarrow$
$\cos x (2 \sin x - 1) = 0 \Rightarrow \cos x = 0$ or
$2 \sin x - 1 = 0 \Rightarrow \sin x = \dfrac{1}{2}$
If $\cos x = 0 \Rightarrow x = \dfrac{\pi}{2}$ or $\dfrac{3\pi}{2}$.
If $\sin x = \dfrac{1}{2} \Rightarrow x = \dfrac{\pi}{6} + 2n\pi$ or
$x = \pi - \left(\dfrac{\pi}{6}\right) + 2n\pi = \dfrac{5\pi}{6} + 2n\pi$.
The solution is $\left\{\dfrac{\pi}{6}, \dfrac{\pi}{2}, \dfrac{5\pi}{6}, \dfrac{3\pi}{2}\right\}$.

33. $\cos\left(2x + \dfrac{\pi}{4}\right) = \cos x \Rightarrow$
$2x + \dfrac{\pi}{4} = x + 2n\pi$ or $2x + \dfrac{\pi}{4} = (2\pi - x) + 2n\pi$
$2x + \dfrac{\pi}{4} = x + 2n\pi \Rightarrow x = -\dfrac{\pi}{4} + 2n\pi$
$2x + \dfrac{\pi}{4} = (2\pi - x) + 2n\pi \Rightarrow$
$3x = \dfrac{7\pi}{4} + 2n\pi \Rightarrow x = \dfrac{7\pi}{12} + \dfrac{2}{3}n\pi$
Now find the values of n that result in solutions in the interval $[0, 2\pi)$.
$n = 0$: $\dfrac{7\pi}{12}$
$n = 1$: $\dfrac{7\pi}{4}$ or $\dfrac{5\pi}{4}$

$n = 2$: $\dfrac{23\pi}{12}$
The solutions are $\left\{\dfrac{7\pi}{12}, \dfrac{5\pi}{4}, \dfrac{7\pi}{4}, \dfrac{23\pi}{12}\right\}$.

35. $\sin 2x = \cos\left(x - \dfrac{\pi}{4}\right) \Rightarrow$
$\sin 2x = \sin\left(\dfrac{\pi}{2} - \left(x - \dfrac{\pi}{4}\right)\right) = \sin\left(\dfrac{3\pi}{4} - x\right) \Rightarrow$
$2x = \dfrac{3\pi}{4} - x + 2n\pi$ or
$2x = \pi - \left(\dfrac{3\pi}{4} - x\right) + 2n\pi$
$2x = \dfrac{3\pi}{4} - x + 2n\pi \Rightarrow 3x = \dfrac{3\pi}{4} + 2n\pi \Rightarrow$
$x = \dfrac{\pi}{4} + \dfrac{2}{3}n\pi$
$2x = \pi - \left(\dfrac{3\pi}{4} - x\right) + 2n\pi = \dfrac{\pi}{4} + x + 2n\pi \Rightarrow$
$x = \dfrac{\pi}{4} + 2n\pi$
Now find the values of n that result in solutions in the interval $[0, 2\pi)$.
$n = 0$: $\dfrac{\pi}{4}$ $\dfrac{\pi}{4}$
$n = 1$: $\dfrac{11\pi}{12}$
$n = 2$: $\dfrac{19\pi}{12}$
The solutions are $\left\{\dfrac{\pi}{4}, \dfrac{11\pi}{12}, \dfrac{19\pi}{12}\right\}$.

37. $\sin\left(2x - \dfrac{\pi}{4}\right) = \cos\left(x + \dfrac{\pi}{4}\right)$
$\sin\left(2x - \dfrac{\pi}{4}\right) = \sin\left(\dfrac{\pi}{2} - \left(x + \dfrac{\pi}{4}\right)\right)$
$= \sin\left(\dfrac{\pi}{4} - x\right)$
$2x - \dfrac{\pi}{4} = \dfrac{\pi}{4} - x + 2n\pi$ or
$2x - \dfrac{\pi}{4} = \pi - \left(\dfrac{\pi}{4} - x\right) + 2n\pi$
$2x - \dfrac{\pi}{4} = \dfrac{\pi}{4} - x + 2n\pi \Rightarrow 3x = \dfrac{\pi}{2} + 2n\pi \Rightarrow$
$x = \dfrac{\pi}{6} + \dfrac{2}{3}n\pi$

(continued on next page)

(continued)

$$2x - \frac{\pi}{4} = \pi - \left(\frac{\pi}{4} - x\right) + 2n\pi$$

$$= \frac{3\pi}{4} + x + 2n\pi \Rightarrow x = \pi + 2n\pi$$

Now find the values of n that result in solutions in the interval $[0, 2\pi)$.

$n = 0: \frac{\pi}{6} \quad \pi$

$n = 1: \frac{5\pi}{6}$

$n = 2: \frac{3\pi}{2}$

The solutions are $\left\{\frac{\pi}{6}, \frac{5\pi}{6}, \pi, \frac{3\pi}{2}\right\}$.

39. $3\cos 2\theta = 1 \Rightarrow \cos 2\theta = \frac{1}{3} \Rightarrow$
    $2\theta \approx 70.5288 + 2n(180°) \Rightarrow \theta \approx 35.3° + 180n$
    or $2\theta \approx 289.4712 + 2n(180°) \Rightarrow$
    $\theta \approx 144.7° + 180n$.
    $\theta \approx 35.3° + 180n \Rightarrow \theta \approx 35.3°, 215.3°$
    $\theta \approx 144.7° + 180n \Rightarrow \theta \approx 144.7°, 324.7°$
    Solution set: $\{35.3°, 144.7°, 215.3°, 324.7°\}$

41. $2\cos 3\theta + 1 = 2 \Rightarrow \cos 3\theta = \frac{1}{2} \Rightarrow$
    $3\theta = 60° + 2n(180°) \Rightarrow \theta = 20° + n(120°)$
    or $3\theta = 300° + 2n(180°) \Rightarrow$
    $\theta = 100° + n(120°)$
    $\theta \approx 20° + n(120°) \Rightarrow \theta = 20°, 140°, 260°$
    $\theta \approx 100° + n(120°) \Rightarrow \theta = 100°, 220°, 340°$
    Solution set: $\{20°, 100°, 140°, 220°, 260°, 340°\}$

43. $3\sin 3\theta + 1 = 0 \Rightarrow \sin 3\theta = -\frac{1}{3} \Rightarrow$
    $3\theta \approx -19.4712 + 2n(180°)$ or
    $3\theta \approx 199.4712 + 2n(180°)$.
    $3\theta \approx -19.4712 + 2n(180°) \Rightarrow$
    $\theta \approx -6.5 + (120°)n \Rightarrow$
    $\theta \approx 113.5°, 233.5°, 353.5°$.
    $3\theta \approx 199.4712 + 2n(180°) \Rightarrow$
    $\theta \approx 66.5° + (120°)n \Rightarrow$
    $\theta \approx 66.5°, 186.5°, 306.5°$
    Solution set: $\{66.5°, 113.5°, 186.5°, 233.5°, 306.5°, 353.5°\}$

45. $2\sin\frac{\theta}{2} + 1 = 0 \Rightarrow \sin\frac{\theta}{2} = -\frac{1}{2} \Rightarrow$
    $\frac{\theta}{2} = 210° + 2n(180°)$ or $\frac{\theta}{2} = 330° + 2n(180°)$
    $\frac{\theta}{2} = 210° + 2n(180°) \Rightarrow \theta = 420° + 4n(180°)$,
    which is outside the desired interval.
    $\frac{\theta}{2} = 330° + 2n(180°) \Rightarrow \theta = 660° + 4n(180°)$,
    which is also outside the desired interval.
    Therefore, the solution set is $\emptyset$.

47. $2\sec 2\theta + 3 = 7 \Rightarrow \sec 2\theta = 2 \Rightarrow$
    $2\theta = 60° + 2n(180°)$ or $2\theta = 300° + 2n(180°)$.
    $2\theta = 60° + 2n(180°) \Rightarrow \theta = 30° + n(180°) \Rightarrow$
    $\theta = 30°, 210°$
    $2\theta = 300° + 2n(180°) \Rightarrow$
    $\theta = 150° + n(180°) \Rightarrow \theta = 150°, 330°$
    Solution set: $\{30°, 150°, 210°, 330°\}$

49. $3\csc\frac{\theta}{2} - 1 = 5 \Rightarrow \csc\frac{\theta}{2} = 2 \Rightarrow$
    $\frac{\theta}{2} = 30° + 2n(180°)$ or $\frac{\theta}{2} = 150° + 2n(180°)$.
    $\frac{\theta}{2} = 30° + 2n(180°) \Rightarrow \theta = 60° + 4n(180°) \Rightarrow$
    $\theta = 60°$.
    $\frac{\theta}{2} = 150° + 2n(180°) \Rightarrow \theta = 300° + 4n(180°) \Rightarrow$
    $\theta = 300°$.
    Solution set: $\{60°, 300°\}$

51. $\sin 2x + \sin x = 0$
    $2\sin\left(\frac{2x+x}{2}\right)\cos\left(\frac{2x-x}{2}\right) = 0$
    $\sin\frac{3x}{2}\cos\frac{x}{2} = 0 \Rightarrow$
    $\sin\frac{3x}{2} = 0$ or $\cos\frac{x}{2} = 0$.
    $\sin\frac{3x}{2} = 0 \Rightarrow \frac{3x}{2} = 0 + 2n\pi$ or $\frac{3x}{2} = 2n\pi$.
    $\frac{3x}{2} = 2n\pi \Rightarrow x = \frac{4n\pi}{3} \Rightarrow x = 0, \frac{4\pi}{3}$.
    $\frac{3x}{2} = \pi \Rightarrow x = \frac{2\pi}{3}$
    $\cos\frac{x}{2} = 0 \Rightarrow \frac{x}{2} = \frac{\pi}{2} + 2n\pi$ or $\frac{x}{2} = \frac{3\pi}{2} + 2n\pi$
    $\frac{x}{2} = \frac{\pi}{2} + 2n\pi \Rightarrow x = \pi + 4n\pi \Rightarrow x = \pi$

(continued on next page)

(*continued*)

$\dfrac{x}{2} = \dfrac{3\pi}{2} + 2n\pi \Rightarrow x = 3\pi + 4n\pi$, which is not in the given interval.

Solution set: $\left\{0, \dfrac{2\pi}{3}, \pi, \dfrac{4\pi}{3}\right\}$

53.
$$\cos 3x - \cos x = 0$$
$$-2\sin\left(\dfrac{3x+x}{2}\right)\sin\left(\dfrac{3x-x}{2}\right) = 0$$
$$\sin 2x \sin x = 0 \Rightarrow$$
$\sin 2x = 0$ or $\sin x = 0$.
$\sin 2x = 0 \Rightarrow 2x = 0 + 2n\pi$ or $2x = \pi + 2n\pi$.
$2x = 0 + 2n\pi \Rightarrow x = 0 + n\pi \Rightarrow x = 0, \pi$.
$2x = \pi + 2n\pi \Rightarrow x = \dfrac{\pi}{2} + n\pi \Rightarrow x = \dfrac{\pi}{2}, \dfrac{3\pi}{2}$
$\sin x = 0 \Rightarrow x = 0 + 2n\pi \Rightarrow x = 0, \pi$

Solution set: $\left\{0, \dfrac{\pi}{2}, \pi, \dfrac{3\pi}{2}\right\}$

55.
$$\cos 3x + \cos 5x = 0$$
$$2\cos\left(\dfrac{3x+5x}{2}\right)\cos\left(\dfrac{3x-5x}{2}\right) = 0$$
$$2\cos(4x)\cos(-x) = 0$$
$$2\cos(4x)\cos x = 0 \Rightarrow$$
$\cos 4x = 0$ or $\cos x = 0$.
$\cos 4x = 0 \Rightarrow 4x = \dfrac{\pi}{2} + 2n\pi$ or
$4x = \dfrac{3\pi}{2} + 2n\pi$.
$4x = \dfrac{\pi}{2} + 2n\pi \Rightarrow x = \dfrac{\pi}{8} + \dfrac{n\pi}{2} \Rightarrow$
$x = \dfrac{\pi}{8}, \dfrac{5\pi}{8}, \dfrac{9\pi}{8}, \dfrac{13\pi}{8}$.
$4x = \dfrac{3\pi}{2} + 2n\pi \Rightarrow x = \dfrac{3\pi}{8} + \dfrac{n\pi}{2} \Rightarrow$
$x = \dfrac{3\pi}{8}, \dfrac{7\pi}{8}, \dfrac{11\pi}{8}, \dfrac{15\pi}{8}$
$\cos x = 0 \Rightarrow x = \dfrac{\pi}{2} + 2n\pi$ or $x = \dfrac{3\pi}{2} + 2n\pi$.
$x = \dfrac{\pi}{2} + 2n\pi \Rightarrow x = \dfrac{\pi}{2}$.
$x = \dfrac{3\pi}{2} + 2n\pi \Rightarrow x = \dfrac{3\pi}{2}$.

Solution set:
$\left\{\dfrac{\pi}{8}, \dfrac{3\pi}{8}, \dfrac{\pi}{2}, \dfrac{5\pi}{8}, \dfrac{7\pi}{8}, \dfrac{9\pi}{8}, \dfrac{11\pi}{8}, \dfrac{3\pi}{2}, \dfrac{13\pi}{8}, \dfrac{15\pi}{8}\right\}$

57.
$$\sin 3x - \sin 5x = 0$$
$$2\sin\left(\dfrac{3x-5x}{2}\right)\cos\left(\dfrac{3x+5x}{2}\right) = 0$$
$$2\sin(-x)\cos(4x) = 0$$
$$-2\sin x\cos(4x) = 0 \Rightarrow$$
$\sin x = 0$ or $\cos 4x = 0$.
$\sin x = 0 \Rightarrow x = 0 + 2n\pi \Rightarrow x = 0, \pi$
$\cos 4x = 0 \Rightarrow 4x = \dfrac{\pi}{2} + 2n\pi$ or
$4x = \dfrac{3\pi}{2} + 2n\pi$.
$4x = \dfrac{\pi}{2} + 2n\pi \Rightarrow x = \dfrac{\pi}{8} + \dfrac{n\pi}{2} \Rightarrow$
$x = \dfrac{\pi}{8}, \dfrac{5\pi}{8}, \dfrac{9\pi}{8}, \dfrac{13\pi}{8}$.
$4x = \dfrac{3\pi}{2} + 2n\pi \Rightarrow x = \dfrac{3\pi}{8} + \dfrac{n\pi}{2} \Rightarrow$
$x = \dfrac{3\pi}{8}, \dfrac{7\pi}{8}, \dfrac{11\pi}{8}, \dfrac{15\pi}{8}$

Solution set:
$\left\{0, \dfrac{\pi}{8}, \dfrac{3\pi}{8}, \dfrac{5\pi}{8}, \dfrac{7\pi}{8}, \pi, \dfrac{9\pi}{8}, \dfrac{11\pi}{8}, \dfrac{13\pi}{8}, \dfrac{15\pi}{8}\right\}$

Use the identity $\sin^{-1} x + \cos^{-1} x = \dfrac{\pi}{2}$, x in $[-1, 1]$ for exercises 59–66.

59.
$$\sin^{-1} x - \cos^{-1} x = \dfrac{\pi}{6}$$
$$\sin^{-1} x - \left(\dfrac{\pi}{2} - \sin^{-1} x\right) = \dfrac{\pi}{6}$$
$$2\sin^{-1} x = \dfrac{2\pi}{3}$$
$$\sin^{-1} x = \dfrac{\pi}{3} \Rightarrow x = \sin\dfrac{\pi}{3} = \dfrac{\sqrt{3}}{2}$$

Solution set: $\left\{\dfrac{\sqrt{3}}{2}\right\}$

61.
$$\sin^{-1} x + \cos^{-1} x = \dfrac{3\pi}{4}$$
$$\sin^{-1} x + \left(\dfrac{\pi}{2} - \sin^{-1} x\right) = \dfrac{3\pi}{4} \Rightarrow \dfrac{\pi}{2} = \dfrac{3\pi}{4}$$

This is a false statement, so there is no solution. Solution set: ∅

63.
$$\sin^{-1} x - \cos^{-1} x = -\frac{\pi}{6}$$
$$\sin^{-1} x - \left(\frac{\pi}{2} - \sin^{-1} x\right) = -\frac{\pi}{6}$$
$$2\sin^{-1} x = \frac{\pi}{3}$$
$$\sin^{-1} x = \frac{\pi}{6} \Rightarrow x = \sin\frac{\pi}{6} = \frac{1}{2}$$

Solution set: $\left\{\frac{1}{2}\right\}$

65. $\sin^{-1} x - \tan^{-1}\frac{2}{3} = \frac{\pi}{4}$
$$\sin^{-1} x = \frac{\pi}{4} + \tan^{-1}\frac{2}{3}$$
$$x = \sin\left(\frac{\pi}{4} + \tan^{-1}\frac{2}{3}\right)$$
$$\approx 0.9806$$

Solution set: {0.9806}
Alternatively,
$$\sin^{-1} x - \tan^{-1}\frac{2}{3} = \frac{\pi}{4}$$

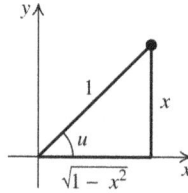 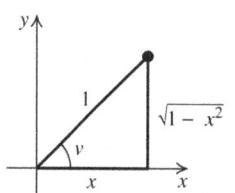

Let $u = \sin^{-1} x$ and $v = \tan^{-1}\frac{2}{3}$. Then $\sin u = x$ and $\tan v = \frac{2}{3}$. From the figure, we have $\cos u = \sqrt{1-x^2}$. Since $\tan v = \frac{2}{3} = \frac{y}{x}$, we know that $r = \sqrt{x^2 + y^2} = \sqrt{3^2 + 2^2} = \sqrt{13}$.

Thus, $\sin v = \frac{2}{\sqrt{13}}$ and $\cos v = \frac{3}{\sqrt{13}}$. Replace the equivalents of the trigonometric functions in the equation.

$$u - v = \frac{\pi}{4} \Rightarrow \sin(u-v) = \sin\left(\frac{\pi}{4}\right) \Rightarrow$$
$$\sin u \cos v - \cos u \sin v = \sin\left(\frac{\pi}{4}\right) \Rightarrow$$
$$x \cdot \frac{3}{\sqrt{13}} - \left(\sqrt{1-x^2}\right)\left(\frac{2}{\sqrt{13}}\right) = \frac{\sqrt{2}}{2} \Rightarrow$$

$$3x - 2\sqrt{1-x^2} = \frac{\sqrt{26}}{2} \Rightarrow$$
$$3x - \frac{\sqrt{26}}{2} = 2\sqrt{1-x^2}$$

Now square both sides.
$$\left(3x - \frac{\sqrt{26}}{2}\right)^2 = \left(2\sqrt{1-x^2}\right)^2 \Rightarrow$$
$$9x^2 - 3\sqrt{26}x + \frac{13}{2} = 4(1-x^2) \Rightarrow$$
$$9x^2 - 3\sqrt{26}x + \frac{13}{2} = 4 - 4x^2 \Rightarrow$$
$$13x^2 - 3\sqrt{26}x + \frac{5}{2} = 0$$

Use the quadratic formula to solve for x.
$$x = \frac{-\left(-3\sqrt{26}\right) \pm \sqrt{\left(-3\sqrt{26}\right)^2 - 4(13)\left(\frac{5}{2}\right)}}{2(13)}$$
$$= \frac{3\sqrt{26} \pm \sqrt{234 - 130}}{26} = \frac{3\sqrt{26} \pm \sqrt{104}}{26}$$
$$= \frac{3\sqrt{26} \pm 2\sqrt{26}}{26} = \frac{5\sqrt{26}}{26} \text{ or } \frac{\sqrt{26}}{26}$$

Use a calculator to check each answer.

```
sin⁻¹(√(26)/26)-t    sin⁻¹(5√(26)/26)-
an⁻¹(2/3)            tan⁻¹(2/3)
         -.3906070437          .7853981634
π/4                  π/4
          .7853981634          .7853981634
```

Solution set: $\left\{\frac{5\sqrt{26}}{26}\right\}$

6.6 Applying the Concepts

67. a. $30 = 60\sin(120\pi t) \Rightarrow \frac{1}{2} = \sin(120\pi t) \Rightarrow$
$$120\pi t = \frac{\pi}{6} \Rightarrow t = \frac{1}{720} \approx 0.0014 \text{ sec or}$$
$$120\pi t = \frac{5\pi}{6} \Rightarrow t = \frac{5}{720} > \frac{1}{720}, \text{ so the smallest}$$
possible positive value is $t \approx 0.0014$ sec.

b. $-20 = 60\sin(120\pi t) \Rightarrow -\frac{1}{3} = \sin(120\pi t) \Rightarrow$
$120\pi t = -0.3398$ (because we are looking for positive values) $\Rightarrow 2\pi - 0.3398 \approx 5.9434 \Rightarrow$
$t \approx 0.0158$ sec or
$120\pi t = -0.3398 = \pi + 0.3398 \approx 3.4814 \Rightarrow$
$t = 0.0092$ sec. The smallest possible positive value is $t \approx 0.0092$ sec.

69. $6\cos\left(\dfrac{\pi}{2}t\right) = 3 \Rightarrow \dfrac{1}{2} = \cos\left(\dfrac{\pi}{2}t\right) \Rightarrow$

$\dfrac{\pi}{2}t = \dfrac{\pi}{3}$ or $\dfrac{\pi}{2}t = \dfrac{5\pi}{3}$. Because we are looking for all positive values, solve

$\dfrac{\pi}{2}t = \dfrac{\pi}{3} + 2\pi n \Rightarrow t = \dfrac{2}{3} + 4n$ sec or

$\dfrac{\pi}{2}t = \dfrac{5\pi}{3} + 2\pi n \Rightarrow t = \dfrac{10}{3} + 4n$ sec

71. a. Note that we use 30 instead of 30,000 for y because y is defined as thousands of people.

$30 = 20 + 10\sin\left(\dfrac{\pi}{26}(x-14)\right) \Rightarrow$

$\sin\left(\dfrac{\pi}{26}(x-14)\right) = 1 \Rightarrow \dfrac{\pi}{26}(x-14) = \dfrac{\pi}{2} \Rightarrow$

$x = 27$

b. $25 = 20 + 10\sin\left(\dfrac{\pi}{26}(x-14)\right) \Rightarrow$

$\sin\left(\dfrac{\pi}{26}(x-14)\right) = \dfrac{1}{2} \Rightarrow \dfrac{\pi}{26}(x-14) = \dfrac{\pi}{6} \Rightarrow$

$x \approx 18.3$ (the 18th week) or

$\dfrac{\pi}{26}(x-14) = \dfrac{5\pi}{6} \Rightarrow x \approx 35.6$ (which is during the 36th week).

c. $15 = 20 + 10\sin\left(\dfrac{\pi}{26}(x-14)\right) \Rightarrow$

$\sin\left(\dfrac{\pi}{26}(x-14)\right) = -\dfrac{1}{2} \Rightarrow \dfrac{\pi}{26}(x-14) = \dfrac{7\pi}{6} \Rightarrow$

$x \approx 44.3 \approx$ the 44th week or $\dfrac{\pi}{26}(x-14)$

$= -\dfrac{\pi}{6} \Rightarrow x \approx 9.7$ (which is during the 10th week).

73. a. $24 = 12 + 12\sin\left(\dfrac{\pi}{4}(x-2)\right) \Rightarrow$

$\sin\left(\dfrac{\pi}{4}(x-2)\right) = 1 \Rightarrow \dfrac{\pi}{4}(x-2) = \dfrac{\pi}{2} \Rightarrow$

$x = 4$ (January)

b. $18 = 12 + 12\sin\left(\dfrac{\pi}{4}(x-2)\right) \Rightarrow$

$\sin\left(\dfrac{\pi}{4}(x-2)\right) = \dfrac{1}{2} \Rightarrow \dfrac{\pi}{4}(x-2) = \dfrac{\pi}{6} \Rightarrow$

$x \approx 2.7$ (which is during December) or

$\dfrac{\pi}{4}(x-2) = \dfrac{5\pi}{6} \Rightarrow x \approx 5.3$ (which is during February).

6.6 Beyond the Basics

75. $\sin^4 2x = 1 \Rightarrow \sin 2x = \pm 1 \Rightarrow 2x = \dfrac{\pi}{2} + 2\pi n \Rightarrow$

$x = \dfrac{\pi}{4} + \pi n$ or $2x = \dfrac{3\pi}{2} + 2\pi n \Rightarrow x = \dfrac{3\pi}{4} + \pi n$

If $x = \dfrac{\pi}{4} + \pi n \Rightarrow x = \dfrac{\pi}{4}$ or $x = \dfrac{5\pi}{4}$

If $x = \dfrac{3\pi}{4} + \pi n \Rightarrow x = \dfrac{3\pi}{4}$ or $x = \dfrac{7\pi}{4}$

Solution set: $\left\{\dfrac{\pi}{4}, \dfrac{3\pi}{4}, \dfrac{5\pi}{4}, \dfrac{7\pi}{4}\right\}$.

77. $\tan^2 2\theta = 1 \Rightarrow \tan 2\theta = \pm 1 \Rightarrow 2\theta = \dfrac{\pi}{4} + \pi n \Rightarrow$

$\theta = \dfrac{\pi}{8} + \dfrac{\pi}{2}n$ or $2\theta = \dfrac{5\pi}{4} + \pi n \Rightarrow \theta = \dfrac{5\pi}{8} + \dfrac{\pi}{2}n$

or $2\theta = \dfrac{3\pi}{4} + \pi n \Rightarrow \theta = \dfrac{3\pi}{8} + \dfrac{\pi}{2}n$ or

$2\theta = \dfrac{7\pi}{4} + \pi n \Rightarrow \theta = \dfrac{7\pi}{8} + \dfrac{\pi}{2}n$

If $\theta = \dfrac{\pi}{8} + \dfrac{\pi}{2}n \Rightarrow \theta = \dfrac{\pi}{8}$ or $\theta = \dfrac{5\pi}{8}$ or $\theta = \dfrac{9\pi}{8}$

or $\theta = \dfrac{13\pi}{8}$. If $\theta = \dfrac{5\pi}{8} + \dfrac{\pi}{2}n \Rightarrow \theta = \dfrac{5\pi}{8}$ or

$\theta = \dfrac{9\pi}{8}$ or $\theta = \dfrac{13\pi}{8}$. If $\theta = \dfrac{3\pi}{8} + \dfrac{\pi}{2}n \Rightarrow$

$\theta = \dfrac{3\pi}{8}$ or $\theta = \dfrac{7\pi}{8}$ or $\theta = \dfrac{11\pi}{8}$ or $\theta = \dfrac{15\pi}{8}$.

If $\theta = \dfrac{7\pi}{8} + \dfrac{\pi}{2}n \Rightarrow \dfrac{7\pi}{8}$ or $\theta = \dfrac{11\pi}{8}$ or $\theta = \dfrac{15\pi}{8}$.

The solution is

$\left\{\dfrac{\pi}{8}, \dfrac{3\pi}{8}, \dfrac{5\pi}{8}, \dfrac{7\pi}{8}, \dfrac{9\pi}{8}, \dfrac{11\pi}{8}, \dfrac{13\pi}{8}, \dfrac{15\pi}{8}\right\}$.

79. Factor by grouping.

$\sin x \cos x - \dfrac{\sqrt{3}}{2}\sin x + \dfrac{1}{2}\cos x - \dfrac{\sqrt{3}}{4} = 0 \Rightarrow$

$\sin x\left(\cos x - \dfrac{\sqrt{3}}{2}\right) + \dfrac{1}{2}\left(\cos x - \dfrac{\sqrt{3}}{2}\right) = 0 \Rightarrow$

$\left(\sin x + \dfrac{1}{2}\right)\left(\cos x - \dfrac{\sqrt{3}}{2}\right) = 0 \Rightarrow \sin x = -\dfrac{1}{2}$

or $\cos x = \dfrac{\sqrt{3}}{2}$.

If $\sin x = -\dfrac{1}{2} \Rightarrow x = \dfrac{7\pi}{6} + 2n\pi \Rightarrow$

$x = \dfrac{7\pi}{6}$ or $x = \dfrac{11\pi}{6} + 2n\pi \Rightarrow x = \dfrac{11\pi}{6}$.

(continued on next page)

374 *Chapter 6 Trigonometric Identities and Equations*

(*continued*)

If $\cos x = \frac{\sqrt{3}}{2} \Rightarrow x = \frac{\pi}{6} + 2n\pi \Rightarrow x = \frac{\pi}{6}$ or $x = \frac{11\pi}{6} + 2n\pi \Rightarrow x = \frac{11\pi}{6}$.

The solution is $\left\{\frac{\pi}{6}, \frac{7\pi}{6}, \frac{11\pi}{6}\right\}$.

81. $\sin^2\theta - \cos\theta = \frac{1}{4} \Rightarrow (1 - \cos^2\theta) - \cos\theta = \frac{1}{4} \Rightarrow -\cos^2\theta - \cos\theta + \frac{3}{4} = 0 \Rightarrow \cos^2\theta + \cos\theta - \frac{3}{4} = 0 \Rightarrow$
$4\cos^2\theta + 4\cos\theta - 3 = 0 \Rightarrow (2\cos\theta + 3)(2\cos\theta - 1) = 0$

$2\cos\theta + 3 \Rightarrow \cos\theta = -\frac{3}{2} \Rightarrow \theta$ is undefined.

$2\cos\theta - 1 \Rightarrow \cos\theta = \frac{1}{2} \Rightarrow \theta = \frac{\pi}{3}, \frac{5\pi}{3}$

The solution is $\left\{\frac{\pi}{3}, \frac{5\pi}{3}\right\}$.

83. $\tan x + \sec x = 2\cos x \Rightarrow \frac{\sin x}{\cos x} + \frac{1}{\cos x} = 2\cos x \Rightarrow \frac{\sin x + 1}{\cos x} = 2\cos x \Rightarrow \sin x + 1 = 2\cos^2 x \Rightarrow$
$\sin x + 1 = 2(1 - \sin^2 x) \Rightarrow 2\sin^2 x + \sin x - 1 = 0 \Rightarrow (2\sin x - 1)(\sin x + 1) = 0 \Rightarrow$

$\sin x = \frac{1}{2} \Rightarrow x = \frac{\pi}{6} + 2n\pi$ or $x = \frac{5\pi}{6} + 2n\pi$

$\sin x = -1 \Rightarrow x = \frac{3\pi}{2}$

Checking each answer shows that $x = \frac{3\pi}{2}$ is extraneous. The solution is $\left\{\frac{\pi}{6}, \frac{5\pi}{6}\right\}$.

85. $\sin 2x + \cos 2x = \sqrt{2}$

Let $u = 2x$. Then, we have $\sin u + \cos u = \sqrt{2}$.

Now use the reduction formula.

$a = 1, b = 1$, so $\sqrt{a^2 + b^2} = \sqrt{2} = A$. θ is any angle in standard position that has (1, 1) on its terminal side,

so $\theta = \frac{\pi}{4}$, and $y = \sin u + \cos u = \sqrt{2}\sin\left(u + \frac{\pi}{4}\right)$

$\sqrt{2}\sin\left(u + \frac{\pi}{4}\right) = \sqrt{2} \Rightarrow \sin\left(u + \frac{\pi}{4}\right) = 1 \Rightarrow u + \frac{\pi}{4} = \frac{\pi}{2} + 2n\pi \Rightarrow u = \frac{\pi}{4} + 2n\pi \Rightarrow$

$2x = \frac{\pi}{4} + 2n\pi \Rightarrow x = \frac{\pi}{8} + n\pi$

The possible solutions in the interval $[0, 2\pi)$ are $x = \frac{\pi}{8}, \frac{9\pi}{8}$. Checking each answer shows that both are

valid. The solution is $\left\{\frac{\pi}{8}, \frac{9\pi}{8}\right\}$.

87. $\cos 2\theta = \cot 2\theta \Rightarrow \cos 2\theta - \cot 2\theta = 0 \Rightarrow \cos 2\theta - \dfrac{\cos 2\theta}{\sin 2\theta} = 0 \Rightarrow \cos 2\theta \left(1 - \dfrac{1}{\sin 2\theta}\right) = 0 \Rightarrow$

$\cos 2\theta = 0$ or $1 - \dfrac{1}{\sin 2\theta} = 0$

$\cos 2\theta = 0 \Rightarrow 2\theta = \dfrac{\pi}{2} + 2n\pi$ or $2\theta = \dfrac{3\pi}{2} + 2n\pi \Rightarrow \theta = \dfrac{\pi}{4} + n\pi$ or $\theta = \dfrac{3\pi}{4} + n\pi$

$1 - \dfrac{1}{\sin 2\theta} = 0 \Rightarrow \dfrac{1}{\sin 2\theta} = 1 \Rightarrow \sin 2\theta = 1 \Rightarrow 2\theta = \dfrac{\pi}{2} + 2n\pi \Rightarrow \theta = \dfrac{\pi}{4} + n\pi$

Now find the values of n that result in solutions in the interval $[0, 2\pi)$:

$n = 0:\ \dfrac{\pi}{4}\ \dfrac{3\pi}{4}\ \dfrac{\pi}{4}$

$n = 1:\ \dfrac{5\pi}{4}\ \dfrac{7\pi}{4}$

Checking each answer shows that all are valid. The solution is $\left\{\dfrac{\pi}{4}, \dfrac{3\pi}{4}, \dfrac{5\pi}{4}, \dfrac{7\pi}{4}\right\}$.

For exercises 89–92, recall that if (a, b) is any point on the terminal side of an angle θ in standard position, then $a \sin x + b \cos x = \sqrt{a^2 + b^2} \sin(x + \theta)$ for any real number x, and $\theta = \tan^{-1}\left(\dfrac{b}{a}\right)$.

89. $3 \sin 2x + 4 \cos 2x = 0$

Using the reduction formula, we have

$3 \sin 2x + 4 \cos 2x = \sqrt{3^2 + 4^2} \sin\left(2x + \tan^{-1}\left(\dfrac{4}{3}\right)\right) = 5 \sin\left(2x + \tan^{-1}\left(\dfrac{4}{3}\right)\right)$.

$5 \sin\left(2x + \tan^{-1}\left(\dfrac{4}{3}\right)\right) = 0 \Rightarrow 2x + \tan^{-1}\left(\dfrac{4}{3}\right) = 0 + 2n\pi$ or $2x + \tan^{-1}\left(\dfrac{4}{3}\right) = \pi + 2n\pi$.

$2x + \tan^{-1}\left(\dfrac{4}{3}\right) = 0 + 2n\pi \Rightarrow 2x = -\tan^{-1}\left(\dfrac{4}{3}\right) + 2n\pi \Rightarrow x = \dfrac{-\tan^{-1}\left(\dfrac{4}{3}\right)}{2} + n\pi \Rightarrow x \approx 2.678, 5.820.$

$2x + \tan^{-1}\left(\dfrac{4}{3}\right) = \pi + 2n\pi \Rightarrow 2x = \pi - \tan^{-1}\left(\dfrac{4}{3}\right) + 2n\pi \Rightarrow x = \dfrac{\pi}{2} - \dfrac{\tan^{-1}\left(\dfrac{4}{3}\right)}{2} + n\pi \Rightarrow x \approx 1.107, 4.249.$

Solution set: $\{1.107, 2.678, 4.249, 5.820\}$

91. $5 \sin 3x - 12 \cos 3x = \dfrac{13\sqrt{3}}{2}$

Using the reduction formula, we have

$5 \sin 3x - 12 \cos 3x = \sqrt{5^2 + (-12)^2} \sin\left(3x + \tan^{-1}\left(-\dfrac{12}{5}\right)\right) = 13 \sin\left(3x + \tan^{-1}\left(-\dfrac{12}{5}\right)\right)$.

$13 \sin\left(3x + \tan^{-1}\left(-\dfrac{12}{5}\right)\right) = \dfrac{13\sqrt{3}}{2} \Rightarrow \sin\left(3x + \tan^{-1}\left(-\dfrac{12}{5}\right)\right) = \dfrac{\sqrt{3}}{2} \Rightarrow$

$3x - \tan^{-1}\left(\dfrac{12}{5}\right) = \dfrac{\pi}{3} + 2n\pi$ or $3x - \tan^{-1}\left(\dfrac{12}{5}\right) = \dfrac{2\pi}{3} + 2n\pi$.

$3x - \tan^{-1}\left(\dfrac{12}{5}\right) = \dfrac{\pi}{3} + 2n\pi \Rightarrow 3x = \dfrac{\pi}{3} + \tan^{-1}\left(\dfrac{12}{5}\right) + 2n\pi \Rightarrow x = \dfrac{\dfrac{\pi}{3} + \tan^{-1}\left(\dfrac{12}{5}\right) + 2n\pi}{3} \Rightarrow$

$x \approx 0.741, 2.835, 4.930.$

$3x - \tan^{-1}\left(\dfrac{12}{5}\right) = \dfrac{2\pi}{3} + 2n\pi \Rightarrow 3x = \dfrac{2\pi}{3} + \tan^{-1}\left(\dfrac{12}{5}\right) + 2n\pi \Rightarrow x = \dfrac{\dfrac{2\pi}{3} + \tan^{-1}\left(\dfrac{12}{5}\right) + 2n\pi}{3} \Rightarrow$

$x \approx 1.090, 3.185, 5.279$

Solution set: $\{0.741, 1.090, 2.835, 3.185, 4.930, 5.279\}$

6.6 Critical Thinking/Discussion/Writing

93. $\sin x \cos x = \frac{1}{2} \Rightarrow \sin^2 x \cos^2 x = \frac{1}{4} \Rightarrow \sin^2 x(1-\sin^2 x) = \frac{1}{4} \Rightarrow \sin^2 x - \sin^4 x - \frac{1}{4} = 0 \Rightarrow$

$\sin^4 x - \sin^2 x + \frac{1}{4} = 0 \Rightarrow \left(\sin^2 x - \frac{1}{2}\right)^2 = 0 \Rightarrow \sin^2 x = \frac{1}{2} \Rightarrow \sin x = \pm\frac{\sqrt{2}}{2} \Rightarrow x = \frac{\pi}{4}$ or $x = \frac{5\pi}{4}$

The solution is $\left\{\frac{\pi}{4}, \frac{5\pi}{4}\right\}$.

94. $\sin x \cos x = 1 \Rightarrow \sin^2 x \cos^2 x = 1 \Rightarrow \sin^2 x(1-\sin^2 x) = 1 \Rightarrow \sin^2 x - \sin^4 x - 1 = 0 \Rightarrow \sin^4 x - \sin^2 x + 1 = 0$.

Letting $u = \sin^2 x$, we have $u^2 - u + 1 = 0 \Rightarrow u = \frac{1 \pm \sqrt{(-1)^2 - 4(1)(1)}}{2(1)} = \frac{1 \pm \sqrt{-3}}{2} = \frac{1}{2} \pm \frac{i\sqrt{3}}{2} \Rightarrow$

$\sin^2 x = \frac{1}{2} \pm \frac{i\sqrt{3}}{2}$.

Similarly, we can show that $\cos^2 x = \frac{1}{2} \pm \frac{i\sqrt{3}}{2}$. So, there are no real roots.

95. $\tan^2 x + (\sec x - 1)^2 + 3 = 0 \Rightarrow \tan^2 x + (\sec x - 1)^2 = -3$

The left side of the equation is positive, while the right side is negative. Thus, there are no real solutions.

96. Answers will vary. Sample answer: $\cos x = 1$.

6.6 Getting Ready for the Next Section

97. $5^2 + 12^2 = c^2 \Rightarrow 169 = c^2 \Rightarrow 13 = c$

99. $a^2 + 6^2 = 10^2 \Rightarrow a^2 + 36 = 100 \Rightarrow a^2 = 64 \Rightarrow a = 8$

101. $a^2 + 3^2 = 6^2 \Rightarrow a^2 + 9 = 36 \Rightarrow a^2 = 27 \Rightarrow a = \sqrt{27} = 3\sqrt{3}$

103. $\dfrac{a}{b} = \dfrac{\frac{1}{2}}{\frac{\sqrt{3}}{2}} = \dfrac{\frac{1}{2} \cdot 2}{\frac{\sqrt{3}}{2} \cdot 2} = \dfrac{1}{\sqrt{3}} \cdot \dfrac{\sqrt{3}}{\sqrt{3}} = \dfrac{\sqrt{3}}{3}$

Chapter 6 Review Exercises

Building Skills

1. $\sin\theta = -\frac{2}{3}$ and $\cos\theta < 0 \Rightarrow \theta$ is in Quadrant III.

$\left(-\frac{2}{3}\right)^2 + (\cos^2\theta) = 1 \Rightarrow \cos\theta = -\frac{\sqrt{5}}{3}$.

$\tan\theta = \dfrac{\sin\theta}{\cos\theta} = \dfrac{-2/3}{-\sqrt{5}/3} = \dfrac{2\sqrt{5}}{5}$, $\cot\theta = \dfrac{1}{\tan\theta} = \dfrac{\sqrt{5}}{2}$, $\sec\theta = \dfrac{1}{\cos\theta} = -\dfrac{3\sqrt{5}}{5}$, $\csc\theta = \dfrac{1}{\sin\theta} = -\dfrac{3}{2}$.

3. $\sec\theta = 3$ and $\tan\theta < 0 \Rightarrow \theta$ is in Quadrant IV.

$\cos\theta = \dfrac{1}{\sec\theta} = \dfrac{1}{3}; \sin^2\theta + \left(\dfrac{1}{3}\right)^2 = 1 \Rightarrow \sin\theta = -\dfrac{2\sqrt{2}}{3}; \tan\theta = \dfrac{\sin\theta}{\cos\theta} = \dfrac{-2\sqrt{2}/3}{1/3} \Rightarrow \tan\theta = -2\sqrt{2};$

$\cot\theta = \dfrac{1}{\tan\theta} = -\dfrac{\sqrt{2}}{4}; \csc\theta = \dfrac{1}{\sin\theta} = -\dfrac{3\sqrt{2}}{4}$

5. $(\sin x + \cos x)^2 + (\sin x - \cos x)^2 = \sin^2 x + 2\sin x \cos x + \cos^2 x + (\sin^2 x - 2\sin x \cos x + \cos^2 x)$
$$= (\sin^2 x + \cos^2 x) + 2\sin x \cos x - 2\sin x \cos x + (\sin^2 x + \cos^2 x)$$
$$= 2(\sin^2 x + \cos^2 x) = 2(1) = 2$$

7. $\dfrac{1-\tan^2\theta}{1+\tan^2\theta} = \dfrac{1-\dfrac{\sin^2\theta}{\cos^2\theta}}{1+\dfrac{\sin^2\theta}{\cos^2\theta}} = \dfrac{\dfrac{\cos^2\theta - \sin^2\theta}{\cos^2\theta}}{\dfrac{\cos^2\theta + \sin^2\theta}{\cos^2\theta}} = \dfrac{\cos^2\theta - \sin^2\theta}{\cos^2\theta + \sin^2\theta} = \dfrac{\cos^2\theta - \sin^2\theta}{1} = \cos^2\theta - \sin^2\theta$

9. $\dfrac{\sin\theta}{1+\cos\theta} + \dfrac{\sin\theta}{1-\cos\theta} = \dfrac{\sin\theta(1-\cos\theta)}{(1+\cos\theta)(1-\cos\theta)} + \dfrac{\sin\theta(1+\cos\theta)}{(1-\cos\theta)(1+\cos\theta)} = \dfrac{\sin\theta - \sin\theta\cos\theta + \sin\theta + \sin\theta\cos\theta}{1-\cos^2\theta}$
$$= \dfrac{2\sin\theta}{\sin^2\theta} = \dfrac{2}{\sin\theta} = 2\csc\theta$$

11. $\dfrac{\sin\theta}{1-\cot\theta} + \dfrac{\cos\theta}{1-\tan\theta} = \dfrac{\sin\theta}{1-\dfrac{\cos\theta}{\sin\theta}} + \dfrac{\cos\theta}{1-\dfrac{\sin\theta}{\cos\theta}} = \dfrac{\sin\theta}{\dfrac{\sin\theta - \cos\theta}{\sin\theta}} + \dfrac{\cos\theta}{\dfrac{\cos\theta - \sin\theta}{\cos\theta}} = \dfrac{\sin^2\theta}{\sin\theta - \cos\theta} + \dfrac{\cos^2\theta}{\cos\theta - \sin\theta}$
$$= \dfrac{\sin^2\theta}{\sin\theta - \cos\theta} - \dfrac{\cos^2\theta}{\sin\theta - \cos\theta} = \dfrac{\sin^2\theta - \cos^2\theta}{\sin\theta - \cos\theta} = \dfrac{(\sin\theta - \cos\theta)(\sin\theta + \cos\theta)}{\sin\theta - \cos\theta}$$
$$= \sin\theta + \cos\theta$$

13. $\dfrac{\tan x}{\sec x - 1} + \dfrac{\tan x}{\sec x + 1} = \dfrac{\tan x(\sec x + 1) + \tan x(\sec x - 1)}{\sec^2 x - 1} = \dfrac{\tan x \sec x + \tan x + \tan x \sec x - \tan x}{\sec^2 x - 1}$
$$= \dfrac{2\tan x \sec x}{\tan^2 x} = \dfrac{2\sec x}{\tan x} = \dfrac{\dfrac{2}{\cos x}}{\dfrac{\sin x}{\cos x}} = \dfrac{2}{\sin x} = 2\csc x$$

15. $\dfrac{1+\sin\theta}{1-\sin\theta} = \dfrac{1+\sin\theta}{1-\sin\theta} \cdot \dfrac{1+\sin\theta}{1+\sin\theta} = \dfrac{(1+\sin\theta)^2}{1-\sin^2\theta} = \dfrac{(1+\sin\theta)^2}{\cos^2\theta} = \left(\dfrac{1+\sin\theta}{\cos\theta}\right)^2 = \left(\dfrac{1}{\cos\theta} + \dfrac{\sin\theta}{\cos\theta}\right)^2 = (\sec\theta + \tan\theta)^2$

17. $\dfrac{\sec x - \tan x}{\sec x + \tan x} = \dfrac{\sec x - \tan x}{\sec x + \tan x} \cdot \dfrac{\sec x - \tan x}{\sec x - \tan x} = \dfrac{(\sec x - \tan x)^2}{\sec^2 x - \tan^2 x} = \dfrac{(\sec x - \tan x)^2}{1} = (\sec x - \tan x)^2$

19. $\dfrac{\sin x - \cos x + 1}{\sin x + \cos x - 1} = \dfrac{\sin x + (1-\cos x)}{\sin x - (1-\cos x)} = \dfrac{\sin x + (1-\cos x)}{\sin x - (1-\cos x)} \cdot \dfrac{\sin x + (1-\cos x)}{\sin x + (1-\cos x)}$
$$= \dfrac{\sin^2 x + 2\sin x(1-\cos x) + (1-\cos x)^2}{\sin^2 x - (1-\cos x)^2} = \dfrac{\sin^2 x + 2\sin x - 2\sin x \cos x + 1 - 2\cos x + \cos^2 x}{\sin^2 x - (1 - 2\cos x + \cos^2 x)}$$
$$= \dfrac{\sin^2 x + \cos^2 x + 2\sin x - 2\sin x \cos x - 2\cos x + 1}{\sin^2 x - 1 + 2\cos x - \cos^2 x} = \dfrac{2 - 2\cos x - 2\sin x \cos x + 2\sin x}{\sin^2 x - 1 + 2\cos x - \cos^2 x}$$
$$= \dfrac{2(1-\cos x) + 2\sin x(1-\cos x)}{1-\cos^2 x - 1 + 2\cos x - \cos^2 x} = \dfrac{(2 + 2\sin x)(1-\cos x)}{-2\cos^2 x - 2\cos x} = \dfrac{2(1+\sin x)(1-\cos x)}{-2\cos x(\cos x - 1)}$$
$$= \dfrac{2(1+\sin x)(1-\cos x)}{2\cos x(1-\cos x)} = \dfrac{1+\sin x}{\cos x}$$

21. $\cos 15° = \cos\dfrac{30°}{2} = \sqrt{\dfrac{1+\cos 30°}{2}} = \sqrt{\dfrac{1+\sqrt{3}/2}{2}} = \dfrac{\sqrt{2+\sqrt{3}}}{2}$

23. $\csc 75° = \dfrac{1}{\sin(45° + 30°)} = \dfrac{1}{\sin 45°\cos 30° + \cos 45°\sin 30°} = \dfrac{1}{\dfrac{\sqrt{2}}{2}\left(\dfrac{\sqrt{3}}{2}\right) + \dfrac{\sqrt{2}}{2}\left(\dfrac{1}{2}\right)} = \dfrac{1}{\dfrac{\sqrt{6}+\sqrt{2}}{4}} = \sqrt{6} - \sqrt{2}$

25. $\sin 41° \cos 49° + \cos 41° \sin 49° = \sin(41° + 49°) = \sin 90° = 1$

27. $\dfrac{\tan 69° + \tan 66°}{1 - \tan 69° \tan 66} = \tan(69° + 66°) = \tan 135° = -1$

In exercises 29–32, $\sin u = \dfrac{4}{5}$ and $0 < u \le \dfrac{\pi}{2} \Rightarrow \cos u = \dfrac{3}{5}$, $\tan u = \dfrac{4}{3}$ and $\cos v = \dfrac{5}{13}$ and $0 < v \le \dfrac{\pi}{2} \Rightarrow \sin v = \dfrac{12}{13}$, $\tan v = \dfrac{12}{5}$.

29. $\sin(u - v) = \sin u \cos v - \cos u \sin v = \dfrac{4}{5} \cdot \dfrac{5}{13} - \dfrac{3}{5} \cdot \dfrac{12}{13} = -\dfrac{16}{65}$

31. $\cos(u - v) = \cos u \cos v + \sin u \sin v = \dfrac{3}{5} \cdot \dfrac{5}{13} + \dfrac{4}{5} \cdot \dfrac{12}{13} = \dfrac{63}{65}$

33. $\sin(x - y)\cos y + \cos(x - y)\sin y = \cos y(\sin x \cos y - \sin y \cos x) + \sin y(\cos x \cos y + \sin x \sin y)$
$= \sin x \cos^2 y - \sin y \cos x \cos y + \sin y \cos x \cos y + \sin x \sin^2 y$
$= \sin x \cos^2 y + \sin x \sin^2 y = \sin x(\cos^2 y + \sin^2 y) = \sin x$

35. Start with the right side.

$\dfrac{\tan u + \tan v}{\tan u - \tan v} = \dfrac{\dfrac{\sin u}{\cos u} + \dfrac{\sin v}{\cos v}}{\dfrac{\sin u}{\cos u} - \dfrac{\sin v}{\cos v}} = \dfrac{\dfrac{\sin u \cos v + \sin v \cos u}{\cos u \cos v}}{\dfrac{\sin u \cos v - \sin v \cos u}{\cos u \cos v}} = \dfrac{\sin u \cos v + \sin v \cos u}{\sin u \cos v - \sin v \cos u} = \dfrac{\sin(u+v)}{\sin(u-v)}$

37. $\dfrac{\sin 4x}{\sin 2x} - \dfrac{\cos 4x}{\cos 2x} = \dfrac{2\sin 2x \cos 2x}{\sin 2x} - \dfrac{2\cos^2 2x - 1}{\cos 2x} = 2\cos 2x - 2\cos 2x + \dfrac{1}{\cos 2x} = \sec 2x$

39. $\sin\left(x - \dfrac{\pi}{6}\right) + \cos\left(x + \dfrac{\pi}{3}\right) = \sin x \cos\dfrac{\pi}{6} - \sin\dfrac{\pi}{6}\cos x + \cos x \cos\dfrac{\pi}{3} - \sin x \sin\dfrac{\pi}{3}$
$= \dfrac{\sqrt{3}}{2}\sin x - \dfrac{1}{2}\cos x + \dfrac{1}{2}\cos x - \dfrac{\sqrt{3}}{2}\sin x = 0$

41. $1 + \tan\theta \tan 2\theta = 1 + \dfrac{\sin\theta}{\cos\theta} \cdot \dfrac{\sin 2\theta}{\cos 2\theta} = 1 + \dfrac{2\sin^2\theta \cos\theta}{\cos\theta(\cos^2\theta - \sin^2\theta)} = 1 + \dfrac{2\sin^2\theta}{\cos^2\theta - \sin^2\theta}$
$= \dfrac{\cos^2\theta - \sin^2\theta + 2\sin^2\theta}{\cos^2\theta - \sin^2\theta} = \dfrac{\cos^2\theta + \sin^2\theta}{\cos^2\theta - \sin^2\theta} = \dfrac{1}{\cos 2\theta} = \sec 2\theta$

43. $\dfrac{\sin\theta + \sin 2\theta}{\cos\theta + \cos 2\theta} = \dfrac{2\sin\dfrac{3\theta}{2}\cos\left(-\dfrac{\theta}{2}\right)}{2\cos\dfrac{3\theta}{2}\cos\left(-\dfrac{\theta}{2}\right)} = \tan\dfrac{3\theta}{2}$

45. $\dfrac{\sin 5x - \sin 3x}{\sin 5x + \sin 3x} = \dfrac{2\sin x \cos 4x}{2\sin 4x \cos x} = \tan x \cot 4x = \dfrac{\tan x}{\tan 4x}$

47. $\dfrac{\tan 3x + \tan x}{\tan 3x - \tan x} = \dfrac{\dfrac{\sin 3x}{\cos 3x} + \dfrac{\sin x}{\cos x}}{\dfrac{\sin 3x}{\cos 3x} - \dfrac{\sin x}{\cos x}} = \dfrac{\dfrac{\sin 3x \cos x + \sin x \cos 3x}{\cos 3x \cos x}}{\dfrac{\sin 3x \cos x - \sin x \cos 3x}{\cos 3x \cos x}} = \dfrac{\sin 3x \cos x + \sin x \cos 3x}{\sin 3x \cos x - \sin x \cos 3x}$
$= \dfrac{\sin 4x}{\sin 2x} = \dfrac{2\sin 2x \cos 2x}{\sin 2x} = 2\cos 2x$

49. $\dfrac{\sin 3x + \sin 5x + \sin 7x + \sin 9x}{\cos 3x + \cos 5x + \cos 7x + \cos 9x} = \dfrac{(\sin 3x + \sin 5x) + (\sin 7x + \sin 9x)}{(\cos 3x + \cos 5x) + (\cos 7x + \cos 9x)} = \dfrac{2\sin 4x \cos x + 2\sin 8x \cos x}{2\cos 4x \cos x + 2\cos 8x \cos x}$

$\qquad\qquad\qquad\qquad\qquad = \dfrac{2\cos x(\sin 4x + \sin 8x)}{2\cos x(\cos 4x + \cos 8x)} = \dfrac{2\sin 6x \cos(-2x)}{2\cos 6x \cos(-2x)} = \dfrac{\sin 6x}{\cos 6x} = \tan 6x$

51. Using the reduction formula, we have $y = \sqrt{3}\sin x + \cos x = a\sin x + b\cos x \Rightarrow$

$a = \sqrt{3}, b = 1 \Rightarrow \sqrt{a^2 + b^2} = 2 = A$. So, θ is any angle in standard position that has $(\sqrt{3}, 1)$ on its terminal

side $\Rightarrow \tan\theta = \dfrac{1}{\sqrt{3}} \Rightarrow \theta = \dfrac{\pi}{6}$. So $y = \sqrt{3}\sin x + \cos x = 2\sin\left(x + \dfrac{\pi}{6}\right)$.

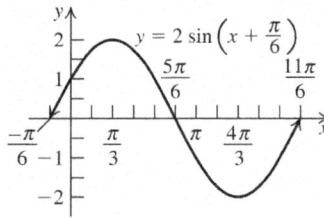

53. $\sin 2A + \sin 2B - \sin 2C = (\sin 2A + \sin 2B) - \sin 2C = 2\sin(A+B)\cos(A-B) - 2\sin C\cos C$
$A + B + C = 180° \Rightarrow C = 180° - (A+B)$ or $A + B = 180° - C$, so
$2\sin(A+B)\cos(A-B) - 2\sin C\cos C = 2\sin(180° - C)\cos(A-B) - 2\sin C\cos C$
$\qquad\qquad\qquad\qquad\qquad = 2\sin C\cos(A-B) - 2\sin C\cos C = 2\sin C[\cos(A-B) - \cos C]$
$\qquad\qquad\qquad\qquad\qquad = 2\sin C[\cos(A-B) - \cos(180° - (A+B))]$
$\qquad\qquad\qquad\qquad\qquad = 2\sin C[\cos(A-B) + \cos(A+B)]$
$\qquad\qquad\qquad\qquad\qquad = 2\sin C[\cos A\cos B + \sin A\sin B + \cos A\cos B - \sin A\sin B]$
$\qquad\qquad\qquad\qquad\qquad = 2\sin C[2\cos A\cos B] = 4\cos A\cos B\sin C$

55. $2\cos^2 x - 1 = 0 \Rightarrow \cos x = \pm\dfrac{\sqrt{2}}{2} \Rightarrow x \in \left\{\dfrac{\pi}{4}, \dfrac{3\pi}{4}, \dfrac{5\pi}{4}, \dfrac{7\pi}{4}\right\}$

57. $2\cos^2 x - \cos x - 1 = 0 \Rightarrow (2\cos x + 1)(\cos x - 1) = 0 \Rightarrow \cos x = -\dfrac{1}{2}$ or $\cos x = 1$.

If $\cos x = -\dfrac{1}{2} \Rightarrow x = \dfrac{2\pi}{3}$ or $x = \dfrac{4\pi}{3}$. If $\cos x = 1 \Rightarrow x = 0$. The solution is $\left\{0, \dfrac{2\pi}{3}, \dfrac{4\pi}{3}\right\}$.

59. $2\sin 3x - 1 = 0 \Rightarrow \sin 3x = \dfrac{1}{2} \Rightarrow 3x \in \left\{\dfrac{\pi}{6}, \dfrac{5\pi}{6}, \dfrac{13\pi}{6}, \dfrac{17\pi}{6}, \dfrac{25\pi}{6}, \dfrac{29\pi}{6}\right\} \Rightarrow x \in \left\{\dfrac{\pi}{18}, \dfrac{5\pi}{18}, \dfrac{13\pi}{18}, \dfrac{17\pi}{18}, \dfrac{25\pi}{18}, \dfrac{29\pi}{18}\right\}$

61. $3\sin\theta - 1 = 0 \Rightarrow \sin\theta = \dfrac{1}{3} \Rightarrow \theta \in \{19.5°, 160.5°\}$

63. $2\sqrt{3}\cos 2\theta - 3 = 0 \Rightarrow \cos 2\theta = \dfrac{\sqrt{3}}{2} \Rightarrow 2\theta \in \{30°, 330°, 390°, 690°\} \Rightarrow \theta \in \{15°, 165°, 195°, 345°\}$

65. $(\sqrt{3}\tan\theta + 1)(2\cos\theta + 1) = 0 \Rightarrow \tan\theta = -\dfrac{\sqrt{3}}{3}$ or $\cos\theta = -\dfrac{1}{2}$. $\tan\theta = -\dfrac{\sqrt{3}}{3} \Rightarrow \theta \in \{150°, 330°\}$.

$\cos\theta = -\dfrac{1}{2} \Rightarrow \theta \in \{120°, 240°\}$. The solution is $\{120°, 150°, 240°, 330°\}$.

Applying the Concepts

67. $\dfrac{2}{3} = \dfrac{4}{3}\cos\alpha \Rightarrow \cos\alpha = \dfrac{1}{2} \Rightarrow \alpha = 60°$

69. $80 = 160\cos 120\pi t \Rightarrow \cos 120\pi t = \dfrac{1}{2} \Rightarrow 120\pi t = \cos^{-1}\dfrac{1}{2} = \dfrac{\pi}{3} \Rightarrow t = \dfrac{1}{360}$ sec

71. $A = 1$ and $x = 2 \Rightarrow 4\sin\dfrac{\theta}{2}\cos\dfrac{\theta}{2} = 1 \Rightarrow 2\left(2\sin\dfrac{\theta}{2}\cos\dfrac{\theta}{2}\right) = 2\sin\theta = 1 \Rightarrow \sin\theta = \dfrac{1}{2} \Rightarrow \theta = \sin^{-1}\left(\dfrac{1}{2}\right) = 30°$

Chapter 6 Practice Test A

1. $\sin\theta = \dfrac{3}{5} = \dfrac{y}{r}$ and $\cos\theta < 0 \Rightarrow \cos\theta = \dfrac{x}{r} = -\dfrac{4}{5} \Rightarrow \tan\theta = -\dfrac{3}{4}$

2. $\tan x = \dfrac{2}{3} = \dfrac{y}{x}$ and $\csc x < 0 \Rightarrow$
   $\sin x < 0$ and $\cos x < 0 \Rightarrow \cos x = \dfrac{x}{r} = -\dfrac{3\sqrt{13}}{13}$

3. $\dfrac{1-\sin^2 x}{\sin^2 x} = \dfrac{\cos^2 x}{\sin^2 x} = \cot^2 x$

4. Let $u = \sin x + \cos x$. Starting with the right side, we have
   $(\sin x + \cos x + 1)(\sin x + \cos x - 1) = (u+1)(u-1) = u^2 - 1 = (\sin x + \cos x)^2 - 1$
   $= \sin^2 x + 2\sin x\cos x + \cos^2 x - 1$
   $= (\sin^2 x + \cos^2 x) - 1 + 2\sin x\cos x = 2\sin x\cos x$

5. $\sin x \sin\left(\dfrac{\pi}{2} - x\right) = \sin x\left(\sin\dfrac{\pi}{2}\cos x - \sin x\cos\dfrac{\pi}{2}\right) = \sin x\cos x = \dfrac{2\sin x\cos x}{2} = \dfrac{\sin 2x}{2}$

6. Start with the right side.
   $\dfrac{\sin x}{1+\tan x} = \dfrac{\sin x}{1+\dfrac{\sin x}{\cos x}} = \dfrac{\sin x}{\dfrac{\cos x + \sin x}{\cos x}} = \dfrac{\sin x\cos x}{\cos x + \sin x} = \dfrac{2\sin x\cos x}{2(\cos x + \sin x)} = \dfrac{\sin 2x}{2(\cos x + \sin x)}$

7. $\dfrac{\sin 2x + \sin 4x}{\cos 2x + \cos 4x} = \dfrac{2\sin\left(\frac{2x+4x}{2}\right)\cos\left(\frac{2x-4x}{2}\right)}{2\cos\left(\frac{2x+4x}{2}\right)\cos\left(\frac{2x-4x}{2}\right)} = \dfrac{2\sin 3x\cos(-x)}{2\cos 3x\cos(-x)} = \dfrac{\sin 3x}{\cos 3x} = \tan 3x$

8. $\cos(x+y)\cos(x-y) = \dfrac{1}{2}\left(\cos\left[(x+y)-(x-y)\right] + \cos\left[(x+y)+(x-y)\right]\right) = \dfrac{1}{2}(\cos 2y + \cos 2x)$
   $= \dfrac{1}{2}\left(2\cos^2 y - 1 + 2\cos^2 x - 1\right) = \cos^2 x + \cos^2 y - 1$

9. $\tan(-x) = 1 \Rightarrow -x = \tan^{-1} 1 \Rightarrow -x = \dfrac{\pi}{4} \Rightarrow x = -\dfrac{\pi}{4} + n\pi$, where n is any integer.

10. $\sin 4x = \dfrac{1}{2} \Rightarrow 4x = \sin^{-1}\left(\dfrac{1}{2}\right) \Rightarrow 4x = \dfrac{\pi}{6} + 2\pi$ or $4x = \dfrac{5\pi}{6} + 2\pi \Rightarrow x = \dfrac{\pi}{24} + \dfrac{\pi}{2}n$ or $x = \dfrac{5\pi}{24} + \dfrac{\pi}{2}n$, where n is any integer.

11. $\cos\dfrac{x}{3} = \dfrac{\sqrt{2}}{2} \Rightarrow \dfrac{x}{3} = \cos^{-1}\left(\dfrac{\sqrt{2}}{2}\right) \Rightarrow \dfrac{x}{3} \in \left\{\dfrac{\pi}{4}, \dfrac{7\pi}{4}\right\} \Rightarrow x = \dfrac{3\pi}{4}$. (The other solution is not in the domain.)

12. $\sin 2x + \cos x = 0 \Rightarrow 2\sin x \cos x + \cos x = 0 \Rightarrow \cos x(2\sin x + 1) = 0 \Rightarrow \cos x = 0$ or $\sin x = -\dfrac{1}{2}$

    If $\cos x = 0 \Rightarrow x = \dfrac{\pi}{2}$ or $x = \dfrac{3\pi}{2}$. If $\sin x = -\dfrac{1}{2} \Rightarrow x = \dfrac{7\pi}{6}$ or $x = \dfrac{11\pi}{6}$.

    The solution is $\left\{\dfrac{\pi}{2}, \dfrac{7\pi}{6}, \dfrac{3\pi}{2}, \dfrac{11\pi}{6}\right\}$.

13. $\sin 56° + \cos 146° = \sin(90° - 34°) + \cos(180° - 34°)$
    $= \sin 90° \cos 34° - \sin 34° \cos 90° + \cos 180° \cos 34° + \sin 180° \sin 34°$
    $= \cos 34° - 0 - \cos 34° + 0 = 0$

14. $\cos 48° \cos 12° - \sin 48° \sin 12° = \cos(48° + 12°) = \cos 60° = \dfrac{1}{2}$

15. $\sin \dfrac{5\pi}{12} = \sin\left(\dfrac{\pi}{6} + \dfrac{\pi}{4}\right) = \sin\dfrac{\pi}{6}\cos\dfrac{\pi}{4} + \cos\dfrac{\pi}{6}\sin\dfrac{\pi}{4} = \dfrac{1}{2}\left(\dfrac{\sqrt{2}}{2}\right) + \dfrac{\sqrt{3}}{2}\left(\dfrac{\sqrt{2}}{2}\right) = \dfrac{\sqrt{6} + \sqrt{2}}{4}$

16. $\cos 2\theta = 1 - 2\sin^2\theta = 1 - 2\left(\dfrac{4}{5}\right)^2 = -\dfrac{7}{25}$

17. $\tan\theta = \dfrac{3}{4} \Rightarrow \sin\theta = \dfrac{3}{5}, \cos\theta = \dfrac{4}{5}$
    $\sin 2\theta = 2\sin\theta\cos\theta = 2\left(\dfrac{3}{5}\right)\left(\dfrac{4}{5}\right) = \dfrac{24}{25}$

18. $y = \cos\left(\dfrac{\pi}{2} - x\right) + \tan x = \sin x + \tan x$
    $f(-x) = \sin(-x) + \tan(-x) = -\sin x - \tan x = -(\sin x + \tan x) \Rightarrow f(x)$ is odd.

19. If $x = y = \dfrac{\pi}{2}, \sin x + \sin y = 1 + 1 = 2 \neq \sin\pi$, so $\sin x + \sin y = \sin(x + y)$ is not an identity.

20. $\dfrac{128^2 \sin 2\theta}{32} = 512 \Rightarrow \sin 2\theta = \dfrac{512 \cdot 32}{128^2} = 1 \Rightarrow 2\theta = 90° \Rightarrow \theta = 45°$

Chapter 6 Practice Test B

1. $\sin\theta = -\dfrac{12}{13}$ and $\pi < \theta < \dfrac{3\pi}{2} \Rightarrow \cos\theta = -\dfrac{5}{13} \Rightarrow \sec\theta = -\dfrac{13}{5}$
   The answer is A.

2. $(\sin x + \cos x)^2 + (\sin x - \cos x)^2 = \sin^2 x + 2\sin x \cos x + \cos^2 x + \sin^2 x - 2\sin x \cos x + \cos^2 x$
   $= 2(\sin^2 x + \cos^2 x) = 2$. The answer is B.

3. $\dfrac{\sin x}{1 + \cos x} + \dfrac{\sin x}{1 - \cos x} = \dfrac{\sin x(1 - \cos x) + \sin x(1 + \cos x)}{1 - \cos^2 x} = \dfrac{2\sin x}{\sin^2 x} = \dfrac{2}{\sin x} = 2\csc x$. The answer is D.

4. $\dfrac{\tan x + \tan y}{\cot x + \cot y} = \dfrac{\tan x + \tan y}{\dfrac{1}{\tan x} + \dfrac{1}{\tan y}} = \dfrac{\tan x + \tan y}{\dfrac{\tan y + \tan x}{\tan x \tan y}} = \tan x \tan y$. The answer is B.

5. $\dfrac{\sin x + \sin y}{\cos x + \cos y} + \dfrac{\cos x - \cos y}{\sin x - \sin y} = \dfrac{(\sin x + \sin y)(\sin x - \sin y) + (\cos x - \cos y)(\cos x + \cos y)}{(\cos x + \cos y)(\sin x - \sin y)}$

$= \dfrac{\sin^2 x - \sin^2 y + \cos^2 x - \cos^2 y}{(\cos x + \cos y)(\sin x - \sin y)} = \dfrac{(\sin^2 x + \cos^2 x) - (\sin^2 y + \cos^2 y)}{(\cos x + \cos y)(\sin x - \sin y)} = 0$

The answer is A.

6. $\dfrac{\sin 2x}{1 + \cos 2x} = \dfrac{2 \sin x \cos x}{1 + 2\cos^2 x - 1} = \dfrac{2 \sin x \cos x}{2 \cos^2 x}$

$= \dfrac{\sin x}{\cos x} = \tan x$. The answer is B.

7. $\dfrac{1 - \cos 2x}{1 + \cos 2x} = \tan^2 x.$

The answer is C.

8. $\dfrac{\sin 3x - \sin x}{\cos x - \cos 3x} = \dfrac{3 \sin x - 4 \sin^3 x - \sin x}{\cos x - (4\cos^3 x - 3 \cos x)}$

$= \dfrac{2 \sin x(1 - 2\sin^2 x)}{4 \cos x(1 - \cos^2 x)}$

$= \dfrac{2 \sin x \cos 2x}{4 \cos x \sin^2 x} = \dfrac{\cos 2x}{2 \sin x \cos x}$

$= \dfrac{\cos 2x}{\sin 2x} = \cot 2x$

The answer is D.

9. $\cot(-x) = -1 \Rightarrow -\cot x = -1 \Rightarrow \cot x = 1 \Rightarrow$

$x = \dfrac{\pi}{4} + \pi n.$

The answer is C.

10. $2 \cos 4x = -1 \Rightarrow \cos 4x = -\dfrac{1}{2} \Rightarrow$

$4x = \dfrac{2\pi}{3} + 2n\pi \Rightarrow x = \dfrac{\pi}{6} + \dfrac{n\pi}{2}$

or $4x = \dfrac{4\pi}{3} + 2n\pi \Rightarrow x = \dfrac{\pi}{3} + \dfrac{n\pi}{2}$, where n is

any integer. The answer is C.

11. $\sin \dfrac{x}{3} = \dfrac{\sqrt{2}}{2} \Rightarrow \dfrac{x}{3} = \dfrac{\pi}{4} + 2n\pi \Rightarrow x = \dfrac{3\pi}{4} + 6n\pi$

or $\dfrac{x}{3} = \dfrac{3\pi}{4} + 2n\pi \Rightarrow x = \dfrac{9\pi}{4} + 6n\pi$, where n is

any integer. The answer is C.

12. $\cos x - \sin 2x = 0 \Rightarrow \cos x - 2 \sin x \cos x = 0 \Rightarrow$

$\cos x(1 - 2 \sin x) = 0 \Rightarrow \cos x = 0$ or $\sin x = \dfrac{1}{2}.$

If $\cos x = 0 \Rightarrow x = \dfrac{\pi}{2}$ or $x = \dfrac{3\pi}{2}.$

If $\sin x = \dfrac{1}{2} \Rightarrow x = \dfrac{\pi}{6}$ or $x = \dfrac{5\pi}{6}.$

The answer is A.

13. $\sin 71° + \cos 161°$
$= \sin(90° - 19°) + \cos(180° - 19°)$
$= \sin 90° \cos 19° - \cos 90° \sin 19°$
$\quad + \cos 180° \cos 19° + \sin 180° \sin 19°$
$= \cos 19° - 0 - \cos 19° + 0 = 0$

The answer is D.

14. $\sin 53° \cos 37° + \cos 53° \sin 37° = \sin(53° + 37°)$
$= \sin 90° = 1$

The answer is D.

15. $\cos 2\theta = 1 - 2 \sin^2 \theta = 1 - 2(2/3)^2 = 1/9$

The answer is B.

16. $\tan \theta = \dfrac{4}{3} = \dfrac{y}{x} \Rightarrow \sin \theta = \dfrac{4}{5}, \cos \theta = \dfrac{3}{5}.$

$\sin 2\theta = 2 \sin \theta \cos \theta = 2(4/5)(3/5) = 24/25$

The answer is A.

17. $f(-x) = \sin\left(\dfrac{\pi}{2} - (-x)\right) + \cot(-x)$

$= \cos(-x) + \cot(-x) = \cos x - \cot x$
$\neq f(x) \Rightarrow f(x)$ is not even (not symmetric
with respect to the y-axis)
$\neq -f(x) \Rightarrow f(x)$ is not odd (not symmetric
with respect to the origin)

$-y = -\left(\sin\left(\dfrac{\pi}{2} - (-x)\right) + \cot(-x)\right)$

$= -(\cos(-x) + \cot(-x))$
$= -(\cos x - \cot x) = -\cos x + \cot x \Rightarrow$

$y = \cos x - \cot x = \sin\left(\dfrac{\pi}{2} - x\right) - \cot x \Rightarrow$

the equation is not symmetric with respect to
the origin. The answer is D.

18. $\tan\left(\dfrac{\pi}{4} + \dfrac{5\pi}{4}\right) = \tan\dfrac{3\pi}{2}$ (undefined)

$\tan\left(\dfrac{\pi}{4}\right) + \tan\left(\dfrac{5\pi}{4}\right) = 1 + 1 = 2$. The answer is C.

19. $2\cos^2 x = \dfrac{3}{2} \Rightarrow \cos^2 x = \dfrac{3}{4} \Rightarrow \cos x = \pm\dfrac{\sqrt{3}}{2} \Rightarrow$

$x \in \left\{\dfrac{\pi}{6}, \dfrac{5\pi}{6}, \dfrac{7\pi}{6}, \dfrac{11\pi}{6}\right\}$. The answer is D.

Copyright © 2019 Pearson Education Inc.

20. $\dfrac{128^2 \sin 2\theta}{32} = 256\sqrt{3} \Rightarrow$

$\sin 2\theta = \dfrac{256\sqrt{3} \cdot 32}{128^2} = \dfrac{\sqrt{3}}{2} \Rightarrow 2\theta = 60°, 120° \Rightarrow$

$\theta = 30°, 60°$

The answer is B.

Cumulative Review Exercises
Chapters P–6

1. $x^2 + x - 1 = 0 \Rightarrow x = \dfrac{-1 \pm \sqrt{1^2 - 4(1)(-1)}}{2(1)} \Rightarrow$

    $x = \dfrac{-1 \pm \sqrt{5}}{2}$

2. $\log_2(x+1) + \log_2(x-1) = 1 \Rightarrow$
    $\log_2\big((x+1)(x-1)\big) = 1 \Rightarrow$

    $x^2 - 1 = 2^1 \Rightarrow x^2 = 3 \Rightarrow x = \sqrt{3}$ (Reject the negative solution; logarithms are not defined for negative numbers.)

3. $5^{-x} = 9 \Rightarrow -x \log 5 = \log 9 \Rightarrow x = -\dfrac{\log 9}{\log 5}$

4. $\sin 2x - \cos x = 0 \Rightarrow 2\sin x \cos x - \cos x = 0 \Rightarrow$
    $\cos x(2\sin x - 1) = 0 \Rightarrow \cos x = 0$ or $\sin x = \dfrac{1}{2}$.

    If $\cos x = 0 \Rightarrow x = \dfrac{\pi}{2}$ or $x = \dfrac{3\pi}{2}$.

    If $\sin x = \dfrac{1}{2} \Rightarrow x = \dfrac{\pi}{6}$ or $x = \dfrac{5\pi}{6}$.

    The solution is $\left\{\dfrac{\pi}{6}, \dfrac{\pi}{2}, \dfrac{5\pi}{6}, \dfrac{3\pi}{2}\right\}$.

5. Solve the associated equation:
    $x^3 - 4x = 0 \Rightarrow x(x-2)(x+2) = 0 \Rightarrow x = 0$ or $x = 2$ or $x = -2$. The intervals to be tested are $(-\infty, -2), (-2, 0), (0, 2),$ and $(2, \infty)$.

Interval	Test point	Value of $x^3 - 4x$	Result
$(-\infty, -2)$	-3	-15	$-$
$(-2, 0)$	-1	3	$+$
$(0, 2)$	1	-3	$-$
$(2, \infty)$	3	15	$+$

The solution set is $(-2, 0) \cup (2, \infty)$.

6. $\dfrac{2x-3}{x+2} - 1 < 0 \Rightarrow \dfrac{2x-3-(x+2)}{x+2} < 0 \Rightarrow$

    $\dfrac{x-5}{x+2} < 0$. Now solve $x - 5 = 0 \Rightarrow x = 5$ and $x + 2 = 0 \Rightarrow x = -2$. The intervals to be tested are $(-\infty, -2), (-2, 5),$ and $(5, \infty)$.

Interval	Test point	Value of $\dfrac{2x-3}{x+2} - 1$	Result
$(-\infty, -2)$	-3	8	$+$
$(-2, 5)$	0	$-5/2$	$-$
$(5, \infty)$	6	$1/8$	$+$

The solution set is $(-2, 5)$.

7. $\dfrac{f(x+h) - f(x)}{h}$

    $= \dfrac{\big(2(x+h)^2 - (x+h) + 3\big) - \big(2x^2 - x + 3\big)}{h}$

    $= \dfrac{2x^2 + 4xh + 2h^2 - x - h + 3 - 2x^2 + x - 3}{h}$

    $= \dfrac{2h^2 + 4xh - h}{h} = \dfrac{h(4x + 2h - 1)}{h} = 4x - 1 + 2h$

8. $f(x) = y = \dfrac{x+2}{2x-1}$. Switch the variables, and then solve for y to find $f^{-1}(x)$:

    $x = \dfrac{y+2}{2y-1} \Rightarrow 2xy - x = y + 2 \Rightarrow$

    $2xy - y = x + 2 \Rightarrow y(2x-1) = x + 2 \Rightarrow$

    $y = f^{-1}(x) = \dfrac{x+2}{2x-1}$

9. First, complete the square to find the transformations needed:

$y = x^2 - 4x - 7 \Rightarrow y + 7 + 4 = x^2 - 4x + 4 \Rightarrow$
$y + 11 = (x-2)^2 \Rightarrow y = (x-2)^2 - 11$

Shift the graph of $y = x^2$ two units to the right and 11 units down.

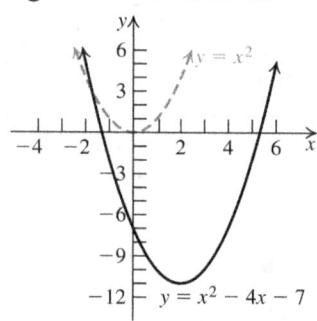

10. Shift the graph of $y = \sqrt{x}$ three units to the right, then stretch the graph vertically by a factor of 2, then shift the resulting graph one unit up.

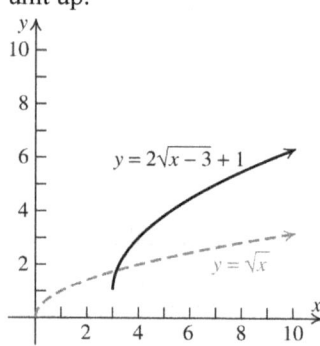

11. Shift the graph of $y = 2^x$ one unit to the right and two units down.

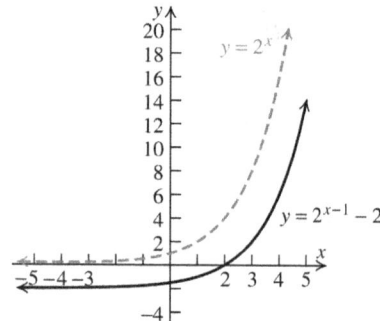

12. Shift the graph of $y = \ln x$ one unit to the left and then stretch the graph vertically by a factor of 2.

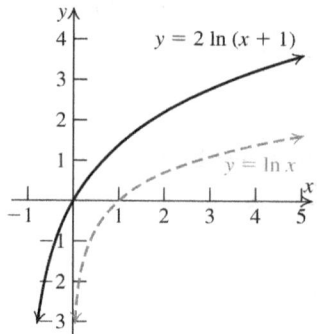

13. The period of $y = \dfrac{1}{2}\sin 2x$ is $\dfrac{2\pi}{2} = \pi$ and its amplitude is $1/2$. Compress the graph of $y = \sin x$ horizontally by a factor of 2 and vertically by a factor of 2.

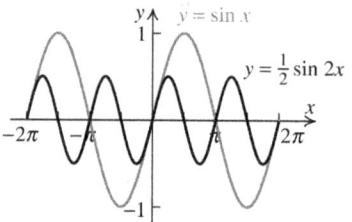

14. The amplitude of

$y = -2\cos(3x - 1) = -2\cos 3\left(x - \dfrac{1}{3}\right)$ is 2, the

period is $2\pi/3$, and the phase shift is $1/3$ to the right.

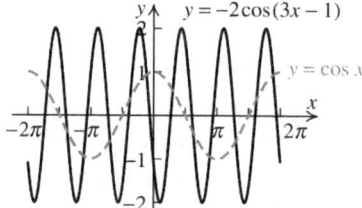

15.

16.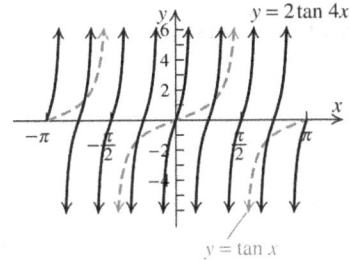

17. $\dfrac{2\csc^2 x - 5\cot x - 5}{2\cot x + 1} = \dfrac{2(1+\cot^2 x) - 5\cot x - 5}{2\cot x + 1}$

$= \dfrac{2\cot^2 x - 5\cot x - 3}{2\cot x + 1}$

$= \dfrac{(\cot x - 3)(2\cot x + 1)}{2\cot x + 1}$

$= \cot x - 3$

18. $\sec 15° = \sec(45° - 30°) = \dfrac{1}{\cos(45° - 30°)}$

$= \dfrac{1}{\cos 45° \cos 30° + \sin 45° \sin 30°}$

$= \dfrac{1}{\left(\frac{\sqrt{2}}{2}\right)\left(\frac{\sqrt{3}}{2}\right) + \left(\frac{\sqrt{2}}{2}\right)\left(\frac{1}{2}\right)} = \dfrac{4}{\sqrt{6} + \sqrt{2}}$

$= \dfrac{4}{\sqrt{6} + \sqrt{2}} \cdot \dfrac{\sqrt{6} - \sqrt{2}}{\sqrt{6} - \sqrt{2}} = \sqrt{6} - \sqrt{2}$

19. $\sin\left(x - \dfrac{3\pi}{2}\right)$

$= \sin x \cos\dfrac{3\pi}{2} - \sin\dfrac{3\pi}{2}\cos x$

$= 0 - (-1)\cos x = \cos x$

20.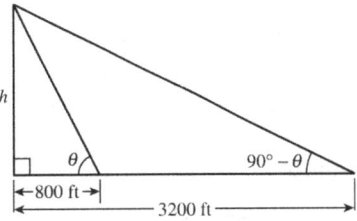

$\tan\theta = \dfrac{h}{800} \Rightarrow h = 800\tan\theta$

$\tan(90° - \theta) = \cot\theta = \dfrac{h}{3200} \Rightarrow h = 3200\cot\theta$

$800\tan\theta = 3200\cot\theta \Rightarrow \dfrac{\tan\theta}{\cot\theta} = 4 \Rightarrow$

$\tan^2\theta = 4 \Rightarrow \tan\theta = \pm 2$

(Reject the negative solution.)

$2 = \dfrac{h}{800} \Rightarrow h = 1600.$

The tower is 1600 feet tall.

Chapter 7 Applications of Trigonometric Functions

7.1 The Law of Sines

7.1 Practice Problems

1. Given: $a = 6$, $A = 30°$, and $C = 72°$
$$\frac{a}{\sin A} = \frac{c}{\sin C} \Rightarrow \frac{6}{\sin 30°} = \frac{c}{\sin 72°} \Rightarrow$$
$$c = \frac{6 \sin 72°}{\sin 30°} \approx 11.4$$

2. Given: $a = 3$, $b = 5$, and $B = 70°$
$$\frac{\sin A}{a} = \frac{\sin B}{b} \Rightarrow \frac{\sin A}{3} = \frac{\sin 70°}{5} \Rightarrow$$
$$\sin A = \frac{3 \sin 70°}{5} \Rightarrow$$
$$A_1 = \sin^{-1}\left(\frac{3 \sin 70°}{5}\right) \approx 34.3° \text{ or}$$
$$A_2 = 180° - \sin^{-1}\left(\frac{3 \sin 70°}{5}\right) \approx 145.7°$$
However,
$A_2 + B = 145.7° + 70° = 215.7° > 180°$, so there is no triangle with angle A_2. Therefore, $A = 34.3°$.

3. Given: $A = 70°$, $B = 65°$, and $a = 16$ in. – an AAS case. $C = 180° - (70° + 65°) = 45°$
$$\frac{a}{\sin A} = \frac{b}{\sin B} \Rightarrow \frac{16}{\sin 70°} = \frac{b}{\sin 65°} \Rightarrow$$
$b \approx 15.4$ in.
$$\frac{a}{\sin A} = \frac{c}{\sin C} \Rightarrow \frac{16}{\sin 70°} = \frac{c}{\sin 45°} \Rightarrow$$
$c \approx 12.0$ in.

4.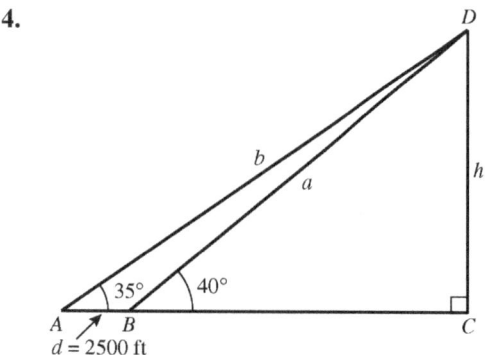

   In triangle ABD, $m\angle DBC = 40° \Rightarrow$
   $m\angle DBA = 180° - 40° = 140°$. Then,
   $m\angle BDA = 180° - (35° + 140°) = 5°$.

$$\frac{2500}{\sin 5°} = \frac{a}{\sin 35°} \Rightarrow$$
$$a = \frac{2500 \sin 35°}{\sin 5°} \approx 16,453 \text{ ft}$$
In triangle BCD,
$$\sin 40° = \frac{h}{a} \Rightarrow h = a \sin 40° \Rightarrow$$
$h \approx 16,453 \sin 40° \approx 10,576$ ft

5. Given: $C = 35°$, $b = 15$ ft, and $c = 12$ ft – an SSA case.
$$\frac{\sin C}{c} = \frac{\sin B}{b} \Rightarrow \frac{\sin 35°}{12} = \frac{\sin B}{15} \Rightarrow$$
$$\sin B = \frac{15 \sin 35°}{12} \approx 0.7170 \Rightarrow B_1 \approx 45.8° \text{ or}$$
$B_2 \approx 180° - 45.8° = 134.2°$
$A_1 \approx 180° - (35° + 45.8°) = 99.2°$
$A_2 \approx 180° - (35° + 134.2°) = 10.8°$
$$\frac{a_1}{\sin A_1} = \frac{c}{\sin C} \Rightarrow \frac{a_1}{\sin 99.2°} = \frac{12}{\sin 35°} \Rightarrow$$
$a_1 \approx 20.7$ ft
$$\frac{a_2}{\sin A_2} = \frac{c}{\sin C} \Rightarrow \frac{a_2}{\sin 10.8°} = \frac{12}{\sin 35°} \Rightarrow$$
$a_2 \approx 3.9$ ft
Solution 1:
$A_1 \approx 99.2°$, $B_1 \approx 45.8°$, $a_1 \approx 20.7$ ft
Solution 2:
$A_2 \approx 10.8°$, $B_2 \approx 134.2°$, $a_2 \approx 3.9$ ft

6. Given: $A = 65°$, $a = 16$ m, and $b = 30$ m – an SSA case.
$$\frac{\sin A}{a} = \frac{\sin B}{b} \Rightarrow \frac{\sin 65°}{16} = \frac{\sin B}{30} \Rightarrow$$
$\sin B \approx 1.6993 > 1$, so no triangle is possible.

7. Given: $C = 60°$, $c = 50$ ft, and $a = 30$ ft – an SSA case.
$$\frac{\sin C}{c} = \frac{\sin A}{a} \Rightarrow \frac{\sin 60°}{50} = \frac{\sin A}{30} \Rightarrow$$
$\sin A \approx 0.5196 \Rightarrow A_1 \approx 31.3°$ or
$A_2 \approx 180° - 31.3° = 148.7°$
$B_1 \approx 180° - (31.3° + 60°) = 88.7°$
$B_2 \approx 180° - (148.7° + 60°) = -28.7°$, which is not possible, so there is one solution.
$$\frac{c}{\sin C} = \frac{b}{\sin B} \Rightarrow \frac{50}{\sin 60°} = \frac{b}{\sin 88.7°} \Rightarrow$$
$b \approx 57.7$ ft

Thus, $A \approx 31.3°$, $B \approx 88.7°$, $b \approx 57.7$ ft.

8. a. In two hours, the ship travels $(2)(25) = 50$ mi.

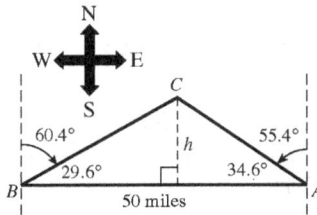

The oil rig (at C), the starting point of the ship (at A) and the position of the ship after two hours (at B) are shown in the figure above. Since the original bearing is N 55.4° W, $m\angle CAB = 90° - 55.4° = 34.6°$. Since the second bearing is 60.4°, $m\angle CBA = 90° - 60.4° = 29.6°$.
$m\angle ACB = 180° - (29.6° + 34.6°) = 115.8°$

Then $\dfrac{BC}{\sin \angle CAB} = \dfrac{AB}{\sin \angle ACB} \Rightarrow$
$\dfrac{BC}{\sin 34.6°} = \dfrac{50}{\sin 115.8°} \Rightarrow BC \approx 31.5$ mi

The ship is approximately 31.5 mi from the oil rig when the second bearing is taken.

b. The shortest distance from the ship to the oil rig is the length of segment h.
$\sin B = \dfrac{h}{BC} \Rightarrow \sin 29.6° \approx \dfrac{h}{31.5} \Rightarrow$
$h \approx 15.6$ mi
The ship passes within 15.6 miles of the oil rig.

7.1 Concepts and Vocabulary

1. If you know two angles of a triangle, then you can determine the third because the sum of all three angles is 180° degrees.

3. If you are given any two angles and one side, then there is exactly one triangle possible.

5. True

7. True

7.1 Building Skills

9. $A = 45°, B = 60°, a = 10$
$\dfrac{a}{\sin A} = \dfrac{b}{\sin B} \Rightarrow \dfrac{10}{\sin 45°} = \dfrac{b}{\sin 60°} \Rightarrow$
$b = \dfrac{10 \sin 60°}{\sin 45°} = \dfrac{10 \cdot \frac{\sqrt{3}}{2}}{\frac{\sqrt{2}}{2}} = \dfrac{10\sqrt{3}}{\sqrt{2}} = \dfrac{10\sqrt{6}}{2}$
$= 5\sqrt{6} \approx 12.2$

11. $B = 45°, C = 75°, b = \sqrt{6}$
$A + B + C = 180° \Rightarrow A + 45° + 75° = 180° \Rightarrow$
$A = 60°$
$\dfrac{a}{\sin A} = \dfrac{b}{\sin B} \Rightarrow \dfrac{a}{\sin 60°} = \dfrac{\sqrt{6}}{\sin 45°} \Rightarrow$
$a = \dfrac{\sqrt{6} \sin 60°}{\sin 45°} = \dfrac{\sqrt{6} \cdot \frac{\sqrt{3}}{2}}{\frac{\sqrt{2}}{2}} = \dfrac{\sqrt{18}}{\sqrt{2}} = \sqrt{9} = 3$

13. $a = 4, b = 6, A = 30°$
$\dfrac{\sin A}{a} = \dfrac{\sin B}{b} \Rightarrow \dfrac{\sin 30°}{4} = \dfrac{\sin B}{6} \Rightarrow$
$\sin B = \dfrac{6 \sin 30°}{4} = \dfrac{6\left(\frac{1}{2}\right)}{4} = \dfrac{3}{4}$
$B_1 = \sin^{-1}\left(\dfrac{3}{4}\right) \approx 48.6°$ or
$B_2 = 180° - \sin^{-1}\left(\dfrac{3}{4}\right) \approx 131.4°$

15. $a = 3, b = 5, C = 60°$
This is the SAS case, so the Law of Sines is not applicable.

17. $a = 2, b = 3, \sin A = \dfrac{2}{3}$
$\dfrac{\sin A}{a} = \dfrac{\sin B}{b} \Rightarrow \dfrac{\frac{2}{3}}{2} = \dfrac{\sin B}{3} \Rightarrow$
$\sin B = 1 \Rightarrow B = 90°$

19. $a = 8, c = 5, C = 60°$
$\dfrac{\sin A}{a} = \dfrac{\sin C}{c} \Rightarrow \dfrac{\sin A}{8} = \dfrac{\sin 60°}{5} \Rightarrow$
$\sin A = \dfrac{8 \sin 60°}{5} \approx 1.4 > 1 \Rightarrow$ no triangle is possible.

21. Given: $A = 61°, B = 56°, c = 100$ ft − an ASA case. $C = 180° - (61° + 56°) = 63°$
$\dfrac{a}{\sin A} = \dfrac{c}{\sin C} \Rightarrow \dfrac{a}{\sin 61°} = \dfrac{100}{\sin 63°} \Rightarrow$
$a \approx 98.2$ ft
$\dfrac{b}{\sin B} = \dfrac{c}{\sin C} \Rightarrow \dfrac{b}{\sin 56°} = \dfrac{100}{\sin 63°} \Rightarrow$
$b \approx 93.0$ ft

23. Given: $A = 110°, C = 43°, c = 47$ m – an AAS case. $B = 180° - (43° + 110°) = 27°$

$\dfrac{a}{\sin A} = \dfrac{c}{\sin C} \Rightarrow \dfrac{a}{\sin 110°} = \dfrac{47}{\sin 43°} \Rightarrow$
$a \approx 64.8$ m

$\dfrac{b}{\sin B} = \dfrac{c}{\sin C} \Rightarrow \dfrac{b}{\sin 27°} = \dfrac{47}{\sin 43°} \Rightarrow$
$b \approx 31.3$ m

25. Given: $A = 40°, B = 35°, a = 100$ m – an AAS case. $C = 180° - (40° + 35°) = 105°$

$\dfrac{a}{\sin A} = \dfrac{b}{\sin B} \Rightarrow \dfrac{100}{\sin 40°} = \dfrac{b}{\sin 35°} \Rightarrow$
$b \approx 89.2$ m

$\dfrac{a}{\sin A} = \dfrac{c}{\sin C} \Rightarrow \dfrac{100}{\sin 40°} = \dfrac{c}{\sin 105°} \Rightarrow$
$c \approx 150.3$ m

27. Given: $A = 46°, C = 55°, a = 75$ cm – an AAS case. $B = 180° - (46° + 55°) = 79°$

$\dfrac{a}{\sin A} = \dfrac{b}{\sin B} \Rightarrow \dfrac{75}{\sin 46°} = \dfrac{b}{\sin 79°} \Rightarrow$
$b \approx 102.3$ cm

$\dfrac{a}{\sin A} = \dfrac{c}{\sin C} \Rightarrow \dfrac{75}{\sin 46°} = \dfrac{c}{\sin 55°} \Rightarrow$
$c \approx 85.4$ cm

29. Given: $A = 35°, C = 47°, c = 60$ ft – an AAS case. $B = 180° - (35° + 47°) = 98°$

$\dfrac{c}{\sin C} = \dfrac{a}{\sin A} \Rightarrow \dfrac{60}{\sin 47°} = \dfrac{a}{\sin 35°} \Rightarrow$
$a \approx 47.1$ ft

$\dfrac{c}{\sin C} = \dfrac{b}{\sin B} \Rightarrow \dfrac{60}{\sin 47°} = \dfrac{b}{\sin 98°} \Rightarrow$
$c \approx 81.2$ ft

31. Given: $B = 43°, C = 67°, b = 40$ in. – an AAS case. $A = 180° - (43° + 67°) = 70°$

$\dfrac{b}{\sin B} = \dfrac{a}{\sin A} \Rightarrow \dfrac{40}{\sin 43°} = \dfrac{a}{\sin 70°} \Rightarrow$
$a \approx 55.1$ in.

$\dfrac{b}{\sin B} = \dfrac{c}{\sin C} \Rightarrow \dfrac{40}{\sin 43°} = \dfrac{c}{\sin 67°} \Rightarrow$
$c \approx 54.0$ in.

33. Given: $B = 110°, C = 46°, c = 23.5$ ft – an AAS case. $A = 180° - (110° + 46°) = 24°$

$\dfrac{c}{\sin C} = \dfrac{a}{\sin A} \Rightarrow \dfrac{23.5}{\sin 46°} = \dfrac{a}{\sin 24°} \Rightarrow$
$a \approx 13.3$ ft

$\dfrac{c}{\sin C} = \dfrac{b}{\sin B} \Rightarrow \dfrac{23.5}{\sin 46°} = \dfrac{b}{\sin 110°} \Rightarrow$
$b \approx 30.7$ ft

35. Given: $A = 35.7°, B = 45.8°, c = 30$ m – an ASA case. $C = 180° - (35.7° + 45.8°) = 98.5°$

$\dfrac{c}{\sin C} = \dfrac{a}{\sin A} \Rightarrow \dfrac{30}{\sin 98.5°} = \dfrac{a}{\sin 35.7°} \Rightarrow$
$a \approx 17.7$ m

$\dfrac{c}{\sin C} = \dfrac{b}{\sin B} \Rightarrow \dfrac{30}{\sin 98.5°} = \dfrac{b}{\sin 45.8°} \Rightarrow$
$b \approx 21.7$ m

37. Given: $A = 65°, a = 20$ ft, $b = 20$ ft – an SSA case. However, the triangle is isosceles, so $m\angle B = m\angle A = 65°$
$C = 180° - (65° + 65°) = 50°$

$\dfrac{a}{\sin A} = \dfrac{c}{\sin C} \Rightarrow \dfrac{20}{\sin 65°} = \dfrac{c}{\sin 50°} \Rightarrow$
$c \approx 16.9$ ft

39. Given: $A = 115°, a = 70$ m, $b = 31$ m – an SSA case.

$\dfrac{\sin A}{a} = \dfrac{\sin B}{b} \Rightarrow \dfrac{\sin 115°}{70} = \dfrac{\sin B}{31} \Rightarrow$
$\sin B \approx 0.4014 \Rightarrow B_1 \approx 23.7°$ or $B_2 \approx 156.3°$
$C_1 = 180° - (115° + 23.7°) = 41.3°$
$C_2 = 180° - (115° + 156.3°) = -91.3°$, so there is one solution, with $B = 23.7°$ and $C = 41.3°$.

$\dfrac{a}{\sin A} = \dfrac{c}{\sin C} \Rightarrow \dfrac{70}{\sin 115°} = \dfrac{c}{\sin 41.3°} \Rightarrow$
$c \approx 51.0$ m

41. $a = 40, b = 70, A = 30°$
$h = b \sin A = 70 \sin 30° = 35$
Since $h < a < b$, there are two triangles with the given measurements.

43. $b = 15, c = 19, B = 58°$
$h = c \sin B = 19 \sin 58° \approx 16.1$
Since $h > b$, there are no triangles with the given measures.

45. $a = 50, b = 70, B = 120°$
Since B is obtuse and $b > a$, there is one triangle with the given measures.

Use this triangle to help solve exercises 47–66.

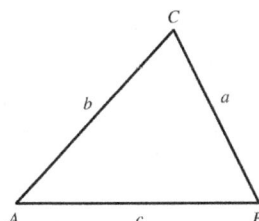

47. Given: $A = 40°, a = 23, b = 20$.
$h = 20 \sin 40° \approx 12.9$; $h < b < a$, so there is one triangle.
$\dfrac{\sin A}{a} = \dfrac{\sin B}{b} \Rightarrow \dfrac{\sin 40°}{23} = \dfrac{\sin B}{20} \Rightarrow$
$B \approx \sin^{-1}\left(\dfrac{20 \sin 40°}{23}\right) \approx 34.0°$
$C = 180° - (40° + 34.0°) \approx 106.0°$
$\dfrac{a}{\sin A} = \dfrac{c}{\sin C} \Rightarrow \dfrac{23}{\sin 40°} = \dfrac{c}{\sin 106.0°} \Rightarrow c \approx 34.4$

49. Given: $A = 30°, a = 25, b = 50$;
$h = b \sin A = 50 \sin 30° = 25$.
$a = h$, so there is one right triangle.
$B = 90°, C = 60°, c = 25\sqrt{3}$ (using the Pythagorean theorem).

51. Given: $A = 40°, a = 10, b = 20$. A is an acute angle. $h = b \sin A = 20 \sin 40° \approx 12.9$; $a < h$, so no triangle exists.

53. Given: $A = 95°, a = 18, b = 21$. A is an obtuse angle and $a < b$, so no triangle exists.

55. Given: $A = 100°, a = 40, b = 34$. $b < a$, so there is one triangle.
$\dfrac{\sin A}{a} = \dfrac{\sin B}{b} \Rightarrow \dfrac{\sin 100°}{40} = \dfrac{\sin B}{34} \Rightarrow$
$B \approx \sin^{-1}\left(\dfrac{34 \sin 100°}{40}\right) \approx 56.8°$
$C = 180° - (100° + 56.8°) \approx 23.2°$
$\dfrac{a}{\sin A} = \dfrac{c}{\sin C} \Rightarrow \dfrac{40}{\sin 100°} = \dfrac{c}{\sin 23.2°} \Rightarrow$
$c \approx 16.0$

57. Given: $B = 50°, b = 22, c = 40$. B is an acute angle. $h = c \sin B = 40 \sin 50° \approx 30.6$; $b < h$, so no triangle exists.

59. Given: $B = 46°, b = 35, c = 40$.
$h = c \sin B = 40 \sin 46° \approx 28.8$; $h < b < c$, so two triangles exist.
$\dfrac{\sin B}{b} = \dfrac{\sin C}{c} \Rightarrow$
$\dfrac{\sin 46°}{35} = \dfrac{\sin C}{40} \Rightarrow C = \sin^{-1}\left(\dfrac{40 \sin 46°}{35}\right) \Rightarrow$
$C_1 \approx 55.3°, C_2 \approx 124.7°$
$A_1 = 180° - (46° + 55.3°) \approx 78.7°$
$\dfrac{b}{\sin B} = \dfrac{a_1}{\sin A_1} \Rightarrow \dfrac{35}{\sin 46°} = \dfrac{a_1}{\sin 78.7°} \Rightarrow$
$a_1 \approx 47.7$
$A_2 = 180° - (46° + 124.7°) \approx 9.3°$
$\dfrac{b}{\sin B} = \dfrac{a_2}{\sin A_2} \Rightarrow \dfrac{35}{\sin 46°} = \dfrac{a_2}{\sin 9.3°} \Rightarrow$
$a_2 \approx 7.9$. The two solutions are: $C_1 \approx 55.3°$, $A_1 \approx 78.7°, a_1 \approx 47.7$ and $C_2 \approx 124.7°$, $A_2 \approx 9.3°, a_2 \approx 7.9$.

61. Given: $B = 97°, b = 27, c = 30$. B is an obtuse angle and $b < c$, so no triangle exists.

63. Given: $A = 42°, a = 55, c = 62$.
$h = c \sin A = 62 \sin 42° \approx 41.5$; $h < a < c$, so two triangles exist. $\dfrac{\sin A}{a} = \dfrac{\sin C}{c} \Rightarrow$
$\dfrac{\sin 42°}{55} = \dfrac{\sin C}{62} \Rightarrow C = \sin^{-1}\left(\dfrac{62 \sin 42°}{55}\right) \Rightarrow$
$C_1 \approx 49.0°, C_2 \approx 131.0°$
$B_1 = 180° - (42° + 49.0°) \approx 89.0°$
$\dfrac{a}{\sin A} = \dfrac{b_1}{\sin B_1} \Rightarrow \dfrac{55}{\sin 42°} = \dfrac{b_1}{\sin 89.0°} \Rightarrow$
$b_1 \approx 82.2$
$B_2 = 180° - (42° + 131.0°) \approx 7.0°$
$\dfrac{a}{\sin A} = \dfrac{b_2}{\sin B_2} \Rightarrow \dfrac{55}{\sin 42°} = \dfrac{b_2}{\sin 7.0°} \Rightarrow$
$b_2 \approx 10.0$. The two solutions are: $C_1 \approx 49.0°$, $B_1 \approx 89.0°, b_1 \approx 82.2$ and $C_2 \approx 131.0°$, $B_2 \approx 7.0°, b_2 \approx 10.0$.

65. Given: $C = 40°, a = 3.3, c = 2.1$.
$h = 3.3 \sin 40° \approx 2.1212 > c$, so no triangle exists.

67. First, use the Law of Sines to find the length of side AD.
$D = 180° - (56° + 37°) = 87°$
$\dfrac{AD}{\sin C} = \dfrac{AC}{\sin D} \Rightarrow \dfrac{AD}{\sin 37°} = \dfrac{156}{\sin 87°} \Rightarrow$
$AD = \dfrac{156 \sin 37°}{\sin 87°} \approx 94.01$
$\sin A = \dfrac{h}{AD} \Rightarrow \sin 56° = \dfrac{h}{94.01} \Rightarrow$
$h = 94.01 \sin 56° \approx 77.9$

69. First, use the Law of Sines to find the length of side AD.
$\angle ACD = 180° - 50° = 130°$
$D = 180° - (31° + 130°) = 19°$
$\dfrac{AD}{\sin \angle ACD} = \dfrac{AC}{\sin D} \Rightarrow \dfrac{AD}{\sin 130°} = \dfrac{35}{\sin 19°} \Rightarrow$
$AD = \dfrac{35 \sin 130°}{\sin 19°} \approx 82.35$
$\sin A = \dfrac{h}{AD} \Rightarrow \sin 31° = \dfrac{h}{82.35} \Rightarrow$
$h = 82.35 \sin 31° \approx 42.4$

71. In right triangle ABD, $\angle A = 90° - 34° = 56°$.
In triangle ADC, $\angle ADC = 34° - 23° = 11°$ and $\angle ACD = 180° - (56° + 11°) = 113°$. Use the Law of Sines in triangle ADC to find the length of side AD.
$\dfrac{AD}{\sin \angle ACD} = \dfrac{AC}{\sin \angle ADC} \Rightarrow$
$\dfrac{AD}{\sin 113°} = \dfrac{20}{\sin 11°} \Rightarrow AD = \dfrac{20 \sin 113°}{\sin 11°} \approx 96.48$
$\sin \angle ADB = \dfrac{h}{AD} \Rightarrow \sin 34° = \dfrac{h}{96.48} \Rightarrow$
$h = 96.48 \sin 34° \approx 54.0$

7.1 Applying the Concepts

73. The length of the side from A to C is b.
$C = 180° - (57° + 46°) = 77°$
$\dfrac{c}{\sin C} = \dfrac{b}{\sin B} \Rightarrow \dfrac{540}{\sin 77°} = \dfrac{b}{\sin 46°} \Rightarrow$
$b \approx 399$ ft

75. Let x represent the length of the palm tree from the base to the top. The measure of the third angle in the triangle is
$180° - (78° + 41°) = 61°$.
$\dfrac{x}{\sin 41°} = \dfrac{60}{\sin 61°} \Rightarrow x = \dfrac{60 \sin 41°}{\sin 61°} \approx 45$
The length of the palm tree is about 45 ft.

77.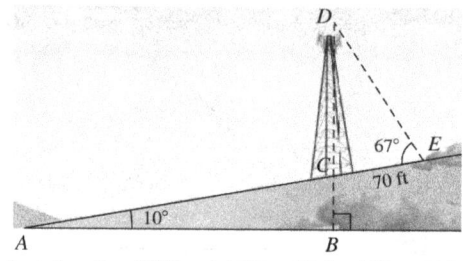

In triangle ABC, $\angle ACB = 90° - 10° = 80°$. So, $\angle DCE = 80°$ because vertical angles are equal. In triangle DCE,
$\angle D = 180° - (80° + 67°) = 33°$.
$\dfrac{DC}{\sin E} = \dfrac{CE}{\sin D} \Rightarrow \dfrac{DC}{\sin 67°} = \dfrac{70}{\sin 33°} \Rightarrow$
$DC = \dfrac{70 \sin 67°}{\sin 33°} \approx 118$
The GPS tower is about 118 ft tall.

79.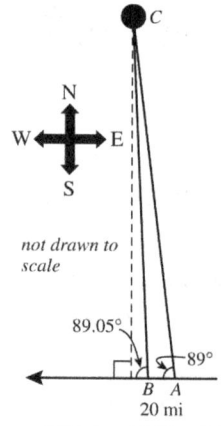

$m\angle CBA = 180° - 89.05° = 90.95°$
$C = 180° - (90.95° + 89°) = 0.05°$
$\dfrac{AC}{\sin 90.95°} = \dfrac{20}{\sin 0.05°} \Rightarrow AC = \dfrac{20 \sin 90.95°}{\sin 0.05°}$
The height of the satellite is the length of the altitude drawn from C to AB: $h = AC \sin 89°$
$= \left(\dfrac{20 \sin 90.95°}{\sin 0.05°}\right) \sin 89° \approx 22{,}912$ mi

81.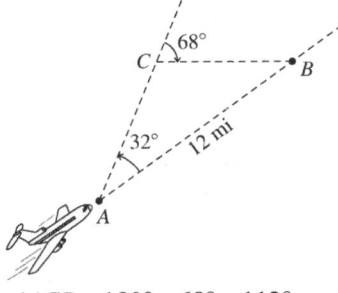

$\angle ACB = 180° - 68° = 112°$ and
$\angle B = 180° - (112° + 32°) = 36°$

Detour = $AC + CB$

$\dfrac{AC}{\sin B} = \dfrac{AB}{\sin \angle ACB} \Rightarrow \dfrac{AC}{\sin 36°} = \dfrac{12}{\sin 112°} \Rightarrow$
$AC = \dfrac{12 \sin 36°}{\sin 112°}$

$\dfrac{CB}{\sin A} = \dfrac{AB}{\sin \angle ACB} \Rightarrow \dfrac{CB}{\sin 32°} = \dfrac{12}{\sin 112°} \Rightarrow$
$CB = \dfrac{12 \sin 32°}{\sin 112°}$

$AC + CB = \dfrac{12 \sin 36°}{\sin 112°} + \dfrac{12 \sin 32°}{\sin 112°} \approx 14.5$

The detour was about 14.5 miles.

83. a.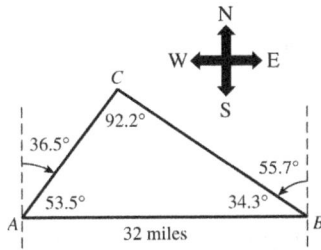

At 16 miles per hour, the ship travels 32 miles in two hours. The interior angles of the triangle are shown in the figure above. The distance from the ship at the second time to the beacon is BC.

$\dfrac{32}{\sin 92.2°} = \dfrac{BC}{\sin 53.5°} \Rightarrow BC \approx 25.7$ mi

b. The closest the ship came to the beacon is the length of the altitude from C to AB:
$h = 25.7 \sin 34.3° \approx 14.5$ mi

85.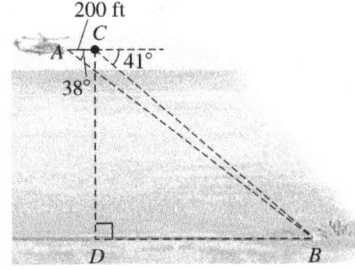

We are looking for the length of DB.
$\angle ACB = 180° - 41° = 139°$
$\angle B = 180° - (139° + 38°) = 3°$
In triangle ABC,
$\dfrac{BC}{\sin A} = \dfrac{AC}{\sin B} \Rightarrow \dfrac{BC}{\sin 38°} = \dfrac{200}{\sin 3°} \Rightarrow$
$BC = \dfrac{200 \sin 38°}{\sin 3°} \approx 2352.73$

In right triangle CDB,
$\angle DCB = 90° - 41° = 49°$.

$\sin \angle DCB = \dfrac{DB}{BC} \Rightarrow \sin 49° = \dfrac{DB}{2352.73} \Rightarrow$
$DB = 2352.73 \sin 49° \approx 1776$

The ground distance from a point directly under the second point (C) to the hospital is about 1776 feet.

87. a.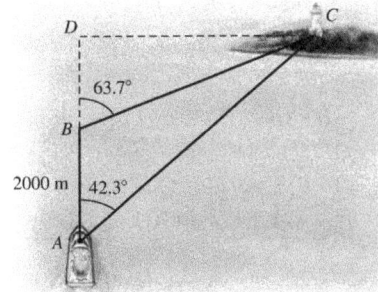

In triangle ABC,
$\angle ABC = 180° - 63.7° = 116.3°$ and
$\angle BCA = 180° - (116.3° + 42.3°) = 21.4°$

$\dfrac{BC}{\sin A} = \dfrac{AB}{\sin \angle BCA} \Rightarrow$
$\dfrac{BC}{\sin 42.3°} = \dfrac{2000}{\sin 21.4°} \Rightarrow$
$BC = \dfrac{2000 \sin 42.3°}{\sin 21.4°} \approx 3689$

The ship is about 3689 meters from the lighthouse at the second sighting.

b. In triangle BDC,
$\sin \angle CBD = \dfrac{DC}{BC} \Rightarrow \sin 63.7° = \dfrac{DC}{3689} \Rightarrow$
$DC = 3689 \sin 63.7° \approx 3307$

If the ship follows the same course, the closest it will be to the lighthouse is about 3307 meter.

In exercises 89 and 90, be sure to carry all decimal places throughout the exercise.

89.

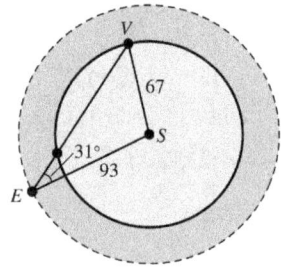

$h = 93 \sin 31° \approx 47.9$
Since $h < VS < ES$, there are two possible solutions.
$$\frac{\sin E}{VS} = \frac{\sin V}{ES} \Rightarrow \frac{\sin 31°}{67} = \frac{\sin V}{93} \Rightarrow$$
$$V_1 = \sin^{-1}\left(\frac{93\sin 31°}{67}\right) \approx 45.6° \text{ and}$$
$V_2 \approx 180° - 45.6352991° \approx 134.3647009°$
$S_1 = 180° - (31° + 45.6352991°)$
$\quad = 103.3647009°$
$S_2 = 180° - (31° + 134.3647009°)$
$\quad = 14.6352991°$
$$\frac{\sin S_1}{EV_1} = \frac{\sin E}{VS} \Rightarrow \frac{\sin 103.3647009°}{EV_1} = \frac{\sin 31°}{67} \Rightarrow$$
$$EV_1 = \frac{67\sin 103.3647009°}{\sin 31°} \approx 126.5645021$$
$$\frac{\sin S_2}{EV_2} = \frac{\sin E}{VS} \Rightarrow$$
$$\frac{\sin 14.6352991°°}{EV_2} = \frac{\sin 31°}{67} \Rightarrow$$
$$EV_2 = \frac{67\sin 14.6352991°}{\sin 31°} \approx 32.86861586$$
The distance between Earth and Venus is approximately 127 million miles or 33 million miles.

91.

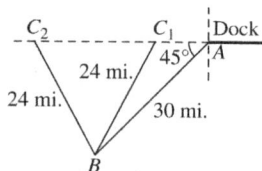

$h = 30 \sin 45° \approx 21.2$
Since $h < BC < AC$, there are two possible solutions.
$$\frac{\sin C_1}{AB} = \frac{\sin A}{BC_1} \Rightarrow \frac{\sin C_1}{30} = \frac{\sin 45°}{24} \Rightarrow$$
$$C_1 = \sin^{-1}\left(\frac{30\sin 45°}{24}\right) \approx 62.1°$$
$C_2 = 180° - C_1 \approx 117.9°$

$B_1 = 180° - (45° + 62.1°) = 72.9°$
$B_2 = 180° - (45° + 117.9°) = 17.1°$
$$\frac{AC_1}{\sin B_1} = \frac{BC_1}{\sin A} \Rightarrow$$
$$AC_1 = \frac{24\sin 72.9°}{\sin 45°} \approx 32.44 \text{ mi}$$
$$\frac{AC_2}{\sin B_2} = \frac{BC_2}{\sin A} \Rightarrow$$
$$AC_2 = \frac{24\sin 17.1°}{\sin 45°} \approx 9.98 \text{ mi}$$
The ship is either 9.98 or 32.44 miles from the dock.

93.

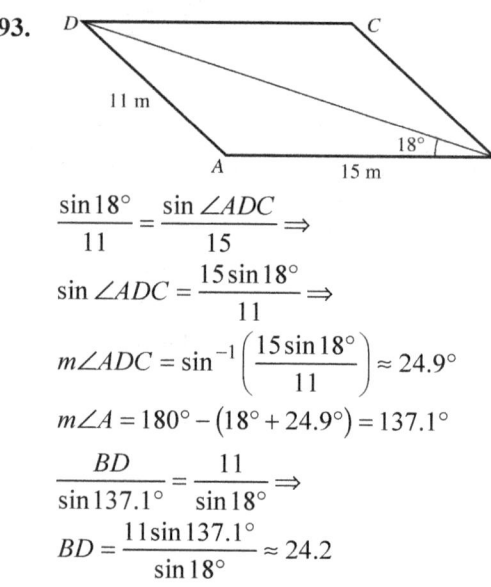

$$\frac{\sin 18°}{11} = \frac{\sin \angle ADC}{15} \Rightarrow$$
$$\sin \angle ADC = \frac{15\sin 18°}{11} \Rightarrow$$
$$m\angle ADC = \sin^{-1}\left(\frac{15\sin 18°}{11}\right) \approx 24.9°$$
$m\angle A = 180° - (18° + 24.9°) = 137.1°$
$$\frac{BD}{\sin 137.1°} = \frac{11}{\sin 18°} \Rightarrow$$
$$BD = \frac{11\sin 137.1°}{\sin 18°} \approx 24.2$$
The diagonal is about 24.2 m.

7.1 Beyond the Basics

95.

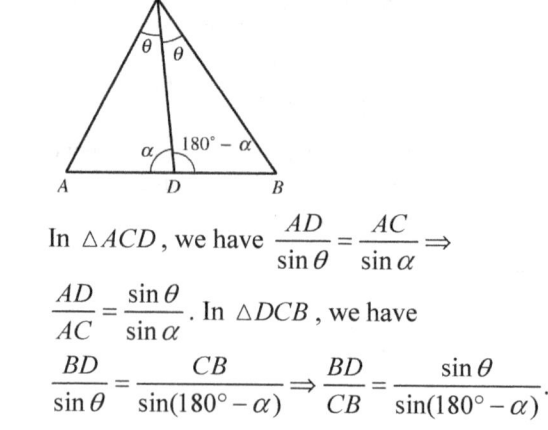

In $\triangle ACD$, we have $\dfrac{AD}{\sin \theta} = \dfrac{AC}{\sin \alpha} \Rightarrow$
$\dfrac{AD}{AC} = \dfrac{\sin \theta}{\sin \alpha}$. In $\triangle DCB$, we have
$\dfrac{BD}{\sin \theta} = \dfrac{CB}{\sin(180° - \alpha)} \Rightarrow \dfrac{BD}{CB} = \dfrac{\sin \theta}{\sin(180° - \alpha)}.$

(continued on next page)

(continued)

$\sin(180° - \alpha) = \sin \alpha$, so
$\dfrac{BD}{CB} = \dfrac{\sin \theta}{\sin(180° - \alpha)} = \dfrac{\sin \theta}{\sin \alpha} = \dfrac{AD}{AC} \Rightarrow$
$\dfrac{AD}{BC} = \dfrac{AC}{CB}$

97.

Given: $A = 45°, a = 20, c = p = 12t, b = 24$.
$h = b \sin A = 24 \sin 45° \approx 17$; $h < a < b$, so two triangles exist.
$\dfrac{\sin A}{a} = \dfrac{\sin B}{b} \Rightarrow \dfrac{\sin 45°}{20} = \dfrac{\sin B}{24} \Rightarrow$
$B = \sin^{-1}\left(\dfrac{24 \sin 45°}{20}\right) \Rightarrow B_1 \approx 58°, B_2 \approx 122°$
$P_1 = 180° - (45° + 58°) \approx 77°$
$\dfrac{a}{\sin A} = \dfrac{p_1}{\sin P_1} \Rightarrow \dfrac{20}{\sin 45°} = \dfrac{p_1}{\sin 77°} \Rightarrow$
$p_1 \approx 27.6 = 12t \Rightarrow t = 2.3$ hr $= 2$ hr 18 min
$P_2 = 180° - (45° + 122°) \approx 13°$
$\dfrac{a}{\sin A} = \dfrac{p_2}{\sin P_2} \Rightarrow \dfrac{20}{\sin 45°} = \dfrac{p_2}{\sin 13°} \Rightarrow$
$p_2 \approx 6.4 = 12t \Rightarrow t = 0.53$ hr ≈ 32 min. The two times are 1:00 P.M. + 2 hr 18 min = 3:18 P.M. and 1:00 P.M. + 32 min = 1:32 P.M.

99. Let $\dfrac{a}{\sin A} = \dfrac{b}{\sin B} = \dfrac{c}{\sin C} = k$, so
$a = k \sin A$, $B = k \sin B$, and $C = k \sin C$.
Then,
$\dfrac{b - c}{a} = \dfrac{k(\sin B - \sin C)}{k \sin A} = \dfrac{\sin B - \sin C}{\sin A}$.
Using the sum-to-product formula, this becomes
$\dfrac{2 \sin\left(\dfrac{B-C}{2}\right) \cos\left(\dfrac{B+C}{2}\right)}{\sin A}$
$\cdot \dfrac{(A + B + C)}{2} = 90° \Rightarrow$

$\dfrac{B+C}{2} = 90° - \dfrac{A}{2}$, so
$\cos\left(\dfrac{B+C}{2}\right) = \cos\left(90° - \dfrac{A}{2}\right) = \sin \dfrac{A}{2}$.
Substituting and using the double angle formula in the denominator, we have
$\dfrac{2 \sin\left(\dfrac{B-C}{2}\right) \cos\left(\dfrac{B+C}{2}\right)}{\sin A}$
$= \dfrac{2 \sin\left(\dfrac{B-C}{2}\right) \sin \dfrac{A}{2}}{2 \sin \dfrac{A}{2} \cos \dfrac{A}{2}} = \dfrac{\sin\left(\dfrac{B-C}{2}\right)}{\cos \dfrac{A}{2}}$.

101. Let $\dfrac{a}{\sin A} = \dfrac{b}{\sin B} = \dfrac{c}{\sin C} = k$, so $a = k \sin A$, $B = k \sin B$, and $C = k \sin C$.
Then $\dfrac{b - c}{b + c} = \dfrac{k(\sin B - \sin C)}{k(\sin B + \sin C)} = \dfrac{\sin B - \sin C}{\sin B + \sin C}$.
Using the sum-to-product formula, this becomes
$\dfrac{b-c}{b+c} = \dfrac{2 \sin\left(\dfrac{B-C}{2}\right) \cos\left(\dfrac{B+C}{2}\right)}{2 \sin\left(\dfrac{B+C}{2}\right) \cos\left(\dfrac{B-C}{2}\right)}$
$= \tan\left(\dfrac{B-C}{2}\right) \cot\left(\dfrac{B-C}{2}\right) = \dfrac{\tan\left(\dfrac{B-C}{2}\right)}{\tan\left(\dfrac{B+C}{2}\right)}$.

7.1 Critical Thinking/Discussion/Writing

103. An isosceles triangle with $m\angle A = m\angle B$ because $a = b \Rightarrow \dfrac{a}{\sin A} = \dfrac{b}{\sin B} \Rightarrow$
$a \sin B = b \sin A \Rightarrow a \sin A = b \sin B$.

104. An isosceles right triangle with right angle C gives
$m\angle A = m\angle B = 45° \Rightarrow$
$\sin A = \sin B = \cos A = \cos B \Rightarrow$
$\dfrac{a}{\sin A} = \dfrac{b}{\sin B} \Rightarrow a \sin B = b \sin A \Rightarrow$
$a \cos A = b \cos B$

Use the triangle to help solve exercises 105 and 106.

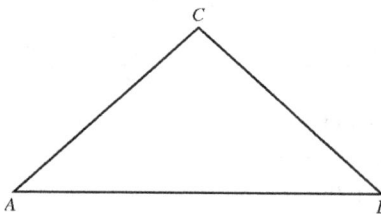

105. Given isosceles triangle ABC with $AB = BC$.
$$\frac{\sin A}{BC} = \frac{\sin B}{AC} \Rightarrow \frac{\sin A}{AC} = \frac{\sin B}{AC} \Rightarrow$$
$$\sin A = \sin B \Rightarrow A = B$$

106. Given triangle ABC with $m\angle A = m\angle B$.
$$\frac{AC}{\sin \angle A} = \frac{BC}{\sin \angle B} \Rightarrow \frac{AC}{\sin \angle A} = \frac{BC}{\sin \angle A} \Rightarrow$$
$$AC = BC$$

7.1 Getting Ready for the Next Section

107. $\cos 135° = -\dfrac{\sqrt{2}}{2}$

109. $\cos^{-1}\left(\dfrac{1}{2}\right) = 60°$

111. $\cos^{-1}\left(-\dfrac{1}{2}\right) = 120°$

113. $\sqrt{2^2 + 3^2 - 2(2)(3)\left(\dfrac{1}{6}\right)} = \sqrt{11}$

115. $\sqrt{3^2 + 1^2 - 2(3)(1)\cos 60°} = \sqrt{10 - 6\left(\dfrac{1}{2}\right)} = \sqrt{7}$

117. $\dfrac{2^2 + 3^2 - 4^2}{2(2)(3)} = -\dfrac{1}{4}$

119. $\cos^{-1}\left(\dfrac{1^2 + \left(\sqrt{3}\right)^2 - 1^2}{2(1)\left(\sqrt{3}\right)}\right) = \cos^{-1}\left(\dfrac{3}{2\sqrt{3}}\right)$
$$= \cos^{-1}\left(\dfrac{\sqrt{3}}{2}\right) = 30°$$

7.2 The Law of Cosines

7.2 Practice Problems

1. Given: $a = 6$, $b = 3\sqrt{3}$, $C = 30°$
$$c^2 = a^2 + b^2 - 2ab\cos C$$
$$c^2 = 6^2 + \left(3\sqrt{3}\right)^2 - 2(6)\left(3\sqrt{3}\right)\left(\dfrac{\sqrt{3}}{2}\right)$$
$$= 36 + 27 - 54 = 9$$
$$c = \sqrt{9} = 3$$

2. Given: $a = 6$, $b = 3$, $c = 4$
$$b^2 = a^2 + c^2 - 2ac\cos B \Rightarrow$$
$$\cos B = \dfrac{a^2 + c^2 - b^2}{2ac} = \dfrac{6^2 + 4^2 - 3^2}{2(6)(4)} \Rightarrow$$
$$B = \cos^{-1}\left(\dfrac{6^2 + 4^2 - 3^2}{2(6)(4)}\right) = \cos^{-1}\dfrac{43}{48} \approx 26.4°$$

3. Given: $c = 25$, $a = 15$, and $B = 60°$ – an SAS case.
$$b = \sqrt{15^2 + 25^2 - 2(15)(25)\cos 60°} \approx 21.8$$
$$\dfrac{\sin A}{a} = \dfrac{\sin B}{b} \Rightarrow \dfrac{\sin A}{15} = \dfrac{\sin 60°}{21.8}$$
$$A = \sin^{-1}\left(\dfrac{15\sin 60°}{21.8}\right) \approx 36.6°$$
$$C = 180° - (60° + 36.6°) \approx 83.4°$$

4.

The measure of $\angle FDB$ is $(90° - 75°) + 90° + 12° = 117°$. Then
$$d^2 = (1375t)^2 + \left[550\left(t + \dfrac{1}{3}\right)\right]^2$$
$$-2(1375t)\left[550\left(t + \dfrac{1}{3}\right)\right]\cos 117°$$

Substitute $t = 3$ and then solve for d:
$$d^2 = (1375 \cdot 3)^2 + \left[550\left(3 + \dfrac{1}{3}\right)\right]^2$$
$$-2(1375 \cdot 3)\left[550\left(3 + \dfrac{1}{3}\right)\right]\cos 117°$$
$$\approx 27,243,342.42$$
$$d \approx 5219.5 \text{ mi}$$

5. Given: $a = 4.5$, $b = 6.7$, and $c = 5.3$ – an SSS case
$5.3^2 = 4.5^2 + 6.7^2 - 2(4.5)(6.7)\cos C \Rightarrow$
$\cos C \approx 0.6144 \Rightarrow C \approx 52.1°$
$\dfrac{\sin 52.1°}{5.3} = \dfrac{\sin B}{6.7} \Rightarrow$
$B \approx \sin^{-1}\left(\dfrac{6.7\sin 52.1°}{5.3}\right) \approx B \approx 86.0°$
$A \approx 180° - (86.0° + 52.1°) \approx 41.9°$
Note that answers will vary slightly if the computations are done in a different order.

6. Given: $a = 2$ in., $b = 3$ in., and $c = 6$ in. – an SSS case
$6^2 = 2^2 + 3^2 - 2(2)(3)\cos A \Rightarrow \cos A \approx -1.9167$
Since $0 \leq \cos\theta \leq 1$, no triangle exists.

7. Given: $C = 35°$, $b = 15$ ft, $c = 12$ ft – an SSA case.
The missing side is a.
$c^2 = a^2 + b^2 - 2ab\cos C$
$12^2 = a^2 + 15^2 - 2a(15)\cos 35°$
$144 = a^2 + 225 - (30\cos 35°)a$
$0 = a^2 - (30\cos 35°)a + 81$
Use the quadratic formula to solve for a.
$a = \dfrac{-(-30\cos 35°) \pm \sqrt{(-30\cos 35°)^2 - 4(1)(81)}}{2(1)}$
$a = \dfrac{24.5746 \pm \sqrt{279.9091}}{2} \Rightarrow$
$a \approx 20.653$ or $a \approx 3.922$
Now use the Law of Cosines to find either of the two remaining angles for each possibility for a.
If $a \approx 20.653$, then
$a^2 = b^2 + c^2 - 2bc\cos A$
$\cos A = \dfrac{b^2 + c^2 - a^2}{2bc}$
$\cos A = \dfrac{15^2 + 12^2 - 20.653^2}{2(15)(12)}$
$A = \cos^{-1}\left(\dfrac{15^2 + 12^2 - 20.653^2}{2(15)(12)}\right) \approx 99.2°$
$B = 180° - (99.2° + 35°) \approx 45.8°$

If $a \approx 3.9$, then
$a^2 = b^2 + c^2 - 2bc\cos A$
$\cos A = \dfrac{b^2 + c^2 - a^2}{2bc}$
$\cos A = \dfrac{15^2 + 12^2 - 3.922^2}{2(15)(12)}$
$A = \cos^{-1}\left(\dfrac{15^2 + 12^2 - 3.922^2}{2(15)(12)}\right) \approx 10.8°$
$B = 180° - (10.8° + 35°) \approx 134.2°$
The solutions are
$\triangle A_1B_1C$: $A_1 \approx 99.2°$, $B_1 \approx 45.8°$, $a_1 \approx 20.7$ ft
$\triangle A_2B_2C$: $A_2 \approx 10.8°$, $B_2 \approx 134.2°$, $a_2 \approx 3.9$ ft
Note that answers will vary slightly depending on the number of decimal places carried through the computations and the order in which the sides and angles are found.

8. Given: $A = 65°$, $a = 16$ m, and $b = 30$ m – an SSA case.
$a^2 = b^2 + c^2 - 2bc\cos A$
$16^2 = 30^2 + c^2 - 2(30)c\cos 65°$
$256 = c^2 - 60\cos 65° c + 900$
$0 = c^2 - 60\cos 65° c + 644$
Use the quadratic formula to solve for c.
$c = \dfrac{60\cos 65° \pm \sqrt{(-60\cos 65°)^2 - 4(1)(644)}}{2(1)}$
$c = \dfrac{25.36 \pm \sqrt{-1933.0177}}{2(1)}$
The discriminant is negative, so there is no solution, and no such triangle exists.

9. Given: $C = 60°$, $c = 50$ ft, and $a = 30$ ft – an SSA case.
$c^2 = a^2 + b^2 - 2ab\cos C$
$50^2 = 30^2 + b^2 - 2(30)b\cos 60°$
$2500 = b^2 - 30b + 900$
$0 = b^2 - 30b - 1600$
Use the quadratic formula to solve for b.
$b = \dfrac{-(-30) \pm \sqrt{(-30)^2 - 4(1)(-1600)}}{2(1)}$
$b = \dfrac{30 \pm \sqrt{7300}}{2} \Rightarrow b \approx 57.72$ or $b \approx -27.72$
Reject the negative solution, so only one triangle is possible.

(*continued on next page*)

(*continued*)

Now find the remaining angles.
$$a^2 = b^2 + c^2 - 2bc\cos A$$
$$\cos A = \frac{b^2 + c^2 - a^2}{2bc}$$
$$\cos A = \frac{57.72^2 + 50^2 - 30^2}{2(57.72)(50)}$$
$$A = \cos^{-1}\left(\frac{57.72^2 + 50^2 - 30^2}{2(57.72)(50)}\right) \approx 31.3°$$
$$B = 180° - (31.3° + 60°) \approx 88.7°$$

7.2 Concepts and Vocabulary

1. One form of the Law of Cosines is $c^2 = a^2 + b^2 - 2ab\cos \underline{C}$.

3. Triangles with SAS given are solve by the Law of Cosines, as are triangles with <u>three</u> sides given.

5. False. Use the Law of Sines when two angles and a side are given.

7. True

Note that intermediate results are carried to one extra decimal place throughout the exercises. This leads to more accurate answers.

7.2 Building Skills

9. Given: $b = 4, c = 5, A = 60°$
$$a^2 = b^2 + c^2 - 2bc\cos A$$
$$a^2 = 4^2 + 5^2 - 2(4)(5)\cos 60° = 21$$
$$a = \sqrt{21} \approx 4.6$$

11. Given: $c = 10, a = 5\sqrt{2}, B = 45°$
$$b^2 = a^2 + c^2 - 2ac\cos B$$
$$b^2 = (5\sqrt{2})^2 + 10^2 - 2(5\sqrt{2})(10)\cos 45° = 50$$
$$b = \sqrt{50} = 5\sqrt{2} \approx 7.1$$

13. Given: $b = 3\sqrt{2}, a = 4, C = 45°$
$$c^2 = a^2 + b^2 - 2ab\cos C$$
$$c^2 = 4^2 + (3\sqrt{2})^2 - 2(4)(3\sqrt{2})\cos 45° = 10$$
$$c = \sqrt{10} \approx 3.2$$

15. Given: $a = 2, b = 3, c = 4$
$$a^2 = b^2 + c^2 - 2bc\cos A \Rightarrow$$
$$\cos A = \frac{b^2 + c^2 - a^2}{2bc} = \frac{3^2 + 4^2 - 2^2}{2(3)(4)} \Rightarrow$$
$$A = \cos^{-1}\left(\frac{3^2 + 4^2 - 2^2}{2(3)(4)}\right) = \cos^{-1}\left(\frac{21}{24}\right)$$
$$= \cos^{-1}\left(\frac{7}{8}\right) \approx 29.0°$$

17. $a = 3, b = 5, c = 10$
$$c^2 = a^2 + b^2 - 2ab\cos C \Rightarrow$$
$$\cos C = \frac{a^2 + b^2 - c^2}{2ab} = \frac{3^2 + 5^2 - 10^2}{2(3)(5)} \Rightarrow$$
$$C = \cos^{-1}\left(\frac{3^2 + 5^2 - 10^2}{2(3)(5)}\right) = \cos^{-1}\left(-\frac{66}{30}\right)$$

Because $\cos^{-1}\left(-\frac{66}{30}\right)$ is not defined, no triangle exists.

19. Given: $B = 106°, a = 14.6, c = 10.5$ – an SAS case.
$$b = \sqrt{10.5^2 + 14.6^2 - 2(10.5)(14.6)\cos 106°}$$
$$\approx 20.2$$
$$A = \sin^{-1}\left(\frac{14.6\sin 106°}{20.2}\right) \approx 44°$$
$$C = 180° - (106° + 44°) \approx 30°$$

21. Given: $a = 30, b = 18, c = 15$ – an SSS case.
$$30^2 = 18^2 + 15^2 - 2(18)(15)\cos A \Rightarrow$$
$$A \approx 130.5°$$
$$18^2 = 30^2 + 15^2 - 2(30)(15)\cos B \Rightarrow$$
$$B \approx 27.1°$$
$$C = 180° - (130.5° + 27.1°) \approx 22.4°$$

23. Given: $a = 15, b = 9, C = 120°$ – an SAS case.
$$c = \sqrt{15^2 + 9^2 - 2(15)(9)\cos 120°} = 21$$
$$\frac{\sin 120°}{21} = \frac{\sin A}{15} \Rightarrow A \approx 38.2°$$
$$B \approx 180° - (120° + 38.2°) \approx 21.8°$$

25. Given: $b = 10, c = 12, A = 62°$ – an SAS case.
$$a = \sqrt{10^2 + 12^2 - 2(10)(12)\cos 62°}$$
$$\approx 11.46 \approx 11.5$$
$$\frac{\sin 62°}{11.46} = \frac{\sin B}{10} \Rightarrow B \approx 50.4°$$
$$C \approx 180° - (62° + 50.2°) \approx 67.6°$$

27. Given: $c = 12, a = 15, b = 11$ – an SSS case
$15^2 = 11^2 + 12^2 - 2(11)(12)\cos A \Rightarrow A \approx 81.3°$
$\dfrac{\sin 81.3}{15} = \dfrac{\sin C}{12} \Rightarrow C \approx 52.3°$
$B \approx 180° - (81.3° + 52.3°) \approx 46.4°$

29. Given: $a = 9, b = 13, c = 18$ – an SSS case
$18^2 = 9^2 + 13^2 - 2(9)(13)\cos C \Rightarrow$
$C \approx 108.44° \approx 108.4°$
$\dfrac{\sin 108.44°}{18} = \dfrac{\sin B}{13} \Rightarrow B \approx 43.3°$
$A \approx 180° - (43.3° + 108.4°) \approx 28.3°$

31. Given: $a = 2.5, b = 3.7, c = 5.4$ – an SSS case
$5.4^2 = 2.5^2 + 3.7^2 - 2(2.5)(3.7)\cos C \Rightarrow$
$C \approx 119.89° \approx 119.9°$
$\dfrac{\sin 119.89°}{5.4} = \dfrac{\sin B}{3.7} \Rightarrow B \approx 36.4°$
$A \approx 180° - (36.4° + 119.9°) \approx 23.7°$

33. Given: $b = 3.2, c = 4.3, A = 97.7°$ – an SAS case
$a = \sqrt{3.2^2 + 4.3^2 - 2(3.2)(4.3)\cos 97.7°}$
$\approx 5.69 \approx 5.7$
$\dfrac{\sin 97.7°}{5.69} = \dfrac{\sin C}{4.3} \Rightarrow C \approx 48.5°$
$B \approx 180° - (97.7° + 48.5°) \approx 33.8°$

35. Given: $c = 4.9, a = 3.9, B = 68.3°$ – an SAS case
$b = \sqrt{3.9^2 + 4.9^2 - 2(3.9)(4.9)\cos 68.3°}$
$\approx 5.01 \approx 5.0$
$\dfrac{\sin 68.3°}{5.01} = \dfrac{\sin A}{3.9} \Rightarrow A \approx 46.3°$
$C \approx 180° - (68.3° + 46.3°) \approx 65.4°$

37. Given: $a = 2.3, b = 2.8, c = 3.7$ – an SSS case
$3.7^2 = 2.3^2 + 2.8^2 - 2(2.3)(2.8)\cos C \Rightarrow$
$C \approx 92.49° \approx 92.5°$
$\dfrac{\sin 92.49°}{3.7} = \dfrac{\sin B}{2.8} \Rightarrow B \approx 49.1°$
$A \approx 180° - (92.5° + 49.1°) \approx 38.4°$

For exercises 39–58, carry all decimal places through the calculations for greater accuracy.

39. Given: $A = 40°, a = 23, b = 20$
$a^2 = b^2 + c^2 - 2bc\cos A$
$23^2 = 20^2 + c^2 - 2(20)c\cos 40°$
$529 = c^2 - 2(20)c\cos 40° + 400$
$0 = c^2 - 40\cos 40° c - 129$

Use the quadratic formula to solve for c.
$c = \dfrac{40\cos 40° \pm \sqrt{(-40\cos 40°)^2 - 4(1)(-129)}}{2(1)}$
$c = \dfrac{30.64 \pm \sqrt{1454.9189}}{2} \Rightarrow$
$c \approx 34.4$ or $c \approx -3.8$
Disregard the negative solution. One triangle exists. Now find the remaining angles
$b^2 = a^2 + c^2 - 2ac\cos B$
$\cos B = \dfrac{a^2 + c^2 - b^2}{2ac}$
$B = \cos^{-1}\left(\dfrac{a^2 + c^2 - b^2}{2ac}\right)$
$= \cos^{-1}\left(\dfrac{23^2 + 34.4^2 - 20^2}{2(23)(34.4)}\right) \approx 34°$
$C = 180° - (40° + 34°) = 106°$

41. Given: $A = 30°, a = 25, b = 50$
$a^2 = b^2 + c^2 - 2bc\cos A$
$25^2 = 50^2 + c^2 - 2(50)c\cos 30°$
$625 = c^2 - 100c\cos 30° + 2500$
$0 = c^2 - 50\sqrt{3}c + 1875$
Use the quadratic formula to solve for c.
$c = \dfrac{50\sqrt{3} \pm \sqrt{(-50\sqrt{3})^2 - 4(1)(1875)}}{2(1)}$
$c = \dfrac{50\sqrt{3} \pm \sqrt{0}}{2} \Rightarrow c = 25\sqrt{3}$
One triangle exists. Now find the remaining angles.
$b^2 = a^2 + c^2 - 2ac\cos B$
$\cos B = \dfrac{a^2 + c^2 - b^2}{2ac}$
$B = \cos^{-1}\left(\dfrac{a^2 + c^2 - b^2}{2ac}\right)$
$= \cos^{-1}\left(\dfrac{25^2 + (25\sqrt{3})^2 - 50^2}{2(25)(25\sqrt{3})}\right) = 90°$
$C = 180° - (30° + 90°) = 60°$

43. Given: $A = 40°$, $a = 10$, and $b = 20$
$$a^2 = b^2 + c^2 - 2bc\cos A$$
$$10^2 = 20^2 + c^2 - 2(20)c\cos 40°$$
$$100 = c^2 - 40\cos 40°c + 400$$
$$0 = c^2 - 40\cos 40°c + 300$$
Use the quadratic formula to solve for c.
$$c = \frac{40\cos 40° \pm \sqrt{(-40\cos 40°)^2 - 4(1)(300)}}{2(1)}$$
$$c = \frac{30.64 \pm \sqrt{-261.08}}{2}$$
The discriminant is negative, so there is no solution, and no such triangle exists.

45. Given: $A = 95°$, $a = 18$, and $b = 21$
$$a^2 = b^2 + c^2 - 2bc\cos A$$
$$18^2 = 21^2 + c^2 - 2(21)c\cos 95°$$
$$324 = c^2 - 42\cos 95°c + 441$$
$$0 = c^2 - 42\cos 95°c + 117$$
Use the quadratic formula to solve for c.
$$c = \frac{42\cos 95° \pm \sqrt{(-42\cos 95°)^2 - 4(1)(117)}}{2(1)}$$
$$c = \frac{-3.66 \pm \sqrt{-454.60}}{2}$$
The discriminant is negative, so there is no solution, and no such triangle exists.

47. Given: $A = 100°$, $a = 40$, $b = 34$
$$a^2 = b^2 + c^2 - 2bc\cos A$$
$$40^2 = 34^2 + c^2 - 2(34)c\cos 100°$$
$$1600 = c^2 - 68\cos 100°c + 1156$$
$$0 = c^2 - 68\cos 100°c - 444$$
Use the quadratic formula to solve for c.
$$c = \frac{68\cos 100° \pm \sqrt{(-68\cos 100°)^2 - 4(1)(-444)}}{2(1)}$$
$$c = \frac{-11.81 \pm \sqrt{1915.43}}{2} \Rightarrow$$
$c \approx 16.0$ or $c \approx -27.8$
Disregard the negative result. Only one triangle is possible.
If $c \approx 16.0$, then
$$b^2 = a^2 + c^2 - 2ac\cos B \Rightarrow$$
$$\cos B = \frac{a^2 + c^2 - b^2}{2ac} = \frac{40^2 + 16.0^2 - 34^2}{2(40)(16.0)} \Rightarrow$$
$$B = \cos^{-1}\left(\frac{40^2 + 16.0^2 - 34^2}{2(40)(16.0)}\right) \approx 56.8°$$
$C = 180° - (100° + 56.8°) = 23.2°$

49. Given: $B = 50°$, $b = 22$, $c = 40$
$$b^2 = a^2 + c^2 - 2ac\cos B$$
$$22^2 = a^2 + 40^2 - 2(40)a\cos 50°$$
$$484 = a^2 - 80\cos 50°a + 1600$$
$$0 = a^2 - 80\cos 50°a + 1116$$
Use the quadratic formula to solve for a.
$$a = \frac{80\cos 50° \pm \sqrt{(-80\cos 50°)^2 - 4(1)(1116)}}{2(1)}$$
$$a = \frac{80\cos 50° \pm \sqrt{-1819.67}}{2}$$
The discriminant is negative, so no triangle exists.

51. Given: $B = 46°$, $b = 35$, $c = 40$
$$b^2 = a^2 + c^2 - 2ac\cos B$$
$$35^2 = a^2 + 40^2 - 2(40)a\cos 46°$$
$$1225 = a^2 - 80\cos 46°a + 1600$$
$$0 = a^2 - 80\cos 46°a + 375$$
Use the quadratic formula to solve for a.
$$a = \frac{80\cos 46° \pm \sqrt{(-80\cos 46°)^2 - 4(1)(375)}}{2(1)}$$
$$a = \frac{80\cos 46° \pm \sqrt{1588.32}}{2}$$
$a \approx 47.7$ or $a \approx 7.9$
If $a \approx 47.7$, then
$$a^2 = b^2 + c^2 - 2bc\cos A$$
$$\cos A = \frac{b^2 + c^2 - a^2}{2bc}$$
$$\cos A = \frac{35^2 + 40^2 - 47.7^2}{2(35)(40)}$$
$$A = \cos^{-1}\left(\frac{35^2 + 40^2 - 47.7^2}{2(35)(40)}\right) \approx 78.7°$$
$C = 180° - (78.7° + 46°) \approx 55.3°$
If $a \approx 7.9$, then
$$a^2 = b^2 + c^2 - 2bc\cos A$$
$$\cos A = \frac{b^2 + c^2 - a^2}{2bc}$$
$$\cos A = \frac{35^2 + 40^2 - 7.9^2}{2(35)(40)}$$
$$A = \cos^{-1}\left(\frac{35^2 + 40^2 - 47.7^2}{2(35)(40)}\right) \approx 9.4°$$
$C = 180° - (9.4° + 46°) \approx 124.6°$

(continued on next page)

(*continued*)

Note: If you carry all decimal places for a, then $A \approx 9.3°, C \approx 124.7°$.

The two solutions are: $C_1 \approx 55.3°$, $A_1 \approx 78.7°, a_1 \approx 47.7$ and $C_2 \approx 124.7°$, $A_2 \approx 9.4°, a_2 \approx 7.9$.

53. Given: $B = 97°, b = 27, c = 30$

$b^2 = a^2 + c^2 - 2ac\cos B$
$27^2 = a^2 + 30^2 - 2(30)a\cos 97°$
$729 = a^2 - 60\cos 97°\, a + 900$
$0 = a^2 - 60\cos 97°\, a + 171$

Use the quadratic formula to solve for a.

$a = \dfrac{60\cos 97° \pm \sqrt{(-60\cos 97°)^2 - 4(1)(171)}}{2(1)}$

$a = \dfrac{60\cos 97° \pm \sqrt{-630.53}}{2}$

The discriminant is negative, so no triangle exists.

55. Given: $A = 42°, a = 55, c = 62$

$a^2 = b^2 + c^2 - 2bc\cos A$
$55^2 = b^2 + 62^2 - 2(62)b\cos 42°$
$3025 = b^2 - 124\cos 42°\, b + 3844$
$0 = b^2 - 124\cos 42°\, b + 819$

Use the quadratic formula to solve for b.

$b = \dfrac{124\cos 42° \pm \sqrt{(-124\cos 42°)^2 - 4(1)(819)}}{2(1)}$

$b = \dfrac{124\cos 42° \pm \sqrt{5215.61}}{2} \Rightarrow$

$b \approx 82.2$ or $b \approx 10.0$

If $b \approx 82.2$, then

$b^2 = a^2 + c^2 - 2ac\cos B \Rightarrow$

$\cos B = \dfrac{a^2 + c^2 - b^2}{2ac} = \dfrac{55^2 + 62^2 - 82.2^2}{2(55)(62)} \Rightarrow$

$B = \cos^{-1}\left(\dfrac{55^2 + 62^2 - 82.2^2}{2(55)(62)}\right)$

$= \cos^{-1}\left(\dfrac{112.16}{6820}\right) \approx 89.1°$

$C = 180° - (42° + 89.1°) = 48.9°$

Note: If you carry all decimal places for b, then $B \approx 89.0°, C = 49.0°$.

If $b \approx 10.0$, then

$b^2 = a^2 + c^2 - 2ac\cos B \Rightarrow$

$\cos B = \dfrac{a^2 + c^2 - b^2}{2ac} = \dfrac{55^2 + 62^2 - 10.0^2}{2(55)(62)} \Rightarrow$

$B = \cos^{-1}\left(\dfrac{55^2 + 62^2 - 10.0^2}{2(55)(62)}\right)$

$= \cos^{-1}\left(\dfrac{6769}{6820}\right) \approx 7.0°$

$C = 180° - (42° + 7.0°) = 131.0°$

The two solutions are: $C_1 \approx 49.0°$, $B_1 \approx 89.0°, b_1 \approx 82.2$ and $C_2 \approx 131.0°$, $B_2 \approx 7.0°, b_2 \approx 10.0$.

57. Given: $C = 40°, a = 3.3, c = 2.1$

$c^2 = a^2 + b^2 - 2ab\cos C$
$2.1^2 = 3.3^2 + b^2 - 2(3.3)b\cos 40°$
$4.41 = b^2 - 6.6\cos 40°\, b + 10.89$
$0 = b^2 - 6.6\cos 40°\, b + 6.48$

Use the quadratic formula to solve for b.

$b = \dfrac{6.6\cos 40° \pm \sqrt{(-6.6\cos 40°)^2 - 4(1)(6.48)}}{2(1)}$

$b = \dfrac{6.6\cos 40° \pm \sqrt{-0.3579}}{2}$

The discriminant is negative, so no triangle exists.

7.2 Applying the Concepts

59. $AB = \sqrt{8^2 + 8^2 - 2(8)(8)\cos 42°} \approx 5.7$ ft

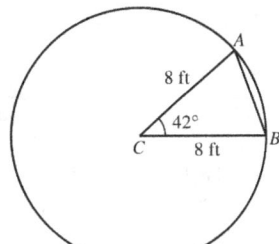

61.

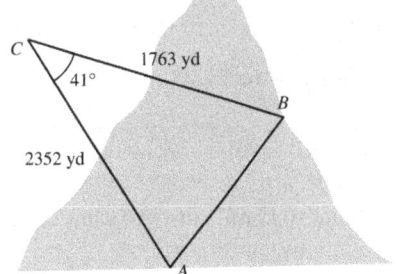

$AB = \sqrt{2352^2 + 1763^2 - 2(2352)(1763)\cos 41°}$
≈ 1543.1 yd

63. $BC = \sqrt{537^2 + 823^2 - 2(537)(823)\cos 130°}$
≈ 1238.5 yd

65.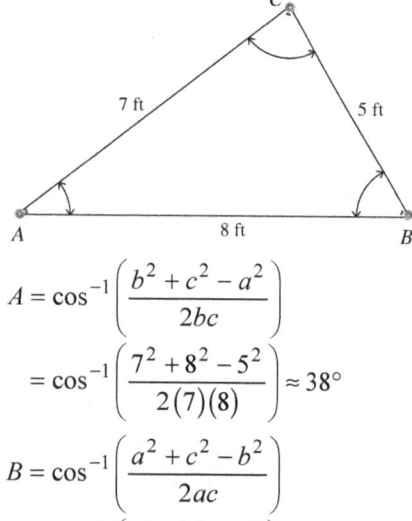

$A = \cos^{-1}\left(\dfrac{b^2 + c^2 - a^2}{2bc}\right)$

$= \cos^{-1}\left(\dfrac{7^2 + 8^2 - 5^2}{2(7)(8)}\right) \approx 38°$

$B = \cos^{-1}\left(\dfrac{a^2 + c^2 - b^2}{2ac}\right)$

$= \cos^{-1}\left(\dfrac{5^2 + 8^2 - 7^2}{2(5)(8)}\right) = \dfrac{1}{2} = 60°$

$C = 180° - (38° + 60°) = 82°$

67. After three hours, Sonia has walked $3(3) = 9$ miles and Tony has walked $3(4.3) = 12.9$ miles.

$d = \sqrt{12.9^2 + 9^2 - 2(12.9)(9)\cos 45°} \approx 9.1$ mi

69.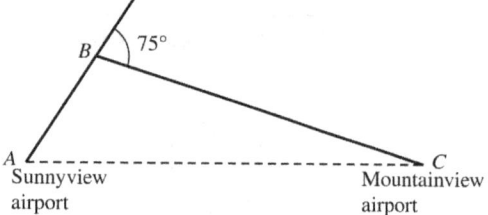

In the triangle, $B = 180° - 75° = 105°$. For the first part of the trip, the plane travels

$(420 \text{ mph})\left(\dfrac{1}{3}\text{ hr}\right) = 140$ mi $= AB$. For the

second part of the trip, the plane travels

$(420 \text{ mph})\left(\dfrac{7}{3}\text{ hr}\right) = 980$ mi $= BC$.

$AC = \sqrt{140^2 + 980^2 - 2(140)(980)\cos 105°}$
≈ 1025 mi

It is about 1025 miles from Sunnyview airport to Mountainview airport.

71. Ship A traveled 7 hours at 18 mph, 126 miles.
Ship B traveled 4.5 hours at 20 mph, 90 miles.

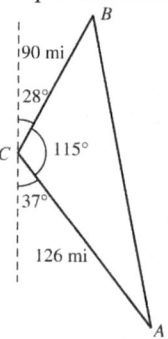

$m\angle ACB = 180° - (37° + 28°) = 115°$

$AB = \sqrt{90^2 + 126^2 - 2(90)(126)\cos 115°}$
≈ 183.2 mi

73. $AB = 1.2 + 2.2 = 3.4$ in.
$BC = 2.2 + 3.1 = 5.3$ in.
$AC = 1.2 + 3.1 = 4.3$ in.
$AB = 3.4 = \sqrt{5.3^2 + 4.3^2 - 2(5.3)(4.3)\cos C} \Rightarrow$
$C \approx 39.8°$
$BC = 5.3 = \sqrt{3.4^2 + 4.3^2 - 2(4.3)(3.4)\cos A} \Rightarrow$
$A \approx 86.2°$
$B \approx 180° - (86.2° + 39.8°) \approx 54.0°$

75.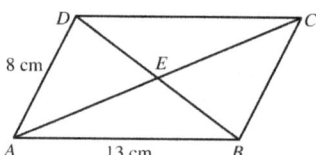

In $\triangle ABD$,
$8^2 = 13^2 + 11^2 - 2(13)(11)\cos \angle ABD \Rightarrow$
$m\angle ABD \approx 37.79°$.

In $\triangle BCD$,
$13^2 = 8^2 + 11^2 - 2(8)(11)\cos \angle CBD \Rightarrow$
$m\angle CBD \approx 84.78°$
$m\angle ABC = m\angle ABD + m\angle CBD \approx 122.57°$

In $\triangle ABC$,
$AC = \sqrt{13^2 + 8^2 - 2(13)(8)\cos \angle ABC}$
$= \sqrt{13^2 + 8^2 - 2(13)(8)\cos 122.57°} \approx 18.6$ cm

77.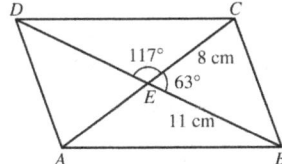

The diagonals of a parallelogram bisect each other, so $BE = ED = 11$ and $AD = EC = 8$. In $\triangle BEC$, $BC = \sqrt{11^2 + 8^2 - 2(11)(8)\cos 63°}$
≈ 10.3 cm

In $\triangle CED$, $CD = \sqrt{11^2 + 8^2 - 2(11)(8)\cos 117°}$
≈ 16.3 cm

The diagonals are 10.3 cm and 16.3 cm.

7.2 Beyond the Basics

79. Using the Law of Cosines and solving for $\cos A$, we have
$a^2 = b^2 + c^2 - 2bc \cos A \Rightarrow$
$\dfrac{a^2 - b^2 - c^2}{-2bc} = \cos A \Rightarrow$
$1 - \cos A = 1 - \left(-\dfrac{a^2 - b^2 - c^2}{2bc}\right)$
$= 1 + \left(\dfrac{a^2 - b^2 - c^2}{2bc}\right).$

Expanding the right side, we have
$\dfrac{(a-b+c)(a+b-c)}{2bc} = \dfrac{a^2 - b^2 - c^2 + 2bc}{2bc}$
$= 1 + \left(\dfrac{a^2 - b^2 - c^2}{2bc}\right).$

Thus, $1 - \cos A = \dfrac{(a-b+c)(a+b-c)}{2bc}$.

81. Using the Law of Cosines and solving for $\cos A$, we have $a^2 = b^2 + c^2 - 2bc \cos A \Rightarrow$
$\dfrac{a^2 - b^2 - c^2}{-2bc} = \cos A \Rightarrow$
$1 + \cos A = 1 - \dfrac{a^2 - b^2 - c^2}{2bc}.$

Expanding the right side of the identity gives
$\dfrac{(b+c+a)(b+c-a)}{2bc} = \dfrac{-a^2 + b^2 + c^2 + 2bc}{2bc}$
$= 1 - \left(\dfrac{a^2 - b^2 - c^2}{2bc}\right).$

Thus, $1 + \cos A = \dfrac{(b+c+a)(b+c-a)}{2bc}$.

83. $\sin\dfrac{A}{2} = \sqrt{\dfrac{1 - \cos A}{2}} = \sqrt{\dfrac{\dfrac{2(s-b)(s-c)}{bc}}{2}}$
$= \sqrt{\dfrac{(s-b)(s-c)}{bc}}$

85. $\sin A = \sin 2\left(\dfrac{A}{2}\right) = 2 \sin\dfrac{A}{2}\cos\dfrac{A}{2}$
$= 2\sqrt{\dfrac{(s-b)(s-c)}{bc}}\sqrt{\dfrac{s(s-a)}{bc}}$
$= \dfrac{2}{bc}\sqrt{s(s-a)(s-b)(s-c)}$

7.2 Critical Thinking/Discussion/Writing

87.

$a = BC = \sqrt{(3-5)^2 + (6-3)^2} = \sqrt{13}$
$b = AC = \sqrt{(3-(-2))^2 + (6-1)^2} = 5\sqrt{2}$
$c = AB = \sqrt{(5-(-2))^2 + (3-1)^2} = \sqrt{53}$
$13 = \left(5\sqrt{2}\right)^2 + \left(\sqrt{53}\right)^2 - 2\left(5\sqrt{2}\right)\left(\sqrt{53}\right)\cos A \Rightarrow$
$A \approx 29.05° \approx 29.1°$
$\dfrac{\sin 29.05°}{\sqrt{13}} = \dfrac{\sin B}{5\sqrt{2}} \Rightarrow B \approx 72.2°$
$C \approx 180° - (29.1° + 72.2°) \approx 78.7°$

88.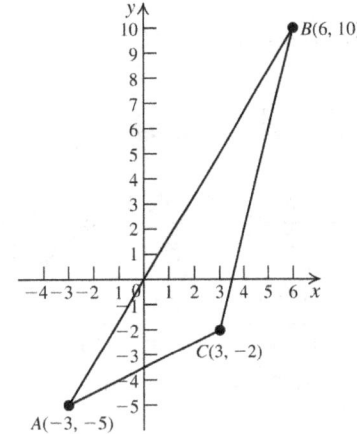

(continued on next page)

(continued)

$a = BC = \sqrt{(6-3)^2 + (10-(-2))^2} = 3\sqrt{17}$

$b = AC = \sqrt{(3-(-3))^2 + (-2-(-5))^2} = 3\sqrt{5}$

$c = AB = \sqrt{(6-(-3))^2 + (10-(-5))^2} = 3\sqrt{34}$

$153 = (3\sqrt{5})^2 + (3\sqrt{34})^2 - 2(3\sqrt{5})(3\sqrt{34})\cos A \Rightarrow$

$A \approx 32.47° \approx 32.5°$

$\dfrac{\sin 32.47°}{3\sqrt{17}} = \dfrac{\sin B}{3\sqrt{5}} \Rightarrow B \approx 16.9°$

$C \approx 180° - (32.5° + 16.9°) \approx 130.6°$

89. $a^2 < b^2 + c^2$ if angle A is an acute angle.

90. If A is an obtuse angle, then $\cos A < 0$.
$a^2 = b^2 + c^2 - 2bc\cos A \Rightarrow$
$a^2 = b^2 + c^2 + 2bc|\cos A| \Rightarrow a^2 > b^2 + c^2$

91. If the sides of a triangle ABC satisfy the condition $b^2 = c^2 + a^2$, then the triangle is a right triangle with $B = 90°$ and A and C are acute angles.

92. If the sides of a triangle ABC satisfy the condition $a^2 = b^2 + c^2 + bc$, then $\cos A = -\tfrac{1}{2}$, and $A = 120°$. B and C are acute angles.

7.2 Getting Ready for the Next Section

For exercises 93–98, $a = 6$, $b = 4$, $c = 3$, $\alpha = 30°$, $\beta = 45°$, and $\theta = 60°$.

93. $b\sin\alpha = 4\sin 30° = 4\left(\dfrac{1}{2}\right) = 2$

95. $\dfrac{1}{2}ab\sin\theta = \dfrac{1}{2}\cdot 6\cdot 4\sin 60° = 6\sqrt{3}$

97. $\dfrac{1}{2}ca\sin\beta = \dfrac{1}{2}\cdot 3\cdot 6\sin 45° = \dfrac{9\sqrt{2}}{2}$

99. a. $s = \dfrac{a+b+c}{2} = \dfrac{3+4+5}{2} = 6$

b. $K = \sqrt{s(s-a)(s-b)(s-c)}$
$= \sqrt{6(6-3)(6-4)(6-5)}$
$= \sqrt{6\cdot 3\cdot 2\cdot 1} = \sqrt{36} = 6$

101. a. $s = \dfrac{a+b+c}{2} = \dfrac{6+6+6}{2} = 9$

b. $K = \sqrt{s(s-a)(s-b)(s-c)}$
$= \sqrt{9(9-6)(9-6)(9-6)}$
$= \sqrt{9\cdot 3\cdot 3\cdot 3} = 9\sqrt{3}$

103. a. $s = \dfrac{a+b+c}{2} = \dfrac{18+10+14}{2} = 21$

b. $K = \sqrt{s(s-a)(s-b)(s-c)}$
$= \sqrt{21(21-18)(21-10)(21-14)}$
$= \sqrt{21\cdot 3\cdot 11\cdot 7} = 21\sqrt{11}$

105. a. $s = \dfrac{a+b+c}{2} = \dfrac{10+13+13}{2} = 18$

b. $K = \sqrt{s(s-a)(s-b)(s-c)}$
$= \sqrt{18(18-10)(18-13)(18-13)}$
$= \sqrt{18\cdot 8\cdot 5\cdot 5} = 60$

7.3 Areas of Polygons Using Trigonometry

7.3 Practice Problems

1.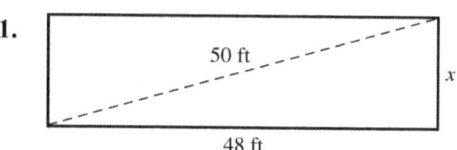

$48^2 + x^2 = 50^2 \Rightarrow x^2 = 50^2 - 48^2 \Rightarrow$
$x = \sqrt{50^2 - 48^2} = \sqrt{2500 - 2304} = \sqrt{196} = 14$
The width of the rectangle is 14 ft.

a. The perimeter is $2(48 + 14) = 2(62) = 124$ ft

b. The area is $(48)(14) = 672$ ft².

2. $K = \dfrac{1}{2}(AC)(BC)\sin\theta = \dfrac{1}{2}(27)(38)\sin 47°$
≈ 375.2 sq ft

3. $K = \dfrac{1}{2}ab\sin\theta$
$6 = \dfrac{1}{2}\cdot 4\cdot 3\sin\theta \Rightarrow 6 = 6\sin\theta \Rightarrow 1 = \sin\theta \Rightarrow$
$\theta = 90°$

4. First, find the third angle of the triangle.
$C = 180° - 63° - 74° = 43°$
We are given side C, so use the formula
$K = \dfrac{c^2 \sin A \sin B}{2\sin C}$ to find the area.

$K = \dfrac{18^2 \sin 63° \sin 74°}{2\sin 43°} \approx 203.4$ sq in.

5. Given: $a = 11$ m, $b = 17$ m, and $c = 20$ m
   Then $s = \dfrac{11+17+20}{2} = 24$.
   $K = \sqrt{24(24-11)(24-17)(24-20)}$
   ≈ 93.5 sq m

6. First find the surface area of the pool.
   $s = \dfrac{25+30+33}{2} = 44$
   $K = \sqrt{44(44-25)(44-30)(44-33)}$
   ≈ 358.8091 sq ft
   The volume of the pool is
   $358.8091 \cdot 5.5 \approx 1973.45$ cu ft.
   One cubic foot contains approximately 7.5 gallons of water, so $1973.45 \times 7.5 \approx 14{,}801$ gal of water will fill the pool.

7.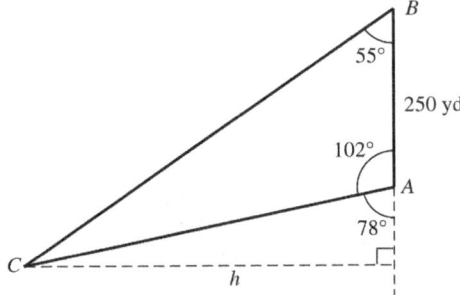

   In triangle ABC, $\angle BAC = 180° - 78° = 102°$ and $C = 180° - (102° + 55°) = 23°$.
   Use the area formula for an AAS triangle that contains side c.
   $K = \dfrac{c^2 \sin A \sin B}{2 \sin C} = \dfrac{250^2 \sin 102° \sin 55°}{2 \sin 23°}$
   ≈ 64082.72 ft^2
   $K = \dfrac{1}{2}ch = \dfrac{1}{2}(250)h = 125h$ ft$^2 \Rightarrow$
   $64082.72 = 125h \Rightarrow h \approx 513$ ft

7.3 Basic Concepts and Skills

1. The area K of a triangle with base b and height h is $K = \dfrac{1}{2}bh$.

3. The area of an SAS triangle ABC with sides a and c is $K = \dfrac{1}{2}ac \sin B$.

5. True

7. True

7.3 Building Skills

9. $K = \dfrac{1}{2}bc \sin A = \dfrac{1}{2} \cdot 6 \cdot 5 \sin 30°$
   $= \dfrac{1}{2} \cdot 6 \cdot 5 \cdot \dfrac{1}{2} = 7.5$

11. $K = \dfrac{1}{2}ab \sin C = \dfrac{1}{2} \cdot 8 \cdot 5 \sin 120°$
    $= \dfrac{1}{2} \cdot 8 \cdot 5 \cdot \dfrac{\sqrt{3}}{2} = 10\sqrt{3}$

13. $K = \dfrac{1}{2}ac \sin B = \dfrac{1}{2} \cdot 12 \cdot 9 \sin 150°$
    $= \dfrac{1}{2} \cdot 12 \cdot 9 \cdot \dfrac{1}{2} = 27$

For exercises 15–22, use one of the formulas
$K = \dfrac{1}{2}bc \sin \theta$, $K = \dfrac{1}{2}ac \sin \theta$, or $K = \dfrac{1}{2}ab \sin \theta$,
where θ is the measure of the angle included between the sides of lengths b and c.

15. $K = \dfrac{1}{2}(30)(52) \sin 57° \approx 654.2$ in.^{2}

17. $K = \dfrac{1}{2}(15)(22) \sin 46° \approx 118.7$ km^2

19. $A = \dfrac{1}{2}(12)(16.7) \sin 38.6° \approx 62.5$ mm^2

21. $A = \dfrac{1}{2}(271)(194.3) \sin 107.3° \approx 25{,}136.6$ ft^2

In exercises 23–30, first find the measure of the third angle, then use one of the formulas listed on page 723 in the text.

23. $A = 180° - 57° - 49° = 74°$
    We are given side a, so use the formula
    $K = \dfrac{a^2 \sin B \sin C}{2 \sin A}$ to find the area.
    $K = \dfrac{16^2 \sin 57° \sin 49°}{2 \sin 74°} \approx 84.3$ sq ft

25. $C = 180° - 64° - 38° = 78°$
    We are given side b, so use the formula
    $K = \dfrac{b^2 \sin C \sin A}{2 \sin B}$ to find the area.
    $K = \dfrac{15.3^2 \sin 78° \sin 64°}{2 \sin 38°} \approx 167.1$ sq yd

27. $A = 180° - 55° - 37.5° = 87.5°$
We are given side c, so use the formula
$K = \dfrac{c^2 \sin A \sin B}{2 \sin C}$ to find the area.
$K = \dfrac{16.3^2 \sin 87.5° \sin 55°}{2 \sin 37.5°} \approx 178.6$ sq cm

29. $C = 180° - 62°15' - 44°30' = 73°15'$
We are given side a, so use the formula
$K = \dfrac{a^2 \sin B \sin C}{2 \sin A}$ to find the area.
$K = \dfrac{65.4^2 \sin 44°30' \sin 73°15'}{2 \sin 62°15'} \approx 1621.9$ sq ft

In exercises 31–38, use Heron's formula.

31. $s = (2+3+4)/2 = 4.5$
$K = \sqrt{4.5(4.5-2)(4.5-3)(4.5-4)} = 2.9$

33. $s = (50+50+75)/2 = 87.5$
$K = \sqrt{87.5(87.5-50)(87.5-50)(87.5-75)}$
$= 1240.2$

35. $s = (7.5+4.5+6.0)/2 = 9$
$K = \sqrt{9(9-7.5)(9-4.5)(9-6)} = 13.5$

37. $s = (3.7+5.1+4.2)/2 = 6.5$
$K = \sqrt{6.5(6.5-3.7)(6.5-5.1)(6.5-4.2)} = 7.7$

For exercises 39–42, use the formula $K = \dfrac{1}{2} ab \sin \theta$, where θ is the angle included between the sides of lengths a and b.

39. $12 = \dfrac{1}{2} \cdot 6 \cdot 8 \sin \theta \Rightarrow 12 = 24 \sin \theta \Rightarrow$
$\sin \theta = \dfrac{1}{2} \Rightarrow \theta = \sin^{-1}\left(\dfrac{1}{2}\right) = 30°$ or since $\sin \theta = \sin(180° - \theta)$,
$\theta = 180° - 30° = 150°$.

41. $30 = \dfrac{1}{2} \cdot 12 \cdot 5 \sin \theta \Rightarrow 30 = 30 \sin \theta \Rightarrow$
$\sin \theta = 1 \Rightarrow \theta = 90°$

43.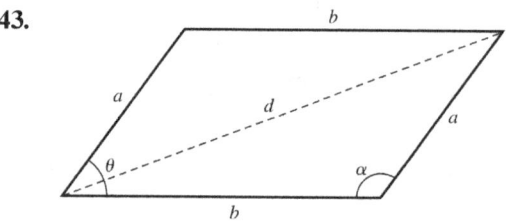

Recall that opposite sides of a parallelogram have equal length. So the area of each triangle formed by the diagonal shown is $\dfrac{1}{2} ab \sin \alpha$, and the area of the parallelogram is
$2 \cdot \dfrac{1}{2} ab \sin \alpha = ab \sin \alpha$.
From the diagram, we know that
$90° < \alpha < 180° \Rightarrow \alpha$ lies in quadrant II, so
$\sin \alpha = \sin(180° - \alpha) = \sin \theta$. Therefore, the area of the parallelogram can also be given by $ab \sin \theta$.

45. Since we are given the lengths of the sides and the diagonal, use Heron's formula to find the area of each of the triangles formed. The area of the parallelogram is twice the area of a triangle.
$s = \dfrac{12+16+22}{2} = 25$
$K_{parallelogram} = 2\sqrt{25(25-12)(25-16)(25-22)}$
$\approx 187.35 \text{ cm}^2$

47.

Because the diagonals of a parallelogram bisect each other, and the vertical angles at E are equal, $\triangle ABE \cong \triangle CDE$ and $\triangle ADE \cong \triangle CBE$ by SAS. The area of
$\triangle ABE = \dfrac{1}{2}\left(\dfrac{d_1}{2} \cdot \dfrac{d_2}{2}\right) \sin \varphi = $ area of $\triangle CDE$.

The area of $\triangle ADE = \dfrac{1}{2}\left(\dfrac{d_1}{2} \cdot \dfrac{d_2}{2}\right) \sin \beta = $ area of $\triangle CBE$. $\beta = 180° - \varphi$, so
$\sin \beta = \sin(180° - \varphi) = \sin \varphi$, and therefore,
$\triangle ADE = \dfrac{1}{2}\left(\dfrac{d_1}{2} \cdot \dfrac{d_2}{2}\right) \sin \varphi = $ area of $\triangle CBE$.

The area of the parallelogram is the sum of the areas of the four triangles, or
$4 \cdot \dfrac{1}{2}\left(\dfrac{d_1}{2} \cdot \dfrac{d_2}{2}\right) \sin \beta = \dfrac{1}{2} d_1 d_2 \sin \beta$
$= \dfrac{1}{2} d_1 d_2 \sin \varphi.$

49. First find β using the Law of Cosines.

$$\beta = \cos^{-1}\left(\frac{4^2 + 5^2 - 2^2}{2(4)(5)}\right) \approx 22.3°$$

$$K = \frac{1}{2}d_1 d_2 \sin \beta$$
$$= \frac{1}{2} \cdot 8 \cdot 10 \sin 22.3° \approx 15.2 \text{ cm}^2$$

7.3 Applying the Concepts

51. Area of the sector $= 64\pi \cdot \frac{105}{360} \approx 58.6 \text{ in.}^2$

$K_{\text{triangle}} = \frac{1}{2}ab\sin C = \frac{1}{2} \cdot 8 \cdot 8 \sin 105°$
$\approx 30.9 \text{ in.}^2$

Area of the segment $= 58.6 - 30.9 = 27.7 \text{ in.}^2$

53.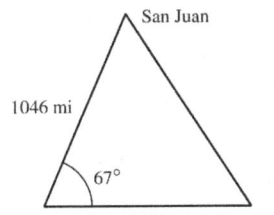

$A = \frac{1}{2}(1026)(1046)\sin 67° \approx 493,941 \text{ mi}^2$

55. $s = (400 + 250 + 274)/2 = 462$
$K = \sqrt{462(462 - 400)(462 - 250)(462 - 274)}$
$\approx 33,788.1 \text{ ft}^2 \approx 0.775668 \text{ acre}$

At $1 million per acre, the lot is worth $775,668.

57. $V = \text{height} \times \text{area of the base}$. Use Heron's formula to find the area of the base:
$s = (11 + 16 + 19)/2 = 23$
$K = \sqrt{23(23-11)(23-16)(23-19)} \approx 87.91 \text{ ft}^2$
$V = 87.91 \times 5 = 439.55 \text{ ft}^3$.
There are $439.55 \times 7.5 \approx 3297 \text{ gal}$.

59. Use Heron's formula.
$s = \frac{23 + 15 + 11}{2} = 24.5$
$K = \sqrt{24.5(24.5-23)(24.5-15)(24.5-11)}$
$\approx 69 \text{ sq ft}$

61. The area of the field is
$\frac{1}{2} \cdot 400 \cdot 275 \sin 51° \approx 42743 \text{ sq yd}$.

$\frac{42743 \text{ sq yd}}{100 \text{ sq yd}} = 427.43 \Rightarrow$ the farmer must pay for 428 applications.

It will cost $428 \cdot \$3.50 = \$1498$ to apply the pesticide.

63.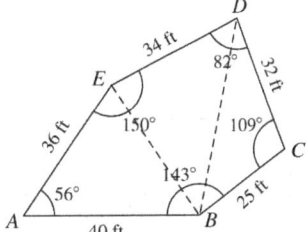

Draw diagonals to split the quadrilateral into 3 triangles.

$K_{\triangle ABE} = \frac{1}{2} \cdot 36 \cdot 40 \sin 56° \approx 596.9$

$K_{\triangle DBC} = \frac{1}{2} \cdot 25 \cdot 32 \sin 109° \approx 378.2$

For $\triangle DBE$, we must first find the lengths of sides DB and EB using the Law of Cosines, and then use Heron's formula.

$EB = \sqrt{40^2 + 36^2 - 2(40)(36)\cos 56°} \approx 35.9$

$DB = \sqrt{32^2 + 25^2 - 2(32)(25)\cos 109°}$
≈ 46.6

$s = \frac{35.9 + 46.6 + 34}{2} = 58.25$

$K_{\triangle DBE}$
$= \sqrt{58.25(58.25 - 35.9)(58.25 - 46.6)(58.25 - 34)}$
≈ 606.5

The area of the pond is about
$596.9 + 378.2 + 606.5 \approx 1581.6 \approx 1600 \text{ sq ft}$.

65.

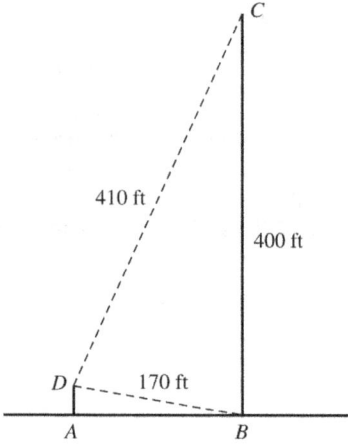

In triangle BCD, use the Law of Cosines to find the measure of angle CBD.

$\angle CBD = \cos^{-1}\left(\dfrac{170^2 + 400^2 - 410^2}{2(170)(400)}\right) \approx 81.2°$

Then, $\angle DBA = 90° - 81.2° = 8.8°$.

$\cos \angle DBA = \dfrac{AB}{170} \Rightarrow \cos 8.8° = \dfrac{AB}{170} \Rightarrow$
$AB = 170 \cos 8.8° \approx 168.0$

The distance between the buildings is about 168 feet.

67.

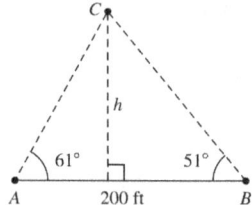

The third angle of the triangle is
$C = 180° - (61° + 51°) = 68°$.

$K = \dfrac{c^2 \sin A \sin B}{2 \sin C}$
$= \dfrac{200^2 \sin 61° \sin 51°}{2 \sin 68°} \approx 14661.76$

$K = \dfrac{1}{2}bh \Rightarrow 14661.76 = \dfrac{1}{2}(200)h \Rightarrow$
$h \approx 146.6$

The width of the river is about 146.6 feet.

7.3 Beyond the Basics

69.

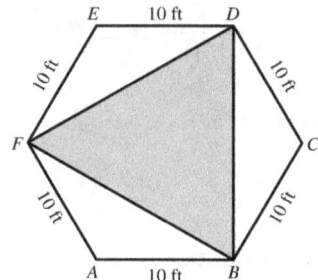

Recall that the measure of an interior angle (in degrees) in a regular n-gon is given by $\dfrac{180(n-2)}{n}$. Using the facts that the interior angles of a regular polygon are equal and the base angles of an isosceles triangle are equal, we can conclude that
$m\angle FBD = m\angle BDF = m\angle DFB$ and that triangle FBD is equilateral.

$\angle A = \dfrac{180(6-2)}{6} = 120$, so

$\angle ABF = \angle AFB = \dfrac{180° - 120°}{2} = 30°$

Find BF using the Law of Sines.

$\dfrac{BF}{\sin 120°} = \dfrac{10}{\sin 30°} \Rightarrow BF = \dfrac{10 \sin 120°}{\sin 30°} = 10\sqrt{3}$

The measure of each angle in triangle BDF is $60°$, so

$K = \dfrac{c^2 \sin A \sin B}{2 \sin C}$

$= \dfrac{\left(10\sqrt{3}\right)^2 \sin 60° \sin 60°}{2 \sin 60°}$

$= \dfrac{300\left(\dfrac{\sqrt{3}}{2}\right)^2}{2\left(\dfrac{\sqrt{3}}{2}\right)} = 75\sqrt{3} \approx 129.9$

The area of the shaded triangle BDF is about 130 sq ft.

Use this figure for exercises 71–74.

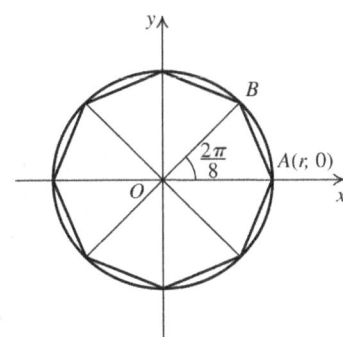

71. a. The coordinates of $A = (r, 0)$. Each central angle $= \dfrac{2\pi}{8} = \dfrac{\pi}{4}$. Thus, the coordinates of

$B = \left(r\cos\dfrac{\pi}{4}, r\sin\dfrac{\pi}{4}\right) = \left(\dfrac{r\sqrt{2}}{2}, \dfrac{r\sqrt{2}}{2}\right)$. Continuing in a counterclockwise direction, the coordinates of the

vertices are $(0, r), \left(-\dfrac{r\sqrt{2}}{2}, \dfrac{r\sqrt{2}}{2}\right), (-r, 0), \left(-\dfrac{r\sqrt{2}}{2}, -\dfrac{r\sqrt{2}}{2}\right), (0, -r), \left(\dfrac{r\sqrt{2}}{2}, -\dfrac{r\sqrt{2}}{2}\right)$.

b. Using the distance formula, we have

$AB = \sqrt{\left(\dfrac{r\sqrt{2}}{2} - r\right)^2 + \left(\dfrac{r\sqrt{2}}{2} - 0\right)^2} = \sqrt{\left(\dfrac{r^2}{2} - r^2\sqrt{2} + r^2\right) + \dfrac{r^2}{2}} = \sqrt{2r^2 - r\sqrt{2}} = r\sqrt{2 - \sqrt{2}}$

c. The perimeter is $8AB = 8r\sqrt{2-\sqrt{2}}$.

73. a. In a regular n-gon, the coordinates of B can be given as $\left(r\cos\dfrac{2\pi}{n}, r\sin\dfrac{2\pi}{n}\right)$. Then

$AB = \sqrt{\left(r\cos\dfrac{2\pi}{n} - r\right)^2 + \left(r\sin\dfrac{2\pi}{n} - 0\right)^2} = \sqrt{r^2\cos^2\dfrac{2\pi}{n} - 2r^2\cos\dfrac{2\pi}{n} + r^2 + r^2\sin^2\dfrac{2\pi}{n}}$

$= \sqrt{r^2\cos^2\dfrac{2\pi}{n} + r^2\sin^2\dfrac{2\pi}{n} - 2r^2\cos\dfrac{2\pi}{n} + r^2} = \sqrt{r^2\left(\cos^2\dfrac{2\pi}{n} + \sin^2\dfrac{2\pi}{n}\right) - 2r^2\cos\dfrac{2\pi}{n} + r^2}$

$= \sqrt{2r^2 - 2r^2\cos\dfrac{2\pi}{n}} = r\sqrt{2 - 2\cos\dfrac{2\pi}{n}}$

Thus, the perimeter $P = rn\sqrt{2 - 2\cos\dfrac{2\pi}{n}}$.

b. $C = 2\pi r = 10\pi \approx 31.4159$

c. $P = 4 \cdot 5\sqrt{2 - 2\cos\dfrac{2\pi}{4}} \approx 28.2843$

$P = 10 \cdot 5\sqrt{2 - 2\cos\dfrac{2\pi}{10}} \approx 30.9017$

$P = 50 \cdot 5\sqrt{2 - 2\cos\dfrac{2\pi}{50}} \approx 31.3953$

$P = 100 \cdot 5\sqrt{2 - 2\cos\dfrac{2\pi}{100}} \approx 31.4108$

The perimeters approach the circumference of the circle.

Use the figure to solve exercises 75–78.

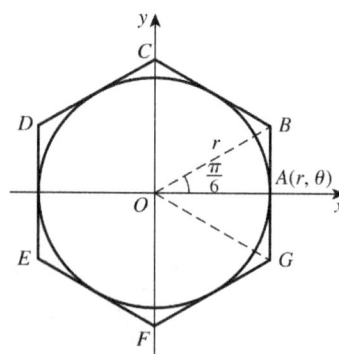

75. **a.** From the figure, it is clear that the x-coordinate of B is r. Find the y-coordinate of B as follows:

$\tan\dfrac{\pi}{6} = \dfrac{AB}{r} \Rightarrow AB = r\tan\dfrac{\pi}{6}$. From geometry, we know that A is the midpoint of the side. Therefore, the length of each side of the hexagon is $2r\tan\dfrac{\pi}{6}$.

b. $P = 6\left(2r\tan\dfrac{\pi}{6}\right) = 12r\tan\dfrac{\pi}{6}$

$= 12r\left(\dfrac{\sqrt{3}}{3}\right) = 4r\sqrt{3}$

77. **a.** Using the same reasoning as in exercise 75, we know that the length of each side of the circumscribed n-gon is $2r\tan\dfrac{\pi}{n}$. Thus, the perimeter of the n-gon is $P = 2nr\tan\dfrac{\pi}{n}$.

b. $C = 2\pi r = 20\pi \approx 62.8319$

c. $P = 2(4)(10)\tan\dfrac{\pi}{4} \approx 80.0$

$P = 2(10)(10)\tan\dfrac{\pi}{10} \approx 64.9839$

$P = 2(50)(10)\tan\dfrac{\pi}{50} \approx 62.9147$

$P = 2(100)(10)\tan\dfrac{\pi}{100} \approx 62.8525$

The perimeters approach the perimeter of the circle.

79. In triangle ABC,

$K = \dfrac{1}{2}ab\sin C$	*Formula for area of a triangle*
$4K^2 = a^2 b^2 (1 - \cos^2 C)$	*Multiply each side by 2 to obtain $2K = ab\sin C$. Square each side to obtain $(2K)^2 = (ab\sin C)^2 \Rightarrow 4K^2 = a^2 b^2 \sin^2 C$. Replace $\sin^2 C$ with its equivalent $1-\cos^2\theta$ to obtain $4K^2 = a^2 b^2(1 - \cos^2 C)$*
$4K^2 = a^2 b^2\left[1 - \left(\dfrac{a^2+b^2-c^2}{2ab}\right)^2\right]$	*Start with the Law of Cosines and solve for $\cos C$.* $c^2 = a^2 + b^2 - 2ab\cos C \Rightarrow$ $2ab\cos C = a^2 + b^2 - c^2 \Rightarrow$ $\cos C = \dfrac{a^2+b^2-c^2}{2ab} \Rightarrow$ $\cos^2\theta = \left(\dfrac{a^2+b^2-c^2}{2ab}\right)^2$

(continued on next page)

(continued)

$16K^2 = 4a^2b^2 - (a^2+b^2-c^2)^2$	On the right side, we have $$a^2b^2\left[1-\left(\frac{a^2+b^2-c^2}{2ab}\right)^2\right]$$ $$= a^2b^2\left[1-\frac{(a^2+b^2-c^2)^2}{4a^2b^2}\right]$$ $$= a^2b^2 - \frac{(a^2+b^2-c^2)^2}{4}$$ Now multiply both sides of the equation by 4. $$4K^2 = a^2b^2 - \frac{(a^2+b^2-c^2)^2}{4} \Rightarrow$$ $$16K^2 = 4a^2b^2 - (a^2+b^2-c^2)^2$$
$16K^2 = \left[2ab+(a^2+b^2-c^2)\right]\left[2ab-(a^2+b^2-c^2)\right]$	Note that the right side of the equation is the difference of squares. Factoring gives $$16K^2 = 4a^2b^2 - (a^2+b^2-c^2)^2$$ $$= \left[2ab+(a^2+b^2-c^2)\right] \cdot$$ $$\left[2ab-(a^2+b^2-c^2)\right]$$
$16K^2 = \left[(a+b)^2 - c^2\right]\left[c^2 - (a-b)^2\right]$	Rearrange terms inside the brackets on the right side. $$16K^2 = \left[a^2+2ab+b^2-c^2\right] \cdot$$ $$\left[-a^2+2ab-b^2+c^2\right]$$ $$= \left[(a+b)^2 - c^2\right]\left[-(a-b)^2 + c^2\right]$$ $$= \left[(a+b)^2 - c^2\right]\left[c^2 - (a-b)^2\right]$$
$K^2 = \left(\frac{a+b+c}{2}\right)\left(\frac{a+b-c}{2}\right)\left(\frac{c+a-b}{2}\right)\left(\frac{c+b-a}{2}\right)$	Again, we have the difference of squares on the right side. $$16K^2 = \left[(a+b)^2-c^2\right]\left[c^2-(a-b)^2\right]$$ $$= \left[(a+b)+c\right]\left[(a+b)-c\right] \cdot$$ $$\left[c+(a-b)\right]\left[c-(a-b)\right]$$ $$= (a+b+c)(a+b-c)(c+a-b)(c+b-a)$$ Now use the fact that $16 = 2^4$ to obtain $$K^2 = \frac{(a+b+c)(a+b-c)(c+a-b)(c+b-a)}{16}$$ $$= \left(\frac{a+b+c}{2}\right)\left(\frac{a+b-c}{2}\right)\left(\frac{c+a-b}{2}\right)\left(\frac{c+b-a}{2}\right)$$

(continued on next page)

(continued)

$$K = \sqrt{s(s-a)(s-b)(s-c)}$$

Note that
$$s = \frac{a+b+c}{2},$$
$$s-a = \frac{a+b+c}{2} - a = \frac{a+b+c-2a}{2} = \frac{b+c-a}{2},$$
$$s-b = \frac{a+b+c}{2} - b = \frac{a+b+c-2b}{2} = \frac{a+c-b}{2},$$
$$s-c = \frac{a+b+c}{2} - c = \frac{a+b+c-2c}{2} = \frac{a+b-c}{2}.$$

Thus, we have
$$K^2 = s\left(\frac{a+b-c}{2}\right)\left(\frac{c+a-b}{2}\right)\left(\frac{c+b-a}{2}\right)$$
$$= s(s-c)(s-b)(s-a) \Rightarrow$$
$$K = \sqrt{s(s-a)(s-b)(s-c)}$$

81. From geometry, we know that the measure of each angle in an equilateral triangle is 60°. Thus, the area of an equilateral triangle whose sides have length a is $K = \frac{1}{2}a^2 \sin 60° = \frac{1}{2}a^2 \cdot \frac{\sqrt{3}}{2} = \frac{\sqrt{3}}{4}a^2$.

83.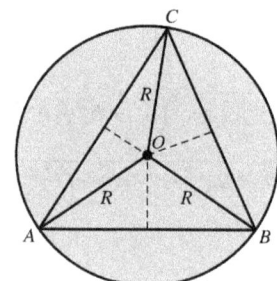

We have $2A = \angle BOC$, $2B = \angle AOC$, and $2C = \angle AOB$. We also have $a = BC$, $b = AC$, and $c = AB$. In $\triangle BOC$, we have

$$a^2 = R^2 + R^2 - 2R \cdot R \cos \angle BOC \Rightarrow \cos \angle BOC = \frac{a^2 - R^2 - R^2}{-2R \cdot R} = \frac{R^2 + R^2 - a^2}{2R \cdot R} = \frac{2R^2 - a^2}{2R^2} = 1 - \frac{a^2}{2R^2} \Rightarrow$$

$$\cos 2A = 1 - \frac{a^2}{2R^2} \Rightarrow 1 - 2\sin^2 A = 1 - \frac{a^2}{2R^2} \Rightarrow \sin^2 A = \frac{a^2}{4R^2} \Rightarrow \sin A = \frac{a}{2R} \Rightarrow R = \frac{a}{2\sin A}$$

Similarly, we can show that $R = \frac{b}{2\sin B}$ and $R = \frac{c}{2\sin C}$. Therefore, $R = \frac{a}{2\sin A} = \frac{b}{2\sin B} = \frac{c}{2\sin C}$.

85. From exercise 80, we have $\sin A = \frac{2}{bc}\sqrt{s(s-a)(s-b)(s-c)}$.

Using exercise 83, we have $R = \frac{a}{2\sin A} = \frac{a}{2\left(\frac{2}{bc}\sqrt{s(s-a)(s-b)(s-c)}\right)} = \frac{abc}{4K}$.

87.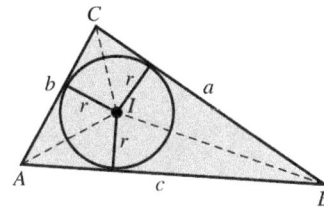

The area of $\triangle AIB$ is $\frac{1}{2}rc$. The area of $\triangle BIC$ is $\frac{1}{2}ra$.

The area of $\triangle AIC$ is $\frac{1}{2}rb$. Thus,

$$K_{\triangle ABC} = \frac{1}{2}ra + \frac{1}{2}rb + \frac{1}{2}rc = \frac{1}{2}r(a+b+c) = rs \Rightarrow$$

$$\sqrt{s(s-a)(s-b)(s-c)} = rs \Rightarrow r = \frac{\sqrt{s(s-a)(s-b)(s-c)}}{s}$$

89.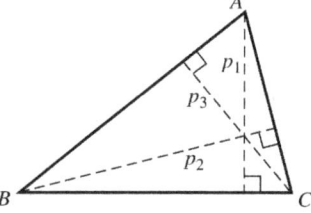

$\frac{1}{2}ap_1 = \frac{1}{2}bp_2 = \frac{1}{2}cp_3 = K \Rightarrow \frac{1}{p_1} = \frac{a}{2K}$, $\frac{1}{p_2} = \frac{b}{2K}$, and $\frac{1}{p_3} = \frac{c}{2K}$.

Adding the three expressions, we have

$$\frac{1}{p_1} + \frac{1}{p_2} + \frac{1}{p_3} = \frac{a}{2K} + \frac{b}{2K} + \frac{c}{2K} = \frac{a+b+c}{2\sqrt{s(s-a)(s-b)(s-c)}} = \frac{2s}{2\sqrt{s(s-a)(s-b)(s-c)}}$$

$$= \frac{s}{\sqrt{s(s-a)(s-b)(s-c)}} \cdot \frac{\sqrt{s}}{\sqrt{s}} = \frac{s\sqrt{s}}{s\sqrt{(s-a)(s-b)(s-c)}} = \sqrt{\frac{s}{(s-a)(s-b)(s-c)}}$$

7.3 Critical Thinking/Discussion/Writing

91.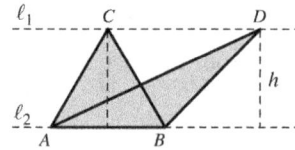

Because the lines are parallel, the height of each triangle = h. The base of triangle ABC is AB, and the base of triangle ABD is also AB. So, the area of triangle ABD = the area of triangle ABC.

92.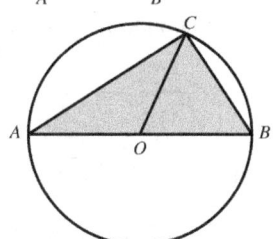

$OA = OB = OC$

For triangle COB, $K = \frac{1}{2}(OC)(OB)\sin \angle BOC$.

For triangle COA, $K = \frac{1}{2}(OC)(OA)\sin \angle COA$.

$\angle COA = 180° - \angle BOC$, so $\sin \angle BOC = \sin \angle COA$. Therefore, the area of triangle COB = the area of triangle COA.

93. A 90° angle has the greatest sine value, so $K = \frac{1}{2}(5)(3)\sin C$ will be the maximum when $C = 90°$.

94. From exercise 47, the area of a parallelogram with diagonals d_1 and d_2 is given by $\frac{1}{2}d_1d_2 \sin \beta$, where β is one of the angles between the diagonals. The maximum sine value occurs when the angle = 90°.

95. Begin with $K = \dfrac{1}{2}ab\sin C$.

Since $C = 180° - (A + B)$, replace C with its equivalent.

$$K = \dfrac{1}{2}ab\sin[180 - (A+B)] = \dfrac{1}{2}ab\sin(A+B) = \dfrac{ab\sin A\cos B + ab\sin B\cos A}{2}$$

$$= \dfrac{(b\sin A)a\cos B + b(a\sin B)\cos A}{2}$$

$$= \dfrac{2(a\sin B)a\cos B + 2b(b\sin A)\cos A}{4} \quad \text{(from the Law of Sines)}$$

$$= \dfrac{a^2\sin 2B + b^2\sin 2A}{4}$$

Similarly, we can prove $K = \dfrac{b^2\sin 2C + c^2\sin 2B}{4}$ and $K = \dfrac{c^2\sin 2A + b^2\sin 2C}{4}$.

96. From exercise 83 in section 7.2, we have

$$\sin\dfrac{A}{2} = \sqrt{\dfrac{(s-b)(s-c)}{bc}},\ \sin\dfrac{B}{2} = \sqrt{\dfrac{(s-a)(s-c)}{ac}},\ \text{and } \sin\dfrac{C}{2} = \sqrt{\dfrac{(s-a)(s-b)}{ab}}.$$

Multiply these three equations to obtain

$$4R\sin\dfrac{A}{2}\sin\dfrac{B}{2}\sin\dfrac{C}{2} = 4R\sqrt{\dfrac{(s-b)(s-c)}{bc}} \cdot \sqrt{\dfrac{(s-a)(s-c)}{ac}} \cdot \sqrt{\dfrac{(s-a)(s-b)}{ab}} = 4R\dfrac{(s-a)(s-b)(s-c)}{abc}$$

$$= 4 \cdot \dfrac{abc}{4K} \cdot \dfrac{(s-a)(s-b)(s-c)}{abc} \quad \text{(from exercise 61)}$$

$$= \dfrac{(s-a)(s-b)(s-c)}{K} = \dfrac{(s-a)(s-b)(s-c)}{\sqrt{s(s-a)(s-b)(s-c)}}$$

$$= \dfrac{\left(\sqrt{(s-a)(s-b)(s-c)}\right)^2}{\sqrt{s(s-a)(s-b)(s-c)}} = \dfrac{\sqrt{(s-a)(s-b)(s-c)}}{\sqrt{s}} = \dfrac{\sqrt{s(s-a)(s-b)(s-c)}}{s} = r$$

(from exercise 87.)

7.3 Getting Ready for the Next Section

97. $2(-x+3y) + 3(x-y) = -2x+6y+3x-3y$
$= x + 3y \Rightarrow a = 1,\ b = 3$

$\sqrt{a^2 + b^2} = \sqrt{1^2 + 3^2} = \sqrt{10}$

99. $\theta' = \tan^{-1}(\sqrt{3}) \Rightarrow \theta' = \dfrac{\pi}{3}$

In quadrant III, $\theta = \theta' + \pi = \dfrac{\pi}{3} + \pi = \dfrac{4\pi}{3}$

101. a. $d(S, T) = \sqrt{(6-1)^2 + (15-3)^2} = \sqrt{25 + 144}$
$= 13$

$d(O, P) = \sqrt{(5-0)^2 + (12-0)^2} = \sqrt{25 + 144}$
$= 13$

b. $m_1 = \dfrac{15-3}{6-1} = \dfrac{12}{5}$

$m_2 = \dfrac{12-0}{5-0} = \dfrac{12}{5}$

c. Since the slopes are equal, the lines are parallel.

103. a. $d(S, T) = \sqrt{(-2-(-1))^2 + (1-2)^2}$
$= \sqrt{(-1)^2 + (-1)^2} = \sqrt{2}$

$d(O, P) = \sqrt{(1-0)^2 + (-1-0)^2} = \sqrt{2}$

b. $m_1 = \dfrac{1-2}{-2-(-1)} = \dfrac{-1}{-1} = 1$

$m_2 = \dfrac{-1-0}{-0} = -1$

c. The product of the slopes is -1, so the lines are perpendicular.

7.4 Vectors

7.4 Practice Problems

1. a.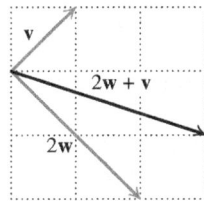

b.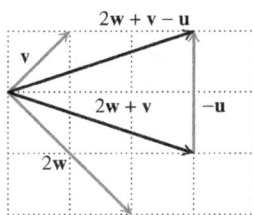

2. $\mathbf{w} = \langle 1-(-2), -3-7 \rangle = \langle 3, -10 \rangle$

3. $\mathbf{v} = \langle -1, 2 \rangle, \mathbf{w} = \langle 2, -3 \rangle$

a. $\mathbf{v} + \mathbf{w} = \langle -1+2, 2+(-3) \rangle = \langle 1, -1 \rangle$

b. $3\mathbf{w} = 3\langle 2, -3 \rangle = \langle 3(2), 3(-3) \rangle = \langle 6, -9 \rangle$

c. $3\mathbf{w} - 2\mathbf{v} = 3\langle 2, -3 \rangle - 2\langle -1, 2 \rangle$
$= \langle 6, -9 \rangle - \langle -2, 4 \rangle$
$= \langle 6-(-2), -9-4 \rangle = \langle 8, -13 \rangle$

d. $\|3\mathbf{w} - 2\mathbf{v}\| = \|\langle 8, -13 \rangle\| = \sqrt{8^2 + (-13)^2}$
$= \sqrt{233}$

4. $\|\mathbf{v}\| = \|\langle -12, 5 \rangle\| = \sqrt{(-12)^2 + 5^2} = \sqrt{169} = 13$

$\mathbf{u} = \dfrac{1}{\|\mathbf{v}\|}\mathbf{v} = \dfrac{1}{13}\langle -12, 5 \rangle = \left\langle -\dfrac{12}{13}, \dfrac{5}{13} \right\rangle$

5. $\mathbf{u} = -3\mathbf{i} + 2\mathbf{j}, \mathbf{v} = \mathbf{i} + 4\mathbf{j}$

a. $3\mathbf{u} + 2\mathbf{v} = 3(-3\mathbf{i} + 2\mathbf{j}) + 2(\mathbf{i} + 4\mathbf{j})$
$= -9\mathbf{i} + 6\mathbf{j} + 2\mathbf{i} + 8\mathbf{j} = -7\mathbf{i} + 14\mathbf{j}$

b. $\|3\mathbf{u} + 2\mathbf{v}\| = \sqrt{(-7)^2 + 14^2} = \sqrt{245} = 7\sqrt{5}$

6. $\mathbf{v} = 2\mathbf{i} - 3\mathbf{j} = \langle 2, -3 \rangle$

$\|\mathbf{v}\| = \sqrt{2^2 + (-3)^2} = \sqrt{13}$

$\tan \theta = \dfrac{-3}{2} = -\dfrac{3}{2}$

The reference angle θ' is given by

$\theta' = \left| \tan^{-1}\left(-\dfrac{3}{2}\right) \right| \approx 56.31°$. Since the point $(2, -3)$ lies in Quadrant IV,
$\theta \approx 360° - 56.31° = 303.69°$.

7. $\mathbf{v} = 2\left(\cos\dfrac{11\pi}{6}\mathbf{i} + \sin\dfrac{11\pi}{6}\mathbf{j}\right) = 2\left(\dfrac{\sqrt{3}}{2}\mathbf{i} - \dfrac{1}{2}\mathbf{j}\right)$
$= \sqrt{3}\mathbf{i} - \mathbf{j}$

8.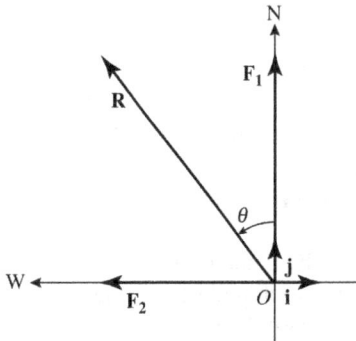

$\mathbf{F}_1 = 40\mathbf{j}$ and $\mathbf{F}_2 = -30\mathbf{i}$. Then
$\mathbf{R} = \mathbf{F}_1 + \mathbf{F}_2 = -30\mathbf{i} + 40\mathbf{j}$.

$\|\mathbf{R}\| = \sqrt{(-30)^2 + 40^2} = 50$

$\tan \theta' = -\dfrac{40}{30} \Rightarrow \theta' = \left| \tan^{-1}\left(-\dfrac{40}{30}\right) \right| \approx 53.13°$

The angle between $\mathbf{R}$ and the y-axis is
$\theta \approx 90° - 53.1° = 36.9°$. Therefore, $\mathbf{R}$ is a force of 50 pounds in the direction N 36.9° W.

9. Let $\mathbf{v}$ be the velocity of the plane in still air. Let $\mathbf{w}$ be the wind speed, and let $\mathbf{r}$ be the resultant ground velocity of the plane. Since the bearing of the plane is N 60° W, the direction angle is 150°, and since the wind direction is S 30° W, its direction angle is 240°.

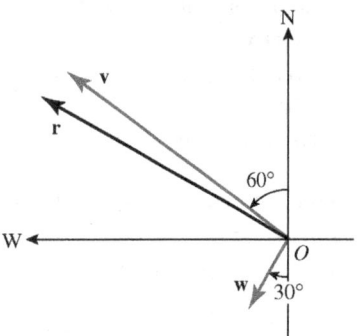

Not drawn to scale

$\mathbf{v} = 800(\cos 150°\mathbf{i} + \sin 150°\mathbf{j})$

$\mathbf{w} = 40(\cos 240°\mathbf{i} + \sin 240°\mathbf{j})$

$\mathbf{r} = \mathbf{v} + \mathbf{w}$
$= 800(\cos 150°\mathbf{i} + \sin 150°\mathbf{j})$
$\quad + 40(\cos 240°\mathbf{i} + \sin 240°\mathbf{j})$
$= (800\cos 150° + 40\cos 240°)\mathbf{i}$
$\quad + (800\sin 150° + 40\sin 240°)\mathbf{j}$

(continued on next page)

(*continued*)

$$\|\mathbf{r}\| = \sqrt{\begin{array}{c}(800\cos 150° + 40\cos 240°)^2 \\ + (800\sin 150° + 40\sin 240°)^2\end{array}}$$
$$= \sqrt{641{,}600} \approx 801.0$$

Now find the direction angle of **r**.

$$\theta' = \left|\tan^{-1}\left(\frac{800\sin 150° + 40\sin 240°}{800\cos 150° + 40\cos 240°}\right)\right|$$
$$\approx 27.1°$$

The angle between **r** and the *y*-axis is $\theta \approx 90° - 27.1° = 62.9°$. Therefore, the actual speed of the plane is approximately 801.0 miles per hour, and its bearing is approximately N 62.9° W.

10. **u** = 200 lb is the direction N 40° E, **v** = 300 lb in the direction N 70° W, and **w** = 400 lb in the direction S 20° E.

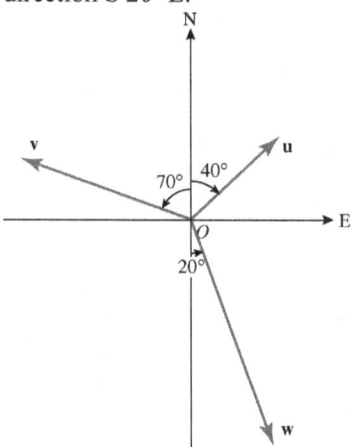

The horizontal component R_1 of **R** is given by
$R_1 = u_1 + v_1 + w_1$
$= 200\cos 50° + 300\cos 160° + 400\cos(-70°)$
≈ -16.5422

The vertical component R_1 of **R** is given by
$R_1 = u_1 + v_1 + w_1$
$= 200\sin 50° + 300\sin 160° + 400\sin(-70°)$
≈ -120.06212

$\mathbf{R} = R_1\mathbf{i} + R_2\mathbf{j} = -16.54\mathbf{i} - 120.06\mathbf{j}$

$\|\mathbf{R}\| = \sqrt{(-16.54)^2 + (-120.06)^2} \approx 121.2$

$\theta = \tan^{-1}\dfrac{-120.06}{-16.54} \approx 82.2°$

From the diagram, we see that the resultant will lie in quadrant III. Thus, the magnitude of the resultant is about 121.2 lb at a bearing of about S 7.8° W.

7.4 Exercises
Concepts and Vocabulary

1. A vector is a quantity that is characterized by a magnitude and a <u>direction</u>.

3. If $\mathbf{v} = \langle a, b \rangle$, then $\|\mathbf{v}\| = \underline{\sqrt{a^2 + b^2}}$, and its direction angle $\theta = \tan^{-1}\underline{\left(\dfrac{b}{a}\right)}$.

5. True

7. False. The position vector equivalent to **v** is $\langle x_2 - x_1, y_2 - y_1 \rangle$.

7.4 Building Skills

9.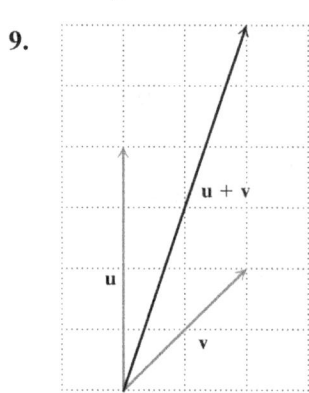

11.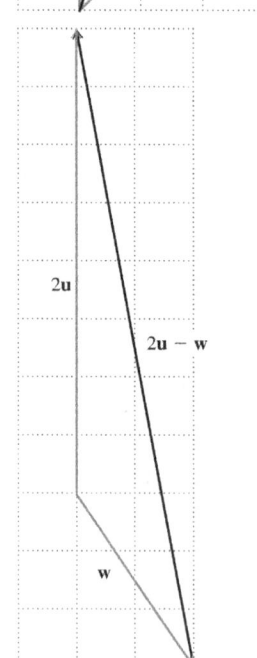

13.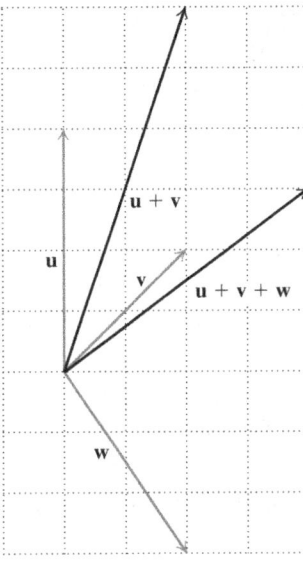

15.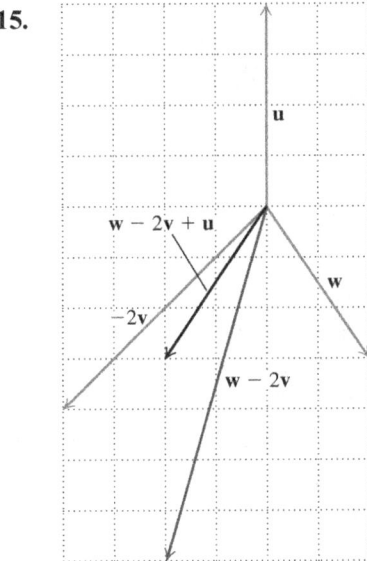

17. $v = \langle 2-3, 9-6 \rangle = \langle -1, 3 \rangle$

19. $v = \langle -3-(-5), -4-(-2) \rangle = \langle 2, -2 \rangle$

21. $v = \langle 2-(-1), -3-4 \rangle = \langle 3, -7 \rangle$

23. $v = \left\langle -\dfrac{1}{2} - \dfrac{1}{2}, -\dfrac{7}{4} - \dfrac{3}{4} \right\rangle = \left\langle -1, -\dfrac{5}{2} \right\rangle$

25. $\overrightarrow{AB} = \langle 3-1, 4-0 \rangle = \langle 2, 4 \rangle$
    $\overrightarrow{CD} = \langle 1-(-1), 6-2 \rangle = \langle 2, 4 \rangle$
    The vectors are equivalent.

27. $\overrightarrow{AB} = \langle 3-2, 5-(-1) \rangle = \langle 1, 6 \rangle$
    $\overrightarrow{CD} = \langle -2-(-1), -3-3 \rangle = \langle -1, -6 \rangle$
    The vectors are not equivalent.

In exercises 29–36, $v = \langle -1, 2 \rangle$ and $w = \langle 3, -2 \rangle$.

29. $\|v\| = \sqrt{(-1)^2 + 2^2} = \sqrt{5}$

31. $v - w = \langle -1-3, 2-(-2) \rangle = \langle -4, 4 \rangle$

33. $2v - 3w = 2\langle -1, 2 \rangle - 3\langle 3, -2 \rangle$
    $= \langle -2, 4 \rangle - \langle 9, -6 \rangle = \langle -11, 10 \rangle$

35. $2v - 3w = 2\langle -1, 2 \rangle - 3\langle 3, -2 \rangle$
    $= \langle -2, 4 \rangle - \langle 9, -6 \rangle = \langle -11, 10 \rangle$
    $\|2v - 3w\| = \sqrt{(-11)^2 + 10^2} = \sqrt{221}$

37. $\|v\| = \sqrt{1^2 + (-1)^2} = \sqrt{2}$
    $u = \dfrac{1}{\sqrt{2}} \langle 1, -1 \rangle = \left\langle \dfrac{\sqrt{2}}{2}, -\dfrac{\sqrt{2}}{2} \right\rangle$

39. $\|v\| = \sqrt{(-4)^2 + 3^2} = 5$
    $u = \dfrac{1}{5} \langle -4, 3 \rangle = \left\langle -\dfrac{4}{5}, \dfrac{3}{5} \right\rangle$

41. $\|v\| = \sqrt{\left(\sqrt{2}\right)^2 + \left(\sqrt{2}\right)^2} = 2$
    $u = \dfrac{1}{2} \langle \sqrt{2}, \sqrt{2} \rangle = \left\langle \dfrac{\sqrt{2}}{2}, \dfrac{\sqrt{2}}{2} \right\rangle$

43. $v = 3i$
    $\|v\| = \sqrt{3^2 + 0^2} = 3$
    $u = \dfrac{1}{3} \cdot 3i = i$

45. $v = -3i + 4j$
    $\|v\| = \sqrt{(-3)^2 + 4^2} = 5$
    $u = \dfrac{1}{5}v = -\dfrac{3}{5}i + \dfrac{4}{5}j$

47. $u + v = (2i - 5j) + (-3i - 2j) = -i - 7j$

49. $2u - 3v = 2(2i - 5j) - 3(-3i - 2j) = 13i - 4j$

51. $\|2u - 3v\| = \|2(2i - 5j) - 3(-3i - 2j)\|$
    $= \|13i - 4j\| = \sqrt{13^2 + (-4)^2} = \sqrt{185}$

53. $\|v\| = 10$; direction angle $= 60°$

55. $\mathbf{v} = -3(\cos 30°\mathbf{i} - \sin 30°\mathbf{j}) = -3\left(\dfrac{\sqrt{3}}{2}\mathbf{i} - \dfrac{1}{2}\mathbf{j}\right)$

$= -\dfrac{3\sqrt{3}}{2}\mathbf{i} + \dfrac{3}{2}\mathbf{j}$

Thus, the vector lies in quadrant II.

$\tan\theta = \dfrac{3/2}{-3\sqrt{3}/2} = -\dfrac{1}{\sqrt{3}} = -\dfrac{\sqrt{3}}{3} \Rightarrow \theta = 150°$

$\|\mathbf{v}\| = 3$; direction angle $= 150°$

57. $\|\mathbf{v}\| = \sqrt{5^2 + 12^2} = 13$

$\theta = \tan^{-1}\dfrac{12}{5} \approx 67.38°$

59. $\|\mathbf{v}\| = \sqrt{(-4)^2 + (-3)^2} = 5$

$\theta' = \left|\tan^{-1}\left(\dfrac{-3}{-4}\right)\right| \approx 36.87°$

Since $(-4, -3)$ lies in Quadrant III,
$\theta \approx 180° + 36.87° = 216.87°$.

61. $\mathbf{v} = 2(\cos 30°\mathbf{i} + \sin 30°\mathbf{j}) = 2\left(\dfrac{\sqrt{3}}{2}\mathbf{i} + \dfrac{1}{2}\mathbf{j}\right)$

$= \sqrt{3}\mathbf{i} + \mathbf{j}$

63. $\mathbf{v} = 4(\cos 120°\mathbf{i} + \sin 120°\mathbf{j}) = 4\left(-\dfrac{1}{2}\mathbf{i} + \dfrac{\sqrt{3}}{2}\mathbf{j}\right)$

$= -2\mathbf{i} + 2\sqrt{3}\mathbf{j}$

65. $\mathbf{v} = 3\left(\cos\dfrac{5\pi}{3}\mathbf{i} + \sin\dfrac{5\pi}{3}\mathbf{j}\right) = 3\left(\dfrac{1}{2}\mathbf{i} - \dfrac{\sqrt{3}}{2}\mathbf{j}\right)$

$= \dfrac{3}{2}\mathbf{i} - \dfrac{3\sqrt{3}}{2}\mathbf{j}$

67. $\mathbf{v} = 7\left(\cos\left(-\dfrac{\pi}{3}\right)\mathbf{i} + \sin\left(-\dfrac{\pi}{3}\right)\mathbf{j}\right)$

$= 7\left(\dfrac{1}{2}\mathbf{i} - \dfrac{\sqrt{3}}{2}\mathbf{j}\right) = \dfrac{7}{2}\mathbf{i} - \dfrac{7\sqrt{3}}{2}\mathbf{j}$

7.4 Applying the Concepts

69.

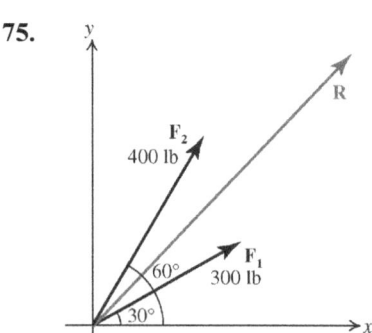

$\mathbf{w} = 25\cos 23°\mathbf{i} + 25\sin 23°\mathbf{j} \approx 23\mathbf{i} + 9.8\mathbf{j}$

71. $\mathbf{F}_1 = 0\mathbf{i} - 25\mathbf{j}$
$\mathbf{F}_2 = -32\mathbf{i} + 0\mathbf{j}$
$\mathbf{R} = \mathbf{F}_1 + \mathbf{F}_2 = -32\mathbf{i} - 25\mathbf{j}$
$\|\mathbf{R}\| = \sqrt{(-32)^2 + (-25)^2} \approx 40.6$

$\theta' = \left|\tan^{-1}\left(\dfrac{-25}{-32}\right)\right| \approx 38.0°$

Since $\mathbf{R}$ lies in quadrant III,
$\theta \approx 180° + 38.0° = 218.0°$. Thus, the resultant is a force of approximately 40.6 lb in the direction $218.0°$.

73. The direction angle of the force is $90° - 65° = 25°$. Thus, the force
$\mathbf{v} = 80(\cos 25°\mathbf{i} + \sin 25°\mathbf{j})$
$\approx 72.5\mathbf{i} + 33.8\mathbf{j} \approx \langle 72.5, 33.8\rangle$.

75.

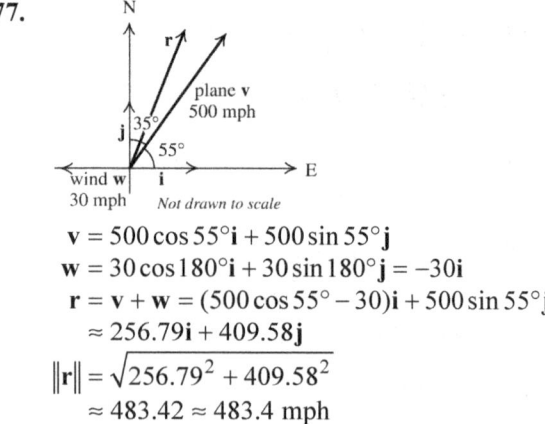

$\mathbf{F}_1 = 300\cos 30°\mathbf{i} + 300\sin 30°\mathbf{j}$
$\mathbf{F}_2 = 400\cos 60°\mathbf{i} + 400\sin 60°\mathbf{j}$
$\mathbf{R} = \mathbf{F}_1 + \mathbf{F}_2 \approx 459.81\mathbf{i} + 496.41\mathbf{j}$
$\|\mathbf{R}\| = \sqrt{(459.81)^2 + (496.41)^2} \approx 676.6$

$\theta' = \left|\tan^{-1}\left(\dfrac{496.41}{459.81}\right)\right| \approx 47.2°$

The resultant is a force of approximately 676.6 lb in the direction $47.2°$.

77.

$\mathbf{v} = 500\cos 55°\mathbf{i} + 500\sin 55°\mathbf{j}$
$\mathbf{w} = 30\cos 180°\mathbf{i} + 30\sin 180°\mathbf{j} = -30\mathbf{i}$
$\mathbf{r} = \mathbf{v} + \mathbf{w} = (500\cos 55° - 30)\mathbf{i} + 500\sin 55°\mathbf{j}$
$\approx 256.79\mathbf{i} + 409.58\mathbf{j}$
$\|\mathbf{r}\| = \sqrt{256.79^2 + 409.58^2}$
$\approx 483.42 \approx 483.4$ mph

(continued on next page)

(*continued*)

$$\theta' = \left|\tan^{-1}\left(\frac{409.58}{256.79}\right)\right| \approx 57.9° \text{ Since } \mathbf{R} \text{ lies in}$$

quadrant I, $\theta \approx 90° - 57.9° = 32.1°$. The plane's ground speed is 483.4 mph, and its bearing is N32.1°E.

79.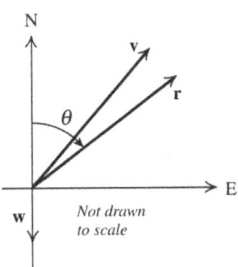

$\mathbf{v} = 15\cos 50°\mathbf{i} + 15\sin 50°\mathbf{j}$
$\mathbf{w} = -4\mathbf{j}$
$\mathbf{r} = \mathbf{v} + \mathbf{w} = 15\cos 50°\mathbf{i} + (15\sin 50° - 4)\mathbf{j}$
$\|\mathbf{r}\| = \sqrt{(15\cos 50°)^2 + (15\sin 50° - 4)^2}$
$\approx 12.21 \approx 12.2$ mph

$$\theta' = \left|\tan^{-1}\left(\frac{15\sin 50° - 4}{15\cos 50°}\right)\right| \approx 37.8°$$

Since **R** lies in quadrant I, $\theta \approx 90° - 37.8° = 52.2°$. The boat's speed and direction are 12.2 mph and N52.2°E.

81. First find the resultant **R** of the system.
$\mathbf{u} = 30(\cos 26°\mathbf{i} + \sin 26°\mathbf{j})$
$\mathbf{v} = 40(\cos 115°\mathbf{i} + \sin 115°\mathbf{j})$
$\mathbf{w} = 50(\cos 270°\mathbf{i} + \sin 270°\mathbf{j})$
$\mathbf{R}_1 = 30\cos 26° + 40\cos 115° + 50\cos 270°$
≈ 10.06
$\mathbf{R}_2 = 30\sin 26° + 40\sin 115° + 50\sin 270°$
≈ -0.60
$\|\mathbf{R}\| = \sqrt{10.06^2 + (-0.60)^2} \approx 10.08$

$$\theta' = \left|\tan^{-1}\frac{\mathbf{R}_2}{\mathbf{R}_1}\right| \approx 3.39°$$

Since **R** lies in Quadrant IV, $\theta \approx 360° - 3.39° = 356.61°$. The force that must be added to the system to obtain equilibrium is $-\mathbf{R}$, a force of 10.08 pounds at an angle of $180° - 3.39° = 176.61°$ with the positive x-axis.

83. To help visualize the system, overlay a coordinate system with the origin at the endpoint of the three vectors.

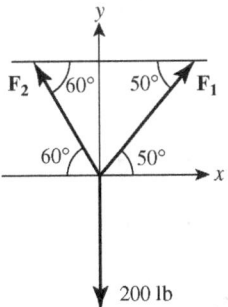

Since the system is in equilibrium, the resultant **R** is $200(\cos 270°\mathbf{i} + \sin 270°\mathbf{j}) = 0\mathbf{i} - 200\mathbf{j}$.
$\mathbf{u} = \|\mathbf{F}_1\|(\cos 50°\mathbf{i} + \sin 50°\mathbf{j})$
$= \|\mathbf{F}_1\|\cos 50°\mathbf{i} + \|\mathbf{F}_1\|\sin 50°\mathbf{j}$
$\mathbf{v} = \|\mathbf{F}_2\|(\cos 120°\mathbf{i} + \sin 120°\mathbf{j})$
$= \|\mathbf{F}_2\|\cos 120°\mathbf{i} + \|\mathbf{F}_2\|\sin 120°\mathbf{j}$

Now solve the system
$\|\mathbf{F}_1\|\cos 50° + \|\mathbf{F}_2\|\cos 120° = 0$
$\|\mathbf{F}_1\|\sin 50° + \|\mathbf{F}_2\|\sin 120° = -200$

We will solve by elimination.
$\|\mathbf{F}_1\|\cos 50° - \frac{1}{2}\|\mathbf{F}_2\| = 0$

$\|\mathbf{F}_1\|\sin 50° + \frac{\sqrt{3}}{2}\|\mathbf{F}_2\| = -200$

$\|\mathbf{F}_1\|\sqrt{3}\cos 50° - \frac{\sqrt{3}}{2}\|\mathbf{F}_2\| = 0$

$\|\mathbf{F}_1\|\sin 50° + \frac{\sqrt{3}}{2}\|\mathbf{F}_2\| = -200$

$\|\mathbf{F}_1\|\sqrt{3}\cos 50° + x\sin 50° = 200$
$\|\mathbf{F}_1\|(\sqrt{3}\cos 50° + \sin 50°) = 200 \Rightarrow$

$$\|\mathbf{F}_1\| = \left|\frac{200}{\sqrt{3}\cos 50° + \sin 50°}\right| \approx 106.42 \text{ lb}$$

$106.42\cos 50° + \|\mathbf{F}_2\|\cos 120° = 0 \Rightarrow$
$\|\mathbf{F}_2\| \approx 136.81$ lb

85. $\mathbf{v} = \langle 4, -3 \rangle; P = (-2, 1)$
$Q_x - (-2) = 4 \Rightarrow Q_x = 2$
$Q_y - 1 = -3 \Rightarrow Q_y = -2$
The terminal point is $(2, -2)$.

87. $\mathbf{u} = \langle -1, 2 \rangle, \mathbf{v} = \langle 3, 5 \rangle$
$2\mathbf{u} - \mathbf{x} = 2\mathbf{x} + 3\mathbf{v} \Rightarrow 3\mathbf{x} = 2\mathbf{u} - 3\mathbf{v} \Rightarrow$
$3\mathbf{x} = 2\langle -1, 2 \rangle - 3\langle 3, 5 \rangle = \langle -11, -11 \rangle \Rightarrow$
$\mathbf{x} = \left\langle -\dfrac{11}{3}, -\dfrac{11}{3} \right\rangle$

89. $\overrightarrow{AB} = \langle 1-2, 4-3 \rangle = \langle -1, 1 \rangle = -\mathbf{i} + \mathbf{j}$
$\overrightarrow{CD} = \langle 1-0, -3-(-2) \rangle = \langle 1, -1 \rangle = \mathbf{i} - \mathbf{j}$
Thus, $\overrightarrow{AB} = -\overrightarrow{CD}$, so $\overrightarrow{AB} \parallel -\overrightarrow{CD}$.
$\|\overrightarrow{AB}\| = \sqrt{(-1)^2 + 1^2} = \sqrt{2}$
$\|\overrightarrow{CD}\| = \sqrt{1^2 + (-1)^2} = \sqrt{2}$

From geometry, we know that if a quadrilateral has two opposite sides that are parallel and equal in length, the quadrilateral is a parallelogram. Alternatively, we could show that $\overrightarrow{BC} = \overrightarrow{AD}$, or $\overrightarrow{BC} \parallel \overrightarrow{AD}$.

91. $\overrightarrow{PQ} = \langle 2-0, 1-(-3) \rangle = \langle 2, 4 \rangle = 2\mathbf{i} + 4\mathbf{j}$
$\overrightarrow{PR} = \langle 3-0, 3-(-3) \rangle = \langle 3, 6 \rangle = 3\mathbf{i} + 6\mathbf{j}$
$\overrightarrow{PQ} = \dfrac{2}{3}\overrightarrow{PR}$, so $P, Q,$ and R are collinear.

93. $\mathbf{v} = \mathbf{i} + 2\mathbf{j}, \mathbf{w} = \mathbf{i} - \mathbf{j}$. The lengths of the diagonals are $\|\mathbf{v} + \mathbf{w}\| = \|2\mathbf{i} + \mathbf{j}\| = \sqrt{5}$ and $\|\mathbf{v} - \mathbf{w}\| = \|3\mathbf{j}\| = 3$.

7.4 Critical Thinking/Discussion/Writing

95. $\overrightarrow{AB} = \langle 4-1, 3-2 \rangle = \langle 3, 1 \rangle$
$\overrightarrow{CP} = \langle 3, 1 \rangle = \langle P_x - 6, P_y - 1 \rangle$
$P_x - 6 = 3 \Rightarrow P_x = 9$
$P_y - 1 = 1 \Rightarrow P_y = 2$
The coordinates of P are $(9, 2)$.

96. $\overrightarrow{CB} = \langle 4-6, 3-1 \rangle = \langle -2, 2 \rangle$
$\overrightarrow{PA} = \langle -2, 2 \rangle = \langle 1 - P_x, 2 - P_y \rangle$
$1 - P_x = -2 \Rightarrow P_x = 3$
$2 - P_y = 2 \Rightarrow P_y = 0$
The coordinates of P are $(3, 0)$.

97. $\overrightarrow{CB} = \langle 4-6, 3-1 \rangle = \langle -2, 2 \rangle$
$\overrightarrow{AP} = \langle -2, 2 \rangle = \langle P_x - 1, P_y - 2 \rangle$
$P_x - 1 = -2 \Rightarrow P_x = -1$
$P_y - 2 = 2 \Rightarrow P_y = 4$
The coordinates of P are $(-1, 4)$.

99. Equivalent vectors have the same length and the same direction. Therefore, two equivalent vectors with the same initial point, by definition, must have the same terminal point.

7.4 Getting Ready for the Next Section

101. False. If $\cos\theta > 0$, then θ lies in quadrant I or in quadrant IV.

103. $\cos^{-1}(-1) = \pi$, $\cos^{-1}(0) = \dfrac{\pi}{2}$, $\cos^{-1}(1) = 0$

For exercises 105–108, $\mathbf{v} = \langle 1, -2 \rangle$ and $\mathbf{w} = \langle 3, 2 \rangle$.

105. $\|\mathbf{v}\| = \sqrt{1^2 + (-2)^2} = \sqrt{5}$

107. $\dfrac{(1)(3) + (-2)(2)}{\|\mathbf{v}\|\|\mathbf{w}\|} = \dfrac{(1)(3) + (-2)(2)}{\sqrt{5}\sqrt{13}} = -\dfrac{1}{\sqrt{65}}$

7.5 The Dot Product

7.5 Practice Problems

1. a. $\mathbf{v} \cdot \mathbf{w} = 1(-2) + (2)(5) = 8$

b. $\mathbf{v} \cdot \mathbf{w} = 4(-3) + (-3)(-4) = 0$

2. $\mathbf{v} \cdot \mathbf{w} = \|\mathbf{v}\|\|\mathbf{w}\|\cos\theta = 4(7)\cos 60° = 14$

3. There is no scalar c such that $\mathbf{v} = c\mathbf{w}$, so the vectors are not parallel. Note that
$\cos\theta = \dfrac{\mathbf{v} \cdot \mathbf{w}}{\|\mathbf{v}\|\|\mathbf{w}\|} = \dfrac{(-1)(2) + (-2)(1)}{\sqrt{5}\sqrt{5}} = -\dfrac{4}{5} \Rightarrow$
$\theta \neq 0$ and $\theta \neq \pi$

4. $\cos\theta = \dfrac{\mathbf{v} \cdot \mathbf{w}}{\|\mathbf{v}\|\|\mathbf{w}\|} = \dfrac{(2)(3) + (-3)(5)}{\sqrt{2^2 + (-3)^2}\sqrt{3^2 + 5^2}}$
$= -\dfrac{9}{\sqrt{13}\sqrt{34}} = -\dfrac{9}{\sqrt{442}} \Rightarrow$
$\theta = \cos^{-1}\left(-\dfrac{9}{\sqrt{442}}\right) \approx 115.3°$

5. $\|\mathbf{v} + 2\mathbf{w}\|^2 = (\mathbf{v} + 2\mathbf{w}) \cdot (\mathbf{v} + 2\mathbf{w})$
$= (\mathbf{v} + 2\mathbf{w}) \cdot \mathbf{v} + (\mathbf{v} + 2\mathbf{w}) \cdot (2\mathbf{w})$
$= \mathbf{v} \cdot \mathbf{v} + 2\mathbf{w} \cdot \mathbf{v} + \mathbf{v} \cdot 2\mathbf{w} + (2\mathbf{w}) \cdot (2\mathbf{w})$
$= \|\mathbf{v}\|^2 + 4(\mathbf{v} \cdot \mathbf{w}) + 4\|\mathbf{w}\|^2$
$= \|\mathbf{v}\|^2 + 4\|\mathbf{v}\|\|\mathbf{w}\|\cos\theta + 4\|\mathbf{w}\|^2$
$= 20^2 + 4(20)(12)\cos 54° + 4(12)^2$
≈ 1540.2738
$\|\mathbf{v} + 2\mathbf{w}\| \approx 39$

6. $\mathbf{v} \cdot \mathbf{w} = 4(2) + (8)(-1) = 0$
   The vectors are orthogonal.

7. The vectors are orthogonal if the dot product equals zero.
   $\mathbf{v} \cdot \mathbf{w} = 4k + (-2)(6) = 0$
   $4k - 12 = 0 \Rightarrow k = 3$

8. a. First compute $\|\mathbf{v}\|, \|\mathbf{w}\|$, and $\mathbf{v} \cdot \mathbf{w}$.
   $\|\mathbf{v}\| = \sqrt{1^2 + (-2)^2} = \sqrt{5}$
   $\|\mathbf{w}\| = \sqrt{2^2 + 5^2} = \sqrt{29}$
   $\mathbf{v} \cdot \mathbf{w} = 1(2) + (-2)(5) = -8$
   The vector projection of $\mathbf{v}$ onto $\mathbf{w}$ is
   $\text{proj}_\mathbf{w} \mathbf{v} = \left(\frac{\mathbf{v} \cdot \mathbf{w}}{\|\mathbf{w}\|^2}\right)\mathbf{w} = \left(-\frac{8}{29}\right)\langle 2, 5 \rangle$
   $= \left\langle -\frac{16}{29}, -\frac{40}{29} \right\rangle.$

   b. The scalar projection of $\mathbf{v}$ onto $\mathbf{w}$ is
   $\frac{\mathbf{v} \cdot \mathbf{w}}{\|\mathbf{w}\|} = -\frac{8}{\sqrt{29}} = -\frac{8\sqrt{29}}{29}.$

9. First compute $\mathbf{v} \cdot \mathbf{w}$ and $\|\mathbf{w}\|^2$.
   $\mathbf{v} \cdot \mathbf{w} = 5(4) + 1(4) = 24$
   $\|\mathbf{w}\|^2 = 4^2 + 4^2 = 32$
   $\mathbf{v}_1 = \text{proj}_\mathbf{w} \mathbf{v} = \left(\frac{\mathbf{v} \cdot \mathbf{w}}{\|\mathbf{w}\|^2}\right)\mathbf{w} = \frac{24}{32}\langle 4, 4 \rangle = \langle 3, 3 \rangle$
   $\mathbf{v}_2 = \mathbf{v} - \mathbf{v}_1 = \langle 5, 1 \rangle - \langle 3, 3 \rangle = \langle 2, -2 \rangle$
   Check that $\mathbf{v}_2$ is orthogonal to $\mathbf{w}$ by showing that $\mathbf{v}_2 \cdot \mathbf{w} = 0$: $2(4) + (-2)(4) = 0.$
   Thus,
   $\mathbf{v} = \mathbf{v}_1 + \mathbf{v}_2$
   $\langle 5, 1 \rangle = \langle 3, 3 \rangle + \langle 2, -2 \rangle$

10. $W = \mathbf{F} \cdot \overrightarrow{PQ} = \|\mathbf{F}\|\|\overrightarrow{PQ}\|\cos\theta$
    $= 40(150)\cos 60° = 3000$ foot-pounds

7.5 Concepts and Vocabulary

1. The dot product of $\mathbf{v} = \langle a_1, a_2 \rangle$ and $\mathbf{w} = \langle b_1, b_2 \rangle$ is defined as $\mathbf{v} \cdot \mathbf{w} = \underline{a_1 b_1 + a_2 b_2}$.

3. If $\mathbf{v}$ and $\mathbf{w}$ are orthogonal, then $\mathbf{v} \cdot \mathbf{w} = \underline{0}$.

5. True. If $\mathbf{v} \cdot \mathbf{w} < 0$, then $\cos\theta$ is negative, and θ is obtuse.

7. False. The dot product is a scalar.

7.5 Building Skills

9. $\mathbf{u} \cdot \mathbf{v} = 1(3) + (-2)(5) = -7$

11. $\mathbf{u} \cdot \mathbf{v} = 2(3) + (-6)(1) = 0$

13. $\mathbf{u} \cdot \mathbf{v} = 1(8) + (-3)(-2) = 14$

15. $\mathbf{u} \cdot \mathbf{v} = 4(0) + (-2)(3) = -6$

17. $\mathbf{u} \cdot \mathbf{v} = 2(5)\cos\frac{\pi}{6} = 5\sqrt{3} \approx 8.7$

19. $\mathbf{u} \cdot \mathbf{v} = 4(5)\cos\frac{2\pi}{3} = -10.0$

21. $\mathbf{u} \cdot \mathbf{v} = 5(4)\cos 65° \approx 8.5$

23. $\mathbf{u} \cdot \mathbf{v} = 3(7)\cos 120° = -10.5$

25. $\cos\theta = \dfrac{2(3) + 3(4)}{\sqrt{2^2 + 3^2}\sqrt{3^2 + 4^2}} = \dfrac{18}{5\sqrt{13}} \Rightarrow$
    $\theta = \cos^{-1}\left(\dfrac{18}{5\sqrt{13}}\right) \approx 3.2°$

27. $\cos\theta = \dfrac{(-2)(1) + (-3)(1)}{\sqrt{(-2)^2 + (-3)^2}\sqrt{1^2 + 1^2}} = -\dfrac{5}{\sqrt{26}} \Rightarrow$
    $\theta = \cos^{-1}\left(-\dfrac{5}{\sqrt{26}}\right) \approx 168.7°$

29. $\cos\theta = \dfrac{2(-5) + 5(2)}{\sqrt{2^2 + 5^2}\sqrt{(-5)^2 + 2^2}} = 0 \Rightarrow$
    $\theta = \cos^{-1}(0) = 90°$

31. $\cos\theta = \dfrac{1(3) + 1(3)}{\sqrt{1^2 + 1^2}\sqrt{3^2 + 3^2}} = 1 \Rightarrow$
    $\theta = \cos^{-1}(1) = 0°$

33. $\cos\theta = \dfrac{(-2)(-4) + (-3)(6)}{\sqrt{(-2)^2 + (-3)^2}\sqrt{(-4)^2 + 6^2}} = -\dfrac{10}{26} \Rightarrow$
    $\theta = \cos^{-1}\left(-\dfrac{10}{26}\right) \approx 112.6°$

In exercises 35–40, use the fact that the angle between two parallel vectors is either 0 or π. Also, two vectors **v** and **w** are parallel if there is a nonzero scalar c such that $\mathbf{v} = c\mathbf{w}$.

35. $\cos\theta = \dfrac{\mathbf{v}\cdot\mathbf{w}}{\|\mathbf{v}\|\|\mathbf{w}\|} = \dfrac{(2)\left(\frac{2}{3}\right)+(3)(1)}{\sqrt{2^2+3^2}\sqrt{\left(\frac{2}{3}\right)^2+1^2}}$

$= \dfrac{\frac{13}{3}}{\sqrt{13}\sqrt{\frac{13}{9}}} = 1 \Rightarrow \theta = 0$

The vectors are parallel.
Alternatively, notice that $\mathbf{v} = 3\mathbf{w}$.

37. $\cos\theta = \dfrac{\mathbf{v}\cdot\mathbf{w}}{\|\mathbf{v}\|\|\mathbf{w}\|} = \dfrac{(3)(-6)+(5)(10)}{\sqrt{3^2+5^2}\sqrt{(-6)^2+10^2}}$

$= \dfrac{32}{\sqrt{34}\sqrt{136}} \Rightarrow \theta \neq 0° \text{ or } \theta \neq \pi$

The vectors are not parallel.

39. $\cos\theta = \dfrac{\mathbf{v}\cdot\mathbf{w}}{\|\mathbf{v}\|\|\mathbf{w}\|} = \dfrac{(6)(2)+(3)(1)}{\sqrt{6^2+3^2}\sqrt{2^2+1^2}}$

$= \dfrac{15}{\sqrt{45}\sqrt{5}} = \dfrac{15}{\sqrt{225}} = \dfrac{15}{15} = 1 \Rightarrow \theta = 0°$

The vectors are parallel.
Alternatively, notice that $\mathbf{v} = \dfrac{1}{3}\mathbf{w}$.

41. Using the Law of Cosines, we have
$\|\mathbf{v}+\mathbf{w}\|^2 = \|\mathbf{v}\|^2 + \|\mathbf{w}\|^2 - 2\|\mathbf{v}\|\|\mathbf{w}\|\cos\theta$
$= 12^2 + 16^2 - 2(12)(16)\cos(180°-50°)$
$\approx 646.83 \Rightarrow \|\mathbf{v}+\mathbf{w}\| \approx 25.433 \approx 25.4$

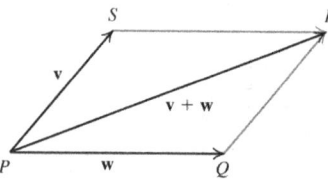

43. Using the Law of Cosines, we have
$\|2\mathbf{v}+\mathbf{w}\|^2 = \|2\mathbf{v}\|^2 + \|\mathbf{w}\|^2 - 2\|2\mathbf{v}\|\|\mathbf{w}\|\cos\theta$
$= ((2)(12))^2 + 16^2$
$\quad - 2(2)(12)(16)\cos(180°-50°)$
≈ 1325.66
$\|2\mathbf{v}+\mathbf{w}\| \approx \sqrt{1325.66} \approx 36.4$

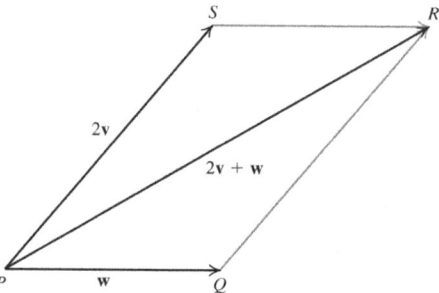

45. Using the Law of Sines and the intermediate result from exercise 41, the angle we want is $\angle RPQ$. So

$\dfrac{\sin\theta}{\|\mathbf{v}\|} = \dfrac{\sin 130°}{\|\mathbf{v}+\mathbf{w}\|} \Rightarrow \dfrac{\sin\theta}{12} = \dfrac{\sin 130°}{25.433} \Rightarrow$

$\theta = \sin^{-1}\left(\dfrac{12\sin 130°}{25.433}\right) \approx 21.2°$

47. Using the Law of Sines and the intermediate result from exercise 43, the angle we want is $\angle SPR$. So

$\dfrac{\sin\theta}{\|\mathbf{w}\|} = \dfrac{\sin 130°}{\|2\mathbf{v}+\mathbf{w}\|} \Rightarrow \dfrac{\sin\theta}{16} = \dfrac{\sin 130°}{36.410} \Rightarrow$

$\theta = \sin^{-1}\left(\dfrac{16\sin 130°}{36.410}\right) \approx 19.7°$

In exercises 49–54, use the fact that the dot product of vectors that are orthogonal equals zero.

49. $\mathbf{v}\cdot\mathbf{w} = 2(-3) + 3(2) = 0$
The vectors are orthogonal.

51. $\mathbf{v}\cdot\mathbf{w} = 2(-7) + 7(-2) = -28$
The vectors are not orthogonal.

53. $\mathbf{v}\cdot\mathbf{w} = 12(2) + 6(-4) = 0$
The vectors are orthogonal.

In exercises 55–62, use the facts that two vectors are orthogonal if the dot product equals zero, and two vectors **v** and **w** are parallel if there is a nonzero scalar n such that $\mathbf{v} = n\mathbf{w}$.

55. a. $\mathbf{v}\cdot\mathbf{w} = 2(3) + 3c = 0$
$6 + 3c = 0 \Rightarrow c = -2$

b. $\mathbf{v} = n\mathbf{w}$
$2 = 3n \Rightarrow \dfrac{2}{3} = n$
$\langle 2,3\rangle = \dfrac{2}{3}\langle 3,c\rangle \Rightarrow 3 = \dfrac{2}{3}c \Rightarrow c = \dfrac{9}{2}$

57. a. $\mathbf{v}\cdot\mathbf{w} = -4c + 3(-6) = 0$
$-4c - 18 = 0 \Rightarrow c = -\dfrac{9}{2}$

b. $\mathbf{v} = n\mathbf{w}$

$3 = -6n \Rightarrow -\dfrac{1}{2} = n$

$\langle -4, 3\rangle = -\dfrac{1}{2}\langle c, -6\rangle \Rightarrow -4 = -\dfrac{1}{2}c \Rightarrow c = 8$

59. a. $\mathbf{v} \cdot \mathbf{w} = 1(2) + 1(-c) = 0$

$2 - c = 0 \Rightarrow c = 2$

b. $\mathbf{v} = n\mathbf{w}$

$1 = 2n \Rightarrow \dfrac{1}{2} = n$

$\mathbf{i} + \mathbf{j} = \dfrac{1}{2}(2\mathbf{i} - c\mathbf{j}) \Rightarrow 1 = \dfrac{1}{2}(-c) \Rightarrow c = -2$

61. a. $\mathbf{v} \cdot \mathbf{w} = c(2) + 3(-1) = 0$

$2c - 3 = 0 \Rightarrow c = \dfrac{3}{2}$

b. $\mathbf{v} = n\mathbf{w}$

$3 = n(-1) \Rightarrow -3 = n$

$c\mathbf{i} + 3\mathbf{j} = -3(2\mathbf{i} - \mathbf{j}) \Rightarrow c = -6$

63. $\text{proj}_{\mathbf{w}} \mathbf{v} = \dfrac{\mathbf{v} \cdot \mathbf{w}}{\|\mathbf{w}\|^2}\mathbf{w} = \dfrac{3(4) + 5(1)}{4^2 + 1^2}\langle 4, 1\rangle$

$= \dfrac{17}{17}\langle 4, 1\rangle = \langle 4, 1\rangle$

65. $\text{proj}_{\mathbf{w}} \mathbf{v} = \dfrac{\mathbf{v} \cdot \mathbf{w}}{\|\mathbf{w}\|^2}\mathbf{w} = \dfrac{0(3) + 2(4)}{3^2 + 4^2}\langle 3, 4\rangle$

$= \dfrac{8}{25}\langle 3, 4\rangle = \left\langle \dfrac{24}{25}, \dfrac{32}{25}\right\rangle$

67. $\text{proj}_{\mathbf{w}} \mathbf{v} = \dfrac{\mathbf{v} \cdot \mathbf{w}}{\|\mathbf{w}\|^2}\mathbf{w} = \dfrac{5(1) + (-3)(0)}{1^2 + 0^2}\langle 1, 0\rangle$

$= 5\langle 1, 0\rangle = \langle 5, 0\rangle$

69. $\text{proj}_{\mathbf{w}} \mathbf{v} = \dfrac{\mathbf{v} \cdot \mathbf{w}}{\|\mathbf{w}\|^2}\mathbf{w} = \dfrac{a(1) + b(1)}{1^2 + 1^2}\langle 1, 1\rangle$

$= \dfrac{a+b}{2}\langle 1, 1\rangle = \left\langle \dfrac{a+b}{2}, \dfrac{a+b}{2}\right\rangle$

71. First compute $\mathbf{v} \cdot \mathbf{w}$ and $\|\mathbf{w}\|^2$.

$\mathbf{v} \cdot \mathbf{w} = 2(1) + 3(0) = 2$

$\|\mathbf{w}\|^2 = 1^2 + 0^2 = 1$

$\mathbf{v}_1 = \text{proj}_{\mathbf{w}} \mathbf{v} = \left(\dfrac{\mathbf{v} \cdot \mathbf{w}}{\|\mathbf{w}\|^2}\right)\mathbf{w} = \dfrac{2}{1}\langle 1, 0\rangle = \langle 2, 0\rangle$

$\mathbf{v}_2 = \mathbf{v} - \mathbf{v}_1 = \langle 2, 3\rangle - \langle 2, 0\rangle = \langle 0, 3\rangle$

Check that $\mathbf{v}_2$ is orthogonal to $\mathbf{w}$ by showing that $\mathbf{v}_2 \cdot \mathbf{w} = 0 : 0(1) + 3(0) = 0$.

$\mathbf{v} = \mathbf{v}_1 + \mathbf{v}_2 \Rightarrow \langle 2, 3\rangle = \langle 2, 0\rangle + \langle 0, 3\rangle$

In exercises 73–78, the check step is omitted for brevity.

73. First compute $\mathbf{v} \cdot \mathbf{w}$ and $\|\mathbf{w}\|^2$.

$\mathbf{v} \cdot \mathbf{w} = 1(1) + 0(1) = 1$

$\|\mathbf{w}\|^2 = 1^2 + 1^2 = 2$

$\mathbf{v}_1 = \text{proj}_{\mathbf{w}} \mathbf{v} = \left(\dfrac{\mathbf{v} \cdot \mathbf{w}}{\|\mathbf{w}\|^2}\right)\mathbf{w} = \dfrac{1}{2}\langle 1, 1\rangle = \left\langle \dfrac{1}{2}, \dfrac{1}{2}\right\rangle$

$\mathbf{v}_2 = \mathbf{v} - \mathbf{v}_1 = \langle 1, 0\rangle - \left\langle \dfrac{1}{2}, \dfrac{1}{2}\right\rangle = \left\langle \dfrac{1}{2}, -\dfrac{1}{2}\right\rangle$

$\mathbf{v} = \mathbf{v}_1 + \mathbf{v}_2 \Rightarrow \langle 1, 0\rangle = \left\langle \dfrac{1}{2}, \dfrac{1}{2}\right\rangle + \left\langle \dfrac{1}{2}, -\dfrac{1}{2}\right\rangle$

75. First compute $\mathbf{v} \cdot \mathbf{w}$ and $\|\mathbf{w}\|^2$.

$\mathbf{v} \cdot \mathbf{w} = 4(-2) + 2(4) = 0$

$\|\mathbf{w}\|^2 = (-2)^2 + 4^2 = 20$

$\mathbf{v}_1 = \text{proj}_{\mathbf{w}} \mathbf{v} = \left(\dfrac{\mathbf{v} \cdot \mathbf{w}}{\|\mathbf{w}\|^2}\right)\mathbf{w} = \dfrac{0}{20}\langle -2, 4\rangle = \langle 0, 0\rangle$

$\mathbf{v}_2 = \mathbf{v} - \mathbf{v}_1 = \langle 4, 2\rangle - \langle 0, 0\rangle = \langle 4, 2\rangle$

$\mathbf{v} = \mathbf{v}_1 + \mathbf{v}_2 \Rightarrow \langle 4, 2\rangle = \langle 0, 0\rangle + \langle 4, 2\rangle$

77. First compute $\mathbf{v} \cdot \mathbf{w}$ and $\|\mathbf{w}\|^2$.

$\mathbf{v} \cdot \mathbf{w} = 4(3) + (-1)(4) = 8$

$\|\mathbf{w}\|^2 = 3^2 + 4^2 = 25$

$\mathbf{v}_1 = \text{proj}_{\mathbf{w}} \mathbf{v} = \left(\dfrac{\mathbf{v} \cdot \mathbf{w}}{\|\mathbf{w}\|^2}\right)\mathbf{w} = \dfrac{8}{25}\langle 3, 4\rangle$

$= \left\langle \dfrac{24}{25}, \dfrac{32}{25}\right\rangle$

$\mathbf{v}_2 = \mathbf{v} - \mathbf{v}_1 = \langle 4, -1\rangle - \left\langle \dfrac{24}{25}, \dfrac{32}{25}\right\rangle = \left\langle \dfrac{76}{25}, -\dfrac{57}{25}\right\rangle$

$\mathbf{v} = \mathbf{v}_1 + \mathbf{v}_2 \Rightarrow \langle 4, -1\rangle = \left\langle \dfrac{24}{25}, \dfrac{32}{25}\right\rangle + \left\langle \dfrac{76}{25}, -\dfrac{57}{25}\right\rangle$

7.5 Applying the Concepts

79.

$\mathbf{v} = 8\cos 0°\mathbf{i} + 8\sin 0°\mathbf{j} = 8\mathbf{i}$

$\mathbf{w} = 6\cos 30°\mathbf{i} + 6\sin 30°\mathbf{j} = 3\sqrt{3}\mathbf{i} + 3\mathbf{j}$

$\mathbf{r} = \mathbf{v} + \mathbf{w} = (8 + 3\sqrt{3})\mathbf{i} + 3\mathbf{j}$

$\|\mathbf{r}\| = \sqrt{(8+3\sqrt{3})^2 + 3^2} \approx 13.53 \approx 13.5$ pounds

$\cos\theta_1 = \dfrac{((8+3\sqrt{3})\mathbf{i} + 3\mathbf{j}) \cdot 8\mathbf{i}}{(13.53)(8)} = \dfrac{8+3\sqrt{3}}{13.53} \Rightarrow$

$\theta_1 = \cos^{-1}\left(\dfrac{8+3\sqrt{3}}{13.53}\right) \approx 12.8°$

$\theta_2 \approx 30° - 12.8° \approx 17.2°$

81. $\mathbf{F} = \langle 64, -20 \rangle$; $\overrightarrow{OP} = \langle 10, 4 \rangle$

$W = \mathbf{F} \cdot \overrightarrow{OP} = (64)(10) + (-20)(4)$
$= 560$ foot-pounds

83. $5220 = \mathbf{F}_1 \cos 18° \approx 5520.2$ lb

85. $\mathbf{F} = \left\langle 10\cos\dfrac{2\pi}{3}, 10\sin\dfrac{2\pi}{3} \right\rangle$; $\overrightarrow{OP} = \langle -6, 3 \rangle$

$W = \mathbf{F} \cdot \overrightarrow{OP}$

$= \left(10\cos\dfrac{2\pi}{3}\right)(-6) + \left(10\sin\dfrac{2\pi}{3}\right)(3)$

≈ 56 foot-pounds

7.5 Beyond the Basics

87. $\overrightarrow{CA} = \langle a-x, -y \rangle$, $\overrightarrow{CB} = \langle -a-x, -y \rangle$

$\overrightarrow{CA} \cdot \overrightarrow{CB} = (a-x)(-a-x) + (-y)(-y)$
$= -a^2 + x^2 + y^2$
$= -a^2 + a^2 = 0$ (using the hint)

Since the dot product equals 0, the vectors are orthogonal (perpendicular).

89. Given $\mathbf{u} = \langle u_1, u_2 \rangle$, $\mathbf{v} = \langle v_1, v_2 \rangle$, and $\mathbf{w} = \langle w_1, w_2 \rangle$. Then

$(\mathbf{v} + \mathbf{w}) = \langle v_1 + w_1, v_2 + w_2 \rangle$.

$\mathbf{u} \cdot (\mathbf{v} + \mathbf{w}) = u_1(v_1 + w_1) + u_2(v_2 + w_2)$
$= u_1 v_1 + u_1 w_1 + u_2 v_2 + u_2 w_2$
$= (u_1 v_1 + u_2 v_2) + (u_1 w_1 + u_2 w_2)$
$= \mathbf{u} \cdot \mathbf{v} + \mathbf{u} \cdot \mathbf{w}$

91. Using the result from exercise 90, we have $\|\mathbf{v}+\mathbf{w}\|^2 = \|\mathbf{v}\|^2 + 2(\mathbf{v} \cdot \mathbf{w}) + \|\mathbf{w}\|^2$. From exercise 88 (the Cauchy-Schwarz inequality), this becomes

$\|\mathbf{v}+\mathbf{w}\|^2 \leq \|\mathbf{v}\|^2 + 2\|\mathbf{v}\|\|\mathbf{w}\| + \|\mathbf{w}\|^2$. Then

$\|\mathbf{v}+\mathbf{w}\|^2 \leq (\|\mathbf{v}\| + \|\mathbf{w}\|)^2 \Rightarrow$
$\|\mathbf{v}+\mathbf{w}\| \leq \|\mathbf{v}\| + \|\mathbf{w}\|$.

93. $\cos\theta = \dfrac{\langle 2,-5 \rangle \cdot \langle 5,2 \rangle}{\|\langle 2,-5 \rangle\|\|\langle 5,2 \rangle\|} = \dfrac{0}{\|\langle 2,-5 \rangle\|\|\langle 5,2 \rangle\|} = 0 \Rightarrow$

$\cos^{-1} 0 = 90°$, so the vectors are perpendicular.

95. Vectors orthogonal to $\mathbf{v} = 3\mathbf{i} - 4\mathbf{j}$ is $\mathbf{w} = 4\mathbf{i} + 3\mathbf{j}$ and $\mathbf{w} = -4\mathbf{i} - 3\mathbf{j}$. (Check by showing that the dot product equals zero.) The unit vectors are

$\dfrac{1}{\|\mathbf{w}\|}\mathbf{w} = \dfrac{1}{\sqrt{4^2+3^2}}\langle 4,3 \rangle = \pm\dfrac{1}{5}\langle 4,3 \rangle = \pm\left\langle \dfrac{4}{5}, \dfrac{3}{5} \right\rangle$.

97. The vectors that make up the sides of the triangles are $\overrightarrow{AB} = \langle 5,0 \rangle$, $\overrightarrow{AC} = \langle 1,-2 \rangle$, and $\overrightarrow{BC} = \langle -4,-2 \rangle$.

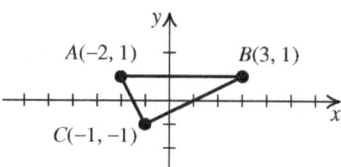

Use the dot product to identify if two vectors are orthogonal.

$\overrightarrow{AC} \cdot \overrightarrow{BC} = (1)(-4) + (-2)(-2) = 0$

Since $\overrightarrow{AC} \cdot \overrightarrow{BC} = 0, \overrightarrow{AC} \perp \overrightarrow{BC}$. Thus $\triangle ABC$ is a right triangle.

99. $\|\mathbf{u}+\mathbf{v}\|^2 - \|\mathbf{u}-\mathbf{v}\|^2$
$= (\mathbf{u}+\mathbf{v}) \cdot (\mathbf{u}+\mathbf{v}) - (\mathbf{u}-\mathbf{v}) \cdot (\mathbf{u}-\mathbf{v})$
$= \mathbf{u} \cdot \mathbf{u} + \mathbf{u} \cdot \mathbf{v} + \mathbf{v} \cdot \mathbf{u} + \mathbf{v} \cdot \mathbf{v}$
$\quad - (\mathbf{u} \cdot \mathbf{u} - \mathbf{u} \cdot \mathbf{v} - \mathbf{v} \cdot \mathbf{u} + \mathbf{v} \cdot \mathbf{v})$
$= \|\mathbf{u}\|^2 + 2(\mathbf{u} \cdot \mathbf{v}) + \|\mathbf{v}\|^2 - \|\mathbf{u}\|^2 + 2(\mathbf{u} \cdot \mathbf{v}) - \|\mathbf{v}\|^2$
$= 4(\mathbf{u} \cdot \mathbf{v})$

101.

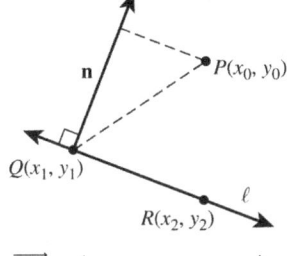

$\overrightarrow{QR} = \langle x_2 - x_1, y_2 - y_1 \rangle$

$ax + by + c = 0 \Rightarrow y = -\dfrac{a}{b}x - \dfrac{c}{b}$

$m = -\dfrac{a}{b} = \dfrac{y_2 - y_1}{x_2 - x_1} \Rightarrow a(x_2 - x_1) = -b(y_2 - y_1)$

$\mathbf{n} \cdot \overrightarrow{QR} = \langle x_2 - x_1, y_2 - y_1 \rangle \cdot \langle a, b \rangle$
$= a(x_2 - x_1) + b(y_2 - y_1)$
$= -b(y_2 - y_1) + b(y_2 - y_1) = 0$

Thus, $\mathbf{n} \perp \overrightarrow{QR}$.

$\mathbf{n} = \langle a, b \rangle = \langle x_0 - x, y_0 - y \rangle$

$\left\| \text{proj}_{\mathbf{n}} \overrightarrow{QP} \right\| = \dfrac{\overrightarrow{QP} \cdot \mathbf{n}}{\|\mathbf{n}\|} = \dfrac{|a(x_0 - x_1) + b(y_0 - y_1)|}{\sqrt{a^2 + b^2}}$

$= \dfrac{|ax_0 + by_0 - (ax_1 + by_1)|}{\sqrt{a^2 + b^2}}$

Since $Q(x_1, y_1)$ is on ℓ, $ax_1 + by_1 = -c$. Thus,

$\left\| \text{proj}_{\mathbf{n}} \overrightarrow{QP} \right\| = \dfrac{|ax_0 + by_0 - (ax_1 + by_1)|}{\sqrt{a^2 + b^2}}$

$= \dfrac{|ax_0 + by_0 + c|}{\sqrt{a^2 + b^2}} = d$

7.5 Critical Thinking/Discussion/Writing

102. False. For example, let $\mathbf{u} = \langle 1, 0 \rangle$, $\mathbf{v} = \langle 0, 2 \rangle$, and $\mathbf{w} = \langle 0, 3 \rangle$. Then $\mathbf{u} \cdot \mathbf{v} = 0$ and $\mathbf{u} \cdot \mathbf{w} = 0$. However, $\mathbf{v} \neq \mathbf{w}$.

103. Since $\cos \theta = \dfrac{\mathbf{v} \cdot \mathbf{w}}{\|\mathbf{v}\| \|\mathbf{w}\|}$, the quotient must be a number between -1 and 1.

104. Yes. If we graph $\mathbf{v}$ and $\mathbf{w}$ along with $-\mathbf{v}$ and $-\mathbf{w}$ on the same coordinate plane, we see that the angles formed are vertical angles. Thus, the angles are equal.

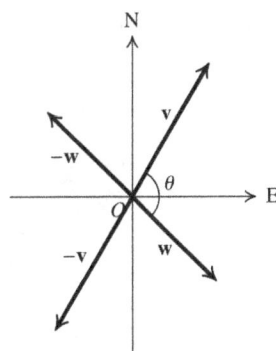

7.5 Getting Ready for the Next Section

105. $x = -1$ is the graph of a vertical line with x-intercept -1 and no y-intercept.

107. $y = x^2 - x - 6$
$0 = x^2 - x - 6 \Rightarrow 0 = (x + 2)(x - 3) \Rightarrow$
$x = -2, 3$
$y = 0^2 - 0 - 6 = -6$

$y = x^2 - x - 6$ is the graph of a parabola with x-intercepts -2 and 3 and y-intercept -6.

109. $x^2 + (y - 1)^2 = 4$
$x^2 + (0 - 1)^2 = 4 \Rightarrow x^2 = 3 \Rightarrow x = \pm\sqrt{3}$
$0^2 + (y - 1)^2 = 4 \Rightarrow (y - 1)^2 = 4 \Rightarrow$
$y - 1 = \pm 2 \Rightarrow y = 1 \pm 2 = -1, 3$

$x^2 + (y - 1)^2 = 4$ is the graph of a circle with center $(0, 1)$ and radius 2. The x-intercepts are $\pm\sqrt{3}$ and the y-intercepts are -1 and 3.

111. $x^2 + y^2 - 2x + 4y + 2 = 0 \Rightarrow$
$x^2 - 2x + y^2 + 4y = -2 \Rightarrow$
$x^2 - 2x + 1 + y^2 + 4y + 4 = -2 + 1 + 4 \Rightarrow$
$(x - 1)^2 + (y + 2)^2 = 3$
$x^2 + 0^2 - 2x + 4(0) + 2 = 0 \Rightarrow$
$x^2 - 2x + 2 = 0 \Rightarrow x^2 - 2x + 1 = -2 + 1 \Rightarrow$
$(x - 1)^2 = -1$, which has no real solution.
$0^2 + y^2 - 2(0) + 4y + 2 = 0 \Rightarrow$
$y^2 + 4y + 4 = -2 + 4 \Rightarrow (y + 2)^2 = 2 \Rightarrow$
$y + 2 = \pm\sqrt{2} \Rightarrow y = -2 \pm \sqrt{2}$

$x^2 + y^2 - 2x + 4y + 2 = 0$ is the graph of a circle with center $(1, -2)$ and radius $\sqrt{3}$. There are no x-intercepts and the y-intercepts are $-2 \pm \sqrt{2}$.

See page 188 in the text to review the tests for symmetry.

113. $x^3 + y^2 = 5$

Test for all three symmetries:

x-axis: $x^3 + (-y)^2 = 5 \Rightarrow x^3 + y^2 = 5$

This is the same as the original equation, so the graph is symmetric with respect to the x-axis.

y-axis: $(-x)^3 + y^2 = 5 \Rightarrow -x^3 + y^2 = 5$, which is not the same as the original equation. So the graph is not symmetric with respect to the y-axis.

origin: $(-x)^3 + (-y)^2 = 5 \Rightarrow -x^3 + y^2 = 5$, which is not the same as the original equation. So the graph is not symmetric with respect to the origin.

115. $2x^2 - 3y^3 = 7$

Test for all three symmetries:

x-axis: $2x^2 - 3(-y)^3 = 7 \Rightarrow 2x^2 + 3y^3 = 7$, which is not the same as the original equation, so the graph is not symmetric with respect to the x-axis.

y-axis: $2(-x)^2 - 3y^3 = 7 \Rightarrow 2x^2 - 3y^3 = 7$, which is the same as the original equation. So the graph is symmetric with respect to the y-axis.

origin:

$2(-x)^2 - 3(-y)^3 = 7 \Rightarrow 2x^2 + 3y^3 = 7$, which is not the same as the original equation. So the graph is not symmetric with respect to the origin.

7.6 Polar Coordinates

7.6 Practice Problems

1. a. $P(2, -150°)$

b. $(-2, 30°)$

c. $(2, 210°)$

d. $(-2, -330°)$

2. a. $(-3, 60°) \Rightarrow x = -3\cos 60° = -\dfrac{3}{2}$,

$y = -3\sin 60° = -\dfrac{3\sqrt{3}}{2}$

The rectangular coordinates of $(-3, 60°)$ are $\left(-\dfrac{3}{2}, -\dfrac{3\sqrt{3}}{2}\right)$.

b. $\left(2, -\dfrac{\pi}{4}\right) \Rightarrow x = 2\cos\left(-\dfrac{\pi}{4}\right) = \sqrt{2}$,

$y = 2\sin\left(-\dfrac{\pi}{4}\right) = -\sqrt{2}$

The rectangular coordinates of $\left(\pi, -\dfrac{\pi}{4}\right)$ are $\left(\sqrt{2}, -\sqrt{2}\right)$.

3. The point $(-1, -1)$ lies in quadrant III with $x = -1$ and $y = -1$.

$(-1, -1) \Rightarrow r = \sqrt{(-1)^2 + (-1)^2} = \sqrt{2}$

$\tan\theta = \dfrac{y}{x} = 1 \Rightarrow \theta = \dfrac{\pi}{4}$ or $\theta = \dfrac{5\pi}{4}$

$(-1, -1)$ is in Quadrant III, so choose $\theta = \dfrac{5\pi}{4}$.

The polar coordinates of $(-1, -1)$ are $\left(\sqrt{2}, \dfrac{5\pi}{4}\right)$.

4. $OE = 15$ in., $EH = 10$ in.
$\overrightarrow{OE} = \langle 15\cos 30°, 15\sin 30°\rangle$, $\overrightarrow{EH} = \langle 10\cos 45°, 10\sin 45°\rangle$
$\overrightarrow{OH} = \overrightarrow{OE} + \overrightarrow{EH} = \langle 15\cos 30° + 10\cos 45°, 15\sin 30° + 10\sin 45°\rangle$
$r = \sqrt{(15\cos 30° + 10\cos 45°)^2 + (15\sin 30° + 10\sin 45°)^2} \approx 24.8$ in.
$\theta = \tan^{-1}\left(\dfrac{15\cos 30° + 10\cos 45°}{15\sin 30° + 10\sin 45°}\right) \approx 36.0°$

The hand is at $(24.8, 36.0°)$ relative to the shoulder.

5.
$$(x+2)^2 + (y-2)^2 = 4$$
$$(r\cos\theta + 2)^2 + (r\sin\theta - 2)^2 = 4$$
$$r^2\cos^2\theta + 4r\cos\theta + 4 + r^2\sin^2\theta - 4r\sin\theta + 4 = 4$$
$$r^2\cos^2\theta + r^2\sin^2\theta + 4r\cos\theta - 4r\sin\theta + 4 = 0$$
$$r^2 + 4r\cos\theta - 4r\sin\theta + 4 = 0$$

6. a. $r = -5 \Rightarrow r^2 = 25 \Rightarrow x^2 + y^2 = 25$.
   A circle with center (0, 0) and radius 5.

   b. $\theta = -\dfrac{\pi}{3} \Rightarrow \tan\theta = -\sqrt{3} = \dfrac{y}{x} \Rightarrow y = -\sqrt{3}x$.
   A line with slope $-\sqrt{3}$ and y-intercept (0, 0).

   c. $r = \sec\theta = \dfrac{1}{\cos\theta} \Rightarrow r\cos\theta = 1 \Rightarrow x = 1$
   A vertical line through $x = 1$.

   d. $r = 2\sin\theta \Rightarrow r^2 = 2r\sin\theta \Rightarrow$
   $x^2 + y^2 = 2y \Rightarrow x^2 + y^2 - 2y = 0 \Rightarrow$
   $x^2 + y^2 - 2y + 1 = 1 \Rightarrow x^2 + (y-1)^2 = 1$
   A circle with center (0, 1) and radius 1.

7.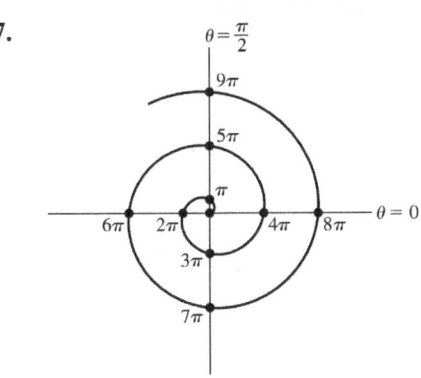

8.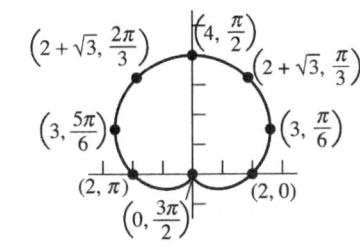

9.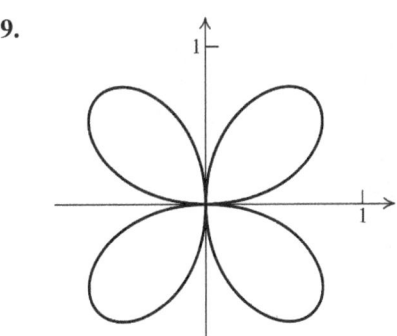

10.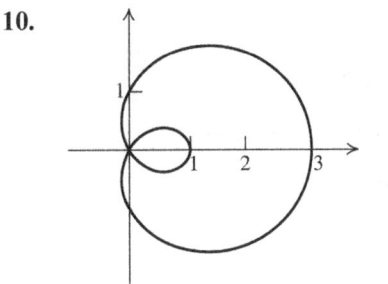

7.6 Concepts and Vocabulary

1. The polar coordinates (r, θ) of a point are the directed distance r to the <u>origin</u> and the directed angle θ from the polar <u>axis</u>.

3. The substitutions $x = \underline{r\cos\theta}$ and $y = \underline{r\sin\theta}$ transform a rectangular equation into a polar equation for the same curve.

5. True

7. False

7.6 Building Skills

9.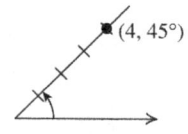

   a. $(-4, 225°)$ **b.** $(-4, -135°)$

   c. $(4, -315°)$

11.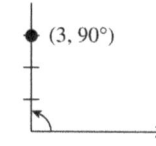

   a. $(-3, 270°)$ **b.** $(-3, -90°)$

   c. $(3, -270°)$

13.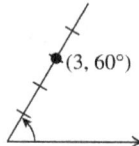

   a. $(-3, 240°)$ **b.** $(-3, -120°)$

   c. $(3, -300°)$

15.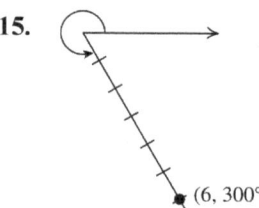

   a. $(-6, 120°)$ **b.** $(-6, -240°)$

   c. $(6, -60°)$

17.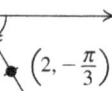

   a. $\left(2, \dfrac{5\pi}{3}\right)$ **b.** $\left(-2, \dfrac{2\pi}{3}\right)$

19.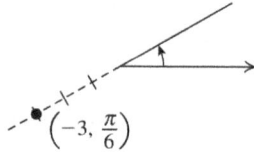

   a. $\left(-3, -\dfrac{11\pi}{6}\right)$ **b.** $\left(3, -\dfrac{5\pi}{6}\right)$

21.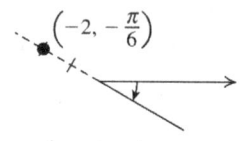

   a. $\left(-2, \dfrac{11\pi}{6}\right)$ **b.** $\left(2, \dfrac{5\pi}{6}\right)$

23.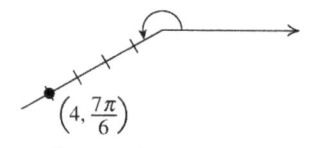

   a. $\left(4, -\dfrac{5\pi}{6}\right)$ **b.** $\left(-4, \dfrac{\pi}{6}\right)$

25. $(3, 60°) \Rightarrow$

$x = 3\cos 60° = \dfrac{3}{2}, y = 3\sin 60° = \dfrac{3\sqrt{3}}{2}$

The rectangular coordinates of $(3, 60°)$ are $\left(\dfrac{3}{2}, \dfrac{3\sqrt{3}}{2}\right)$.

27. $(5, -60°) \Rightarrow x = 5\cos(-60°) = \dfrac{5}{2}$,

$y = 5\sin(-60°) = -\dfrac{5\sqrt{3}}{2}$

The rectangular coordinates of $(5, -60°)$ are $\left(\dfrac{5}{2}, -\dfrac{5\sqrt{3}}{2}\right)$.

29. $(3, \pi) \Rightarrow x = 3\cos\pi = -3, y = 3\sin\pi = 0$

The rectangular coordinates of $(3, \pi)$ are $(-3, 0)$.

31. $\left(-2, -\dfrac{5\pi}{6}\right) \Rightarrow x = -2\cos\left(-\dfrac{5\pi}{6}\right) = \sqrt{3}$,

$y = -2\sin\left(-\dfrac{5\pi}{6}\right) = 1$

The rectangular coordinates of $\left(-2, -\dfrac{5\pi}{6}\right)$ are $\left(\sqrt{3}, 1\right)$.

33. $(1,-1) \Rightarrow r = \sqrt{1^2 + (-1)^2} = \sqrt{2}$
$\tan\theta = \frac{y}{x} = -1 \Rightarrow \theta = \frac{3\pi}{4}$ or $\theta = \frac{7\pi}{4}$
$(1,-1)$ is in Quadrant IV, so choose $\theta = \frac{7\pi}{4}$.
The polar coordinates of $(1,-1)$ are $\left(\sqrt{2}, \frac{7\pi}{4}\right)$.

35. $(3,3) \Rightarrow r = \sqrt{3^2 + 3^2} = 3\sqrt{2}$
$\tan\theta = \frac{y}{x} = 1 \Rightarrow \theta = \frac{\pi}{4}$ or $\theta = \frac{5\pi}{4}$
$(3,3)$ is in Quadrant I, so choose $\theta = \frac{\pi}{4}$.
The polar coordinates of $(3,3)$ are $\left(3\sqrt{2}, \frac{\pi}{4}\right)$.

37. $(3,-3) \Rightarrow r = \sqrt{3^2 + (-3)^2} = 3\sqrt{2}$
$\tan\theta = \frac{y}{x} = -1 \Rightarrow \theta = \frac{3\pi}{4}$ or $\theta = \frac{7\pi}{4}$
$(3,-3)$ is in Quadrant IV, so choose $\theta = \frac{7\pi}{4}$.
The polar coordinates of $(3,-3)$ are $\left(3\sqrt{2}, \frac{7\pi}{4}\right)$.

39. $\left(-1, \sqrt{3}\right) \Rightarrow r = \sqrt{(-1)^2 + \left(\sqrt{3}\right)^2} = 2$
$\tan\theta = \frac{y}{x} = -\sqrt{3} \Rightarrow \theta = \frac{2\pi}{3}$ or $\theta = \frac{5\pi}{3}$
$\left(-1, \sqrt{3}\right)$ is in Quadrant II, so choose $\theta = \frac{2\pi}{3}$.
The polar coordinates of $\left(-1, \sqrt{3}\right)$ are $\left(2, \frac{2\pi}{3}\right)$.

41. $x^2 + y^2 = 16 \Rightarrow$
$(r\cos\theta)^2 + (r\sin\theta)^2 = 16 \Rightarrow$
$r^2(\cos^2\theta + \sin^2\theta) = 16 \Rightarrow r^2 = 16 \Rightarrow r = 4$

43. $y = -2 \Rightarrow r\sin\theta = -2 \Rightarrow r = -\frac{2}{\sin\theta} = -2\csc\theta$

45. $y^2 = 4x \Rightarrow (r\sin\theta)^2 = 4(r\cos\theta) \Rightarrow$
$r^2\sin^2\theta = 4r\cos\theta \Rightarrow$
$r = \frac{4\cos\theta}{\sin^2\theta} = 4\cot\theta\csc\theta$

47. $x^3 = 3y^2 \Rightarrow (r\cos\theta)^3 = 3(r\sin\theta)^2 \Rightarrow$
$r = \frac{3\sin^2\theta}{\cos^3\theta} = 3\tan^2\theta\sec\theta$

49. $y^2 = 6y - x^2 \Rightarrow x^2 + y^2 = 6y \Rightarrow$
$(r\cos\theta)^2 + (r\sin\theta)^2 = 6r\sin\theta \Rightarrow$
$r^2 = 6r\sin\theta \Rightarrow r = 6\sin\theta$

51. $x^2 + y^2 - 4x + 6y = 12$
$(r\cos\theta)^2 + (r\sin\theta)^2 - 4r\cos\theta + 6r\sin\theta = 12$
$r^2(\cos^2\theta + \sin^2\theta) - 4r\cos\theta + 6r\sin\theta = 12$
$r^2 - 4r\cos\theta + 6r\sin\theta = 12$

53. $r = 2 \Rightarrow r^2 = 4 \Rightarrow x^2 + y^2 = 4$.
A circle with center (0, 0) and radius 2.

54. $r = -3 \Rightarrow r^2 = 9 \Rightarrow x^2 + y^2 = 9$.
A circle with center (0, 0) and radius 3.

55. $\theta = \frac{3\pi}{4} \Rightarrow \tan\theta = \tan\frac{3\pi}{4} = -1 \Rightarrow \frac{y}{x} = -1 \Rightarrow$
$y = -x$. A line with slope -1 and y-intercept 0.

57. $r\cos\theta = -2 \Rightarrow x = -2$

59. $r = 4\cos\theta \Rightarrow r^2 = 4r\cos\theta \Rightarrow x^2 + y^2 = 4x \Rightarrow$
$x^2 - 4x + y^2 = 0 \Rightarrow x^2 - 4x + 4 + y^2 = 4 \Rightarrow$
$(x-2)^2 + y^2 = 4$.
A circle with center (2, 0) and radius 2.

61. $r = -2\sin\theta \Rightarrow r^2 = -2r\sin\theta \Rightarrow$
$x^2 + y^2 = -2y \Rightarrow x^2 + y^2 + 2y = 0 \Rightarrow$
$x^2 + y^2 + 2y + 1 = 1 \Rightarrow x^2 + (y+1)^2 = 1$
A circle with center $(0, -1)$ and radius 1.

63. $r = -2 \Rightarrow r^2 = 4 \Rightarrow x^2 + y^2 = 4$

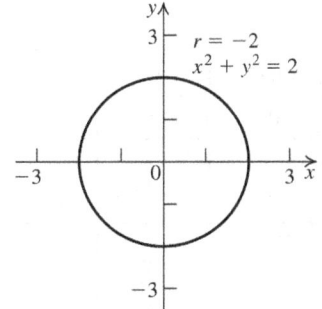

65. $r \sin \theta = 2 \Rightarrow y = 2$

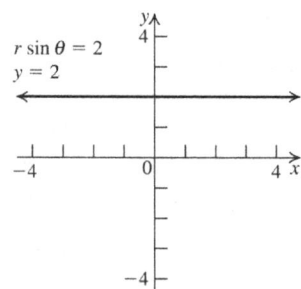

67. $r = 2 \sec \theta = \dfrac{2}{\cos \theta} \Rightarrow r \cos \theta = 2 \Rightarrow x = 2$

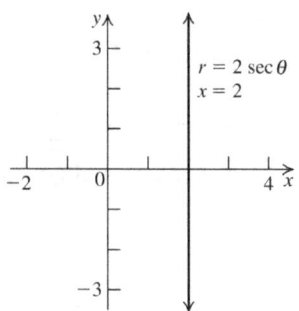

69. $r = \dfrac{1}{\cos \theta + \sin \theta} \Rightarrow r \cos \theta + r \sin \theta = 1 \Rightarrow$
$x + y = 1$

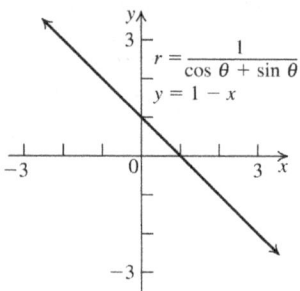

71.

θ	$\sin \theta$	θ	$\sin \theta$
0	0	π	0
$\dfrac{\pi}{6}$	$\dfrac{1}{2}$	$\dfrac{5\pi}{6}$	$\dfrac{1}{2}$
$\dfrac{\pi}{4}$	$\dfrac{\sqrt{2}}{2} \approx 0.7071$	$\dfrac{3\pi}{4}$	$\dfrac{\sqrt{2}}{2} \approx 0.7071$
$\dfrac{\pi}{3}$	$\dfrac{\sqrt{3}}{2} \approx 0.8660$	$\dfrac{2\pi}{3}$	$\dfrac{\sqrt{3}}{2} \approx 0.8660$
$\dfrac{\pi}{2}$	1		

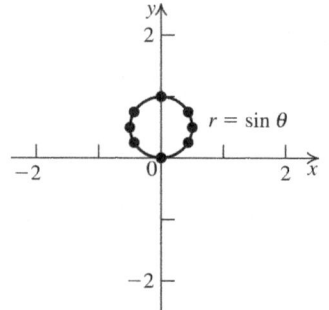

73.

θ	$1 - \cos \theta$	θ	$1 - \cos \theta$
0	0	π	2
$\dfrac{\pi}{6}$	$1 - \dfrac{\sqrt{3}}{2} \approx 0.1340$	$-\dfrac{\pi}{6}$	$1 - \dfrac{\sqrt{3}}{2} \approx 0.1340$
$\pi/3$	$1/2$	$-\pi/3$	$1/2$
$\pi/2$	1	$-\pi/2$	1
$\dfrac{2\pi}{3}$	$\dfrac{3}{2}$	$-\dfrac{2\pi}{3}$	$\dfrac{3}{2}$
$\dfrac{5\pi}{6}$	$1 + \dfrac{\sqrt{3}}{2} \approx 1.8660$	$-\dfrac{5\pi}{6}$	$1 + \dfrac{\sqrt{3}}{2} \approx 1.8660$

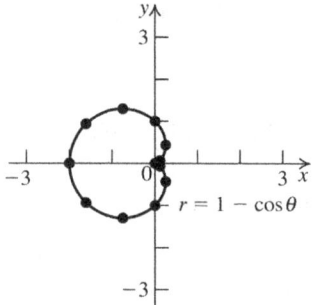

75.

θ	$\cos 3\theta$	θ	$\cos 3\theta$
0	1	π	1
$\dfrac{\pi}{6}$	0	$-\dfrac{\pi}{6}$	0
$\pi/3$	-1	$-\pi/3$	-1
$\pi/2$	0	$-\pi/2$	0
$\dfrac{2\pi}{3}$	1	$-\dfrac{2\pi}{3}$	1
$\dfrac{5\pi}{6}$	0	$-\dfrac{5\pi}{6}$	0

(*continued on next page*)

(continued)

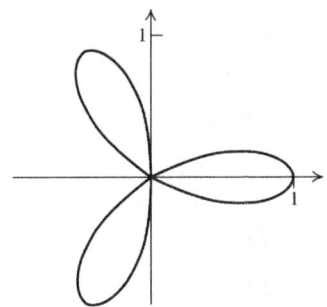

77.

θ	sin 4θ	θ	sin 4θ
0	0	π	0
$\frac{\pi}{6}$	$\frac{\sqrt{3}}{2} \approx 0.8660$	$-\frac{\pi}{6}$	$-\frac{\sqrt{3}}{2} \approx -0.8660$
π/3	$-\frac{\sqrt{3}}{2} \approx -0.8660$	$-\pi/3$	$\frac{\sqrt{3}}{2} \approx 0.8660$
π/2	0	$-\pi/2$	0
$\frac{2\pi}{3}$	$\frac{\sqrt{3}}{2} \approx 0.8660$	$-\frac{2\pi}{3}$	$-\frac{\sqrt{3}}{2} \approx -0.8660$
$\frac{5\pi}{6}$	$-\frac{\sqrt{3}}{2} \approx -0.8660$	$-\frac{5\pi}{6}$	$\frac{\sqrt{3}}{2} \approx 0.8660$

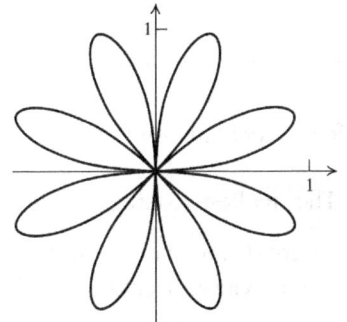

79.

θ	1 − 2 sin θ	θ	1 − 2 sin θ
0	1	π	1
$\frac{\pi}{6}$	0	$-\frac{\pi}{6}$	2
π/3	$1-\sqrt{3}$	$-\pi/3$	$1+\sqrt{3}$
π/2	−1	$-\pi/2$	3
$\frac{2\pi}{3}$	$1-\sqrt{3}$	$-\frac{2\pi}{3}$	$1+\sqrt{3}$
$\frac{5\pi}{6}$	0	$-\frac{5\pi}{6}$	2

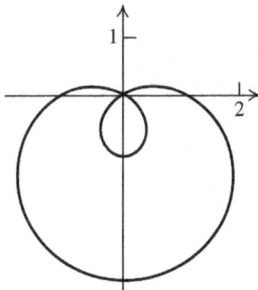

81.

θ	2 + 4 cos θ	θ	2 + 4 cos θ
0	6	π	−2
$\frac{\pi}{6}$	$2+2\sqrt{3}$	$-\frac{\pi}{6}$	$2+2\sqrt{3}$
π/3	4	$-\pi/3$	4
π/2	2	$-\pi/2$	2
$\frac{2\pi}{3}$	0	$-\frac{2\pi}{3}$	0
$\frac{5\pi}{6}$	$2-2\sqrt{3}$	$-\frac{5\pi}{6}$	$2-2\sqrt{3}$

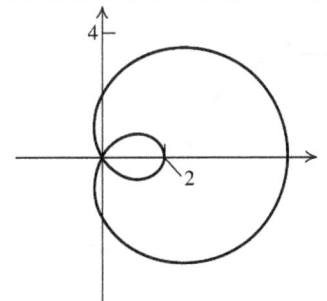

83.

θ	$3-2\sin\theta$	θ	$3-2\sin\theta$
0	3	π	3
$\dfrac{\pi}{6}$	2	$-\dfrac{\pi}{6}$	4
$\pi/3$	$3-\sqrt{3}$	$-\pi/3$	$3+\sqrt{3}$
$\pi/2$	1	$-\pi/2$	5
$\dfrac{2\pi}{3}$	$3-\sqrt{3}$	$-\dfrac{2\pi}{3}$	$3+\sqrt{3}$
$\dfrac{5\pi}{6}$	2	$-\dfrac{5\pi}{6}$	4

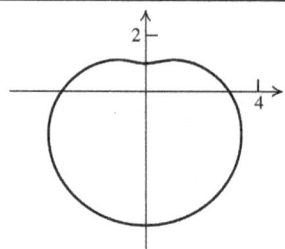

85.

θ	$4+3\cos\theta$	θ	$4+3\cos\theta$
0	7	π	1
$\dfrac{\pi}{6}$	$4+\dfrac{3\sqrt{3}}{2}$	$-\dfrac{\pi}{6}$	$4+\dfrac{3\sqrt{3}}{2}$
$\pi/3$	$\dfrac{11}{2}$	$-\pi/3$	$\dfrac{11}{2}$
$\pi/2$	4	$-\pi/2$	4
$\dfrac{2\pi}{3}$	$\dfrac{5}{2}$	$-\dfrac{2\pi}{3}$	$\dfrac{5}{2}$
$\dfrac{5\pi}{6}$	$4-\dfrac{3\sqrt{3}}{2}$	$-\dfrac{5\pi}{6}$	$4-\dfrac{3\sqrt{3}}{2}$

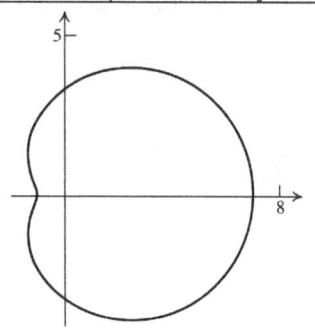

7.6 Applying the Concepts

87. $\alpha = 45°, \beta = 30° \Rightarrow$
$\overrightarrow{OE} = \langle 14\cos 45°, 14\sin 45°\rangle = \langle 7\sqrt{2}, 7\sqrt{2}\rangle$
$\overrightarrow{EH} = \langle 8\cos 30°, 8\sin 30°\rangle = \langle 4\sqrt{3}, 4\rangle$
$\overrightarrow{OH} = \overrightarrow{OE} + \overrightarrow{EH} = \langle 7\sqrt{2}+4\sqrt{3}, 7\sqrt{2}+4\rangle$
$r = \sqrt{\left(7\sqrt{2}+4\sqrt{3}\right)^2 + \left(7\sqrt{2}+4\right)^2} \approx 21.8$ in.
$\theta = \tan^{-1}\left(\dfrac{7\sqrt{2}+4}{7\sqrt{2}+4\sqrt{3}}\right) \approx 39.6°$

The hand is at $(21.8, 39.6°)$ relative to the shoulder.

89. $\alpha = -70°, \beta = 0° \Rightarrow$
$\overrightarrow{OE} = \langle 14\cos(-70°), 14\sin(-70°)\rangle$
$\overrightarrow{EH} = \langle 8\cos 0°, 8\sin 0°\rangle = \langle 8, 0\rangle$
$\overrightarrow{OH} = \overrightarrow{OE} + \overrightarrow{EH}$
$\phantom{\overrightarrow{OH}} = \langle 14\cos(-70°)+8, 14\sin(-70°)\rangle$
$r = \sqrt{\left(14\cos(-70°)+8\right)^2 + \left(14\sin(-70°)\right)^2}$
$ \approx 18.3$ in.
$\theta = \tan^{-1}\left(\dfrac{14\sin(-70)}{14\cos(-70°)+8}\right) \approx -45.8°$

The hand is at $(18.3, -45.8°)$ relative to the shoulder.

91. $r = 40, \ \theta = \dfrac{1}{3}(2\pi) = \dfrac{2\pi}{3}$

Her polar coordinates are $\left(40, \dfrac{2\pi}{3}\right)$.

For Exercises 93–96, apply the formula
$r = \dfrac{a(1-e^2)}{1-e\cos\theta}$. The smallest distance occurs at the minimum value of $\cos\theta$, and the largest distance occurs at the maximum value of $\cos\theta$.

93. a. $r = \dfrac{1(1-0.017^2)}{1-0.017\cos\pi} = 0.983$ au

   b. $r = \dfrac{1(1-0.017^2)}{1-0.017\cos 0} = 1.017$ au

95. a. $r = \dfrac{30.1(1-0.009^2)}{1-0.009\cos\pi} = 29.8291$ au

   b. $r = \dfrac{30.1(1-0.009^2)}{1-0.009\cos 0} = 30.3709$ au

Copyright © 2019 Pearson Education Inc.

7.6 Beyond the Basics

97. $y = x\tan\left(\sqrt{x^2+y^2}\right) \Rightarrow \dfrac{y}{x} = \tan r \Rightarrow \dfrac{\sin\theta}{\cos\theta} = \tan r \Rightarrow r = \theta$

99. $r(1-\sin\theta) = 3 \Rightarrow r - r\sin\theta = 3 \Rightarrow \sqrt{x^2+y^2} - y = 3 \Rightarrow \sqrt{x^2+y^2} = y+3 \Rightarrow$
$x^2 + y^2 = (y+3)^2 = y^2 + 6y + 9 \Rightarrow x^2 - 9 = 6y \Rightarrow y = \dfrac{x^2}{6} - \dfrac{3}{2}$

101. $r\left(1 + \dfrac{1}{2}\cos\theta\right) = 1 \Rightarrow r + \dfrac{r\cos\theta}{2} = 1 \Rightarrow \sqrt{x^2+y^2} + \dfrac{x}{2} = 1 \Rightarrow \sqrt{x^2+y^2} = 1 - \dfrac{x}{2} \Rightarrow$
$x^2 + y^2 = \left(1 - \dfrac{x}{2}\right)^2 = 1 - x + \dfrac{x^2}{4} \Rightarrow y^2 = 1 - x - \dfrac{3x^2}{4}$

103. $r(1 - 3\cos\theta) = 5 \Rightarrow r - 3r\cos\theta = 5 \Rightarrow \sqrt{x^2+y^2} - 3x = 5 \Rightarrow \sqrt{x^2+y^2} = 3x+5 \Rightarrow$
$x^2 + y^2 = (3x+5)^2 = 9x^2 + 30x + 25 \Rightarrow y^2 - 8x^2 - 30x - 25 = 0$

105. $r = \dfrac{6}{1 - e\cos\theta}$

   a. $r = \dfrac{6}{1 - \dfrac{1}{2}\cos\theta} \Rightarrow r - \dfrac{1}{2}r\cos\theta = 6 \Rightarrow \sqrt{x^2+y^2} - \dfrac{1}{2}x = 6 \Rightarrow \sqrt{x^2+y^2} = \dfrac{1}{2}x + 6 \Rightarrow$
$2\sqrt{x^2+y^2} = x + 12 \Rightarrow 4(x^2+y^2) = x^2 + 24x + 144 \Rightarrow 4x^2 + 4y^2 = x^2 + 24x + 144 \Rightarrow$
$3x^2 + 4y^2 - 24x - 144 = 0$

   b. $r = \dfrac{6}{1 - 1\cdot\cos\theta} \Rightarrow r - r\cos\theta = 6 \Rightarrow \sqrt{x^2+y^2} - x = 6 \Rightarrow \sqrt{x^2+y^2} = x+6 \Rightarrow$
$x^2 + y^2 = x^2 + 12x + 36 \Rightarrow y^2 - 12x - 36 = 0$

   c. $r = \dfrac{6}{1 - 2\cos\theta} \Rightarrow r - 2r\cos\theta = 6 \Rightarrow \sqrt{x^2+y^2} - 2x = 6 \Rightarrow \sqrt{x^2+y^2} = 2x+6 \Rightarrow$
$x^2 + y^2 = 4x^2 + 24x + 36 \Rightarrow 3x^2 - y^2 + 24x + 36 = 0$

107. The points (r_1, θ_1) and (r_2, θ_2) have rectangular coordinates $(r_1\cos\theta_1, r_1\sin\theta_1)$ and $(r_2\cos\theta_2, r_2\sin\theta_2)$.
The midpoint M of the line segment joining the points is $\left(\dfrac{r_1\cos\theta_1 + r_2\cos\theta_2}{2}, \dfrac{r_1\sin\theta_1 + r_2\sin\theta_2}{2}\right)$.

109. $D = \sqrt{(x_2-x_1)^2 + (y_2-y_1)^2} = \sqrt{(r_2\cos\theta_2 - r_1\cos\theta_1)^2 + (r_2\sin\theta_2 - r_1\sin\theta_1)^2}$
$= \sqrt{r_2^2\cos^2\theta_2 - 2r_1r_2\cos\theta_1\cos\theta_2 + r_1^2\cos^2\theta_1 + r_2^2\sin^2\theta_2 - 2r_1r_2\sin\theta_1\sin\theta_2 + r_1^2\sin^2\theta_1}$
$= \sqrt{r_2^2(\cos^2\theta_2 + \sin^2\theta_2) - 2r_1r_2(\cos\theta_1\cos\theta_2 + \sin\theta_1\sin\theta_2) + r_1^2(\cos^2\theta_1 + \sin^2\theta_1)}$
$= \sqrt{r_1^2 + r_2^2 - 2r_1r_2(\cos\theta_1\cos\theta_2 + \sin\theta_1\sin\theta_2)}$
$= \sqrt{r_1^2 + r_2^2 - 2r_1r_2\cos(\theta_1 - \theta_2)}$

111.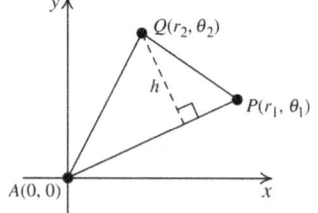

$h = d(A,Q)\sin A, d(A,Q) = r_2, d(A,P) = r_1$, and $m\angle A = \theta_2 - \theta_1$. So, $K = \frac{1}{2}r_1 r_2 \sin(\theta_2 - \theta_1)$.

113.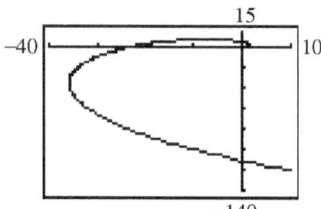

115.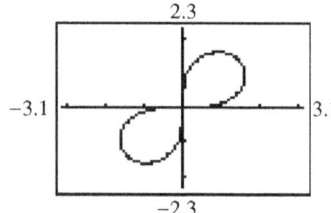

117.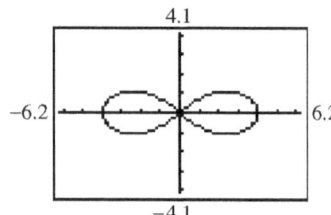

119.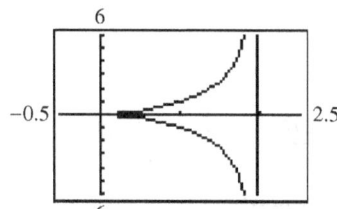

121.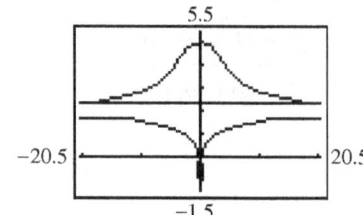

7.6 Critical Thinking/Discussion/Writing

123. No. Let $u = 2\theta$. Then,
$r = \sin(-2\theta) = \sin(-u) = -\sin u = -\sin 2\theta$
This indicates that the graph is symmetric with respect to the line $\theta = 0$.
Replacing (r, θ) with $(r, \pi + \theta)$ leads to
$r = \sin 2(\pi + \theta) = 2\sin(\pi + \theta)\cos(\pi + \theta)$
$= 2(\sin \pi \cos \theta + \cos \pi \sin \theta)$
$\quad \cdot (\cos \pi \cos \theta - \sin \pi \sin \theta)$
$= 2(-\sin \theta)(-\cos \theta)$
$= 2\sin \theta \cos \theta = \sin 2\theta$
This indicates that the graph is symmetric with respect to the pole. If a polar equation has two of these symmetries, then it also has the third symmetry, so $r = 2\sin\theta$ is symmetric about the y-axis.

124. Replacing (r, θ) with $(r, -\theta)$ leads to
$-r = 2 + 3\sin(-\theta) = 2 - 3\sin\theta$, which is inconclusive. However, replacing (r, θ) with $(r, \pi - \theta)$ leads to
$r = 2 + 3\sin(\pi - \theta)$
$= 2 + 3(\sin \pi \cos \theta - \cos \pi \sin \theta)$
$= 2 + 3(-(-1)\sin\theta) = 2 + 3\sin\theta$
Therefore, the graph is symmetric about the y-axis.

125. Replacing (r, θ) with $(-r, -\theta)$ leads to
$-r = \cos\left(-\frac{\theta}{4}\right) = \cos\left(\frac{\theta}{4}\right)$, which is inconclusive.
Now, examine the graph of $r = \cos\left(\frac{\theta}{4}\right)$. If the relation is graphed for $0 \leq \theta \leq 2\pi$, then it is clear that the graph has no symmetries.

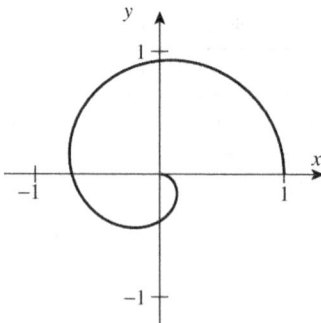

(continued on next page)

(*continued*)

However, if the relation is graphed for $0 \le \theta \le 4\pi$, then the graph appears to be symmetric with respect to the y-axis.

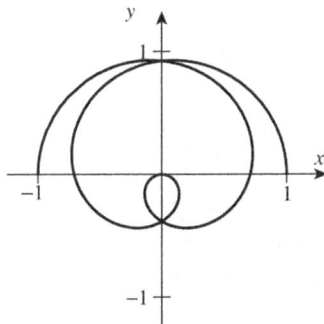

Replace (r, θ) with $(-r, 4\pi - \theta)$ yields
$$-r = \cos\left(\frac{4\pi - \theta}{4}\right) = \cos\left(\pi - \frac{\theta}{4}\right)$$
$$= \cos \pi \cos\left(-\frac{\theta}{4}\right) + \sin \pi \sin\left(-\frac{\theta}{4}\right)$$
$$= -\cos\left(-\frac{\theta}{4}\right) = -\cos\left(\frac{\theta}{4}\right),$$
showing that the graph is symmetric about the y-axis.

Graphing $r = \cos\left(\frac{\theta}{4}\right)$ over $0 \le \theta \le 8\pi$ shows that the relation actually has all three symmetries.

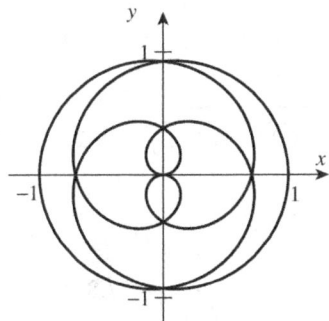

126.

$r = 2\cos\theta \Rightarrow r^2 = 2r\cos\theta \Rightarrow x^2 + y^2 = 2x \Rightarrow$
$x^2 - 2x + y^2 = 0 \Rightarrow x^2 - 2x + 1 + y^2 = 1 \Rightarrow$
$(x-1)^2 + y^2 = 1$
This is a circle of radius 1, centered at (1, 0).
$r = 2\cos\left(\theta - \frac{\pi}{3}\right)$
$r = 2\left(\cos\theta\cos\frac{\pi}{3} + \sin\theta\sin\frac{\pi}{3}\right)$
$r = 2\left(\frac{1}{2}\cos\theta + \frac{\sqrt{3}}{2}\sin\theta\right) = \cos\theta + \sqrt{3}\sin\theta$
$r^2 = r\cos\theta + r\sqrt{3}\sin\theta \Rightarrow$
$x^2 + y^2 = x + \sqrt{3}y \Rightarrow$
$x^2 - x + y^2 - \sqrt{3}y = 0 \Rightarrow$
$\left(x^2 - x + \frac{1}{4}\right) + \left(y^2 - \sqrt{3}y + \frac{3}{4}\right) = 1 \Rightarrow$
$\left(x - \frac{1}{2}\right)^2 + \left(y - \frac{\sqrt{3}}{2}\right)^2 = 1$
This is a circle of radius 1, centered at $\left(\frac{1}{2}, \frac{\sqrt{3}}{2}\right)$. This is the first circle rotated through an angle of $\frac{\pi}{3}$ about the origin. The graph of $r = f(\theta - \alpha)$ is the graph of $r = f(\theta)$ rotated through angle α about the origin.

7.6 Getting Ready for the Next Section

127. $(2 - 3i) + (-5 + i) = -3 - 2i$

129. $i(2 + 5i) = 2i + 5i^2 = 2i - 5 = -5 + 2i$

131. $(2 + i)(3 - i) = 6 + i - i^2 = 6 + i - (-1)$
$= 7 + i$

133. $(2 - 3i)^2 = 4 - 12i + 9i^2 = 4 - 12i - 9$
$= -5 - 12i$

135. $\frac{1}{3 - 4i} = \frac{1}{3 - 4i} \cdot \frac{3 + 4i}{3 + 4i} = \frac{3 + 4i}{9 - 16i^2} = \frac{3 + 4i}{9 + 16}$
$= \frac{3 + 4i}{25} = \frac{3}{25} + \frac{4}{25}i$

137. $\frac{2 - 3i}{3 + 4i} = \frac{2 - 3i}{3 + 4i} \cdot \frac{3 - 4i}{3 - 4i} = \frac{6 - 17i + 12i^2}{9 - 16i^2}$
$= \frac{6 - 17i - 12}{9 + 16} = \frac{-6 - 17i}{25} = -\frac{6}{25} - \frac{17}{25}i$

139. $\mathbf{v} = \mathbf{i} + \mathbf{j}$

$\|\mathbf{v}\| = \sqrt{1^2 + 1^2} = \sqrt{2}$

$\theta' = \left|\tan^{-1}\left(\frac{1}{1}\right)\right| = \left|\frac{\pi}{4}\right| = \frac{\pi}{4}$

Since $(1, 1)$ lies in quadrant I, we have $\theta = \frac{\pi}{4}$.

Thus, $\mathbf{v} = \mathbf{i} + \mathbf{j} = \sqrt{2}\left(\cos\frac{\pi}{4}\mathbf{i} + \sin\frac{\pi}{4}\mathbf{j}\right)$

141. $\mathbf{v} = -\sqrt{3}\mathbf{i} + \mathbf{j}$

$\|\mathbf{v}\| = \sqrt{\left(-\sqrt{3}\right)^2 + 1^2} = 2$

$\theta' = \tan^{-1}\left(\frac{1}{-\sqrt{3}}\right) = \left|\tan^{-1}\left(-\frac{\sqrt{3}}{3}\right)\right|$

$= \left|-\frac{\pi}{6}\right| = \frac{\pi}{6}$

Since $\left(-\sqrt{3}, 1\right)$ lies in quadrant II, we have

$\theta = \pi - \frac{\pi}{6} = \frac{5\pi}{6}$. Thus,

$\mathbf{v} = -\sqrt{3}\mathbf{i} + \mathbf{j} = 2\left(\cos\frac{5\pi}{6}\mathbf{i} + \sin\frac{5\pi}{6}\mathbf{j}\right)$

7.7 Polar Form of Complex Numbers; DeMoivre's Theorem

7.7 Practice Problems

1.

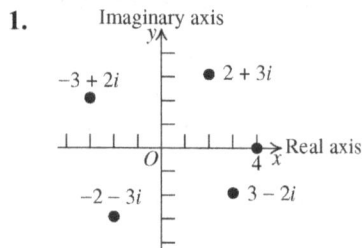

2. a. $|-5 + 12i| = \sqrt{(-5)^2 + 12^2} = 13$

b. $|-7| = 7$

c. $|i| = 1$

d. $|a - bi| = \sqrt{a^2 + (-b)^2} = \sqrt{a^2 + b^2}$

3. $z = -1 - i \Rightarrow a = -1, b = -1$

$r = \sqrt{(-1)^2 + (-1)^2} = \sqrt{2}$

$\theta = \tan^{-1}\left(\frac{-1}{-1}\right) = \tan^{-1} 1 \Rightarrow \theta = \frac{\pi}{4}$ or $\theta = \frac{5\pi}{4}$

Since $(-1, -1)$ lies in quadrant III, $\theta = \frac{5\pi}{4}$.

Thus, $z = -1 - i = \sqrt{2}\left(\cos\frac{5\pi}{4} + i\sin\frac{5\pi}{4}\right)$.

4. $z = 4\left(\cos\frac{5\pi}{3} + i\sin\frac{5\pi}{3}\right) = 4\cos\frac{5\pi}{3} + 4i\sin\frac{5\pi}{3}$

$= 4\left(\frac{1}{2}\right) + 4i\left(-\frac{\sqrt{3}}{2}\right) = 2 - 2i\sqrt{3}$

5. $z_1 = 5(\cos 75° + i\sin 75°)$

$z_2 = 2(\cos 60° + i\sin 60°)$

$z_1 z_2 = 5 \cdot 2\left(\cos(75° + 60°) + i\sin(75° + 60°)\right)$

$= 10(\cos 135° + i\sin 135°)$

$\frac{z_1}{z_2} = \frac{5}{2}\left(\cos(75° - 60°) + i\sin(75° - 60°)\right)$

$= \frac{5}{2}(\cos 15° + i\sin 15°)$

6. $Z_1 = 4(\cos 45° + i\sin 45°)$

$Z_2 = 6(\cos 0° + i\sin 0°)$

To find $Z_1 + Z_2$, first convert each to rectangular form. Then convert the sum back to polar form:

$Z_1 = 4(\cos 45° + i\sin 45°) = 4\left(\frac{\sqrt{2}}{2} + i\frac{\sqrt{2}}{2}\right)$

$= 2\sqrt{2} + 2\sqrt{2}i$

$Z_2 = 6(\cos 0° + i\sin 0°) = 6$

$Z_1 + Z_2 = 6 + 2\sqrt{2} + 2\sqrt{2}i \Rightarrow$

$a = 6 + 2\sqrt{2}, b = 2\sqrt{2}$

$r = \sqrt{\left(6 + 2\sqrt{2}\right)^2 + \left(2\sqrt{2}\right)^2} \approx 9.27$

$\tan\theta = \frac{2\sqrt{2}}{6 + 2\sqrt{2}}$

$\theta = \tan^{-1}\left(\frac{2\sqrt{2}}{6 + 2\sqrt{2}}\right) \approx 17.8°$

The polar form of $Z_1 + Z_2$ is approximately $9.27(\cos 17.8° + i\sin 17.8°)$.

Now find $Z_1 Z_2$.

$Z_1 Z_2 = 4 \cdot 6\left(\cos(45° + 0°) + i\sin(45° + 0°)\right)$

$= 24(\cos 45° + i\sin 45°)$

$\frac{Z_1 Z_2}{Z_1 + Z_2} \approx \frac{24(\cos 45° + i\sin 45°)}{9.27(\cos 17.8° + i\sin 17.8°)}$

$\approx 2.6\left(\cos(45° - 17.8°) + i\sin(45° - 17.8°)\right)$

$\approx 2.6(\cos 27.2° + i\sin 27.2°)$

7. First convert $-1+i$ to polar form:
$r = \sqrt{(-1)^2 + 1^2} = \sqrt{2}; \theta = \tan^{-1}(-1) = \dfrac{3\pi}{4}$
(Note that $-1+i$ lies in Quadrant II.)

   a. $(-1+i)^8 = \left[\sqrt{2}\left(\cos\dfrac{3\pi}{4} + i\sin\dfrac{3\pi}{4}\right)\right]^8$
   $= \left(\sqrt{2}\right)^8 \left(\cos\left(8 \cdot \dfrac{3\pi}{4}\right) + i\sin\left(8 \cdot \dfrac{3\pi}{4}\right)\right)$
   $= 2^4 (\cos 6\pi + i\sin 6\pi)$
   $= 16 \cdot 1 + 16 \cdot 0i = 16$

   b. $(-1+i)^{-12}$
   $= \left[\sqrt{2}\left(\cos\dfrac{3\pi}{4} + i\sin\dfrac{3\pi}{4}\right)\right]^{-12}$
   $= \left(\sqrt{2}\right)^{-12}\left(\cos\left(-12 \cdot \dfrac{3\pi}{4}\right) + i\sin\left(-12 \cdot \dfrac{3\pi}{4}\right)\right)$
   $= \dfrac{1}{2^6}(\cos(-9\pi) + i\sin(-9\pi))$
   $= \dfrac{1}{64}(-1) + \dfrac{1}{64}i \cdot 0 = -\dfrac{1}{64}$

8. First convert $-1+i$ to polar form:
$r = \sqrt{(-1)^2 + 1^2} = \sqrt{2}; \theta = \tan^{-1}(-1) = 135°$
(Note that $-1+i$ lies in Quadrant II.)
$z_k = \left(\sqrt{2}\right)^{1/3}\left(\cos\dfrac{135° + 360° \cdot k}{3} + i\sin\dfrac{135° + 360° \cdot k}{3}\right)$ for $k = 0, 1, 2$.

$z_0 = \left(\sqrt{2}\right)^{1/3}\left(\cos\dfrac{135°}{3} + i\sin\dfrac{135°}{3}\right)$
$= 2^{1/6}(\cos 45° + i\sin 45°)$

$z_1 = \left(\sqrt{2}\right)^{1/3}\left(\cos\dfrac{495°}{3} + i\sin\dfrac{495°}{3}\right)$p
$= 2^{1/6}(\cos 165° + i\sin 165°)$

$z_2 = \left(\sqrt{2}\right)^{1/3}\left(\cos\dfrac{855°}{3} + i\sin\dfrac{855°}{3}\right)$
$= 2^{1/6}(\cos 285° + i\sin 285°)$

9. The polar form for 1 is $1 + 0i = \cos 0° + i\sin 0°$
$z_k = \cos\dfrac{0° + 360° \cdot k}{4} + i\sin\dfrac{0° + 360° \cdot k}{4}$ for $k = 0, 1, 2, 3$.

$z_0 = \cos\dfrac{0°}{4} + i\sin\dfrac{0°}{4} = \cos 0° + i\sin 0°$

$z_1 = \cos\dfrac{360°}{4} + i\sin\dfrac{360°}{4} = \cos 90° + i\sin 90°$

$z_2 = \cos\dfrac{720°}{4} + i\sin\dfrac{720°}{4}$
$= \cos 180° + i\sin 180°$

$z_3 = \cos\dfrac{1080°}{4} + i\sin\dfrac{1080°}{4}$
$= \cos 270° + i\sin 270°$

7.7 Concepts and Vocabulary

1. If $z = a + bi$, then $|z| = \sqrt{a^2 + b^2}$ and $\arg z = \theta = \tan^{-1}\left(\dfrac{b}{a}\right)$.

3. To multiply two complex numbers in polar form, multiply their moduli and add their arguments.

5. True.
$z^{-1} = r^{-1}(\cos(-\theta) + i\sin(-\theta))$
$= \dfrac{1}{r}(\cos\theta - i\sin\theta)$

7. False. The complex number $z = a + bi$ is written in rectangular form.

9–15.

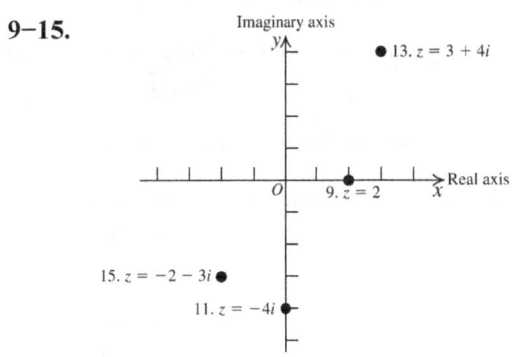

9. $|2| = 2$

11. $|-4i| = \sqrt{(-4)^2} = 4$

13. $|3 + 4i| = \sqrt{3^2 + 4^2} = 5$

15. $|-2 - 3i| = \sqrt{(-2)^2 + (-3)^2} = \sqrt{13}$

17. $r = \sqrt{1^2 + \left(\sqrt{3}\right)^2} = 2; \theta = \tan^{-1}\left(\dfrac{\sqrt{3}}{1}\right) = 60°$ or $\theta = 240°$. $1 + \sqrt{3}i$ is in Quadrant I, so $\theta = 60°$
$1 + \sqrt{3}i = 2(\cos 60° + i\sin 60°)$

19. $r = \sqrt{(-1)^2 + 1^2} = \sqrt{2}; \theta = \tan^{-1}(-1) = 135°$ or $\theta = 315°$. Note that $-1+i$ lies in Quadrant II, so $\theta = 135°$
$-1 + i = \sqrt{2}(\cos 135° + i \sin 135°)$

21. $r = \sqrt{0^2 + 1^2} = 1; \tan\theta = \dfrac{1}{0} \Rightarrow \tan\theta$ is undefined $\Rightarrow \theta = 90°$ or $\theta = 270°$
$i = \cos 90° + i \sin 90°$

23. $r = \sqrt{1^2 + 0^2} = 1; \theta = \tan^{-1} 0 = 0°$ or $\theta = 180°$
$1 = \cos 0° + i \sin 0°$

25. $r = \sqrt{3^2 + (-3)^2} = 3\sqrt{2}$
$\theta = \tan^{-1}\left(-\dfrac{3}{3}\right) = \tan^{-1}(-1) = 135°$ or $\theta = 315°$
$3 - 3i$ is in Quadrant IV, so $\theta = 315°$
$3 - 3i = 3\sqrt{2}(\cos 315° + i \sin 315°)$

27. $r = \sqrt{2^2 + \left(-2\sqrt{3}\right)^2} = 4$
$\theta = \tan^{-1}\left(-\dfrac{2\sqrt{3}}{2}\right) = 120°$ or $\theta = 300°$.
$2 - 2\sqrt{3}i$ is in Quadrant IV, so $\theta = 300°$.
$2 - 2\sqrt{3}i = 4(\cos 300° + i \sin 300°)$

29. $2(\cos 60° + i \sin 60°) = 2\cos 60° + 2i \sin 60°$
$= 1 + \sqrt{3}i$

31. $3(\cos \pi + i \sin \pi) = 3\cos\pi + 3i\sin\pi = -3$

33. $5(\cos 240° + i \sin 240°) = 5\cos 240° + 5i\sin 240°$
$= -\dfrac{5}{2} - \dfrac{5\sqrt{3}}{2}i$

35. $8(\cos 0° + i \sin 0°) = 8\cos 0° + 8i \sin 0° = 8$

37. $6\left(\cos\dfrac{5\pi}{6} + i \sin\dfrac{5\pi}{6}\right) = 6\cos\dfrac{5\pi}{6} + 6i\sin\dfrac{5\pi}{6}$
$= -3\sqrt{3} + 3i$

39. $3\left(\cos\left(-\dfrac{\pi}{3}\right) + i \sin\left(-\dfrac{\pi}{3}\right)\right)$
$= 3\cos\left(-\dfrac{\pi}{3}\right) + 3i\sin\left(-\dfrac{\pi}{3}\right) = \dfrac{3}{2} - \dfrac{3\sqrt{3}}{2}i$

41. $z_1 = 4(\cos 75° + i \sin 75°)$
$z_2 = 2(\cos 15° + i \sin 15°)$
$z_1 z_2 = 4 \cdot 2(\cos(75° + 15°) + i \sin(75° + 15°))$
$= 8(\cos 90° + i \sin 90°) = 8i$

43. $z_1 = 5(\cos 240° + i \sin 240°)$
$z_2 = 2(\cos 60° + i \sin 60°)$
$z_1 z_2 = 5 \cdot 2(\cos(240° + 60°) + i \sin(240° + 60°))$
$= 10(\cos 300° + i \sin 300°)$
$= 10\left(\dfrac{1}{2} - \dfrac{\sqrt{3}}{2}i\right) = 5 - 5\sqrt{3}i$

45. $z_1 = 3(\cos 40° + i \sin 40°)$
$z_2 = 5(\cos 20° + i \sin 20°)$
$z_1 z_2 = 3 \cdot 5(\cos(40° + 20°) + i \sin(40° + 20°))$
$= 15(\cos 60° + i \sin 60°)$
$= 15\left(\dfrac{1}{2} + \dfrac{\sqrt{3}}{2}i\right) = \dfrac{15}{2} + \dfrac{15\sqrt{3}}{2}i$

47. $z_1 = 3(\cos 140° + i \sin 140°)$
$z_2 = 2(\cos 100° + i \sin 100°)$
$z_1 z_2 = 3 \cdot 2(\cos(140° + 100°) + i \sin(140° + 100°))$
$= 6(\cos 240° + i \sin 240°)$
$= 6\left(-\dfrac{1}{2} - \dfrac{\sqrt{3}}{2}i\right) = -3 - 3\sqrt{3}i$

49. $z_1 = 8(\cos 72° + i \sin 72°),$
$z_2 = 4(\cos 12° + i \sin 12°)$
$\dfrac{z_1}{z_2} = \dfrac{8}{4}\left(\cos(72° - 12°) + i \sin(72° - 12°)\right)$
$= 2(\cos 60° + i \sin 60°)$
$= 2\left(\dfrac{1}{2} + \dfrac{\sqrt{3}}{2}i\right) = 1 + \sqrt{3}i$

51. $z_1 = 5(\cos 102° + i \sin 102°)$
$z_2 = 2.5(\cos 12° + i \sin 12°)$
$\dfrac{z_1}{z_2} = \dfrac{5}{2.5}\left(\cos(102° - 12°) + i \sin(102° - 12°)\right)$
$= 2(\cos 90° + i \sin 90°) = 2(0 + i) = 2i$

53. $z_1 = 6(\cos 70° + i \sin 70°)$
$z_2 = 3(\cos 190° + i \sin 190°)$
$\dfrac{z_1}{z_2} = \dfrac{6}{3}\left(\cos(70° - 190°) + i \sin(70° - 190°)\right)$
$= 2(\cos(-120°) + i \sin(-120°))$
$= 2(\cos 240° + i \sin 240°)$
$= 2\left(-\dfrac{1}{2} - \dfrac{\sqrt{3}}{2}i\right) = -1 - \sqrt{3}i$

55. $z_1 = 3\left(\cos\dfrac{\pi}{12} + i\sin\dfrac{\pi}{12}\right)$

$z_2 = 2\left(\cos\dfrac{5\pi}{12} + i\sin\dfrac{5\pi}{12}\right)$

$\dfrac{z_1}{z_2} = \dfrac{3}{2}\left(\cos\left(\dfrac{\pi}{12} - \dfrac{5\pi}{12}\right) + i\sin\left(\dfrac{\pi}{12} - \dfrac{5\pi}{12}\right)\right)$

$= \dfrac{3}{2}\left(\cos\left(-\dfrac{\pi}{3}\right) + i\sin\left(-\dfrac{\pi}{3}\right)\right)$

$= \dfrac{3}{2}\left(\cos\dfrac{5\pi}{3} + i\sin\dfrac{5\pi}{3}\right)$

$= \dfrac{3}{2}\left(\dfrac{1}{2} - \dfrac{\sqrt{3}}{2}i\right) = \dfrac{3}{4} - \dfrac{3\sqrt{3}}{4}i$

57. $\left[2\left(\cos\dfrac{\pi}{3} + i\sin\dfrac{\pi}{3}\right)\right]^{12}$

$= 2^{12}\left(\cos 12\left(\dfrac{\pi}{3}\right) + i\sin 12\left(\dfrac{\pi}{3}\right)\right)$

$= 2^{12}(\cos 4\pi + i\sin 4\pi) = 2^{12}$

59. $\left[2\left(\cos\left(-\dfrac{3\pi}{4}\right) + i\sin\left(-\dfrac{3\pi}{4}\right)\right)\right]^6$

$= 2^6\left(\cos 6\left(-\dfrac{3\pi}{4}\right) + i\sin 6\left(-\dfrac{3\pi}{4}\right)\right)$

$= 2^6\left(\cos\left(-\dfrac{9\pi}{2}\right) + i\sin\left(-\dfrac{9\pi}{2}\right)\right)$

$= -2^6 i = -64i$

61. $\left[2\left(\cos\left(\dfrac{3\pi}{4}\right) + i\sin\left(\dfrac{3\pi}{4}\right)\right)\right]^{-6}$

$= 2^{-6}\left(\cos\left((-6)\left(\dfrac{3\pi}{4}\right)\right) + i\sin\left((-6)\left(\dfrac{3\pi}{4}\right)\right)\right)$

$= 2^{-6}\left(\cos\left(-\dfrac{9\pi}{2}\right) + i\sin\left(-\dfrac{9\pi}{2}\right)\right)$

$= -2^{-6}i = -\dfrac{1}{64}i$

63. $\left[2\left(\cos\left(-\dfrac{\pi}{4}\right) + i\sin\left(-\dfrac{\pi}{4}\right)\right)\right]^{-10}$

$= 2^{-10}\left(\cos\left(-10\left(-\dfrac{\pi}{4}\right)\right) + i\sin\left(-10\left(-\dfrac{\pi}{4}\right)\right)\right)$

$= 2^{-10}\left(\cos\dfrac{5\pi}{2} + i\sin\dfrac{5\pi}{2}\right) = \dfrac{1}{2^{10}}i$

65. First convert $1 - i$ to polar form:

$r = \sqrt{1^2 + (-1)^2} = \sqrt{2}; \theta = \tan^{-1}(-1) = 315°$

(Note that $1 - i$ lies in Quadrant III.)

So,

$(1-i)^{12} = \left[\sqrt{2}(\cos 315° + i\sin 315°)\right]^{12}$

$= (\sqrt{2})^{12}(\cos(12 \cdot 315°) + i\sin(12 \cdot 315°))$

$= 2^6(\cos° 3780 + i\sin 3780°) = -64$

67. $i^{25} = i$

69. First write $\dfrac{1}{2} - \dfrac{\sqrt{3}}{2}i$ in polar form:

$r = \sqrt{\left(\dfrac{1}{2}\right)^2 + \left(-\dfrac{\sqrt{3}}{2}\right)^2} = 1$

$\theta = \tan^{-1}(-\sqrt{3}) = 300°$

(Note that $\dfrac{1}{2} - \dfrac{\sqrt{3}}{2}i$ lies in Quadrant III.) So

$\left(\dfrac{1}{2} - \dfrac{\sqrt{3}}{2}i\right)^{-8} = (\cos 300° + i\sin 300°)^{-8}$

$= \cos(-8 \cdot 300°) + i\sin(-8 \cdot 300°)$

$= \cos(-2400°) + i\sin(-2400°)$

$= -\dfrac{1}{2} + \dfrac{\sqrt{3}}{2}i$

71. The polar form for 64 is

$64 + 0i = 64(\cos 0° + i\sin 0°)$

$z_k = 64^{1/3}\left[\cos\dfrac{0° + 360° \cdot k}{3} + i\sin\dfrac{0° + 360° \cdot k}{3}\right]$

for $k = 0, 1, 2$.

$z_0 = 64^{1/3}\left(\cos\dfrac{0°}{3} + i\sin\dfrac{0°}{3}\right)$

$= 4(\cos 0° + i\sin 0°)$

$z_1 = 4\left(\cos\dfrac{360°}{3} + i\sin\dfrac{360°}{3}\right)$

$= 4(\cos 120° + i\sin 120°)$

$z_2 = 4\left(\cos\dfrac{720°}{3} + i\sin\dfrac{720°}{3}\right)$

$= 4(\cos 240° + i\sin 240°)$

73. The polar form for i is

$0 + i = \cos 90° + i\sin 90°$

$z_k = \cos\dfrac{90° + 360° \cdot k}{2} + i\sin\dfrac{90° + 360° \cdot k}{2}$

for $k = 0, 1$.

$z_0 = \cos\dfrac{90°}{2} + i\sin\dfrac{90°}{2} = \cos 45° + i\sin 45°$

$z_1 = \cos\dfrac{450°}{2} + i\sin\dfrac{450°}{2}$

$= \cos 225° + i\sin 225°$

75. The polar form for -1 is
$-1+0i = \cos 180° + i \sin 180°$
$$z_k = \cos\frac{180° + 360° \cdot k}{6} + i\sin\frac{180° + 360° \cdot k}{6}$$ for
$k = 0, 1, 2, 3, 4, 5$.

$z_0 = \cos\frac{180°}{6} + i\sin\frac{180°}{6} = \cos 30° + i\sin 30°$

$z_1 = \cos\frac{540°}{6} + i\sin\frac{540°}{6} = \cos 90° + i\sin 90°$

$z_2 = \cos\frac{900°}{6} + i\sin\frac{900°}{6}$
$= \cos 150° + i\sin 150°$

$z_3 = \cos\frac{1260°}{6} + i\sin\frac{1260°}{6}$
$= \cos 210° + i\sin 210°$

$z_4 = \cos\frac{1620°}{6} + i\sin\frac{1620°}{6}$
$= \cos 270° + i\sin 270°$

$z_5 = \cos\frac{1980°}{6} + i\sin\frac{1980°}{6}$
$= \cos 330° + i\sin 330°$

77. Write $1 - \sqrt{3}i$ in polar form:
$r = \sqrt{1^2 + \left(-\sqrt{3}\right)^2} = 2; \theta = \tan^{-1}\left(-\sqrt{3}\right) = 300°$
(Note that $1 - \sqrt{3}i$ lies in Quadrant III.)

$z_k = 2^{1/2}\left(\cos\frac{300° + 360° \cdot k}{2} + i\sin\frac{300° + 360° \cdot k}{2}\right)$

for $k = 0, 1$.

$z_0 = \sqrt{2}\left(\cos\frac{300°}{2} + i\sin\frac{300°}{2}\right)$
$= \sqrt{2}\left(\cos 150° + i\sin 150°\right)$

$z_1 = \sqrt{2}\left(\cos\frac{660°}{2} + i\sin\frac{660°}{2}\right)$
$= \sqrt{2}\left(\cos 330° + i\sin 330°\right)$

7.7 Applying the Concepts

79. By DeMoivre's theorem,
$\cos 3\theta + i\sin 3\theta = (\cos\theta + i\sin\theta)^3$.
Expanding, we have
$(\cos\theta + i\sin\theta)^3$
$= \cos^3\theta + 3\cos^2\theta\, i\sin\theta - 3\cos\theta\sin^2\theta - i\sin^3\theta$

a. To solve for $\sin 3\theta$, gather the even-powered cosine terms and simplify:
$i\sin 3\theta = 3i\cos^2\theta\sin\theta - i\sin^3\theta$
$= i\sin\theta(3\cos^2\theta - \sin^2\theta) \Rightarrow$

$\sin 3\theta = \sin\theta(3\cos^2\theta - \sin^2\theta)$
$= \sin\theta(3(1 - \sin^2\theta) - \sin^2\theta)$
$= \sin\theta(3 - 4\sin^2\theta)$
$= 3\sin\theta - 4\sin^3\theta$

b. To solve for $\cos 3\theta$, gather the odd-powered cosine terms and simplify:
$\cos 3\theta = \cos^3\theta - 3\cos\theta\sin^2\theta$
$= \cos\theta(\cos^2\theta - 3\sin^2\theta) \Rightarrow$
$\cos 3\theta = \cos\theta(\cos^2\theta - 3\sin^2\theta)$
$= \cos\theta(\cos^2\theta - 3(1 - \cos^2\theta))$
$= \cos\theta(4\cos^2\theta - 3)$
$= 4\cos^3\theta - 3\cos\theta$

81. $I = \dfrac{V}{Z} = \dfrac{120(\cos 60° + i\sin 60°)}{8(\cos 30° + i\sin 30°)}$
$= 15(\cos 30° + i\sin 30°)$

83. To find $z_1 + z_2$, first convert each to rectangular form. Then convert the sum back to polar form:
$z_1 = 16(\cos 180° + i\sin 180°) = -16$
$z_2 = 2(\cos 150° + i\sin 150°) = -\sqrt{3} + i$
$z_1 + z_2 = -16 - \sqrt{3} + i$
$r = \sqrt{\left(-16 - \sqrt{3}\right)^2 + 1^2} \approx 17.76$
$\theta = \tan^{-1}\left(\dfrac{1}{-16 - \sqrt{3}}\right) \approx -3.2° = 176.8°$.
So, $z_1 + z_2 \approx 17.76(\cos(176.8°) + i\sin(176.8°))$
$z_1 z_2 = 32(\cos 330° + i\sin 330°)$
$\dfrac{z_1 z_2}{z_1 + z_2} \approx \dfrac{32(\cos 330° + i\sin 330°)}{17.76(\cos(176.8°) + i\sin(176.8°))}$
$\approx 1.8(\cos 153.2° + i\sin 153.2°)$

7.7 Beyond the Basics

85. First convert each factor to polar form:
$z_1 = 1 - \sqrt{3}i = 2(\cos 300° + i \sin 300°)$
$z_2 = -2 + 2i = 2\sqrt{2}(\cos 135° + i \sin 135°)$
Now use DeMoivre's theorem to raise each factor to the indicated power:
$z_1^{10} = (1 - \sqrt{3}i)^{10} = [2(\cos 300° + i \sin 300°)]^{10} = 2^{10}(\cos 3000° + i \sin 3000°)$
$z_2^{-6} = (-2 + 2i)^{-6} = [2\sqrt{2}(\cos 135° + i \sin 135°)]^{-6} = 2^{-9}(\cos(-810°) + i \sin(-810°))$
Multiply the results:
$z_1^{10} z_2^{-6} = 2^{10}(\cos 3000° + i \sin 3000°) \cdot 2^{-9}(\cos(-810°) + i \sin(-810°))$
$= 2(\cos 2190° + i \sin 2190°) = 2(\cos 30° + i \sin 30°)$

87. $\left(\sin \dfrac{\pi}{6} + i \cos \dfrac{\pi}{6}\right)^{10} = (\sin 30° + i \cos 30°)^{10} = (\cos 60° + i \sin 60°)^{10}$
$= \cos 600° + i \sin 600° = \cos 240° + i \sin 240°$

89. $\left(\sin \dfrac{5\pi}{3} + i \cos \dfrac{5\pi}{3}\right)^{-8} = (\sin 300° + i \cos 300°)^{-8} = (\cos 150° + i \sin 150°)^{-8}$
$= \cos(-1200°) + i \sin(-1200°) = \cos 240° + i \sin 240°$

91. $z^4 = 1 \Rightarrow z^4 = \cos 0° + i \sin 0° \Rightarrow z = (\cos 0° + i \sin 0°)^{1/4}$
$z_k = \cos \dfrac{0° + 360° \cdot k}{4} + i \sin \dfrac{0° + 360° \cdot k}{4}$, for $k = 0, 1, 2, 3$
$z_0 = \cos \dfrac{0°}{4} + i \sin \dfrac{0°}{4} = \cos 0° + i \sin 0°$
$z_1 = \cos \dfrac{360°}{4} + i \sin \dfrac{360°}{4} = \cos 90° + i \sin 90°$
$z_2 = \cos \dfrac{720°}{4} + i \sin \dfrac{720°}{4} = \cos 180° + i \sin 180°$
$z_3 = \cos \dfrac{1080°}{4} + i \sin \dfrac{1080°}{4} = \cos 270° + i \sin 270°$

93. $z^3 = 1 + i \Rightarrow z^3 = \sqrt{2}(\cos 45° + i \sin 45°) \Rightarrow z = \left(\sqrt{2}(\cos 45° + i \sin 45°)\right)^{1/3}$
$z_k = 2^{1/6}\left(\cos \dfrac{45° + 360° \cdot k}{3} + i \sin \dfrac{45° + 360° \cdot k}{3}\right)$, for $k = 0, 1, 2$
$z_0 = 2^{1/6}\left(\cos \dfrac{45°}{3} + i \sin \dfrac{45°}{3}\right) = 2^{1/6}(\cos 15° + i \sin 15°)$
$z_1 = 2^{1/6}\left(\cos \dfrac{405°}{3} + i \sin \dfrac{405°}{3}\right) = 2^{1/6}(\cos 135° + i \sin 135°)$
$z_2 = 2^{1/6}\left(\cos \dfrac{765°}{3} + i \sin \dfrac{765°}{3}\right) = 2^{1/6}(\cos 255° + i \sin 255°)$

95. $z^4 - 2z^2 + 4 = 0$

Let $u = z^2$. Then $z^4 - 2z^2 + 4 = 0 \Rightarrow u^2 - 2u + 4 = 0 \Rightarrow u^2 - 2u + 1 = -3 \Rightarrow (u-1)^2 = -3 \Rightarrow$

$u - 1 = \pm\sqrt{3}i \Rightarrow u = 1 \pm \sqrt{3}i$

Now find the roots of $u = z^2$.

$u = z^2 = 1 + \sqrt{3}i \Rightarrow z^2 = 2(\cos 60° + i\sin 60°)$

$z = 2^{1/2}(\cos 60° + i\sin 60°)^{1/2}$

$z_k = \sqrt{2}\left(\cos\dfrac{60° + 360° \cdot k}{2} + i\sin\dfrac{60° + 360° \cdot k}{2}\right)$, for $k = 0, 1$

$z_0 = \sqrt{2}\left(\cos\dfrac{60°}{2} + i\sin\dfrac{60°}{2}\right) = \sqrt{2}(\cos 30° + i\sin 30°) = \sqrt{2}\left(\dfrac{\sqrt{3}}{2} + \dfrac{1}{2}i\right) = \dfrac{\sqrt{6}}{2} + \dfrac{\sqrt{2}}{2}i$

$z_1 = \sqrt{2}\left(\cos\dfrac{60° + 360°}{2} + i\sin\dfrac{60° + 360°}{2}\right) = \sqrt{2}(\cos 210° + i\sin 210°) = \sqrt{2}\left(-\dfrac{\sqrt{3}}{2} - \dfrac{1}{2}i\right) = -\dfrac{\sqrt{6}}{2} - \dfrac{\sqrt{2}}{2}i$

$u = z^2 = 1 - \sqrt{3}i \Rightarrow z^2 = 2(\cos(-60°) + i\sin(-60°))$

$z = 2^{1/2}(\cos(-60°) + i\sin(-60°))^{1/2}$

$z_k = \sqrt{2}\left(\cos\dfrac{-60° + 360° \cdot k}{2} + i\sin\dfrac{-60° + 360° \cdot k}{2}\right)$, for $k = 0, 1$

$z_0 = \sqrt{2}\left(\cos\dfrac{-60°}{2} + i\sin\dfrac{-60°}{2}\right) = \sqrt{2}(\cos(-30°) + i\sin(-30°)) = \sqrt{2}\left(\dfrac{\sqrt{3}}{2} - \dfrac{1}{2}i\right) = \dfrac{\sqrt{6}}{2} - \dfrac{\sqrt{2}}{2}i$

$z_1 = \sqrt{2}\left(\cos\dfrac{-60° + 360°}{2} + i\sin\dfrac{-60° + 360°}{2}\right) = \sqrt{2}(\cos 150° + i\sin 150°) = \sqrt{2}\left(-\dfrac{\sqrt{3}}{2} + \dfrac{1}{2}i\right) = -\dfrac{\sqrt{6}}{2} + \dfrac{\sqrt{2}}{2}i$

97. If $z = \cos\theta + i\sin\theta, z \neq 0$, then $\dfrac{1}{z} = z^{-1} = (\cos\theta + i\sin\theta)^{-1} = \cos(-\theta) + i\sin(-\theta)$

$\cos(-\theta) = \cos\theta$ and $\sin(-\theta) = -\sin\theta$, so this becomes $\cos\theta - i\sin\theta$.

99. $z - \dfrac{1}{z} = (\cos\theta + i\sin\theta) - (\cos\theta - i\sin\theta) = 2i\sin\theta$

101. $z^n - \dfrac{1}{z^n} = (\cos\theta + i\sin\theta)^n - (\cos\theta + i\sin\theta)^{-n} = (\cos n\theta + i\sin n\theta) - (\cos(-n\theta) + i\sin(-n\theta))$

$= (\cos n\theta + i\sin n\theta) - (\cos n\theta - i\sin n\theta) = 2i\sin n\theta$

103. Let $z_1 = r_1(\cos\theta_1 + i\sin\theta_1)$ and $z_2 = r_2(\cos\theta_2 + i\sin\theta_2)$. Then

$z_1 z_2 = r_1(\cos\theta_1 + i\sin\theta_1) \cdot r_2(\cos\theta_2 + i\sin\theta_2)$

$= r_1 r_2(\cos\theta_1 \cos\theta_2 + i\cos\theta_1 \sin\theta_2 + i\sin\theta_1 \cos\theta_2 - \sin\theta_1 \sin\theta_2)$

$= r_1 r_2\left[(\cos\theta_1 \cos\theta_2 - \sin\theta_1 \sin\theta_2) + i(\cos\theta_1 \sin\theta_2 + \sin\theta_1 \cos\theta_2)\right]$

$= r_1 r_2\left[\cos(\theta_1 + \theta_2) + i\sin(\theta_1 + \theta_2)\right]$

105. $(3 + 2i)(5 + i) = 15 + 13i - 2 = 13 + 13i$

We want to prove that $\tan^{-1}\left(\dfrac{2}{3}\right) + \tan^{-1}\left(\dfrac{1}{5}\right) = \dfrac{\pi}{4}$. Since complex multiplication adds arguments, multiply the complex numbers with arguments of $\tan^{-1}\left(\dfrac{2}{3}\right)$ and $\tan^{-1}\left(\dfrac{1}{5}\right)$, namely, $(3 + 2i)(5 + i)$. The product has an argument of $\tan^{-1}\left(\dfrac{13}{13}\right) = \tan^{-1}(1) = \dfrac{\pi}{4}$.

107. DeMoivre's theorem states that for any integer n, $z^n = r^n(\cos n\theta + i\sin n\theta)$, so the argument is $n\tan^{-1}\theta$. Thus, the argument of $(2+3i)^4$ is $4\tan^{-1}\left(\frac{3}{2}\right)$.

$$(2+3i)^4 = \left((2+3i)^2\right)^2 = (4+12i-9)^2$$
$$= (-5+12i)^2 = 25-120i-144$$
$$= -119-120i$$

The argument of $(-119-120i)$ is $\tan^{-1}\left(\frac{120}{119}\right)$. Since complex division subtracts arguments, divide the complex numbers to find the difference of the arguments.

$$4\tan^{-1}\left(\frac{3}{2}\right) - \tan^{-1}\left(\frac{120}{119}\right) = \arg\left(\frac{(2+3i)^4}{-119-120i}\right)$$
$$= \arg(-1+0i)$$
$$= \tan^{-1}(0) = \pi$$

109.
$x^3 + x^2 + x + 1 = 0 \Rightarrow$
$(x-1)(x^3 + x^2 + x + 1) = 0 \Rightarrow$
$x^4 - 1 = 0 \Rightarrow (x^2+1)(x^2-1) = 0 \Rightarrow$
$x = \pm i, \pm 1$

We reject the root $x = 1$ since it was introduced when the equation was multiplied by $x - 1$.
Solution set: $\{-i, i, -1\}$

7.7 Critical Thinking/Discussion/Writing

111. The exercises that are false are incorrect usages of DeMoivre's theorem.
  a. True   b. False
  c. True   d. False
  e. True   f. True
  g. True   h. True
  i. False   j. True

7.7 Getting Ready for the Next Section

113. $-2x + 9y = 5$, $y = 3$
Substituting, we have
$-2x + 9(3) = 5 \Rightarrow -2x + 27 = 5 \Rightarrow$
$-2x = -22 \Rightarrow x = 11$

115. $23x - 14y = -5$, $x = 1$
Substituting, we have
$23(1) - 14y = -5 \Rightarrow 23 - 14y = -5 \Rightarrow$
$-14y = -28 \Rightarrow y = 2$

117. Rearrange the equation to put it in slope-intercept form.
$$15x + 5y = 2 \Rightarrow 5y = -15x + 2 \Rightarrow y = -3x + \frac{2}{5}$$
The slope is -3.

119. Parallel lines have the same slope. Rearrange the equation to put it in slope-intercept form.
$$x + 2y = 12 \Rightarrow 2y = -x + 12 \Rightarrow y = -\frac{1}{2}x + 6$$
The slope is $-\frac{1}{2}$. Now use the point-slope form to write the equation of the parallel line.
$$y - 3 = -\frac{1}{2}(x-3) \Rightarrow 2y - 6 = -x + 3 \Rightarrow$$
$$x + 2y = 9$$

Chapter 7 Review Exercises

Building Skills

1. $B = 90° - 30° = 60°$
$\sin A = \frac{a}{c} \Rightarrow \sin 30° = \frac{6}{c} \Rightarrow c = 12$
$b = \sqrt{c^2 - a^2} = \sqrt{12^2 - 6^2} = \sqrt{108} = 6\sqrt{3}$
≈ 10.4

3. $B = 90° - 37° = 53°$
$\sin B = \frac{b}{c} \Rightarrow \sin 53° = \frac{4}{c} \Rightarrow c = 5.0$
$a = \sqrt{c^2 - b^2} = \sqrt{5.0^2 - 4^2} \approx 3.0$

5. $B = 90° - 40° = 50°$
$\sin A = \frac{a}{c} \Rightarrow \sin 40° = \frac{a}{12} \Rightarrow a \approx 7.7$
$b = \sqrt{c^2 - a^2} = \sqrt{5.0^2 - 4^2} \approx 9.2$

7. $c = \sqrt{a^2 + b^2} = \sqrt{3^2 + 5^2} = \sqrt{34} \approx 5.8$
$B = \tan^{-1}\left(\frac{b}{a}\right) = \tan^{-1}\left(\frac{5}{3}\right) \approx 59.0°$
$A \approx 90° - 59.0° = 31.0°$

9. $b = \sqrt{c^2 - a^2} = \sqrt{6^2 - 4^2} = \sqrt{20} \approx 4.5$
$A = \sin^{-1}\left(\frac{a}{c}\right) = \sin^{-1}\left(\frac{4}{6}\right) \approx 41.8°$
$B \approx 90° - 41.8° = 48.2°$

Use this triangle to help solve exercises 11–18.

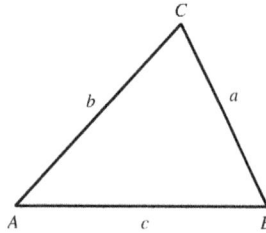

11. Given: $A = 40°, B = 35°, c = 100$ – an ASA case. $C = 180° - (40° + 35°) = 105°$
$$\frac{a}{\sin A} = \frac{c}{\sin C} \Rightarrow \frac{a}{\sin 40°} = \frac{100}{\sin 105°} \Rightarrow a \approx 66.5$$
$$\frac{b}{\sin B} = \frac{c}{\sin C} \Rightarrow \frac{b}{\sin 35°} = \frac{100}{\sin 105°} \Rightarrow b \approx 59.4$$

13. Given: $A = 45°, a = 25, b = 75$ – an SSA case. A is an acute angle, so examine h:
$h = b \sin A = 75 \sin 45° \approx 43.3$
$a < h$, so no triangle exists.

15. Given: $A = 48.5°, C = 57.3°, b = 47.3$ – an ASA case. $B = 180° - (48.5° + 57.3°) = 74.2°$
$$\frac{b}{\sin B} = \frac{a}{\sin A} \Rightarrow \frac{47.3}{\sin 74.2°} = \frac{a}{\sin 48.5°} \Rightarrow$$
$a \approx 36.8$
$$\frac{b}{\sin B} = \frac{c}{\sin A} \Rightarrow \frac{47.3}{\sin 74.2°} = \frac{c}{\sin 57.3°} \Rightarrow$$
$c \approx 41.4$

17. Given: $A = 65.2°, a = 21.3, b = 19$ – an SSA case. A is an acute angle, so examine h:
$h = b \sin A = 19 \sin 65.2° \approx 17.2$. $h < b < a$, so there is one triangle. $\frac{\sin 65.2°}{21.3} = \frac{\sin B}{19} \Rightarrow$
$B = \sin^{-1}\left(\frac{19 \sin 65.2°}{21.3}\right) \approx 54.1°$
$C \approx 180° - (65.2° + 54.1°) \approx 60.7°$
$\frac{21.3}{\sin 65.2°} = \frac{c}{\sin 60.7°} \Rightarrow c \approx 20.5$

19. Answers will vary.
  (i) $h = c \sin B = 20 \sin 60° = 10\sqrt{3}$. For C to have two possible values, $h < b < c$, so choose a value of b such that $10\sqrt{3} < b < 20$.
  (ii) C has exactly one value if $b = 10\sqrt{3}$.
  (iii) C has no value if $b < c \sin B \Rightarrow b < 10\sqrt{3}$.

21. Given: $a = 60, b = 90, c = 125$ – an SSS case.
$60^2 = 90^2 + 125^2 - 2(90)(125) \cos A \Rightarrow$
$A \approx 26.6°$
$90^2 = 60^2 + 125^2 - 2(60)(125) \cos B \Rightarrow$
$B \approx 42.1°$
$C = 180° - (26.6° + 42.1°) \approx 111.3°$

23. Given: $a = 40, c = 38, B = 80°$ – an SAS case.
$b^2 = 40^2 + 38^2 - 2(40)(38) \cos 80° \Rightarrow b \approx 50.2$
$\frac{\sin 80°}{50.2} = \frac{\sin A}{40} \Rightarrow$
$A = \sin^{-1}\left(\frac{40 \sin 80°}{50.2}\right) \approx 51.7°$
$C \approx 180° - (80° + 51.7°) \approx 48.3°$

25. Given: $a = 2.6, b = 3.7, c = 4.8$ – an SSS case.
$2.6^2 = 3.7^2 + 4.8^2 - 2(3.7)(4.8) \cos A \Rightarrow$
$A \approx 32.462° \approx 32.5°$
$\frac{\sin 32.462°}{2.6} = \frac{\sin B}{3.7} \Rightarrow B \approx 49.8°$
$C \approx 180° - (32.5° + 49.8°) \approx 97.7°$

27. Given: $a = 12, b = 7, C = 130°$ – an SAS case.
$c^2 = 12^2 + 7^2 - 2(12)(7) \cos 130° \Rightarrow$
$c \approx 17.349 \approx 17.3$
$\frac{\sin A}{12} = \frac{\sin 130°}{17.349} \Rightarrow A \approx 32.0°$
$B \approx 180° - (130° + 32.0°) \approx 18.0°$

29. $s = \frac{5 + 7 + 10}{2} = 11$
$k = \sqrt{11(11-5)(11-7)(11-10)} \approx 16 \text{ m}^2$

31. $k = \frac{1}{2} bc \sin A = \frac{1}{2}(6)(4) \sin 65° \approx 11 \text{ ft}^2$

33. $\mathbf{v} = \langle 2-3, 7-5 \rangle = \langle -1, 2 \rangle = -\mathbf{i} + 2\mathbf{j}$

35. $5\mathbf{v} = 5\langle -2, 3 \rangle = \langle -10, 15 \rangle$

37. $3\mathbf{v} - 2\mathbf{w} = 3\langle -2, 3 \rangle - 2\langle 5, -6 \rangle = \langle -16, 21 \rangle$

39. $\mathbf{v} = \mathbf{i} + \mathbf{j} \Rightarrow \|\mathbf{v}\| = \sqrt{1^2 + 1^2} = \sqrt{2}$
$\mathbf{u} = \frac{1}{\sqrt{2}}(\mathbf{i} + \mathbf{j}) = \frac{\sqrt{2}}{2}\mathbf{i} + \frac{\sqrt{2}}{2}\mathbf{j}$

41. $\mathbf{v} = \langle 3, -5 \rangle \Rightarrow \|\mathbf{v}\| = \sqrt{3^2 + (-5)^2} = \sqrt{34}$
$\mathbf{u} = \frac{1}{\sqrt{34}}\langle 3, -5 \rangle = \left\langle \frac{3\sqrt{34}}{34}, -\frac{5\sqrt{34}}{34} \right\rangle$

43. Given: $\|\mathbf{v}\| = 6, \theta = 30°$. Then
$\mathbf{v} = 6(\cos 30°\mathbf{i} + \sin 30°\mathbf{j})$
$= 6\left(\dfrac{\sqrt{3}}{2}\mathbf{i} + \dfrac{1}{2}\mathbf{j}\right) = 3\sqrt{3}\mathbf{i} + 3\mathbf{j}$

45. Given: $\|\mathbf{v}\| = 12, \theta = 225°$. Then
$\mathbf{v} = 12(\cos 225°\mathbf{i} + \sin 225°\mathbf{j})$
$= 12\left(-\dfrac{\sqrt{2}}{2}\mathbf{i} - \dfrac{\sqrt{2}}{2}\mathbf{j}\right) = -6\sqrt{2}\mathbf{i} - 6\sqrt{2}\mathbf{j}$

47. $\mathbf{v} \cdot \mathbf{w} = \langle 2, -3 \rangle \cdot \langle 3, 4 \rangle = 2(3) + (-3)(4) = -6$
$\text{proj}_{\mathbf{w}} \mathbf{v} = \dfrac{\mathbf{v} \cdot \mathbf{w}}{\|\mathbf{w}\|^2}\mathbf{w} = \dfrac{-6}{3^2 + 4^2}\langle 3, 4 \rangle$
$= -\dfrac{6}{25}\langle 3, 4 \rangle = \left\langle -\dfrac{18}{25}, -\dfrac{24}{25} \right\rangle$

49. $\mathbf{v} \cdot \mathbf{w} = \langle 2, -5 \rangle \cdot \langle 5, 2 \rangle = 2(5) + (-5)(2) = 0$
$\text{proj}_{\mathbf{w}} \mathbf{v} = \dfrac{\mathbf{v} \cdot \mathbf{w}}{\|\mathbf{w}\|^2}\mathbf{w} = \dfrac{0}{5^2 + 2^2}\langle 5, 0 \rangle$
$= 0\langle 5, 0 \rangle = \langle 0, 0 \rangle$

For exercises 51–54, use the formula
$\mathbf{v} \cdot \mathbf{w} = \|\mathbf{v}\|\|\mathbf{w}\|\cos\theta$ to find the value of θ.

51. Given: $\mathbf{v} = 2\mathbf{i} + 3\mathbf{j}, \mathbf{w} = -\mathbf{i} + 2\mathbf{j}$.
$\|\mathbf{v}\| = \sqrt{2^2 + 3^2} = \sqrt{13}; \|\mathbf{w}\| = \sqrt{(-1)^2 + 2^2} = \sqrt{5}$
$\mathbf{v} \cdot \mathbf{w} = (2)(-1) + (3)(2) = 4$
$\mathbf{v} \cdot \mathbf{w} = \|\mathbf{v}\|\|\mathbf{w}\|\cos\theta \Rightarrow 4 = \sqrt{13}\sqrt{5}\cos\theta \Rightarrow$
$\theta = \cos^{-1}\left(\dfrac{4}{\sqrt{65}}\right) \approx 60.3°$

53. Given: $\mathbf{v} = \langle 1, 1 \rangle, \mathbf{w} = \langle -3, 2 \rangle$.
$\|\mathbf{v}\| = \sqrt{1^2 + 1^2} = \sqrt{2}; \|\mathbf{w}\| = \sqrt{(-3)^2 + 2^2} = \sqrt{13}$
$\mathbf{v} \cdot \mathbf{w} = (1)(-3) + (1)(2) = -1$
$\mathbf{v} \cdot \mathbf{w} = \|\mathbf{v}\|\|\mathbf{w}\|\cos\theta \Rightarrow -1 = \sqrt{2}\sqrt{13}\cos\theta \Rightarrow$
$\theta = \cos^{-1}\left(-\dfrac{1}{\sqrt{26}}\right) \approx 101.3°$

55. $x = 24\cos 30° = 12\sqrt{3}; y = 24\sin 30° = 12$ The rectangular coordinates of $(24, 30°)$ are $(12\sqrt{3}, 12)$

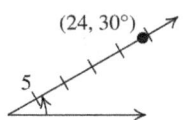

57. $x = -2\cos\left(-\dfrac{\pi}{4}\right) = -2\left(\dfrac{\sqrt{2}}{2}\right) = -\sqrt{2}$
$y = -2\sin\left(-\dfrac{\pi}{4}\right) = -2\left(-\dfrac{\sqrt{2}}{2}\right) = \sqrt{2}$
The rectangular coordinates of $(-2, -\pi/4)$ are $\left(-\sqrt{2}, \sqrt{2}\right)$.

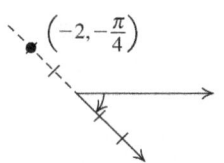

59. $r = \sqrt{(-2)^2 + 2^2} = 2\sqrt{2}$
$\tan\theta = \dfrac{y}{x} = -\dfrac{2}{2} \Rightarrow \theta = \dfrac{3\pi}{4}$
The polar form of $(-2, 2)$ is $\left(2\sqrt{2}, \dfrac{3\pi}{4}\right)$.

61. $r = \sqrt{\left(2\sqrt{3}\right)^2 + (-2)^2} = 4$
$\tan\theta = \dfrac{y}{x} = -\dfrac{2}{2\sqrt{3}} \Rightarrow \theta = \dfrac{\pi}{6}$ or $\theta = \dfrac{11\pi}{6}$
$\left(2\sqrt{3}, -2\right)$ lies in Quadrant IV, so $\theta = \dfrac{11\pi}{6}$.
The polar form of $\left(2\sqrt{3}, -2\right)$ is $\left(4, \dfrac{11\pi}{6}\right)$.

63. $3x + 2y = 12 \Rightarrow 3r\cos\theta + 2r\sin\theta = 12 \Rightarrow$
$r(3\cos\theta + 2\sin\theta) = 12 \Rightarrow r = \dfrac{12}{3\cos\theta + 2\sin\theta}$

65. $x^2 + y^2 = 8x \Rightarrow r^2 = 8r\cos\theta \Rightarrow r = 8\cos\theta$

67. $r = -3 \Rightarrow r^2 = 9 \Rightarrow x^2 + y^2 = 9$

69. $r = 3\csc\theta \Rightarrow r = \dfrac{3}{\sin\theta} \Rightarrow r\sin\theta = 3 \Rightarrow y = 3$

71. $r = 1 - 2\sin\theta \Rightarrow r + 2\sin\theta = 1 \Rightarrow$
$r^2 + 2r\sin\theta = r \Rightarrow (r^2 + 2r\sin\theta)^2 = r^2 \Rightarrow$
$(x^2 + y^2 + 2y)^2 = x^2 + y^2$

73. $-3i = 3\left(\cos\dfrac{3\pi}{2} + i\sin\dfrac{3\pi}{2}\right)$

Chapter 7 Applications of Trigonometric Functions

75. $5\sqrt{3} - 5i \Rightarrow r = \sqrt{(5\sqrt{3})^2 + (-5)^2} = 10$

$\tan\theta = -\dfrac{5}{5\sqrt{3}} \Rightarrow \theta = \dfrac{5\pi}{6}$ or $\theta = \dfrac{11\pi}{6}$

$5\sqrt{3} - 5i$ lies in Quadrant IV, so $\theta = \dfrac{11\pi}{6}$.

The polar form of $5\sqrt{3} - 5i$ is $10\left(\cos\dfrac{11\pi}{6} + i\sin\dfrac{11\pi}{6}\right)$.

77. $2(\cos 45° + i\sin 45°) = 2\left(\dfrac{\sqrt{2}}{2} + \dfrac{\sqrt{2}}{2}i\right) = \sqrt{2} + \sqrt{2}i$

79. $6\left(\cos\dfrac{3\pi}{4} + i\sin\dfrac{3\pi}{4}\right) = 6\left(-\dfrac{\sqrt{2}}{2} + \dfrac{\sqrt{2}}{2}i\right) = -3\sqrt{2} + 3\sqrt{2}i$

81. $z_1 = 3(\cos 25° + i\sin 25°);\ z_2 = 2(\cos 10° + i\sin 10°)$
$z_1 z_2 = 6(\cos 35° + i\sin 35°)$
$\dfrac{z_1}{z_2} = \dfrac{3}{2}(\cos 15° + i\sin 15°)$

83. $z_1 = 2\left(\cos\dfrac{5\pi}{6} + i\sin\dfrac{5\pi}{6}\right);\ z_2 = 3\left(\cos\dfrac{\pi}{3} + i\sin\dfrac{\pi}{3}\right)$
$z_1 z_2 = 6\left(\cos\dfrac{7\pi}{6} + i\sin\dfrac{7\pi}{6}\right)$
$\dfrac{z_1}{z_2} = \dfrac{2}{3}\left(\cos\dfrac{\pi}{2} + i\sin\dfrac{\pi}{2}\right)$

85. $[3(\cos 40° + i\sin 40°)]^3 = 3^3(\cos(3 \cdot 40°) + i\sin(3 \cdot 40°)) = 27(\cos 120° + i\sin 120°)$

87. First, convert $2 - 2\sqrt{3}i$ to polar form: $r = \sqrt{2^2 + (-2\sqrt{3})^2} = 4;\ \theta' = \left|\tan^{-1}\left(\dfrac{-2\sqrt{3}}{2}\right)\right| = \left|-\dfrac{\pi}{3}\right| = \dfrac{\pi}{3}$

$2 - 2\sqrt{3}i$ lies in quadrant IV, so $\theta = 2\pi - \dfrac{\pi}{3} = \dfrac{5\pi}{3}$.

The polar form of $2 - 2\sqrt{3}i$ is $4\left(\cos\dfrac{5\pi}{3} + i\sin\dfrac{5\pi}{3}\right)$.

$\left[4\left(\cos\dfrac{5\pi}{3} + i\sin\dfrac{5\pi}{3}\right)\right]^6 = 4^6\left(\cos\left(6 \cdot \dfrac{5\pi}{3}\right) + i\sin\left(6 \cdot \dfrac{5\pi}{3}\right)\right)$
$= 4096(\cos 10\pi + i\sin 10\pi) = 4096(\cos 0 + i\sin 0)$

89. The polar form for -125 is $125(\cos 180° + i\sin 180°)$.

$z_k = 125^{1/3}\left(\cos\dfrac{180° + 360° \cdot k}{3} + i\sin\dfrac{180° + 360° \cdot k}{3}\right)$ for $k = 0, 1, 2$.

$z_0 = 125^{1/3}\cos\dfrac{180°}{3} + i\sin\dfrac{180°}{3} = 5\cos 60° + i\sin 60°$

$z_1 = 125^{1/3}\cos\dfrac{540°}{3} + i\sin\dfrac{540°}{3} = 5\cos 180° + i\sin 180°$

$z_2 = 125^{1/3}\cos\dfrac{900°}{3} + i\sin\dfrac{900°}{3} = 5\cos 300° + i\sin 300°$

91. Write $-1+\sqrt{3}i$ in polar form: $r = \sqrt{(-1)^2 + (\sqrt{3})^2} = 2$; $\theta' = \left|\tan^{-1}(-\sqrt{3})\right| = |-60°| = 60°$

    Since $-1+\sqrt{3}i$ lies in Quadrant II, $\theta = 180° - 60° = 120°$

    $z_k = 2^{1/5}\left(\cos\dfrac{120° + 360°\cdot k}{5} + i\sin\dfrac{120° + 360°\cdot k}{5}\right)$ for $k = 0, 1, 2, 3, 4$.

    $z_0 = 2^{1/5}\left(\cos\dfrac{120°}{5} + i\sin\dfrac{120°}{5}\right) = 2^{1/5}(\cos 24° + i\sin 24°)$

    $z_1 = 2^{1/5}\left(\cos\dfrac{480°}{5} + i\sin\dfrac{480°}{5}\right) = 2^{1/5}(\cos 96° + i\sin 96°)$

    $z_2 = 2^{1/5}\left(\cos\dfrac{840°}{5} + i\sin\dfrac{840°}{5}\right) = 2^{1/5}(\cos 168° + i\sin 168°)$

    $z_3 = 2^{1/5}\left(\cos\dfrac{1200°}{5} + i\sin\dfrac{1200°}{5}\right) = 2^{1/5}(\cos 240° + i\sin 240°)$

    $z_4 = 2^{1/5}\left(\cos\dfrac{1560°}{5} + i\sin\dfrac{1560°}{5}\right) = 2^{1/5}(\cos 312° + i\sin 312°)$

Applying the Concepts

93.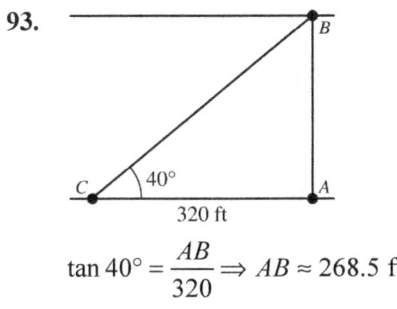

    $\tan 40° = \dfrac{AB}{320} \Rightarrow AB \approx 268.5$ ft

95. After 80 minutes, the cars have traveled
    $55\left(\dfrac{80}{60}\right) \approx 73.33$ mi and $65\left(\dfrac{80}{60}\right) \approx 86.67$ mi,
    respectively. The distance between the two cars is the length of the third side of the triangle that is formed by the roads. Using the Law of Cosines, we have
    $c = \sqrt{73.33^2 + 86.67^2 - 2(73.33)(86.67)\cos 72°}$
    ≈ 94.7 miles

97. Using Heron's Formula, we have
    $s = \dfrac{310 + 415 + 175}{2} = 450$
    $k = \sqrt{450(450-310)(450-415)(450-175)}$
    $\approx 24{,}625$ ft^2

99. $W = F \cdot \overrightarrow{PQ} = \|F\|\|\overrightarrow{PQ}\|\cos\theta$
    $= 30(60)\cos 48° \approx 1204.4$ foot-pounds

Chapter 7 Practice Test A

1. Given: $A = 115°, C = 35°, c = 15$ – an AAS case. $B = 180° - (115° + 35°) = 30°$
   $\dfrac{a}{\sin 115°} = \dfrac{15}{\sin 35°} \Rightarrow a \approx 23.7$
   $\dfrac{b}{\sin 30°} = \dfrac{15}{\sin 35°} \Rightarrow b \approx 13.1$

2. Given: $A = 42°, B = 37°, a = 50$ m – an AAS case. $C = 180° - (42° + 37°) = 101°$
   $\dfrac{50}{\sin 42°} = \dfrac{b}{\sin 37°} \Rightarrow b \approx 45.0$ m
   $\dfrac{50}{\sin 42°} = \dfrac{c}{\sin 101°} \Rightarrow c \approx 73.4$ m

3. Given: $a = 35, B = 106°, c = 53$ – an SAS case.
   $b^2 = 35^2 + 53^2 - 2(35)(53)\cos 106° \Rightarrow$
   $b \approx 71.11 \approx 71.1$
   $\dfrac{\sin 106°}{71.11} = \dfrac{\sin A}{35} \Rightarrow A \approx 28.2°$
   $C \approx 180° - (106° + 28.2°) \approx 45.8°$

4. Given: $a = 30$ ft, $b = 20$ ft, $c = 25$ ft – an SSS case. $30^2 = 20^2 + 25^2 - 2(20)(25)\cos A \Rightarrow$
   $A \approx 82.819° \approx 82.8°$
   $\dfrac{\sin 82.819°}{30} = \dfrac{\sin B}{20} \Rightarrow B \approx 41.4°$
   $C \approx 180° - (82.8° + 41.4°) \approx 55.8°$

446 Chapter 7 Applications of Trigonometric Functions

5. Given: $a = 5.6, b = 4.1$.
   Since the triangle is a right triangle, use the Pythagorean theorem to find c:
   $c = \sqrt{a^2 + b^2} = \sqrt{5.6^2 + 4.1^2} \approx 6.9$
   $A = \tan^{-1}\left(\dfrac{a}{b}\right) = \tan^{-1}\left(\dfrac{5.6}{4.1}\right) \approx 53.8°$
   $B = 90° - A = 90 - 53.8° = 36.2°$

6.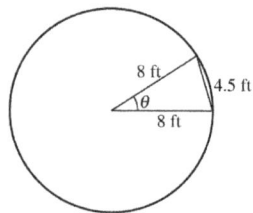

   $4.5^2 = 8^2 + 8^2 - 2(8)(8)\cos\theta \Rightarrow \theta \approx 32.7°$

7.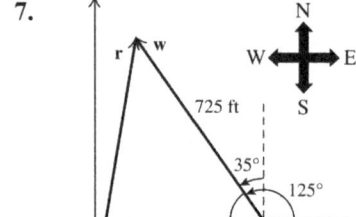

   $\mathbf{v} = 580(\cos 20°\mathbf{i} + \sin 20°\mathbf{j}) \approx 545.0\mathbf{i} + 198.4\mathbf{j}$
   $\mathbf{w} = 725(\cos 125°\mathbf{i} + \sin 125°\mathbf{j})$
   $\approx -415.8\mathbf{i} + 593.9\mathbf{j}$
   $\mathbf{r} = \mathbf{v} + \mathbf{w} = 129.2\mathbf{i} + 792.3\mathbf{j}$
   $\|\mathbf{r}\| = \sqrt{129.2^2 + 792.3^2} \approx 803$ feet

8. $\mathbf{v} = \langle 2-3, -7-5\rangle = \langle -1, -12\rangle$

9. $\mathbf{v} - \mathbf{w} = \langle -2-1, 3-5\rangle = \langle -3, -2\rangle$

10. $2\mathbf{v} - 3\mathbf{w} = 2(-2\mathbf{i}+3\mathbf{j}) - 3(\mathbf{i}+5\mathbf{j}) = -7\mathbf{i}-9\mathbf{j}$

11. $v_1 = \|\mathbf{v}\|\cos\theta = 3\cos(-30°) = \dfrac{3\sqrt{3}}{2}$
    $v_2 = \|\mathbf{v}\|\sin\theta = 3\sin(-30°) = -\dfrac{3}{2}$
    $\mathbf{v} = \dfrac{3\sqrt{3}}{2}\mathbf{i} - \dfrac{3}{2}\mathbf{j}$

12. $\mathbf{v}\cdot\mathbf{w} = (4\mathbf{i}+3\mathbf{j})\cdot(-\mathbf{i}+7\mathbf{j}) = -4 + 3(7) = 17$

13. Given: $\mathbf{v} = 3\mathbf{i} - 4\mathbf{j}, \mathbf{w} = -2\mathbf{i} + 5\mathbf{j}$
    $\cos\theta = \dfrac{\mathbf{v}\cdot\mathbf{w}}{\|\mathbf{v}\|\|\mathbf{w}\|} = \dfrac{(3)(-2)+(-4)(5)}{\left(\sqrt{3^2+(-4)^2}\right)\left(\sqrt{(-2)^2+5^2}\right)}$
    $= -\dfrac{26}{5\sqrt{29}} \Rightarrow \theta = \cos^{-1}\left(-\dfrac{26}{5\sqrt{29}}\right) \approx 164.9°$

14. $x = -2\cos(-45°) = -\sqrt{2}; y = -2\sin(-45°) = \sqrt{2}$
    The rectangular coordinates for $(-2, -45°)$ are $\left(-\sqrt{2}, \sqrt{2}\right)$.

15. $r = \sqrt{\left(-\sqrt{3}\right)^2 + (-1)^2} = 2; \tan\theta = \dfrac{1}{\sqrt{3}} \Rightarrow$
    $\theta = \dfrac{\pi}{6}$ or $\theta = \dfrac{7\pi}{6}$. $\left(-\sqrt{3}, -1\right)$ is in Quadrant III, so $\theta = \dfrac{7\pi}{6}$. The polar coordinates for $\left(-\sqrt{3}, -1\right)$ are $\left(2, \dfrac{7\pi}{6}\right)$.

16. $r = -3 \Rightarrow r^2 = 9 \Rightarrow x^2 + y^2 = 9$. The curve is a circle with center $(0, 0)$ and radius 3.

17. $3\left(\cos\left(\dfrac{2\pi}{3}\right) + i\sin\left(\dfrac{2\pi}{3}\right)\right) = 3\left(-\dfrac{1}{2} + \dfrac{\sqrt{3}}{2}i\right)$
    $= -\dfrac{3}{2} + \dfrac{3\sqrt{3}}{2}i$

18. First write both expressions in polar form:
    $z_1 = 1 - \sqrt{3}i \Rightarrow r = \sqrt{1^2 + \left(-\sqrt{3}\right)^2} = 2$
    $\tan\theta = -\sqrt{3} \Rightarrow \theta = 120°$ or $\theta = 300°$
    $1 - \sqrt{3}i$ lies in Quadrant IV, so $\theta = 300°$
    $z_1 = 2(\cos 300° + i\sin 300°)$
    $z_2 = -1 + i \Rightarrow r = \sqrt{(-1)^2 + 1^2} = \sqrt{2}$
    $\theta' = \left|\tan^{-1}(-1)\right| \Rightarrow \theta' = 45°$
    $-1 + i$ lies in Quadrant II, so
    $\theta = 180° - 45° = 135°$.
    $z_2 = \sqrt{2}(\cos 135° + i\sin 135°)$
    $\dfrac{z_1}{z_2} = \dfrac{2(\cos 300° + i\sin 300°)}{\sqrt{2}(\cos 135° + i\sin 135°)}$
    $= \sqrt{2}(\cos 165° + i\sin 165°)$

19. $\left(\sqrt{5}\left(\cos\dfrac{\pi}{4} + i\sin\dfrac{\pi}{4}\right)\right)^{-6}$
    $= 5^{-3}\left(\cos\left(-\dfrac{3\pi}{2}\right) + i\sin\left(-\dfrac{3\pi}{2}\right)\right) = \dfrac{1}{125}i$

20. First write $1+i$ in polar form:
$r = \sqrt{1^2 + 1^2} = \sqrt{2}$; $\theta' = \tan^{-1}(1) \Rightarrow \theta' = 45°$
$1+i$ lies in Quadrant I, so $\theta = 45°$.
The polar coordinates for $1+i$ are
$\sqrt{2}(\cos 45° + i \sin 45°)$.

$z_k = (\sqrt{2})^{1/4} \left(\cos \dfrac{45° + 360° \cdot k}{4} + i \sin \dfrac{45° + 360° \cdot k}{4} \right)$

$= 2^{1/8} \left(\cos \dfrac{45° + 360° \cdot k}{4} + i \sin \dfrac{45° + 360° \cdot k}{4} \right)$

for $k = 0, 1, 2, 3$.

$z_0 = 2^{1/8} \left(\cos \dfrac{45°}{4} + i \sin \dfrac{45°}{4} \right)$
$= 2^{1/8} (\cos 11.25° + i \sin 11.25°)$

$z_1 = 2^{1/8} \left(\cos \dfrac{405°}{4} + i \sin \dfrac{405°}{4} \right)$
$= 2^{1/8} (\cos 101.25° + i \sin 101.25°)$

$z_2 = 2^{1/8} \left(\cos \dfrac{765°}{4} + i \sin \dfrac{765°}{4} \right)$
$= 2^{1/8} (\cos 191.25° + i \sin 191.25°)$

$z_3 = 2^{1/8} \left(\cos \dfrac{1125°}{4} + i \sin \dfrac{1125°}{4} \right)$
$= 2^{1/8} (\cos 281.25° + i \sin 281.25°)$

Chapter 7 Practice Test B

1. Given: $A = 105°, C = 35°, c = 12$ — an AAS case. $B = 180° - (105° + 35°) = 40°$

$\dfrac{b}{\sin 40°} = \dfrac{12}{\sin 35°} \Rightarrow b \approx 13.4$

The answer is A.

2. Given: $A = 27°, B = 50°, a = 25$ m — an AAS case. $C = 180° - (27° + 50°) = 103°$

$\dfrac{c}{\sin 103°} = \dfrac{25}{\sin 27°} \Rightarrow c \approx 53.7$ m

The answer is C.

3. Given: $A = 25°, a = 25$ ft, $b = 30$ ft — an SSA case. A is acute, so examine h:
$h = b \sin A = 30 \sin 25° \approx 12.7 \Rightarrow h < a < b$, so there are two triangles.

$\dfrac{\sin 25°}{25} = \dfrac{\sin B}{30} \Rightarrow B_1 \approx 30.473°, B_2 \approx 149.527°$

$C_1 \approx 180° - (25° + 30.5°) \approx 124.527°$

$\dfrac{25}{\sin 25°} = \dfrac{c_1}{\sin 124.527°} \Rightarrow c_1 \approx 48.7$

$C_2 \approx 180° - (25° + 149.527°) \approx 5.473°$

$\dfrac{25}{\sin 25°} = \dfrac{c_2}{\sin 5.473°} \Rightarrow c_2 \approx 5.6$

The answer is C.

4. Given: $B = 110°, a = 37, c = 21$ — an SAS case.
$b^2 = 37^2 + 21^2 - 2(37)(21) \cos 110° \Rightarrow b \approx 48.4$
The answer is D.

5.

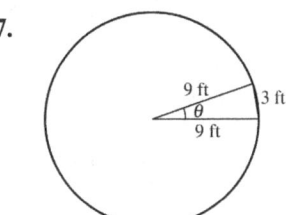

$\tan \theta = \dfrac{12}{45 + x} = \dfrac{3}{x} \Rightarrow x = 1.5$ ft

The answer is A.

6. Given: $a = 12$ ft, $b = 13$ ft, $c = 7$ ft — an SSS case. $12^2 = 13^2 + 7^2 - 2(13)(7) \cos A \Rightarrow A \approx 66°$. The answer is B.

7.

$3^2 = 9^2 + 9^2 - 2(9)(9) \cos \theta \Rightarrow \theta \approx 19.2°$.
The answer is D.

8.

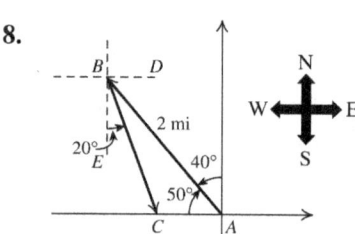

$m\angle ABD = m\angle BAC = 50°$
$m\angle BAC + m\angle ABC + m\angle CBE = 90° \Rightarrow$
$m\angle ABC = 90° - 20° - 50° = 20°$
$m\angle BCA = 180° - (20° + 50°) = 110°$
$\dfrac{2}{\sin 110°} = \dfrac{BC}{\sin 50°} \Rightarrow BC \approx 1.6 \text{ mi}$.
The answer is C.

9. $\mathbf{v} = \langle 3-(-2), -1-5 \rangle = \langle 5, -6 \rangle$
The answer is A.

10. $\mathbf{v} - \mathbf{w} = \langle 1-3, -4-(-2) \rangle = \langle -2, -2 \rangle$
The answer is C.

11. $2\mathbf{v} - 3\mathbf{w} = 2(7\mathbf{i}+3\mathbf{j}) - 3(-2\mathbf{i}+\mathbf{j}) = 20\mathbf{i}+3\mathbf{j}$
The answer is A.

12. $v_1 = 6\cos(-60°) = 3$; $v_2 = 6\sin(-60°) = -3\sqrt{3}$
$\mathbf{v} = 3\mathbf{i} - 3\sqrt{3}\mathbf{j}$. The answer is D.

13. $\mathbf{v} \bullet \mathbf{w} = (-2)(9) + (-5)(-4) = 2$.
The answer is D.

14. $\cos\theta = \dfrac{\mathbf{v}\bullet\mathbf{w}}{\|\mathbf{v}\|\|\mathbf{w}\|} = \dfrac{7-\sqrt{2}}{\left(\sqrt{7^2+(-1)^2}\right)\left(\sqrt{1^2+\left(\sqrt{2}\right)^2}\right)}$

$= \dfrac{7-\sqrt{2}}{\left(5\sqrt{2}\right)\sqrt{3}} = \dfrac{7-\sqrt{2}}{5\sqrt{6}} \Rightarrow$

$\theta = \cos^{-1}\left(\dfrac{7-\sqrt{2}}{5\sqrt{6}}\right) \approx 62.9° \approx 63°$

The answer is B.

15. $x = -4\cos(-30°) = -2\sqrt{3}$
$y = -4\sin(-30°) = 2$
The rectangular coordinates for $(-4, -30°)$ are $\left(-2\sqrt{3}, 2\right)$. The answer is C.

16. $r = \sqrt{3^2 + \left(-\sqrt{3}\right)^2} = 2\sqrt{3}$

$\tan\theta = -\dfrac{\sqrt{3}}{3} \Rightarrow \theta = 150°$ or $\theta = 330°$

$\left(3, -\sqrt{3}\right)$ is in Quadrant IV, so $\theta = 330°$.

The polar coordinates for $\left(3, -\sqrt{3}\right)$ are $\left(2\sqrt{3}, 330°\right)$. The answer is B.

17. The answer is B.
$r = -2 \Rightarrow r^2 = 4 \Rightarrow x^2 + y^2 = 4$

18. $5\left(\cos\dfrac{7\pi}{3} + i\sin\dfrac{7\pi}{3}\right) = 5\left(\cos\dfrac{\pi}{3} + i\sin\dfrac{\pi}{3}\right)$

$= 5\left(\dfrac{1}{2} + \dfrac{\sqrt{3}}{2}i\right)$

$= \dfrac{5}{2} + \dfrac{5\sqrt{3}}{2}i$

The answer is C.

19. $z_1 = i \Rightarrow z_1 = \cos 90° + i\sin 90°$
$z_2 = 1 - i \Rightarrow r = \sqrt{1^2 + (-1)^2} = \sqrt{2}$
$\theta' = \left|\tan^{-1}(-1)\right| = 45°$
$1 - i$ is in Quadrant IV, so
$\theta = 360° - 45° = 315°$.
$z_2 = \sqrt{2}(\cos 315° + i\sin 315°)$
$\dfrac{z_1}{z_2} = \dfrac{\cos 90° + i\sin 90°}{\sqrt{2}(\cos 315° + i\sin 315°)}$

$= \dfrac{\sqrt{2}}{2}(\cos(-225°) + i\sin(-225°))$

$= \dfrac{\sqrt{2}}{2}(\cos 135° + i\sin 135°)$

The answer is C.

20. $\left[\sqrt{2}\left(\cos\dfrac{\pi}{12} + i\sin\dfrac{\pi}{12}\right)\right]^4$

$= \left(\sqrt{2}\right)^4 \left(\cos\left(4\cdot\dfrac{\pi}{12}\right) + i\sin\left(4\cdot\dfrac{\pi}{12}\right)\right)$

$= 4\left(\cos\dfrac{\pi}{3} + i\sin\dfrac{\pi}{3}\right) = 4\left(\dfrac{1}{2} + \dfrac{\sqrt{3}}{2}i\right)$

$= 2 + 2\sqrt{3}i$
The answer is C.

Cumulative Review Exercises
Chapters P-7

1. $f(x) = \sqrt{x}$; $g(x) = \dfrac{1}{x-2}$

   $\dfrac{g}{f} = \dfrac{\frac{1}{x-2}}{\sqrt{x}} = \dfrac{1}{(x-2)\sqrt{x}}$

   f is not defined for $x < 0$, while g is not defined for $x = 2$. $\dfrac{g}{f}$ is not defined for $x \leq 0$ or $x = 2$, so the domain of $\dfrac{g}{f}$ is $(0, 2) \cup (2, \infty)$.

2. $f(x) = 2x + 7 \Rightarrow y = 2x + 7$
   Interchange x and y, then solve for y:
   $x = 2y + 7 \Rightarrow y = \dfrac{x-7}{2} \Rightarrow f^{-1}(x) = \dfrac{x}{2} - \dfrac{7}{2}$.

3. The graph of $f(x) = -2x^2 + 8x + 10$ is a parabola with $a = -2$, $b = 8$ and $c = 10$. The parabola opens downward because $a < 0$. Now, find the vertex:

   $h = -\dfrac{b}{2a} = -\dfrac{8}{2(-2)} = 2$

   $k = f(h) = f(2) = -2(2)^2 + 8(2) + 10 = 18$ Thus, the vertex (h, k) is $(2, 18)$.
   Next, find the x-intercepts:
   $-2x^2 + 8x + 10 = 0 \Rightarrow -2(x^2 - 4x - 5) = 0 \Rightarrow -2(x-5)(x+1) = 0 \Rightarrow x = 5$ or $x = -1$

   Now, find the y-intercept: $f(0) = -2(0)^2 + 8(0) + 10 = 10$.

   Thus, the intercepts are $(-1, 0)$, $(5, 0)$ and $(0, 10)$. Use the fact that the parabola is symmetric with respect to its axis, $x = 2$, to locate additional points. Plot the vertex, the x-intercepts, the y-intercept, and any additional points, and join them with a parabola.

   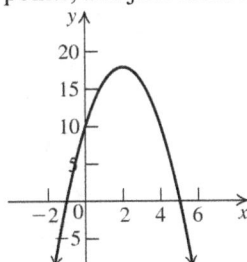

4. First solve the associated equation $x^2 - 6x + 8 = 0$.
   $x^2 - 6x + 8 = 0 \Rightarrow (x-4)(x-2) = 0 \Rightarrow x = 4$ or $x = 2$.
   Choose a value of x between 2 and 4 (i.e., 3) to test.
   $3^2 - 6(3) + 8 \overset{?}{>} 0$
   $-1 < 0$
   Thus, 3 is not in the solution set. The solution set is $(-\infty, 2) \cup (4, \infty)$.

5. $\sin\theta\cos\theta = -\dfrac{\sqrt{3}}{4} \Rightarrow 2\sin\theta\cos\theta = -\dfrac{\sqrt{3}}{2} \Rightarrow \sin 2\theta = -\dfrac{\sqrt{3}}{2} \Rightarrow 2\theta \in \left\{\dfrac{4\pi}{3},\dfrac{5\pi}{3},\dfrac{10\pi}{3},\dfrac{11\pi}{3}\right\} \Rightarrow$
$\theta \in \left\{\dfrac{2\pi}{3},\dfrac{5\pi}{6},\dfrac{5\pi}{3},\dfrac{11\pi}{6}\right\}$

6. $4\sin^2\theta - 1 = 0 \Rightarrow (2\sin\theta - 1)(2\sin\theta + 1) = 0 \Rightarrow \sin\theta = \dfrac{1}{2}$ or $\sin\theta = -\dfrac{1}{2}$. If $\sin\theta = \dfrac{1}{2}$, $\theta \in \left\{\dfrac{\pi}{6},\dfrac{5\pi}{6}\right\}$.
If $\sin\theta = -\dfrac{1}{2}$, $\theta \in \left\{\dfrac{7\pi}{6},\dfrac{11\pi}{6}\right\}$. The solution set is $\left\{\dfrac{\pi}{6},\dfrac{5\pi}{6},\dfrac{7\pi}{6},\dfrac{11\pi}{6}\right\}$.

7. Write $\sqrt{3}+i$ in polar form:
$r = \sqrt{\left(\sqrt{3}\right)^2+1} = 2;\ \theta' = \left|\tan^{-1}\left(\dfrac{1}{\sqrt{3}}\right)\right| \Rightarrow \theta' = 30°$. $\sqrt{3}+i$ lies in Quadrant I, so $\theta = 30°$.
The polar coordinates for $\sqrt{3}+i$ are $2(\cos 30° + i\sin 30°)$.

$z_k = 2^{1/4}\left(\cos\dfrac{30°+360°\cdot k}{4} + i\sin\dfrac{30°+360°\cdot k}{4}\right)$ for $k = 0, 1, 2, 3$.

$z_0 = 2^{1/4}\left(\cos\dfrac{30°}{4} + i\sin\dfrac{30°}{4}\right) = 2^{1/4}\left(\cos 7.5° + i\sin 7.5°\right)$

$z_1 = 2^{1/4}\left(\cos\dfrac{390°}{4} + i\sin\dfrac{390°}{4}\right) = 2^{1/4}\left(\cos 97.5° + i\sin 97.5°\right)$

$z_2 = 2^{1/4}\left(\cos\dfrac{750°}{4} + i\sin\dfrac{750°}{4}\right) = 2^{1/4}\left(\cos 187.5° + i\sin 187.5°\right)$

$z_3 = 2^{1/4}\left(\cos\dfrac{1110°}{4} + i\sin\dfrac{1110°}{4}\right) = 2^{1/4}\left(\cos 277.5° + i\sin 277.5°\right)$

8. Write -64 in polar form: $-64 = 64(\cos 180° + i\sin 180°)$

$z_k = 64^{1/6}\left(\cos\dfrac{180°+360°\cdot k}{6} + i\sin\dfrac{180°+360°\cdot k}{6}\right) = 2\left(\cos\dfrac{180°+360°\cdot k}{6} + i\sin\dfrac{180°+360°\cdot k}{6}\right)$ for $k = 0, 1, 2, 3, 4, 5$.

$z_0 = 2\left(\cos\dfrac{180°}{6} + i\sin\dfrac{180°}{6}\right) = 2(\cos 30° + i\sin 30°)$

$z_1 = 2\left(\cos\dfrac{540°}{6} + i\sin\dfrac{540°}{6}\right) = 2(\cos 90° + i\sin 90°)$

$z_2 = 2\left(\cos\dfrac{900°}{6} + i\sin\dfrac{900°}{6}\right) = 2(\cos 150° + i\sin 150°)$

$z_3 = 2\left(\cos\dfrac{1260°}{6} + i\sin\dfrac{1260°}{6}\right) = 2(\cos 210° + i\sin 210°)$

$z_4 = 2\left(\cos\dfrac{1620°}{6} + i\sin\dfrac{1620°}{6}\right) = 2(\cos 270° + i\sin 270°)$

$z_5 = 2\left(\cos\dfrac{1980°}{6} + i\sin\dfrac{1980°}{6}\right) = 2(\cos 330° + i\sin 330°)$

9. First find the slope of the line $2x+3y+6=0$: $2x+3y+6=0 \Rightarrow y = -\dfrac{2}{3}x - 2 \Rightarrow m = -\dfrac{2}{3}$. So the slope of the perpendicular line is $\dfrac{3}{2}$. The equation of the perpendicular line is $y - 5 = \dfrac{3}{2}(x+1) \Rightarrow y = \dfrac{3}{2}x + \dfrac{13}{2}$.

10. $2\log x + \dfrac{1}{2}\log(y+1) - \log(3x+1) = \log\dfrac{x^2\sqrt{y+1}}{3x+1}$

11. Shift the graph of $y = \sqrt{x}$ one unit right, stretch it vertically by a factor of 2, then shift the graph three units down.

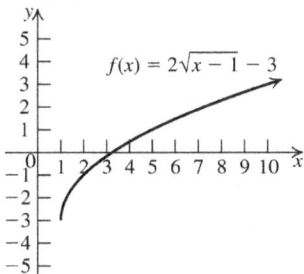

12. Shift the graph of $y = |x|$ one unit left, reflect it in the x-axis, then shift it up two units.

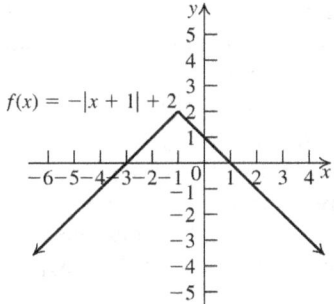

13. Reflect the graph of $y = \cos x$ in the x-axis. The amplitude is 2, the period is $2\pi/3$, and the phase shift is $-\pi/3$.

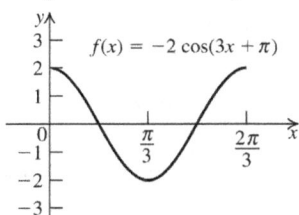

14. Stretch the graph of $y = \sec x$ vertically by a factor of 2. The period is $2\pi/3$.

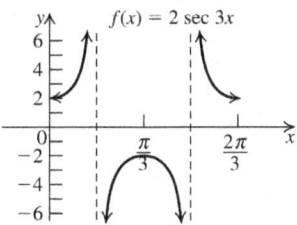

15. $\cos\theta = -\dfrac{5}{13} = \dfrac{x}{r} \Rightarrow r = 13, x = -5 \Rightarrow y = 12 \Rightarrow$
$\tan\theta = -\dfrac{12}{5}$

16. $\cos 65° \cos 35° + \sin 65° \sin 35°$
$= \cos(65° - 35°) = \cos 30° = \dfrac{\sqrt{3}}{2}$

17. $\dfrac{1}{1-\cos x} + \dfrac{1}{1+\cos x} = \dfrac{1+\cos x + 1 - \cos x}{1-\cos^2 x}$
$= \dfrac{2}{\sin^2 x} = 2\csc^2 x$
$= 2(1 + \cot^2 x)$

18. $\dfrac{1+\cos 2x}{\sin 2x} = \dfrac{1+(2\cos^2 x - 1)}{2\sin x \cos x} = \dfrac{2\cos^2 x}{2\sin x \cos x}$
$= \dfrac{\cos x}{\sin x} = \cot x$

19. Given: $A = 65°, B = 46°, c = 60$ m — an ASA case. $C = 180° - (65° + 46°) = 69°$
$\dfrac{60}{\sin 69°} = \dfrac{a}{\sin 65°} \Rightarrow a \approx 58.2$ m
$\dfrac{60}{\sin 69°} = \dfrac{b}{\sin 46°} \Rightarrow b \approx 46.2$ m

20.

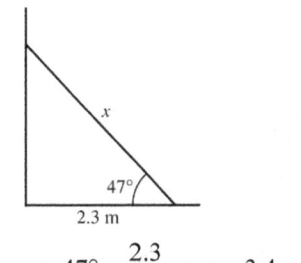

$\cos 47° = \dfrac{2.3}{x} \Rightarrow x \approx 3.4$ m

Chapter 8 Systems of Equations and Inequalities

8.1 Systems of Linear Equations in Two Variables

8.1 Practice Problems

1. **a.** Check (2, 2).

    Equation (1) Equation (2)
    $x + y = 4$ $3x - y = 0$
    $2 + 2 \stackrel{?}{=} 4$ $3(2) - 2 \stackrel{?}{=} 0$
    $4 = 4$ ✓ $4 = 0$ ✗

    Because (2, 2) does not satisfy both equations, it is not a solution of the system.

   b. Check (1, 3).

    Equation (1) Equation (2)
    $x + y = 4$ $3x - y = 0$
    $1 + 3 \stackrel{?}{=} 4$ $3(1) - 3 \stackrel{?}{=} 0$
    $4 = 4$ ✓ $0 = 0$ ✓

    Because (1, 3) satisfies both equations, it is a solution of the system.

2. The solution is $\{(-1, 3)\}$.
    $\begin{cases} x + y = 2 \\ 4x + y = -1 \end{cases}$

    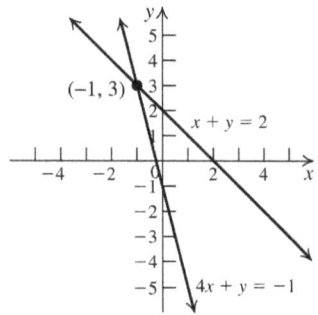

3. $\begin{cases} x - y = 5 & (1) \\ 2x + y = 7 & (2) \end{cases}$

    Solve the equation (1) for x, then substitute that expression into equation (2) and solve for y.
    $x - y = 5 \Rightarrow x = y + 5$
    $2x + y = 7 \Rightarrow 2(y + 5) + y = 7 \Rightarrow$
    $3y + 10 = 7 \Rightarrow 3y = -3 \Rightarrow y = -1$
    Now substitute the value for y into equation (1) and solve for x.
    $x - y = 5 \Rightarrow x - (-1) = 5 \Rightarrow x = 4$
    The solution is $\{(4, -1)\}$.

4. $\begin{cases} x - 3y = 1 & (1) \\ -2x + 6y = 3 & (2) \end{cases}$

    Solve the equation (1) for x, then substitute that expression into the equation (2) and solve for y.
    $x - 3y = 1 \Rightarrow x = 1 + 3y$
    $-2x + 6y = 3 \Rightarrow -2(1 + 3y) + 6y = 3 \Rightarrow$
    $-2 - 6y + 6y = 3 \Rightarrow -2 = 3$
    Since the equation $-2 = 3$ is false, the system is inconsistent. The solution set is $\varnothing$.

5. $\begin{cases} -2x + y = -3 & (1) \\ 4x - 2y = 6 & (2) \end{cases}$

    Solve the equation (1) for y, then substitute that expression into equation (2) and solve for x.
    $-2x + y = -3 \Rightarrow y = 2x - 3$
    $4x - 2y = 6 \Rightarrow 4x - 2(2x - 3) = 6 \Rightarrow$
    $4x - 4x + 6 = 6 \Rightarrow 6 = 6$
    The equation $6 = 6$ is true for every value of x. Thus, any value of x can be used in the equation $y = 2x - 3$. The solutions of the system are of the form $\{(x, 2x - 3)\}$.

6. $\begin{cases} 3x + 2y = 3 & (1) \\ 9x - 4y = 4 & (2) \end{cases}$

    Multiply the equation (1) by 2, then add the two equations.
    $\begin{cases} 3x + 2y = 3 \\ 9x - 4y = 4 \end{cases} \Rightarrow \begin{cases} 6x + 4y = 6 \\ 9x - 4y = 4 \end{cases} \Rightarrow 15x = 10 \Rightarrow$
    $x = \dfrac{2}{3}$

    Now substitute $x = \dfrac{2}{3}$ into the equation (2) and solve for y:

    $9x - 4y = 4 \Rightarrow 9\left(\dfrac{2}{3}\right) - 4y = 4 \Rightarrow 6 - 4y = 4 \Rightarrow$
    $-4y = -2 \Rightarrow y = \dfrac{1}{2}$

    The solution is $\left\{\left(\dfrac{2}{3}, \dfrac{1}{2}\right)\right\}$.

7. $\begin{cases} \dfrac{4}{x} + \dfrac{3}{y} = 1 \\ \dfrac{2}{x} - \dfrac{6}{y} = 3 \end{cases}$

Let $u = \dfrac{1}{x}$ and $v = \dfrac{1}{y}$. Then the equations become $\begin{cases} 4u + 3v = 1 \\ 2u - 6v = 3 \end{cases}$. Multiply the first equation by 2, then add the equations:

$\begin{cases} 4u + 3v = 1 \\ 2u - 6v = 3 \end{cases} \Rightarrow \begin{cases} 8u + 6v = 2 \\ 2u - 6v = 3 \end{cases} \Rightarrow 10u = 5 \Rightarrow u = \dfrac{1}{2}$

Substitute this value into the first equation to solve for v:

$4\left(\dfrac{1}{2}\right) + 3v = 1 \Rightarrow 3v = -1 \Rightarrow v = -\dfrac{1}{3}$.

Now substitute these values into the expressions for u and v to solve for x and y.

$u = \dfrac{1}{x} \Rightarrow \dfrac{1}{2} = \dfrac{1}{x} \Rightarrow x = 2$.

$v = \dfrac{1}{y} \Rightarrow -\dfrac{1}{3} = \dfrac{1}{y} \Rightarrow y = -3$

The solution is $\{(2, -3)\}$.

8. $\begin{cases} p = 20 + 0.002x & (1) \\ p = 77 - 0.008x & (2) \end{cases}$

Solve by substitution.
$20 + 0.002x = 77 - 0.008x \Rightarrow 0.010x = 57 \Rightarrow$
$x = 5700$
Substitute this value into equation (1) and solve for p.

$p = 20 + 0.002x \Rightarrow p = 20 + 0.002(5700) \Rightarrow$
$p = 31.4$
The equilibrium point is (5700, 31.4).

9. Let x = the amount invested at 12%.
Let y = the amount invested at 8%.
Then $0.12x$ = the income from the 12% investment and $0.08y$ = the income from the 8% investment. The system of equations is

$\begin{cases} x + y = 150{,}000 & (1) \\ 0.12x + 0.08y = 15{,}400 & (2) \end{cases}$

We will use the elimination method to solve the system.

$\begin{array}{ll} -12x - 12y = -1{,}800{,}000 & \text{Multiply by } -12. \\ \underline{12x + 8y = 1{,}540{,}000} & \text{Multiply by 100.} \\ -4y = -260{,}000 & \text{Add.} \end{array}$

$y = \dfrac{-260{,}000}{-4} = 65{,}000 \qquad$ Solve for y.

Back-substitute $y = 65{,}000$ into equation (1) and solve for x.
$x + 65{,}000 = 150{,}000$
$x = 85{,}000$
Check:
12% of 85,000 = 10,200
8% of 65,000 = 5200
85,000 + 65,000 = 150,000 and
10,200 + 5200 = 15,400, as given.
Solution: $85,000 was invested at 12% and $65,000 was invested at 8%.

8.1 Concepts and Vocabulary

1. The ordered pair (a, b) is a solution of a system of equations in x and y provided that when x is replaced by a and y is replace by b, the resulting equations are true.

3. If, in the process of solving a system of equations, you get an equation of the form $0 = k$, where k is not zero, then the system is inconsistent.

5. False. A system consisting of two identical equations has an infinite number of solutions.

7. True

8.1 Building Skills

9. Substituting each ordered pair into the system $\begin{cases} 2x + 3y = 3 \\ 3x - 4y = 13 \end{cases}$, we find that $(3, -1)$ is a solution.

$\begin{cases} 2(3) + 3(-1) = 6 - 3 = 3 \\ 3(3) - 4(-1) = 9 + 4 = 13 \end{cases}$

11. Substituting each ordered pair into the system $\begin{cases} 5x - 2y = 7 \\ -10x + 4y = 11 \end{cases}$, we find that none of the ordered pairs are solutions.

13. Substituting each ordered pair into the system $\begin{cases} x + y = 1 \\ \dfrac{1}{2}x + \dfrac{1}{3}y = 2 \end{cases}$, we find that $(10, -9)$ is a solution.

$\begin{cases} 10 - 9 = 1 \\ \dfrac{1}{2}(10) + \dfrac{1}{3}(-9) = 5 - 3 = 2 \end{cases}$

15. The solution is $\{(2, 1)\}$.
$$\begin{cases} 2+1=3 \\ 2-1=1 \end{cases}$$

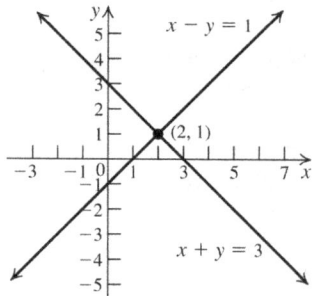

17. The solution is $\{(2, 2)\}$.
$$\begin{cases} 2+2(2)=2+4=6 \\ 2(2)+2=4+2=6 \end{cases}$$

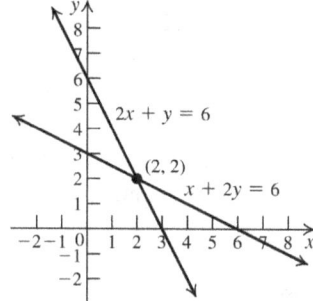

19. The system is inconsistent.

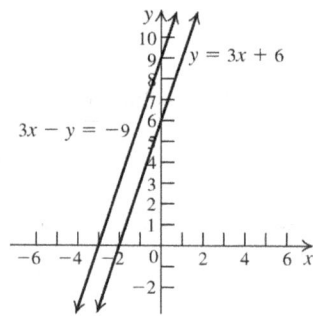

21. The solution is $\left\{\left(\dfrac{7}{3}, \dfrac{14}{3}\right)\right\}$.
$$\begin{cases} \dfrac{7}{3}+\dfrac{14}{3}=\dfrac{21}{3}=7 \\ \dfrac{14}{3}=2\left(\dfrac{7}{3}\right) \end{cases}$$

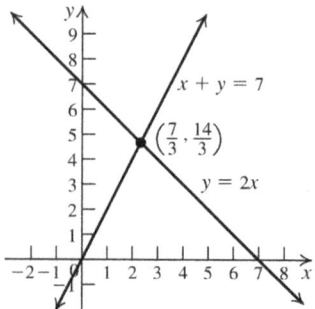

23. The system is dependent. The general solution is $\{(x, 12-3x)\}$.

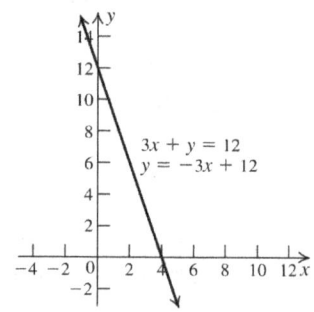

25. The slopes of the two lines are different, so the system is independent.

27. The slopes of the two lines are different, so the system is independent.

29. The slopes and y-intercepts of the two lines are the same, so they coincide, and the system is dependent.

31. The slopes of the two lines are different, so the system is independent.

33. The slopes of the two lines are the same while the y-intercepts are different, so the system is inconsistent.

35. The slopes of the two lines are the same while the y-intercepts are different, so the system is inconsistent.

37. The slopes of the two lines are different, so the system is independent.

In exercises 39–48, check your answers by substituting the values of x and y into both of the original equations.

39. Substitute $y = 2x+1$ into
$5x + 2y = 9 \Rightarrow 5x + 2(2x+1) = 9 \Rightarrow$
$5x + 4x + 2 = 9 \Rightarrow 9x = 7 \Rightarrow x = 7/9$

Use this value of x to find y: $y = 2\left(\dfrac{7}{9}\right)+1 \Rightarrow$

$y = \dfrac{23}{9}$. The solution is $\left\{\left(\dfrac{7}{9}, \dfrac{23}{9}\right)\right\}$.

41. Solve the second equation for y, and then substitute $y = 7 - x$ into the first equation:
$3x - (7 - x) = 5 \Rightarrow 3x - 7 + x = 5 \Rightarrow 4x = 12 \Rightarrow$
$x = 3$. Substitute this value into the second equation to find y: $3 + y = 7 \Rightarrow y = 4$
The solution is $\{(3, 4)\}$.

43. Solve the first equation for y, and then substitute $y = 2x - 5$ into the second equation:
$-4x + 2(2x - 5) = 7 \Rightarrow -4x + 4x - 10 = 7 \Rightarrow$
$-10 \neq 7 \Rightarrow$ there is no solution.
Solution set: $\varnothing$

45. Solve the first equation for y, and then substitute $y = 3 - \frac{2}{3}x$ into the second equation:
$3x + 2\left(3 - \frac{2}{3}x\right) = 1 \Rightarrow 3x + 6 - \frac{4}{3}x = 1 \Rightarrow$
$\frac{5}{3}x = -5 \Rightarrow x = -3$. Substitute this value into the first equation to find y:
$\frac{2}{3}(-3) + y = 3 \Rightarrow y = 5$.
The solution is $\{(-3, 5)\}$.

47. Solve the first equation for y, and then substitute $y = \frac{1}{2}(x - 5)$ into the second equation:
$-3x + 6\left(\frac{1}{2}(x - 5)\right) = -15 \Rightarrow -3x + 3x - 15 = -15 \Rightarrow$
$-15 = -15 \Rightarrow$ the system is dependent.
The general form of the solution is
$\left\{\left(x, \frac{1}{2}(x - 5)\right)\right\}$.

49. Add the two equations and solve for x:
$\begin{cases} x - y = 1 \\ x + y = 5 \end{cases} \Rightarrow 2x = 6 \Rightarrow x = 3$. Now substitute $x = 3$ into the second equation and solve for y:
$3 + y = 5 \Rightarrow y = 2$. The solution is $\{(3, 2)\}$.

51. Multiply the first equation by -2:
$\begin{cases} x + y = 0 \\ 2x + 3y = 3 \end{cases} \Rightarrow \begin{cases} -2x - 2y = 0 \\ 2x + 3y = 3 \end{cases}$
Add the equations and solve for y:
$\begin{cases} -2x - 2y = 0 \\ 2x + 3y = 3 \end{cases} \Rightarrow y = 3$.
Substitute this value into the first equation and solve for x: $x + 3 = 0 \Rightarrow x = -3$.
The solution is $\{(-3, 3)\}$.

53. Multiply the first equation by 2:
$\begin{cases} 5x - y = 5 \\ 3x + 2y = -10 \end{cases} \Rightarrow \begin{cases} 10x - 2y = 10 \\ 3x + 2y = -10 \end{cases}$
Add the equations and solve for x:
$\begin{cases} 10x - 2y = 10 \\ 3x + 2y = -10 \end{cases} \Rightarrow 13x = 0 \Rightarrow x = 0$.
Substitute this value into the first equation and solve for y: $5(0) - y = 5 \Rightarrow y = -5$. The solution is $\{(0, -5)\}$.

55. Multiply the first equation by 2:
$\begin{cases} x - y = 2 \\ -2x + 2y = 5 \end{cases} \Rightarrow \begin{cases} 2x - 2y = 4 \\ -2x + 2y = 5 \end{cases}$. Adding the equations gives $0 = 9 \Rightarrow$ the equations are inconsistent, and there is no solution.
Solution set: $\varnothing$

57. Multiply the second equation by -2:
$\begin{cases} 4x + 6y = 12 \\ 2x + 3y = 6 \end{cases} \Rightarrow \begin{cases} 4x + 6y = 12 \\ -4x - 6y = -12 \end{cases}$. Adding the equations gives $0 = 0 \Rightarrow$ the equations are dependent. Solving the second equation for y, we have $2x + 3y = 6 \Rightarrow y = -\frac{2x}{3} + 2$.
The general form of the solution is
$\left\{\left(x, -\frac{2x}{3} + 2\right)\right\}$.

59. Solve the first equation for y, and then substitute this value into the second equation to solve for x: $2x + y = 9 \Rightarrow y = -2x + 9$.
$2x - 3(-2x + 9) = 5 \Rightarrow 8x - 27 = 5 \Rightarrow x = 4$.
Substitute this value into the first equation and solve for y: $2(4) + y = 9 \Rightarrow y = 1$.
The solution is $\{(4, 1)\}$.

61. Multiply the second equation by -2 and then add the equations:
$\begin{cases} 2x + 5y = 2 \\ x + 3y = 2 \end{cases} \Rightarrow \begin{cases} 2x + 5y = 2 \\ -2x - 6y = -4 \end{cases} \Rightarrow -y = -2 \Rightarrow$
$y = 2$. Substitute this value into the second equation and solve for x: $x + 3(2) = 2 \Rightarrow x = -4$.
The solution is $\{(-4, 2)\}$.

63. Multiply the second equation by -3 and then add the equations:
$\begin{cases} 2x + 3y = 7 \\ 3x + y = 7 \end{cases} \Rightarrow \begin{cases} 2x + 3y = 7 \\ -9x - 3y = -21 \end{cases} \Rightarrow$
$-7x = -14 \Rightarrow x = 2$. Substitute this value into the second equation and solve for y:
$3(2) + y = 7 \Rightarrow y = 1$.
The solution is $\{(2, 1)\}$.

65. Multiply the first equation by -2 and the second equation by 3, then add the equations to solve for x:
$$\begin{cases} 2x+3y=9 \\ 3x+2y=11 \end{cases} \Rightarrow \begin{cases} -4x-6y=-18 \\ 9x+6y=33 \end{cases} \Rightarrow$$
$5x=15 \Rightarrow x=3$.
Substitute this value into the first equation to solve for y: $2(3)+3y=9 \Rightarrow 3y=3 \Rightarrow y=1$.
The solution is $\{(3, 1)\}$.

67. First, simplify both equations:
$$\begin{cases} \dfrac{x}{4}+\dfrac{y}{6}=1 \\ x+2(x-y)=7 \end{cases} \Rightarrow \begin{cases} 6x+4y=24 \\ 3x-2y=7 \end{cases}.$$
Now multiply the second equation by 2, and then add the equations:
$$\begin{cases} 6x+4y=24 \\ 3x-2y=7 \end{cases} \Rightarrow \begin{cases} 6x+4y=24 \\ 6x-4y=14 \end{cases} \Rightarrow$$
$12x=38 \Rightarrow x=\dfrac{19}{6}$.
Substitute this value into the second equation and solve for y:
$\dfrac{19}{6}+2\left(\dfrac{19}{6}-y\right)=7 \Rightarrow \dfrac{57}{6}-2y=7 \Rightarrow$
$-2y=-\dfrac{5}{2} \Rightarrow y=\dfrac{5}{4}$.
The solution is $\left\{\left(\dfrac{19}{6},\dfrac{5}{4}\right)\right\}$.

69. Simplify the first equation:
$3x=2(x+y) \Rightarrow 3x=2x+2y \Rightarrow x=2y$.
Substitute this expression for x into the second equation: $3(2y)-5y=2 \Rightarrow y=2$. Substitute this value into the first equation to solve for x:
$3x=2(x+2) \Rightarrow 3x=2x+4 \Rightarrow x=4$.
The solution is $\{(4, 2)\}$.

71. Multiply the first equation by -2, then add the equations:
$$\begin{cases} 0.2x+0.7y=1.5 \\ 0.4x-0.3y=1.3 \end{cases} \Rightarrow \begin{cases} -0.4x-1.4y=-3 \\ 0.4x-0.3y=1.3 \end{cases} \Rightarrow$$
$-1.7y=-1.7 \Rightarrow y=1$. Substitute this value into the first equation to solve for x:
$0.2x+0.7(1)=1.5 \Rightarrow 0.2x=0.8 \Rightarrow x=4$.
The solution is $\{(4, 1)\}$.

73. $\begin{cases} \dfrac{x}{2}+\dfrac{y}{3}=1 & (1) \\ \dfrac{3x}{4}+\dfrac{y}{2}=1 & (2) \end{cases}$

Clear the fractions by multiplying equation (1) by 6 and equation (2) by 8.
$$\begin{cases} 3x+2y=6 & (1) \\ 6x+4y=8 & (2) \end{cases}$$
Now multiply equation (1) by -2, then add the two equations.
$$\begin{array}{r} -6x-4y=-24 \\ 6x+4y=8 \\ \hline 0=-16 \quad \text{False} \end{array}$$
The system is inconsistent.
The solution set is $\varnothing$.

75. $\begin{cases} \dfrac{x}{3}-\dfrac{y}{2}=1 & (1) \\ \dfrac{3y}{8}-\dfrac{x}{4}=-\dfrac{3}{4} & (2) \end{cases}$

Clear the fractions by multiplying equation (1) by 6 and equation (2) by 8.
$$\begin{cases} 2x-3y=6 & (1) \\ 3y-2x=-6 & (2) \end{cases}$$
Now add the two equations.
$$\begin{array}{r} 2x-3y=6 \\ -2x+3y=-6 \\ \hline 0=0 \ \checkmark \end{array}$$
The system is dependent.
Solve equation (1) for y in terms of x.
$\dfrac{x}{3}-\dfrac{y}{2}=1 \Rightarrow 2x-3y=6 \Rightarrow -3y=-2x+6 \Rightarrow$
$3y=2(x-3) \Rightarrow y=\dfrac{2}{3}(x-3)$
The solution set is $\left\{\left(x,\dfrac{2}{3}(x-3)\right)\right\}$.

77. $\begin{cases} \dfrac{2}{x}+\dfrac{5}{y}=-5 & (1) \\ \dfrac{3}{x}-\dfrac{2}{y}=-17 & (2) \end{cases}$

Let $u=\dfrac{1}{x}$ and $v=\dfrac{1}{y}$. Then the system becomes $\begin{cases} 2u+5v=-5 \\ 3u-2v=-17 \end{cases}$.

(*continued on next page*)

(continued)

Multiply the first equation by 2 and the second equation by 5, then add the equations:
$\begin{cases} 2u + 5v = -5 \\ 3u - 2v = -17 \end{cases} \Rightarrow$

$\begin{cases} 4u + 10v = -10 \\ 15u - 10v = -85 \end{cases} \Rightarrow 19u = -95 \Rightarrow u = -5$.

Substitute this value into the first equation to solve for v: $2(-5) + 5v = -5 \Rightarrow 5v = 5 \Rightarrow v = 1$.

$u = \dfrac{1}{x} \Rightarrow -5 = \dfrac{1}{x} \Rightarrow x = -\dfrac{1}{5}$.

$v = \dfrac{1}{y} \Rightarrow 1 = \dfrac{1}{y} \Rightarrow y = 1$.

The solution is $\left\{\left(-\dfrac{1}{5}, 1\right)\right\}$.

79. $\begin{cases} \dfrac{3}{x} + \dfrac{1}{y} = 4 & (1) \\ \dfrac{6}{x} - \dfrac{1}{y} = 2 & (2) \end{cases}$

Let $u = \dfrac{1}{x}$ and $v = \dfrac{1}{y}$. Then the system becomes $\begin{cases} 3u + v = 4 \\ 6u - v = 2 \end{cases}$. Add the equations and solve for u: $9u = 6 \Rightarrow u = \dfrac{2}{3}$. Substitute this value into the first equation and solve for v:

$3\left(\dfrac{2}{3}\right) + v = 4 \Rightarrow v = 2$.

$u = \dfrac{1}{x} \Rightarrow \dfrac{2}{3} = \dfrac{1}{x} \Rightarrow x = \dfrac{3}{2}$.

$v = \dfrac{1}{y} \Rightarrow 2 = \dfrac{1}{y} \Rightarrow y = \dfrac{1}{2}$.

The solution is $\left\{\left(\dfrac{3}{2}, \dfrac{1}{2}\right)\right\}$.

81. $\begin{cases} \dfrac{5}{x} + \dfrac{10}{y} = 3 & (1) \\ \dfrac{2}{x} - \dfrac{12}{y} = -2 & (2) \end{cases}$

Let $u = \dfrac{1}{x}$ and $v = \dfrac{1}{y}$. Then the system becomes $\begin{cases} 5u + 10v = 3 \\ 2u - 12v = -2 \end{cases}$. Multiply the first equation by 2 and the second equation by -5, then add the equation to solve for v:

$\begin{cases} 5u + 10v = 3 \\ 2u - 12v = -2 \end{cases} \Rightarrow \begin{cases} 10u + 20v = 6 \\ -10u + 60v = 10 \end{cases} \Rightarrow$

$80v = 16 \Rightarrow v = \dfrac{1}{5}$. Substitute this value into the first equation to solve for u:

$5u + 10\left(\dfrac{1}{5}\right) = 3 \Rightarrow 5u = 1 \Rightarrow u = \dfrac{1}{5}$.

$u = \dfrac{1}{x} \Rightarrow \dfrac{1}{5} = \dfrac{1}{x} \Rightarrow x = 5$.

$v = \dfrac{1}{y} \Rightarrow \dfrac{1}{5} = \dfrac{1}{y} \Rightarrow y = 5$.

The solution is $\{(5, 5)\}$.

83. $\begin{cases} \dfrac{2}{x} + \dfrac{1}{y} = 4 & (1) \\ x + 2y = 6xy & (2) \end{cases}$

Divide equation (2) by $6xy$.

$\begin{cases} \dfrac{2}{x} + \dfrac{1}{y} = 4 \\ \dfrac{1}{6y} + \dfrac{1}{3x} = 1 \end{cases} \Rightarrow \begin{cases} \dfrac{2}{x} + \dfrac{1}{y} = 4 & (1) \\ \dfrac{1}{3x} + \dfrac{1}{6y} = 1 & (2) \end{cases}$

Let $u = \dfrac{1}{x}$ and $v = \dfrac{1}{y}$. Then the system becomes $\begin{cases} 2u + v = 4 \\ \dfrac{1}{3}u + \dfrac{1}{6}v = 1 \end{cases}$.

Clear the fractions in equation (2) by multiplying by 6.

$\begin{cases} 2u + v = 4 \\ \dfrac{1}{3}u + \dfrac{1}{6}v = 1 \end{cases} \Rightarrow \begin{cases} 2u + v = 4 \\ 2u + v = 6 \end{cases}$

Multiply equation (2) by -1, then add the equations.
$\quad 2u + v = 4$
$\underline{-2u - v = -6}$
$\qquad 0 = -2 \quad$ False

The system is inconsistent.
The solution set is $\varnothing$.

8.1 Applying the Concepts

In exercises 85–88, the equilibrium point is the point that satisfies both the demand equation and the supply equation.

85. Add the equations to solve for p:
$\begin{cases} 2p + x = 140 \\ 12p - x = 280 \end{cases} \Rightarrow 14p = 420 \Rightarrow p = 30$.

Substitute this value into the first equation to solve for x: $2(30) + x = 140 \Rightarrow x = 80$.

The equilibrium point is $(80, 30)$.

87. Multiply the second equation by −1, then add the equations to solve for x:
$$\begin{cases} 2p + x = 25 \\ x - p = 13 \end{cases} \Rightarrow \begin{cases} 2p + x = 25 \\ p - x = -13 \end{cases} \Rightarrow 3p = 12 \Rightarrow$$
$p = 4$. Substitute this value into the second equation to solve for x: $x - 4 = 13 \Rightarrow x = 17$. The equilibrium point is $(17, 4)$.

89. Let x = the diameter of the largest pizza, and let y = the diameter of the smallest pizza. Then
$$\begin{cases} x + y = 29 \\ x - y = 13 \end{cases} \Rightarrow 2x = 42 \Rightarrow x = 21.$$
$21 + y = 29 \Rightarrow y = 8$.
The largest pizza has diameter 21 inches, and the smallest pizza has diameter 8 inches.

91. Let x = the percentage of paper trash, and let y = the percentage of plastic trash. If the total amount of trash is t, then
$$\begin{cases} x + y = 48 \\ x = 5y \end{cases}.$$
Substitute the expression for x from the second equation into the first equation to solve for y:
$5y + y = 48 \Rightarrow y = 8$. Substitute this value into the first equation to solve for x:
$x + 8 = 48 \Rightarrow x = 40$. So, 8% of the trash is plastic and 40% is paper.

93. Let x = the number of beads, and let y = the number of doubloons. Then
$$\begin{cases} 0.4x + 0.3y = 265 \\ x + y = 770 \end{cases}.$$ Multiply the second equation by −0.3 and then add the equations to solve for x: $\begin{cases} 0.4x + 0.3y = 265 \\ x + y = 770 \end{cases} \Rightarrow$
$\begin{cases} 0.4x + 0.3y = 265 \\ -0.3x - 0.3y = -231 \end{cases} \Rightarrow 0.1x = 34 \Rightarrow x = 340.$
Substitute this value into the second equation to solve for y: $340 + y = 770 \Rightarrow y = 430$. Levon bought 340 beads and 430 doubloons.

95. Let x = the number of Egg McMuffins, and let y = the number of Breakfast Burritos. Then
$$\begin{cases} 27x + 21y = 123 \\ 17x + 13y = 77 \end{cases}.$$ Multiply the first equation by 13 and the second equation by −21, then add the equations to solve for x:
$\begin{cases} 27x + 21y = 123 \\ 17x + 13y = 77 \end{cases} \Rightarrow \begin{cases} 351x + 273y = 1599 \\ -357x - 273y = -1617 \end{cases} \Rightarrow$
$-6x = -18 \Rightarrow x = 3$.

Substitute this value into the first equation to solve for y:
$27(3) + 21y = 123 \Rightarrow 21y = 42 \Rightarrow y = 2$. You will need to eat three Egg McMuffins and two Breakfast Burritos.

97. Let x = the amount invested at 7.5%, and let y = the amount invested at 12%. Then
$$\begin{cases} x + y = 50{,}000 \\ 0.075x + 0.12y = 5190 \end{cases}.$$ Solve the first equation for x and substitute this value into the second equation to solve for y:
$x + y = 50{,}000 \Rightarrow x = 50{,}000 - y$.
$0.075(50{,}000 - y) + 0.12y = 5190 \Rightarrow$
$3750 - 0.075y + 0.12y = 5190 \Rightarrow$
$0.045y = 1440 \Rightarrow y = \$32{,}000$.
Substitute this value into the first equation to solve for x:
$x + 32{,}000 = 50{,}000 \Rightarrow x = 18{,}000$.
Mrs. García invested $18,000 at 7.5% and $32,000 at 12%.

99. Let x = the amount earned tutoring, and let y = the amount earned working at McDougal's.
Then $\begin{cases} x = 2y \\ \dfrac{x + y}{2} = 11.25 \end{cases}.$
Substitute the expression for x from the first equation into the second equation and solve for y. $\dfrac{2y + y}{2} = 11.25 \Rightarrow 3y = 22.50 \Rightarrow y = 7.50$.
Substitute this value into the first equation to solve for x: $x = 2(7.50) = 15$. She earned $15 tutoring and $7.50 working at McDougal's.

101. Let x = the number of pounds of the herb, and let y = the number of pounds of tea. Then
$$\begin{cases} x + y = 100 \\ 5.50x + 3.20y = 3.66(100) \end{cases}.$$ Solve the first equation for x, and substitute this expression into the second equation to solve for y:
$x + y = 100 \Rightarrow x = 100 - y$.
$5.5(100 - y) + 3.20y = 3.66(100) \Rightarrow$
$550 - 5.5y + 3.2y = 366 \Rightarrow -2.3y = -184 \Rightarrow$
$y = 80$. Substitute this value into the first equation to solve for x: $x + 80 = 100 \Rightarrow x = 20$. There are 20 pounds of the herb and 80 pounds of tea in the mixture.

103. Let $x =$ the speed of the plane in still air, and let $y =$ the wind speed. So, the speed of the plane with the wind is $x + y$, and the speed of the plane against the wind is $x - y$. Then, using the fact that rate × time = distance, we have
$$\begin{cases} 5(x+y) = 3000 \\ 6(x-y) = 3000 \end{cases} \Rightarrow \begin{cases} x+y = 600 \\ x-y = 500 \end{cases}.$$ Add the equations to solve for x: $2x = 1100 \Rightarrow x = 550$. Substitute this value into the first equation to solve for y: $5(550+y) = 3000 \Rightarrow$ $550 + y = 600 \Rightarrow y = 50$. The speed of the plane is 550 kph, and the wind speed is 50 kph.

105. a. $C(x) = y = 2x + 30,000$; $R(x) = y = 3.50x$

b.

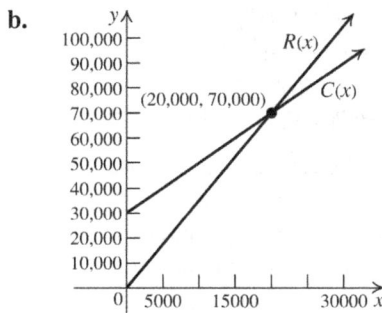

c. $\begin{cases} y = 2x + 30,000 \\ y = 3.5x \end{cases}$.

Substitute the expression for y from the second equation into the first equation to solve for x:
$2x + 30,000 = 3.5x \Rightarrow 1.5x = 30,000 \Rightarrow$
$x = 20,000$ magazines

107. Let $x =$ Shanaysha's weekly sales. Then her salary at store B is $150 + 0.04x$.
$150 + 0.04x > 400 \Rightarrow 0.04x > 250 \Rightarrow x > 6250$.
Her weekly sales should be more than \$6250.

8.1 Beyond the Basics

109. $\begin{cases} 2\log_3 x + 3\log_3 y = 8 & (1) \\ 3\log_3 x - \log_3 y = 1 & (2) \end{cases}$

Multiply equation (2) by 3. Then add the two equations.
$2\log_3 x + 3\log_3 y = 8$
$9\log_3 x - 3\log_3 y = 3$
$\overline{ 11\log_3 x = 11}$
Solve for x.
$11\log_3 x = 11 \Rightarrow \log_3 x = 1 \Rightarrow x = 3$

Substitute $x = 3$ into equation (2) and solve for y:
$3\log_3 3 - \log_3 y = 1 \Rightarrow 3(1) - \log_3 y = 1 \Rightarrow$
$-\log_3 y = -2 \Rightarrow \log_3 y = 2 \Rightarrow y = 3^2 = 9$
Be sure to check your answer in the original system of equations. The solution is $\{(3, 9)\}$.

111. $\begin{cases} 3e^x - 4e^y = 4 & (1) \\ 2e^x + 5e^y = 18 & (2) \end{cases}$

Multiply equation (1) by 5 and equation (2) by 4. Then add the equations.
$15e^x - 20e^y = 20$
$8e^x + 20e^y = 72$
$\overline{ 23e^x = 92}$
Now solve for x.
$23e^x = 92 \Rightarrow e^x = \dfrac{92}{23} \Rightarrow x = \ln 4$.
Substitute this value in equation (1) and solve for y.
$3e^{\ln 4} - 4e^y = 4 \Rightarrow 3(4) - 4e^y = 4 \Rightarrow$
$-4e^y = -8 \Rightarrow e^y = 2 \Rightarrow y = \ln 2$
The solution is $(\ln 4, \ln 2)$.

113. First solve the system $\begin{cases} x + 2y = 7 \\ 3x + 5y = 11 \end{cases}$.

Multiply the first equation by –3, then add the equations to solve for y:
$\begin{cases} x + 2y = 7 \\ 3x + 5y = 11 \end{cases} \Rightarrow \begin{cases} -3x - 6y = -21 \\ 3x + 5y = 11 \end{cases} \Rightarrow -y = -10 \Rightarrow$
$y = 10$. Substitute this value into the first equation to solve for x: $x + 2(10) = 7 \Rightarrow$
$x = -13$. Now substitute $(-13, 10)$ into the third equation to solve for c:
$-13c + 3(10) = 4 \Rightarrow -13c = -26 \Rightarrow c = 2$.

115. $\begin{cases} 2x + cy = 11 \\ 5x - 7y = 5 \end{cases}$

The system is inconsistent if the lines are parallel, i.e., if the coefficients of one line are multiples of the coefficients of the other line, but the constant in the first line is not the same multiple of the constant in the second line.
$2x = \dfrac{2}{5}(5x)$, so $c = \dfrac{2}{5}(-7) = -\dfrac{14}{5}$

117. $\ell_1 = a_1 x + b_1 y + c_1 = 0$
$\ell_2 = a_2 x + b_2 y + c_2 = 0 \Rightarrow \ell_1 + k\ell_2 = 0$ (1)
If $P(h, k)$ is the point of intersection of ℓ_1 and ℓ_2, then $a_1 h + b_1 k + c_1 = 0$ and $a_2 h + b_2 k + c_2 = 0$. Since $\ell_1 + k\ell_2 = 0$, P lies on this line also. Therefore,
$\ell_1 + k\ell_2 = a_1 x + b_1 y + c_1 + k(a_2 x + b_2 y + c_2)$
$= 0$
is an equation of the line passing through P.

119. To find the intersection of the two lines, solve the system $\begin{cases} 5x + 2y = 7 \\ 6x - 5y = 38 \end{cases}$. Multiply the first equation by 5 and the second equation by 2, then add the equations to solve for x:
$\begin{cases} 5x + 2y = 7 \\ 6x - 5y = 38 \end{cases} \Rightarrow \begin{cases} 25x + 10y = 35 \\ 12x - 10y = 76 \end{cases} \Rightarrow 37x = 111 \Rightarrow$
$x = 3$. Substitute this value into the first equation to solve for y: $5(3) + 2y = 7 \Rightarrow$
$2y = -8 \Rightarrow y = -4$. So the line passes through the point $(3, -4)$. Because the line also passes through the point $(1, 3)$, the slope of the line is
$\dfrac{3 - (-4)}{1 - 3} = -\dfrac{7}{2}$. The equation of the line is
$y - 3 = -\dfrac{7}{2}(x - 1) \Rightarrow y = -\dfrac{7}{2}x + \dfrac{13}{2}$. Verify graphically:

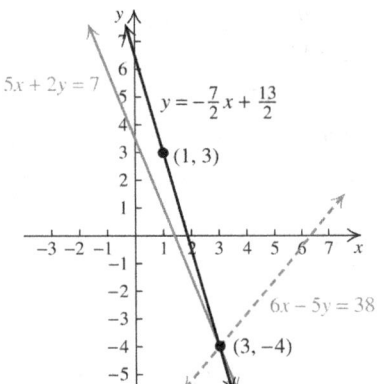

8.1 Critical Thinking/Discussion/Writing

121. a. There is only one solution if the equations are independent. Solve for x by multiplying the first equation by b_2 and multiplying the second equation by $-b_1$, then adding the equations:

$\begin{cases} a_1 x + b_1 y = c_1 \\ a_2 x + b_2 y = c_2 \end{cases} \Rightarrow$
$\begin{cases} b_2 a_1 x + b_2 b_1 y = b_2 c_1 \\ -b_1 a_2 x - b_1 b_2 y = -b_1 c_2 \end{cases} \Rightarrow$
$b_2 a_1 x - b_1 a_2 x = b_2 c_1 - b_1 c_2 \Rightarrow$
$x(b_2 a_1 - b_1 a_2) = b_2 c_1 - b_1 c_2 \Rightarrow$
$x = \dfrac{b_2 c_1 - b_1 c_2}{a_1 b_2 - a_2 b_1}$

Solve for y by multiplying the first equation by a_2 and multiplying the second equation by $-a_1$, then adding the equations.

$\begin{cases} a_1 x + b_1 y = c_1 \\ a_2 x + b_2 y = c_2 \end{cases} \Rightarrow$
$\begin{cases} a_1 a_2 x + a_2 b_1 y = a_2 c_1 \\ -a_1 a_2 x - a_1 b_2 y = -a_1 c_2 \end{cases} \Rightarrow$
$a_2 b_1 y - a_1 b_2 y = a_2 c_1 - a_1 c_2 \Rightarrow$
$y(a_2 b_1 - a_1 b_2) = a_2 c_1 - a_1 c_2 \Rightarrow$
$y = \dfrac{a_2 c_1 - a_1 c_2}{a_2 b_1 - a_1 b_2} = \dfrac{a_1 c_2 - a_2 c_1}{a_1 b_2 - a_2 b_1}$.

Note that the denominator of x and y, $a_1 b_2 - a_2 b_1$, cannot equal zero.

b. There is no solution if the equations are inconsistent. They are inconsistent if they are parallel. The slopes of the lines are equal and $\dfrac{b_1}{a_1} = \dfrac{b_2}{a_2} \Rightarrow \dfrac{a_1}{a_2} = \dfrac{b_1}{b_2}$. The intercepts are different. If $c_1 = c_2$, then
$\dfrac{c_1}{a_1} \neq \dfrac{c_2}{a_2}$.

c. There are infinitely many solutions if the equations are dependent. From part (b), we have $\dfrac{a_1}{a_2} = \dfrac{b_1}{b_2}$. The intercepts of the lines must be the same, so $\dfrac{a_1}{a_2} = \dfrac{b_1}{b_2} = \dfrac{c_1}{c_2}$.

122. a. The slope of ℓ_1 is $-\dfrac{3}{4}$, so the slope of the perpendicular is $\dfrac{4}{3}$. The equation of ℓ_2 is
$y - 8 = \dfrac{4}{3}(x - 5) \Rightarrow 3y - 24 = 4x - 20 \Rightarrow$
$3y - 4x = 4$.

b. To find Q, solve the system $\begin{cases} 3x+4y=12 \\ -4x+3y=4 \end{cases}$.

Multiply the first equation by 4 and the second equation by 3, then add the equations to solve for x: $\begin{cases} 3x+4y=12 \\ -4x+3y=4 \end{cases} \Rightarrow$

$\begin{cases} 12x+16y=48 \\ -12x+9y=12 \end{cases} \Rightarrow 25y=60 \Rightarrow y=\dfrac{12}{5}$.

Substitute this value into the first equation to solve for x: $3x+4\left(\dfrac{12}{5}\right)=12 \Rightarrow 3x=\dfrac{12}{5} \Rightarrow$

$x=\dfrac{4}{5}$. The coordinates of Q are $\left(\dfrac{4}{5},\dfrac{12}{5}\right)$.

c. $d(P,Q)=\sqrt{\left(5-\dfrac{4}{5}\right)^2+\left(8-\dfrac{12}{5}\right)^2}=\sqrt{49}=7$

123. a. The slope of ℓ_1 is $-\dfrac{a}{b}$, so the slope of the perpendicular is $\dfrac{b}{a}$. The equation of ℓ_2 is

$y-y_1=\dfrac{b}{a}(x-x_1) \Rightarrow$

$ay-ay_1=bx-bx_1 \Rightarrow ay-bx=ay_1-bx_1$ or $b(x-x_1)-a(y-y_1)=0$.

b. The intersection of ℓ_1 and ℓ_2 is the solution of the system

$\begin{cases} ax+by=-c \\ -bx+ay=ay_1-bx_1 \end{cases} \Rightarrow$

$\begin{cases} abx+b^2y=-bc \\ -abx+a^2y=a^2y_1-abx_1 \end{cases} \Rightarrow$

$y(a^2+b^2)=-bc+a^2y_1-abx_1 \Rightarrow$

$y=\dfrac{-bc+a^2y_1-abx_1}{a^2+b^2}$

$\begin{cases} ax+by=-c \\ -bx+ay=ay_1-bx_1 \end{cases} \Rightarrow$

$\begin{cases} -a^2x-aby=ac \\ -b^2x+aby=aby_1-b^2x_1 \end{cases} \Rightarrow$

$-x(a^2+b^2)=ac+aby_1-b^2x_1 \Rightarrow$

$x=\dfrac{-ac-aby_1+b^2x_1}{a^2+b^2}$

$=-\dfrac{ac+aby_1-b^2x_1}{a^2+b^2}$

The intersection of the two lines is

$\left(-\dfrac{ac+aby_1-b^2x_1}{a^2+b^2},\dfrac{-abx_1+a^2y_1-bc}{a^2+b^2}\right)$.

c. Now find the distance between (x_1,y_1) and

$\left(-\dfrac{ac+aby_1-b^2x_1}{a^2+b^2},\dfrac{-bc+a^2y_1-abx_1}{a^2+b^2}\right)$:

$\sqrt{\left(x_1+\dfrac{ac+aby_1-b^2x_1}{a^2+b^2}\right)^2+\left(y_1-\dfrac{-bc+a^2y_1-abx_1}{a^2+b^2}\right)^2}$

$=\sqrt{\left(\dfrac{a(ax_1+by_1+c)}{a^2+b^2}\right)^2+\left(\dfrac{b(ax_1+by_1+c)}{a^2+b^2}\right)^2}$

$=\sqrt{\dfrac{a^2(ax_1+by_1+c)^2}{(a^2+b^2)^2}+\dfrac{b^2(ax_1+by_1+c)^2}{(a^2+b^2)^2}}$

$=\sqrt{\dfrac{(a^2+b^2)(ax_1+by_1+c)^2}{(a^2+b^2)^2}}$

$=\sqrt{\dfrac{(ax_1+by_1+c)^2}{a^2+b^2}}=\dfrac{|ax_1+by_1+c|}{\sqrt{a^2+b^2}}$.

124. a. $a=1, b=1, c=-7, x_1=2, y_1=3$

$\dfrac{|ax_1+by_1+c|}{\sqrt{a^2+b^2}}=\dfrac{|1(2)+1(3)-7|}{\sqrt{1^2+1^2}}=\dfrac{2}{\sqrt{2}}=\sqrt{2}$

b. $a=2, b=-1, c=3, x_1=-2, y_1=5$

$\dfrac{|ax_1+by_1+c|}{\sqrt{a^2+b^2}}=\dfrac{|2(-2)-1(5)+3|}{\sqrt{2^2+(-1)^2}}$

$=\dfrac{6}{\sqrt{5}}=\dfrac{6\sqrt{5}}{5}$

c. $a=5, b=-2, c=-7, x_1=3, y_1=4$

$\dfrac{|ax_1+by_1+c|}{\sqrt{a^2+b^2}}=\dfrac{|5(3)-2(4)-7|}{\sqrt{3^2+4^2}}=0$

d. $a=a, b=b, c=c, x_1=0, y_1=0$

$\dfrac{|ax_1+by_1+c|}{\sqrt{a^2+b^2}}=\dfrac{|a(0)+b(0)+c|}{\sqrt{a^2+b^2}}$

$=\dfrac{|c|}{\sqrt{a^2+b^2}}$

125.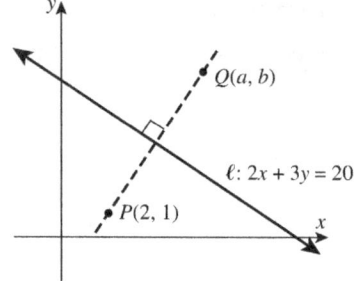

Let $Q(a, b)$ be the reflection point. Using the hint, we have two equations, one developed from the fact that the midpoint of $\overline{PQ}$ lies on line ℓ, and the other developed from the fact that ℓ is perpendicular to the line containing the segment $\overline{PQ}$.

Equation (1): The midpoint of $\overline{PQ}$ is $M\left(\dfrac{a+2}{2}, \dfrac{b+1}{2}\right)$. Since M lies on line ℓ, we have $2\left(\dfrac{a+2}{2}\right) + 3\left(\dfrac{b+1}{2}\right) = 20$. We can simplify this to
$(a+2) + \dfrac{3}{2}b + \dfrac{3}{2} = 20 \Rightarrow a + \dfrac{3}{2}b = \dfrac{33}{2} \Rightarrow$
$2a + 3b = 33$

Equation (2): The slope of ℓ is $-\dfrac{2}{3}$, so the slope of the line perpendicular to line ℓ is $\dfrac{3}{2}$.

Using the formula for the slope of $\overline{PQ}$, we have
$\dfrac{3}{2} = \dfrac{b-1}{a-2} \Rightarrow 3a - 6 = 2b - 2 \Rightarrow 3a - 2b = 4$.

So, the system is
$\begin{cases} 2a + 3b = 33 & (1) \\ 3a - 2b = 4 & (2) \end{cases}$

Multiply the first equation by 2 and the second equation by 3. Then add the two equations, then solve for a.
$4a + 6b = 66$
$\underline{9a - 6b = 12}$
$\quad 13a = 78 \Rightarrow a = 6$

Substitute $a = 6$ into equation (2) and solve for b.
$3(6) - 2b = 4 \Rightarrow -2b = -14 \Rightarrow b = 7$
Be sure to check your answer.
The reflection point is $Q(6, 7)$.

126.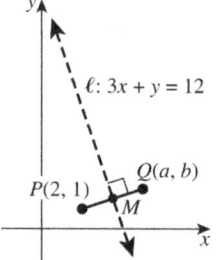

Let $Q(a, b)$ be the reflection point. Using the hint from exercise 122, we have two equations, one developed from the fact that the midpoint of $\overline{PQ}$ lies on line ℓ, and the other developed from the fact that ℓ is perpendicular to the line containing the segment $\overline{PQ}$.

Equation (1): The midpoint of $\overline{PQ}$ is $M\left(\dfrac{a+2}{2}, \dfrac{b+1}{2}\right)$. Since M lies on line ℓ, we have $3\left(\dfrac{a+2}{2}\right) + \left(\dfrac{b+1}{2}\right) = 12$. We can simplify this to
$3(a+2) + (b+1) = 24 \Rightarrow$
$3a + 6 + b + 1 = 24 \Rightarrow 3a + b = 17$

Equation (2): The slope of ℓ is -3, so the slope of the line perpendicular to line ℓ is $\dfrac{1}{3}$.

Using the formula for the slope of $\overline{PQ}$, we have
$\dfrac{1}{3} = \dfrac{b-1}{a-2} \Rightarrow 3b - 3 = a - 2 \Rightarrow a - 3b = -1$.

So, the system is
$\begin{cases} 3a + b = 17 & (1) \\ a - 3b = -1 & (2) \end{cases}$

Multiply the first equation by 3. Then add the two equations, then solve for a.
$9a + 3b = 51$
$\underline{a - 3b = -1}$
$\quad 10a = 50 \Rightarrow a = 5$

Substitute $a = 5$ into equation (1) and solve for b.
$3(5) + b = 17 \Rightarrow b = 2$
Be sure to check your answer.
The reflection point is $Q(5, 2)$.

8.1 Getting Ready for the Next Section

127. $y + 2z = 7;\ z = 2$
$y + 2(2) = 7 \Rightarrow y + 4 = 7 \Rightarrow y = 3$

129. $2x + 3y + 5z = 21;\ y = 1;\ z = 2$
$2x + 3(1) + 5(2) = 21 \Rightarrow 2x + 13 = 21 \Rightarrow$
$2x = 8 \Rightarrow x = 4$

131. $2x + 3y - 2z = 5;\ y - z = 4,\ z = -3$
$y - (-3) = 4 \Rightarrow y + 3 = 4 \Rightarrow y = 1$
$2x + 3(1) - 2(-3) = 5 \Rightarrow 2x + 9 = 5 \Rightarrow$
$2x = -4 \Rightarrow x = -2$

8.2 Systems of Linear Equations in Three Variables

8.2 Practice Problems

1. Substituting the ordered triple into the system
$$\begin{cases} x + y + z = 1 \\ 3x + 4y + z = -4, \\ 2x + y + 2z = 5 \end{cases}$$
we find that $(2, -3, 2)$ is a solution.
$$\begin{cases} 2 + (-3) + 2 = 1 \\ 3(2) + 4(-3) + 2 = -4 \\ 2(2) + (-3) + 2(2) = 5 \end{cases}$$

2. $\begin{cases} 2x + 5y = 1 & (1) \\ x - 3y + 2z = 1 & (2) \\ -x + 2y + z = 7 & (3) \end{cases}$

Interchange equations (1) and (2).
$$\begin{cases} x - 3y + 2z = 1 & (2) \\ 2x + 5y = 1 & (1) \\ -x + 2y + z = 7 & (3) \end{cases}$$

Multiply equation (2) by -2, then add the resulting equation and equation (1) to eliminate x, then replace equation (1) with the resulting equation.
$-2x + 6y - 4z = -2$ (2)
$\underline{2x + 5y\ \ \ \ \ \ \ \ = 1}$ (1)
$\ \ \ \ \ 11y - 4z = -1$ (4)

$$\begin{cases} x - 3y + 2z = 1 & (2) \\ 11y - 4z = -1 & (4) \\ -x + 2y + z = 7 & (3) \end{cases}$$

Now add equations (2) and (3), then replace equation (3) with the resulting equation.
$x - 3y + 2z = 1$ (2)
$\underline{-x + 2y + z = 7}$ (3)
$\ \ \ \ \ \ -y + 3z = 8$ (5)

$$\begin{cases} x - 3y + 2z = 1 & (2) \\ 11y - 4z = -1 & (4) \\ -y + 3z = 8 & (5) \end{cases}$$

Now multiply equation (5) by 11, add the resulting equation to equation (4), then solve for z.
$11y - 4z = -1$ (4)
$\underline{-11y + 33z = 88}$ (5)
$\ \ \ \ \ \ \ \ \ \ \ 29z = 87$
$\ \ \ \ \ \ \ \ \ \ \ \ \ z = 3$ (6)

The system becomes
$$\begin{cases} x - 3y + 2z = 1 & (2) \\ 11y - 4z = -1 & (4). \\ z = 3 & (6) \end{cases}$$

Now substitute $z = 3$ into equation (4) and solve for y.
$11y - 4(3) = -1 \Rightarrow 11y = 11 \Rightarrow y = 1$

Substitute the values of y and z into equation (2) and solve for x.
$x - 3(1) + 2(3) = 1 \Rightarrow x + 3 = 1 \Rightarrow x = -2$

So, we have $x = -2, y = 1$, and $z = 3$. Now check in the original system.
$$\begin{cases} 2(-2) + 5(1) = 1 \ \checkmark \\ -2 - 3(1) + 2(3) = 1 \ \checkmark \\ -(-2) + 2(1) + 3 = 7 \ \checkmark \end{cases}$$

The solution is $\{(-2, 1, 3)\}$.

3. $\begin{cases} 2x + 2y + 2z = 12 & (1) \\ -3x + y - 11z = -6 & (2) \\ 2x + y + 4z = -8 & (3) \end{cases}$

Divide equation (1) by 2 and replace equation (1) with the resulting equation.
$$\begin{cases} x + y + z = 6 & (4) \\ -3x + y - 11z = -6 & (2) \\ 2x + y + 4z = -8 & (3) \end{cases}$$

Multiply equation (4) by 3, add the resulting equation to equation (2), and replace equation (2) with the result. Similarly, multiply equation (4) by -2, add the resulting equation to equation (3), and replace equation (3) with the result.
$$\begin{cases} x + y + z = 6 & (4) \\ 4y - 8z = 12 & (5) \\ -y + 2z = -20 & (6) \end{cases}$$

Multiply equation (6) by 4, add the resulting equation to equation (5), and replace equation (6) with the result.
$4y - 8z = 12$
$\underline{-4y + 8z = -80}$
$\ \ \ \ \ \ \ \ \ \ 0 = -68$

$$\begin{cases} x + y + z = 6 & (4) \\ 4y - 8z = 12 & (5) \\ 0 = -68 & (7) \end{cases}$$

Equation (7) is false, so the solution set of the system is $\varnothing$.

4. $\begin{cases} x + y + z = 5 & (1) \\ -4x - y - 8z = -29 & (2) \\ 2x + 5y - 2z = 1 & (3) \end{cases}$

   Multiply equation (1) by 4, add the resulting equation to equation (2), and replace equation (2) with the result. Similarly, multiply equation (1) by -2, add the resulting equation to equation (3), and replace equation (3) with the resulting equation.
   $\begin{cases} x + y + z = 5 & (1) \\ 3y - 4z = -9 & (4) \\ 3y - 4z = -9 & (5) \end{cases}$

   Now multiply equation (4) by -1, add the result to equation (5), and replace equation (5) with the resulting equation.
   $\begin{cases} x + y + z = 5 & (1) \\ 3y - 4z = -9 & (4) \\ 0 = 0 & (6) \end{cases}$

   Solve equation (4) for y.
   $3y - 4z = -9 \Rightarrow y = \dfrac{-9 + 4z}{3} = -3 + \dfrac{4}{3}z$

   Substitute this expression into equation (1) and solve for x.
   $x + y + z = 5 \Rightarrow x + \left(-3 + \dfrac{4}{3}z\right) + z = 5 \Rightarrow$
   $x + \dfrac{7}{3}z - 3 = 5 \Rightarrow x = -\dfrac{7}{3}z + 8$

   The solution set is $\left\{\left(8 - \dfrac{7}{3}z, -3 + \dfrac{4}{3}z, z\right)\right\}$.

5. $\begin{cases} x + 3y + 2z = 4 & (1) \\ 2x + 7y - z = 5 & (2) \end{cases}$

   Eliminate x.
   $\begin{array}{r} -2x - 6y - 4z = -8 \\ 2x + 7y - z = 5 \\ \hline y - 5z = -3 \end{array}$

   $\begin{cases} x + 3y + 2z = 4 & (1) \\ y - 5z = -3 & (3) \end{cases}$

   Solve equation (3) for y in terms of z.
   $y - 5z = -3 \Rightarrow y = 5z - 3$

   Substitute this expression into equation (1) and then solve for x in terms of z.
   $x + 3(5z - 3) + 2z = 4 \Rightarrow$
   $x + 15z - 9 + 2z = 4 \Rightarrow x = -17z + 13$

   The solution set is $\{(-17z + 13, 5z - 3, z)\}$.

6. The system is
   $\begin{cases} x + y = 0.65 & \text{(beam 1)} \\ x + z = 0.70 & \text{(beam 2)} \\ y + z = 0.55 & \text{(beam 3)} \end{cases}$

   Multiply the first equation by -1, add the result to the second equation, and replace the second equation with the new equation:
   $-1(x + y = 0.65) \Rightarrow -x - y = -0.65$
   $\begin{cases} -x - y = -0.65 \\ x + z = 0.70 \end{cases} \Rightarrow -y + z = 0.05$

   $\begin{cases} x + y = 0.65 \\ x + z = 0.70 \\ y + z = 0.55 \end{cases} \Rightarrow \begin{cases} x + y = 0.65 \\ -y + z = 0.05 \\ y + z = 0.55 \end{cases}$

   Add the second and third equations and solve for z: $2z = 0.60 \Rightarrow z = 0.30$. Substituting this value into the original second equation, we have $x + 0.30 = 0.70 \Rightarrow x = 0.40$. Substituting this value into the original first equation, we have $0.40 + y = 0.65 \Rightarrow y = 0.25$. Referring to table 8.1, we see that cell A contains bone (since $x = 0.40$), cell B contains healthy tissue (since $y = 0.25$), and cell C contains tumorous tissue (since $z = 0.30$).

8.2 Concepts and Vocabulary

1. Systems of equations that have the same solution sets are called <u>equivalent</u> systems.

3. If any of the equations in a system has no solution, then the system is <u>inconsistent</u>.

5. True

7. False. The system will have no solution.

8.2 Building Skills

9. Substituting the ordered triple into the system
   $\begin{cases} 2x - 2y - 3z = 1 \\ 3y + 2z = -1 \\ y + z = 0 \end{cases}$
   we find that $(1, -1, 1)$ *is* a solution.
   $\begin{cases} 2(1) - 2(-1) - 3(1) = 1 \\ 3(-1) + 2(1) = -1 \\ -1 + 1 = 0 \end{cases}$

11. Substituting the ordered triple into the system
    $\begin{cases} x + 3y - 2z = 0 \\ 2x - y + 4z = 5 \\ x - 11y + 14z = 0 \end{cases}$
    we find that $(-10, 8, 7)$ *is not* a solution.
    $\begin{cases} -10 + 3(8) - 2(7) = 0 \\ 2(-10) - 8 + 4(7) = 0 \neq 5 \\ (-10) - 11(8) + 14(7) = 0 \end{cases}$

13. To solve the system $\begin{cases} x+y+z=4 \\ y-2z=4 \\ z=-1 \end{cases}$,

substitute $z=-1$ into the second equation, and solve for y: $y-2(-1)=4 \Rightarrow y=2$.
Now substitute the values for y and z into the first equation, and solve for x:
$x+2-1=4 \Rightarrow x=3$.
The solution is $\{(3, 2, -1)\}$.

15. To solve the system $\begin{cases} x-5y+3z=-1 \\ y-2z=-6 \\ z=4 \end{cases}$,

substitute $z=4$ into the second equation, and solve for y: $y-2(4)=-6 \Rightarrow y=2$.
Now substitute the values for y and z into the first equation, and solve for x:
$x-5(2)+3(4)=-1 \Rightarrow x=-3$. The solution is $\{(-3, 2, 4)\}$.

17. $\begin{cases} 2x-2y-3z=1 \\ 3y+2z=-1 \\ y+z=0 \end{cases} \Rightarrow \begin{cases} 2x-2y-3z=1 \\ y+z=0 \\ 3y+2z=-1 \end{cases}$.

Multiply the first equation by $1/2$:
$\frac{1}{2}(2x-2y-3z=1) \Rightarrow x-y-\frac{3}{2}z=\frac{1}{2}$
Now multiply the second equation by -3, add the result to the third equation, and replace the third equation with the new equation:
$-3(y+z=0) \Rightarrow -3y-3z=0$.
$\begin{cases} -3y-3z=0 \\ 3y+2z=-1 \end{cases} \Rightarrow -z=-1 \Rightarrow z=1$.

So the system becomes
$\begin{cases} x-y-\frac{3}{2}z=\frac{1}{2} \\ y+z=0 \\ z=1 \end{cases}$

19. Multiply the first equation by -2, add the result to the second equation, and replace the second equation with the new equation:
$-2(x+3y-2z=0) \Rightarrow -2x-6y+4z=0$
$\begin{cases} -2x-6y+4z=0 \\ 2x-y+4z=5 \end{cases} \Rightarrow -7y+8z=5$.

The system becomes:
$\begin{cases} x+3y-2z=0 \\ 2x-y+4z=5 \\ x-11y+14z=0 \end{cases} \Rightarrow \begin{cases} x+3y-2z=0 \\ -7y+8z=5 \\ x-11y+14z=0 \end{cases}$

Multiply the first equation by -1, add the result to the third equation, and replace the third equation with the new equation:
$-1(x+3y-2z=0) \Rightarrow -x-3y+2z=0$
$\begin{cases} -x-3y+2z=0 \\ x-11y+14z=5 \end{cases} \Rightarrow -14y+16z=5$.

The system becomes:
$\begin{cases} x+3y-2z=0 \\ -7y+8z=5 \\ x-11y+14z=0 \end{cases} \Rightarrow \begin{cases} x+3y-2z=0 \\ -7y+8z=5 \\ -14y+16z=5 \end{cases}$

Multiply the second equation by -2, add the result to the third equation, and replace the third equation with the new equation:
$-2(-7y+8z=5) \Rightarrow 14y-16z=-10$
$\begin{cases} 14y-16z=-10 \\ -14y+16z=5 \end{cases} \Rightarrow 0=-5$.

The system becomes:
$\begin{cases} x+3y-2z=0 \\ -7y+8z=5 \\ x-11y+14z=0 \end{cases} \Rightarrow \begin{cases} x+3y-2z=0 \\ y-\frac{8}{7}z=\frac{5}{7} \\ 0=-5 \end{cases}$

21. $\begin{cases} x-y+z=6 \\ 2y+3z=5 \\ 2z=6 \end{cases} \Rightarrow \begin{cases} x-y+z=6 \\ 2y+3z=5 \\ z=3 \end{cases}$.

Substitute $z=3$ into the second equation to solve for y: $2y+3(3)=5 \Rightarrow y=-2$. Now substitute the values for y and z into the first equation to solve for x: $x-(-2)+3=6 \Rightarrow x=1$. The solution is $\{(1, -2, 3)\}$.

23. Multiply the first equation by $-3/4$ add the result to the second equation, then replace the second equation with the new equation:

$-\frac{3}{4}(4x+4y+4z=7) \Rightarrow$

$-3x-3y-3z=-\frac{21}{4}$

$\begin{cases} -3x-3y-3z=-\frac{21}{4} \\ 3x-8y=14 \end{cases} \Rightarrow -11y-3z=\frac{35}{4} \Rightarrow$

$y+\frac{3}{11}z=-\frac{35}{44}$

$\begin{cases} 4x+4y+4z=7 \\ 3x-8y=14 \\ 4z=-1 \end{cases} \Rightarrow \begin{cases} x+y+z=\frac{7}{4} \\ y+\frac{3}{11}z=-\frac{35}{44} \\ z=-\frac{1}{4} \end{cases}$

(continued on next page)

Chapter 8 Systems of Equations and Inequalities

(*continued*)

Substitute $z = -1/4$ into the second equation to solve for y:

$$y + \frac{3}{11}\left(-\frac{1}{4}\right) = -\frac{35}{44} \Rightarrow y = -\frac{8}{11}.$$

Now substitute the values for y and z into the first equation to solve for x:

$$4x + 4\left(-\frac{8}{11}\right) + 4\left(-\frac{1}{4}\right) = 7 \Rightarrow x = \frac{30}{11}.$$

The solution is $\left\{\left(\dfrac{30}{11}, -\dfrac{8}{11}, -\dfrac{1}{4}\right)\right\}$.

In exercises 25–46, be sure to check the answers by substituting the values into the original system of equations.

25. Multiply the first equation by -1, add the result to the second equation, and replace the second equation with the new equation:

$-1(x + y + z = 6) \Rightarrow -x - y - z = -6$

$\begin{cases} -x - y - z = -6 \\ x - y + z = 2 \end{cases} \Rightarrow -2y = -4$

$\begin{cases} x + y + z = 6 \\ x - y + z = 2 \\ 2x + y - z = 1 \end{cases} \Rightarrow \begin{cases} x + y + z = 6 \\ -2y = -4 \\ 2x + y - z = 1 \end{cases}$

Simplify the new second equation, and then multiply the first equation by -2, add the result to the third equation, and replace the third equation with the new equation:

$-2(x + y + z = 6) \Rightarrow -2x - 2y - 2z = -12$

$\begin{cases} -2x - 2y - 2z = -12 \\ 2x + y - z = 1 \end{cases} \Rightarrow -y - 3z = -11$

$\begin{cases} x + y + z = 6 \\ -2y = -4 \\ 2x + y - z = 1 \end{cases} \Rightarrow \begin{cases} x + y + z = 6 \\ y = 2 \\ -y - 3z = -11 \end{cases}$

Add the second and third equations, replacing the third equation with the new equation:

$\begin{cases} x + y + z = 6 \\ y = 2 \\ -y - 3z = -11 \end{cases} \Rightarrow \begin{cases} x + y + z = 6 \\ y = 2 \\ -3z = -9 \end{cases} \Rightarrow$

$\begin{cases} x + y + z = 6 \\ y = 2 \\ z = 3 \end{cases}$

Substitute the values for y and z into the original first equation to solve for x:
$x + 2 + 3 = 6 \Rightarrow x = 1$.
The solution is $\{(1, 2, 3)\}$.

27. Switch the first and second equations.

$\begin{cases} 2x + 3y + z = 9 \\ x + 2y + 3z = 6 \\ 3x + y + 2z = 8 \end{cases} \Rightarrow \begin{cases} x + 2y + 3z = 6 \\ 2x + 3y + z = 9 \\ 3x + y + 2z = 8 \end{cases}$

Multiply the first equation by -2, add the result to the second equation, then replace the second equation with the new equation:
$-2(x + 2y + 3z = 6) = -2x - 4y - 6z = -12$

$\begin{cases} -2x - 4y - 6z = -12 \\ 2x + 3y + z = 9 \end{cases} \Rightarrow -y - 5z = -3 \Rightarrow$

$y + 5z = 3$

$\begin{cases} x + 2y + 3z = 6 \\ 2x + 3y + z = 9 \\ 3x + y + 2z = 8 \end{cases} \Rightarrow \begin{cases} x + 2y + 3z = 6 \\ y + 5z = 3 \\ 3x + y + 2z = 8 \end{cases}$

Multiply the first equation by -3, add the result to the third equation, then replace the third equation with the new equation:
$-3(x + 2y + 3z = 6) = -3x - 6y - 9z = -18$

$\begin{cases} -3x - 6y - 9z = -18 \\ 3x + y + 2z = 8 \end{cases} \Rightarrow -5y - 7z = -10 \Rightarrow$

$5y + 7z = 10$

$\begin{cases} x + 2y + 3z = 6 \\ y + 5z = 3 \\ 3x + y + 2z = 8 \end{cases} \Rightarrow \begin{cases} x + 2y + 3z = 6 \\ y + 5z = 3 \\ 5y + 7z = 10 \end{cases}$

Multiply the second equation by -5, and add the result to the third equation, then replace the third equation with the new equation:
$-5(y + 5z = 3) \Rightarrow -5y - 25z = -15$

$\begin{cases} -5y - 25z = -15 \\ 5y + 7z = 10 \end{cases} \Rightarrow -18z = -5 \Rightarrow z = \dfrac{5}{18}$

$\begin{cases} x + 2y + 3z = 6 \\ y + 5z = 3 \\ 5y + 7z = 10 \end{cases} \Rightarrow \begin{cases} x + 2y + 3z = 6 \\ y + 5z = 3 \\ z = \dfrac{5}{18} \end{cases}$

Substitute the value of z into the second equation to solve for y:

$y + 5\left(\dfrac{5}{18}\right) = 3 \Rightarrow y = \dfrac{29}{18}$.

Substitute the values for y and z into the first equation to solve for x:

$x + 2\left(\dfrac{29}{18}\right) + 3\left(\dfrac{5}{18}\right) = 6 \Rightarrow x = \dfrac{35}{18}$.

The solution is $\left\{\left(\dfrac{35}{18}, \dfrac{29}{18}, \dfrac{5}{18}\right)\right\}$.

29. $\begin{cases} x - 4y + 7z = 14 \\ 3x + 8y - 2z = 13 \\ 7x - 8y + 26z = 5 \end{cases}$

Multiply the first equation by -3, add the result to the second equation, then replace the second equation with the new equation:
$-3(x - 4y + 7z = 14) \Rightarrow -3x + 12y - 21z = -42$
$\begin{cases} -3x + 12y - 21z = -42 \\ 3x + 8y - 2z = 3 \end{cases} \Rightarrow 20y - 23z = -39$

$\begin{cases} x - 4y + 7z = 14 \\ 3x + 8y - 2z = 13 \\ 7x - 8y + 26z = 5 \end{cases} \Rightarrow \begin{cases} x - 4y + 7z = 14 \\ 20y - 23z = -39 \\ 7x - 8y + 26z = 5 \end{cases}$

Multiply the first equation by -7, add the result to the third equation, then replace the third equation:
$-7(x - 4y + 7z = 14) \Rightarrow -7x + 28y - 49z = -98$
$\begin{cases} -7x + 28y - 49z = -98 \\ 7x - 8y + 26z = 5 \end{cases} \Rightarrow 20y - 23z = -93$

$\begin{cases} x - 4y + 7z = 14 \\ 20y - 23z = -39 \\ 7x - 8y + 26z = 5 \end{cases} \Rightarrow \begin{cases} x - 4y + 7z = 14 \\ 20y - 23z = -39 \\ 20y - 23z = -93 \end{cases}$

Multiplying the second equation by -1, and adding the result to the third equation gives:
$-1(20y - 23z = -39) \Rightarrow -20y + 23z = 39$
$\begin{cases} -20y + 23z = 39 \\ 20y - 23z = -93 \end{cases} \Rightarrow 0 = -54$ False

Thus, the system is inconsistent and the solution set is $\varnothing$.

31. Multiply the second equation by -2, add the result to the first equation, then replace the second equation with the new equation:
$-2(x + 3y - z = -2) \Rightarrow -2x - 6y + 2z = 4$
$\begin{cases} -2x - 6y + 2z = 4 \\ 2x + 3y + 2z = 7 \end{cases} \Rightarrow -3y + 4z = 11$

$\begin{cases} 2x + 3y + 2z = 7 \\ x + 3y - z = -2 \\ x - y + 2z = 8 \end{cases} \Rightarrow \begin{cases} 2x + 3y + 2z = 7 \\ -3y + 4z = 11 \\ x - y + 2z = 8 \end{cases}$

Multiply the third equation by -2, add the result to the first equation, then replace the third equation with the new equation:
$-2(x - y + 2z = 8) \Rightarrow -2x + 2y - 4z = -16$
$\begin{cases} -2x + 2y - 4z = -16 \\ 2x + 3y + 2z = 7 \end{cases} \Rightarrow 5y - 2z = -9$

$\begin{cases} 2x + 3y + 2z = 7 \\ x + 3y - z = -2 \\ x - y + 2z = 8 \end{cases} \Rightarrow \begin{cases} 2x + 3y + 2z = 7 \\ -3y + 4z = 11 \\ 5y - 2z = -9 \end{cases}$

Multiply the third equation by 2, and add the result to the second equation to solve for y:
$2(5y - 2z = -9) \Rightarrow 10y - 4z = -18$
$\begin{cases} 10y - 4z = -18 \\ -3y + 4z = 11 \end{cases} \Rightarrow 7y = -7 \Rightarrow y = -1$

$\begin{cases} 2x + 3y + 2z = 7 \\ -3y + 4z = 11 \\ 5y - 2z = -9 \end{cases} \Rightarrow \begin{cases} 2x + 3y + 2z = 7 \\ y = -1 \\ 5y - 2z = -9 \end{cases}$

Multiply the second equation by -5, then add the result to the third equation and replace the third equation with the new equation:
$\begin{cases} -5y = 5 \\ 5y - 2z = -9 \end{cases} \Rightarrow -2z = -4 \Rightarrow z = 2$

$\begin{cases} 2x + 3y + 2z = 7 \\ y = -1 \\ 5y - 2z = -9 \end{cases} \Rightarrow \begin{cases} 2x + 3y + 2z = 7 \\ y = -1 \\ z = 2 \end{cases}$

Substitute the values for y and z into the first equation to solve for x:
$2x + 3(-1) + 2(2) = 7 \Rightarrow x = 3$.
The solution is $\{(3, -1, 2)\}$.

33. Switch the first and third equations:
$\begin{cases} 4x - 2y + z = 5 \\ 2x + y - 2z = 4 \\ x + 3y - 2z = 6 \end{cases} \Rightarrow \begin{cases} x + 3y - 2z = 6 \\ 2x + y - 2z = 4 \\ 4x - 2y + z = 5 \end{cases}$

Multiply the first equation by -2, add the result to the second equation, and replace the second equation with the new equation:
$-2(x + 3y - 2z = 6) \Rightarrow -2x - 6y + 4z = -12$
$\begin{cases} -2x - 6y + 4z = -12 \\ 2x + y - 2z = 4 \end{cases} \Rightarrow -5y + 2z = -8$

$\begin{cases} x + 3y - 2z = 6 \\ 2x + y - 2z = 4 \\ 4x - 2y + z = 5 \end{cases} \Rightarrow \begin{cases} x + 3y - 2z = 6 \\ -5y + 2z = -8 \\ 4x - 2y + z = 5 \end{cases}$

Multiply the first equation by -4, add the result to the third equation, and replace the third equation with the new equation:
$-4(x + 3y - 2z = 6) \Rightarrow -4x - 12y + 8z = -24$
$\begin{cases} -4x - 12y + 8z = -24 \\ 4x - 2y + z = 5 \end{cases} \Rightarrow -14y + 9z = -19$

$\begin{cases} x + 3y - 2z = 6 \\ -5y + 2z = -8 \\ 4x - 2y + z = 5 \end{cases} \Rightarrow \begin{cases} x + 3y - 2z = 6 \\ -5y + 2z = -8 \\ -14y + 9z = -19 \end{cases}$

Multiply the second equation by $-\frac{14}{5}$ and add the result to the third equation and replace the third equation with the new equation:

(continued on next page)

(continued)

$$-\frac{14}{5}(-5y+2z=-8) \Rightarrow 14y-\frac{28}{5}z=\frac{112}{5}$$

$$\begin{cases} 14y-\frac{28}{5}z=\frac{112}{5} \\ -14y+9z=-19 \end{cases} \Rightarrow \frac{17}{5}z=\frac{17}{5} \Rightarrow z=1$$

$$\begin{cases} x+3y-2z=6 \\ -5y+2z=-8 \\ -14y+9z=-19 \end{cases} \Rightarrow \begin{cases} x+3y-2z=6 \\ -5y+2z=-8 \\ z=1 \end{cases}$$

Substitute $z=1$ into the second equation to obtain $-5y+2(1)=-8 \Rightarrow y=2$. Substitute the values for y and z into the first equation to solve for x: $x+3(2)-2(1)=6 \Rightarrow x=2$. The solution is $\{(2, 2, 1)\}$.

35. Multiply the second equation by -1 and then add the result to the third equation:
$-(x+y-z=1) \Rightarrow -x-y+z=-1$

$$\begin{cases} -x-y+z=-1 \\ x+y+2z=4 \end{cases} \Rightarrow 3z=3 \Rightarrow z=1$$

Multiply the second equation by -2, add the result to the first equation, and replace the second equation with the new equation:
$-2(x+y-z=1) \Rightarrow -2x-2y+2z=-2$

$$\begin{cases} 2x+y+z=6 \\ -2x-2y+2z=-2 \end{cases} \Rightarrow -y+3z=4$$

$$\begin{cases} 2x+y+z=6 \\ x+y-z=1 \\ x+y+2z=4 \end{cases} \Rightarrow \begin{cases} 2x+y+z=6 \\ -y+3z=4 \\ z=1 \end{cases}$$

Substitute $z=1$ into the second equation to solve for y: $-y+3(1)=4 \Rightarrow y=-1$.

Substitute the values for y and z into the first equation to solve for x: $2x-1+1=6 \Rightarrow x=3$. The solution is $(3, -1, 1)$.

37. $\begin{cases} x-y-z=3 \\ x+9y+z=3 \\ 2x+3y-z=6 \end{cases}$

Multiply the first equation by -1, add the result to the second equation, and replace the second equation with the new equation:
$-1(x-y-z=3) \Rightarrow -x+y+z=-3$

$$\begin{cases} -x+y+z=-3 \\ x+9y+z=3 \end{cases} \Rightarrow 10y+2z=0$$

$$\begin{cases} x-y-z=3 \\ x+9y+z=3 \\ 2x+3y-z=6 \end{cases} \Rightarrow \begin{cases} x-y-z=3 \\ 10y+2z=0 \\ 2x+3y-z=6 \end{cases}$$

Multiply the first equation by -2, add the result to the third equation, and replace the third equation with the new equation:

$-2(x-y-z=3) \Rightarrow -2x+2y+2z=-6$

$$\begin{cases} -2x+2y+2z=-6 \\ 2x+3y-z=6 \end{cases} \Rightarrow 5y+z=0$$

$$\begin{cases} x-y-z=3 \\ 10y+2z=0 \\ 2x+3y-z=6 \end{cases} \Rightarrow \begin{cases} x-y-z=3 \\ 10y+2z=0 \\ 5y+z=0 \end{cases}$$

Multiply the third equation by -2, then add the result to the second equation:
$-2(5y+z=0) \Rightarrow -10y-2z=0$

$$\begin{cases} 10y+2z=0 \\ -10y-2z=0 \end{cases} \Rightarrow 0=0$$

The equation $0=0$ is equivalent to $0z=0$, which is true for every value of z. Solving the second equation for y, we have

$10y+2z=0 \Rightarrow y=-\frac{1}{5}z$. Substituting this into the first equation, we have
$x-y-z=3 \Rightarrow$

$x-\left(-\frac{1}{5}z\right)-z=3 \Rightarrow x=3+\frac{4}{5}z$

Thus, the solution is $\left\{\left(3+\frac{4}{5}z, -\frac{1}{5}z, z\right)\right\}$.

39. Rearrange the equations as shown:

$$\begin{cases} x+y=0 \\ y+2z=-4 \\ y+z=4-x \end{cases} \Rightarrow \begin{cases} x+y=0 \\ y+2z=-4 \\ x+y+z=4 \end{cases} \Rightarrow$$

$$\begin{cases} x+y+z=4 \\ y+2z=-4 \\ x+y=0 \end{cases}$$

Subtract the third equation from the first equation, and replace the third equation:

$$\begin{cases} x+y+z=4 \\ y+2z=-4 \\ x+y=0 \end{cases} \Rightarrow \begin{cases} x+y+z=4 \\ y+2z=-4 \\ z=4 \end{cases}$$

Substitute $z=4$ into the second equation to solve for y: $y+2(4)=-4 \Rightarrow y=-12$.

Substitute the values for x and y into the first equation to solve for x:
$x-12+4=4 \Rightarrow x=12$.
The solution is $\{(12, -12, 4)\}$.

41. Add the first and second equations, and replace the second equation:
$$\begin{cases} 2y - z = -4 \\ x + z = 3 \end{cases} \Rightarrow x + 2y = -1.$$

$$\begin{cases} 2y - z = -4 \\ x + z = 3 \\ 2x + 3y = -1 \end{cases} \Rightarrow \begin{cases} 2y - z = -4 \\ x + 2y = -1 \\ 2x + 3y = -1 \end{cases}$$

Switch the first and third equations, multiply the second equation by -2, add the result to the first equation, and replace the second equation with the new equation:
$-2(x + 2y = -1) \Rightarrow -2x - 4y = 2$

$$\begin{cases} -2x - 4y = 2 \\ 2x + 3y = -1 \end{cases} \Rightarrow -y = 1 \Rightarrow y = -1$$

$$\begin{cases} 2x + 3y = -1 \\ y = -1 \\ 2y - z = -4 \end{cases}$$

Multiply the second equation by -2, and add the result to the third equation.

$$\begin{cases} -2y = 2 \\ 2y - z = -4 \end{cases} \Rightarrow -z = -2 \Rightarrow z = 2$$

$$\begin{cases} 2x + 3y = -1 \\ y = -1 \\ z = 2 \end{cases}$$

Substitute the value for y into the first equation to solve for x:
$2x + 3(-1) = -1 \Rightarrow x = 1.$
The solution is $\{(1, -1, 2)\}$.

43. Multiply the second equation by -2, add the result to the third equation, and replace the second equation with the result:
$-2(2x + y = 8) \Rightarrow -4x - 2y = -16$

$$\begin{cases} 2y + 3z = 1 \\ -4x - 2y = -16 \end{cases} \Rightarrow -4x + 3z = -15$$

$$\begin{cases} 3x - 2z = 11 \\ 2x + y = 8 \\ 2y + 3z = 1 \end{cases} \Rightarrow \begin{cases} 3x - 2z = 11 \\ -4x + 3z = -15 \\ 2y + 3z = 1 \end{cases}$$

Multiply the first equation by 4 and the second equation by 3, add the two result and replace the second equation:
$4(3x - 2z = 11) \Rightarrow 12x - 8z = 44$
$3(-4x + 3z = -15) \Rightarrow -12x + 9x = -45$

$$\begin{cases} 12x - 8z = 44 \\ -12x + 9z = -45 \end{cases} \Rightarrow z = -1$$

$$\begin{cases} 3x - 2z = 11 \\ -4x + 3z = -15 \\ 2y + 3z = 1 \end{cases} \Rightarrow$$

$$\begin{cases} 3x - 2z = 11 \\ z = -1 \\ 2y + 3z = 1 \end{cases} \Rightarrow \begin{cases} 3x - 2z = 11 \\ 2y + 3z = 1 \\ z = -1 \end{cases}$$

Substitute $z = -1$ into the first and second equations to solve for x and y:
$2y + 3(-1) = 1 \Rightarrow y = 2.$
$3x - 2(-1) = 11 \Rightarrow x = 3.$
The solution is $\{(3, 2, -1)\}$.

45. Multiplying the second equation by -1, and then adding the two equations, we have:
$$\begin{cases} 2x + 6y + 11 = 0 \\ -6y + 18z - 1 = 0 \end{cases} \Rightarrow 2x + 18z + 10 = 0 \Rightarrow$$
$2x + 18z = -10 \Rightarrow x = -5 - 9z$

$6y - 18z + 1 = 0 \Rightarrow y = -\dfrac{1}{6} + 3z.$

The solution is $\left\{\left(-5 - 9z, -\dfrac{1}{6} + 3z, z\right)\right\}$.

The system is dependent.

8.2 Applying the Concepts

47. Let x = the amount invested at 4%, y = the amount invested at 5%, and z = the amount invested at 6%. Then, we have the system:
$$\begin{cases} x + y + z = 20{,}000 \\ 0.04x + 0.05y + 0.06z = 1060 \\ 0.06z = 2(0.05y) \end{cases} \Rightarrow$$

$$\begin{cases} x + y + z = 20{,}000 \\ 4x + 5y + 6z = 106{,}000 \\ 6z = 10y \end{cases} \Rightarrow$$

$$\begin{cases} x + y + z = 20{,}000 \\ 4x + 5y + 6z = 106{,}000 \\ -10y + 6z = 0 \end{cases}$$

Multiply the first equation by -4, add the result to the second equation, and replace the second equation with the new equation:
$-4(x + y + z = 20{,}000) \Rightarrow$
$-4x - 4y - 4z = -80{,}000$

$$\begin{cases} -4x - 4y - 4z = -80{,}000 \\ 4x + 5y + 6z = 106{,}000 \end{cases} \Rightarrow y + 2z = 26{,}000$$

$$\begin{cases} x + y + z = 20{,}000 \\ y + 2z = 26{,}000 \\ -10y + 6z = 0 \end{cases}$$

(*continued on next page*)

(*continued*)

Multiply the second equation by 10, add the result to the third equation, and solve for z:
$10(y + 2z = 26,000) \Rightarrow 10y + 20z = 260,000$
$\begin{cases} 10y + 20z = 260,000 \\ -10y + 6z = 0 \end{cases} \Rightarrow 26z = 260,000 \Rightarrow$
$z = 10,000$

$\begin{cases} x + y + z = 20,000 \\ y + 2z = 26,000 \\ -10y + 6z = 0 \end{cases} \Rightarrow$

$\begin{cases} x + y + z = 20,000 \\ y + 2z = 26,000 \\ z = 10,000 \end{cases}$

Substitute $z = 10,000$ into the second equation to solve for y:
$y + 2(10,000) = 26,000 \Rightarrow y = 6000$.
Substitute the values for y and z into the first equation to solve for x:
$x + 6000 + 10,000 = 20,000 \Rightarrow x = 4000$.
Miguel invested $4000 at 4%, $6000 at 5%, and $10,000 at 6%.

49. Let a = the number of hours Alex worked, b = the number of hours Becky worked, and c = the number of hours Courtney worked. Then, we have the system:
$\begin{cases} a + b + c = 6 \\ 124a + 118b + 132c = 741 \\ b + c = 2a \end{cases} \Rightarrow$

$\begin{cases} a + b + c = 6 \\ 124a + 118b + 132c = 741 \\ -2a + b + c = 0 \end{cases}$

Subtract the third equation from the first, and replace the first equation with the result:
$\begin{cases} a + b + c = 6 \\ 124a + 118b + 132c = 741 \Rightarrow \\ -2a + b + c = 0 \end{cases}$

$\begin{cases} 3a = 6 \\ 124a + 118b + 132c = 741 \Rightarrow \\ -2a + b + c = 0 \end{cases}$

$\begin{cases} a = 2 \\ 124a + 118b + 132c = 741 \Rightarrow \\ -2a + b + c = 0 \end{cases}$

Substitute $a = 2$ into the second and third equations and simplify:
$\begin{cases} a = 2 \\ 124(2) + 118b + 132c = 741 \Rightarrow \\ -2(2) + b + c = 0 \end{cases}$

$\begin{cases} a = 2 \\ 118b + 132c = 493 \\ b + c = 4 \end{cases}$

Multiply the third equation by −118, and add the result to the second equation to solve for c:
$-118(b + c = 4) \Rightarrow -118b - 118c = -472$
$\begin{cases} 118b + 132c = 493 \\ -118b - 118c = -472 \end{cases} \Rightarrow 14c = 21 \Rightarrow c = 1.5$

Substitute $c = 1.5$ into the third equation to solve for b: $b + 1.5 = 4 \Rightarrow b = 2.5$.
Alex worked 2 hours, Becky worked 2.5 hours, and Courtney worked 1.5 hours.

51. Let n = the number of nickels, d = the number of dimes, and q = the number of quarters. Then, we have the system
$\begin{cases} n + d + q = 300 \\ d = 3(n + q) \\ 0.05n + 0.1d + 0.25q = 30.05 \end{cases} \Rightarrow$

$\begin{cases} n + d + q = 300 \\ -3n + d - 3q = 0 \\ 5n + 10d + 25q = 3005 \end{cases}$

Multiply the first equation by 3, add the result to the second equation, and replace the second equation with the new equation:
$3(n + d + q = 300) \Rightarrow 3n + 3d + 3q = 900$
$\begin{cases} 3n + 3d + 3q = 900 \\ -3n + d - 3q = 0 \end{cases} \Rightarrow 4d = 900 \Rightarrow d = 225$

$\begin{cases} n + d + q = 300 \\ d = 225 \\ 5n + 10d + 25q = 3005 \end{cases}$

Multiply the first equation by −5, add the result to the third equation, and replace the third equation with the new equation:
$-5(n + d + q = 300) \Rightarrow -5n - 5d - 5q = -1500$
$\begin{cases} -5n - 5d - 5q = -1500 \\ 5n + 10d + 25q = 3005 \end{cases} \Rightarrow 5d + 20q = 1505$

$\begin{cases} n + d + q = 300 \\ d = 225 \\ 5n + 10d + 25q = 3005 \end{cases} \Rightarrow$

$\begin{cases} n + d + q = 300 \\ d = 225 \\ 5d + 20q = 1505 \end{cases}$

Substitute $d = 225$ into the third equation to solve for q, then substitute the values for d and q into the first equation to solve for n:
$5(225) + 20q = 1505 \Rightarrow q = 19$.
$n + 225 + 19 = 300 \Rightarrow n = 56$.
There are 56 nickels, 225 dimes, and 19 quarters.

53. Let x = the number of daytime hours Amy worked, y = the number of night hours Amy worked, and z = the number of holiday hours Amy worked. Then, we have the system
$$\begin{cases} x + y + z = 53 \\ 7.40x + 9.20y + 11.75z = 452.20 \\ x = y + z + 9 \end{cases} \Rightarrow$$

$$\begin{cases} x + y + z = 53 \\ 7.40x + 9.20y + 11.75z = 452.20 \\ x - y - z = 9 \end{cases}$$

Add the first and third equations, and replace the first equation with the result:
$$\begin{cases} x + y + z = 53 \\ 7.40x + 9.20y + 11.75z = 452.20 \\ x - y - z = 9 \end{cases} \Rightarrow$$

$$\begin{cases} 2x = 62 \\ 7.40x + 9.20y + 11.75z = 452.20 \\ x - y - z = 9 \end{cases} \Rightarrow$$

$$\begin{cases} x = 31 \\ 7.40x + 9.20y + 11.75z = 452.20 \\ x - y - z = 9 \end{cases}$$

Substitute $x = 31$ into the second and third equations and simplify:
$$\begin{cases} x = 31 \\ 7.40x + 9.20y + 11.75z = 452.20 \\ x - y - z = 9 \end{cases} \Rightarrow$$

$$\begin{cases} x = 31 \\ 7.40(31) + 9.20y + 11.75z = 452.20 \\ 31 - y - z = 9 \end{cases} \Rightarrow$$

$$\begin{cases} x = 31 \\ 9.20y + 11.75z = 222.80 \\ -y - z = -22 \end{cases}$$

Multiply the third equation by 9.2, add the result to the second equation, and solve for z:
$9.2(-y - z = -22) \Rightarrow -9.2y - 9.2z = 202.4$
$$\begin{cases} 9.20y + 11.75z = 222.80 \\ -9.20y - 9.20z = -202.40 \end{cases} \Rightarrow$$
$2.55z = 20.40 \Rightarrow z = 8$
$$\begin{cases} x = 31 \\ 9.20y + 11.75z = 222.80 \\ -y - z = -22 \end{cases} \Rightarrow$$

$$\begin{cases} x = 31 \\ 9.20y + 11.75z = 222.80 \\ z = 8 \end{cases}$$

$9.20y + 11.75(8) = 222.80 \Rightarrow y = 14$

Amy worked 31 daytime hours, 14 night hours, and 8 holiday hours.

For exercises 55–58, refer to Table 8.1 in the text.

55. The system is
$$\begin{cases} x + y = 0.54 \text{ (beam 1)} \\ x + z = 0.40 \text{ (beam 2)} \\ y + z = 0.52 \text{ (beam 3)} \end{cases}$$

Multiply the first equation by -1, add the result to the second equation, and replace the second equation with the new equation:
$-1(x + y = 0.54) \Rightarrow -x - y = -0.54$
$$\begin{cases} -x - y = -0.54 \\ x + z = 0.40 \end{cases} \Rightarrow -y + z = -0.14$$

$$\begin{cases} x + y = 0.54 \\ x + z = 0.40 \\ y + z = 0.52 \end{cases} \Rightarrow \begin{cases} x + y = 0.54 \\ -y + z = -0.14 \\ y + z = 0.52 \end{cases}$$

Add the second and third equations and solve for z: $2z = 0.38 \Rightarrow z = 0.19$. Substituting this value into the original second equation, we have $x + 0.19 = 0.40 \Rightarrow x = 0.21$. Substituting this value into the original first equation, we have $0.21 + y = 0.54 \Rightarrow y = 0.33$.

Cell A contains healthy tissue (because $x = 0.21$), cell B contains tumorous tissue (because $y = 0.33$), and cell C contains healthy tissue (because $z = 0.19$).

57. The system is
$$\begin{cases} x + y = 0.51 \text{ (beam 1)} \\ x + z = 0.49 \text{ (beam 2)} \\ y + z = 0.44 \text{ (beam 3)} \end{cases}$$

Multiply the first equation by -1, add the result to the second equation, and replace the second equation with the new equation:
$-1(x + y = 0.51) \Rightarrow -x - y = -0.51$
$$\begin{cases} -x - y = -0.51 \\ x + z = 0.49 \end{cases} \Rightarrow -y + z = -0.02$$

$$\begin{cases} x + y = 0.51 \\ x + z = 0.49 \\ y + z = 0.44 \end{cases} \Rightarrow \begin{cases} x + y = 0.51 \\ -y + z = -0.02 \\ y + z = 0.44 \end{cases}$$

Add the second and third equations and solve for z: $2z = 0.42 \Rightarrow z = 0.21$. Substituting this value into the original second equation, we have $x + 0.21 = 0.49 \Rightarrow x = 0.28$. Substituting this value into the original first equation, we have $0.28 + y = 0.51 \Rightarrow y = 0.23$.

Cell A contains healthy tissue (because $x = 0.28$), cell B contains healthy tissue (because $y = 0.23$), and cell C contains healthy tissue (because $z = 0.21$).

8.2 Beyond the Basics

59. Because each of the ordered triples satisfies the given equation, we can find the values of the coefficients by solving the system

$$\begin{cases} 1 + 0b + 0c = d \\ 0 + 1b + 0c = d \\ 0 + 0b + 1c = d \end{cases} \Rightarrow \begin{cases} 1 = d \\ b = d \\ c = d \end{cases} \Rightarrow b = c = d = 1.$$

The equation is $x + y + z = 1$.

61. Because each of the ordered triples satisfies the given equation, we can find the values of the coefficients by solving the system

$$\begin{cases} 3 - 4b + 0c = d \\ 0 + \frac{1}{4}b + \frac{1}{2}c = d \\ 1 + 1b - 4c = d \end{cases} \Rightarrow \begin{cases} -4b - d = -3 \\ b + 2c - 4d = 0 \\ b - 4c - d = -1 \end{cases}$$

Multiplying the second equation by 2, adding it to the third equation, and replacing the second equation with the new equation, we have

$2(b + 2c - 4d = 0) \Rightarrow 2b + 4c - 8d = 0$

$\begin{cases} 2b + 4c - 8d = 0 \\ b - 4c - d = -1 \end{cases} \Rightarrow 3b - 9d = -1$

$\begin{cases} -4b - d = -3 \\ b + 2c - 4d = 0 \\ b - 4c - d = -1 \end{cases} \Rightarrow \begin{cases} -4b - d = -3 \\ 3b - 9d = -1 \\ b - 4c - d = 1 \end{cases}$

From the first equation, we have $d = -3 - 4b$. Substituting this into the second equation, we have

$3b - 9(-3 - 4b) = -1 \Rightarrow 39b = 26 \Rightarrow b = \frac{2}{3}$.

Then $d = 3 - 4\left(\frac{2}{3}\right) - 3 = \frac{1}{3}$.

$\frac{1}{4}\left(\frac{2}{3}\right) + \frac{1}{2}c = \frac{1}{3} \Rightarrow \frac{1}{2}c = \frac{1}{6} \Rightarrow c = \frac{1}{3}$.

The equation is $x + \frac{2}{3}y + \frac{1}{3}z = \frac{1}{3}$.

63. Because each of the ordered triples satisfies the given equation, we can find the values of the coefficients by solving the system

$$\begin{cases} a(0)^2 + b(0) + c = 1 \\ a(-1)^2 + b(-1) + c = 0 \\ a(1)^2 + b(1) + c = 4 \end{cases} \Rightarrow \begin{cases} c = 1 \\ a - b + c = 0 \\ a + b + c = 4 \end{cases}$$

Substituting $c = 1$ into the second and third equations and then solving for b and c, we have $\begin{cases} a - b + 1 = 0 \\ a + b + 1 = 4 \end{cases} \Rightarrow 2a = 2 \Rightarrow a = 1$

$1 - b + 1 = 0 \Rightarrow b = 2$

The equation is $y = x^2 + 2x + 1$.

65. Because each of the ordered triples satisfies the given equation, we can find the values of the coefficients by solving the system

$$\begin{cases} a(1)^2 + b(1) + c = 2 \\ a(-1)^2 + b(-1) + c = 4 \\ a(2)^2 + b(2) + c = 4 \end{cases} \Rightarrow \begin{cases} c = 2 \\ a - b + c = 4 \\ 4a + 2b + c = 4 \end{cases}$$

Substituting $c = 2$ into the second and third equations and then solving for b and c, we have

$\begin{cases} a - b + 2 = 4 \\ 4a + 2b + 2 = 4 \end{cases} \Rightarrow \begin{cases} 2a - 2b + 4 = 8 \\ 4a + 2b + 2 = 4 \end{cases} \Rightarrow$

$6a = 6 \Rightarrow a = 1$

$1 - b + 2 = 4 \Rightarrow b = -1$

The equation is $y = x^2 - x + 2$.

67. Because each of the ordered triples satisfies the given equation, we can find the values of the coefficients by solving the system

$$\begin{cases} 0^2 + 4^2 + a(0) + b(4) + c = 0 \\ (2\sqrt{2})^2 + (2\sqrt{2})^2 + a(2\sqrt{2}) + b(2\sqrt{2}) + c = 0 \\ (-4)^2 + 0^2 + a(-4) + b(0) + c = 0 \end{cases} \Rightarrow$$

$$\begin{cases} 4b + c = -16 \\ (2\sqrt{2})a + (2\sqrt{2})b + c = -16 \\ -4a + c = -16 \end{cases}$$

From the third equation, we have $c = 4a - 16$. Substituting this expression into the first and second equations, we have

$\begin{cases} 4b + 4a - 16 = -16 \\ (2\sqrt{2})a + (2\sqrt{2})b + 4a - 16 = -16 \end{cases} \Rightarrow$

$\begin{cases} 4a + 4b = 0 \\ (2\sqrt{2} + 4)a + (2\sqrt{2})b = 0 \end{cases} \Rightarrow$

$\begin{cases} a + b = 0 \\ (2\sqrt{2} + 4)a + (2\sqrt{2})b = 0 \end{cases} \Rightarrow$

$\begin{cases} a = -b \\ (2\sqrt{2} + 4)a + (2\sqrt{2})b = 0 \end{cases} \Rightarrow$

$(2\sqrt{2} + 4)(-b) + (2\sqrt{2})b = 0 \Rightarrow$

$-4b = 0 \Rightarrow b = 0$. Substituting this value into the first equation, we have $4(0) + c = -16 \Rightarrow c = -16$. Substituting into the third equation, we have $-4a - 16 = -16 \Rightarrow a = 0$. The equation is $x^2 + y^2 - 16 = 0$.

69. Because each of the ordered triples satisfies the given equation, we can find the values of the coefficients by solving the system
$$\begin{cases} 1^2 + 2^2 + a(1) + b(2) + c = 0 \\ 6^2 + (-3)^2 + a(6) + b(-3) + c = 0 \\ (4)^2 + 1^2 + a(4) + b(1) + c = 0 \end{cases} \Rightarrow$$
$$\begin{cases} a + 2b + c = -5 \\ 6a - 3b + c = -45 \\ 4a + b + c = -17 \end{cases}$$

Multiply the first equation by –6, add the result to the second equation, then replace the second equation with the new equation. Similarly, multiply the first equation by –4, add that result to the third equation, then replace the third equation with the new equation.
$$-6(a + 2b + c = -5) \Rightarrow -6a - 12b - 6c = 30$$
$$\begin{cases} -6a - 12b - 6c = 30 \\ 6a - 3b + c = -45 \end{cases} \Rightarrow -15b - 5c = -15 \Rightarrow$$
$$3b + c = 3$$
$$-4(a + 2b + c = -5) \Rightarrow -4a - 8b - 4c = 20$$
$$\begin{cases} -4a - 8b - 4c = 20 \\ 4a + b + c = -17 \end{cases} \Rightarrow -7b - 3c = 3$$
$$\begin{cases} a + 2b + c = -5 \\ 6a - 3b + c = -45 \\ 4a + b + c = -17 \end{cases} \Rightarrow \begin{cases} a + 2b + c = -5 \\ 3b + c = 3 \\ -7b - 3c = 3 \end{cases}$$

From the second equation, we have $c = -3b + 3$. Substitute this expression into the third equation, and solve for b.

$-7b - 3(-3b + 3) = 3 \Rightarrow 2b = 12 \Rightarrow b = 6$.

So, $c = -3(6) + 3 \Rightarrow c = -15$. Substituting into the original first equation, we have $a + 2(6) - 15 = -5 \Rightarrow a = -2$.

The equation is $x^2 + y^2 - 2x + 6y - 15 = 0$.

71. Letting $u = 1/x, v = 1/y, w = 1/z$, we have
$$\begin{cases} \dfrac{1}{x} + \dfrac{3}{y} - \dfrac{1}{z} = 5 \\ \dfrac{2}{x} + \dfrac{4}{y} + \dfrac{6}{z} = 4 \\ \dfrac{2}{x} + \dfrac{3}{y} + \dfrac{1}{z} = 3 \end{cases} \Rightarrow \begin{cases} u + 3v - w = 5 \\ 2u + 4v + 6w = 4 \\ 2u + 3v + w = 3 \end{cases}$$

Multiplying the first equation by –2, adding the result to the second and third equations, and replacing those equations with the new equations, we have
$-2(u + 3v - w = 5) \Rightarrow -2u - 6v + 2w = -10$
$$\begin{cases} -2u - 6v + 2w = -10 \\ 2u + 4v + 6w = 4 \end{cases} \Rightarrow -2v + 8w = -6 \Rightarrow$$
$v - 4w = 3$
$$\begin{cases} -2u - 6v + 2w = -10 \\ 2u + 3v + w = 3 \end{cases} \Rightarrow -3v + 3w = -7$$
$$\begin{cases} u + 3v - w = 5 \\ 2u + 4v + 6w = 4 \\ 2u + 3v + w = 3 \end{cases} \Rightarrow \begin{cases} u + 3v - w = 5 \\ v - 4w = 3 \\ -3v + 3w = -7 \end{cases}$$

Multiplying the second equation by 3 and adding the result to the third equation, we have $3(v - 4w = 3) \Rightarrow 3v - 12w = 9$
$$\begin{cases} 3v - 12w = 9 \\ -3v + 3w = -7 \end{cases} \Rightarrow -9w = 2 \Rightarrow w = -\dfrac{2}{9}.$$

Substituting, we have
$$v - 4\left(-\dfrac{2}{9}\right) = 3 \Rightarrow v = \dfrac{19}{9} \text{ and}$$
$$u + 3\left(\dfrac{19}{9}\right) - \left(-\dfrac{2}{9}\right) = 5 \Rightarrow u = -\dfrac{14}{9}. \text{ So,}$$
$$x = -\dfrac{9}{14}, y = \dfrac{9}{19}, z = -\dfrac{9}{2}.$$

73. Solve the first three equations in the system for x, y, and z:
$-3(x + 2y - 5z = 9) \Rightarrow -3x - 6y + 15z = -27$
$$\begin{cases} -3x - 6y + 15z = -27 \\ 3x - y + 2z = 14 \end{cases} \Rightarrow -7y + 17z = -13$$
$-2(x + 2y - 5z = 9) \Rightarrow -2x - 4y + 10z = -18$
$$\begin{cases} -2x - 4y + 10z = -18 \\ 2x + 3y - z = 3 \end{cases} \Rightarrow -y + 9z = -15$$
$$\begin{cases} x + 2y - 5z = 9 \\ 3x - y + 2z = 14 \\ 2x + 3y - z = 3 \end{cases} \Rightarrow \begin{cases} x + 2y - 5z = 9 \\ -7y + 17z = -13 \\ -y + 9z = -15 \end{cases}$$
$-7(-y + 9z = -15) \Rightarrow 7y - 63z = 105$
$$\begin{cases} -7y + 17z = -13 \\ 7y - 63z = 105 \end{cases} \Rightarrow -46z = 92 \Rightarrow z = -2$$
$-7y + 17(-2) = -13 \Rightarrow -7y = 21 \Rightarrow y = -3$
$x + 2(-3) - 5(-2) = 9 \Rightarrow x = 5$

Substituting $x = 5$, $y = -3$, and $z = -2$ into the fourth equation of the original system, we have
$5c - 5(-3) + (-2) + 3 = 0 \Rightarrow$
$5c = -16 \Rightarrow c = -\dfrac{16}{5}.$

75. Each of the ordered triples satisfies the given equation, so find the values of the coefficients by solving the system

$$\begin{cases} a(-1)^2 + b(-1) + c = -1 \\ a(0)^2 + b(0) + c = 5 \\ a(2)^2 + b(2) + c = 5 \end{cases} \Rightarrow \begin{cases} a - b + c = -1 \\ c = 5 \\ 4a + 2b + c = 5 \end{cases}$$

Substitute $c = 5$ into the first and third equations and then solve those equations for a and b

$$\begin{cases} a - b + 5 = -1 \\ 4a + 2b + 5 = 5 \end{cases} \Rightarrow \begin{cases} a - b = -6 \\ 2a + b = 0 \end{cases} \Rightarrow$$

$3a = -6 \Rightarrow a = -2$
$-2 - b + 5 = -1 \Rightarrow b = 4.$

The equation is $y = -2x^2 + 4x + 5$.

8.2 Critical Thinking/Discussion/Writing

Answers may vary for exercises 77 and 78.

77. $\begin{cases} 1 + (-1) - 2 = -2 \\ 2(1) - (-1) + 3(2) = 9 \\ 1 + (-1) + 2 = 2 \end{cases} \Rightarrow \begin{cases} x + y - z = -2 \\ 2x - y + 3z = 9 \\ x + y + z = 2 \end{cases}$

78. a. $\begin{cases} x + 3y + 3z = 15 \\ x + 2y + z = 1 \\ 2y + 4z = 11 \end{cases}$ **b.** $\begin{cases} x + 2y - z = 4 \\ x + 3y + 2z = 5 \\ y + 3z = 1 \end{cases}$

8.2 Getting Ready for the Next Section

79. $\dfrac{1}{2 \cdot 3} = \dfrac{1}{2} + \dfrac{B}{3} \Rightarrow 1 = 3 + 2B \Rightarrow -2 = 2B \Rightarrow$
$B = -1$

81. $\dfrac{1}{n \cdot (n+1)} = \dfrac{A}{n} - \dfrac{1}{n+1} \Rightarrow 1 = A(n+1) - n \Rightarrow$
$1 + n = A(n+1) \Rightarrow 1 = A$

83. $2x + 3 = A(x + 3) + B(x - 1)$
$2x + 3 = Ax + 3A + Bx - B = (A + B)x + 3A - B$
Equating the coefficients gives the system
$\begin{cases} A + B = 2 \\ 3A - B = 3 \end{cases} \Rightarrow 4A = 5 \Rightarrow A = \dfrac{5}{4}$
$\dfrac{5}{4} + B = 2 \Rightarrow B = \dfrac{3}{4}$

85. $x^2 + 5x + 6 = (x + 2)(x + 3)$

87. $2x^2 + 5x - 3 = (2x - 1)(x + 3)$

89. $4x^2 - 9 = (2x - 3)(2x + 3)$

8.3 Partial-Fraction Decomposition

8.3 Practice Problems

1. $\dfrac{2x - 7}{(x+1)(x-2)} = \dfrac{A}{x+1} + \dfrac{B}{x-2} \Rightarrow$
$2x - 7 = A(x - 2) + B(x + 1) \Rightarrow$
$2x - 7 = (A + B)x + (-2A + B) \Rightarrow$
$\begin{cases} A + B = 2 \\ -2A + B = -7 \end{cases} \Rightarrow \begin{cases} -A - B = -2 \\ -2A + B = -7 \end{cases} \Rightarrow$
$-3A = -9 \Rightarrow A = 3$
$3 + B = 2 \Rightarrow B = -1$
$\dfrac{2x - 7}{(x+1)(x-2)} = \dfrac{3}{x+1} - \dfrac{1}{x-2}$

2. $\dfrac{3x^2 + 4x + 3}{x^3 - x} = \dfrac{3x^2 + 4x + 3}{x(x-1)(x+1)}$
$= \dfrac{A}{x} + \dfrac{B}{x-1} + \dfrac{C}{x+1} \Rightarrow$
$3x^2 + 4x + 3$
$= A(x+1)(x-1) + Bx(x+1) + Cx(x-1)$
$= Ax^2 - A + Bx^2 + Bx + Cx^2 - Cx$
$= (A + B + C)x^2 + (B - C)x - A$
$\begin{cases} A + B + C = 3 \\ B - C = 4 \Rightarrow A = -3 \\ -A = 3 \end{cases}$
$\begin{cases} -3 + B + C = 3 \\ B - C = 4 \end{cases} \Rightarrow -3 + 2B = 7 \Rightarrow B = 5$
$5 - C = 4 \Rightarrow C = 1$
$\dfrac{3x^2 + 4x + 3}{x^3 - x} = -\dfrac{3}{x} + \dfrac{5}{x-1} + \dfrac{1}{x+1}$

3. $\dfrac{x + 5}{x(x-1)^2} = \dfrac{A}{x} + \dfrac{B}{(x-1)} + \dfrac{C}{(x-1)^2} \Rightarrow$
$x + 5 = A(x - 1)^2 + Bx(x - 1) + Cx \Rightarrow$
$x + 5 = (A + B)x^2 + (-2A - B + C)x + A$
$\begin{cases} A + B = 0 \\ -2A - B + C = 1 \Rightarrow A = 5, B = -5 \\ A = 5 \end{cases}$
$-2A - B + C = 1 \Rightarrow -2(5) + 5 + C = 1 \Rightarrow C = 6$
$\dfrac{x + 5}{x(x-1)^2} = \dfrac{5}{x} - \dfrac{5}{(x-1)} + \dfrac{6}{(x-1)^2}$

4. $\dfrac{2x+1}{(x+3)^2} = \dfrac{A}{(x+3)} + \dfrac{B}{(x+3)^2} \Rightarrow$
$2x+1 = A(x+3) + B \Rightarrow 2x+1 = Ax + 3A + B$
$\begin{cases} A = 2 \\ 3A+B = 1 \end{cases} \Rightarrow A = 2$
$3A + B = 1 \Rightarrow 3(2) + B = 1 \Rightarrow B = -5$
$\dfrac{2x+1}{(x+3)^2} = \dfrac{2}{x+3} - \dfrac{5}{(x+3)^2}$

5. $\dfrac{3x^2+5x-2}{x(x^2+2)} = \dfrac{A}{x} + \dfrac{Bx+C}{x^2+2} \Rightarrow$
$3x^2 + 5x - 2 = A(x^2+2) + (Bx+C)x \Rightarrow$
$3x^2 + 5x - 2 = (A+B)x^2 + Cx + 2A \Rightarrow$
$\begin{cases} A+B = 3 \\ C = 5 \\ 2A = -2 \end{cases} \Rightarrow A = -1, C = 5$
$A+B = 3 \Rightarrow -1 + B = 3 \Rightarrow B = 4$
$\dfrac{3x^2+5x-2}{x(x^2+2)} = -\dfrac{1}{x} + \dfrac{4x+5}{x^2+2}$

6. $\dfrac{x^2+3x+1}{(x^2+1)^2} = \dfrac{Ax+B}{x^2+1} + \dfrac{Cx+D}{(x^2+1)^2} \Rightarrow$
$x^2 + 3x + 1 = (Ax+B)(x^2+1) + Cx + D \Rightarrow$
$x^2 + 3x + 1 = Ax^3 + Bx^2 + (A+C)x + (B+D) \Rightarrow$
$\begin{cases} A = 0 \\ B = 1 \\ A+C = 3 \\ B+D = 1 \end{cases} \Rightarrow A = 0, B = 1, C = 3, D = 0$
$\dfrac{x^2+3x+1}{(x^2+1)^2} = \dfrac{1}{x^2+1} + \dfrac{3x}{(x^2+1)^2}$

7. $\dfrac{1}{4\cdot 5} + \dfrac{1}{5\cdot 6} + \dfrac{1}{6\cdot 7} + \cdots + \dfrac{1}{3111\cdot 3112}$
$= \left(\dfrac{1}{4} - \dfrac{1}{5}\right) + \left(\dfrac{1}{5} - \dfrac{1}{6}\right) + \left(\dfrac{1}{6} - \dfrac{1}{7}\right) + \cdots$
$\quad + \left(\dfrac{1}{3111} - \dfrac{1}{3112}\right)$
$= \dfrac{1}{4} - \dfrac{1}{3112} = \dfrac{778-1}{3112} = \dfrac{777}{3112}$

8. $R = \dfrac{(x+1)(x+2)}{4x+7} \Rightarrow \dfrac{1}{R} = \dfrac{4x+7}{(x+1)(x+2)}$
$\dfrac{4x+7}{(x+1)(x+2)} = \dfrac{A}{x+1} + \dfrac{B}{x+2} \Rightarrow$
$4x+7 = A(x+2) + B(x+1) \Rightarrow$
$4x+7 = (A+B)x + (2A+B) \Rightarrow$

$\begin{cases} A+B = 4 \\ 2A+B = 7 \end{cases} \Rightarrow A = 3, B = 1$
$\dfrac{4x+7}{(x+1)(x+2)} = \dfrac{3}{x+1} + \dfrac{1}{x+2}$
$= \dfrac{1}{\dfrac{x+1}{3}} + \dfrac{1}{x+2} = \dfrac{1}{R}$

This means that if two resistances $R_1 = \dfrac{x+1}{3}$ and $R_2 = x+2$ are connected in parallel, they will produce a total resistance given by $\dfrac{(x+1)(x+2)}{4x+7}$.

8.3 Concepts and Vocabulary

1. In a rational expression, if the degree of the numerator, $P(x)$, is less than the degree of the denominator, $Q(x)$, then the expression is a proper fraction.

3. In a rational expression, if $(x-8)$ is a linear factor that is repeated three times in the denominator, then the portion of the expression's partial-fraction decomposition that corresponds to $(x-8)^3$ has three terms.

5. False. The factors are repeated linear factors.

7. True

8.3 Building Skills

9. $\dfrac{1}{(x-1)(x+2)} = \dfrac{A}{x-1} + \dfrac{B}{x+2}$

11. $\dfrac{1}{x^2+7x+6} = \dfrac{1}{(x+6)(x+1)} = \dfrac{A}{x+6} + \dfrac{B}{x+1}$

13. $\dfrac{2}{x^3-x^2} = \dfrac{2}{x^2(x-1)} = \dfrac{A}{x^2} + \dfrac{B}{x} + \dfrac{C}{x-1}$

15. $\dfrac{x^2-3x+3}{(x+1)(x^2-x+1)} = \dfrac{A}{x+1} + \dfrac{Bx+C}{x^2-x+1}$

17. $\dfrac{3x-4}{(x^2+1)^2} = \dfrac{Ax+B}{x^2+1} + \dfrac{Cx+D}{(x^2+1)^2}$

19. $\dfrac{2x+1}{(x+1)(x+2)} = \dfrac{A}{x+1} + \dfrac{B}{x+2} \Rightarrow 2x+1 = A(x+2) + B(x+1) \Rightarrow 2x+1 = (A+B)x + (2A+B) \Rightarrow$

$\begin{cases} A+B=2 \\ 2A+B=1 \end{cases} \Rightarrow \begin{cases} -A-B=-2 \\ 2A+B=1 \end{cases} \Rightarrow A=-1, B=3$

$\dfrac{2x+1}{(x+1)(x+2)} = -\dfrac{1}{x+1} + \dfrac{3}{x+2} = \dfrac{3}{x+2} - \dfrac{1}{x+1}$

21. $\dfrac{1}{x^2+4x+3} = \dfrac{1}{(x+3)(x+1)} = \dfrac{A}{x+3} + \dfrac{B}{x+1} \Rightarrow 1 = A(x+1) + B(x+3) = (A+B)x + (A+3B) \Rightarrow$

$\begin{cases} A+B=0 \\ A+3B=1 \end{cases} \Rightarrow \begin{cases} -A-B=0 \\ A+3B=1 \end{cases} \Rightarrow 2B=1 \Rightarrow B=\dfrac{1}{2}, A=-\dfrac{1}{2}$

$\dfrac{1}{x^2+4x+3} = -\dfrac{1}{2(x+3)} + \dfrac{1}{2(x+1)}$

23. $\dfrac{2}{x^2+2x} = \dfrac{2}{x(x+2)} = \dfrac{A}{x} + \dfrac{B}{x+2} \Rightarrow 2 = A(x+2) + Bx = (A+B)x + 2A \Rightarrow \begin{cases} A+B=0 \\ 2A=2 \end{cases} \Rightarrow A=1, B=-1 \Rightarrow$

$\dfrac{2}{x^2+2x} = \dfrac{1}{x} - \dfrac{1}{x+2}$

25. $\dfrac{x}{(x+1)(x+2)(x+3)} = \dfrac{A}{x+1} + \dfrac{B}{x+2} + \dfrac{C}{x+3} \Rightarrow x = A(x+2)(x+3) + B(x+1)(x+3) + C(x+1)(x+2) \Rightarrow$

$x = A(x^2+5x+6) + B(x^2+4x+3) + C(x^2+3x+2) = (A+B+C)x^2 + (5A+4B+3C)x + (6A+3B+2C)$

$\begin{cases} A+B+C=0 \\ 5A+4B+3C=1 \\ 6A+3B+2C=0 \end{cases}$

$-5(A+B+C=0) = -5A-5B-5C=0$

$\begin{cases} -5A-5B-5C=0 \\ 5A+4B+3C=1 \end{cases} \Rightarrow -B-2C=1$

$-6(A+B+C=0) = -6A-6B-6C=0$

$\begin{cases} -6A-6B-6C=0 \\ 6A+3B+2C=0 \end{cases} \Rightarrow -3B-4C=0$

$\begin{cases} A+B+C=0 \\ 5A+3B+3C=1 \\ 6A+3B+2C=0 \end{cases} \Rightarrow \begin{cases} A+B+C=0 \\ -B-2C=1 \\ -3B-4C=0 \end{cases}$

$-3(-B-2C=1) = 3B+6C=-3$

$\begin{cases} 3B+6C=-3 \\ -3B-4C=0 \end{cases} \Rightarrow 2C=-3 \Rightarrow C=-\dfrac{3}{2}$

$-B-2\left(-\dfrac{3}{2}\right) = 1 \Rightarrow -B=-4 \Rightarrow B=2$

$A+2-\dfrac{3}{2}=0 \Rightarrow A=-\dfrac{1}{2}$

$\dfrac{x}{(x+1)(x+2)(x+3)} = -\dfrac{1}{2(x+1)} + \dfrac{2}{x+2} - \dfrac{3}{2(x+3)}$

Section 8.3 Partial-Fraction Decomposition 477

27. $\dfrac{x^2}{(x-1)(x^2+5x+4)} = \dfrac{x^2}{(x-1)(x+1)(x+4)} = \dfrac{A}{x-1} + \dfrac{B}{x+1} + \dfrac{C}{x+4} \Rightarrow$

$x^2 = A(x+1)(x+4) + B(x-1)(x+4) + C(x-1)(x+1) \Rightarrow$

$x^2 = A(x^2+5x+4) + B(x^2+3x-4) + C(x^2-1) \Rightarrow$

$x^2 = (A+B+C)x^2 + (5A+3B)x + (4A-4B-C) \Rightarrow$

$\begin{cases} A+B+C = 1 \\ 5A+3B = 0 \\ 4A-4B-C = 0 \end{cases} \Rightarrow \begin{cases} A+B+C = 1 \\ 4A-4B-C = 0 \end{cases} \Rightarrow 5A - 3B = 1$

$\begin{cases} A+B+C = 1 \\ 5A+3B = 0 \\ 4A-4B-C = 0 \end{cases} \Rightarrow \begin{cases} A+B+C = 1 \\ 5A+3B = 0 \\ 5A-3B = 1 \end{cases} \Rightarrow 10A = 1 \Rightarrow A = \dfrac{1}{10}$

$5\left(\dfrac{1}{10}\right) + 3B = 0 \Rightarrow 3B = -\dfrac{1}{2} \Rightarrow B = -\dfrac{1}{6}$

$\dfrac{1}{10} - \dfrac{1}{6} + C = 1 \Rightarrow C = \dfrac{16}{15}$

$\dfrac{x^2}{(x-1)(x^2+5x+4)} = \dfrac{1}{10(x-1)} - \dfrac{1}{6(x+1)} + \dfrac{16}{15(x+4)}$

29. $\dfrac{3x+1}{(x-5)(x+3)^2} = \dfrac{A}{(x+3)^2} + \dfrac{B}{x+3} + \dfrac{C}{x-5}$

$3x+1 = A(x-5) + B(x+3)(x-5) + C(x+3)^2$

$3x+1 = A(x-5) + B(x^2 - 2x - 15) + C(x^2 + 6x + 9)$

$3x+1 = x^2(B+C) + x(A - 2B + 6C) + (-5A - 15B + 9C) \Rightarrow$

$\begin{cases} B+C = 0 \\ A - 2B + 6C = 3 \\ -5A - 15B + 9C = 1 \end{cases} \Rightarrow \begin{cases} A - 2B + 6C = 3 \\ -5A - 15B + 9C = 1 \end{cases} \Rightarrow \begin{cases} 5A - 10B + 30C = 15 \\ -5A - 15B + 9C = 1 \end{cases} \Rightarrow -25B + 39C = 16$

$\begin{cases} B+C = 0 \\ -25B + 39C = 16 \end{cases} \Rightarrow \begin{cases} 25B + 25C = 0 \\ -25B + 39C = 16 \end{cases} \Rightarrow 64C = 16 \Rightarrow C = \dfrac{1}{4}$

$B + \dfrac{1}{4} = 0 \Rightarrow B = -\dfrac{1}{4}$

$A - 2\left(-\dfrac{1}{4}\right) + 6\left(\dfrac{1}{4}\right) = 3 \Rightarrow A + \dfrac{1}{2} + \dfrac{3}{2} = 3 \Rightarrow A = 1$

$\dfrac{3x+1}{(x-5)(x+3)^2} = \dfrac{1}{(x+3)^2} - \dfrac{1}{4(x+3)} + \dfrac{1}{4(x-5)}$

31. $\dfrac{8x+1}{(x+2)^2(x-3)} = \dfrac{A}{(x+2)^2} + \dfrac{B}{x+2} + \dfrac{C}{x-3}$

$8x+1 = A(x-3) + B(x+2)(x-3) + C(x+2)^2$

$8x+1 = A(x-3) + B(x^2 - x - 6) + C(x^2 + 4x + 4)$

$8x+1 = x^2(B+C) + x(A - B + 4C) + (-3A - 6B + 4C) \Rightarrow$

$\begin{cases} B+C = 0 \\ A - B + 4C = 8 \\ -3A - 6B + 4C = 1 \end{cases} \Rightarrow \begin{cases} A - B + 4C = 8 \\ -3A - 6B + 4C = 1 \end{cases} \Rightarrow \begin{cases} 3A - 3B + 12C = 24 \\ -3A - 6B + 4C = 1 \end{cases} \Rightarrow -9B + 16C = 25$

$\begin{cases} B+C = 0 \\ -9B + 16C = 25 \end{cases} \Rightarrow \begin{cases} 9B + 9C = 0 \\ -9B + 16C = 25 \end{cases} \Rightarrow 25C = 25 \Rightarrow C = 1$

$B + 1 = 0 \Rightarrow B = -1$

(continued on next page)

(*continued*)

$$A - (-1) + 4(1) = 8 \Rightarrow A = 3$$
$$\frac{8x+1}{(x+2)^2(x-3)} = \frac{3}{(x+2)^2} - \frac{1}{x+2} + \frac{1}{x-3}$$

33. $\dfrac{x-1}{(x+1)^2} = \dfrac{A}{(x+1)^2} + \dfrac{B}{(x+1)} \Rightarrow x - 1 = A + B(x+1) = Bx + (A+B) \Rightarrow \begin{cases} B = 1 \\ A + B = -1 \end{cases} \Rightarrow B = 1, A = -2$

$$\frac{x-1}{(x+1)^2} = -\frac{2}{(x+1)^2} + \frac{1}{(x+1)}$$

35. $\dfrac{x-1}{(2x-3)^2} = \dfrac{A}{2x-3} + \dfrac{B}{(2x-3)^2} \Rightarrow x - 1 = A(2x-3) + B \Rightarrow x - 1 = 2Ax + (-3A + B)$

$$\begin{cases} 2A = 1 \\ -3A + B = -1 \end{cases} \Rightarrow A = \frac{1}{2}$$

$$-3\left(\frac{1}{2}\right) + B = -1 \Rightarrow B = \frac{1}{2}$$

$$\frac{x-1}{(2x-3)^2} = \frac{1}{2(2x-3)} + \frac{1}{2(2x-3)^2}$$

37. $\dfrac{2x^2+x}{(x+1)^3} = \dfrac{A}{(x+1)^3} + \dfrac{B}{(x+1)^2} + \dfrac{C}{x+1} \Rightarrow$

$$2x^2 + x = A + B(x+1) + C(x+1)^2 = A + Bx + B + C(x^2 + 2x + 1) = Cx^2 + (B + 2C)x + (A + B + C) \Rightarrow$$

$$\begin{cases} A + B + C = 0 \\ B + 2C = 1 \\ C = 2 \end{cases} \Rightarrow C = 2, B = -3, A = 1$$

$$\frac{2x^2+x}{(x+1)^3} = \frac{2}{x+1} - \frac{3}{(x+1)^2} + \frac{1}{(x+1)^3}$$

39. $\dfrac{2x+3}{(x+2)^3} = \dfrac{A}{x+2} + \dfrac{B}{(x+2)^2} + \dfrac{C}{(x+2)^3} \Rightarrow$

$$2x + 3 = A(x+2)^2 + B(x+2) + C = A(x^2 + 4x + 4) + Bx + 2B + C = Ax^2 + x(4A + B) + 2B + C$$

$$\begin{cases} A = 0 \\ 4A + B = 2 \Rightarrow 4(0) + B = 2 \Rightarrow B = 2 \\ 2B + C = 3 \end{cases}$$

$$2(2) + C = 3 \Rightarrow C = -1$$

$$\frac{2x+3}{(x+2)^3} = \frac{2}{(x+2)^2} - \frac{1}{(x+2)^3}$$

41. $\dfrac{-x^2+3x+1}{x^3+2x^2+x} = \dfrac{-x^2+3x+1}{x(x+1)^2} = \dfrac{A}{(x+1)^2} + \dfrac{B}{x+1} + \dfrac{C}{x} \Rightarrow$

$$-x^2 + 3x + 1 = Ax + Bx(x+1) + C(x+1)^2 = Ax + Bx^2 + Bx + Cx^2 + 2Cx + C = (B+C)x^2 + (A+B+2C)x + C$$

$$\begin{cases} B + C = -1 \\ A + B + 2C = 3 \\ C = 1 \end{cases} \Rightarrow C = 1, B = -2$$

$$A + 2(-2) + 3(1) = 2 \Rightarrow A = 3$$

$$\frac{-x^2+3x+1}{x^3+2x^2+x} = \frac{1}{x} - \frac{2}{x+1} + \frac{3}{(x+1)^2}$$

43. $\dfrac{3x+1}{(x+1)(x^2-1)} = \dfrac{3x+1}{(x+1)(x+1)(x-1)} = \dfrac{A}{x-1} + \dfrac{B}{x+1} + \dfrac{C}{(x+1)^2} \Rightarrow$

$3x+1 = A(x+1)^2 + B(x+1)(x-1) + C(x-1) = A(x^2+2x+1) + B(x^2-1) + C(x-1)$

$3x+1 = x^2(A+B) + x(2A+C) + (A-B-C) \Rightarrow$

$\begin{cases} A+B = 0 \\ 2A+C = 3 \\ A-B-C = 1 \end{cases} \Rightarrow \begin{cases} A+B = 0 \\ A-B-C = 1 \end{cases} \Rightarrow 2A - C = 1$

$\begin{cases} 2A+C = 3 \\ 2A-C = 1 \end{cases} \Rightarrow A = 1, C = 1$

$1 + B = 0 \Rightarrow B = -1$

$\dfrac{3x+1}{(x+1)(x^2-1)} = \dfrac{1}{x-1} - \dfrac{1}{x+1} + \dfrac{1}{(x+1)^2}$

45. $\dfrac{1}{x^2(x+1)^2} = \dfrac{A}{x} + \dfrac{B}{x^2} + \dfrac{C}{x+1} + \dfrac{D}{(x+1)^2} \Rightarrow 1 = Ax(x+1)^2 + B(x+1)^2 + Cx^2(x+1) + Dx^2$

Letting $x = -1$, we have $D = 1$. Letting $x = 0$, we have $B = 1$. Substitute the values for B and D, expand the equation, and simplify:

$1 = Ax(x+1)^2 + 1(x+1)^2 + Cx^2(x+1) + 1x^2 = (A+C)x^3 + (2 + 2A + C)x^2 + (2+A)x + 1 \Rightarrow$

$\begin{cases} A+C = 0 \\ 2A+C = -2 \Rightarrow A = -2, C = 2 \\ A = -2 \end{cases}$

$\dfrac{1}{x^2(x+1)^2} = -\dfrac{2}{x} + \dfrac{1}{x^2} + \dfrac{2}{x+1} + \dfrac{1}{(x+1)^2}$

47. $\dfrac{1}{(x^2-1)^2} = \dfrac{1}{((x-1)(x+1))^2} = \dfrac{A}{x-1} + \dfrac{B}{(x-1)^2} + \dfrac{C}{x+1} + \dfrac{D}{(x+1)^2} \Rightarrow$

$1 = A(x-1)(x+1)^2 + B(x+1)^2 + C(x-1)^2(x+1) + D(x-1)^2$

Letting $x = 1$, we have $1 = B(2)^2 \Rightarrow B = 1/4$. Letting $x = -1$, we have $1 = D(-2)^2 \Rightarrow D = 1/4$. Substitute the values for B and D, expand the equation, and simplify:

$1 = A(x-1)(x+1)^2 + \dfrac{1}{4}(x+1)^2 + C(x-1)^2(x+1) + \dfrac{1}{4}(x-1)^2$

$1 = (A+C)x^3 + \left(\dfrac{1}{2} + A - C\right)x^2 + (-A-C)x + \left(\dfrac{1}{2} - A + C\right) \Rightarrow$

$\begin{cases} A+C = 0 \\ A - C = -\dfrac{1}{2} \\ -A - C = 0 \\ -A + C = \dfrac{1}{2} \end{cases} \Rightarrow A = -\dfrac{1}{4}, C = \dfrac{1}{4}$

$\dfrac{1}{(x^2-1)^2} = -\dfrac{1}{4(x-1)} + \dfrac{1}{4(x-1)^2} + \dfrac{1}{4(x+1)} + \dfrac{1}{4(x+1)^2}$

480 Chapter 8 Systems of Equations and Inequalities

49. $\dfrac{x^2-5}{(x-2)(x^2-2x+3)} = \dfrac{A}{x-2} + \dfrac{Bx+C}{x^2-2x+3}$

$x^2 - 5 = A(x^2 - 2x + 3) + (Bx + C)(x - 2) = A(x^2 - 2x + 3) + (Bx^2 - 2Bx + Cx - 2C)$

$\qquad = A(x^2 - 2x + 3) + (Bx^2 + (-2B + C)x - 2C) = x^2(A + B) + x(-2A - 2B + C) + (3A - 2C)$

$\begin{cases} A + B = 1 \\ -2A - 2B + C = 0 \\ 3A - 2C = -5 \end{cases} \Rightarrow \begin{cases} A + B = 1 \\ -2A - 2B + C = 0 \end{cases} \Rightarrow \begin{cases} 2A + 2B = 2 \\ -2A - 2B + C = 0 \end{cases} \Rightarrow C = 2$

$3A - 2(2) = -5 \Rightarrow 3A = -1 \Rightarrow A = -\dfrac{1}{3}$

$-\dfrac{1}{3} + B = 1 \Rightarrow B = \dfrac{4}{3}$

$\dfrac{x^2 - 5}{(x-2)(x^2 - 2x + 3)} = -\dfrac{1}{3(x-2)} + \dfrac{\frac{4}{3}x + 2}{x^2 - 2x + 3} = -\dfrac{1}{3(x-2)} + \dfrac{4x + 6}{3(x^2 - 2x + 3)} = -\dfrac{1}{3(x-2)} + \dfrac{2(2x + 3)}{3(x^2 - 2x + 3)}$

51. $\dfrac{x-1}{(x+1)^2(x^2 + 2x + 2)} = \dfrac{A}{x+1} + \dfrac{B}{(x+1)^2} + \dfrac{Cx + D}{x^2 + 2x + 2}$

$x - 1 = A(x+1)(x^2 + 2x + 2) + B(x^2 + 2x + 2) + (Cx + D)(x+1)^2$

Substitute $x = -1$ into the equation to obtain

$-2 = B\left((-1)^2 + 2(-1) + 2\right) \Rightarrow B = -2$

Substitute the value for B, expand the equation, and simplify:

$x - 1 = A(x^3 + 3x^2 + 4x + 2) - 2(x^2 + 2x + 2) + (Cx^3 + 2Cx^2 + Cx + Dx^2 + 2Dx + D)$

$\qquad = x^3(A + C) + x^2(3A + 2C + D - 2) + x(4A + C + 2D - 4) + (2A + D - 4)$

$\begin{cases} A + C = 0 \\ 3A + 2C + D - 2 = 0 \\ 4A + C + 2D - 4 = 1 \\ 2A + D - 4 = -1 \end{cases} \Rightarrow \begin{cases} A + C = 0 \\ 3A + 2C + D = 2 \\ 4A + C + 2D = 5 \\ 2A + D = 3 \end{cases} \Rightarrow \begin{cases} 3A + 2C + D = 2 \\ 4A + C + 2D = 5 \end{cases} \Rightarrow \begin{cases} 3A + 2C + D = 2 \\ -8A - 2C - 4D = -10 \end{cases} \Rightarrow$

$-5A - 3D = -8$

$\begin{cases} -5A - 3D = -8 \\ 2A + D = 3 \end{cases} \Rightarrow \begin{cases} -5A - 3D = -8 \\ 6A + 3D = 9 \end{cases} \Rightarrow A = 1,\ D = 1$

$A + C = 0 \Rightarrow C = -1$

$\dfrac{x-1}{(x+1)^2(x^2 + 2x + 2)} = \dfrac{1}{x+1} - \dfrac{2}{(x+1)^2} + \dfrac{-x + 1}{x^2 + 2x + 2}$

53. $\dfrac{x^2 + 1}{(x-2)(x^2 - 3x + 2)} = \dfrac{x^2 + 1}{(x-2)(x-2)(x-1)} = \dfrac{A}{x-2} + \dfrac{B}{(x-2)^2} + \dfrac{C}{x-1}$

$x^2 + 1 = A(x - 2)(x - 1) + B(x - 1) + C(x - 2)^2$

$\qquad = A(x^2 - 3x + 2) + B(x - 1) + C(x^2 - 4x + 4) = x^2(A + C) + x(-3A + B - 4C) + (2A - B + 4C)$

$\begin{cases} A + C = 1 \\ -3A + B - 4C = 0 \\ 2A - B + 4C = 1 \end{cases} \Rightarrow \begin{cases} -3A + B - 4C = 0 \\ 2A - B + 4C = 1 \end{cases} \Rightarrow -A = 1 \Rightarrow A = -1,\ C = 2$

$3 + B - 8 = 0 \Rightarrow B = 5$

$\dfrac{x^2 + 1}{(x-2)(x^2 - 3x + 2)} = -\dfrac{1}{x-2} + \dfrac{5}{(x-2)^2} + \dfrac{2}{x-1}$

Section 8.3 Partial-Fraction Decomposition   481

55. $\dfrac{6x+7}{4x^2+12x+9} = \dfrac{6x+7}{(2x+3)^2} = \dfrac{A}{2x+3} + \dfrac{B}{(2x+3)^2}$

$6x+7 = A(2x+3) + B = 2Ax + (3A+B)$

$\begin{cases} 2A = 6 \\ 3A+B = 7 \end{cases} \Rightarrow A = 3, B = -2$

$\dfrac{6x+7}{4x^2+12x+9} = \dfrac{3}{2x+3} - \dfrac{2}{(2x+3)^2}$

57. $\dfrac{x-3}{x^3+x^2} = \dfrac{x-3}{x^2(x+1)} = \dfrac{A}{x} + \dfrac{B}{x^2} + \dfrac{C}{x+1}$

$x-3 = Ax(x+1) + B(x+1) + Cx^2 = (A+C)x^2 + (A+B)x + B$

$\begin{cases} A + C = 0 \\ A + B = 1 \\ B = -3 \end{cases} \Rightarrow B = -3, A = 4, C = -4$

$\dfrac{x-3}{x^3+x^2} = \dfrac{4}{x} - \dfrac{3}{x^2} - \dfrac{4}{x+1}$

59. $\dfrac{x^2+2x+4}{x^3+x^2} = \dfrac{x^2+2x+4}{x^2(x+1)} = \dfrac{A}{x} + \dfrac{B}{x^2} + \dfrac{C}{x+1}$

$x^2+2x+4 = Ax(x+1) + B(x+1) + Cx^2 = (A+C)x^2 + (A+B)x + B$

$\begin{cases} A + C = 1 \\ A + B = 2 \\ B = 4 \end{cases} \Rightarrow B = 4, A = -2, C = 3$

$\dfrac{x^2+2x+4}{x^3+x^2} = -\dfrac{2}{x} + \dfrac{4}{x^2} + \dfrac{3}{x+1}$

61. $\dfrac{x}{x^4-1} = \dfrac{x}{(x-1)(x+1)(x^2+1)} = \dfrac{A}{x-1} + \dfrac{B}{x+1} + \dfrac{Cx+D}{x^2+1}$

$x = A(x+1)(x^2+1) + B(x-1)(x^2+1) + (Cx+D)(x^2-1)$

$x = (A+B+C)x^3 + (A-B+D)x^2 + (A+B-C)x + (A-B-D)$

$\begin{cases} A+B+C = 0 \\ A-B+D = 0 \\ A+B-C = 1 \\ A-B-D = 0 \end{cases} \Rightarrow A = \dfrac{1}{4}, B = \dfrac{1}{4}, C = -\dfrac{1}{2}, D = 0$

$\dfrac{x}{x^4-1} = \dfrac{1}{4(x-1)} + \dfrac{1}{4(x+1)} - \dfrac{x}{2(x^2+1)}$

63. $\dfrac{1}{x(x^2+1)^2} = \dfrac{A}{x} + \dfrac{Bx+C}{x^2+1} + \dfrac{Dx+E}{(x^2+1)^2} \Rightarrow$

$1 = A(x^2+1)^2 + (Bx+C)x(x^2+1) + (Dx+E)x = (A+B)x^4 + Cx^3 + (2A+B+D)x^2 + (C+E)x + A$

$\begin{cases} A + B = 0 \\ C = 0 \\ 2A + B + D = 0 \\ C + E = 0 \\ A = 1 \end{cases} \Rightarrow A = 1, B = -1, C = 0, D = -1, E = 0$

$\dfrac{1}{x(x^2+1)^2} = \dfrac{1}{x} - \dfrac{x}{x^2+1} - \dfrac{x}{(x^2+1)^2}$

65. $\dfrac{2x^2+3x}{(x^2+1)(x^2+2)} = \dfrac{Ax+B}{x^2+1} + \dfrac{Cx+D}{x^2+2}$

$2x^2+3x = (Ax+B)(x^2+2)+(Cx+D)(x^2+1) = (A+C)x^3+(B+D)x^2+(2A+C)x+(2B+D)$

$\begin{cases} A+C=0 \\ B+D=2 \\ 2A+C=3 \\ 2B+D=0 \end{cases} \Rightarrow A=3, B=-2, C=-3, D=4$

$\dfrac{2x^2+3x}{(x^2+1)(x^2+2)} = \dfrac{3x-2}{x^2+1} + \dfrac{-3x+4}{x^2+2}$

8.3 Applying the Concepts

67. $\dfrac{1}{1\cdot 2} + \dfrac{1}{2\cdot 3} + \dfrac{1}{3\cdot 4} + \cdots + \dfrac{1}{n(n+1)} = \left(1-\dfrac{1}{2}\right)+\left(\dfrac{1}{2}-\dfrac{1}{3}\right)+\left(\dfrac{1}{3}-\dfrac{1}{4}\right)+\cdots+\left(\dfrac{1}{n}-\dfrac{1}{n+1}\right) = 1-\dfrac{1}{n+1} = \dfrac{n}{n+1}$

69. Each term is of the form $\dfrac{2}{(2n-1)(2n+1)}$. Decomposing $\dfrac{2}{(2n-1)(2n+1)}$ we have

$\dfrac{2}{(2n-1)(2n+1)} = \dfrac{A}{2n-1} + \dfrac{B}{2n+1} \Rightarrow$

$2 = A(2n+1)+B(2n-1) = (2A+2B)n+(A-B) \Rightarrow$

$\begin{cases} 2A+2B=0 \\ A-B=2 \end{cases} \Rightarrow A=1, B=-1$

$\dfrac{2}{(2n-1)(2n+1)} = \dfrac{1}{2n-1} - \dfrac{1}{2n+1}.$

So,

$\dfrac{2}{1\cdot 3} + \dfrac{2}{3\cdot 5} + \dfrac{2}{5\cdot 7} + \cdots + \dfrac{2}{(2n-1)(2n+1)} = \left(\dfrac{1}{1}-\dfrac{1}{3}\right)+\left(\dfrac{1}{3}-\dfrac{1}{5}\right)+\left(\dfrac{1}{5}-\dfrac{1}{7}\right)+\cdots+\dfrac{1}{2n-1}-\dfrac{1}{2n+1}$

$= 1-\dfrac{1}{2n+1} = \dfrac{2n}{2n+1}$

71. $\dfrac{1}{R} = \dfrac{2x+4}{(x+1)(x+3)}$

$\dfrac{2x+4}{(x+1)(x+3)} = \dfrac{A}{x+1} + \dfrac{B}{x+3} \Rightarrow 2x+4 = A(x+3)+B(x+1) = (A+B)x+(3A+B)$

$\begin{cases} A+B=2 \\ 3A+B=4 \end{cases} \Rightarrow A=1, B=1$

$\dfrac{2x+4}{(x+1)(x+3)} = \dfrac{1}{x+1} + \dfrac{1}{x+3} = \dfrac{1}{R}.$

This means that if two resistances $R_1 = x+1$ and $R_2 = x+3$ are connected in parallel, they will produce a total resistance given by $\dfrac{(x+1)(x+3)}{2x+4}.$

73. $\dfrac{1}{R} = \dfrac{R_1R_2 + R_2R_3 + R_3R_1}{R_1R_2R_3} = \dfrac{A}{R_1} + \dfrac{B}{R_2} + \dfrac{C}{R_3}$

$R_1R_2 + R_2R_3 + R_3R_1 = AR_2R_3 + BR_1R_3 + CR_1R_2$

$\begin{cases} A = 1 \\ B = 1 \\ C = 1 \end{cases}$

$\dfrac{R_1R_2 + R_2R_3 + R_3R_1}{R_1R_2R_3} = \dfrac{1}{R_1} + \dfrac{1}{R_2} + \dfrac{1}{R_3} = \dfrac{1}{R}$

This means that if three resistances $R_1, R_2,$ and R_3 are connected in parallel, they will produce a total resistance given by $\dfrac{R_1R_2R_3}{R_1R_2 + R_2R_3 + R_3R_1}$.

8.3 Beyond the Basics

75. $\begin{cases} A + B + C = 0 \\ 2B - 2C = 10 \\ -4A = -4 \end{cases} \Rightarrow \begin{cases} A + B + C = 0 \\ B - C = 5 \\ A = 1 \end{cases}$

Subtract the third equation from the first equation, and replace the first equation with the new equation:

$\begin{cases} A + B + C = 0 \\ B - C = 5 \\ A = 1 \end{cases} \Rightarrow \begin{cases} B + C = -1 \\ B - C = 5 \\ A = 1 \end{cases}$

Add the first and second equations to solve for B:

$\begin{cases} B + C = -1 \\ B - C = 5 \\ A = 1 \end{cases} \Rightarrow \begin{cases} B + C = -1 \\ 2B = 4 \\ A = 1 \end{cases} \Rightarrow \begin{cases} B + C = -1 \\ B = 2 \\ A = 1 \end{cases}$

Substitute the value for B into the first equation to solve for C:

$\begin{cases} B + C = -1 \\ B = 2 \\ A = 1 \end{cases} \Rightarrow \begin{cases} 2 + C = -1 \\ B = 2 \\ A = 1 \end{cases} \Rightarrow \begin{cases} C = -3 \\ B = 2 \\ A = 1 \end{cases}$

77. $\dfrac{4x}{(x^2-1)^2} = \dfrac{4x}{(x-1)^2(x+1)^2} = \dfrac{A}{x-1} + \dfrac{B}{(x-1)^2} + \dfrac{C}{x+1} + \dfrac{D}{(x+1)^2}$

$4x = A(x-1)(x+1)^2 + B(x+1)^2 + C(x-1)^2(x+1) + D(x-1)^2$

$= A(x^3 + x^2 - x - 1) + B(x^2 + 2x + 1) + C(x^3 - x^2 - x + 1) + D(x^2 - 2x + 1)$

$4x = (A+C)x^3 + (A+B-C+D)x^2 + (-A+2B-C-2D)x + (-A+B+C+D)$

$\begin{cases} A + C = 0 \quad (1) \\ A + B - C + D = 0 \quad (2) \\ -A + 2B - C - 2D = 4 \quad (3) \\ -A + B + C + D = 0 \quad (4) \end{cases}$

Add equations (1) and (3), and (2) and (4), and replace equations (3) and (4):

$\begin{cases} A + C = 0 \\ A + B - C + D = 0 \\ -A + 2B - C - 2D = 4 \\ -A + B + C + D = 0 \end{cases} \Rightarrow \begin{cases} A + C = 0 \quad (1) \\ A + B - C + D = 0 \quad (2) \\ 2B - 2D = 4 \quad (3) \\ 2B + 2D = 0 \quad (4) \end{cases} \Rightarrow \begin{cases} A + C = 0 \\ A + B - C + D = 0 \\ 4B = 4 \\ 2B + 2D = 0 \end{cases} \Rightarrow$

$B = 1, D = -1, A = 0, C = 0$

$\dfrac{4x}{(x^2-1)^2} = \dfrac{1}{(x-1)^2} - \dfrac{1}{(x+1)^2}$

79. $\dfrac{x+1}{(x^2+1)(x-1)^2} = \dfrac{A}{x-1} + \dfrac{B}{(x-1)^2} + \dfrac{Cx+D}{x^2+1}$

$x+1 = A(x-1)(x^2+1) + B(x^2+1) + (Cx+D)(x-1)^2$
$= (A+C)x^3 + (-A+B-2C+D)x^2 + (A+B+C-2D)x + (-A+B+D)$

$\begin{cases} A + C = 0 \\ -A+B-2C+D = 0 \\ A+B+C-2D = 1 \\ -A+B+D = 1 \end{cases}$

From the first equation, we have $C = -A$. Substitute this into the last three equations and simplify:

$\begin{cases} A+C=0 \\ -A+B-2C+D=0 \\ A+C-2D=1 \\ -A+B+D=1 \end{cases} \Rightarrow \begin{cases} C=-A \\ -A+B-2(-A)+D=0 \\ A+(-A)-2D=1 \\ -A+B+D=1 \end{cases} \Rightarrow \begin{cases} C=-A \\ A+B+D=0 \\ -2D=1 \\ -A+B+D=1 \end{cases} \Rightarrow \begin{cases} C=-A \\ A+B+D=0 \\ D=-\dfrac{1}{2} \\ -A+B+D=1 \end{cases}$

Substitute the value for D into the second and fourth equations, and simplify:

$\begin{cases} C=-A \\ A+B=\dfrac{1}{2} \\ D=-\dfrac{1}{2} \\ -A+B=\dfrac{3}{2} \end{cases} \Rightarrow B=1, A=-\dfrac{1}{2}, C=\dfrac{1}{2}$

$\dfrac{x+1}{(x^2+1)(x-1)^2} = -\dfrac{1}{2(x-1)} + \dfrac{1}{(x-1)^2} + \dfrac{x-1}{2(x^2+1)}$

81. $\dfrac{x^3}{(x+1)^2(x+2)^2} = \dfrac{A}{x+1} + \dfrac{B}{(x+1)^2} + \dfrac{C}{x+2} + \dfrac{D}{(x+2)^2}$

$x^3 = A(x+1)(x+2)^2 + B(x+2)^2 + C(x+2)(x+1)^2 + D(x+1)^2$

Letting $x = -2$, we have $(-2)^3 = D(-1)^2 \Rightarrow D = -8$. Letting $x = -1$, we have $(-1)^3 = B(1)^2 \Rightarrow B = -1$.
Expand the equation and substitute the values for B and D:

$x^3 = (A+C)x^3 + (-9+5A+4C)x^2 + (-20+8A+5C)x + (-12+4A+2C)$

$\begin{cases} A+C=1 \\ 5A+4C=9 \\ 8A+5C=20 \\ 4A+2C=12 \end{cases} \Rightarrow A=5, C=-4$

$\dfrac{x^3}{(x+1)^2(x+2)^2} = \dfrac{5}{x+1} - \dfrac{1}{(x+1)^2} - \dfrac{4}{x+2} - \dfrac{8}{(x+2)^2}$

83.

$\begin{array}{r} 1 \\ x^2+3x+2 \overline{)x^2+4x+5} \\ \underline{x^2+3x+2} \\ x+3 \end{array}$

$\dfrac{x^2+4x+5}{x^2+3x+2} = 1 + \dfrac{x+3}{x^2+3x+2}$

Decompose $\dfrac{x+3}{x^2+3x+2}$:

$\dfrac{x+3}{x^2+3x+2} = \dfrac{x+3}{(x+1)(x+2)} = \dfrac{A}{x+1} + \dfrac{B}{x+2} \Rightarrow$

$x+3 = A(x+2) + B(x+1) = (A+B)x + (2A+B)$

$\begin{cases} A+B=1 \\ 2A+B=3 \end{cases} \Rightarrow A=2, B=-1$

$\dfrac{x^2+4x+5}{x^2+3x+2} = 1 + \dfrac{2}{x+1} - \dfrac{1}{x+2}$

85. Each term is of the form
$$\frac{1}{\sqrt{2n+1}+\sqrt{2n+3}} \cdot \frac{\sqrt{2n+1}-\sqrt{2n+3}}{\sqrt{2n+1}-\sqrt{2n+3}} = \frac{\sqrt{2n+1}-\sqrt{2n+3}}{2n+1-(2n+3)} = \frac{\sqrt{2n+1}-\sqrt{2n+3}}{-2} = \frac{\sqrt{2n+3}-\sqrt{2n+1}}{2}.$$
So,
$$\frac{1}{\sqrt{3}+\sqrt{5}} + \frac{1}{\sqrt{5}+\sqrt{7}} + \frac{1}{\sqrt{7}+\sqrt{9}} + \cdots + \frac{1}{\sqrt{2n-1}+\sqrt{2n+1}} + \frac{1}{\sqrt{2n+1}+\sqrt{2n+3}}$$
$$= \frac{\sqrt{5}-\sqrt{3}}{2} + \frac{\sqrt{7}-\sqrt{5}}{2} + \frac{\sqrt{9}-\sqrt{7}}{2} + \cdots + \frac{\sqrt{2n+1}-\sqrt{2n-1}}{2} + \frac{\sqrt{2n+3}-\sqrt{2n+1}}{2}$$
$$= \frac{\sqrt{2n+3}-\sqrt{3}}{2}$$

8.3 Critical Thinking/Discussion/Writing

87. Two equal polynomials have equal corresponding coefficients.

88. $f(x) = \dfrac{x+8}{x^2+x-2} = \dfrac{x+8}{(x+2)(x-1)}$
$= \dfrac{A}{x+2} + \dfrac{B}{x-1} \Rightarrow$
$x+8 = A(x-1) + B(x+2)$
$= (A+B)x + (-A+2B)$
$\begin{cases} A+B = 1 \\ -A+2B = 8 \end{cases} \Rightarrow 3B = 9 \Rightarrow B = 3, A = -2$
$\dfrac{x+8}{x^2+x-2} = -\dfrac{2}{x+2} + \dfrac{3}{x-1}$
$g(x) = -\dfrac{2}{x+2}, \; h(x) = \dfrac{3}{x-1}$

a.

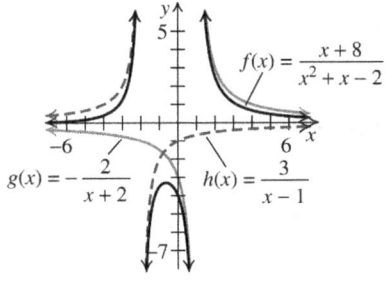

b. The graphs of f and g have the same vertical asymptote, $x = -2$, and the same horizontal asymptote, $y = 0$, or the x-axis. Also g is an asymptote of f as x approaches 1 from the right or from the left.

c. The graphs of f and h have the same vertical asymptote, $x = 1$, and the same horizontal asymptote, $y = 0$, or the x-axis. Also h is an asymptote of f as x approaches -2 from the right or from the left.

8.3 Getting Ready for the Next Section

89. $x^2 + 10x + 21 = 0 \Rightarrow (x+7)(x+3) = 0 \Rightarrow$
$x = -7, x = -3$

91. $2x^2 + x - 10 = 0 \Rightarrow (2x+5)(x-2) = 0 \Rightarrow$
$x = -\dfrac{5}{2}, x = 2$

In exercises 93–100, be sure to check the solution in both equations.

93. $\begin{cases} x+y = 3 & (1) \\ 3x+2y = 7 & (2) \end{cases}$

From equation (1), we have $y = 3 - x$.
Substituting in equation (2), we have
$3x + 2(3-x) = 7 \Rightarrow x + 6 = 7 \Rightarrow x = 1$
Substituting $x = 1$ in equation (1) gives
$1 + y = 3 \Rightarrow y = 2$.
Solution set: $\{(1, 2)\}$

95. $\begin{cases} x - 2y = 1 & (1) \\ 2x + 3y = 16 & (2) \end{cases}$

From equation (1), we have $x = 2y + 1$.
Substituting in equation (2), we have
$2(2y+1) + 3y = 16 \Rightarrow 7y + 2 = 16 \Rightarrow$
$7y = 14 \Rightarrow y = 2$
Substituting $y = 2$ in equation (1) gives
$x - 2(2) = 1 \Rightarrow x - 4 = 1 \Rightarrow x = 5$
Solution set: $\{(5, 2)\}$

97. $\begin{cases} 2x - y = 4 & (1) \\ 3x + 2y = 13 & (2) \end{cases}$

Multiply equation (1) by 2, then add the two equations and solve for x.
$4x - 2y = 8$
$3x + 2y = 13$
$\overline{7x = 21} \Rightarrow x = 3$
Substitute $x = 3$ in equation (1), then solve for y.
$2(3) - y = 4 \Rightarrow 6 - y = 4 \Rightarrow y = 2$
Solution set: $\{(3, 2)\}$

99. $\begin{cases} 2x + 5y = 1 & (1) \\ 3x - 2y = -8 & (2) \end{cases}$

Multiply equation (1) by 2 and equation (2) by 5 to eliminate y. Add the resulting equations and solve for x.
$4x + 10y = 2$
$15x - 10y = -40$
$\overline{19x = -38} \Rightarrow x = -2$
Substitute $x = -2$ in equation (1), then solve for y.
$2(-2) + 5y = 1 \Rightarrow -4 + 5y = 1 \Rightarrow 5y = 5 \Rightarrow y = 1$
Solution set: $\{(-2, 1)\}$

8.4 Systems of Nonlinear Equations

8.4 Practice Problems

1. $\begin{cases} x^2 + y = 2 \\ 2x + y = 3 \end{cases} \Rightarrow \begin{cases} y = 2 - x^2 \\ 2x + y = 3 \end{cases} \Rightarrow$
$2x + (2 - x^2) = 3 \Rightarrow -x^2 + 2x - 1 = 0 \Rightarrow$
$x^2 - 2x + 1 = 0 \Rightarrow (x - 1)^2 = 0 \Rightarrow x = 1$
$2(1) + y = 3 \Rightarrow y = 1$
The solution is $\{(1, 1)\}$.

2. $\begin{cases} x^2 + 2y^2 = 34 \\ x^2 - y^2 = 7 \end{cases} \Rightarrow 3y^2 = 27 \Rightarrow y^2 = 9 \Rightarrow$
$y = \pm 3$
$x^2 - (-3)^2 = 7 \Rightarrow x^2 = 16 \Rightarrow x = \pm 4$
$x^2 - (3)^2 = 7 \Rightarrow x^2 = 16 \Rightarrow x = \pm 4$
The solution is $\{(-4, -3), (-4, 3), (4, -3), (4, 3)\}$.

3. Let x = the number of shares received as dividends, and let p = the selling price per share. Then $xp = 1950$. The number of shares she sold at p dollars per share is $240 + x$. Thus, revenue = $(240 + x)p$ and the cost of her stock was $(240)(40) + 100 = 9700$.
Since revenue − cost = profit, we have
$(240 + x)p - 9700 = 7850 \Rightarrow$
$240p + xp = 17,550$
Thus, the system of equations is
$\begin{cases} xp = 1950 \\ 240p + xp = 17,550 \end{cases}.$

$\begin{cases} xp = 1950 \\ 240p + xp = 17,550 \end{cases} \Rightarrow$
$240p + 1950 = 17,550 \Rightarrow 240p = 15,600 \Rightarrow$
$p = 65$
$65x = 1950 \Rightarrow x = 30$
Danielle received 30 shares as dividends and sold her stock at $65 per share.

8.4 Concepts and Vocabulary

1. In a system of nonlinear equations, <u>at least one</u> equation must be nonlinear.

3. The solutions of the system
$\begin{cases} ax + by = c & (1) \\ (x - h)^2 + (y - k)^2 = r^2 & (2) \end{cases}$
represent the points of <u>intersection</u> of the graphs of equations (1) and (2).

5. True

7. True

8.4 Building Skills

9. Substituting each ordered pair into the system
$\begin{cases} 2x + 3y = 3 \\ x - y^2 = 2 \end{cases}$
we find that $(3, -1)$ is a solution.
$\begin{cases} 2(3) + 3(-1) = 6 - 3 = 3 \\ 3 - (-1)^2 = 3 - 1 = 2 \end{cases}$

11. Substituting each ordered pair into the system
$\begin{cases} 5x - 2y = 7 \\ x^2 + y^2 = 2 \end{cases},$
we find that $(1, -1)$ is a solution:
$\begin{cases} 5(1) - 2(-1) = 5 + 2 = 7 \\ (1)^2 + (-1)^2 = 1 + 1 = 2 \end{cases}$

13. Substituting each ordered pair into the system
$$\begin{cases} 4x^2 + 5y^2 = 180 \\ x^2 - y^2 = 9 \end{cases}$$
we find that (5, 4), (−5, 4), and (−5, −4) are solutions.
$$\begin{cases} 4(5)^2 + 5(4)^2 = 100 + 80 = 180 \\ 5^2 - 4^2 = 25 - 16 = 9 \end{cases}$$
$$\begin{cases} 4(-5)^2 + 5(4)^2 = 100 + 80 = 180 \\ (-5)^2 - (4)^2 = 25 - 16 = 9 \end{cases}$$
$$\begin{cases} 4(-5)^2 + 5(-4)^2 = 100 + 80 = 180 \\ (-5)^2 - (-4)^2 = 25 - 16 = 9 \end{cases}$$

15. Substituting each ordered pair into the system
$$\begin{cases} y = e^{x-1} \\ y = 2x - 1 \end{cases}$$
we find that (1, 1) is a solution:
$$\begin{cases} y = e^{1-1} = e^0 = 1 \\ y = 2(1) - 1 = 1 \end{cases}$$

17. $\begin{cases} y = x^2 \\ y = x + 2 \end{cases} \Rightarrow x^2 = x + 2 \Rightarrow x^2 - x - 2 = 0 \Rightarrow$
$(x - 2)(x + 1) = 0 \Rightarrow x = 2$ or $x = -1$
$y = 2^2 = 4$ or $y = (-1)^2 = 1$
The solution is $\{(2, 4), (-1, 1)\}$.

19. $\begin{cases} x^2 - y = 6 \\ x - y = 0 \end{cases} \Rightarrow \begin{cases} x^2 - y = 6 \\ x = y \end{cases} \Rightarrow x^2 - x = 6 \Rightarrow$
$x^2 - x - 6 = 0 \Rightarrow (x - 3)(x + 2) = 0 \Rightarrow$
$x = 3$ or $x = -2$
$3 - y = 0 \Rightarrow y = 3$ or $-2 - y = 0 \Rightarrow y = -2$
The solution is $\{(3, 3), (-2, -2)\}$.

21. $\begin{cases} x^2 + y^2 = 9 \\ x = 3 \end{cases} \Rightarrow 3^2 + y^2 = 9 \Rightarrow y = 0$
The solution is $\{(3, 0)\}$.

23. $\begin{cases} x^2 + y^2 = 5 \\ x - y = -3 \end{cases} \Rightarrow \begin{cases} x^2 + y^2 = 5 \\ x = y - 3 \end{cases} \Rightarrow$
$(y - 3)^2 + y^2 = 5 \Rightarrow 2y^2 - 6y + 4 = 0 \Rightarrow$
$y^2 - 3y + 2 = 0 \Rightarrow (y - 2)(y - 1) = 0 \Rightarrow$
$y = 2$ or $y = 1$
$x - 2 = -3 \Rightarrow x = -1$
$x - 1 = -3 \Rightarrow x = -2$
The solution is $\{(-2, 1), (-1, 2)\}$.

25. $\begin{cases} x^2 - 4x + y^2 = -2 \\ x - y = 2 \end{cases} \Rightarrow \begin{cases} x^2 - 4x + y^2 = -2 \\ x = y + 2 \end{cases} \Rightarrow$
$(y + 2)^2 - 4(y + 2) + y^2 = -2 \Rightarrow 2y^2 = 2 \Rightarrow$
$y = \pm 1$
$x - (-1) = 2 \Rightarrow x = 1$
$x - 1 = 2 \Rightarrow x = 3$
The solution is $\{(1, -1), (3, 1)\}$.

27. $\begin{cases} x - y = -2 \\ xy = 3 \end{cases} \Rightarrow \begin{cases} x = y - 2 \\ xy = 3 \end{cases} \Rightarrow y(y - 2) = 3 \Rightarrow$
$y^2 - 2y - 3 = 0 \Rightarrow (y - 3)(y + 1) = 0 \Rightarrow$
$y = 3$ or $y = -1$
$x - 3 = -2 \Rightarrow x = 1$
$x - (-1) = -2 \Rightarrow x = -3$
The solution is $\{(1, 3), (-3, -1)\}$.

29. $\begin{cases} 4x^2 + y^2 = 25 \\ x + y = 5 \end{cases} \Rightarrow \begin{cases} 4x^2 + y^2 = 25 \\ x = 5 - y \end{cases} \Rightarrow$
$4(5 - y)^2 + y^2 = 25 \Rightarrow 5y^2 - 40y + 75 = 0 \Rightarrow$
$y^2 - 8y + 15 = 0 \Rightarrow (y - 5)(y - 3) = 0 \Rightarrow$
$y = 5$ or $y = 3$
$x + 5 = 5 \Rightarrow x = 0$
$x + 3 = 5 \Rightarrow x = 2$
The solution is $\{(0, 5), (2, 3)\}$.

31. $\begin{cases} x^2 - y^2 = 24 \\ 5x - 7y = 0 \end{cases} \Rightarrow \begin{cases} x^2 - y^2 = 24 \\ x = \dfrac{7}{5}y \end{cases} \Rightarrow$
$\left(\dfrac{7}{5}y\right)^2 - y^2 = 24 \Rightarrow \dfrac{24}{25}y^2 = 24 \Rightarrow y^2 = 25 \Rightarrow$
$y = \pm 5$
$5x - 7(-5) = 0 \Rightarrow x = -7$
$5x - 7(5) = 0 \Rightarrow x = 7$
The solution is $\{(7, 5), (-7, -5)\}$.

33. $\begin{cases} x^2 + y^2 = 20 \\ x^2 - y^2 = 12 \end{cases} \Rightarrow 2x^2 = 32 \Rightarrow x = \pm 4$
$(-4)^2 + y^2 = 20 \Rightarrow y^2 = 4 \Rightarrow y = \pm 2$
$(4)^2 + y^2 = 20 \Rightarrow y^2 = 4 \Rightarrow y = \pm 2$
The solution is $\{(-4, -2), (-4, 2), (4, -2), (4, 2)\}$.

35. $\begin{cases} x^2 + 2y^2 = 12 \\ 7y^2 - 5x^2 = 8 \end{cases} \Rightarrow \begin{cases} 5x^2 + 10y^2 = 60 \\ 7y^2 - 5x^2 = 8 \end{cases} \Rightarrow$
$17y^2 = 68 \Rightarrow y^2 = 4 \Rightarrow y = \pm 2$
$x^2 + 2(-2)^2 = 12 \Rightarrow x^2 = 4 \Rightarrow x = \pm 2$
$x^2 + 2(2)^2 = 12 \Rightarrow x^2 = 4 \Rightarrow x = \pm 2$
The solution is $\{(-2, -2), (-2, 2), (2, -2), (2, 2)\}$.

37. $\begin{cases} x^2 - y = 2 \\ 2x - y = 4 \end{cases} \Rightarrow \begin{cases} x^2 - y = 2 \\ -2x + y = -4 \end{cases} \Rightarrow$

$x^2 - 2x = -2 \Rightarrow x^2 - 2x + 2 = 0 \Rightarrow$

$x = \dfrac{2 \pm \sqrt{4 - 8}}{2} = 1 \pm i \Rightarrow$ there are no real solutions. Solution set: $\varnothing$

39. $\begin{cases} x^2 + y^2 = 5 \\ 3x^2 - 2y^2 = -5 \end{cases} \Rightarrow \begin{cases} 2x^2 + 2y^2 = 10 \\ 3x^2 - 2y^2 = -5 \end{cases} \Rightarrow$

$5x^2 = 5 \Rightarrow x = \pm 1$

$(1)^2 + y^2 = 5 \Rightarrow y = \pm 2$

$(-1)^2 + y^2 = 5 \Rightarrow y = \pm 2$

The solution is $\{(-1, -2), (-1, 2), (1, -2), (1, 2)\}$.

41. $\begin{cases} x^2 + y^2 + 2x = 9 \\ x^2 + 4y^2 + 3x = 14 \end{cases} \Rightarrow$

$\begin{cases} 4x^2 + 4y^2 + 8x = 36 \\ x^2 + 4y^2 + 3x = 14 \end{cases} \Rightarrow 3x^2 + 5x - 22 = 0 \Rightarrow$

$(x - 2)(3x + 11) = 0 \Rightarrow x = 2$ or $x = -\dfrac{11}{3}$

$(2)^2 + y^2 + 2(2) = 9 \Rightarrow y = \pm 1$

$\left(-\dfrac{11}{3}\right)^2 + y^2 + 2\left(-\dfrac{11}{3}\right) = 9 \Rightarrow y^2 + \dfrac{55}{9} = 9 \Rightarrow$

$y = \pm \dfrac{\sqrt{26}}{3}$

The solution is $\left\{(2, -1), (2, 1), \left(-\dfrac{11}{3}, -\dfrac{\sqrt{26}}{3}\right), \left(-\dfrac{11}{3}, \dfrac{\sqrt{26}}{3}\right)\right\}$.

43. Using substitution, we have

$\begin{cases} x + y = 8 \\ xy = 15 \end{cases} \Rightarrow \begin{cases} x = 8 - y \\ xy = 15 \end{cases} \Rightarrow$

$y(8 - y) = 15 \Rightarrow -y^2 + 8y - 15 = 0 \Rightarrow$

$y^2 - 8y + 15 = 0 \Rightarrow (y - 3)(y - 5) = 0 \Rightarrow$

$y = 3$ or $y = 5$

$x + 3 = 8 \Rightarrow x = 5$

$x + 5 = 8 \Rightarrow x = 3$

The solution is $\{(3, 5), (5, 3)\}$.

45. Using elimination, we have

$\begin{cases} x^2 + y^2 = 2 \\ 3x^2 + 3y^2 = 9 \end{cases} \Rightarrow \begin{cases} -3x^2 - 3y^2 = -6 \\ 3x^2 + 3y^2 = 9 \end{cases} \Rightarrow$

$0 = 3 \Rightarrow$ there is no solution. Solution set: $\varnothing$

47. Using substitution, we have

$\begin{cases} y^2 = 4x + 4 \\ y = 2x - 2 \end{cases} \Rightarrow (2x - 2)^2 = 4x + 4 \Rightarrow$

$4x^2 - 12x = 0 \Rightarrow 4x(x - 3) = 0 \Rightarrow x = 0$ or $x = 3$

$y = 2(0) - 2 = -2$

$y = 2(3) - 2 = 4$

The solution is $\{(0, -2), (3, 4)\}$.

49. Using substitution, we have

$\begin{cases} x^2 + 4y^2 = 25 \\ x - 2y + 1 = 0 \end{cases} \Rightarrow \begin{cases} x^2 + 4y^2 = 25 \\ x = 2y - 1 \end{cases} \Rightarrow$

$(2y - 1)^2 + 4y^2 = 25 \Rightarrow 8y^2 - 4y - 24 = 0 \Rightarrow$

$4(y - 2)(2y + 3) = 0 \Rightarrow y = 2$ or $y = -\dfrac{3}{2}$

$x - 2(2) + 1 = 0 \Rightarrow x = 3$

$x - 2\left(-\dfrac{3}{2}\right) + 1 = 0 \Rightarrow x = -4$

The solution is $\left\{(3, 2), \left(-4, -\dfrac{3}{2}\right)\right\}$.

51. Using elimination, we have

$\begin{cases} x^2 - 3y^2 = 1 \\ x^2 + 4y^2 = 8 \end{cases} \Rightarrow -7y^2 = -7 \Rightarrow y = \pm 1$

$x^2 - 3(-1)^2 = 1 \Rightarrow x = \pm 2$

$x^2 - 3(1)^2 = 1 \Rightarrow x = \pm 2$

The solution is $\{(-2, -1), (-2, 1), (2, -1), (2, 1)\}$.

53. Using elimination, we have

$\begin{cases} x^2 - xy + 5x = 4 \\ 2x^2 - 3xy + 10x = -2 \end{cases} \Rightarrow$

$\begin{cases} -3x^2 + 3xy - 15x = -12 \\ 2x^2 - 3xy + 10x = -2 \end{cases} \Rightarrow -x^2 - 5x = -14 \Rightarrow$

$x^2 + 5x - 14 = 0 \Rightarrow (x + 7)(x - 2) = 0 \Rightarrow$

$x = -7$ or $x = 2$

$(-7)^2 - (-7)y + 5(-7) = 4 \Rightarrow y = -\dfrac{10}{7}$

$(2)^2 - 2y + 5(2) = 4 \Rightarrow y = 5$

The solution is $\left\{\left(-7, -\dfrac{10}{7}\right), (2, 5)\right\}$.

55. Using elimination, we have
$$\begin{cases} x^2 + y^2 - 8x = -8 \\ x^2 - 4y^2 + 6x = 0 \end{cases} \Rightarrow$$
$$\begin{cases} 4x^2 + 4y^2 - 32x = -32 \\ x^2 - 4y^2 + 6x = 0 \end{cases} \Rightarrow 5x^2 - 26x = -32 \Rightarrow$$
$5x^2 - 26x + 32 = 0 \Rightarrow (x-2)(5x-16) = 0 \Rightarrow$
$x = 2$ or $x = \dfrac{16}{5}$
$2^2 + y^2 - 8(2) = -8 \Rightarrow y = \pm 2$
$\left(\dfrac{16}{5}\right)^2 + y^2 - 8\left(\dfrac{16}{5}\right) = -8 \Rightarrow y^2 = \dfrac{184}{25} \Rightarrow$
$y = \pm \dfrac{2\sqrt{46}}{5}$

The solution is
$\left\{(2,-2),(2,2),\left(\dfrac{16}{5},-\dfrac{2\sqrt{46}}{5}\right),\left(\dfrac{16}{5},\dfrac{2\sqrt{46}}{5}\right)\right\}$.

8.4 Applying the Concepts

57. Using elimination, we have
$$\begin{cases} p + 2x^2 = 96 \\ p - 13x = 39 \end{cases} \Rightarrow 2x^2 + 13x = 57 \Rightarrow$$
$2x^2 + 13x - 57 = 0 \Rightarrow (x-3)(2x+19) = 0 \Rightarrow$
$x = 3$ or $x = -\dfrac{19}{2}$ (reject this)
$p - 13(3) = 39 \Rightarrow p = 78$

Market equilibrium occurs when 3 (hundred) units are sold and the price is $78/unit.

59. Let x = the first positive number and let y = the second positive number. Then
$$\begin{cases} x + y = 24 \\ xy = 143 \end{cases} \Rightarrow \begin{cases} y = 24 - x \\ xy = 143 \end{cases} \Rightarrow$$
$x(24 - x) = 143 \Rightarrow -x^2 + 24x - 143 = 0 \Rightarrow$
$-(x-13)(x-11) = 0 \Rightarrow x = 11$ or $x = 13$
$11 + y = 24 \Rightarrow y = 13$
$13 + y = 24 \Rightarrow y = 11$

The numbers are 11 and 13.

61. a. Let x = the length of the two equal sides and let y = the length of the third side.

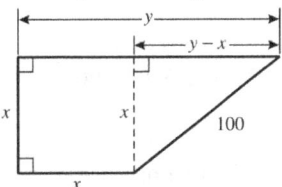

Using the Pythagorean theorem, we have
$x^2 + (y - x)^2 = 100^2$. So,
$$\begin{cases} 2x + y + 100 = 360 \\ x^2 + (y-x)^2 = 100^2 \end{cases} \Rightarrow$$
$$\begin{cases} y = 260 - 2x \\ x^2 + (y-x)^2 = 100^2 \end{cases} \Rightarrow$$
$x^2 + ((260 - 2x) - x)^2 = 10,000 \Rightarrow$
$10x^2 - 1560x + 57,600 = 0 \Rightarrow$
$10(x - 60)(x - 96) = 0 \Rightarrow x = 60$ or $x = 96$
$2(60) + y + 100 = 360 \Rightarrow y = 140$
$2(96) + y + 100 = 360 \Rightarrow y = 68$

Because $y > x$ (from the diagram), we reject $x = 96$ and $y = 68$.
So, $x = 60$ m and $y = 140$ m

b. The area is $60^2 + \dfrac{1}{2}(60)(80) = 6000$ m^2.

63. Let x = the original number of students in the group and let y = the original cost per student. Then
$$\begin{cases} xy = 960 \\ (x+8)(y-6) = 960 \end{cases} \Rightarrow$$
$$\begin{cases} y = \dfrac{960}{x} \\ (x+8)(y-6) = 960 \end{cases} \Rightarrow$$
$(x+8)\left(\dfrac{960}{x} - 6\right) = 960 \Rightarrow$
$-6x + \dfrac{7680}{x} + 912 = 960 \Rightarrow$
$-6x + \dfrac{7680}{x} - 48 = 0 \Rightarrow$
$-6x^2 - 48x + 7680 = 0 \Rightarrow$
$-6(x - 32)(x + 40) = 0 \Rightarrow x = 32$ or $x = -40$ (reject this)
$32y = 960 \Rightarrow y = 30$

There were originally 32 students at a cost of $30 each.

490 Chapter 8 Systems of Equations and Inequalities

65. Let x = the number of shares of stock she bought and let y = the original price per share. Then

$\begin{cases} xy + 100 = 10,000 \\ (x+30)(y+3) = 11,900 + 100 \end{cases} \Rightarrow$

$\begin{cases} y = \dfrac{9900}{x} \\ (x+30)(y+3) = 12,000 \end{cases} \Rightarrow$

$(x+30)\left(\dfrac{9900}{x} + 3\right) = 12,000 \Rightarrow$

$3x + \dfrac{297,000}{x} - 2010 = 0 \Rightarrow$

$3x^2 - 2010x + 297,000 = 0 \Rightarrow$
$3(x-450)(x-220) = 0 \Rightarrow x = 450$ or $x = 220$

The problem says that she bought more than 400 shares, so reject $x = 220$.
$450y + 100 = 10,000 \Rightarrow y = 22$
She bought 450 shares at \$22 per share.

8.4 Beyond the Basics

67. $\begin{cases} x^2 + y^2 - 7x + 5y + 6 = 0 \\ y = 0 \end{cases} \Rightarrow$

$x^2 - 7x + 6 = 0 \Rightarrow (x-6)(x-1) = 0 \Rightarrow$
$x = 6$ or $x = 1$
The circle intersects the x-axis at $A(1, 0)$ and $B(6, 0)$. $d(A, B) = \sqrt{(6-1)^2 + (0-0)^2} = 5$.

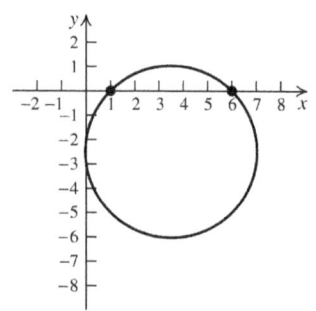

69. $\begin{cases} x^2 + y^2 + 2x - 4y - 5 = 0 \\ x - y + 1 = 0 \end{cases} \Rightarrow$

$\begin{cases} x^2 + y^2 + 2x - 4y - 5 = 0 \\ x = y - 1 \end{cases} \Rightarrow$

$(y-1)^2 + y^2 + 2(y-1) - 4y - 5 = 0 \Rightarrow$
$2y^2 - 4y - 6 = 0 \Rightarrow 2(y-3)(y+1) = 0 \Rightarrow$
$y = 3$ or $y = -1$
$x - 3 + 1 = 0 \Rightarrow x = 2$
$x - (-1) + 1 \Rightarrow x = -2$
The circle and the line intersect at $A(-2, -1)$ and $B(2, 3)$.
$d(A, B) = \sqrt{(-2-2)^2 + (-1-3)^2} = \sqrt{16+16} = \sqrt{32} = 4\sqrt{2}$.

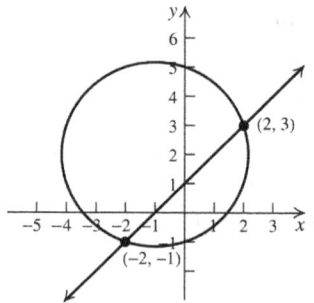

71. $\begin{cases} x^2 + y^2 - 2x + 2y - 3 = 0 \\ x + 2y + 6 = 0 \end{cases} \Rightarrow$

$\begin{cases} x^2 + y^2 - 2x + 2y - 3 = 0 \\ x = -2y - 6 \end{cases} \Rightarrow$

$(-2y-6)^2 + y^2 - 2(-2y-6) + 2y - 3 = 0 \Rightarrow$
$5y^2 + 30y + 45 = 0 \Rightarrow 5(y+3)^2 = 0 \Rightarrow y = -3$
$x + 2(-3) + 6 = 0 \Rightarrow x = 0$
The line and the circle intersect at only one point $(0, -3)$, so the line is tangent to the circle.

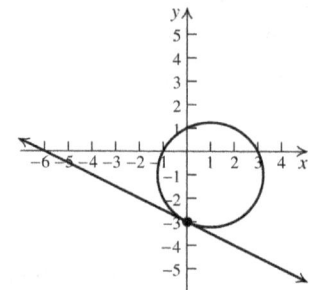

73. $\begin{cases} x + y = 2 \quad (1) \\ x^2 + y^2 = c^2 \quad (2) \end{cases}$

From equation (1), we have $y = 2 - x$.
Substituting in equation (2) and then solving for x, we have
$x^2 + (2-x)^2 = c^2 \Rightarrow x^2 + 4 - 4x + x^2 = c^2 \Rightarrow$
$2x^2 - 4x + 4 - c^2 = 0 \Rightarrow$

$x = \dfrac{-(-4) \pm \sqrt{(-4)^2 - 4(2)(4-c^2)}}{2(2)}$

Since the equations are tangent, there is only one solution. Recall that the value of the discriminant is 0 when there is exactly one real solution, so solve
$(-4)^2 - 4(2)(4-c^2) = 0 \Rightarrow$
$16 - 32 + 8c^2 = 0 \Rightarrow c^2 = 2 \Rightarrow c = \pm\sqrt{2}$

75. $\begin{cases} 4x^2 + y^2 = 25 & (1) \\ 8x + 3y = c & (2) \end{cases}$

From equation (2), we have $y = \dfrac{c - 8x}{3}$.

Substituting in equation (1) and then solving for x, we have

$$4x^2 + \left(\dfrac{c - 8x}{3}\right)^2 = 25$$

$$4x^2 + \dfrac{c^2 - 16cx + 64x^2}{9} = 25$$

$$36x^2 + c^2 - 16cx + 64x^2 = 225$$

$$100x^2 - 16cx + c^2 - 225 = 0$$

$$x = \dfrac{-(-16c) \pm \sqrt{(-16c)^2 - 4(100)(c^2 - 225)}}{2(100)}$$

Since the equations are tangent, there is only one solution. Recall that the value of the discriminant is 0 when there is exactly one real solution, so solve

$$(-16c)^2 - 4(100)(c^2 - 225) = 0 \Rightarrow$$
$$256c^2 - 400c^2 + 90,000 = 0 \Rightarrow$$
$$90,000 - 144c^2 = 0 \Rightarrow$$
$$(300 - 12c)(300 + 12c) = 0 \Rightarrow c = \pm 25$$

In exercises 77–80, let $u = \dfrac{1}{x}$ and $v = \dfrac{1}{y}$.

77. $\begin{cases} \dfrac{1}{x} - \dfrac{1}{y} = 5 \\ \dfrac{1}{xy} - 24 = 0 \end{cases} \Rightarrow \begin{cases} u - v = 5 & (1) \\ uv - 24 = 0 & (2) \end{cases}$

From equation (1), we have $u = v + 5$. Substitute this into equation (2) and solve for v.

$v(v + 5) - 24 = 0 \Rightarrow v^2 + 5v - 24 = 0 \Rightarrow$
$(v + 8)(v - 3) = 0 \Rightarrow v = -8, 3$
$v = -8 \Rightarrow u + 8 = 5 \Rightarrow u = -3 \Rightarrow$
$x = -\dfrac{1}{3}, \ y = -\dfrac{1}{8}$

$v = 3 \Rightarrow u - 3 = 5 \Rightarrow u = 8 \Rightarrow x = \dfrac{1}{8}, \ y = \dfrac{1}{3}$

Solution set: $\left\{\left(-\dfrac{1}{3}, -\dfrac{1}{8}\right), \left(\dfrac{1}{8}, \dfrac{1}{3}\right)\right\}$

79. $\begin{cases} \dfrac{5}{x^2} - \dfrac{3}{y^2} = 2 \\ \dfrac{6}{x^2} + \dfrac{1}{y^2} = 7 \end{cases} \Rightarrow \begin{cases} 5u^2 - 3v^2 = 2 & (1) \\ 6u^2 + v^2 = 7 & (2) \end{cases}$

Solve equation (2) for v^2, then substitute this expression in equation (1) and solve for u.

$v^2 = 7 - 6u^2$
$5u^2 - 3(7 - 6u^2) = 2 \Rightarrow 23u^2 - 21 = 2 \Rightarrow$
$u^2 = 1 \Rightarrow u = \pm 1 \Rightarrow x = \pm 1$

If $u = 1$, then
$6(1^2) + v^2 = 7 \Rightarrow v^2 = 1 \Rightarrow v = \pm 1 \Rightarrow y = \pm 1$

If $u = -1$, then
$6(-1^2) + v^2 = 7 \Rightarrow v^2 = 1 \Rightarrow v = \pm 1 \Rightarrow y = \pm 1$

Solution set: $\{(-1, -1), (-1, 1), (1, -1), (1, 1)\}$

81. $(a + bi)^2 = -5 + 12i \Rightarrow$
$a^2 + 2abi - b^2 = -5 + 12i \Rightarrow$
$\begin{cases} a^2 - b^2 = -5 \\ 2ab = 12 \end{cases} \Rightarrow \begin{cases} a^2 - b^2 = -5 & (1) \\ ab = 6 & (2) \end{cases}$

From equation (2), we have $b = \dfrac{6}{a}$. Substitute this into equation (1) and solve for a.

$a^2 - \left(\dfrac{6}{a}\right)^2 = -5 \Rightarrow a^4 + 5a^2 - 36 = 0 \Rightarrow$
$(a^2 + 9)(a^2 - 4) = 0 \Rightarrow$
$(a^2 + 9)(a - 2)(a + 2) = 0 \Rightarrow a = \pm 2, \ \pm 3i$

Since a is real, reject $\pm 3i$. Using equation (2), if $a = -2$, then $b = -3$. If $a = 2$, then $b = 3$. Thus, the square roots of $w = -5 + 12i$ are $-2 - 3i$ and $2 + 3i$.

83. $(a + bi)^2 = 7 - 24i \Rightarrow$
$a^2 + 2abi - b^2 = 7 - 24i \Rightarrow$
$\begin{cases} a^2 - b^2 = 7 \\ 2ab = -24 \end{cases} \Rightarrow \begin{cases} a^2 - b^2 = 7 & (1) \\ ab = -12 & (2) \end{cases}$

From equation (2), we have $b = -\dfrac{12}{a}$. Substitute this into equation (1) and solve for a.

$a^2 - \left(-\dfrac{12}{a}\right)^2 = 7 \Rightarrow a^4 - 7a^2 - 144 = 0 \Rightarrow$
$(a^2 - 16)(a^2 + 9) = 0 \Rightarrow$
$(a^2 + 9)(a - 4)(a + 4) = 0 \Rightarrow a = \pm 4, \ \pm 3i$

Since a is real, reject $\pm 3i$.
Using equation (2), if $a = -4$, then $b = 3$. If $a = 4$, then $b = -3$. Thus, the square roots of $w = 7 - 24i$ are $-4 + 3i$ and $4 - 3i$.

85. $(a+bi)^2 = 13 + 8\sqrt{3}i \Rightarrow$
$a^2 + 2abi - b^2 = 13 + 8\sqrt{3}i \Rightarrow$
$\begin{cases} a^2 - b^2 = 13 \\ 2ab = 8\sqrt{3} \end{cases} \Rightarrow \begin{cases} a^2 - b^2 = 13 & (1) \\ ab = 4\sqrt{3} & (2) \end{cases}$

From equation (2), we have $b = \dfrac{4\sqrt{3}}{a}$.

Substitute this into equation (1) and solve for a.

$a^2 - \left(\dfrac{4\sqrt{3}}{a}\right)^2 = 13 \Rightarrow a^4 - 13a^2 - 48 = 0 \Rightarrow$
$(a^2 - 16)(a^2 + 3) = 0 \Rightarrow$
$(a^2 + 3)(a - 4)(a + 4) = 0 \Rightarrow a = \pm 4, \pm\sqrt{3}i$

Since a is real, reject $\pm\sqrt{3}i$. Using equation (2), if $a = -4$, then $b = -\sqrt{3}$. If $a = 4$, then $b = \sqrt{3}$. Thus, the square roots of $w = 13 + 8\sqrt{3}i$ are $-4 - \sqrt{3}i$ and $4 + \sqrt{3}i$.

87. Using substitution, we have
$\begin{cases} y = 3^x + 4 \\ y = 3^{2x} - 2 \end{cases} \Rightarrow 3^{2x} - 2 = 3^x + 4 \Rightarrow$
$3^{2x} - 3^x - 6 = 0$
Let $u = 3^x$. Then $u^2 - u - 6 = 0 \Rightarrow$
$(u - 3)(u + 2) = 0 \Rightarrow u = 3$ or $u = -2$
(Reject the negative solution.)
$3 = 3^x \Rightarrow x = 1$
$y = 3^1 + 4 = 7$
The solution is $\{(1, 7)\}$.

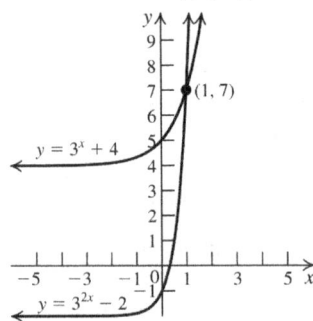

89. Using substitution, we have
$\begin{cases} y = 2^x + 3 \\ y = 2^{2x} + 1 \end{cases} \Rightarrow 2^x + 3 = 2^{2x} + 1 \Rightarrow$
$2^{2x} - 2^x - 2 = 0.$

Let $u = 2^x$. Then $u^2 - u - 2 = 0 \Rightarrow$
$(u - 2)(u + 1) = 0 \Rightarrow u = 2$ or $u = -1$
(reject the negative solution.)
$2 = 2^x \Rightarrow x = 1;\ y = 2^1 + 3 = 5$
The solution is $\{(1, 5)\}$.

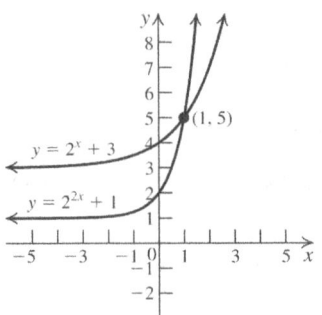

91. $\begin{cases} x = 3^y \\ 3^{2y} = 3x - 2 \end{cases} \Rightarrow 3^{2y} = 3 \cdot 3^y - 2 \Rightarrow$
$3^{2y} - 3 \cdot 3^y + 2 = 0.$ Let $u = 3^y$. Then
$u^2 - 3u + 2 = 0 \Rightarrow (u - 2)(u - 1) = 0 \Rightarrow$
$u = 2$ or $u = 1$
$2 = 3^y \Rightarrow \ln 2 = y \ln 3 \Rightarrow \dfrac{\ln 2}{\ln 3} = y;$
$x = 3^{\ln 2/\ln 3} = 2$ or $1 = 3^y \Rightarrow y = 0$
$x = 3^0 = 1$

The solution is $\left\{(1, 0), \left(2, \dfrac{\ln 2}{\ln 3}\right)\right\}$.

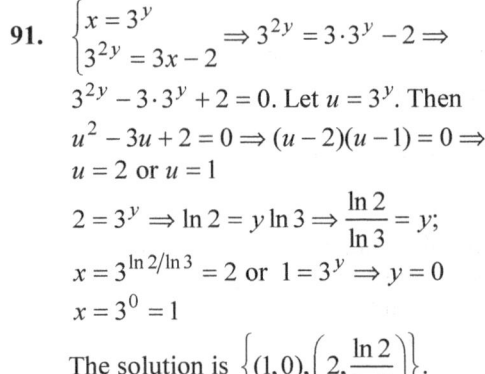

8.4 Critical Thinking/Discussion/Writing

93. a. Not possible   **b.** Possible
    c. Possible   **d.** Possible
    e. Not possible   **f.** Not possible.

94. a. Possible   **b.** Possible
    c. Possible   **d.** Possible
    e. Possible   **f.** Not possible.

8.4 Getting Ready for the Next Section

95.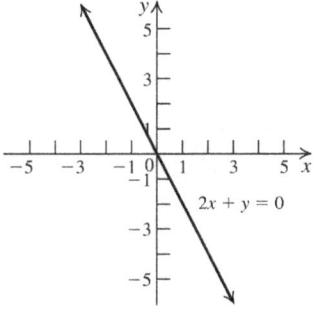

97.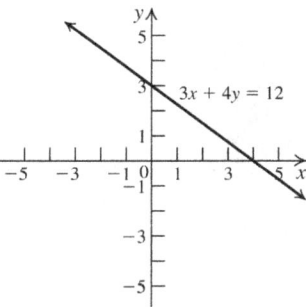

99.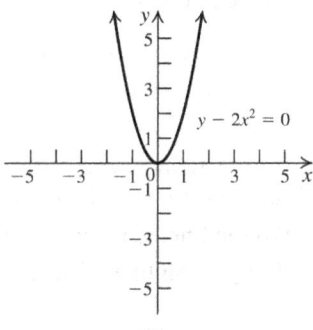

101.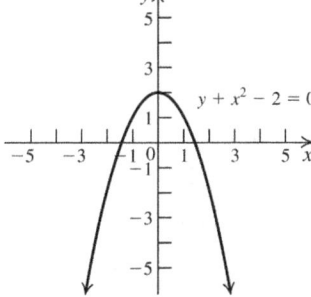

103.

Point	$5x + 3y = 0?$
(0, 0)	$5(0) + 3(0) \stackrel{?}{=} 0$ $0 = 0$ ✓
(−3, 5)	$5(-3) + 3(5) \stackrel{?}{=} 0$ $0 = 0$ ✓
(−1, 1)	$5(-1) + 3(1) \stackrel{?}{=} 0$ $-2 = 0$ ✗
(1, −1)	$5(1) + 3(-1) \stackrel{?}{=} 0$ $2 = 0$ ✗

(0, 0) and (−3, 5) lie on the graph of the equation $5x + 3y = 0$.

105.

Point	$y = x^2 + 3?$
(0, 0)	$0 \stackrel{?}{=} 0^2 + 3$ $0 = 3$ ✗
(1, 4)	$4 \stackrel{?}{=} 1^2 + 3$ $3 = 3$ ✓
(−1, 4)	$4 \stackrel{?}{=} (-1)^2 + 3$ $4 = 4$ ✓
(2, 5)	$5 \stackrel{?}{=} 2^2 + 3$ $5 = 7$ ✗

(1, 4) and (−1, 4) lie on the graph of the equation $y = x^2 + 3$.

8.5 Systems of Inequalities

8.5 Practice Problems

1.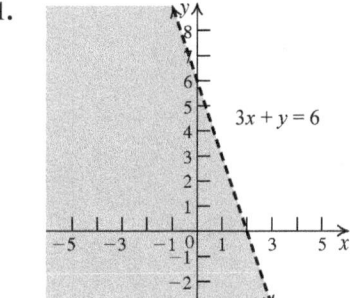

2.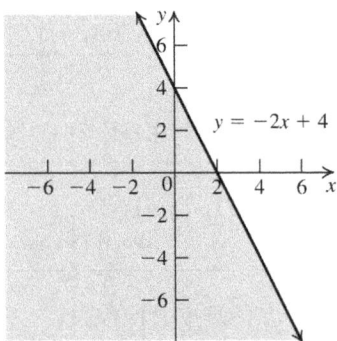

3.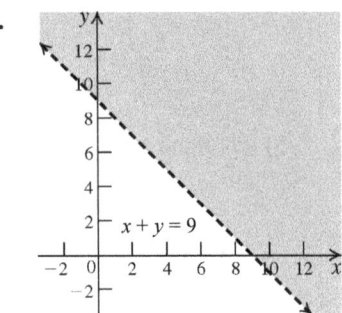

4.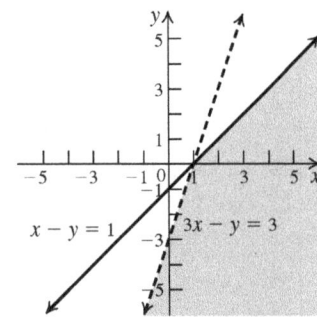

5. Solve the systems $\begin{cases} 2x+3y=16 \\ 4x+2y=16 \end{cases}$,

   $\begin{cases} 4x+2y=16 \\ 2x-y=8 \end{cases}$, and $\begin{cases} 2x+3y=16 \\ 2x-y=8 \end{cases}$ to find the

   corner points: (4, 0), (5, 2), and (2, 4).

   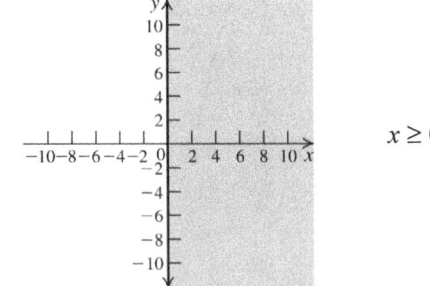

6.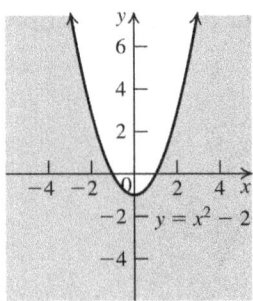

7.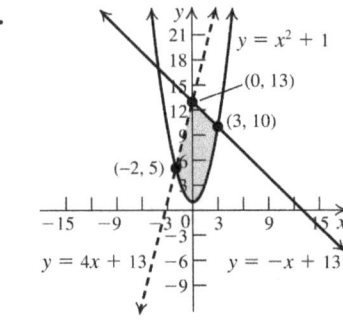

8.5 Concepts and Vocabulary

1. In the graph of $x - 3y > 1$, the corresponding equation $x - 3y = 1$ is graphed as a <u>dashed</u> line.

3. In a system of inequalities containing both $2x + y > 5$ and $2x - y \le 3$, the point of intersection of the lines $2x + y = 5$ and $2x - y = 3$ is <u>not a</u> solution of the system.

5. True

7. True

8.5 Building Skills

9.

$x \ge 0$

11. 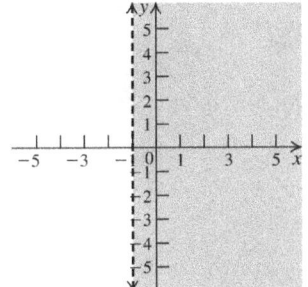 $x > -1$

13. 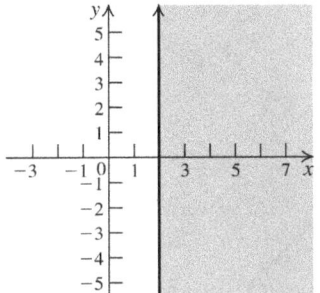 $x \geq 2$

15. 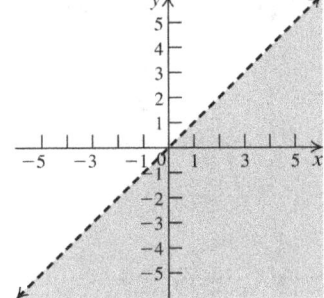 $y - x < 0$

17. 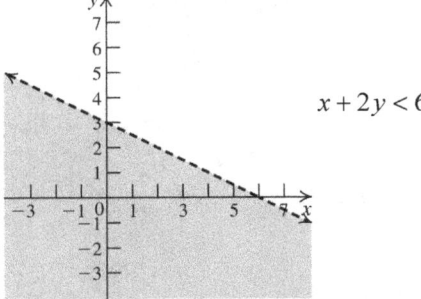 $x + 2y < 6$

19. 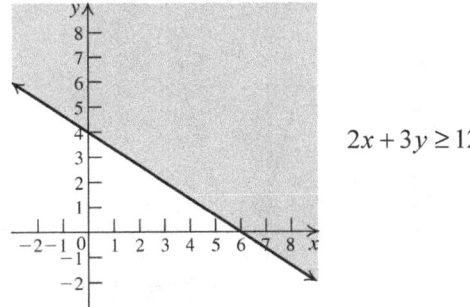 $2x + 3y \geq 12$

21. 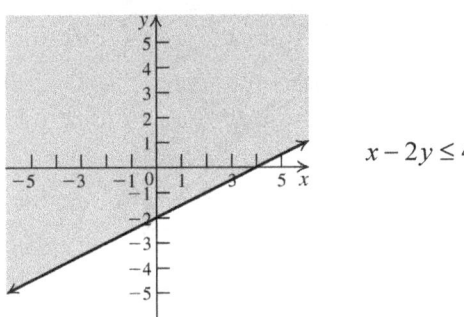 $x - 2y \leq 4$

23. 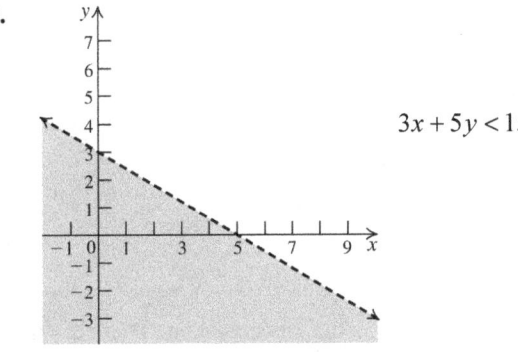 $3x + 5y < 15$

25. Substitute each ordered pair into the system:
$$\begin{cases} 0 + 0 < 2 \\ 2(0) + 0 \geq 6 \end{cases} \quad \begin{cases} -4 - 1 < 2 \\ 2(-4) - 1 \geq 6 \end{cases}$$
$$\begin{cases} 3 + 0 < 2 \\ 2(3) + 0 \geq 6 \end{cases} \quad \begin{cases} 0 + 3 < 2 \\ 2(0) + 3 \geq 6 \end{cases}$$
None of the ordered pairs are a solution.

27. Substitute each ordered pair into the system:
$$\begin{cases} 0 < 0 + 2 \\ 0 + 0 \leq 4 \end{cases} \begin{cases} 0 < 1 + 2 \\ 1 + 0 \leq 4 \end{cases} \begin{cases} 1 < 0 + 2 \\ 0 + 1 \leq 4 \end{cases} \begin{cases} 1 < 1 + 2 \\ 1 + 1 \leq 4 \end{cases}$$
All of the ordered pairs are solutions.

29. Substitute each ordered pair into the system:
$$\begin{cases} 3(0) - 4(0) \leq 12 \\ 0 + 0 \leq 4 \\ 5(0) - 2(0) \geq 6 \end{cases} \quad \begin{cases} 3(2) - 4(0) \leq 12 \\ 2 + 0 \leq 4 \\ 5(2) - 2(0) \geq 6 \end{cases}$$
$$\begin{cases} 3(3) - 4(1) \leq 12 \\ 3 + 1 \leq 4 \\ 5(3) - 2(1) \geq 6 \end{cases} \quad \begin{cases} 3(2) - 4(2) \leq 12 \\ 2 + 2 \leq 4 \\ 5(2) - 2(2) \geq 6 \end{cases}$$
The solutions are (2, 0), (3, 1) and (2, 2).

31.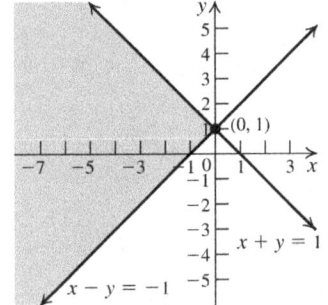

33.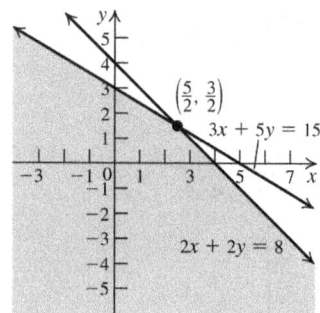

35.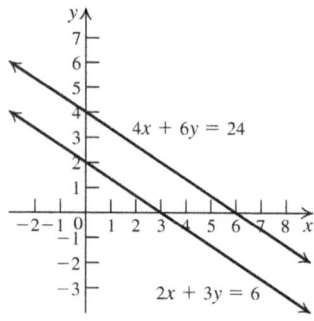

The system is inconsistent. There are no vertices of the solution set.

37.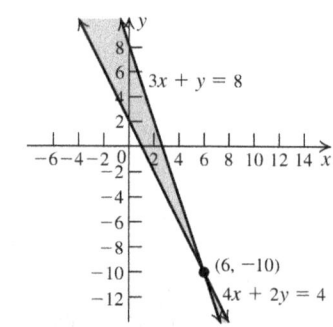

39.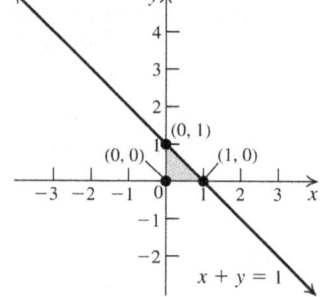

41.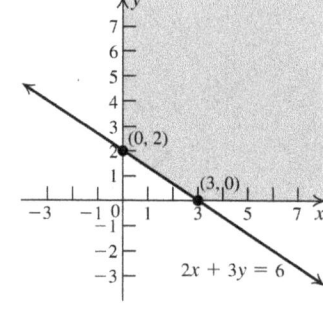

43.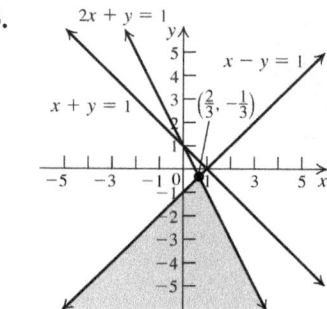

45.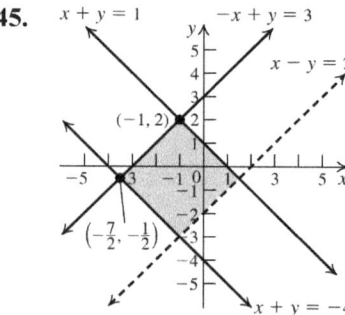

47. A **49.** B **51.** E

53.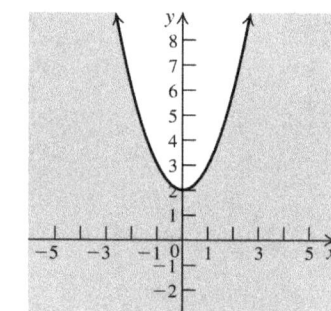

55.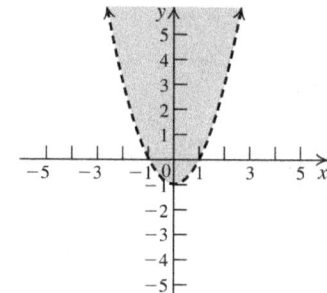

57.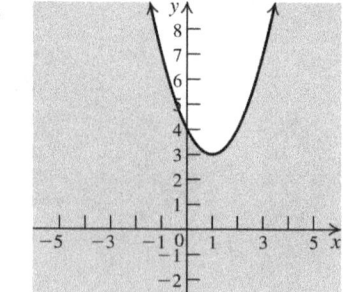

59.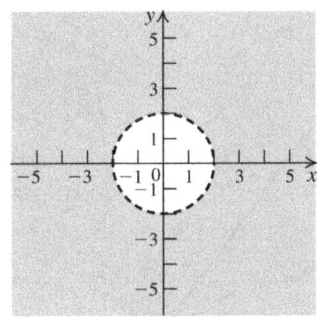

61.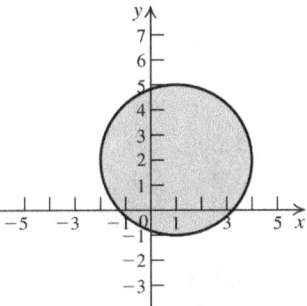

63.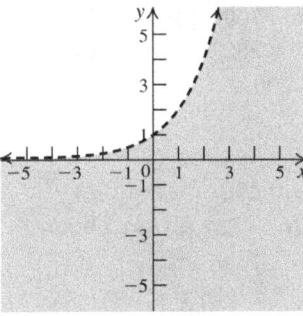

65.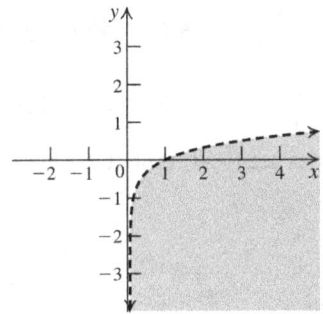

67. D, K

69. E, L

71. E, B

73.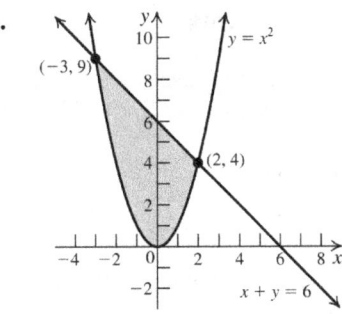

75.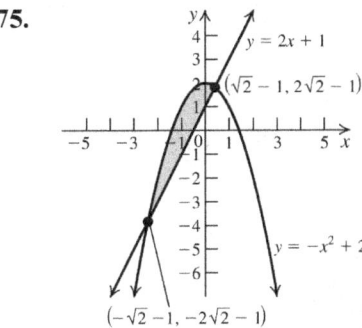

77.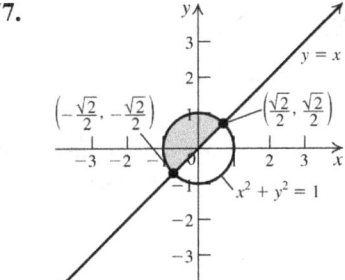

79.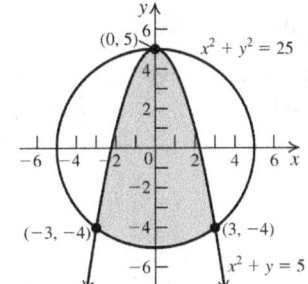

8.5 Applying the Concepts

81. Let $x =$ the number of cases of Coke and $y =$ the number of cases of Sprite. Then we have
$$\begin{cases} x + y \leq 40 & (1) \\ x \geq 10 & (2) \\ y \geq 5 & (3) \end{cases}$$

To find the vertices, solve the systems
$$\begin{cases} x + y = 40 & (1) \\ x = 10 & (2) \end{cases}, \begin{cases} x + y = 40 & (1) \\ y = 5 & (3) \end{cases}, \text{ and}$$
$$\begin{cases} x = 10 & (2) \\ y = 5 & (3) \end{cases}.$$

The vertices are (10, 30), (35, 5), and (10, 5).

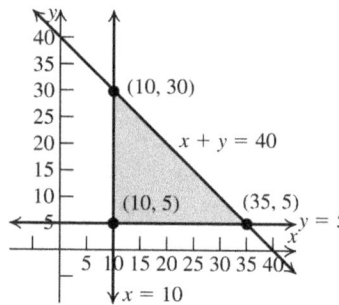

83. Let $x =$ the number of two story houses and $y =$ the number of one story houses. Then we have
$$\begin{cases} 7x + 5y \leq 43 \\ 4x + 3y \leq 25 \\ x \geq 0 \\ y \geq 0 \end{cases}$$

To find the vertices, solve the systems
$$\begin{cases} 7x + 5y = 43 \\ 4x + 3y = 25 \end{cases}, \begin{cases} 7x + 5y = 43 \\ y = 0 \end{cases}, \begin{cases} 4x + 3y = 25 \\ x = 0 \end{cases},$$
and $\begin{cases} x = 0 \\ y = 0 \end{cases}$.

$\begin{cases} 7x + 5y = 43 \\ 4x + 3y = 25 \end{cases} \Rightarrow \begin{cases} 28x + 20y = 172 \\ -28x - 21y = -175 \end{cases} \Rightarrow$
$-y = -3 \Rightarrow y = 3$
$7x + 15 = 43 \Rightarrow x = 4$
$\begin{cases} 7x + 5y = 43 \\ y \geq 0 \end{cases} \Rightarrow x = \dfrac{43}{7}$
$\begin{cases} 4x + 3y = 25 \\ x = 0 \end{cases} \Rightarrow y = \dfrac{25}{3}$

The vertices are $(0, 0), \left(0, \dfrac{25}{3}\right), \left(\dfrac{43}{7}, 0\right),$ and (4, 3).

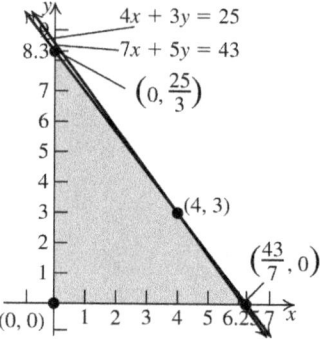

85. Let $x =$ number of 3 hp engines and $y =$ the number of 5 hp engines. Then we have
$$\begin{cases} 3x + 4.5y \leq 360 & (1) \\ 2x + y \leq 200 & (2) \\ 0.5x + 0.75y \leq 60 & (3) \\ x \geq 0 \\ y \geq 0 \end{cases}$$

Equations (1) and (3) coincide. To find the vertices, solve the systems
$\begin{cases} 3x + 4.5y = 360 \\ 2x + y = 200 \end{cases}, \begin{cases} 3x + 4.5y = 360 \\ x = 0 \end{cases},$ and
$\begin{cases} 2x + y = 200 \\ y = 0 \end{cases}.$

$\begin{cases} 3x + 4.5y = 360 \\ 2x + y = 200 \end{cases} \Rightarrow$
$3x + 4.5(200 - 2x) = 360 \Rightarrow 900 - 6x = 360 \Rightarrow$
$x = 90$
$2(90) + y = 200 \Rightarrow y = 20$
$\begin{cases} 3x + 4.5y = 360 \\ x = 0 \end{cases} \Rightarrow 4.5y = 360 \Rightarrow y = 80$
$\begin{cases} 2x + y = 200 \\ y = 0 \end{cases} \Rightarrow 2x = 200 \Rightarrow x = 100$

The vertices are (0, 0) (100, 0), (0, 80), and (90, 20).

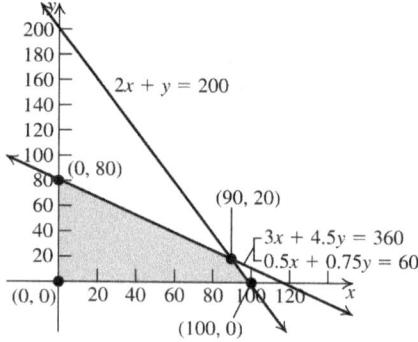

8.5 Beyond the Basics

87. The equation of the line connecting (0, 2) and (3, 0) is $y = -\frac{2}{3}x + 2$. The system is
$$\begin{cases} y \leq -\frac{2}{3}x + 2 \\ x \geq 0 \\ y \geq 0 \end{cases}.$$

89. $\begin{cases} x \geq -1 \\ x \leq 3 \end{cases}$

91. The equation of the line connecting (0, 4) and (2, 3) is $y = -\frac{1}{2}x + 4$. The equation of the line connecting (2, 3) and (4, 0) is $y = -\frac{3}{2}x + 6$. The system is
$$\begin{cases} x \geq 0 \\ y \geq 0 \\ y \leq -\frac{1}{2}x + 4 \\ y \leq -\frac{3}{2}x + 6 \end{cases}.$$

93. The equation of the line connecting (0, 16) and (10, 6) is $y = -x + 16$. The equation of the line connecting (10, 6) and (5, 1) is $y = x - 4$. The equation of the line connecting (5, 1) and (0, 6) is $y = -x + 6$. The system is
$$\begin{cases} x \geq 0 \\ y \leq -x + 16 \\ y \geq x - 4 \\ y \geq -x + 6 \end{cases}.$$

95.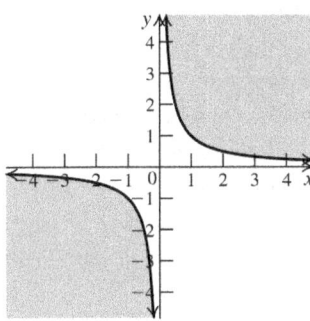

97.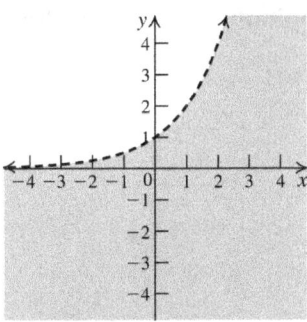

99.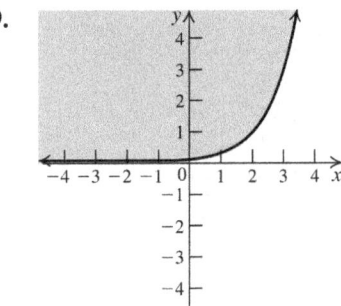

101.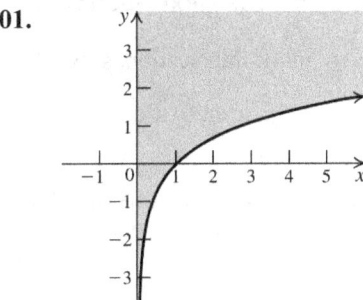

103.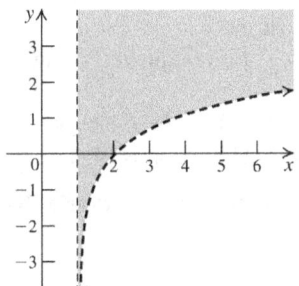

105.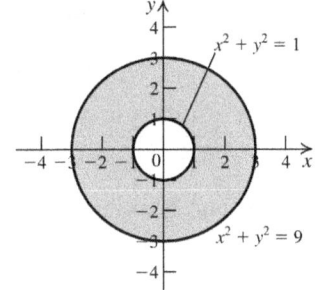

8.5 Critical Thinking/Discussion/Writing

107.

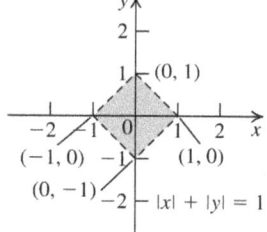

108.

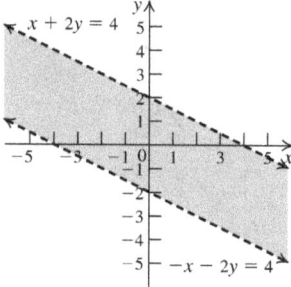

8.5 Getting Ready for the Next Section

109. To find the point of intersection, solve the system
$$\begin{cases} x + 2y = 40 & (1) \\ 3x + y = 30 & (2) \end{cases}$$
Solve equation (2) for y in terms of x, then substitute the expression in equation (1) and solve for x.
$3x + y = 30 \Rightarrow y = -3x + 30$
$x + 2(-3x + 30) = 40 \Rightarrow -5x + 60 = 40 \Rightarrow$
$-5x = -20 \Rightarrow x = 4$
Substitute $x = 4$ into equation (1) and solve for y.
$4 + 2y = 40 \Rightarrow 2y = 36 \Rightarrow y = 18$
The point of intersection is (4, 18).

111. To find the point of intersection, solve the system
$$\begin{cases} x - 2y = 2 & (1) \\ 3x + 2y = 12 & (2) \end{cases}$$
Add the two equations, then solve for x.
$x - 2y = 2$
$3x + 2y = 12$
$\overline{}$
$4x = 14 \Rightarrow x = \dfrac{14}{4} = \dfrac{7}{2}$
Substitute $x = \dfrac{7}{2}$ into equation (1) and solve for y.

$\dfrac{7}{2} - 2y = 2 \Rightarrow -2y = -\dfrac{3}{2} \Rightarrow y = \dfrac{3}{4}$

The point of intersection is $\left(\dfrac{7}{2}, \dfrac{3}{4}\right)$.

113.

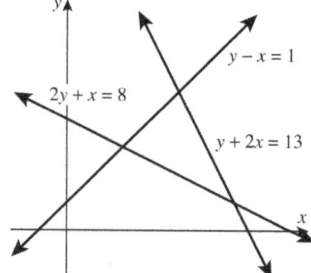

Solve each pair of equations to find the points of intersection.

I: $\begin{cases} 2y + x = 8 \\ y - x = 1 \end{cases}$

II: $\begin{cases} y - x = 1 \\ y + 2x = 13 \end{cases}$

III: $\begin{cases} y + 2x = 13 \\ 2y + x = 8 \end{cases}$

For system I, we will use substitution.
$y = x + 1$
$2(x + 1) + x = 8 \Rightarrow 3x + 2 = 8 \Rightarrow 3x = 6 \Rightarrow$
$x = 2$
$y - 2 = 1 \Rightarrow y = 3$
The point of intersection is (2, 3).
For system II, we will use substitution.
$y = x + 1$
$(x + 1) + 2x = 13 \Rightarrow 3x + 1 = 13 \Rightarrow 3x = 12 \Rightarrow$
$x = 4;\quad y - 4 = 1 \Rightarrow y = 5$
The point of intersection is (4, 5).
For system III, we will use substitution.
$y = -2x + 13$
$2(-2x + 13) + x = 8 \Rightarrow -3x + 26 = 8 \Rightarrow$
$-3x = -18 \Rightarrow x = 6$
$y + 2(6) = 13 \Rightarrow y + 12 = 13 \Rightarrow y = 1$
The point of intersection is (6, 1).
The vertices of the triangle are (2, 3), (4, 5), and (6, 1).

115. a.

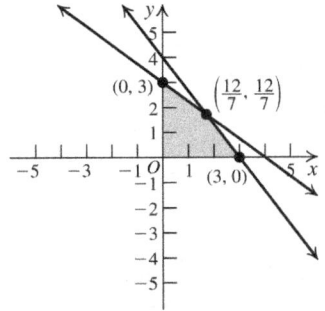

b. Solve the systems

$$\begin{cases} x = 0 \\ y = 0 \end{cases}, \begin{cases} 3x + 4y = 12 \\ y = 0 \end{cases}, \begin{cases} 3x + 4y = 12 \\ 4x + 3y = 12 \end{cases},$$

and $\begin{cases} 4x + 3y = 12 \\ x = 0 \end{cases}$

to find the vertices $(0, 0)$, $(0, 3)$, $\left(\dfrac{12}{7}, \dfrac{12}{7}\right)$, and $(3, 0)$.

8.6 Linear Programming

8.6 Practice Problems

1. Maximize $f = 4x + 5y$ subject to the constraints
$$\begin{cases} 5x + 7y \le 35 \\ x \ge 0 \\ y \ge 0 \end{cases}.$$

First, graph the solution set of the constraints.

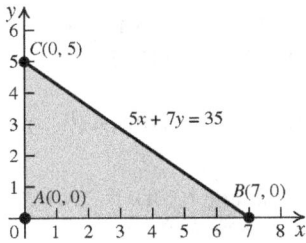

Solve the systems $\begin{cases} x = 0 \\ y = 0 \end{cases}, \begin{cases} 5x + 7y = 35 \\ y = 0 \end{cases}$, and

$\begin{cases} 5x + 7y = 35 \\ x = 0 \end{cases}$ to find the vertices: $(0, 0)$,

$(7, 0)$ and $(0, 5)$. Now find the values of the objective function at each vertex:

Ordered pair	$f = 4x + 5y$
(0, 0)	0
(7, 0)	28
(0, 5)	25

The maximum is 28 at $(7, 0)$.

2. Minimize $f = \dfrac{1}{2}x + y$ subject to the constraints
$$\begin{cases} x + y \le 8 \\ x + 2y \ge 4 \\ 3x + 2y \ge 6 \\ x \ge 0 \\ y \ge 0 \end{cases}.$$

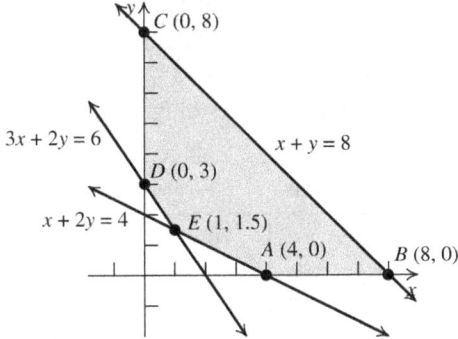

The graph and vertices of the set of feasible solutions are as given in Example 2. The vertices are: $(0, 8)$, $(0, 3)$, $(1, 1.5)$, $(4, 0)$, and $(8, 0)$. Now find the values of the objective function at each vertex:

Ordered pair	$f = \dfrac{1}{2}x + y$
(0, 8)	8
(0, 3)	3
(1, 1.5)	2
(4, 0)	2
(8, 0)	4

The minimum is 2 at $(1, 1.5)$ and $(4, 0)$. In fact, the unique minimum value occurs at every point on the line segment AE. So the linear programming problem has infinitely many solutions, each of which satisfies the equation $x + 2y = 4$.

3. Let $x =$ the number of ounces of soup and let $y =$ the number of ounces of salad. The number of calories in the two items is $f = 30x + 60y$. (This is the objective function.) The constraints are as given in Example 3.
$$\begin{cases} x + y \ge 10 \\ 3x + 2y \ge 24 \\ x \ge 0 \\ y \ge 0 \end{cases}.$$

The graph and vertices of the set of feasible solutions are as given in Example 3. The vertices are $(0, 12)$, $(4, 6)$, and $(10, 0)$.

(continued on next page)

(continued)

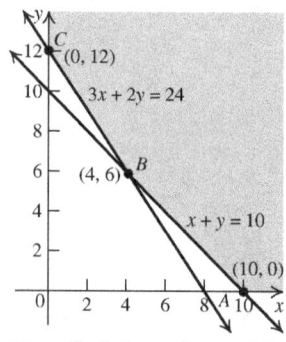

Now find the values of the objective function at each vertex:

Ordered pair	$f = 30x + 60y$
(0, 12)	720
(4, 6)	480
(10, 0)	300

The minimum is 300 at (10, 0). This means that the lunch menu for Peter Griffin should contain 10 ounces of soup and 0 ounces of salad.

8.6 Concepts and Vocabulary

1. The process of finding the maximum or minimum value of a quantity is called optimization.

3. The inequalities that determine the region S in a linear programming problem are called constraints, and S is called the set of feasible solutions.

5. True

7. True

8.6 Building Skills

9.
Ordered pair	$f = x + y$
(10, 0)	10
(3, 2)	5
(1, 4)	5
(0, 8)	8
(10, 8)	18

Maximum: 18; minimum: 5

11.
Ordered pair	$f = x + 2y$
(10, 0)	10
(3, 2)	7
(1, 4)	9
(0, 8)	16
(10, 8)	26

Maximum: 26; minimum: 7

13.
Ordered pair	$f = 5x + 2y$
(10, 0)	50
(3, 2)	19
(1, 4)	13
(0, 8)	16
(10, 8)	66

Maximum: 66; minimum: 13

15.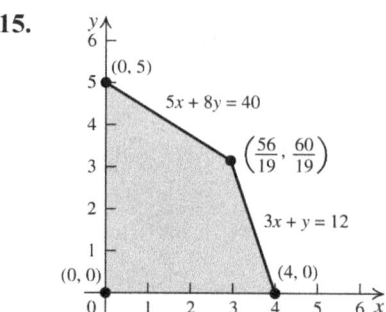

Solve the systems
$\begin{cases} x = 0 \\ 5x + 8y = 40 \end{cases}$, $\begin{cases} 5x + 8y = 40 \\ 3x + y = 12 \end{cases}$, $\begin{cases} 3x + y = 12 \\ y = 0 \end{cases}$,

and $\begin{cases} x = 0 \\ y = 0 \end{cases}$ to find the vertices: (0, 5), $\left(\frac{56}{19}, \frac{60}{19}\right)$, (4, 0), and (0, 0). Now find the values of the objective function at each vertex:

Ordered pair	$f = 9x + 13y$
(0, 5)	65
$\left(\frac{56}{19}, \frac{60}{19}\right)$	$\frac{1284}{19}$
(4, 0)	36
(0, 0)	0

The maximum is $\frac{1284}{19}$ at $\left(\frac{56}{19}, \frac{60}{19}\right)$.

17.

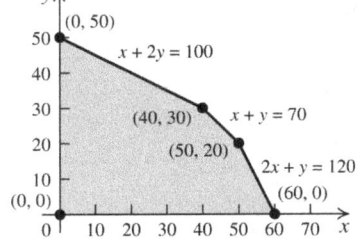

Solve the systems
$\begin{cases} x=0 \\ x+2y=100 \end{cases}$, $\begin{cases} x+2y=100 \\ x+y=70 \end{cases}$, $\begin{cases} x+y=70 \\ 2x+y=120 \end{cases}$,

$\begin{cases} 2x+y=120 \\ y=0 \end{cases}$, and $\begin{cases} x=0 \\ y=0 \end{cases}$ to find the

vertices: (0, 50), (40, 30), (50, 20), (60, 0), and (0, 0). Now find the values of the objective function at each vertex:

Ordered pair	$f = 5x + 7y$
(0, 50)	350
(40, 30)	410
(50, 20)	390
(60, 0)	300
(0, 0)	0

The maximum is 410 at (40, 30).

19.

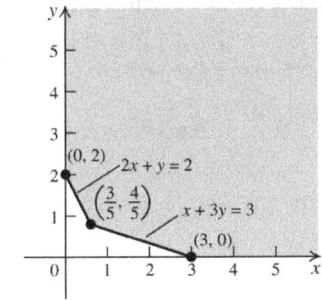

Solve the systems
$\begin{cases} x=0 \\ 2x+y=2 \end{cases}$, $\begin{cases} 2x+y=2 \\ x+3y=3 \end{cases}$, and $\begin{cases} x+3y=3 \\ y=0 \end{cases}$ to

find the vertices: (0, 2), $\left(\dfrac{3}{5}, \dfrac{4}{5}\right)$, and (3, 0).

Now find the values of the objective function at each vertex:

Ordered pair	$f = x + 4y$
(0, 2)	8
$\left(\dfrac{3}{5}, \dfrac{4}{5}\right)$	$\dfrac{19}{5}$
(3, 0)	3

The minimum is 3 at (3, 0).

21.

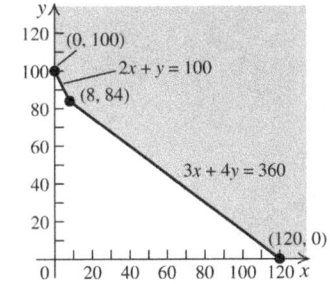

Solve the systems
$\begin{cases} x=0 \\ 2x+y=100 \end{cases}$, $\begin{cases} 2x+y=100 \\ 3x+4y=360 \end{cases}$, and

$\begin{cases} 3x+4y=360 \\ y=0 \end{cases}$ to find the vertices: (0, 100),

(8, 84), and (120, 0). Now find the values of the objective function at each vertex.

Ordered pair	$f = 13x + 15y$
(0, 100)	1500
(8, 84)	1364
(120, 0)	1560

The minimum is 1364 at (8, 84).

23.

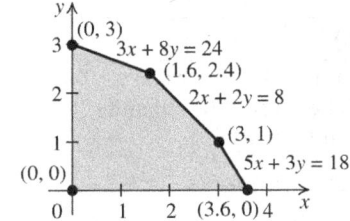

Solve the systems $\begin{cases} x=0 \\ y=0 \end{cases}$, $\begin{cases} 5x+3y=18 \\ y=0 \end{cases}$,

$\begin{cases} 2x+2y=8 \\ 5x+3y=18 \end{cases}$, $\begin{cases} 3x+8y=24 \\ 2x+2y=8 \end{cases}$, and

$\begin{cases} 3x+8y=24 \\ x=0 \end{cases}$ to find the vertices: (0, 0),

(3.6, 0), (3, 1), (1.6, 2.4), and (0, 3). Now find the values of the objective function at each vertex:

Ordered pair	$f = 15x + 7y$
(0, 0)	0
(3.6, 0)	54
(3, 1)	52
(1.6, 2.4)	40.8
(0, 3)	21

The maximum is 54 at (3.6, 0).

25.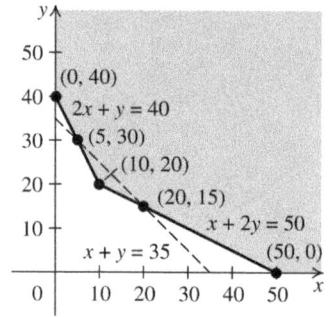

Solve the systems
$$\begin{cases} y=0 \\ x+2y=50 \end{cases}, \begin{cases} x+2y=50 \\ x+y=35 \end{cases}, \begin{cases} x+y=35 \\ 2x+y=40 \end{cases},$$
and $\begin{cases} 2x+y=40 \\ x=0 \end{cases}$ to find the vertices: (50, 0), (20, 15), (30, 5), and (0, 40). Now find the values of the objective function at each vertex:

Ordered pair	$f = 8x + 16y$
(50, 0)	400
(20, 15)	400
(10, 20)	400
(5, 30)	520
(0, 40)	640

The minimum is 400 for all (x, y) on the line segments between (10, 20) and (20, 15), and (20, 15) and (50, 0).

8.6 Applying the Concepts

27. Let x = the number of corn acres, and let y = the number of soybean acres. Then, the profit $p = 50x + 40y$. The constraints are $x \geq 0$, $y \geq 0$, $x + y \leq 240$ (number of acres) and $2x + y \leq 320$ (number of labor hours).
Solve the systems
$$\begin{cases} x=0 \\ x+y=240 \end{cases}, \begin{cases} x+y=240 \\ 2x+y=320 \end{cases}, \text{ and }$$
$\begin{cases} 2x+y=320 \\ y=0 \end{cases}$ to find the vertices: (0, 240), (80, 160), and (160, 0). Note that (0, 0) is also a vertex.

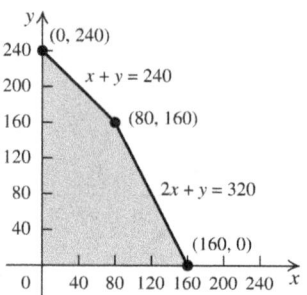

Now find the values of the profit function at each vertex:

Ordered pair	$p = 50x + 40y$
(0, 240)	9600
(80, 160)	10,400
(160, 0)	8000

The maximum profit is $10,400 when 80 acres of corn and 160 acres of soybeans are planted.

29. Let x = the number of hours machine I operates, and let y = the number hours machine II operates. Then, the cost is $c = 50x + 80y$. The constraints are $x \geq 0, y \geq 0, 20x + 30y \geq 1400$ (number of units of Grade A plywood) and $10x + 40y \geq 1200$ (number of units of Grade B plywood.) Solve the systems
$$\begin{cases} x=0 \\ 20x+30y=1400 \end{cases}, \begin{cases} 20x+30y=1400 \\ 10x+40y=1200 \end{cases}, \text{ and }$$
$\begin{cases} 10x+40y=1200 \\ y=0 \end{cases}$ to find the vertices $\left(0, \dfrac{140}{3}\right)$, (40, 20), and (120, 0).

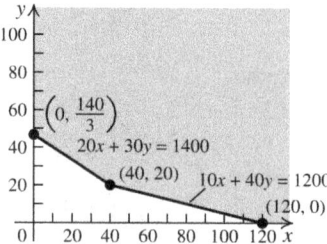

Now find the values of the cost function at each vertex:

Ordered pair	$c = 50x + 80y$
$\left(0, \dfrac{140}{3}\right)$	≈ 3733.33
(40, 20)	3600
(120, 0)	6000

The minimum cost is $3600 when machine I operates for 40 hours and machine II operates for 20 hours.

31. Let $x =$ the number of minutes of television time, and let $y =$ the number of pages of newspaper advertising. Then, the exposure is $f = 60{,}000x + 20{,}000y$. The constraints are $x \geq 1, y \geq 2,$ and $1000x + 500y \leq 6000$ (budget.) Solve the systems
$$\begin{cases} x = 1 \\ 1000x + 500y = 6000 \end{cases}, \begin{cases} 1000x + 500y = 6000 \\ y = 2 \end{cases},$$
and $\begin{cases} x = 1 \\ y = 2 \end{cases}$ to find the vertices (1, 10), (5, 2), and (1, 2).

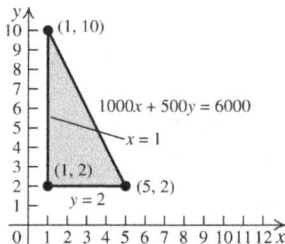

Now find the values of the exposure function at each vertex:

Ordered pair	$f = 60{,}000x + 20{,}000y$
(1, 10)	260,000
(5, 2)	340,000
(1, 2)	100,000

There are a maximum of 340,000 viewers if the company buys 5 minutes of television time and 2 pages of newspaper advertising.

33. Let $x =$ the number of orange acres, and let $y =$ the number of grapefruit acres. Then, the profit is $p = 40x + 30y - 3000$. The constraints are $x \geq 0, y \geq 0, x + y \leq 480$ (number of acres) and $2x + y \leq 800$ (number of labor hours). Solve the systems
$$\begin{cases} x = 0 \\ x + y = 480 \end{cases}, \begin{cases} x + y = 480 \\ 2x + y = 800 \end{cases}, \text{ and}$$
$\begin{cases} 2x + y = 800 \\ y = 0 \end{cases}$ to find the vertices: (0, 480), (320, 160), and (400, 0). Note that (0, 0) is also a vertex.

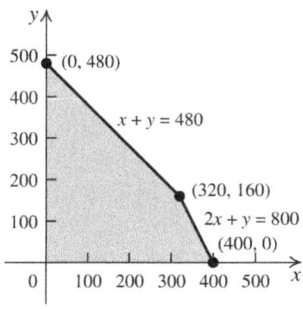

Now find the values of the profit function at each vertex:

Ordered pair	$40x + 30y - 3000$
(0, 480)	11,400
(320, 160)	14,600
(400, 0)	13,000

The maximum profit is $14,600 when 320 acres of oranges and 160 acres of grapefruits are planted.

35. Let $x =$ the number of rectangular tables, and let $y =$ the number of circular tables. The profit is $p = 3x + 4y$. The number of hours to assemble the rectangular tables is x and the number of hours to assemble the circular tables is y. The number of hours to finish the rectangular tables is x and the number of hours to finish the circular tables is $2y$. The 20 assemblers work a total of $(20)(40) = 800$ hours, and the 30 finishers work a total of $(30)(40) = 1200$ hours. So, the constraints are $x \geq 0, y \geq 0, x + y \leq 800,$ and $x + 2y \leq 1200$.

Solve the systems
$$\begin{cases} x = 0 \\ x + 2y = 1200 \end{cases}, \begin{cases} x + 2y = 1200 \\ x + y = 800 \end{cases}, \text{ and}$$
$\begin{cases} x + y = 800 \\ y = 0 \end{cases}$ to find the vertices: (0, 600), (400, 400), and (800, 0). Note that (0, 0) is also a vertex.

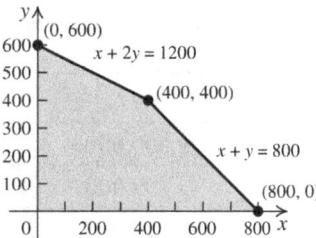

Now find the values of the profit function at each vertex:

Ordered pair	$p = 3x + 4y$.
(0, 600)	2400
(400, 400)	2800
(800, 0)	2400

The maximum profit is $2800 for 400 rectangular tables and 400 circular tables.

37. Let x = the number of terraced houses, and let y = the number of cottages. Then, the revenue is $r = 40{,}000x + 45{,}000y$. The number of units of concrete for the two house types is $x + y$; the number of units of wood for the two house types is $2x + y$; and the number of units of glass for the two house types is $2x + 5y$. The constraints are $x \geq 0$, $y \geq 0$, $x + y \leq 100$, $2x + y \leq 160$, and $2x + 5y \leq 400$. Solve the systems
$\begin{cases} x = 0 \\ 2x + 5y = 400 \end{cases}$, $\begin{cases} 2x + 5y = 400 \\ x + y = 100 \end{cases}$,
$\begin{cases} x + y = 100 \\ 2x + y = 160 \end{cases}$, and $\begin{cases} 2x + y = 160 \\ y = 0 \end{cases}$ to find the
vertices: $(0, 80)$, $\left(\dfrac{100}{3}, \dfrac{200}{3}\right)$, $(60, 40)$, and $(80, 0)$. Note that $(0, 0)$ is also a vertex.

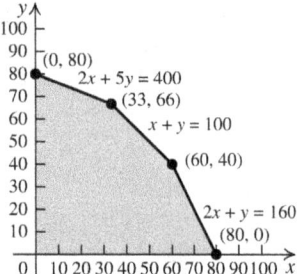

Now find the values of the profit function at each vertex. Note that we cannot use $\left(\dfrac{100}{3}, \dfrac{200}{3}\right)$ because there cannot be a fraction of a house. This value becomes $(33, 66)$.

Ordered pair	$r = 40{,}000x + 45{,}000y$
$(0, 80)$	3,600,000
$(33, 66)$	4,290,000
$(60, 40)$	4,200,000
$(80, 0)$	3,200,000

The maximum profit is $4,290,000 when 33 terraced houses and 66 cottages are built.

39. Let x = the number of female guests, and let y = the number of male guests. Then, the number of guests who eat in the restaurant is $0.5x + 0.3y$, and the number of guests who do not eat in the restaurant is $0.5x + 0.7y$. The profit $p = 15(0.5x + 0.3y) - 2(0.5x + 0.7y)$ $= 6.5x + 3.1y$. The constraints are $x \geq 0$, $y \geq 0$, $x \leq 125$, $y \leq 220$, $x + y \geq 100$ and $x + y \leq 300$. Solve the systems $\begin{cases} x = 0, \\ y = 220, \end{cases}$
$\begin{cases} y = 220 \\ x + y = 300 \end{cases}$, $\begin{cases} x + y = 300 \\ x = 125 \end{cases}$, $\begin{cases} x = 125 \\ y = 0 \end{cases}$,
$\begin{cases} y = 0 \\ x + y = 100 \end{cases}$, and $\begin{cases} x + y = 100 \\ x = 0 \end{cases}$ to find the
vertices: $(0, 220)$, $(80, 220)$, $(125, 175)$, $(125, 0)$, $(100, 0)$, and $(0, 100)$.

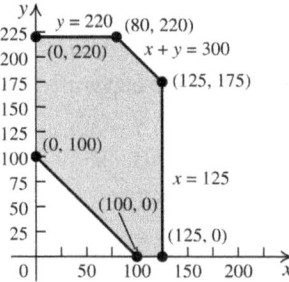

Now find the values of the profit function at each vertex:

Ordered pair	$p = 6.5x + 3.1y$
$(0, 220)$	682
$(80, 220)$	1202
$(125, 175)$	1355
$(125, 0)$	812.50
$(100, 0)$	650
$(0, 100)$	310

The maximum profit is $1355 when there are 125 female guests and 175 male guests.

41. Let x = the number of pounds of enchiladas, and let y = the number of pounds of vegetable loaf. The cost $c = 3.50x + 2.25y$. The number of units of carbohydrate is $5x + 2y$. The number of units of protein is $3x + 2y$. The number of units of fat is $5x + y$.
The constraints are
$x \geq 0$, $y \geq 0$, $5x + 2y \geq 60$, $3x + 2y \geq 40$, and $5x + y \geq 35$. Solve the systems
$\begin{cases} x = 0, \\ 5x + y = 35 \end{cases}$, $\begin{cases} 5x + y = 35 \\ 5x + 2y = 60 \end{cases}$, $\begin{cases} 5x + 2y = 60 \\ 3x + 2y = 40 \end{cases}$,
and $\begin{cases} 3x + 2y = 40 \\ x = 0 \end{cases}$ to find the vertices:
$(0, 35)$, $(2, 25)$, $(10, 5)$, and $(40/3, 0)$.

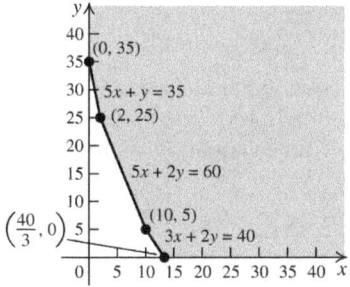

Now find the values of the profit function at each vertex:

Ordered pair	$c = 3.50x + 2.25y$
$(0, 35)$	78.75
$(2, 25)$	63.25
$(10, 5)$	46.25
$(40/3, 0)$.	46.67

The minimum cost is $46.25 when Elisa buys 10 enchilada meals and 5 vegetable loafs.

8.6 Beyond the Basics

43.

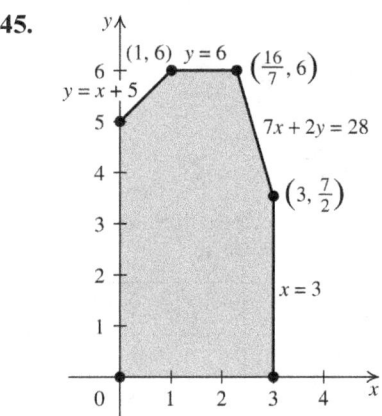

Solve the systems
$\begin{cases} x + 2y = 10 \\ 4x - y = 40 \end{cases}$, $\begin{cases} 4x - y = 40 \\ x + y = 20 \end{cases}$, $\begin{cases} x + y = 20 \\ 8y - 7x = 40 \end{cases}$,
and $\begin{cases} 8y - 7x = 40 \\ x + 2y = 10 \end{cases}$ to find the vertices:
$(10, 0)$, $(12, 8)$, $(8, 12)$, and $(0, 5)$. Now find the values of the objective function at each vertex:

Ordered pair	$f = 8x + 7y$
$(10, 0)$	80
$(12, 8)$	152
$(8, 12)$	148
$(0, 5)$	35

The maximum is 152 at $(12, 8)$.

45.

Solve the systems
$\begin{cases} x = 0 \\ y = 0 \end{cases}$, $\begin{cases} y = 0 \\ x = 3 \end{cases}$, $\begin{cases} x = 3 \\ 7x + 2y = 28 \end{cases}$, $\begin{cases} 7x + 2y = 28 \\ y = 6 \end{cases}$,
and $\begin{cases} y = 6 \\ y = x + 5 \end{cases}$ to find the vertices: $(0, 0)$, $(3, 0)$, $(3, \frac{7}{2})$, $(\frac{16}{7}, 6)$, $(1, 6)$, and $(0, 5)$.

Now find the values of the objective function at each vertex:

Ordered pair	$f = 5x + 6y$
$(0, 0)$	0
$(3, 0)$	15
$(3, \frac{7}{2})$	36
$(\frac{16}{7}, 6)$	$\frac{332}{7}$
$(1, 6)$	41
$(0, 5)$	30

The minimum is 0 at $(0, 0)$.

47. $ax_1 + by_1 = M = ax_2 + by_2 \Rightarrow$
$ax_1 - ax_2 = by_2 - by_1 \Rightarrow$
$a(x_1 - x_2) = b(y_2 - y_1) \Rightarrow$
$\dfrac{a}{b} = \dfrac{y_2 - y_1}{x_1 - x_2} = -\dfrac{y_2 - y_1}{x_2 - x_1} = -m$, where m is the slope of $\overline{PQ}$.

Since the slope is the same between any two points on a given line segment, we can conclude that f has the same value M at every point of the line segment PQ.

8.6 Critical Thinking/Discussion/Writing

49.

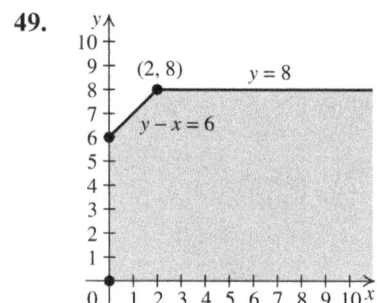

a. S is unbounded because there is no upper limit on the values for x.

b.

Ordered pair	$f = 2x + 3y$
$(0, 0)$	0
$(0, 6)$	18
$(2, 8)$	28
$(x, 8)$	$2x + 24$
$(x, 0)$	$2x$

As x increases, the value of the objective function $f = 2x + 3y$ increases. Since S is unbounded, there is no solution.

50. Minimize $f = 3.2x + 2.5y + 1.3z$ subject to the constraints
$x + 2y + 5 \geq 1$
$10x + 5y + 4z \geq 50$
$6x + 20y + 7z \geq 10$
$x \geq 0, y \geq 0, z \geq 0$

8.6 Getting Ready for the Next Section

51. $\begin{cases} x - 2y = 4 & (1) \\ -3x + 5y = -7 & (2) \end{cases}$

The first equation already has a leading coefficient of 1. To eliminate x from equation (2), add 3 times equation (1) to equation (2). Then solve for y.

$\begin{array}{ll} 3x - 6y = 12 & 3(x - 2y = 12) \\ -3x + 5y = -7 & (2) \\ \hline -y = 5 \\ y = -5 & (3) \end{array}$

The equivalent system in triangular form is
$\begin{cases} x - 2y = 4 & (1) \\ y = -5 & (3) \end{cases}$

Back-substitute the value of y into equation (1) and solve for x.
$x - 2(-5) = 4 \Rightarrow x + 10 = 4 \Rightarrow x = -6$

The solution set is $\{(-6, -5)\}$. Be sure to check the solution in each of the original equations.

53. $\begin{cases} x - 4y - z = 11 & (1) \\ 2x - 5y + 2z = 39 & (2) \\ -3x + 2y + z = 1 & (3) \end{cases}$

The first equation already has a leading coefficient of 1. To eliminate x from equation (2), add -2 times equation (1) to equation (2).

$\begin{array}{ll} -2x + 8y + 2z = -22 & (-2)(x - 4y - z = 11) \\ 2x - 5y + 2z = 39 & (2) \\ \hline 3y + 4z = 17 & (4) \end{array}$

Multiply equation (4) by $\tfrac{1}{3}$.
$\begin{cases} x - 4y - z = 11 & (1) \\ y + \dfrac{4}{3}z = \dfrac{17}{4} & (4) \\ -3x + 2y + z = 1 & (3) \end{cases}$

To eliminate x from equation (3), add 3 times equation (1) to equation (3).

$\begin{array}{ll} 3x - 12y - 3z = 33 & (3)(x - 4y - z = 11) \\ -3x + 2y + z = 1 & (3) \\ \hline -10y - 2z = 34 & (5) \end{array}$

Multiply equation (5) by $-\tfrac{1}{10}$.

$\begin{cases} x - 4y - z = 11 & (1) \\ y + \dfrac{4}{3}z = \dfrac{17}{3} & (4) \\ y + \dfrac{1}{5}z = -\dfrac{17}{5} & (5) \end{cases}$

Eliminate y in equation (5) by adding -1 times equation (2). Then solve for z.

$\begin{array}{ll} -y - \dfrac{4}{3}z = -\dfrac{17}{3} & (-1)\left(y + \tfrac{4}{3}z = \tfrac{17}{4}\right) \\ y + \dfrac{1}{5}z = -\dfrac{17}{5} & (5) \\ \hline -\dfrac{17}{15}z = -\dfrac{136}{15} \\ z = 8 & (6) \end{array}$

(continued on next page)

(*continued*)

The equivalent system in triangular form is
$$\begin{cases} x - 4y - z = 11 & (1) \\ y + \dfrac{4}{3}z = \dfrac{17}{3} & (4) \\ z = 8 & (6) \end{cases}$$

Back-substitute the value of z into equation (4) and solve for y.

$$y + \dfrac{4}{3}(8) = \dfrac{17}{3} \Rightarrow y + \dfrac{32}{3} = \dfrac{17}{3} \Rightarrow y = -5$$

Back-substitute $z = 8$ and $y = -5$ into equation (1) and solve for x.

$$x - 4(-5) - 8 = 11 \Rightarrow x + 12 = 11 \Rightarrow x = -1$$

The solution set is $\{(-1, -5, 8)\}$. Be sure to check the solution in each of the original equations.

For exercises 55–58, refer to the following array.

$$\begin{bmatrix} 2 & 4 & -6 & 10 \\ 3 & -2 & 4 & 12 \end{bmatrix} \begin{matrix} \leftarrow \text{row 1} \\ \leftarrow \text{row 2} \end{matrix}$$

55. Multiply each number in row 1 by $\frac{1}{2}$.

$$\begin{bmatrix} 1 & 2 & -3 & 5 \\ 3 & -2 & 4 & 12 \end{bmatrix}$$

57. Using the rectangular array in exercise 56, multiply each number in row 2 by $-\frac{1}{8}$.

$$\begin{bmatrix} 1 & 2 & -3 & 5 \\ 0 & 1 & -\dfrac{13}{8} & \dfrac{3}{8} \end{bmatrix}$$

Chapter 8 Review Exercises

Building Skills

1. Using elimination, we have
$$\begin{cases} 3x - y = -5 \\ x + 2y = 3 \end{cases} \Rightarrow \begin{cases} 6x - 2y = -10 \\ x + 2y = 3 \end{cases} \Rightarrow 7x = -7 \Rightarrow$$
$x = -1$
$3(-1) - y = -5 \Rightarrow y = 2$
The solution is $\{(-1, 2)\}$.

3. Using elimination, we have
$$\begin{cases} 2x + 4y = 3 \\ 3x + 6y = 10 \end{cases} \Rightarrow \begin{cases} -6x - 12y = -9 \\ 6x + 12y = 20 \end{cases} \Rightarrow 0 = 11 \Rightarrow$$
there is no solution. Solution set: $\varnothing$

5. Using elimination, we have
$$\begin{cases} 3x - y = 3 \\ \dfrac{1}{2}x + \dfrac{1}{3}y = 2 \end{cases} \Rightarrow \begin{cases} x - \dfrac{1}{3}y = 1 \\ \dfrac{1}{2}x + \dfrac{1}{3}y = 2 \end{cases} \Rightarrow \dfrac{3}{2}x = 3 \Rightarrow$$
$x = 2$
$3(2) - y = 3 \Rightarrow y = 3$
The solution is $\{(2, 3)\}$.

7. Multiply the first equation by -2, add the result to the second equation, and replace the second equation with the new equation:
$$\begin{cases} x + 3y + z = 0 \\ 2x - y + z = 5 \\ 3x - 3y + 2z = 10 \end{cases} \Rightarrow \begin{cases} x + 3y + z = 0 \\ -7y - z = 5 \\ 3x - 3y + 2z = 10 \end{cases}$$

Multiply the first equation by -3, add the result to the third equation, and replace the third equation with the new equation:
$$\begin{cases} x + 3y + z = 0 \\ -7y - z = 5 \\ 3x - 3y + 2z = 10 \end{cases} \Rightarrow \begin{cases} x + 3y + z = 0 \\ -7y - z = 5 \\ -12y - z = 10 \end{cases}$$

Subtract the third equation from the second equation, and replace the second equation with the result:
$$\begin{cases} x + 3y + z = 0 \\ -7y - z = 5 \\ -12y - z = 10 \end{cases} \Rightarrow \begin{cases} x + 3y + z = 0 \\ 5y = -5 \\ -12y - z = 10 \end{cases} \Rightarrow$$
$$\begin{cases} x + 3y + z = 0 \\ y = -1 \\ -12y - z = 10 \end{cases}$$

Substitute $y = -1$ into the third equation to solve for z and then substitute the values for z and y into the first equation to solve for x:
$$\begin{cases} x + 3y + z = 0 \\ y = -1 \\ -12y - z = 10 \end{cases} \Rightarrow \begin{cases} x + 3y + z = 0 \\ y = -1 \\ -12(-1) - z = 10 \end{cases} \Rightarrow$$
$$\begin{cases} x + 3y + z = 0 \\ y = -1 \\ z = 2 \end{cases} \Rightarrow \begin{cases} x + 3(-1) + 2 = 0 \\ y = -1 \\ z = 2 \end{cases} \Rightarrow$$
$$\begin{cases} x = 1 \\ y = -1 \\ z = 2 \end{cases}$$

The solution is $\{(1, -1, 2)\}$.

9. Switch the second and third equations, multiply the first equation by -2, add the result to the second equation, and replace the second equation with the result.
$$\begin{cases} x + y = 1 \\ 3y + 2z = 0 \\ 2x - 3z = 7 \end{cases} \Rightarrow \begin{cases} x + y = 1 \\ 2x - 3z = 7 \\ 3y + 2z = 0 \end{cases} \Rightarrow$$

(*continued on next page*)

(*continued*)

$$\begin{cases} x + y = 1 \\ -2y - 3z = 5 \\ 3y + 2z = 0 \end{cases}$$

Multiply the second equation by 3, multiply the third equation by 2, add the results, replace the third equation with the new equation, and solve for z:

$$\begin{cases} x + y = 1 \\ -2y - 3z = 5 \\ 3y + 2z = 0 \end{cases} \Rightarrow \begin{cases} x + y = 1 \\ -2y - 3z = 5 \\ -5z = 15 \end{cases} \Rightarrow$$

$$\begin{cases} x + y = 1 \\ -2y - 3z = 5 \\ z = -3 \end{cases}$$

Substitute $z = -3$ into the second equation to solve for y, and then substitute the value for y into the first equation to solve for x:

$$\begin{cases} x + y = 1 \\ -2y - 3z = 5 \\ z = -3 \end{cases} \Rightarrow \begin{cases} x + y = 1 \\ -2y - 3(-3) = 5 \\ z = -3 \end{cases} \Rightarrow$$

$$\begin{cases} x + y = 1 \\ y = 2 \\ z = -3 \end{cases} \Rightarrow \begin{cases} x + 2 = 1 \\ y = 2 \\ z = -3 \end{cases} \Rightarrow \begin{cases} x = -1 \\ y = 2 \\ z = -3 \end{cases}$$

The solution is $\{(-1, 2, -3)\}$.

11. Multiply the first equation by −5, add the result to the second equation, and replace the second equation with the result, then switch the first and second equations:

$$\begin{cases} x + y + z = 1 \\ x + 5y + 5z = -1 \\ 3x - y - z = 4 \end{cases} \Rightarrow \begin{cases} x + y + z = 1 \\ -4x = -6 \\ 3x - y - z = 4 \end{cases} \Rightarrow$$

$$\begin{cases} x = \dfrac{3}{2} \\ x + y + z = 1 \\ 3x - y - z = 4 \end{cases}$$

Add the second and third equations, and replace the second equation with the result.

$$\begin{cases} x = \dfrac{3}{2} \\ x + y + z = 1 \\ 3x - y - z = 4 \end{cases} \Rightarrow \begin{cases} x = \dfrac{3}{2} \\ 4x = 5 \\ 3x - y - z = 4 \end{cases} \Rightarrow$$

$$\begin{cases} x = 3/2 \\ x = 5/4 \\ 3x - y - z = 4 \end{cases} \Rightarrow \text{There is no solution.}$$

Solution set: ∅

13. The system is a dependent system because there are two equations in three unknowns. Multiply the first equation by −5, and multiply the second equation by 4, then add the resulting equations and solve for z:

$$\begin{cases} x + 4y + 3z = 1 \\ 2x + 5y + 4z = 4 \end{cases} \Rightarrow \begin{cases} -5x - 20y - 15z = -5 \\ 8x + 20y + 16z = 16 \end{cases} \Rightarrow$$

$3x + z = 11 \Rightarrow z = -3x + 11$

Substitute the expression for z into the first equation and solve for y:

$x + 4y + 3(-3x + 11) = 1 \Rightarrow -8x + 4y + 33 = 1 \Rightarrow$
$y = 2x - 8$.

The solution is $\{(x, 2x - 8, -3x + 11)\}$.

15. Add the first and third equations, and replace the first equation with the result:

$$\begin{cases} 3x - y = 2 \\ x + 2y = 9 \\ 3x + y = 10 \end{cases} \Rightarrow \begin{cases} 6x = 12 \\ x + 2y = 9 \\ 3x + y = 10 \end{cases} \Rightarrow \begin{cases} x = 2 \\ x + 2y = 9 \\ 3x + y = 10 \end{cases}$$

Substitute $x = 2$ into the second equation and solve for y:

$$\begin{cases} x = 2 \\ x + 2y = 9 \\ 3x + y = 10 \end{cases} \Rightarrow \begin{cases} x = 2 \\ 2 + 2y = 9 \\ 3x + y = 10 \end{cases} \Rightarrow \begin{cases} x = 2 \\ y = \dfrac{7}{2} \\ 3x + y = 10 \end{cases}$$

$3(2) + \dfrac{7}{2} = \dfrac{19}{2} \neq 10 \Rightarrow$ there is no solution.

Solution set: ∅

17. $\begin{cases} x + y + z = 3 \quad (1) \\ 2x - y + z = 4 \quad (2) \\ x + 4y + 2z = 5 \quad (3) \end{cases}$

Multiply equation (1) by −2, then add the result to equation (2), and replace equation (2) with the new equation:

$$\begin{cases} x + y + z = 3 \quad (1) \\ 2x - y + z = 4 \quad (2) \\ x + 4y + 2z = 5 \quad (3) \end{cases} \Rightarrow$$

$$\begin{cases} x + y + z = 3 \quad (1) \\ -3y - z = -2 \quad (4) \\ x + 4y + 2z = 5 \quad (3) \end{cases}$$

Multiply equation (1) by −1, then add the result to equation (3), and replace equation (3) with the new equation:

$$\begin{cases} x + y + z = 3 \quad (1) \\ 2x - y + z = 4 \quad (2) \\ x + 4y + 2z = 5 \quad (3) \end{cases} \Rightarrow \begin{cases} x + y + z = 3 \quad (1) \\ -3y - z = -2 \quad (4) \\ 3y + z = 2 \quad (5) \end{cases}$$

(*continued on next page*)

(*continued*)

Add equation (4) to equation (5) to solve for *z*:
$$\begin{cases} x+y+z=3 & (1) \\ -3y-z=-2 & (4) \\ 3y+z=2 & (5) \end{cases} \Rightarrow 0=0$$

The equation $0 = 0$ is equivalent to $0z = 0$, which is true for every value of *z*. Solving the second equation for *y*, we have

$-3y - z = -2 \Rightarrow y = \frac{2}{3} - \frac{1}{3}z$. Substituting this into the first equation, we have $x + y + z = 3 \Rightarrow$

$x + \left(\frac{2}{3} - \frac{1}{3}z\right) - z = 3 \Rightarrow x = \frac{7}{3} - \frac{2}{3}z$

Thus, the solution is $\left\{\left(\frac{7}{3} - \frac{2}{3}z, \frac{2}{3} - \frac{1}{3}z, z\right)\right\}$.

19.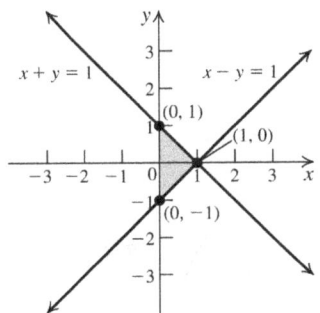

21.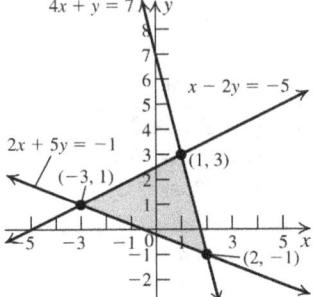

23.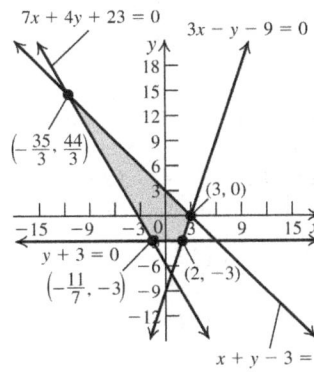

25.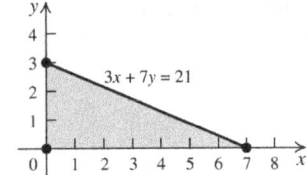

Solve the systems
$\begin{cases} x=0 \\ 3x+7y=21 \end{cases}$, $\begin{cases} 3x+7y=21 \\ y=0 \end{cases}$, and $\begin{cases} y=0 \\ x=0 \end{cases}$ to
find the vertices: (0, 3), (7, 0), and (0, 0). Now find the values of the objective function at each vertex:

Ordered pair	$z = 2x + 3y$
(0, 3)	9
(7, 0)	14
(0, 0)	0

The maximum is 14 at (7, 0).

27.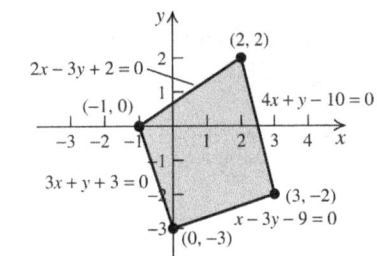

Solve the systems
$\begin{cases} 2x-3y=-2 \\ 4x+y=10 \end{cases}$, $\begin{cases} 4x+y=10 \\ x-3y=9 \end{cases}$, $\begin{cases} x-3y=9 \\ 3x+y=-3 \end{cases}$,

and $\begin{cases} 3x+y=-3 \\ 2x-3y=-2 \end{cases}$ to find the vertices: (2, 2),

(3, −2), (0, −3), and (−1, 0). Now find the values of the objective function at each vertex:

Ordered pair	$z = x + 3y$
(2, 2)	8
(3, −2)	−3
(0, −3)	−9
(−1, 0)	−1

The minimum is −9 at (0, −3).

29. Using substitution, we have
$\begin{cases} x + 3y = 1 \\ x^2 - 3x = 7y + 3 \end{cases} \Rightarrow \begin{cases} x = 1 - 3y \\ x^2 - 3x = 7y + 3 \end{cases} \Rightarrow$
$(1 - 3y)^2 - 3(1 - 3y) = 7y + 3 \Rightarrow$
$9y^2 - 4y - 5 = 0 \Rightarrow (y - 1)(9y + 5) = 0 \Rightarrow$
$y = 1 \text{ or } y = -\dfrac{5}{9}$
$x + 3(1) = 1 \Rightarrow x = -2$
$x + 3\left(-\dfrac{5}{9}\right) = 1 \Rightarrow x = \dfrac{24}{9} = \dfrac{8}{3}$
The solution is $\left\{(-2, 1), \left(\dfrac{8}{3}, -\dfrac{5}{9}\right)\right\}$.

31. Using substitution, we have
$\begin{cases} x - y = 4 \\ 5x^2 + y^2 = 24 \end{cases} \Rightarrow \begin{cases} y = x - 4 \\ 5x^2 + y^2 = 24 \end{cases} \Rightarrow$
$5x^2 + (x - 4)^2 = 24 \Rightarrow 6x^2 - 8x - 8 = 0 \Rightarrow$
$2(x - 2)(3x + 2) = 0 \Rightarrow x = 2 \text{ or } x = -\dfrac{2}{3}$
$2 - y = 4 \Rightarrow y = -2$
$-\dfrac{2}{3} - y = 4 \Rightarrow y = -\dfrac{14}{3}$
The solution is $\left\{(2, -2), \left(-\dfrac{2}{3}, -\dfrac{14}{3}\right)\right\}$.

33. Using substitution, we have
$\begin{cases} xy = 2 \\ x^2 + 2y^2 = 9 \end{cases} \Rightarrow \begin{cases} y = \dfrac{2}{x} \\ x^2 + 2y^2 = 9 \end{cases} \Rightarrow$
$x^2 + 2\left(\dfrac{2}{x}\right)^2 = 9 \Rightarrow x^2 - 9 + \dfrac{8}{x^2} = 0 \Rightarrow$
$x^4 - 9x^2 + 8 = 0$
Let $u = x^2$. Then $u^2 - 9u + 8 = 0 \Rightarrow$
$(u - 8)(u - 1) = 0 \Rightarrow u = 8 \Rightarrow x^2 = 8 \Rightarrow$
$x = \pm 2\sqrt{2}$ or $u = 1 \Rightarrow x^2 = 1 \Rightarrow x^2 = \pm 1$
$(2\sqrt{2})y = 2 \Rightarrow y = \dfrac{\sqrt{2}}{2}$
$(-2\sqrt{2})y = 2 \Rightarrow y = -\dfrac{\sqrt{2}}{2}$
$(1)y = 2 \Rightarrow y = 2; (-1)y = 2 \Rightarrow y = -2$
None of the values are extraneous, so the
solution is $\left\{(1, 2), (-1, -2), \left(2\sqrt{2}, \dfrac{\sqrt{2}}{2}\right), \left(-2\sqrt{2}, -\dfrac{\sqrt{2}}{2}\right)\right\}$.

35.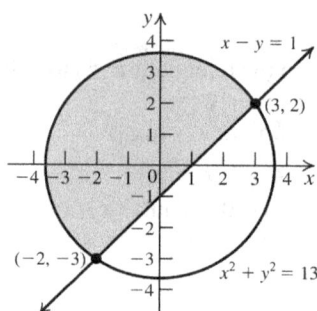

37.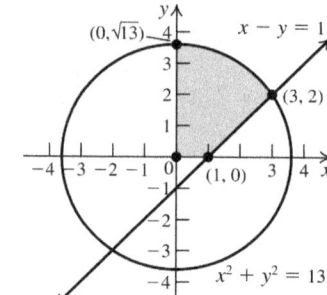

39.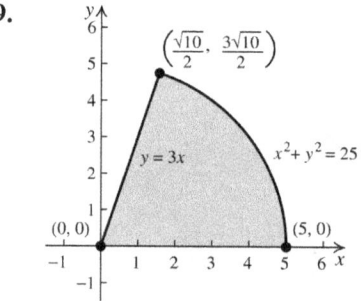

41. $\dfrac{x + 4}{x^2 + 5x + 6} = \dfrac{x + 4}{(x + 2)(x + 3)} = \dfrac{A}{x + 2} + \dfrac{B}{x + 3} \Rightarrow$
$x + 4 = A(x + 3) + B(x + 2) \Rightarrow$
$x + 4 = (A + B)x + (3A + 2B) \Rightarrow$
$\begin{cases} A + B = 1 \\ 3A + 2B = 4 \end{cases} \Rightarrow \begin{cases} -2A - 2B = -2 \\ 3A + 2B = 4 \end{cases} \Rightarrow$
$A = 2, B = -1$
$\dfrac{x + 4}{x^2 + 5x + 6} = \dfrac{2}{x + 2} - \dfrac{1}{x + 3}$

43. $\dfrac{3x^2 + x + 1}{x(x - 1)^2} = \dfrac{A}{x} + \dfrac{B}{x - 1} + \dfrac{C}{(x - 1)^2} \Rightarrow$
$3x^2 + x + 1 = A(x - 1)^2 + Bx(x - 1) + Cx \Rightarrow$
$3x^2 + x + 1$
$= (A + B)x^2 + (-2A - B + C)x + A \Rightarrow$
$\begin{cases} A + B = 3 \\ -2A - B + C = 1 \Rightarrow A = 1, B = 2, C = 5 \\ A = 1 \end{cases}$
$\dfrac{3x^2 + x + 1}{x(x - 1)^2} = \dfrac{1}{x} + \dfrac{2}{x - 1} + \dfrac{5}{(x - 1)^2}$

45. $\dfrac{x^2+2x+3}{(x^2+4)^2} = \dfrac{Ax+B}{x^2+4} + \dfrac{Cx+D}{(x^2+4)^2} \Rightarrow$

$x^2+2x+3 = (Ax+B)(x^2+4) + Cx+D \Rightarrow$

x^2+2x+3
$= Ax^3 + Bx^2 + (4A+C)x + (4B+D) \Rightarrow$

$\begin{cases} A=0 \\ B=1 \\ 4A+C=2 \\ 4B+D=3 \end{cases} \Rightarrow A=0, B=1, C=2, D=-1$

$\dfrac{x^2+2x+3}{(x^2+4)^2} = \dfrac{1}{x^2+4} + \dfrac{2x-1}{(x^2+4)^2}$

Applying the Concepts

47. Let x = the amount invested at high-risk. Let y = the amount invested at 4% interest. Then

$\begin{cases} x + y = 15{,}000 \\ 0.12x + 0.04y = 1300 \end{cases} \Rightarrow$

$\begin{cases} -0.04x - 0.04y = -600 \\ 0.12x + 0.04y = 1300 \end{cases} \Rightarrow 0.08x = 700 \Rightarrow$

$x = 8750$

$8750 + y = 15{,}000 \Rightarrow y = 6250$

The speculator invested $8750 at 12% and $6250 at 4%.

49. Let x = the length of the rectangle. Let y = the width of the rectangle. Then

$\begin{cases} xy = 63 \\ 2x+2y = 33 \end{cases} \Rightarrow \begin{cases} y = \dfrac{63}{x} \\ 2x+2y = 33 \end{cases} \Rightarrow$

$2x + 2\left(\dfrac{63}{x}\right) = 33 \Rightarrow 2x - 33 + \dfrac{126}{x} = 0 \Rightarrow$

$2x^2 - 33x + 126 = 0 \Rightarrow (x-6)(2x-21) = 0 \Rightarrow$

$x = 6$ or $x = \dfrac{21}{2} = 10.5$; $6y = 63 \Rightarrow y = \dfrac{21}{2} = 10.5$

$\dfrac{21}{2}y = 63 \Rightarrow y = 6$

The rectangle is 10.5 feet by 6 feet.

51. Let x = the length of one leg. Let y = the length of the other leg. Then

$\begin{cases} x^2 + y^2 = 17^2 \\ (x+1)^2 + (y+4)^2 = 20^2 \end{cases} \Rightarrow$

$\begin{cases} y^2 = 289 - x^2 \\ x^2 + 2x + y^2 + 8y - 383 = 0 \end{cases} \Rightarrow$

$x^2 + 2x + 289 - x^2 + 8\sqrt{289-x^2} - 383 = 0 \Rightarrow$

$8\sqrt{289-x^2} = -2x + 94 \Rightarrow$

$64(289 - x^2) = (-2x+94)^2 \Rightarrow$
$18{,}496 - 64x^2 = 4x^2 - 376x + 8836 \Rightarrow$
$68x^2 - 376x - 9660 = 0 \Rightarrow$
$4(x-15)(17x+161) = 0 \Rightarrow x = 15$ or $x = -\dfrac{161}{7}$

(Reject the negative solution.)

$15^2 + y^2 = 17^2 \Rightarrow y = 8$

The legs of the triangle are 15 and 8.

53. Let x = the width of the smaller pastures. Let y = the length of the pastures.

x	$2x$	x
y		

Then,

$\begin{cases} 4xy = 6400 \\ 4x + 2y = 240 \end{cases} \Rightarrow \begin{cases} y = \dfrac{1600}{x} \\ 4x + 2y = 240 \end{cases} \Rightarrow$

$4x + 2\left(\dfrac{1600}{x}\right) = 240 \Rightarrow 4x - 240 + \dfrac{3200}{x} = 0 \Rightarrow$

$4x^2 - 240x + 3200 = 0 \Rightarrow$
$4(x-40)(x-20) = 0 \Rightarrow x = 40$ or $x = 20$
$4(40)y = 6400 \Rightarrow y = 40$
$4(20)y = 6400 \Rightarrow y = 80$

The dimensions of the pasture are 40 meters by 160 meters.

55. a. $C(x) = 12x + 60{,}000$; $R(x) = 20x$

b.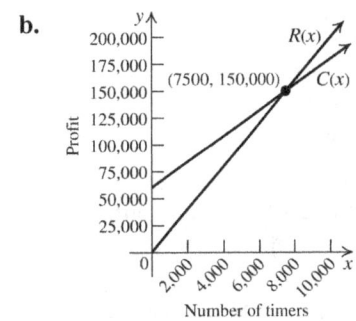

c. $C(x) = R(x) \Rightarrow 12x + 60{,}000 = 20x \Rightarrow$
$x = 7500$
$R(12) = 20(7500) = 150{,}000$
The breakeven point is (7500, 150,000).

d. $P(x) = R(x) - C(x) = 0.15 C(x) \Rightarrow$
$20x - (12x + 60{,}000)$
$\qquad = 0.15(12x + 60{,}000) \Rightarrow$
$8x - 60{,}000 = 1.8x + 9000 \Rightarrow$
$6.2x = 69{,}000 \Rightarrow x \approx 11{,}129$
11,129 timers should be sold.

57. Let x = the double occupancy rate. Let y = the single occupancy rate. Each pays half of the double occupancy rate, so Alisha paid $6\left(\dfrac{x}{2}\right) + 3(x-y) = 6x - 3y$ for the nine months, and Sunita paid $6\left(\dfrac{x}{2}\right) + 3y = 3x + 3y$ for the nine months.
$\begin{cases} 6x - 3y = 2250 \\ 3x + 3y = 3150 \end{cases} \Rightarrow 9x = 5400 \Rightarrow x = 600$
$3(600) + 3y = 3150 \Rightarrow y = 450$
The double occupancy rate is $600 per month, and the single occupancy rate is $450 per month.

59. Let x = the number of students who passed the exam. Let y = the number of students who failed the exam. The total number of points scored for the class is $26 \times 72 = 1872$. Then
$\begin{cases} x + y = 26 \\ 78x + 26y = 1872 \end{cases} \Rightarrow \begin{cases} -26x - 26y = -676 \\ 78x + 26y = 1872 \end{cases} \Rightarrow$
$52x = 1196 \Rightarrow x = 23, y = 3$
Three students failed the exam.

61. Let x = Steve's age now. Let y = Janet's age now. Then $x - y$ = the difference in their ages.
$\begin{cases} y = 2(y - (x-y)) \\ x + ((x-y) + x) = 119 \end{cases} \Rightarrow \begin{cases} y = 4y - 2x \\ 3x - y = 119 \end{cases} \Rightarrow$
$\begin{cases} -2x + 3y = 0 \\ 3x - y = 119 \end{cases} \Rightarrow \begin{cases} -2x + 3y = 0 \\ 9x - 3y = 357 \end{cases} \Rightarrow$
$7x = 357 \Rightarrow x = 51$
$51 + (51 - y) + 51 = 119 \Rightarrow y = 34$
So, Steve is 51 years old now and Janet is 34 years old now.

63. Let x = Butch's amount. Let y = Sundance's amount. Let z = Billy's amount. Then
$\begin{cases} x = 0.75(y + z) \\ y = 0.5(x + z) + 500 \\ z = 3(x - y) - 1000 \end{cases} \Rightarrow$
$\begin{cases} x - 0.75y - 0.75z = 0 \\ -0.5x + y - 0.5z = 500 \\ -3x + 3y + z = -1000 \end{cases}$
Multiply the first equation by 0.5, add the result to the second equation, and replace the second equation with the new equation. Then, multiply the first equation by 3, add the result to the third equation, and replace the third equation with the new equation.

$\begin{cases} x - 0.75y - 0.75z = 0 \\ -0.5x + y - 0.5z = 500 \\ -3x + 3y + z = -1000 \end{cases} \Rightarrow$
$\begin{cases} x - 0.75y - 0.75z = 0 \\ 0.625y - 0.875z = 500 \\ 0.75y - 1.25z = -1000 \end{cases}$
Multiply the second equation by -1.2, add the result to the third equation, replace the third equation with the new equation, then solve for z:
$\begin{cases} x - 0.75y - 0.75z = 0 \\ 0.625y - 0.875z = 500 \\ 0.75y - 1.25z = -1000 \end{cases} \Rightarrow$
$\begin{cases} x - 0.75y - 0.75z = 0 \\ 0.625y - 0.875z = 500 \\ -0.2z = -1600 \end{cases} \Rightarrow$
$\begin{cases} x - 0.75y - 0.75z = 0 \\ 0.625y - 0.875z = 500 \\ z = 8000 \end{cases}$
Substitute $z = 8000$ into the second equation to solve for y, then substitute the values for y and z into the first equation to solve for x:
$\begin{cases} x - 0.75y - 0.75z = 0 \\ 0.625y - 0.875z = 500 \\ z = 8000 \end{cases} \Rightarrow$
$\begin{cases} x - 0.75y - 0.75z = 0 \\ 0.625y - 0.875(8000) = 500 \\ z = 8000 \end{cases} \Rightarrow$
$\begin{cases} x - 0.75y - 0.75z = 0 \\ y = 12,000 \\ z = 8000 \end{cases} \Rightarrow$
$\begin{cases} x - 0.75(12,000) - 0.75(8000) = 0 \\ y = 12,000 \\ z = 8000 \end{cases} \Rightarrow$
$\begin{cases} x = 15,000 \\ y = 12,000 \\ z = 8000 \end{cases}$
So, they stole $35,000. Butch received $15,000, Sundance received $12,000, and Billy received $8000.

65. $2^x = 8^y \Rightarrow 2^x = 2^{3y} \Rightarrow x = 3y$
$9^y = 3^{x-2} \Rightarrow 3^{2y} = 3^{x-2} \Rightarrow 2y = x - 2$
$\begin{cases} x = 3y \\ 2y = x - 2 \end{cases} \Rightarrow 2y = 3y - 2 \Rightarrow y = 2, x = 6$

67. Let x = the number of two-story houses. Let y = the number of one-story houses. The profit $p = 10,000x + 4000y$. The two-story houses use $7x$ units of material, while the one-story houses use y units of material. The two-story houses use x units of labor, while the one-story houses use $2y$ units of labor. So, the constraints are $x \geq 0, y \geq 0, 7x + y \leq 42$, and $x + 2y \leq 32$.

Solve the systems $\begin{cases} x = 0 \\ x + 2y = 32 \end{cases}, \begin{cases} x + 2y = 32 \\ 7x + y = 42 \end{cases}$,

and $\begin{cases} 7x + y = 42 \\ y = 0 \end{cases}$ to find the vertices: (0, 16), (4, 14), and (6, 0).

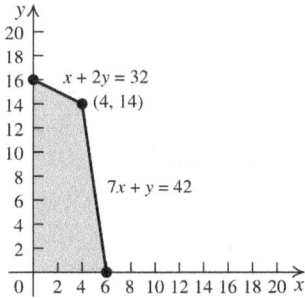

Now find the values of the profit function at each vertex:

Ordered pair	$p = 10,000x + 4000y$.
(0, 16)	64,000
(4, 14)	96,000
(6, 0)	60,000

The builder should build 4 two-story houses and 14 one-story houses for a maximum profit of $96,000.

69. Let x = the cost of the small lobster, y = the cost of the medium lobster, and z = the cost of the large lobster. Then we have

$\begin{cases} 4x + 2y + z = 344 & (1) \\ 3x + 2z = 255 & (2) \\ 5x + 2y + 2z = 449 & (3) \end{cases}$

Interchange equations (2) and (3).

$\begin{cases} 4x + 2y + z = 344 & (1) \\ 5x + 2y + 2z = 449 & (3) \\ 3x + 2z = 255 & (2) \end{cases}$

Multiply equation (1) by −1, add the result to equation (3), then replace equation (1) with the result (4).

$\begin{cases} 4x + 2y + z = 344 \\ 5x + 2y + 2z = 449 \\ 3x + 2z = 255 \end{cases} \Rightarrow$

$\begin{cases} -4x - 2y - z = -344 \\ 5x + 2y + 2z = 449 \\ 3x + 2z = 255 \end{cases} \Rightarrow$

$\begin{cases} x + z = 105 & (4) \\ 5x + 2y + 2z = 449 & (3) \\ 3x + 2z = 255 & (2) \end{cases}$

Multiply equation (4) by −3, add the result to equation (2), then replace equation (2) with the result (5).

$\begin{cases} x + z = 105 & (4) \\ 5x + 2y + 2z = 499 & (3) \\ 3x + 2z = 255 & (2) \end{cases} \Rightarrow$

$\begin{cases} -3x - 3z = -315 \\ 5x + 2y + 2z = 449 \\ 3x + 2z = 255 \end{cases} \Rightarrow$

$\begin{cases} x + z = 105 & (4) \\ 5x + 2y + 2z = 449 & (3) \\ -z = -60 \Rightarrow z = 60 & (5) \end{cases}$

Substitute $z = 60$ in equation (4) and solve for x: $x + 60 = 105 \Rightarrow x = 45$

Now substitute $x = 45$ and $z = 60$ into equation (3) and solve for y:
$5(45) + 2y + 2(60) = 449 \Rightarrow 2y = 104 \Rightarrow y = 52$

A small lobster costs $45, a medium lobster costs $52, and a large lobster costs $60.

Chapter 8 Practice Test A

1. Using elimination, we have
$\begin{cases} 2x - y = 4 \\ 2x + y = 4 \end{cases} \Rightarrow 4x = 8 \Rightarrow x = 2$
$2(2) + y = 4 \Rightarrow y = 0$
The solution is $\{(2, 0)\}$.

2. Using elimination, we have
$\begin{cases} x + 2y = 8 \\ 3x + 6y = 24 \end{cases} \Rightarrow \begin{cases} 3x + 6y = 24 \\ 3x + 6y = 24 \end{cases} \Rightarrow$ the system

is consistent. The solution is $\left\{\left(x, 4 - \dfrac{x}{2}\right)\right\}$.

3. Using substitution, we have
$\begin{cases} -2x + y = 4 \\ 4x - 2y = 4 \end{cases} \Rightarrow \begin{cases} y = 2x + 4 \\ 4x - 2y = 4 \end{cases} \Rightarrow$
$4x - 2(2x + 4) = 4 \Rightarrow -8 = 4 \Rightarrow$ there is no solution.
Solution set: ∅

4. $\begin{cases} 3x+3y=-15 \\ 2x-2y=-10 \end{cases} \Rightarrow \begin{cases} x+y=-5 \\ x-y=-5 \end{cases} \Rightarrow 2x=-10 \Rightarrow$
   $x=-5$
   $3(-5)+3y=-15 \Rightarrow y=0$
   The solution is $\{(-5, 0)\}$.

5. Using elimination, we have
   $\begin{cases} \dfrac{5}{3}x+\dfrac{y}{2}=14 \\ \dfrac{2}{3}x-\dfrac{y}{8}=3 \end{cases} \Rightarrow \begin{cases} 10x+3y=84 \\ 16x-3y=72 \end{cases} \Rightarrow$
   $26x=156 \Rightarrow x=6$
   $\dfrac{5}{3}(6)+\dfrac{y}{2}=14 \Rightarrow y=8$
   The solution is $\{(6, 8)\}$.

6. Using substitution, we have
   $\begin{cases} y=x^2 \\ 3x-y+4=0 \end{cases} \Rightarrow 3x-x^2+4=0 \Rightarrow$
   $x^2-3x-4=0 \Rightarrow (x-4)(x+1)=0 \Rightarrow$
   $x=4$ or $x=-1$
   $y=4^2=16$
   $y=(-1)^2=1$
   The solution is $\{(4, 16), (-1, 1)\}$.

7. Using elimination, we have
   $\begin{cases} x-3y=-4 \\ 2x^2+3x-3y=8 \end{cases} \Rightarrow -2x^2-2x=-12 \Rightarrow$
   $x^2+x-6=0 \Rightarrow (x+3)(x-2)=0 \Rightarrow$
   $x=-3$ or $x=2$
   $-3-3y=-4 \Rightarrow y=\dfrac{1}{3}$
   $2-3y=-4 \Rightarrow y=2$
   The solution is $\left\{\left(-3,\dfrac{1}{3}\right),(2,2)\right\}$.

8. a. Let $x=$ the weight of one bar. Let $y=$ the weight of the other bar. Then
   $\begin{cases} x+y=485 \\ x-y=15 \end{cases}$

   b. $\begin{cases} x+y=485 \\ x-y=15 \end{cases} \Rightarrow 2x=500 \Rightarrow x=250$
   $250+y=485 \Rightarrow y=235$
   One bar weight 250 pounds and the other weighs 235 pounds.

9. Multiply the first equation by -2, add the result to the second equation, and replace the second equation with the new equation. Then divide the second equation by 4.
   $\begin{cases} 2x+y+2z=4 \\ 4x+6y+z=15 \\ 2x+2y+7z=-1 \end{cases} \Rightarrow \begin{cases} 2x+y+2z=4 \\ y-\dfrac{3}{4}z=\dfrac{7}{4} \\ 2x+2y+7z=-1 \end{cases}$
   Subtract the third equation from the first equation and replace the third equation with the new equation:
   $\begin{cases} 2x+y+2z=4 \\ y-\dfrac{3}{4}z=\dfrac{7}{4} \\ 2x+2y+7z=-1 \end{cases} \Rightarrow \begin{cases} 2x+y+2z=4 \\ y-\dfrac{3}{4}z=\dfrac{7}{4} \\ -y-5z=5 \end{cases}$
   Add the second and third equations, replace the third equation with the new equation
   $\begin{cases} 2x+y+2z=4 \\ y-\dfrac{3}{4}z=\dfrac{7}{4} \\ -y-5z=5 \end{cases} \Rightarrow \begin{cases} 2x+y+2z=4 \\ y-\dfrac{3}{4}z=\dfrac{7}{4} \\ -\dfrac{23}{4}z=\dfrac{27}{4} \end{cases} \Rightarrow$
   Solve for the third equation for z. Divide the first equation by 2:
   $\begin{cases} x+\dfrac{1}{2}y+z=2 \\ y-\dfrac{3}{4}z=\dfrac{7}{4} \\ z=-\dfrac{27}{23} \end{cases}$

10. $\begin{cases} x-6y+3z=-2 \\ 9y-5z=2 \\ 2z=10 \end{cases} \Rightarrow \begin{cases} x-6y+3z=-2 \\ 9y-5z=2 \\ z=5 \end{cases}$
    $9y-5(5)=2 \Rightarrow y=3$
    $x-6(3)+3(5)=-2 \Rightarrow x=1$
    The solution is $\{(1, 3, 5)\}$.

11. Subtract the third equation from the first equation, replace the third equation with the new equation, and solve for z.
    $\begin{cases} x+y+z=8 \\ 2x-2y+2z=4 \\ x+y-z=12 \end{cases} \Rightarrow \begin{cases} x+y+z=8 \\ 2x-2y+2z=4 \\ z=-2 \end{cases}$
    Multiply the first equation by -2, add the result to the second equation, and replace the second equation with the new equation. Solve for y:
    $\begin{cases} x+y+z=8 \\ 2x-2y+2z=4 \\ z=-2 \end{cases} \Rightarrow \begin{cases} x+y+z=8 \\ y=3 \\ z=-2 \end{cases}$

    (continued on next page)

(*continued*)

Substitute the values for y and z into the first equation to solve for x:
$$\begin{cases} x+3-2=8 \\ y=3 \\ z=-2 \end{cases} \Rightarrow \begin{cases} x=7 \\ y=3 \\ z=-2 \end{cases}$$
The solution is $\{(7, 3, -2)\}$.

12. Add the second and third equations, and replace the third equation with the result:
$$\begin{cases} x+z=-1 \\ 3y+2z=5 \\ 3x-3y+z=-8 \end{cases} \Rightarrow \begin{cases} x+z=-1 \\ 3y+2z=5 \\ 3x+3z=-3 \end{cases} \Rightarrow$$
$$\begin{cases} x+z=-1 \\ 3y+2z=5 \\ x+z=-1 \end{cases}$$
The first and third equations are the same, so we have two equations in three unknowns. Solve the first equation for z, then substitute that expression into the second equation to solve for y:
$z=-x-1;\ 3y+2(-x-1)=5 \Rightarrow$
$3y=2x+7 \Rightarrow y=\dfrac{2}{3}x+\dfrac{7}{3}$
The solution is $\left\{\left(x, \dfrac{2}{3}x+\dfrac{7}{3}, -x-1\right)\right\}$.

13. Add the first and second equations, replace the first equation with the solution, then solve for x:
$$\begin{cases} 2x-y+z=2 \\ x+y-z=-1 \\ x-5y+5z=7 \end{cases} \Rightarrow \begin{cases} x=\dfrac{1}{3} \\ x+y-z=-1 \\ x-5y+5z=7 \end{cases}$$
Substitute $x=1/3$ into the second and third equations, then solve the system
$$\begin{cases} \dfrac{1}{3}+y-z=-1 \\ \dfrac{1}{3}-5y+5z=7 \end{cases} \Rightarrow \begin{cases} 5y-5z=-\dfrac{20}{3} \\ -5y+5z=\dfrac{20}{3} \end{cases} \Rightarrow 0=0$$
Now, solve for y: $\dfrac{1}{3}+y-z=-1 \Rightarrow z=z-\dfrac{4}{3}$.
The solution is $\left\{\left(\dfrac{1}{3}, -\dfrac{4}{3}+z, z\right)\right\}$.

14. Let $n =$ the number of nickels. Let $d =$ the number of dimes. Let $q =$ the number of quarters. Then
$$\begin{cases} n+d+q=300 \\ d=3(n+q) \\ 0.05n+0.1d+0.25q=30.65 \end{cases} \Rightarrow$$
$$\begin{cases} n+d+q=300 \\ -3n+d-3q=0 \\ 5n+10d+25q=3065 \end{cases}$$
Multiply the first equation by 3, add the result to the second equation, and replace the second equation with the new equation
$$\begin{cases} n+d+q=300 \\ -3n+d-3q=0 \\ 5n+10d+25q=3065 \end{cases} \Rightarrow$$
$$\begin{cases} n+d+q=300 \\ 4d=900 \\ 5n+10d+25q=3065 \end{cases} \Rightarrow$$
$$\begin{cases} n+d+q=300 \\ d=225 \\ 5n+10d+25q=3065 \end{cases}$$
Multiply the first equation by -5, add the result to the third equation, and replace the third equation with the new equation:
$$\begin{cases} n+d+q=300 \\ d=225 \\ 5n+10d+25q=3065 \end{cases} \Rightarrow$$
$$\begin{cases} n+d+q=300 \\ d=225 \\ 5d+20q=1565 \end{cases}$$
Substitute $d=225$ into the third equation, solve for q, then substitute the values for d and q into the first equation to solve for n:
$5(225)+20q=1565 \Rightarrow q=22$
$n+225+22=300 \Rightarrow n=53$
There are 53 nickels, 225 dimes, and 22 quarters.

15. $\dfrac{2x}{(x-5)(x+1)}=\dfrac{A}{x-5}+\dfrac{B}{x+1}$

16. $\dfrac{-5x^2+x-8}{(x-2)(x^2+1)^2}$
$=\dfrac{A}{x-2}+\dfrac{Bx+C}{x^2+1}+\dfrac{Dx+E}{(x^2+1)^2}$

17. $\dfrac{x+3}{(x+4)^2(x-7)} = \dfrac{A}{x-7} + \dfrac{B}{x+4} + \dfrac{C}{(x+4)^2} \Rightarrow$

$x+3 = A(x+4)^2 + B(x-7)(x+4) + C(x-7)$

Letting $x = 7$, we have $10 = 121A \Rightarrow A = \dfrac{10}{121}$.

Letting $x = -4$, we have

$-1 = -11C \Rightarrow \dfrac{1}{11} = C$. Substitute the values for A and C into the equation and expand.

$x+3 = \left(B + \dfrac{10}{121}\right)x^2 + \left(\dfrac{91}{121} - 3B\right)x$

$\qquad + \left(\dfrac{83}{121} - 28B\right) \Rightarrow$

$B + \dfrac{10}{121} = 0 \Rightarrow B = -\dfrac{10}{121}$

$\dfrac{x+3}{(x+4)^2(x-7)}$

$= \dfrac{10}{121(x-7)} - \dfrac{10}{121(x+4)} + \dfrac{1}{11(x+4)^2}$

18.

19.

20. Solve the systems $\begin{cases} x = 0 \\ x+3y = 3 \end{cases}$, $\begin{cases} x+3y = 3 \\ y = 0 \end{cases}$,

and $\begin{cases} y = 0 \\ x = 0 \end{cases}$ to find the vertices: (0, 1), (3, 0), and (0, 0).

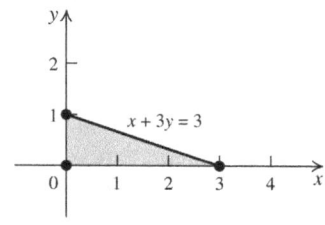

Now find the values of the objective function at each vertex:

Ordered pair	$z = 2x + y$
(0, 1)	1
(3, 0)	6
(0, 0)	0

The maximum value is 6 at (3, 0).

Chapter 8 Practice Test B

1. Substitute each of the ordered pairs into the system.
$\begin{cases} 2(5) - 3(-2) = 16 \\ 5 - (-2) = 7 \end{cases}$ The answer is A.

2. Substitute each of the ordered pairs into the system.
$\begin{cases} 6\left(\dfrac{44}{3}\right) - 9(10) = -2 \\ 3\left(\dfrac{44}{3}\right) - 5(10) = -6 \end{cases}$

The answer is C.

3. Substitute each of the ordered pairs into the system.
$\begin{cases} 2\left(\dfrac{5}{2}y + \dfrac{9}{2}\right) - 5y = 9 \\ 4\left(\dfrac{5}{2}y + \dfrac{9}{2}\right) - 10y = 18 \end{cases}$

The answer is B.

4. Substituting each of the ordered pairs into the system, we find that none are solutions. The answer is C.

5. Substitute each of the ordered pairs into the system.
$\begin{cases} \dfrac{1}{5}(1) + \dfrac{2}{5}(2) = 1 \\ \dfrac{1}{4}(1) - \dfrac{1}{3}(2) = -\dfrac{5}{12} \end{cases}$

The answer is B.

6. Substitute each of the ordered pairs into the system.
$\begin{cases} \left(3y + \dfrac{1}{2}\right) - 3y = \dfrac{1}{2} \\ -2\left(3y + \dfrac{1}{2}\right) + 6y = -1 \end{cases}$

The answer is D.

7. Substitute each of the ordered pairs into the system.
$$\begin{cases} 3(4)+4(0)=12 \\ 3(4)^2+16(0)^2=48 \end{cases} \begin{cases} 3(2)+4\left(\dfrac{3}{2}\right)=12 \\ 3(2)^2+16\left(\dfrac{3}{2}\right)^2=48 \end{cases}$$
The answer is D.

8. Let x = the number of student tickets and y = the number of nonstudent tickets. Then
$$\begin{cases} x+y=300 \\ 2x+5y=975 \end{cases} \Rightarrow \begin{cases} -2x-2y=-600 \\ 2x+5y=975 \end{cases} \Rightarrow$$
$3y=375 \Rightarrow y=125$
$x+125=300 \Rightarrow x=175$
The answer is D.

9. Multiply the first equation by 2, subtract the second equation from that result, then replace the second equation with the new equation:
$$\begin{cases} x+3y+3z=4 \\ 2x+5y+4z=5 \\ x+2y+2z=6 \end{cases} \Rightarrow \begin{cases} x+3y+3z=4 \\ y+2z=3 \\ x+2y+2z=6 \end{cases}$$
Subtract the third equation from the first equation, then replace the third equation with the new equation:
$$\begin{cases} x+3y+3z=4 \\ y+2z=3 \\ x+2y+2z=6 \end{cases} \Rightarrow \begin{cases} x+3y+3z=4 \\ y+2z=3 \\ y+z=-2 \end{cases}$$
Subtract the third equation from the second equation and replace the third equation with the new equation:
$$\begin{cases} x+3y+3z=4 \\ y+2z=3 \\ y+z=-2 \end{cases} \Rightarrow \begin{cases} x+3y+3z=4 \\ y+2z=3 \\ z=5 \end{cases}$$
The answer is C.

10. From the third equation, $z=3$.
$10y-2(3)=4 \Rightarrow y=1; 2x+1+3=0 \Rightarrow$
$x=-2$
The answer is A.

11. Substitute each of the ordered pairs into the system.
$$\begin{cases} 2(1)+13(-1)+6(2)=1 \\ 3(1)+10(-1)+11(2)=15 \\ 2(1)+10(-1)+8(2)=8 \end{cases}$$
The answer is B.

12. Substituting each of the ordered pairs into the system, we find that none are solutions.
The answer is D.

13. Substitute each of the ordered pairs into the system.
$$\begin{cases} x-2(5x+1)+3(3x+2)=4 \\ 2x-(5x+1)+(3x+2)=1 \\ x+(5x+1)-2(3x+2)=-3 \end{cases}$$
The answer is C.

14. Let x = the number of students going to France. Let y = the number of students going to Italy. Let z = the number of students going to Spain. Then
$$\begin{cases} x+y+z=46 \\ x+y=4+z \\ x=z-2 \end{cases} \Rightarrow \begin{cases} x+y+z=46 \\ x+y-z=4 \\ x-z=-2 \end{cases} \text{ Subtract}$$
the second equation from the first equation, replace the second equation with the new equation, and solve for z:
$$\begin{cases} x+y+z=46 \\ x+y-z=4 \\ x-z=-2 \end{cases} \Rightarrow \begin{cases} x+y+z=46 \\ 2z=42 \\ x-z=-2 \end{cases} \Rightarrow$$
$$\begin{cases} x+y+z=46 \\ z=21 \\ x-z=-2 \end{cases}$$
Substitute $z=21$ into the third equation to solve for x. Then substitute the values for x and z into the first equation to solve for y:
$x-21=-2 \Rightarrow x=19; 19+y+21=46 \Rightarrow$
$y=6$
The answer is B.

15. C   16. A

17. $\dfrac{x^2+15x+18}{x^3-9x} = \dfrac{x^2+15x+18}{x(x-3)(x+3)}$
$= \dfrac{A}{x}+\dfrac{B}{x-3}+\dfrac{C}{x+3} \Rightarrow$
$x^2+15x+18$
$= A(x^2-9)+Bx(x+3)+Cx(x-3)$
Letting $x=-3$, we have
$(-3)^2+15(-3)+18=-3C(-3-3) \Rightarrow$
$-18=18C \Rightarrow -1=C$
Letting $x=3$, we have
$(3)^2+15(3)+18=3B(3+3) \Rightarrow$
$72=18B \Rightarrow 4=B$
Substituting the values for B and C and expanding the equation, we have
$x^2+15x+18$
$=A(x^2-9)+4x(x+3)-x(x-3)$
$=(A+3)x^2+15x-9A \Rightarrow$
$$\begin{cases} A+3=1 \\ -9A=18 \end{cases} \Rightarrow A=-2$$

(continued on next page)

(continued)

$$\frac{x^2+15x+18}{x^3-9x} = -\frac{2}{x}+\frac{4}{x-3}-\frac{1}{x+3}$$

The answer is C.

18. B   19. A

20. Solve the systems
$$\begin{cases} x=0 \\ 2x+3y=16 \end{cases} \begin{cases} 2x+3y=16 \\ 2x+y=8 \end{cases}, \begin{cases} 2x+y=8 \\ y=0 \end{cases} \text{ and}$$
$$\begin{cases} y=0 \\ x=0 \end{cases} \text{ to find the vertices: } \left(0,\frac{16}{3}\right),$$
$(2, 4)$, $(4, 0)$, and $(0, 0)$.

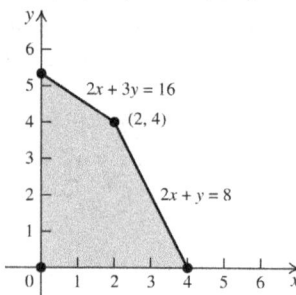

Now find the values of the objective function at each vertex:

Ordered pair	$z = 3x + 21y$
$\left(0, \frac{16}{3}\right)$	112
$(2, 4)$	90
$(4, 0)$	12
$(0, 0)$	0

The maximum value is 112 at $\left(0, \frac{16}{3}\right)$.

The answer is B.

Cumulative Review Exercises (Chapters P–8)

1. $\frac{1}{x-1}+\frac{4}{x-4}=\frac{5}{x-5} \Rightarrow$
$(x-4)(x-5)+4(x-1)(x-5)$
$\qquad = 5(x-1)(x-4) \Rightarrow$
$5x^2-33x+40 = 5x^2-25x+20 \Rightarrow$
$-8x = -20 \Rightarrow x = \frac{5}{2}$

2. Let $u = x + \frac{1}{x}$. Then
$2u^2 - 7u + 5 = 0 \Rightarrow (u-1)(2u-5) = 0 \Rightarrow$
$u = 1$ or $u = 5/2$
$x + \frac{1}{x} = 1 \Rightarrow x^2 - x + 1 = 0 \Rightarrow$
$x = \frac{1 \pm \sqrt{1-4}}{2} = \frac{1 \pm i\sqrt{3}}{2} = \frac{1}{2} \pm \frac{i\sqrt{3}}{2}$
$x + \frac{1}{x} = \frac{5}{2} \Rightarrow 2x^2 - 5x + 2 = 0 \Rightarrow$
$(x-2)(2x-1) = 0 \Rightarrow x = 2$ or $x = 1/2$
The solution is $\left\{\frac{1}{2} \pm \frac{i\sqrt{3}}{2}, 2, \frac{1}{2}\right\}$.

3. $\sqrt{3x-5} = x-3 \Rightarrow 3x-5 = (x-3)^2 \Rightarrow$
$x^2 - 9x + 14 = 0 \Rightarrow (x-7)(x-2) = 0 \Rightarrow$
$x = 7$ or $x = 2$.
Check each answer to see if either is extraneous:
$\sqrt{3(7)-5} = \sqrt{16} = 4 = 7-3$
$\sqrt{3(2)-5} = \sqrt{1} = 1 \ne 2-3 \Rightarrow 2$ is extraneous.
The solution is $\{7\}$.

4. $\frac{x-1}{x+3} = 0 \Rightarrow x = 1$ is the x-intercept. The vertical asymptote is $x = -3$. The intervals to be tested are $(-\infty, -3)$, $(-3, 1]$, and $[1, \infty)$.

Interval	Test Point	Value of $\frac{x-1}{x+3}$	Result
$(-\infty, -3)$	-4	5	$+$
$(-3, 1]$	0	$-\frac{1}{3}$	$-$
$[1, \infty)$	2	$\frac{1}{5}$	$+$

The solution is $(-3, 1]$.

5. Solve the associated equation to find the test intervals:
$x^2 - 9x + 20 = 0 \Rightarrow (x-4)(x-5) = 0 \Rightarrow$
$x = 4$ or $x = 5$.
The intervals to be tested are $(-\infty, 4), (4, 5)$, and $(5, \infty)$.

Interval	Test Point	Value of $x^2 - 9x + 20$	Result
$(-\infty, 4)$	0	20	+
$(4, 5)$	4.5	-0.25	$-$
$(5, \infty)$	10	30	+

The solution is $(-\infty, 4) \cup (5, \infty)$.

6. $2^{x-1} = 5 \Rightarrow (x-1)\ln 2 = \ln 5 \Rightarrow$
$x \ln 2 - \ln 2 = \ln 5 \Rightarrow x = \dfrac{\ln 5 + \ln 2}{\ln 2}$

7. $\log_x 16 = 4 \Rightarrow x^4 = 16 \Rightarrow x^4 = 2^4 \Rightarrow x = 2$

8. $\log(x-3) + \log(x-1) = \log(2x-5) \Rightarrow$
$\log((x-3)(x-1)) = \log(2x-5) \Rightarrow$
$x^2 - 4x + 3 = 2x - 5 \Rightarrow x^2 - 6x + 8 = 0 \Rightarrow$
$(x-4)(x-2) = 0 \Rightarrow x = 4$ or $x = 2$
Reject $x = 2$ because it makes $\log(2-3) = \log(-1)$, which is not possible.
The solution is $\{4\}$.

9. Shift the graph of $y = |x|$ one unit left and two units down.

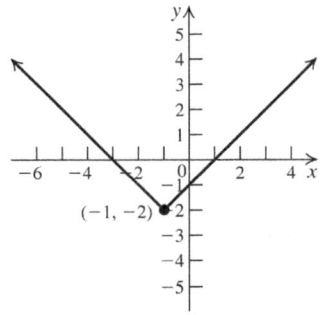

10. Shift the graph of $y = x^2$ two units right, reflect it across the x-axis, then shift it three units up.

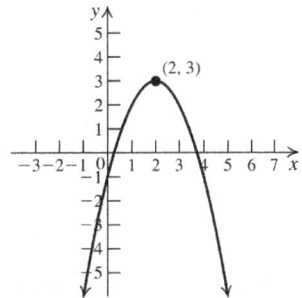

11. Shift the graph of $y = 2^x$ one unit right and three units down.

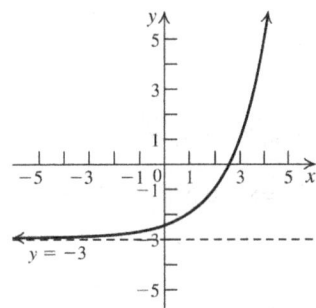

12. Shift the graph of $y = \sqrt{x}$ one unit right and three units up.

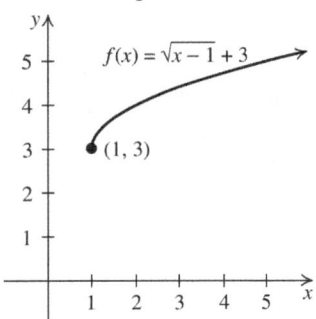

13. a. Switch the variables and solve for y:
$y = 2x - 2 \Rightarrow x = 2y - 2 \Rightarrow y = f^{-1} = \dfrac{x+2}{2}$

b.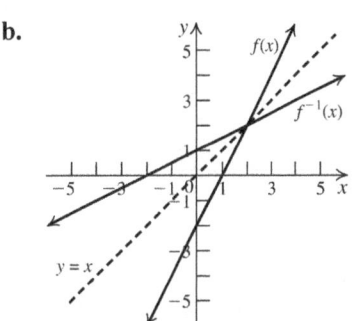

14. a. The factors of the constant term are $\{\pm 1, \pm 2, \pm 3, \pm 6\}$. The factors of the leading coefficient are $\{\pm 1\}$. The possible rational zeros are $\{\pm 1, \pm 2, \pm 3, \pm 6\}$.

b.
$$\underline{2|}\ \begin{array}{cccc} 1 & -1 & 1 & -6 \\ & 2 & 2 & 6 \\ \hline 1 & 1 & 3 & 0 \end{array}$$
A real zero is 2.

c. $x^3 - x^2 + x - 6 = (x-2)(x^2 + x + 3)$

$x^2 + x + 3 = 0 \Rightarrow x = \dfrac{-1 \pm \sqrt{1-12}}{2} \Rightarrow$

$x = \dfrac{-1 \pm i\sqrt{11}}{2}$

The zeros are $\left\{2, -\dfrac{1}{2} \pm \dfrac{i\sqrt{11}}{2}\right\}$.

15. $\log_3(9x^4) = \log_3 9 + \log_3 x^4 = 2 + 4\log_3 x$

16. a. $0.5 = e^{-0.05t} \Rightarrow \ln 0.5 = -0.05t \Rightarrow$

$t = -\dfrac{\ln 0.5}{0.05} \approx 13.86$ years

b. $0.5 = e^{-0.0002t} \Rightarrow \ln 0.5 = -0.0002t \Rightarrow$

$t = -\dfrac{\ln 0.5}{0.0002} \approx 3465.74$ years

17. $f^{-1}(9) = 2$

18. $A = Pe^{rt} \Rightarrow 2P = Pe^{0.075t} \Rightarrow 2 = e^{0.075t} \Rightarrow$

$\ln 2 = 0.075t \Rightarrow t = \dfrac{\ln 2}{0.075} \approx 9.24$

It will take about 9.24 years to double the money.

19. $\begin{cases} 5x - 2y + 25 = 0 \\ 4y - 3x - 29 = 0 \end{cases} \Rightarrow \begin{cases} 10x - 4y = -50 \\ -3x + 4y = 29 \end{cases} \Rightarrow$

$7x = -21 \Rightarrow x = -3$

$4y - 3(-3) - 29 = 0 \Rightarrow y = 5$

The solution is $\{(-3, 5)\}$.

20. Multiply the second equation by -2, add the result to the first equation, and replace the second equation with the new equation:

$\begin{cases} 2x - y + z = 3 \\ x + 3y - 2z = 11 \\ 3x - 2y + z = 4 \end{cases} \Rightarrow \begin{cases} 2x - y + z = 3 \\ -7y + 5z = -19 \\ 3x - 2y + z = 4 \end{cases}$

Multiply the first equation by 3, multiply the third equation by -2, add the two and replace the third equation with the new equation:

$\begin{cases} 2x - y + z = 3 \\ -7y + 5z = -19 \\ 3x - 2y + z = 4 \end{cases} \Rightarrow \begin{cases} 2x - y + z = 3 \\ -7y + 5z = -19 \\ y + z = 1 \end{cases}$

Multiply the third equation by 7, add the result to the second equation, replace the third equation and solve for z:

$\begin{cases} 2x - y + z = 3 \\ -7y + 5z = -19 \\ y + z = 1 \end{cases} \Rightarrow \begin{cases} 2x - y + z = 3 \\ -7y + 5z = -19 \\ 12z = -12 \end{cases} \Rightarrow$

$z = -1$

$-7y + 5(-1) = -19 \Rightarrow y = 2$

$2x - 2 - 1 = 3 \Rightarrow x = 3$

The solution is $\{(3, 2, -1)\}$.

Chapter 9 Matrices and Determinants

9.1 Matrices and Systems of Equations

9.1 Practice Problems

1. a. 3×2 b. 1×2

2. $\begin{bmatrix} 0 & 3 & -1 & | & 8 \\ 1 & 4 & 0 & | & 14 \\ 0 & -2 & 9 & | & 0 \end{bmatrix}$

3. $\begin{bmatrix} 3 & 4 & 5 \\ 2 & 4 & 6 \end{bmatrix} \xrightarrow{R_1 \leftrightarrow R_2} \begin{bmatrix} 2 & 4 & 6 \\ 3 & 4 & 5 \end{bmatrix} \xrightarrow{\frac{1}{2}R_1}$
$\begin{bmatrix} 1 & 2 & 3 \\ 3 & 4 & 5 \end{bmatrix} \xrightarrow{-3R_1 + R_2 \to R_2} \begin{bmatrix} 1 & 2 & 3 \\ 0 & -2 & -4 \end{bmatrix}$

4. First write the system as an augmented matrix.
$\begin{cases} x - 6y + 3z = -2 \\ 3x + 3y - 2z = -2 \\ 2x - 3y + z = -2 \end{cases} \Rightarrow \begin{bmatrix} 1 & -6 & 3 & | & -2 \\ 3 & 3 & -2 & | & -2 \\ 2 & -3 & 1 & | & -2 \end{bmatrix}$

   Now perform the row operations.
$\begin{bmatrix} 1 & -6 & 3 & | & -2 \\ 3 & 3 & -2 & | & -2 \\ 2 & -3 & 1 & | & -2 \end{bmatrix}$
$\xrightarrow[2R_1 - R_3 \to R_3]{3R_1 - R_2 \to R_2} \begin{bmatrix} 1 & -6 & 3 & | & -2 \\ 0 & -21 & 11 & | & -4 \\ 0 & -9 & 5 & | & -2 \end{bmatrix}$
$\xrightarrow{9R_2 - 21R_3 \to R_3} \begin{bmatrix} 1 & -6 & 3 & | & -2 \\ 0 & -21 & 11 & | & -4 \\ 0 & 0 & -6 & | & 6 \end{bmatrix}$
$\xrightarrow[-\frac{1}{6}R_3 \to R_3]{-\frac{1}{21}R_2 \to R_2} \begin{bmatrix} 1 & -6 & 3 & | & -2 \\ 0 & 1 & -\frac{11}{21} & | & \frac{4}{21} \\ 0 & 0 & 1 & | & -1 \end{bmatrix}$

   Thus, $z = -1$.
   Using back-substitution, we have
   $y - \frac{11}{21}(-1) = \frac{4}{21} \Rightarrow y = -\frac{7}{21} = -\frac{1}{3}$. Then,
   $x - 6\left(-\frac{1}{3}\right) + 3(-1) = -2 \Rightarrow x - 1 = -2 \Rightarrow x = -1$

   The solution is $\left\{\left(-1, -\frac{1}{3}, -1\right)\right\}$.

5. First write the system as an augmented matrix.
$\begin{cases} x - 2y = 1 \\ 2x + 3y = 16 \end{cases} \Rightarrow \begin{bmatrix} 1 & -2 & | & 1 \\ 2 & 3 & | & 16 \end{bmatrix}$

   Now perform the row operations.
$\begin{bmatrix} 1 & -2 & | & 1 \\ 2 & 3 & | & 16 \end{bmatrix} \xrightarrow{-2R_1 + R_2 \to R_2} \begin{bmatrix} 1 & -2 & | & 1 \\ 0 & 7 & | & 14 \end{bmatrix}$
$\xrightarrow{\frac{1}{7}R_2 \to R_2} \begin{bmatrix} 1 & -2 & | & 1 \\ 0 & 1 & | & 2 \end{bmatrix}$

   Thus, $y = 2$. Using back-substitution, we have
   $x - 2(2) = 1 \Rightarrow x = 5$. The solution is $\{(5, 2)\}$.

6. First write the system as an augmented matrix.
$\begin{cases} 2x + y - z = 7 \\ x - 3y - 3z = 4 \\ 4x + y + z = 3 \end{cases} \Rightarrow \begin{bmatrix} 2 & 1 & -1 & | & 7 \\ 1 & -3 & -3 & | & 4 \\ 4 & 1 & 1 & | & 3 \end{bmatrix}$

   Now perform the row operations.
$\begin{bmatrix} 2 & 1 & -1 & | & 7 \\ 1 & -3 & -3 & | & 4 \\ 4 & 1 & 1 & | & 3 \end{bmatrix}$
$\xrightarrow{R_1 \leftrightarrow R_2} \begin{bmatrix} 1 & -3 & -3 & | & 4 \\ 2 & 1 & -1 & | & 7 \\ 4 & 1 & 1 & | & 3 \end{bmatrix}$
$\xrightarrow[-4R_1 + R_3 \to R_3]{-2R_1 + R_2 \to R_2} \begin{bmatrix} 1 & -3 & -3 & | & 4 \\ 0 & 7 & 5 & | & -1 \\ 0 & 13 & 13 & | & -13 \end{bmatrix}$
$\xrightarrow[\frac{1}{13}R_3 \to R_3]{\frac{1}{7}R_2 \to R_2} \begin{bmatrix} 1 & -3 & -3 & | & 4 \\ 0 & 1 & \frac{5}{7} & | & -\frac{1}{7} \\ 0 & 1 & 1 & | & -1 \end{bmatrix}$
$\xrightarrow{\frac{7}{2}(-R_2 + R_3) \to R_3} \begin{bmatrix} 1 & -3 & -3 & | & 4 \\ 0 & 1 & \frac{5}{7} & | & -\frac{1}{7} \\ 0 & 0 & 1 & | & -3 \end{bmatrix}$

   Thus, $z = -3$. Using back-substitution, we
   have $y + \frac{5}{7}(-3) = -\frac{1}{7} \Rightarrow y = 2$ and
   $x - 3(2) - 3(-3) = 4 \Rightarrow x = 1$.
   The solution is $\{(1, 2, -3)\}$.

7. $\begin{cases} 6x+8y-14z=3 \\ 3x+4y-7z=12 \\ 6x+3y+z=0 \end{cases} \Rightarrow \begin{bmatrix} 6 & 8 & -14 & | & 3 \\ 3 & 4 & -7 & | & 12 \\ 6 & 3 & 1 & | & 0 \end{bmatrix}$

Now perform the row operations.

$\begin{bmatrix} 6 & 8 & -14 & | & 3 \\ 3 & 4 & -7 & | & 12 \\ 6 & 3 & 1 & | & 0 \end{bmatrix}$

$\xrightarrow{R_1 \leftrightarrow R_2} \begin{bmatrix} 3 & 4 & -7 & | & 12 \\ 6 & 8 & -14 & | & 3 \\ 6 & 3 & 1 & | & 0 \end{bmatrix}$

$\xrightarrow{\frac{1}{3}R_1 \to R_1} \begin{bmatrix} 1 & \frac{4}{3} & -\frac{7}{3} & | & 4 \\ 6 & 8 & -14 & | & 3 \\ 6 & 3 & 1 & | & 0 \end{bmatrix}$

$\xrightarrow{-6R_1 + R_2 \to R_2} \begin{bmatrix} 1 & \frac{4}{3} & -\frac{7}{3} & | & 4 \\ 0 & 0 & 0 & | & -21 \\ 6 & 3 & 1 & | & 0 \end{bmatrix}$

Since the second row is equivalent to $0 = -21$, the system is inconsistent and the solution is $\varnothing$.

8. $\begin{cases} 2x-3y-2z=0 \\ x+y-2z=7 \\ 3x-5y-5z=3 \end{cases} \Rightarrow \begin{bmatrix} 2 & -3 & -2 & | & 0 \\ 1 & 1 & -2 & | & 7 \\ 3 & -5 & -5 & | & 3 \end{bmatrix}$

$\begin{bmatrix} 2 & -3 & -2 & | & 0 \\ 1 & 1 & -2 & | & 7 \\ 3 & -5 & -5 & | & 3 \end{bmatrix} \xrightarrow{R_1 \leftrightarrow R_2} \begin{bmatrix} 1 & 1 & -2 & | & 7 \\ 2 & -3 & -2 & | & 0 \\ 3 & -5 & -5 & | & 3 \end{bmatrix}$

$\xrightarrow[-3R_1+R_3 \to R_3]{-2R_1+R_2 \to R_2} \begin{bmatrix} 1 & 1 & -2 & | & 7 \\ 0 & -5 & 2 & | & -14 \\ 0 & -8 & 1 & | & -18 \end{bmatrix}$

$\xrightarrow{-\frac{1}{5}R_2 \to R_2} \begin{bmatrix} 1 & 1 & -2 & | & 7 \\ 0 & 1 & -\frac{2}{5} & | & \frac{14}{5} \\ 0 & -8 & 1 & | & -18 \end{bmatrix}$

$\xrightarrow[8R_2+R_3 \to R_3]{R_1-R_2 \to R_1} \begin{bmatrix} 1 & 0 & -\frac{8}{5} & | & \frac{21}{5} \\ 0 & 1 & -\frac{2}{5} & | & \frac{14}{5} \\ 0 & 0 & -\frac{11}{5} & | & \frac{22}{5} \end{bmatrix}$

$\xrightarrow{-\frac{5}{11}R_3 \to R_3} \begin{bmatrix} 1 & 0 & -\frac{8}{5} & | & \frac{21}{5} \\ 0 & 1 & -\frac{2}{5} & | & \frac{14}{5} \\ 0 & 0 & 1 & | & -2 \end{bmatrix}$

$\xrightarrow[R_2+\frac{2}{5}R_3 \to R_2]{R_1+\frac{8}{5}R_3 \to R_1} \begin{bmatrix} 1 & 0 & 0 & | & 1 \\ 0 & 1 & 0 & | & 2 \\ 0 & 0 & 1 & | & -2 \end{bmatrix}$

The solution is $\{(1, 2, -2)\}$.

9. $\begin{cases} x+z=-1 \\ 3y+2z=5 \\ 3x-3y+z=-8 \end{cases} \Rightarrow \begin{bmatrix} 1 & 0 & 1 & | & -1 \\ 0 & 3 & 2 & | & 5 \\ 3 & -3 & 1 & | & -8 \end{bmatrix}$

$\begin{bmatrix} 1 & 0 & 1 & | & -1 \\ 0 & 3 & 2 & | & 5 \\ 3 & -3 & 1 & | & -8 \end{bmatrix}$

$\xrightarrow{-3R_1+R_3 \to R_3} \begin{bmatrix} 1 & 0 & 1 & | & -1 \\ 0 & 3 & 2 & | & 5 \\ 0 & -3 & -2 & | & -5 \end{bmatrix}$

$\xrightarrow[R_2+R_3 \to R_3]{\frac{1}{3}R_2 \to R_2} \begin{bmatrix} 1 & 0 & 1 & | & -1 \\ 0 & 1 & \frac{2}{3} & | & \frac{5}{3} \\ 0 & 0 & 0 & | & 0 \end{bmatrix}$

The matrix is in row echelon form. The equivalent system is $\begin{cases} x+z=-1 \\ y+\frac{2}{3}z=\frac{5}{3} \end{cases}$. Solving for x and y in terms of z, we have $x=-z-1$ and $y=\frac{5}{3}-\frac{2}{3}z$.

The solution set is $\left\{\left(-z-1, \frac{5}{3}-\frac{2}{3}z, z\right)\right\}$.

10. a. $\begin{cases} s = 0.01s + 0.1c + 0.1t + 12.46 \\ c = 0.2s + 0.02c + 0.01t + 3 \\ t = 0.25s + 0.3c + 2.7 \end{cases}$ or
$\begin{cases} 0.99s - 0.1c - 0.1t = 12.46 \\ -0.2s + 0.98c - 0.01t = 3 \\ -0.25s - 0.3c + t = 2.7 \end{cases}$

b. $0.99(14) - 0.1(6) - 0.1(8) = 12.46$ ✓
$-0.2(14) + 0.98(6) - 0.01(8) = 3$ ✓
$-0.25(14) - 0.3(6) + (8) = 2.7$ ✓

9.1 Concepts and Vocabulary

1. A matrix is any rectangular array of <u>numbers</u>.

3. Two matrices are row equivalent if one can be obtained form the other by a sequence of <u>row operations</u>.

5. False. A 3×4 matrix has three rows with four entries in each row. In other words, the matrix has three rows and four columns.

7. False. The augmented matrix of the system is
$$\begin{bmatrix} 2 & 3 & 0 & | & 5 \\ 0 & 3 & -4 & | & -1 \\ 1 & 0 & 2 & | & 3 \end{bmatrix}.$$

9.1 Building Skills

9. 1×1 11. 2×4 13. 2×3

15. $a_{13} = 3, a_{31} = 9, a_{33} = 11, a_{34} = 12$

17. No, the array is not a matrix because it is not a rectangular array of numbers.

19. $\begin{bmatrix} 2 & 4 & | & 2 \\ 1 & -3 & | & 1 \end{bmatrix}$

21. $\begin{bmatrix} 5 & -7 & | & 11 \\ -13 & 17 & | & 19 \end{bmatrix}$

23. $\begin{bmatrix} -1 & 2 & 3 & | & 8 \\ 2 & -3 & 9 & | & 16 \\ 4 & -5 & -6 & | & 32 \end{bmatrix}$

25. $\begin{cases} x + 2y - 3z = 4 \\ -2x - 3y + z = 5 \\ 3x - 3y + 2z = 7 \end{cases}$

27. $\begin{cases} x - y + z = 2 \\ 2x + y - 3z = 6 \end{cases}$

29.(i) $\begin{bmatrix} 2 & 3 & 5 \\ 1 & 2 & 3 \end{bmatrix} \xrightarrow{R_1 \leftrightarrow R_2} \begin{bmatrix} 1 & 2 & 3 \\ 2 & 3 & 5 \end{bmatrix}$

(ii) $\begin{bmatrix} 1 & 2 & 3 \\ 2 & 3 & 5 \end{bmatrix} \xrightarrow{-2R_1 + R_2 \to R_2} \begin{bmatrix} 1 & 2 & 3 \\ 0 & -1 & -1 \end{bmatrix}$

(iii) $\begin{bmatrix} 1 & 2 & 3 \\ 0 & -1 & -1 \end{bmatrix} \xrightarrow{-R_2} \begin{bmatrix} 1 & 2 & 3 \\ 0 & 1 & 1 \end{bmatrix}$

31.(i) $\begin{bmatrix} 1 & 2 & 3 & 4 \\ 0 & 4 & -3 & 11 \\ 0 & 1 & 5 & -3 \end{bmatrix} \xrightarrow{R_2 \leftrightarrow R_3}$
$\begin{bmatrix} 1 & 2 & 3 & 4 \\ 0 & 1 & 5 & -3 \\ 0 & 4 & -3 & 11 \end{bmatrix}$

(ii) $\begin{bmatrix} 1 & 2 & 3 & 4 \\ 0 & 1 & 5 & -3 \\ 0 & 4 & -3 & 11 \end{bmatrix} \xrightarrow{-4R_2 + R_3 \to R_3}$
$\begin{bmatrix} 1 & 2 & 3 & 4 \\ 0 & 1 & 5 & -3 \\ 0 & 0 & -23 & 23 \end{bmatrix}$

(iii) $\begin{bmatrix} 1 & 2 & 3 & 4 \\ 0 & 1 & 5 & -3 \\ 0 & 0 & -23 & 23 \end{bmatrix} \xrightarrow{-\frac{1}{23}R_3}$
$\begin{bmatrix} 1 & 2 & 3 & 4 \\ 0 & 1 & 5 & -3 \\ 0 & 0 & 1 & -1 \end{bmatrix}$

33. $\begin{bmatrix} 4 & 5 & -7 \\ 5 & 4 & -2 \end{bmatrix} \xrightarrow{\frac{1}{4}R_1} \begin{bmatrix} 1 & \frac{5}{4} & -\frac{7}{4} \\ 5 & 4 & -2 \end{bmatrix}$

$\xrightarrow{-5R_1 + R_2 \to R_2} \begin{bmatrix} 1 & \frac{5}{4} & -\frac{7}{4} \\ 0 & -\frac{9}{4} & \frac{27}{4} \end{bmatrix}$

$\xrightarrow{-\frac{4}{9}R_2 \to R_2} \begin{bmatrix} 1 & \frac{5}{4} & -\frac{7}{4} \\ 0 & 1 & -3 \end{bmatrix}$

35. $\begin{bmatrix} 1 & 4 & 3 & 1 \\ 0 & -3 & -2 & 0 \\ 0 & 7 & 5 & -3 \end{bmatrix} \xrightarrow{-\frac{1}{3}R_2} \begin{bmatrix} 1 & 4 & 3 & 1 \\ 0 & 1 & \frac{2}{3} & 0 \\ 0 & 7 & 5 & -3 \end{bmatrix}$

$\xrightarrow{-7R_2 + R_3 \to R_3} \begin{bmatrix} 1 & 4 & 3 & 1 \\ 0 & 1 & \frac{2}{3} & 0 \\ 0 & 0 & \frac{1}{3} & -3 \end{bmatrix}$

$\xrightarrow{3R_3} \begin{bmatrix} 1 & 4 & 3 & 1 \\ 0 & 1 & \frac{2}{3} & 0 \\ 0 & 0 & 1 & -9 \end{bmatrix}$

37. No. The matrix does not have a step-like pattern that moves down and to the right. (Property 2)

39. Yes. The matrix is in reduced row-echelon form.

41. Yes. The matrix is in reduced row-echelon form.

43. Yes. The matrix is in reduced row-echelon form.

45. $\begin{cases} x + 2y = 1 \\ y = -2 \end{cases} \Rightarrow x + 2(-2) = 1 \Rightarrow x = 5$

The solution is $\{(5, -2)\}$.

47. $\begin{cases} x + 4y + 2z = 2 \\ z = 3 \end{cases} \Rightarrow x + 4y = -4 \Rightarrow x = -4y - 4$

The solution is $\{(-4y - 4, y, 3)\}$.

49. $\begin{cases} x + 2y + 3z = 2 \\ y - 2z = 4 \\ z = -1 \end{cases} \Rightarrow y - 2(-1) = 4 \Rightarrow y = 2$

$x + 2(2) + 3(-1) = 2 \Rightarrow x = 1$
The solution is $\{(1, 2, -1)\}$.

51. $\begin{cases} x = 2 \\ y = -5 \\ z + 2w = 3 \end{cases} \Rightarrow z = -2w + 3$

The solution is $\{(2, -5, -2w + 3, w)\}$.

53. $\begin{cases} x = -5 \\ y = 4 \\ z + 2w = 3 \\ w = 0 \end{cases} \Rightarrow z + 2(0) = 3 \Rightarrow z = 3$

The solution is $\{(-5, 4, 3, 0)\}$.

55. $\begin{bmatrix} 1 & -2 & | & 11 \\ 2 & -1 & | & 13 \end{bmatrix} \xrightarrow{-2R_1 + R_2 \to R_2} \begin{bmatrix} 1 & -2 & | & 11 \\ 0 & 3 & | & -9 \end{bmatrix}$

$\xrightarrow{\frac{1}{3}R_2 \to R_2} \begin{bmatrix} 1 & -2 & | & 11 \\ 0 & 1 & | & -3 \end{bmatrix} \Rightarrow$

$y = -3; x - 2(-3) = 11 \Rightarrow x = 5$
The solution is $\{(5, -3)\}$.

57. $\begin{bmatrix} 2 & -3 & | & 3 \\ 4 & -1 & | & 11 \end{bmatrix} \xrightarrow{\frac{1}{2}R_1 \to R_1} \begin{bmatrix} 1 & -\frac{3}{2} & | & \frac{3}{2} \\ 4 & -1 & | & 11 \end{bmatrix}$

$\xrightarrow{-4R_1 + R_2 \to R_2} \begin{bmatrix} 1 & -\frac{3}{2} & | & \frac{3}{2} \\ 0 & 5 & | & 5 \end{bmatrix}$

$\xrightarrow{\frac{1}{5}R_2 \to R_2} \begin{bmatrix} 1 & -\frac{3}{2} & | & \frac{3}{2} \\ 0 & 1 & | & 1 \end{bmatrix} \Rightarrow$

$y = 1; x - \frac{3}{2}(1) = \frac{3}{2} \Rightarrow x = 3$
The solution is $\{(3, 1)\}$.

59. $\begin{bmatrix} 3 & -5 & | & 4 \\ 4 & -15 & | & 13 \end{bmatrix} \xrightarrow{\frac{1}{3}R_1 \to R_1} \begin{bmatrix} 1 & -\frac{5}{3} & | & \frac{4}{3} \\ 4 & -15 & | & 13 \end{bmatrix}$

$\xrightarrow{-4R_1 + R_2 \to R_2} \begin{bmatrix} 1 & -\frac{5}{3} & | & \frac{4}{3} \\ 0 & -\frac{25}{3} & | & \frac{23}{3} \end{bmatrix}$

$\xrightarrow{-\frac{3}{25}R_2 \to R_2} \begin{bmatrix} 1 & -\frac{5}{3} & | & \frac{4}{3} \\ 0 & 1 & | & -\frac{23}{25} \end{bmatrix} \Rightarrow$

$y = -\frac{23}{25}; x - \left(\frac{5}{3}\right)\left(-\frac{23}{25}\right) = \frac{4}{3} \Rightarrow x = -\frac{1}{5}$

The solution is $\left\{\left(-\frac{1}{5}, -\frac{23}{25}\right)\right\}$.

61. $\begin{bmatrix} 1 & -1 & | & 1 \\ 2 & 1 & | & 5 \\ 3 & -4 & | & 2 \end{bmatrix} \xrightarrow[-3R_1 + R_3 \to R_3]{-2R_1 + R_2 \to R_2} \begin{bmatrix} 1 & -1 & | & 1 \\ 0 & 3 & | & 3 \\ 0 & -1 & | & -1 \end{bmatrix}$

$\xrightarrow[-R_3 \to R_3]{\frac{1}{3}R_2 \to R_2} \begin{bmatrix} 1 & -1 & | & 1 \\ 0 & 1 & | & 1 \\ 0 & 1 & | & 1 \end{bmatrix} \Rightarrow$

$y = 1; x - 1 = 1 \Rightarrow x = 2$
The solution is $\{(2, 1)\}$.

63. $\begin{bmatrix} 1 & 1 & 1 & | & 6 \\ 1 & -1 & 1 & | & 2 \\ 2 & 1 & -1 & | & 1 \end{bmatrix}$

$\xrightarrow[-2R_1 + R_3 \to R_3]{R_1 - R_2 \to R_2} \begin{bmatrix} 1 & 1 & 1 & | & 6 \\ 0 & 2 & 0 & | & 4 \\ 0 & -1 & -3 & | & -11 \end{bmatrix}$

$\xrightarrow{\frac{1}{2}R_2 \to R_2} \begin{bmatrix} 1 & 1 & 1 & | & 6 \\ 0 & 1 & 0 & | & 2 \\ 0 & -1 & -3 & | & -11 \end{bmatrix}$

$\xrightarrow{R_2 + R_3 \to R_3} \begin{bmatrix} 1 & 1 & 1 & | & 6 \\ 0 & 1 & 0 & | & 2 \\ 0 & 0 & -3 & | & -9 \end{bmatrix}$

$\xrightarrow{-\frac{1}{3}R_3 \to R_3} \begin{bmatrix} 1 & 1 & 1 & | & 6 \\ 0 & 1 & 0 & | & 2 \\ 0 & 0 & 1 & | & 3 \end{bmatrix} \Rightarrow$

$z = 3; y = 2; x + 2 + 3 = 6 \Rightarrow x = 1$
The solution is $\{(1, 2, 3)\}$.

65. $\begin{bmatrix} 2 & 3 & -1 & | & 9 \\ 1 & 1 & 1 & | & 9 \\ 3 & -1 & -1 & | & -1 \end{bmatrix}$

$\xrightarrow{R_1 \leftrightarrow R_2} \begin{bmatrix} 1 & 1 & 1 & | & 9 \\ 2 & 3 & -1 & | & 9 \\ 3 & -1 & -1 & | & -1 \end{bmatrix}$

$\xrightarrow[-3R_1+R_3 \to R_3]{-2R_1+R_2 \to R_2} \begin{bmatrix} 1 & 1 & 1 & | & 9 \\ 0 & 1 & -3 & | & -9 \\ 0 & -4 & -4 & | & -28 \end{bmatrix}$

$\xrightarrow{4R_2+R_3 \to R_3} \begin{bmatrix} 1 & 1 & 1 & | & 9 \\ 0 & 1 & -3 & | & -9 \\ 0 & 0 & -16 & | & -64 \end{bmatrix}$

$\xrightarrow{-\frac{1}{16}R_3 \to R_3} \begin{bmatrix} 1 & 1 & 1 & | & 9 \\ 0 & 1 & -3 & | & -9 \\ 0 & 0 & 1 & | & 4 \end{bmatrix} \Rightarrow$

$z = 4; y - 3(4) = -9 \Rightarrow y = 3$
$x + 3 + 4 = 9 \Rightarrow x = 2$
The solution is $\{(2, 3, 4)\}$.

67. $\begin{bmatrix} 3 & 2 & 4 & | & 19 \\ 2 & -1 & 1 & | & 3 \\ 6 & 7 & -1 & | & 17 \end{bmatrix}$

$\xrightarrow{R_1 \leftrightarrow \frac{1}{2}R_2} \begin{bmatrix} 1 & -\frac{1}{2} & \frac{1}{2} & | & \frac{3}{2} \\ 3 & 2 & 4 & | & 19 \\ 6 & 7 & -1 & | & 17 \end{bmatrix}$

$\xrightarrow[-6R_1+R_3 \to R_3]{-3R_1+R_2 \to R_2} \begin{bmatrix} 1 & -\frac{1}{2} & \frac{1}{2} & | & \frac{3}{2} \\ 0 & \frac{7}{2} & \frac{5}{2} & | & \frac{29}{2} \\ 0 & 10 & -4 & | & 8 \end{bmatrix}$

$\xrightarrow{\frac{2}{7}R_2 \to R_2} \begin{bmatrix} 1 & -\frac{1}{2} & \frac{1}{2} & | & \frac{3}{2} \\ 0 & 1 & \frac{5}{7} & | & \frac{29}{7} \\ 0 & 10 & -4 & | & 8 \end{bmatrix}$

$\xrightarrow{-10R_2+R_3 \to R_3} \begin{bmatrix} 1 & -\frac{1}{2} & \frac{1}{2} & | & \frac{3}{2} \\ 0 & 1 & \frac{5}{7} & | & \frac{29}{7} \\ 0 & 0 & -\frac{78}{7} & | & -\frac{234}{7} \end{bmatrix}$

$\xrightarrow{-\frac{7}{78}R_3 \to R_3} \begin{bmatrix} 1 & -\frac{1}{2} & \frac{1}{2} & | & \frac{3}{2} \\ 0 & 1 & \frac{5}{7} & | & \frac{29}{7} \\ 0 & 0 & 1 & | & 3 \end{bmatrix} \Rightarrow$

$z = 3$
$y + \left(\frac{5}{7}\right)(3) = \frac{29}{7} \Rightarrow y = 2$
$x - \frac{1}{2}(2) + \frac{1}{2}(3) = \frac{3}{2} \Rightarrow x = 1$
The solution is $\{(1, 2, 3)\}$.

69. $\begin{bmatrix} 1 & -1 & 0 & | & 1 \\ 1 & 0 & -1 & | & -1 \\ 2 & 1 & -1 & | & 3 \end{bmatrix}$

$\xrightarrow[-2R_1+R_3 \to R_3]{R_1-R_2 \to R_2} \begin{bmatrix} 1 & -1 & 0 & | & 1 \\ 0 & -1 & 1 & | & 2 \\ 0 & 3 & -1 & | & 1 \end{bmatrix}$

$\xrightarrow[3R_2+R_3 \to R_3]{-R_2+R_1 \to R_1} \begin{bmatrix} 1 & 0 & -1 & | & -1 \\ 0 & -1 & 1 & | & 2 \\ 0 & 0 & 2 & | & 7 \end{bmatrix}$

$\xrightarrow{\frac{1}{2}R_3 \to R_3} \begin{bmatrix} 1 & 0 & -1 & | & -1 \\ 0 & -1 & 1 & | & 2 \\ 0 & 0 & 1 & | & \frac{7}{2} \end{bmatrix}$

$\xrightarrow[-R_2+R_3 \to R_2]{R_3+R_1 \to R_1} \begin{bmatrix} 1 & 0 & 0 & | & \frac{5}{2} \\ 0 & 1 & 0 & | & \frac{3}{2} \\ 0 & 0 & 1 & | & \frac{7}{2} \end{bmatrix} \Rightarrow$

$x = \frac{5}{2}, y = \frac{3}{2}, z = \frac{7}{2}$
The solution is $\left\{\left(\frac{5}{2}, \frac{3}{2}, \frac{7}{2}\right)\right\}$.

71. $\begin{bmatrix} 1 & 1 & -1 & | & 4 \\ 1 & 3 & 5 & | & 10 \\ 3 & 5 & 3 & | & 18 \end{bmatrix}$

$\xrightarrow[-3R_1+R_3 \to R_3]{R_1-R_2 \to R_2} \begin{bmatrix} 1 & 1 & -1 & | & 4 \\ 0 & -2 & -6 & | & -6 \\ 0 & 2 & 6 & | & 6 \end{bmatrix}$

$\xrightarrow{R_2+R_3 \to R_3} \begin{bmatrix} 1 & 1 & -1 & | & 4 \\ 0 & -2 & -6 & | & -6 \\ 0 & 0 & 0 & | & 0 \end{bmatrix}$

(*continued on next page*)

(continued)

$$\xrightarrow{-\frac{1}{2}R_2 \to R_2} \begin{bmatrix} 1 & 1 & -1 & | & 4 \\ 0 & 1 & 3 & | & 3 \\ 0 & 0 & 0 & | & 0 \end{bmatrix}$$

$$\xrightarrow{R_1 - R_2 \to R_1} \begin{bmatrix} 1 & 0 & -4 & | & 1 \\ 0 & 1 & 3 & | & 3 \\ 0 & 0 & 0 & | & 0 \end{bmatrix} \Rightarrow$$

$x - 4z = 1 \Rightarrow x = 1 + 4z$
$y + 3z = 3 \Rightarrow y = 3 - 3z$
The solution is $\{(1+4z, 3-3z, z)\}$.

73. $\begin{bmatrix} 1 & 2 & -1 & | & 6 \\ 3 & 1 & 2 & | & 3 \\ 2 & 5 & 3 & | & 9 \end{bmatrix}$

$$\xrightarrow[-2R_1 + R_3 \to R_3]{-3R_1 + R_2 \to R_2} \begin{bmatrix} 1 & 2 & -1 & | & 6 \\ 0 & -5 & 5 & | & -15 \\ 0 & 1 & 5 & | & -3 \end{bmatrix}$$

$$\xrightarrow{-\frac{1}{5}R_2 \to R_2} \begin{bmatrix} 1 & 2 & -1 & | & 6 \\ 0 & 1 & -1 & | & 3 \\ 0 & 1 & 5 & | & -3 \end{bmatrix}$$

$$\xrightarrow[R_2 - R_3 \to R_3]{-2R_2 + R_1 \to R_1} \begin{bmatrix} 1 & 0 & 1 & | & 0 \\ 0 & 1 & -1 & | & 3 \\ 0 & 0 & -6 & | & 6 \end{bmatrix}$$

$$\xrightarrow{-\frac{1}{6}R_3 \to R_3} \begin{bmatrix} 1 & 0 & 1 & | & 0 \\ 0 & 1 & -1 & | & 3 \\ 0 & 0 & 1 & | & -1 \end{bmatrix}$$

$$\xrightarrow[R_2 + R_3 \to R_2]{R_1 - R_3 \to R_1} \begin{bmatrix} 1 & 0 & 0 & | & 1 \\ 0 & 1 & 0 & | & 2 \\ 0 & 0 & 1 & | & -1 \end{bmatrix} \Rightarrow$$

$x = 1, y = 2, z = -1$
The solution is $\{(1, 2, -1)\}$.

75. **a.** There is one solution: $\{(2, 3)\}$.

b. There are infinitely many solutions:
$\{(-2y + 2, y, 2)\}$.

c. There is no solution.

77. False. Each row of matrix A has seven entries.

9.1 Applying the Concepts

79. **a.** $\begin{cases} a = 0.1b + 1000 \\ b = 0.2a + 780 \end{cases}$

b. Rewrite the system as
$\begin{cases} a - 0.1b = 1000 \\ -0.2a + b = 780 \end{cases} \Rightarrow \begin{bmatrix} 1 & -0.1 & | & 1000 \\ -0.2 & 1 & | & 780 \end{bmatrix}$

c. $\begin{bmatrix} 1 & -0.1 & | & 1000 \\ -0.2 & 1 & | & 780 \end{bmatrix}$

$$\xrightarrow{0.2R_1 + R_2 \to R_2} \begin{bmatrix} 1 & -0.1 & | & 1000 \\ 0 & 0.98 & | & 980 \end{bmatrix}$$

$$\xrightarrow{\frac{100}{98} R_2 \to R_2} \begin{bmatrix} 1 & -0.1 & | & 1000 \\ 0 & 1 & | & 1000 \end{bmatrix}$$

$$\xrightarrow{0.1R_2 + R_1 \to R_1} \begin{bmatrix} 1 & 0 & | & 1100 \\ 0 & 1 & | & 1000 \end{bmatrix} \Rightarrow$$

$a = 1100, b = 1000$

81. **a.** $\begin{cases} l = 0.4t + 0.2f + 10,000 \\ t = 0.5l + 0.3t + 20,000 \\ f = 0.5l + 0.05t + 0.35f + 10,000 \end{cases}$

b. Rewrite the system as
$\begin{cases} l - 0.4t - 0.2f = 10,000 \\ -0.5l + 0.7t - 0f = 20,000 \\ -0.5l - 0.05t + 0.65f = 10,000 \end{cases} \Rightarrow$

$\begin{bmatrix} 1 & -0.4 & -0.2 & | & 10,000 \\ -0.5 & 0.7 & 0 & | & 20,000 \\ -0.5 & -0.05 & 0.65 & | & 10,000 \end{bmatrix}$

c. $\begin{bmatrix} 1 & -0.4 & -0.2 & | & 10,000 \\ -0.5 & 0.7 & 0 & | & 20,000 \\ -0.5 & -0.05 & 0.65 & | & 10,000 \end{bmatrix}$

$$\xrightarrow[R_1 + 2R_3 \to R_3]{R_1 + 2R_2 \to R_2} \begin{bmatrix} 1 & -0.4 & -0.2 & | & 10,000 \\ 0 & 1 & -0.2 & | & 50,000 \\ 0 & -0.5 & 1.1 & | & 30,000 \end{bmatrix}$$

$$\xrightarrow{R_2 + 2R_3 \to R_3} \begin{bmatrix} 1 & -0.4 & -0.2 & | & 10,000 \\ 0 & 1 & -0.2 & | & 50,000 \\ 0 & 0 & 2 & | & 110,000 \end{bmatrix}$$

$$\xrightarrow[-\frac{1}{2}R_3 \to R_3]{R_1 + 0.4R_2 \to R_1} \begin{bmatrix} 1 & 0 & -0.28 & | & 30,000 \\ 0 & 1 & -0.2 & | & 50,000 \\ 0 & 0 & 1 & | & 55,000 \end{bmatrix}$$

$$\xrightarrow[R_2 + 0.2R_3 \to R_2]{R_1 + 0.28R_3 \to R_1} \begin{bmatrix} 1 & 0 & 0 & | & 45,400 \\ 0 & 1 & 0 & | & 61,000 \\ 0 & 0 & 1 & | & 55,000 \end{bmatrix}$$

$l = \$45,400, t = \$61,000, f = \$55,000$

83. $T_1 = \dfrac{0 + 0 + 300 + T_2}{4} \Rightarrow 4T_1 - T_2 = 300$

$T_2 = \dfrac{T_1 + T_3 + 190 + 200}{4} \Rightarrow -T_1 + 4T_2 - T_3 = 390$

$T_3 = \dfrac{0 + 0 + 100 + T_2}{4} \Rightarrow -T_2 + 4T_3 = 100$

(continued on next page)

(continued)

$$\begin{bmatrix} 4 & -1 & 0 & | & 300 \\ -1 & 4 & -1 & | & 390 \\ 0 & -1 & 4 & | & 100 \end{bmatrix}$$

$\xrightarrow{R_1 \leftrightarrow -R_2}$ $\begin{bmatrix} 1 & -4 & 1 & | & -390 \\ 4 & -1 & 0 & | & 300 \\ 0 & -1 & 4 & | & 100 \end{bmatrix}$

$\xrightarrow{-4R_1 + R_2 \to R_2}$ $\begin{bmatrix} 1 & -4 & 1 & | & -390 \\ 0 & 15 & -4 & | & 1860 \\ 0 & -1 & 4 & | & 100 \end{bmatrix}$

$\xrightarrow{R_2 + R_3 \to R_2}$ $\begin{bmatrix} 1 & -4 & 1 & | & -390 \\ 0 & 14 & 0 & | & 1960 \\ 0 & -1 & 4 & | & 100 \end{bmatrix}$

$\xrightarrow{\frac{1}{14}R_2 \to R_2}$ $\begin{bmatrix} 1 & -4 & 1 & | & -390 \\ 0 & 1 & 0 & | & 140 \\ 0 & -1 & 4 & | & 100 \end{bmatrix}$

$\xrightarrow{R_2 + R_3 \to R_3}$ $\begin{bmatrix} 1 & -4 & 1 & | & -390 \\ 0 & 1 & 0 & | & 140 \\ 0 & 0 & 4 & | & 240 \end{bmatrix}$

$\xrightarrow[\frac{1}{4}R_3 \to R_3]{R_1 + 4R_2 \to R_1}$ $\begin{bmatrix} 1 & 0 & 1 & | & 170 \\ 0 & 1 & 0 & | & 140 \\ 0 & 0 & 1 & | & 60 \end{bmatrix}$

$\xrightarrow{R_1 - R_3 \to R_1}$ $\begin{bmatrix} 1 & 0 & 0 & | & 110 \\ 0 & 1 & 0 & | & 140 \\ 0 & 0 & 1 & | & 60 \end{bmatrix} \Rightarrow$

$T_1 = 110, T_2 = 140, T_3 = 60$

85. a. $\begin{cases} x - y & = 270 - 200 \\ -x + z = 180 - 300 \\ y - z = 40 + 70 - 60 \end{cases}$

$\Rightarrow \begin{cases} x - y = 70 \\ -x + z = -120 \\ y - z = 50 \end{cases}$

b. $\begin{bmatrix} 1 & -1 & 0 & | & 70 \\ -1 & 0 & 1 & | & -120 \\ 0 & 1 & -1 & | & 50 \end{bmatrix}$

$\xrightarrow{R_1 + R_2 \to R_2}$ $\begin{bmatrix} 1 & -1 & 0 & | & 70 \\ 0 & -1 & 1 & | & -50 \\ 0 & 1 & -1 & | & 50 \end{bmatrix}$

$\xrightarrow[R_2 + R_3 \to R_3]{R_1 - R_2 \to R_1}$ $\begin{bmatrix} 1 & 0 & -1 & | & 120 \\ 0 & -1 & 1 & | & -50 \\ 0 & 0 & 0 & | & 0 \end{bmatrix} \Rightarrow$

$x - z = 120 \Rightarrow x = z + 120$
$-y + z = -50 \Rightarrow y = z + 50$
The solution is $\{(z + 120, z + 50, z)\}$.
There are $300 + 200 + 60 = 560$ cars entering the system, so
$0 \le (z + 120) + (z + 50) + z \le 560$
$0 \le 3z + 170 \le 560$
$0 \le 3z \le 390$
$0 \le z \le 130$
Thus, the system has 131 solutions.

87. $9 = a(-1)^2 + b(-1) + c = a - b + c$
$3 = a(1)^2 + b(1) + c = a + b + c$
$6 = a(2)^2 + b(2) + c = 4a + 2b + c$

$\begin{bmatrix} 1 & -1 & 1 & | & 9 \\ 1 & 1 & 1 & | & 3 \\ 4 & 2 & 1 & | & 6 \end{bmatrix}$

$\xrightarrow[-4R_1 + R_3 \to R_3]{-R_1 + R_2 \to R_2}$ $\begin{bmatrix} 1 & -1 & 1 & | & 9 \\ 0 & 2 & 0 & | & -6 \\ 0 & 6 & -3 & | & -30 \end{bmatrix}$

$\xrightarrow[-\frac{1}{3}R_3 \to R_3]{\frac{1}{2}R_2 \to R_2}$ $\begin{bmatrix} 1 & -1 & 1 & | & 9 \\ 0 & 1 & 0 & | & -3 \\ 0 & -2 & 1 & | & 10 \end{bmatrix}$

$\xrightarrow{2R_2 + R_3 \to R_3}$ $\begin{bmatrix} 1 & -1 & 1 & | & 9 \\ 0 & 1 & 0 & | & -3 \\ 0 & 0 & 1 & | & 4 \end{bmatrix}$

$\xrightarrow{R_1 + R_2 \to R_1}$ $\begin{bmatrix} 1 & 0 & 1 & | & 6 \\ 0 & 1 & 0 & | & -3 \\ 0 & 0 & 1 & | & 4 \end{bmatrix}$

$\xrightarrow{R_1 - R_3 \to R_1}$ $\begin{bmatrix} 1 & 0 & 0 & | & 2 \\ 0 & 1 & 0 & | & -3 \\ 0 & 0 & 1 & | & 4 \end{bmatrix}$

Thus, $a = 2$, $b = -3$, and $c = 4$. The equation is $y = 2x^2 - 3x + 4$.

89. $-3 = a + b\sin\left(\frac{\pi}{4}\right) + c\sin\left(2\left(\frac{\pi}{4}\right)\right) \Rightarrow$

$-3 = a + \frac{\sqrt{2}}{2}b + c \Rightarrow -6 = 2a + \sqrt{2}b + 2c$

$2 = a + b\sin\left(\frac{3\pi}{4}\right) + c\sin\left(2\left(\frac{3\pi}{4}\right)\right) \Rightarrow$

$2 = a + \frac{\sqrt{2}}{2}b - c \Rightarrow 4 = 2a + \sqrt{2}b - 2c$

(continued on next page)

(continued)

$5 = a + b\sin\left(\frac{7\pi}{4}\right) + c\sin\left(2\left(\frac{7\pi}{4}\right)\right) \Rightarrow$

$5 = a - \frac{\sqrt{2}}{2}b - c \Rightarrow 10 = 2a - \sqrt{2}b - 2c$

The system of equations is
$\begin{cases} 2a + \sqrt{2}b + 2c = -6 \\ 2a + \sqrt{2}b - 2c = 4 \\ 2a - \sqrt{2}b - 2c = 10 \end{cases}$

$\begin{bmatrix} 2 & \sqrt{2} & 2 & | & -6 \\ 2 & \sqrt{2} & -2 & | & 4 \\ 2 & -\sqrt{2} & -2 & | & 10 \end{bmatrix}$

$\xrightarrow[R_1 + R_2 \to R_2]{\frac{(R_1 + R_3)}{4} \to R_1} \begin{bmatrix} 1 & 0 & 0 & | & 1 \\ 4 & 2\sqrt{2} & 0 & | & -2 \\ 2 & -\sqrt{2} & -2 & | & 10 \end{bmatrix}$

$\xrightarrow[-2R_1 + R_3 \to R_3]{-4R_1 + R_2 \to R_2} \begin{bmatrix} 1 & 0 & 0 & | & 1 \\ 0 & 2\sqrt{2} & 0 & | & -6 \\ 0 & -\sqrt{2} & -2 & | & 8 \end{bmatrix}$

$\xrightarrow{\frac{R_2}{2\sqrt{2}} \to R_2} \begin{bmatrix} 1 & 0 & 0 & | & 1 \\ 0 & 1 & 0 & | & -\frac{3\sqrt{2}}{2} \\ 0 & -\sqrt{2} & -2 & | & 8 \end{bmatrix}$

$\xrightarrow{\frac{\sqrt{2}R_2 + R_3}{-2} \to R_3} \begin{bmatrix} 1 & 0 & 0 & | & 1 \\ 0 & 1 & 0 & | & -\frac{3\sqrt{2}}{2} \\ 0 & 0 & 1 & | & -\frac{5}{2} \end{bmatrix}$

The equation is $f(x) = 1 - \frac{3\sqrt{2}}{2}\sin x - \frac{5}{2}\sin(2x)$.

91. $\begin{cases} a(1)^3 + b(1)^2 + c(1) + d = 5 \\ a(-1)^3 + b(-1)^2 + c(-1) + d = 1 \\ a(2)^3 + b(2)^2 + c(2) + d = 7 \\ a(-2)^3 + b(-2)^2 + c(-2) + d = 11 \end{cases} \Rightarrow$

$\begin{cases} a + b + c + d = 5 \\ -a + b - c + d = 1 \\ 8a + 4b + 2c + d = 7 \\ -8a + 4b - 2c + d = 11 \end{cases}$

Using a graphing calculator, we have

```
[A]
[[1   1   1   1   5 ]
 [-1  1  -1   1   1 ]
 [8   4   2   1   7 ]
 [-8  4  -2   1  11]]
```
```
rref([A])
[[1  0  0  0  -1]
 [0  1  0  0   2]
 [0  0  1  0   3]
 [0  0  0  1   1]]
```

The equation is $y = -x^3 + 2x^2 + 3x + 1$.

93. For $1 \le x \le 4$, the system to be solved is
$\begin{cases} 4a + b = 4 \\ a + b = 2 \end{cases}$.

$\begin{bmatrix} 4 & 1 & | & 4 \\ 1 & 1 & | & 2 \end{bmatrix} \xrightarrow{\frac{R_1 - R_2}{3} \to R_1} \begin{bmatrix} 1 & 0 & | & \frac{2}{3} \\ 1 & 1 & | & 2 \end{bmatrix}$

$\xrightarrow{R_2 - R_1 \to R_2} \begin{bmatrix} 1 & 0 & | & \frac{2}{3} \\ 0 & 1 & | & \frac{4}{3} \end{bmatrix}$

The equation is $f(x) = \frac{2}{3}x + \frac{4}{3}$.

For $4 \le x \le 6$, the system to be solved is
$\begin{cases} 4a + b = 4 \\ 6a + b = 1 \end{cases}$.

$\begin{bmatrix} 4 & 1 & | & 4 \\ 6 & 1 & | & 1 \end{bmatrix} \xrightarrow{\frac{R_1 - R_2}{-2} \to R_1} \begin{bmatrix} 1 & 0 & | & -\frac{3}{2} \\ 6 & 1 & | & 1 \end{bmatrix}$

$\xrightarrow{R_2 - 6R_1 \to R_2} \begin{bmatrix} 1 & 0 & | & -\frac{3}{2} \\ 0 & 1 & | & 10 \end{bmatrix}$

The equation is $f(x) = -\frac{3}{2}x + 10$.

The piecewise linear function is

$f(x) = \begin{cases} \frac{2}{3}x + \frac{4}{3}, & 1 \le x \le 4 \\ -\frac{3}{2}x + 10, & 4 \le x \le 6 \end{cases}$

9.1 Beyond the Basics

95. $\begin{bmatrix} 1 & 1 & 1 & 1 & | & 0 \\ 1 & 3 & 2 & 4 & | & 0 \\ 2 & 0 & 1 & -1 & | & 0 \end{bmatrix}$

$\xrightarrow[-2R_1 + R_3 \to R_3]{R_1 - R_2 \to R_2} \begin{bmatrix} 1 & 1 & 1 & 1 & | & 0 \\ 0 & -2 & -1 & -3 & | & 0 \\ 0 & -2 & -1 & -3 & | & 0 \end{bmatrix}$

$\xrightarrow[R_2 - R_3 \to R_3]{\frac{1}{2}R_2 + R_1 \to R_1} \begin{bmatrix} 1 & 0 & \frac{1}{2} & -\frac{1}{2} & | & 0 \\ 0 & 1 & \frac{1}{2} & \frac{3}{2} & | & 0 \\ 0 & 0 & 0 & 0 & | & 0 \end{bmatrix}$

$\xrightarrow{-\frac{1}{2}R_2 \to R_2} \begin{bmatrix} 1 & 0 & \frac{1}{2} & -\frac{1}{2} & | & 0 \\ 0 & 1 & \frac{1}{2} & \frac{3}{2} & | & 0 \\ 0 & 0 & 0 & 0 & | & 0 \end{bmatrix} \Rightarrow$

(continued on next page)

(continued)

$y + \frac{1}{2}z + \frac{3}{2}w = 0 \Rightarrow z = -2y - 3w$

$x + \frac{1}{2}z - \frac{1}{2}w = 0 \Rightarrow z = w - 2x$

$-2y - 3w = w - 2x \Rightarrow x = 2w + y$

The solution is $\{(y + 2w, y, -2y - 3w, w)\}$.

97. a. $\begin{bmatrix} 1 & 2 & -3 & 1 \\ 1 & 0 & -3 & 2 \\ 0 & 1 & 1 & 0 \\ 2 & 3 & 0 & -2 \end{bmatrix}$

$\xrightarrow{R_1 \leftrightarrow R_4} \begin{bmatrix} 2 & 3 & 0 & -2 \\ 1 & 0 & -3 & 2 \\ 0 & 1 & 1 & 0 \\ 1 & 2 & -3 & 1 \end{bmatrix}$

$\xrightarrow[R_1 - 2R_4 \to R_4]{R_2 - R_4 \to R_2} \begin{bmatrix} 2 & 3 & 0 & -2 \\ 0 & -2 & 0 & 1 \\ 0 & 1 & 1 & 0 \\ 0 & -1 & 6 & -4 \end{bmatrix}$

$\xrightarrow[R_2 + 2R_3 \to R_3]{R_2 - 2R_4 \to R_4} \begin{bmatrix} 2 & 3 & 0 & -2 \\ 0 & -2 & 0 & 1 \\ 0 & 0 & 2 & 1 \\ 0 & 0 & -12 & 9 \end{bmatrix}$

$\xrightarrow{6R_3 + R_4 \to R_4} \begin{bmatrix} 2 & 3 & 0 & -2 \\ 0 & -2 & 0 & 1 \\ 0 & 0 & 2 & 1 \\ 0 & 0 & 0 & 15 \end{bmatrix}$

$\xrightarrow[\substack{\frac{1}{2}R_1 \to R_1 \\ -\frac{1}{2}R_2 \to R_2 \\ \frac{1}{2}R_3 \to R_3 \\ \frac{1}{15}R_4 \to R_4}]{} \begin{bmatrix} 1 & \frac{3}{2} & 0 & -1 \\ 0 & 1 & 0 & -\frac{1}{2} \\ 0 & 0 & 1 & \frac{1}{2} \\ 0 & 0 & 0 & 1 \end{bmatrix} = B$

$\begin{bmatrix} 1 & 2 & -3 & 1 \\ 1 & 0 & -3 & 2 \\ 0 & 1 & 1 & 0 \\ 2 & 3 & 0 & -2 \end{bmatrix}$

$\xrightarrow[2R_1 - R_4 \to R_4]{R_1 - R_2 \to R_2} \begin{bmatrix} 1 & 2 & -3 & 1 \\ 0 & 2 & 0 & -1 \\ 0 & 1 & 1 & 0 \\ 0 & 1 & -6 & 4 \end{bmatrix}$

$\xrightarrow{R_2 - 2R_4 \to R_4} \begin{bmatrix} 1 & 2 & -3 & 1 \\ 0 & 2 & 0 & -1 \\ 0 & 1 & 1 & 0 \\ 0 & 0 & 12 & -9 \end{bmatrix}$

$\xrightarrow[\frac{1}{12}R_4 \to R_4]{\frac{1}{2}R_2 \to R_2} \begin{bmatrix} 1 & 2 & -3 & 1 \\ 0 & 1 & 0 & -\frac{1}{2} \\ 0 & 1 & 1 & 0 \\ 0 & 0 & 1 & -\frac{3}{4} \end{bmatrix}$

$\xrightarrow{R_2 - R_3 \to R_3} \begin{bmatrix} 1 & 2 & -3 & 1 \\ 0 & 1 & 0 & -\frac{1}{2} \\ 0 & 0 & -1 & -\frac{1}{2} \\ 0 & 0 & 1 & -\frac{3}{4} \end{bmatrix}$

$\xrightarrow{R_3 + R_4 \to R_4} \begin{bmatrix} 1 & 2 & -3 & 1 \\ 0 & 1 & 0 & -\frac{1}{2} \\ 0 & 0 & -1 & -\frac{1}{2} \\ 0 & 0 & 0 & -\frac{5}{4} \end{bmatrix}$

$\xrightarrow[-\frac{4}{5}R_4 \to R_4]{-R_3 \to R_3} \begin{bmatrix} 1 & 2 & -3 & 1 \\ 0 & 1 & 0 & -\frac{1}{2} \\ 0 & 0 & 1 & \frac{1}{2} \\ 0 & 0 & 0 & 1 \end{bmatrix} = C$

b. Use a calculator to show that the reduced row-echelon form of all three matrices is

$\begin{bmatrix} 1 & 0 & 0 & 0 \\ 0 & 1 & 0 & 0 \\ 0 & 0 & 1 & 0 \\ 0 & 0 & 0 & 1 \end{bmatrix}$.

532 Chapter 9 Matrices and Determinants

99. a. $\begin{bmatrix} a & b & m \\ c & d & n \end{bmatrix} \xrightarrow[\frac{1}{c}R_2 \to R_2]{\frac{1}{a}R_1 \to R_1} \begin{bmatrix} 1 & \frac{b}{a} & \frac{m}{a} \\ 1 & \frac{d}{c} & \frac{n}{c} \end{bmatrix}$

$\xrightarrow{R_1 - R_2 \to R_2} \begin{bmatrix} 1 & \frac{b}{a} & \frac{m}{a} \\ 0 & \frac{b}{a} - \frac{d}{c} & \frac{m}{a} - \frac{n}{c} \end{bmatrix}$

$= \begin{bmatrix} 1 & \frac{b}{a} & \frac{m}{a} \\ 0 & \frac{bc - ad}{ac} & \frac{cm - an}{ac} \end{bmatrix}$

$\xrightarrow{\frac{ac}{bc-da}R_2 \to R_2} \begin{bmatrix} 1 & \frac{b}{a} & \frac{m}{a} \\ 0 & 1 & \frac{cm - an}{bc - ad} \end{bmatrix}$

$\xrightarrow{-\frac{b}{a}R_2 + R_1 \to R_1} \begin{bmatrix} 1 & 0 & \frac{dm - bn}{ad - bc} \\ 0 & 1 & \frac{cm - an}{bc - ad} \end{bmatrix}$

The solution is $\left(\dfrac{dm - bn}{ad - bc}, \dfrac{cm - an}{bc - ad} \right)$.

b. (i) There is a unique solution if $bc \neq ad$.

   (ii) There is no solution if $bc = ad$ and $\dfrac{m}{b} \neq \dfrac{n}{d}$.

   (iii) There are infinitely many solutions if $bc = ad$ and $\dfrac{m}{b} = \dfrac{n}{d}$.

101. Using the hint suggested, we have
$\begin{bmatrix} 1 & 1 & 1 & 6 \\ 3 & -1 & 3 & 10 \\ 5 & 5 & -4 & 3 \end{bmatrix}$

$\xrightarrow[-5R_1 + R_3 \to R_3]{-3R_1 + R_2 \to R_2} \begin{bmatrix} 1 & 1 & 1 & 6 \\ 0 & -4 & 0 & -8 \\ 0 & 0 & -9 & -27 \end{bmatrix}$

$\xrightarrow[-\frac{1}{9}R_3 \to R_3]{-\frac{1}{4}R_2 \to R_2} \begin{bmatrix} 1 & 1 & 1 & 6 \\ 0 & 1 & 0 & 2 \\ 0 & 0 & 1 & 3 \end{bmatrix} \Rightarrow$

$w = 3, v = 2, u + 2 + 3 = 6 \Rightarrow u = 1$
$\log x = 1 \Rightarrow x = 10;$
$\log y = 2 \Rightarrow y = 100$
$\log z = 3 \Rightarrow y = 1000$
The solution is $\{(10, 100, 1000)\}$.

103. Substitute the given points into the given system, subtract c from both sides of each equation, and then write and solve the augmented matrix.

$\begin{bmatrix} 2 & 1 & -1 & 0 \\ \frac{1}{3} & -\frac{1}{9} & -1 & 0 \\ -3 & -\frac{7}{3} & -1 & 0 \end{bmatrix}$

$\xrightarrow{\frac{R_1}{2} \to R_1} \begin{bmatrix} 1 & \frac{1}{2} & -\frac{1}{2} & 0 \\ \frac{1}{3} & -\frac{1}{9} & -1 & 0 \\ -3 & -\frac{7}{3} & -1 & 0 \end{bmatrix}$

$\xrightarrow[3R_1 + R_3 \to R_3]{R_1 - 3R_2 \to R_2} \begin{bmatrix} 1 & \frac{1}{2} & -\frac{1}{2} & 0 \\ 0 & \frac{5}{6} & \frac{5}{2} & 0 \\ 0 & -\frac{5}{6} & -\frac{5}{2} & 0 \end{bmatrix}$

$\xrightarrow[\frac{5}{3}R_1 + R_3 \to R_3]{\substack{R_1 + \frac{1}{5}R_2 \to R_1 \\ R_2 + R_3 \to R_2}} \begin{bmatrix} 1 & \frac{2}{3} & 0 & 0 \\ 0 & 0 & 0 & 0 \\ \frac{5}{3} & 0 & -\frac{10}{3} & 0 \end{bmatrix}$

$\xrightarrow{\frac{3}{5}R_3 \to R_3} \begin{bmatrix} 1 & \frac{2}{3} & 0 & 0 \\ 0 & 0 & 0 & 0 \\ 1 & 0 & -2 & 0 \end{bmatrix}$

This yields the system
$\begin{cases} a + \frac{2}{3}b = 0 \\ a - 2c = 0 \end{cases} \Rightarrow \begin{cases} b = -\frac{3}{2}a \\ c = \frac{1}{2}a \end{cases}$.

Because this holds for values other than $a = b = c = 0$, the points are collinear.

105. Substitute the given points into the given system, subtract d from both sides of each equation, and then write and solve the augmented matrix.

$\begin{bmatrix} 2 & 1 & -\frac{5}{4} & -1 & 0 \\ -1 & 0 & -\frac{3}{2} & -1 & 0 \\ 4 & 3 & -\frac{7}{4} & -1 & 0 \\ 1 & -2 & 0 & -1 & 0 \end{bmatrix}$

$\xrightarrow[\frac{-3R_1 + R_3}{-2} \to R_3]{2R_1 + R_4 \to R_4} \begin{bmatrix} 2 & 1 & -\frac{5}{4} & -1 & 0 \\ -1 & 0 & -\frac{3}{2} & -1 & 0 \\ 1 & 0 & -1 & -1 & 0 \\ 5 & 0 & -\frac{5}{2} & -3 & 0 \end{bmatrix}$

(continued on next page)

(continued)

$$\xrightarrow{2\left(-\frac{5}{2}R_3+R_4\right)\to R_4} \begin{bmatrix} 2 & 1 & -\frac{5}{4} & -1 & 0 \\ -1 & 0 & -\frac{3}{2} & -1 & 0 \\ 1 & 0 & -1 & -1 & 0 \\ 5 & 0 & 0 & -1 & 0 \end{bmatrix}$$

$$\xrightarrow[\frac{-R_3+R_4}{4}\to R_3]{R_1-R_2\to R_2} \begin{bmatrix} 2 & 1 & -\frac{5}{4} & -1 & 0 \\ 3 & 1 & \frac{1}{4} & 0 & 0 \\ 1 & 0 & \frac{1}{4} & 0 & 0 \\ 5 & 0 & 0 & -1 & 0 \end{bmatrix}$$

$$\xrightarrow{R_2-R_3\to R_2} \begin{bmatrix} 2 & 1 & -\frac{5}{4} & -1 & 0 \\ 2 & 1 & 0 & 0 & 0 \\ 1 & 0 & \frac{1}{4} & 0 & 0 \\ 5 & 0 & 0 & -1 & 0 \end{bmatrix}$$

$$\xrightarrow{R_1-R_2\to R_1} \begin{bmatrix} 0 & 0 & -\frac{5}{4} & -1 & 0 \\ 2 & 1 & 0 & 0 & 0 \\ 1 & 0 & \frac{1}{4} & 0 & 0 \\ 5 & 0 & 0 & -1 & 0 \end{bmatrix}$$

$$\xrightarrow{R_1+5R_3\to R_1} \begin{bmatrix} 5 & 0 & 0 & -1 & 0 \\ 2 & 1 & 0 & 0 & 0 \\ 1 & 0 & \frac{1}{4} & 0 & 0 \\ 5 & 0 & 0 & -1 & 0 \end{bmatrix}$$

$$\xrightarrow{R_1-R_4\to R_1} \begin{bmatrix} 0 & 0 & 0 & 0 & 0 \\ 2 & 1 & 0 & 0 & 0 \\ 1 & 0 & \frac{1}{4} & 0 & 0 \\ 5 & 0 & 0 & -1 & 0 \end{bmatrix} \Rightarrow$$

$2a+b=0 \Rightarrow b=-2a$
$a+\frac{1}{4}c=0 \Rightarrow c=-4a$
$5a-d=0 \Rightarrow d=5a$
Because this holds for values other than $a=b=c=d=0$, the points are collinear.

107. Three points that satisfy the function are $(1, 1)$, $(2, 5)$, and $(3, 12)$. Find the equation by solving the system
$$\begin{cases} a(1)^2+b(1)+c=1 \\ a(2)^2+b(2)+c=5 \\ a(3)^2+b(3)+c=12 \end{cases} \Rightarrow \begin{cases} a+b+c=1 \\ 4a+2b+c=5 \\ 9a+3b+c=12 \end{cases}.$$

$$\begin{bmatrix} 1 & 1 & 1 & 1 \\ 4 & 2 & 1 & 5 \\ 9 & 3 & 1 & 12 \end{bmatrix}$$

$$\xrightarrow[-9R_1+R_3\to R_3]{-4R_1+R_2\to R_2} \begin{bmatrix} 1 & 1 & 1 & 1 \\ 0 & -2 & -3 & 1 \\ 0 & -6 & -8 & 3 \end{bmatrix}$$

$$\xrightarrow{-3R_2+R_3\to R_3} \begin{bmatrix} 1 & 1 & 1 & 1 \\ 0 & -2 & -3 & 1 \\ 0 & 0 & 1 & 0 \end{bmatrix}$$

$$\xrightarrow[\frac{R_2+3R_3}{-2}\to R_2]{R_1-R_3\to R_1} \begin{bmatrix} 1 & 1 & 0 & 1 \\ 0 & 1 & 0 & -\frac{1}{2} \\ 0 & 0 & 1 & 0 \end{bmatrix}$$

$$\xrightarrow{R_1-R_2\to R_1} \begin{bmatrix} 1 & 0 & 0 & \frac{3}{2} \\ 0 & 1 & 0 & -\frac{1}{2} \\ 0 & 0 & 1 & 0 \end{bmatrix}$$

The equation is $P(n)=\frac{3}{2}n^2-\frac{1}{2}n$.

9.1 Critical Thinking/Discussion/Writing

109. If $a \neq 0$,
$$\begin{bmatrix} a \\ b \end{bmatrix} \xrightarrow[\frac{1}{b}R_2\to R_2]{\frac{1}{a}R_1\to R_1} \begin{bmatrix} 1 \\ 1 \end{bmatrix} \xrightarrow{R_1-R_2\to R_2} \begin{bmatrix} 1 \\ 0 \end{bmatrix}.$$

If $a=0$, $b=0$, $\begin{bmatrix} a \\ b \end{bmatrix} = \begin{bmatrix} 0 \\ 0 \end{bmatrix}$.

110. If the matrix is $\begin{bmatrix} 0 & 0 \\ 0 & 0 \end{bmatrix}$, then it is in reduced row-echelon form already. If the matrix is $\begin{bmatrix} 1 & k \\ 0 & 0 \end{bmatrix}$ (for k any real number), then it is in reduced row-echelon form already.

If the matrix is $\begin{bmatrix} 0 & 1 \\ 0 & 0 \end{bmatrix}$ then it is in reduced row-echelon form already. If the matrix is $\begin{bmatrix} a & b \\ c & d \end{bmatrix}$, then we have

$$\begin{bmatrix} a & b \\ c & d \end{bmatrix} \xrightarrow[\frac{1}{c}R_2\to R_2]{\frac{1}{a}R_1\to R_1} \begin{bmatrix} 1 & \frac{b}{a} \\ 1 & \frac{d}{c} \end{bmatrix}$$

(continued on next page)

(*continued*)

$$\xrightarrow{R_1-R_2 \to R_2} \begin{bmatrix} 1 & \dfrac{b}{a} \\ 0 & \dfrac{b}{a}-\dfrac{d}{c} \end{bmatrix} = \begin{bmatrix} 1 & \dfrac{b}{a} \\ 0 & \dfrac{bc-ad}{ac} \end{bmatrix}$$

$$\xrightarrow{\frac{ac}{bc-ad} R_2 \to R_2} \begin{bmatrix} 1 & \dfrac{b}{a} \\ 0 & 1 \end{bmatrix}$$

$$\xrightarrow{-\frac{b}{a} R_2 + R_1 \to R_1} \begin{bmatrix} 1 & 0 \\ 0 & 1 \end{bmatrix}$$

111. Yes. Use the inverse of the operation used to transform A to B.

112. a. True. For example,
$$\begin{bmatrix} 1 & 0 & 0 & 6 \\ 0 & 1 & 0 & 3 \\ 0 & 0 & 1 & 2 \end{bmatrix} \to \begin{bmatrix} 1 & 0 & 0 & 6 \\ 0 & 0 & 1 & 2 \end{bmatrix}$$

b. False. For example,
$$\begin{bmatrix} 1 & 0 & 0 & 6 \\ 0 & 1 & 0 & 3 \\ 0 & 0 & 1 & 2 \end{bmatrix} \to \begin{bmatrix} 1 & 0 & 6 \\ 0 & 0 & 3 \\ 0 & 1 & 2 \end{bmatrix}$$

9.1 Maintaining Skills

113. $3(x-1) = 5 - x \Rightarrow 3x - 3 = 5 - x \Rightarrow$
$4x = 8 \Rightarrow x = 2$
Solution set: $\{2\}$

115. $-3(x+4) + 2 = 8 - x \Rightarrow -3x - 12 + 2 = 8 - x \Rightarrow$
$-3x - 10 = 8 - x \Rightarrow -2x = 18 \Rightarrow x = -9$
Solution set: $\{-9\}$

Be sure to check your solutions in the original equations in exercises 117–120.

117. $\begin{cases} 2x - y = 5 & (1) \\ x + 2y = 25 & (2) \end{cases}$

From equation (1) we have $y = 2x - 5$. Substituting this expression in equation (2) gives
$x + 2(2x - 5) = 25 \Rightarrow x + 4x - 10 = 25 \Rightarrow$
$5x = 35 \Rightarrow x = 7$
Substitute $x = 7$ in equation (1) and solve for y.
$2(7) - y = 5 \Rightarrow 14 - y = 5 \Rightarrow y = 9$
Solution set: $\{(7, 9)\}$

119. $\begin{cases} x + 3y = 6 & (1) \\ 2x + 6y = 8 & (2) \end{cases}$

Multiply equation (1) by -2, then add the two equations.

$-2x - 6y = -12$
$\underline{2x + 6y = 8}$
$0 = -4$ False
Solution set: $\varnothing$

121. True

123. True. This is an example of the distributive property.

9.2 Matrix Algebra

9.2 Practice Problems

1. $\begin{bmatrix} 1 & 2x - y \\ x + y & 5 \end{bmatrix} = \begin{bmatrix} 1 & 1 \\ 2 & 5 \end{bmatrix} \Rightarrow$
$\begin{cases} 2x - y = 1 \\ x + y = 2 \end{cases} \Rightarrow x = 1, \ y = 1$

2. $\begin{bmatrix} 2 & -1 & 4 \\ 5 & 0 & 9 \end{bmatrix} + \begin{bmatrix} -8 & 2 & 9 \\ 7 & 3 & 6 \end{bmatrix} = \begin{bmatrix} -6 & 1 & 13 \\ 12 & 3 & 15 \end{bmatrix}$

3. $2A - 3B = 2\begin{bmatrix} 7 & -4 \\ 3 & 6 \\ 0 & -2 \end{bmatrix} - 3\begin{bmatrix} 1 & -3 \\ 2 & 2 \\ 5 & 8 \end{bmatrix}$

$= \begin{bmatrix} 14 & -8 \\ 6 & 12 \\ 0 & -4 \end{bmatrix} - \begin{bmatrix} 3 & -9 \\ 6 & 6 \\ 15 & 24 \end{bmatrix}$

$= \begin{bmatrix} 11 & 1 \\ 0 & 6 \\ -15 & -28 \end{bmatrix}$

4. $5\begin{bmatrix} 1 & -1 \\ 3 & 5 \end{bmatrix} + 3X = 2\begin{bmatrix} 2 & 7 \\ -3 & -5 \end{bmatrix} \Rightarrow$

$\begin{bmatrix} 5 & -5 \\ 15 & 25 \end{bmatrix} + 3X = \begin{bmatrix} 4 & 14 \\ -6 & -10 \end{bmatrix} \Rightarrow$

$3X = \begin{bmatrix} 4 & 14 \\ -6 & -10 \end{bmatrix} - \begin{bmatrix} 5 & -5 \\ 15 & 25 \end{bmatrix} = \begin{bmatrix} -1 & 19 \\ -21 & -35 \end{bmatrix} \Rightarrow$

$X = \begin{bmatrix} -\dfrac{1}{3} & \dfrac{19}{3} \\ -7 & -\dfrac{35}{3} \end{bmatrix}$

5. Yes, the product matrix AB is defined. A is a 2×3 matrix, while B is a 3×1 matrix. The product matrix has order 2×1.

6. $N = \begin{bmatrix} S & C & M \end{bmatrix} = \begin{bmatrix} 10 & 30 & 45 \end{bmatrix}; P = \begin{bmatrix} 41 \\ 26 \\ 19 \end{bmatrix}$

$NP = \begin{bmatrix} 10 & 30 & 45 \end{bmatrix} \begin{bmatrix} 41 \\ 26 \\ 19 \end{bmatrix} = 10(41) + 30(26) + 45(19) = \$2045 \text{ thousand}$

7. $\begin{bmatrix} 3 & -1 & 2 & 7 \end{bmatrix} \begin{bmatrix} -2 \\ 0 \\ 1 \\ 5 \end{bmatrix} = \begin{bmatrix} 3(-2) - 1(0) + 2(1) + 7(5) \end{bmatrix} = \begin{bmatrix} 31 \end{bmatrix}$

8. AB is not defined because A is a 2×2 matrix and B is a 3×2 matrix.

$BA = \begin{bmatrix} 8 & 1 \\ -2 & 6 \\ 0 & 4 \end{bmatrix} \begin{bmatrix} 5 & 0 \\ 2 & -1 \end{bmatrix} = \begin{bmatrix} 8(5)+1(2) & 8(0)+1(-1) \\ -2(5)+6(2) & -2(0)+6(-1) \\ 0(5)+4(2) & 0(0)+4(-1) \end{bmatrix} = \begin{bmatrix} 42 & -1 \\ 2 & -6 \\ 8 & -4 \end{bmatrix}$

9. $AB = \begin{bmatrix} 7 & 1 \\ 0 & 3 \end{bmatrix} \begin{bmatrix} 2 & -1 \\ 4 & 4 \end{bmatrix} = \begin{bmatrix} 7(2)+1(4) & 7(-1)+1(4) \\ 0(2)+3(4) & 0(-1)+3(4) \end{bmatrix} = \begin{bmatrix} 18 & -3 \\ 12 & 12 \end{bmatrix}$

$BA = \begin{bmatrix} 2 & -1 \\ 4 & 4 \end{bmatrix} \begin{bmatrix} 7 & 1 \\ 0 & 3 \end{bmatrix} = \begin{bmatrix} 2(7)-1(0) & 2(1)-1(3) \\ 4(7)+4(0) & 4(1)+4(3) \end{bmatrix} = \begin{bmatrix} 14 & -1 \\ 28 & 16 \end{bmatrix}$

10. $AD = \begin{bmatrix} 0 & 1 \\ 1 & 0.25 \end{bmatrix} \begin{bmatrix} 0 & 4 & 4 & 1 & 1 & 0 \\ 0 & 0 & 1 & 1 & 6 & 6 \end{bmatrix}$

$= \begin{bmatrix} 0(0)+1(0) & 0(4)+1(0) & 0(4)+1(1) & 0(1)+1(1) & 0(1)+1(6) & 0(0)+1(6) \\ 1(0)+0.25(0) & 1(4)+0.25(0) & 1(4)+0.25(1) & 1(1)+0.25(1) & 1(1)+0.25(6) & 1(0)+0.25(6) \end{bmatrix}$

$= \begin{bmatrix} 0 & 0 & 1 & 1 & 6 & 6 \\ 0 & 4 & 4.25 & 1.25 & 2.5 & 1.5 \end{bmatrix}$

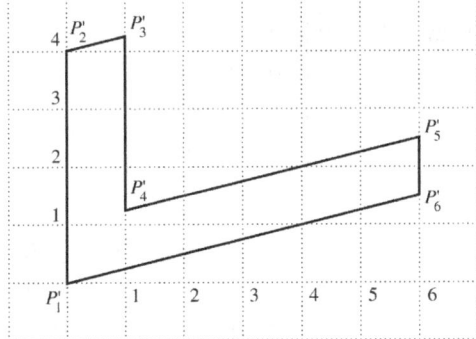

9.2 Concepts and Vocabulary

1. Two $m \times n$ matrices $A = \begin{bmatrix} a_{ij} \end{bmatrix}$ and $B = \begin{bmatrix} b_{ij} \end{bmatrix}$ are equal if $\underline{a_{ij} = b_{ij}}$ for all i and j.

3. The product of a $1 \times n$ matrix A and $n \times 1$ matrix B is $\underline{1 \times 1}$ matrix.

5. False. It is possible that $AB = BA$, but not necessarily true.

7. False. For example, suppose $A = \begin{bmatrix} 0 & 1 \\ 0 & 0 \end{bmatrix}$. Then $A^2 = \begin{bmatrix} 0(0)+1(0) & 0(1)+1(0) \\ 0(0)+0(0) & 0(0)+0(0) \end{bmatrix} = \begin{bmatrix} 0 & 0 \\ 0 & 0 \end{bmatrix}$

9.2 Building Skills

9. $\begin{bmatrix} 2 \\ -3 \end{bmatrix} = \begin{bmatrix} x \\ u \end{bmatrix} \Rightarrow x = 2, u = -3$

11. $\begin{bmatrix} 2 & x \\ y & -3 \end{bmatrix} = \begin{bmatrix} 2 & -1 \\ 3 & -3 \end{bmatrix} \Rightarrow x = -1, y = 3$

13. $\begin{bmatrix} 2x-3y & -4 \\ 5 & 3x+y \end{bmatrix} = \begin{bmatrix} 1 & -4 \\ 5 & 7 \end{bmatrix} \Rightarrow$

$\begin{cases} 2x - 3y = 1 \\ 3x + y = 7 \end{cases} \Rightarrow x = 2, y = 1$

15. $\begin{bmatrix} x-y & 1 & 2 \\ 4 & 3x-2y & 3 \\ 5 & 6 & 5x-10y \end{bmatrix} = \begin{bmatrix} -1 & 1 & 2 \\ 4 & -1 & 3 \\ 5 & 6 & 6 \end{bmatrix} \Rightarrow$

$\begin{cases} x - y = -1 \\ 3x - 2y = -1 \\ 5x - 10y = 6 \end{cases} \Rightarrow \begin{bmatrix} 1 & -1 & | & -1 \\ 3 & -2 & | & -1 \\ 5 & -10 & | & 6 \end{bmatrix}$

$\xrightarrow[-5R_1+R_3 \to R_3]{-3R_1+R_2 \to R_2} \begin{bmatrix} 1 & -1 & | & -1 \\ 0 & 1 & | & 2 \\ 0 & -5 & | & 11 \end{bmatrix}$

$\xrightarrow{5R_2+R_3 \to R_3} \begin{bmatrix} 1 & -1 & | & -1 \\ 0 & 1 & | & 2 \\ 0 & 0 & | & 21 \end{bmatrix} \Rightarrow 0 = 21$

There is no solution.

17. a. $A + B = \begin{bmatrix} 1 & 2 \\ 3 & 4 \end{bmatrix} + \begin{bmatrix} -1 & 0 \\ 2 & -3 \end{bmatrix} = \begin{bmatrix} 0 & 2 \\ 5 & 1 \end{bmatrix}$

b. $A - B = \begin{bmatrix} 1 & 2 \\ 3 & 4 \end{bmatrix} - \begin{bmatrix} -1 & 0 \\ 2 & -3 \end{bmatrix} = \begin{bmatrix} 2 & 2 \\ 1 & 7 \end{bmatrix}$

c. $-3A = -3\begin{bmatrix} 1 & 2 \\ 3 & 4 \end{bmatrix} = \begin{bmatrix} -3 & -6 \\ -9 & -12 \end{bmatrix}$

d. $3A - 2B = 3\begin{bmatrix} 1 & 2 \\ 3 & 4 \end{bmatrix} - 2\begin{bmatrix} -1 & 0 \\ 2 & -3 \end{bmatrix}$

$= \begin{bmatrix} 3 & 6 \\ 9 & 12 \end{bmatrix} - \begin{bmatrix} -2 & 0 \\ 4 & -6 \end{bmatrix} = \begin{bmatrix} 5 & 6 \\ 5 & 18 \end{bmatrix}$

e. $(A+B)^2 = \left(\begin{bmatrix} 1 & 2 \\ 3 & 4 \end{bmatrix} + \begin{bmatrix} -1 & 0 \\ 2 & -3 \end{bmatrix}\right)^2$

$= \left(\begin{bmatrix} 0 & 2 \\ 5 & 1 \end{bmatrix}\right)^2 = \begin{bmatrix} 0 & 2 \\ 5 & 1 \end{bmatrix}\begin{bmatrix} 0 & 2 \\ 5 & 1 \end{bmatrix}$

$= \begin{bmatrix} 0(0)+2(5) & 0(2)+2(1) \\ 5(0)+1(5) & 5(2)+1(1) \end{bmatrix}$

$= \begin{bmatrix} 10 & 2 \\ 5 & 11 \end{bmatrix}$

f. $A^2 = \begin{bmatrix} 1 & 2 \\ 3 & 4 \end{bmatrix}\begin{bmatrix} 1 & 2 \\ 3 & 4 \end{bmatrix}$

$= \begin{bmatrix} 1(1)+2(3) & 1(2)+2(4) \\ 3(1)+4(3) & 3(2)+4(4) \end{bmatrix} = \begin{bmatrix} 7 & 10 \\ 15 & 22 \end{bmatrix}$

$B^2 = \begin{bmatrix} -1 & 0 \\ 2 & -3 \end{bmatrix}\begin{bmatrix} -1 & 0 \\ 2 & -3 \end{bmatrix}$

$= \begin{bmatrix} -1(-1)+0(2) & -1(0)+0(-3) \\ 2(-1)-3(2) & 2(0)-3(-3) \end{bmatrix}$

$= \begin{bmatrix} 1 & 0 \\ -8 & 9 \end{bmatrix}$

$A^2 - B^2 = \begin{bmatrix} 7 & 10 \\ 15 & 22 \end{bmatrix} - \begin{bmatrix} 1 & 0 \\ -8 & 9 \end{bmatrix} = \begin{bmatrix} 6 & 10 \\ 23 & 13 \end{bmatrix}$

19. a. $A + B$ is not defined.

b. $A - B$ is not defined.

c. $-3A = -3\begin{bmatrix} 2 & 3 \\ -4 & 5 \end{bmatrix} = \begin{bmatrix} -6 & -9 \\ 12 & -15 \end{bmatrix}$

d. $3A - 2B$ is not defined

e. $(A+B)^2$ is not defined.

f. $A^2 - B^2$ is not defined.

21. a. $A + B = \begin{bmatrix} 4 & 0 & -1 \\ -2 & 5 & 2 \\ 0 & 0 & 1 \end{bmatrix} + \begin{bmatrix} 3 & 1 & 0 \\ 1 & -4 & 2 \\ 2 & 1 & 3 \end{bmatrix}$

$= \begin{bmatrix} 7 & 1 & -1 \\ -1 & 1 & 4 \\ 2 & 1 & 4 \end{bmatrix}$

b. $A - B = \begin{bmatrix} 4 & 0 & -1 \\ -2 & 5 & 2 \\ 0 & 0 & 1 \end{bmatrix} - \begin{bmatrix} 3 & 1 & 0 \\ 1 & -4 & 2 \\ 2 & 1 & 3 \end{bmatrix}$

$= \begin{bmatrix} 1 & -1 & -1 \\ -3 & 9 & 0 \\ -2 & -1 & -2 \end{bmatrix}$

c. $-3A = -3\begin{bmatrix} 4 & 0 & -1 \\ -2 & 5 & 2 \\ 0 & 0 & 1 \end{bmatrix} = \begin{bmatrix} -12 & 0 & 3 \\ 6 & -15 & -6 \\ 0 & 0 & -3 \end{bmatrix}$

d. $3A - 2B = 3\begin{bmatrix} 4 & 0 & -1 \\ -2 & 5 & 2 \\ 0 & 0 & 1 \end{bmatrix} - 2\begin{bmatrix} 3 & 1 & 0 \\ 1 & -4 & 2 \\ 2 & 1 & 3 \end{bmatrix} = \begin{bmatrix} 12 & 0 & -3 \\ -6 & 15 & 6 \\ 0 & 0 & 3 \end{bmatrix} - \begin{bmatrix} 6 & 2 & 0 \\ 2 & -8 & 4 \\ 4 & 2 & 6 \end{bmatrix} = \begin{bmatrix} 6 & -2 & -3 \\ -8 & 23 & 2 \\ -4 & -2 & -3 \end{bmatrix}$

e. $(A+B)^2 = \begin{bmatrix} 7 & 1 & -1 \\ -1 & 1 & 4 \\ 2 & 1 & 4 \end{bmatrix}\begin{bmatrix} 7 & 1 & -1 \\ -1 & 1 & 4 \\ 2 & 1 & 4 \end{bmatrix} = \begin{bmatrix} 7(7)+1(-1)-1(2) & 7(1)+1(1)-1(1) & 7(-1)+1(4)-1(4) \\ -1(7)+1(-1)+4(2) & -1(1)+1(1)+4(1) & -1(-1)+1(4)+4(4) \\ 2(7)+1(-1)+4(2) & 2(1)+1(1)+4(1) & 2(-1)+1(4)+4(4) \end{bmatrix}$

$= \begin{bmatrix} 46 & 7 & -7 \\ 0 & 4 & 21 \\ 21 & 7 & 18 \end{bmatrix}$

f. $A^2 = \begin{bmatrix} 4 & 0 & -1 \\ -2 & 5 & 2 \\ 0 & 0 & 1 \end{bmatrix}\begin{bmatrix} 4 & 0 & -1 \\ -2 & 5 & 2 \\ 0 & 0 & 1 \end{bmatrix} = \begin{bmatrix} 4(4)+0(-2)-1(0) & 4(0)+0(5)-1(0) & 4(-1)+0(2)-1(1) \\ -2(4)+5(-2)+2(0) & -2(0)+5(5)+2(0) & -2(-1)+5(2)+2(1) \\ 0(4)+0(-2)+1(0) & 0(0)+0(5)+1(0) & 0(-1)+0(2)+1(1) \end{bmatrix}$

$= \begin{bmatrix} 16 & 0 & -5 \\ -18 & 25 & 14 \\ 0 & 0 & 1 \end{bmatrix}$

$B^2 = \begin{bmatrix} 3 & 1 & 0 \\ 1 & -4 & 2 \\ 2 & 1 & 3 \end{bmatrix}\begin{bmatrix} 3 & 1 & 0 \\ 1 & -4 & 2 \\ 2 & 1 & 3 \end{bmatrix} = \begin{bmatrix} 3(3)+1(1)+0(2) & 3(1)+1(-4)+0(1) & 3(0)+1(2)+0(3) \\ 1(3)-4(1)+2(2) & 1(1)-4(-4)+2(1) & 1(0)-4(2)+2(3) \\ 2(3)+1(1)+3(2) & 2(1)+1(-4)+3(1) & 2(0)+1(2)+3(3) \end{bmatrix} = \begin{bmatrix} 10 & -1 & 2 \\ 3 & 19 & -2 \\ 13 & 1 & 11 \end{bmatrix}$

$A^2 - B^2 = \begin{bmatrix} 16 & 0 & -5 \\ -18 & 25 & 14 \\ 0 & 0 & 1 \end{bmatrix} - \begin{bmatrix} 10 & -1 & 2 \\ 3 & 19 & -2 \\ 13 & 1 & 11 \end{bmatrix} = \begin{bmatrix} 6 & 1 & -7 \\ -21 & 6 & 16 \\ -13 & -1 & -10 \end{bmatrix}$

23. a. $A + B = \begin{bmatrix} 1 & 2 & -3 \\ 3 & 4 & 5 \\ 2 & -1 & 0 \end{bmatrix} + \begin{bmatrix} 3 & 1 & 0 \\ 1 & -4 & 2 \\ 2 & 1 & 3 \end{bmatrix} = \begin{bmatrix} 4 & 3 & -3 \\ 4 & 0 & 7 \\ 4 & 0 & 3 \end{bmatrix}$

b. $A - B = \begin{bmatrix} 1 & 2 & -3 \\ 3 & 4 & 5 \\ 2 & -1 & 0 \end{bmatrix} - \begin{bmatrix} 3 & 1 & 0 \\ 1 & -4 & 2 \\ 2 & 1 & 3 \end{bmatrix} = \begin{bmatrix} -2 & 1 & -3 \\ 2 & 8 & 3 \\ 0 & -2 & -3 \end{bmatrix}$

c. $-3A = -3\begin{bmatrix} 1 & 2 & -3 \\ 3 & 4 & 5 \\ 2 & -1 & 0 \end{bmatrix} = \begin{bmatrix} -3 & -6 & 9 \\ -9 & -12 & -15 \\ -6 & 3 & 0 \end{bmatrix}$

d. $3A - 2B = 3\begin{bmatrix} 1 & 2 & -3 \\ 3 & 4 & 5 \\ 2 & -1 & 0 \end{bmatrix} - 2\begin{bmatrix} 3 & 1 & 0 \\ 1 & -4 & 2 \\ 2 & 1 & 3 \end{bmatrix} = \begin{bmatrix} 3 & 6 & -9 \\ 9 & 12 & 15 \\ 6 & -3 & 0 \end{bmatrix} - \begin{bmatrix} 6 & 2 & 0 \\ 2 & -8 & 4 \\ 4 & 2 & 6 \end{bmatrix} = \begin{bmatrix} -3 & 4 & -9 \\ 7 & 20 & 11 \\ 2 & -5 & -6 \end{bmatrix}$

e. $(A+B)^2 = \begin{bmatrix} 4 & 3 & -3 \\ 4 & 0 & 7 \\ 4 & 0 & 3 \end{bmatrix}\begin{bmatrix} 4 & 3 & -3 \\ 4 & 0 & 7 \\ 4 & 0 & 3 \end{bmatrix} = \begin{bmatrix} 4(4)+3(4)-3(4) & 4(3)+3(0)-3(0) & 4(-3)+3(7)-3(3) \\ 4(4)+0(4)+7(4) & 4(3)+0(0)+7(0) & 4(-3)+0(7)+7(3) \\ 4(4)+0(4)+3(4) & 4(3)+0(0)+3(0) & 4(-3)+0(7)+3(3) \end{bmatrix}$

$= \begin{bmatrix} 16 & 12 & 0 \\ 44 & 12 & 9 \\ 28 & 12 & -3 \end{bmatrix}$

f. $A^2 = \begin{bmatrix} 1 & 2 & -3 \\ 3 & 4 & 5 \\ 2 & -1 & 0 \end{bmatrix}\begin{bmatrix} 1 & 2 & -3 \\ 3 & 4 & 5 \\ 2 & -1 & 0 \end{bmatrix} = \begin{bmatrix} 1(1)+2(3)-3(2) & 1(2)+2(4)-3(-1) & 1(-3)+2(5)-3(0) \\ 3(1)+4(3)+5(2) & 3(2)+4(4)+5(-1) & 3(-3)+4(5)+5(0) \\ 2(1)-1(3)+0(2) & 2(2)-1(4)+0(-1) & 2(-3)-1(5)+0(0) \end{bmatrix}$

$= \begin{bmatrix} 1 & 13 & 7 \\ 25 & 17 & 11 \\ -1 & 0 & -11 \end{bmatrix}$

$B^2 = \begin{bmatrix} 3 & 1 & 0 \\ 1 & -4 & 2 \\ 2 & 1 & 3 \end{bmatrix}\begin{bmatrix} 3 & 1 & 0 \\ 1 & -4 & 2 \\ 2 & 1 & 3 \end{bmatrix} = \begin{bmatrix} 3(3)+1(1)+0(2) & 3(1)+1(-4)+0(1) & 3(0)+1(2)+0(3) \\ 1(3)-4(1)+2(2) & 1(1)-4(-4)+2(1) & 1(0)-4(2)+2(3) \\ 2(3)+1(1)+3(2) & 2(1)+1(-4)+3(1) & 2(0)+1(2)+3(3) \end{bmatrix} = \begin{bmatrix} 10 & -1 & 2 \\ 3 & 19 & -2 \\ 13 & 1 & 11 \end{bmatrix}$

$A^2 - B^2 = \begin{bmatrix} 1 & 13 & 7 \\ 25 & 17 & 11 \\ -1 & 0 & -11 \end{bmatrix} - \begin{bmatrix} 10 & -1 & 2 \\ 3 & 19 & -2 \\ 13 & 1 & 11 \end{bmatrix} = \begin{bmatrix} -9 & 14 & 5 \\ 22 & -2 & 13 \\ -14 & -1 & -22 \end{bmatrix}$

25. $\begin{bmatrix} 2 & 3 & -1 \\ 1 & -2 & 4 \end{bmatrix} + X = \begin{bmatrix} -2 & 1 & 0 \\ 2 & 3 & 4 \end{bmatrix} \Rightarrow X = \begin{bmatrix} -2 & 1 & 0 \\ 2 & 3 & 4 \end{bmatrix} - \begin{bmatrix} 2 & 3 & -1 \\ 1 & -2 & 4 \end{bmatrix} = \begin{bmatrix} -4 & -2 & 1 \\ 1 & 5 & 0 \end{bmatrix}$

27. $2X - \begin{bmatrix} 2 & 3 & -1 \\ 1 & -2 & 4 \end{bmatrix} = \begin{bmatrix} -2 & 1 & 0 \\ 2 & 3 & 4 \end{bmatrix} \Rightarrow 2X = \begin{bmatrix} -2 & 1 & 0 \\ 2 & 3 & 4 \end{bmatrix} + \begin{bmatrix} 2 & 3 & -1 \\ 1 & -2 & 4 \end{bmatrix} \Rightarrow$

$2X = \begin{bmatrix} 0 & 4 & -1 \\ 3 & 1 & 8 \end{bmatrix} \Rightarrow X = \begin{bmatrix} 0 & 2 & -\frac{1}{2} \\ \frac{3}{2} & \frac{1}{2} & 4 \end{bmatrix}$

29. $2X + 3\begin{bmatrix} 2 & 3 & -1 \\ 1 & -2 & 4 \end{bmatrix} = \begin{bmatrix} -2 & 1 & 0 \\ 2 & 3 & 4 \end{bmatrix} \Rightarrow 2X + \begin{bmatrix} 6 & 9 & -3 \\ 3 & -6 & 12 \end{bmatrix} = \begin{bmatrix} -2 & 1 & 0 \\ 2 & 3 & 4 \end{bmatrix} \Rightarrow$

$2X = \begin{bmatrix} -2 & 1 & 0 \\ 2 & 3 & 4 \end{bmatrix} - \begin{bmatrix} 6 & 9 & -3 \\ 3 & -6 & 12 \end{bmatrix} \Rightarrow 2X = \begin{bmatrix} -8 & -8 & 3 \\ -1 & 9 & -8 \end{bmatrix} \Rightarrow X = \begin{bmatrix} -4 & -4 & \frac{3}{2} \\ -\frac{1}{2} & \frac{9}{2} & -4 \end{bmatrix}$

31. $2\begin{bmatrix} 2 & 3 & -1 \\ 1 & -2 & 4 \end{bmatrix} + 3\begin{bmatrix} -2 & 1 & 0 \\ 2 & 3 & 4 \end{bmatrix} + 4X = 0 \Rightarrow \begin{bmatrix} 4 & 6 & -2 \\ 2 & -4 & 8 \end{bmatrix} + \begin{bmatrix} -6 & 3 & 0 \\ 6 & 9 & 12 \end{bmatrix} = -4X \Rightarrow$

$\begin{bmatrix} -2 & 9 & -2 \\ 8 & 5 & 20 \end{bmatrix} = -4X \Rightarrow X = \begin{bmatrix} \frac{1}{2} & -\frac{9}{4} & \frac{1}{2} \\ -2 & -\frac{5}{4} & -5 \end{bmatrix}$

33. a. $\begin{bmatrix} 1 & 2 \\ 3 & 4 \end{bmatrix}\begin{bmatrix} -2 & 1 \\ 3 & 5 \end{bmatrix} = \begin{bmatrix} 1(-2)+2(3) & 1(1)+2(5) \\ 3(-2)+4(3) & 3(1)+4(5) \end{bmatrix} = \begin{bmatrix} 4 & 11 \\ 6 & 23 \end{bmatrix}$

b. $\begin{bmatrix} -2 & 1 \\ 3 & 5 \end{bmatrix}\begin{bmatrix} 1 & 2 \\ 3 & 4 \end{bmatrix} = \begin{bmatrix} -2(1)+1(3) & -2(2)+1(4) \\ 3(1)+5(3) & 3(2)+5(4) \end{bmatrix} = \begin{bmatrix} 1 & 0 \\ 18 & 26 \end{bmatrix}$

35. a. $\begin{bmatrix} 2 & -1 & 0 \\ -3 & 1 & 2 \end{bmatrix}\begin{bmatrix} 1 & 5 \\ -2 & 3 \\ 4 & 0 \end{bmatrix} = \begin{bmatrix} 2(1)-1(-2)+0(4) & 2(5)-1(3)+0(0) \\ -3(1)+1(-2)+2(4) & -3(5)+1(3)+2(0) \end{bmatrix} = \begin{bmatrix} 4 & 7 \\ 3 & -12 \end{bmatrix}$

b. $\begin{bmatrix} 1 & 5 \\ -2 & 3 \\ 4 & 0 \end{bmatrix} \begin{bmatrix} 2 & -1 & 0 \\ -3 & 1 & 2 \end{bmatrix} = \begin{bmatrix} 1(2)+5(-3) & 1(-1)+5(1) & 1(0)+5(2) \\ -2(2)+3(-3) & -2(-1)+3(1) & -2(0)+3(2) \\ 4(2)+0(-3) & 4(-1)+0(1) & 4(0)+0(2) \end{bmatrix} = \begin{bmatrix} -13 & 4 & 10 \\ -13 & 5 & 6 \\ 8 & -4 & 0 \end{bmatrix}$

37. a. $\begin{bmatrix} 2 & 3 & 5 \end{bmatrix} \begin{bmatrix} 1 \\ -2 \\ 4 \end{bmatrix} = \begin{bmatrix} 2(1)+3(-2)+5(4) \end{bmatrix} = \begin{bmatrix} 16 \end{bmatrix}$

b. $\begin{bmatrix} 1 \\ -2 \\ 4 \end{bmatrix} \begin{bmatrix} 2 & 3 & 5 \end{bmatrix} = \begin{bmatrix} 1(2) & 1(3) & 1(5) \\ -2(2) & -2(3) & -2(5) \\ 4(2) & 4(3) & 4(5) \end{bmatrix} = \begin{bmatrix} 2 & 3 & 5 \\ -4 & -6 & -10 \\ 8 & 12 & 20 \end{bmatrix}$

39. a. $\begin{bmatrix} 1 & 2 & 3 \end{bmatrix} \begin{bmatrix} 1 & 2 & -1 \\ 0 & 3 & 1 \\ 2 & 0 & -3 \end{bmatrix} = \begin{bmatrix} 1(1)+2(0)+3(2) & 1(2)+2(3)+3(0) & 1(-1)+2(1)+3(-3) \end{bmatrix} = \begin{bmatrix} 7 & 8 & -8 \end{bmatrix}$

b. The product BA is not defined.

41. a. $\begin{bmatrix} 2 & 0 & 1 \\ 1 & 4 & 2 \\ 3 & -1 & 0 \end{bmatrix} \begin{bmatrix} 3 & 1 & 0 \\ -1 & 2 & 0 \\ 4 & 5 & 2 \end{bmatrix} = \begin{bmatrix} 2(3)+0(-1)+1(4) & 2(1)+0(2)+1(5) & 2(0)+0(0)+1(2) \\ 1(3)+4(-1)+2(4) & 1(1)+4(2)+2(5) & 1(0)+4(0)+2(2) \\ 3(3)-1(-1)+0(4) & 3(1)-1(2)+0(5) & 3(0)-1(0)+0(2) \end{bmatrix} = \begin{bmatrix} 10 & 7 & 2 \\ 7 & 19 & 4 \\ 10 & 1 & 0 \end{bmatrix}$

b. $\begin{bmatrix} 3 & 1 & 0 \\ -1 & 2 & 0 \\ 4 & 5 & 2 \end{bmatrix} \begin{bmatrix} 2 & 0 & 1 \\ 1 & 4 & 2 \\ 3 & -1 & 0 \end{bmatrix} = \begin{bmatrix} 3(2)+1(1)+0(3) & 3(0)+1(4)+0(-1) & 3(1)+1(2)+0(0) \\ -1(2)+2(1)+0(3) & -1(0)+2(4)+0(-1) & -1(1)+2(2)+0(0) \\ 4(2)+5(1)+2(3) & 4(0)+5(4)+2(-1) & 4(1)+5(2)+2(0) \end{bmatrix} = \begin{bmatrix} 7 & 4 & 5 \\ 0 & 8 & 3 \\ 19 & 18 & 14 \end{bmatrix}$

43. $AB = \begin{bmatrix} 3 & 1 & 2 \\ 0 & 4 & 3 \\ 1 & -2 & 2 \end{bmatrix} \begin{bmatrix} 2 & 5 & 0 \\ 1 & 2 & -1 \\ 3 & 0 & 2 \end{bmatrix} = \begin{bmatrix} 3(2)+1(1)+2(3) & 3(5)+1(2)+2(0) & 3(0)+1(-1)+2(2) \\ 0(2)+4(1)+3(3) & 0(5)+4(2)+3(0) & 0(0)+4(-1)+3(2) \\ 1(2)-2(1)+2(3) & 1(5)-2(2)+2(0) & 1(0)-2(-1)+2(2) \end{bmatrix} = \begin{bmatrix} 13 & 17 & 3 \\ 13 & 8 & 2 \\ 6 & 1 & 6 \end{bmatrix}$

$BA = \begin{bmatrix} 2 & 5 & 0 \\ 1 & 2 & -1 \\ 3 & 0 & 2 \end{bmatrix} \begin{bmatrix} 3 & 1 & 2 \\ 0 & 4 & 3 \\ 1 & -2 & 2 \end{bmatrix} = \begin{bmatrix} 2(3)+5(0)+0(1) & 2(1)+5(4)+0(-2) & 2(2)+5(3)+0(2) \\ 1(3)+2(0)-1(1) & 1(1)+2(4)-1(-2) & 1(2)+2(3)-1(2) \\ 3(3)+0(0)+2(1) & 3(1)+0(4)+2(-2) & 3(2)+0(3)+2(2) \end{bmatrix} = \begin{bmatrix} 6 & 22 & 19 \\ 2 & 11 & 6 \\ 11 & -1 & 10 \end{bmatrix}$

So, $AB \ne BA$.

45. $(AB)C = \left(\begin{bmatrix} 1 & 2 \\ 3 & 4 \end{bmatrix} \begin{bmatrix} 2 & -3 \\ 3 & 5 \end{bmatrix} \right) \begin{bmatrix} 0 & 1 \\ 2 & 4 \end{bmatrix} = \begin{bmatrix} 1(2)+2(3) & 1(-3)+2(5) \\ 3(2)+4(3) & 3(-3)+4(5) \end{bmatrix} \begin{bmatrix} 0 & 1 \\ 2 & 4 \end{bmatrix} = \begin{bmatrix} 8 & 7 \\ 18 & 11 \end{bmatrix} \begin{bmatrix} 0 & 1 \\ 2 & 4 \end{bmatrix}$

$= \begin{bmatrix} 8(0)+7(2) & 8(1)+7(4) \\ 18(0)+11(2) & 18(1)+11(4) \end{bmatrix} = \begin{bmatrix} 14 & 36 \\ 22 & 62 \end{bmatrix}$

$A(BC) = \begin{bmatrix} 1 & 2 \\ 3 & 4 \end{bmatrix} \left(\begin{bmatrix} 2 & -3 \\ 3 & 5 \end{bmatrix} \begin{bmatrix} 0 & 1 \\ 2 & 4 \end{bmatrix} \right) = \begin{bmatrix} 1 & 2 \\ 3 & 4 \end{bmatrix} \begin{bmatrix} 2(0)-3(2) & 2(1)-3(4) \\ 3(0)+5(2) & 3(1)+5(4) \end{bmatrix} = \begin{bmatrix} 1 & 2 \\ 3 & 4 \end{bmatrix} \begin{bmatrix} -6 & -10 \\ 10 & 23 \end{bmatrix}$

$= \begin{bmatrix} 1(-6)+2(10) & 1(-10)+2(23) \\ 3(-6)+4(10) & 3(-10)+4(23) \end{bmatrix} = \begin{bmatrix} 14 & 36 \\ 22 & 62 \end{bmatrix} \Rightarrow (AB)C = A(BC)$

47. $(A+B)C = \left(\begin{bmatrix} 1 & 2 \\ 3 & 4 \end{bmatrix} + \begin{bmatrix} 2 & -3 \\ 3 & 5 \end{bmatrix}\right)\begin{bmatrix} 0 & 1 \\ 2 & 4 \end{bmatrix} = \begin{bmatrix} 3 & -1 \\ 6 & 9 \end{bmatrix}\begin{bmatrix} 0 & 1 \\ 2 & 4 \end{bmatrix} = \begin{bmatrix} 3(0)-1(2) & 3(1)-1(4) \\ 6(0)+9(2) & 6(1)+9(4) \end{bmatrix} = \begin{bmatrix} -2 & -1 \\ 18 & 42 \end{bmatrix}$

$AC + BC = \begin{bmatrix} 1 & 2 \\ 3 & 4 \end{bmatrix}\begin{bmatrix} 0 & 1 \\ 2 & 4 \end{bmatrix} + \begin{bmatrix} 2 & -3 \\ 3 & 5 \end{bmatrix}\begin{bmatrix} 0 & 1 \\ 2 & 4 \end{bmatrix} = \begin{bmatrix} 1(0)+2(2) & 1(1)+2(4) \\ 3(0)+4(2) & 3(1)+4(4) \end{bmatrix} + \begin{bmatrix} 2(0)-3(2) & 2(1)-3(4) \\ 3(0)+5(2) & 3(1)+5(4) \end{bmatrix}$

$= \begin{bmatrix} 4 & 9 \\ 8 & 19 \end{bmatrix} + \begin{bmatrix} -6 & -10 \\ 10 & 23 \end{bmatrix} = \begin{bmatrix} -2 & -1 \\ 18 & 42 \end{bmatrix} \Rightarrow (A+B)C = AC + BC$

9.2 Applying the Concepts

49.

$$A + B = \begin{bmatrix} 7 & 3 & 18 \\ 4 & 1 & 3 \end{bmatrix} + \begin{bmatrix} 6 & 2 & 20 \\ 3 & 1 & 4 \end{bmatrix} = \begin{array}{c} \text{Steel} \quad \text{Glass} \quad \text{Wood} \\ \begin{bmatrix} 13 & 5 & 38 \\ 7 & 2 & 7 \end{bmatrix} \begin{array}{l} \text{Material} \\ \text{Transportation} \end{array} \end{array}$$

51. $\begin{bmatrix} 100 & 300 & 400 \end{bmatrix}\begin{bmatrix} 60 \\ 38 \\ 17 \end{bmatrix} = [100(60) + 300(38) + 400(17)] = [24,200]$

The total cost is $24,200.

53. a.

$$\begin{array}{c} \quad\quad\quad \text{Chairman} \quad \text{President} \quad \text{Vice President} \\ \begin{array}{l} \text{Salary} \\ \text{Bonus} \\ \text{Stock} \end{array} \begin{bmatrix} 2,500,000 & 1,250,000 & 100,000 \\ 1,500,000 & 750,000 & 150,000 \\ 50,000 & 25,000 & 5000 \end{bmatrix} \end{array}$$

b. $\begin{bmatrix} 1 \\ 1 \\ 4 \end{bmatrix} \begin{array}{l} \text{Chairman} \\ \text{President} \\ \text{Vice President} \end{array}$

c. $\begin{bmatrix} 2,500,000 & 1,250,000 & 100,000 \\ 1,500,000 & 750,000 & 150,000 \\ 50,000 & 25,000 & 5000 \end{bmatrix}\begin{bmatrix} 1 \\ 1 \\ 4 \end{bmatrix} = \begin{bmatrix} 2,500,000(1)+1,250,000(1)+100,000(4) \\ 1,500,000(1)+750,000(1)+150,000(4) \\ 50,000(1)+25,000(1)+5000(4) \end{bmatrix} = \begin{array}{c} \text{Totals} \\ \begin{bmatrix} 4,150,000 \\ 2,850,000 \\ 95,000 \end{bmatrix} \begin{array}{l} \text{Salary} \\ \text{Bonus} \\ \text{Stock} \end{array} \end{array}$

55. $AD = \begin{bmatrix} 1 & 0 \\ 0 & -1 \end{bmatrix}\begin{bmatrix} 0 & 4 & 4 & 1 & 1 & 0 \\ 0 & 0 & 1 & 1 & 6 & 6 \end{bmatrix}$

$= \begin{bmatrix} 1(0)+0(0) & 1(4)+0(0) & 1(4)+0(1) & 1(1)+0(1) & 1(1)+0(6) & 1(0)+0(6) \\ 0(0)-1(0) & 0(4)-1(0) & 0(4)-1(1) & 0(1)-1(1) & 0(1)-1(6) & 0(0)-1(6) \end{bmatrix}$

$= \begin{bmatrix} 0 & 4 & 4 & 1 & 1 & 0 \\ 0 & 0 & -1 & -1 & -6 & -6 \end{bmatrix}$

57. $AD = \begin{bmatrix} 1 & 0 \\ 0.25 & 1 \end{bmatrix}\begin{bmatrix} 0 & 4 & 4 & 1 & 1 & 0 \\ 0 & 0 & 1 & 1 & 6 & 6 \end{bmatrix}$

$= \begin{bmatrix} 1(0)+0(0) & 1(4)+0(0) & 1(4)+0(1) & 1(1)+0(1) & 1(1)+0(6) & 1(0)+0(6) \\ 0.25(0)+1(0) & 0.25(4)+1(0) & 0.25(4)+1(1) & 0.25(1)+1(1) & 0.25(1)+1(6) & 0.25(0)+1(6) \end{bmatrix}$

$= \begin{bmatrix} 0 & 4 & 4 & 1 & 1 & 0 \\ 0 & 1 & 2 & 1.25 & 6.25 & 6 \end{bmatrix}$

In exercises 59–62, the matrix representation of the quadrilateral $PQRS = D = \begin{bmatrix} 0 & 2 & 1 & 0 \\ 0 & 0 & 2 & 1 \end{bmatrix}$, where row 1 contains the x-coordinates of P, Q, R, S, and row 2 contains the y-coordinates.

59. $AD = \begin{bmatrix} 2 & 0 \\ 0 & 1 \end{bmatrix} \begin{bmatrix} 0 & 2 & 1 & 0 \\ 0 & 0 & 2 & 1 \end{bmatrix} = \begin{bmatrix} 0 & 4 & 2 & 0 \\ 0 & 0 & 2 & 1 \end{bmatrix}$

The coordinates of the vertices are $(0, 0)$, $(4, 0)$, $(2, 2)$, and $(0, 1)$.

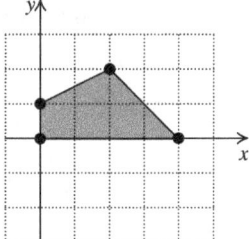

61. $AD = \begin{bmatrix} 0.6 & 0.8 \\ -0.8 & 0.6 \end{bmatrix} \begin{bmatrix} 0 & 2 & 1 & 0 \\ 0 & 0 & 2 & 1 \end{bmatrix} = \begin{bmatrix} 0 & 1.2 & 2.2 & 0.8 \\ 0 & -1.6 & 0.4 & 0.6 \end{bmatrix}$

The coordinates of the vertices are $(0, 0)$, $(1.2, -1.6)$, $(2.2, 0.4)$, and $(0.8, 0.6)$.

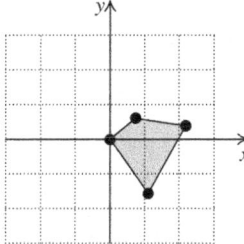

In exercises 63–66, the matrix representation of the quadrilateral $PQRS = D = \begin{bmatrix} 0 & 1 & 2 & 1 \\ 0 & 0 & 1 & 2 \end{bmatrix}$, where row 1 contains the x-coordinates of P, Q, R, S, and row 2 contains the y-coordinates.

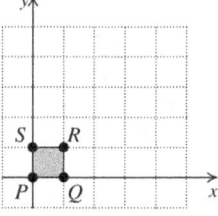

63.

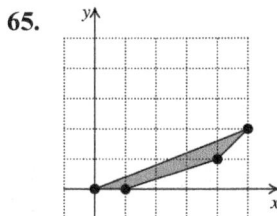

$AD = \begin{bmatrix} 0 & 2 & 4 & 2 \\ 0 & 0 & 1 & 2 \end{bmatrix} = \begin{bmatrix} a & b \\ c & d \end{bmatrix} \begin{bmatrix} 0 & 1 & 2 & 1 \\ 0 & 0 & 1 & 2 \end{bmatrix}$

$= \begin{bmatrix} a \cdot 0 + b \cdot 0 & a \cdot 1 + b \cdot 0 & a \cdot 2 + b \cdot 1 & a \cdot 1 + b \cdot 2 \\ c \cdot 0 + d \cdot 0 & c \cdot 1 + d \cdot 0 & c \cdot 2 + d \cdot 1 & c \cdot 1 + d \cdot 2 \end{bmatrix}$

$= \begin{bmatrix} 0 & a & 2a+b & a+2b \\ 0 & c & 2c+d & c+2d \end{bmatrix} \Rightarrow a = 2, c = 0$

$\begin{aligned} 2a + b = 4 &\Rightarrow 2(2) + b = 4 \Rightarrow b = 0 \\ 2c + d = 1 &\Rightarrow 2(0) + d = 1 \Rightarrow d = 1 \end{aligned} \Rightarrow A = \begin{bmatrix} 2 & 0 \\ 0 & 1 \end{bmatrix}$

65.

$AD = \begin{bmatrix} 0 & 1 & 4 & 5 \\ 0 & 0 & 1 & 2 \end{bmatrix} = \begin{bmatrix} a & b \\ c & d \end{bmatrix} \begin{bmatrix} 0 & 1 & 2 & 1 \\ 0 & 0 & 1 & 2 \end{bmatrix}$

$= \begin{bmatrix} a \cdot 0 + b \cdot 0 & a \cdot 1 + b \cdot 0 & a \cdot 2 + b \cdot 1 & a \cdot 1 + b \cdot 2 \\ c \cdot 0 + d \cdot 0 & c \cdot 1 + d \cdot 0 & c \cdot 2 + d \cdot 1 & c \cdot 1 + d \cdot 2 \end{bmatrix}$

$= \begin{bmatrix} 0 & a & 2a+b & a+2b \\ 0 & c & 2c+d & c+2d \end{bmatrix} \Rightarrow a = 1, c = 0$

$\begin{aligned} 2a + b = 4 &\Rightarrow 2(1) + b = 4 \Rightarrow b = 2 \\ 2c + d = 1 &\Rightarrow 2(0) + d = 1 \Rightarrow d = 1 \end{aligned} \Rightarrow A = \begin{bmatrix} 1 & 2 \\ 0 & 1 \end{bmatrix}$

Copyright © 2019 Pearson Education Inc.

9.2 Beyond the Basics

67. $AB = \begin{bmatrix} 0 & 3 \\ 0 & 0 \end{bmatrix}\begin{bmatrix} 2 & 1 \\ 3 & 0 \end{bmatrix} = \begin{bmatrix} 0(2)+3(3) & 0(1)+3(0) \\ 0(2)+0(3) & 0(1)+0(0) \end{bmatrix} = \begin{bmatrix} 9 & 0 \\ 0 & 0 \end{bmatrix}$

$AC = \begin{bmatrix} 0 & 3 \\ 0 & 0 \end{bmatrix}\begin{bmatrix} 5 & 4 \\ 3 & 0 \end{bmatrix} = \begin{bmatrix} 0(5)+3(3) & 0(4)+3(0) \\ 0(5)+0(3) & 0(4)+0(0) \end{bmatrix} = \begin{bmatrix} 9 & 0 \\ 0 & 0 \end{bmatrix}$

So, $AB = AC$ does not imply $B = C$.

69. Answers will vary. Let $A = \begin{bmatrix} 1 & 2 \\ 3 & 4 \end{bmatrix}$ and $B = \begin{bmatrix} -1 & 0 \\ 2 & -3 \end{bmatrix}$. Then $A + B = \begin{bmatrix} 1 & 2 \\ 3 & 4 \end{bmatrix} + \begin{bmatrix} -1 & 0 \\ 2 & -3 \end{bmatrix} = \begin{bmatrix} 0 & 2 \\ 5 & 1 \end{bmatrix}$

$(A+B)^2 = \begin{bmatrix} 0 & 2 \\ 5 & 1 \end{bmatrix}\begin{bmatrix} 0 & 2 \\ 5 & 1 \end{bmatrix} = \begin{bmatrix} 0(0)+2(5) & 0(2)+2(1) \\ 5(0)+1(5) & 5(2)+1(1) \end{bmatrix} = \begin{bmatrix} 10 & 2 \\ 5 & 11 \end{bmatrix}$

$A^2 = \begin{bmatrix} 1 & 2 \\ 3 & 4 \end{bmatrix}\begin{bmatrix} 1 & 2 \\ 3 & 4 \end{bmatrix} = \begin{bmatrix} 1(1)+2(3) & 1(2)+2(4) \\ 3(1)+4(3) & 3(2)+4(4) \end{bmatrix} = \begin{bmatrix} 7 & 10 \\ 15 & 22 \end{bmatrix}$

$2AB = \begin{bmatrix} 1 & 2 \\ 3 & 4 \end{bmatrix}\begin{bmatrix} -1 & 0 \\ 2 & -3 \end{bmatrix} = 2\begin{bmatrix} 1(-1)+2(2) & 1(0)+2(-3) \\ 3(-1)+4(2) & 3(0)+4(-3) \end{bmatrix} = 2\begin{bmatrix} 3 & -6 \\ 5 & -12 \end{bmatrix} = \begin{bmatrix} 6 & -12 \\ 10 & -24 \end{bmatrix}$

$B^2 = \begin{bmatrix} -1 & 0 \\ 2 & -3 \end{bmatrix}\begin{bmatrix} -1 & 0 \\ 2 & -3 \end{bmatrix} = \begin{bmatrix} -1(-1)+0(2) & -1(0)+0(-3) \\ 2(-1)-3(2) & 2(0)-3(-3) \end{bmatrix} = \begin{bmatrix} 1 & 0 \\ -8 & 9 \end{bmatrix}$

$A^2 + 2AB + B^2 = \begin{bmatrix} 7 & 10 \\ 15 & 22 \end{bmatrix} + \begin{bmatrix} 6 & -12 \\ 10 & -24 \end{bmatrix} + \begin{bmatrix} 1 & 0 \\ -8 & 9 \end{bmatrix} = \begin{bmatrix} 14 & -2 \\ 17 & 7 \end{bmatrix}$

So, $(A+B)^2 \neq A^2 + 2AB + B^2$.

71.

```
3[A]²                2[A]              [I]              3[A]²-2[A]+[I]
[[18  -27  21]    [[2  -4   6]     [[1  0  0]         [[17   -23  15]
 [-9   21  12]     [4  -8   2]      [0  1  0]          [-13   30  10]
 [-3   12  24]]    [6  -10  4]]     [0  0  1]]         [-9    22  21]]
```

73. Let $B = \begin{bmatrix} x & y \\ z & w \end{bmatrix}$

$\begin{bmatrix} 2 & 3 \\ 1 & 2 \end{bmatrix}\begin{bmatrix} x & y \\ z & w \end{bmatrix} = \begin{bmatrix} 1 & 0 \\ 0 & 1 \end{bmatrix} \Rightarrow \begin{bmatrix} 2x+3z & 2y+3w \\ x+2z & y+2w \end{bmatrix} = \begin{bmatrix} 1 & 0 \\ 0 & 1 \end{bmatrix} \Rightarrow \begin{cases} 2x+3z = 1 \\ 2y+3w = 0 \\ x+2z = 0 \\ y+2w = 1 \end{cases} \Rightarrow$

$\begin{bmatrix} 2 & 0 & 3 & 0 & | & 1 \\ 0 & 2 & 0 & 3 & | & 0 \\ 1 & 0 & 2 & 0 & | & 0 \\ 0 & 1 & 0 & 2 & | & 1 \end{bmatrix} \xrightarrow{\begin{array}{c}-(R_1-2R_3)\to R_3\\ -(R_2-2R_4)\to R_4\end{array}} \begin{bmatrix} 2 & 0 & 3 & 0 & | & 1 \\ 0 & 2 & 0 & 3 & | & 0 \\ 0 & 0 & 1 & 0 & | & -1 \\ 0 & 0 & 0 & 1 & | & 2 \end{bmatrix} \xrightarrow{\begin{array}{c}\frac{1}{2}(R_1-3R_3)\to R_1\\ \frac{1}{2}(R_2-3R_4)\to R_2\end{array}} \begin{bmatrix} 1 & 0 & 0 & 0 & | & 2 \\ 0 & 1 & 0 & 0 & | & -3 \\ 0 & 0 & 1 & 0 & | & -1 \\ 0 & 0 & 0 & 1 & | & 2 \end{bmatrix} \Rightarrow$

$x = 2, y = -3, z = -1, w = 2 \Rightarrow B = \begin{bmatrix} 2 & -3 \\ -1 & 2 \end{bmatrix}$

75. $\left(4\begin{bmatrix} 2 & -1 & 3 \\ 1 & 0 & 2 \end{bmatrix} - \begin{bmatrix} 1 & 2 & -1 \\ 2 & -3 & 4 \end{bmatrix}\right)\begin{bmatrix} 2 \\ -1 \\ 1 \end{bmatrix} = \begin{bmatrix} x \\ y \end{bmatrix} \Rightarrow \left(\begin{bmatrix} 8 & -4 & 12 \\ 4 & 0 & 8 \end{bmatrix} - \begin{bmatrix} 1 & 2 & -1 \\ 2 & -3 & 4 \end{bmatrix}\right)\begin{bmatrix} 2 \\ -1 \\ 1 \end{bmatrix} = \begin{bmatrix} x \\ y \end{bmatrix} \Rightarrow$

$\begin{bmatrix} 7 & -6 & 13 \\ 2 & 3 & 4 \end{bmatrix}\begin{bmatrix} 2 \\ -1 \\ 1 \end{bmatrix} = \begin{bmatrix} x \\ y \end{bmatrix} \Rightarrow \begin{bmatrix} 7(2)-6(-1)+13(1) \\ 2(2)+3(-1)+4(1) \end{bmatrix} = \begin{bmatrix} x \\ y \end{bmatrix} \Rightarrow \begin{bmatrix} 33 \\ 5 \end{bmatrix} = \begin{bmatrix} x \\ y \end{bmatrix} \Rightarrow x = 33, y = 5$

77. $AB = \begin{bmatrix} 0 & 1 \\ 0 & 0 \end{bmatrix}\begin{bmatrix} 1 & 0 \\ 0 & 0 \end{bmatrix} = \begin{bmatrix} 0(1)+1(0) & 0(0)+1(0) \\ 0(1)+0(0) & 0(0)+0(0) \end{bmatrix} = \begin{bmatrix} 0 & 0 \\ 0 & 0 \end{bmatrix} = 0$

$BA = \begin{bmatrix} 1 & 0 \\ 0 & 0 \end{bmatrix}\begin{bmatrix} 0 & 1 \\ 0 & 0 \end{bmatrix} = \begin{bmatrix} 1(0)+0(0) & 1(1)+0(0) \\ 0(0)+0(0) & 0(1)+0(0) \end{bmatrix} = \begin{bmatrix} 0 & 1 \\ 0 & 0 \end{bmatrix} \neq 0$

79. $2A - B = \begin{bmatrix} 6 & -6 & 0 \\ -4 & 2 & 1 \end{bmatrix}$ $\quad A + B = \begin{bmatrix} 3 & 0 & 3 \\ -2 & 1 & -4 \end{bmatrix}$

$(2A-B)+(A+B) = 3A = \begin{bmatrix} 6 & -6 & 0 \\ -4 & 2 & 1 \end{bmatrix} + \begin{bmatrix} 3 & 0 & 3 \\ -2 & 1 & -4 \end{bmatrix} = \begin{bmatrix} 9 & -6 & 3 \\ -6 & 3 & -3 \end{bmatrix} \Rightarrow A = \begin{bmatrix} 3 & -2 & 1 \\ -2 & 1 & -1 \end{bmatrix}$

$A + B = \begin{bmatrix} 3 & -2 & 1 \\ -2 & 1 & -1 \end{bmatrix} + B = \begin{bmatrix} 3 & 0 & 3 \\ -2 & 1 & -4 \end{bmatrix} \Rightarrow B - \begin{bmatrix} 3 & 0 & 3 \\ -2 & 1 & -4 \end{bmatrix} - \begin{bmatrix} 3 & -2 & 1 \\ -2 & 1 & -1 \end{bmatrix} = \begin{bmatrix} 0 & 2 & 2 \\ 0 & 0 & -3 \end{bmatrix}$

81. $AB = \begin{bmatrix} 2 & -3 & -5 \\ -1 & 4 & 5 \\ 1 & -3 & -4 \end{bmatrix}\begin{bmatrix} 2 & -2 & -4 \\ -1 & 3 & 4 \\ 1 & -2 & -3 \end{bmatrix}$

$= \begin{bmatrix} 2(2)+(-3)(-1)+(-5)(1) & 2(-2)+(-3)(3)+(-5)(-2) & 2(-4)+(-3)(4)+(-5)(-3) \\ (-1)(2)+4(-1)+5(1) & (-1)(-2)+4(3)+5(-2) & (-1)(-4)+4(4)+5(-3) \\ 1(2)+(-3)(-1)+(-4)(1) & 1(-2)+(-3)(3)+(-4)(-2) & 1(-4)+(-3)(4)+(-4)(-3) \end{bmatrix}$

$= \begin{bmatrix} 2 & -3 & -5 \\ -1 & 4 & 5 \\ 1 & -3 & -4 \end{bmatrix} = A$

$BA = \begin{bmatrix} 2 & -2 & -4 \\ -1 & 3 & 4 \\ 1 & -2 & -3 \end{bmatrix}\begin{bmatrix} 2 & -3 & -5 \\ -1 & 4 & 5 \\ 1 & -3 & -4 \end{bmatrix}$

$= \begin{bmatrix} 2(2)+(-2)(-1)+(-4)(1) & 2(-3)+(-2)(4)+(-4)(-3) & 2(-5)+(-2)(5)+(-4)(-4) \\ (-1)(2)+3(-1)+4(1) & (-1)(-3)+3(4)+4(-3) & (-1)(-5)+3(5)+4(-4) \\ 1(2)+(-2)(-1)+(-3)(1) & 1(-3)+(-2)(4)+(-3)(-3) & 1(-5)+(-2)(5)+(-3)(-4) \end{bmatrix}$

$= \begin{bmatrix} 2 & -2 & -4 \\ -1 & 3 & 4 \\ 1 & -2 & -3 \end{bmatrix} = B$

$A^2 = \begin{bmatrix} 2 & -3 & -5 \\ -1 & 4 & 5 \\ 1 & -3 & -4 \end{bmatrix}\begin{bmatrix} 2 & -3 & -5 \\ -1 & 4 & 5 \\ 1 & -3 & -4 \end{bmatrix}$

$= \begin{bmatrix} 2(2)+(-3)(-1)+(-5)(1) & 2(-3)+(-3)(4)+(-5)(-3) & 2(-5)+(-3)(5)+(-5)(-4) \\ (-1)(2)+4(-1)+5(1) & (-1)(-3)+4(4)+5(-3) & (-1)(-5)+4(5)+5(-4) \\ 1(2)+(-3)(-1)+(-4)(1) & 1(-3)+(-3)(4)+(-4)(-3) & 1(-5)+(-3)(5)+(-4)(-4) \end{bmatrix}$

$= \begin{bmatrix} 2 & -3 & -5 \\ -1 & 4 & 5 \\ 1 & -3 & -4 \end{bmatrix} = A$

83. $A^2 = \begin{bmatrix} 4 & -1 & -4 \\ 3 & 0 & -4 \\ 3 & -1 & -3 \end{bmatrix}\begin{bmatrix} 4 & -1 & -4 \\ 3 & 0 & -4 \\ 3 & -1 & -3 \end{bmatrix}$

$= \begin{bmatrix} 4(4)+(-1)(3)+(-4)(3) & 4(-1)+(-1)(0)+(-4)(-1) & 4(-4)+(-1)(-4)+(-4)(-3) \\ 3(4)+0(3)+(-4)(3) & 3(-1)+0(0)+(-4)(-1) & 3(-4)+0(-4)+(-4)(-3) \\ 3(4)+(-1)(3)+(-3)(3) & 3(-1)+(-1)(0)+(-3)(-1) & 3(-4)+(-1)(-4)+(-3)(-3) \end{bmatrix}$

$= \begin{bmatrix} 1 & 0 & 0 \\ 0 & 1 & 0 \\ 0 & 0 & 1 \end{bmatrix} = I$

9.2 Critical Thinking/Discussion/Writing

84. a. AB is defined when $n = 5$. The order of AB when this product is defined is $3 \times m$.

b. BA is defined when $m = 3$. The order of BA when this product is defined is $5 \times n$.

85. $(CA)B$

86. a. $P^2 = \begin{bmatrix} 0.4 & 0.3 & 0.3 \\ 0.6 & 0.3 & 0.1 \\ 0.6 & 0.1 & 0.3 \end{bmatrix} \begin{bmatrix} 0.4 & 0.3 & 0.3 \\ 0.6 & 0.3 & 0.1 \\ 0.6 & 0.1 & 0.3 \end{bmatrix} = \begin{bmatrix} 0.52 & 0.24 & 0.24 \\ 0.48 & 0.28 & 0.24 \\ 0.48 & 0.24 & 0.28 \end{bmatrix} = \begin{bmatrix} \dfrac{13}{25} & \dfrac{6}{25} & \dfrac{6}{25} \\ \dfrac{12}{25} & \dfrac{7}{25} & \dfrac{6}{25} \\ \dfrac{12}{25} & \dfrac{6}{25} & \dfrac{7}{25} \end{bmatrix}$

$XP^2 = \begin{bmatrix} \dfrac{1}{3} & \dfrac{1}{3} & \dfrac{1}{3} \end{bmatrix} \begin{bmatrix} \dfrac{13}{25} & \dfrac{6}{25} & \dfrac{6}{25} \\ \dfrac{12}{25} & \dfrac{7}{25} & \dfrac{6}{25} \\ \dfrac{12}{25} & \dfrac{6}{25} & \dfrac{7}{25} \end{bmatrix} = \begin{bmatrix} \dfrac{37}{75} & \dfrac{19}{75} & \dfrac{19}{75} \end{bmatrix}$

b. $P^3 = P^2 P = \begin{bmatrix} 0.4 & 0.3 & 0.3 \\ 0.6 & 0.3 & 0.1 \\ 0.6 & 0.1 & 0.3 \end{bmatrix} \begin{bmatrix} 0.52 & 0.24 & 0.24 \\ 0.48 & 0.28 & 0.24 \\ 0.48 & 0.24 & 0.28 \end{bmatrix} = \begin{bmatrix} 0.496 & 0.252 & 0.252 \\ 0.504 & 0.252 & 0.244 \\ 0.504 & 0.244 & 0.252 \end{bmatrix} = \begin{bmatrix} \dfrac{62}{125} & \dfrac{63}{250} & \dfrac{63}{250} \\ \dfrac{63}{125} & \dfrac{63}{250} & \dfrac{61}{250} \\ \dfrac{63}{125} & \dfrac{61}{250} & \dfrac{63}{250} \end{bmatrix}$

$XP^3 = \begin{bmatrix} \dfrac{1}{3} & \dfrac{1}{3} & \dfrac{1}{3} \end{bmatrix} \begin{bmatrix} \dfrac{62}{125} & \dfrac{63}{250} & \dfrac{63}{250} \\ \dfrac{63}{125} & \dfrac{63}{250} & \dfrac{61}{250} \\ \dfrac{63}{125} & \dfrac{61}{250} & \dfrac{63}{250} \end{bmatrix} = \begin{bmatrix} \dfrac{188}{375} & \dfrac{187}{750} & \dfrac{187}{750} \end{bmatrix}$

$P^4 = P^3 P = \begin{bmatrix} 0.496 & 0.252 & 0.252 \\ 0.504 & 0.252 & 0.244 \\ 0.504 & 0.244 & 0.252 \end{bmatrix} \begin{bmatrix} 0.4 & 0.3 & 0.3 \\ 0.6 & 0.3 & 0.1 \\ 0.6 & 0.1 & 0.3 \end{bmatrix} = \begin{bmatrix} 0.5008 & 0.2496 & 0.2496 \\ 0.4992 & 0.2512 & 0.2496 \\ 0.4992 & 0.2496 & 0.2512 \end{bmatrix} = \begin{bmatrix} \dfrac{313}{625} & \dfrac{156}{625} & \dfrac{156}{625} \\ \dfrac{312}{625} & \dfrac{157}{625} & \dfrac{156}{625} \\ \dfrac{312}{625} & \dfrac{156}{625} & \dfrac{157}{625} \end{bmatrix}$

$XP^4 = \begin{bmatrix} \dfrac{1}{3} & \dfrac{1}{3} & \dfrac{1}{3} \end{bmatrix} \begin{bmatrix} \dfrac{313}{625} & \dfrac{156}{625} & \dfrac{156}{625} \\ \dfrac{312}{625} & \dfrac{157}{625} & \dfrac{156}{625} \\ \dfrac{312}{625} & \dfrac{156}{625} & \dfrac{157}{625} \end{bmatrix} = \begin{bmatrix} \dfrac{937}{1875} & \dfrac{469}{1875} & \dfrac{469}{1875} \end{bmatrix}$

(continued on next page)

(*continued*)

$$P^5 = P^4 P = \begin{bmatrix} 0.5008 & 0.2496 & 0.2496 \\ 0.4992 & 0.2512 & 0.2496 \\ 0.4992 & 0.2496 & 0.2512 \end{bmatrix} \begin{bmatrix} 0.4 & 0.3 & 0.3 \\ 0.6 & 0.3 & 0.1 \\ 0.6 & 0.1 & 0.3 \end{bmatrix} = \begin{bmatrix} 0.49984 & 0.25008 & 0.25008 \\ 0.50016 & 0.25008 & 0.24976 \\ 0.50016 & 0.24976 & 0.25008 \end{bmatrix}$$

$$= \begin{bmatrix} \dfrac{1562}{3125} & \dfrac{1563}{6250} & \dfrac{1563}{6250} \\ \dfrac{1563}{3125} & \dfrac{1563}{6250} & \dfrac{1561}{6250} \\ \dfrac{1563}{3125} & \dfrac{1561}{6250} & \dfrac{1563}{6250} \end{bmatrix}$$

$$XP^5 = \begin{bmatrix} \dfrac{1}{3} & \dfrac{1}{3} & \dfrac{1}{3} \end{bmatrix} \begin{bmatrix} \dfrac{1562}{3125} & \dfrac{1563}{6250} & \dfrac{1563}{6250} \\ \dfrac{1563}{3125} & \dfrac{1563}{6250} & \dfrac{1561}{6250} \\ \dfrac{1563}{3125} & \dfrac{1561}{6250} & \dfrac{1563}{6250} \end{bmatrix} = \begin{bmatrix} \dfrac{4688}{9375} & \dfrac{4687}{18,750} & \dfrac{4687}{18,750} \end{bmatrix} \Rightarrow$$

$$XP^n = \begin{bmatrix} \dfrac{1}{2} & \dfrac{1}{4} & \dfrac{1}{4} \end{bmatrix}$$

In the long term, the market shares for each company are: A, 50%; B, 25%; C, 25%.

9.2 Getting Ready for the Next Section

87. $\left(\dfrac{1}{2}\right)^{-1} = 2$

89. $x^{-1} = \dfrac{1}{8} \Rightarrow x = 8$

91. $\dfrac{1}{4}x - \dfrac{7}{12} = \dfrac{11}{12} - \dfrac{5}{4}x \Rightarrow \dfrac{6}{4}x = \dfrac{18}{12} \Rightarrow$
$x = \dfrac{18}{12} \cdot \dfrac{4}{6} = 1$

93. $\dfrac{2}{4x-1} = \dfrac{3}{4x+1} \Rightarrow 2(4x+1) = 3(4x-1) \Rightarrow$
$8x + 2 = 12x - 3 \Rightarrow -4x = -5 \Rightarrow x = \dfrac{5}{4}$

95. $\begin{cases} 3x + 5y = 2 & (1) \\ 6x + 10y = 4 & (2) \end{cases}$

Multiply equation (1) by −2, then add the two equations.
$-6x - 10y = -4$
$\underline{6x + 10y = 4}$
$0 = 0$

The equations are dependent and the system has infinitely many solutions. Solve equation (1) for y in terms of x to find the general solution.

$3x + 5y = 2 \Rightarrow 5y = -3x + 2 \Rightarrow$
$y = -\dfrac{3}{5}x + \dfrac{2}{5} = \dfrac{2-3x}{5}$

Thus, the solution set can be written as
$\left\{\left(x, \dfrac{2-3x}{5}\right)\right\}.$

97. $\begin{cases} 3x + 2y = 6 & (1) \\ 6x + 4y = -13 & (2) \end{cases}$

Multiply equation (1) by −2, then add the equations.
$-6x - 4y = -12$
$\underline{6x + 4y = -13}$
$0 = -25$ False

The system is inconsistent and the solution set is $\emptyset$.

9.3 The Matrix Inverse

9.3 Practice Problems

1. We need to verify that $AB = BA = I$.
$AB = \begin{bmatrix} 3 & 2 \\ 2 & 1 \end{bmatrix} \begin{bmatrix} -1 & 2 \\ 2 & -3 \end{bmatrix}$
$= \begin{bmatrix} 3(-1) + 2(2) & 3(2) + 2(-3) \\ 2(-1) + 1(2) & 2(2) + 1(-3) \end{bmatrix} = \begin{bmatrix} 1 & 0 \\ 0 & 1 \end{bmatrix}$

(*continued on next page*)

(*continued*)

$$BA = \begin{bmatrix} -1 & 2 \\ 2 & -3 \end{bmatrix}\begin{bmatrix} 3 & 2 \\ 2 & 1 \end{bmatrix} = \begin{bmatrix} -1(3)+2(2) & -1(2)+2(1) \\ 2(3)-3(2) & 2(2)-3(1) \end{bmatrix} = \begin{bmatrix} 1 & 0 \\ 0 & 1 \end{bmatrix}$$

Thus B is the inverse of A.

2. Suppose A has an inverse B, where $B = \begin{bmatrix} x & y \\ z & w \end{bmatrix}$. Then

$$\begin{bmatrix} 3 & 1 \\ 3 & 1 \end{bmatrix}\begin{bmatrix} x & y \\ z & w \end{bmatrix} = \begin{bmatrix} 1 & 0 \\ 0 & 1 \end{bmatrix} \Rightarrow \begin{bmatrix} 3x+z & 3y+w \\ 3x+z & 3y+w \end{bmatrix} = \begin{bmatrix} 1 & 0 \\ 0 & 1 \end{bmatrix}.$$

Since the matrices are equal, the entries must be equal. So $3x+z=1$ (in the 1, 1 position) and $3x+z=0$ (in the 2, 1) position. This is a contradiction, so A does not have an inverse.

3. $[A\mid I] = \begin{bmatrix} 1 & 4 & -2 & | & 1 & 0 & 0 \\ -1 & 1 & 2 & | & 0 & 1 & 0 \\ 3 & 7 & -6 & | & 0 & 0 & 1 \end{bmatrix} \xrightarrow[-3R_1+R_3 \to R_3]{R_1+R_2 \to R_2} \begin{bmatrix} 1 & 4 & -2 & | & 1 & 0 & 0 \\ 0 & 5 & 0 & | & 1 & 1 & 0 \\ 0 & -5 & 0 & | & -3 & 0 & 1 \end{bmatrix}$

$\xrightarrow{R_2+R_3 \to R_3} \begin{bmatrix} 1 & 4 & -2 & | & 1 & 0 & 0 \\ 0 & 5 & 0 & | & 1 & 1 & 0 \\ 0 & 0 & 0 & | & -2 & 1 & 1 \end{bmatrix}$

The inverse of A does not exist.

4. $[A\mid I] = \begin{bmatrix} 1 & 2 & 3 & | & 1 & 0 & 0 \\ -2 & 3 & 1 & | & 0 & 1 & 0 \\ 4 & 5 & -2 & | & 0 & 0 & 1 \end{bmatrix} \xrightarrow[-4R_1+R_3 \to R_3]{2R_1+R_2 \to R_2} \begin{bmatrix} 1 & 2 & 3 & | & 1 & 0 & 0 \\ 0 & 7 & 7 & | & 2 & 1 & 0 \\ 0 & -3 & -14 & | & -4 & 0 & 1 \end{bmatrix}$

$\xrightarrow{\frac{1}{11}(2R_2+R_3) \to R_2} \begin{bmatrix} 1 & 2 & 3 & | & 1 & 0 & 0 \\ 0 & 1 & 0 & | & 0 & \frac{2}{11} & \frac{1}{11} \\ 0 & -3 & -14 & | & -4 & 0 & 1 \end{bmatrix} \xrightarrow[\left(-\frac{1}{14}\right)(3R_2+R_3) \to R_3]{R_1-2R_2 \to R_1} \begin{bmatrix} 1 & 0 & 3 & | & 1 & -\frac{4}{11} & -\frac{2}{11} \\ 0 & 1 & 0 & | & 0 & \frac{2}{11} & \frac{1}{11} \\ 0 & 0 & 1 & | & \frac{2}{7} & -\frac{3}{77} & -\frac{1}{11} \end{bmatrix}$

$\xrightarrow{R_1-3R_3 \to R_1} \begin{bmatrix} 1 & 0 & 3 & | & \frac{1}{7} & -\frac{19}{77} & \frac{1}{11} \\ 0 & 1 & 0 & | & 0 & \frac{2}{11} & \frac{1}{11} \\ 0 & 0 & 1 & | & \frac{2}{7} & -\frac{3}{77} & -\frac{1}{11} \end{bmatrix} \Rightarrow A^{-1} = \begin{bmatrix} \frac{1}{7} & -\frac{19}{77} & \frac{1}{11} \\ 0 & \frac{2}{11} & \frac{1}{11} \\ \frac{2}{7} & -\frac{3}{77} & -\frac{1}{11} \end{bmatrix}$

5. a. $ad - bc = (8)(1) - (2)(4) = 0$

   The inverse of matrix A does not exist.

   b. $ad - bc = (8)(1) - (-2)(3) = 14$

   Thus, matrix B is invertible.

   $$B^{-1} = \frac{1}{14}\begin{bmatrix} 1 & 2 \\ -3 & 8 \end{bmatrix} = \begin{bmatrix} \frac{1}{14} & \frac{1}{7} \\ -\frac{3}{14} & \frac{4}{7} \end{bmatrix}$$

6. $\begin{cases} 3x+2y+3z = 9 \\ 3x+ y = 12 \\ x+ z = 6 \end{cases} \Rightarrow AX = B \Rightarrow \begin{bmatrix} 3 & 2 & 3 \\ 3 & 1 & 0 \\ 1 & 0 & 1 \end{bmatrix} \begin{bmatrix} x \\ y \\ z \end{bmatrix} = \begin{bmatrix} 9 \\ 12 \\ 6 \end{bmatrix} \Rightarrow X = A^{-1}B$

Using a graphing calculator, we find that $A^{-1} = \begin{bmatrix} -\frac{1}{6} & \frac{1}{3} & \frac{1}{2} \\ \frac{1}{2} & 0 & -\frac{3}{2} \\ \frac{1}{6} & -\frac{1}{3} & \frac{1}{2} \end{bmatrix}$.

$X = A^{-1}B \Rightarrow \begin{bmatrix} x \\ y \\ z \end{bmatrix} = \begin{bmatrix} -\frac{1}{5} & \frac{2}{5} & \frac{2}{5} \\ \frac{3}{5} & -\frac{1}{5} & -\frac{6}{5} \\ \frac{1}{5} & -\frac{2}{5} & \frac{3}{5} \end{bmatrix} \begin{bmatrix} 9 \\ 12 \\ 6 \end{bmatrix} \Rightarrow X = \begin{bmatrix} \frac{11}{2} \\ -\frac{9}{2} \\ \frac{1}{2} \end{bmatrix}$

Solution set: $\left\{ \left(\frac{11}{2}, -\frac{9}{2}, \frac{1}{2} \right) \right\}$

7. Following the discussion given in Example 7, we have $X = (I - A)^{-1} D = \frac{1}{19} \begin{bmatrix} 36 & 24 \\ 16 & 36 \end{bmatrix} \begin{bmatrix} 800 \\ 3400 \end{bmatrix} = \begin{bmatrix} \frac{110,400}{19} \\ \frac{135,200}{19} \end{bmatrix}$

   To meet the consumer demand for 800 units of energy and 3400 units of food, the energy produced must be $\frac{110,400}{19}$ units and food production must be $\frac{135,200}{19}$ units.

8. Associate each letter in the phrase with the number representing its position in the alphabet, and partition the numbers into groups of three (inserting two zeros at the right end), forming a 3 × 6 matrix:

   J   A   C   K   I   S   N   O   W   S   A   F   E
   [10 1   3]   [11 0   9]   [19 0   14] [15 23   0] [19 1   6] [5   0   0]

   $M = \begin{bmatrix} 10 & 11 & 19 & 15 & 19 & 5 \\ 1 & 0 & 0 & 23 & 1 & 0 \\ 3 & 9 & 14 & 0 & 6 & 0 \end{bmatrix}$

   Let $A = \begin{bmatrix} 1 & 2 & 3 \\ 1 & 3 & 3 \\ 1 & 2 & 4 \end{bmatrix}$. Then, $AM = \begin{bmatrix} 1 & 2 & 3 \\ 1 & 3 & 3 \\ 1 & 2 & 4 \end{bmatrix} \begin{bmatrix} 10 & 11 & 19 & 15 & 19 & 5 \\ 1 & 0 & 0 & 23 & 1 & 0 \\ 3 & 9 & 14 & 0 & 6 & 0 \end{bmatrix} = \begin{bmatrix} 21 & 38 & 61 & 61 & 39 & 5 \\ 22 & 38 & 61 & 84 & 40 & 5 \\ 24 & 47 & 75 & 61 & 45 & 5 \end{bmatrix}$

   The cryptogram is 21 22 24 38 38 47 61 61 75 61 84 61 39 40 45 5 5 5.

9.3 Concepts and Vocabulary

1. For $n \times n$ matrices A and B, if $AB = I$, then B is called the <u>inverse</u> of A.

3. To find the inverse of an invertible matrix A, we transform $[A | I]$ by a sequence of row operations into $[I | B]$, where $B = \underline{A^{-1}}$.

5. False.

7. True

9.3 Building Skills

9. $AB = \begin{bmatrix} 1 & 2 \\ 1 & 3 \end{bmatrix} \begin{bmatrix} 3 & -2 \\ -1 & 1 \end{bmatrix} = \begin{bmatrix} 1 & 0 \\ 0 & 1 \end{bmatrix}$

   $BA = \begin{bmatrix} 3 & -2 \\ -1 & 1 \end{bmatrix} \begin{bmatrix} 1 & 2 \\ 1 & 3 \end{bmatrix} = \begin{bmatrix} 1 & 0 \\ 0 & 1 \end{bmatrix} \Rightarrow$

   B is the inverse of A.

11. $AB = \begin{bmatrix} 3 & 2 \\ 1 & 4 \end{bmatrix} \begin{bmatrix} \frac{2}{5} & -\frac{1}{5} \\ -\frac{1}{10} & \frac{3}{10} \end{bmatrix} = \begin{bmatrix} 1 & 0 \\ 0 & 1 \end{bmatrix}$

$BA = \begin{bmatrix} \frac{2}{5} & -\frac{1}{5} \\ -\frac{1}{10} & \frac{3}{10} \end{bmatrix} \begin{bmatrix} 3 & 2 \\ 1 & 4 \end{bmatrix} = \begin{bmatrix} 1 & 0 \\ 0 & 1 \end{bmatrix} \Rightarrow$

B is the inverse of A.

13. $AB = \begin{bmatrix} 1 & 0 & -1 \\ -1 & 1 & 1 \end{bmatrix} \begin{bmatrix} 2 & 1 \\ 1 & 1 \\ 1 & 1 \end{bmatrix} = \begin{bmatrix} 1 & 0 \\ 0 & 1 \end{bmatrix}$

$BA = \begin{bmatrix} 2 & 1 \\ 1 & 1 \\ 1 & 1 \end{bmatrix} \begin{bmatrix} 1 & 0 & -1 \\ -1 & 1 & 1 \end{bmatrix} = \begin{bmatrix} 1 & 1 & -1 \\ 0 & 1 & 0 \\ 0 & 1 & 0 \end{bmatrix} \Rightarrow$

B is not the inverse of A.

15. $AB = \begin{bmatrix} 1 & -2 & 1 \\ -8 & 6 & -2 \\ 5 & -3 & 1 \end{bmatrix} \left(\frac{1}{2} \begin{bmatrix} 0 & 1 & 2 \\ 1 & 2 & 3 \\ 3 & 1 & 1 \end{bmatrix} \right)$

$= \begin{bmatrix} \frac{1}{2} & -1 & -\frac{3}{2} \\ 0 & 1 & 0 \\ 0 & 0 & 1 \end{bmatrix} \Rightarrow B$ is not the inverse of A.

17. $AB = \begin{bmatrix} 1 & 1 & 1 \\ 1 & 2 & 3 \\ -1 & 1 & -1 \end{bmatrix} \left(\frac{1}{4} \begin{bmatrix} 5 & -2 & -1 \\ 2 & 0 & 2 \\ -3 & 2 & -1 \end{bmatrix} \right)$

$= \begin{bmatrix} 1 & 0 & 0 \\ 0 & 1 & 0 \\ 0 & 0 & 1 \end{bmatrix}$

$BA = \left(\frac{1}{4} \begin{bmatrix} 5 & -2 & -1 \\ 2 & 0 & 2 \\ -3 & 2 & -1 \end{bmatrix} \right) \begin{bmatrix} 1 & 1 & 1 \\ 1 & 2 & 3 \\ -1 & 1 & -1 \end{bmatrix}$

$= \begin{bmatrix} 1 & 0 & 0 \\ 0 & 1 & 0 \\ 0 & 0 & 1 \end{bmatrix} \Rightarrow B$ is the inverse of A.

In exercises 19–28, the steps to find the inverse may vary. Be sure to verify that $AA^{-1} = I$.

19. $[A\,|\,I] = \begin{bmatrix} 2 & 0 & | & 1 & 0 \\ 1 & 3 & | & 0 & 1 \end{bmatrix} \xrightarrow{\frac{1}{2}R_1} \begin{bmatrix} 1 & 0 & | & \frac{1}{2} & 0 \\ 1 & 3 & | & 0 & 1 \end{bmatrix}$

$\xrightarrow{R_1 - R_2 \to R_2} \begin{bmatrix} 1 & 0 & | & \frac{1}{2} & 0 \\ 0 & -3 & | & \frac{1}{2} & -1 \end{bmatrix}$

$\xrightarrow{-\frac{1}{3}R_2} \begin{bmatrix} 1 & 0 & | & \frac{1}{2} & 0 \\ 0 & 1 & | & -\frac{1}{6} & \frac{1}{3} \end{bmatrix} \Rightarrow$

$A^{-1} = \begin{bmatrix} \frac{1}{2} & 0 \\ -\frac{1}{6} & \frac{1}{3} \end{bmatrix}$

21. $[A\,|\,I] = \begin{bmatrix} 2 & 4 & | & 1 & 0 \\ 3 & 6 & | & 0 & 1 \end{bmatrix}$

$\xrightarrow{\frac{1}{2}R_1 \to R_1} \begin{bmatrix} 1 & 2 & | & \frac{1}{2} & 0 \\ 3 & 6 & | & 0 & 1 \end{bmatrix}$

$\xrightarrow{-3R_1 + R_2 \to R_2} \begin{bmatrix} 1 & 2 & | & \frac{1}{2} & 0 \\ 0 & 0 & | & -\frac{3}{2} & 1 \end{bmatrix} \Rightarrow$

The inverse of A does not exist.

23. $[A\,|\,I] = \begin{bmatrix} 1 & 6 & 4 & | & 1 & 0 & 0 \\ 0 & 2 & 3 & | & 0 & 1 & 0 \\ 0 & 1 & 2 & | & 0 & 0 & 1 \end{bmatrix}$

$\xrightarrow{2R_3 - R_2 \to R_3} \begin{bmatrix} 1 & 6 & 4 & | & 1 & 0 & 0 \\ 0 & 2 & 3 & | & 0 & 1 & 0 \\ 0 & 0 & 1 & | & 0 & -1 & 2 \end{bmatrix}$

$\xrightarrow{\frac{1}{2}(-3R_3 + R_2) \to R_2} \begin{bmatrix} 1 & 6 & 4 & | & 1 & 0 & 0 \\ 0 & 1 & 0 & | & 0 & 2 & -3 \\ 0 & 0 & 1 & | & 0 & -1 & 2 \end{bmatrix}$

$\xrightarrow[R_1 - 6R_2 \to R_1]{R_1 - 4R_3 \to R_1} \begin{bmatrix} 1 & 0 & 0 & | & 1 & -8 & 10 \\ 0 & 1 & 0 & | & 0 & 2 & -3 \\ 0 & 0 & 1 & | & 0 & -1 & 2 \end{bmatrix}$

$A^{-1} = \begin{bmatrix} 1 & -8 & 10 \\ 0 & 2 & -3 \\ 0 & -1 & 2 \end{bmatrix}$

Section 9.3 The Matrix Inverse   549

25. $[A|I] = \begin{bmatrix} 2 & 3 & 1 & | & 1 & 0 & 0 \\ 2 & 4 & 1 & | & 0 & 1 & 0 \\ 3 & 7 & 2 & | & 0 & 0 & 1 \end{bmatrix}$

$\xrightarrow{R_2 - R_1 \to R_2} \begin{bmatrix} 2 & 3 & 1 & | & 1 & 0 & 0 \\ 0 & 1 & 0 & | & -1 & 1 & 0 \\ 3 & 7 & 2 & | & 0 & 0 & 1 \end{bmatrix}$

$\xrightarrow{2R_1 - R_3 \to R_1} \begin{bmatrix} 1 & -1 & 0 & | & 2 & 0 & -1 \\ 0 & 1 & 0 & | & -1 & 1 & 0 \\ 3 & 7 & 2 & | & 0 & 0 & 1 \end{bmatrix}$

$\xrightarrow[R_3 - 7R_2 \to R_3]{R_1 + R_2 \to R_1} \begin{bmatrix} 1 & 0 & 0 & | & 1 & 1 & -1 \\ 0 & 1 & 0 & | & -1 & 1 & 0 \\ 3 & 0 & 2 & | & 7 & -7 & 1 \end{bmatrix}$

$\xrightarrow{\frac{1}{2}(R_3 - 3R_1) \to R_3} \begin{bmatrix} 1 & 0 & 0 & | & 1 & 1 & -1 \\ 0 & 1 & 0 & | & -1 & 1 & 0 \\ 0 & 0 & 1 & | & 2 & -5 & 2 \end{bmatrix}$

$A^{-1} = \begin{bmatrix} 1 & 1 & -1 \\ -1 & 1 & 0 \\ 2 & -5 & 2 \end{bmatrix}$

27. $[A|I] = \begin{bmatrix} 3 & -3 & 4 & | & 1 & 0 & 0 \\ 2 & -3 & 4 & | & 0 & 1 & 0 \\ 0 & -1 & 1 & | & 0 & 0 & 1 \end{bmatrix}$

$\xrightarrow{R_1 - R_2 \to R_1} \begin{bmatrix} 1 & 0 & 0 & | & 1 & -1 & 0 \\ 2 & -3 & 4 & | & 0 & 1 & 0 \\ 0 & -1 & 1 & | & 0 & 0 & 1 \end{bmatrix}$

$\xrightarrow{2R_1 - R_2 \to R_2} \begin{bmatrix} 1 & 0 & 0 & | & 1 & -1 & 0 \\ 0 & 3 & -4 & | & 2 & -3 & 0 \\ 0 & -1 & 1 & | & 0 & 0 & 1 \end{bmatrix}$

$\xrightarrow{-(R_2 + 4R_3) \to R_2} \begin{bmatrix} 1 & 0 & 0 & | & 1 & -1 & 0 \\ 0 & 1 & 0 & | & -2 & 3 & -4 \\ 0 & -1 & 1 & | & 0 & 0 & 1 \end{bmatrix}$

$\xrightarrow{R_2 + R_3 \to R_3} \begin{bmatrix} 1 & 0 & 0 & | & 1 & -1 & 0 \\ 0 & 1 & 0 & | & -2 & 3 & -4 \\ 0 & 0 & 1 & | & -2 & 3 & -3 \end{bmatrix}$

$A^{-1} = \begin{bmatrix} 1 & -1 & 0 \\ -2 & 3 & -4 \\ -2 & 3 & -3 \end{bmatrix}$

29. $ad - bc = (1)(2) - (0)(3) = 2$

$A^{-1} = \frac{1}{2}\begin{bmatrix} 2 & 0 \\ -3 & 1 \end{bmatrix} = \begin{bmatrix} 1 & 0 \\ -\frac{3}{2} & \frac{1}{2} \end{bmatrix}$

31. $ad - bc = (2)(5) - (-3)(-3) = 1$

$A^{-1} = \begin{bmatrix} 5 & 3 \\ 3 & 2 \end{bmatrix}$

33. $ad - bc = (a)(-a) - (b)(-b) = -a^2 + b^2$

$A^{-1} = \frac{1}{-a^2 + b^2}\begin{bmatrix} -a & b \\ -b & a \end{bmatrix}$ where $a^2 \neq b^2$.

35. $\begin{cases} 2x + 3y = -9 \\ x - 3y = 13 \end{cases} \Rightarrow \begin{bmatrix} 2 & 3 \\ 1 & -3 \end{bmatrix}\begin{bmatrix} x \\ y \end{bmatrix} = \begin{bmatrix} -9 \\ 13 \end{bmatrix}$

37. $\begin{cases} 3x + 2y + z = 8 \\ 2x + y + 3z = 7 \\ x + 3y + 2z = 9 \end{cases} \Rightarrow \begin{bmatrix} 3 & 2 & 1 \\ 2 & 1 & 3 \\ 1 & 3 & 2 \end{bmatrix}\begin{bmatrix} x \\ y \\ z \end{bmatrix} = \begin{bmatrix} 8 \\ 7 \\ 9 \end{bmatrix}$

39. $\begin{bmatrix} 1 & -2 \\ 2 & 1 \end{bmatrix}\begin{bmatrix} x \\ y \end{bmatrix} = \begin{bmatrix} 0 \\ 5 \end{bmatrix} \Rightarrow \begin{cases} x - 2y = 0 \\ 2x + y = 5 \end{cases}$

41. $\begin{bmatrix} 2 & 3 & 1 \\ 5 & 7 & -1 \\ 4 & 3 & 0 \end{bmatrix}\begin{bmatrix} x_1 \\ x_2 \\ x_3 \end{bmatrix} = \begin{bmatrix} -1 \\ 5 \\ 5 \end{bmatrix} \Rightarrow \begin{cases} 2x_1 + 3x_2 + x_3 = -1 \\ 5x_1 + 7x_2 - x_3 = 5 \\ 4x_1 + 3x_2 = 5 \end{cases}$

43. $\begin{bmatrix} 1 & 2 & 5 \\ 2 & 3 & 8 \\ -1 & 1 & 2 \end{bmatrix}\begin{bmatrix} 2 & -1 & -1 \\ 12 & -7 & -2 \\ -5 & 3 & 1 \end{bmatrix} = \begin{bmatrix} 1 & 0 & 0 \\ 0 & 1 & 0 \\ 0 & 0 & 1 \end{bmatrix}$

45. $\begin{cases} x_1 + 2x_2 + 5x_3 = 1 \\ 2x_1 + 3x_2 + 8x_3 = 3 \\ -x_1 + x_2 + 2x_3 = -3 \end{cases} \Rightarrow AX = B \Rightarrow$

$\begin{bmatrix} 1 & 2 & 5 \\ 2 & 3 & 8 \\ -1 & 1 & 2 \end{bmatrix}\begin{bmatrix} x \\ y \\ z \end{bmatrix} = \begin{bmatrix} 1 \\ 3 \\ -3 \end{bmatrix} \Rightarrow X = A^{-1}B \Rightarrow$

$\begin{bmatrix} x \\ y \\ z \end{bmatrix} = \begin{bmatrix} 2 & -1 & -1 \\ 12 & -7 & -2 \\ -5 & 3 & 1 \end{bmatrix}\begin{bmatrix} 1 \\ 3 \\ -3 \end{bmatrix} \Rightarrow \begin{bmatrix} x \\ y \\ z \end{bmatrix} = \begin{bmatrix} 2 \\ -3 \\ 1 \end{bmatrix}$

The solution is $\{(2, -3, 1)\}$.

47. a. $[A|I] = \begin{bmatrix} 1 & 1 & 1 & | & 1 & 0 & 0 \\ 1 & 2 & 3 & | & 0 & 1 & 0 \\ 1 & 4 & 9 & | & 0 & 0 & 1 \end{bmatrix}$

$\xrightarrow[R_3 - R_1 \to R_3]{R_2 - R_1 \to R_2} \begin{bmatrix} 1 & 1 & 1 & | & 1 & 0 & 0 \\ 0 & 1 & 2 & | & -1 & 1 & 0 \\ 0 & 3 & 8 & | & -1 & 0 & 1 \end{bmatrix}$

$\xrightarrow[\frac{1}{2}(R_3 - 3R_2) \to R_3]{R_1 - R_2 \to R_1} \begin{bmatrix} 1 & 0 & -1 & | & 2 & -1 & 0 \\ 0 & 1 & 2 & | & -1 & 1 & 0 \\ 0 & 0 & 1 & | & 1 & -\frac{3}{2} & \frac{1}{2} \end{bmatrix}$

(*continued on next page*)

(continued)

$$\xrightarrow[R_2-2R_3 \to R_2]{R_1+R_3 \to R_1} \begin{bmatrix} 1 & 0 & 0 & | & 3 & -\frac{5}{2} & \frac{1}{2} \\ 0 & 1 & 0 & | & -3 & 4 & -1 \\ 0 & 0 & 1 & | & 1 & -\frac{3}{2} & \frac{1}{2} \end{bmatrix}$$

$$A^{-1} = \begin{bmatrix} 3 & -\frac{5}{2} & \frac{1}{2} \\ -3 & 4 & -1 \\ 1 & -\frac{3}{2} & \frac{1}{2} \end{bmatrix}$$

b. $\begin{cases} x+y+z=6 \\ x+2y+3z=14 \\ x+4y+9z=36 \end{cases} \Rightarrow AX=B \Rightarrow$

$$\begin{bmatrix} 1 & 1 & 1 \\ 1 & 2 & 3 \\ 1 & 4 & 9 \end{bmatrix} \begin{bmatrix} x \\ y \\ z \end{bmatrix} = \begin{bmatrix} 6 \\ 14 \\ 36 \end{bmatrix} \Rightarrow X = A^{-1}B \Rightarrow$$

$$\begin{bmatrix} x \\ y \\ z \end{bmatrix} = \begin{bmatrix} 3 & -\frac{5}{2} & \frac{1}{2} \\ -3 & 4 & -1 \\ 1 & -\frac{3}{2} & \frac{1}{2} \end{bmatrix} \begin{bmatrix} 6 \\ 14 \\ 36 \end{bmatrix} \Rightarrow \begin{bmatrix} x \\ y \\ z \end{bmatrix} = \begin{bmatrix} 1 \\ 2 \\ 3 \end{bmatrix}$$

The solution is $\{(1,2,3)\}$.

9.3 Applying the Concepts

49. $\begin{cases} 3x+7y=17 \\ -5x+4y=13 \end{cases} \Rightarrow \begin{bmatrix} 3 & 7 \\ -5 & 4 \end{bmatrix} \begin{bmatrix} x \\ y \end{bmatrix} = \begin{bmatrix} 11 \\ 13 \end{bmatrix}$

$$[A\,|\,I] = \begin{bmatrix} 3 & 7 & | & 1 & 0 \\ -5 & 4 & | & 0 & 1 \end{bmatrix}$$

$$\xrightarrow{5R_1+3R_2 \to R_2} \begin{bmatrix} 3 & 7 & | & 1 & 0 \\ 0 & 47 & | & 5 & 3 \end{bmatrix}$$

$$\xrightarrow{\frac{1}{47}R_2 \to R_2} \begin{bmatrix} 3 & 7 & | & 1 & 0 \\ 0 & 1 & | & \frac{5}{47} & \frac{3}{47} \end{bmatrix}$$

$$\xrightarrow{\frac{1}{3}(R_1-7R_2) \to R_1} \begin{bmatrix} 1 & 0 & | & \frac{4}{47} & -\frac{7}{47} \\ 0 & 1 & | & \frac{5}{47} & \frac{3}{47} \end{bmatrix}$$

$$A^{-1} = \begin{bmatrix} \frac{4}{47} & -\frac{7}{47} \\ \frac{5}{47} & \frac{3}{47} \end{bmatrix}$$

$$X = A^{-1}B \Rightarrow \begin{bmatrix} x \\ y \end{bmatrix} = \begin{bmatrix} \frac{4}{47} & -\frac{7}{47} \\ \frac{5}{47} & \frac{3}{47} \end{bmatrix} \begin{bmatrix} 11 \\ 13 \end{bmatrix} = \begin{bmatrix} -1 \\ 2 \end{bmatrix}$$

The solution is $\{(-1,2)\}$.

51. $\begin{cases} x+y+2z=7 \\ x-y-3z=-6 \\ 2x+3y+z=4 \end{cases} \Rightarrow \begin{bmatrix} 1 & 1 & 2 \\ 1 & -1 & -3 \\ 2 & 3 & 1 \end{bmatrix} \begin{bmatrix} x \\ y \\ z \end{bmatrix} = \begin{bmatrix} 7 \\ -6 \\ 4 \end{bmatrix}$

$$[A\,|\,I] = \begin{bmatrix} 1 & 1 & 2 & | & 1 & 0 & 0 \\ 1 & -1 & -3 & | & 0 & 1 & 0 \\ 2 & 3 & 1 & | & 0 & 0 & 1 \end{bmatrix}$$

$$\xrightarrow[R_3-2R_1 \to R_3]{R_1-R_2 \to R_2} \begin{bmatrix} 1 & 1 & 2 & | & 1 & 0 & 0 \\ 0 & 2 & 5 & | & 1 & -1 & 0 \\ 0 & 1 & -3 & | & -2 & 0 & 1 \end{bmatrix}$$

$$\xrightarrow[\frac{1}{11}(R_2-2R_3) \to R_3]{2R_1-R_2 \to R_1} \begin{bmatrix} 2 & 0 & -1 & | & 1 & 1 & 0 \\ 0 & 2 & 5 & | & 1 & -1 & 0 \\ 0 & 0 & 1 & | & \frac{5}{11} & -\frac{1}{11} & -\frac{2}{11} \end{bmatrix}$$

$$\xrightarrow[\frac{1}{2}(R_2-5R_3) \to R_2]{\frac{1}{2}(R_1+R_3) \to R_1} \begin{bmatrix} 1 & 0 & 0 & | & \frac{8}{11} & \frac{5}{11} & -\frac{1}{11} \\ 0 & 1 & 0 & | & -\frac{7}{11} & -\frac{3}{11} & \frac{5}{11} \\ 0 & 0 & 1 & | & \frac{5}{11} & -\frac{1}{11} & -\frac{2}{11} \end{bmatrix}$$

$$A^{-1} = \begin{bmatrix} \frac{8}{11} & \frac{5}{11} & -\frac{1}{11} \\ -\frac{7}{11} & -\frac{3}{11} & \frac{5}{11} \\ \frac{5}{11} & -\frac{1}{11} & -\frac{2}{11} \end{bmatrix}$$

$X = A^{-1}B \Rightarrow$

$$\begin{bmatrix} x \\ y \\ z \end{bmatrix} = \begin{bmatrix} \frac{8}{11} & \frac{5}{11} & -\frac{1}{11} \\ -\frac{7}{11} & -\frac{3}{11} & \frac{5}{11} \\ \frac{5}{11} & -\frac{1}{11} & -\frac{2}{11} \end{bmatrix} \begin{bmatrix} 7 \\ -6 \\ 4 \end{bmatrix} = \begin{bmatrix} 2 \\ -1 \\ 3 \end{bmatrix}$$

The solution is $\{(2, -1, 3)\}$.

53. $\begin{cases} 2x+2y+3x=7 \\ 5x+3y+5z=3 \\ 3x+5y+z=-5 \end{cases} \Rightarrow \begin{bmatrix} 2 & 2 & 3 \\ 5 & 3 & 5 \\ 3 & 5 & 1 \end{bmatrix} \begin{bmatrix} x \\ y \\ z \end{bmatrix} = \begin{bmatrix} 7 \\ 3 \\ -5 \end{bmatrix}$

$[A \mid I] = \begin{bmatrix} 2 & 2 & 3 & | & 1 & 0 & 0 \\ 5 & 3 & 5 & | & 0 & 1 & 0 \\ 3 & 5 & 1 & | & 0 & 0 & 1 \end{bmatrix} \xrightarrow[3R_1 - 2R_3 \to R_3]{5R_1 - 2R_2 \to R_2} \begin{bmatrix} 2 & 2 & 3 & | & 1 & 0 & 0 \\ 0 & 4 & 5 & | & 5 & -2 & 0 \\ 0 & -4 & 7 & | & 3 & 0 & -2 \end{bmatrix}$

$\xrightarrow[\frac{1}{12}(R_2+R_3) \to R_3]{R_2 - 2R_1 \to R_1} \begin{bmatrix} -4 & 0 & -1 & | & 3 & -2 & 0 \\ 0 & 4 & 5 & | & 5 & -2 & 0 \\ 0 & 0 & 1 & | & \frac{2}{3} & -\frac{1}{6} & -\frac{1}{6} \end{bmatrix} \xrightarrow[\frac{1}{4}(R_2 - 5R_3) \to R_2]{-\frac{1}{4}(R_1+R_3) \to R_1} \begin{bmatrix} 1 & 0 & 0 & | & -\frac{11}{12} & \frac{13}{24} & \frac{1}{24} \\ 0 & 1 & 0 & | & \frac{5}{12} & -\frac{7}{24} & \frac{5}{24} \\ 0 & 0 & 1 & | & \frac{2}{3} & -\frac{1}{6} & -\frac{1}{6} \end{bmatrix} \Rightarrow$

$A^{-1} = \begin{bmatrix} -\frac{11}{12} & \frac{13}{24} & \frac{1}{24} \\ \frac{5}{12} & -\frac{7}{24} & \frac{5}{24} \\ \frac{2}{3} & -\frac{1}{6} & -\frac{1}{6} \end{bmatrix}$

$X = A^{-1}B \Rightarrow \begin{bmatrix} x \\ y \\ z \end{bmatrix} = \begin{bmatrix} -\frac{11}{12} & \frac{13}{24} & \frac{1}{24} \\ \frac{5}{12} & -\frac{7}{24} & \frac{5}{24} \\ \frac{2}{3} & -\frac{1}{6} & -\frac{1}{6} \end{bmatrix} \begin{bmatrix} 7 \\ 3 \\ -5 \end{bmatrix} = \begin{bmatrix} -5 \\ 1 \\ 5 \end{bmatrix}$

The solution is $\{(-5, 1, 5)\}$.

55. Let x = the amount invested in a treasury bill, y = the amount invested in bonds, and z = the amount invested in a mutual fund. Then we have

$\begin{cases} x + y + z = 90{,}000 \\ 0.03x + 0.07y - 0.11z = 980 \\ 0.03x + 0.07y + 0.08z = 5920 \end{cases} \Rightarrow \begin{bmatrix} 1 & 1 & 1 \\ 0.03 & 0.07 & -0.11 \\ 0.03 & 0.07 & 0.08 \end{bmatrix} \begin{bmatrix} x \\ y \\ z \end{bmatrix} = \begin{bmatrix} 90{,}000 \\ 980 \\ 5920 \end{bmatrix}$

$[A \mid I] = \begin{bmatrix} 1 & 1 & 1 & | & 1 & 0 & 0 \\ \frac{3}{100} & \frac{7}{100} & -\frac{11}{100} & | & 0 & 1 & 0 \\ \frac{3}{100} & \frac{7}{100} & \frac{8}{100} & | & 0 & 0 & 1 \end{bmatrix} \xrightarrow[-\frac{3}{100}R_3 + R_3 \to]{-\frac{3}{100}R_1 + R_2 \to R_2} \begin{bmatrix} 1 & 1 & 1 & | & 1 & 0 & 0 \\ 0 & \frac{1}{25} & -\frac{7}{50} & | & -\frac{3}{100} & 1 & 0 \\ 0 & \frac{1}{25} & \frac{1}{20} & | & -\frac{3}{100} & 0 & 1 \end{bmatrix}$

$\xrightarrow{-\frac{100}{19}(R_2 - R_3) \to R_3} \begin{bmatrix} 1 & 1 & 1 & | & 1 & 0 & 0 \\ 0 & \frac{1}{25} & -\frac{7}{50} & | & -\frac{3}{100} & 1 & 0 \\ 0 & 0 & 1 & | & 0 & -\frac{100}{19} & \frac{100}{19} \end{bmatrix} \xrightarrow{25R_2 \to R_2} \begin{bmatrix} 1 & 1 & 1 & | & 1 & 0 & 0 \\ 0 & 1 & -\frac{7}{2} & | & -\frac{3}{4} & 25 & 0 \\ 0 & 0 & 1 & | & 0 & -\frac{100}{19} & \frac{100}{19} \end{bmatrix}$

(continued on next page)

(*continued*)

$$\xrightarrow[R_2 + \frac{7}{2}R_3 \to R_2]{R_1 - R_3 \to R_1} \begin{bmatrix} 1 & 1 & 0 & 1 & \frac{100}{19} & -\frac{100}{19} \\ 0 & 1 & 0 & -\frac{3}{4} & \frac{125}{19} & \frac{350}{19} \\ 0 & 0 & 1 & 0 & -\frac{100}{19} & \frac{100}{19} \end{bmatrix} \xrightarrow{R_1 - R_2 \to R_1} \begin{bmatrix} 1 & 1 & 0 & \frac{7}{4} & -\frac{25}{19} & -\frac{450}{19} \\ 0 & 1 & 0 & -\frac{3}{4} & \frac{125}{19} & \frac{350}{19} \\ 0 & 0 & 1 & 0 & -\frac{100}{19} & \frac{100}{19} \end{bmatrix}$$

$$A^{-1} = \begin{bmatrix} \frac{7}{4} & -\frac{25}{19} & -\frac{450}{19} \\ -\frac{3}{4} & \frac{125}{19} & \frac{350}{19} \\ 0 & -\frac{100}{19} & \frac{100}{19} \end{bmatrix} \Rightarrow X = A^{-1}B \Rightarrow \begin{bmatrix} x \\ y \\ z \end{bmatrix} = \begin{bmatrix} \frac{7}{4} & -\frac{25}{19} & -\frac{450}{19} \\ -\frac{3}{4} & \frac{125}{19} & \frac{350}{19} \\ 0 & -\frac{100}{19} & \frac{100}{19} \end{bmatrix} \begin{bmatrix} 90,000 \\ 980 \\ 5920 \end{bmatrix} = \begin{bmatrix} 16,000 \\ 48,000 \\ 26,000 \end{bmatrix}$$

Liz invested $16,000 in a treasury bill, $48,000 in bonds, and $26,000 in a mutual fund.

57. $\begin{cases} V - E + R = 2 \\ 2E = 3V \\ 2R = E - 1 \end{cases} \Rightarrow \begin{cases} V - E + R = 2 \\ -3V + 2E = 0 \\ -E + 2R = -1 \end{cases} \Rightarrow \begin{bmatrix} 1 & -1 & 1 \\ -3 & 2 & 0 \\ 0 & -1 & 2 \end{bmatrix} \begin{bmatrix} x \\ y \\ z \end{bmatrix} = \begin{bmatrix} 2 \\ 0 \\ 1 \end{bmatrix}$

$$[A \mid I] = \begin{bmatrix} 1 & -1 & 1 & 1 & 0 & 0 \\ -3 & 2 & 0 & 0 & 1 & 0 \\ 0 & -1 & 2 & 0 & 0 & 1 \end{bmatrix} \xrightarrow{3R_1 + R_2 \to R_2} \begin{bmatrix} 1 & -1 & 1 & 1 & 0 & 0 \\ 0 & -1 & 3 & 3 & 1 & 0 \\ 0 & -1 & 2 & 0 & 0 & 1 \end{bmatrix}$$

$$\xrightarrow{-R_2 \to R_2} \begin{bmatrix} 1 & -1 & 1 & 1 & 0 & 0 \\ 0 & 1 & -3 & -3 & -1 & 0 \\ 0 & -1 & 2 & 0 & 0 & 1 \end{bmatrix} \xrightarrow[-(R_3 + R_2) \to R_3]{R_1 + R_2 \to R_1} \begin{bmatrix} 1 & 0 & -2 & -2 & -1 & 0 \\ 0 & 1 & -3 & -3 & -1 & 0 \\ 0 & 0 & 1 & 3 & 1 & -1 \end{bmatrix}$$

$$\xrightarrow[R_2 + 3R_3 \to R_2]{R_1 + 2R_3 \to R_1} \begin{bmatrix} 1 & 0 & 0 & 4 & 1 & -2 \\ 0 & 1 & 0 & 6 & 2 & -3 \\ 0 & 0 & 1 & 3 & 1 & -1 \end{bmatrix}$$

$$A^{-1} = \begin{bmatrix} 4 & 1 & -2 \\ 6 & 2 & -3 \\ 3 & 1 & -1 \end{bmatrix} \Rightarrow X = A^{-1}B \Rightarrow \begin{bmatrix} x \\ y \\ z \end{bmatrix} = \begin{bmatrix} 4 & 1 & -2 \\ 6 & 2 & -3 \\ 3 & 1 & -1 \end{bmatrix} \begin{bmatrix} 2 \\ 0 \\ 1 \end{bmatrix} = \begin{bmatrix} 10 \\ 15 \\ 7 \end{bmatrix}$$

There are 10 vertices, 15 edges, and 7 regions.

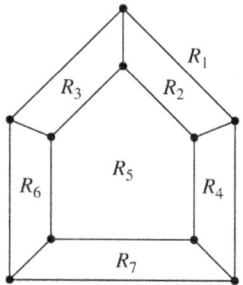

59. $X = \left(\begin{bmatrix} 1 & 0 \\ 0 & 1 \end{bmatrix} - \begin{bmatrix} 0.1 & 0.4 \\ 0.5 & 0.2 \end{bmatrix}\right)^{-1} \begin{bmatrix} 50 \\ 30 \end{bmatrix}$; $(I-A)^{-1} = \begin{bmatrix} \dfrac{9}{10} & -\dfrac{2}{5} \\ -\dfrac{1}{2} & \dfrac{4}{5} \end{bmatrix}^{-1}$

$\left[\begin{array}{cc|cc} \dfrac{9}{10} & -\dfrac{2}{5} & 1 & 0 \\ -\dfrac{1}{2} & \dfrac{4}{5} & 0 & 1 \end{array}\right] \xrightarrow{\frac{1}{2}R_1 + \frac{9}{10}R_2 \to R_2} \left[\begin{array}{cc|cc} \dfrac{9}{10} & -\dfrac{2}{5} & 1 & 0 \\ 0 & \dfrac{13}{25} & \dfrac{1}{2} & \dfrac{9}{10} \end{array}\right] \xrightarrow{\frac{13}{25}R_1 + \frac{2}{5}R_2 \to R_1} \left[\begin{array}{cc|cc} \dfrac{117}{250} & 0 & \dfrac{18}{25} & \dfrac{9}{25} \\ 0 & \dfrac{13}{25} & \dfrac{1}{2} & \dfrac{9}{10} \end{array}\right]$

$\xrightarrow[\frac{25}{13}R_2 \to R_2]{\frac{250}{117}R_1 \to R_1} \left[\begin{array}{cc|cc} 1 & 0 & \dfrac{20}{13} & \dfrac{10}{13} \\ 0 & 1 & \dfrac{25}{26} & \dfrac{45}{26} \end{array}\right]$

$(I-A)^{-1} = \begin{bmatrix} \dfrac{20}{13} & \dfrac{10}{13} \\ \dfrac{25}{26} & \dfrac{45}{26} \end{bmatrix} \Rightarrow X = (I-A)^{-1}D = \begin{bmatrix} \dfrac{20}{13} & \dfrac{10}{13} \\ \dfrac{25}{26} & \dfrac{45}{26} \end{bmatrix}\begin{bmatrix} 50 \\ 30 \end{bmatrix} = \begin{bmatrix} 100 \\ 100 \end{bmatrix}$

61. $X = \left(\begin{bmatrix} 1 & 0 & 0 \\ 0 & 1 & 0 \\ 0 & 0 & 1 \end{bmatrix} - \begin{bmatrix} 0.2 & 0.2 & 0 \\ 0.1 & 0.1 & 0.3 \\ 0.1 & 0 & 0.2 \end{bmatrix}\right)^{-1}\begin{bmatrix} 400 \\ 600 \\ 800 \end{bmatrix}$

$(I-A)^{-1} = \begin{bmatrix} \dfrac{4}{5} & -\dfrac{1}{5} & 0 \\ -\dfrac{1}{10} & \dfrac{9}{10} & -\dfrac{3}{10} \\ -\dfrac{1}{10} & 0 & \dfrac{4}{5} \end{bmatrix}^{-1} \Rightarrow \left[\begin{array}{ccc|ccc} \dfrac{4}{5} & -\dfrac{1}{5} & 0 & 1 & 0 & 0 \\ -\dfrac{1}{10} & \dfrac{9}{10} & -\dfrac{3}{10} & 0 & 1 & 0 \\ -\dfrac{1}{10} & 0 & \dfrac{4}{5} & 0 & 0 & 1 \end{array}\right] \xrightarrow[\frac{1}{10}R_1 + \frac{4}{5}R_3 \to R_3]{\frac{1}{10}R_1 + \frac{4}{5}R_2 \to R_2} \left[\begin{array}{ccc|ccc} \dfrac{4}{5} & -\dfrac{1}{5} & 0 & 1 & 0 & 0 \\ 0 & \dfrac{7}{10} & -\dfrac{6}{25} & \dfrac{1}{10} & \dfrac{4}{5} & 0 \\ 0 & -\dfrac{1}{50} & \dfrac{16}{25} & \dfrac{1}{10} & 0 & \dfrac{4}{5} \end{array}\right]$

$\xrightarrow{\frac{625}{277}\left(\frac{1}{50}R_2 + \frac{7}{10}R_3\right) \to R_3} \left[\begin{array}{ccc|ccc} \dfrac{4}{5} & -\dfrac{1}{5} & 0 & 1 & 0 & 0 \\ 0 & \dfrac{7}{10} & -\dfrac{6}{25} & \dfrac{1}{10} & \dfrac{4}{5} & 0 \\ 0 & 0 & 1 & \dfrac{45}{277} & \dfrac{10}{277} & \dfrac{350}{277} \end{array}\right]$

$\xrightarrow[\frac{10}{7}\left(\frac{6}{25}R_3 + R_2\right) \to R_2]{5R_1 \to R_1} \left[\begin{array}{ccc|ccc} 4 & -1 & 0 & 5 & 0 & 0 \\ 0 & 1 & 0 & \dfrac{55}{277} & \dfrac{320}{277} & \dfrac{120}{277} \\ 0 & 0 & 1 & \dfrac{45}{277} & \dfrac{10}{277} & \dfrac{350}{277} \end{array}\right] \xrightarrow{\frac{1}{4}(R_1 + R_2) \to R_1} \left[\begin{array}{ccc|ccc} 1 & 0 & 0 & \dfrac{360}{277} & \dfrac{80}{277} & \dfrac{30}{277} \\ 0 & 1 & 0 & \dfrac{55}{277} & \dfrac{320}{277} & \dfrac{120}{277} \\ 0 & 0 & 1 & \dfrac{45}{277} & \dfrac{10}{277} & \dfrac{350}{277} \end{array}\right] \Rightarrow$

$(I-A)^{-1} = \begin{bmatrix} \dfrac{360}{277} & \dfrac{80}{277} & \dfrac{30}{277} \\ \dfrac{55}{277} & \dfrac{320}{277} & \dfrac{120}{277} \\ \dfrac{45}{277} & \dfrac{10}{277} & \dfrac{350}{277} \end{bmatrix}$; $X = (I-A)^{-1}D = \begin{bmatrix} \dfrac{360}{277} & \dfrac{80}{277} & \dfrac{30}{277} \\ \dfrac{55}{277} & \dfrac{320}{277} & \dfrac{120}{277} \\ \dfrac{45}{277} & \dfrac{10}{277} & \dfrac{350}{277} \end{bmatrix}\begin{bmatrix} 400 \\ 600 \\ 800 \end{bmatrix} = \begin{bmatrix} \dfrac{216{,}000}{277} \\ \dfrac{310{,}000}{277} \\ \dfrac{304{,}000}{277} \end{bmatrix}$

Chapter 9 Matrices and Determinants

Answers will vary in exercises 63−66.

63. a. Associate each letter in the phrase with the number representing its position in the alphabet, and partition the numbers into groups of two, forming a 2×9 matrix:

$$M = \begin{bmatrix} 3 & 14 & 15 & 6 & 14 & 3 & 12 & 13 & 15 \\ 1 & 14 & 20 & 9 & 4 & 15 & 15 & 2 & 0 \end{bmatrix} \text{ Let } A = \begin{bmatrix} 1 & 2 \\ 2 & 1 \end{bmatrix} \text{ Then,}$$

$$AM = \begin{bmatrix} 1 & 2 \\ 2 & 1 \end{bmatrix}\begin{bmatrix} 3 & 14 & 15 & 6 & 14 & 3 & 12 & 13 & 15 \\ 1 & 14 & 20 & 9 & 4 & 15 & 15 & 2 & 0 \end{bmatrix} = \begin{bmatrix} 5 & 42 & 55 & 24 & 22 & 33 & 42 & 17 & 15 \\ 7 & 42 & 50 & 21 & 32 & 21 & 39 & 28 & 30 \end{bmatrix}$$

The cryptogram is 5 7 42 42 55 50 24 21 22 32 33 21 42 39 17 28 15 30.

b. $A^{-1} = \begin{bmatrix} 1 & 2 & | & 1 & 0 \\ 2 & 1 & | & 0 & 1 \end{bmatrix} \xrightarrow{\frac{1}{3}(2R_1 - R_2) \to R_2} \begin{bmatrix} 1 & 2 & | & 1 & 0 \\ 0 & 1 & | & \frac{2}{3} & -\frac{1}{3} \end{bmatrix} \xrightarrow{R_1 - 2R_2 \to R_1} \begin{bmatrix} 1 & 0 & | & -\frac{1}{3} & \frac{2}{3} \\ 0 & 1 & | & \frac{2}{3} & -\frac{1}{3} \end{bmatrix} = \begin{bmatrix} -\frac{1}{3} & \frac{2}{3} \\ \frac{2}{3} & -\frac{1}{3} \end{bmatrix}$

$$M = A^{-1}(AM) = \begin{bmatrix} -\frac{1}{3} & \frac{2}{3} \\ \frac{2}{3} & -\frac{1}{3} \end{bmatrix}\begin{bmatrix} 5 & 42 & 55 & 24 & 22 & 33 & 42 & 17 & 15 \\ 7 & 42 & 50 & 21 & 32 & 21 & 39 & 28 & 30 \end{bmatrix}$$

$$= \begin{bmatrix} 3 & 14 & 15 & 6 & 14 & 3 & 12 & 13 & 15 \\ 1 & 14 & 20 & 9 & 4 & 15 & 15 & 2 & 0 \end{bmatrix}$$

65. a. Associate each letter in the phrase with the number representing its position in the alphabet, and partition the numbers into groups of three, forming a 3×6 matrix:

$$M = \begin{bmatrix} 3 & 14 & 6 & 4 & 12 & 2 \\ 1 & 15 & 9 & 3 & 15 & 15 \\ 14 & 20 & 14 & 15 & 13 & 0 \end{bmatrix} \text{ Let } A = \begin{bmatrix} 2 & 1 & 1 \\ 1 & 2 & 1 \\ 1 & 1 & 2 \end{bmatrix} \text{ Then,}$$

$$AM = \begin{bmatrix} 2 & 1 & 1 \\ 1 & 2 & 1 \\ 1 & 1 & 2 \end{bmatrix}\begin{bmatrix} 3 & 14 & 6 & 4 & 12 & 2 \\ 1 & 15 & 9 & 3 & 15 & 15 \\ 14 & 20 & 14 & 15 & 13 & 0 \end{bmatrix} = \begin{bmatrix} 21 & 63 & 35 & 26 & 52 & 19 \\ 19 & 64 & 38 & 25 & 55 & 32 \\ 32 & 69 & 43 & 37 & 53 & 17 \end{bmatrix}$$

The cryptogram is 21 39 32 63 64 69 35 38 43 26 25 37 52 55 53 19 32 17.

b. $A^{-1} = \begin{bmatrix} 2 & 1 & 1 & | & 1 & 0 & 0 \\ 1 & 2 & 1 & | & 0 & 1 & 0 \\ 1 & 1 & 2 & | & 0 & 0 & 1 \end{bmatrix} \xrightarrow[R_1 - 2R_3 \to R_3]{R_1 - 2R_2 \to R_2} \begin{bmatrix} 2 & 1 & 1 & | & 1 & 0 & 0 \\ 0 & -3 & -1 & | & 1 & -2 & 0 \\ 0 & -1 & -3 & | & 1 & 0 & -2 \end{bmatrix}$

$\xrightarrow{\frac{1}{8}(R_2 - 3R_3) \to R_3} \begin{bmatrix} 2 & 1 & 1 & | & 1 & 0 & 0 \\ 0 & -3 & -1 & | & 1 & -2 & 0 \\ 0 & 0 & 1 & | & -\frac{1}{4} & -\frac{1}{4} & \frac{3}{4} \end{bmatrix} \xrightarrow[-\frac{1}{3}(R_3 + R_2) \to R_2]{R_1 - R_3 \to R_1} \begin{bmatrix} 2 & 1 & 0 & | & \frac{5}{4} & \frac{1}{4} & -\frac{3}{4} \\ 0 & 1 & 0 & | & -\frac{1}{4} & \frac{3}{4} & -\frac{1}{4} \\ 0 & 0 & 1 & | & -\frac{1}{4} & -\frac{1}{4} & \frac{3}{4} \end{bmatrix}$

$\xrightarrow{\frac{1}{2}(R_1 - R_2) \to R_1} \begin{bmatrix} 1 & 0 & 0 & | & \frac{3}{4} & -\frac{1}{4} & -\frac{1}{4} \\ 0 & 1 & 0 & | & -\frac{1}{4} & \frac{3}{4} & -\frac{1}{4} \\ 0 & 0 & 1 & | & -\frac{1}{4} & -\frac{1}{4} & \frac{3}{4} \end{bmatrix} = \begin{bmatrix} \frac{3}{4} & -\frac{1}{4} & -\frac{1}{4} \\ -\frac{1}{4} & \frac{3}{4} & -\frac{1}{4} \\ -\frac{1}{4} & -\frac{1}{4} & \frac{3}{4} \end{bmatrix}$

$$M = A^{-1}(AM) = \begin{bmatrix} \frac{3}{4} & -\frac{1}{4} & -\frac{1}{4} \\ -\frac{1}{4} & \frac{3}{4} & -\frac{1}{4} \\ -\frac{1}{4} & -\frac{1}{4} & \frac{3}{4} \end{bmatrix} \begin{bmatrix} 21 & 63 & 35 & 26 & 52 & 19 \\ 19 & 64 & 38 & 25 & 55 & 32 \\ 32 & 69 & 43 & 37 & 53 & 17 \end{bmatrix} = \begin{bmatrix} 3 & 14 & 6 & 4 & 12 & 2 \\ 1 & 15 & 9 & 3 & 15 & 15 \\ 14 & 20 & 14 & 15 & 13 & 0 \end{bmatrix}$$

9.3 Beyond the Basics

67. a. Yes, by the definition of *inverse* if $AB = I$, then $BA = I$. So $B = A^{-1}$ and $A = B^{-1}$.

  b. $\left(A^{-1}\right)^{-1} = A$

69. $I = A^2 B = A(AB)$, so AB is the inverse of A.

71. $ABB^{-1}A^{-1} = AIA^{-1} = AA^{-1} = I$
$B^{-1}A^{-1}AB = B^{-1}IB = B^{-1}B = I \Rightarrow AB$ is invertible and $(AB)^{-1} = B^{-1}A^{-1}$.

73. If A is invertible, then there exists a matrix D such that $AD = I$ and $DA = I$. Then $AB = AC \Leftrightarrow DAB = DAC \Leftrightarrow IB = IC \Leftrightarrow B = C$.

75. If A is invertible, then $I = A^{-1}A = A^{-1}(AB) = \left(A^{-1}A\right)B = IB = B$.

77. a. $\begin{bmatrix} 3 & 4 \\ 2 & 3 \end{bmatrix}^2 - 6\begin{bmatrix} 3 & 4 \\ 2 & 3 \end{bmatrix} + \begin{bmatrix} 1 & 0 \\ 0 & 1 \end{bmatrix} = \begin{bmatrix} 17 & 24 \\ 12 & 17 \end{bmatrix} - \begin{bmatrix} 18 & 24 \\ 12 & 18 \end{bmatrix} + \begin{bmatrix} 1 & 0 \\ 0 & 1 \end{bmatrix} = \begin{bmatrix} 0 & 0 \\ 0 & 0 \end{bmatrix}$

  b. $A^2 - 6A + I = 0 \Rightarrow I = 6A - A^2 \Rightarrow AA^{-1} = 6A - A^2 \Rightarrow A^{-1} = 6I - A$

  c. $A^{-1} = 6I - A = \begin{bmatrix} 6 & 0 \\ 0 & 6 \end{bmatrix} - \begin{bmatrix} 3 & 4 \\ 2 & 3 \end{bmatrix} = \begin{bmatrix} 3 & -4 \\ -2 & 3 \end{bmatrix}$

79. $\begin{cases} x - y - z - w = -4 \\ x + y + z - w = 2 \\ 2x + y + z - w = 3 \\ x - y + z - w = -2 \end{cases} \Rightarrow \begin{bmatrix} 1 & -1 & -1 & -1 \\ 1 & 1 & 1 & -1 \\ 2 & 1 & 1 & -1 \\ 1 & -1 & 1 & -1 \end{bmatrix} \begin{bmatrix} x \\ y \\ z \\ w \end{bmatrix} = \begin{bmatrix} -4 \\ 2 \\ 3 \\ -2 \end{bmatrix} \Rightarrow \begin{bmatrix} 1 & -1 & -1 & -1 \\ 1 & 1 & 1 & -1 \\ 2 & 1 & 1 & -1 \\ 1 & -1 & 1 & -1 \end{bmatrix}^{-1} \begin{bmatrix} -4 \\ 2 \\ 3 \\ -2 \end{bmatrix} = \begin{bmatrix} x \\ y \\ z \\ w \end{bmatrix}$

Using a graphing calculator, we have

$\begin{bmatrix} 1 & -1 & -1 & -1 \\ 1 & 1 & 1 & -1 \\ 2 & 1 & 1 & -1 \\ 1 & -1 & 1 & -1 \end{bmatrix}^{-1} = \begin{bmatrix} 0 & -1 & 1 & 0 \\ 0 & \frac{1}{2} & 0 & -\frac{1}{2} \\ -\frac{1}{2} & 0 & 0 & \frac{1}{2} \\ -\frac{1}{2} & -\frac{3}{2} & 1 & 0 \end{bmatrix}$ and then $\begin{bmatrix} 0 & -1 & 1 & 0 \\ 0 & \frac{1}{2} & 0 & -\frac{1}{2} \\ -\frac{1}{2} & 0 & 0 & \frac{1}{2} \\ -\frac{1}{2} & -\frac{3}{2} & 1 & 0 \end{bmatrix} \begin{bmatrix} -4 \\ 2 \\ 3 \\ -2 \end{bmatrix} = \begin{bmatrix} 1 \\ 2 \\ 1 \\ 2 \end{bmatrix}$

Solution set: $\{(1, 2, 1, 2)\}$

81. a. $ABB^{-1} = AI = A \Rightarrow A = \begin{bmatrix} -1 & 3 \\ 0 & 2 \end{bmatrix} \begin{bmatrix} 1 & 2 \\ 3 & 4 \end{bmatrix} = \begin{bmatrix} 8 & 10 \\ 6 & 8 \end{bmatrix}$

  b. $B^{-1}BA = IA = A \Rightarrow A = \begin{bmatrix} 1 & 2 \\ 3 & 4 \end{bmatrix} \begin{bmatrix} -1 & 3 \\ 0 & 2 \end{bmatrix} = \begin{bmatrix} -1 & 7 \\ -3 & 17 \end{bmatrix}$

c. $B^{-1}BAB^{-1} = IAB^{-1} = AB^{-1}$; $\quad AB^{-1} = \begin{bmatrix} 1 & 2 \\ 3 & 4 \end{bmatrix}\begin{bmatrix} 1 & 1 \\ 1 & 2 \end{bmatrix} = \begin{bmatrix} 3 & 5 \\ 7 & 11 \end{bmatrix}$

$\begin{bmatrix} a & b \\ c & d \end{bmatrix}\begin{bmatrix} 1 & 2 \\ 3 & 4 \end{bmatrix} = \begin{bmatrix} 3 & 5 \\ 7 & 11 \end{bmatrix} \Rightarrow \begin{bmatrix} a+3b & 2a+4b \\ c+3d & 2c+4d \end{bmatrix} = \begin{bmatrix} 3 & 5 \\ 7 & 11 \end{bmatrix} \Rightarrow \begin{cases} a+3b=3 \\ 2a+4b=5 \\ c+3d=7 \\ 2c+4d=11 \end{cases} \Rightarrow$

$a = \dfrac{3}{2}, b = \dfrac{1}{2}, c = \dfrac{5}{2}, d = \dfrac{3}{2} \Rightarrow A = \begin{bmatrix} \dfrac{3}{2} & \dfrac{1}{2} \\ \dfrac{5}{2} & \dfrac{3}{2} \end{bmatrix}$

d. $B^{-1}ABB^{-1} = B^{-1}AI = B^{-1}A \Rightarrow B^{-1}A = \begin{bmatrix} 1 & 1 \\ 1 & 2 \end{bmatrix}\begin{bmatrix} 1 & 2 \\ 3 & 4 \end{bmatrix} = \begin{bmatrix} 4 & 6 \\ 7 & 10 \end{bmatrix}$

$\begin{bmatrix} 1 & 2 \\ 3 & 4 \end{bmatrix}\begin{bmatrix} a & b \\ c & d \end{bmatrix} = \begin{bmatrix} 4 & 6 \\ 7 & 10 \end{bmatrix} \Rightarrow \begin{bmatrix} a+2c & b+2d \\ 3a+4c & 3b+4d \end{bmatrix} = \begin{bmatrix} 4 & 6 \\ 7 & 10 \end{bmatrix} \Rightarrow \begin{cases} a+2c=4 \\ b+2d=6 \\ 3a+4c=7 \\ 3b+4d=10 \end{cases} \Rightarrow$

$\begin{bmatrix} 1 & 0 & 2 & 0 \\ 0 & 1 & 0 & 2 \\ 3 & 0 & 4 & 0 \\ 0 & 3 & 0 & 4 \end{bmatrix}\begin{bmatrix} a \\ b \\ c \\ d \end{bmatrix} = \begin{bmatrix} 4 \\ 6 \\ 7 \\ 10 \end{bmatrix} \Rightarrow \begin{bmatrix} a \\ b \\ c \\ d \end{bmatrix} = \begin{bmatrix} 1 & 0 & 2 & 0 \\ 0 & 1 & 0 & 2 \\ 3 & 0 & 4 & 0 \\ 0 & 3 & 0 & 4 \end{bmatrix}^{-1}\begin{bmatrix} 4 \\ 6 \\ 7 \\ 10 \end{bmatrix} = \begin{bmatrix} -2 & 0 & 1 & 0 \\ 0 & -2 & 0 & 1 \\ \dfrac{3}{2} & 0 & -\dfrac{1}{2} & 0 \\ 0 & \dfrac{3}{2} & 0 & -\dfrac{1}{2} \end{bmatrix}\begin{bmatrix} 4 \\ 6 \\ 7 \\ 10 \end{bmatrix} = \begin{bmatrix} -1 \\ -2 \\ \dfrac{5}{2} \\ 4 \end{bmatrix} \Rightarrow$

$A = \begin{bmatrix} -1 & -2 \\ \dfrac{5}{2} & 4 \end{bmatrix}$

9.3 Critical Thinking/Discussion/Writing

83. a. True. $A^2B = AAB = ABA = BAA = BA^2$

    b. False. For example, if $A = I$ and $B = -I$,
    then $A + B = \begin{bmatrix} 1 & 0 \\ 0 & 1 \end{bmatrix} + \begin{bmatrix} -1 & 0 \\ 0 & -1 \end{bmatrix} = \begin{bmatrix} 0 & 0 \\ 0 & 0 \end{bmatrix}$,
    which is not invertible.

84. Let $A = \begin{bmatrix} 0 & -1 \\ 0 & 0 \end{bmatrix}$. Then $A^2 = \begin{bmatrix} 0 & 0 \\ 0 & 0 \end{bmatrix}$.

    $I + A = \begin{bmatrix} 1 & 0 \\ 0 & 1 \end{bmatrix} + \begin{bmatrix} 0 & -1 \\ 0 & 0 \end{bmatrix} = \begin{bmatrix} 1 & -1 \\ 0 & 1 \end{bmatrix}$

    $(I+A)^{-1} = \begin{bmatrix} 1 & 1 \\ 0 & 1 \end{bmatrix}$

    $I - A = \begin{bmatrix} 1 & 0 \\ 0 & 1 \end{bmatrix} - \begin{bmatrix} 0 & -1 \\ 0 & 0 \end{bmatrix} = \begin{bmatrix} 1 & 1 \\ 0 & 1 \end{bmatrix}$

85. True. A matrix has an inverse if and only if it is square.

9.3 Getting Ready for the Next Section

87. 1

89. -1

91. $\begin{bmatrix} -1 \\ -3 \end{bmatrix}[5 \ 0] = \begin{bmatrix} -1(5) & -1(0) \\ -3(5) & -3(0) \end{bmatrix} = \begin{bmatrix} -5 & 0 \\ -15 & 0 \end{bmatrix}$

93. $(-1)\begin{bmatrix} -6 & 7 \\ 4 & -1 \end{bmatrix} = \begin{bmatrix} 6 & -7 \\ -4 & 1 \end{bmatrix}$

In exercises 94–97, be sure to check the solution in the original equations.

95. $\begin{cases} 16x - 9y = -5 \quad (1) \\ 10x + 18y = -11 \quad (2) \end{cases}$

    Multiply equation (1) by 2, then add the resulting equation and equation (2).
    $32x - 18y = -10$
    $10x + 18y = -11$
    $\overline{}$
    $42x = -21 \Rightarrow x = -\dfrac{1}{2}$

    Substitute $x = -\dfrac{1}{2}$ in equation (2) and solve for y.

    (*continued on next page*)

(continued)

$10\left(-\dfrac{1}{2}\right) + 18y = -11 \Rightarrow -5 + 18y = -11 \Rightarrow$

$18y = -6 \Rightarrow y = -\dfrac{1}{3}$

Solution set: $\left\{\left(-\dfrac{1}{2}, -\dfrac{1}{3}\right)\right\}$

97. $\begin{cases} 2x - y + 2z = 3 & (1) \\ 2x + 2y - z = 0 & (2) \\ -x + 2y + 2z = -12 & (3) \end{cases}$

Write the augmented matrix.

$\begin{bmatrix} 2 & -1 & 2 & | & 3 \\ 2 & 2 & -1 & | & 0 \\ -1 & 2 & 2 & | & -12 \end{bmatrix}$

Now use Gauss-Jordan elimination to solve the system.

$\begin{bmatrix} 2 & -1 & 2 & | & 3 \\ 2 & 2 & -1 & | & 0 \\ -1 & 2 & 2 & | & -12 \end{bmatrix}$

$\xrightarrow[R_1 + 2R_3 \to R_3]{R_2 - R_1 \to R_2} \begin{bmatrix} 2 & -1 & 2 & | & 3 \\ 0 & 3 & -3 & | & -3 \\ 0 & 3 & 6 & | & -21 \end{bmatrix}$

$\xrightarrow{\frac{1}{9}(R_3 - R_2) \to R_3} \begin{bmatrix} 2 & -1 & 2 & | & 3 \\ 0 & 3 & -3 & | & -3 \\ 0 & 0 & 1 & | & -2 \end{bmatrix}$

$\xrightarrow[\frac{1}{3}(R_2 + 3R_3) \to R_2]{R_1 - 2R_3 \to R_1} \begin{bmatrix} 2 & -1 & 0 & | & 7 \\ 0 & 1 & 0 & | & -3 \\ 0 & 0 & 1 & | & -2 \end{bmatrix}$

$\xrightarrow{\frac{1}{2}(R_1 + R_2) \to R_1} \begin{bmatrix} 1 & 0 & 0 & | & 2 \\ 0 & 1 & 0 & | & -3 \\ 0 & 0 & 1 & | & -2 \end{bmatrix}$

Solution set: $\{(2, -3, -2)\}$

9.4 Determinants and Cramer's Rule

9.4 Practice Problems

1. a. $\begin{vmatrix} 5 & 1 \\ 3 & -7 \end{vmatrix} = 5(-7) - 1(3) = -38$

b. $\begin{vmatrix} 2 & -9 \\ -4 & 18 \end{vmatrix} = 2(18) - (-9)(-4) = 0$

2. a. $M_{11} = \begin{bmatrix} 3 & -1 & 2 \\ 4 & 5 & 6 \\ 7 & 1 & 2 \end{bmatrix} = \begin{vmatrix} 5 & 6 \\ 1 & 2 \end{vmatrix}$

$= (5)(2) - 6(1) = 4$

$M_{23} = \begin{bmatrix} 3 & -1 & 2 \\ 4 & 5 & 6 \\ 7 & 1 & 2 \end{bmatrix} = \begin{vmatrix} 3 & -1 \\ 7 & 1 \end{vmatrix}$

$= 3(1) - (-1)(7) = 10$

$M_{32} = \begin{bmatrix} 3 & -1 & 2 \\ 4 & 5 & 6 \\ 7 & 1 & 2 \end{bmatrix} = \begin{vmatrix} 3 & 2 \\ 4 & 6 \end{vmatrix}$

$= 3(6) - 2(4) = 10$

b. $A_{11} = (-1)^{1+1} M_{11} = 4$

$A_{23} = (-1)^{2+3} M_{23} = -10$

$A_{32} = (-1)^{3+2} M_{32} = -10$

3. Expand by the third row:

$\begin{vmatrix} 2 & -3 & 7 \\ -2 & -1 & 9 \\ 0 & 2 & -9 \end{vmatrix}$

$= a_{31} A_{31} + a_{32} A_{32} + a_{33} A_{33}$

$= 0 + 2(-1)^{3+2} \begin{vmatrix} 2 & 7 \\ -2 & 9 \end{vmatrix} - 9(-1)^{3+3} \begin{vmatrix} 2 & -3 \\ -2 & -1 \end{vmatrix}$

$= 0 - 2(32) - 9(-8) = 8$

4. $D = \begin{vmatrix} 2 & 3 \\ 5 & 9 \end{vmatrix} = 3, \ D_x = \begin{vmatrix} 7 & 3 \\ 4 & 9 \end{vmatrix} = 51$

$D_y = \begin{vmatrix} 2 & 7 \\ 5 & 4 \end{vmatrix} = -27$

$x = \dfrac{D_x}{D} = \dfrac{51}{3} = 17; \ y = \dfrac{D_y}{D} = \dfrac{-27}{3} = -9$

The solution is $\{(17, -9)\}$.

5. $D = \begin{vmatrix} 3 & 2 & 1 \\ 4 & 3 & 1 \\ 5 & 1 & 1 \end{vmatrix} = -3, \ D_x = \begin{vmatrix} 4 & 2 & 1 \\ 5 & 3 & 1 \\ 9 & 1 & 1 \end{vmatrix} = -6$

$D_y = \begin{vmatrix} 3 & 4 & 1 \\ 4 & 5 & 1 \\ 5 & 9 & 1 \end{vmatrix} = 3, \ D_z = \begin{vmatrix} 3 & 2 & 4 \\ 4 & 3 & 5 \\ 5 & 1 & 9 \end{vmatrix} = 0$

$x = \dfrac{D_x}{D} = \dfrac{-6}{-3} = 2; \ y = \dfrac{D_y}{D} = \dfrac{3}{-3} = -1$

$z = \dfrac{D_z}{D} = \dfrac{0}{-3} = 0$

Solution set: $\{(2, -1, 0)\}$

6.

$$AD = \begin{bmatrix} -1 & 2 \\ 2 & -1 \end{bmatrix} \begin{matrix} P & Q & R & S \\ \begin{bmatrix} 0 & 4 & 2 & 1 \\ 0 & 0 & 2 & 2 \end{bmatrix} \end{matrix}$$

$$= \begin{matrix} P' & Q' & R' & S' \\ \begin{bmatrix} 0 & -4 & 2 & 3 \\ 0 & 6 & 2 & 0 \end{bmatrix} \end{matrix}$$

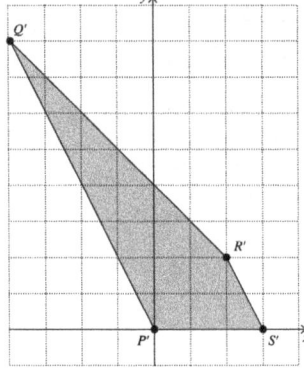

From Example 6, $A_{PQRS} = 5$ units^2.

$$A_{P'Q'R'S'} = |\det A| \cdot A_{PQRS}$$
$$= |-1(-1) - 2(2)| \cdot 5 \text{ units}^2$$
$$= |-3| \cdot 5 \text{ units}^2$$
$$= 15 \text{ units}^2$$

9.4 Concepts and Vocabulary

1. The determinant of $\begin{bmatrix} a & b \\ c & d \end{bmatrix}$ is $\underline{ad - bc}$.

3. To expand an $n \times n$ determinant, you multiply each element of some row (or column) by its <u>cofactor</u> and add the result.

5. False.

7. True

9.4 Building Skills

9. $\begin{vmatrix} 2 & 3 \\ 4 & 5 \end{vmatrix} = 2(5) - 4(3) = -2$

11. $\begin{vmatrix} 4 & -2 \\ 3 & -3 \end{vmatrix} = 4(-3) - (-2)(3) = -6$

13. $\begin{vmatrix} -1 & -3 \\ -4 & -5 \end{vmatrix} = (-1)(-5) - (-3)(-4) = -7$

15. $\begin{vmatrix} \frac{3}{8} & \frac{1}{2} \\ -\frac{1}{9} & 5 \end{vmatrix} = \left(\frac{3}{8}\right)(5) - \left(\frac{1}{2}\right)\left(-\frac{1}{9}\right) = \frac{139}{72}$

17. $\begin{vmatrix} \sqrt{a} & \sqrt{b} \\ \sqrt{b} & \sqrt{a} \end{vmatrix} = a - b$

19. a. $M_{21} = \begin{bmatrix} 2 & -3 & 4 \\ 1 & -1 & 2 \\ 0 & 1 & 2 \end{bmatrix} = \begin{vmatrix} -3 & 4 \\ 1 & 2 \end{vmatrix}$
$= (-3)(2) - 4(1) = -10$

b. $M_{23} = \begin{bmatrix} 2 & -3 & 4 \\ 1 & -1 & 2 \\ 0 & 1 & 2 \end{bmatrix} = \begin{vmatrix} 2 & -3 \\ 0 & 1 \end{vmatrix} = 2(1) - 0 = 2$

c. $M_{32} = \begin{bmatrix} 2 & -3 & 4 \\ 1 & -1 & 2 \\ 0 & 1 & 2 \end{bmatrix} = \begin{vmatrix} 2 & 4 \\ 1 & 2 \end{vmatrix} = 2(2) - 4(1) = 0$

21. a. $M_{11} = \begin{bmatrix} 2 & -3 & 4 \\ 1 & -1 & 2 \\ 0 & 1 & 2 \end{bmatrix} = \begin{vmatrix} -1 & 2 \\ 1 & 2 \end{vmatrix}$
$= (-1)(2) - 2(1) = -4$

b. $M_{22} = \begin{bmatrix} 2 & -3 & 4 \\ 1 & -1 & 2 \\ 0 & 1 & 2 \end{bmatrix} = \begin{vmatrix} 2 & 4 \\ 0 & 2 \end{vmatrix} = 4$

c. $M_{31} = \begin{bmatrix} 2 & -3 & 4 \\ 1 & -1 & 2 \\ 0 & 1 & 2 \end{bmatrix} = \begin{vmatrix} -3 & 4 \\ -1 & 2 \end{vmatrix}$
$= (-3)(2) - 4(-1) = -2$

23. Expand by the third row:

$\begin{vmatrix} 1 & 0 & -1 \\ 0 & 2 & 2 \\ -1 & 0 & 0 \end{vmatrix} = a_{31}A_{31} + a_{32}A_{32} + a_{33}A_{33}$

$= -1(-1)^{3+1}\begin{vmatrix} 0 & -1 \\ 2 & 2 \end{vmatrix} + 0 + 0 = -2$

25. Expand by the third row:

$\begin{vmatrix} 1 & 2 & 3 \\ 0 & 3 & 4 \\ 0 & 0 & 4 \end{vmatrix} = a_{31}A_{31} + a_{32}A_{32} + a_{33}A_{33}$

$= 0 + 0 + 4(-1)^{3+3}\begin{vmatrix} 1 & 2 \\ 0 & 3 \end{vmatrix} = 12$

27. Expand by the first row:
$$\begin{vmatrix} 1 & 0 & 0 \\ 2 & 0 & 0 \\ 3 & 4 & 5 \end{vmatrix} = a_{11}A_{11} + a_{12}A_{12} + a_{13}A_{13}$$
$$= 1(-1)^{1+1}\begin{vmatrix} 0 & 0 \\ 4 & 5 \end{vmatrix} + 0 + 0 = 0$$

29. Expand by the first row:
$$\begin{vmatrix} 1 & 6 & 0 \\ 2 & 5 & 3 \\ 3 & 4 & 0 \end{vmatrix} = a_{11}A_{11} + a_{12}A_{12} + a_{13}A_{13}$$
$$= 1(-1)^{1+1}\begin{vmatrix} 5 & 3 \\ 4 & 0 \end{vmatrix} + 6(-1)^{1+2}\begin{vmatrix} 2 & 3 \\ 3 & 0 \end{vmatrix} + 0$$
$$= -12 - 6(-9) = 42$$

31. Expand by the first row:
$$\begin{vmatrix} 3 & 4 & 1 \\ 1 & 4 & 3 \\ 4 & 3 & 1 \end{vmatrix} = a_{11}A_{11} + a_{12}A_{12} + a_{13}A_{13}$$
$$= 3(-1)^{1+1}\begin{vmatrix} 4 & 3 \\ 3 & 1 \end{vmatrix} + 4(-1)^{1+2}\begin{vmatrix} 1 & 3 \\ 4 & 1 \end{vmatrix}$$
$$+ 1(-1)^{1+3}\begin{vmatrix} 1 & 4 \\ 4 & 3 \end{vmatrix}$$
$$= 3(-5) - 4(-11) + 1(-13) = 16$$

33. Expand by the first row:
$$\begin{vmatrix} 0 & 1 & 6 \\ 1 & 0 & 4 \\ 8 & 3 & 1 \end{vmatrix} = a_{11}A_{11} + a_{12}A_{12} + a_{13}A_{13}$$
$$= 0 + 1(-1)^{1+2}\begin{vmatrix} 1 & 4 \\ 8 & 1 \end{vmatrix} + 6(-1)^{1+3}\begin{vmatrix} 1 & 0 \\ 8 & 3 \end{vmatrix}$$
$$= -1(-31) + 6(3) = 49$$

35. Expand by the first row:
$$\begin{vmatrix} a & b & 0 \\ 0 & a & b \\ b & 0 & a \end{vmatrix} = a_{11}A_{11} + a_{12}A_{12} + a_{13}A_{13}$$
$$= a(-1)^{1+1}\begin{vmatrix} a & b \\ 0 & a \end{vmatrix} + b(-1)^{1+2}\begin{vmatrix} 0 & b \\ b & a \end{vmatrix} + 0$$
$$= a(a^2) - b(-b^2) = a^3 + b^3$$

37. Expand by the first row:
$$\begin{vmatrix} a & b & c \\ c & a & b \\ b & c & a \end{vmatrix} = a_{11}A_{11} + a_{12}A_{12} + a_{13}A_{13}$$
$$= a(-1)^{1+1}\begin{vmatrix} a & b \\ c & a \end{vmatrix} + b(-1)^{1+2}\begin{vmatrix} c & b \\ b & a \end{vmatrix}$$
$$+ c(-1)^{1+3}\begin{vmatrix} c & a \\ b & c \end{vmatrix}$$
$$= a(a^2 - bc) - b(ac - b^2) + c(c^2 - ab)$$
$$= a^3 + b^3 + c^3 - 3abc$$

39. $D = \begin{vmatrix} 1 & 1 \\ 1 & -1 \end{vmatrix} = -2, D_x = \begin{vmatrix} 8 & 1 \\ -2 & -1 \end{vmatrix} = -6$

$D_y = \begin{vmatrix} 1 & 8 \\ 1 & -2 \end{vmatrix} = -10$

$x = \dfrac{-6}{-2} = 3, y = \dfrac{-10}{-2} = 5$

The solution is $\{(3, 5)\}$.

41. $D = \begin{vmatrix} 5 & 3 \\ 2 & 1 \end{vmatrix} = -1, D_x = \begin{vmatrix} 11 & 3 \\ 4 & 1 \end{vmatrix} = -1$

$D_y = \begin{vmatrix} 5 & 11 \\ 2 & 4 \end{vmatrix} = -2$

$x = \dfrac{-1}{-1} = 1, y = \dfrac{-2}{-1} = 2$

The solution is $\{(1, 2)\}$.

43. $D = \begin{vmatrix} 2 & 9 \\ 3 & -2 \end{vmatrix} = -31, D_x = \begin{vmatrix} 4 & 9 \\ 6 & -2 \end{vmatrix} = -62$

$D_y = \begin{vmatrix} 2 & 4 \\ 3 & 6 \end{vmatrix} = 0$

$x = \dfrac{-62}{-31} = 2, y = \dfrac{0}{-31} = 0$

The solution is $\{(2, 0)\}$.

45. $D = \begin{vmatrix} 2 & -3 \\ 4 & -6 \end{vmatrix} = 0 \Rightarrow$ there is not a unique

solution. $2x - 3y = 4 \Rightarrow x = \dfrac{3}{2}y + 2$.

The solution is $\left\{\left(\dfrac{3}{2}y + 2, y\right)\right\}$.

47. Let $u = \dfrac{1}{x}$ and $v = \dfrac{1}{y}$. Then

$$\begin{cases} \dfrac{2}{x} + \dfrac{3}{y} = 2 \\ \dfrac{5}{x} + \dfrac{8}{y} = \dfrac{31}{6} \end{cases} \Rightarrow \begin{cases} 2u + 3v = 2 \\ 5u + 8v = \dfrac{31}{6} \end{cases} \Rightarrow$$

$$D = \begin{vmatrix} 2 & 3 \\ 5 & 8 \end{vmatrix} = 1,\ D_u = \begin{vmatrix} 2 & 3 \\ \frac{31}{6} & 8 \end{vmatrix} = \dfrac{1}{2},$$

$$D_v = \begin{vmatrix} 2 & 2 \\ 5 & \frac{31}{6} \end{vmatrix} = \dfrac{1}{3} \Rightarrow u = \dfrac{1}{2}, v = \dfrac{1}{3} \Rightarrow$$

$x = 2,\ y = 3$
The solution is $\{(2, 3)\}$.

49. $D = \begin{vmatrix} 1 & -2 & 1 \\ 3 & 1 & -1 \\ 0 & 1 & 1 \end{vmatrix} = 11,\ D_x = \begin{vmatrix} -1 & -2 & 1 \\ 4 & 1 & -1 \\ 1 & 1 & 1 \end{vmatrix} = 11,$

$D_y = \begin{vmatrix} 1 & -1 & 1 \\ 3 & 4 & -1 \\ 0 & 1 & 1 \end{vmatrix} = 11,\ D_z = \begin{vmatrix} 1 & -2 & -1 \\ 3 & 1 & 4 \\ 0 & 1 & 1 \end{vmatrix} = 0$

$x = \dfrac{11}{11} = 1,\ y = \dfrac{11}{11} = 1,\ z = \dfrac{0}{11} = 0$
The solution is $\{(1, 1, 0)\}$.

51. $D = \begin{vmatrix} 1 & 1 & -1 \\ 2 & 3 & 1 \\ 0 & 2 & 1 \end{vmatrix} = -5,\ D_x = \begin{vmatrix} -3 & 1 & -1 \\ 2 & 3 & 1 \\ 1 & 2 & 1 \end{vmatrix} = -5,$

$D_y = \begin{vmatrix} 1 & -3 & -1 \\ 2 & 2 & 1 \\ 0 & 1 & 1 \end{vmatrix} = 5,\ D_z = \begin{vmatrix} 1 & 1 & -3 \\ 2 & 3 & 2 \\ 0 & 2 & 1 \end{vmatrix} = -15$

$x = \dfrac{-5}{-5} = 1,\ y = \dfrac{5}{-5} = -1,\ z = \dfrac{-15}{-5} = 3$
The solution is $\{(1, -1, 3)\}$.

53. $D = \begin{vmatrix} 1 & 1 & 1 \\ 2 & -3 & 5 \\ 1 & 2 & -4 \end{vmatrix} = 22,\ D_x = \begin{vmatrix} 3 & 1 & 1 \\ 4 & -3 & 5 \\ -1 & 2 & -4 \end{vmatrix} = 22,$

$D_y = \begin{vmatrix} 1 & 3 & 1 \\ 2 & 4 & 5 \\ 1 & -1 & -4 \end{vmatrix} = 22,\ D_z = \begin{vmatrix} 1 & 1 & 3 \\ 2 & -3 & 4 \\ 1 & 2 & -1 \end{vmatrix} = 22$

$x = \dfrac{22}{22} = 1,\ y = \dfrac{22}{22} = 1,\ z = \dfrac{22}{22} = 1$
The solution is $\{(1, 1, 1)\}$.

55. $D = \begin{vmatrix} 2 & -3 & 5 \\ 3 & 5 & -2 \\ 1 & 2 & -3 \end{vmatrix} = -38,$

$D_x = \begin{vmatrix} 11 & -3 & 5 \\ 7 & 5 & -2 \\ -4 & 2 & -3 \end{vmatrix} = -38,$

$D_y = \begin{vmatrix} 2 & 11 & 5 \\ 3 & 7 & -2 \\ 1 & -4 & -3 \end{vmatrix} = -76,$

$D_z = \begin{vmatrix} 2 & -3 & 11 \\ 3 & 5 & 7 \\ 1 & 2 & -4 \end{vmatrix} = -114$

$x = \dfrac{-38}{-38} = 1,\ y = \dfrac{-76}{-38} = 2,\ z = \dfrac{-114}{-38} = 3$
The solution is $\{(1, 2, 3)\}$.

57. $D = \begin{vmatrix} 5 & 2 & 1 \\ 2 & 1 & 3 \\ 3 & 2 & 4 \end{vmatrix} = -7,\ D_x = \begin{vmatrix} 12 & 2 & 1 \\ 13 & 1 & 3 \\ 19 & 2 & 4 \end{vmatrix} = -7,$

$D_y = \begin{vmatrix} 5 & 12 & 1 \\ 2 & 13 & 3 \\ 3 & 19 & 4 \end{vmatrix} = -14,\ D_z = \begin{vmatrix} 5 & 2 & 12 \\ 2 & 1 & 13 \\ 3 & 2 & 19 \end{vmatrix} = -21$

$x = \dfrac{-7}{-7} = 1,\ y = \dfrac{-14}{-7} = 2,\ z = \dfrac{-21}{-7} = 3$
The solution is $\{(1, 2, 3)\}$.

9.4 Applying the Concepts

59. $A = |D| = \dfrac{1}{2} \begin{vmatrix} 1 & 2 & 1 \\ -3 & 4 & 1 \\ 4 & 6 & 1 \end{vmatrix} = 11$

61. $A = |D| = \dfrac{1}{2} \begin{vmatrix} -2 & 1 & 1 \\ -3 & -5 & 1 \\ 2 & 4 & 1 \end{vmatrix} = \dfrac{21}{2} = 10.5$

63. $\begin{vmatrix} 0 & 3 & 1 \\ -1 & 1 & 1 \\ 2 & 7 & 1 \end{vmatrix} = 0 \Rightarrow$ the points are collinear.

65. $\begin{vmatrix} 0 & -4 & 1 \\ 3 & -2 & 1 \\ 1 & -4 & 1 \end{vmatrix} = -2 \Rightarrow$ the points are not collinear.

67. $\begin{vmatrix} x & y & 1 \\ -1 & -1 & 1 \\ 1 & 3 & 1 \end{vmatrix} = 0 \Rightarrow x(-1)^{1+1} \begin{vmatrix} -1 & 1 \\ 3 & 1 \end{vmatrix} + y(-1)^{1+2} \begin{vmatrix} -1 & 1 \\ 1 & 1 \end{vmatrix} + 1(-1)^{1+3} \begin{vmatrix} -1 & -1 \\ 1 & 3 \end{vmatrix} = 0 \Rightarrow -4x + 2y - 2 = 0 \Rightarrow y = 2x + 1$

69. $\begin{vmatrix} x & y & 1 \\ 0 & \frac{1}{3} & 1 \\ 1 & 1 & 1 \end{vmatrix} = 0 \Rightarrow x(-1)^{1+1} \begin{vmatrix} \frac{1}{3} & 1 \\ 1 & 1 \end{vmatrix} + y(-1)^{1+2} \begin{vmatrix} 0 & 1 \\ 1 & 1 \end{vmatrix} + 1(-1)^{1+3} \begin{vmatrix} 0 & \frac{1}{3} \\ 1 & 1 \end{vmatrix} = 0 \Rightarrow -\frac{2}{3}x + y - \frac{1}{3} = 0 \Rightarrow y = \frac{2}{3}x + \frac{1}{3}$

For exercise 71–74, *PQRS* is a trapezoid. Its area is $\frac{1}{2}(\text{base}_1 + \text{base}_2) \cdot \text{altitude} = \frac{1}{2}(4+2) \cdot 2 = 6$ units².

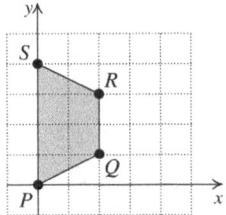

71. $A = \begin{bmatrix} 1 & -2 \\ 2 & 1 \end{bmatrix}$

$A_{P'Q'R'S'} = |\det A| \cdot A_{PQRS} = |1(1) - (-2)(2)| \cdot 6 \text{ units}^2 = |5| \cdot 6 \text{ units}^2 = 30 \text{ units}^2$

73. $A = \begin{bmatrix} -2 & 1 \\ 1 & 1 \end{bmatrix}$

$A_{P'Q'R'S'} = |\det A| \cdot A_{PQRS} = |-2(1) - 1(1)| \cdot 6 \text{ units}^2 = |-3| \cdot 6 \text{ units}^2 = 18 \text{ units}^2$

9.4 Beyond the Basics

75. a. $\begin{vmatrix} 0 & 0 \\ 2 & 5 \end{vmatrix} = 0(5) - 2(0) = 0$

    b. $\begin{vmatrix} 1 & 0 \\ 3 & 0 \end{vmatrix} = 1(0) - 3(0) = 0$

    c. Expand by the second row: $\begin{vmatrix} 1 & -2 & -3 \\ 0 & 0 & 0 \\ 4 & 4 & -7 \end{vmatrix} = 0(-1)^{2+1} \begin{vmatrix} -2 & -3 \\ 4 & -7 \end{vmatrix} + 0(-1)^{2+2} \begin{vmatrix} 1 & -3 \\ 4 & -7 \end{vmatrix} + 0(-1)^{2+3} \begin{vmatrix} 1 & -2 \\ 4 & 4 \end{vmatrix} = 0$

    d. Expand by the third column: $\begin{vmatrix} 4 & 5 & 0 \\ 6 & -7 & 0 \\ 8 & 15 & 0 \end{vmatrix} = 0(-1)^{1+3} \begin{vmatrix} 6 & -7 \\ 8 & 15 \end{vmatrix} + 0(-1)^{2+3} \begin{vmatrix} 4 & 5 \\ 8 & 15 \end{vmatrix} + 0(-1)^{3+3} \begin{vmatrix} 4 & 5 \\ 6 & -7 \end{vmatrix} = 0$

77. a. Expand by the second row: $\begin{vmatrix} 2 & 3 & 5 \\ -1 & 4 & 8 \\ 2 & 3 & 5 \end{vmatrix} = -1(-1)^{2+1} \begin{vmatrix} 3 & 5 \\ 3 & 5 \end{vmatrix} + 4(-1)^{2+2} \begin{vmatrix} 2 & 5 \\ 2 & 5 \end{vmatrix} + 8(-1)^{2+3} \begin{vmatrix} 2 & 5 \\ 2 & 5 \end{vmatrix} = 0$

    b. Expand by the second column: $\begin{vmatrix} 3 & 4 & 3 \\ 1 & 2 & 1 \\ -1 & 6 & -1 \end{vmatrix} = 4(-1)^{1+2} \begin{vmatrix} 1 & 1 \\ -1 & -1 \end{vmatrix} + 2(-1)^{2+2} \begin{vmatrix} 3 & 3 \\ -1 & -1 \end{vmatrix} + 6(-1)^{3+2} \begin{vmatrix} 3 & 3 \\ 1 & 1 \end{vmatrix} = 0$

79. $\begin{vmatrix} 2 & -3 & 4 \\ -4 & 7 & -8 \\ 5 & -1 & 3 \end{vmatrix} = -14, \quad \begin{vmatrix} 2 & -3 & 4 \\ 0 & 1 & 0 \\ 5 & -1 & 3 \end{vmatrix} = -14$

81. $\begin{vmatrix} 2 & 0 & -3 & 4 \\ 0 & 1 & 0 & 5 \\ 5 & 0 & -9 & 8 \\ 1 & 2 & 0 & 7 \end{vmatrix} \xrightarrow[R_3 - 5R_4 \to R_3]{R_1 - 2R_4 \to R_1} \begin{vmatrix} 0 & -4 & -3 & -10 \\ 0 & 1 & 0 & 5 \\ 0 & -10 & -9 & -27 \\ 1 & 2 & 0 & 7 \end{vmatrix}$

Expand by column 1:
$\begin{vmatrix} 0 & -4 & -3 & -10 \\ 0 & 1 & 0 & 5 \\ 0 & -10 & -9 & -27 \\ 1 & 2 & 0 & 7 \end{vmatrix} = 0 + 0 + 0 + 1(-1)^{4+1} \begin{vmatrix} -4 & -3 & -10 \\ 1 & 0 & 5 \\ -10 & -9 & -27 \end{vmatrix} = 21$

83. $\begin{vmatrix} 5 & 7 & 1 & 2 \\ 6 & 8 & 9 & 3 \\ 24 & 22 & 6 & 10 \\ 21 & 17 & 7 & 10 \end{vmatrix} \xrightarrow[\substack{-6R_1 + R_3 \to R_3 \\ -7R_1 + R_4 \to R_4}]{-9R_1 + R_2 \to R_2} \begin{vmatrix} 5 & 7 & 1 & 2 \\ -39 & -55 & 0 & -15 \\ -6 & -20 & 0 & -2 \\ -14 & -32 & 0 & -4 \end{vmatrix}$

Expand by column 3: $\begin{vmatrix} 5 & 7 & 1 & 2 \\ -39 & -55 & 0 & -15 \\ -6 & -20 & 0 & -2 \\ -14 & -32 & 0 & -4 \end{vmatrix} = 1(-1)^{1+3} \begin{vmatrix} -39 & -55 & -15 \\ -6 & -20 & -2 \\ -14 & -32 & -4 \end{vmatrix} = 476$

85. $\begin{vmatrix} 3 & 2 \\ 6 & x \end{vmatrix} = 0 \Rightarrow 3x - 12 = 0 \Rightarrow x = 4$

87. $\begin{vmatrix} x & -2 \\ 1 & x-1 \end{vmatrix} = 0 \Rightarrow x^2 - x + 2 = 0 \Rightarrow x = \dfrac{1 \pm \sqrt{1-8}}{2} = \dfrac{1 \pm i\sqrt{7}}{2}$

89. Expand by the second row:
$\begin{vmatrix} 1 & -3 & 1 \\ 4 & 7 & x \\ 0 & 2 & 2 \end{vmatrix} = 0 \Rightarrow 4(-1)^{2+1} \begin{vmatrix} -3 & 1 \\ 2 & 2 \end{vmatrix} + 7(-1)^{2+2} \begin{vmatrix} 1 & 1 \\ 0 & 2 \end{vmatrix} + x(-1)^{2+3} \begin{vmatrix} 1 & -3 \\ 0 & 2 \end{vmatrix} = 0 \Rightarrow$
$-4(-8) + 7(2) - 2x = 46 - 2x = 0 \Rightarrow x = 23$

91. Expand by the second row:
$\begin{vmatrix} x & 0 & 1 \\ 0 & x & 0 \\ 1 & 0 & x \end{vmatrix} = 0 \Rightarrow 0 + x(-1)^{2+2} \begin{vmatrix} x & 1 \\ 1 & x \end{vmatrix} + 0 \Rightarrow x(x^2 - 1) = 0 \Rightarrow x(x-1)(x+1) = 0 \Rightarrow x = 0 \text{ or } x = 1 \text{ or } x = -1$

93. $\dfrac{1}{2} \begin{vmatrix} -4 & 2 & 1 \\ 0 & k & 1 \\ -2 & k & 1 \end{vmatrix} = 28 \Rightarrow 0 + k(-1)^{2+2} \begin{vmatrix} -4 & 1 \\ -2 & 1 \end{vmatrix} + 1(-1)^{2+3} \begin{vmatrix} -4 & 2 \\ -2 & k \end{vmatrix} = 56 \Rightarrow -2k - (-4k + 4) = 2k - 4 = 56 \Rightarrow k = 30$

9.4 Critical Thinking/Discussion/Writing

95. Let $u = \dfrac{1}{x-2}$ and $v = \dfrac{1}{y+1}$. Then $\begin{cases} \dfrac{1}{x-2} + \dfrac{3}{y+1} = 13 \\ \dfrac{4}{x-2} - \dfrac{5}{y+1} = 1 \end{cases} \Rightarrow \begin{cases} u + 3v = 13 \\ 4u - 5v = 1 \end{cases}$.

$D = \begin{vmatrix} 1 & 3 \\ 4 & -5 \end{vmatrix} = -17, \; D_u = \begin{vmatrix} 13 & 3 \\ 1 & -5 \end{vmatrix} = -68, \; D_v = \begin{vmatrix} 1 & 13 \\ 4 & 1 \end{vmatrix} = -51 \Rightarrow u = \dfrac{-68}{-14}, \; v = \dfrac{-51}{-17} = 3$

$4 = \dfrac{1}{x-2} \Rightarrow 4x - 8 = 1 \Rightarrow x = \dfrac{9}{4}; \quad 3 = \dfrac{1}{y+1} \Rightarrow 3y + 3 = 1 \Rightarrow y = -\dfrac{2}{3}$

The solution is $\left\{\left(\dfrac{9}{4}, -\dfrac{2}{3}\right)\right\}$.

96. Let $u = \dfrac{1}{x+1}$ and $v = \dfrac{1}{y-1}$. Then $\begin{cases} \dfrac{6}{x+1} + \dfrac{4}{y-1} = 7 \\ \dfrac{8}{x+1} + \dfrac{5}{y-1} = 9 \end{cases} \Rightarrow \begin{cases} 6u + 4v = 7 \\ 8u + 5v = 9 \end{cases}$

$D = \begin{vmatrix} 6 & 4 \\ 8 & 5 \end{vmatrix} = -2, \; D_u = \begin{vmatrix} 7 & 4 \\ 9 & 5 \end{vmatrix} = -1, \; D_v = \begin{vmatrix} 6 & 7 \\ 8 & 9 \end{vmatrix} = -2, \; u = \dfrac{-1}{-2} = \dfrac{1}{2}, \; v = \dfrac{-2}{-2} = 1$

$\dfrac{1}{2} = \dfrac{1}{x+1} \Rightarrow x + 1 = 2 \Rightarrow x = 1; \quad 1 = \dfrac{1}{y-1} \Rightarrow y - 1 = 1 \Rightarrow y = 2$

The solution is $\{(1, 2)\}$.

97. Let $u = \dfrac{1}{3^x}$ and $v = \dfrac{1}{4^y}$. Then $\begin{cases} \dfrac{1}{3^x} + \dfrac{4}{4^y} = 25 \\ \dfrac{2}{3^x} - \dfrac{1}{4^y} = 14 \end{cases} \Rightarrow \begin{cases} u + 4v = 25 \\ 2u - v = 14 \end{cases}$

$D = \begin{vmatrix} 1 & 4 \\ 2 & -1 \end{vmatrix} = -9, \; D_u = \begin{vmatrix} 25 & 4 \\ 14 & -1 \end{vmatrix} = -81, \; D_v = \begin{vmatrix} 1 & 25 \\ 2 & 14 \end{vmatrix} = -36, \; u = \dfrac{-81}{-9} = 9, \; v = \dfrac{-36}{-9} = 4$

$9 = \dfrac{1}{3^x} \Rightarrow 3^x = \dfrac{1}{9} \Rightarrow x = -2; \quad 4 = \dfrac{1}{4^y} \Rightarrow 4^y = \dfrac{1}{4} \Rightarrow y = -1$

The solution is $\{(-2, -1)\}$.

98. We use the change of base formula along with the power rule of logarithms. See section 4.3.

  a. Recall that $\log 8 = \log(2^3) = 3\log 2$.

  $\begin{vmatrix} \log_2 3 & \log_8 3 \\ \log_3 4 & \log_3 4 \end{vmatrix} = \log_2 3 \cdot \log_3 4 - \log_8 3 \cdot \log_3 4 = \dfrac{\log 3 \cdot \log 4}{\log 2 \cdot \log 3} - \dfrac{\log 3 \cdot \log 4}{\log 8 \cdot \log 3} = \dfrac{\log 3 \cdot \log 4}{\log 2 \cdot \log 3} - \dfrac{\log 3 \cdot \log 4}{3\log 2 \cdot \log 3}$

  $= \dfrac{\log 4}{\log 2} - \dfrac{\log 4}{3\log 2} = \dfrac{3\log 4 - \log 4}{3\log 2} = \dfrac{2\log 4}{3\log 2} = \dfrac{2\log(2^2)}{3\log 2} = \dfrac{4\log 2}{3\log 2} = \dfrac{4}{3}$

  b. $\begin{vmatrix} \log_3 512 & \log_4 3 \\ \log_3 8 & \log_4 9 \end{vmatrix} = \log_3 512 \cdot \log_4 9 - \log_4 3 \cdot \log_3 8 = \dfrac{\log 512 \cdot \log 9}{\log 3 \cdot \log 4} - \dfrac{\log 3 \cdot \log 8}{\log 4 \cdot \log 3}$

  $= \dfrac{\log(2 \cdot 16^2) \cdot 2\log 3}{\log 3 \cdot \log 4} - \dfrac{3\log 2}{\log 4} = \dfrac{(\log 2 + 2\log 16) \cdot 2}{2\log 2} - \dfrac{3\log 2}{2\log 2} = \dfrac{\log 2 + 2\log(2^4)}{\log 2} - \dfrac{3}{2}$

  $= \dfrac{\log 2 + 8\log 2}{\log 2} - \dfrac{3}{2} = \dfrac{9\log 2}{\log 2} - \dfrac{3}{2} = 9 - \dfrac{3}{2} = \dfrac{15}{2}$

99. A system has no solution when $D = 0$.

$$D = \begin{vmatrix} k & 3 & -1 \\ 1 & 2 & 1 \\ -k & 1 & 2 \end{vmatrix}$$

$$= -1\begin{vmatrix} 3 & -1 \\ 1 & 2 \end{vmatrix} + 2\begin{vmatrix} k & -1 \\ -k & 2 \end{vmatrix} - 1\begin{vmatrix} k & 3 \\ -k & 1 \end{vmatrix}$$

$$= -7 + 2(2k - k) - (k + 3k)$$

$$= -7 + 2k - 4k = -7 - 2k \Rightarrow -7 - 2k = 0 \Rightarrow$$

$$k = -\frac{7}{2}$$

The system has no solution for $k = -\frac{7}{2}$.

100. $D = \begin{vmatrix} 2 & a & 6 \\ 1 & 2 & b \\ 1 & 1 & 3 \end{vmatrix} = \begin{vmatrix} a & 6 \\ 2 & b \end{vmatrix} - \begin{vmatrix} 2 & 6 \\ 1 & b \end{vmatrix} + 3\begin{vmatrix} 2 & a \\ 1 & 2 \end{vmatrix}$

$$= (ab - 12) - (2b - 6) + 3(4 - a)$$

$$= ab - 3a - 2b + 6$$

The system of equations has a unique solution when $D \neq 0$.

$ab - 3a - 2b + 6 = 0 \Rightarrow ab - 3a = 2b - 6 \Rightarrow$
$a(b - 3) = 2(b - 3) \Rightarrow a = 2, b \neq 3$

So, the system has a unique solution if $a \neq 2$ and $b \neq 3$.

If $a = 2$, then row 1 = 2 row 3, so there are infinitely many solutions.
If $a = b = 3$, then $D = 0$, but all rows are different, so there is no solution.

9.4 Getting Ready for the Next Section

You may want to refer to sections 1.2, 8.3, and 8.4.

101. $x^2 = 8y \Rightarrow \frac{1}{8}x^2 = y$

103. $(x - 2)^2 = 6(y - 3) \Rightarrow x^2 - 4x + 4 = 6y - 18 \Rightarrow$
$x^2 - 4x + 22 = 6y \Rightarrow \frac{1}{6}x^2 - \frac{2}{3}x + \frac{11}{3} = y$

105. $-2 = a(-2)^2 \Rightarrow -2 = 4a \Rightarrow -\frac{1}{2} = a$

The function is $y = -\frac{1}{2}x^2$.

107. The formula for the x-value of the vertex gives

$x = -\frac{b}{2a} = -3 \Rightarrow b = 6a$

Using $(0, 10)$ gives
$10 = a(0^2) + b(0) + c \Rightarrow c = 10$.

Substituting $(-3, -8)$, $b = 6a$, and $c = 10$ gives

$-8 = a(-3)^2 + 6a(-3) + 10 \Rightarrow$
$-8 = 9a - 18a + 10 \Rightarrow -18 = -9a \Rightarrow 2 = a \Rightarrow$
$b = 12$

The function is $y = 2x^2 + 12x + 10$.

109. $f(x) = 2x^2 + 4x - 6$

a. Opens up since $a > 0$.

b. $x = -\frac{b}{2a} = -\frac{4}{2(2)} = -1$

$f(-1) = 2(-1)^2 + 4(-1) - 6 = -8$

The vertex is $(-1, -8)$.

c. The axis of symmetry is $x = -1$.

d. To find the x-intercepts, let $y = 0$ and solve for x.

$2x^2 + 4x - 6 = 0 \Rightarrow 2(x^2 + 2x - 3) = 0 \Rightarrow$
$(x + 3)(x - 1) = 0 \Rightarrow x = -3, x = 1$

The x-intercepts are -3 and 1.

e. To find the y-intercept, let $x = 0$ and solve for y.

$y = 2(0)^2 + 4(0) - 6 = -6$

The y-intercept is -6.

Chapter 9 Review Exercises

Building Skills

1. 1×4 **3.** 3×2

5. $a_{12} = -1, a_{14} = -4, a_{23} = 3, a_{21} = 5$

7. $\begin{bmatrix} 2 & -3 & | & 7 \\ 3 & 1 & | & 6 \end{bmatrix}$

9. Answers may vary.

$\begin{bmatrix} 0 & 1 & 2 & 1 \\ 2 & 0 & 3 & 4 \\ 1 & -2 & 1 & 7 \end{bmatrix} \xrightarrow{R_1 \leftrightarrow R_3} \begin{bmatrix} 1 & -2 & 1 & 7 \\ 2 & 0 & 3 & 4 \\ 0 & 1 & 2 & 1 \end{bmatrix}$

$\xrightarrow{2R_1 - R_2 \to R_2} \begin{bmatrix} 1 & -2 & 1 & 7 \\ 0 & -4 & -1 & 10 \\ 0 & 1 & 2 & 1 \end{bmatrix}$

$\xrightarrow{R_2 \leftrightarrow R_3} \begin{bmatrix} 1 & -2 & 1 & 7 \\ 0 & 1 & 2 & 1 \\ 0 & -4 & -1 & 10 \end{bmatrix}$

$\xrightarrow{\frac{1}{7}(4R_2 + R_3) \to R_3} \begin{bmatrix} 1 & -2 & 1 & 7 \\ 0 & 1 & 2 & 1 \\ 0 & 0 & 1 & 2 \end{bmatrix}$

11. $\begin{bmatrix} 3 & 1 & 3 & 1 \\ 2 & 1 & 1 & 1 \\ 1 & -1 & -1 & 0 \end{bmatrix} \xrightarrow{R_1 \leftrightarrow R_3} \begin{bmatrix} 1 & -1 & -1 & 0 \\ 2 & 1 & 1 & 1 \\ 3 & 1 & 3 & 1 \end{bmatrix}$

$\xrightarrow[-3R_1+R_3 \to R_3]{-2R_1+R_2 \to R_2} \begin{bmatrix} 1 & -1 & -1 & 0 \\ 0 & 3 & 3 & 1 \\ 0 & 4 & 6 & 1 \end{bmatrix}$

$\xrightarrow{-\frac{1}{2}(-2R_2+R_3)\to R_2} \begin{bmatrix} 1 & -1 & -1 & 0 \\ 0 & 1 & 0 & \frac{1}{2} \\ 0 & 4 & 6 & 1 \end{bmatrix}$

$\xrightarrow[\frac{1}{6}(R_3-4R_2)\to R_3]{R_1+R_2 \to R_1} \begin{bmatrix} 1 & 0 & -1 & \frac{1}{2} \\ 0 & 1 & 0 & \frac{1}{2} \\ 0 & 0 & 1 & -\frac{1}{6} \end{bmatrix}$

$\xrightarrow{R_1+R_3 \to R_1} \begin{bmatrix} 1 & 0 & 0 & \frac{1}{3} \\ 0 & 1 & 0 & \frac{1}{2} \\ 0 & 0 & 1 & -\frac{1}{6} \end{bmatrix}$

13. $\begin{cases} x+y-z=0 \\ 2x+y-2z=3 \\ 3x-2y+3z=9 \end{cases} \Rightarrow \begin{bmatrix} 1 & 1 & -1 & | & 0 \\ 2 & 1 & -2 & | & 3 \\ 3 & -2 & 3 & | & 9 \end{bmatrix}$

$\xrightarrow[3R_1-R_3 \to R_3]{2R_1-R_2 \to R_2} \begin{bmatrix} 1 & 1 & -1 & | & 0 \\ 0 & 1 & 0 & | & -3 \\ 0 & 5 & -6 & | & -9 \end{bmatrix}$

$\xrightarrow[-\frac{1}{6}(R_3-5R_2)\to R_3]{R_1-R_2\to R_1} \begin{bmatrix} 1 & 0 & -1 & | & 3 \\ 0 & 1 & 0 & | & -3 \\ 0 & 0 & 1 & | & -1 \end{bmatrix}$

$\xrightarrow{R_1+R_3 \to R_1} \begin{bmatrix} 1 & 0 & 0 & | & 2 \\ 0 & 1 & 0 & | & -3 \\ 0 & 0 & 1 & | & -1 \end{bmatrix}$

The solution is $\{(2,-3,-1)\}$.

15. $\begin{cases} x-2y+3z=-2 \\ 2x-3y+z=9 \\ 3x-y+2z=5 \end{cases} \Rightarrow \begin{bmatrix} 1 & -2 & 3 & | & -2 \\ 2 & -3 & 1 & | & 9 \\ 3 & -1 & 2 & | & 5 \end{bmatrix}$

$\xrightarrow[R_3-3R_1 \to R_3]{R_2-2R_1 \to R_2} \begin{bmatrix} 1 & -2 & 3 & | & -2 \\ 0 & 1 & -5 & | & 13 \\ 0 & 5 & -7 & | & 11 \end{bmatrix}$

$\xrightarrow{\frac{1}{18}(R_3-5R_2)\to R_3} \begin{bmatrix} 1 & -2 & 3 & | & -2 \\ 0 & 1 & -5 & | & 13 \\ 0 & 0 & 1 & | & -3 \end{bmatrix}$

$z=-3; y-5(-3)=13 \Rightarrow y=-2$
$x-2(-2)+3(-3)=-2 \Rightarrow x=3$
The solution is $\{(3,-2,-3)\}$.

17. $\begin{cases} x-2y-2z=11 \\ 3x+4y-z=-2 \\ 4x+5y+7z=7 \end{cases} \Rightarrow \begin{bmatrix} 1 & -2 & -2 & | & 1 \\ 3 & 4 & -1 & | & -2 \\ 4 & 5 & 7 & | & 7 \end{bmatrix}$

$\xrightarrow[R_3-4R_1 \to R_3]{\frac{1}{5}(R_2-3R_1)\to R_2} \begin{bmatrix} 1 & -2 & -2 & | & 1 \\ 0 & 2 & 1 & | & -7 \\ 0 & 13 & 15 & | & -37 \end{bmatrix}$

$\xrightarrow[\frac{2}{17}(R_3-\frac{13}{2}R_2)\to R_3]{R_1+R_2 \to R_1} \begin{bmatrix} 1 & 0 & -1 & | & 4 \\ 0 & 2 & 1 & | & -7 \\ 0 & 0 & 1 & | & 1 \end{bmatrix}$

$\xrightarrow[\frac{1}{2}(R_2-R_3)\to R_2]{R_1+R_3 \to R_1} \begin{bmatrix} 1 & 0 & 0 & | & 5 \\ 0 & 1 & 0 & | & -4 \\ 0 & 0 & 1 & | & 1 \end{bmatrix}$

The solution is $\{(5,-4,1)\}$.

19. $\begin{cases} 2x-y+3z=4 \\ x+3y+3z=-2 \\ 3x+2y-6z=6 \end{cases} \Rightarrow \begin{bmatrix} 2 & -1 & 3 & | & 4 \\ 1 & 3 & 3 & | & -2 \\ 3 & 2 & -6 & | & 6 \end{bmatrix}$

$\xrightarrow{R_1 \leftrightarrow R_2} \begin{bmatrix} 1 & 3 & 3 & | & -2 \\ 2 & -1 & 3 & | & 4 \\ 3 & 2 & -6 & | & 6 \end{bmatrix}$

$\xrightarrow[R_3-3R_1 \to R_3]{2R_1-R_2 \to R_2} \begin{bmatrix} 1 & 3 & 3 & | & -2 \\ 0 & 7 & 3 & | & -8 \\ 0 & -7 & -15 & | & 12 \end{bmatrix}$

$\xrightarrow{-\frac{1}{12}(R_2+R_3)\to R_3} \begin{bmatrix} 1 & 3 & 3 & | & -2 \\ 0 & 7 & 3 & | & -8 \\ 0 & 0 & 1 & | & -\frac{1}{3} \end{bmatrix}$

$\xrightarrow[\frac{1}{7}(R_2-3R_3)\to R_2]{R_1-3R_3 \to R_1} \begin{bmatrix} 1 & 3 & 0 & | & -1 \\ 0 & 1 & 0 & | & -1 \\ 0 & 0 & 1 & | & -\frac{1}{3} \end{bmatrix}$

$\xrightarrow{R_1-3R_2 \to R_1} \begin{bmatrix} 1 & 0 & 0 & | & 2 \\ 0 & 1 & 0 & | & -1 \\ 0 & 0 & 1 & | & -\frac{1}{3} \end{bmatrix}$

The solution is $\left\{\left(2,-1,-\frac{1}{3}\right)\right\}$.

21. $\begin{bmatrix} x-y & 0 \\ 1 & x+y \end{bmatrix} = \begin{bmatrix} 1 & 0 \\ 1 & 3 \end{bmatrix} \Rightarrow \begin{cases} x-y=1 \\ x+y=3 \end{cases} \Rightarrow$
$2x=4 \Rightarrow x=2, y=1$

23. a. $\begin{bmatrix} 1 & 2 \\ -3 & 4 \end{bmatrix} + \begin{bmatrix} 2 & -3 \\ -5 & 6 \end{bmatrix} = \begin{bmatrix} 3 & -1 \\ -8 & 10 \end{bmatrix}$

b. $\begin{bmatrix} 1 & 2 \\ -3 & 4 \end{bmatrix} - \begin{bmatrix} 2 & -3 \\ -5 & 6 \end{bmatrix} = \begin{bmatrix} -1 & 5 \\ 2 & -2 \end{bmatrix}$

c. $2\begin{bmatrix} 1 & 2 \\ -3 & 4 \end{bmatrix} = \begin{bmatrix} 2 & 4 \\ -6 & 8 \end{bmatrix}$

d. $-3\begin{bmatrix} 2 & -3 \\ -5 & 6 \end{bmatrix} = \begin{bmatrix} -6 & 9 \\ 15 & -18 \end{bmatrix}$

e. $2\begin{bmatrix} 1 & 2 \\ -3 & 4 \end{bmatrix} - 3\begin{bmatrix} 2 & -3 \\ -5 & 6 \end{bmatrix}$

$= \begin{bmatrix} 2 & 4 \\ -6 & 8 \end{bmatrix} - \begin{bmatrix} 6 & -9 \\ -15 & 18 \end{bmatrix} = \begin{bmatrix} -4 & 13 \\ 9 & -10 \end{bmatrix}$

25. $3A + 2B - 3X = 0 \Rightarrow$

$3\begin{bmatrix} 1 & 2 \\ -3 & 4 \end{bmatrix} + 2\begin{bmatrix} 2 & -3 \\ -5 & 6 \end{bmatrix} = 3X \Rightarrow$

$\begin{bmatrix} 3 & 6 \\ -9 & 12 \end{bmatrix} + \begin{bmatrix} 4 & -6 \\ -10 & 12 \end{bmatrix} = 3X \Rightarrow$

$\begin{bmatrix} 7 & 0 \\ -19 & 24 \end{bmatrix} = 3X \Rightarrow X = \begin{bmatrix} \frac{7}{3} & 0 \\ -\frac{19}{3} & 8 \end{bmatrix}$

27. a. $AB = \begin{bmatrix} 0 & 1 \\ 2 & 3 \end{bmatrix}\begin{bmatrix} -1 & -1 \\ -3 & 4 \end{bmatrix}$

$= \begin{bmatrix} 0(-1)+1(-3) & 0(-1)+1(4) \\ 2(-1)+3(-3) & 2(-1)+3(4) \end{bmatrix}$

$= \begin{bmatrix} -3 & 4 \\ -11 & 10 \end{bmatrix}$

b. $BA = \begin{bmatrix} -1 & -1 \\ -3 & 4 \end{bmatrix}\begin{bmatrix} 0 & 1 \\ 2 & 3 \end{bmatrix}$

$= \begin{bmatrix} -1(0)-1(2) & -1(1)-1(3) \\ -3(0)+4(2) & -3(1)+4(3) \end{bmatrix}$

$= \begin{bmatrix} -2 & -4 \\ 8 & 9 \end{bmatrix}$

29. a. $AB = \begin{bmatrix} 1 & 2 & -1 \end{bmatrix}\begin{bmatrix} 2 \\ 3 \\ 1 \end{bmatrix} = [1(2) + 2(3) - 1(1)]$

$= [7]$

b. $BA = \begin{bmatrix} 2 \\ 3 \\ 1 \end{bmatrix}\begin{bmatrix} 1 & 2 & -1 \end{bmatrix}$

$= \begin{bmatrix} 2(1) & 2(2) & 2(-1) \\ 3(1) & 3(2) & 3(-1) \\ 1(1) & 1(2) & 1(-1) \end{bmatrix} = \begin{bmatrix} 2 & 4 & -2 \\ 3 & 6 & -3 \\ 1 & 2 & -1 \end{bmatrix}$

31. $\begin{bmatrix} 1 & 2 \\ 3 & -1 \end{bmatrix}A = \begin{bmatrix} 5 & 6 \\ 1 & -3 \end{bmatrix} \Rightarrow A = \begin{bmatrix} 1 & 2 \\ 3 & -1 \end{bmatrix}^{-1}\begin{bmatrix} 5 & 6 \\ 1 & -3 \end{bmatrix}$

$\begin{bmatrix} 1 & 2 \\ 3 & -1 \end{bmatrix}^{-1} = \frac{1}{-1-6}\begin{bmatrix} -1 & -2 \\ -3 & 1 \end{bmatrix} = \begin{bmatrix} \frac{1}{7} & \frac{2}{7} \\ \frac{3}{7} & -\frac{1}{7} \end{bmatrix}$

$A = \begin{bmatrix} \frac{1}{7} & \frac{2}{7} \\ \frac{3}{7} & -\frac{1}{7} \end{bmatrix}\begin{bmatrix} 5 & 6 \\ 1 & -3 \end{bmatrix} = \begin{bmatrix} 1 & 0 \\ 2 & 3 \end{bmatrix}$

33. $\begin{bmatrix} 5 & 3 \\ 4 & 2 \end{bmatrix}A = \begin{bmatrix} 1 & 0 \\ 0 & 1 \end{bmatrix} \Rightarrow A = \begin{bmatrix} 5 & 3 \\ 4 & 2 \end{bmatrix}^{-1}\begin{bmatrix} 1 & 0 \\ 0 & 1 \end{bmatrix}$

$\begin{bmatrix} 5 & 3 \\ 4 & 2 \end{bmatrix}^{-1} = \frac{1}{10-12}\begin{bmatrix} 2 & -3 \\ -4 & 5 \end{bmatrix} = \begin{bmatrix} -1 & \frac{3}{2} \\ 2 & -\frac{5}{2} \end{bmatrix}$

$A = \begin{bmatrix} -1 & \frac{3}{2} \\ 2 & -\frac{5}{2} \end{bmatrix}\begin{bmatrix} 1 & 0 \\ 0 & 1 \end{bmatrix} = \begin{bmatrix} -1 & \frac{3}{2} \\ 2 & -\frac{5}{2} \end{bmatrix}$

35. a. $A^{-1} = \frac{1}{0-(-1)}\begin{bmatrix} 3 & -1 \\ 1 & 0 \end{bmatrix} = \begin{bmatrix} 3 & -1 \\ 1 & 0 \end{bmatrix}$

b. $(A^2)^{-1} = \left(\begin{bmatrix} 0 & 1 \\ -1 & 3 \end{bmatrix}^2\right)^{-1} = \begin{bmatrix} -1 & 3 \\ -3 & 8 \end{bmatrix}^{-1}$

$= \frac{1}{-8+9}\begin{bmatrix} 8 & -3 \\ 3 & -1 \end{bmatrix} = \begin{bmatrix} 8 & -3 \\ 3 & -1 \end{bmatrix}$

c. $(A^{-1})^2 = \begin{bmatrix} 3 & -1 \\ 1 & 0 \end{bmatrix}\begin{bmatrix} 3 & -1 \\ 1 & 0 \end{bmatrix} = \begin{bmatrix} 8 & -3 \\ 3 & -1 \end{bmatrix}$

d. $(A^2)(A^{-1})^2 = \begin{bmatrix} -1 & 3 \\ -3 & 8 \end{bmatrix}\begin{bmatrix} 8 & -3 \\ 3 & -1 \end{bmatrix} = \begin{bmatrix} 1 & 0 \\ 0 & 1 \end{bmatrix}$

So $(A^{-1})^2$ is the inverse of A^2.

For exercises 37–40, we show that $AB = I$. You should also show that $BA = I$ in order to show that B is the inverse of A.

37. $AB = \begin{bmatrix} 7 & 6 \\ 6 & 5 \end{bmatrix} \begin{bmatrix} -5 & 6 \\ 6 & -7 \end{bmatrix} = \begin{bmatrix} 1 & 0 \\ 0 & 1 \end{bmatrix}$

39. $AB = \begin{bmatrix} 2 & -1 & 0 \\ 1 & 0 & 4 \\ 1 & -1 & 1 \end{bmatrix} \left(\dfrac{1}{5}\right) \begin{bmatrix} 4 & 1 & -4 \\ 3 & 2 & -8 \\ -1 & 1 & 1 \end{bmatrix} = \left(\dfrac{1}{5}\right) \begin{bmatrix} 2 & -1 & 0 \\ 1 & 0 & 4 \\ 1 & -1 & 1 \end{bmatrix} \begin{bmatrix} 4 & 1 & -4 \\ 3 & 2 & -8 \\ -1 & 1 & 1 \end{bmatrix} = \left(\dfrac{1}{5}\right) \begin{bmatrix} 5 & 0 & 0 \\ 0 & 5 & 0 \\ 0 & 0 & 5 \end{bmatrix} = \begin{bmatrix} 1 & 0 & 0 \\ 0 & 1 & 0 \\ 0 & 0 & 1 \end{bmatrix}$

41. $\begin{bmatrix} 3 & 1 \\ 2 & 4 \end{bmatrix}^{-1} = \dfrac{1}{12-2} \begin{bmatrix} 4 & -1 \\ -2 & 3 \end{bmatrix} = \begin{bmatrix} \dfrac{2}{5} & -\dfrac{1}{10} \\ -\dfrac{1}{5} & \dfrac{3}{10} \end{bmatrix}$

43. $\left[\begin{array}{ccc|ccc} 1 & 2 & -2 & 1 & 0 & 0 \\ -1 & 3 & 0 & 0 & 1 & 0 \\ 0 & -2 & 1 & 0 & 0 & 1 \end{array}\right] \xrightarrow{R_1+R_2 \to R_2} \left[\begin{array}{ccc|ccc} 1 & 2 & -2 & 1 & 0 & 0 \\ 0 & 5 & -2 & 1 & 1 & 0 \\ 0 & -2 & 1 & 0 & 0 & 1 \end{array}\right] \xrightarrow[R_2+2R_3 \to R_2]{R_1+R_3 \to R_1} \left[\begin{array}{ccc|ccc} 1 & 0 & -1 & 1 & 0 & 1 \\ 0 & 1 & 0 & 1 & 1 & 2 \\ 0 & -2 & 1 & 0 & 0 & 1 \end{array}\right]$

$\xrightarrow{2R_2+R_3 \to R_3} \left[\begin{array}{ccc|ccc} 1 & 0 & -1 & 1 & 0 & 1 \\ 0 & 1 & 0 & 1 & 1 & 2 \\ 0 & 0 & 1 & 2 & 2 & 5 \end{array}\right] \xrightarrow{R_1+R_3 \to R_1} \left[\begin{array}{ccc|ccc} 1 & 0 & 0 & 3 & 2 & 6 \\ 0 & 1 & 0 & 1 & 1 & 2 \\ 0 & 0 & 1 & 2 & 2 & 5 \end{array}\right]$

Thus, $\begin{bmatrix} 1 & 2 & -2 \\ -1 & 3 & 0 \\ 0 & -2 & 1 \end{bmatrix}^{-1} = \begin{bmatrix} 3 & 2 & 6 \\ 1 & 1 & 2 \\ 2 & 2 & 5 \end{bmatrix}$

45. $\begin{cases} x+3y = 7 \\ 2x+5y = 4 \end{cases} \Rightarrow \begin{bmatrix} 1 & 3 \\ 2 & 5 \end{bmatrix} \begin{bmatrix} x \\ y \end{bmatrix} = \begin{bmatrix} 7 \\ 4 \end{bmatrix} \Rightarrow \begin{bmatrix} x \\ y \end{bmatrix} = \begin{bmatrix} 1 & 3 \\ 2 & 5 \end{bmatrix}^{-1} \begin{bmatrix} 7 \\ 4 \end{bmatrix} = \begin{bmatrix} -5 & 3 \\ 2 & -1 \end{bmatrix} \begin{bmatrix} 7 \\ 4 \end{bmatrix} = \begin{bmatrix} -23 \\ 10 \end{bmatrix}$

The solution is $\{(-23, 10)\}$.

47. $\begin{cases} x+3y+3z = 3 \\ x+4y+3z = 5 \\ x+3y+4z = 6 \end{cases} \Rightarrow \begin{bmatrix} 1 & 3 & 3 \\ 1 & 4 & 3 \\ 1 & 3 & 4 \end{bmatrix} \begin{bmatrix} x \\ y \\ z \end{bmatrix} = \begin{bmatrix} 3 \\ 5 \\ 6 \end{bmatrix} \Rightarrow \begin{bmatrix} x \\ y \\ z \end{bmatrix} = \begin{bmatrix} 1 & 3 & 3 \\ 1 & 4 & 3 \\ 1 & 3 & 4 \end{bmatrix}^{-1} \begin{bmatrix} 3 \\ 5 \\ 6 \end{bmatrix} = \begin{bmatrix} 7 & -3 & -3 \\ -1 & 1 & 0 \\ -1 & 0 & 1 \end{bmatrix} \begin{bmatrix} 3 \\ 5 \\ 6 \end{bmatrix} = \begin{bmatrix} -12 \\ 2 \\ 3 \end{bmatrix}$

The solution is $\{(-12, 2, 3)\}$.

49. $\begin{cases} x-y+z = 3 \\ 4x+2y = 5 \\ 7x-y-z = 6 \end{cases} \Rightarrow \begin{bmatrix} 1 & -1 & 1 \\ 4 & 2 & 0 \\ 7 & -1 & -1 \end{bmatrix} \begin{bmatrix} x \\ y \\ z \end{bmatrix} = \begin{bmatrix} 3 \\ 5 \\ 6 \end{bmatrix} \Rightarrow \begin{bmatrix} x \\ y \\ z \end{bmatrix} = \begin{bmatrix} 1 & -1 & 1 \\ 4 & 2 & 0 \\ 7 & -1 & -1 \end{bmatrix}^{-1} \begin{bmatrix} 3 \\ 5 \\ 6 \end{bmatrix} = \begin{bmatrix} \dfrac{1}{12} & \dfrac{1}{12} & \dfrac{1}{12} \\ -\dfrac{1}{6} & \dfrac{1}{3} & -\dfrac{1}{6} \\ \dfrac{3}{4} & \dfrac{1}{4} & -\dfrac{1}{4} \end{bmatrix} \begin{bmatrix} 3 \\ 5 \\ 6 \end{bmatrix} = \begin{bmatrix} \dfrac{7}{6} \\ \dfrac{1}{6} \\ 2 \end{bmatrix}$

The solution is $\left\{\left(\dfrac{7}{6}, \dfrac{1}{6}, 2\right)\right\}$.

51. $\begin{vmatrix} 2 & -5 \\ 3 & 4 \end{vmatrix} = 2(4) - (-5)(3) = 23$

53. $\begin{vmatrix} 12 & 21 \\ 4 & 7 \end{vmatrix} = 12(7) - 21(4) = 0$

55. a. $A = \begin{bmatrix} 4 & -1 & 2 \\ -2 & -3 & 5 \\ 0 & 2 & -4 \end{bmatrix}$

$M_{12} = \begin{vmatrix} -2 & 5 \\ 0 & -4 \end{vmatrix} = (-2)(-4) - 5(0) = 8$ $\qquad M_{23} = \begin{vmatrix} 4 & -1 \\ 0 & 2 \end{vmatrix} = 4(2) - (-1)(0) = 8$

$M_{22} = \begin{vmatrix} 4 & 2 \\ 0 & -4 \end{vmatrix} = 4(-4) - 2(0) = -16$

b. $A_{12} = (-1)^{1+2} M_{12} = -8 \qquad A_{23} = (-1)^{2+3} M_{23} = -8 \qquad A_{22} = (-1)^{2+2} M_{22} = -16$

57. a. $\begin{vmatrix} 1 & -2 & 3 \\ 4 & -1 & -2 \\ -2 & 1 & 5 \end{vmatrix} = a_{21}A_{21} + a_{22}A_{22} + a_{23}A_{23} = 4(-1)^{2+1}\begin{vmatrix} -2 & 3 \\ 1 & 5 \end{vmatrix} - 1(-1)^{2+2}\begin{vmatrix} 1 & 3 \\ -2 & 5 \end{vmatrix} - 2(-1)^{2+3}\begin{vmatrix} 1 & -2 \\ -2 & 1 \end{vmatrix}$

$= -4(-13) - 1(11) + 2(-3) = 35$

b. $\begin{vmatrix} 1 & -2 & 3 \\ 4 & -1 & -2 \\ -2 & 1 & 5 \end{vmatrix} = a_{13}A_{13} + a_{23}A_{23} + a_{33}A_{33} = 3(-1)^{1+3}\begin{vmatrix} 4 & -1 \\ -2 & 1 \end{vmatrix} - 2(-1)^{2+3}\begin{vmatrix} 1 & -2 \\ -2 & 1 \end{vmatrix} + 5(-1)^{3+3}\begin{vmatrix} 1 & -2 \\ 4 & -1 \end{vmatrix}$

$= 3(2) + 2(-3) + 5(7) = 35$

59. Expand by the first column: $\begin{vmatrix} 1 & 2 & 0 \\ 0 & 1 & 2 \\ 1 & 0 & 2 \end{vmatrix} = 1(-1)^{1+1}\begin{vmatrix} 1 & 2 \\ 0 & 2 \end{vmatrix} + 0 + 1(-1)^{3+1}\begin{vmatrix} 2 & 0 \\ 1 & 2 \end{vmatrix} = 2 + 4 = 6$

61. $D = \begin{vmatrix} 5 & 3 \\ 2 & 1 \end{vmatrix} = -1, \ D_x = \begin{vmatrix} 11 & 3 \\ 4 & 1 \end{vmatrix} = -1, \ D_y = \begin{vmatrix} 5 & 11 \\ 2 & 4 \end{vmatrix} = -2$

$x = \dfrac{-1}{-1} = 1, y = \dfrac{-2}{-1} = 2$

The solution is $\{(1, 2)\}$.

63. $D = \begin{vmatrix} 3 & 1 & -1 \\ 1 & 3 & -1 \\ 1 & 1 & -3 \end{vmatrix} = -20, \ D_x = \begin{vmatrix} 14 & 1 & -1 \\ 16 & 3 & -1 \\ -10 & 1 & -3 \end{vmatrix} = -100, \ D_y = \begin{vmatrix} 3 & 14 & -1 \\ 1 & 16 & -1 \\ 1 & -10 & -3 \end{vmatrix} = -120, \ D_z = \begin{vmatrix} 3 & 1 & 14 \\ 1 & 3 & 16 \\ 1 & 1 & -10 \end{vmatrix} = -140$

$x = \dfrac{-100}{-20} = 5, y = \dfrac{-120}{-20} = 6, z = \dfrac{-140}{-20} = 7$

The solution is $\{(5, 6, 7)\}$.

65. Expand by the third row:

$\begin{vmatrix} 1 & 2 & 4 \\ -3 & 5 & 7 \\ 1 & x & 4 \end{vmatrix} = 0 \Rightarrow 1(-1)^{3+1}\begin{vmatrix} 2 & 4 \\ 5 & 7 \end{vmatrix} + x(-1)^{3+2}\begin{vmatrix} 1 & 4 \\ -3 & 7 \end{vmatrix} + 4(-1)^{3+3}\begin{vmatrix} 1 & 2 \\ -3 & 5 \end{vmatrix} = 0 \Rightarrow -6 - 19x + 4(11) = 0 \Rightarrow x = 2$

67. $\begin{vmatrix} x & y & 1 \\ 2 & 3 & 1 \\ 1 & -4 & 1 \end{vmatrix} = 0 \Rightarrow x(-1)^{1+1}\begin{vmatrix} 3 & 1 \\ -4 & 1 \end{vmatrix} + y(-1)^{1+2}\begin{vmatrix} 2 & 1 \\ 1 & 1 \end{vmatrix} + 1(-1)^{1+3}\begin{vmatrix} 2 & 3 \\ 1 & -4 \end{vmatrix} = 0 \Rightarrow 7x - y - 11 = 0 \Rightarrow y = 7x - 11$

Using the given points to find the slope and y-intercept, we have $m = \dfrac{3 - (-4)}{2 - 1} = 7$ and $3 = 7(2) + b \Rightarrow b = -11$.

Applying the Concepts

69.
$$\begin{array}{c} \text{Method 1} \\ \text{Method 2} \\ \text{Method 3} \end{array} \begin{array}{c} \text{A B C} \\ \begin{bmatrix} 4 & 8 & 2 \\ 5 & 7 & 1 \\ 5 & 4 & 8 \end{bmatrix} \end{array} \begin{bmatrix} 10 \\ 4 \\ 6 \end{bmatrix} = \begin{bmatrix} 84 \\ 84 \\ 114 \end{bmatrix} \Rightarrow \text{Method 3 is the most profitable.}$$

71. Let x = the speed of the plane. Let y = the velocity of the wind. Using Gauss-Jordan elimination, we have
$$\begin{cases} 3(x+y) = 1680 \\ 3.5(x-y) = 1680 \end{cases} \Rightarrow \begin{cases} 3x + 3y = 1680 \\ 3.5x - 3.5y = 1680 \end{cases} \Rightarrow$$

$$\begin{bmatrix} 3 & 3 & | & 1680 \\ 3.5 & -3.5 & | & 1680 \end{bmatrix} \xrightarrow[\frac{1}{3.5}R_2 \to R_2]{\frac{1}{3}R_1 \to R_1} \begin{bmatrix} 1 & 1 & | & 560 \\ 1 & -1 & | & 480 \end{bmatrix} \xrightarrow[\frac{1}{2}(R_1-R_2) \to R_2]{\frac{1}{2}(R_1+R_2) \to R_1} \begin{bmatrix} 1 & 0 & | & 520 \\ 0 & 1 & | & 40 \end{bmatrix} \Rightarrow x = 520, y = 40$$

The plane is traveling at 520 mph and the wind velocity is 40 mph.

73. Let x = Andrew's amount. Let y = Bonnie's amount. Let z = Chauncie's amount. Using Gaussian elimination,
we have $\begin{cases} x+y+z = 320 \\ x = 2z \\ y+z = x-20 \end{cases} \Rightarrow \begin{cases} x+y+z = 320 \\ x -2z = 0 \\ -x+y+z = -20 \end{cases} \Rightarrow$

$$\begin{vmatrix} 1 & 1 & 1 & | & 320 \\ 1 & 0 & -2 & | & 0 \\ -1 & 1 & 1 & | & -20 \end{vmatrix} \xrightarrow[\frac{1}{2}(R_1+R_3) \to R_3]{R_1 - R_2 \to R_2} \begin{vmatrix} 1 & 1 & 1 & | & 320 \\ 0 & 1 & 3 & | & 320 \\ 0 & 1 & 1 & | & 150 \end{vmatrix} \xrightarrow{\frac{1}{2}(R_2-R_3) \to R_3} \begin{vmatrix} 1 & 1 & 1 & | & 320 \\ 0 & 1 & 3 & | & 320 \\ 0 & 0 & 1 & | & 85 \end{vmatrix} \Rightarrow$$

$z = 85, y + 3(85) = 320 \Rightarrow y = 65$
$x + 65 + 85 = 320 \Rightarrow x = 170$
Andrew has \$170, Bonnie has \$65, and Chauncie has \$85.

75. Let x = the number of registered nurses. Let y = the number of licensed practical nurses. Let z = the number of nurse's aids. Using Gaussian elimination, we have $\begin{cases} 75x + 20y + 30z = 4850 \\ 3x + 5y + 4z = 530 \\ x + y + z = 130 \end{cases} \Rightarrow$

$$\begin{bmatrix} 75 & 20 & 30 & | & 4850 \\ 3 & 5 & 4 & | & 530 \\ 1 & 1 & 1 & | & 130 \end{bmatrix} \xrightarrow{R_1 \leftrightarrow R_3} \begin{bmatrix} 1 & 1 & 1 & | & 130 \\ 3 & 5 & 4 & | & 530 \\ 75 & 20 & 30 & | & 4850 \end{bmatrix} \xrightarrow[75R_1 - R_3 \to R_3]{\frac{1}{2}(R_2 - 3R_1) \to R_2} \begin{bmatrix} 1 & 1 & 1 & | & 130 \\ 0 & 1 & \frac{1}{2} & | & 70 \\ 0 & 55 & 45 & | & 4900 \end{bmatrix}$$

$$\xrightarrow{\frac{2}{35}(R_3 - 55R_2) \to R_3} \begin{bmatrix} 1 & 1 & 1 & | & 130 \\ 0 & 1 & \frac{1}{2} & | & 70 \\ 0 & 0 & 1 & | & 60 \end{bmatrix} \Rightarrow z = 60, y + \frac{1}{2}(60) = 70 \Rightarrow y = 40, \; x + 60 + 40 = 130 \Rightarrow x = 30$$

There are 30 registered nurses, 40 licensed practical nurses, and 60 nurses aides.

Chapter 9 Practice Test A

1. $5 \times 4, \; a_{43} = 9$

2. $\begin{bmatrix} 7 & -3 & 9 & | & 5 \\ -2 & 4 & 3 & | & -12 \\ 8 & -5 & 1 & | & -9 \end{bmatrix}$

3. $\begin{cases} 4x -z = -3 \\ x + 3y = 9 \\ 2x + 7y + 5z = 8 \end{cases}$

4. $x - 5 = 5 \Rightarrow x = 10; \; y + 3 = 13 \Rightarrow y = 10; \; z = 0$
The solution is $\{(10, 10, 0)\}$.

5. $\begin{cases} x+2y+z=6 \\ x+y-z=7 \\ 2x-y+2z=-3 \end{cases} \Rightarrow \begin{bmatrix} 1 & 2 & 1 & | & 6 \\ 1 & 1 & -1 & | & 7 \\ 2 & -1 & 2 & | & -3 \end{bmatrix}$

$\xrightarrow[\frac{1}{5}(2R_1-R_3)\to R_3]{R_1-R_2\to R_2} \begin{bmatrix} 1 & 2 & 1 & | & 6 \\ 0 & 1 & 2 & | & -1 \\ 0 & 1 & 0 & | & 3 \end{bmatrix}$

$\xrightarrow{\frac{1}{2}(R_2-R_3)\to R_3} \begin{bmatrix} 1 & 2 & 1 & | & 6 \\ 0 & 1 & 2 & | & -1 \\ 0 & 0 & 1 & | & -2 \end{bmatrix}$

$\xrightarrow[R_2-2R_3\to R_2]{R_1-R_3\to R_1} \begin{bmatrix} 1 & 2 & 0 & | & 8 \\ 0 & 1 & 0 & | & 3 \\ 0 & 0 & 1 & | & -2 \end{bmatrix}$

$\xrightarrow{R_1-2R_2\to R_1} \begin{bmatrix} 1 & 0 & 0 & | & 2 \\ 0 & 1 & 0 & | & 3 \\ 0 & 0 & 1 & | & -2 \end{bmatrix}$

The solution is $\{(2, 3, -2)\}$.

6. $\begin{cases} 2x+y-4z=6 \\ -x+3y-z=-2 \\ 2x-6y+2z=4 \end{cases} \Rightarrow \begin{bmatrix} 2 & 1 & -4 & | & 6 \\ -1 & 3 & -1 & | & -2 \\ 2 & -6 & 2 & | & 4 \end{bmatrix}$

$\xrightarrow[R_1-R_3\to R_3]{R_1+2R_2\to R_2} \begin{bmatrix} 2 & 1 & -4 & | & 6 \\ 0 & 7 & -6 & | & 2 \\ 0 & 7 & -6 & | & 2 \end{bmatrix}$

$\xrightarrow[R_2-R_3\to R_3]{\frac{1}{14}(7R_1-R_2)\to R_1} \begin{bmatrix} 1 & 0 & -\frac{11}{7} & | & \frac{20}{7} \\ 0 & 7 & -6 & | & 2 \\ 0 & 0 & 0 & | & 0 \end{bmatrix}$

$\xrightarrow{\frac{1}{7}R_2\to R_2} \begin{bmatrix} 1 & 0 & -\frac{11}{7} & | & \frac{20}{7} \\ 0 & 1 & -\frac{6}{7} & | & \frac{2}{7} \\ 0 & 0 & 0 & | & 0 \end{bmatrix} \Rightarrow$

$x-\frac{11}{7}z=\frac{20}{7} \Rightarrow x=\frac{11}{7}z+\frac{20}{7}$

$y-\frac{6}{7}z=\frac{2}{7} \Rightarrow y=\frac{6}{7}z+\frac{2}{7}$

The solution is $\left\{\left(\frac{11}{7}z+\frac{20}{7}, \frac{6}{7}z+\frac{2}{7}, z\right)\right\}$.

7. $A-B = \begin{bmatrix} 5 & -2 \\ 4 & 0 \\ 7 & 6 \end{bmatrix} - \begin{bmatrix} -3 & -1 \\ 0 & -8 \\ -4 & 6 \end{bmatrix} = \begin{bmatrix} 8 & -1 \\ 4 & 8 \\ 11 & 0 \end{bmatrix}$

8. $AB = \begin{bmatrix} 3 & -7 & 2 \end{bmatrix} \begin{bmatrix} 0 \\ 1 \\ 4 \end{bmatrix} = [3(0)-7(1)+2(4)] = [1]$

9. The product AB is not defined.

10. $2A = 2\begin{bmatrix} -3 & 1 & 0 \\ 5 & 7 & 2 \end{bmatrix} = \begin{bmatrix} -6 & 2 & 0 \\ 10 & 14 & 4 \end{bmatrix}$

11. $A+BA = \begin{bmatrix} -3 & 1 & 0 \\ 5 & 7 & 2 \end{bmatrix} + \begin{bmatrix} -1 & 4 \\ 8 & 2 \end{bmatrix}\begin{bmatrix} -3 & 1 & 0 \\ 5 & 7 & 2 \end{bmatrix}$

$= \begin{bmatrix} -3 & 1 & 0 \\ 5 & 7 & 2 \end{bmatrix} + \begin{bmatrix} 23 & 27 & 8 \\ -14 & 22 & 4 \end{bmatrix}$

$= \begin{bmatrix} 20 & 28 & 8 \\ -9 & 29 & 6 \end{bmatrix}$

12. $C^2 = \begin{bmatrix} 1 & 5 \\ 0 & 4 \end{bmatrix}\begin{bmatrix} 1 & 5 \\ 0 & 4 \end{bmatrix} = \begin{bmatrix} 1 & 25 \\ 0 & 16 \end{bmatrix}$

13. $C^{-1} = \begin{bmatrix} 1 & 5 \\ 0 & 4 \end{bmatrix} = \frac{1}{4}\begin{bmatrix} 4 & -5 \\ 0 & 1 \end{bmatrix} = \begin{bmatrix} 1 & -\frac{5}{4} \\ 0 & \frac{1}{4} \end{bmatrix}$

14. $\begin{bmatrix} 2 & 1 & 3 \\ 1 & 2 & -1 \\ 3 & 1 & 5 \end{bmatrix}^{-1} = \begin{bmatrix} 2 & 1 & 3 & | & 1 & 0 & 0 \\ 1 & 2 & -1 & | & 0 & 1 & 0 \\ 3 & 1 & 5 & | & 0 & 0 & 1 \end{bmatrix}$

$\xrightarrow{R_1\leftrightarrow R_2} \begin{bmatrix} 1 & 2 & -1 & | & 0 & 1 & 0 \\ 2 & 1 & 3 & | & 1 & 0 & 0 \\ 3 & 1 & 5 & | & 0 & 0 & 1 \end{bmatrix}$

$\xrightarrow[3R_1-R_3\to R_3]{2R_1-R_2\to R_2} \begin{bmatrix} 1 & 2 & -1 & | & 0 & 1 & 0 \\ 0 & 3 & -5 & | & -1 & 2 & 0 \\ 0 & 5 & -8 & | & 0 & 3 & -1 \end{bmatrix}$

$\xrightarrow[3R_3-5R_2\to R_3]{5R_1-R_2\to R_1} \begin{bmatrix} 5 & 7 & 0 & | & 1 & 3 & 0 \\ 0 & 3 & -5 & | & -1 & 2 & 0 \\ 0 & 0 & 1 & | & 5 & -1 & -3 \end{bmatrix}$

$\xrightarrow{\frac{1}{3}(R_2+5R_3)\to R_2} \begin{bmatrix} 5 & 7 & 0 & | & 1 & 3 & 0 \\ 0 & 1 & 0 & | & 8 & -1 & -5 \\ 0 & 0 & 1 & | & 5 & -1 & -3 \end{bmatrix}$

$\xrightarrow{\frac{1}{5}(R_1-7R_2)\to R_1} \begin{bmatrix} 1 & 0 & 0 & | & -11 & 2 & 7 \\ 0 & 1 & 0 & | & 8 & -1 & -5 \\ 0 & 0 & 1 & | & 5 & -1 & -3 \end{bmatrix}$

$\begin{bmatrix} 2 & 1 & 3 \\ 1 & 2 & -1 \\ 3 & 1 & 5 \end{bmatrix}^{-1} = \begin{bmatrix} -11 & 2 & 7 \\ 8 & -1 & -5 \\ 5 & -1 & -3 \end{bmatrix}$

15. $\begin{bmatrix} 5 & 2 \\ 3 & 1 \end{bmatrix} \begin{bmatrix} x \\ y \end{bmatrix} = \begin{bmatrix} 32 \\ 18 \end{bmatrix}$

16. $\begin{cases} 12x - 3y = 5 \\ -2x + 7y = -9 \end{cases}$

17. $\begin{bmatrix} 1 & 5 & -2 \\ 4 & -2 & 7 \end{bmatrix} - 5X = 2\begin{bmatrix} 2 & 5 & -11 \\ 18 & 8 & 11 \end{bmatrix} \Rightarrow$

    $\begin{bmatrix} 1 & 5 & -2 \\ 4 & -2 & 7 \end{bmatrix} - 5X = \begin{bmatrix} 4 & 10 & -22 \\ 36 & 16 & 22 \end{bmatrix} \Rightarrow$

    $\begin{bmatrix} 1 & 5 & -2 \\ 4 & -2 & 7 \end{bmatrix} - \begin{bmatrix} 4 & 10 & -22 \\ 36 & 16 & 22 \end{bmatrix} = 5X \Rightarrow$

    $\begin{bmatrix} -3 & -5 & 20 \\ -32 & -18 & -15 \end{bmatrix} = 5X \Rightarrow$

    $\begin{bmatrix} -\frac{3}{5} & -1 & 4 \\ -\frac{32}{5} & -\frac{18}{5} & -3 \end{bmatrix} = X$

18. $\begin{vmatrix} \frac{1}{2} & -\frac{1}{4} \\ \frac{1}{2} & \frac{3}{4} \end{vmatrix} = \left(\frac{1}{2}\right)\left(\frac{3}{4}\right) - \left(-\frac{1}{4}\right)\left(\frac{1}{2}\right) = \frac{1}{2}$

19. Expand by the second row:

    $\begin{vmatrix} 1 & 3 & 5 \\ 2 & 0 & 10 \\ -3 & 1 & -15 \end{vmatrix}$

    $= 2(-1)^{2+1} \begin{vmatrix} 3 & 5 \\ 1 & -15 \end{vmatrix} + 0 + 10(-1)^{2+3} \begin{vmatrix} 1 & 3 \\ -3 & 1 \end{vmatrix}$

    $= -2(-50) - 10(10) = 0$

20. $\begin{cases} 2x - y + z = 3 \\ x + y + z = 6 \\ 4x + 3y - 2z = 4 \end{cases} \Rightarrow D = \begin{vmatrix} 2 & -1 & 1 \\ 1 & 1 & 1 \\ 4 & 3 & -2 \end{vmatrix}$

    $D_x = \begin{vmatrix} 3 & -1 & 1 \\ 6 & 1 & 1 \\ 4 & 3 & -2 \end{vmatrix}, D_y = \begin{vmatrix} 2 & 3 & 1 \\ 1 & 6 & 1 \\ 4 & 4 & -2 \end{vmatrix},$

    $D_z = \begin{vmatrix} 2 & -1 & 3 \\ 1 & 1 & 6 \\ 4 & 3 & 4 \end{vmatrix}$

    $x = \dfrac{\begin{vmatrix} 3 & -1 & 1 \\ 6 & 1 & 1 \\ 4 & 3 & -2 \end{vmatrix}}{\begin{vmatrix} 2 & -1 & 1 \\ 1 & 1 & 1 \\ 4 & 3 & -2 \end{vmatrix}}, y = \dfrac{\begin{vmatrix} 2 & 3 & 1 \\ 1 & 6 & 1 \\ 4 & 4 & -2 \end{vmatrix}}{\begin{vmatrix} 2 & -1 & 1 \\ 1 & 1 & 1 \\ 4 & 3 & -2 \end{vmatrix}},$

    $z = \dfrac{\begin{vmatrix} 2 & -1 & 3 \\ 1 & 1 & 6 \\ 4 & 3 & 4 \end{vmatrix}}{\begin{vmatrix} 2 & -1 & 1 \\ 1 & 1 & 1 \\ 4 & 3 & -2 \end{vmatrix}}$

Chapter 9 Practice Test B

1. D 2. D 3. D

4. $x + 3 = 9 \Rightarrow x = 6; y + 4 = 2 \Rightarrow y = -2; z = 3$
   The answer is C.

5. $\begin{cases} 2x + y = 15 \\ 2y + z = 25 \\ 2z + x = 26 \end{cases} \Rightarrow \begin{bmatrix} 2 & 1 & 0 & | & 15 \\ 0 & 2 & 1 & | & 25 \\ 1 & 0 & 2 & | & 26 \end{bmatrix}$

   $\xrightarrow{R_1 - 2R_3 \to R_3} \begin{bmatrix} 2 & 1 & 0 & | & 15 \\ 0 & 2 & 1 & | & 25 \\ 0 & 1 & -4 & | & -37 \end{bmatrix}$

   $\xrightarrow{\frac{1}{9}(R_2 - 2R_3) \to R_3} \begin{bmatrix} 2 & 1 & 0 & | & 15 \\ 0 & 2 & 1 & | & 25 \\ 0 & 0 & 1 & | & 11 \end{bmatrix} \Rightarrow z = 11$

   The answer is C.

6. $\begin{cases} 2x + y = 17 \\ y + 2z = 15 \\ x + z = 9 \end{cases} \Rightarrow \begin{bmatrix} 2 & 1 & 0 & | & 17 \\ 0 & 1 & 2 & | & 15 \\ 1 & 0 & 1 & | & 9 \end{bmatrix}$

   $\xrightarrow{2R_3 - R_1 \to R_3} \begin{bmatrix} 2 & 1 & 0 & | & 17 \\ 0 & 1 & 2 & | & 15 \\ 0 & -1 & 2 & | & 1 \end{bmatrix}$

   $\xrightarrow[\frac{1}{4}(R_2 + R_3) \to R_3]{\frac{1}{2}(R_1 + R_3) \to R_1} \begin{bmatrix} 1 & 0 & 1 & | & 9 \\ 0 & 1 & 2 & | & 15 \\ 0 & 0 & 1 & | & 4 \end{bmatrix}$

   $\xrightarrow[R_2 - 2R_3 \to R_2]{R_1 - R_3 \to R_1} \begin{bmatrix} 1 & 0 & 0 & | & 5 \\ 0 & 1 & 0 & | & 7 \\ 0 & 0 & 1 & | & 4 \end{bmatrix} \Rightarrow$

   $x = 5, y = 7, z = 4 \Rightarrow 4(5) + 3(7) + 4 = 45$
   The answer is D.

7. $\begin{bmatrix} -1 & 4 \\ 0 & 4 \\ 8 & -4 \end{bmatrix} - \begin{bmatrix} 7 & 2 \\ 17 & 4 \\ 2 & 2 \end{bmatrix} = \begin{bmatrix} -8 & 2 \\ -17 & 0 \\ 6 & -6 \end{bmatrix}$

   The answer is C.

8. $AB = \begin{bmatrix} -8 & 2 & 9 \end{bmatrix} \begin{bmatrix} 3 \\ 0 \\ -3 \end{bmatrix} = [-8(3) + 2(0) + 9(-3)]$
$= [-51]$. The answer is B.

9. AB is not defined. The answer is A.

10. $2A = 2\begin{bmatrix} 2 & 1 & -3 \\ -5 & 2 & 1 \end{bmatrix} = \begin{bmatrix} 4 & 2 & -6 \\ -10 & 4 & 2 \end{bmatrix}$
The answer is B.

11. $A + BA = \begin{bmatrix} 2 & 1 & -3 \\ -5 & 2 & 1 \end{bmatrix} + \begin{bmatrix} -3 & 7 \\ 2 & 4 \end{bmatrix} \begin{bmatrix} 2 & 1 & -3 \\ -5 & 2 & 1 \end{bmatrix}$
$= \begin{bmatrix} 2 & 1 & -3 \\ -5 & 2 & 1 \end{bmatrix} + \begin{bmatrix} -41 & 11 & 16 \\ -16 & 10 & -2 \end{bmatrix}$
$= \begin{bmatrix} -39 & 12 & 13 \\ -21 & 12 & -1 \end{bmatrix}$
The answer is C.

12. $C^2 = \begin{bmatrix} 5 & 4 \\ 1 & 0 \end{bmatrix} \begin{bmatrix} 5 & 4 \\ 1 & 0 \end{bmatrix} = \begin{bmatrix} 29 & 20 \\ 5 & 4 \end{bmatrix}$
The answer is C.

13. $C^{-1} = \begin{bmatrix} 5 & 4 \\ 1 & 0 \end{bmatrix}^{-1} = -\frac{1}{4}\begin{bmatrix} 0 & -4 \\ -1 & 5 \end{bmatrix} = \begin{bmatrix} 0 & 1 \\ \frac{1}{4} & -\frac{5}{4} \end{bmatrix}$
The answer is B.

14. $\begin{bmatrix} 1 & 0 & 0 \\ 2 & 1 & 0 \\ 3 & -4 & 1 \end{bmatrix}^{-1} = \begin{bmatrix} 1 & 0 & 0 & | & 1 & 0 & 0 \\ 2 & 1 & 0 & | & 0 & 1 & 0 \\ 3 & -4 & 1 & | & 0 & 0 & 1 \end{bmatrix}$
$\xrightarrow[3R_1 - R_3 \to R_3]{R_2 - 2R_1 \to R_2} \begin{bmatrix} 1 & 0 & 0 & | & 1 & 0 & 0 \\ 0 & 1 & 0 & | & -2 & 1 & 0 \\ 0 & 4 & -1 & | & 3 & 0 & -1 \end{bmatrix}$
$\xrightarrow{4R_2 - R_3 \to R_3} \begin{bmatrix} 1 & 0 & 0 & | & 1 & 0 & 0 \\ 0 & 1 & 0 & | & -2 & 1 & 0 \\ 0 & 0 & 1 & | & -11 & 4 & 1 \end{bmatrix} \Rightarrow$
$\begin{bmatrix} 1 & 0 & 0 \\ 2 & 1 & 0 \\ 3 & -4 & 1 \end{bmatrix}^{-1} = \begin{bmatrix} 1 & 0 & 0 \\ -2 & 1 & 0 \\ -11 & 4 & 1 \end{bmatrix}$
The answer is D.

15. A   16. A

17. $\begin{bmatrix} -2 & -3 & 1 \\ -5 & 3 & -2 \end{bmatrix} - 3X = -5\begin{bmatrix} -1 & -2 & -1 \\ 1 & 0 & 1 \end{bmatrix} \Rightarrow$
$\begin{bmatrix} -2 & -3 & 1 \\ -5 & 3 & -2 \end{bmatrix} - 3X = \begin{bmatrix} 5 & 10 & 5 \\ -5 & 0 & -5 \end{bmatrix} \Rightarrow$
$\begin{bmatrix} -2 & -3 & 1 \\ -5 & 3 & -2 \end{bmatrix} - \begin{bmatrix} 5 & 10 & 5 \\ -5 & 0 & -5 \end{bmatrix} = 3X \Rightarrow$
$3X = \begin{bmatrix} -7 & -13 & -4 \\ 0 & 3 & 3 \end{bmatrix} \Rightarrow$
$X = \begin{bmatrix} -\frac{7}{3} & -\frac{13}{3} & -\frac{4}{3} \\ 0 & 1 & 1 \end{bmatrix}$
The answer is B.

18. $\begin{vmatrix} -8 & 5 \\ -4 & -1 \end{vmatrix} = (-8)(-1) - 5(-4) = 28$
The answer is D.

19. $\begin{vmatrix} 2 & 3 & -2 \\ 3 & 0 & -3 \\ -3 & 0 & -5 \end{vmatrix} = 3(-1)^{1+2} \begin{vmatrix} 3 & -3 \\ -3 & -5 \end{vmatrix} = -3(-24) = 72$
The answer is C.

20. $\begin{cases} x + y + z = -6 \\ x - y + 3z = -22 \\ 2x + y + z = -10 \end{cases} \Rightarrow D = \begin{vmatrix} 1 & 1 & 1 \\ 1 & -1 & 3 \\ 2 & 1 & 1 \end{vmatrix} = 4$

$D_x = \begin{vmatrix} -6 & 1 & 1 \\ -22 & -1 & 3 \\ -10 & 1 & 1 \end{vmatrix} = -16, D_y = \begin{vmatrix} 1 & -6 & 1 \\ 1 & -22 & 3 \\ 2 & -10 & 1 \end{vmatrix} = 12$

$D_z = \begin{vmatrix} 1 & 1 & -6 \\ 1 & -1 & -22 \\ 2 & 1 & -10 \end{vmatrix} = -20$

$x = \frac{-16}{4} = -4, y = \frac{12}{4} = 3, z = \frac{-20}{4} = -5$
The answer is B.

Cumulative Review Exercises (Chapters P–9)

1. $\sqrt{(-1-2)^2 + (y-(-3))^2} = 5 \Rightarrow$
$9 + (y+3)^2 = 25 \Rightarrow y^2 + 6y - 7 = 0 \Rightarrow$
$(y+7)(y-1) = 0 \Rightarrow y = -7$ or $y = 1$

2. $\frac{4}{x-1} - \frac{3}{x+2} = \frac{18}{(x+2)(x-1)} \Rightarrow$
$4(x+2) - 3(x-1) = 18 \Rightarrow x + 11 = 18 \Rightarrow x = 7$

3. $|2x - 5| = 3 \Rightarrow 2x - 5 = 3$ or $2x - 5 = -3 \Rightarrow$
$x = 4$ or $x = 1$

4. $4x^2 = 8x - 13 \Rightarrow 4x^2 - 8x + 13 = 0 \Rightarrow$
$x = \dfrac{8 \pm \sqrt{64 - 4(4)(13)}}{2(4)} = \dfrac{8 \pm \sqrt{-144}}{8} = 1 \pm \dfrac{3}{2}i$

5. $\left(\dfrac{3x-1}{x+5}\right)^2 - 3\left(\dfrac{3x-1}{x+5}\right) - 28 = 0$

   Let $u = \dfrac{3x-1}{x+5}$. Then we have $u^2 - 3u - 28 = 0 \Rightarrow$
   $(u+4)(u-7) = 0 \Rightarrow u = -4$ or $u = 7$
   $\dfrac{3x-1}{x+5} = -4 \Rightarrow 3x - 1 = -4x - 20 \Rightarrow x = -\dfrac{19}{7}$
   $\dfrac{3x-1}{x+5} = 7 \Rightarrow 3x - 1 = 7x + 35 \Rightarrow x = -9$
   The solution is $\left\{-\dfrac{19}{7}, -9\right\}$.

6. $\log_2 |x| + \log_2 |x+6| = 4 \Rightarrow$
   $\log_2 (|x| \cdot |x+6|) = 4 \Rightarrow 2^4 = |x^2 + 6x| \Rightarrow$
   $x^2 + 6x = 16$ or $x^2 + 6x = -16$
   $x^2 + 6x - 16 = 0 \Rightarrow (x+8)(x-2) = 0 \Rightarrow$
   $x = -8$ or $x = 2$
   $x^2 + 6x + 16 = 0 \Rightarrow$
   $x = \dfrac{-6 \pm \sqrt{36 - 64}}{2} = -3 \pm i\sqrt{7}$ (reject this)
   The solution is $\{-8, 2\}$.

7. Solve $x + 2 = 0 \Rightarrow x = -2$ and
   $2x - 1 = 0 \Rightarrow x = \dfrac{1}{2}$. The intervals to be tested
   are $(-\infty, -2), \left(-2, \dfrac{1}{2}\right)$, and $\left(\dfrac{1}{2}, \infty\right)$.

Interval	Test point	Value of $\dfrac{x+2}{2x-1}$	Result
$(-\infty, -2)$	-3	$1/7$	$+$
$(-2, 1/2)$	0	-2	$-$
$(1/2, \infty)$	1	3	$+$

   The solution set is $(-\infty, -2) \cup \left(\dfrac{1}{2}, \infty\right)$.

8. Solve the associated equation:
   $x^2 - 7x + 6 = 0 \Rightarrow (x-6)(x-1) = 0 \Rightarrow x = 6$
   or $x = 1$. The intervals are $(-\infty, 1], [1, 6]$, and $[6, \infty)$.

Interval	Test point	Value of $x^2 - 7x + 6$	Result
$(-\infty, 1]$	0	6	$+$
$[1, 6]$	2	-4	$-$
$(6, \infty)$	7	6	$+$

   The solution set is $[1, 6]$.

9. The factors of the constant term are $\{\pm 1, \pm 3\}$.
   The factors of the leading coefficient are
   $\{\pm 1, \pm 2, \pm 4\}$. The possible rational zeros are
   $\left\{\pm 1, \pm \dfrac{1}{2}, \pm \dfrac{1}{4}, \pm 3, \pm \dfrac{3}{2}, \pm \dfrac{3}{4}\right\}$.

10. Using synthetic division, we have

    $\begin{array}{r|rrrr} \frac{1}{2} & 4 & 8 & -11 & 3 \\ & & 2 & 5 & -3 \\ \hline & 4 & 10 & -6 & 0 \end{array}$

    $4x^3 + 8x^2 - 11x + 3 = \left(x - \dfrac{1}{2}\right)(4x^2 + 10x - 6)$

    The zeros of the depressed function
    $4x^2 + 10x - 6$ are also zeros of the original function.
    $4x^2 + 10x - 6 = 0 \Rightarrow 2(x+3)(2x-1) = 0 \Rightarrow$
    $x = -3$ or $x = \dfrac{1}{2}$.
    So $\dfrac{1}{2}$ is a zero of multiplicity 2.

11. $h = kr^3 \Rightarrow 10,125 = k(15^3) \Rightarrow k = 3$
    $h = 3(20^3) = 24,000$ horsepower

12. Let $x =$ the number of acres to be annexed.
    Then
    $0.12(400 + x) = 0.02(400) + 0.2x \Rightarrow$
    $48 + 0.12x = 8 + 0.2x \Rightarrow x = 500$ square miles

13. Using substitution, we have
    $\begin{cases} 3x + y = 2 \\ 4x + 5y = -1 \end{cases} \Rightarrow y = -3x + 2$
    $4x + 5(-3x + 2) = -1 \Rightarrow -11x + 10 = -1 \Rightarrow$
    $x = 1; y = -3(1) + 2 = -1$
    The solution is $\{(1, -1)\}$.

574 Chapter 9 *Matrices and Determinants*

14. Using elimination, we have
$$\begin{cases} 2x + y = 5 \\ y^2 - 2y = -3x + 5 \end{cases} \Rightarrow \begin{cases} 2x + y = 5 \\ 3x + y^2 - 2y = 5 \end{cases} \Rightarrow$$
$$\begin{cases} -6x - 3y = -15 \\ 6x + 2y^2 - 4y = 10 \end{cases} \Rightarrow 2y^2 - 7y = -5 \Rightarrow$$
$$2y^2 - 7y + 5 = 0 \Rightarrow (y-1)(2y-5) = 0 \Rightarrow$$
$$y = 1 \text{ or } y = \frac{5}{2}$$
$$2x + 1 = 5 \Rightarrow x = 2$$
$$2x + \frac{5}{2} = 5 \Rightarrow x = \frac{5}{4}$$
The solution is $\left\{ (2,1), \left(\frac{5}{4}, \frac{5}{2} \right) \right\}$.

15. Shift the graph of $f(x) = |x|$ one unit left, stretch by a factor of 3, then shift the resulting graph two units up.

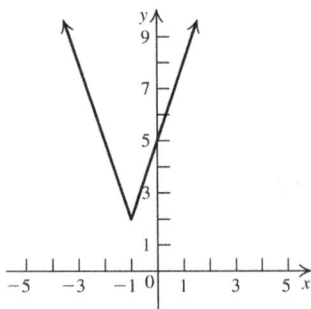

16. $\begin{bmatrix} 1 & 2 & -2 \\ -1 & 3 & 0 \\ 0 & -2 & 1 \end{bmatrix}^{-1} = \begin{bmatrix} 1 & 2 & -2 & | & 1 & 0 & 0 \\ -1 & 3 & 0 & | & 0 & 1 & 0 \\ 0 & -2 & 1 & | & 0 & 0 & 1 \end{bmatrix}$

$\xrightarrow{R_1 + R_2 \to R_2} \begin{bmatrix} 1 & 2 & -2 & | & 1 & 0 & 0 \\ 0 & 5 & -2 & | & 1 & 1 & 0 \\ 0 & -2 & 1 & | & 0 & 0 & 1 \end{bmatrix}$

$\xrightarrow[2R_2 + 5R_3 \to R_3]{R_1 + R_3 \to R_1} \begin{bmatrix} 1 & 0 & -1 & | & 1 & 0 & 1 \\ 0 & 5 & -2 & | & 1 & 1 & 0 \\ 0 & 0 & 1 & | & 2 & 2 & 5 \end{bmatrix}$

$\xrightarrow[\frac{1}{5}(R_2 + 2R_3) \to R_2]{R_1 + R_3 \to R_1} \begin{bmatrix} 1 & 0 & 0 & | & 3 & 2 & 6 \\ 0 & 1 & 0 & | & 1 & 1 & 2 \\ 0 & 0 & 1 & | & 2 & 2 & 5 \end{bmatrix} \Rightarrow$

$\begin{bmatrix} 1 & 2 & -2 \\ -1 & 3 & 0 \\ 0 & -2 & 1 \end{bmatrix}^{-1} = \begin{bmatrix} 3 & 2 & 6 \\ 1 & 1 & 2 \\ 2 & 2 & 5 \end{bmatrix}$

17. $\begin{bmatrix} 1 & 2 & -2 \\ -1 & 3 & 0 \\ 0 & -2 & 1 \end{bmatrix} \begin{bmatrix} x \\ y \\ z \end{bmatrix} = \begin{bmatrix} 5 \\ 2 \\ -3 \end{bmatrix} \Rightarrow$

$\begin{bmatrix} x \\ y \\ z \end{bmatrix} = \begin{bmatrix} 1 & 2 & -2 \\ -1 & 3 & 0 \\ 0 & -2 & 1 \end{bmatrix}^{-1} \begin{bmatrix} 5 \\ 2 \\ -3 \end{bmatrix}$

$= \begin{bmatrix} 3 & 2 & 6 \\ 1 & 1 & 2 \\ 2 & 2 & 5 \end{bmatrix} \begin{bmatrix} 5 \\ 2 \\ -3 \end{bmatrix} = \begin{bmatrix} 1 \\ 1 \\ -1 \end{bmatrix}$

The solution is $\{(1, 1, -1)\}$.

18. a. $F(x) = (x+2)^2 + 3(x+2) - 1 = x^2 + 7x + 9$

    b. $F(4) = 4^2 + 7(4) + 9 = 53$

19. $y = \dfrac{x}{x+4}$. Switch the variables, and then solve for y to find $f^{-1}(x)$:

$x = \dfrac{y}{y+4} \Rightarrow$
$xy + 4x = y \Rightarrow xy - y = -4x \Rightarrow$
$y(x-1) = -4x \Rightarrow y = -\dfrac{4x}{x-1} \Rightarrow$
$f^{-1}(x) = -\dfrac{4x}{x-1}$

20. Domain: $(-\infty, -4) \cup (-4, \infty)$
    Range: $(-\infty, 1) \cup (1, \infty)$

Chapter 10 Conic Sections

10.2 The Parabola

10.2 Practice Problems

1. a.

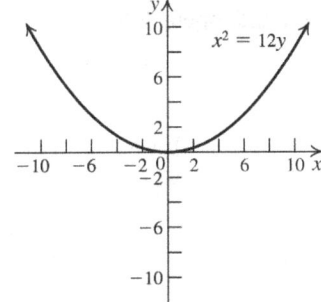

Vertex: (0, 0)
$x^2 = 12y = 4py \Rightarrow p = 3$, so the focus is (0, 3).
The directrix is $y = -3$.
The axis is the y-axis.

b.

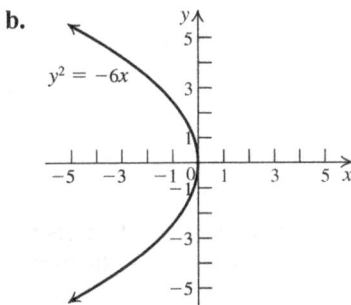

Vertex: (0, 0)
$y^2 = -6x = 4px \Rightarrow p = -\dfrac{3}{2}$, so the focus is $\left(-\dfrac{3}{2}, 0\right)$.
The directrix is $x = \dfrac{3}{2}$.
The axis is the x-axis.

2. $y^2 = -8x = 4px \Rightarrow p = -2$
Vertex: (0, 0)
Focus: $(-2, 0)$
Directrix: $x = 2$
Focal diameter: $-4p = -4(-2) = 8$

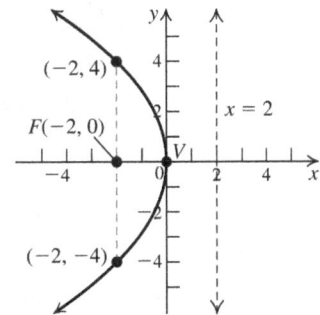

3. a. The vertex is (0, 0) and the focus is (0, 2), so the graph opens up. The general form is $x^2 = 4py$.
$p = 2 \Rightarrow x^2 = 4(2)y \Rightarrow x^2 = 8y$.

b. Since the vertex is (0, 0), the axis of the parabola is the x-axis, and the parabola passes through (1, 2) which is to the right of the vertex, the parabola opens to the right and the general form is $y^2 = 4px$.
We need to solve for p:
$2^2 = 4p(1) \Rightarrow 4 = 4p \Rightarrow 1 = p$.
The equation is $y^2 = 4x$.

4. a. Focus: (0, 2), directrix: $y = -10$
The distance between the focus (0, 2) and the directrix $y = -10$ is $2 - (-10) = 12$ units.
Then the vertex is $\dfrac{1}{2}(12) = 6$ units below the focus, so $p = 6$, and the vertex is $(0, 2 - 6) = (0, -4) = (h, k)$. The standard equation when the directrix is below the vertex is $(x - h)^2 = 4p(y - k)$, so the equation is
$x^2 = 4(6)(y - (-4)) = 24(y + 4)$.
The focal diameter is 24.

b. Vertex: (1, 2), directrix: $x = 5$
The distance between the vertex (1, 2) and the directrix $x = 5$ is $p = 5 - 1 = 4$ units.
The vertex $(1, 2) = (h, k)$. The standard equation when the directrix is to the right of the vertex is $(y - k)^2 = -4p(x - h)$, so the equation is
$(y - 2)^2 = -4(4)(x - 1) = -16(x - 1)$.
The focal diameter is 16.

576 Chapter 10 Conic Sections

c. Vertex: (1, 7), focus: (2, 7)
The distance between the vertex and the focus is $p = 2 - 1 = 1$ unit. The vertex $(1, 7) = (h, k)$. The standard equation when the vertex to the left of the focus is
$(y - k)^2 = 4p(x - h)$, so the equation is
$(y - 7)^2 = 4(1)(x - 1) = 4(x - 1)$.
The focal diameter is 4.

5. Rewrite the equation as $2(x^2 - 4x) = y - 7$, then complete the square to put the equation into standard form:
$$2x^2 - 8x - y + 7 = 0$$
$$2x^2 - 8x = y - 7$$
$$2(x^2 - 4x) = y - 7$$
$$2(x^2 - 4x + 4) = y - 7 + 2(4)$$
$$2(x - 2)^2 = y + 1$$
$$(x - 2)^2 = \frac{1}{2}(y + 1)$$
$4p = \frac{1}{2} \Rightarrow p = \frac{1}{8}$; vertex $= (h, k) = (2, -1)$
The parabola opens up, so the focus is at
$(h, k + p) = \left(2, -1 + \frac{1}{8}\right) = \left(2, -\frac{7}{8}\right)$. The directrix is located at
$y = k - p = -1 - \frac{1}{8} = -\frac{9}{8}$.

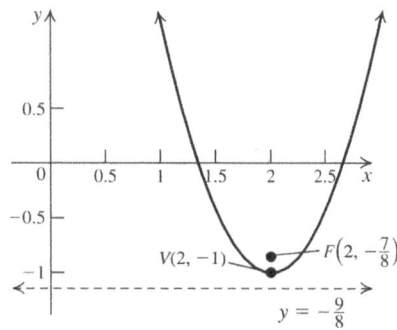

6. a. The vertex $(h, k) = (0, 0)$, so $h = 0$ and $k = 0$. The distance between the vertex $(0, 0)$ and the focus $(0, -2)$ is 2 units, so $p = 2$. The equation for a parabola that opens down is $(x - h)^2 = -4p(y - k)$, so the equation of the graph is
$x^2 = -4(2)y = -8y$.

b. The vertex $(h, k) = (-2, -1)$, so $h = -2$ and $k = -1$. The distance between the vertex and the focus is the same as the distance between the vertex and the directrix. So p is the distance between the vertex $(-2, -1)$ and the point $(3, -1)$ on the directrix $x = 3$. So, $p = 3 - (-2) = 5$ units. The equation for a parabola that opens left is
$(y - k)^2 = -4p(x - h)$, so the equation of the graph is
$(y - (-1))^2 = -4(5)(x - (-2))$, or
$(y + 1)^2 = -20(x + 2)$.

7. The equation is $x^2 = 4py \Rightarrow x^2 = 4(7.3)y \Rightarrow$
$x^2 = 29.2y$. To find the thickness y of the mirror at the edge, substitute $x = 1.5$ (half the diameter) in the equation and solve for y.
$1.5^2 = 29.2y \Rightarrow y \approx 0.0771$
The mirror is about 0.0771 in. thick at the edge.

10.2 Concepts and Vocabulary

1. A parabola is the set of all points P in the plane that are equidistant from a fixed line called the <u>directrix</u> and a fixed point not on the line called the <u>focus</u>.

3. The point at which the axis intersects the parabola is called the <u>vertex</u> of the parabola.

5. True

7. False

10.2 Building Skills

9. $x^2 = 2y = 4py \Rightarrow p = \frac{1}{2}$
Focus: $\left(0, \frac{1}{2}\right)$, directrix: $y = -\frac{1}{2}$, graph (e)

11. $16x^2 = -9y \Rightarrow x^2 = -\frac{9}{16}y = 4py \Rightarrow p = -\frac{9}{64}$
Focus: $\left(0, -\frac{9}{64}\right)$, directrix: $y = \frac{9}{64}$, graph (d)

13. $y^2 = 2x = 4px \Rightarrow p = \frac{1}{2}$
Focus: $\left(\frac{1}{2}, 0\right)$, directrix: $x = -\frac{1}{2}$, graph (g)

15. $9y^2 = -16x \Rightarrow y^2 = -\dfrac{16}{9}x = 4px \Rightarrow p = -\dfrac{4}{9}$

   Focus: $\left(-\dfrac{4}{9}, 0\right)$, directrix: $x = \dfrac{4}{9}$, graph (f)

For exercises 17–28, the vertex is (0, 0).

17. The vertex and the focus (2, 0) lie on the x-axis, so the parabola opens to the right. The equation is of the form $y^2 = 4px$ with $p = 2$. The equation of the parabola is
$y^2 = 4(2)x = 8x$.

19. The vertex and the focus (0, –3) lie on the y-axis, so the parabola opens down. The equation is of the form $x^2 = -4py$ with $p = 3$. The equation of the parabola is
$x^2 = -4(3)y = -12y$.

21. The vertex is (0, 0), the axis of the parabola is the y-axis, and the parabola passes through (–6, –3) which is below the x-axis, so the parabola opens down and the standard form is $x^2 = 4py$. We need to solve for p:
$(-6)^2 = -4p(-3) \Rightarrow 36 = 12p \Rightarrow 3 = p$.
The equation is $x^2 = -4(3)y = -12y$.

23. The vertex is (0, 0), the axis of the parabola is the x-axis, and the parabola passes through (3, 6) which is to the right of the y-axis, so the parabola opens to the right and the standard form is $y^2 = 4px$. We need to solve for p:
$(6)^2 = 4p(3) \Rightarrow 36 = 12p \Rightarrow 3 = p$.
The equation is $y^2 = 4(3)x = 12x$.

25. The endpoints of the latus rectum are (4, 8) and (4, –8), so the focus is (4, 0), the axis of the parabola is the x-axis, $p = 4$, and the parabola opens to the right.
The equation is $y^2 = 4(4)y = 16x$.

27. The endpoints of the latus rectum are (4, 2) and (–4, 2), so the focus is (0, 2), the axis of the parabola is the y-axis, $p = 2$, and the parabola opens up.
The equation is $x^2 = 4(2)y = 8y$.

29. The focus is (0, 2) and the directrix is $y = 4$, so the vertex is (0, 3), and the graph opens down. The general form is $(x - h)^2 = -4p(y - k)$.
$p = 3 - 2 = 1 \Rightarrow x^2 = -4(y - 3)$. The focal diameter = 4(1) = 4.

31. The focus is (–2, 0) and the directrix is $x = 3$, so the vertex is $(1/2, 0)$, and the graph opens to the left. The general form is
$(y - k)^2 = -4a(x - h)$. $a = \dfrac{1}{2} - (-2) = \dfrac{5}{2} \Rightarrow$
$y^2 = -4\left(\dfrac{5}{2}\right)\left(x - \dfrac{1}{2}\right) \Rightarrow y^2 = -10\left(x - \dfrac{1}{2}\right)$.
The focal diameter = $4\left(\dfrac{5}{2}\right) = 10$.

33. The vertex is (1, 1) and the directrix is $x = 3$, so the graph opens to the left. The general form is $(y - k)^2 = -4p(x - h)$.
$p = 3 - 1 = 2 \Rightarrow$
$(y - 1)^2 = -4(2)(x - 1) \Rightarrow (y - 1)^2 = -8(x - 1)$.
The focal diameter = 4(2) = 8.

35. The vertex is (1, 1) and the directrix is $y = -3$, so the graph opens up. The general form is $(x - h)^2 = 4p(y - k)$. $p = 1 - (-3) = 4 \Rightarrow$
$(x - 1)^2 = 4(4)(y - 1) \Rightarrow (x - 1)^2 = 16(y - 1)$.
The focal diameter = 4(4) = 16.

37. The vertex is (1, 0) and the focus is (3, 0), so the graph opens to the right. The general form is $(y - k)^2 = 4p(x - h)$. $p = 3 - 1 = 2 \Rightarrow$
$(y - 0)^2 = 4(2)(x - 1) \Rightarrow y^2 = 8(x - 1)$.
The focal diameter = 4(2) = 8.

39. The vertex is (0, 1) and the focus is (0, –2), so the graph opens down. The general form is $(x - h)^2 = -4p(y - k)$. $p = 1 - (-2) = 3 \Rightarrow$
$(x - 0)^2 = -4(3)(y - 1) \Rightarrow x^2 = -12(y - 1)$.
The focal diameter = 4(3) = 12.

41. The vertex is (2, 3) and the directrix is $x = 4$, so the graph opens to the left. The general form is $(y - k)^2 = -4p(x - h)$.
$p = 4 - 2 = 2 \Rightarrow$
$(y - 3)^2 = -4(2)(x - 2) \Rightarrow (y - 3)^2 = -8(x - 2)$. The focal diameter = 4(2) = 8.

43. The vertex is (2, 3) and the directrix is $y = 1$ so the graph opens up. The general form is $(x-h)^2 = 4p(y-k)$. $p = 3-1 = 2 \Rightarrow$ $(x-2)^2 = 4(2)(y-3) \Rightarrow (x-2)^2 = 8(y-3)$. The focal diameter = $4(2) = 8$.

45. $4p = 2 \Rightarrow p = 1/2$. The graph opens to the right, so the focus is $(h + p, k)$, and the directrix is $x = h - p$. Vertex: $(-1, 1)$, focus $(-1/2, 1)$, directrix: $x = -3/2$.

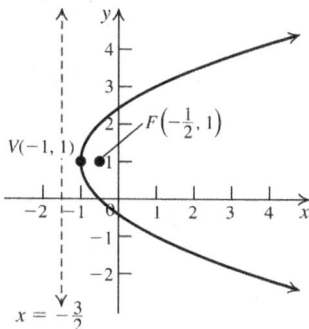

47. $4p = 3 \Rightarrow p = 3/4$. The graph opens up, so the focus is $(h, k + p)$, and the directrix is $y = k - p$. Vertex: $(-2, 2)$, focus $(-2, 11/4)$, directrix: $y = 5/4$.

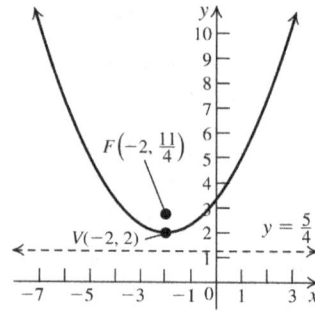

49. $4p = 6 \Rightarrow p = 3/2$. The graph opens to the left, so the focus is $(h - p, k)$, and the directrix is $x = h + p$. Vertex: $(2, -1)$, focus $(1/2, -1)$, directrix: $x = 7/2$.

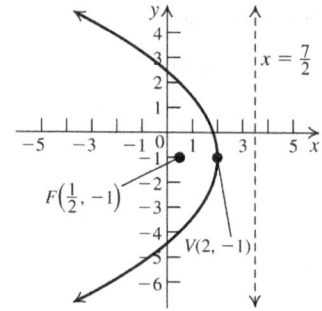

51. $4p = 10 \Rightarrow p = \frac{5}{2}$. The graph opens down, so the focus is $(h, k - p)$, and the directrix is $y = k + p$. Vertex: $(1, 3)$, focus $\left(1, \frac{1}{2}\right)$, directrix: $y = \frac{11}{2}$.

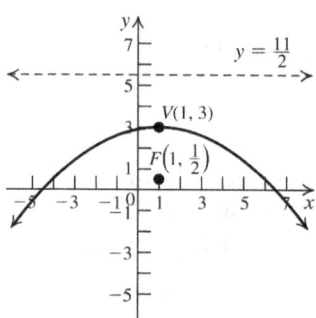

53. Rewrite the equation as $y - 2 = x^2 + 2x$, then complete the square to put the equation into standard form: $y - 2 + 1 = x^2 + 2x + 1 \Rightarrow$ $y - 1 = (x + 1)^2 \Rightarrow 4a = 1 \Rightarrow a = 1/4$. The graph opens up, so the focus is $(h, k + a)$, and the directrix is $y = k - a$. Vertex: $(-1, 1)$, focus $(-1, 5/4)$, directrix: $y = 3/4$.

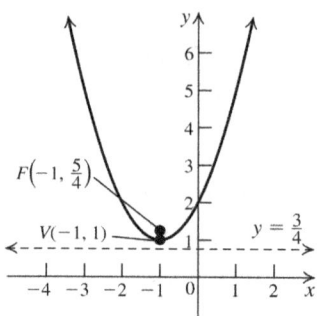

55. Rewrite the equation as $2(y^2 + 2y) = 2x - 1$, then complete the square to put the equation into standard form:
$2(y^2 + 2y + 1) = 2x - 1 + 2 \Rightarrow$
$2(y + 1)^2 = 2\left(x + \frac{1}{2}\right) \Rightarrow (y + 1)^2 = x + \frac{1}{2} \Rightarrow$
$4p = 1 \Rightarrow p = \frac{1}{4}$.
The graph opens to the right, so the focus is $(h + p, k)$, and the directrix is $x = h - p$.

(continued on next page)

(continued)

Vertex: $\left(-\frac{1}{2}, -1\right)$, focus $\left(-\frac{1}{4}, -1\right)$,

directrix: $x = -\frac{3}{4}$.

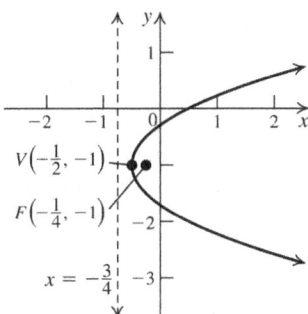

57. Complete the square to put the equation into standard form:

$x + 2 - 48 = -3\left(y^2 - 8y + 16\right) \Rightarrow$

$x - 46 = -3(y-4)^2 \Rightarrow -\frac{1}{3}(x-46) = (y-4)^2 \Rightarrow$

$4p = \frac{1}{3} \Rightarrow p = \frac{1}{12}$. The graph opens to the left, so the focus is $(h - p, k)$, and the directrix is $x = h + p$. Vertex: (46, 4), focus $\left(\frac{551}{12}, 4\right)$,

directrix: $x = \frac{553}{12}$.

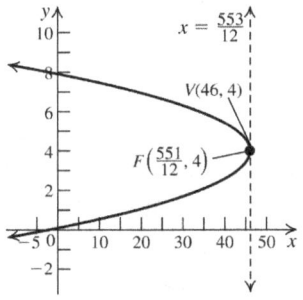

59. The vertex is $(0, 0) = (h, k)$, the focus is $(0, 3)$, and the graph opens up. The general form is $(x - h)^2 = 4p(y - k)$. $p = 3 - 0 = 3$, so the equation of the graph is

$(x - 0)^2 = 4(3)(y - 0)$, or $x^2 = 12y$.

61. The vertex $(h, k) = (-1, 2)$, the directrix is $x = 2$, and the graph opens to the left. The general form for a parabola that opens left is $(y - k)^2 = -4p(x - h)$. $p = 2 - (-1) = 3$, so the equation of the graph is

$(y - 2)^2 = -4(3)(x - (-1))$, or

$(y - 2)^2 = -12(x + 1)$.

63. The vertex is $(1, -1) = (h, k)$, the focal diameter $= 4p = 4$, and the graph opens down. The general form is $(x - h)^2 = -4p(y - k)$, so the equation of the graph is

$(x - 1)^2 = -4(y - (-1)) \Rightarrow (x - 1)^2 = -4(y + 1)$.

65. If the axis of symmetry is the y-axis, point P is above the vertex, and the parabola opens up. Substitute the coordinates of the vertex and P into the standard equation to find a:

$(x - h)^2 = 4p(y - k) \Rightarrow (1 - 0)^2 = 4p(2 - 0) \Rightarrow$

$p = \frac{1}{8} \Rightarrow$ the equation is $x^2 = \frac{1}{2}y$. If the axis of symmetry is the x-axis, point P is to the right of the vertex, and the parabola opens to the right. Substitute the coordinates of the vertex and P into the standard equation to find p:

$(y - k)^2 = 4a(x - h) \Rightarrow (2 - 0)^2 = 4a(1 - 0) \Rightarrow$

$a = 1 \Rightarrow y^2 = 4x$.

67. If the axis of symmetry is parallel to the y-axis, point P is above the vertex, and the parabola opens up. Substitute the coordinates of the vertex and P into the standard equation to find p: $(x - h)^2 = 4p(y - k) \Rightarrow$

$(2 - 0)^2 = 4p(3 - 1) \Rightarrow p = \frac{1}{2} \Rightarrow$ the equation is $x^2 = 2(y - 1)$. If the axis of symmetry is parallel to the x-axis, point P is to the right of the vertex, and the parabola opens to the right. Substitute the coordinates of the vertex and P into the standard equation to find p:

$(y - k)^2 = 4p(x - h) \Rightarrow$

$(3 - 1)^2 = 4p(2 - 0) \Rightarrow p = \frac{1}{2} \Rightarrow$ the equation is $(y - 1)^2 = 2x$.

10.2 Applying the Concepts

69. If we sketch the parabola so that the focus is on the *y*-axis, and the vertex is at (0, 0), then the points (20, 20) and (−20, 20) must lie on the parabola.

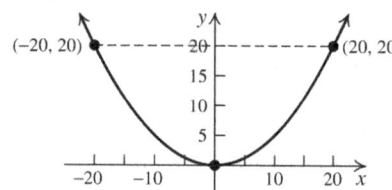

$x^2 = 4py \Rightarrow 20^2 = 4p(20) \Rightarrow p = 5 \Rightarrow$ the receptor should be placed at the focus 5 inches from the vertex.

71. If we sketch the parabola so that the focus is on the *y*-axis, and the vertex is at (0, 0), then the points (9, 6) and (−9, 6) must lie on the parabola.

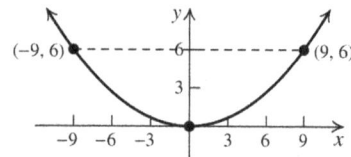

$x^2 = 4py \Rightarrow 9^2 = 4p(6) \Rightarrow p = \dfrac{81}{24} = \dfrac{27}{8} \Rightarrow$ the heating element should be placed at the focus $3\dfrac{3}{8}$ feet from the vertex.

73. $x = 4y^2 \Rightarrow \dfrac{x}{4} = y^2 \Rightarrow 4p = \dfrac{1}{4} \Rightarrow p = \dfrac{1}{16} \Rightarrow$ the bulb should be placed at the focus, $\left(\dfrac{1}{16}, 0\right)$.

75. $y = 4x^2 \Rightarrow \dfrac{1}{4}y = x^2 \Rightarrow 4p = \dfrac{1}{4} \Rightarrow p = \dfrac{1}{16} \Rightarrow$ the microphone should be placed at the focus, $\left(0, \dfrac{1}{16}\right)$.

77. If we sketch the parabola so that the focus is on the *y*-axis, the roadbed is the *x*-axis, and the vertex is at (0, 0), then the points (400, 120) and (−400, 120) must lie on the parabola.

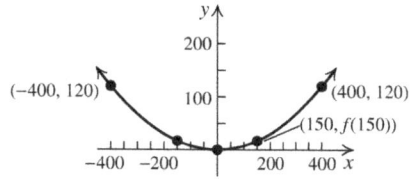

$x^2 = 4py \Rightarrow 400^2 = 4p(120) \Rightarrow p = \dfrac{1000}{3} \Rightarrow$ the equation of the parabola is $x^2 = \dfrac{4000}{3}y \Rightarrow y = \dfrac{3}{4000}x^2$. The point on the cable 250 feet from the tower has coordinates $(150, f(150))$. The height of the cable is $f(150) = \dfrac{135}{8} = 16.875$ feet.

79. The vertex of the parabola occurs at (35, 30). The parabola also passes through the origin.

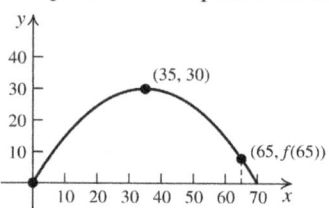

The standard form of the equation is $(x - 35)^2 = -4p(y - 30)$. Substitute (0, 0) for (*x*, *y*) and solve for *a*:

$(0 - 35)^2 = -4p(0 - 30) \Rightarrow p = \dfrac{245}{24}$. The equation of the parabola is

$(x - 35)^2 = -\dfrac{245}{6}(y - 30) \Rightarrow$

$y = -\dfrac{6}{245}(x - 35)^2 + 30$.

Now find $f(65) = \dfrac{390}{49} \approx 7.96$ yards.

81. Rewrite the equation as $10y - 500 = x^2 - 30x$, then complete the square to put the equation in standard form:

$10y - 500 + 225 = x^2 - 30x + 225 \Rightarrow$

$10(y - 27.5) = (x - 15)^2$

So, the vertex is (15, 27.5).
The output is 15 tons at a cost of $27.50.

10.2 Beyond the Basics

83. Solve the system $\begin{cases} 2x - 3y = -16 \\ y^2 = 16x \end{cases}$ using substitution: $2x - 3y = -16 \Rightarrow x = \frac{3}{2}y - 8 \Rightarrow$

$y^2 = 16\left(\frac{3}{2}y - 8\right) \Rightarrow y^2 = 24y - 128 \Rightarrow$

$y^2 - 24y + 128 = 0 \Rightarrow (y - 8)(y - 16) = 0 \Rightarrow$
$y = 8$ or $y = 16$.
$2x - 3(8) + 16 = 0 \Rightarrow x = 4$
$2x - 3(16) + 16 = 0 \Rightarrow x = 16$. The line and the parabola intersect at $(16, 16)$ and $(4, 8)$. Verify using a graphing calculator.

85. The parabola opens up or down, so the equation is of the form $(x - h)^2 = \pm 4p(y - k)$. Substitute the coordinates of the given points into the equation $ax^2 + bx + c = y$, and solve the system

$\begin{cases} a(0^2) + b(0) + c = 5 \\ a(1^2) + b(1) + c = 4 \\ a(2^2) + b(2) + c = 7 \end{cases} \Rightarrow a = 2, b = -3, c = 5$.

Now rewrite the equation $2x^2 - 3x + 5 = y$ as

$2\left(x^2 - \frac{3}{2}x\right) = y - 5$ and complete the square:

$2\left(x^2 - \frac{3}{2}x + \frac{9}{16}\right) = y - 5 + \frac{9}{8} \Rightarrow$

$\left(x - \frac{3}{4}\right)^2 = \frac{1}{2}\left(y - \frac{31}{8}\right)$.

87. Rewrite the equation as $x^2 - 8x = -2y - 4$, and then complete the square to put the equation in standard form:
$x^2 - 8x + 16 = -2y - 4 + 16 \Rightarrow$
$(x - 4)^2 = -2(y - 6)$. The vertex is $(4, 6)$.

$4a = -2 \Rightarrow p = -\frac{1}{2}$, so the focus is $\left(4, \frac{11}{2}\right)$.

The directrix is $y = \frac{13}{2}$. The axis of the parabola is $x = 4$.

89. Rewrite the equation as $y^2 - 6y = -3x - 15$, and then complete the square to put the equation in standard form:
$y^2 - 6y + 9 = -3x - 15 + 9 \Rightarrow$
$(y - 3)^2 = -3(x + 2)$. The vertex is $(-2, 3)$.

$4p = -3 \Rightarrow p = -\frac{3}{4}$, so the focus is

$\left(-\frac{11}{4}, 3\right)$. The directrix is $x = -\frac{5}{4}$. The axis is $y = 3$.

91. Step 1: Let m = the slope of the tangent line. Then the equation of the tangent line is
$y - 3 = m(x - 1) \Rightarrow y = mx - m + 3$.

Step 2: $y = 3x^2 \Rightarrow mx - m + 3 = 3x^2$.

Step 3:
$mx - m + 3 = 3x^2 \Rightarrow 3x^2 - mx + m - 3 = 0 \Rightarrow$
$a = 3, b = -m, c = m - 3$
$b^2 - 4ac = (-m)^2 - 4(3)(m - 3)$, so
$b^2 - 4ac = 0 \Rightarrow (-m)^2 - 4(3)(m - 3) = 0 \Rightarrow$
$m^2 - 12m + 36 = 0 \Rightarrow (m - 6)^2 = 0 \Rightarrow m = 6$

Step 4: The equation of the tangent line is
$y - 3 = 6(x - 1) \Rightarrow y = 6x - 3$.

93. Step 1: Let m = the slope of the tangent line. Then the equation of the tangent line is

$y - y_1 = m(x - x_1) \Rightarrow \frac{y - y_1}{m} + x_1 = x$.

Step 2:

$x = 4py^2 \Rightarrow \frac{y - y_1}{m} + x_1 = 4py^2$.

$x_1 = 4py_1^2 \Rightarrow \frac{y - y_1}{m} + x_1 = 4py^2 \Rightarrow$

$\frac{y - y_1}{m} + 4py_1^2 = 4py^2$

Step 3:

$\frac{y - y_1}{m} + 4py_1^2 = 4py^2 \Rightarrow$

$4py^2 - \frac{y}{m} + \left(\frac{y_1}{m} - 4py_1^2\right) = 0 \Rightarrow$

$4py^2 - \frac{y}{m} + \left(\frac{y_1 - 4mpy_1^2}{m}\right) = 0$

(continued on next page)

(continued)

$a = 4p$, $b = -\dfrac{1}{m}$, $c = \dfrac{y_1}{m} - 4py_1^2$

$b^2 - 4ac = \left(-\dfrac{1}{m}\right)^2 - 4(4p)\left(\dfrac{y_1}{m} - 4py_1^2\right)$, so

$b^2 - 4ac = 0 \Rightarrow \dfrac{1}{m^2} - \dfrac{16py_1}{m} + 64p^2y_1^2 = 0 \Rightarrow$

$\left(\dfrac{1}{m} - 8py_1\right)^2 = 0 \Rightarrow \dfrac{1}{m} = 8py_1 \Rightarrow m = \dfrac{1}{8py_1}$

Step 4: The equation of the tangent line is

$y - y_1 = \dfrac{1}{8py_1}(x - x_1) \Rightarrow$

$y - y_1 = \dfrac{1}{8py_1}x - \dfrac{1}{8py_1}x_1 \Rightarrow$

$y - y_1 = \dfrac{1}{8py_1}x - \dfrac{1}{8py_1}(4py_1^2) \Rightarrow$

$y - y_1 = \dfrac{1}{8py_1}x - \dfrac{1}{2}y_1 \Rightarrow y = \dfrac{1}{8py_1}x + \dfrac{1}{2}y_1$

95. Rewrite the equation in standard form to find the vertex and focus.

$x^2 - 2x - 12y + 25 = 0$
$x^2 - 2x = 12y - 25$
$x^2 - 2x + 1 = 12y - 24$
$(x - 1)^2 = 12(y - 2) = 4(3)(y - 2)$

The vertex is (1, 2), and $p = 3$, so the focus is $(1, 2 + 3) = (1, 5)$. Find the endpoints of the latus rectum by substituting $y = 5$ into the equation and solving for x.

$x^2 - 2x - 12(5) + 25 = 0$
$x^2 - 2x - 35 = 0$
$(x + 5)(x - 7) = 0 \Rightarrow x = -5, x = 7$

The endpoints of the latus rectum are $A(-5, 5)$ and $B(7, 5)$.

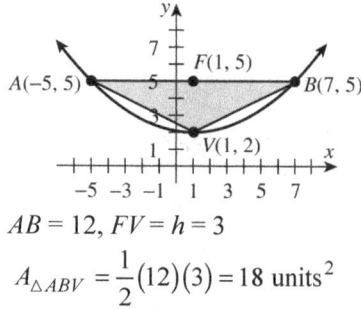

$AB = 12$, $FV = h = 3$

$A_{\triangle ABV} = \dfrac{1}{2}(12)(3) = 18$ units^2

10.2 Critical Thinking/Discussion/Writing

97. a. No. A parabola that opens to the right or left is not the graph of a function because it does not pass the vertical line test.

b. A parabola is always a function if the directrix is parallel to the x-axis, so the slope of the directrix is 0.

98.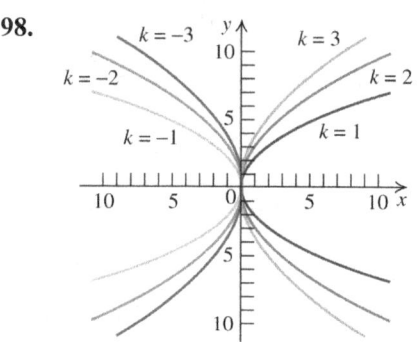

The parabola widens as $|k|$ increases.

99. For a point $P(x_1, y_1)$ on a parabola, the distance from P to the focus is the same as the distance from P to the directrix. Using the given formula, the distance from P to the directrix is $\dfrac{|3x_1 + 4y_1 - 7|}{\sqrt{3^2 + 4^2}} = \dfrac{|3x_1 + 4y_1 - 7|}{5}$.

The distance from P to the focus is $\sqrt{(x_1 - 4)^2 + (y_1 - 5)^2}$. Find the equation of the parabola by setting

$\dfrac{|3x_1 + 4y_1 - 7|}{5} = \sqrt{(x_1 - 4)^2 + (y_1 - 5)^2}$ and simplifying:

$16x^2 + 9y^2 - 24xy - 158x - 194y + 976 = 0$

The axis is perpendicular to the directrix and passes through the focus. The slope of the directrix is $-\dfrac{3}{4}$, so the slope of the axis is $\dfrac{4}{3}$.

The equation of the line through (4, 5) with slope $\dfrac{4}{3}$ is $y - 5 = \dfrac{4}{3}(x - 4) \Rightarrow y = \dfrac{4}{3}x - \dfrac{1}{3}$.

100. Using the formula given in exercise 88, the distance from the vertex to the directrix is

$\dfrac{|3(6) - 5(-3) + 1|}{\sqrt{3^2 + (-5)^2}} = \dfrac{34}{\sqrt{34}} = \sqrt{34}$. The distance from the focus (x, y) to the vertex is

(1) $\sqrt{(x - 6)^2 + (y + 3)^2} = \sqrt{34}$.

The distance from the focus (x, y) to the directrix is twice the distance from the focus to the vertex:

(2) $\dfrac{|3x - 5y + 1|}{\sqrt{3^2 + (-5)^2}} = 2\sqrt{34} \Rightarrow |3x - 5y + 1| = 68$.

(continued on next page)

(*continued*)

Solve the system consisting of equations (1) and (2) to find the coordinates of the focus:

$\begin{cases}(x-6)^2+(y+3)^2=34\\3x-5y=67\end{cases}\Rightarrow$

$\begin{cases}x^2+y^2-12x+6y+11=0\\x=\dfrac{5}{3}y+\dfrac{67}{3}\end{cases}\Rightarrow$

$y^2+16y+64=0\Rightarrow y=-8$
$3x-5(-8)=67\Rightarrow x=9$

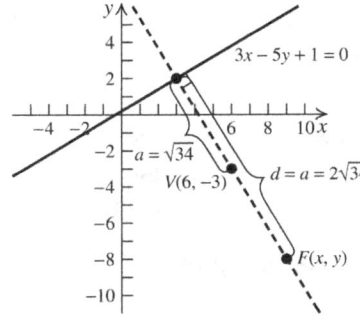

The focus is at $(9,-8)$.

10.2 Getting Ready for the Next Section

101. $d=\sqrt{(x_2-x_1)^2+(y_2-y_1)^2}$
$=\sqrt{(3-(-1))^2+(-2-4)^2}=\sqrt{52}=2\sqrt{13}$

103. $M=\left(\dfrac{x_1+x_2}{2},\dfrac{y_1+y_2}{2}\right)=\left(\dfrac{3+(-7)}{2},\dfrac{5+5}{2}\right)$
$=(-2,5)$

105. $\dfrac{x^2}{9}+\dfrac{y^2}{25}=1$

x-intercepts: $\dfrac{x^2}{9}+\dfrac{0^2}{25}=1\Rightarrow x^2=9\Rightarrow x=\pm 3$

y-intercepts:
$\dfrac{0^2}{9}+\dfrac{y^2}{25}=1\Rightarrow y^2=25\Rightarrow y=\pm 5$

107. $x^2+6x+1=0\Rightarrow x^2+6x=-1\Rightarrow$
$x^2+6x+9=-1+9\Rightarrow (x+3)^2=8$

109. $y-y_1=m(x-x_1)$
$y-(-4)=-2(x-3)$
$y+4=-2x+6$
$y=-2x+2$

10.3 The Ellipse

10.3 Practice Problems

1. Since the foci are $(0,-8)$ and $(0,8)$, the major axis is on the y-axis, and $c=8$. One vertex is $(0,10)$, so the other vertex is $(0,-10)$, and $a=10$.
$b^2=a^2-c^2\Rightarrow b^2=10^2-8^2\Rightarrow b=6$.
Thus, the equation is $\dfrac{x^2}{36}+\dfrac{y^2}{100}=1$.

2. $4x^2+y^2=16\Rightarrow\dfrac{x^2}{4}+\dfrac{y^2}{16}=1\Rightarrow$
$a^2=16\Rightarrow a=4;\quad b^2=4\Rightarrow b=2$

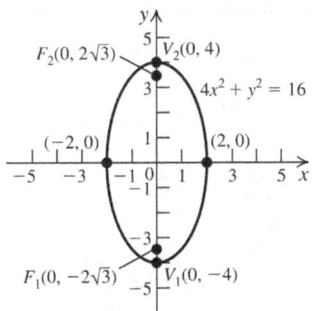

The length of the major axis $=2a=8$.
The length of the minor axis $=2b=4$.

3. **a.** Since the foci $(2,-3)$ and $(2,5)$ lie on the vertical line $x=2$, the ellipse is a vertical ellipse. The center of the ellipse is at $\left(2,\dfrac{-3+5}{2}\right)=(2,1)=(h,k)$. Since the major axis has length 10, the vertices are 5 units from the center, so $a=5$. The foci are 4 units from the center, so $c=4$.
$b^2=a^2-c^2\Rightarrow b^2=5^2-4^2\Rightarrow b^2=9$.
Thus, the equation is
$\dfrac{(x-2)^2}{9}+\dfrac{(y-1)^2}{25}=1$.

b. Since the vertices $(0,0)$ and $(0,10)$ lie on the y-axis, the ellipse is a vertical axis. The center is $(0,5)$. The distance a from the center to a vertex is $a=5$. The distance c from the center to a focus $(0,8)$ is $c=3$.
$c^2=a^2-b^2\Rightarrow 3^2=5^2-b^2\Rightarrow$
$9=25-b^2\Rightarrow b^2=16$
Therefore, the equation is
$\dfrac{x^2}{16}+\dfrac{(y-5)^2}{25}=1$.

4. Rewrite the equation as $\left(x^2 - 6x\right) + 4\left(y^2 + 2y\right) = 29$, then complete both squares and write the equation in standard form:

$$\left(x^2 - 6x\right) + 4\left(y^2 + 2y\right) = 29$$
$$\left(x^2 - 6x + 9\right) + 4\left(y^2 + 2y + 1\right) = 29 + 9 + 4$$
$$(x-3)^2 + 4(y+1)^2 = 42$$
$$\frac{(x-3)^2}{42} + \frac{2(y+1)^2}{21} = 1$$
$$\frac{(x-3)^2}{42} + \frac{(y+1)^2}{21/2} = 1$$

The center is $(3, -1)$.
$a^2 = 42 \Rightarrow a = \sqrt{42}$, so the vertices are $\left(3 - \sqrt{42}, -1\right)$ and $\left(3 + \sqrt{42}, -1\right)$.

$b^2 = a^2 - c^2 \Rightarrow \frac{21}{2} = 42 - c^2 \Rightarrow c^2 = \frac{63}{2} \Rightarrow$

$c = \sqrt{\frac{63}{2}} = \frac{\sqrt{126}}{2} = \frac{3\sqrt{14}}{2}$, the foci are

$\left(3 - \frac{3\sqrt{14}}{2}, -1\right)$ and $\left(3 + \frac{3\sqrt{14}}{2}, -1\right)$.

5. a. The center is $(1, -2)$, so $h = 1$ and $k = -2$. The distance a between the center $(1, -2)$ and the vertex $(1, 1)$ is $a = 1 - (-2) = 3$. The distance b between the center and $(-1, -2)$ is $b = 1 - (-1) = 2$. The equation for a vertical ellipse is $\dfrac{(x-h)^2}{b^2} + \dfrac{(y-k)^2}{a^2} = 1$, so the equation of the ellipse is

$$\frac{(x-1)^2}{4} + \frac{(y+2)^2}{9} = 1.$$

b. The center is $(-1, 1)$ so $h = -1$ and $k = 1$. The distance c between the center and the focus is $c = \left(-1 + \sqrt{3}\right) - (-1) = \sqrt{3}$. The focus and the center lie on the horizontal line $y = 1$, so the ellipse is horizontal, and $(-1, 0)$ is an endpoint of the minor axis. The distance b between the center and $(-1, 0)$ is $b = 1 - 0 = 1$.

$c^2 = a^2 - b^2 \Rightarrow \left(\sqrt{3}\right)^2 = a^2 - 1^2 \Rightarrow$
$a^2 = 3 + 1 = 4$

The equation for a horizontal ellipse is $\dfrac{(x-h)^2}{a^2} + \dfrac{(y-k)^2}{b^2} = 1$, so the equation of the ellipse is

$$\frac{(x+1)^2}{4} + \frac{(y-1)^2}{1} = \frac{(x+1)^2}{4} + (y-1)^2 = 1.$$

6. Since the length of the major axis of the ellipse is 8 feet, $a = 4$. Since the length of the minor axis is 4 feet, $b = 2$.
$2^2 = 4^2 - c^2 \Rightarrow c^2 = 12 \Rightarrow c = 2\sqrt{3}$. If we position the center of the ellipse at $(0, 0)$ and the major axis along the x-axis, the foci of the ellipse are $\left(-2\sqrt{3}, 0\right)$ and $\left(2\sqrt{3}, 0\right)$. The distance between the two foci is $4\sqrt{3} \approx 6.9282$ feet. Thus the stone should be positioned 6.9282 feet from the source.

10.3 Concepts and Vocabulary

1. An ellipse is the set of all points in the plane, the <u>sum</u> of whose distances from two fixed points is a constant.

3. The standard equation of an ellipse with center $(0, 0)$, vertices $(\pm a, 0)$, foci $(\pm c, 0)$ is $\dfrac{x^2}{a^2} + \dfrac{y^2}{b^2} = 1$, where $b^2 = \underline{a^2 - c^2}$.

5. True

7. True

10.3 Building Skills

9. The major axis is on the x-axis.
$b^2 = a^2 - c^2 \Rightarrow b^2 = 3^2 - 1^2 = 8$.
The equation is $\dfrac{x^2}{9} + \dfrac{y^2}{8} = 1$.

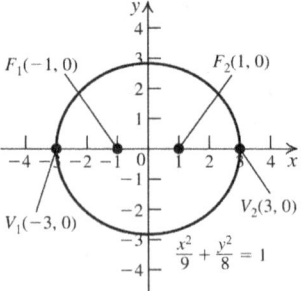

11. The major axis is on the y-axis.
$b^2 = a^2 - c^2 \Rightarrow b^2 = 4^2 - 2^2 = 12$.
The equation is $\dfrac{x^2}{12} + \dfrac{y^2}{16} = 1$.

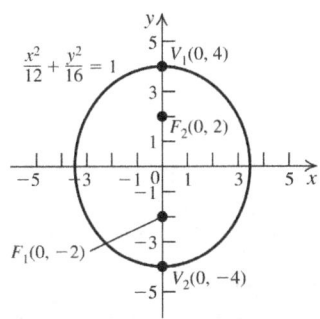

13. The major axis is on the x-axis.
$b^2 = a^2 - c^2 \Rightarrow 3^2 = a^2 - 4^2 \Rightarrow a^2 = 25$.
The equation is $\dfrac{x^2}{25} + \dfrac{y^2}{9} = 1$.

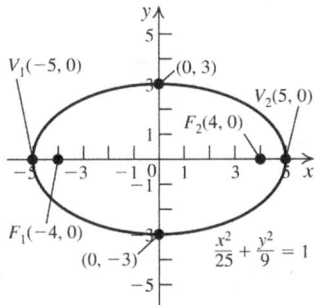

15. The major axis is on the y-axis.
$b^2 = a^2 - c^2 \Rightarrow 4^2 = a^2 - 2^2 \Rightarrow a^2 = 20$.
The equation is $\dfrac{x^2}{16} + \dfrac{y^2}{20} = 1$.

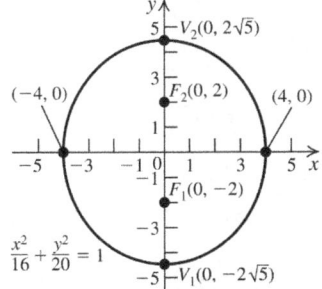

17. The major axis is on the y-axis.
Major axis length = 10 $\Rightarrow a = 5$ and minor axis length = 6 $\Rightarrow b = 3$. The equation is
$\dfrac{x^2}{9} + \dfrac{y^2}{25} = 1$.

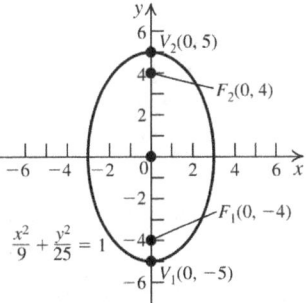

19. The major axis is on the x-axis. $a = 6$.
$b^2 = a^2 - c^2 \Rightarrow b^2 = 6^2 - 3^2 = 27$.
The equation is $\dfrac{x^2}{36} + \dfrac{y^2}{27} = 1$.

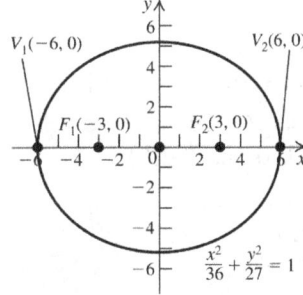

21. The major axis is on the y-axis. $c = 2$.
$b^2 = a^2 - c^2 \Rightarrow 3^2 = a^2 - 2^2 \Rightarrow a^2 = 13$.
The equation is $\dfrac{x^2}{9} + \dfrac{y^2}{13} = 1$.

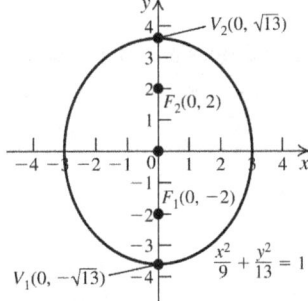

23. $a^2 = 16 \Rightarrow$ the vertices are $(4, 0)$ and $(-4, 0)$.
$b^2 = a^2 - c^2 \Rightarrow 4 = 16 - c^2 \Rightarrow c = 2\sqrt{3} \Rightarrow$
the foci are $\left(2\sqrt{3}, 0\right)$ and $\left(-2\sqrt{3}, 0\right)$.

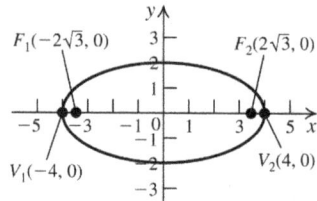

25. $a^2 = 9 \Rightarrow$ the vertices are $(3, 0)$ and $(-3, 0)$.
$b^2 = a^2 - c^2 \Rightarrow 1 = 9 - c^2 \Rightarrow c = 2\sqrt{2} \Rightarrow$
the foci are $\left(2\sqrt{2}, 0\right)$ and $\left(-2\sqrt{2}, 0\right)$.

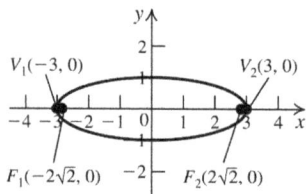

27. $a^2 = 25 \Rightarrow$ the vertices are $(5, 0)$ and $(-5, 0)$.
$b^2 = a^2 - c^2 \Rightarrow 16 = 25 - c^2 \Rightarrow c = 3 \Rightarrow$
the foci are $(3, 0)$ and $(-3, 0)$.

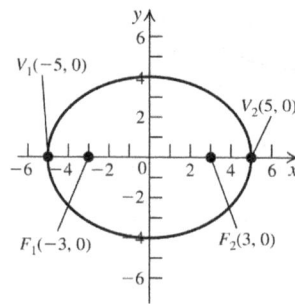

29. $a^2 = 36 \Rightarrow$ the vertices are $(0, 6)$ and $(0, -6)$.
$b^2 = a^2 - c^2 \Rightarrow 16 = 36 - c^2 \Rightarrow c = 2\sqrt{5} \Rightarrow$
the foci are $\left(0, 2\sqrt{5}\right)$ and $\left(0, -2\sqrt{5}\right)$.

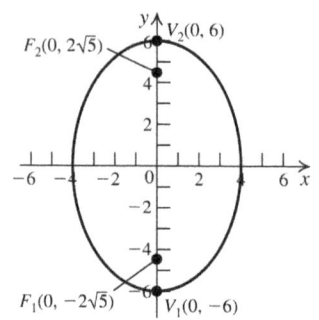

31. A circle with radius 2, centered at the origin.

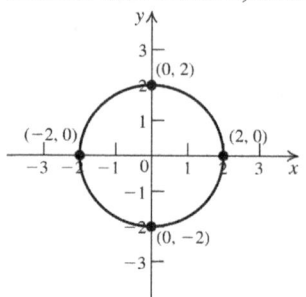

33. $x^2 + 4y^2 = 4 \Rightarrow \dfrac{x^2}{4} + y^2 = 1$. $a^2 = 4 \Rightarrow$ the vertices are $(2, 0)$ and $(-2, 0)$.
$b^2 = a^2 - c^2 \Rightarrow 1 = 4 - c^2 \Rightarrow c = \sqrt{3} \Rightarrow$
the foci are $\left(\sqrt{3}, 0\right)$ and $\left(-\sqrt{3}, 0\right)$.

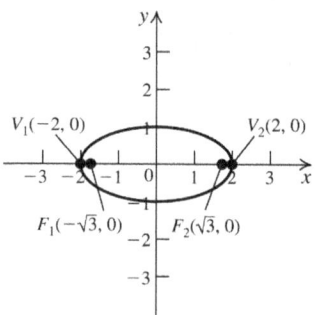

35. $9x^2 + 4y^2 = 36 \Rightarrow \dfrac{x^2}{4} + \dfrac{y^2}{9} = 1$. $a^2 = 9 \Rightarrow$
the vertices are $(0, 3)$ and $(0, -3)$.
$b^2 = a^2 - c^2 \Rightarrow 4 = 9 - c^2 \Rightarrow c = \pm\sqrt{5} \Rightarrow$
the foci are $\left(0, \sqrt{5}\right)$ and $\left(0, -\sqrt{5}\right)$.

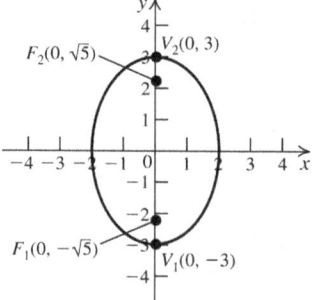

37. $3x^2 + 4y^2 = 12 \Rightarrow \dfrac{x^2}{4} + \dfrac{y^2}{3} = 1$. $a^2 = 4 \Rightarrow$
the vertices are $(2, 0)$ and $(-2, 0)$.
$b^2 = a^2 - c^2 \Rightarrow 3 = 4 - c^2 \Rightarrow c = 1 \Rightarrow$
the foci are $(1, 0)$ and $(-1, 0)$.

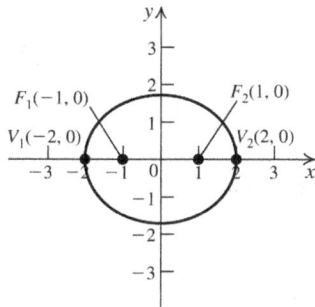

39. $5x^2 - 10 = -2y^2 \Rightarrow 5x^2 + 2y^2 = 10 \Rightarrow$
$\dfrac{x^2}{2} + \dfrac{y^2}{5} = 1$. $a^2 = 5 \Rightarrow$ the vertices are
$\left(0, \sqrt{5}\right)$ and $\left(0, -\sqrt{5}\right)$.
$b^2 = a^2 - c^2 \Rightarrow 2 = 5 - c^2 \Rightarrow c = \sqrt{3} \Rightarrow$
the foci are $\left(0, \sqrt{3}\right)$ and $\left(0, -\sqrt{3}\right)$.

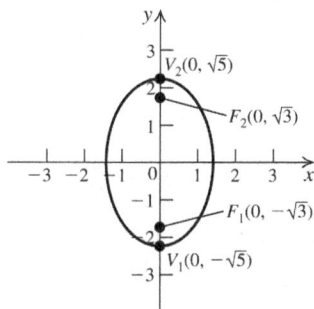

41. $2x^2 + 3y^2 = 7 \Rightarrow \dfrac{x^2}{7/2} + \dfrac{y^2}{7/3} = 1 \Rightarrow a^2 = \dfrac{7}{2} \Rightarrow$
the vertices are $\left(\dfrac{\sqrt{14}}{2}, 0\right)$ and $\left(-\dfrac{\sqrt{14}}{2}, 0\right)$.
$b^2 = a^2 - c^2 \Rightarrow \dfrac{7}{3} = \dfrac{7}{2} - c^2 \Rightarrow c = \dfrac{\sqrt{42}}{6} \Rightarrow$
the foci are $\left(\dfrac{\sqrt{42}}{6}, 0\right)$ and $\left(-\dfrac{\sqrt{42}}{6}, 0\right)$.

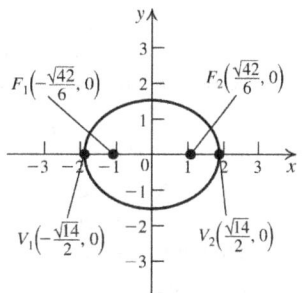

43. The major axis has length 10, so $a = 5$. The minor axis has length 6, so $b = 3$. The ellipse is oriented vertically with center $(-2, 3)$, so the vertices are $(-2, 3 - 5) = (-2, -2)$ and $(-2, 3 + 5) = (-2, 8)$
$b^2 = a^2 - c^2 \Rightarrow 9 = 25 - c^2 \Rightarrow c^2 = 16 \Rightarrow c = 4$
Thus, the foci are located at
$(-2, 3 - 4) = (-2, -1)$ and
$(-2, 3 + 4) = (-2, 7)$.

45. The vertices are located at $(1, 0)$ and $(9, 0)$, so the major axis has length 8 and $a = 4$. The center of the ellipse is $\left(\dfrac{1+9}{2}, 0\right) = (5, 0)$.
Since a focus is located at $(7, 0)$, $c = 7 - 5 = 2$.
$b^2 = a^2 - c^2 \Rightarrow b^2 = 16 - 4 \Rightarrow b^2 = 12$
Thus, the equation of the ellipse is
$\dfrac{(x-5)^2}{16} + \dfrac{y^2}{12} = 1$.

47. The vertices are at $(2, -5)$ and $(2, 3)$, so the ellipse is positioned vertically and the length of the major axis is 8. Therefore, $a = 4$. The minor axis has length 2, so $b = 1$. The center of the ellipse is $\left(\dfrac{2+2}{2}, \dfrac{-5+3}{2}\right) = (2, -1)$.
Thus, the equation of the ellipse is
$(x - 2)^2 + \dfrac{(y+1)^2}{16} = 1$.

49. The foci are located at (−2, 3) and (4, 3) and a vertex is located at (−5, 3). Therefore, the ellipse is positioned horizontally. The center of the ellipse is $\left(\dfrac{-2+4}{2}, \dfrac{3+3}{2}\right) = (1, 3)$.

The distance from the given vertex (−5, 3) to the center is 6, so $a = 6$. The distance from the focus (−2, 3) to the center is 3, so $c = 3$.
$b^2 = a^2 - c^2 \Rightarrow b^2 = 36 - 9 \Rightarrow b^2 = 27$
Thus, the equation of the ellipse is
$\dfrac{(x-1)^2}{36} + \dfrac{(y-3)^2}{27} = 1.$

51. Center: (1, 1); $a = 3 \Rightarrow$ vertices: (1, 4) and (1, −2).
$b^2 = a^2 - c^2 \Rightarrow 4 = 9 - c^2 \Rightarrow c = \sqrt{5}$. The foci are $\left(1, 1+\sqrt{5}\right)$ and $\left(1, 1-\sqrt{5}\right)$.

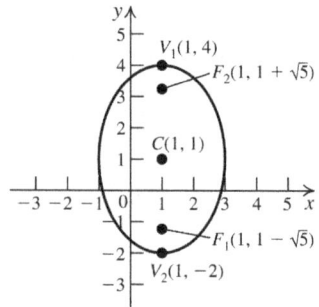

53. $4(x+3)^2 + 5(y-1)^2 = 20 \Rightarrow$
$\dfrac{(x+3)^2}{5} + \dfrac{(y-1)^2}{4} = 1 \Rightarrow$ center: (−3, 1).
$a = \sqrt{5} \Rightarrow$ vertices: $\left(-3+\sqrt{5}, 1\right)$ and $\left(-3-\sqrt{5}, 1\right)$. $b^2 = a^2 - c^2 \Rightarrow 4 = 5 - c^2 \Rightarrow c = \sqrt{5}$

 The foci are $(-2, 1)$ and $(-4, 1)$.

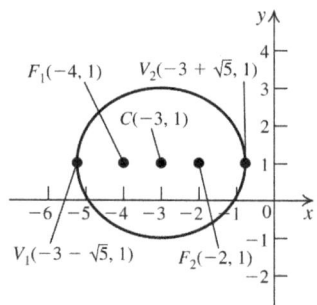

55. Rewrite the equation as
$5(x^2 + 2x) + 9(y^2 - 4y) = 4$, then complete both squares:
$5(x^2 + 2x + 1) + 9(y^2 - 4y + 4) = 4 + 5 + 36 \Rightarrow$
$5(x+1)^2 + 9(y-2)^2 = 45 \Rightarrow$
$\dfrac{(x+1)^2}{9} + \dfrac{(y-2)^2}{5} = 1 \Rightarrow$ center: $(-1, 2)$.
$a = 3 \Rightarrow$ vertices: $(2, 2)$ and $(-4, 2)$.
$b^2 = a^2 - c^2 \Rightarrow 5 = 9 - c^2 \Rightarrow c = 2$.
The foci are $(1, 2)$ and $(-3, 2)$.

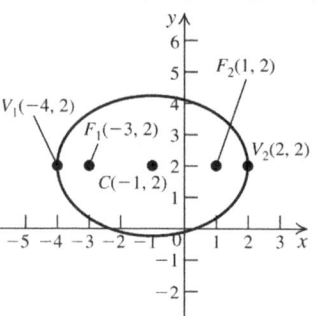

57. Rewrite the equation as
$9(x^2 + 4x) + 5(y^2 - 8y) = -71$, then complete both squares: $9(x^2 + 4x + 4) + 5(y^2 - 8y + 16)$
$= -71 + 36 + 80 \Rightarrow$
$9(x+2)^2 + 5(y-4)^2 = 45 \Rightarrow$
$\dfrac{(x+2)^2}{5} + \dfrac{(y-4)^2}{9} = 1 \Rightarrow$ center: $(-2, 4)$.
$a = 3 \Rightarrow$ vertices: $(-2, 7)$ and $(-2, 1)$.
$b^2 = a^2 - c^2 \Rightarrow 5 = 9 - c^2 \Rightarrow c = 2$.
The foci are $(-2, 2)$ and $(-2, 6)$.

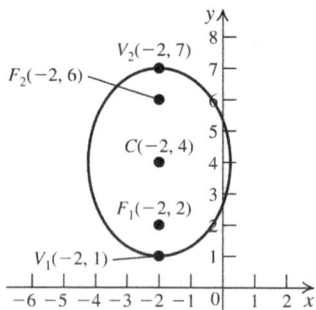

59. Rewrite the equation as
$(x^2 - 2x) + 2(y^2 + 2y) = -1$, then complete both squares:
$(x^2 - 2x + 1) + 2(y^2 + 2y + 1) = -1 + 1 + 2 \Rightarrow$
$(x-1)^2 + 2(y+1)^2 = 2 \Rightarrow$
$\dfrac{(x-1)^2}{2} + (y+1)^2 = 1 \Rightarrow$ center: $(1, -1)$.
$a = \sqrt{2} \Rightarrow$ vertices: $\left(1 + \sqrt{2}, -1\right)$ and $\left(1 - \sqrt{2}, -1\right)$.
$b^2 = a^2 - c^2 \Rightarrow 1 = 2 - c^2 \Rightarrow c = 1$. The foci are $(2, -1)$ and $(0, -1)$.

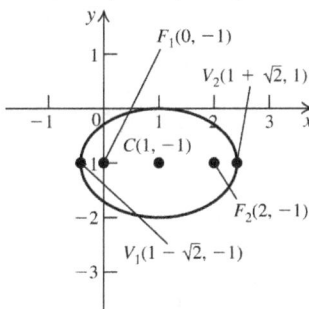

61. Rewrite the equation as
$2(x^2 - 2x) + 9(y^2 + 2y) = -12$, then complete both squares:
$2(x^2 - 2x + 1) + 9(y^2 + 2y + 1) = -12 + 2 + 9 \Rightarrow$
$2(x-1)^2 + 9(y+1)^2 = -1 \Rightarrow$ there is no graph.

63. The major axis is on the y-axis. Major axis length $= 6 \Rightarrow a = 3$ and minor axis length $= 2$
$\Rightarrow b = 1$. The equation is $x^2 + \dfrac{y^2}{9} = 1$.

65. The major axis is on the x-axis. $c = 4$, $b = 2$.
$b^2 = a^2 - c^2 \Rightarrow 2^2 = a^2 - 4^2 \Rightarrow a^2 = 20$.
The equation is $\dfrac{x^2}{20} + \dfrac{y^2}{4} = 1$.

67. The center is $(0, 1) = (h, k)$. The distance a between the center and the vertex $(3, 1)$ is $3 - 0 = 3$. The distance b between the center and the vertex of the minor axis $(0, 3)$ is $3 - 1 = 2$.
The equation is $\dfrac{x^2}{9} + \dfrac{(y-1)^2}{4} = 1$.

10.3 Applying the Concepts

69. Sketch the ellipse so that its center is at the origin and the major axis lies on the x-axis.

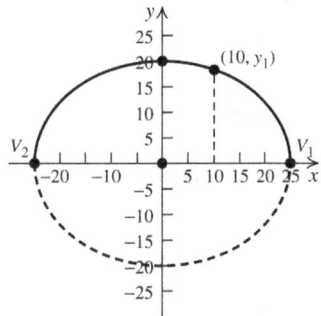

Then $a = 25$ and $b = 20$. The equation of the ellipse is $\dfrac{x^2}{25^2} + \dfrac{y^2}{20^2} = 1$. Let $x = 10$, then solve for y:
$\dfrac{10^2}{25^2} + \dfrac{y^2}{20^2} = 1 \Rightarrow y = 4\sqrt{21} \approx 18.3$ ft.

71. Sketch the ellipse so that its center is at the origin and the major axis lies on the x-axis.

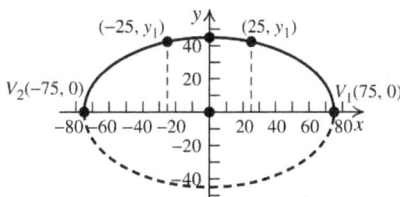

Then $a = 75$ and $b = 45$. The equation of the ellipse is $\dfrac{x^2}{75^2} + \dfrac{y^2}{45^2} = 1$. Let $x = 25$, then solve for y:
$\dfrac{25^2}{75^2} + \dfrac{y^2}{45^2} = 1 \Rightarrow y = 30\sqrt{2} \approx 42.4$ m.

73. $a = 48$ and $b = 23$, so $b^2 = a^2 - c^2 \Rightarrow$
$23^2 = 48^2 - c^2 \Rightarrow c = 5\sqrt{71} \approx 42.13$ feet from the center or $48 - 42.13 \approx 5.9$ feet from the wall on the opposite side, along the major axis.

75. Using the hints given, we have
$a = \dfrac{128.49 + 154.83}{2} = 141.66 \Rightarrow$
$a^2 = 20{,}067.5556$ and
$c = 141.66 - 128.49 = 13.17$.
$b^2 = 141.66^2 - 13.17^2 = 19{,}894.1067$. The equation is $\dfrac{x^2}{20{,}067.5556} + \dfrac{y^2}{19{,}894.1067} = 1$.

77. Using the hints given in exercise 75, we have
$a = \dfrac{28.56 + 43.88}{2} = 36.22 \Rightarrow a^2 = 1311.8884$
and $c = 36.22 - 28.56 = 7.66$.
$b = 36.22^2 - 7.66^2 = 1253.2128$. The equation is $\dfrac{x^2}{1311.8884} + \dfrac{y^2}{1253.2128} = 1$.

79. $a = \dfrac{768,806}{2} = 384,403$ and
$b = \dfrac{767,746}{2} = 383,873$. So $b^2 = a^2 - c^2 \Rightarrow$
$383,873^2 = 384,403^2 - c^2 \Rightarrow c \approx 20,179$.
The perihelion = $a - c \approx 364,224$ km.
The aphelion = $a + c \approx 404,582$ km.

81. The perihelion of the satellite is 4070 miles and the aphelion is 4124 miles.

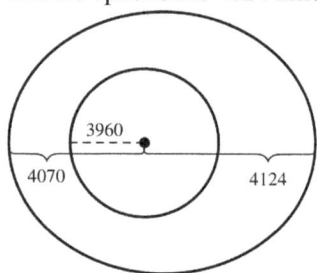

Not drawn to scale

$a = \dfrac{4070 + 4124}{2} = 4097 \Rightarrow a^2 = 16,785,409$
and $c = 4097 - 4070 = 27$. Then
$b^2 = 16,784,680$. The equation is
$\dfrac{x^2}{16,785,409} + \dfrac{y^2}{16,784,680} = 1$.

10.3 Beyond the Basics

83. The center is (0, 0) and $a = 3$, so the equation is either $\dfrac{x^2}{3^2} + \dfrac{y^2}{b^2} = 1$ or $\dfrac{x^2}{b^2} + \dfrac{y^2}{3^2} = 1$.

Substitute the coordinates of the point for x and y, then solve for b:
$\dfrac{(1)^2}{5^2} + \dfrac{\left(3\sqrt{5}/2\right)^2}{b^2} = 1 \Rightarrow \dfrac{1}{25} + \dfrac{45}{4b^2} = 1 \Rightarrow$
$b^2 = \dfrac{1125}{96} \Rightarrow b \approx 3.4$. Because the exercise states that the length of the major axis is 6 there is no solution.

85. The general form of the equation is
$\dfrac{(x-2)^2}{a^2} + \dfrac{(y-1)^2}{b^2} = 1$. Substitute the coordinates of the points into the equation of the ellipse and then solve the system:
$\begin{cases} \dfrac{(10-2)^2}{a^2} + \dfrac{(1-1)^2}{b^2} = 1 \\ \dfrac{(6-2)^2}{a^2} + \dfrac{(2-1)^2}{b^2} = 1 \end{cases} \Rightarrow \begin{cases} \dfrac{8^2}{a^2} + \dfrac{0}{b^2} = 1 \\ \dfrac{16}{a^2} + \dfrac{1}{b^2} = 1 \end{cases} \Rightarrow$
$a^2 = 64, b^2 = 4/3$. Then the equation is
$\dfrac{(x-2)^2}{64} + \dfrac{(y-1)^2}{4/3} = 1 \Rightarrow$
$(x-2)^2 + 48(y-1)^2 = 64$

87. $a = 4, b^2 = a^2 - c^2 \Rightarrow 9 = 16 - c^2 \Rightarrow c = \sqrt{7}$
$e = \sqrt{7}/4$

89. $x^2 + 4y^2 = 1 \Rightarrow x^2 + \dfrac{y^2}{1/4} = 1$
$a = 1, b^2 = a^2 - c^2 \Rightarrow \dfrac{1}{4} = 1 - c^2 \Rightarrow c = \dfrac{\sqrt{3}}{2}$
$e = \sqrt{3}/2$

91. Rewrite the equation as
$(x^2 - 2x) + 2(y^2 + 2y) = -1$, then complete both squares:
$(x^2 - 2x + 1) + 2(y^2 + 2y + 1) = -1 + 1 + 2 \Rightarrow$
$(x+1)^2 + 2(y+1)^2 = 2 \Rightarrow \dfrac{(x+1)^2}{2} + (y+1)^2 = 1$
$a = \sqrt{2}, b^2 = a^2 - c^2 \Rightarrow 1 = 2 - c^2 \Rightarrow c = 1$
$e = \dfrac{1}{\sqrt{2}} = \dfrac{\sqrt{2}}{2}$

93. $a = 3; e = \dfrac{c}{a} \Rightarrow \dfrac{1}{3} = \dfrac{c}{3} \Rightarrow c = 1$
$b^2 = a^2 - c^2 \Rightarrow b^2 = 9 - 1 \Rightarrow b^2 = 8$
The equation is $\dfrac{x^2}{9} + \dfrac{y^2}{8} = 1$.

95. Using substitution, we have
$\begin{cases} x + 3y = -2 \\ 4x^2 + 3y^2 = 7 \end{cases} \Rightarrow \begin{cases} x = -3y - 2 \\ 4x^2 + 3y^2 = 7 \end{cases} \Rightarrow 4(-3y-2)^2 + 3y^2 = 7 \Rightarrow 39y^2 + 48y + 9 = 0 \Rightarrow$
$y = -1$ or $y = -\dfrac{3}{13}$
$x + 3(-1) = -2 \Rightarrow x = 1$
$x + 3\left(-\dfrac{3}{13}\right) = -2 \Rightarrow x = -\dfrac{17}{13}$

The points of intersection are $\left(-\dfrac{17}{13}, -\dfrac{3}{13}\right)$ and $(1, -1)$.

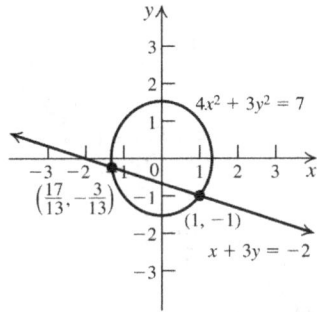

97. Using elimination, we have $\begin{cases} x^2 + y^2 = 20 \\ 9x^2 + y^2 = 36 \end{cases} \Rightarrow 8x^2 = 16 \Rightarrow x = \pm\sqrt{2}$

$2 + y^2 = 20 \Rightarrow y = \pm 3\sqrt{2}$

The points of intersection are $\left(-\sqrt{2}, -3\sqrt{2}\right), \left(-\sqrt{2}, 3\sqrt{2}\right), \left(\sqrt{2}, -3\sqrt{2}\right),$ and $\left(\sqrt{2}, 3\sqrt{2}\right)$.

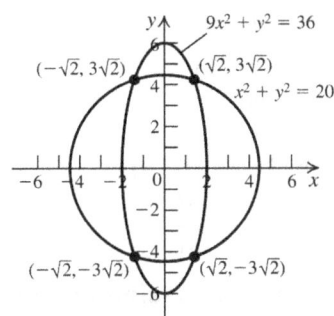

99. $\dfrac{x^2}{a^2} + \dfrac{y^2}{b^2} = 1 \Rightarrow \dfrac{x^2}{(5/2)^2} + \dfrac{y^2}{5^2} = 1 \Rightarrow \dfrac{4x^2}{25} + \dfrac{y^2}{25} = 1 \Rightarrow 4x^2 + y^2 = 25 \Rightarrow y^2 = 25 - 4x^2$

The tangent line equation is $y = m(x - 2) + 3$. Solve $25 - 4x^2 = [m(x-2)+3]^2$ to find m.

$25 - 4x^2 = m^2(x-2)^2 + 6m(x-2) + 9 \Rightarrow$

$(4 + m^2)x^2 + (6m - 4m^2)x + (4m^2 - 12m - 16) = 0 \Rightarrow (6m - 4m^2)^2 - 4(4 + m^2)(4m^2 - 12m - 16) = 0 \Rightarrow$

$36m^2 + 192m + 256 = 0 \Rightarrow 9m^2 + 48m + 64 = 0 \Rightarrow (3m + 8)^2 = 0 \Rightarrow m = -\dfrac{8}{3}$

For a unique solution, the discriminant must be zero. The tangent line as equation
$y = -\dfrac{8}{3}(x-2) + 3 = -\dfrac{8}{3}x + \dfrac{25}{3}$.

101. $9(x+1)^2 + 16(y-2)^2 = 144 \Rightarrow$
$\dfrac{(x+1)^2}{16} + \dfrac{(y-2)^2}{9} = 1 \Rightarrow a = 4, b = 3$
$A = \pi ab = 12\pi$

103. $16(x-1)^2 + 9(y+2)^2 = 144 \Rightarrow$
$\dfrac{(x-1)^2}{9} + \dfrac{(y+2)^2}{16} = 1$
This ellipse has center $(1, -2)$ and is oriented vertically. $a = 3$, $b = 4$, so $A = 12\pi$.
$16(x-1)^2 + 4(y+2)^2 = 64 \Rightarrow$
$\dfrac{(x-1)^2}{4} + \dfrac{(y+2)^2}{16} = 1$
This ellipse has center $(1, -2)$ and is oriented vertically. $a = 2$, $b = 4$, so $A = 8\pi$.
The area between the two ellipses is $12\pi - 8\pi = 4\pi$.

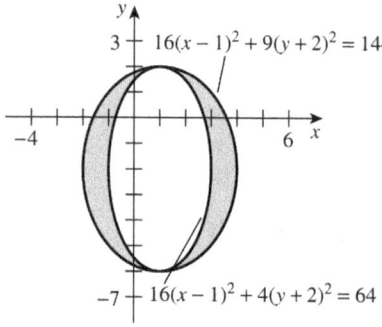

105. Let the length of $P_1F_1 = s$. Then, the coordinates of P_1 are (c, s). But, $ae = c$, so the coordinates of P_1 are (ae, s). Since P_1 lies on the ellipse
$\dfrac{x^2}{a^2} + \dfrac{y^2}{b^2} = 1 \Rightarrow \dfrac{c^2}{a^2} + \dfrac{s^2}{b^2} = 1 \Rightarrow$
$\dfrac{a^2 - b^2}{a^2} + \dfrac{s^2}{b^2} = 1 \Rightarrow$
$b^2(a^2 - b^2) + a^2 s^2 = a^2 b^2 \Rightarrow$
$s^2 = \dfrac{a^2 b^2 - b^2(a^2 - b^2)}{a^2}$
$= \dfrac{b^2(a^2 - a^2 + b^2)}{a^2} \Rightarrow s = \dfrac{b^2}{a}$
The length of the ellipse is symmetric with respect to the x-axis, so the length of the latus rectum, P_1P_2, is $\dfrac{2b^2}{a}$.

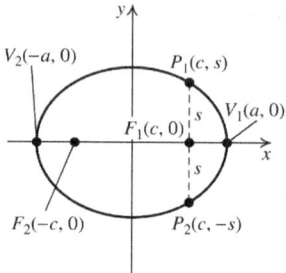

107. $\dfrac{x^2}{16} + \dfrac{y^2}{8} = 1 \Rightarrow a^2 = 16 \Rightarrow a = 4, b^2 = 8$
L.R. $= \dfrac{2b^2}{a} = \dfrac{2(8)}{4} = 4$

109. $5x^2 + 9y^2 = 45 \Rightarrow \dfrac{x^2}{9} + \dfrac{y^2}{5} = 1 \Rightarrow$
$a^2 = 9 \Rightarrow a = 3, b^2 = 5$
L.R. $= \dfrac{2b^2}{a} = \dfrac{2(5)}{3} = \dfrac{10}{3}$

111. The length of the major axis
$2a = 3(2b) \Rightarrow a = 3b$.
L.R. $= 2 = \dfrac{2b^2}{a} = \dfrac{2b^2}{3b} \Rightarrow 2 = \dfrac{2b}{3} \Rightarrow b = 3$ and $a = 9$.
The center of the ellipse is at $(0, 0)$, so an equation of the ellipse is
$\dfrac{x^2}{9^2} + \dfrac{y^2}{3^2} = 1 \Rightarrow \dfrac{x^2}{81} + \dfrac{y^2}{9} = 1$.
Another ellipse with major axis $3b$, latus rectum = 2, and center $(0, 0)$ is
$\dfrac{x^2}{3^2} + \dfrac{y^2}{9^2} = 1 \Rightarrow \dfrac{x^2}{9} + \dfrac{y^2}{81} = 1$.

10.3 Critical Thinking/Discussion/Writing

113. There are four circles, one in each quadrant:
$(x-3)^2 + (y-3)^2 = 9, (x+3)^2 + (y-3)^2 = 9$,
$(x-3)^2 + (y+3)^2 = 9$, and
$(x+3)^2 + (y+3)^2 = 9$.

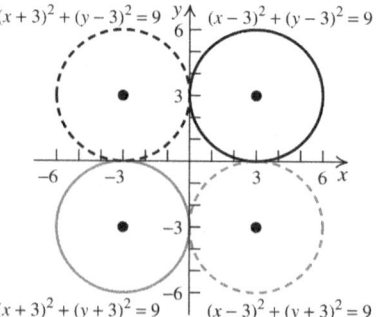

114. There are eight ellipses, two in each quadrant:
$\frac{(x-3)^2}{9}+\frac{(y-2)^2}{4}=1, \frac{(x-2)^2}{4}+\frac{(y-3)^2}{9}=1,$
$\frac{(x+3)^2}{9}+\frac{(y-2)^2}{4}=1, \frac{(x+2)^2}{4}+\frac{(y-3)^2}{9}=1,$
$\frac{(x+3)^2}{9}+\frac{(y+2)^2}{4}=1, \frac{(x+2)^2}{4}+\frac{(y+3)^2}{9}=1,$
$\frac{(x-3)^2}{9}+\frac{(y+2)^2}{4}=1, \frac{(x-2)^2}{4}+\frac{(y+3)^2}{9}=1$

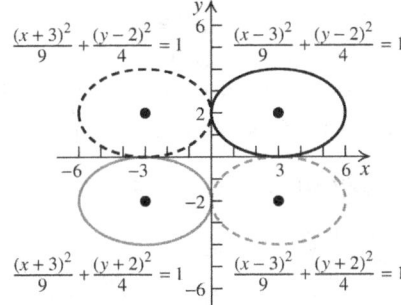

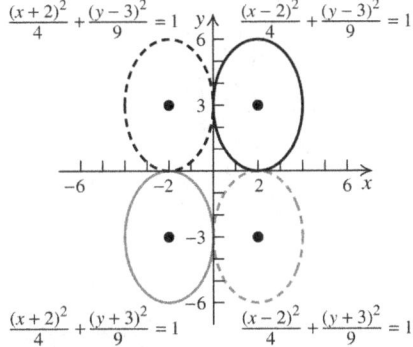

10.3 Getting Ready for the Next Section

115. $y-y_1=m(x-x_1) \Rightarrow y-5=-(x+3) \Rightarrow$
$y=-x-3+5=-x+2$

117. $\begin{cases} y=3x+6 \\ y=-3x \end{cases} \Rightarrow 2y=6 \Rightarrow y=3$
$y=-3x \Rightarrow 3=-3x \Rightarrow x=-1$
The point of intersection is $(-1, 3)$.

119. $4y^2-x^2=1$
x-intercepts:
$4(0)^2-x^2=1 \Rightarrow x^2=-1 \Rightarrow$ there are no x-intercepts.
y-intercepts:
$4y^2-(0)^2=1 \Rightarrow y^2=\frac{1}{4} \Rightarrow y=\pm\frac{1}{2}$

121. $\left(\sqrt{29}\right)^2=a^2+4^2 \Rightarrow a^2=29-16=13 \Rightarrow$
$a=\pm\sqrt{13}$

123. $f(x)=\frac{x^2+1}{x-1}$
$x-1=0 \Rightarrow x=1$ is the vertical asymptote.

$$\begin{array}{r} x+1 \\ x-1\overline{\smash{\big)}\,x^2+0x+1} \\ \underline{x^2-x} \\ x+1 \\ \underline{x-1} \\ 2 \end{array}$$

$\frac{x^2+1}{x-1}=x+1+\frac{2}{x-1} \Rightarrow y=x+1$ is the slant asymptote.

10.4 The Hyperbola

10.4 Practice Problems

1. The transverse axis is on the y-axis.

2. Rewrite the equation in standard form to determine a and b. $x^2-4y^2=8 \Rightarrow$
$\frac{x^2}{8}-\frac{y^2}{2}=1 \Rightarrow a^2=8 \Rightarrow a=2\sqrt{2}$ and
$b^2=2 \Rightarrow b=\sqrt{2}$.
The transverse axis of the hyperbola is along the x-axis, so the vertices are $\left(-2\sqrt{2},0\right)$ and $\left(2\sqrt{2},0\right)$.
To find the foci, we need c:
$c^2=a^2+b^2 \Rightarrow c^2=8+2 \Rightarrow c=\sqrt{10}$. The foci are $\left(-\sqrt{10},0\right)$ and $\left(\sqrt{10},0\right)$.

3. a. Since the foci of the hyperbola, $(0, -6)$ and $(0, 6)$ lie on the y-axis, the transverse axis also lies on the y-axis, and $c = 6$. The center of the hyperbola is $(0, 0)$, so the standard form of this hyperbola is
$\frac{y^2}{a^2}-\frac{x^2}{b^2}=1$.
The vertices are $(0, -3)$ and $(0, 3)$, so $a = 3$.
$c^2=a^2+b^2 \Rightarrow 6^2=3^2+b^2 \Rightarrow b^2=27$.
The equation is $\frac{y^2}{9}-\frac{x^2}{27}=1$.

b. The center (0, 0) and the vertex (5, 0) are on the x-axis, so the transverse axis is on the x-axis. The distance between the center and the vertex is 5, so $a = 5$. The distance between the center and the focus (7, 0) is 7, so $c = 7$.
$$c^2 = a^2 + b^2 \Rightarrow 7^2 = 5^2 + b^2 \Rightarrow$$
$$b^2 = 49 - 25 = 24$$
The equation is $\dfrac{x^2}{25} - \dfrac{y^2}{24} = 1$.

c. The foci are on the y-axis, so the tranverse axis is also on the y-axis. The length of the transverse axis is 10, so $2a = 10 \Rightarrow a = 5$. The center is (0, 0) and the distance between the center and the focus (0, 6) is $c = 6$.
$$c^2 = a^2 + b^2 \Rightarrow 6^2 = 5^2 + b^2 \Rightarrow$$
$$b^2 = 36 - 25 = 11$$
The equation is $\dfrac{y^2}{25} - \dfrac{x^2}{11} = 1$.

4. The hyperbola $\dfrac{y^2}{4} - \dfrac{x^2}{9} = 1$ is of the form $\dfrac{y^2}{a^2} - \dfrac{x^2}{b^2} = 1$, so $b = 3$ and $a = 2$. The asymptotes are of the form $y = \dfrac{a}{b}x$ and $y = -\dfrac{a}{b}x$. They are $y = \dfrac{2}{3}x$ and $y = -\dfrac{2}{3}x$.

5. a. The foci are on the x-axis, so the transverse axis is on the x-axis and the asymptotes are $y = \pm\dfrac{b}{a}x$. The asymptotes for this hyperbola are $y = \pm x$, so $\dfrac{b}{a} = 1 \Rightarrow a = b$. For this hyperbola, $c = 2\sqrt{2}$.
$$c^2 = a^2 + b^2 \Rightarrow (2\sqrt{2})^2 = a^2 + b^2 \Rightarrow$$
$$8 = a^2 + a^2 \Rightarrow 8 = 2a^2 \Rightarrow a^2 = 4$$
$$a = b \Rightarrow a^2 = b^2 = 4$$
The equation is $\dfrac{x^2}{4} - \dfrac{y^2}{4} = 1$.

b. The vertices are on the y-axis, so the so the transverse axis is on the y-axis and the asymptotes are $y = \pm\dfrac{a}{b}x$.

The asymptotes for this hyperbola are $y = \pm\dfrac{1}{3}x$, so $\dfrac{a}{b} = \dfrac{1}{3} \Rightarrow a = \dfrac{1}{3}b$. For this hyperbola, $a = 1$, so $b = 3$. The equation is
$$y^2 - \dfrac{x^2}{9} = 1.$$

6. a. $25x^2 - 4y^2 = 100 \Rightarrow \dfrac{x^2}{4} - \dfrac{y^2}{25} = 1 \Rightarrow a = 2$, $b = 5$.
$$c^2 = a^2 + b^2 \Rightarrow c^2 = 4 + 25 \Rightarrow c = \sqrt{29}.$$
The vertices are (2, 0) and (−2, 0). The endpoints of the conjugate axis are (0, 5) and (0, −5), and the foci are $\left(-\sqrt{29}, 0\right)$ and $\left(\sqrt{29}, 0\right)$. The asymptotes are
$$y = \dfrac{b}{a}x = \dfrac{5}{2}x \text{ and } y = -\dfrac{b}{a}x = -\dfrac{5}{2}x.$$

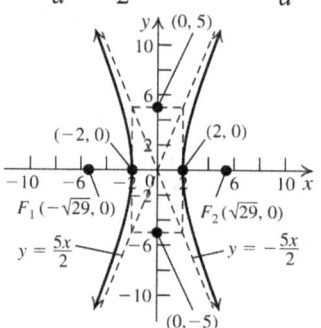

b. $9y^2 - x^2 = 1 \Rightarrow \dfrac{y^2}{1/9} - \dfrac{x^2}{1} = 1 \Rightarrow a^2 = \dfrac{1}{9} \Rightarrow$
$a = \dfrac{1}{3}$ and $b = 1$.
$$c^2 = a^2 + b^2 \Rightarrow c^2 = \dfrac{1}{9} + 1 \Rightarrow c = \dfrac{\sqrt{10}}{3}.$$
The vertices are $\left(0, \dfrac{1}{3}\right)$ and $\left(0, -\dfrac{1}{3}\right)$.
The endpoints of the conjugate axis are (1, 0) and (−1, 0), and the foci are $\left(0, \dfrac{\sqrt{10}}{3}\right)$ and $\left(0, -\dfrac{\sqrt{10}}{3}\right)$.

The asymptotes are $y = \dfrac{a}{b}x = \dfrac{1/3}{1}x = \dfrac{1}{3}x$ and $y = -\dfrac{a}{b}x = -\dfrac{1/3}{1}x = -\dfrac{1}{3}x$.

(*continued on next page*)

(*continued*)

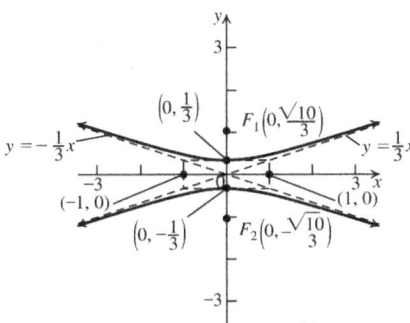

7. Complete the square to put the equation in standard form:

$$(x^2 - 2x) - 4(y^2 - 4y) = 20$$
$$(x^2 - 2x + 1) - 4(y^2 - 4y + 4) = 20 + 1 - 16$$
$$(x-1)^2 - 4(y-2)^2 = 5$$
$$\frac{(x-1)^2}{5} - \frac{4(y-2)^2}{5} = 1$$
$$\frac{(x-1)^2}{5} - \frac{(y-2)^2}{5/4} = 1$$

The center is (1, 2), $a = \sqrt{5}$ and $b = \frac{\sqrt{5}}{2}$.

The vertices are $(1+\sqrt{5}, 2)$ and $(1-\sqrt{5}, 2)$.

The endpoints of the conjugate axis are $\left(1, 2+\frac{\sqrt{5}}{2}\right)$ and $\left(1, 2-\frac{\sqrt{5}}{2}\right)$.

The asymptotes are $y - k = \pm \frac{b}{a}(x-h) \Rightarrow$

$y - 2 = \pm \frac{\sqrt{5}/2}{\sqrt{5}}(x-1) \Rightarrow y - 2 = \pm \frac{1}{2}(x-1)$.

$c^2 = a^2 + b^2 \Rightarrow c^2 = 5 + \frac{5}{4} = \frac{25}{4} \Rightarrow c = \frac{5}{2}$. The

foci are $\left(1+\frac{5}{2}, 2\right) = \left(\frac{7}{2}, 2\right)$ and

$\left(1-\frac{5}{2}, 2\right) = \left(-\frac{3}{2}, 2\right)$.

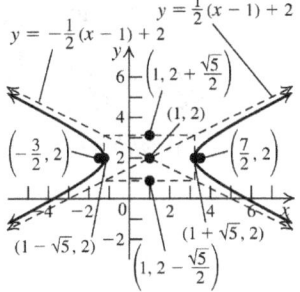

8. **a.** The center is (0, –2), so $h = 0$ and $k = -2$. The distance from the center to the vertex is $a = 1$. The distance between the center and the focus is $c = 3$. The transverse axis lies on the horizontal line $y = -2$.
$c^2 = a^2 + b^2 \Rightarrow 3^2 = 1^2 + b^2 \Rightarrow b^2 = 8$.
The equation is
$$\frac{x^2}{1} - \frac{(y+2)^2}{8} = x^2 - \frac{(y+2)^2}{8} = 1.$$

b. The center is (–1, 0), so $h = -1$ and $k = 0$. The distance between the center and the vertex (–1, 1) is $a = 1$. The slope of the asymptote with positive slope is $\frac{1}{2}$ and the transverse axis is on the vertical line $x = -1$, so $\frac{a}{b} = \frac{1}{2} \Rightarrow b = 2a = 2(1) = 2$.
The equation is
$$\frac{y^2}{1} - \frac{(x+1)^2}{4} = y^2 - \frac{(x+1)^2}{4} = 1.$$

9. The hyperbola has foci (–150, 0) and (150, 0), so $c = 150$. The distance from A to B is 260 miles, so $a = 130$. $c^2 = a^2 + b^2 \Rightarrow 150^2 = 130^2 + b^2 \Rightarrow b^2 = 5600$.
The equation of the hyperbola is
$$\frac{x^2}{130^2} - \frac{y^2}{5600} = 1 \Rightarrow \frac{x^2}{16,900} - \frac{y^2}{5600} = 1.$$

10.4 Concepts and Vocabulary

1. A hyperbola is a set of all points in the plane, the <u>difference</u> of whose distances form two fixed points is constant.

3. The standard equation of the hyperbola with center (0, 0), vertices ($\pm a$, 0), and foci ($\pm c$, 0) is $\frac{x^2}{a^2} - \frac{y^2}{b^2} = 1$, where $b^2 = \underline{c^2 - a^2}$.

5. True

7. False. If one asymptote of a hyperbola has equation $y = -2x$, then the other has the equation $y = 2x$.

10.4 Building Skills

9. G 11. H

13. D 15. B

13. D   **15.** B

17. $a^2 = 1 \Rightarrow$ the vertices are $(1, 0)$ and $(-1, 0)$.
$c^2 = a^2 + b^2 \Rightarrow c^2 = 1 + 4 \Rightarrow c = \sqrt{5} \Rightarrow$
the foci are $(\sqrt{5}, 0)$ and $(-\sqrt{5}, 0)$.

Transverse axis: x-axis. Hyperbola opens left and right. Vertices of the fundamental rectangle: $(1, 2), (-1, 2), (-1, -2), (1, -2)$. Asymptotes: $y = \pm 2x$.

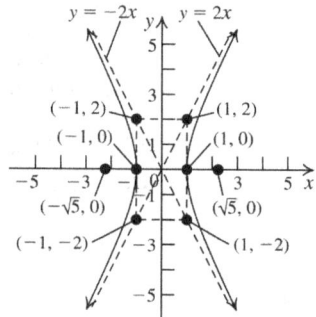

19. $x^2 - y^2 = -1 \Rightarrow y^2 - x^2 = 1 \Rightarrow a^2 = 1 \Rightarrow$ the vertices are $(0, 1)$ and $(0, -1)$.
$c^2 = a^2 + b^2 \Rightarrow c^2 = 1 + 1 \Rightarrow c = \sqrt{2} \Rightarrow$
the foci are $(0, \sqrt{2})$ and $(0, -\sqrt{2})$.

Transverse axis: y-axis. Hyperbola opens up and down. Vertices of the fundamental rectangle: $(1, 1), (-1, 1), (-1, -1), (1, -1)$. Asymptotes: $y = \pm x$.

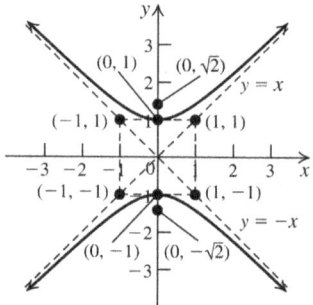

21. $9y^2 - x^2 = 36 \Rightarrow \dfrac{y^2}{4} - \dfrac{x^2}{36} = 1 \Rightarrow a^2 = 4 \Rightarrow$ the vertices are $(0, 2)$ and $(0, -2)$.
$c^2 = a^2 + b^2 \Rightarrow c^2 = 4 + 36 \Rightarrow c = 2\sqrt{10} \Rightarrow$
the foci are $(0, 2\sqrt{10})$ and $(0, -2\sqrt{10})$.

Transverse axis: y-axis. Hyperbola opens up and down. Vertices of the fundamental rectangle: $(6, 2), (-6, 2), (-6, -2), (6, -2)$. Asymptotes: $y = \pm \dfrac{a}{b} x = \pm \dfrac{1}{3} x$.

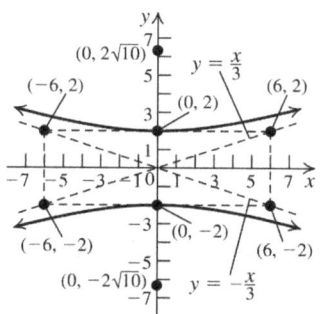

23. $4x^2 - 9y^2 - 36 = 0 \Rightarrow \dfrac{x^2}{9} - \dfrac{y^2}{4} = 1 \Rightarrow$
$a^2 = 9 \Rightarrow$ the vertices are $(3, 0)$ and $(-3, 0)$.
$c^2 = a^2 + b^2 \Rightarrow c^2 = 9 + 4 \Rightarrow c = \sqrt{13} \Rightarrow$
the foci are $(\sqrt{13}, 0)$ and $(-\sqrt{13}, 0)$.

Transverse axis: x-axis. Hyperbola opens left and right. Vertices of the fundamental rectangle: $(3, 2), (-3, 2), (-3, -2), (3, -2)$.
Asymptotes: $y = \pm \dfrac{b}{a} x = \pm \dfrac{2}{3} x$.

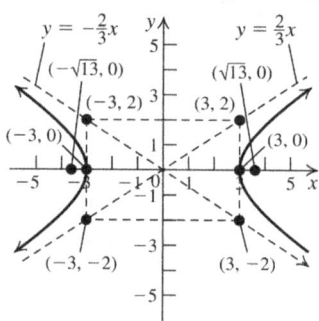

25. $y = \pm \sqrt{4x^2 + 1} \Rightarrow y^2 - \dfrac{x^2}{1/4} = 1 \Rightarrow a^2 = 1 \Rightarrow$
the vertices are $(0, 1)$ and $(0, -1)$.
$c^2 = a^2 + b^2 \Rightarrow c^2 = 1 + \dfrac{1}{4} \Rightarrow c = \dfrac{\sqrt{5}}{2} \Rightarrow$ the foci are $\left(0, \dfrac{\sqrt{5}}{2}\right)$ and $\left(0, -\dfrac{\sqrt{5}}{2}\right)$. Transverse axis: y-axis. Hyperbola opens up and down. Vertices of the fundamental rectangle: $\left(\dfrac{1}{2}, 1\right), \left(-\dfrac{1}{2}, 1\right), \left(-\dfrac{1}{2}, -1\right)$, and $\left(\dfrac{1}{2}, -1\right)$.

Asymptotes: $y = \pm \dfrac{a}{b} x = \pm 2x$.

(continued on next page)

(*continued*)

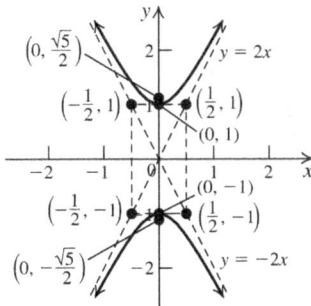

27. $y = \pm\sqrt{9x^2 - 1} \Rightarrow 9x^2 - y^2 = 1 \Rightarrow$

$\dfrac{x^2}{1/9} - y^2 = 1 \Rightarrow a^2 = \dfrac{1}{9} \Rightarrow$ the vertices are

$\left(\dfrac{1}{3}, 0\right)$ and $\left(-\dfrac{1}{3}, 0\right)$. $c^2 = a^2 + b^2 \Rightarrow$

$c^2 = \dfrac{1}{9} + 1 \Rightarrow c = \dfrac{\sqrt{10}}{3} \Rightarrow$ the foci are $\left(\dfrac{\sqrt{10}}{3}, 0\right)$

and $\left(-\dfrac{\sqrt{10}}{3}, 0\right)$. Transverse axis:

x-axis. Hyperbola opens left and right. Vertices of the fundamental rectangle:

$\left(\dfrac{1}{3}, 1\right), \left(-\dfrac{1}{3}, 1\right), \left(-\dfrac{1}{3}, -1\right)$, and $\left(\dfrac{1}{3}, -1\right)$.

Asymptotes: $y = \pm\dfrac{b}{a}x \Rightarrow y = \pm 3x$.

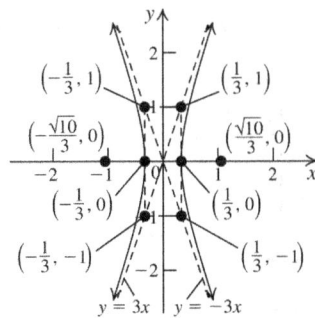

29. The transverse axis is the x-axis and the center is $(0, 0)$, so the equation is of the form

$\dfrac{x^2}{a^2} - \dfrac{y^2}{b^2} = 1$. $a = 2, c = 3 \Rightarrow 9 = 4 + b^2 \Rightarrow$

$b^2 = 5$. The equation is $\dfrac{x^2}{4} - \dfrac{y^2}{5} = 1$.

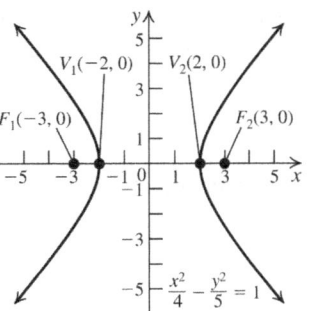

31. The transverse axis is the y-axis and the center is $(0, 0)$, so the equation is of the form

$\dfrac{y^2}{a^2} - \dfrac{x^2}{b^2} = 1$. $a = 4, c = 6 \Rightarrow 36 = 16 + b^2 \Rightarrow$

$b^2 = 20$. The equation is $\dfrac{y^2}{16} - \dfrac{x^2}{20} = 1$.

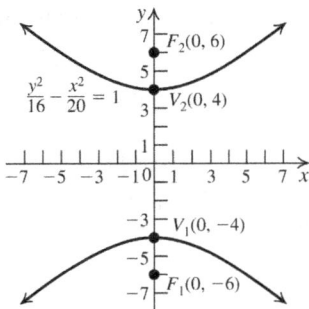

33. The transverse axis is the y-axis and the center is $(0, 0)$, so the equation is of the form

$\dfrac{y^2}{a^2} - \dfrac{x^2}{b^2} = 1$. $a = 2, c = 5 \Rightarrow 25 = 4 + b^2 \Rightarrow$

$b^2 = 21$. The equation is $\dfrac{y^2}{4} - \dfrac{x^2}{21} = 1$.

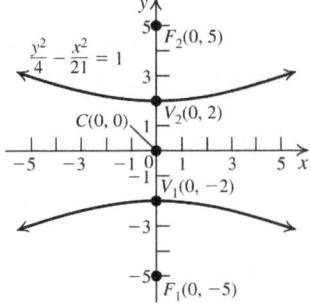

35. The transverse axis is the x-axis and the center is (0, 0), so the equation is of the form
$\dfrac{x^2}{a^2} - \dfrac{y^2}{b^2} = 1$. $a = 1, c = 5 \Rightarrow 25 = 1 + b^2 \Rightarrow$
$b^2 = 24$. The equation is $x^2 - \dfrac{y^2}{24} = 1$.

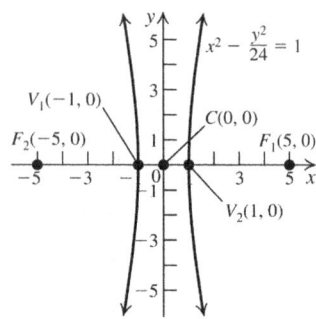

37. The transverse axis is the y-axis and the center is (0, 0), so the equation is of the form
$\dfrac{y^2}{a^2} - \dfrac{x^2}{b^2} = 1$. Length of the transverse axis = 6,
so, $a = 3$. $c = 5 \Rightarrow 25 = 9 + b^2 \Rightarrow b^2 = 16$.
The equation is $\dfrac{y^2}{9} - \dfrac{x^2}{16} = 1$.

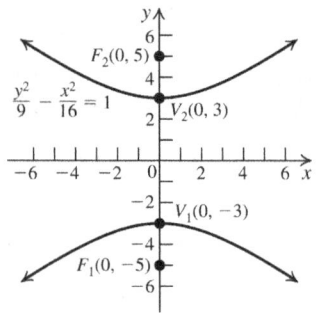

39. The transverse axis is the x-axis and the center is (0, 0), so the equation is of the form
$\dfrac{x^2}{a^2} - \dfrac{y^2}{b^2} = 1$. $y = 2x = \dfrac{b}{a}x \Rightarrow b = 2a$.
$c^2 = a^2 + b^2 \Rightarrow 5 = a^2 + (2a)^2 \Rightarrow a = 1, b = 2$.
The equation is $x^2 - \dfrac{y^2}{4} = 1$.

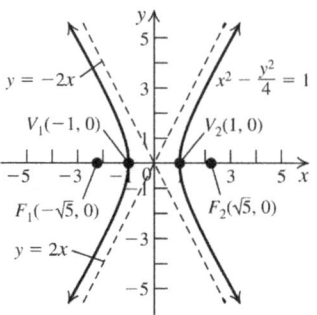

41. The transverse axis is the y-axis and the center is (0, 0), so the equation is of the form
$\dfrac{y^2}{a^2} - \dfrac{x^2}{b^2} = 1$. $y = x = \dfrac{a}{b}x \Rightarrow a = b$.
$a = 4 \Rightarrow b = 4$. The equation is $\dfrac{y^2}{16} - \dfrac{x^2}{16} = 1$.

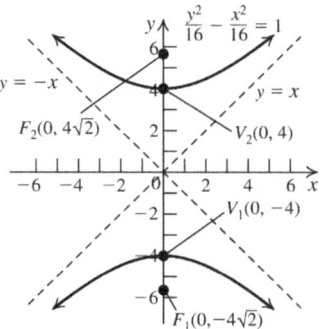

43. Center: $(1, -1)$; vertices: $(4, -1), (-2, -1)$; transverse axis: $y = -1$; asymptotes:
$y - k = \pm\dfrac{b}{a}(x - h) \Rightarrow y + 1 = \pm\dfrac{4}{3}(x - 1)$

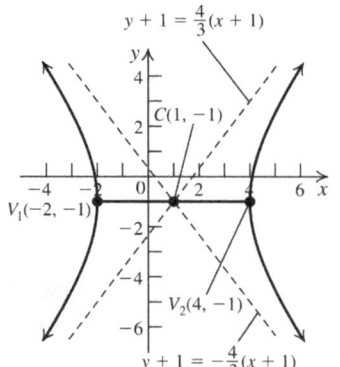

45. Center: (−2, 0); vertices: (3, 0), (−7, 0); transverse axis: x-axis (y = 0); asymptotes:

$$y - k = \pm \frac{b}{a}(x-h) \Rightarrow y = \pm \frac{7}{5}(x+2)$$

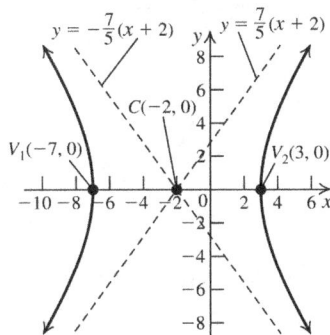

47. Center: (−4, −3); vertices: (1, −3), (−9, −3); transverse axis: y = −3; asymptotes:

$$y - k = \pm \frac{b}{a}(x-h) \Rightarrow y + 3 = \pm \frac{7}{5}(x+4).$$

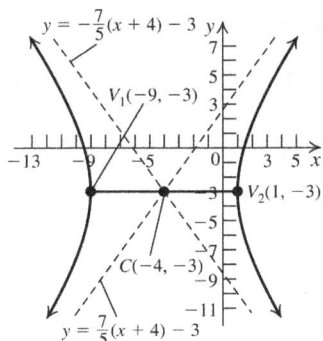

49. $4x^2 - (y+1)^2 = 25 \Rightarrow \dfrac{x^2}{25/4} - \dfrac{(y+1)^2}{25} = 1 \Rightarrow$

Center: (0, −1); vertices: $\left(\dfrac{5}{2}, -1\right), \left(-\dfrac{5}{2}, -1\right)$;

transverse axis: y = −1; asymptotes:

$$y - k = \pm \frac{b}{a}(x-h) \Rightarrow y + 1 = \pm 2x.$$

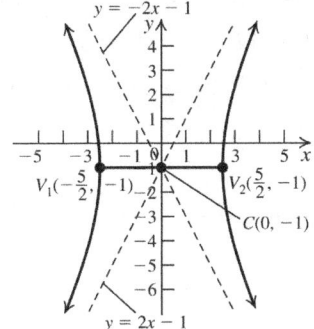

51. $(y+1)^2 - 9(x-2)^2 = 25 \Rightarrow$

$\dfrac{(y+1)^2}{25} - \dfrac{(x-2)^2}{25/9} = 1 \Rightarrow$ center: (2, −1);

vertices: (2, 4), (2, −6);
transverse axis: x = 2;

asymptotes: $y - k = \pm \dfrac{a}{b}(x-h) \Rightarrow$

$y + 1 = \pm 3(x-2)$.

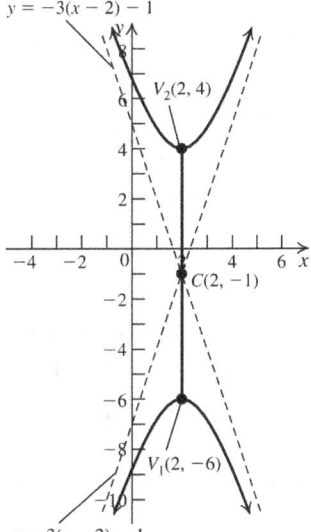

53. Complete the square to put the equation in standard form: $x^2 - y^2 + 6x = 36 \Rightarrow$

$x^2 + 6x + 9 - y^2 = 36 + 9 \Rightarrow$

$\dfrac{(x+3)^2}{45} - \dfrac{y^2}{45} = 1 \Rightarrow$ center: (−3, 0);

vertices: $\left(-3 + 3\sqrt{5}, 0\right), \left(-3 - 3\sqrt{5}, 0\right)$;

transverse axis:
x-axis (y = 0); asymptotes:

$$y - k = \pm \frac{b}{a}(x-h) \Rightarrow y = \pm(x+3).$$

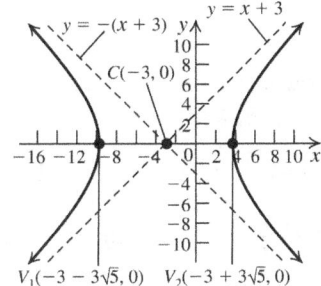

55. Complete the square to put the equation in standard form: $x^2 + 4x - 4y^2 = 12 \Rightarrow$
$x^2 + 4x + 4 - 4y^2 = 12 + 4 \Rightarrow$
$\dfrac{(x+2)^2}{16} - \dfrac{y^2}{4} = 1 \Rightarrow$ center: $(-2, 0)$; vertices: $(2, 0), (-6, 0)$; transverse axis: x-axis ($y = 0$);
asymptotes: $y - k = \pm\dfrac{b}{a}(x - h) \Rightarrow$
$y = \pm\dfrac{1}{2}(x + 2)$.

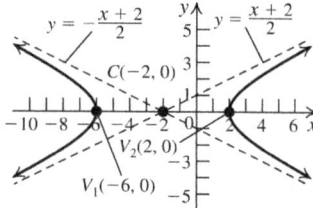

57. Complete the square to put the equation in standard form: $2x^2 - y^2 + 12x - 8y + 3 = 0 \Rightarrow$
$2(x^2 + 6x + 9) - (y^2 + 8y + 16) = -3 + 18 - 16 \Rightarrow$
$(y + 4)^2 - \dfrac{(x+3)^2}{1/2} = 1 \Rightarrow$ center: $(-3, -4)$;
vertices: $(-3, -3), (-3, -5)$; transverse axis: $x = -3$; asymptotes: $y - k = \pm\dfrac{a}{b}(x - h) \Rightarrow$
$y + 4 = \pm\sqrt{2}(x + 3)$.

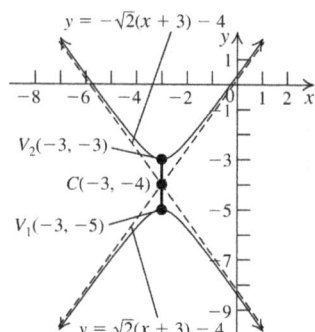

59. Complete the square to put the equation in standard form: $3x^2 - 18x - 2y^2 - 8y + 1 = 0 \Rightarrow$
$3(x^2 - 6x + 9) - 2(y^2 + 4y + 4) = -1 + 27 - 8 \Rightarrow$
$\dfrac{(x-3)^2}{6} - \dfrac{(y+2)^2}{9} = 1 \Rightarrow$ center: $(3, -2)$;
vertices: $(3 + \sqrt{6}, -2), (3 - \sqrt{6}, -2)$; transverse axis: $y = -2$; asymptotes:
$y - k = \pm\dfrac{b}{a}(x - h) \Rightarrow y + 2 = \pm\dfrac{\sqrt{6}}{2}(x - 3)$

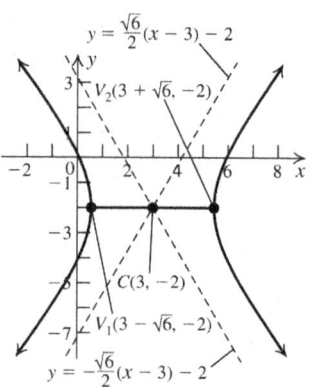

61. Complete the square to put the equation in standard form: $y^2 + 2\sqrt{2} - x^2 + 2\sqrt{2}x = 1 \Rightarrow$
$y^2 - \left(x^2 - 2\sqrt{2}x + 2\right) = 1 - 2\sqrt{2} - 2 \Rightarrow$
$\dfrac{(x - \sqrt{2})^2}{1 + 2\sqrt{2}} - \dfrac{y^2}{1 + 2\sqrt{2}} = 1 \Rightarrow$ center: $(\sqrt{2}, 0)$;
vertices: $\left(\sqrt{2} + \sqrt{1 + 2\sqrt{2}}, 0\right)$,
$\left(\sqrt{2} - \sqrt{1 + 2\sqrt{2}}, 0\right)$; transverse axis: x-axis ($y = 0$); asymptotes: $y - k = \pm\dfrac{b}{a}(x - h) \Rightarrow$
$y = \pm(x - \sqrt{2})$

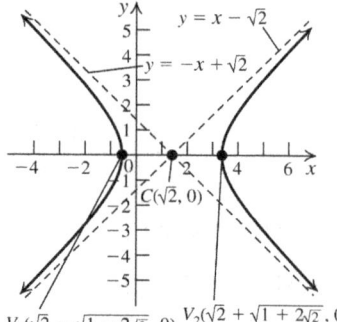

63. $x^2 - 6x + 12y + 33 = 0 \Rightarrow$
$x^2 - 6x + 9 = -12y - 33 + 9 \Rightarrow$
$(x - 3)^2 = -12(y + 2) \Rightarrow$ the conic is a parabola.

65. $y^2 - 9x^2 = -1 \Rightarrow \dfrac{x^2}{1/9} - y^2 = 1 \Rightarrow$ the conic is a hyperbola.

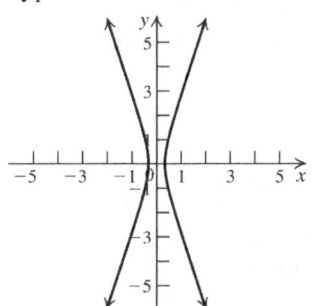

67. $x^2 + y^2 - 4x + 8y = 16 \Rightarrow$
$x^2 - 4x + 4 + y^2 + 8y + 16 = 16 + 4 + 16 \Rightarrow$
$(x-2)^2 + (y+4)^2 = 36 \Rightarrow$ the conic is a circle.

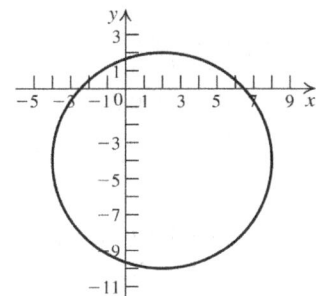

69. $2x^2 - 4x + 3y + 8 = 0 \Rightarrow$
$2(x^2 - 2x + 1) = -3y - 8 + 2 \Rightarrow$
$2(x-1)^2 = -3(y+2) \Rightarrow$ the conic is a parabola.

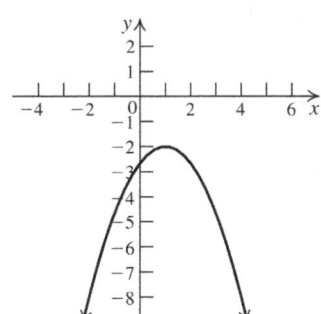

71. $4x^2 + 9y^2 + 8x - 54y + 49 = 0 \Rightarrow$
$4(x^2 + 2x + 1) + 9(y^2 - 6y + 9) = -49 + 4 + 81 \Rightarrow$
$\dfrac{(x+1)^2}{9} + \dfrac{(y-3)^2}{4} = 1 \Rightarrow$ the conic is an ellipse.

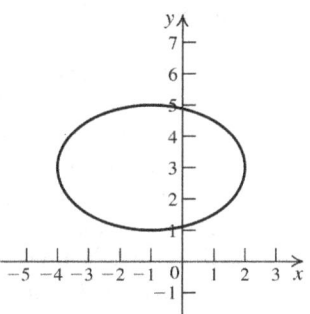

73. The distance from the center to the vertex is $a = 3$. The distance from the center to the focus is $c = \sqrt{17}$.
$c^2 = a^2 + b^2 \Rightarrow 17 = 9 + b^2 \Rightarrow b^2 = 8$.
The transverse axis is on the y-axis, so the equation is $\dfrac{y^2}{9} - \dfrac{x^2}{8} = 1$.

75. The distance from the center to the vertex is $a = 3$. The distance from the center to the focus is $c = 4$.
$c^2 = a^2 + b^2 \Rightarrow 16 = 9 + b^2 \Rightarrow b^2 = 7$.
The transverse axis is on the horizontal line $y = 1$, so the equation is $\dfrac{(x+2)^2}{9} - \dfrac{(y-1)^2}{7} = 1$.

77. The vertex and center are on the x-axis, so the transverse axis is on the x-axis and the asymptotes are $y = \pm \dfrac{b}{a} x$. The asymptote with positive slope has slope $\dfrac{5}{4}$, so
$\dfrac{b}{a} = \dfrac{5}{4} \Rightarrow b = \dfrac{5}{4} a$. For this hyperbola, $a = 2$, so
$b = \dfrac{5}{4}(2) = \dfrac{5}{2}$.
The equation is
$\dfrac{(x+1)^2}{4} - \dfrac{y^2}{(5/2)^2} = 1 \Rightarrow \dfrac{(x+1)^2}{4} - \dfrac{4y^2}{25} = 1$.

10.4 Applying the Concepts

79. A hyperbola is the set of all points in the plane whose distances from two fixed points have a constant difference. The difference in the distances of the location of the explosion from point A to point B is 600 meters, so A and B are the foci of the hyperbola. Let the coordinates of A be $(-500, 0)$ and those of B be $(500, 0)$, so $c = 500$. The distance between V_1 and V_2 is 600, so $a = 300$.
$$c^2 = a^2 + b^2 \Rightarrow$$
$$500^2 = 300^2 + b^2 \Rightarrow b^2 = 160{,}000$$
The equation is $\dfrac{x^2}{90{,}000} - \dfrac{y^2}{160{,}000} = 1$.

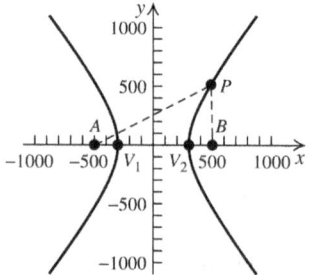

81. Using the same reasoning as in exercises 79 and 80, Nicole and Juan are located at the foci of the hyperbola, $(-4000, 0)$ and $(4000, 0)$. The distance between the vertices is 3300, so $a = 1650$.
$$c^2 = a^2 + b^2 \Rightarrow 4000^2 = 1650^2 + b^2 \Rightarrow$$
$$b^2 = 13{,}277{,}500.$$ The equation is
$$\dfrac{x^2}{2{,}722{,}500} - \dfrac{y^2}{13{,}277{,}500} = 1.$$
Because Nicole hears the thunder first, the graph consists only of the part of the hyperbola closest to Nicole.

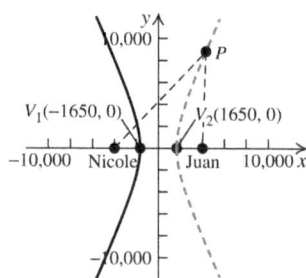

83. Let the coordinates of A and B be $(-150, 0)$ and $(150, 0)$, respectively. At 300,000 km per sec, the difference in the distances the signals travel is 150 km, so $a = 75$. $c^2 = a^2 + b^2 \Rightarrow$
$150^2 = 75^2 + b^2 \Rightarrow b^2 = 16{,}875$. The equation is $\dfrac{x^2}{5625} - \dfrac{y^2}{16{,}875} = 1$.

85. Let the coordinates of A be $(-125, 0)$ and those of B be $(125, 0)$. The difference in the distances of the location of the plane from point A to point B is 500 microseconds × 980 feet per microsecond = 490,000 feet = 92.8 miles. So $a = 245{,}000$ feet ≈ 46.4015 miles and $c = 125$ miles.
$$c^2 = a^2 + b^2 \Rightarrow 125^2 = 46.4015^2 + b^2 \Rightarrow$$
$$b^2 = 13{,}471.9008.$$ The equation is
$$\dfrac{x^2}{2153.0992} - \dfrac{y^2}{13{,}471.9008} = 1.$$
The plane is flying along the line $y = 50$, so
$$\dfrac{x^2}{2153.0992} - \dfrac{50^2}{13{,}471.9008} = 1 \Rightarrow$$
$x = \pm 50.5238$ miles
Since the signal from point B arrives sooner than the signal from point A, the plane is located closer to B. It is at approximately $(50.5238, 50)$.

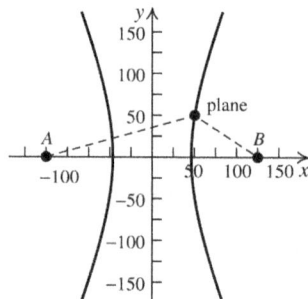

87. The fishing boat is located at the intersection of two hyperbolas, one with foci at A and B, and the other with foci at A and C. For the first hyperbola, $c = 200$ and $a = 150$.
$c^2 = a^2 + b^2 \Rightarrow 200^2 = 150^2 + b^2 \Rightarrow$
$b^2 = 17{,}500$. The equation of the first hyperbola is $\dfrac{x^2}{22{,}500} - \dfrac{y^2}{17{,}500} = 1$. For the second hyperbola, $c = 500$ and $a = 200$.
$500^2 = 200^2 + b^2 \Rightarrow b^2 = 210{,}000$. The center of the second hyperbola is $(200, 500)$, so the equation of the second hyperbola is
$\dfrac{(y-500)^2}{40{,}000} - \dfrac{(x-200)^2}{210{,}000} = 1$. Solve the system
$\begin{cases} \dfrac{x^2}{22{,}500} - \dfrac{y^2}{17{,}500} = 1 \\ \dfrac{(y-500)^2}{40{,}000} - \dfrac{(x-200)^2}{210{,}000} = 1 \end{cases}$
or use a graphing calculator to find the intersections (the possible locations of the boat).

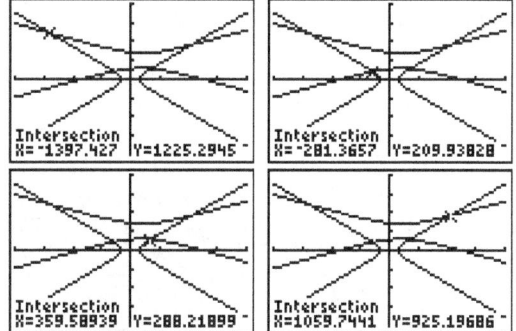

10.4 Beyond the Basics

89.

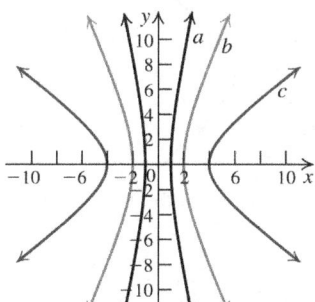

91.

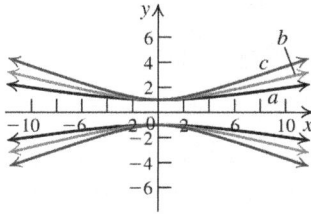

93. Assume that the transverse axis is the x-axis. Then the equation of the hyperbola is $\dfrac{x^2}{a^2} - \dfrac{y^2}{b^2} = 1$. The hyperbola is equilateral, so $a = b$, and the equations of the asymptotes are $y = \pm x$. These lines are perpendicular to each other. Similarly, it can be shown that the asymptotes are perpendicular to each other if the transverse axis is the y-axis.

95. Since $b > 0$ and $c^2 = a^2 + b^2$, it follows that $c^2 > a^2$ and $c > a$. So, $e > 1$. When $e = 1$, the hyperbola becomes the union of two rays.

97. $36x^2 - 25y^2 = 900 \Rightarrow \dfrac{x^2}{25} - \dfrac{y^2}{36} = 1$
$a = 5, b = 6 \Rightarrow c^2 = 25 + 36 \Rightarrow c = \sqrt{61}$
$e = \dfrac{c}{a} = \dfrac{\sqrt{61}}{5}$.
Length of latus rectum $= \dfrac{2b^2}{a} = \dfrac{72}{5}$.

99. $8(x-1)^2 - (y+2)^2 = 2 \Rightarrow$
$\dfrac{(x-1)^2}{1/4} - \dfrac{(y+2)^2}{2} = 1$
$a = \dfrac{1}{2}, b = \sqrt{2} \Rightarrow c^2 = \dfrac{1}{4} + 2 \Rightarrow c = \dfrac{3}{2}$
$e = \dfrac{c}{a} = 3$.
Length of latus rectum $= \dfrac{2b^2}{a} = 8$.

101. $e = \dfrac{c}{a} \Rightarrow e^2 = \dfrac{c^2}{a^2} \Rightarrow$
$e^2 = \dfrac{a^2 + b^2}{a^2} = 1 + \dfrac{b^2}{a^2} \Rightarrow$
$e^2 - 1 = \dfrac{b^2}{a^2} \Rightarrow a^2(e^2 - 1) = b^2$

103. The foci are $(\pm 6, 0) \Rightarrow c = 6$ and the center is $(0, 0)$. The transverse axis lies along the x-axis.
Length of the latus rectum =
$\dfrac{2b^2}{a} = 10 \Rightarrow b^2 = 5a$
$c^2 = a^2 + b^2 \Rightarrow 36 = a^2 + 5a \Rightarrow$
$a^2 + 5a - 36 = 0 \Rightarrow (a-4)(a+9) = 0 \Rightarrow$
$a = 4$, $a = -9$ (disregard)
$b^2 = 5(4) = 20$, so an equation of the hyperbola is $\dfrac{x^2}{4^2} - \dfrac{y^2}{20} = 1 \Rightarrow \dfrac{x^2}{16} - \dfrac{y^2}{20} = 1$.

105. The foci are $(0, \pm 12) \Rightarrow c = 12$ and the center is $(0, 0)$. The transverse axis lies along the y-axis.
Length of the latus rectum =
$\dfrac{2b^2}{a} = 36 \Rightarrow b^2 = 18a$
$c^2 = a^2 + b^2 \Rightarrow 144 = a^2 + 18a \Rightarrow$
$a^2 + 18a - 144 = 0 \Rightarrow (a-6)(a+24) = 0 \Rightarrow$
$a = 6$, $a = -24$ (disregard)
$b^2 = 18(6) = 108$, so an equation of the hyperbola is $\dfrac{y^2}{6^2} - \dfrac{x^2}{108} = 1 \Rightarrow \dfrac{y^2}{36} - \dfrac{x^2}{108} = 1$.

107. The foci are $(0, \pm\sqrt{10}) \Rightarrow c = \sqrt{10} \Rightarrow c^2 = 10$.
The transverse axis lies along the y-axis.
$a^2 + b^2 = c^2 \Rightarrow a^2 + b^2 = 10 \Rightarrow$
$b^2 = 10 - a^2$
The graph passes through $(2, 3)$ so
$\dfrac{y^2}{a^2} - \dfrac{x^2}{b^2} = 1 \Rightarrow \dfrac{3^2}{a^2} - \dfrac{2^2}{10-a^2} = 1 \Rightarrow$
$9(10 - a^2) - 4a^2 = a^2(10 - a^2)$
$90 - 9a^2 - 4a^2 = 10a^2 - a^4$
$a^4 - 23a^2 + 90 = 0 \Rightarrow$
$(a^2 - 18)(a^2 - 5) = 0 \Rightarrow a^2 = 18$, $a^2 = 5$
If $a^2 = 18$, then $a^2 + b^2 = 10 \Rightarrow$
$18 + b^2 = 10 \Rightarrow b^2 = -8$, which is impossible.
If $a^2 = 5$, then $a^2 + b^2 = 10 \Rightarrow$
$5 + b^2 = 10 \Rightarrow b^2 = 5$.
The equation is $\dfrac{y^2}{5} - \dfrac{x^2}{5} = 1$.

109. Solve using substitution:
$\begin{cases} y - 2x - 20 = 0 \\ y^2 - 4x^2 = 36 \end{cases} \Rightarrow \begin{cases} y = 2x + 20 \\ y^2 - 4x^2 = 36 \end{cases} \Rightarrow$
$(2x + 20)^2 - 4x^2 = 36 \Rightarrow$
$80x + 400 = 36 \Rightarrow x = -\dfrac{91}{20}$
$y = 2\left(-\dfrac{91}{20}\right) + 20 = \dfrac{109}{10}$
The only point of intersection is $\left(-\dfrac{91}{20}, \dfrac{109}{10}\right)$.

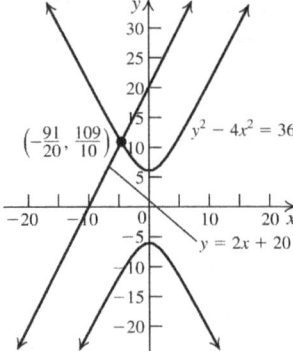

111. Solve using elimination:
$\begin{cases} x^2 + y^2 = 15 \\ x^2 - y^2 = 1 \end{cases} \Rightarrow 2x^2 = 16 \Rightarrow x = \pm 2\sqrt{2}$
$(\pm 2\sqrt{2})^2 + y^2 = 15 \Rightarrow y^2 = 7 \Rightarrow y = \pm\sqrt{7}$
The points of intersection are
$(2\sqrt{2}, \sqrt{7})$, $(2\sqrt{2}, -\sqrt{7})$, $(-2\sqrt{2}, -\sqrt{7})$, and $(-2\sqrt{2}, \sqrt{7})$.

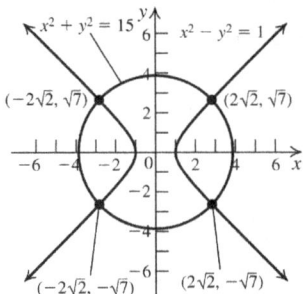

113. The equation of the hyperbola is

$\dfrac{x^2}{5/3} - \dfrac{y^2}{5/7} = 1 \Rightarrow \dfrac{3x^2}{5} - \dfrac{7y^2}{5} = 1 \Rightarrow 3x^2 - 7y^2 = 5 \Rightarrow 7y^2 = 3x^2 - 5$

Step 1: Let m = the slope of the tangent line. Then the equation of the tangent line is
$y - 1 = m(x - 2) \Rightarrow y = m(x - 2) + 1$

Step 2: $7y^2 = 7\left[m(x-2)+1\right]^2 = 7\left[m^2(x-2)^2 + 2m(x-2) + 1\right] = 7m^2(x-2)^2 + 14m(x-2) + 7$

Step 3:

$$3x^2 - 5 = 7m^2(x-2)^2 + 14m(x-2) + 7$$
$$3x^2 - 5 = 7m^2(x^2 - 4x + 4) + 14mx - 28m + 7$$
$$3x^2 - 5 = 7m^2x^2 - 28m^2x + 28m^2 + 14mx - 28m + 7$$

$3x^2 - 7m^2x^2 + 28m^2x - 14mx - 28m^2 + 28m - 12 = 0$
$(3 - 7m^2)x^2 + (28m^2 - 14m)x - (28m^2 - 28m + 12) = 0$

Now set the discriminant equal to 0 and solve for x:

$$(28m^2 - 14m)^2 - 4(3 - 7m^2)(28m^2 - 28m + 12) = 0$$
$$784m^4 - 784m^3 + 196m^2 - 784m^4 + 784m^3 - 336m + 144 = 0$$
$$196m^2 - 336m + 144 = 0$$
$$49m^2 - 84m + 36 = 0$$
$$(7m - 6)^2 = 0$$
$$m = \dfrac{6}{7}$$

Step 4: The equation of the tangent line is $y - 1 = \dfrac{6}{7}(x - 2) \Rightarrow y = \dfrac{6}{7}x - \dfrac{5}{7}$.

10.4 Critical Thinking/Writing/Discussion

115. a. $4x^2 - 9y^2 = 0 \Rightarrow (2x + 3y)(2x - 3y) = 0$.

The graph consists of two lines, $y = \pm\dfrac{2}{3}x$.

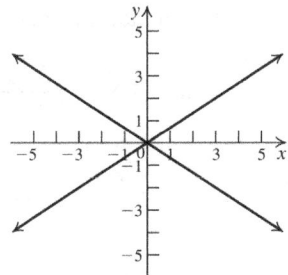

b. All the variable terms on the left side of the equation are always nonnegative, so there are no x or y values that would satisfy the equation. There is no graph.

c. Both x^2 and y^2 are nonnegative, so the only solution of the equation is $(0, 0)$.

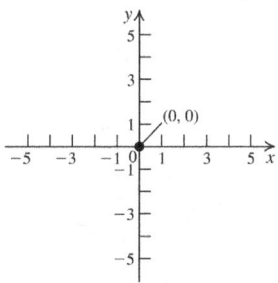

d. Complete the square:
$x^2 + y^2 - 4x + 8y = -20 \Rightarrow$
$(x^2 - 4x + 4) + (y^2 + 8y + 16) = -20 + 4 + 16 \Rightarrow$
$(x - 2)^2 + (y + 4)^2 = 0 \Rightarrow x = 2$ and $y = -4$.

(continued on next page)

(*continued*)

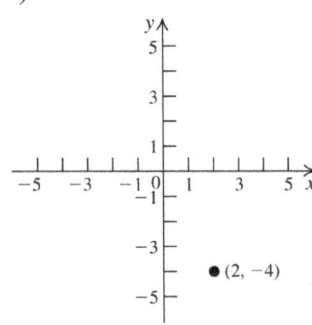

e. $y^2 - 2x^2 = 0 \Rightarrow y^2 = 2x^2 \Rightarrow y = \pm x\sqrt{2}$.
The graph consists of two lines.

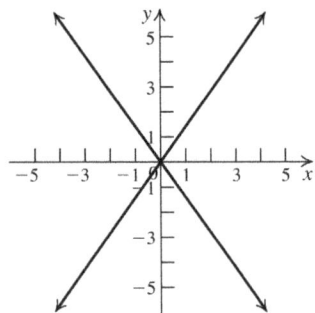

116. $Ax^2 + Cy^2 + Dx + Ey + F = 0$, $AC \neq 0$
Rewriting the equation, we have
$$A\left(x^2 + \frac{D}{A}x + \frac{D^2}{4A^2}\right) + C\left(y^2 + \frac{E}{C}y + \frac{E^2}{4C^2}\right)$$
$$= \frac{D^2}{4A} + \frac{E^2}{4C} - F \Rightarrow$$
$$A\left(x + \frac{D}{2A}\right)^2 + C\left(y + \frac{E}{2C}\right)^2 = \frac{D^2}{4A} + \frac{E^2}{4C} - F.$$
If the term on the right side is not zero, then the equation can be rewritten as
$$\frac{A\left(x + \frac{D}{2A}\right)^2}{\frac{D^2}{4A} + \frac{E^2}{4C} - F} + \frac{C\left(y + \frac{E}{2C}\right)^2}{\frac{D^2}{4A} + \frac{E^2}{4C} - F} = 1 \Rightarrow$$
$$\frac{\left(x + \frac{D}{2A}\right)^2}{\frac{D^2}{4A^2} + \frac{E^2}{4AC} - \frac{F}{A}} + \frac{\left(y + \frac{E}{2C}\right)^2}{\frac{D^2}{4AC} + \frac{E^2}{4C^2} - \frac{F}{C}} = 1$$

(i) If $A > 0$ and $C > 0$, and if $\frac{D^2}{4A} + \frac{E^2}{4C} - F = 0$, the graph consists of one point, $\left(-\frac{D}{2A}, -\frac{E}{2C}\right)$. If $\frac{D^2}{4A} + \frac{E^2}{4C} - F > 0$, the graph is an ellipse. If $\frac{D^2}{4A} + \frac{E^2}{4C} - F < 0$, then there is no graph.

(ii) If $A > 0$ and $C < 0$, and if $\frac{D^2}{4A} + \frac{E^2}{4C} - F = 0$, the graph consists of two lines, $\sqrt{A}\left(x + \frac{D}{2A}\right) = \pm\sqrt{-C}\left(y + \frac{E}{2C}\right)$. If $\frac{D^2}{4A} + \frac{E^2}{4C} - F > 0$, the graph is a hyperbola with a horizontal transverse axis. If $\frac{D^2}{4A} + \frac{E^2}{4C} - F < 0$, then the graph is a hyperbola with a vertical transverse axis.

(iii) If $A < 0$ and $C > 0$, and if $\frac{D^2}{4A} + \frac{E^2}{4C} - F = 0$, the graph consists of two lines, $\sqrt{-A}\left(x + \frac{D}{2A}\right) = \pm\sqrt{C}\left(y + \frac{E}{2C}\right)$. If $\frac{D^2}{4A} + \frac{E^2}{4C} - F > 0$, the graph is a hyperbola with a vertical transverse axis. If $\frac{D^2}{4A} + \frac{E^2}{4C} - F < 0$, then the graph is a hyperbola with a horizontal transverse axis.

(iv) If $A < 0$ and $C < 0$, and if $\frac{D^2}{4A} + \frac{E^2}{4C} - F = 0$, the graph consists of one point, $\left(-\frac{D}{2A}, -\frac{E}{2C}\right)$. If $\frac{D^2}{4A} + \frac{E^2}{4C} - F > 0$, there is no graph. If $\frac{D^2}{4A} + \frac{E^2}{4C} - F < 0$, then the graph is an ellipse.

10.4 Getting Ready for the Next Section

117. $f(3) = \frac{1}{1+3} = \frac{1}{4}$

119. $h(5) = (-1)^3 3^{5-1} = -3^4 = -81$

121. $(-1)^{17} = -1$

123. $(-1)^5 2^5 = 32$

125. $\dfrac{3 \cdot \pi \cdot e^2 \cdot 7 \cdot 11 \cdot 19 \cdot a}{a \cdot 3 \cdot \pi \cdot e^2 \cdot 19} = 7 \cdot 11 = 77$

127. $c(1 - 5a + 3b) = c - 5ac + 3bc$
True

Chapter 10 Review Exercises

1. $y^2 = -6x \Rightarrow -6 = 4a \Rightarrow a = -\dfrac{3}{2}$

Vertex: (0, 0), focus: $\left(-\dfrac{3}{2}, 0\right)$, axis: x-axis,

directrix: $x = \dfrac{3}{2}$.

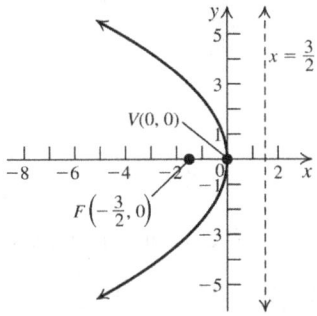

3. $x^2 = 7y \Rightarrow 7 = 4a \Rightarrow \dfrac{7}{4} = a$.

Vertex: (0, 0), focus: $\left(0, \dfrac{7}{4}\right)$, axis: y-axis,

directrix: $y = -\dfrac{7}{4}$.

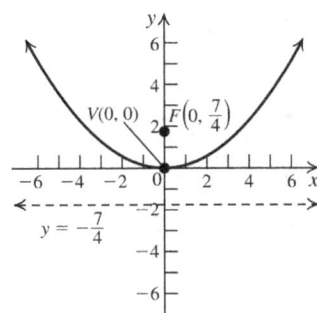

5. $(x-2)^2 = -(y+3) \Rightarrow -1 = 4a \Rightarrow -\dfrac{1}{4} = a$.

Vertex: (2, −3), focus: $\left(2, -\dfrac{13}{4}\right)$, axis: x = 2,

directrix: $y = -\dfrac{11}{4}$.

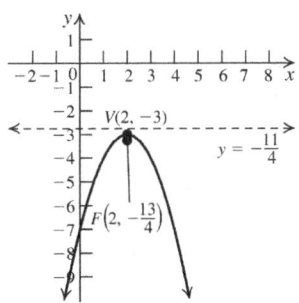

7. Rearrange the equation and complete the square to put the equation in standard form:

$y^2 = -4y + 2x + 1 \Rightarrow y^2 + 4y + 4 = 2x + 1 + 4 \Rightarrow$

$(y+2)^2 = 2\left(x + \dfrac{5}{2}\right) \Rightarrow 2 = 4a \Rightarrow a = \dfrac{1}{2}$.

Vertex: $\left(-\dfrac{5}{2}, -2\right)$, focus: (−2, −2),

axis: y = −2, directrix: x = −3.

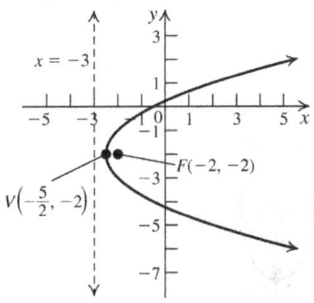

9. The vertex is (0, 0) and the focus is (−3, 0), so the graph opens to the left. The general form is $(y-k)^2 = -4a(x-h)$.

$a = -3 \Rightarrow y^2 = -12x$.

11. The focus is (0, 4) and the directrix is $y = -4$, so the vertex is (0, 0), and the graph opens up. The general form is $(x-h)^2 = 4a(y-k)$.

$a = 4 \Rightarrow x^2 = 16y$.

13. $a^2 = 25 \Rightarrow$ the vertices are (5, 0) and (−5, 0). $b^2 = a^2 - c^2 \Rightarrow 4 = 25 - c^2 \Rightarrow c = \pm\sqrt{21} \Rightarrow$ the foci are $\left(\sqrt{21}, 0\right)$ and $\left(-\sqrt{21}, 0\right)$.

Endpoints of the minor axis: (0, −2) and (0, 2).

(*continued on next page*)

608 Chapter 10 Conic Sections

(continued)

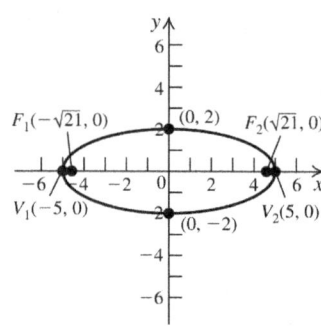

15. $4x^2 + y^2 = 4 \Rightarrow x^2 + \dfrac{y^2}{4} = 1$.

$a^2 = 4 \Rightarrow$ the vertices are (0, 2) and (0, −2).
$b^2 = a^2 - c^2 \Rightarrow 1 = 4 - c^2 \Rightarrow c = \pm\sqrt{3} \Rightarrow$
the foci are $\left(0, \sqrt{3}\right)$ and $\left(0, -\sqrt{3}\right)$.
Endpoints of the minor axis: (1, 0) and (−1, 0).

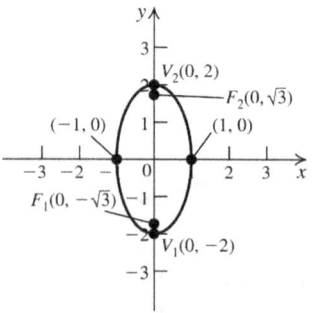

17. $16(x+1)^2 + 9(y+4)^2 = 144 \Rightarrow$

$\dfrac{(x+1)^2}{9} + \dfrac{(y+4)^2}{16} = 1.$

The center is (−1, −4) and $a^2 = 16 \Rightarrow$ the vertices are (−1, −8) and (−1, 0).
$b^2 = a^2 - c^2 \Rightarrow 9 = 16 - c^2 \Rightarrow c = \pm\sqrt{7}.$
The foci are $\left(-1, -4 - \sqrt{7}\right)$ and $\left(-1, -4 + \sqrt{7}\right)$.
Endpoints of the minor axis: (−4, −4) and (2, −4).

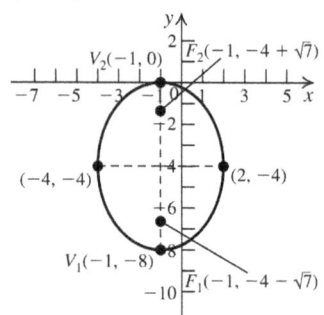

19. Rearrange the equation and complete the square to put the equation in standard form:
$x^2 + 9y^2 + 2x - 18y + 1 = 0 \Rightarrow$
$x^2 + 2x + 1 + 9(y^2 - 2y + 1) = -1 + 1 + 9 \Rightarrow$
$(x+1)^2 + 9(y-1)^2 = 9 \Rightarrow \dfrac{(x+1)^2}{9} + (y-1)^2 = 1$

The center is (−1, 1) and $a^2 = 9 \Rightarrow$ the vertices are (−4, 1) and (2, 1).
$b^2 = a^2 - c^2 \Rightarrow 1 = 9 - c^2 \Rightarrow c = \pm 2\sqrt{2}$
The foci are $\left(-1 - 2\sqrt{2}, 1\right)$ and $\left(-1 + 2\sqrt{2}, 1\right)$.
Endpoints of the minor axis: (−1, 2) and (−1, 0).

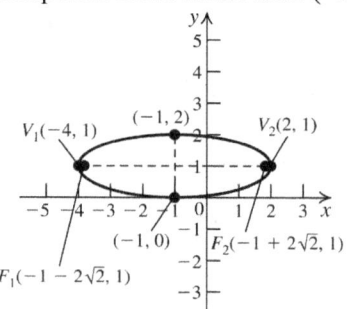

21. $a = 4, b = 2$, major axis: x-axis. $\dfrac{x^2}{16} + \dfrac{y^2}{4} = 1$

23. $a = 10$, center (0, 0), major axis: x-axis, $c = 5$, so $b^2 = 100 - 25 = 75$. $\dfrac{x^2}{100} + \dfrac{y^2}{75} = 1$

25. $a^2 = 16 \Rightarrow$ the vertices are (0, 4) and (0, −4).
$c^2 = a^2 + b^2 \Rightarrow c^2 = 16 + 4 \Rightarrow c = \pm 2\sqrt{5} \Rightarrow$
the foci are $\left(0, 2\sqrt{5}\right)$ and $\left(0, -2\sqrt{5}\right)$.

Asymptotes: $y = \pm \dfrac{a}{b}x = \pm 2x$.

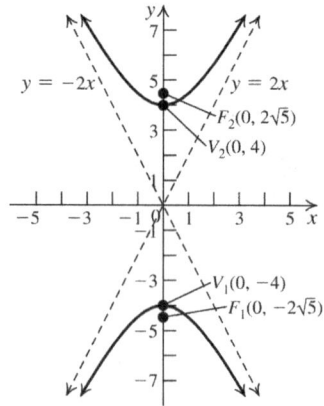

Copyright © 2019 Pearson Education Inc.

27. $8x^2 - y^2 = 8 \Rightarrow x^2 - \dfrac{y^2}{8} = 1 \Rightarrow a^2 = 1 \Rightarrow$ the vertices are $(-1, 0)$ and $(1, 0)$.
$c^2 = a^2 + b^2 \Rightarrow c^2 = 1 + 8 \Rightarrow c = \pm 3 \Rightarrow$ the foci are $(3, 0)$ and $(-3, 0)$. Asymptotes: $y = \pm \dfrac{b}{a} x = \pm 2\sqrt{2} x$.

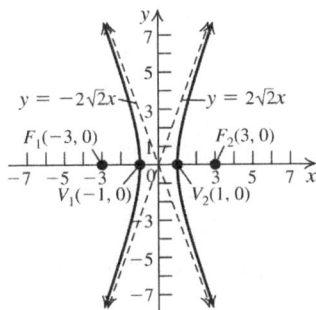

29. The center is $(-2, 3)$. $a^2 = 9 \Rightarrow$ the vertices are $(1, 3)$ and $(-5, 3)$.
$c^2 = a^2 + b^2 \Rightarrow c^2 = 9 + 4 \Rightarrow c = \pm\sqrt{13} \Rightarrow$ the foci are $\left(-2 + \sqrt{13}, 3\right)$ and $\left(-2 - \sqrt{13}, 3\right)$. Asymptotes:
$y - k = \pm \dfrac{b}{a}(x - h) \Rightarrow y - 3 = \pm \dfrac{2}{3}(x + 2)$.

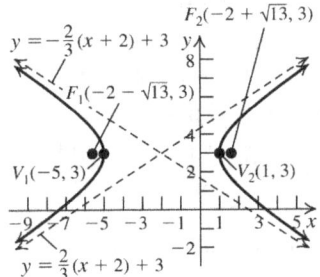

31. Rearrange the equation and complete the square to put the equation in standard form:
$4y^2 - x^2 + 40y - 4x + 60 = 0 \Rightarrow$
$4(y^2 + 10y + 25) - (x^2 + 4x + 4)$
$\qquad = -60 + 100 - 4$
$4(y+5)^2 - (x+2)^2 = 36 \Rightarrow$
$\dfrac{(y+5)^2}{9} - \dfrac{(x+2)^2}{36} = 1$.
The center is $(-2, -5)$. $a^2 = 9 \Rightarrow$ the vertices are $(-2, -5 - 3) = (-2, -8)$ and $(-2, -5 + 3) = (-2, -2)$.

$c^2 = a^2 + b^2 \Rightarrow c^2 = 9 + 36 \Rightarrow c = \pm 3\sqrt{5} \Rightarrow$ the foci are $\left(-2, -5 + 3\sqrt{5}\right)$ and $\left(-2, -5 - 3\sqrt{5}\right)$. Asymptotes:
$y - k = \pm \dfrac{a}{b}(x - h) \Rightarrow y + 5 = \pm \dfrac{1}{2}(x + 2)$.

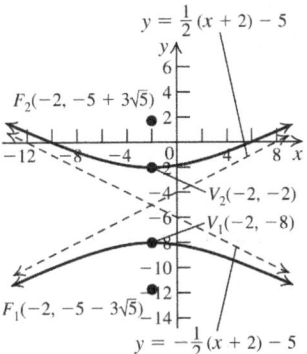

33. The vertices are $(\pm 1, 0)$ and the foci are $(\pm 2, 0)$, so the center is $(0, 0)$, $a = 1$, $c = 2$, and the transverse axis is the x-axis.
$c^2 = a^2 + b^2 \Rightarrow 4 = 1 + b^2 \Rightarrow b^2 = 3$. The equation is $x^2 - \dfrac{y^2}{3} = 1$.

35. The vertices are $(\pm 2, 0)$ so the center is $(0, 0)$, $a = 2$, and the transverse axis is the x-axis. The asymptotes are $y = \pm 3x \Rightarrow \pm \dfrac{b}{a} = \pm 3 \Rightarrow \dfrac{b}{2} = 3 \Rightarrow b = 6$. The equation is $\dfrac{x^2}{4} - \dfrac{y^2}{36} = 1$.

37. Hyperbola. In standard form, the equation is $\dfrac{x^2}{4} - \dfrac{y^2}{5} = 1$.

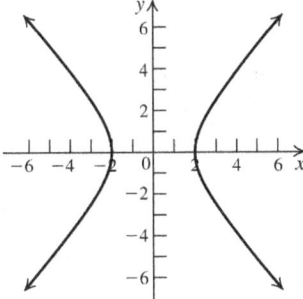

39. Ellipse. In standard form, the equation is
$$\frac{(x-2)^2}{22/3}+\frac{(y+1)^2}{11/2}=1.$$

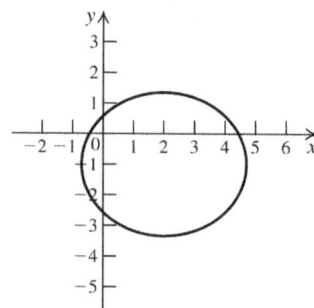

41. Circle. In standard form, the equation is
$(x+1)^2+y^2=4$.

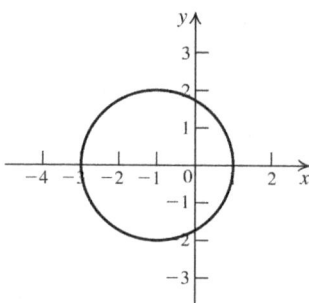

43. Hyperbola. In standard form, the equation is
$y^2-x^2=3$.

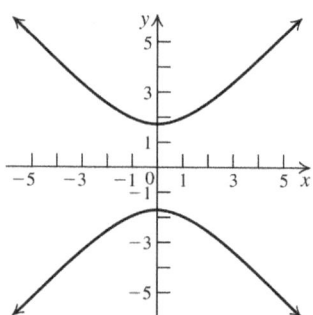

45. Circle. In standard form, the equation is
$$(x-1)^2+(y+2)^2=\frac{10}{3}$$

47. Ellipse. In standard form, the equation is
$$\frac{x^2}{4}+\frac{y^2}{9/2}=1.$$

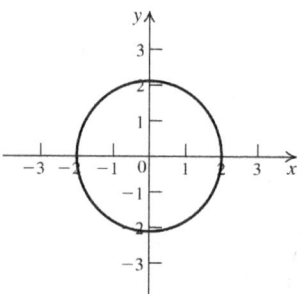

49. $4x^2+9y^2=36 \Rightarrow \frac{x^2}{9}+\frac{y^2}{4}=1 \Rightarrow$ the vertices of the ellipse are (3, 0) and (−3, 0). $b^2=a^2-c^2 \Rightarrow 4=9-c^2 \Rightarrow c=\pm\sqrt{5} \Rightarrow$ the foci are $(\sqrt{5},0)$ and $(-\sqrt{5},0)$. For the hyperbola, the transverse axis is the x-axis, $a=\sqrt{5}$, and $c=3$, so $c^2=a^2+b^2 \Rightarrow 9=5+b^2 \Rightarrow b^2=4$. The equation of the hyperbola is $\frac{x^2}{5}-\frac{y^2}{4}=1$.

51. Solve using substitution:
$$\begin{cases} x^2-4y^2=36 \\ x-2y-20=0 \end{cases} \Rightarrow \begin{cases} x^2-4y^2=36 \\ x=2y+20 \end{cases} \Rightarrow$$
$(2y+20)^2-4y^2=36 \Rightarrow 80y+400=36 \Rightarrow$
$y=-\frac{91}{20}; \ x=2\left(-\frac{91}{20}\right)+20=\frac{109}{10}$

The only point of intersection is $\left(\frac{109}{10},-\frac{91}{20}\right)$.

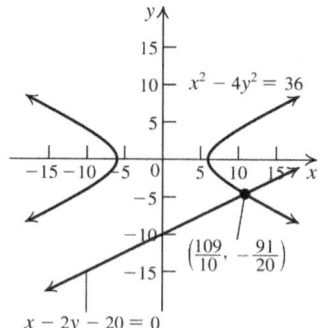

53. Solve using elimination:
$\begin{cases} 3x^2 - 7y^2 = 5 \\ 9y^2 - 2x^2 = 1 \end{cases} \Rightarrow \begin{cases} 6x^2 - 14y^2 = 10 \\ -6x^2 + 27y^2 = 3 \end{cases} \Rightarrow$
$13y^2 = 13 \Rightarrow y = \pm 1$
$3x^2 - 7 = 5 \Rightarrow x^2 = 4 \Rightarrow x = \pm 2$
The points of intersection are $(-2, -1)$, $(-2, 1)$, $(2, -1)$, and $(2, 1)$.

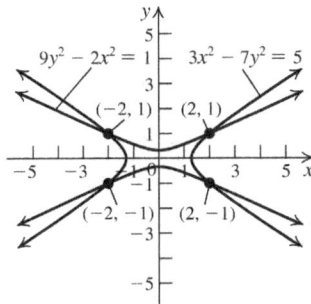

55. Let the end of the pipe, and therefore the vertex of the parabola, be at $(0, 0)$. Then the ground is at $y = -20$. The water goes through the point $(8, -6)$.

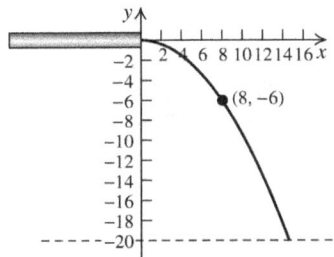

The equation is of the form $y = -4ax^2$.
Substitute $(8, -6)$ into the equation and solve for a: $-6 = -4a(64) \Rightarrow a = \dfrac{3}{128}$ and the equation of the parabola is $y = -\dfrac{3}{32}x^2$.
Let $y = -20$ and solve for x:
$-20 = -\dfrac{3}{32}x^2 \Rightarrow x^2 = \dfrac{640}{3} \Rightarrow x = \pm 14.61$.
We are looking for the positive solution. The water hits the ground 14.6 feet from the end of the pipe.

57. The equation of the ellipse is $\dfrac{x^2}{36} + \dfrac{y^2}{49/4} = 1$.

At $x = 4$, $y = \dfrac{7\sqrt{5}}{6} \approx 2.61$. (Reject the negative solution.) This is the radius of the cross section. The circumference $= 2\pi(2.61) \approx 16.4$ inches.

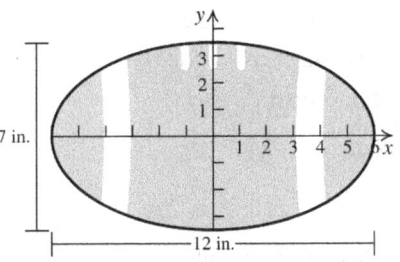

Chapter 10 Practice Test A

1. The focus is at $(0, 12)$ and the directrix is $x = -12$, so the vertex is at $(-6, 12)$ and the parabola opens to the right. The general form is $(y - k)^2 = 4a(x - h)$. $a = 0 - (-6) = 6 \Rightarrow$
$(y - 12)^2 = 4(6)(x + 6) = 24(x + 6)$.

2. $y^2 - 2y + 8x + 25 = 0 \Rightarrow$
$y^2 - 2y + 1 = -8x - 25 + 1 \Rightarrow$
$(y - 1)^2 = -8(x + 3)$

3. $x^2 = -9y = 4ay \Rightarrow a = -\dfrac{9}{4}$
Focus: $\left(0, -\dfrac{9}{4}\right)$, directrix: $y = \dfrac{9}{4}$.

4.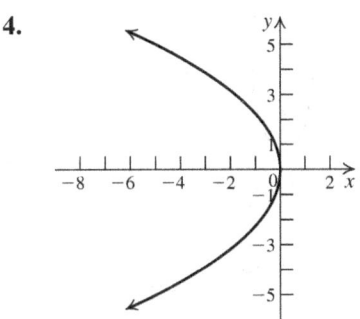

5. $(x + 2)^2 = -8(y - 1) = 4a(y - 1) \Rightarrow a = -2$.
Vertex: $(-2, 1)$, focus: $(-2, -1)$, directrix: $y = 3$.

6. Let the vertex be at $(0, 0)$. Then the parabola goes through the point $(3, 2)$. The general form of the equation is $(x - h)^2 = 4a(y - k)$. Substitute the values for $x, y, h,$ and k, then solve for a: $3^2 = 4a(2) \Rightarrow a = \dfrac{9}{8}$. The equation of the parabola is $x^2 = \dfrac{9}{2}y$. Now find the value of y for $x = 1$:
$1 = \dfrac{9}{2}y \Rightarrow y = \dfrac{2}{9} \approx 0.22$ feet.

7. Vertex (3, −1) and directrix $x = -3 \Rightarrow a = 6$. The parabola opens to the right, so the general form of the equation is $(y-k)^2 = 4a(x-h)$. The equation is $(y+1)^2 = 24(x-3)$.

8. Foci (0, −2) and (0, 2) $\Rightarrow c = 2$.
Vertices (0, −4) and (0, 4) $\Rightarrow a = 4$.
$b^2 = a^2 - c^2 \Rightarrow b^2 = 4^2 - 2^2 = 12$.
The equation is $\dfrac{x^2}{12} + \dfrac{y^2}{16} = 1$.

9. Foci (−2, 0) and (2, 0) $\Rightarrow c = 2$.
y-intercepts −5 and 5 $\Rightarrow b = 5$.
$b^2 = a^2 - c^2 \Rightarrow 25 = a^2 - 2^2 \Rightarrow a^2 = 29$.
The equation is $\dfrac{x^2}{29} + \dfrac{y^2}{25} = 1$.

10. $a = 9$, $b = 2$. The equation is $\dfrac{x^2}{81} + \dfrac{y^2}{4} = 1$.

11.

12. $49y^2 - x^2 = 49 \Rightarrow y^2 - \dfrac{x^2}{49} = 1 \Rightarrow$
$a^2 = 1, b^2 = 49$. The vertices are at (0, 1) and (0, −1).
$c^2 = a^2 + b^2 \Rightarrow c^2 = 1 + 49 \Rightarrow c = \pm 5\sqrt{2}$. The foci are at $\left(0, 5\sqrt{2}\right)$ and $\left(0, -5\sqrt{2}\right)$.

13.

14. The center is (0, 0) and the transverse axis is the y-axis, so the equation is of the form
$\dfrac{y^2}{a^2} - \dfrac{x^2}{b^2} = 1$.
$a = 6, c = \sqrt{45} \Rightarrow 45 = 36 + b^2 \Rightarrow b^2 = 9$.
The equation is $\dfrac{y^2}{36} - \dfrac{x^2}{9} = 1$.

15. $y^2 - x^2 + 2x = 2 \Rightarrow$
$y^2 - (x^2 - 2x + 1) = 2 - 1 \Rightarrow y^2 - (x-1)^2 = 1$

16. $x^2 - 2x - 4y^2 - 16y = 19 \Rightarrow$
$(x^2 - 2x + 1) - 4(y^2 + 4y + 4) = 19 + 1 - 16 \Rightarrow$
$(x-1)^2 - 4(y+2)^2 = 4 \Rightarrow$
$\dfrac{(x-1)^2}{4} - (y+2)^2 = 1$

17. Hyperbola

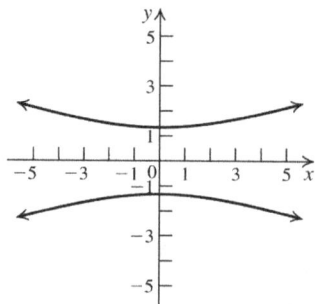

18. Parabola

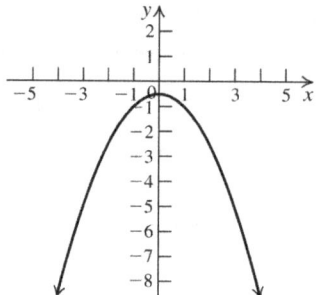

19. Circle

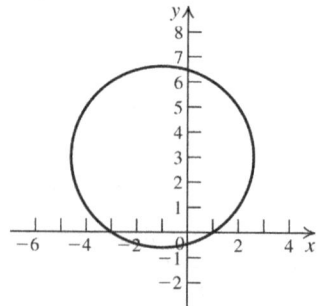

20. Ellipse

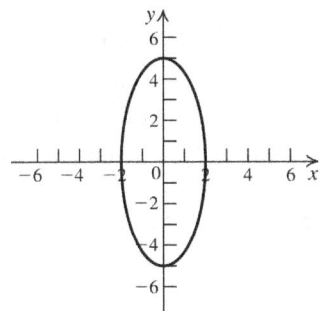

Chapter 10 Practice Test B

1. The focus is at (−10, 0) and the directrix is $x = 10$, so the vertex is at (0, 0) and the parabola opens to the left. The general form is $(y-k)^2 = -4a(x-h)$.
$a = 0 - (-10) = 10 \Rightarrow y^2 = -4(10)x = -40x$.
The answer is B.

2. $y^2 - 4y - 5x + 24 = 0 \Rightarrow$
$y^2 - 4y + 4 = 5x - 24 + 4 \Rightarrow (y-2)^2 = 5(x-4)$
The answer is B.

3. $x = 7y^2 \Rightarrow \dfrac{x}{7} = y^2 \Rightarrow \dfrac{1}{7} = 4a \Rightarrow \dfrac{1}{28} = a \Rightarrow$ the focus at $\left(\dfrac{1}{28}, 0\right)$ and the directrix is $x = -\dfrac{1}{28}$.
The answer is A.

4. The answer is A.

5. $(x-3)^2 = 12(y-1) \Rightarrow$ the vertex is (3, 1).
$12 = 4a \Rightarrow a = 3 \Rightarrow$ the focus is at (3, 4), and the directrix is $y = -2$. The answer is A.

6. Let the vertex be located at (0, 25) and one end of the base is at (90, 0). The general form of the equation is $(x-h)^2 = -4a(y-k)$. Substitute the values for x, y, h, and k, then solve for a:
$(90-0)^2 = -4a(0-25) \Rightarrow a = 81$
The equation of the parabola is
$x^2 = -4(81)(y-25) \Rightarrow x^2 = -324(y-25)$.
Find the value of y for
$x = 45$: $45^2 = -324(y-25) \Rightarrow y = 18.75$.
The answer is D.

7. Vertex (−2, 1) and directrix $x = 2 \Rightarrow a = 4$.
The parabola opens to the left, so the general form of the equation is
$(y-k)^2 = -4a(x-h)$.
The equation is $(y-1)^2 = -16(x+2)$.
The answer is A.

8. Foci (−3, 0) and (3, 0) $\Rightarrow c = 3$.
Vertices (−5, 0) and (5, 0) $\Rightarrow a = 5$.
$b^2 = a^2 - c^2 \Rightarrow b^2 = 5^2 - 3^2 = 16$
The equation is $\dfrac{x^2}{25} + \dfrac{y^2}{16} = 1$.
The answer is B.

9. Foci (0, −3) and (0, 3) $\Rightarrow c = 3$. y-intercepts −7 and 7 $\Rightarrow a = 7$. $b^2 = a^2 - c^2 \Rightarrow$
$49 = 3^2 - b^2 \Rightarrow b^2 = 40$. The equation is
$\dfrac{x^2}{40} + \dfrac{y^2}{49} = 1$. The answer is D.

10. $a = 8, b = 4$. The equation is $\dfrac{x^2}{16} + \dfrac{y^2}{64} = 1$.
The answer is A.

11. $9(x-1)^2 + 4(y-2)^2 = 36 \Rightarrow$
$\dfrac{(x-1)^2}{4} + \dfrac{(y-2)^2}{9} = 1$. The answer is A.

12. $a = 11, b = 2$. The vertices are at (11, 0) and (−11, 0). $c^2 = a^2 + b^2 \Rightarrow c^2 = 121 + 4 \Rightarrow$
$c = 5\sqrt{5}$. The foci are at $\left(-5\sqrt{5}, 0\right)$ and $\left(5\sqrt{5}, 0\right)$. The answer is C.

13. The answer is B.

14. The center is (0, 0) and the transverse axis is the y-axis, so the equation is of the form
$\dfrac{y^2}{a^2} - \dfrac{x^2}{b^2} = 1$.
$a = 5, c = 10 \Rightarrow 100 = 25 + b^2 \Rightarrow b^2 = 75$.
The equation is $\dfrac{y^2}{25} - \dfrac{x^2}{75} = 1$.
The answer is B.

15. $y^2 - 4x^2 - 2y - 16x - 19 = 0 \Rightarrow$
$y^2 - 2y + 1 - 4(x^2 + 4x + 4) = 19 + 1 - 16 \Rightarrow$
$(y-1)^2 - 4(x+2)^2 = 4 \Rightarrow$
$\dfrac{(y-1)^2}{4} - (x+2)^2 = 1$. The answer is C.

16. $4y^2 - 9x^2 - 16y - 36x - 56 = 0 \Rightarrow$
$4(y^2 - 4y + 4) - 9(x^2 + 4x + 4)$
$\qquad = 56 + 16 - 36 \Rightarrow$
$4(y-2)^2 - 9(x+2)^2 = 36 \Rightarrow$
$\dfrac{(y-2)^2}{9} - \dfrac{(x+2)^2}{4} = 1$. The answer is C.

17. C   **18.** A   **19.** D   **20.** D

Cumulative Review Exercises (Chapters P–10)

1. $\dfrac{\big((x+h)^2 - 3(x+h) + 2\big) - (x^2 - 3x + 2)}{h}$

$= \dfrac{x^2 + 2xh + h^2 - 3x - 3h + 2 - x^2 + 3x - 2}{h}$

$= \dfrac{h^2 + 2xh - 3h}{h} = h + 2x - 3$

2.
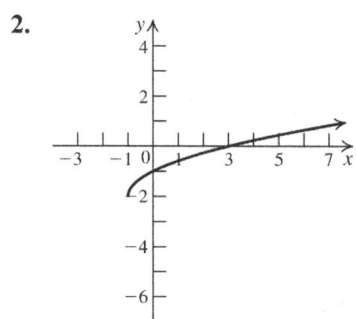

3. Switch the variables, then solve for y:
$y = 2x - 3 \Rightarrow x = 2y - 3 \Rightarrow \dfrac{x+3}{2} = y = f^{-1}(x)$
$f(f^{-1}(x)) = 2\left(\dfrac{x+3}{2}\right) - 3 = x$

4.
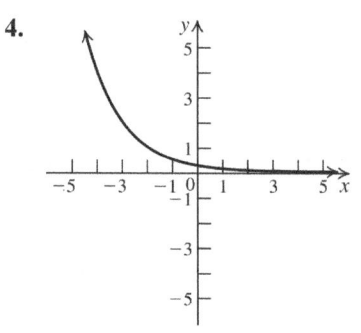

5. $\log_5(x-1) + \log_5(x-2) = 3\log_5\sqrt[3]{6} \Rightarrow$
$\log_5\big((x-1)(x-2)\big) = \log_5\left(\sqrt[3]{6}\right)^3 \Rightarrow$
$x^2 - 3x + 2 = 6 \Rightarrow x^2 - 3x - 4 = 0 \Rightarrow$
$(x-4)(x+1) = 0 \Rightarrow x = 4$ or $x = -1$. (Reject the negative solution.) The solution is $\{4\}$.

6. $\log_a \sqrt[3]{x\sqrt{yz}} = \log_a \left(x\sqrt{yz}\right)^{1/3}$
$= \dfrac{1}{3}\log_a x + \dfrac{1}{3}\log_a \sqrt{yz}$
$= \dfrac{1}{3}\log_a x + \dfrac{1}{3}\log_a (yz)^{1/2}$
$= \dfrac{1}{3}\log_a x + \dfrac{1}{3} \cdot \dfrac{1}{2}\log_a (yz)$
$= \dfrac{1}{3}\log_a x + \dfrac{1}{6}\log_a y + \dfrac{1}{6}\log_a z$

7. $\dfrac{x}{x-2} \geq 1 \Rightarrow \dfrac{x}{x-2} - 1 \geq 0 \Rightarrow \dfrac{2}{x-2} \geq 0$.
$x - 2 > 0 \Rightarrow x > 2$. The solution is $(2, \infty)$.

8. $I = \dfrac{V}{R}(1 - e^{-0.3t}) \Rightarrow 1 - \dfrac{IR}{V} = e^{-0.3t} \Rightarrow$
$\ln\left(1 - \dfrac{IR}{V}\right) = -0.3t \Rightarrow t = -\dfrac{\ln\left(1 - \dfrac{IR}{V}\right)}{0.3}$

9. Solve using elimination:
$\begin{cases} 1.4x - 0.5y = 1.3 \\ 0.4x + 1.1y = 4.1 \end{cases} \Rightarrow \begin{cases} 0.56x - 0.2y = 0.52 \\ -0.56x - 1.54y = -5.74 \end{cases} \Rightarrow$
$-1.74y = -5.22 \Rightarrow y = 3$
$0.4x + 1.1(3) = 4.1 \Rightarrow x = 2$
The solution is $\{(2, 3)\}$.

10. Switch the first and second equations:
$\begin{cases} 2x + y - 4z = 3 \\ x - 2y + 3z = 4 \\ -3x + 4y - z = -2 \end{cases} \Rightarrow \begin{cases} x - 2y + 3z = 4 \\ 2x + y - 4z = 3 \\ -3x + 4y - z = -2 \end{cases}$.

Multiply the first equation by -2, add the result to the second equation, and replace the second equation with the new equation:
$\begin{cases} x - 2y + 3z = 4 \\ 2x + y - 4z = 3 \\ -3x + 4y - z = -2 \end{cases} \Rightarrow \begin{cases} x - 2y + 3z = 4 \\ 5y - 10z = -5 \\ -3x + 4y - z = -2 \end{cases}$

Divide the second equation by 5 and replace the equation with the result. Then multiply the first equation by 3, add the result to the third equation, and replace the third equation with the new equation.
$\begin{cases} x - 2y + 3z = 4 \\ 5y - 10z = -5 \\ -3x + 4y - z = -2 \end{cases} \Rightarrow \begin{cases} x - 2y + 3z = 4 \\ y - 2z = -1 \\ -2y + 8z = 10 \end{cases}$.

(continued on next page)

(*continued*)

Divide the third equation by −2 and replace the equation with the result:
$$\begin{cases} x-2y+3z=4 \\ y-2z=-1 \\ -2y+8z=10 \end{cases} \Rightarrow \begin{cases} x-2y+3z=4 \\ y-2z=-1 \\ y-4z=-5 \end{cases}.$$
Subtract the third equation from the second equation and solve for *z*:
$$\begin{cases} x-2y+3z=4 \\ y-2z=-1 \\ y-4z=-5 \end{cases} \Rightarrow \begin{cases} x-2y+3z=4 \\ y-2z=-1 \\ z=2 \end{cases}$$
Substitute $z=2$ into the second equation and solve for *y*: $y-2(2)=-1 \Rightarrow y=3$. Substitute the values for *y* and *z* into the first equation and solve for *x*: $x-2(3)+3(2)=4 \Rightarrow x=4$.
The solution is $\{(4,3,2)\}$.

11. Solve using substitution:
$$\begin{cases} y=2-\log x \\ y-\log(x+3)=1 \end{cases} \Rightarrow$$
$2-\log x - \log(x+3)=1 \Rightarrow$
$\log x + \log(x+3)=1 \Rightarrow \log(x(x+3))=1 \Rightarrow$
$x^2+3x=10 \Rightarrow x^2+3x-10=0 \Rightarrow$
$(x+5)(x-2)=0 \Rightarrow x=-5$ or $x=2$
(Reject the negative solution.)
The solution is $\{(2, 2-\log 2)\}$.

12. Solve using substitution:
$$\begin{cases} y=x^2-1 \\ 3x^2+8y^2=8 \end{cases} \Rightarrow \begin{cases} y+1=x^2 \\ 3x^2+8y^2=8 \end{cases} \Rightarrow$$
$3(y+1)+8y^2=8 \Rightarrow 8y^2+3y-5=0 \Rightarrow$
$(y+1)(8y-5)=0 \Rightarrow y=-1$ or $y=\dfrac{5}{8}$
$-1=x^2-1 \Rightarrow x^2=0 \Rightarrow x=0$
$\dfrac{5}{8}=x^2-1 \Rightarrow \dfrac{13}{8}=x^2 \Rightarrow x=\pm\dfrac{\sqrt{26}}{4}$
The solutions are $\left\{(0,-1),\left(\dfrac{\sqrt{26}}{4},\dfrac{5}{8}\right),\left(-\dfrac{\sqrt{26}}{4},\dfrac{5}{8}\right)\right\}$.

13. Expand by the first column:
$$\begin{vmatrix} 1 & 4 & 7 \\ 2 & 5 & 8 \\ 3 & 6 & 9 \end{vmatrix} = a_{11}A_{11}+a_{21}A_{21}+a_{31}A_{31}$$
$$= 1(-1)^{1+1}\begin{vmatrix} 5 & 8 \\ 6 & 9 \end{vmatrix} + 2(-1)^{2+1}\begin{vmatrix} 4 & 7 \\ 6 & 9 \end{vmatrix}$$
$$+ 3(-1)^{3+1}\begin{vmatrix} 4 & 7 \\ 5 & 8 \end{vmatrix}$$
$$= -3 - 2(-6) + 3(-3) = 0$$

14. $\begin{cases} 2x-3y=-4 \\ 5x+7y=1 \end{cases} \Rightarrow D = \begin{vmatrix} 2 & -3 \\ 5 & 7 \end{vmatrix} = 29$
$D_x = \begin{vmatrix} -4 & -3 \\ 1 & 7 \end{vmatrix} = -25,\ D_y = \begin{vmatrix} 2 & -4 \\ 5 & 1 \end{vmatrix} = 22$
$x = -\dfrac{25}{29},\ y = \dfrac{22}{29}$

15. $A = \begin{vmatrix} 3 & -2 \\ -5 & 4 \end{vmatrix} \Rightarrow A^{-1} = \dfrac{1}{12-10}\begin{vmatrix} 4 & 2 \\ 5 & 3 \end{vmatrix} = \begin{vmatrix} 2 & 1 \\ \dfrac{5}{2} & \dfrac{3}{2} \end{vmatrix}$

16. To find the point of intersection, solve the system
$$\begin{cases} x+2y-3=0 \\ 3x+4y-5=0 \end{cases} \Rightarrow \begin{cases} -2x-4y+6=0 \\ 3x+4y-5=0 \end{cases} \Rightarrow$$
$x=-1; -1+2y-3=0 \Rightarrow y=2$
The point of intersection is $(-1, 2)$. The line $x-3y+5=0 \Rightarrow y=\dfrac{1}{3}x+\dfrac{5}{3} \Rightarrow$ its slope is $1/3$. The slope of the perpendicular is -3. The equation of the line through $(-1, 2)$ with slope -3 is $y-2=-3(x+1) \Rightarrow y=-3x-1$.

17. Let $u=x^2$. Then the equation becomes
$2u^2-5u+3=0 \Rightarrow (2u-3)(u-1)=0 \Rightarrow$
$u=\dfrac{3}{2}$ or $u=1$. Now solve for *x*:
$x^2=\dfrac{3}{2} \Rightarrow x=\pm\dfrac{\sqrt{6}}{2}; x^2=1 \Rightarrow x=\pm1$.
The solutions are $\left\{\pm\dfrac{\sqrt{6}}{2}, \pm 1\right\}$.

18. Hyperbola

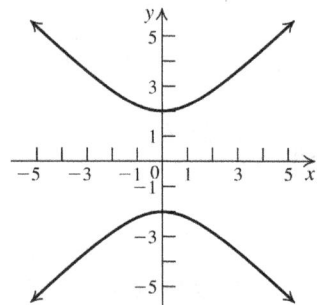

19. Circle

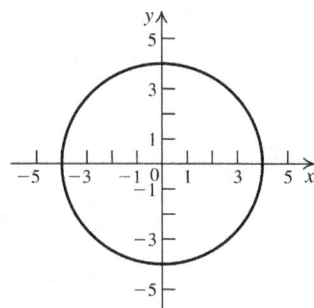

20.

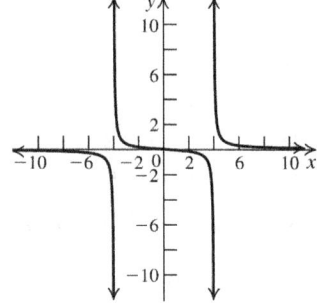

Chapter 11 Further Topics in Algebra

11.1 Sequences and Series

11.1 Practice Problems

1. $a_n = -2^n \Rightarrow a_1 = -2^1 = -2;\; a_2 = -2^2 = -4;$
   $a_3 = -2^3 = -8;\; a_4 = -2^4 = -16$

2. $a_n = 1 + \left(\dfrac{1}{2}\right)^n$

   $a_1 = 1 + \left(\dfrac{1}{2}\right)^1 = \dfrac{3}{2}$

   $a_2 = 1 + \left(\dfrac{1}{2}\right)^2 = \dfrac{5}{4}$

   $a_3 = 1 + \left(\dfrac{1}{2}\right)^3 = \dfrac{9}{8}$

   $a_4 = 1 + \left(\dfrac{1}{2}\right)^4 = \dfrac{17}{16}$

   $a_5 = 1 + \left(\dfrac{1}{2}\right)^5 = \dfrac{33}{32}$

   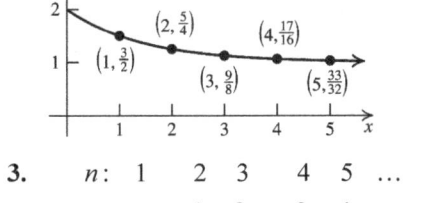

3. n: 1, 2, 3, 4, 5, ..., n

   term: $0,\; -\dfrac{1}{2},\; \dfrac{2}{3},\; -\dfrac{3}{4},\; \dfrac{4}{5},\; \ldots,\; a_n$

   The terms alternate, and the negative terms are the even-numbered terms, so use the factor $(-1)^{n+1}$ to alternate the signs. The absolute value of each term is $\dfrac{n-1}{n}$, so the general term is $(-1)^{n+1}\left(\dfrac{n-1}{n}\right)$ or $(-1)^{n+1}\left(1 - \dfrac{1}{n}\right)$.

4. $a_1 = -3,\; a_{n+1} = 2a_n + 5$
   $a_2 = 2a_1 + 5 = 2(-3) + 5 = -1$
   $a_3 = 2a_2 + 5 = 2(-1) + 5 = 3$
   $a_4 = 2a_3 + 5 = 2(3) + 5 = 11$
   $a_5 = 2a_4 + 5 = 2(11) + 5 = 27$

5. The female bee is a_0. Since female bees have two parents, $a_1 = 2$. The female parent has two parents and the male parent has one, so $a_2 = 3$. There are two females and one male in this generation, so the next generation has three females and two males. We can use a tree to represent this:

   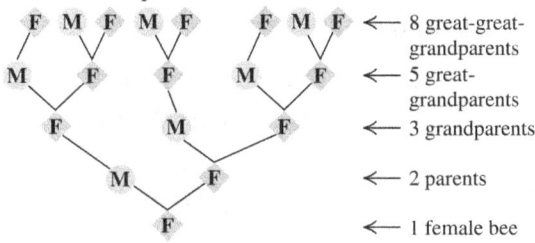

   $a_k = a_{k-2} + a_{k-1}$ for $k \geq 3$.

6. a. $\dfrac{13!}{12!} = \dfrac{13 \cdot 12!}{12!} = 13$

   b. $\dfrac{n!}{(n-3)!} = \dfrac{n(n-1)(n-2)(n-3)!}{(n-3)!}$
   $= n(n-1)(n-2)$

7. $a_n = \dfrac{(-1)^n 2^n}{n!} \Rightarrow a_1 = \dfrac{(-1)^1 2^1}{1!} = -2$

   $a_2 = \dfrac{(-1)^2 2^2}{2!} = \dfrac{4}{2 \cdot 1} = 2$

   $a_3 = \dfrac{(-1)^3 2^3}{3!} = \dfrac{-8}{3 \cdot 2 \cdot 1} = -\dfrac{8}{6} = -\dfrac{4}{3}$

   $a_4 = \dfrac{(-1)^4 2^4}{4!} = \dfrac{16}{4 \cdot 3 \cdot 2 \cdot 1} = \dfrac{16}{24} = \dfrac{2}{3}$

   $a_5 = \dfrac{(-1)^5 2^5}{5!} = \dfrac{-32}{5 \cdot 4 \cdot 3 \cdot 2 \cdot 1} = -\dfrac{32}{120} = -\dfrac{4}{15}$

8. $\displaystyle\sum_{k=0}^{3} (-1)^k k!$
   $= (-1)^0(0!) + (-1)^1(1!) + (-1)^2(2!) + (-1)^3(3!)$
   $= 1 + (-1) + (2 \cdot 1) + (-1)(3 \cdot 2 \cdot 1) = -4$

9. $2 - 4 + 6 - 8 + 10 - 12 + 14$ alternates addition and subtraction of consecutive even integers from 2 to 14. The even integers can be represented as $2k$ with k ranging from 1 to 7. We alternate the signs using the factor $(-1)^{k+1}$. Thus, the expression is
   $\displaystyle\sum_{k=1}^{7} (-1)^{k+1}(2k).$

618 Chapter 11 Further Topics in Algebra

11.1 Concepts and Vocabulary

1. An infinite sequence is a function whose domain is the set of <u>positive integers</u>.

3. By definition, 0! = <u>1</u>.

5. False. $a_1 = 3$, $a_2 = a_1^2 = 3^2 = 9$, $a_3 = a_2^2 = 9^2 = 81$

7. True

11.1 Building Skills

9. $a_1 = 3(1) - 2 = 1, a_2 = 3(2) - 2 = 4,$
$a_3 = 3(3) - 2 = 7, a_4 = 3(4) - 2 = 10$

11. $a_1 = 1 - \frac{1}{1} = 0, a_2 = 1 - \frac{1}{2} = \frac{1}{2},$
$a_3 = 1 - \frac{1}{3} = \frac{2}{3}, a_4 = 1 - \frac{1}{4} = \frac{3}{4}$

13. $a_1 = -1^2 = -1, a_2 = -2^2 = -4,$
$a_3 = -3^2 = -9, a_4 = -4^2 = -16$

15. $a_1 = \frac{2(1)}{1+1} = 1, a_2 = \frac{2(2)}{2+1} = \frac{4}{3}, a_3 = \frac{2(3)}{3+1} = \frac{3}{2},$
$a_4 = \frac{2(4)}{4+1} = \frac{8}{5}$

17. $a_1 = (-1)^{1+1} = 1, a_2 = (-1)^{2+1} = -1,$
$a_3 = (-1)^{3+1} = 1, a_4 = (-1)^{4+1} = -1$

19. $a_1 = 3 - \frac{1}{2^1} = \frac{5}{2}, a_2 = 3 - \frac{1}{2^2} = \frac{11}{4},$
$a_3 = 3 - \frac{1}{2^3} = \frac{23}{8}, a_4 = 3 - \frac{1}{2^4} = \frac{47}{16}$

21. $a_1 = a_2 = a_3 = a_4 = 0.6$

23. $a_1 = \frac{(-1)^1}{1!} = -1, a_2 = \frac{(-1)^2}{2!} = \frac{1}{2},$
$a_3 = \frac{(-1)^3}{3!} = -\frac{1}{6}, a_4 = \frac{(-1)^4}{4!} = \frac{1}{24}$

25. $a_1 = (-1)^1 3^{-1} = -\frac{1}{3}, a_2 = (-1)^2 3^{-2} = \frac{1}{9},$
$a_3 = (-1)^3 3^{-3} = -\frac{1}{27}, a_4 = (-1)^4 3^{-4} = \frac{1}{81}$

27. $a_1 = \frac{e^1}{2(1)} = \frac{e}{2}, a_2 = \frac{e^2}{2(2)} = \frac{e^2}{4}, a_3 = \frac{e^3}{2(3)} = \frac{e^3}{6},$
$a_4 = \frac{e^4}{2(4)} = \frac{e^4}{8}$

29. $a_n = 2n - 3$
$a_1 = 2(1) - 3 = -1 \quad a_2 = 2(2) - 3 = 1$
$a_3 = 2(3) - 3 = 3 \quad a_4 = 2(4) - 3 = 5$
The points to be plotted are (1, –1), (2, 1), (3, 3), and (4, 5).

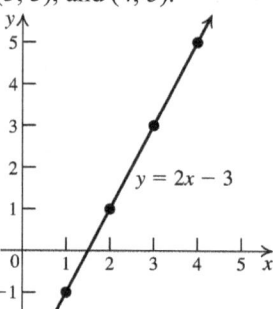

31. $a_n = \frac{1}{2}n^2 - n$
$a_1 = \frac{1}{2}(1)^2 - 1 = -\frac{1}{2} \quad a_2 = \frac{1}{2}(2)^2 - 2 = 0$
$a_3 = \frac{1}{2}(3)^2 - 3 = \frac{3}{2} \quad a_4 = \frac{1}{2}(4)^2 - 4 = 4$
The points to be plotted are $\left(1, -\frac{1}{2}\right)$, (2, 0), $\left(3, \frac{3}{2}\right)$, (4, 4).

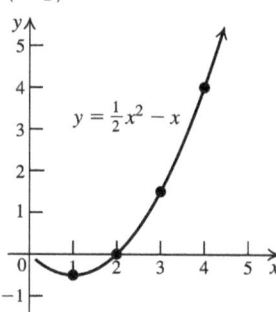

33. $a_n = (n-2)^2 - 1$
$a_1 = (1-2)^2 - 1 = 0 \quad a_2 = (2-2)^2 - 1 = -1$
$a_3 = (3-2)^2 - 1 = 0 \quad a_4 = (4-2)^2 - 1 = 3$
The points to be plotted are (1, 0), (2, –1), (3, 0), (4, 3).

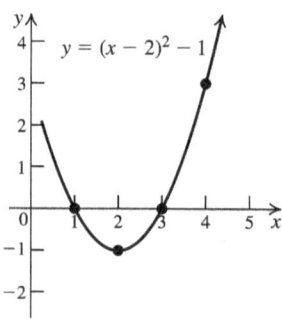

Copyright © 2019 Pearson Education Inc.

35. $a_n = 3 - \dfrac{1}{n}$

$a_1 = 3 - \dfrac{1}{1} = 2 \quad a_2 = 3 - \dfrac{1}{2} = \dfrac{5}{2}$

$a_3 = 3 - \dfrac{1}{3} = \dfrac{8}{3} \quad a_4 = 3 - \dfrac{1}{4} = \dfrac{11}{4}$

The points to be plotted are $(1, 2)$, $\left(2, \tfrac{5}{2}\right)$, $\left(3, \tfrac{8}{3}\right)$, $\left(4, \tfrac{11}{4}\right)$.

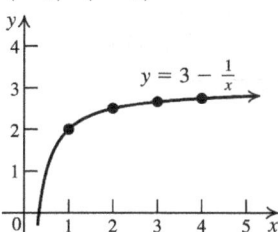

37. $a_n = 2^{n-1} - 3$

$a_1 = 2^{1-1} - 3 = -2 \quad a_2 = 2^{2-1} - 3 = -1$

$a_3 = 2^{3-1} - 3 = 1 \quad a_4 = 2^{4-1} - 3 = 5$

The points to be plotted are $(1, -2)$, $(2, -1)$, $(3, 1)$, and $(4, 5)$.

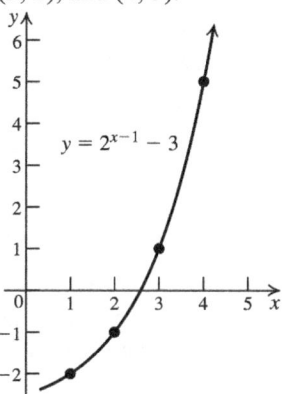

39. $a_n = 3n - 2$

41. $a_n = \dfrac{1}{n+1}$

43. $a_n = (-1)^{n-1}(2)$

45. $a_n = \dfrac{3^{n-1}}{2^n}$

47. $a_n = n(n+1)$

49. $a_n = 2 - \dfrac{(-1)^n}{n+1}$

51. $a_n = \dfrac{3^{n+1}}{n+1}$

53. $a_1 = 2, a_2 = 2 + 3 = 5, a_3 = 5 + 3 = 8,$
$a_4 = 8 + 3 = 11, a_5 = 11 + 3 = 14$

55. $a_1 = 3, a_2 = 2(3) = 6, a_3 = 2(6) = 12,$
$a_4 = 2(12) = 24, a_5 = 2(24) = 48$

57. $a_1 = 7, a_2 = -2(7) + 3 = -11,$
$a_3 = -2(-11) + 3 = 25, a_4 = -2(25) + 3 = -47,$
$a_5 = -2(-47) + 3 = 97$

59. $a_1 = 2, a_2 = \dfrac{1}{2}, a_3 = \dfrac{1}{1/2} = 2, a_4 = \dfrac{1}{2},$
$a_5 = \dfrac{1}{1/2} = 2$

61. $a_1 = 25, a_2 = \dfrac{(-1)^1}{5(25)} = -\dfrac{1}{125},$
$a_3 = \dfrac{(-1)^2}{5(-1/125)} = -25, a_4 = \dfrac{(-1)^3}{5(-25)} = \dfrac{1}{125},$
$a_5 = \dfrac{(-1)^4}{5(1/125)} = 25$

In exercises 63–68, we show only the first screen to determine the terms of the sequence before we list the ten terms.

63. a.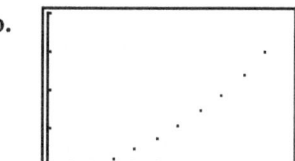

2, 11, 26, 47, 74, 107, 146, 191, 242, 299

b.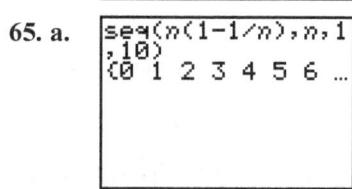

65. a.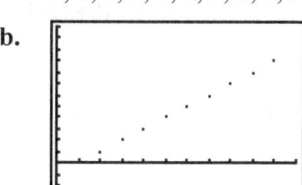

0, 1, 2, 3, 4, 5, 6, 7, 8, 9

b.

67. a.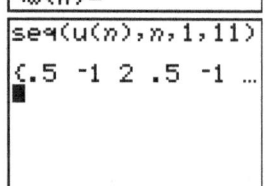

0.5, −1, 2, 0.5, −1, 2, 0.5, −1, 2, 0.5

b.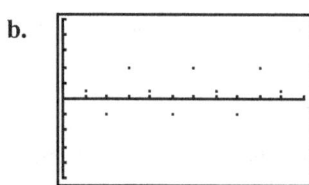

69. $\dfrac{3!}{5!} = \dfrac{3!}{5 \cdot 4 \cdot 3!} = \dfrac{1}{20}$

71. $\dfrac{12!}{11!} = \dfrac{12 \cdot 11!}{11!} = 12$

73. $\dfrac{n!}{(n+1)!} = \dfrac{n!}{(n+1) \cdot n!} = \dfrac{1}{n+1}$

75. $\dfrac{(2n+1)!}{(2n)!} = \dfrac{(2n+1) \cdot (2n)!}{(2n)!} = 2n+1$

77. $\sum\limits_{k=1}^{7} 5 = 5+5+5+5+5+5+5 = 35$

79. $\sum\limits_{j=0}^{5} j^2 = 0^2+1^2+2^2+3^2+4^2+5^2 = 55$

81. $\sum\limits_{i=1}^{5}(2i-1) = (2(1)-1)+(2(2)-1)+(2(3)-1)$
$\qquad\qquad\qquad\qquad + (2(4)-1)+(2(5)-1) = 25$

83. $\sum\limits_{j=3}^{7} \dfrac{j+1}{j} = \dfrac{4}{3}+\dfrac{5}{4}+\dfrac{6}{5}+\dfrac{7}{6}+\dfrac{8}{7} = \dfrac{853}{140}$

85. $\sum\limits_{i=2}^{6}(-1)^i 3^{i-1}$
$= (-1)^2(3^1)+(-1)^3(3^2)+(-1)^4(3^3)$
$\qquad + (-1)^5(3^4)+(-1)^6(3^5)$
$= 183$

87. $\sum\limits_{k=4}^{7}(2-k^2)$
$= (2-4^2)+(2-5^2)+(2-6^2)+(2-7^2)$
$= -118$

89. $\sum\limits_{k=1}^{51}(2k-1)$ **91.** $\sum\limits_{k=1}^{11} \dfrac{1}{5k}$

93. $\sum\limits_{k=1}^{50} \dfrac{(-1)^{k+1}}{k}$ **95.** $\sum\limits_{k=1}^{10} \dfrac{k}{k+1}$

97. sum(seq(12n²,n,1,10)) = 4620

99. sum(seq(7/(1−n²),n,5,30)) = −1.345430108

101. sum(seq((−1)^n(n),n,1,100)) = 50

11.1 Applying the Concepts

103. a. Use a table to determine the pattern. It appears that each second, the body falls 32 feet per second faster than it did in the second before.

Time	Distance each second
1	16
2	48
3	80
4	112
5	144
6	176
7	208

b. $16 + 32(n-1) = 32n - 16$ feet

105. The fine is $50 plus $25 for each additional day the work is not done. So on the ninth day, she will be fined $50 + 8($25) = $250.

107. In three years, there are 6 six-month periods. For the first six-month period, cell phone usage was 600 minutes per month. For the next 5 six-month periods, the usage doubled during each period. At the end of three years, cell phone use was $600 \cdot 2^5 = 19,200$ minutes per month.

109. a. $A_1 = 10,000\left(1+\dfrac{0.06}{2}\right)^1 = \$10,300$

$A_2 = 10,000\left(1+\dfrac{0.06}{2}\right)^2 = \$10,609$

$A_3 = 10,000\left(1+\dfrac{0.06}{2}\right)^3 = \$10,927.27$

$A_4 = 10,000\left(1+\dfrac{0.06}{2}\right)^4 = \$11,255.09$

$A_5 = 10,000\left(1+\dfrac{0.06}{2}\right)^5 = \$11,592.74$

$A_6 = 10,000\left(1+\dfrac{0.06}{2}\right)^6 = \$11,940.52$

b. The interest is compounded semiannually, so in eight years, there are 16 compounding periods.

$A_{16} = 100\left(1+\dfrac{0.08}{4}\right)^{16} = \$16,047.10$

111. $A_1 = (100,000)(1.05) = \$105,000$
$A_2 = (105,000)(1.05) = \$110,250$
$A_3 = (110,250)(1.05) = \$115,762.50$
$A_4 = (115,762.50)(1.05) = \$121,550.63$
$A_5 = (121,550.63)(1.05) = \$127,628.16$
$A_6 = (127,628.16)(1.05) = \$134,009.56$
$A_7 = (134,009.56)(1.05) = \$140,710.04$

The formula is $100,000\left(1.05^n\right)$.

11.1 Beyond the Basics

113. $a_1 = 2^{1/2}, a_2 = \left(2 \cdot 2^{1/2}\right)^{1/2} = 2^{3/4}$,

$a_3 = \left(2 \cdot 2^{3/4}\right)^{1/2} = 2^{7/8}$,

$a_4 = \left(2 \cdot 2^{7/8}\right)^{1/2} = 2^{15/16}$,

$a_5 = \left(2 \cdot 2^{15/16}\right)^{1/2} = 2^{31/32}$

$a_n = 2^{(2^n-1)/2^n} = 2^{1-1/2^n}$

115. a. $a_1 = 1, a_2 = \dfrac{1}{2}, a_3 = \dfrac{1}{4}, a_4 = \dfrac{1}{8}, a_5 = \dfrac{1}{16}$

b. $a_n = \dfrac{1}{2^{n-1}}$

117. $a_1 = a_2 = 1$,
$a_3 = a_{a_2} + a_{3-a_2} = a_1 + a_2 = 1+1 = 2$,
$a_4 = a_{a_3} + a_{4-a_3} = a_2 + a_2 = 1+1 = 2$,
$a_5 = a_{a_4} + a_{5-a_4} = a_2 + a_3 = 1+2 = 3$,
$a_6 = a_{a_5} + a_{6-a_5} = a_3 + a_3 = 2+2 = 4$,
$a_7 = a_{a_6} + a_{7-a_6} = a_4 + a_3 = 2+2 = 4$,
$a_8 = a_{a_7} + a_{8-a_7} = a_4 + a_4 = 2+2 = 4$,
$a_9 = a_{a_8} + a_{9-a_8} = a_4 + a_5 = 2+3 = 5$,
$a_{10} = a_{a_9} + a_{10-a_9} = a_5 + a_5 = 3+3 = 6$

119. $\displaystyle\sum_{n=1}^{20} n^2 = \sum_{m=0}^{19} a_m$

Write the first five terms to determine the pattern: $\displaystyle\sum_{n=1}^{1} n^2 = 1 = \sum_{m=0}^{0} a_m = a_0 \Rightarrow a_0 = 1$,

$\displaystyle\sum_{n=1}^{2} n^2 = 1+2^2 = 5 = \sum_{m=0}^{1} a_m = a_0 + a_1 \Rightarrow$
$5 = 1+a_1 \Rightarrow a_1 = 4 = 2^2$.

$\displaystyle\sum_{n=1}^{3} n^2 = 1+2^2+3^2 = 14$

$= \displaystyle\sum_{m=0}^{2} a_m = a_0 + a_1 + a_2 \Rightarrow$
$14 = 1+4+a_2 \Rightarrow a_2 = 9 = 3^2$

$\displaystyle\sum_{n=1}^{4} n^2 = 1+2^2+3^2+4^2 = 30$

$= \displaystyle\sum_{m=0}^{3} a_m = a_0 + a_1 + a_2 + a_3 \Rightarrow$
$30 = 1+4+9+a_3 \Rightarrow a_3 = 16 = 4^2$

$\displaystyle\sum_{n=1}^{5} n^2 = 1+2^2+3^2+4^2+5^2 = 55$

$= \displaystyle\sum_{m=0}^{4} a_m = a_0 + a_1 + a_2 + a_3 + a_4 \Rightarrow$
$55 = 1+4+9+16+a_4 \Rightarrow a_4 = 25 = 5^2$

Following the pattern, we see that
$a_m = (m+1)^2$.

121. $\displaystyle\sum_{n=0}^{10} n^3 = \sum_{m=p}^{q} (m-2)^3 \Rightarrow n^3 = (m-2)^3 \Rightarrow$
$n = m-2 \Rightarrow m = n+2$. So, $p = 2$ and $q = 12$.

123. $7! - 2(5!) = (7 \cdot 6)(5!) - 2(5!)$
$= (42-2)(5!) = 40(5!)$

125. $\dfrac{1}{5!}+\dfrac{1}{6!}=\dfrac{x}{7!}\Rightarrow \dfrac{1}{5!}+\dfrac{1}{6\cdot 5!}=\dfrac{x}{7\cdot 6\cdot 5!}$

Multiply both sides of the equation by $7!=7\cdot 6\cdot 5!$ and simplify. Then solve for x.

$\dfrac{7\cdot 6\cdot 5!}{5!}+\dfrac{7\cdot 6\cdot 5!}{6\cdot 5!}=\dfrac{7\cdot 6\cdot 5!\,x}{7\cdot 6\cdot 5!}\Rightarrow 42+7=x\Rightarrow x=49$

11.1 Critical Thinking/Discussion/Writing

127.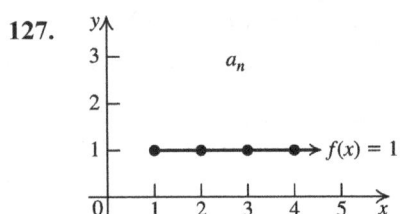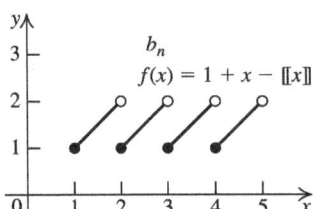

The general term formula $a_n=1$ better represents the fact that the given sequence is constant.

128. $a_n=1+\dfrac{(-1)^n}{n}$

$a_1=1+\dfrac{(-1)^1}{1}=0 \qquad a_3=1+\dfrac{(-1)^3}{3}=\dfrac{2}{3} \qquad a_5=1+\dfrac{(-1)^5}{5}=\dfrac{4}{5} \qquad a_7=1+\dfrac{(-1)^7}{7}=\dfrac{6}{7} \qquad a_9=1+\dfrac{(-1)^9}{9}=\dfrac{8}{9}$

$a_2=1+\dfrac{(-1)^2}{2}=\dfrac{3}{2} \qquad a_4=1+\dfrac{(-1)^4}{4}=\dfrac{5}{4} \qquad a_6=1+\dfrac{(-1)^6}{6}=\dfrac{7}{6} \qquad a_8=1+\dfrac{(-1)^8}{8}=\dfrac{9}{8} \qquad a_{10}=1+\dfrac{(-1)^{10}}{10}=\dfrac{11}{10}$

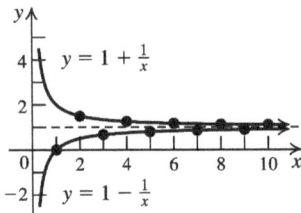

The points representing a_n alternate between $f(x)=1-\dfrac{1}{x}$ and $g(x)=1+\dfrac{1}{x}$. As $n\to\infty$, the values of a_n approach 1.

129. $n^2+n=(n-1)(n-2)(n-3)(n-4)(n-5)+n^2+n\Rightarrow 0=(n-1)(n-2)(n-3)(n-4)(n-5)\Rightarrow n=1,2,3,4,5$, so $k=5$ is the upper limit.

130. If $a_0=13$, then $a_1=3(13)+1=40,\ a_2=a_1/2=40/2=20,\ a_3=a_2/2=20/2=10,\ a_4=a_3/2=10/2=5$,
$a_5=3a_4+1=3(5)+1=16,\ a_6=a_5/2=16/2=8,\ a_7=a_6/2=8/2=4,\ a_8=a_7/2=4/2=2,\,.$
$a_9=a_8/2=2/2=1$
So the conjecture is true for $a_0=13$.

131. $\dfrac{(2n)!}{n!}=\dfrac{(2n)(2n-1)(2n-2)(2n-3)(2n-4)(2n-5)\cdots}{n(n-1)(n-2)\cdots}=\dfrac{2n(2n-1)2(n-1)(2n-3)2(n-2)(2n-5)\cdots}{n(n-1)(n-2)\cdots}$
$=2^n(2n-1)(2n-3)(2n-5)\cdots$

Note that the odd numbers are represented by $2n-1,\ 2n-3,\ 2n-5$, etc. Therefore,

$\dfrac{(2n)!}{n!}=2^n(2n-1)(2n-3)(2n-5)\cdots=2^n\left[1\cdot 3\cdot 5\cdot (2n-1)\right].$

132. $n!$ is divisible by all integers between 2 and n (from the definition of $n!$). So, if we divide $(n!+1)$ by any integer between 2 and n, we will end up with remainder 1. Hence, $(n!+1)$ is not divisible by any integer between 2 and n.

11.1 Getting Ready for the Next Section

133. $a_2 - a_1 = 5 - 1 = 4$
$a_3 - a_2 = 9 - 5 = 4$
$a_4 - a_3 = 13 - 9 = 4$
$a_5 - a_4 = 17 - 13 = 4$

135. $a_2 - a_1 = 1 - 6 = -5$
$a_3 - a_2 = -4 - 1 = -5$
$a_4 - a_3 = -9 - (-4) = -5$
$a_5 - a_4 = -14 - (-9) = -5$

137. $a_2 = a_1 + d = 3 + 4 = 7$
$a_3 = a_2 + d = 7 + 4 = 11$
$a_4 = a_3 + d = 11 + 4 = 15$
$a_5 = a_4 + d = 15 + 4 = 19$

139. $a_2 = a_1 + d = 7 + (-3) = 4$
$a_3 = a_2 + d = 4 + (-3) = 1$
$a_4 = a_3 + d = 1 + (-3) = -2$
$a_5 = a_4 + d = -2 + (-3) = -5$

141. $a_n = 3n + 5$
$a_1 = 3(1) + 5 = 8$
$a_2 = 3(2) + 5 = 11$
$a_3 = 3(3) + 5 = 14$
$a_4 = 3(4) + 5 = 17$
$a_5 = 3(5) + 5 = 20$
$a_2 - a_1 = 11 - 8 = 3$
$a_3 - a_2 = 14 - 11 = 3$
$a_4 - a_3 = 17 - 14 = 3$
$a_5 - a_4 = 20 - 17 = 3$
Note that
$a_n - a_{n-1} = (3n + 5) - (3(n-1) + 5)$
$= (3n + 5) - (3n - 3 + 5)$
$= (3n + 5) - (3n + 2) = 3$

143. $a_n = -2n - 6$
$a_1 = -2(1) - 6 = -8$
$a_2 = -2(2) - 6 = -10$
$a_3 = -2(3) - 6 = -12$
$a_4 = -2(4) - 6 = -14$
$a_5 = -2(5) - 6 = -16$
$a_2 - a_1 = -10 - (-8) = -2$
$a_3 - a_2 = -12 - (-10) = -2$
$a_4 - a_3 = -14 - (-12) = -2$
$a_5 - a_4 = -16 - (-14) = -2$

Note that
$a_n - a_{n-1} = (-2n - 6) - (-2(n-1) - 6)$
$= (-2n - 6) - (-2n + 2 - 6)$
$= (-2n - 6) - (-2n - 4) = -2$

145. $a_n = 2n - 5$
$a_{n+1} = 2(n+1) - 5 = 2n + 2 - 5 = 2n - 3$
$a_{n-1} = 2(n-1) - 5 = 2n - 2 - 5 = 2n - 7$

11.2 Arithmetic Sequences; Partial Sums

11.2 Practice Problems

1. $(-2) - 3 = -5$, $-7 - (-2) = -5$, etc. The common difference is -5.

2. $a_1 = -3$ and $d = 1 - (-3) = 4$, so the expression is
$a_n = -3 + 4(n-1) = -3 + 4n - 4 = 4n - 7$.

3. $a_n = 7, 4, 1, -2, -5, \ldots$

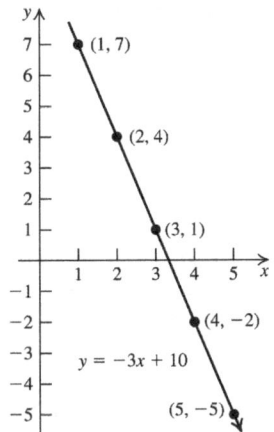

We will use the points $(1, 7)$ and $(2, 4)$ to find the linear function that produces the sequence.
$m = \dfrac{7 - 4}{1 - 2} = -3$
$y - 7 = -3(x - 1) \Rightarrow y = -3x + 10$
The general term for this sequence is
$a_n = f(n) = -3n + 10$.

4. $a_4 = a_1 + d(4 - 1) \Rightarrow 41 = a_1 + 3d$
$a_{15} = a_1 + d(15 - 1) \Rightarrow 8 = a_1 + 14d$
$\begin{cases} a_1 + 3d = 41 \\ a_1 + 14d = 8 \end{cases} \Rightarrow -11d = 33 \Rightarrow d = -3, a_1 = 50$
$a_n = 50 - 3(n - 1) = -3n + 53$

5. $d = \frac{5}{6} - \frac{2}{3} = \frac{1}{6}; n = 10; a_1 = \frac{2}{3}; a_{10} = \frac{13}{6}$

$S_{10} = 10\left(\frac{\frac{2}{3} + \frac{13}{6}}{2}\right) = \frac{85}{6}$

6. $d = 32; n = 5; a_{11} = 32(11) - 16 = 336;$
$a_{15} = 32(15) - 16 = 464$
$S = 5\left(\frac{336 + 464}{2}\right) = 2000$

The object fell 2000 feet during the 11th through 15th seconds.

11.2 Concepts and Vocabulary

1. If 5 is the common difference of an arithmetic sequence with general term a_n, then $a_{17} - a_{16} = \underline{5}$.

3. If 14 is the term immediately following the sequence term 17 is an arithmetic sequence, then the common difference is $\underline{-3}$.

5. False. The common difference of an arithmetic sequence may be positive or negative.

7. False. $a_n = a_1 + (n-1)d$

11.2 Building Skills

9. The sequence is arithmetic. $a_1 = 1, d = 1$

11. The sequence is arithmetic. $a_1 = 2, d = 3$

13. The sequence is not arithmetic.

15. The sequence is not arithmetic.

17. The sequence is arithmetic. $a_1 = 0.6, d = -0.4$

19. Write the first four terms of the sequence:
$a_1 = 2(1) + 6 = 8, a_2 = 2(2) + 6 = 10,$
$a_3 = 2(3) + 6 = 12, a_4 = 2(4) + 6 = 14$.
It appears that the sequence is arithmetic, with a common difference of 2. Verify as follows:
$d = a_{n+1} - a_n = (2(n+1) + 6) - (2n + 6) = 2$.

21. Write the first four terms of the sequence:
$a_1 = 1 - 1^2 = 0, a_2 = 1 - 2^2 = -3,$
$a_3 = 1 - 3^2 = -8, a_4 = 1 - 4^2 = -15$.
There is no common difference, so the sequence is not arithmetic.

23. $d = 3, a_1 = 5 \Rightarrow a_n = 5 + 3(n-1) = 3n + 2$

25. $d = -5, a_1 = 11 \Rightarrow a_n = 11 - 5(n-1) = 16 - 5n$

27. $d = -\frac{1}{4}, a_1 = \frac{1}{2} \Rightarrow a_n = \frac{1}{2} - \frac{1}{4}(n-1) = \frac{3-n}{4}$

29. $d = -\frac{2}{5}, a_1 = -\frac{3}{5} \Rightarrow$
$a_n = -\frac{3}{5} - \frac{2}{5}(n-1) = \frac{-1-2n}{5} = -\frac{2n+1}{5}$

31. $d = 3, a_1 = e \Rightarrow a_n = e + 3(n-1) = e + 3n - 3$

33. $a_n = -1, -3, -5, -7, -9, \ldots$

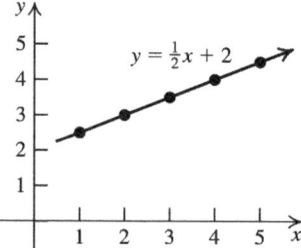

$y = -2x + 1$

We will use the points $(1, -1)$ and $(2, -3)$ to find the linear function that produces the sequence.
$m = \frac{-3 - (-1)}{2 - 1} = -2$
$y - (-1) = -2(x - 1) \Rightarrow y = -2x + 1$
The general term for this sequence is
$a_n = f(n) = -2n + 1$.

35. $a_n = \frac{5}{2}, 3, \frac{7}{2}, 4, \frac{9}{2}, \ldots$

$y = \frac{1}{2}x + 2$

We will use the points $(2, 3)$ and $(4, 4)$ to find the linear function that produces the sequence.
$m = \frac{4 - 3}{4 - 2} = \frac{1}{2}$
$y - 3 = \frac{1}{2}(x - 2) \Rightarrow y = \frac{1}{2}x + 2$
The general term for this sequence is
$a_n = f(n) = \frac{1}{2}n + 2$.

37. $a_n = -\dfrac{5}{4}, -\dfrac{1}{2}, \dfrac{1}{4}, 1, \dfrac{7}{4}, \ldots$

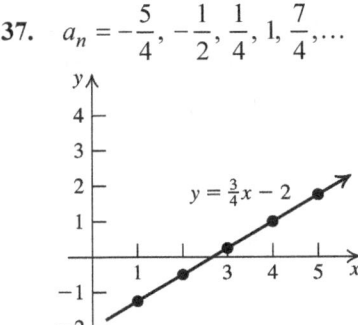

We will use the points $\left(3, \dfrac{1}{4}\right)$ and $\left(5, \dfrac{7}{4}\right)$ to find the linear function that produces the sequence.

$m = \dfrac{\frac{7}{4} - \frac{1}{4}}{5 - 3} = \dfrac{\frac{3}{2}}{2} = \dfrac{3}{4}$

$y - \dfrac{1}{4} = \dfrac{3}{4}(x - 3) \Rightarrow y = \dfrac{3}{4}x - 2$

The general term for this sequence is

$a_n = f(n) = \dfrac{3}{4}n - 2.$

39. $a_4 = a_1 + d(4-1) \Rightarrow 21 = a_1 + 3d$
$a_{10} = a_1 + d(10-1) \Rightarrow 60 = a_1 + 9d$

$\begin{cases} a_1 + 3d = 21 \\ a_1 + 9d = 60 \end{cases} \Rightarrow -6d = -39 \Rightarrow d = \dfrac{13}{2}, a_1 = \dfrac{3}{2}$

$a_n = \dfrac{3}{2} + \dfrac{13}{2}(n-1) = \dfrac{13}{2}n - 5$

41. $a_7 = a_1 + d(7-1) \Rightarrow 8 = a_1 + 6d$
$a_{15} = a_1 + d(15-1) \Rightarrow -8 = a_1 + 14d$

$\begin{cases} a_1 + 6d = 8 \\ a_1 + 14d = -8 \end{cases} \Rightarrow 8d = -16 \Rightarrow d = -2, a_1 = 20$

$a_n = 20 - 2(n-1) = 22 - 2n$

43. $a_3 = a_1 + d(3-1) \Rightarrow 7 = a_1 + 2d$
$a_{23} = a_1 + d(23-1) \Rightarrow 17 = a_1 + 22d$

$\begin{cases} a_1 + 2d = 7 \\ a_1 + 22d = 17 \end{cases} \Rightarrow 20d = 10 \Rightarrow d = \dfrac{1}{2}, a_1 = 6$

$a_n = 6 + \dfrac{1}{2}(n-1) = \dfrac{n+11}{2}$

45. $n = 50; S = 50\left(\dfrac{1+50}{2}\right) = 1275$

47. $d = 2, a_1 = 1, a_n = 99 \Rightarrow 99 = 1 + 2(n-1) \Rightarrow$
$n = 50; S = 50\left(\dfrac{1+99}{2}\right) = 2500$

49. $d = 3, a_1 = 3, a_n = 300 \Rightarrow 300 = 3 + 3(n-1) \Rightarrow$
$n = 100; S = 100\left(\dfrac{3+300}{2}\right) = 15,150$

51. $d = -3, a_1 = 2, a_n = -34 \Rightarrow$
$-34 = 2 - 3(n-1) \Rightarrow n = 13$
$S = 13\left(\dfrac{2-34}{2}\right) = -208$

53. $d = \dfrac{2}{3}, a_1 = \dfrac{1}{3}, a_n = 7 \Rightarrow$
$7 = \dfrac{1}{3} + \dfrac{2}{3}(n-1) \Rightarrow n = 11$
$S = 11\left(\dfrac{1/3 + 7}{2}\right) = \dfrac{121}{3}$

55. $d = 5, n = 50, a_1 = 2 \Rightarrow a_n = 2 + 5(50-1) = 247$
$S = 50\left(\dfrac{2+247}{2}\right) = 6225$

57. $d = 4, n = 20, a_1 = -15 \Rightarrow$
$a_n = -15 + 4(20-1) = 61$
$S = 20\left(\dfrac{-15+61}{2}\right) = 460$

59. $d = 0.2, n = 100, a_1 = 3.5 \Rightarrow$
$a_n = 3.5 + 0.2(100-1) = 23.3$
$S = 100\left(\dfrac{3.5+23.3}{2}\right) = 1340$

In exercises 61–66, use the formula for finding the terms in an arithmetic sequence, $a_n = a_1 + d(n-1)$.

61. $a_n = 75, a_1 = 1, d = 2$
$75 = 1 + 2(n-1) \Rightarrow 37 = n - 1 \Rightarrow n = 38$

63. $a_n = 95, a_1 = -1, d = 4$
$95 = -1 + 4(n-1) \Rightarrow 24 = n - 1 \Rightarrow n = 25$

65. $a_n = 50\sqrt{3}, a_1 = 2\sqrt{3}, d = 2\sqrt{3}$
$50\sqrt{3} = 2\sqrt{3} + 2\sqrt{3}(n-1) \Rightarrow$
$48\sqrt{3} = 2\sqrt{3}(n-1) \Rightarrow 24 = n - 1 \Rightarrow n = 25$

11.2 Applying the Concepts

67. $d = 2, n = 30, a_1 = 10 \Rightarrow a_n = 10 + 2(30-1) = 68$
$S = 30\left(\dfrac{10+68}{2}\right) = 1170$

69. $d = 25, n = 30, a_1 = 50 \Rightarrow$
$a_{30} = 50 + 25(30-1) = 775$
$S = 30\left(\dfrac{50+775}{2}\right) = \$12,375$

71. Antonio has 17 different hourly wages over the four years (his original wage plus 16 raises), so $n = 17$. $a_1 = \$12.75, d = 0.25 \Rightarrow$
$a_{13} = 12.75 + 0.25(17 - 1) = \$16.75$.

73. $n = 25, a_1 = 20, d = 2 \Rightarrow$
$a_{25} = 20 + 2(25 - 1) = 68$
$S = 25\left(\dfrac{20 + 68}{2}\right) = 1100$.
There are 1100 seats in the theater.

75. $n = 17, a_1 = 40, a_{28} = 8 \Rightarrow$
$S = 17\left(\dfrac{40 + 8}{2}\right) = 408$
There are 408 boxes in the stack.

77. Using the formula for the sum of an arithmetic sequence with $a_1 = 3$, we have
$192 = n\left(\dfrac{3 + a_n}{2}\right) \Rightarrow 384 = 3n + na_n$. The last term in the sequence, given $d = 6$, is
$a_n = 3 + 6(n - 1) = 6n - 3$. Solve the system:
$\begin{cases} 3n + na_n = 384 \\ a_n = 6n - 3 \end{cases} \Rightarrow 3n + n(6n - 3) = 384 \Rightarrow$
$6n^2 = 384 \Rightarrow n^2 = 64 \Rightarrow n = \pm 8$ (Reject the negative solution.) The flea market was open for 8 days.

79. 1, 3, 5, ...
The common difference is 2.
$a_{20} = 1 + 2(20 - 1) = 39$
At the 20th stage, there will be 39 dots.

81. 13, 16, 19, ...
The common difference is 3.
$a_{20} = 13 + 3(20 - 1) = 70$
At the 20th stage, there will be 70 dots.

11.2 Beyond the Basics

83. $a_n = \log\left(\dfrac{a^n}{b^{n-1}}\right)$, $a_{n+1} = \log\left(\dfrac{a^{n+1}}{b^n}\right)$

$a_{n+1} - a_n = \log\left(\dfrac{a^{n+1}}{b^n}\right) - \log\left(\dfrac{a^n}{b^{n-1}}\right)$

$= \log\left(\dfrac{\frac{a^{n+1}}{b^n}}{\frac{a^n}{b^{n-1}}}\right) = \log\left(\dfrac{a^{n+1}}{b^n} \cdot \dfrac{b^{n-1}}{a^n}\right)$

$= \log\dfrac{a}{b}$

85. $S_n = 3n^2 + 4n$
$S_{n-1} = 3(n-1)^2 + 4(n-1)$
$= 3(n^2 - 2n + 1) + 4n - 4$
$= 3n^2 - 6n + 3 + 4n - 4$
$= 3n^2 - 2n - 1$
$a_n = S_n - S_{n-1} = (3n^2 + 4n) - (3n^2 - 2n - 1)$
$= 6n + 1$
$d = a_n - a_{n-1} = (6n + 1) - (6(n-1) + 1)$
$= (6n + 1) - (6n - 5) = 6$
Since $d = 6$ is a constant, the sequence a_n is an arithmetic sequence.

87. 45 is divisible by 3 and the largest number less than 100 that is divisible by 3 is 99, so $a_1 = 45$, $d = 3, a_n = 99 \Rightarrow 99 = 45 + 3(n-1) \Rightarrow n = 19$.
$S = 19\left(\dfrac{45 + 99}{2}\right) = 1368$.

89. The sequence of the reciprocals is $2, \dfrac{5}{3}, \dfrac{4}{3}, 1, \ldots$. The common difference is $-\dfrac{1}{3}$, so the sequence is arithmetic, and the original sequence is harmonic.

91. $m = a + d$ and $b = a + 2d$. So
$\dfrac{a + b}{2} = \dfrac{a + a + 2d}{2} = \dfrac{2a + 2d}{2} = a + d = m$

11.2 Critical Thinking/Discussion/Writing

93. $a_1 = 10, a_{21} = a_1 + d(21 - 1) \Rightarrow 0 = 10 + 20d \Rightarrow$
$d = -\dfrac{1}{2} \Rightarrow a_n = 10 - \dfrac{1}{2}(n - 1) = \dfrac{21 - n}{2}$

94. First term $= -a_1$; difference $= -d$.

95. There are $100 - 22 + 1 = 79$ terms in the series.

96. In the set of counting numbers, $a_n = n$.
$12{,}403 = n\left(\dfrac{1 + n}{2}\right) \Rightarrow 24{,}806 = n^2 + n \Rightarrow$
$n^2 + n - 24{,}806 = 0 \Rightarrow (n - 157)(n + 158) = 0 \Rightarrow$
$n = -158$ (reject this) or $n = 157$.
The sum of the first 157 counting number is 12,403.

11.2 Getting Ready for the Next Section

97. $\dfrac{a_2}{a_1} = \dfrac{6}{3} = 2$

$\dfrac{a_3}{a_2} = \dfrac{12}{6} = 2$

$\dfrac{a_4}{a_3} = \dfrac{24}{12} = 2$

$\dfrac{a_5}{a_4} = \dfrac{48}{24} = 2$

99. $\dfrac{a_2}{a_1} = \dfrac{-12}{4} = -3$

$\dfrac{a_3}{a_2} = \dfrac{36}{-12} = -3$

$\dfrac{a_4}{a_3} = \dfrac{-108}{36} = -3$

$\dfrac{a_5}{a_4} = \dfrac{324}{-108} = -3$

101. $a_2 = a_1 r = 2 \cdot 3 = 6$
$a_3 = a_2 r = 6 \cdot 3 = 18$
$a_4 = a_3 r = 18 \cdot 3 = 54$
$a_5 = a_4 r = 54 \cdot 3 = 162$

103. $a_2 = a_1 r = 1(-2) = -2$
$a_3 = a_2 r = -2(-2) = 4$
$a_4 = a_3 r = 4(-2) = -8$
$a_5 = a_4 r = -8(-2) = 16$

105. $a_n = 5 \cdot 2^n$
$a_1 = 5 \cdot 2^1 = 10$
$a_2 = 5 \cdot 2^2 = 20$
$a_3 = 5 \cdot 2^3 = 40$
$a_4 = 5 \cdot 2^4 = 80$
$a_5 = 5 \cdot 2^5 = 160$

$\dfrac{a_2}{a_1} = \dfrac{20}{10} = 2$

$\dfrac{a_3}{a_2} = \dfrac{40}{20} = 2$

$\dfrac{a_4}{a_3} = \dfrac{80}{40} = 2$

$\dfrac{a_5}{a_4} = \dfrac{160}{80} = 2$

Note that

$\dfrac{a_{n+1}}{a_n} = \dfrac{5 \cdot 2^{n+1}}{5 \cdot 2^n} = \dfrac{5 \cdot 2^n \cdot 2}{5 \cdot 2^n} = 2$

107. $a_n = 2(-3)^n$
$a_1 = 2(-3)^1 = -6$
$a_2 = 2(-3)^2 = 18$
$a_3 = 2(-3)^3 = -54$
$a_4 = 2(-3)^4 = 162$
$a_5 = 2(-3)^5 = -486$

$\dfrac{a_2}{a_1} = \dfrac{18}{-6} = -3$

$\dfrac{a_3}{a_2} = \dfrac{-54}{18} = -3$

$\dfrac{a_4}{a_3} = \dfrac{162}{-54} = -3$

$\dfrac{a_5}{a_4} = \dfrac{-486}{162} = -3$

Note that

$\dfrac{a_{n+1}}{a_n} = \dfrac{2(-3)^{n+1}}{2(-3)^n} = \dfrac{2(-3)^n(-3)}{2(-3)^n} = -3$

109. $a_n = -3 \cdot 2^n$
$a_{n+1} = -3 \cdot 2^{n+1}$
$a_{n-1} = -3 \cdot 2^{n-1}$

11.3 Geometric Sequences and Series

11.3 Practice Problems

1. $r = \dfrac{18}{6} = 3$

2. The first four terms of the sequence are

$\dfrac{3}{2}, \left(\dfrac{3}{2}\right)^2 = \dfrac{9}{4}, \left(\dfrac{3}{2}\right)^3 = \dfrac{27}{8}, \left(\dfrac{3}{2}\right)^4 = \dfrac{81}{16}.$

Since the common ratio is $\dfrac{3}{2}$, the sequence is geometric.

3. a. $a_1 = 2$ b. $r = \dfrac{6/5}{2} = \dfrac{3}{5}$

   c. $a_n = 2 \cdot \left(\dfrac{3}{5}\right)^{n-1}$

4. $a_n = \dfrac{1}{3}, \dfrac{2}{3}, \dfrac{4}{3}, \dfrac{8}{3}, \dfrac{16}{3}, \ldots$

We will use the points $\left(1, \tfrac{1}{3}\right)$ and $\left(2, \tfrac{2}{3}\right)$ to find the exponential function that produces the sequence. First, find the common ratio B.

$B = \dfrac{\tfrac{2}{3}}{\tfrac{1}{3}} = 2$

Use one of the points to find the constant A.

$y = A \cdot 2^x \Rightarrow \dfrac{1}{3} = A \cdot 2^1 \Rightarrow \dfrac{1}{6} = A$

The function that produces this sequence is

$y = \dfrac{1}{6} \cdot 2^x$ and the general term is $a_n = \dfrac{1}{6} \cdot 2^n$.

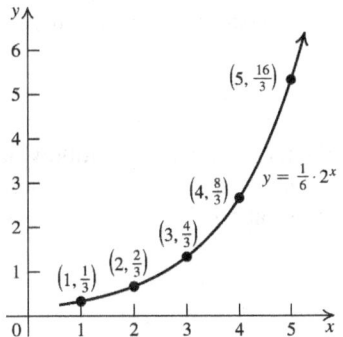

5. $a_{18} = a_1 r^{18-1} = 7(1.5^{17}) \approx 6896.8288$

6. $a_1 = \frac{1}{9}; r = \frac{1/3}{1/9} = 3$

$a_n = a_1 r^{n-1} \Rightarrow \frac{1}{9} \cdot 3^{n-1} = 243 \Rightarrow$
$3^{n-1} = 2187 \Rightarrow 3^{n-1} = 3^7 \Rightarrow$
$n - 1 = 7 \Rightarrow n = 8$
There are 8 terms in the sequence.

7. $a_1 = 3(0.4)^1 = 1.2; r = 0.4$

$S_{17} = \sum_{i=1}^{17} 3(0.4)^i = 1.2 \left(\frac{1 - 0.4^{17}}{1 - 0.4} \right) \approx 2$

8. $A = 1500 \left[\frac{\left(1 + \frac{0.045}{1}\right)^{(1)(30)} - 1}{\frac{0.045}{1}} \right]$
$= \$91,510.60$

9. $a_1 = 3, r = \frac{6/3}{3} = \frac{2}{3}$

Since $|r| < 1$, we can use the formula for the sum of an infinite geometric series.

$S = \frac{a_1}{1-r} = \frac{3}{1 - \frac{2}{3}} = 9$

10. $\sum_{i=1}^{\infty} (10,000,000)(0.85)^{i-1} = \frac{a_1}{1-r}$
$= \frac{10,000,000}{1 - 0.85}$
$\approx \$66,666,666.67$

11.3 Basic Concepts and Skills

1. If 5 is the common ratio of a geometric sequence with general term a_n, then $\frac{a_{63}}{a_{62}} = \underline{5}$.

3. If 24 is the term immediately following the sequence term 8 is a geometric sequence, then the common ratio is $\underline{3}$.

5. True

7. False. $a_n = a_1 r^{n-1}$

11.3 Building Skills

9. The sequence is geometric. $a_1 = 3, r = 2$

11. The sequence is not geometric.

13. The sequence is geometric. $a_1 = 1, r = -3$

15. The sequence is geometric. $a_1 = 7, r = -1$

17. The sequence is geometric. $a_1 = 9, r = \frac{1}{3}$

19. The sequence is geometric. $a_1 = -\frac{1}{2}, r = -\frac{1}{2}$

21. The sequence is geometric. $a_1 = 1, r = 2$

23. The sequence is not geometric.

25. The sequence is geometric. $a_1 = \frac{1}{3}, r = \frac{1}{3}$

27. The sequence is geometric. $a_1 = \sqrt{5}, r = \sqrt{5}$

39. $a_1 = 2, r = 5, a_n = 2 \cdot 5^{n-1}$

31. $a_1 = 5, r = \frac{2}{3}, a_n = 5\left(\frac{2}{3}\right)^{n-1}$

33. $a_1 = 0.2, r = -3, a_n = 0.2(-3)^{n-1}$

35. $a_1 = \pi^4, r = \pi^2, a_n = \pi^4(\pi^2)^{n-1} = \pi^{2n+2}$

37. $a_n = 6, 12, 24, 48, 96, \ldots$

We will use the points (1, 6) and (2, 12) to find the exponential function that produces the sequence. First, find the common ratio B.

$B = \frac{12}{6} = 2$

Use one of the points to find the constant A.
$y = A \cdot 2^x \Rightarrow 6 = A \cdot 2^1 \Rightarrow 3 = A$
The function that produces this sequence is
$y = 3 \cdot 2^x$ and the general term is $a_n = 3 \cdot 2^n$.

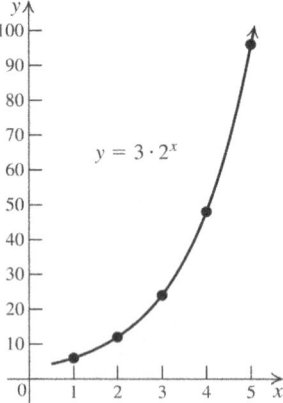

39. $a_n = 36, 12, 4, \dfrac{4}{3}, \dfrac{4}{9}, \ldots$

We will use the points (1, 36) and (2, 12) to find the exponential function that produces the sequence. First, find the common ratio B.

$B = \dfrac{12}{36} = \dfrac{1}{3}$

Use one of the points to find the constant A.

$y = A \cdot \left(\dfrac{1}{3}\right)^x \Rightarrow 36 = A \cdot \left(\dfrac{1}{3}\right)^1 \Rightarrow 108$

The function that produces this sequence is

$y = 108 \cdot \left(\dfrac{1}{3}\right)^x$ and the general term is

$a_n = 108 \cdot \left(\dfrac{1}{3}\right)^n$.

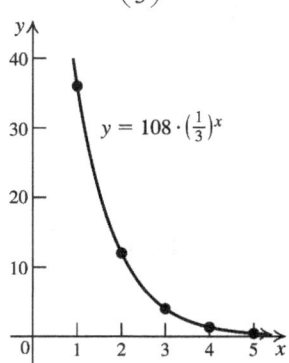

41. $a_n = -\dfrac{4}{5}, -\dfrac{16}{5}, -\dfrac{64}{5}, -\dfrac{256}{5}, -\dfrac{1024}{5}, \ldots$

We will use the points $\left(1, -\dfrac{4}{5}\right)$ and $\left(2, -\dfrac{16}{5}\right)$ to find the exponential function that produces the sequence. First, find the common ratio B.

$B = \dfrac{-\frac{16}{5}}{-\frac{4}{5}} = 4$

Use one of the points to find the constant A.

$y = A \cdot 4^x \Rightarrow -\dfrac{4}{5} = A \cdot 4^1 \Rightarrow -\dfrac{1}{5} = A$

The function that produces this sequence is

$y = -\dfrac{1}{5} \cdot 4^x$ and the general term is

$a_n = -\dfrac{1}{5} \cdot 4^n$.

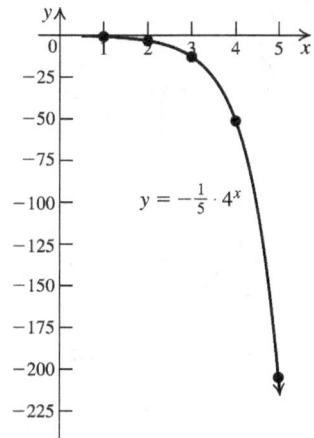

43. $a_7 = a_1 r^{7-1} = 5(2^6) = 320$

45. $a_{10} = a_1 r^{10-1} = 3(-2)^9 = -1536$

47. $a_6 = a_1 r^{6-1} = \dfrac{1}{16}(3)^5 = \dfrac{243}{16}$

49. $a_9 = a_1 r^{9-1} = -1\left(\dfrac{5}{2}\right)^8 = -\dfrac{390,625}{256}$

51. $a_{20} = a_1 r^{20-1} = 500\left(-\dfrac{1}{2}\right)^{19} = -\dfrac{125}{131,072}$

53. $a_1 = 5;\ r = \dfrac{10}{5} = 2$

$a_n = a_1 r^{n-1} \Rightarrow 5 \cdot 2^{n-1} = 5120 \Rightarrow$

$2^{n-1} = 1024 \Rightarrow 2^{n-1} = 2^{10} \Rightarrow$

$n - 1 = 10 \Rightarrow n = 11$

There are 11 terms in the sequence.

55. $a_1 = 16;\ r = \dfrac{8}{16} = \dfrac{1}{2}$

$a_n = a_1 r^{n-1} \Rightarrow 16 \cdot \left(\dfrac{1}{2}\right)^{n-1} = \dfrac{1}{32} \Rightarrow$

$\left(\dfrac{1}{2}\right)^{n-1} = \dfrac{1}{16 \cdot 32} \Rightarrow \left(\dfrac{1}{2}\right)^{n-1} = \dfrac{1}{2^4 \cdot 2^5} = \dfrac{1}{2^9} \Rightarrow$

$n - 1 = 9 \Rightarrow n = 10$

There are 10 terms in the sequence.

57. $a_1 = 18; r = \dfrac{-12}{18} = -\dfrac{2}{3}$

$a_n = a_1 r^{n-1} \Rightarrow 18 \cdot \left(-\dfrac{2}{3}\right)^{n-1} = \dfrac{512}{729} \Rightarrow$

$\left(-\dfrac{2}{3}\right)^{n-1} = \dfrac{512}{729 \cdot 18} \Rightarrow$

$\left(-\dfrac{2}{3}\right)^{n-1} = \dfrac{2^9}{3^6 \cdot 3^2 \cdot 2} = \dfrac{2^8}{3^8} = \left(\dfrac{2}{3}\right)^8 = \left(-\dfrac{2}{3}\right)^8 \Rightarrow$

$n - 1 = 8 \Rightarrow n = 9$

There are 9 terms in the sequence.

59. $r = 5, S_{10} = \dfrac{(1/10)(1 - 5^{10})}{1 - 5} = \dfrac{1{,}220{,}703}{5}$

61. $r = -5, S_{12} = \dfrac{(1/25)(1 - 5^{12})}{1 - (-5)} = -\dfrac{40{,}690{,}104}{25}$

63. $r = \dfrac{1}{4}, S_8 = \dfrac{5(1 - (1/4)^8)}{1 - 1/4} = \dfrac{109{,}225}{16{,}384}$

65. $\sum_{i=1}^{5} \left(\dfrac{1}{2}\right)^{i-1} = \dfrac{(1)(1 - (1/2)^5)}{1 - 1/2} = \dfrac{31}{16}$

67. $a_1 = 3, r = \dfrac{2}{3}$

$\sum_{i=1}^{8} 3\left(\dfrac{2}{3}\right)^{i-1} = \dfrac{3(1 - (2/3)^8)}{1 - (2/3)} = \dfrac{6305}{729}$

69. $a_1 = \dfrac{1}{4}, r = 2$

$\sum_{i=3}^{10} \dfrac{2^{i-1}}{4} = \sum_{i=1}^{10} \dfrac{2^{i-1}}{4} - \sum_{i=1}^{2} \dfrac{2^{i-1}}{4}$

$= \dfrac{(1/4)(1 - 2^{10})}{1 - 2} - \dfrac{(1/4)(1 - 2^2)}{1 - 2}$

$= \dfrac{1023}{4} - \dfrac{3}{4} = 255$

71. $a_1 = -\dfrac{3}{5}, r = -\dfrac{5}{2}$

$\sum_{i=1}^{20} \left(-\dfrac{3}{5}\right)\left(-\dfrac{5}{2}\right)^{i-1} = \dfrac{(-3/5)(1 - (-5/2)^{20})}{1 - (-5/2)}$

$= \dfrac{3(5^{20} - 2^{20})}{35(2^{19})}$

73. $a_1 = \dfrac{1}{5}; r = \dfrac{1}{2}$

$\dfrac{1}{320} = \dfrac{1}{5}\left(\dfrac{1}{2}\right)^{n-1} \Rightarrow \dfrac{1}{64} = \left(\dfrac{1}{2}\right)^{n-1} \Rightarrow$

$\left(\dfrac{1}{2}\right)^6 = \left(\dfrac{1}{2}\right)^{n-1} \Rightarrow 6 = n - 1 \Rightarrow n = 7$

Thus, the sum is $\sum_{n=1}^{7} \dfrac{1}{5}\left(\dfrac{1}{2}\right)^{n-1}$.

```
sum(seq((1/5)(1/
2)^(n-1),n,1,7)
           .396875
Ans▶Frac
          127/320
```

75.
```
sum(seq(3^(n-2),
n,1,10)
       9841.333333
```

77.
```
sum(seq((-1)^(n-
1)*3^(2-n),n,1,1
0)
       2.249961896
```

79. $a_1 = \dfrac{1}{3}, r = \dfrac{1}{3}$

$\dfrac{1}{3} + \dfrac{1}{9} + \dfrac{1}{27} + \dfrac{1}{81} + \cdots = \dfrac{1/3}{1 - (1/3)} = \dfrac{1}{2}$

81. $a_1 = -\dfrac{1}{2}, r = -\dfrac{1}{2}$

$-\dfrac{1}{2} + \dfrac{1}{4} - \dfrac{1}{8} + \dfrac{1}{16} + \cdots = \dfrac{-1/2}{1 - (-1/2)} = -\dfrac{1}{3}$

83. $a_1 = 8, r = -\dfrac{1}{4}$

$8 - 2 + \dfrac{1}{2} - \dfrac{1}{8} + \cdots = \dfrac{8}{1 - (-1/4)} = \dfrac{32}{5}$

85. $a_1 = 5, r = \dfrac{1}{3}; \sum_{n=0}^{\infty} 5\left(\dfrac{1}{3}\right)^n = \dfrac{5}{1 - (1/3)} = \dfrac{15}{2}$

87. $a_1 = 1, r = -\dfrac{1}{4}; \sum_{n=0}^{\infty} \left(-\dfrac{1}{4}\right)^n = \dfrac{1}{1 - (-1/4)} = \dfrac{4}{5}$

11.3 Applying the Concepts

89. $a_5 = 20{,}000(1.03^5) = 23{,}185$

91. $a_{36} = 100\left(\dfrac{\left(1 + \dfrac{0.06}{12}\right)^{36} - 1}{\dfrac{0.06}{12}}\right) = \$3933.61$

93. $a_{10} = 2^{10} = 1024$.

A person has 1024 ancestors in the tenth generation back.

95. $a_1 = \$36,000, r = 1.05$

$S_{20} = \dfrac{36,000(1-1.05^{20})}{1-1.05} = \$1,190,374.35$

97. $a_1 = 56.25, r = 0.8$

$S_5 = \dfrac{56.25(1-0.8^5)}{1-0.8} = 189.09$ cm

99. $D = 5 + 2(5)\left(\dfrac{3}{5}\right) + 2(5)\left(\dfrac{3}{5}\right)^2 + 2(5)\left(\dfrac{3}{5}\right)^3 + \ldots$

$= 5 + 10\left(\dfrac{3}{5}\right) + 10\left(\dfrac{3}{5}\right)^2 + 10\left(\dfrac{3}{5}\right)^3 + \ldots$

$= 5 + \dfrac{6}{1-(3/5)} = 20$ m

101. Number of squares: 1, 2, 4, …

Common ratio $r = \dfrac{2}{1} = 2$

$a_{10} = a_1 \cdot r^9 = 1 \cdot 2^9 = 512$

At the 10th stage, there will be 512 squares.

103. The ratio of the sides of the triangles is $\dfrac{1}{2}$, so the ratio of the areas of the triangles is $\left(\dfrac{1}{2}\right)^2 = \dfrac{1}{4}$. The areas of the shaded triangles are $\dfrac{1}{4}, \dfrac{1}{16}, \dfrac{1}{64}, \dfrac{1}{256}, \ldots$

So, $a_n = \dfrac{1}{4} \cdot \left(\dfrac{1}{4}\right)^{n-1} = \left(\dfrac{1}{4}\right)^n$. Because $|r| = \dfrac{1}{4} < 1$, we can use the formula for the sum of an infinite geometric series.

$S = \dfrac{a_1}{1-r} = \dfrac{\frac{1}{4}}{1-\frac{1}{4}} = \dfrac{\frac{1}{4}}{\frac{3}{4}} = \dfrac{1}{3}$

This sum of the areas of the shaded triangles is $\dfrac{1}{3}$.

11.3 Beyond the Basics

105. The area of each square is 1/2 the area of the next larger square. However, we are including the shaded regions only, so $r = -1/2$. (This will eliminate the white squares.)

$\displaystyle\sum_{i=1}^{\infty} \left(-\dfrac{1}{2}\right)^{i-1} = \dfrac{1}{1-(-1/2)} = \dfrac{2}{3}$.

107. Since $a_1, a_2, a_3, \ldots$ is a geometric sequence, $a_n = a_{n-1} r$ for every $n \geq 1$, where r is the common ratio. Taking the logarithm of both sides, we have $\ln a_n = \ln(a_{n-1} r)$
$= \ln a_{n-1} + \ln r$. So the sequence $\ln a_1, \ln a_2, \ln a_3, \ldots$ is arithmetic with common difference $\ln r$.

109. Since $a_1, a_2, a_3, \ldots$ is a geometric sequence, $a_n = a_{n-1} r$ for every $n \geq 1$, where r is the common ratio. Taking the reciprocal of both sides, we have $\dfrac{1}{a_n} = \dfrac{1}{a_{n-1} r} = \dfrac{1}{a_{n-1}} \cdot \dfrac{1}{r}$. So the sequence $\dfrac{1}{a_1}, \dfrac{1}{a_2}, \dfrac{1}{a_3}, \ldots$ is geometric with common ratio $1/r$.

111. Since $a_n = \dfrac{2a_{n-1}}{x}$ for every $n \geq 1$, the sequence is geometric with common ratio $2/x$.

113. Let the general term be represented by b_n. Since $b_n = \dfrac{b_{n-1}}{y}$, the sequence is geometric with common ratio $1/y$.

11.3 Critical Thinking/Discussion/Writing

115. $a_n = 2^{4-n} = \dfrac{2^4}{2^n} = 16 \cdot \left(\dfrac{1}{2}\right)^n$, which is a geometric sequence.

$b_n = \dfrac{4}{2^{n-2}} = \dfrac{2^2}{\frac{2^n}{2^2}} = \dfrac{2^4}{2^n} = 16 \cdot \left(\dfrac{1}{2}\right)^n$

The sequences are the same.

116. $4, -2, 1, -\dfrac{1}{2}, \dfrac{1}{4}, \ldots$

$r = \dfrac{-2}{4} = -\dfrac{1}{2}$

$y = A \cdot \left(-\dfrac{1}{2}\right)^n \Rightarrow 4 = A \cdot \left(-\dfrac{1}{2}\right)^1 \Rightarrow A = -8$

So, $y = -8 \cdot \left(-\dfrac{1}{2}\right)^x$ is a function that

produces this sequence and $a_n = -8 \cdot \left(-\dfrac{1}{2}\right)^n$

is the general term for the sequence.

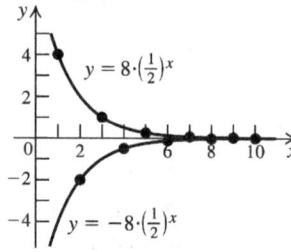

Another way of looking at this is as follows:

Examine just the positive values: $4, 1, \dfrac{1}{4}, \ldots$

These values are generated by

$f(x) = 8 \cdot \left(\dfrac{1}{2}\right)^x$.

Now, look at just the negative values:

$-2, -\dfrac{1}{2}, -\dfrac{1}{8}, \ldots$

These values are generated by

$g(x) = -8 \cdot \left(\dfrac{1}{2}\right)^x$.

The points representing a_n alternate between $f(x)$ and $g(x)$. As $x \to \infty$, $f(x) \to 0$, and $g(x) \to 0$.

117. $a_{1001} = 10 + 1000(2.7) = 2710$.

$b_{1001} = 10(2^{1000})$, so b_{1001} is larger.

118. $\dfrac{a_1\left(1 - \left(\dfrac{1}{2}\right)^{10}\right)}{1 - \dfrac{1}{2}} = 1023 \Rightarrow a_1 = 512$

119. $5 + 5(2) + 5(2^2) + \cdots + 5(2^{15}) = \displaystyle\sum_{i=0}^{15} 5(2^i)$ or

$\displaystyle\sum_{i=1}^{16} 5(2^{i-1})$.

120. Let $\dfrac{a}{r}, a,$ and ar be the three terms of the geometric sequence. Then we have

$\begin{cases} \dfrac{a}{r} + a + ar = 35 \\ \left(\dfrac{a}{r}\right)a(ar) = 1000 \end{cases} \Rightarrow \begin{cases} a + ar + ar^2 = 35r \\ a^3 = 1000 \end{cases} \Rightarrow$

$a = 10$

$10 + 10r + 10r^2 = 35r \Rightarrow 2r^2 - 5r + 2 = 0 \Rightarrow$

$(2r - 1)(r - 2) \Rightarrow r = \dfrac{1}{2}$ or $r = 2$.

If $r = \dfrac{1}{2}$, then the three numbers are 20, 10, and 5. If $r = 2$, then the numbers are also 5, 10, and 20.

121. Let $a, a+d,$ and $a+2d$ be the three terms of the arithmetic sequence. Then
$a + (a+d) + (a+2d) = 3a + 3d = 15 \Rightarrow$
$a + d = 5$.
The geometric sequence is $a+1, a+d+4,$
and $a+2d+19$, and
$\dfrac{a+d+4}{a+1} = \dfrac{a+2d+19}{a+d+4}$.

Use substitution to solve the system:
$\begin{cases} a + d = 5 \\ \dfrac{a+d+4}{a+1} = \dfrac{a+2d+19}{a+d+4} \end{cases} \Rightarrow$

$\begin{cases} d = 5 - a \\ \dfrac{a+d+4}{a+1} = \dfrac{a+2d+19}{a+d+4} \end{cases} \Rightarrow$

$\dfrac{a+(5-a)+4}{a+1} = \dfrac{a+2(5-a)+19}{a+(5-a)+4} \Rightarrow$

$\dfrac{9}{a+1} = \dfrac{29-a}{9} \Rightarrow 29 + 28a - a^2 = 81 \Rightarrow$

$-(a^2 - 28a + 52) = 0 \Rightarrow (a-2)(a-26) = 0 \Rightarrow$

$a = 2$ or $a = 26$.
If $a = 2$, then $d = 3$, and the arithmetic sequence is 2, 5, 8. If $a = 26$, then $d = -21$, and the sequence is 26, 5, -16.

122. Let $a, ar,$ and ar^2 be the three terms of the geometric sequence. So,

$a \cdot ar \cdot ar^2 = a^3 r^3 = 1000 \Rightarrow ar = 10$.

The arithmetic sequence is a, $ar + 6$, and

$ar^2 + 7$. Then,

$(ar + 6) - a = (ar^2 + 7) - (ar + 6)$.

(continued on next page)

(*continued*)

Substituting $ar = 10$ gives
$(10+6) - a = (10r+7) - (10+6) \Rightarrow$
$16 - a = 10r - 9 \Rightarrow 25 = 10r + a$
$ar = 10 \Rightarrow r = \dfrac{10}{a}$

Substituting, we have
$25 = 10\left(\dfrac{10}{a}\right) + a \Rightarrow 25a = 100 + a^2 \Rightarrow$
$a^2 - 25a + 100 = 0 \Rightarrow (a-5)(a-20) = 0 \Rightarrow$
$a = 5, a = 20$

If $a = 5$, then we have $5r = 10 \Rightarrow r = 2$, and the geometric sequence is 5, 10, 20.

If $a = 20$, then we have $20r = 10 \Rightarrow r = \dfrac{1}{2}$, and the geometric sequence is 20, 10, 5.

11.3 Getting Ready for the Next Section

123. $P_n : n(n+1)$ is even.
$P_3 : 3(3+1)$ is even.   True
$P_4 : 4(4+1)$ is even.   True

125. $P_n : 1 + 2 + 3 + \cdots + n = \dfrac{n(n+1)}{2}$

$P_3 : 1 + 2 + 3 = \dfrac{3(3+1)}{2}$ True

$P_4 : 1 + 2 + 3 + 4 = \dfrac{4(4+1)}{2}$ True

127. $P_n : n^2 + n$ is divisible by 2.
$P_{n+1} : (n+1)^3 + (n+1)$ is divisible by 2.

129. $P_n : 3^n > 5n$
$P_{n+1} : 3^{n+1} > 5(n+1)$

131. $\dfrac{k(k+1)}{2} + (k+1) = \dfrac{k(k+1)}{2} + \dfrac{2(k+1)}{2}$
$= \dfrac{k^2 + k + 2k + 2}{2}$
$= \dfrac{k^2 + 3k + 2}{2}$
$= \dfrac{(k+1)(k+2)}{2}$

11.4 Mathematical Induction

11.4 Practice Problems

1. $P_{k+1} : \left[(k+1)+3\right]^2 > (k+1)^2 + 9 \Rightarrow$
$(k+4)^2 > (k+1)^2 + 9$

2. For $n = 1 : 1 = \dfrac{1(1+1)}{2}$ is true. Assume that it is true for $n = k : 1 + 2 + 3 + \ldots + k = \dfrac{k(k+1)}{2}$.
Then for $n = k+1 : 1 + 2 + 3 + \ldots + k + (k+1)$
$= \dfrac{k(k+1)}{2} + (k+1) = \dfrac{k^2 + k + 2k + 2}{2}$
$= \dfrac{(k+1)(k+2)}{2}$, which is exactly the statement for $n = k+1$. Therefore the formula is true for all natural numbers.

3. For $n = 1 : 3^1 > 1$ is true. Assume that the inequality is true for $n = k : 3^k > k$. Then, we must use this fact to prove that for $n = k+1 : 3^{k+1} > k+1$.
$3^k > k \Rightarrow 3^k \cdot 3 > 3 \cdot k \Rightarrow 3^{k+1} > 3k > k+1$, which is exactly the statement for $n = k+1$. Therefore the formula is true for all natural numbers.

11.4 Concepts and Vocabulary

1. Mathematical induction can only be used to prove statements about the <u>natural numbers</u>.

3. The second step in a mathematical induction proof is to assume that P_k is true for a natural number k, and then show that <u>P_{k+1} is true</u>.

5. True

7. False. The number e is a constant. The statement $e^n \geq n$ can be proven by mathematical induction for all natural numbers n.

11.4 Building Skills

9. $P_{k+1} : ((k+1)+1)^2 - 2(k+1)$
$= k^2 + 4k + 4 - 2k - 2$
$= k^2 + 2k + 2 = (k+1)^2 + 1$

11. $P_{k+1} : 2^{k+1} > 5(k+1)$

13. For $n=1: 2=2$ is true. Assume that it is true for $n=k$:
$$2+4+6+\cdots+2k = k^2+k = k(k+1)$$
Then for $n=k+1$:
$$2+4+6+\ldots+2k+2(k+1) = k(k+1)+2(k+1) = k^2+3k+2 = (k+1)(k+2),$$
which is exactly the statement for $n=k+1$. Therefore the formula is true for all natural numbers.

15. For $n=1: 4=2(1)(1+1)$ is true. Assume that it is true for $n=k$:
$$4+8+12+\cdots+4k = 2k(k+1)$$
Then for $n=k+1$:
$$4+8+12+\cdots+4k+(4k+4) = 2k(k+1)+(4k+4) = 2k^2+2k+4k+4$$
$$= 2(k^2+3k+2) = 2(k+1)(k+2),$$
which is exactly the statement for $n=k+1$. Therefore the formula is true for all natural numbers.

17. For $n=1: 1=1(2-1)$ is true. Assume that it is true for $n=k$:
$$1+5+9+\cdots+(4k-3) = k(2k-1)$$
Then for $n=k+1$:
$$1+5+9+\cdots+(4k-3)+(4k+1) = k(2k-1)+(4k+1) = 2k^2-k+4k+1$$
$$= 2k^2+3k+1 = (k+1)(2k+1) = (k+1)(2(k+1)-1),$$
which is exactly the statement for $n=k+1$. Therefore the formula is true for all natural numbers.

19. For $n=1: 3=\dfrac{3(3-1)}{2}$ is true. Assume that it is true for $n=k$:
$$3+9+27+\cdots+3^k = \frac{3(3^k-1)}{2}$$
Then for $n=k+1$:
$$3+9+27+\cdots+3^k+3^{k+1} = \frac{3(3^k-1)}{2}+3^{k+1} = \frac{3^{k+1}-3+2(3^{k+1})}{2} = \frac{3(3^{k+1})-3}{2} = \frac{3(3^{k+1}-1)}{2},$$
which is exactly the statement for $n=k+1$. Therefore the formula is true for all natural numbers.

21. For $n=1: 1\cdot 2 = \dfrac{1}{3}\cdot(1+1)(4\cdot 1-1)$ is true. Assume that the statement is true for $n=k$:
$$1\cdot 2+3\cdot 4+5\cdot 6+\cdots+(2k-1)(2k) = \frac{1}{3}k(k+1)(4k-1)$$
Then, for $n=k+1$:
$$\frac{1}{3}k(k+1)(4k-1)+(2(k+1)-1)(2(k+1)) = \frac{1}{3}k(k+1)(4k-1)+(2k+1)(2k+2)$$
$$= \frac{4}{3}k^3+k^2-\frac{1}{3}k+4k^2+6k+2$$
$$= \frac{4}{3}k^3+5k^2+\frac{17}{3}k+2$$
$$= \frac{4k^3+15k^2+17k+6}{3}$$
$$= \frac{(k+1)(k+2)(4k+3)}{3}$$
$$= \frac{1}{3}(k+1)(k+2)(4(k+1)-1)$$
This is exactly the statement for $n=k+1$. Therefore, the formula is true for all natural numbers.

23. For $n=1: \dfrac{1}{(1)(2)} = \dfrac{1}{1+1}$ is true. Assume that it is true for $n = k$:

$\dfrac{1}{(1)(2)} + \dfrac{1}{(2)(3)} + \dfrac{1}{(3)(4)} + \cdots + \dfrac{1}{k(k+1)} = \dfrac{k}{k+1}$. Then for $n = k+1$:

$\dfrac{1}{(1)(2)} + \dfrac{1}{(2)(3)} + \dfrac{1}{(3)(4)} + \cdots + \dfrac{1}{k(k+1)} + \dfrac{1}{(k+1)(k+2)} = \dfrac{k}{k+1} + \dfrac{1}{(k+1)(k+2)} = \dfrac{k^2+2k+1}{(k+1)(k+2)}$

$= \dfrac{(k+1)^2}{(k+1)(k+2)} = \dfrac{k+1}{k+2}$

which is exactly the statement for $n = k+1$. Therefore the formula is true for all natural numbers.

25. For $n = 1: 2 \le 2$ is true. Assume that it is true for $n = k: 2 \le 2^k$. Then for $n = k+1$:
$2 \le 2^k \Rightarrow 2 \cdot 2 \le 2(2^k) \Rightarrow 4 \le 2^{k+1} \Rightarrow 2 \le 2^{k+1}$ Therefore the formula is true for all natural numbers.

27. For $n = 1: 1(1+2) < (1+1)^2$ is true. Assume that it is true for $n = k: k(k+2) < (k+1)^2$. Then for $n = k+1$:
$(k+1)(k+1+2) = (k+1)(k+3) < (k+1+1)^2 \Rightarrow k^2 + 4k + 3 < k^2 + 4k + 4 \Rightarrow$
$k^2 + 2k + 2k + 3 < k^2 + 2k + 1 + 2k + 3 \Rightarrow k(k+2) + (2k+3) < (k+1)^2 + (2k+3) \Rightarrow k(k+2) < (k+1)^2$,
which is exactly the statement for $n = k+1$. Therefore the formula is true for all natural numbers.

29. For $n = 1: \dfrac{1!}{1} = (1-1)!$ is true. Assume that it is true for $n = k: \dfrac{k!}{k} = (k-1)!$. Then for $n = k+1$:

$\dfrac{(k+1)!}{k+1} = \dfrac{(k+1)k!}{k+1} = k!$, which is exactly the statement for $n = k+1$. Therefore the formula is true for all natural numbers.

31. For $n = 1: 1^2 = \dfrac{1(1+1)(2+1)}{6}$ is true. Assume that it is true for $n = k$:

$1^2 + 2^2 + 3^2 + \cdots + k^2 = \dfrac{k(k+1)(2k+1)}{6}$.

Then for $n = k+1$:

$1^2 + 2^2 + 3^2 + \cdots + k^2 + (k+1)^2 = \dfrac{k(k+1)(2k+1)}{6} + (k+1)^2 = \dfrac{k(k+1)(2k+1) + 6(k+1)^2}{6}$

$= \dfrac{(k+1)(k(2k+1) + 6(k+1))}{6} = \dfrac{(k+1)(2k^2 + 7k + 6)}{6}$

$= \dfrac{(k+1)(k+2)(2k+3)}{6}$

which is exactly the statement for $n = k+1$. Therefore the formula is true for all natural numbers.

33. For $n = 1$: 2 is a factor of $1^2 + 1 = 2$.

For $n = k$: assume that 2 is a factor of $n^2 + n \Rightarrow k^2 + k = 2p$ for some integer p. Then for $n = k+1$:
$(k+1)^2 + (k+1) = k^2 + 3k + 2 = (k^2 + k) + (2k+2) = 2p + 2(k+1) = 2(p+k+1)$
$(p+k+1)$ is an integer, so $2(p+k+1)$ is divisible by 2. Therefore the formula is true for all natural numbers.

35. For $n=1$: 6 is a factor of $1(1+1)(1+2)=6$. For $n=k$: Assume that 6 is a factor of $n(n+1)(n+2) \Rightarrow k(k+1)(k+2) = 6p$ for some integer p.
Then for $n = k+1$:
$(k+1)(k+2)(k+3) = k(k+1)(k+2) + 3(k+1)(k+2) = 6p + 3(k+1)(k+2)$
Either $k+1$ or $k+2$ is even, so $3(k+1)(k+2)$ is divisible by 6 and, thus, $6p + 3(k+1)(k+2)$ is divisible by 6. Therefore the formula is true for all natural numbers.

37. For $n=1: (1+a)^1 \geq 1 \cdot a$, $a > 1$ is true. Assume that the statement is true for $n = k$,:
$(1+a)^k \geq ka$, $a > 1$. Then for $n = k+1$,
$(1+a)^k \geq ka \Rightarrow (1+a)(1+a)^k \geq ka(1+a) \Rightarrow (1+a)^{k+1} \geq ka + ka^2 \geq (k+1)a$, $a > 1$
Therefore the formula is true for all natural numbers $a > 1$.

39. For $n=1$: $\sum_{i=1}^{1} 3 \cdot 4^i = 4(4^1 - 1) = 12$ is true. Assume that the statement is true for $n = k$: $\sum_{i=1}^{k} 3 \cdot 4^i = 4(4^k - 1)$.
Then, for $n = k+1$,
$\sum_{i=1}^{k+1} 3 \cdot 4^i = 4(4^k - 1) + 3 \cdot 4^{k+1} = 4(4^k - 1) + 4(4^k \cdot 3) = 4(4^k - 1 + 4^k \cdot 3)$
$= 4((4^k + 4^k \cdot 3) - 1) = 4(4^k(1+3) - 1) = 4(4 \cdot 4^k - 1) = 4(4^{k+1} - 1)$,
which is exactly the statement for $n = k+1$. Therefore the formula is true for all natural numbers.

41. For $n=1$: $1 + \frac{1}{1} = 1+1$ is true. Assume that the statement is true for $n = k$:
$\left(1+\frac{1}{1}\right)\left(1+\frac{1}{2}\right)\left(1+\frac{1}{3}\right) + \cdots + \left(1+\frac{1}{k}\right) = k+1$.
Then, for $n = k+1$,
$(k+1)\left(1+\frac{1}{k+1}\right) = (k+1)\left(\frac{k+2}{k+1}\right) = k+2 = (k+1)+1$
This is exactly the statement for $n = k+1$. Therefore the formula is true for all natural numbers.

43. For $n=1: (ab)^1 = a^1 b^1$ is true. Assume that it is true for $n = k$: $(ab)^k = a^k b^k$.
Then for $n = k+1$: $(ab)^{k+1} = (ab)^k (ab) = a^k b^k ab = a^{k+1} b^{k+1}$, which is exactly the statement for $n = k+1$.
Therefore the formula is true for all natural numbers.

11.4 Applying the Concepts

45. For $n=2$: $\frac{2^2 - 2}{2} = 1$, which is the correct number of hugs for two people. Assume that the number of hugs for k people is $\frac{k^2 - k}{2}$. If there are $k+1$ people, then we can separate one of them. The remaining k have $\frac{k^2 - k}{2}$ hugs by the hypothesis. The $k+1$ st person has to hug each of the other k people. So the number of hugs is $\frac{k^2 - k}{2} + k = \frac{k^2 - k + 2k}{2} = \frac{k^2 + 2k + 1 - (k+1)}{2} = \frac{(k+1)^2 - (k+1)}{2}$, which is exactly the statement for $n = k+1$. Therefore the formula is true for all natural numbers $n \geq 2$.

47. a. The number of sides of the nth figure is $3(4^{n-1})$. For $n=1: 3(4^0)=3$, which is the number of sides of the triangle. Assume the number of sides of the kth figure is $3(4^{k-1})$. When making the $k+1$st figure, each new small triangle splits the original side into two sides, thus doubling the number of sides. In addition, each new triangle adds on two new sides per original side. So, the number of sides of the $k+1$st figure is four times the number of sides of the kth figure, i.e., $4(3)(4^{k-1}) = 3(4^k)$, which is exactly the formula for $n=k+1$. Therefore the formula is true for all natural numbers.

b. The perimeter of the nth figure is $3\left(\frac{4}{3}\right)^{n-1}$.

For $n=1: 3\left(\frac{4}{3}\right)^0 = 3$, which is the perimeter of the equilateral triangle with side length 1. Assume that the perimeter of the kth figure is $3\left(\frac{4}{3}\right)^{k-1}$. When making the $k+1$st figure, each new small triangle adds on two new sides and deletes one small side, each with length $1/3$ of the original side. So, each new triangle increases the perimeter by $1/3$ of an original side. Therefore the perimeter of the $k+1$st figure is $4/3$ times that of the kth figure, i.e., $\frac{4}{3}(3)\left(\frac{4}{3}\right)^{k-1} = 3\left(\frac{4}{3}\right)^k$, which is exactly the formula for $n=k+1$. Therefore the formula is true for all natural numbers.

49. The smallest number of moves to accomplish the transfer is $2^n - 1$. Prove as follows: for $n=1: 2^1 - 1 = 1$ is the number of necessary moves for one ring. Assume that the smallest number of moves for k rings is $2^k - 1$. When we move the biggest peg from the bottom to another peg, all the other rings must be stacked on the third peg in decreasing order from bottom to top. This requires $2^k - 1$ moves by the hypothesis. It takes one move to put the biggest ring onto the other peg. Then we have to move all the other rings on top of it, with again requires $2^k - 1$ moves. Altogether, then, there are
$(2^k - 1) + 1 + (2^k - 1) = 2(2^k) - 1 = 2^{k+1} - 1$
moves, which is exactly the formula for $n = k+1$. Therefore the formula is true for all natural numbers.

11.4 Beyond the Basics

51. For $n=1$, 5 is a factor of $8^1 - 3^1 = 5$. For $n=k$, assume that 5 is a factor of $8^n - 3^n \Rightarrow$ $8^k - 3^k = 5p$ for some integer p. Then for $n=k+1$: $8^{k+1} - 3^{k+1} = 8(8^k) - 3(3^k)$
$= 8(8^k) - 8(3^k) + 5(3^k) = 8(8^k - 3^k) + 5(3^k)$
$8(5p) + 5(3^k) = 5(8p + 3^k)$. Since $8p + 3^k$ is an integer, $5(8p + 3^k)$ is divisible by 5. Therefore the formula is true for all natural numbers.

53. For $n=1$, 64 is a factor of $3^{2(1)+2} - 8(1) - 9 = 64$. For $n=k$, assume that 64 is a factor of $3^{2n+2} - 8n - 9 \Rightarrow$ $3^{2k+2} - 8k - 9 = 64p$ for some integer p. Then for $n=k+1$:
$3^{2(k+1)+2} - 8(k+1) - 9$
$= 3^{2k+4} - 8(k+1) - 9$
$= 3^2(3^{2k+2}) - 8k - 8 - 9$
$= 9(3^{2k+2}) - 9(8k) - 9(9) + 8(8k) + 64$
$= 9(3^{2k+2} - 8k - 9) + 64(k+1)$
$= 9(64p) + 64(k+1) = 64(9p + k + 1)$
Since $9p + k + 1$ is an integer, $64(9p + k + 1)$ is divisible by 64. Therefore the formula is true for all natural numbers.

55. For $n=1$, 3 is a factor of $2^{2(1)+1} + 1 = 9$. For $n=k$, assume that 3 is a factor of $2^{2n+1} + 1 \Rightarrow 2^{2k+1} + 1 = 3p$ for some integer p. Then for $n=k+1$:
$2^{2(k+1)+1} + 1 = 2^{2k+3} + 1 = 2^2 2^{2k+1} + 1$
$= 2^2 2^{2k+1} + 4 - 3$
$= 4(2^{2k+1} + 1) - 3$
$= 4(3p) - 3 = 3(4p - 1)$
Since $4p - 1$ is an integer, $3(4p - 1)$ is divisible by 3. Therefore the formula is true for all natural numbers.

57. For $n = 1$, $a-b$ is a factor of $a^1 - b^1 = a - b$. For $n = k$, assume that $a - b$ is a factor of
$a^n - b^n \Rightarrow a^k - b^k = (a-b)P(a,b)$ where $P(a,b)$ is a polynomial of a and b. Then for $n = k+1$:
$$a^{k+1} - b^{k+1} = a\left(a^k - b^k\right) + b^k(a-b) = a(a-b)P(a,b) + b^k(a-b) = (a-b)\left(aP(a,b) + b^k\right)$$
Since $aP(a,b) + b^k$ is a polynomial of a and b, $a-b$ is a factor of $(a-b)\left(aP(a,b) + b^k\right)$. Therefore the statement is true for all natural numbers.

59. For $n=1$: $\sum_{k=1}^{2} \frac{1}{k+1} = \frac{1}{2} + \frac{1}{3} \leq \frac{5}{6}$ is true. Assume that it is true for $n = m$: $\sum_{k=1}^{m+1} \frac{1}{k+m} \leq \frac{5}{6}$.

Then for $n = m+1$: $\sum_{k=1}^{m+2} \frac{1}{k+m+1} = \sum_{k=2}^{m+3} \frac{1}{k+m} = \left(\sum_{k=1}^{m+1} \frac{1}{k+m}\right) + \frac{1}{2m+2} + \frac{1}{2m+3} - \frac{1}{m+1}$.

In the last expression, $\frac{1}{2m+2} + \frac{1}{2m+3} - \frac{1}{m+1} = \frac{(2m+3) + (2m+2) - 2(2m+3)}{2(m+1)(2m+3)} = -\frac{1}{2(m+1)(2m+3)} < 0$

So, $\sum_{k=1}^{m+2} \frac{1}{k+m+1} = \left(\sum_{k=1}^{m+1} \frac{1}{k+m}\right) + \frac{1}{2m+2} + \frac{1}{2m+3} - \frac{1}{m+1} \leq \left(\sum_{k=1}^{m+1} \frac{1}{k+m}\right) \leq \frac{5}{6}$

This is exactly the formula for $n = k+1$. Therefore the formula is true for all natural numbers.

11.4 Critical Thinking/Discussion/Writing

60. The proof is not valid because the first step of the mathematical induction, which is to show that the statement is true for P_1, is missing from this proof.

61. For $n = m = 4$: $2^4 \geq 4^2 \Rightarrow 16 \geq 16$ is true. Now assume that the statement is true for $n = k$: $2^k \geq k^2$. Now we must use P_k to prove that P_{k+1} is true. That is, $P_{k+1}: 2^{k+1} \geq (k+1)^2$.
$2 \cdot 2^k \geq 2k^2 \Rightarrow 2^{k+1} \geq 2k^2 \Rightarrow 2^{k+1} \geq k^2 + k^2$
Since $k \geq 4$, we have $2k \geq 8 \Rightarrow 2k > 1$. Also, since $k \geq 4$, we have $k^2 \geq 4k$.
$2^{k+1} > k^2 + k^2 > k^2 + 4k \Rightarrow 2^{k+1} > k^2 + 4k \Rightarrow 2^{k+1} > k^2 + 2k + 2k \Rightarrow$
$2^{k+1} > k^2 + 2k + 2k > k^2 + 2k + 1 \Rightarrow 2^{k+1} > k^2 + 2k + 1 \Rightarrow 2^{k+1} > (k+1)^2$
Thus, $P_k \Rightarrow P_{k+1}$, and the statement is true for all natural numbers.

62. For $n = m = 6$, $6! > 6^3 \Rightarrow 720 > 216$ is true. Now assume that the statement is true for $n = k$: $k! > k^3$.
Now we must use P_k to prove that P_{k+1} is true. That is, $P_{k+1}: (k+1)! \geq (k+1)^3$.
$k! > k^3 \Rightarrow (k+1) \cdot k! \geq (k+1) \cdot k^3 \Rightarrow (k+1)! \geq (k+1) \cdot k^3$.
Since $k + 1 \geq 7$, we have $(k+1)! \geq (k+1) \cdot k^3 \Rightarrow (k+1)! \geq 7k^3$.
We need to show that $7k^3 \geq (k+1)^3$ for $k \geq 6$:
$(k+1)^3 = k^3 + 3k^2 + 3k + 1 \Rightarrow 7k^3 \geq k^3 + 3k^2 + 3k + 1 \Rightarrow 6k^3 \geq 3k^2 + 3k + 1 \Rightarrow$
$k^3 \geq \frac{1}{2}k^2 + \frac{1}{2}k + \frac{1}{6} \Rightarrow k^3 \geq \frac{1}{2}\left(k^2 + k + \frac{1}{3}\right) \Rightarrow 7k^3 \geq \frac{7}{2}\left(k^2 + k + \frac{1}{3}\right)$
$k \geq 6 \Rightarrow 7k^2 \cdot k \geq 7k^2 \cdot 6 \Rightarrow 7k^3 \geq 42k^2$

(continued on next page)

(*continued*)

Now compare $\frac{7}{2}\left(k^2 + k + \frac{1}{3}\right)$ with $42k^2$:

$12\left[\frac{7}{2}\left(k^2 + k + \frac{1}{3}\right)\right] = 42k^2 + 42k + 14 \Rightarrow 42k^2 \geq \frac{7}{2}\left(k^2 + k + \frac{1}{3}\right)$, so

$7k^3 \geq \frac{7}{2}\left(k^2 + k + \frac{1}{3}\right) \Rightarrow (k+1)! \geq \frac{7}{2}\left(k^2 + k + \frac{1}{3}\right) \Rightarrow (k+1)! \geq (k+1)^3$.

11.4 Getting Ready for the Next Section

63. $(x+y)^2 = (x+y)(x+y) = x^2 + 2xy + y^2$

65. $(x+y)^4 = (x+y)(x+y)^3 = (x+y)\left(x^3 + 3x^2y + 3xy^2 + y^3\right)$
$= x^4 + 3x^3y + 3x^2y^2 + xy^3 + yx^3 + 3x^2y^2 + 3xy^3 + y^4 = x^4 + 4x^3y + 6x^2y^2 + 4xy^3 + y^4$

67. The exponents of x decrease from n to 1 when expanding $(x+y)^n$.

69. In exercise 63, we see that the sum of the exponents in each term is 2. In exercise 64, the sum of the exponents in each term is 3. In exercise 65, the sum of the exponents in each term is 4, and in exercise 66, the sum is 5. Thus, we can say that the sum of the exponents on x and y in each term in the expansion of $(x+y)^n$ is n.

11.5 The Binomial Theorem

11.5 Practice Problems

1. $(3y-x)^6 = 1(3y)^6 + 6(3y)^5(-x) + 15(3y)^4(-x)^2 + 20(3y)^3(-x)^3 + 15(3y)^2(-x)^4 + 6(3y)(-x)^5 + 1(-x)^6$
$= 729y^6 - 1458y^5x + 1215y^4x^2 - 540y^3x^3 + 135y^2x^4 - 18yx^5 + x^6$

2. a. $\binom{6}{2} = \frac{6!}{2!(6-2)!} = \frac{6 \cdot 5 \cdot 4!}{2!4!} = 15$

   b. $\binom{12}{9} = \frac{12!}{9!(12-9)!} = \frac{12 \cdot 11 \cdot 10 \cdot 9!}{9!3!} = \frac{12 \cdot 11 \cdot 10}{3 \cdot 2 \cdot 1} = 220$

3. $(3x-y)^4 = \binom{4}{0}(3x)^4 + \binom{4}{1}(3x)^3(-y) + \binom{4}{2}(3x)^2(-y)^2 + \binom{4}{3}(3x)(-y)^3 + \binom{4}{4}(-y)^4$
$= \frac{4!}{0!4!}(81x^4) - \frac{4!}{1!3!}(27x^3y) + \frac{4!}{2!2!}(9x^2y^2) - \frac{4!}{3!1!}(3xy^3) + \frac{4!}{4!0!}y^4$
$= 81x^4 - 108x^3y + 54x^2y^2 - 12xy^3 + y^4$

4. The term x^3y^9 is the tenth term in the expansion of $(x+y)^{12}$. So its coefficient is $\binom{12}{9}$.

   $\binom{12}{9} = \frac{12!}{9!(12-9)!} = \frac{12 \cdot 11 \cdot 10 \cdot 9!}{9!3!} = \frac{12 \cdot 11 \cdot 10}{3 \cdot 2 \cdot 1} = 220$

5. $\binom{n}{n-r}x^r(2a)^{n-r} = \binom{15}{15-3}x^3(2a)^{15-3} = \binom{15}{12}x^3(2a)^{12} = \frac{15!}{12!(15-12)!}x^3(2a)^{12} = \frac{15!}{12!3!}x^3(2a)^{12}$
$= \frac{15 \cdot 14 \cdot 13 \cdot 12!}{12!3!}x^3(2a)^{12} = \frac{15 \cdot 14 \cdot 13}{3 \cdot 2 \cdot 1}x^3a^{12} = 455x^3(2a)^{12} = 1{,}863{,}680x^3a^{12}$

6. The fourth term in the expansion of $(x-3)^{12}$ is
$$\binom{12}{3}x^{12-4+1}(-3)^{4-1} = \frac{12!}{3!(12-3)!}x^9(-3)^3$$
$$= 220(-3)^3 x^9$$
$$= -5940x^9$$

11.5 Concepts and Vocabulary

1. The expansion of $(x+y)^n$ has $\underline{n+1}$ terms.

3. Expanding a difference, such as $(2x-y)^{10}$, results in <u>alternating signs</u> between terms.

5. False. If n is even, then there are an odd number of terms in the expansion, so one coefficient appears just once.

7. True

11.5 Building Skills

9. $\dfrac{6!}{3!} = \dfrac{6 \cdot 5 \cdot 4 \cdot 3!}{3!} = 120$

11. $\dfrac{12!}{11!} = \dfrac{12 \cdot 11!}{11!} = 12$

13. $\binom{6}{4} = \dfrac{6!}{4!(6-4)!} = \dfrac{6 \cdot 5 \cdot 4!}{4!2!} = 15$

15. $\binom{9}{0} = \dfrac{9!}{0!(9-0)!} = \dfrac{9!}{0!9!} = 1$

17. $\binom{7}{1} = \dfrac{7!}{1!(7-1)!} = \dfrac{7 \cdot 6!}{1!6!} = 7$

19. $\binom{45}{45} = \dfrac{45!}{45!(45-45)!} = 1$

21. $\binom{100}{98} = \dfrac{100!}{98!(100-98)!} = \dfrac{100 \cdot 99 \cdot 98!}{98!2!} = 4950$

23. $\binom{21}{19} = \dfrac{21!}{19!(21-19)!} = \dfrac{21 \cdot 20 \cdot 19!}{19!2!} = 210$

---

25. $(x+2)^4 = x^4 + 4(2x^3) + 6(2^2 x^2) + 4(2^3 x) + 2^4 = x^4 + 8x^3 + 24x^2 + 32x + 16$

27. $(x-2)^5 = x^5 + 5(-2)x^4 + 10(-2)^2 x^3 + 10(-2)^3 x^2 + 5(-2)^4 x + (-2)^5 = x^5 - 10x^4 + 40x^3 - 80x^2 + 80x - 32$

29. $(2-3x)^3 = 2^3 + 3(2^2)(-3x) + 3(2)(-3x)^2 + (-3x)^3 = 8 - 36x + 54x^2 - 27x^3$

31. $(2x+3y)^4 = (2x)^4 + 4(2x)^3(3y) + 6(2x)^2(3y)^2 + 4(2x)(3y)^3 + (3y)^4$
$= 16x^4 + 96x^3 y + 216x^2 y^2 + 216xy^3 + 81y^4$

33. $(x+1)^4 = x^4 + 4(1x^3) + 6(1^2 x^2) + 4(1^3 x) + 1^4 = x^4 + 4x^3 + 6x^2 + 4x + 1$

35. $(x-1)^5 = x^5 + 5(-1)x^4 + 10(-1)^2 x^3 + 10(-1)^3 x^2 + 5(-1)^4 x + (-1)^5 = x^5 - 5x^4 + 10x^3 - 10x^2 + 5x - 1$

37. $(y-3)^3 = y^3 + 3(-3)y^2 + 3(-3)^2 y + (-3)^3 = y^3 - 9y^2 + 27y - 27$

39. $(x+y)^6$
$= \binom{6}{0}x^6 + \binom{6}{1}x^5 y + \binom{6}{2}x^4 y^2 + \binom{6}{3}x^3 y^3 + \binom{6}{4}x^2 y^4 + \binom{6}{5}xy^5 + \binom{6}{6}y^6$
$= \dfrac{6!}{0!(6-0)!}x^6 + \dfrac{6!}{1!(6-1)!}x^5 y + \dfrac{6!}{2!(6-2)!}x^4 y^2 + \dfrac{6!}{3!(6-3)!}x^3 y^3 + \dfrac{6!}{4!(6-4)!}x^2 y^4$
$+ \dfrac{6!}{5!(6-5)!}xy^5 + \dfrac{6!}{6!(6-6)!}y^6$
$= x^6 + 6x^5 y + 15x^4 y^2 + 20x^3 y^3 + 15x^2 y^4 + 6xy^5 + y^6$

41. $(1+3y)^5 = 1^5 + 5(1^4)(3y) + (10)(1^3)(3y)^2 + (10)(1^2)(3y)^3 + (5)(1)(3y)^4 + (3y)^5$
$= 1 + 15y + 90y^2 + 270y^3 + 405y^4 + 243y^5$

43. $(2x+1)^4 = (2x)^4 + 4(2x)^3(1) + 6(2x)^2(1)^2 + 4(2x)(1)^3 + 1^4 = 16x^4 + 32x^3 + 24x^2 + 8x + 1$

45. $(x-2y)^3 = x^3 + 3(x^2)(-2y) + 3(x)(-2y)^2 + (-2y)^3 = x^3 - 6x^2y + 12xy^2 - 8y^3$

47. $(2x+y)^4 = (2x)^4 + 4(2x)^3 y + 6(2x)^2 y^2 + 4(2x)y^3 + y^4 = 16x^4 + 32x^3 y + 24x^2 y^2 + 8xy^3 + y^4$

49. $\left(\dfrac{x}{2}+2\right)^7 = \binom{7}{0}\left(\dfrac{x}{2}\right)^7 + \binom{7}{1}\left(\dfrac{x}{2}\right)^6 (2) + \binom{7}{2}\left(\dfrac{x}{2}\right)^5 (2^2) + \binom{7}{3}\left(\dfrac{x}{2}\right)^4 (2^3) + \binom{7}{4}\left(\dfrac{x}{2}\right)^3 (2^4) + \binom{7}{5}\left(\dfrac{x}{2}\right)^2 (2^5)$

$\qquad + \binom{7}{6}\left(\dfrac{x}{2}\right)(2^6) + \binom{7}{7}(2^7)$

$= \dfrac{7!}{0!(7-0)!}\left(\dfrac{x}{2}\right)^7 + \dfrac{7!}{1!(7-1)!}\left(\dfrac{x}{2}\right)^6 (2) + \dfrac{7!}{2!(7-2)!}\left(\dfrac{x}{2}\right)^5 (2^2) + \dfrac{7!}{3!(7-3)!}\left(\dfrac{x}{2}\right)^4 (2^3)$

$\qquad + \dfrac{7!}{4!(7-4)!}\left(\dfrac{x}{2}\right)^3 (2^4) + \dfrac{7!}{5!(7-5)!}\left(\dfrac{x}{2}\right)^2 (2^5) + \dfrac{7!}{6!(7-6)!}\left(\dfrac{x}{2}\right)(2^6) + \dfrac{7!}{7!(7-7)!}(2^7)$

$= \dfrac{1}{128}x^7 + \dfrac{7}{32}x^6 + \dfrac{21}{8}x^5 + \dfrac{35}{2}x^4 + 70x^3 + 168x^2 + 224x + 128$

51. $\left(a^2 - \dfrac{1}{3}\right)^4 = (a^2)^4 + 4(a^2)^3\left(-\dfrac{1}{3}\right) + 6(a^2)^2\left(-\dfrac{1}{3}\right)^2 + 4(a^2)\left(-\dfrac{1}{3}\right)^3 + \left(-\dfrac{1}{3}\right)^4 = a^8 - \dfrac{4}{3}a^6 + \dfrac{2}{3}a^4 - \dfrac{4}{27}a^2 + \dfrac{1}{81}$

53. $\left(\dfrac{1}{x}+y\right)^3 = \left(\dfrac{1}{x}\right)^3 + 3\left(\dfrac{1}{x}\right)^2 y + 3\left(\dfrac{1}{x}\right)y^2 + y^3 = \dfrac{1}{x^3} + \dfrac{3y}{x^2} + \dfrac{3y^2}{x} + y^3$

55. The coefficient of the $x^2 y^7$ term in $(x+y)^9$ is $\binom{9}{9-2} = \dfrac{9!}{2!7!} = \dfrac{9 \cdot 8 \cdot 7!}{2!7!} = 36$

57. The x^5 in $(x+3)^8$ has a factor of $3^3 = 27$ in its coefficient. The coefficient is

$\binom{8}{8-5}3^3 = \left(\dfrac{8!}{5!3!}\right)3^3 = \left(\dfrac{8 \cdot 7 \cdot 6 \cdot 5!}{5!3!}\right)3^3 = 1512.$

59. The $x^2 y^7$ term in $(2x+3y)^9$ has a factor of $2^2 \cdot 3^7 = 8748$ in its coefficient. The coefficient is

$\binom{9}{9-2}(8748) = \dfrac{9!}{7!2!}(8748) = \dfrac{9 \cdot 8 \cdot 7!}{7!2!}(8748) = 314,928.$

61. The $x^{10} y^{12}$ term in $(x^2+y^3)^9$ is the $(x^2)^5 (y^3)^4$ term in the expansion. The coefficient is

$\binom{9}{9-5} = \dfrac{9!}{4!5!} = \dfrac{9 \cdot 8 \cdot 7 \cdot 6 \cdot 5!}{4!5!} = 126.$

63. The term containing x^7 is the fourth term in the expansion: $\binom{10}{7}x^7 y^3 = \dfrac{10!}{7!(10-7)!}x^7 y^3 = 120x^7 y^3$

65. The term containing x^3 is the tenth term in the expansion:

$\binom{12}{3}x^3(-2)^9 = \dfrac{12!}{3!(12-3)!}(-512)x^3 = -112,640x^3.$

67. The term containing x^6 is the third term in the expansion:

$\binom{8}{2}(2x)^6 (3y)^2 = \dfrac{8!}{2!(8-2)!}(64x^6)(9y^2) = 16,128x^6 y^2.$

69. The term containing y^9 is the tenth term in the expansion:
$$\binom{11}{9}(5x)^2(-2y)^9 = \frac{11!}{2!(11-2)!}(25x^2)(-512y^9) = -704,000x^2y^9.$$

71. The seventh term in the expansion of $(x-1)^9$ is $\binom{9}{7-1}x^{9-7+1}(-1)^{7-1} = \frac{9!}{6!3!}x^3(-1)^6 = 84x^3.$

73. The fifth term in the expansion of $(2x-y)^7$ is $\binom{7}{5-1}(2x)^{7-5+1}(-y)^{5-1} = \frac{7!}{4!3!}(2x)^3(-y)^4 = 280x^3y^4.$

75. The third term in the expansion of $(x+3y)^{10}$ is $\binom{10}{3-1}x^{10-3+1}(3y)^{3-1} = \frac{10!}{2!8!}x^8(3y)^2 = 405x^8y^2$

77. $(1.2)^5 = (1+0.2)^5 = 1^5 + 5(1)^4(0.2) + 10(1)^3(0.2)^2 + 10(1)^2(0.2)^3 + 5(1)(0.2)^4 + (0.2)^5 = 2.48832$

11.5 Beyond the Basics

79. The middle term is the sixth term in the expansion:
$$\binom{10}{5}(\sqrt{x})^5\left(-\frac{2}{x^2}\right)^5 = \frac{10!}{5!(10-5)!}x^2\sqrt{x}\left(-\frac{32}{x^{10}}\right) = -\frac{8064\sqrt{x}}{x^8}$$

81. The middle term is the seventh term in the expansion: $\binom{12}{6}(1^6)(-x^2y^{-3})^6 = \frac{12!}{6!(12-6)!}\left(\frac{x^{12}}{y^{18}}\right) = \frac{924x^{12}}{y^{18}}$

83. By the Binomial Theorem, we have
$$2^n = (1+1)^n = \binom{n}{0}(1)^n(1)^0 + \binom{n}{1}(1)^{n-1}(1)^1 + \binom{n}{2}(1)^{n-2}(1)^2 + \cdots + \binom{n}{n}(1)^0(1)^n$$
$$= \binom{n}{0} + \binom{n}{1} + \binom{n}{2} + \cdots + \binom{n}{n}$$

85. $\binom{k}{j} + \binom{k}{j-1} = \frac{k!}{j!(k-j)!} + \frac{k!}{(j-1)!(k-j+1)!} = \frac{k!(j-1)!(k-j+1)! + k!j!(k-j)!}{j!(k-j)!(j-1)!(k-j+1)!}$
$= \frac{[k!(j-1)!(k-j)!][j+(k-j+1)]}{j!(k-j)!(j-1)!(k-j+1)!} = \frac{k!(j+(k-j+1))}{j!(k+1-j)!} = \frac{k!(k+1)}{j!(k+1-j)!} = \frac{(k+1)!}{j!(k+1-j)!} = \binom{k+1}{j}$

87. $(2x-1)^4 + 4(2x-1)^3(3-2x) + 6(2x-1)^2(3-2x)^2 + 4(2x-1)(3-2x)^3 + (3-2x)^4$
$= ((2x-1)+(3-2x))^4 = 2^4 = 16$

89. $(3x-1)^5 + 5(3x-1)^4(1-2x) + 10(3x-1)^3(1-2x)^2 + 10(3x-1)^2(1-2x)^3 + 5(3x-1)(1-2x)^4 + (1-2x)^5$
$= ((3x-1)+(1-2x))^5 = x^5$

91. The constant term is the third term in the expansion:
$$240 = \binom{6}{2}(kx)^4\left(-\frac{1}{x^2}\right)^2 = \frac{6!}{2!(6-2)!}k^4x^4\left(\frac{1}{x^4}\right) = 15k^4 \Rightarrow k^4 = 16 \Rightarrow k = \pm 2$$

93. The general term in the expansion has the form $\binom{11}{k}(2x^2)^k\left(-\frac{1}{4x}\right)^{11-k} = \binom{11}{k}(2^k)\left(-\frac{1}{4}\right)^{11-k}\frac{x^{2k}}{x^{11-k}}.$

In order to get a constant term, $2k$ must equal $11-k$. However, the solution of this equation is not an integer, so there is no constant term in the expansion.

11.5 Critical Thinking/Discussion/Writing

94. $2^6 = (1+1)^6$

$= \binom{6}{0}(1)^6(1)^0 + \binom{6}{1}(1)^5(1)^1 + \binom{6}{2}(1)^4(1)^2 + \binom{6}{3}(1)^3(1)^3 + \binom{6}{4}(1)^2(1)^4 + \binom{6}{5}(1)^1(1)^5 + \binom{6}{6}(1)^0(1)^6$

$= \binom{6}{0} + \binom{6}{1} + \binom{6}{2} + \binom{6}{3} + \binom{6}{4} + \binom{6}{5} + \binom{6}{6}$

95. $0 = (1-1)^{10}$

$= \binom{10}{0}(1)^{10}(1)^0 - \binom{10}{1}(1)^9(1)^1 + \binom{10}{2}(1)^8(1)^2 - \binom{10}{3}(1)^7(1)^3 + \binom{10}{4}(1)^6(1)^4 - \binom{10}{5}(1)^5(1)^5 + \binom{10}{6}(1)^4(1)^6$

$\quad - \binom{10}{7}(1)^3(1)^7 + \binom{10}{8}(1)^2(1)^8 - \binom{10}{9}(1)^1(1)^9 + \binom{10}{10}(1)^9(1)^{10}$

$= \binom{10}{0} - \binom{10}{1} + \binom{10}{2} - \binom{10}{3} + \binom{10}{4} - \binom{10}{5} + \binom{10}{6} - \binom{10}{7} + \binom{10}{8} - \binom{10}{9} + \binom{10}{10}$

96. $(x+y)^2 = x^2 + 2xy + y^2$. If $x > 0$ and $y > 0$, then $2xy > 0$ and $x^2 + 2xy + y^2 > x^2 + y^2$. Therefore, $(x+y)^2 > x^2 + y^2$.

97. $(x+y)^n = x^n + \binom{n}{1}x^{n-1}y + \binom{n}{2}x^{n-2}y^2 + \cdots + \binom{n}{n-1}xy^{n-1} + y^n$. If $x > 0$ and $y > 0$, then all the intermediate terms in the expansion are positive. Therefore $(x+y)^n > x^n + y^n$ if $n > 1$.

98. $(1+x)^n = 1 + \binom{n}{1}x + \binom{n}{2}x^2 + \ldots + \binom{n}{n}x^n = 1 + nx + \binom{n}{1}x + \binom{n}{2}x^2 + \ldots + \binom{n}{n}x^n$. If $x > 0$, then all the terms in the sum are positive, and $1 + nx + \binom{n}{1}x + \binom{n}{2}x^2 + \ldots + \binom{n}{n}x^n > 1 + nx$. Therefore, $(1+x)^n > 1 + nx$ if $n > 1$.

11.5 Getting Ready for the Next Section

99. $5 \cdot 6 \cdot 7 \cdot 8 = \dfrac{8!}{4!}$

101. $2 \cdot 4 \cdot 6 \cdot 8 \cdot 10 \cdot 12$
$= 2 \cdot (2 \cdot 2) \cdot (2 \cdot 3) \cdot (2 \cdot 4) \cdot (2 \cdot 5) \cdot (2 \cdot 6)$
$= 2^6 \cdot 6!$

103. $\dfrac{12!}{10!} = 12 \cdot 11 = 132$

105. $\dfrac{8!}{5!3!} = \dfrac{8 \cdot 7 \cdot 6}{3 \cdot 2 \cdot 1} = 56$

11.6 Counting Principles

11.6 Practice Problems

1.

There are 9 ways to choose a course.

2. There are 7 restaurants and 5 movies, so there are $7 \cdot 5 = 35$ different choices.

3. There are $10^6 = 1,000,000$ possibilities.

4. There are $7! = 7 \cdot 6 \cdot 5 \cdot 4 \cdot 3 \cdot 2 \cdot 1 = 5040$ possible arrangements.

5. There are $9 \cdot 8 \cdot 7 \cdot 6 = 3024$ possibilities.

6. a. $P(9,2) = \dfrac{9!}{(9-2)!} = \dfrac{9 \cdot 8 \cdot 7!}{7!} = 72$

   b. $P(n,0) = \dfrac{n!}{(n-0)!} = \dfrac{n!}{n!} = 1$

7. {bears, bulls, lions}, {bears, bulls, tigers}, {bears, lions, tigers}, {bulls, lions, tigers}
   $C(4,3) = 4$

8. a. $C(9,2) = \dfrac{9!}{(9-2)!2!} = \dfrac{9 \cdot 8 \cdot 7!}{7!2!} = 36$

   b. $C(n,0) = \dfrac{n!}{(n-0)!0!} = \dfrac{n!}{n!} = 1$

   c. $C(n,n) = \dfrac{n!}{(n-n)!n!} = \dfrac{n!}{n!} = 1$

9. $C(12,3) = \dfrac{12!}{(12-3)!3!} = \dfrac{12 \cdot 11 \cdot 10 \cdot 9!}{9!3!}$
   $= \dfrac{12 \cdot 11 \cdot 10}{3 \cdot 2 \cdot 1} = 220$
   There are 220 ways that three of the twelve chocolates can be chosen.

10. $\dfrac{6!}{2!2!2!} = 90$
    There are 90 ways to send six counselors in pairs to three different locations.

11.6 Concepts and Vocabulary

1. Any arrangement of n distinct objects in a fixed order in which no object is used more than once is called a permutation.

3. When r objects are chosen from n distinct objects, the set of r objects is called a combination of n objects taken r at a time.

5. True

7. False. Three Coke, two Sprite, and three Pepsi cans can be arranged in $\dfrac{8!}{3!2!3!} = 560$ ways.

11.6 Building Skills

9. $P(6,1) = \dfrac{6!}{(6-1)!} = 6$

11. $P(8,2) = \dfrac{8!}{(8-2)!} = 56$

13. $P(9,9) = \dfrac{9!}{(9-9)!} = 362,880$

15. $P(7,0) = \dfrac{7!}{(7-0!)} = 1$

17. $C(8,3) = \dfrac{8!}{(8-3)!3!} = 56$

19. $C(9,4) = \dfrac{9!}{(9-4)!(4!)} = 126$

21. $C(5,5) = \dfrac{5!}{(5-5)!(5!)} = 1$

23. $C(3,0) = \dfrac{3!}{(3-0)!0!} = 1$

25. a. The first letter can be chosen in 26 different ways. The second letter can also be chosen in 26 different ways, so the number of different 2-letter codes is $26 \cdot 26 = 676$.

    b. The first letter can be chosen in 26 different ways. The second letter can be chosen in 25 different ways, so the number of different 2-letter codes is $26 \cdot 25 = 650$.

27. The president can be chosen in 50 ways, while the vice-president can be chosen in 49 ways. So the number of possibilities is $50 \cdot 49 = 2450$.

29. There are four people to be seated in a row of four chairs, so there are
    $P(4,4) = \dfrac{4!}{(4-4)!} = 24$ different ways to arrange the people.

31. There are five letters in the word ANGLE. Using three letters at a time, there are
    $P(5,3) = \dfrac{5!}{(5-3)!} = 5 \cdot 4 \cdot 3 = 60$ ways to arrange the letters.

33. There are five different toppings. So there are
    $C(5,2) = \dfrac{5!}{(5-2)!2!} = 10$ different pizzas with two additional toppings.

35. There are eleven problems, so it is possible to choose nine of them in
$$C(11,9) = \frac{11!}{(11-9)!(9!)} = 55 \text{ different ways.}$$

37. Order is not important, so find the number of combinations. There are $C(11,4) = 330$ different combinations.

39. Order is important, so find the number of permutations. There are 8 seats and six people will be seated, so there are
$$P(8,6) = \frac{8!}{(8-6)!} = 20,160 \text{ different ways to}$$
seat 6 people.

41. Order is important, so find the number of permutations or use the Fundamental Counting Principle. There are
$$P(5,5) = \frac{5!}{(5-5)!} = 120 \text{ ways to visit the five}$$
colleges.

11.6 Applying the Concepts

43. a. The first digit cannot be a zero or a one, so there are eight possibilities for the first digit. There are two possibilities for the second digit and ten possibilities for the third digit. So there are $8 \cdot 2 \cdot 10 = 160$ possible three-digit area codes.

b. There are eight possibilities for the first digit, ten possibilities for the second digit, and ten possibilities for the third digit. So there are $8 \cdot 10 \cdot 10 = 800$ possible three-digit area codes.

45. a. If no repetitions are allowed, there are $24 \cdot 23 \cdot 22 = 12,144$ three-letter fraternity names.

b. If repetitions are allowed, there are $24^3 = 13,824$ three-letter fraternity names.

47. To reach a majority decision, five, six, seven, eight or all nine of the nine justices must agree. Since order is not important, find the sum of the combinations
$$C(9,5) + C(9,6) + C(9,7) + C(9,8) + C(9,9)$$
$$= \frac{9!}{(9-5)!(5!)} + \frac{9!}{(9-6)!(6!)} + \frac{9!}{(9-7)!(7!)}$$
$$+ \frac{9!}{(9-8)!(8!)} + \frac{9!}{(9-9)!(9!)}$$
$$= 126 + 84 + 36 + 9 + 1 = 256$$

There are 256 ways to form a majority of five from the nine justices.

49. Al can choose from $8 \cdot 8 \cdot 8 = 512$ different outfits. Since there are 365 days in a year, he can wear a different outfit every day.

51. a. There are $7! = 5040$ different ways to arrange the houses since any arrangement of the seven houses could be made to correspond to a permutation of the seven designs by agreeing that the first four go on one side of the street and the remaining three go on the other side of the street.

b. There are $7! = 5040$ different ways to arrange the houses since any arrangement of the seven houses could be made to correspond to a permutation of the seven designs by agreeing that the first five go on one side of the street and the remaining two go on the other side of the street.

53. a. There are seven letters in the word CHARITY. If two letters in each four-letter group must be an A and an R, then there are five letters that can be chosen in groups of two: $C(5,2) = \frac{5!}{(5-2)!(2)!} = 10$ possibilities.

b. If one letter in each four-letter group must be an A and none of the letters in the group can be an R, then there are five letters that can be chosen in groups of three:
$$C(5,3) = \frac{5!}{(5-3)!(3)!} = 10 \text{ possibilities.}$$

c. If none of the letters in each four-letter group can be an A or an R, there there are five letters that can be chosen in groups of four: $C(5,4) = \frac{5!}{(5-4)!(4)!} = 5$ possibilities.

55. The store will need to stock $5 \cdot 7 \cdot 3 = 105$ shirts to have one of each type.

57. The maximum number of folders is necessary if all of the questionnaires have different rankings. There are $10! = 3,628,800$ possible rankings, so they would need that many folders.

59. In each group of four, if one child is "constant", then the other three children can be chosen in $C(6,3) = \dfrac{6!}{(6-3)!3!} = 20$ ways. So the maximum number of times any one child will go to the circus is 20.

    There are $C(7,4) = \dfrac{7!}{(7-4)!4!} = 35$ combinations of four children, so the father will go to the circus 35 times.

61. Because order is not important, we use combinations. Under the "B", there are $C(15,5) = \dfrac{15!}{(15-5)!5!} = 3003$ combinations of numbers. There are the same number under the "I", "G", and "O". Under the "N", there are $C(15,4) = \dfrac{15!}{(15-4)!4!} = 1365$. So there are $3003^4 \cdot 1365 = 111{,}007{,}923{,}832{,}370{,}565$ different bingo cards.

63. There are 10 people and four are needed for the committee. Once the first person is chosen, that person's spouse is eliminated, so there are 8 ways to choose the next person. That person's spouse cannot be chosen next, so there are 6 ways to choose the third person, and 4 ways to choose the fourth person. That gives 1920 possible permutations. However, the order that people are chosen for the committee is not important, so the number of combinations is $\dfrac{1920}{4!} = 80$.

65. If Cory wants to give his sister one coin, there are 4 different combinations of one coin. If he wants to give his sister two coins, there are $C(4,2) = 6$ different combinations. If he wants to give his sister three coins, there are 4 combinations, and if he wants to give his sister four coins, there is one combination. So there are $4 + 6 + 4 + 1 = 15$ different combinations of coins.

67. Out of six letters, there are two "A"s and the rest are singles. So, there are $\dfrac{6!}{2!1!1!1!1!} = 360$ distinguishable ways to arrange the letters.

69. Out of seven letters, there are three "S"s, two "C"s, and the rest are singles. So, there are $\dfrac{7!}{3!2!1!1!} = 420$ distinguishable ways to arrange the letters.

11.6 Beyond the Basics

71. The company needs n workers in combinations of three. So
    $$C(n,3) = \dfrac{n!}{(n-3)!3!} \geq 20 \Rightarrow$$
    $$\dfrac{n(n-1)(n-2)(n-3)!}{(n-3)!3!} \geq 20 \Rightarrow$$
    $$n^3 - 3n^2 + 2n \geq 120 \Rightarrow$$
    $$n^3 - 3n^2 + 2n - 120 \geq 0$$
    The possible rational zeros are
    $\pm 1, \pm 2, \pm 3, \pm 4, \pm 5, \pm 6, \pm 8, \pm 10, \pm 12,$
    $\pm 15, \pm 20, \pm 24, \pm 30, \pm 60, \pm 120.$
    Using synthetic division, we find that 6 is a zero:

    ```
    6| 1   -3    2   -120
           6    18    120
       ─────────────────────
       1    3    20     0
    ```

    and $(n-6)(n^2 + 3n + 20) = 0$.

    The two zeros of $n^2 + 3n + 20 = 0$ are complex, so we reject them. Therefore, six different workers are necessary.

73. There are $C(30,2) = \dfrac{30!}{(30-2)!2!} = 435$ lines.

75. $\dbinom{m+1}{m-1} = 3! \Rightarrow$
    $$\dfrac{(m+1)!}{(m-1)!((m+1)-(m-1))!} = 6 \Rightarrow$$
    $$\dfrac{(m+1)!}{(m-1)!2!} = 6 \Rightarrow \dfrac{(m+1)m(m-1)!}{(m-1)!} = 12 \Rightarrow$$
    $m^2 + m - 12 = 0 \Rightarrow (m+4)(m-3) = 0 \Rightarrow$
    $m = 3$ or $m = -4$
    Reject the negative solution because factorials are not defined for negative numbers. The solution is $\{3\}$.

77. $2\dbinom{n-1}{2} = \dbinom{n}{3} \Rightarrow \dfrac{2(n-1)!}{2!(n-3)!} = \dfrac{n!}{3!(n-3)!} \Rightarrow$
    $$\dfrac{(n-1)(n-2)(n-3)!}{(n-3)!} = \dfrac{n(n-1)(n-2)(n-3)!}{6(n-3)!} \Rightarrow$$
    $6(n-1)(n-2) = n(n-1)(n-2) \Rightarrow n = 6$

79. Using the Binomial Theorem,
$$\binom{n}{0}+\binom{n}{1}+\binom{n}{2}+\cdots+\binom{n}{n}=64$$ is the sum of the coefficients in the nth row of the expansion of $(x+y)^n$. If $x=1$ and $y=1$, then
$$(1+1)^n = \binom{n}{0}+\binom{n}{1}+\binom{n}{2}+\cdots+\binom{n}{n}=64 \Rightarrow$$
$2^n = 64 \Rightarrow n = 6$.

81. $\binom{m}{4}=\binom{m}{5} \Rightarrow \dfrac{m!}{(m-4)!4!} = \dfrac{m!}{(m-5)!5!} \Rightarrow$
$\dfrac{5!}{4!} = \dfrac{(m-4)!}{(m-5)!} \Rightarrow 5 = \dfrac{(m-4)(m-5)!}{(m-5)!} \Rightarrow$
$5 = m-4 \Rightarrow m = 9$

11.6 Critical Thinking/Discussion/Writing

83. There are $5 \cdot 5 = 25$ terms of the form $x^n y^m$ if n and m are any integers such that $1 \le n \le 5$ and $1 \le m \le 5$.

84. There are two terms in the first factor and three terms in the second factor, so there are $2 \cdot 3 = 6$ terms.

11.6 Preparing for the Next Section

85. $E = \{2, 5, 7\} \Rightarrow n(E) = 3$

87. $E = \{0\} \Rightarrow n(E) = 1$

For exercises 89–92, $A = \{1, 2, 3, 5\}$ and $B = \{3, 4, 5, 6, 7\}$.

89. $n(A) = 4$, $n(B) = 5$

91. $A \cap B = \{3, 5\}$
$n(A \cap B) = 2$

For exercises 93–96, $A = \{a, b, c\}$ and $B = \{2, 4, 6, 8\}$.

93. $n(A) = 3$, $n(B) = 4$

95. $A \cap B = \emptyset$
$n(A \cap B) = 0$

11.7 Probability

11.7 Practice Problems

1. a. $E_1 = \{2, 4, 6\}$, $P(E_1) = \dfrac{3}{6} = \dfrac{1}{2}$

b. $E_2 = \{5, 6\}$, $P(E_2) = \dfrac{2}{6} = \dfrac{1}{3}$

2. $S = \{(H, H), (H, T), (T, H), (T, T)\}$
$E = \{(H, T), (T, H)\}$
$P(E) = \dfrac{2}{4} = \dfrac{1}{2}$

3. $E = \{(6, 1), (5, 2), (4, 3), (3, 4), (2, 5), (6, 1)\}$
There are 36 possible outcomes, so
$P(E) = \dfrac{6}{36} = \dfrac{1}{6}$

4. Since the order in which the numbers are selected is not important, the sample space S consists of all sets of 6 numbers that can be selected from 50 numbers. So,
$n(S) = C(50, 6) = \dfrac{50!}{(50-6)!6!}$
$= \dfrac{50 \cdot 49 \cdot 48 \cdot 47 \cdot 46 \cdot 45}{6 \cdot 5 \cdot 4 \cdot 3 \cdot 2 \cdot 1} = 15{,}890{,}700$
$P(\text{winning the lottery}) = \dfrac{1}{15{,}890{,}700}$

5. $P(\text{jack or king}) = P(\text{jack}) + P(\text{king})$
$= \dfrac{4}{52} + \dfrac{4}{52} = \dfrac{8}{52} = \dfrac{2}{13}$

6. $P(\text{Brandy is not selected}) = \dfrac{C(9,3)}{C(10,3)}$
$= \dfrac{9!}{3!6!} \cdot \dfrac{3!7!}{10!} = \dfrac{7}{10}$
$P(\text{Brandy is selected})$
$= 1 - P(\text{Brandy is not selected}) = 1 - \dfrac{7}{10} = \dfrac{3}{10}$

7. The probability that a student selected at random is a male is $\frac{45,143}{82,074} \approx 0.55$. The probability that a student selected at random attends Sacramento College is $\frac{20,878}{82,074} \approx 0.25$. The probability that a student selected at random is a male attending Sacramento College is $\frac{9040}{82,074} \approx 0.11$. So, the probability that a student selected at random is a male or attends Sacramento College is approximately $0.55 + 0.25 - 0.11 = 0.69$.

11.7 Concepts and Vocabulary

1. If no outcome of an experiment results more often than any other outcome, then the outcomes are said to be <u>equally likely</u>.

3. If it is impossible for two events to occur simultaneously, then the events are said to be <u>mutually exclusive</u>.

5. True (if "between 0 and 1" includes 0 and 1.)

7. False. The probability will be $\frac{1}{10}$ only if each outcome is equally likely.

11.7 Building Skills

9. $S = \{$(Wendy's, McDonald's, Burger King), (Wendy's, Burger King, McDonald's), (McDonald's, Wendy's, Burger King), (McDonald's, Burger King, Wendy's), (Burger King, McDonald's, Wendy's), (Burger King, Wendy's, McDonald's)$\}$

11. $S = \{\{Thriller, The Wall\}\}, \{Thriller, Eagles: Their Greatest Hits\}, \{Thriller, Led Zeppelin IV\}, \{The Wall, Eagles: Their Greatest Hits\}, \{The Wall, Led Zeppelin IV\}, \{Eagles: Their Greatest Hits, Led Zeppelin IV\}\}$

13. $S = \{$(white, male), (white, female), (African-American, male), (African-American, female), (Native American, male), (Native American, female), (Asian, male), (Asian, female), (other, male), (other, female), (multiracial, male), (multiracial, female)$\}$

15. $P = 0$.

17. $P = 1$

19. experimental

21. theoretical

23. experimental

25. theoretical

27. experimental

29. An event that is very likely to happen has probability 0.999. An event that will surely happen has probability 1. An event that is a rare event has probability 0.001. An event that is equally likely to happen or not happen has probability 0.5. An event that will never happen has probability 0.

31. $E = \{1, 6\}, P(E) = \frac{1}{3}$

33. $E = \{5, 6\}, P(E) = \frac{1}{3}$

35. $E = \{2, 4, 6\}, P(E) = \frac{1}{2}$

37. $E = \{$club queen, spade queen, heart queen, diamond queen$\}$, $P(E) = \frac{1}{13}$

39. $E = \{$heart ace, heart 2, heart 3, heart 4, heart 5, heart 6, heart 7, heart 8, heart 9, heart 10, heart jack, heart queen, heart king$\}$, $P(E) = \frac{1}{4}$

41. $E = \{$heart jack, heart queen, heart king, diamond jack, diamond queen, diamond king, club jack, club queen, club king, spade jack, spade queen, spade king$\}$, $P(E) = \frac{3}{13}$

43. $E = \{8 + 3\}, P(E) = \frac{1}{10}$

45. $E = \{0 + 3\}, P(E) = \frac{1}{10}$

47. $E = \{1 + 3, 3 + 3, 5 + 3, 7 + 3, 9 + 3\}, P(E) = \frac{1}{2}$

11.7 Applying the Concepts

49. The probability the Tony will not like his blind date is $1 - 0.3 = 0.7$

51. $P = \frac{20}{100} = \frac{1}{5}$

53. 6% of the Marines were women, so 94% were men. The probability that a Marine chosen at random is a man is $\frac{94}{100} = \frac{47}{50}$.

55. If you have 10 tickets and there is only one prize, then the probability of winning is $\frac{10}{125} = \frac{2}{25}$. If there are two prizes, then there are $C(125, 2) = 7750$ possible winning combinations, and you have $C(10, 2) = 45$ possible winning tickets. So the probability of winning is $\frac{45}{7750} = \frac{9}{1550}$.

57. a. There are $2^2 = 4$ possible combinations (bb, bg, gb, gg). The probability of having two boys is $1/4$.

  b. The probability of having two girls is $1/4$.

  c. The probability of having one boy and one girl is $1/2$.

59. a. Out of the 4440 families, 2370 families own two cars, so the probability of a family owning two cars is $\frac{2370}{4440} = \frac{79}{148}$.

  b. Out of the 4440 families, 37 own no cars, 1256 own one car, and 2370 own two cars. So the probability of a family owning at most two cars is $\frac{3663}{4440} = \frac{33}{40}$.

  c. Out of the 4440 families, none own six cars, so the probability of a family owning six cars is 0.

  d. All of the families own fewer than six cars, so the probability of a family owning at most six cars is 1.

  e. Of the 4440 families, all but 37 own at least one car, so the probability of a family owning at least one car is $\frac{4403}{4340} = \frac{119}{120}$.

61. The probability that a non-Hispanic white American has no lactose intolerance is $1 - 0.2 = 0.8 = \frac{4}{5}$. The probability that an African-American, Asian, or Native American has no lactose intolerance is $1 - 0.75 = 0.25 = 1/4$.

63. a. There are $C(10, 2) = 10$ different combinations of people for the committee. There is only one combination of the two oldest members, so the probability of the committee consisting of the two oldest members is $1/10$.

  b. There is only one combination of the oldest and youngest members, so the probability is $1/10$.

65.

Number of people who are	Depressed according to the test	Normal according to the test
Actually depressed	90	10
Actually normal	135	765

If 10% of the population suffers from depression, then we expect that 10% of the 1000 people tested = 100 people will suffer from depression and 900 will not. Of the 100 people, the test will work for 90% of them, or 90 people, and the remaining 10 people are normal according to the test. Of the 900 people who don't suffer from depression, the test works for 85% of them, or 765 people. The remaining 135 are depressed according to the test.

67. The possible combinations are {bbbb, bbbg, bbgg, bggg, gggg}. (Birth order is not important.) So, there are two ways to have three children of one gender and one of the other gender, while there is only one way to have two children of each gender. Therefore, it is more likely to have three children of one gender and one of the other.

69. a. Out of 180 people, 50 people received the placebo, a total of 56 people had no pain relief, while 34 people received the placebo and had no pain relief. So the probability that a person either received the placebo or had no pain relief is $\frac{50}{180} + \frac{56}{180} - \frac{34}{180} = \frac{2}{5}$.

  b. Out of 180 people, 70 people received the new medicine, a total of 69 people had complete pain relief, while 40 people who received the new medicine also had complete pain relief. So, the probability that a person either received the new medicine or had complete pain relief is $\frac{70}{180} + \frac{69}{180} - \frac{40}{180} = \frac{99}{180}$.

11.7 Beyond the Basics

71. To find the probability that a number between 1 and 1,000,000 does not contain the digit 3, note that each place value through the hundred-thousand's place can contain one of nine digits. So the probability that a number between 1 and 1,000,000 will not contain the digit 3 is $\frac{9^6}{1,000,000} = \frac{531,441}{1,000,000}$. So the probability that a number between 1 and 1,000,000 will contain the digit 3 is $1 - \frac{531,441}{1,000,000} = \frac{468,559}{1,000,000}$.

73. Because one of the digits is repeated, there are $\frac{4!}{2!1!1!} = 12$ different ways to arrange the digits. The probability that Maggie will dial the correct number is $1/12$.

11.7 Critical Thinking/Discussion/Writing

74. The probability that a person chosen at random in the survey would report his or her highest level of education as high school graduation is $\frac{59,840,000}{187,000,000} = \frac{8}{25}$.

75. If there are x dark caramels in the box, then there are $2x$ light caramels, and a total of $3x$ caramels in the box. So the probability of choosing a dark caramel is $\frac{x}{3x} = \frac{1}{3}$.

76. If a single die is rolled two times, then there are $6 \cdot 6 = 36$ possible outcomes. If the first number rolled is n, then there are $6 - n$ possible rolls in which the second roll will be greater than the first roll. So there are $\sum_{n=1}^{5} 6 - n = 15$ possible rolls. The probability that the second roll will result in a number larger than the first roll is $\frac{15}{36} = \frac{5}{12}$.

Chapter 11 Review Exercises

1. $a_1 = 2(1) - 3 = -1, a_2 = 2(2) - 3 = 1,$
$a_3 = 2(3) - 3 = 3, a_4 = 2(4) - 3 = 5,$
$a_5 = 2(5) - 3 = 7$

3. $a_1 = \frac{1}{2(1)+1} = \frac{1}{3}, a_2 = \frac{2}{2(2)+1} = \frac{2}{5},$
$a_3 = \frac{3}{2(3)+1} = \frac{3}{7}, a_4 = \frac{4}{2(4)+1} = \frac{4}{9},$
$a_5 = \frac{5}{2(5)+1} = \frac{5}{11}$

5. This is an arithmetic sequence with a common difference of -2. $a_n = 32 - 2n$ for $n \geq 1$.

7. $\frac{9!}{8!} = 9$

9. $\frac{(n+1)!}{n!} = n+1$

11. $\sum_{k=1}^{4} k^3 = 1^3 + 2^3 + 3^3 + 4^3 = 100$

13. $\sum_{k=1}^{7} \frac{k+1}{k}$
$= \frac{1+1}{1} + \frac{2+1}{2} + \frac{3+1}{3} + \frac{4+1}{4} + \frac{5+1}{5} + \frac{6+1}{6} + \frac{7+1}{7} = \frac{1343}{140}$

15. $\sum_{k=1}^{50} \frac{1}{k}$

17. The sequence is arithmetic. $a_1 = 11, d = -5$

19. The sequence is not arithmetic.

21. $a_n = 3n$

23. $a_n = x + n - 1$

25. $a_3 = a_1 + d(3-1) \Rightarrow 7 = a_1 + 2d$
$a_8 = a_1 + d(8-1) \Rightarrow 17 = a_1 + 7d$
$\begin{cases} a_1 + 2d = 7 \\ a_1 + 7d = 17 \end{cases} \Rightarrow -5d = -10 \Rightarrow d = 2, a_1 = 3$
$a_n = 3 + 2(n-1) = 2n + 1$

27. $d = 2, a_1 = 7, a_n = 37 \Rightarrow$
$37 = 7 + 2(n-1) \Rightarrow n = 16$
$S = 16\left(\frac{7+37}{2}\right) = 352$

29. $d = 5, a_1 = 3, n = 40 \Rightarrow$
$a_n = 3 + 5(40-1) \Rightarrow a_n = 198$
$S = 40\left(\frac{3+198}{2}\right) = 4020$

31. The sequence is geometric. $a_1 = 4, r = -2$

33. The sequence is not geometric.

35. $a_1 = 16, r = -\frac{1}{4}, a_n = 16 \cdot \left(-\frac{1}{4}\right)^{n-1} = \frac{(-1)^{n-1}}{4^{n-3}}$

37. $a_{10} = a_1 r^{10-1} = 2 \cdot 3^9 = 39,366$

39. $r = \frac{1/5}{1/10} = 2; S_{12} = \frac{(1/10)(1 - 2^{12})}{1 - 2} = \frac{819}{2}$

41. $r = \frac{1/6}{1/2} = \frac{1}{3}; S = \frac{1/2}{1 - (1/3)} = \frac{3}{4}$

43. $a_1 = \frac{3}{5}, r = \frac{3}{5}; \sum_{i=1}^{\infty} \left(\frac{3}{5}\right)^i = \frac{3/5}{1 - 3/5} = \frac{3}{2}$

45. For $n = 1, \sum_{k=1}^{1} 2^k = 2 = 2^{1+1} - 2$ is true.

    Assume that it is true for
    $n = m: \sum_{k=1}^{m} 2^k = 2^{m+1} - 2$.

    Then for $n = m + 1$,
    $\sum_{k=1}^{m+1} 2^k = \left(\sum_{k=1}^{m} 2^k\right) + 2^{m+1} = 2^{m+1} - 2 + 2^{m+1}$
    $= 2(2^{m+1}) - 2 = 2^{m+2} - 2$

    which is exactly the statement for $n = m + 1$. Therefore the formula is true for all natural numbers.

47. $\binom{12}{7} = \frac{12!}{7!(12-7)!} = \frac{12 \cdot 11 \cdot 10 \cdot 9 \cdot 8 \cdot 7!}{7!5!}$
    $= \frac{12 \cdot 11 \cdot 10 \cdot 9 \cdot 8}{5 \cdot 4 \cdot 3 \cdot 2 \cdot 1} = 792$

49. $(x - 3)^4$
    $= x^4 + 4(-3)x^3 + 6(-3)^2 x^2 + 4(-3)^3 x + (-3)^4$
    $= x^4 - 12x^3 + 54x^2 - 108x + 81$

51. The term containing x^5 is the eighth term.
    $\binom{12}{5} x^5 (2^7) = \frac{12!}{5!(12-5)!} x^5 (2^7) = 101,376 x^5$

53. If no repetitions are allowed, using the Fundamental Counting Principle, there are $4 \cdot 3 \cdot 2 \cdot 1 = 24$ different numbers that can be written.

55. Since order is important, find the number of permutations of 7 taken 7 at a time:
    $P(7,7) = \frac{7!}{(7-7)!} = 5040$.
    There are 5040 different ways that seven people can line up.

57. The five movies can be listed in $5! = 120$ different ways.

59. Order is not important, so find the number of combinations of three candies from a group of ten. There $C(10, 3) = \frac{10!}{3!(10-3)!} = 120$ ways to choose three candies from a box of ten candies.

61. There are $C(12, 2) = 66$ ways to choose two shirts from 12, and $C(8, 3) = 56$ ways to choose three pairs of pants from eight. So, there are $66 \cdot 56 = 3696$ ways to choose two shirts and three pairs of pants.

63. There are nine letters, so $n = 9$. There are two R's, three E's, and two T's. So there are
    $\frac{9!}{2!3!2!1!1!} = 15,120$ distinguishable ways to arrange the letters.

65. Pair the two coins that appear together as one so that there are three positions to fill. Then, (with the paired coins counted as one) there are $3! = 6$ ways to fill these positions. Since the paired coins can be inserted into any position in $2! = 2$ ways, there are $3!2! = 6 \times 2 = 12$ allowable ways to arrange the coins. The total number of ways to arrange the coins is $4! = 24$, so the probability of arranging the coins so that the two most recent dates will be next to each other is $\frac{12}{24} = \frac{1}{2}$.

67. There are six letters, so there are $C(6, 2) = 15$ combinations of two letters. There are four consonants, so there are $C(4, 2) = 6$ combinations of two consonants. So the probability of choosing two consonants is $\frac{6}{15} = \frac{2}{5}$.

69. There are $C(8, 2) = 28$ ways to choose two people from the group of either. There are five ways to choose one man and three ways to choose one woman, so there are $5 \cdot 3 = 15$ ways to choose one man and one woman. So the probability of choosing one man and one woman is $\dfrac{15}{28}$.

71. There are 13 clubs, so there are $C(13, 2) = 78$ ways to choose two clubs. There are $C(52, 2) = 1326$ ways to choose two cards from the entire deck, so the probability of choosing two clubs is $\dfrac{78}{1326} = \dfrac{1}{17}$.

73. a. $\dfrac{1}{9}$ **b.** $\dfrac{4}{9}$ **c.** $\dfrac{5}{9}$

Chapter 11 Practice Test A

1. $a_1 = 3(5 - 4(1)) = 3, a_2 = 3(5 - 4(2)) = -9,$
$a_3 = 3(5 - 4(3)) = -21, a_4 = 3(5 - 4(4)) = -33,$
$a_5 = 3(5 - 4(5)) = -45$.
The sequence is arithmetic.

2. $a_1 = -3(2^1) = -6, a_2 = -3(2^2) = -12,$
$a_3 = -3(2^3) = -24, a_4 = -3(2^4) = -48,$
$a_5 = -3(2^5) = -96$. The sequence is geometric.

3. $a_1 = -2, a_2 = 3a_1 + 5 = 3(-2) + 5 = -1,$
$a_3 = 3a_2 + 5 = 3(-1) + 5 = 2,$
$a_4 = 3a_3 + 5 = 3(2) + 5 = 11,$
$a_5 = 3a_4 + 5 = 3(11) + 5 = 38$

4. $\dfrac{(n-1)!}{n!} = \dfrac{(n-1)!}{n(n-1)!} = \dfrac{1}{n}$

5. $a_7 = -3 + 6(4) = 21$

6. $a_8 = 13\left(-\dfrac{1}{2}\right)^7 = -\dfrac{13}{128}$

7. $a_1 = 1, a_{20} = 58$
$\displaystyle\sum_{k=1}^{20}(3k - 2) = 20\left(\dfrac{1+58}{2}\right) = 590$

8. $a_1 = \dfrac{3}{8}, a_5 = \dfrac{3}{128}$
$\displaystyle\sum_{k=1}^{5}\left(\dfrac{3}{4}\right)(2^{-k}) = \dfrac{\dfrac{3}{8}\left(1 - \left(\dfrac{1}{2}\right)^5\right)}{1 - \dfrac{1}{2}} = \dfrac{93}{128}$

9. $a_1 = \dfrac{9}{50}; \displaystyle\sum_{k=1}^{\infty}18\left(\dfrac{1}{100}\right)^k = \dfrac{9/50}{1 - 1/100} = \dfrac{2}{11}$

10. $\dbinom{13}{0} = \dfrac{13!}{0!(13-0)!} = 1$

11. $(1 - 2x)^4$
$= 1^4 + 4(1^3)(-2x) + 6(1^2)(-2x)^2$
$\quad + 4(1)(-2x)^3 + (-2x)^4$
$= 16x^4 - 32x^3 + 24x^2 - 8x + 1$

12. The term containing x^1 is the fourth term.
$\dbinom{4}{1}(2x)^1(1)^3 = \dfrac{4!}{1!(4-1)!}(2x)^1(1)^3 = 8x$

13. There are $2^{10} = 1024$ different ways to answer every question on a ten-question true-false test.

14. $P(9, 2) = \dfrac{9!}{(9-2)!} = 72$

15. $C(7, 5) = \dfrac{7!}{(7-5)!5!} = 21$

16. There are $6 \cdot 10 = 60$ ways to fill the positions.

17. There are $26 \cdot 26 \cdot 10^4 = 6,760,000$ possible license plate numbers.

18. $\dfrac{5}{12}$ **19.** $\dfrac{5}{6}$

20. There are 18 coins and $C(18, 2) = 153$ ways to choose two coins. There are $C(7, 2) = 21$ ways to choose two quarters. So the probability of choosing two quarters is $\dfrac{21}{153} = \dfrac{7}{51}$.

Chapter 11 Practice Test B

1. $a_1 = 4(2(1) - 3) = -4, a_2 = 4(2(2) - 3) = 4,$
$a_3 = 4(2(3) - 3) = 12, a_4 = 4(2(4) - 3) = 20,$
$a_5 = 4(2(5) - 3) = 28$. The sequence is arithmetic. The answer is C.

2. $a_1 = 2(4^1) = 8, a_2 = 2(4^2) = 32,$
   $a_3 = 2(4^3) = 128, a_4 = 2(4^4) = 512,$
   $a_5 = 2(4^5) = 2048$. The sequence is geometric. The answer is D.

3. $a_1 = 5, a_2 = 2a_1 + 4 = 2(5) + 4 = 14,$
   $a_3 = 2a_2 + 4 = 2(14) + 4 = 32,$
   $a_4 = 2a_3 + 4 = 2(32) + 4 = 68,$
   $a_5 = 2a_4 + 4 = 2(68) + 4 = 140.$
   The answer is A.

4. $\dfrac{(n+2)!}{(n+2)} = \dfrac{(n+2)(n+1)!}{(n+2)} = (n+1)!$.
   The answer is C.

5. $a_8 = a_1 + 7d = -6 + 7(3) = 15$
   The answer is A.

6. $a_{10} = 247\left(\dfrac{1}{3}\right)^9 = \dfrac{247}{19,683}$
   The answer is B.

7. $a_1 = -3, a_{45} = 4(45) - 7 = 173$
   $\displaystyle\sum_{k=1}^{45}(4k - 7) = 45\left(\dfrac{-3 + 173}{2}\right) = 3825$
   The answer is D.

8. $a_1 = \dfrac{8}{3}, r = 2; \displaystyle\sum_{k=1}^{5}\left(\dfrac{4}{3}\right)(2^k) = \dfrac{\dfrac{8}{3}(1 - 2^5)}{1 - 2} = \dfrac{248}{3}$.
   The answer is A.

9. $a_1 = 8; \displaystyle\sum_{k=1}^{\infty}8(-0.3)^{k-1} = \dfrac{8}{1 - (-0.3)} \approx 6.15$.
   The answer is B.

10. $\dbinom{12}{11} = \dfrac{12!}{11!(12-11)!} = 12$
    The answer is B.

11. $(3x - 1)^4$
    $= (3x)^4 + 4(3x)^3(-1) + 6(3x)^2(-1)^2$
    $\quad + 4(3x)(-1)^3 + (-1)^4$
    $= 81x^4 - 108x^3 + 54x^2 - 12x + 1$
    The answer is A.

12. The term containing x is the third term.
    $\dbinom{3}{1}(2x)^1(3)^2 = \dfrac{3!}{1!(3-1)!}(2x)^1(9) = 54x$
    The answer is D.

13. There are $4 \cdot 10 \cdot 3 = 120$ ways to choose a necklace, a pair of earrings, and a bracelet. The answer is C.

14. $P(7, 3) = \dfrac{7!}{(7-3)!} = 210$
    The answer is A.

15. $C(10, 7) = \dfrac{10!}{7!(10-7)!} = 120$
    The answer is D.

16. The first digit can be chosen in nine ways (note that the first digit cannot be zero), and the second digit can be chosen in nine ways. So there are $9 \cdot 9 = 81$ two-digit numbers that can be formed without using a digit more than once. The answer is B.

17. Since one CD is a "must buy", four CDs must be chosen from the remaining seven. There are $C(7, 4) = \dfrac{7!}{4!(7-4)!} = 35$ ways to choose the CDs. The answer is D.

18. The answer is C.

19. There are $6 \cdot 6 = 36$ possible outcomes. The outcomes in which the sum is greater than 9 are $\{(4, 6), (6, 4), (5, 5), (5, 6), (6, 5) (6, 6)\}$, so the probability of getting a sum greater than 9 when two die are rolled is $\dfrac{6}{36} = \dfrac{1}{6}$.
    The answer is A.

20. There are $52 - 13 = 39$ non-spades in the deck, so the probability of choosing a card that is not a spade is $\dfrac{39}{52} = \dfrac{3}{4}$.
    The answer is D.

Cumulative Review Exercises (Chapters P–11)

1. $|3x - 8| = |x| \Rightarrow 3x - 8 = x$ or $3x - 8 = -x \Rightarrow$
   $x = 4$ or $x = 2$. The solution is $\{2, 4\}$.

2. $x^2(x^2 - 5) = -4 \Rightarrow x^4 - 5x^2 + 4 = 0 \Rightarrow$
   $(x^2 - 4)(x^2 - 1) = 0 \Rightarrow$
   $(x - 2)(x + 2)(x - 1)(x + 1) = 0 \Rightarrow$
   $x = 2$ or $x = -2$ or $x = 1$ or $x = -1$
   The solution is $\{-2, -1, 1, 2\}$.

3. $\log_2(3x-5) + \log_2 x = 1 \Rightarrow$
$\log_2 x(3x-5) = 1 \Rightarrow 3x^2 - 5x = 2 \Rightarrow$
$3x^2 - 5x - 2 = 0 \Rightarrow (3x+1)(x-2) = 0 \Rightarrow$
$x = -\frac{1}{3}$ or $x = 2$
Reject the negative answer. The solution is $\{2\}$.

4. Let $u = e^x$. Then $e^{2x} - e^x - 2 = 0 \Rightarrow$
$u^2 - u - 2 = 0 \Rightarrow (u-2)(u+1) = 0 \Rightarrow$
$u = 2$ or $u = -1$. If $u = -1$, then
$-1 = e^x \Rightarrow \ln(-1) = x$, which is impossible,
so reject -1. If $u = 2$, then $2 = e^x \Rightarrow x = \ln 2$.
The solution is $\{\ln 2\}$.

5. $\dfrac{6}{x+2} = \dfrac{4}{x} \Rightarrow 6x = 4x + 8 \Rightarrow x = 4$

6. $2(x-8)^{-1} = (x-2)^{-1} \Rightarrow \dfrac{2}{x-8} = \dfrac{1}{x-2} \Rightarrow$
$2x - 4 = x - 8 \Rightarrow x = -4$

7. $\sqrt{x-3} = \sqrt{x} - 1 \Rightarrow \left(\sqrt{x-3}\right)^2 = \left(\sqrt{x}-1\right)^2 \Rightarrow$
$x - 3 = x - 2\sqrt{x} + 1 \Rightarrow 2 = \sqrt{x} \Rightarrow x = 4$

8. $x^2 + 4x \geq 0 \Rightarrow x(x+4) \geq 0$
Solving the associated equation, we have $x = 0$ or $x = -4$. The intervals to be tested are $(-\infty, -4], [-4, 0], [0, \infty)$.

Interval	Test point	Value of $x(x+4)$	Result
$(-\infty, -4]$	-5	5	$+$
$[-4, 0]$	-1	-3	$-$
$[0, \infty)$	1	5	$+$

The solution is $(-\infty, -4] \cup [0, \infty)$.

9. $18 \leq x^2 + 6x \Rightarrow x^2 + 6x - 18 \geq 0$
Solving the associated equation, we have
$x = \dfrac{-6 \pm \sqrt{36 - 4(-18)}}{2} = \dfrac{-6 \pm \sqrt{108}}{2}$
$= \dfrac{-6 \pm 6\sqrt{3}}{2} = -3 \pm 3\sqrt{3}$
The intervals to be tested are $\left(-\infty, -3 - 3\sqrt{3}\right]$, $\left[-3 - 3\sqrt{3}, -3 + 3\sqrt{3}\right], \left[-3 + 3\sqrt{3}, \infty\right)$.

Interval	Test point	Value of $x^2 + 6x - 18$	Result
$\left(-\infty, -3 - 3\sqrt{3}\right]$	-10	22	$+$
$\left[-3 - 3\sqrt{3}, -3 + 3\sqrt{3}\right]$	0	-18	$-$
$\left[-3 + 3\sqrt{3}, \infty\right)$	3	9	$+$

The solution is $\left(-\infty, -3 - 3\sqrt{3}\right] \cup \left[-3 + 3\sqrt{3}, \infty\right)$.

10. $\begin{cases} 5x + 4y = 6 \\ 4x - 3y = 11 \end{cases} \Rightarrow \begin{cases} 15x + 12y = 18 \\ 16x - 12y = 44 \end{cases} \Rightarrow$
$31x = 62 \Rightarrow x = 2$
$5(2) + 4y = 6 \Rightarrow y = -1$

11. $\begin{cases} x - 3y + 6z = -8 \\ 5x - 6y - 2z = 7 \\ 3x - 2y - 10z = 11 \end{cases} \Rightarrow \begin{bmatrix} 1 & -3 & 6 & | & -8 \\ 5 & -6 & -2 & | & 7 \\ 3 & -2 & -10 & | & 11 \end{bmatrix}$

$\xrightarrow[R_3 - 3R_1 \to R_3]{R_2 - 5R_1 \to R_2} \begin{bmatrix} 1 & -3 & 6 & | & -8 \\ 0 & 9 & -32 & | & 47 \\ 0 & 7 & -28 & | & 35 \end{bmatrix}$

$\xrightarrow{-\frac{1}{28}(9R_3 - 7R_2) \to R_3} \begin{bmatrix} 1 & -3 & 6 & | & -8 \\ 0 & 9 & -32 & | & 47 \\ 0 & 0 & 1 & | & \frac{1}{2} \end{bmatrix}$

$\xrightarrow{\frac{1}{9}(R_2 + 32R_3) \to R_2} \begin{bmatrix} 1 & -3 & 6 & | & -8 \\ 0 & 1 & 0 & | & 7 \\ 0 & 0 & 1 & | & \frac{1}{2} \end{bmatrix} \Rightarrow$

$y = 7, z = \dfrac{1}{2}; x - 3(7) + 6\left(\dfrac{1}{2}\right) = -8 \Rightarrow x = 10$

The solution is $\left\{\left(10, 7, \dfrac{1}{2}\right)\right\}$.

12.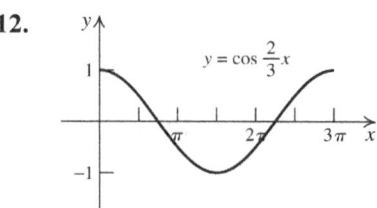

13. $\sin\left(-\dfrac{7\pi}{12}\right) = -\sin\left(\dfrac{7\pi}{12}\right) = -\sin\left(\dfrac{\pi}{4}+\dfrac{\pi}{3}\right)$

$= -\left(\sin\dfrac{\pi}{4}\cos\dfrac{\pi}{3}+\cos\dfrac{\pi}{4}\sin\dfrac{\pi}{3}\right)$

$= -\left(\dfrac{\sqrt{2}}{2}\cdot\dfrac{1}{2}+\dfrac{\sqrt{2}}{2}\cdot\dfrac{\sqrt{3}}{2}\right)$

$= -\left(\dfrac{\sqrt{2}+\sqrt{6}}{4}\right) = \dfrac{-\sqrt{2}-\sqrt{6}}{4}$

14. $r = \sqrt{2^2+\left(2\sqrt{3}\right)^2} = 4$

$\tan\theta = \sqrt{3} \Rightarrow \theta = \dfrac{\pi}{3}$ or $\theta = \dfrac{4\pi}{3}$

$2+2\sqrt{3}i$ lies in Quadrant I, so $\theta = \dfrac{\pi}{3}$

The polar coordinates for $(2+2\sqrt{3}i)$ are

$4\left(\cos\dfrac{\pi}{3}+i\sin\dfrac{\pi}{3}\right)$.

15. Shift the graph of $y=\sqrt{x}$ three units to the right.

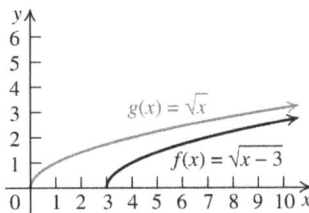

16. The domain is $(-\infty,-1)\cup(-1,0)\cup(0,\infty)$.

17. $7\ln x - \ln(x-5) = \ln\left(\dfrac{x^7}{x-5}\right)$

18. $[A\,|\,I] = \begin{bmatrix} 1 & 0 & -2 & | & 1 & 0 & 0 \\ 4 & 1 & 0 & | & 0 & 1 & 0 \\ 1 & 1 & 7 & | & 0 & 0 & 1 \end{bmatrix}$

$\xrightarrow[R_3-R_1\to R_3]{-4R_1+R_2\to R_2} \begin{bmatrix} 1 & 0 & -2 & | & 1 & 0 & 0 \\ 0 & 1 & 8 & | & -4 & 1 & 0 \\ 0 & 1 & 9 & | & -1 & 0 & 1 \end{bmatrix}$

$\xrightarrow{R_3-R_2\to R_3} \begin{bmatrix} 1 & 0 & -2 & | & 1 & 0 & 0 \\ 0 & 1 & 8 & | & -4 & 1 & 0 \\ 0 & 0 & 1 & | & 3 & -1 & 1 \end{bmatrix}$

$\xrightarrow[R_2-8R_3\to R_2]{R_1+2R_3\to R_1} \begin{bmatrix} 1 & 0 & 0 & | & 7 & -2 & 2 \\ 0 & 1 & 0 & | & -28 & 9 & -8 \\ 0 & 0 & 1 & | & 3 & -1 & 1 \end{bmatrix} \Rightarrow$

$A^{-1} = \begin{bmatrix} 7 & -2 & 2 \\ -28 & 9 & -8 \\ 3 & -1 & 1 \end{bmatrix}$

19. Since order is important, find the number of permutations of four people taken two at a time: $P(4,2) = \dfrac{4!}{(4-2)!} = 12$.

There are 12 arrangements.

20. There are 18 people in total. Three people who are 25 years or older have received speeding tickets, and six people whose ages are between 17 and 24 have received a speeding ticket. The probability that a person chosen at random will be 25 years old or older is $\dfrac{6}{18}$. The probability that a person chosen at random will have received a speeding ticket is $\dfrac{9}{18}$. The probability that a person 25 years old or older will have received a speeding ticket is $\dfrac{3}{18}$. So, the probability that a person chosen at random from the room will be a person who is 25 years or older or has gotten a speeding ticket is $\dfrac{6}{18}+\dfrac{9}{18}-\dfrac{3}{18}=\dfrac{12}{18}=\dfrac{2}{3}$.